ANTHROPOLOGY
THE STUDY OF MAN

(Physical Anthropology, Prehistory Anthropology and Social-Cultural Anthropology)

For Students of Anthropology

Dr. (Mrs.) INDRANI BASU ROY

Consultant Anthropologist

Kolkata

S Chand And Company Limited

(ISO 9001 Certified Company)

S Chand And Company Limited

(ISO 9001 Certified Company)

Head Office: D-92, Sector–2, Noida – 201301, U.P. (India), Ph. 91-120-4682700

Registered Office: A-27, 2nd Floor, Mohan Co-operative Industrial Estate, New Delhi – 110 044, Phone: 011-49731800

www.**schandpublishing.com**; e-mail: **info@schandpublishing.com**

Marketing Offices:

Chennai : Ph: 23632120; chennai@schandpublishing.com
Guwahati : Ph: 2738811, 2735640; guwahati@schandpublishing.com
Hyderabad : Ph: 40186018; hyderabad@schandpublishing.com
Jalandhar : Ph: 4645630; jalandhar@schandpublishing.com
Kolkata : Ph: 23357458, 23353914; kolkata@schandpublishing.com
Lucknow : Ph: 4003633; lucknow@schandpublishing.com
Mumbai : Ph: 25000297; mumbai@schandpublishing.com
Patna : Ph: 4011400; patna@schandpublishing.com

First Edition 2003
Reprints 2005, 2009, 2010, 2011, 2012
Revised Edition 2013;Reprint 2021 (Twice), 2022 (Twice), 2024

Reprint 2025

ISBN: 978-81-219-2259-3 **Product Code:** H5ANP41ANTH10ENZX13O

PRINTED IN INDIA

By Vikas Publishing House Private Limited, Plot 20/4, Site-IV, Industrial Area Sahibabad, Ghaziabad – 201 010 and Published by S Chand And Company Limited, A-27, 2nd Floor, Mohan Co-operative Industrial Estate, New Delhi – 110 044.

PREFACE

Anthropology, the science of man, is the most interesting subject in the way to know ourselves —how man and his culture has developed and perpetuated. It reveals the history of continuous bio-social adjustment of man under bare nature. It also asserts the present position of man with cultural dynamics.

Although the subject Anthropology is linked to other specialized science like Geology, Archaeology, Biology, Zoology, Physiology. Psychology, Ethnology, Sociology, Political Science, etc. but none of those sciences could explain the whole tale of man from the beginning, from the days of struggle and confrontation to the present state of advancement. Rather, the survival of man contributed in the existence of those sciences.

This is a valuable text book of Anthropology which aims to serve all students of Anthropology, specially in Indian Universities. It is complete with three distinct parts—Physical Anthropology, Prehistory or Archaeological Anthropology and Social-Cultural Anthropology. Each of these parts deals with specific portion of the subject matter and corresponds to the major branches of Anthropology.

The book has been written lucidly in simple language with plenty of examples. It offers a blueprint for the subject Anthropology as such to satisfy the general readers also who are enthusiastic to know more and more about Man.

I am really thankful to Shri Ravindra Kumar Gupta, Managing Director, M/s S. Chand and Company for his constant interest in my book. I am also thankful to Shri Navin Joshi, General Manager, Shri R.S. Saxena, Advisor of this prestigious publishing house. The total editorial staff deserves special thanks whose co-operation has made my work very easy.

Suggestions for further improvement will be highly appreciated.

DR. (MRS.) INDRANI BASU ROY

CONTENTS

ARCHAEOLOGICAL ANTHROPOLOGY OR PRE-HISTORY

SOCIAL CULTURAL ANTHROPOLOGY

INTRODUCTION

MEANING AND SCOPE OF ANTHROPOLOGY

Anthropology is a discipline, which serves the infinite curiosity about human beings. Etymologically the term is derived from two distinct Greek words —'Anthropos', the meaning of which is man and the 'logos' refers to science or study. Therefore, we define anthropology as a discipline, which studies the human beings, scientifically. But this definition is incomplete for the reason that there are also several disciplines, which are concerned with man; they study one aspect of man or the other. Sociology, psychology, political science, economics, history, human biology and even the humanistic disciplines like philosophy, literature, etc. form this group. Each of those disciplines is specialized to deal with a typical aspect of different groups of man.They may also cling to specific cultures and their moorings. Therefore, none of these disciplines can cover the whole jurisdiction of anthropology. Rather anthropology is a larger whole where different disciplines unite together despite the diversity of their interest. It possesses its own distinctiveness in the study of man. It is the only discipline, which strives to understand man and his actions in totality. Anthropologists believe in the integration of knowledge and realize the harmful effects of compartmentalization.

The index of anthropologists is man—wherever may he be whether on land, air or sea. They study the human beings in all climates and times. Men of the prehistoric as well as the historic past, men of the present generation and also of coming future come within the purview of anthropologists. But obviously they are not concerned with a particular man as such; their attention centre on 'men in group'. They perceive man not only as animal but also a social human having a history. People irrespective of their genders, ages and occupations are considered. Anthropologists deal with both male and female—old, middle-aged and young. Doctors, lawyers, students, agriculturists, public administrators, bureaucrats, etc. all are taken into account. People with different ideologies (democrat, communist, socialist etc.) or different creeds (Hindu, Muslim, Christian, Jain, Zoroastrian, pagan, ancestor worshipper, atheist, etc) appear to them with same importance. Even, the village folk and the city people are investigated with equal attention.

Man has been conceived as the creator of his cultural destiny. Therefore, anthropology is concerned with a rounded study of man—it studies men at all levels of culture. None of the other disciplines can be so pervasive. For example, the economists who are interested in economic behaviour of man , study man solely from the economic point of view. Political scientists work with that human behaviour, which are related only to the political affairs. A historian while is concerned with the past events of men, a geographer wants to project man in relation to his habitat and environment. A human biologist or a physiologist similarly involves him for the determination of biological or physiological configuration of a body where a psychologist wholly deals with the mental behaviour ot an individual. Thus, each of these disciplines segregates some of the aspects instead of studying them all at a time. Approach of anthropology is therefore unique in the study of man. It never analyses human behaviour in peace meal manner. Rather it tries to cover all aspects; all possible range of human behaviour.

By dint of the very nature, anthropology is holistic and comparative. It is a holistic one because it offers a total study of all aspects of culture and society in an integrated and comprehensive manner. All aspects of culture, say for example, religion, politics, social life, family, kinship, economics, aesthetics, health, technology, etc. are combined into one whole. It is believed that each aspect of culture, directly or indirectly, affects on the other aspects of culture, for better or for worse. Anthropology is said to be comparative because it takes an account of all human groups, all types of culture and society throughout the world for working out the similarities and differences in human body, behaviour and values. The ultimate goal is to evolve certain generalizations, which can be applied more or less to all human kind. The whole world is an anthropological laboratory; it is possible to deduce certain rules of human conduct.

Since the field of anthropology is vast and complicated, it is impossible for any scholar to acquire mastery over whole of the discipline. On the other hand, though specialization take place, discipline of anthropology does not at all fail to retain its holistic orientation. It remains entangled with the organic factors in one side and on the other side it reacts with social factors. Both types of factors are equally relevant to the subject. In practice, anthropology accepts and uses the general principles of biology and proceeds further to formulate a scientific concept of culture. Its field of investigation is extremely dynamic. It intends to understand the whole development of man and the wide variation of culture as a result of change over long periods of time.

POSITION OF ANTHROPOLOGY

Anthropology deals with man who is not merely a part on nature but also a dynamic creature in terms of biological and social features. It is a theoretical problem to determine the position of anthropology—where the discipline has to be put—whether in the fold of sciences or in the fold of humanities. A group of anthropologists took it as a natural science whereas some other anthropologists placed it as a subject under humanities.

In nineteenth century some German idealists and before that in eighteenth century a few French humanists considered anthropology as a branch of history and therefore they placed the discipline strictly under humanity. According to them man is a social creature as they live in a society and lead a social life. Although the biopsychic nature of man is of prime importance, but as man behaves within an organized group of social relatives, it enters into a new level, which is more or less super-psychic and super-organic. Therefore, in this level he is guided very little by his natural instinct; rather the norms of the particular group dictate him. Starting from the food-habit (what type of food should be taken and the very way to eat them), everything in a man's life - the dress-pattern, family structure, marriage form, religious belief and so on are decided by the social norm. Within a social system, man is thus more social creature than biological organism. This school of thought also held that the social relations are essentially the products of history, bound together by the moral values and not by the natural forces. Anthropology was viewed as a part of history and the anthropologist's role lay in social reconstruction. Kroeber, Bidney and Evans pritchard were in favour of this ideology,

In fact, there is a close relationship between history and anthropology for which controversies are found for a long time. Everything in this world offer a history as their existence is counted by time factor. A sort of historical investigation is essentially required in order to understand the factors and processes of change. Since human is the subject of anthropological investigation, we can not proceed at all without the consideration of temporal dimension. Both the disciplines aim to unveil the unexplored events of human life situation but differ from one another in tackling the problems. Each of them has developed its own methodological principles. History is chiefly concerned with the events. They count actions and interactions of human, both in individual and group perspectives. Whereas, anthropology takes interest in determination of culture; biological evolution terminates in cultural evolution.

Anthropology and more particularly the social anthropology is indebted to history. Earlier scholars like August Comte, Herbert Spencer, Emile Durkheim and Max Weber in studying social phenomena deliberately drew facts from history. Sir J.G.Frazer being first chairman in the school of Social anthropology in Britain gave emphasis on the historical analysis of the anthropological facts. In 1899, Franz Boas as a founder of the First University department of Anthropology at Columbia tried to highlight the life-ways of the primitive communities through historical methods. A.L.Kroeber in his two important papers, 'History and Science in Anthropology' (1935) and 'Anthropologist looks at History' (1966) attempted to establish the logical ground that the study of prliterate people would be more meaningful if the facts could be analyzed in historical perspective. According to him, anthropology is not wholly a historical science but its large areas are historical in interest. Moreover, he believed that the difference between the two disciplines was for the difference of the nature of insight but they were complimentary to each other. In a lecture at the University of Manchester in 1961, E.E.Evans Pritchard said, "the main differences between history and anthropology are not aim or method, for fundamentally both are trying to do the same thing". There is no doubt in this point that the continuity of a social process can be clearly estimated if historical methods are applied side by side with anthropological methods.

The relation between anthropology and history can be established in three distinct ways:

1. The subject matter of anthropology is basically historical in character. Anthropologists select different aspects of human culture derived from a common matrix. Since human cultures are not eternal like the subject matters of physics and chemistry, it changes with time. Each and every institutionalized organization viz., technological organization; economic organization, political organization, religious organization etc. are subjected to change. They remain largely relative and restricted to the particular situations. Therefore, all phenomena need a historical analysis.

2. Many of the institutions studied by the anthropologists deal with such a structure, which is essentially temporal or historical. For example, to study any development anthropologists have to trace the event from the beginning. Naturally such a study gets associated with history. Again, some of the problems have to be understood in the light of early stages, which are completely different from the present form. We can illustrate this point with the structure of feudalism, capitalism or socialism.

3. Anthropology often employs methods of Historical analysis, which is not always sufficient to deal with any problem of anthropology, but there are different types of historical analysis appropriate to different kinds of problems in anthropological science. In majority of cases historians have accepted the idea that each age will tend to view the past in the light of its own cultural milieu and stress upon the aspects of the past which provide an explanation of the existing problem.

The common features between history and anthropology are, both the disciplines depend for their materials on the actual happenings or occurrences in the natural course of human life. Teamwork is Suitable for both. Both of them differ from other scientists who make and get their data by experiments as per their needs. It is true that traditionally the historians differed from the anthropologists; historians were interested in past periods while the anthropologists-were concerned with the primitive people. But now both are inclined to study the contemporary problems of the modern civilizations of the world. Both of them have been able to account for the whole of a society. They do not remain satisfied after knowing what happened and what happens, their interests have also extended to find out the nature of social processes and associated regulations.

With the advent of the Darwinian theory of biological evolution and also with the introduction of new archaeological evidences, the quest in study of man got a new dimension. Unlike the seventeenth and eighteenth century thinkers, the nineteenth century historians and ethnologists became interested in the natural history of cultural development. Tylor, Lubbock, Maine and Morgan took anthropology as a historical discipline concerned with the culture of preliterate people.

The group of thinkers who believed anthropology as a subject of science includes Malinowski, Radcliffe-Brown, Fortes, Nadel and other eminent anthropologists. They pointed out that the subject studies human society following the methods of natural sciences. Science is the systematic investigation of phenomenon in the universe for the search of universal truth. It applies the logic, order and precision to identify laws, principles and generalizations. Anthropology proceeds like science. Its task is not to keep long reports about the individual cultures but through a comparative analysis of all such specific reports, it tries to arrive at 'social laws' regarding the emergence, growth, functioning and change of human societies. The believers of this school suggest that there are some regularities in social life which remain unaffected by the variation of time and place, so anthropologists can build a body of scientific laws by dealing with the repetitive, non-variant relations and events. In fact, anthropologists follow the scientific law to discover the rules of behaviour, conduct and organization. Their research methods and techniques strive for validity and reliability.

In 1920, B. Malinowski pointed out the importance of field work. He believed that the participant observation (fieldwork) was only method to go deep into the social forces of human society. According to him, anthropologists should not fully depend on the recorded materials like the historians; they must meet people and through long-term intimate contact data will come out. Although the early workers like Franz Boas and A.R.Radcliffe Brown realized the importance of direct contact with the people in the field, but Malinowski categorically pointed out fieldwork as a method of establishing scientific facts and laws.

At present the scope of anthropology as a scientific discipline has been established. Though the subject utilizes the historical method and draw data from history and other subjects of humanities, it is more meaningful as a science. Because its orientation is much more towards the science than the humanities. The next point is the determination of its exact position. In science, there are four divisions as per their nature and the field of operation. They are, Physical science comprising the subjects like Physics, Chemistry, etc.; Natural science embracing Geology, Astronomy, etc.; Biological science containing Zoology, Botany, etc.; and lastly the Social science consisting of Economics, Political science, Sociology, etc. Ambiguity appeared with the point that in which of the divisions 'anthropology' should be placed.

Anthropology has polarity within the subject itself. One of its branches is concerned with the anatomical structure and physical features of the man. This branch is known as physical anthropology, which is more or less akin to the biological science. It shares many materials with zoology, physiology, embryology, etc. In this branch man is predominantly an animal rather than having a history and social qualities. The other branch of anthropology is concerned with the social and historical factors mainly. It shares the concept from different social sciences like economics, history, jurisprudence, sociology, etc. Here anthropology is non-organic or more than organic having a closer relation to humanities. The subject, thus possesses a holistic tendency to explain man from all respect— biological and social point of view. It is superior to all other disciplines, even from which it borrows ideas and theoretical concepts. In fact, different disciplines are held together in an invisible fine thread from which anthropology harvests the essence of life. Man being the greatest wonder of the world when deserves to study himself, it surpasses all other disciplines of own creation. It becomes both a scientific and humanistic study. Its methodologies are highly abstract and sophisticated as of science. In one way, it perceives human beings as a product of socio-cultural process, and compels human feelings and sentiments to lead a group life demanding cooperation, competition. accommodation and adjustment. At the same time, it initiates human imagination to find expression in arts, artefacts and other mental faculties. On the whole, the subject offers both biological and social dimensions to be a master-science.

BRANCHES OF ANTHROPOLOGY

In the beginning, anthropology was divided into two major branches according to the scope of study. They were Physical anthropology and Cultural anthropology. The biological aspect of man was dealt solely under physical anthropology and the cultural anthropology included the whole of mental, rational and material-technological processes and products of the human being in an integrated pattern. But with time development of more divisions in the subject has been noted. Physical anthropology continues to be one of the major fields of anthropology. The cultural anthropology has given rise three major sub-fields—Archaeological anthropology, Linguistic anthropology and Social cultural anthropology. (Figure 1.1).

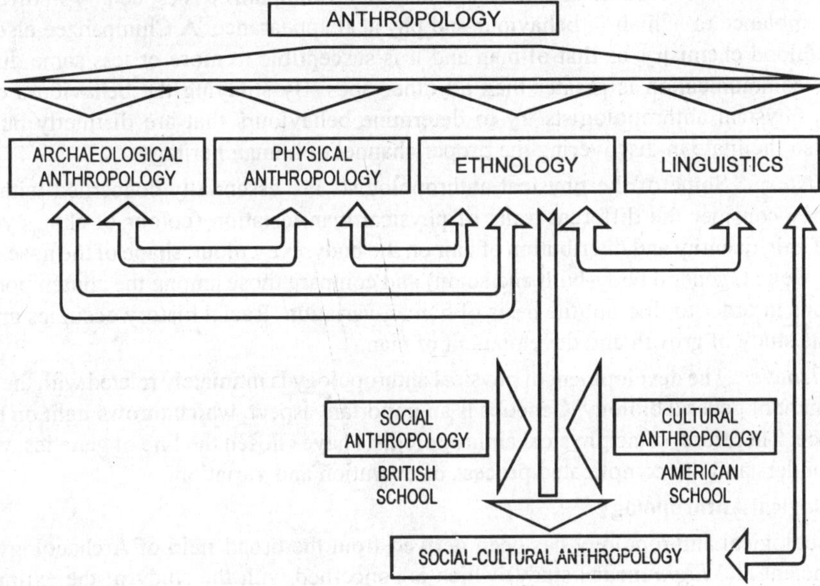

Fig. 1.1. Anthropoiogy and Its Sub-Divisions

Physical Anthropology

It is the oldest branch of anthropology, which was established much earlier than the other branches. As the name indicates, it studies the physical characteristics of man. It uses the general principles of biology and utilizes the findings of anatomy, physiology, embryology, zoology, paleontology and so on, Paul Broca (1871), the famous biologist defined physical anthropology as the "science whose objective is the study of humanity considered as a whole, in its parts and in relationship to the rest of nature". Although it is related to the biological sciences like anatomy, physiology etc., it does not restrict itself in the study of "contemporary average man". Rather it is interested in the comparative study of man considering the past, present and even future. Physical anthropology starts with the study of origin and evolution of man's physical characters and goes with the diversity of forms. It also analyses the biosocial adaptation of different human populations living in different geographical and ecological zones. The major areas of interest in physical anthropology have been classified as Human Paleontology or Paleoanthropology, Ethnology or Primate Behaviour, Racial History and Human Genetics.

Human Paleontology or Paleoanthropology : This field deals with the human origin. In order to trace out the phases of human evolution, some of the physical anthropologists search for and study the skeletal remains of human, pre-human and related animals. They also record the other geological information on the succession of climate, environment, and plant and animal populations. For this, they have to take the help of other scientists like Geo-chemist, Paleo-botanist, Paleo-geologist and Paleo-zoologists. The main objective is to study the extinct remains of man and to relate them with geological stratification. They also study the comparative anatomy of the primates i.e. the classificatory group to which man belongs with the apes and monkey. The comparative study of the anatomy reveals the physical relationship between the large variety of types that constitute the group. Therefore, the knowledge of comparative anatomy is essentially required for proper understanding of fossil evidences in relation to evolution.

Ethnology or Primate Behaviour : A comparative study of primate behaviour is also important in order to find out man's actual position in primate kingdom. The likeness and differences must be discovered before arriving at any conclusion. Most of the species included under the group primate may be observed in the wildlife as well as in laboratory. Chimpanzee is especially studied as it bears close resemblance to human in behaviour and physical appearance. A Chimpanzee also possesses the same blood chemistry, as that of man and it is susceptible to more or less same diseases. His power of communication is greater than all other apes. By studying the behaviours of different primates, physical anthropologists try to determine behaviours that are distinctly human. Such studies also facilitate in discovering the proper channel of human heritage.

Racial History : Some of the physical anthropologists are essentially concerned with the living races. They consider the different traits of physical manifestation (colour of skin, eye and hair, texture of hair, quantity and distribution of hair on the body, eye-colour, shape of the nose, lip, eyelid, etc. body weight, general body-built and so on) and compare those among the contemporary human populations in order to find out the basis of human variation. Racial history occupies an important rank in the study of growth and development of man.

Human Genetics: The development of physical anthropology is intimately related with the theoretical development of general biology. Genetics is an important aspect, which throws light on the ways of inheritance. Therefore, some physical anthropologists have chosen the line of genetics, which helps them to understand the complicated process of evolution and variation.

Archaeological Anthropology

Archaeological anthropology has been derived from the broad field of Archaeology *(archaios* means ancient and *logia* means study) which is concerned with the study of the extinct cultures. Man, the central figure of anthropology existed long before the development of written record. Therefore, archaeology is able to supplement anthropology by recovering the remains of ancient men of bygone days along with the material evidences of his culture.

Classical archaeology is a combination of fine arts, history and classics. It seeks the antiquities of the past. So it can not be an exclusive domain of anthropologists, but anthropologists have to depend on archaeologists in describing the human of the past and to find out the ancient cultures which were flourished before5000 years from now. Archaeologists often work with the paleontologists, geologist, and chemists to reconstruct the days of prehistory. For many parts of the world like Australia, Melanesia, Polynesia and most of the New World and Africa, knowledge of writing is fairly recent. Naturally to discover the prehistoric man and his cultural activities, anthropologists have found no way other than to rely on the archaeologist' work. Archaeology, thus, has become an indispensable part of anthropology. Without archaeology, physical anthropologists could not have been successful in determining the place of Homo sapiens in nature; the long process of human development would very little to be understood.

Cultural anthropology also depends on archaeology. Cultural anthropologists deal with the social behaviour of man; the past and the present are equally important to them. They trace the emergence and development of customs and social behaviour from the prehistoric level and go up to the contemporary level where both the primitive and civilized people form the social counterpart. Since most of the evidences of human life in prehistoric days are intangible and perishable, they leave no permanent imprint behind. Past life-ways and cultural processes can only be understood on the basis of a few tools, which have been dug out and interpreted by the archaeologists. Archaeology considers multifarious events of past, dates them and builds up the sequences. In this respect the prehistoric part of the cultural anthropology is akin to archaeology. Not only it attempts to describe the early men and their way of living; it also discusses the genesis of cultural capacity in order to explain the cultural development and diversities through time and space. Cultural anthropology would stand meaningless and incomplete without the aid of archaeology.

Linguistic Anthropology

It is another sub-field of cultural anthropology, which is concerned solely with language. But the study of languages as a discipline (linguistics) arose much before the birth of anthropology. It became close to anthropology for sometime during the first half of twentieth century but diverted afterwards. At present, some of the cultural anthropologists specialize themselves in linguistics because language is an important aspect of human behaviour. Transmission of culture from one generation to other has been possible only for language. Language enables man to preserve the traditions of the past and to make provisions of future. However, the anthropologists who study language in order to understand a culture help to form this distinct wing. They deal with the emergence and divergence of languages over time. Contemporary languages also belong to their attention.

There were a number of anthropologists whose interest revolved round this essential component of human culture. Edward Sapir (1921) defined language as a purely human and non-instinctive way of communicating ideas, emotions and desires by means of a system of voluntarily produced symbols. Leslie A. White in his book 'The Science of Culture' (1949) mentioned that all behaviours originated and based on man's capacity to use symbols. In fact, use of symbols paved the way for communicable speech. Nineteenth century linguists were engaged in describing the languages and classifying them into families and sub-families on the basis of their similarities and dissimilarities. The subject was then known as philology, not the linguistics. Anthropologists contributed to philology by studying languages spoken by the aboriginals. They broadened the horizon of philology by the reconstruction of lexical forms, grammar, sound patterns and in finding out the ancestral language. Thus, the field of linguists was closely linked with anthropology for several decades. But from 1950 onwards, a conceptual revolution took place; the two disciplines turned from each other.

At present anthropologists and linguists differ in their theme though both of them study languages. Linguists have always kept language isolated from the other activities of an ethnic group whereas anthropologists have never estimated language as an autonomous phenomenon; they treat it as a part of culture. However, linguistic anthropology has two parts. The part, which deals with emergence and divergence of languages, is called historical linguistics and the other part, which offers the description of sound units, is called structural linguistics. Structural linguistics also discovers the rules that reveal how sounds and words are incorporated in actual speech. This aspect is known as socio-linguistics.

The pattern of speech varies from society to society on the basis of action, behaviour and communication. Cognitive anthropology is the outcome of linguistic anthropology, which employs the principles on which speakers of a particular language classify and conceptualize the phenomena. Anthropology in one way has learnt from the linguistics: on the other way has contributed to it.

Social-Cultural Anthropology

Social anthropology is a sub-field of cultural anthropology because of its intensive interest in social behaviour and the organization of social groups. The term social anthropology is popular in Great Britain and other Commonwealth countries. It is also extensively used in France,-Netherlands and the Scandinavian countries being supported by Prof. Clude Levi-Strauss. In the United States of America the term cultural anthropology is preferred still now. According to the cultural anthropologists, as culture subsumes society, cultural anthropology incorporates social anthropology. But in nineteenth century neither the term social anthropology, not the term cultural anthropology was popular. Instead, the term ethnology was used. The Greek term *ethnos* means the race or indigenous people and *logia* means study. Therefore, ethnology is the study of living peoples and their culture; it is sharply distinguished from archaeology, which is the study of extinct cultures. Ethnology as a subject is concerned with the customary ways of thought and behaviour of ethnic groups, and tries to discover the fact for which a culture differs from the other. Ethnology also deals with dynamics of the culture i.e. it attempts to point out how cultures develop and change. The subject ethnology had been derived from ethnography which offers a first hand account of a particular people at a particular time.

Ethnographic facts are interpreted by ethnology through classification, analysis and formulation of the principles. Thus, broad anthropological theories and hypotheses are achieved in terms of the nature of human behaviour, the evolution and functioning of culture.

Ethnography is the basis of the cultural anthropology. Some cultural anthropologists essentially treat with ethnography and ethnology. They seek to understand the relations and patterns of life among the different types of people as revealed through their established institutions and groups such as kinship, marriage, family, economic transaction, religious practices, social control, political behaviour, art forms, music consciousness, etc. The facts of such study are collected through firsthand and face-to-face investigation. The meaning of any item of culture is attained only in reference to the context and by relating the pattern of relationship with other items within the totality of culture. Most of these studies are conducted in small-scale and relatively isolated societies, called primitive societies.

During the period between 1840 to 1870, there was a great quarrel with the two words—ethnology and anthropology. On the one side there were historians and philosophers and on the other side there were supporters of science, particularly of biology. In fact, ethnology as a sub-field of cultural anthropology gave rise to social-cultural anthropology. But at present, social-cultural anthropology has come a long way from the premises of early 'descriptive ethnology'. Moreover, the subtle distinction between the two terms, social anthropology and cultural anthropology is not considered today. These two concepts have been amalgamated and the new term social-cultural anthropology has been adopted.

The British anthropologists in using the term social anthropology have emphasized on the concept of society, which is an aggregate of individuals who live in face-to-face association and share some common sentiments. Different social interrelationships and interactions are their object of study. On the other hand, the American anthropologists preferring the term cultural anthropology have concentrated on the concept of culture which is the sum-total of human behaviour, verbal or nonverbal and their products—material and non-material. Cultural anthropologists try to analyze each and every interaction and interrelationship by judging the value behind it. But in reality, the concept of society and culture are keenly related. A culture can not be understood without the reference of human society. Similarly the understanding regarding a society remains incomplete without the consideration of its culture. The relationship between these two is like that of a container and content or like a body and mind. None of them can exist without the association of other. Therefore, normally, the social and cultural phenomena co-occur; within a larger socio-cultural field they are not always be well distinguished. Hence the term social-cultural anthropology is justified, instead of the term social anthropology or cultural anthropology.

The study of change is an integral part of anthropology. No culture and no society, regardless of circumstances, is beyond change. Whether it is a remote and isolated village or large industrialized city, everywhere people experience a variety of changes in their pattern of living, which is manifested with the passing of time. They occur when some new ideas are accepted to the norms of the society. It may be in the form of movement or communication, conflict and peace, scientific discovery, philosophical insight and so on. For example, introduction of a new medicine or a new mode of transport may bring a great change in social as well as cultural context. Thus, instances of social and cultural change are virtually countless which take place over time. According to De Fleur, D'Antonio and DeFleur (1976), whenever a transformation occurs in a structure, frame, pattern or function of a social system, or social organization (within a single unit of a society or in the society as a whole), it is designated as a social change. In an opposite way, when an alteration takes place in the prescribed standard of existence or in the content of those standards is called a culture change. Despite conceptual distinctions, the relationship between the two is very close. So, the term 'social change' and 'cultural change' are often used interchangeably. Since change is an unavoidable aspect of human dimension, a multiplicity of anthropological specialization has been derived from the study of change.

Each episode of change is responsible for socio-cultural alteration.

Social cultural anthropology is indebted to a number of social sciences such as sociology, economics, political science, law and jurisprudence. But its greatest resemblance is found with sociology. Both of these two disciplines share a common historical origin with the contribution of some intellectual forebears like Marx, Weber and Durkheim. As a result, they enjoy a common terminology as well as a general theory of society and culture. In spite of the commonalties, each of these two disciplines has been developed in its own way with considerable differentiation from one another. Separate academic interest had grown in them. Naturally the ways of approach, areas of concentration, methods of investigation and techniques of considerations vary, although both of the disciplines study society and social relationships. While social-cultural anthropology is prone to work in terms of culture and deals with whole of a society, sociology tends to work in small compartments of complex Western society. Therefore, anthropologists concentrate chiefly on small and isolate societies of primitive people whereas sociologists prefer own contemporary civilization for study. Anthropologists' study is more intensive and analytical than the sociologists. Sociologists' approach is objective as well as extensive; they depend more on statistical formulation. For a long time social anthropology used to be regarded as a branch of sociological study that indulged in the study of primitive societies. In earlier time primitive groups was the exclusive concern of anthropology. Today, although the field of social anthropology has been much enlarged to include all human societies and culture, a lot of anthropologists still devote themselves in the study of primitive society.

Even at the end of twentieth century, an ambiguity existed among the anthropologists regarding the theoretical and methodological issues of the subject anthropology. One group of scholars still looks upon anthropology solely as 'social anthropology', where there is no concern of 'bones and stones'; they believe that anthropology is interchangeable with sociology. But the other group has not agreed with this point. They conceive anthropology as a comprehensive and unified study of man and his culture. They also recognize the sub-divisions like archaeological anthropology, physical anthropology, linguistic anthropology and social-cultural anthropology. The first group is numerically weaker than the second group who believes in specialization. But all specialization do not hold the same importance as evident in the grade of popularity, and prospect of teaching and research in different university departments of our country. The scope of specialization goes on a descending order as social-cultural anthropology, physical anthropology and archaeological anthropology. So far as India is concerned, no anthropology department is found which pursues specialization in linguistic anthropology. However, the situation is different in America where we find a balanced unified development of the subject with the said four wings of specialization.

The concluding point regarding anthropology is that, from the very beginning his is a discipline with tremendous diversity. It relates the past and present of man and tries to anticipate his future. Therefore, the sub-disciplines of anthropology have been designed to explore human biology, culture and language from the origin. The Archaeological anthropology studies the prehistoric and historic human societies. The physical anthropology explains the biological base of mankind which gave them the capacity of culture. The linguistic anthropology analyses the languages cross-culturally from their inception and the social-cultural anthropology studies the similarities and differences in the organizations of the societies of the world. In case of no other discipline such a diversity of interest has been noticed. While examining the nature of anthropology, Kroeber compared it with a forum where different types of people came as being interested in minute things and ultimately find connections between the small things to understand a bigger issue. This means anthropology keeps together several sub-disciplines that work jointly inspite of their differences. But, the integration of the subject is not lost at all; it continues to be a distinct discipline and tries to understand a man in totality.

Human beings share ideas, experiences and biological endowments as an order of the society. Anthropology does not feel ashamed in inviting knowledgeable persons from various other disciplines whose expertise stand indispensable for the growth of the subject. Anthropologists are the most skilled technocrats who can readily assimilate all sorts of knowledge to enrich their own discipline.

APPLIED IMPORTANCE

A pursuit of knowledge without any practical ends has no justification. Therefore every scientific discipline have been bifurcated into pure and applied wings. The former is justified in terms of the latter. Anthropology plays a central role in the integration of the human sciences. Different subjects have been fused together to make a general comprehensive science of mankind. It corresponds to the concept of 'general anthropology' as advocated by Paul Broca, the leading anthropologist in France in 1859. Like any other science, anthropology has its practical applications. Society is the soil from which anthropologists collect the materials to flourish their discipline; they put back their knowledge in the same field for social use. An anthropologist never dissociates himself in the dilemma of pure and applied research. Usually, the pure research is considered as the most valuable and prestigious work—an ideal model of scientific inquiry. It does not bother about the practical result. But the applied research is always conducted in reference to the practical field. It aims at solving various problems and therefore it often uses the concepts, methods, theories and findings of pure science as per the specific purpose. In anthropology, since practical problems of humanity give rise to the subject, the opposition between the pure and applied aspect is not very clearly manifested. In the ultimate analysis, either the human problems converge in understanding of the principles of human relation or the solution comes down to combat against the problems. An anthropologist may learn and teach at the same time. Obviously this learning or teaching does not limit to a classroom; it may take place anywhere—where people live and work. In fact, anthropology is a part of life. For the wide breadth of its subject matter, it encompasses the whole of the mankind. By identifying a set of values, it holds different subjects together as well as provides a special character for which the experts can come to a consensus to pass down the synthesized knowledge from generation to generation.

The original anthropologists included the anatomists, philologists, geographers, and antiquarians who met together to solve an all-inclusive problem. They had developed mutual tolerance and dependence among themselves. Sol Tax (1964) pointed out; "The anatomist might come to read papers on craniometry to philologists and to students of customs, and to listen in turn to papers on chipped-stone industries, on grammar or on folklore. Those first anthropologists surely contributed not only breadth, but also a great tolerance for variety in subject matter and in techniques of study. They established these as values for which anthropology ever since tended to select experts". * 1. He also said, "It is not surprising that anthropology should characteristically form a society of scholars open to new techniques, tools, ideas and men. The tools brought in range from the 'Law of uniformity' from geology to a Freudian model, or a carbon-fourteen dating from chemistry. We have adopted, reinterpreted and made our own what ever has appealed useful to our varied problems. People of other fields are drawn in because they are wanted and needed". *2.

All sub-fields of this discipline have the applied scope. They serve the mankind with the specialized anthropological knowledge. The success depends upon the effectiveness of the anthropologists in understanding the laws of cultural variation and change. For instance, during the Second World War, anthropological linguists improved the methods and materials for intensive instruction in a large number of foreign languages. Similarly, insights of linguists are used at present to develop the instructional programs intending the teaching of the Native American children. Indeed, most of the linguistic researches are funded by the missionary societies, which need the knowledge of native languages in order to facilitate instruction in Christian dogma. Thus, linguists serve many practical purposes related with exotic languages.

* 1 & 2. Tax Sol (ed.) - Horizons of Anthropology, Aldine Pub. Co., Great Britain, p. 252.

In a similar way, physical anthropologists are involved in various applied tasks. They cover a large number of spheres namely, designing the garments and equipment, sport activities, criminology and penology, medico-legal aspects, health-hygiene and diseases, genetic counselling, eugenics, medicine and public welfare, education, race relation, etc. As physical anthropology offers a rich understanding about the biological foundation of culture and human living, it debunks the myth of racism and race superiority. Physical anthropologists have used their knowledge of anthropometry in designing machinery, artificial limbs and clothing. Data on human osteology, serology and genetics have proved immensely beneficial for medical applications. They also contribute to disease research, diet and nutrition, health planning, reconstructive surgery, growth and development. Inputs of physical anthropology even help in forensic tests to identify criminals and crimes.

Considering the applied use of archaeological anthropology we can mention the museum work where reconstruction of history is found through the excavated materials and those are kept open for the public visit. Historical sites in different states attract tourists and serve educational functions. Archeological anthropology thus offers general information and education to the people. It enhances the knowledge about past on which our existence depends.

Among all sub-fields, socio-cultural anthropology is the most deeply involved arena for direct application. It undertakes the study of all human problems that take place for the technological changes particularly in underdeveloped areas. It can find out causes of labour management, friction in industry to minimize the tensions. It also deals with minority problems, community development projects, economic development schemes etc. It has become the therapeutic science of human relations. In a world torn by racial misunderstanding and conflict, the contribution of the anthropologists are very important. They understand the psychological attributes of the people along with the values of culture, so can suggest a way of solution very easily. There are also a large number of external agents which continuously threaten to modify a customary way of life; an anthropologist with his professional expertise study the nature, extent and consequences of these exposures and tries to restore the equilibrium of a social system.

The potentiality of cultural anthropology was fully recognized by E.B.Tylor and Sir John Lubbock as early as nineteenth century. The earliest anthropological organizations of Great Britain and the United States strived for the elimination of human slavery and amelioration of poverty. They also chalked development programmes of the colonized people. At that beginning, the major field of application was the administration of native people. Franz Boas (1928) in his book 'Anthropology and Modern life' had pointed out that anthropologists could make themselves useful by undertaking the studies like hereditary factors in crime, the significance of race, the role of education in human society etc. Boas gave a lead to American anthropology. He established the fact that there is no scope of secluded 'pure' interest; rather human problems themselves made anthropology the science of man. Like Boas, Malinowski and some other functionalists ascertained the applied importance in British anthropology. Contemporary writers namely Kluckhohn and Nadel supported Boas's statement with a list of numerous contributions made by anthropologists in America after the winning of the Second World War. They discussed the contributions in terms of capacity, efficiency, intelligence and training.

Anthropology is devoted to the service of man and the society. Raymond Firth compared the role of anthropology with 'social engineering' as the anthropologists often try to bridge the gulf between the primitive and the modern. Firth could not think so far that an anthropologist might be able to influence in matters of social policy in its development or in execution. But he understood that an anthropologist is essentially required in diagnosis and prediction of the social trend. Kluckhohn in later years regarded, the role of an anthropologist as 'social doctor', instead of 'social engineer', who can prescribe the medicines for the removal of social ill. In reality, anthropology really possesses a higher degree of predictive precision and therefore is able to manage a social situation most scientifically. Recently two specific terms have gained popularity in the context of application of anthropological knowledge—applied anthropology and action anthropology.

Applied Anthropology

Daniel G. Brinton (1895) in his paper 'The aims of Anthropology' first put forward the concept of applied anthropology. According to him "Applied anthropology aims accurately to ascertain what are the criteria of civilization, what individual or social elements have in the past contributed most to it. how these can be continued and strengthened and what new forces, if any may be called in to hasten the progress". *

The concept of 'Applied Anthropology' was developed in United States particularly after the Second World War. It contributed in administration and development policy in the third world. Applied anthropologists in general tried to improve the lives of the people who were in a disadvantageous position in the modern world of colonialism or imperialism. They realized the need for change and so undertook the challenging task of development in the sphere of colonial administration. They also made themselves involved in monitoring the efforts of others in changing people's lives. The most famous test case of 'applied anthropology' is Cornell University's Vicos project in Peru, where an anthropological team under the guidance of A.Holmberg handled the role of 'patron' in a large estate. The team carried out a basically paternalistic reform plan but aimed at developing power to the producer. Another successful case may be cited with the tribe 'Ashanti' in western coast of Africa. The Ashantis traditionally possess a gold-decorated stool, which is believed to have descended from the sky. It is extremely sacred to the members of the tribe; they never put their stool down on the bare ground. To maintain the sanctity they always cover it with an elephant's skin and the whole thing is wrapped within a special cloth. In 1896, when British came in contact with Ashanti, they tried to snatch the stool. But that was not possible as the Ashantis could hide the stool very carefully. In 1921, suddenly it was discovered that the gold decorations on the stool had been stolen out. The Ashantis became furious and turbulent; they demanded the offenders to be killed. Day by day as the situation was getting worse, a government anthropologist was recruited to manage the whole. The anthropologist stepped into the problem, considered the pros and cons of it and lastly allowed the Ashantis to keep the stool with them and punished the offenders with the order of banishment.

In Indian situation, the renowned anthropologist S.C.Roy among the Oraons of Chotonagpur provided a happy solution. It was a dispute with flag. Traditionally each Oraon village was in the possession of a flag, which were used in their inter-village dancing program called *yatra.* Once a contractor, at the time of construction of a new bridge over a river, presented one flag containing a picture of a railway engine to the Oraons of the adjacent village. It was not a mere presentation, it was associated with some superstitious belief by which the spirit dwelling in the river under the bridge was appeased as the previous bridges were washed away due to rain and flood. However, the flag became the symbol of power. This created resentment among the Oraons of another village and they painted a railway train on a large flag in order to increase their respective power. Naturally the original possessors of painted flag did not tolerate the imitation and protested vigorously. Heavy quarrel and fight was followed for which police had to intervene. A large number of people were prosecuted for the breach of peace. S.C.Roy tackled the dramatic situation by making a flag with the picture of an aeroplane. He presented that to the Oraons of the first village who got the flag from contractor. At this time he called the village elders in a meeting and explained the superiority of aeroplane. They became satisfied with the explanation: their anger subsided and peace came back in the region.

Applied anthropology did not acquire the same importance in all countries. Even in United States where it first developed, could not build a positive image. Non-applied anthropology was considered more prestigious than the applied anthropology. But some developing countries like Mexico, Latin America etc. appreciated the dignity of applied anthropologists. Again, Dutch, British

* *Brinton D. G. - The Aims of Anthropology, published in the Proceeding of the A.A.A.S. XLIV, 1895)*

and later French put great value to anthropology or anthropological training, as they were interested in developing markets for European industrial goods and enlarging the production of raw materials in different colonial setting. They felt the necessity to study the languages, customs and practices of the people for getting them under the stronghold. It should be remembered that though Americans realized the utility of anthropology before the British, but in practice Dutch and French were the forerunners.

Writings about anthropology in human welfare first appeared in the documents of American anthropological association. It was depicted that all human problems involve changes in behaviour, attitude, institution and relationship. Most of the scientists, even who devoted themselves in pure research, cherished this idea. But until Second World War majority of the anthropologists in America used to work in colleges, universities and museums and applied scope was absolutely unknown. In 1933, Commissioner John Collier tried to associate anthropologists with the Bureau of Indian Affairs. Soil Conservation Service also asked the help of the anthropologists in their programmes to check the depletion of natural resources as well as to assist the Indians in managing their own affairs. During the Second World War, Government of United States hired many anthropologists in order to use their knowledge of culture in predicting the behaviour of enemy.

There were at least two reasons for which applied anthropology bore a negative image and the general field of applied anthropologists stood outside the colleges, universities and museums. Firstly, for a long time applied anthropology was the monopoly of exploitative colonial administrators of some European nations, therefore a sort of apathy was grown among the anthropologists to select this chore for professional involvement. Secondly, anthropologists who lacked the academic job opportunities were prone to accept the unusual jobs outside the traditional academic field under the public and private organizations. The agencies that employed the anthropologists in different programmes were highly placed and amply funded. The anthropologists became the servants of those powers; their freedom in work was lost. Though the applied knowledge of the anthropologists was highly appreciated to bring desired solution to difficult problems, but.they did not get the due respect. They were never allowed in formulation of new social policy; their roles were largely confined to the execution of policies framed by others. Naturally applied anthropology was not viewed favourably in anthropological profession,

However, the British and the American both utilized applied anthropology in different purposes. British Government employed anthropologists in colonial administration. American Government used the knowledge of anthropology in their own home. A number of native races of North America were uprooted and exterminated from their own territory due to want of a sensible policy. By the application of anthropological knowledge they were brought to the mainstream. Dying Red Indian communities were also saved. In the post-war years, the United Nations and many agencies of the United States Government consciously inaugurated programmes of planned change throughout the world, particularly in the so-called underdeveloped nations. American anthropologists were entrusted with the task of administrating the Pacific Trust territories. The National Indian Institute of Mexico was not only advised but also administered by the anthropologists. In Mexico most of the anthropologists were absorbed in Government agencies and institutes to work on social problems. Anthropologists began to serve in different development organizations like the United States Agency for International Development (AID), the Peace corps, the United Nations Economic, Social and Cultural Organization (UNESCO), United Nations Children's Emergency Fund (UNICEF) etc. Since backward nations or under developed areas were found to face enormous problems, many American and British anthropologists were sent forth to develop economic potential of the people. They also tried to minimize the various concomitants of poverty such as illiteracy, high infant mortality, substandard diet and housing, inadequate public health and so on. They handled the areas like education, public health, urban affairs, industry, etc. All human problems that required intercultural understanding demanded the presence of anthropologists. Anthropologists were also able to show their excellent performance in the field of training and research.

With the technical specialization, anthropologists are expected to study the local situations, make recommendation for an action for the desired result. An applied anthropologist should possess the skill to train up the personnel in order to run rehabilitation or a resettlement program successfully. They must learn to identify the needs of the people and to innovate the ways of change, anticipating the sources of resistance. In the realm of administration, they have to interpret the ends and interests of'the employer. Sometimes the administration wants an anthropologist in creating an atmosphere where a dominant subordinate relationship can be maintained but sometimes he is directed towards aiding of the people. According to Sol Tax, an anthropologist can show his best performance as an administrator. As the work of an administrator involves policy making as well as policy implementing, it requires a clear understanding about the people in concern. The non-anthropologist officials often fail to apprehend the needs and aspiration of the native people that in turn hampers the success of the plan and threaten the credibility of the officials. Unlike the members of the bureaucracy, the anthropologists analyze the basic characteristics of the society and culture along with the pattern of interaction. They try to locate the cultural barriers that inhibit change, stimulants to change and other motivations of the people for the overall good in a society. G.M.Foster (1962) had rightly pointed out the limitations of an ordinary administrator or a technician who makes a decision without weighing the evidence furnished by an anthropologist. In Foster's words, "it is not so easy for a technician or an administrator to accept the fact that an understanding of his attitudes, values and motivations is just as important in successfully bringing about a change. It is painful to realize that one implicitly accepts assumptions that have little validity beyond tradition and that one's professional outlook has been uncritically acquired. It is not easy for either the technician or administrator to admit that the way he views his assignment and how his works are conditioned by such things as how he perceives his role in the bureaucracy to which he belongs, how he reacts to his supervisor, and how he deals with his colleagues and those who work under his direction" * 1.
Eliot D. Chapple (1955) had commented in the same context that "by using anthropological methods, the administrator can attain a control in the field of human relations comparable to that which he already had in the field of cost and production. He can can understood and estimate the effects of change and see what steps have to be taken to modify his organization or to restore it to a state of balance. He can do this both through acquiring a knowledge of anthropological principles and by using anthropologists to make analyses of existing situations" *2. Ordinarily anthropologists are called only after the administrator has decided the policy. Such a policy is not usually devised to utilize the existing system of relations and most of the problems are left undefined. Therefore, every administrator should learn to formulate his or her objectives in terms of the principles of anthropology.

Applied anthropology may bring planned culture change. It requires pushing of new alternatives to a society in such a way that its members would accept those. But if those new alternatives violate the existing norms or deep-seated taboos in a society, a great chaos is created. People either reject or resist the novel ideas as the traditional customs and institutions come in conflict with the changes. It is not possible to know definitely that whether a proposed change will be truly beneficial for a target population or not. Sometimes, in spite of delivering best efforts, the long-range developments may appear detrimental, even those which initially seemed to be advantageous. Therefore, for the successful implementation of a plan, some expert anthropologists are wanted who can develop an all-round idea about a culture within a short period or can reach the root of a problem very easily. He is the best performer among all other social scientists. His judgements of good or bad, right or wrong are not governed by the sentiments; they are wholly concerned with the disposition of recipient

---

*1. G. M. Foster *Traditional Societies and Technological Change,* Second Edition, Allied Publishers Pvt. Ltd. New Delhi, p. 198.

*2. Eliot D.Chapple *'Anthropological Engineering: It's use to Administration'* in *Readings in Anthropology* edited by Hoebel and others, McGraw-Hill Book Co. Inc, New York.

people. For this very reason, values of western civilization can never be applied to a peasant society. Further, anthropologists believe in the functional relatedness among the different aspects of culture, so they remain alert while proceeding from one step to another. For example, the introduction of any new concept, whether an agricultural technique or a new health practice eventually affects on other areas of life.

Applied anthropology has been criticized particularly for its apolitical philosophy. Some scholars lik.e to designate it 'developmental anthropology' where greater political awareness is involved. According to the critics, applied anthropology focuses on cultural differences and thus obscures the structures of social and politico-economic dominance for which developmental problems arise out. Van Willigen (1986) traced the development of applied anthropology through different stages, which he named as the 'applied ethnology stage', the 'Federal Service stage', and the 'role extension, value—explicit stage' and the 'policy research stage'. He also reviewed the offshoots of applied anthropology, which have been resulted from various theoretical and ideological attitudes. For example, the 'action anthropology' as proposed by Sol Tax, 'research and development anthropology' as shown by the Cornell-Peru project, the community development approach and the more recent approaches like advocacy anthropology and cultural brokerage.

Action Anthropology

Sol Tax proposed the term 'action anthropology' in 1958 through his paper 'Values in Action' published in the journal "Human Organization', Vol. 17, No. 1. Though it is an offshoot development from applied anthropology, it does not stop with the humanistic study as an applied anthropologist does with the natives and minority peoples. Rather, the action anthropologists involve themselves intimately with anthropological problems and pursue their studies in a context of action. In such a study, the distinction between the pure research and the applied research generally disappears. The anthropologist accepts a problem as his own and proceeds through trial and error method. In a first exposure he may not be successful, but he never feels disappointed or frustrated. Also he is not in the habit of blaming others. Rather he rectifies his own strategy and procedures; carries on the same task with fresh vigor. He does not forget to follow up the whole from time to time. The method of action anthropology is thus wholly clinical or experimental. Action anthropologists seldom keep themselves as mere observers or catalyzers. They recognize their own responsibility in solving human problems. Therefore, they stick on the problems until they are solved. By the way of solving action anthropologists may generate new theories and findings, acceptable to the general anthropology.

Action anthropology as a new method of research did not appear in the imagination of Paul Broca. In modem anthropology, the 'action anthropology' and 'applied anthropology are the parallel developments belonging to two different schools of thought. According to E.J.Jay (1987), applied anthropologists are the experts on the culture, which they study. So on the basis of their knowledge they are able to make proper recommendations for administration. But those recommendations may not always cent percent true. Because, it is impossible for an anthropologist to know every minute things of a society, other than his own, no matter how closely he is associated with it. So his recommendations are liable to a certain degree of errors; complete and accurate prediction of human behaviour is not always possible. In this respect, action anthropologists can show the ideal performance, as their duty does not end with formulation of concrete recommendations. They remain constantly associated with a project until and unless the goal is achieved. As a program of action proceeds, the anthropologists revise their judgement and recommend for farther action according to the reactions of the recipient groups.

The difference between the applied and action anthropology lies in the modes of approach. In the words of L.R.Peattie (1987), "Applied anthropology tries to move back and forth between value-interest and disinterested consideration of relevant fact" while "Anthropology in action is suspended between these two poles and swings between them". Both the applied anthropology and the action anthropology depend on 'value infused observation' and therefore bear practical utility. When an applied anthropologist feels the urge for a course of prolonged action to solve a problem, action anthropology is initiated. But in reality, the scope of action anthropology is quite limited. In the notion of Sol Tax (1964) "Action anthropology requires the intellectual and the political independence that one associates with a pure researcher, it depends upon university and foundation connections and support rather than those of a client or government. But also requires that the anthropologist leave his ivory tower and that without losing his objectivity he enters into some world of affairs which becomes for the time being his laboratory".

The field for the application of anthropological knowledge was widened at the beginning of twentieth century according to the circumstances of global welfare. The efforts of the anthropologists have become mature and effective, as a steady development is evident in the resources of the discipline. The availability of data have increased, methods have been refined and above all the number of anthropologists and their potentialities for research have been enhanced a lot. Acculturation studies and culture-contact studies are now closely linked with applied anthropology. They have attempted to resolve the contradictions between traditional socio-cultural pattern and the needs of economic and technological development.

New inventions of science and changes in old technology tend to disturb the equilibrium between the individuals and the groups. Applied knowledge of anthropology shows its usefulness in controlling the Change variables. Moreover, this knowledge has proved itself worthy in successful manipulation of human beings to achieve some particular end. Despite the humanitarian motivation of human welfare, some of the anthropologists had rejected the applied outlook. They restrained themselves from the applied field because of the problems of ethics. As a social doctor they could not negate the responsibility of begetting a beneficial change and at the same time could not surpass the limitation of anticipating the end result of a change. As a result, they suffered from a phobia—whether their scheme would really yield any benefit to the target group or not. The question of ethics appeared as a great gulf and the anthropologists did not find a moral support when a plan turned unsuccessful.

Nevertheless, the use of the professional skill of anthropologists was quite legitimate in studying the processes of culture change in action programmes or in introducing the advanced programmes on health, agriculture, education, etc. For the benefit of developing countries, a society for applied anthropology was formed in 1949 who formulated the codes of professional ethics to guide the confused anthropologists. The British social anthropologists working on the problems of colonial administration in Africa and other parts of the British Empire shared these ethical views with their American colleagues. In 1963, the codes were again revitalized. As a matter of fact, the large number of anthropologists who happily accepted the long-term and short-term assignments after the end of Second World War and hoped to deliver effective technical assistance, were found to shrink abruptly. A growing frustration was noticed among the anthropologists regarding the war in Vietnam, which could not be controlled despite the anthropologists had enough potentiality. Consequently, in 1970's American Anthropological Association had to face repeated agitation, acquisition and denial on a massive scale. But the old-fashioned codes were ultimately replaced by the newly developed keen sense of the anthropologists. The enlightened anthropologists themselves got a broadened understanding of the ethical dimension. They continue in their judgement, which is based on the sensitivity to the human sufferings, but comply with the cultural situation. The activity shows a peculiar blending of science with humanities. Many renowned anthropologists namely, F. Boas, B.Malinowski, A.R.Radcliffe Brown, R.Redfield, S. Tax, S.F.Nadel, G.M.Foster, M.J.Herskovits, E.E.Evans-Pritchard, A.L. Kroeber, C.Kluckhohn, R. Linton, D.G.Mandelbum, E.Sapir, L.A.White, etc. devoted their attention in applied field.

A Few Instances on the use of Anthropology

Since the days of colonialism, anthropology has served various interests of people. The earliest use was found in the political administration in England. The expertise of the anthropologists was utilized in tackling certain piecemeal problems in British colonies. But as the employer solely decided the objectives, the anthropologists hardly had any scope to understand the situation with empathy. The needs and consequences of the target-group community did not get any importance. However, the duty of the anthropologists were chiefly concerned with the recording of the behaviours of the native people which prevented the administrators to take crude decisions and hasty political actions so,that the whole situation would not go out of their hand. Some anthropological training centers were established for future colonial rulers where ethnology and comparative linguistics occupied a prominent place. Anthropological surveys on colonial people were instituted from 1929, under the enthusiasm of B. Malinowski. Malinowski was able to use his concept of culture in this respect. To Malinowski culture was entirely humanistic which functioned to secure human survival. He chalked out a design for its application in human welfare.

In the United States anthropology began to be used in the office of Indian Affairs by 1933. The most significant use was noticed with agriculture.It helped in understanding the way of life of American farmers. However, the knowledge of anthropology as used in the office of Indian Affairs was almost similar to the colonial administration in England. None of them were conscious about the ethics. France, Belgium and Netherlands also started employing anthropologists to facilitate the administration of their colonies.

During the Second World War, United States Government hired many anthropologists to help in solving their military problems. The situation got complicated with the Japanese prisoners who did not behave in the usual way. All of the Japanese soldiers who were captured in the battle showed a practice of self-killing before they were taken to the prison. Studies by the anthropologists on Japanese culture and character made the American military leaders aware that the Japanese soldiers preferred an honourable death in own hand because they considered capture as a matter of great disgrace. Anthropologists also informed that the Japanese regarded their emperor almost like god, his word was sufficient to cause all Japanese soldiers to lay down their arms. After working with the anthropologists, Americans not only learnt about the Japanese thought, they also acted as per the advice of the anthropologists to win the situations*. Since anthropologists solved many political and military matters, after the end of the world war, a considerable number of applied anthropologists were called upon to upgrade different other projects associated with health, economy, education etc. They also contributed in the fields of religion, art, system of value and personality structure. They were no longer restricted to their colonialist and political image. A list furnished by Clyde Kluckhohn showed that the anthropologists of America worked in the various departments of state, in the office of strategic services, in the Board of economic Welfare, in the Military and Naval services, in the office of war information, in Peace Corps, in Foreign Economic Administration, in Federal Security Administration, in the medical branch of Army-Air Forces, in Chemical Welfare Division etc. The expertise of anthropologists was recognized in a variety of situations. Their roles were basically like advisers. Sometimes they advised the government of an emergent nation about building of roads, bridges, dams etc. and sometimes helped a public health specialist in epidemiology by advising regarding the control of Malaria, Leprosy, Tuberculosis, etc. He might act as an economist, an educator or an agronomist. Many applied anthropologists conducted rehabilitation programs for the handicapped. Not only that, some anthropologists also made them associated with the private charitable and philanthropic institutions. They have tried to develop cooperative groups, ultimately to produce a self-supporting and integrated community.

---

* Ruth Benedict was asked to compile a report on Japanese collective attitudes. Benedict collected her data from those Japanese living in America and also she explored some documentary evidences. Her report stood as a ground stone for the study of National character on the basis of personality analysis.

In India, both the applied anthropology and action anthropology has great scopes. The British East India Company made the first application of anthropological knowledge in 1807. Francis Buchanan was appointed to conduct ethnographical surveys in order to project a clear picture on the life-style and religion of the local inhabitants. Anthropological training was provided to some officials who performed administrative jobs as well as prepared some handbooks, gazetteers and monographs on tribes and castes of India. In this context we can profess the names like Risley, Dalton, Grigson, Thruston, etc. A few British anthropologists viz. Rivers, Radcliffe-Brown and Hutton also took enough initiative to know the Indian situation. Their critical examinations and recommendations are the assets to the Indian anthropology. The officials who worked in health services and the educators, who were in missionary schools, utilized this information in typical British style. At this time, colonial ethnography and applied anthropology were synonymous with tribal studies. Verrier Elwin and C.V. Furer-Haimendorf came forward with the concept of a tribe. They also framed the tribal policies. The first applied anthropologist of Indian National was Rai Bahadur Sarat Chandra Roy who made extensive studies on the tribes of Chotonagpur and Orissa. S.C.Roy and Ananthakrishna lyer fought for the interest of the tribals. They wanted to establish the 'land right' for the tribals and provided an ideological framework for their movement of autonomy in the pre-independence period. In the postcolonial period, we get a few responsible anthropologists namely D.N.Majumdar, A.Aiyappan, N.K.Bose and others who consumed their energy to understand the tribal situation of India.

Some of the anthropologists being motivated by the imperialist idealism did not understand the dynamics of freedom struggle. Therefore, they uttered a humanistic philosophy and tried to incorporate all sections of Indian people under a united political system. They evolved constitution of India. These anthropologists kept themselves engaged in formulating development programmes for the weaker section of the country and showed further interest in the implementation and evaluation of the programs. The knowledge of anthropology helped both the union government and the states in framing policies of socio-economic amelioration. Anthropology-oriented approach was proved useful for internal reconstruction of the country.

The popularity of applied anthropology increased greatly after independence. This was the time from when Indian scholars started interacting with the American anthropologist very actively; the term 'applied anthropology' or 'action anthropology was coined at this time. A great change of outlook has been noticed between the earlier anthropologists and the modern anthropologists. The anthropologists of independent India are more concerned with the welfare of tribals and other downtrodden people unlike the earlier .anthropologists who acted totally in favour of colonial rulers. The nature of approach has now been totalistic, instead of 'piece-meal' one. Not only that, anthropologists by themselves have formed a forum at present to raise their voice against various social injustices and exploitations. On the one hand, they are diagnosing social evils and prescribing appropriate remedies; on the other hand they have been vocal in pointing out the flaws of implementation. It has been found that in past most of the administrators in development programs had sought quick and easy solution without having sufficient awareness of local situations, so majority of the projects fell flat. Social workers and administrators of present day have been eager to get the advice of applied anthropologists to ensure the success of the project.

Today's anthropologists do not limit themselves with the tribal life or rural people as like past. They have extended their knowledge and views for overall development of the nation. They have provided attention on health and nutritional status, family welfare and childcare, care of the aged and disabled, and so on. Physical anthropologists over the years have studied the localized incidence of diseases like goiter, anemia, marusmus, kwashiorkor. sickle cell traits, mental retardation, juvenile delinquency etc. and tried to correlate them with genetic and environmental factors. The work of the anthropologists ranges from tribal rehabilitation to population growth, malnutrition to adult education, vocational training to labour unrest. The fields include administration, medicine and public health, education, industry, economic development, community development and also the areas of race and ethnic relations with social policy issues. The opinions of anthropologists have even found valuable in the needs of the athletics, defence services, forensic science and criminology, eugenics, national integration and international relations. Their views are also respected in museums and art gallery. Whether in the applied or action field, anthropologists maintain their fundamental values and ethics of anthropological science and never deviate from the ultimate aim. The eminent Indian anthropologists in applied field are K.P.Chattopadhaya, S.C.Dube, T.C.Das, L.P.Vidyarthi, B.K. Roy Burman, L.R.N. Srivastva, Yogendra Singh, M.N. Srinivas, M.N.Basu, P.K. Bhowmick, A.K. Das and many others, whose valued thoughts and actions have accelerated the progress of India. These anthropologists have found out the resources and organized them for social ends. Some of these anthropologists also succeeded to tie up the welfare measures with the International Programmes, which yielded the maximum benefit.

Although it is true that a large number of anthropologists have been able to associate themselves with the various plans and projects of modern India, but it is a matter of great regret that, still in most of the cases, their opinions do not get due weightage in the formulation of the plan or the strategy of its implementation. Because, those development projects are entirely operated by the influential section of the members who possess little or no anthropological guidance. But when such plans fail to achieve a desired goal, anthropologists at work are discredited despite the fact that they have no fault of their own. It is absolutely a fault of the power structure and negative attitude of the society towards the anthropologists. Usually the non-anthropologists with their limited conception and stubbornness stick to the old ideas and can not perceive varied dimensions and depth of anthropological pursuit. Naturally the success is delayed.

ORIGIN AND DEVELOPMENT OF ANTHROPOLOGY

APPEARANCE OF ANTHROPOLOGY AS A DISCIPLINE

Man and his surroundings have always been a perennial source of wonder and reflection for himself. This consciousness instigated him in searching the realities. Therefore, it is futile to talk about the beginning of the study of man. For the genesis of systematic thinking we usually refer back to the classical Greek Civilization especially to the writings of Herodotus in fifth century BC. Not only Herodotus, many other Greek and Roman historians namely Socrates, Aristotle, Hippocrates, Plato, etc. are considered as pioneer social thinkers. They first expressed their significant interest in man's affairs considering the perspective of Universe. Their approach was purely humanistic and they postulated a social theory from organismic point of view.

Anthropology is the youngest of all scientific disciplines. The contribution of Greeks and Romans in the rise of anthropology is beyond any doubt. The Greco-Roman interpretations of man and universe were based on the concept of natural order and natural law. The Greek scholars perceived the Universe and its components within one whole where change can not break the essential relationships. The Roman view was just the opposite. They believed that all things remain in a constant state of change. The common feature between those two approaches was that, both emphasized on the mutual interrelationship of different entities to form the total structure of the society. Both of them possessed a didactic interest and so they compared the great personalities and the nations, particularly in reference to their customs, religion and political organization.

Nevertheless, the emergence of anthropology as a distinct discipline occured only recently in nineteenth century. Sydney Slotkin in his book 'Readings in Early Anthropology' traced the history of many anthropological sub-disciplines from seventeenth and eighteenth centuries and in his opinion most of the fields of anthropology was developed by the end of eighteenth century. But he also agreed that the real professional interest of the subject did not appear until nineteenth century. In fact, the intellectual heritage of Greeks and Romans was muted considerably by the Christian Theology. Again it was revived at a later date when voyages were undertaken to discover the new territories. The unusual peoples and their unknown way of life evoked interest of navigators and other explorers. As a result in 1800 a society named as 'Observers of Man' was founded in Paris by the union of naturalists and medical-men. This society promoted the study of natural history by providing guidance to the travelers and explorers of far places. But meanwhile, for the long series of Napoleonic wars, the commerce and the foreign travel were interrupted. Naturally the study of natural history was neglected and instead, the questions of philosophy, ethnology and politics came forward. The society could not stand long and in 1838 another society for the protection of aborigines was established in London. Eminent scholars joined in that society whose aim was political and social, rather than scientific. Again, within a very short period need of a scientific society was realized. One of the influential members, Mr. Hodgkin in collaboration with several other distinguished persons, in 1839 inaugurated an 'Ethnological Society' in Berlin. Eminent naturalist Milne-Edwards took there an active part. In 1841, a similar type of society was formed in London and soon after that in 1842 the third 'Ethnological Society' was founded in New York. The establishment of ethnological societies can be taken as an important landmark in the emergence of anthropology.

Anthropology is therefore considered as the product of scientific developments in western world. The tradition of social philosophies continued till the advent of industrialization in west and it emerged as a distinct discipline in the nineteenth century; Charles Darwin's Origin of Species (1859) perhaps boosted the zeal of all scientists in different fields. Darwin showed that life had evolved from the unicellular organism and went to the way of complex multicellular organism, through the process of evolution. This idea not only opened the new avenues for zoology, anatomy, physiology, philology, palaeontology, archaeology and geology; it also accelerated the pace of socio-cultural studies. Being influenced by Darwin, a group of intellectuals namely Spencer, Morgan, Tylor reached to the conclusion that evolution did not operate only in case of physical aspect of mankind, but also in cultural life. Accordingly, the year 1859 may be taken as the date of birth of anthropology; R. R. Marret (1912) termed *anthropology* as 'child of Darwin'. In the same year 1859, Paul Broca founded an 'Anthropological Society' in Paris. Broca himself was an anatomist and human biologist. He advocated the idea of general biology by synthesizing all specialized studies in order to understand a man. Anthropology made a significant progress in America following Broca's light.

On the other hand, in 1863, James Hunt withdrew himself from the British Ethnological Society and established an Anthropological Society in London with the dissenting members of Ethnological Society. Hunt declared anthropology as ' a whole science of man' which deals with the origin and development of humanity. According to him, the work of Ethnological Society was too narrow in scope. In 1868, Thomas Huxley was elected the President of the Anthropological Society. Despite his biological orientation, he belonged to the Ethnological Society in London for a considerable span of time. However, this was the time from when the work of Ethnological Society and the Anthropological Society merged together. The difference was maintained only in names.

Nearly thirty years, from 1840 to 1870, a great debate continued with the two words —ethnology and anthropology. This was also a debate between the humanitarians and the pure scientists. But the progress of anthropology did not dwindle down for its ambiguity in the name. France, Germany and England highly appreciated the subject. In fact, anthropology earned a great popularity over whole of Europe. International Congresses of Anthropology and prehistoric archaeology were held in 1866, 1867 and 1868 in different parts of Europe. Anthropological Institute of Great Britain and Ireland was formed in 1871. But in 1873 there was again a split; a new 'London Anthropological Society' came out. This new society presented a journal named 'Anthropologia'. International communication, research and publication were the main objectives of this society. By this time, the names such as anthropology, ethnology, ethnography, archaeology, prehistory, philology and linguistics were firmly established. These names although differed with time and place, but they certainly got the recognition as the branches of the study of man.

Paul Broca in his address 'The Progress of Anthropology' (1869) had pointed out that anatomy along with Biology formed the pivotal base of anthropology upon which the subject could extract the ultimate ideas of general anthropology by means of a rigorous synthesis. He also believed that the general anthropology had the potential to evolve out with crown and glory in the scientific field, sooner or later. After a few years, anthropology really acquired a synthetic character and honoured both in Europe and America. In Europe, the different names as anthropology, ethnology, prehistory and linguistics are still in vogue; they are complementary to each other to cover the whole study of man. But in America and most of the Asia, the word anthropology itself is sufficient, to connote the entire meaning. Before the discovery of America by Columbus, American aborigines had their indigenous ideas about the nature of man. Later, European tradition of Science and scholarship touched them as happened in case of Africa, Oceania and part of Asia. It is said that the English, French, German and other Europeans provided anthropology with the tradition of scholarship, the books and the theories while the Americans provided a nice laboratory close by.

The Europeans not only uncovered the long prehistory of America; they also utilized their rich field in studying the non-European cultures. But, the anthropology of America was not merely an offshoot of European thought; it became an independent center for the development of anthropology in nineteenth century. The empirical field work of European anthropologists for training and practice continued upto the twentieth century. Thomas Jefferson, the third President of United States is said to be the first anthropologist in America. He was the pioneer empirical researcher in archaeology. He also studied Indian tribes and their languages. Jefferson inspired nineteenth century statesmans and scholars to investigate among the American Indians. Such works were essential for them because history already pushed them into the isolated corners of the country. Gallatin, Cass, Schoolcraft etc. were the names of the best American Indian students. Lewis Henry Morgan was one of the leading personalities of the world who combined his personal intensive fieldwork in a native culture with comparative work and general theory. The missionaries and the others, who used to live at that time though endeavoured to publish their observations, Morgan's status was different from them as his observations had a worldwide perspective. In fact, Morgan founded the great branch of anthropology, known as social-cultural anthropology through the comparative analysis of family and kinship structure.

It is therefore clear that the philanthropic ideas mixed with scientific attitude led to the formation of ethnological societies in Europe around 1840 and then it was carried to America. Since the scholars in America got special field opportunities, they were able to raise up the discipline within a short period. On the other hand, being deprived of field-situations, England exhibited a great development only in social theories during the latter half of the nineteenth century. Spencer, Tylor, Frazer were the renowned names but they are known as the arm-chair anthropologists as they had little scope to test their hypothesis in actual field and depended largely on the studies incurred by the other persons. However, in later period we get the scholars like Durkheim, Mauss and others who deliberately tried to avoid the arm-chair speculation.

In the last half of nineteenth century, some anthropologists became interested in the study of racial stock and also in the biological evolution of man. France contributed a good deal in prehistory and physical anthropology. Germany, at first established a psychological and later a geographical tradition of cultural anthropology. Theodore Waitz developed the basic physical anthropology, which embraced the peoples of the whole world. Adolf Bastain by surveying the cultures of people all-over the world inferred about the basic psychological configuration in man. Friedrich Ratzel blended geography with anthropology and created a new sub-field, anthropogeography. All these were empirical studies. In the context of empirical study we must mention the contribution of Franz Boas (1853-1942) who was born in Germany and was educated at the Universities of Heidelberg, Bonn and Keil. Basically Boas was a student of pure science, proficient in physics and mathematics. Later he studied geography and got the Doctoral Degree in 1881, on the colour of sea-water. At the late phase of his academic career, he moved to American University where he became interested in the study of people; he lost his faith in geographical determinants. He successfully delineated the task of ethnology as the study of the total range of phenomena in social life. The field work experience initiated him in developing a new historical method. He opposed the conjectural evolutionary theories of arm-chair anthropologists in the Metropolitan Universities, most of whom had never seen a tribe. Boas believed evolution as having no scientific foundation for anthropology. He also criticized the comparative method and recommended for replacing it by his 'historical method'. The historical method, proposed by Boas was based upon the principle that like phenomena do not always occur due to the same causes, but there must be one common cause which has to be empirically established. Therefore, when Franz Boas went to America with his philosophy, he could easily qualify himself with the American tradition. Not only that, he also adjusted the tools and theories of Americans with his own ways and thus American anthropology got the influence of European anthropology.

The concept of culture given by Sir Edward Burnett Tylor (1832-1917) established anthropology as an academically recognized discipline in Europe. Tylor has been recognized as the father of modern anthropology. The anthropologists of America also believe this. Originally the word 'culture' was used in the field of biology; its German synonym 'kultur' was applied to human societies in eighteenth century (Kroeber and Kluckhohn: 1963) to correlate diversities of behaviour with racial differences. In his epoch-making book 'Primitive Culture' (1871), Tylor first defined culture in the following words: "culture or civilization taken in its wide ethnographic sense, is that complex whole, which included knowledge, belief, art, morals, law, custom and any other capabilities and habits acquired by man as a member of society". He wanted to mean that culture, as a totally learned behaviour in social setting can never be inherited through biological traits. Tylor was definitely one of the principal founders of anthropology who forwarded the idea of cultural evolution. He also stressed the need for first-hand data collection. Like Tylor, Lewis Henry Morgan (1818-1881) of America had also shown the growth of human culture with time. Both of them believed in 'the psychic unity of mankind'.

Increasing complexity of western world with its multiplying social problems as well as the concern for understanding them gradually led to the fissioning of social philosophies and gave birth to various social sciences. The discipline of sociology and anthropology appeared as twin sisters particularly after industrial revolution and colonial expansion respectively. Industrialization suddenly increased man's productivity. As a result, traditional agrarian feudal order collapsed down and this impact rolled on the structure of family system. Extensive kin bonds were cracked; emotional stress appeared on interpersonal relations. At the sametime inhuman exploitation centering the unorganized labour took a violent shape. Innumerable social problems like broken homes, industrial diseases, juvenile delinquency, prostitution, alcoholism, gambling, sprouting of slums etc. appeared in alarming proportion. Sensitive, humane and liberal thinkers started to seek solutions to these severe problems. The concern came to be known as Sociology.

On the other hand, when the industrializing western nations like Britain, France, Holland, Portugal and Spain started to colonize in non-western world in order to find raw materials for their industries and to get markets for the finished goods, anthropology came out as a product of their ventures. In the colonial situation, the white people with the Christian ideology tried to understand the different beliefs, habits, customs and practices of the natives. It was essential for them as they came with a view to trade, preach or to rule. They were in the habit of publishing their queer notes about the exotic people after returning back to home. The materials used to be sold like hot cakes; they were highly valued because based on those facts, the anthropologists made most of their propositions. This is the reason for which those anthropologists are designated as arm-chair anthropologist and since then anthropologists gained the 'other-culture' perspective as the most fundamental tenets of methods and inspiration. This is also the reason for which anthropology is often referred as the study of primitives or tribes, although later, it became equipped in studying complex societies, ranging from peasants' institution to urban situations.

After the First World War, outlook of anthropology changed greatly. Nineteenth century anthropologists were totally unacquainted with the peoples with whom they were concerned. They used to depend on the stale and fabricated data collected by other non-anthropologists like explorers, missionaries, administrators etc. As the anthropologists put forward their propositions sitting merely in their libraries, many speculations were involved there. They leaned chiefly on Comparison. Bronislaw Malinowski (1884 -1942) first broke that trend. He taught the importance of field-study in contrast to speculation about the primitive people. Malinowski was a Polish by birth but later became a British citizen who passed a long time among the natives of Trobriand Islands during the First World War. He laid the foundation of the functional theory in culture following the ideas of Emile Durkhiem. A new trend was brought by Malinowski for which the course of development in anthropology was changed since 1925.

During the second world war, the American anthropologists became attentive to the psychological problems and issues of whole nations in order to understand the basic features of developed civilizations like Japanese, the Chinese, the Russians etc. National character studies became very popular in this period. Moreover, some interesting development took place in the arena of 'language and culture'. At the end of the world war, anthropology got the French scholar Claude Levi-Strauss whose emphasis was restricted to the formal aspect of culture. Levi-Strauss received a vast influence from structural linguistics and in turn influenced the social anthropologists of the Old World by injecting linguistic consciousness in them. Claude Levi-Strauss has been recognized as a structuralist; his role seems to be indispensable for the discipline of anthropology.

By the end of Second World War, the physical aspect of anthropology also took a new turn. It was no longer a study limited to different kinds of measurements; study of growth and development became prominent due to the rediscovery of genetics. The progress in the study of human genetics provided a firm basis of integration between physical anthropology and social anthropology. Anthropologists' basic interest in prehistory is more or less akin to primitive ethnography. But the New World prehistorians did not proceed in the same way as like the prehistorians of the Old World. The Old World prehistorians used to dig the remains for knowing the material base of the people's life but the New World prehistorians were not satisfied after knowing the material base of the ancient people, they tried to know even the kinship pattern. Changes were also become evident in the field of linguistic anthropology. Topics like classification of languages, working on sound-patterns and grammars of unwritten languages became old-fashioned in the socio-cultural and psychological setting. Instead, socio-linguistics and ethnography of communication became pervasive to project the field of linguistic anthropology.

We can, however, conclude in the way that, at the formative phase, anthropology was linked with the exploration of life and culture of colonial countries. Anthropologists' task was to interpret the materials collected by the non-anthropologists. British scholars not only gave a conceptual lead to this field, they first organized the 'International Congress of Anthropological and Ethnological Sciences' (ICAES) in 1934 where nine hundred anthropologists, ethnologists and other scientists of allied field from forty-three countries had participated. By that time, the British Evolutionary and Diffusionist theories met a setback but the structural-functional theory emerged as the most important school. The period can be said as the formative phase of institutionalization of anthropology. It was further strengthened by British social anthropology, which acquired a definite recognition on the global plane. The impact of British anthropology was dominant in the next three Congresses at Copenhagen (1938), Brussels (1948) and Vienna (1952), but after the ravages of Second World War it became almost dead and deserved revitalization.

The fifth International Congress of Anthropological and Ethnological Sciences held in Philadelphia (USA) in 1956 showed the post-war dominance of American anthropology. It was a remarkable anthropological conference organized by Wenner Gren Foundation for Anthropological Research Inc. at New York that was established under the chairmanship of A.L.Kroeber in 1952. It was also significant because Soviet delegates for the first time attended this ICAES. American model of anthropology was followed in this Congress instead of British model. In the introductory lecture Kroeber said, "the subject of anthropology is limited only by man. It is not restricted by time; it goes back into geology as far as men can be traced. It is not restricted by the region but it is world-wide in scope. It has specialized on the primitives because no other science would deal seriously with them, it has never renounced its intent to understand the higher civilization also". Levi-Strauss did not differ with the fact that anthropology maintains a close relationship with the humanities, the social sciences and the natural sciences, but he further stated that "anthropology tries to jump to a deep level, while social sciences have been so far, unable to do so, and the result in an extremely difficult situation for cooperation between anthropology and the social sciences (sociology)". Margaret Mead clarified Levi-Strauss's view emphatically. She said, 'Anthropology uses methods of analysis from different levels (Embryology, Geology) with different units and different time scales, cross-cutting all disciplines relating man to nature, on the one hand, and in history, on the other, while sociology is an analysis on a single level, leaving psychological and biological analysis to other disciplines'. According to L.P.Vidyarthi (1979) the integrated image of anthropology emerged in the Symposium of Vienna in 1952 but the approaches of planing and deliberation in the study of 'integrated man' was worked out in the fifth ICAES in 1956. A book entitled 'Current Anthropology' by William Thomas (1956) and also a journal named as 'Current Anthropology' edited by Sol. Tax (since 1960) further ascertained the image of integration. As a result, in the Paris Congress (1960), the Structural-Functional School was found to be deeply entrenched and American model of anthropology dominantly occupied the ground.

A large number of associates of the Current Anthropology under the dynamic guidance of Prof. Sol. Tax were able to link the whole world with anthropology. The Soviet ethnography was found to be developed on the line of evolutionary theory and later that became modified by the writings of Marx, Engles and Lenin. Thus, new dimensions were added to anthropology, which was rather well balanced and free from Anglo-American influence. The concept of colonial anthropology or neo-imperialism in anthropology is comparatively a recent achievement. The scholars of the third-world countries were neglected for a long time. Now they are also getting due importance; an increased rate of participation has been noticed in different sessions of ICAES. But in spite of all these recent developments, we must not forget the beginning of anthropology that was blessed with the European and American courage.

GROWTH OF INDIAN ANTHROPOLOGY

Anthropology is a young discipline in India. By the term 'Indian Anthropology', Andre Beteille (1996) wanted to mean the study of Society and Culture in India, by the anthropologists, irrespective of their nationality. There were many anthropologists inside or outside of India who took interest in the study of Indian society and culture. However, anthropology owes its origin in the later half of the nineteenth century with the ethnographic compilation of tradition, custom and belief of different tribes and caste in various provinces of India. Prof. D.N.Majumdar found the beginning of Indian Anthropology in the establishment of Asiatic Society of Bengal that was inaugurated by Sir William Jones in 1774. But there is no convincing evidence for an emergence of anthropology in India during eighteenth century. It is true that the Asiatic Society began to publish essays of anthropological and antiquarian interest in its journal and proceedings, but all those were written by government Officials and missionaries who had no academic interest. British administrators undertook such studies in order to know the people of India whom they wanted to bring under their control. Therefore, the contributors were all foreigners who were trying to understand alien cultures. Genuine cultural work started in India around the middle of nineteenth century when Asiatic Society served as a forum for organizing ethnographic and ethnological studies as well as started publishing the results of those studies.

In the world scenario, anthropological work in true sense, except Tylor's pioneering work, started with twentieth century. Tylor in his book 'Anthropology' (1881) considered language, race, physical features, customs and practices of primitive people as well as the old remains of people as parts of anthropology. His idea was derived chiefly from the reports furnished by the administrators and missionaries who penetrated into different parts of the world at the wake of trade and commerce, and subsequent colonization. The objectives of missionaries were different from the trading people or administrators; they wanted to covert the natives to their own religion, Christianity. However, both the administrators and missionaries wondered when they came across various types of people having entirely different kinds of life. So they tried to communicate their strange experience through writing, by describing the people and their facts. Not only Tylor, around the mid-nineteenth century, some notable persons of England—lawyers, doctors, businessmen, men of letters and scholars became interested in non-European customs and practices. They wanted to explain the variegated life-style of the peoples in perspective of their own societal model. Most of the thinkers of this period came to the conclusion that the social institutions essentially underwent a gradual development i.e. they had evolved from a lower stage to a higher stage with time which can otherwise be called a development from simple to complex form. Since the thinkers themselves were not the data-collectors, they did not meet the primitive people in reality and their perceptions were built only upon the available accounts. Accounts also came from some navigators and casual visitors who traveled from place to place in fifteenth and sixteenth centuries. But those early records were carefully abandoned for their ethnocentric and prejudiced nature.

However, at the end of nineteenth century, the administrators and missionaries in India, like other parts of the world, wrote a lot of about the Indian people and their way of life. After colonialization, the administrators took more interest in the issues of colonial people for the sake of good administration in newly acquired territories. Trained British personnel namely Risley, Dalton, Thurston, O'Malley, Russel, Crooke, Blunt, Mills and others who were posted in different parts of India, wrote compendia on tribes and castes on India. Later, some scholars of Indian national joined them. At the same time some British anthropologists like Rivers, Seligman, Radcliffe-Brown, Hutton came in India with a view to work on the native people. They influenced a few missionaries like Bodding and Hoffman. All those works provided the basic information about the life and culture of the peoples of India.

Throughout a whole Century after this, anthropology in India proceeded successfully. Indian anthropologists borrowed the ideas, framework and procedures of work from western anthropologists and practiced 'self-study' instead of studying 'other culture'. But their pattern of work became unique with regard to assumptions, choice of data, criteria of relevance and some other matters.

Different scholars at different times were interested in finding out the genesis and development of anthropology in India. In this context we can put the names of seven eminent Indian scholars—S.C.Roy, D.N.Majumdar, G.S.Ghurye, S.C.Dube, N.K.Bose L.P.Vidyarthi and S.Sinha. Earliest attempt was made by S.C.Roy who in his paper 'Anthropological Researches in India' (1921) furnished a bibliographical account of the publications on tribal and caste studies as published before 1921 *1. He counted all materials published in magazines, all compilations in handbooks of different regions and all monographs on tribes. This work of S.C.Roy appeared as a testimony where the dominance of British administrators, foreign missionaries and travelers have been noted in conducting anthropological researches in India. D.N.Majumdar made the second attempt after a gap of twenty-five years. He reviewed the development of anthropology in a Memorial lecture at Nagpur University and showed a poor progress of Indian anthropology under continual efforts of British administrator and anthropologists. None of the two scholars mentioned the influence of Americans. But in another review Majumdar tried to relate the developing discipline anthropology with the theories of culture that originated in England and America. This view was published in the journal of Anthropological Society of Bombay with the title 'Anthropology under glass' (1950).

Next to them was G.S.Ghurye. In his article 'The Teaching of Sociology, Social psychology and Social anthropology' (1956) he wrote, "Social anthropology in India has not kept pace with the development in England, in the European continent or in America. Although social anthropologists in India are to some extent familiar with the work of important British anthropologists, or some continental scholars, their knowledge of American Social anthropology is not inadequate" *2. It is known that G.S.Ghurye with K.P.Chattopadhaya took research guidance under W.H.R.Rivers in 1923 and afterward they became the head of the departments of Indian Universities *3. Naturally it is expected that the initial stimuli for the development of Sociology and Anthropology in India came from Cambridge.

---

*1. S.C.Roy '*Anthropological Researches in India*' in *Man-in-India*, Vol. 1,1921 pp. 11-56.

*2. G.S. Ghurye "The teaching of Sociology, Social Psychology and Social Anthropology' in *The Teaching of Social Sciences in India*. UNESCO Publication, pp. 148-53.

*3. G.S.Ghurye became head of the Sociology in Bombay University and K.P.Chattopadhaya became Head of Anthropology in Calcutta University.

S.C.Dube in 1952 discussed the same issue in IVth International Congress of Anthropological and Ethnological Sciences (ICAES) held in Vienna. He criticized the unfortunate prejudice and distrust of Social workers and popular political leaders towards anthropologists and suggested an active cooperation among the Government, the Universities and individual scholars to form a central organization at National level for solving the disadvantageous situations of social anthropological research in India. After ten years, in anothert paper Dube highlighted the weakness of the contemporary Indian Social anthropology and recommended for more refined techniques of research and methodology *.

N.K.Bose (1963) published a booklet entitled "Fifty years of Sciences in India, Progress of Anthropology and Archaeology" which was prepared under the auspices of the Indian Science Congress Association from Calcutta. He discussed the progress of anthropology in India under headings—Prehistoric archaeology, Physical anthropology and Cultural anthropology. Next to Bose, came L.P. Vidyarthi who tried to focus the major trend in Anthropology in course of its growth in India. His paper was presented in the VIIth ICAES at Moscow in 1964. In another paper published in the same year, Vidyarthi specifically referred a few recent trends like village studies, caste studies, study of leadership and power structure, religion, kinship and social organization of tribal village, even the applied anthropology. He felt the need of an integrated effort from various disciplines for a proper understanding of man and society. His main stress was laid on 'Indianness' which appeared as a distinct cultural milieu as well as an attitude oriented value system of the Indians. According to him, ideas of Indian thinkers as reflected in ancient scriptures were full of social facts and so those could be explored in understanding of cultural process and civilization history of India. Surajit Sinha (1968) in a conference organized by Wenner Gren Foundation for anthropological research at New York supported the view of L. P. Vidyarthi. He happily announced that the Indian anthropologists readily responded to the latest developments of the west but they had given logical priority to the Indian situation. Because, they did not want to imitate the west blindly; rather they went on a search to establish Indian tradition in anthropology.

In addition to these valuable thoughts, we get some fall-length articles and introductory observations that delineate the development of anthropology in India. Some western scholars also took interest in the study of the development of Indian anthropology. Louis Dumont and D. Pocock in a series of publication have attempted to highlight the significant research areas in Indian social anthropology. Thus, the growth of anthropology in Indian context has been monitored time to time by both the Indian and foreign scholars. However, in the light of Vidyarthi and Sinha , we can divide the growth of Indian anthropology into the following historical phases.

I. The Formative Phase (1774-1919)

The establishment of Asiatic society of Bengal in 1774 is considered as beginning of scientific study of 'nature and man' in India. A number of anthropological studies were initiated by the efforts of Asiatic society under the leadership of the founder—President Sir William Zones. The Society gave birth of a Journal in which scholarly interests on the diversity of Indian customs were reflected. British administrators, missionaries, travelers and other writers used to get a scope to publish their collected information on tribal culture and rural life. Not only a single journal of Asiatic Society had made its appearance in 1784, a number of journals came one by one like Indian Antiquary in 1872, journal of Bihar and Orissa Research Society in 1915 and Man in India in 1921. Also a series of district gazetteers and handbooks appeared containing ethnographic notes on tribes and castes. Some monographs were especially designed with the tribes of Assam. The purpose of these publications was to acquaint the Government Officials and private personnel with the nature of regional tribes and castes in order to ensure effective Colonial Administration.

---

* S.C.Dube "Anthropology in India" in *Indian Anthropology: Essays in Memory* of D.N.Majumdar (Ed.) by T.N.Madan and Gopala Sarana, Bombay, Asia Publishing House, 1962.

The ground stones of anthropology were thus laid in an orderly way in the form of ethnographical mapping. Hence the phase is considered as the formative phase in the history of Indian anthropology. The administrators like Campbell, Latham and Risley published a few general books on ethnology while they were on duty. Several missionaries including Hoffman and Bodding undertook linguistic studies along with ethnographic search. The inspiration was drawn from British anthropologists who came to work in India. For example, W.H.R.Riverse put his attention on the Todas of Nilgiri Hills (present day Tamil Nadu); A.R.Radcliffe-Brown dealt with the Andaman Islanders, G.H-Seligman and B.G.Seligman concentrated on the Veddas of Ceylon, etc. Extensive monographs were prepared on Lushai kukis by J.Shakespeare. P.R.T.Gurdon wrote a monograph on Khasis, J.P.Mills on Lotha Nagas, N.E.Perry on Lakhers and T.V.Grigson on Maria Gonds of Baster. Among the Indian scholars, Hiralal and L.K.Ananthakrishna lyer though did some independent work, first credit crowned on Rai Bahadur Sarat Chandra Roy who was basically a lawyer. He used to live in Ranchi and there tribal people were his clients. Coming in contact with several tribes S.C.Roy became interested in knowing them. He was the first Indian who was able to put Indian anthropology on the world map. He wrote exhaustive monographs on different tribes of Chotonagpur like Munda, Oraon, Ho, Birhor, Hill Bhuiya and Kharia. British anthropologists of his time appreciated his work. Hutton in 1938 designated him as the Father of Indian Ethnology. S.C.Roy maintained contacts with many leading western anthropologists namely J.G.Frazer, R.R.Marett and R.B.Dixon. Next to S.C.Roy we can recall the name of R.P.Chanda who published a book on Indo-Aryan race in 1916. The book provides invaluable information on the cultural history of India. British Governor of Bihar, Sir Edward Gait also deserves mention for his outstanding works in Bihar.

II. The Constructive Phase (1920-1949)

The 'Formative phase' predominated by ethnographic studies took a new turn when social anthropology was included in the Post-graduate curriculum of the University of Calcutta in 1920. No discipline is supposed to be established unless it is recognized as a subject of study and research at the University level. Although anthropology found a berth in the University of Calcutta in 1918 as a subsidery subject, it required two more years to get proper recognition. An independent Anthropology department came out in 1920 which was a great achievement for the subject itself. Except a few Universities in the world (Cambridge, Oxford, London Universities in the U.K. and a few other Universities in the U.S.A.) nowhere anthropology got such recognition at that time. However, K.P.Chattopadhaya was the first Professor in Anthropology at Calcutta who was trained at Cambridge by W.H.R, Rivers and A.C.Haddon. R.P.Chanda became the first lecturer there. He was famous for his hypothesis, Brachycephalizatilon in Eastern India. L.K.Ananthakrishna lyer also came more or less at the same time. Sir Asutosh Mukhopadhaya, the then vice-chancellor of Calcutta University invited L.K.Ananthakrishna lyer and R.P.Chanda to take the charge of the newly opened department of anthropology. As an academic discipline anthropology got its foothold in India. lyer had in his credit the ethnography of Travancore and Cochin (Kerala). He became the first Head in the Department of Anthropology, Calcutta University. There were also other professional anthropologists who received their academic training in the discipline from abroad. Among the first generation of the anthropologists in the University of Calcutta we find N.K.Bose, D.N.Majumdar, B.S.Guha, P.C.Biswas, T.C.Das, S.S.Sarkar, Dharani Sen and Andre Beteille who made their names in the anthropological world.

Department of anthropology in the Universities of Delhi, Lucknow and Guwahati were built up in 1947, 1950 and 1952 respectively. Thereafter, a series of Universities namely Saugarh, Madras, Pune, Ranchi, Dibrugarh, Utkal, Ravi Shankar, Karnataka, North Bengal, North-East Hills, etc. started to include the wing of anthropology in their academic setup. But it is a matter of disgrace that although India advanced highly in the researches of anthropology, the academic circle did not flourish with the same pace. The number of Universities that teach this subject is relatively insignificant in comparison to the number of the total universities in India.

The year 1938 stood as a landmark because a joint session of the Indian Science Congress Association and the British Association was held where eminent anthropologists from abroad discussed their plans for future anthropological researches in India. Moreover, this was the period when some Indian anthropologists presented very remarkable outstanding work. D.N.Majumdar conducted a number of racial or ethnic surveys in Bengal, Gujrat and Uttar Pradesh. He made notable contribution both in physical and social anthropology. K.P.Chattopadhay's attention was on kinship. T.C. Das's work on Purum Kukis was internationally appreciated. Srinivas's publication on marriage and family in Mysore also achieved a great fame. N.K.Bose was interested in the temple-art and architecture. So he had conducted some excavations on prehistoric sites in Orissa. Besides, his ideas on caste dynamics as that reflected in the article 'Hindu Methods of Tribal Absorption' stood unparallel in Indian anthropological literature. Iravati Karve contributed both in socio-cultural anthropology and prehistory. Writings of these scholars brought anthropology to the attention of public, extending beyond the circle of professional anthropology.

In 1945, a full-fledged research institution, the 'Anthropological Survey of India' was established by the effort of B.S.Guha. Dr. Guha was an Indian anthropologist who received his professional training from the University of Harvard. In 1924 he joined the Zoological Survey of India as an anthropologist but felt the need of a separate department for anthropology under the Government of India. He conducted anthropometric studies in different parts of India and worked out a racial classification basing on ethnic differences. Dr. Guha's monumental work during the Census operation of 1931 strengthened his position as a Government anthropologist. In 1946 when Government of India started a separate Department of anthropology in Indian Museum, he was appointed as its director. Anthropological survey of India became the biggest anthropological research body of its kind in the world where hundreds of professional anthropologists got their employment. Beside this important achievement, the period was further marked by the entrance of Verrier Elwin, a missionary who came to India with the intention of converting the Indians, particularly tribals to Christianity. But with time Elwin was entangled with anthropological work and his motive changed. He became totally involved among the tribes of Central India. Although he was not a professional anthropologist like Majumdar, Chattopadhaya or Guha, he wrote some excellent problem-oriented ethnographic accounts of the tribal people of M.P., Orissa and Arunachal Pradesh. His books entitled, "The Baiga (1939), The Agaria (1943), Maria Murder and Suicide (1943), The Muria and their Ghotul (1947), Religion of the Indian Tribe (1955)" are regarded as classics in anthropological literature. He was also appointed as a Governor of Assam, especially on tribal affairs. C.Von. Furer-Haimendorf also provided some unique publications on the tribes of Hyderabad. His two books, "The Chenchus (1943) and The Reddis of the Bison Hills (1945)" deserve special mention as models for future work in India. In reference, to the general trend of work, we can quote the words of L.P.Vidyarthi and conclude that Indian anthropology was born and brought up under the predominant influence of British and matured during the constructive phase on the line of British anthropology. Indian anthropologists, like the anthropologists at Cambridge, Oxford and London made themselves involved in ethnological and monographic studies with a special emphasis on researches in kinship and social organization.

III. Analytical Period (1950-1990)

Contact of Indian anthropologists with American anthropologists occurred after the World War II and especially after India's independence. A shift in approach was noted with the intervention of the American scholars. The influence of British anthropology with its emphasis on preliterate isolate society was gradually replaced by the analytical study of the complex societies. Study of Indian Village became very fashionable .The American anthropologists viz. Morris Opler of Cornell University, Oscar Lewis of the University of Illinois, David Mandelbaum of the University of California and a lot of their students came in India with research team. Their objective was three fold. Firstly, they wanted to make a systematic study of Indian villages for testing some of their own hypotheses. Secondly, they tried to refine their already established methodological framework and thirdly, they urged to assist the community development programmes in Indian villages.

The modern Bengal has been the principal centre of anthropological investigation from the very inception of anthropology in India. Calcutta is said to be the cradle of Indian anthropologists as different anthropological ideas and talents have always been diffused from his spot by enhancing its glory. Later, when the fields of anthropology was extended to include the complex society, Indian anthropologists also made them involved in the studies like acculturation, culture-contacts, technological change, sociopolitical organization, economic activities etc. Gradually the study of civilization, applied and action research also gained a ground in India. In this phase of anthropological development, apart from the impact of British and American anthropologists, influence of French Structural anthropologists were also found. Approach of Levi-Strauss was followed in this field of kinship and folklore studies. Dumont and Leach influenced considerably in the study of Caste system.

Among the Indian anthropologists D.N.Majumdar, M.N.Srinivas and S.C.Dube made notable contributions to community and village studies. American anthropologists namely, R. Redfield, M. Singer, M. Marriott and Bernard S.Cohn devoted themselves in the study of the dimensions of Indian civilization. Redfield's 'Great traditions and Little traditions' as well as 'Folk-urban Continuum' were the universal propositions. K.Gaugh, E. Leach, N.K.Bose and A.Beteille were busy for unveiling the socio-economic basis of Indian society. Beteille conducted his studies on caste and stratification in Tamil Nadu. Srinivas's concept of sanskritization and westernization are the important tools to understand the caste dynamics. American anthropologist G.P.Steed and British anthropologist G.M. Carstairs had spent huge time to relate anthropology with psychology. Uma Chowdhury and P.C.Roy tried to relate personality structure with the cultural tradition. Thus horizon of Indian anthropology gradually expanded with time.

Indian scholars felt that the application of western model would be totally unwise, as those seemed to be inadequate to explain the complexities of Indian society and civilization. Moreover, the spirit of nationalism was very prominent among the anthropologists of this generation. Therefore many of them devised their own approach to study their own society. For example, N.K.Bose developed his indigenous 'Swaraj model' which was an in-depth approach on tribe and caste to know the process of modernization. A.K.Danda developed a composite culture model. He conceived civilization as an amalgamation of number of Traditions—both great and small. S.C.Dube looked at the Indian civilization with a six-fold classification of Traditions—classical, national, regional, local, western and local sub-cultural traditions of social groups. G.S.Ghurye explained Indian social system on the basis of classical writings. I. Karve tried to understand Indian civilization with four distinct variables—historical, linguistic, cultural and structural. B.K. RoyBurman introduced a concept of nation and sub-nation; he wanted to understand the social system of India in terms of socio-political process. S. Sinha projected the tribe and caste at two extreme poles of a lineal continuum to identify the social structure of Indian society. M.N.Srinivas provided a mobility model for the analysis and interpretation of social and cultural change in Indian context. L.P.Vidyarthi leaned on India's ancient scriptures and epics for social facts and he perceived Indian civilization in terms of sacred centres and their networks.

Folklore studies have been the other interesting arena in modern anthropology in India. It is believed that huge social elements remain hidden in folk songs and folk-tales, which need to be unearthed. Significant contributions to this field came from the anthropologists like Durga Bhagat, L.P.Vidyarthi, K.V.Upadhaya, P.D.Goswami, Sankar Sengupta, Dulal Chowdhury, R.M.Sarkar, Pasupati Mahato and others. Anthropological interest in the study of primitive religion did not come to an end with the concept of Bongaism as forwarded by Prof. D.N.Majumdar (1950) or Srinivas's work in South India, 'Religion and Society among the Coorgs' (1952). Rather it became clothed with the idea of civilization and reflected in the works of L.P.Vidyarthi, Baidyanath Saraswati, Makhan

Jha and others. Vidyarthi himself developed the concept of sacred complex for studying the places of pilgrimage as a dimension of Indian civilization. Studies in power structure and leadership as stimulated by the ideas of Oscar Lewis and H.S.Dhillon influenced a number of Indian scholars. As a result, studies on tribal and rural leadership was undertaken in different parts of India. In this respect we can utter the names of R. Kothari, K.S.Singh, B.K.Roy Burman and others.

Most of the anthropologists of the present day have been conscious with the applied field. They want to direct their knowledge in the welfare and development of the society. Tribal as well as rural community development programmes have been benefited a lot by the advice of eminent anthropologists like S.C. Dube, L.P. Vidyarthi, Prof. Sachchidananda, T.B. Nayek, T.N. Madan and others. In many cases anthropologists themselves have chosen the path of action anthropology being guided by the idea of American anthropologist Sol. Tax. In this context, the Lodha project of P. K. Bhowmick deserves special mention. Lodha is an ex-criminal tribe in the district of Midnapore, West Bengal. Prof. Bhowmick had combined his research with the action programmes and earnestly tried to ameliorate the conditions of Lodhas. As an action anthropologist he also devised some ways to rehabilitate a section of Lodhas, economically as well as psychologically in the midst of their advanced neighbours.

At the end of the twentieth century, after hundred years from Risley, we again found an all India study on the Tribes and Caste to assess the aspect of change. The study was conducted by Anthropological Survey of India. Apart from this, a lot of in-depth and analytical studies on various communities were encouraged which were absolutely free from the bias of western theoretical model. Here we can cite the examples of B.K.Roy Burman's study on Totos (1967), A.K.Das's study on Lepchas (1969) and so on. Thus, fresh efforts to undertake researches on the unexplored areas and communities, publication of new bulletins and journals, establishment of more and more research centres characterize the 'analytic phase' in the growth and development of Indian anthropology.

III. Evaluative Phase (1990 onwards)

Recently we have entered silently into a phase of evaluation. Since western anthropology under the influence of British and American failed to explain the complexity of Indian society, a critical appraisal and reorientation of the discipline was needed for Indian situation. Indian scholars had developed indigenous models intending to apprehend the cultural matrix of India. The alternative methodological framework did not merely helped in establishing a refined concept; it also aimed at 'Indianness' for maintaining the quality of national life. In fact, Indian anthropology demands for an active, humanistic and critical outlook towards the subject matter in order to overcome the barrier of intellectual colonialism and neo-colonialism.

It has been understood that the Indian society on account of its vastness has to be dealt from three different perspectives:

(1) Examining of isolable wholes such as the village, the tribe, etc.

(2) To note the change in those units following the intrusion of various factors from outside.

(3) Minute study within the complex societies itself.

Indian anthropology from 1990's has been much concerned with problems of own society, both empirical and normative. New types of data are encountered; the concepts, methods and theories are continually shaped and reshaped. New ways of looking at new types of data have made Indian anthropology much more distinctive ever before. Unlike western countries, in India, a close relationship between sociology and social anthropology prevails from the very beginning. The great size and density of Indian population have facilitated such closeness between the two disciplines. The present phase of anthropology in India has brought sociology much closer; both the disciplines go on investigating the tribal, agrarian and industrial socio-cultural systems. Many renowned anthropologists namely M.N.Srinivas, S.C.Dube and others were found to penetrate into the field of sociologists to combine the two disciplines successfully, for yielding better result.

LIFE AND WORK OF SOME INDIAN ANTHROPOLOGISTS

Sarat Chandra Roy (1871-1942)

Sarat Chandra Roy was one of the pioneers of anthropological studies in India, although he had no formal degree on anthropology. He was born in the year 1871. He passed Matriculation in 1888 from the City Collegiate School in Calcutta. After this he was graduated with Honours in English from the General Assembly Institute (Scottish Church College) in 1892. He also obtained M.A. degree from the University of Calcutta in 1893. His academic career did not stop at this stage. He proceeded for B.L. degree which he achieved in 1895 from the Ripon College, Calcutta. Since 1897 he started practice in the District Court of 24-Paraganas (Alipore, Calcutta) but within a year he shifted to Ranchi where he could successfully built up his professional career and earned a good reputation in the bar of the Judicial Commissioner's Court. For his outstanding literary work and public services the then Government awarded him with a silver medal, 'Kaiser-I-Hind' in 1913 and the title 'Rai Bahadur' in 1919. Besides, he also owned two other certificates of honour and two more medals.

In 1920, he was elected as an Honorary Member of the Folklore Society of London. He was the only Indian who was crowned with such an honour. In the same year he became the President of the Anthropological Section of the Indian Science Congress. He was also elected the President for the Section of Anthropology and Folklore of the All India Oriental Conference in the years 1932 and 1933. Besides, Sri Roy was selected as a member of Council d' honour of the International Congress of the Anthropological and Ethnological Sciences. National Institute of Sciences in India and Patna felt proud getting S.C.Roy as a foundation fellow in their institutes. Sri Roy secured his seat in the Legislative Council of Bihar and Orissa for successive terms. His presence as a member of Simon Commission was much appreciated.

Sarat Chandra Roy was the first man of India who delivered a course of lectures on anthropology in any Indian University and first to make an effort to publish a quarterly journal for anthropology. He brought 'Man-in-India' as an exclusive journal in anthropology, which is still serving the students and scholars of anthropology. Sri Roy devoted himself earnestly in the anthropological and related ethnological studies in India. His studies laid the foundation of anthropological knowledge of the tribal population of Bihar and adjacent areas. He started his investigation amongst the Mundas as early as the beginning of twentieth century and his findings began to be published right from the year 1907. The journal 'The Modern Review' took interest in the work of Sri Roy from its very inception in 1907. However, the first book came to the scene in 1912 with the title 'The Mundas and their Country'. This was followed by a number of voluminous monographs—"The Oraons of Chotonagpur' (1915), 'The Birhors' (1925), 'Oraon Religion and Customs' (1928), 'The Hill Bhuiyas of Orissa' (1935), and 'The Kharias' (1937). Sri Roy kept himself associated with the 'Bihar and Orissa Research Society' in Patna since its origin. His articles enriched the journal of this society and the museum of this society was filled with his huge collection—ethnographic and archaeological.

The arena of S.C. Roy's interest was very wide. He wrote on a variety of topics namely caste, Hindu religion, racial migration, cultural ethnography and so on. A number of articles on the emergence of Marathas in the Indian sub-continent (as Maratha dominated over a vast area extending over northern and southern India) were written before the British came into power. He was able to throw considerable light on the socio-historical perspective of the country.

While dealing with a particular tribe, he tried to cover different aspects of their life like arts and crafts, traditions and customs etc. which were very important for ethnology. Despite lots of information, his monographs were not confined in presenting a rounded picture of the tribes; he aimed at a concise information of the life of tribes as "universal categories of culture". These works were praised by the eminent anthropologists of that time like J.G.Frazer, W.H.R.Rivers, R.R. Marett and R.B. Dixon. S.C. Roy contributed in strengthening the integrated character of anthropology in India. He proceeded in the same line as followed by the anthropologists of pre-Malinowskian period.

Initially his writings were much influenced by historical facts but soon he started analyzing various factors, which enlived the studies. Altogether he had more than a hundred of original papers and seven books in English. In addition, he wrote seventeen articles in Bengali, sixteen of which was published in the Bengali journal 'Prabasi'. His bibliography was published twice. First one was compiled by W.G. Archer and published in *Man in India* (Vol. XXII, No. 4, 1973). The second one was an exhaustive bibliography published in the bulletin of the Anthropological Survey of India (Vol. XVII, No. 2, 1968).

Sarat Chandra Roy is considered as the Father of Indian Ethnography. He not only made valuable contribution in the study of human society and culture; he tried to popularize the subject of anthropology among the general readers. Anthropology is also indebted to his encyclopedic scholarship and analytic appraisal; otherwise anthropology could not have reached the degree of precision or objectivity to fulfil its demand as a science. Sri Roy also pleaded for the application of anthropology in finding out the causes of sufferings of the people. The great personality passed away in 1942 causing an irrevocable loss.

Kshitish Prasad Chattopadhaya (1897-1963)

Prof. Kshitish Prasad Chattopadyaya was born in Calcutta on 1897 and died on 1963 at the age of sixty-six years. Taking an early training in Physics, he went to Cambridge for advanced studies in that subject, but later his mind was changed and he took interest to research work in anthropology. Soon he stood as a prominent figure in the field of Indian Anthropology. He held the position of professor as well as Head of the Department of Anthropology in the University of Calcutta, since 1937. He was in-charge of this chair for twenty-five years at a stretch until his retirement in 1962. During his working span he trained up a large number of scholars in the field of anthropology and he himself kept valuable contribution for the future generation.

Prof. Chattopadhaya had a brilliant academic career from his childhood. In 1915 he stood first in I.Sc. Examination from the University of Calcutta. In 1917 he passed B.Sc. from the same University with first class Honours in Physics. After this he was sent to Cambridge for further study in Physics. But as his mind changed, he began to study anthropology and secured M.Sc. degree (anthropology) in 1922. University of Cambridge awarded him the Anthony Wilkin studentship in 1923. During his stay in Cambridge, he got the scope to work with renowned anthropologists like W.H.R. Rivers , A.C. Haddon, M.S. Berkit and others.

K.P. Chattopadhaya was lean towards the nationalistic activities from the early days of his life. He took part in the freedom movement of the country in association with Deshbandhu Chittaranjan Das and Subhash Chandra Bose. Returning from Cambridge in 1924 he joined the Corporation of Calcutta and became the first 'Education Officer'. There he kept himself involved in organizing the educational system of the country till 1937 and thereafter joined the Calcutta University.

Apart from the job of teaching in the University, Prof. Chattopadhaya made himself engaged with different other activities in other institutions. He went to the United Kingdom in 1934 to represent India in the First International Congress of the anthropologists. He also participated in other Sessions of the Congress—in Vienna and Paris in 1952 and 1960 respectively. He was a fellow of the National Institute of Sciences (India) and also an Honorary Fellow of the Sanskrit College, Calcutta. The Sanskrit scholars (Pandit Samaj) of Calcutta conferred him the title 'Sarvabhauma'. Retirement could not bring depression to Prof. Chattopadhaya. He continued his research under the Retired Scientists Scheme of the Council in Scientific and Industrial Research.

K.P.Chattopadhaya had done extensive fieldwork among the Korku (Melghat), the Khasi (Maphlong) and the Santal. His interests included both rural and urban surveys and biometry. He wrote a large number of papers on anthropology, education and allied subjects. Some of his important books are as follows:

1. *A Plan for Rehabilitation in Bengal* (1946)
2. *Reports on Santals in Bengal* (1947)
3. *Municipal Labour in Calcutta* (1947)
4. *Our Education, Calcutta* (1948)
5. *A Socio-economic Survey of Jute Labour* (1952)
6. *Under-graduate Students in Calcutta* (1954)
7. *Study of Changes in Traditional Culture* (1957)
8. *How Gurdians plan Education* (1962)

Biraja Sankar Guha (1894 -1961)

Dr. B.S.Guha is a pioneer physical anthropologist in India who completed his professional training at the University of Harvard in 1924. Coming to own country, at first-he made himself associated with the anthropology department of Calcutta University as a part-time lecturer. Thereafter in 1927, he joined the Zoological Survey of India as an Anthropologist in its anthropological section. Dr. Guha was an active as well as an enthusiastic anthropologist, who undertook immense trouble in anthropological work during the period from 1927 to 1945. His monumental work during 1931 Census Operation strengthened his position as a Government anthropologist. In 1938, he was elected as Vice-President in physical anthropology in the International Congress at Copenhagen. In 1946, he succeeded in implementing his long cherished dreams by establishing the Anthropological Survey of India. In fact, after the Second World War, he found a scope to increase and improve the study of anthropology in India under the programme of post-war reconstruction.

In 1945 Dr. Guha submitted a proposal to the Government for the establishment of a separate Survey of India in favour of fast anthropological research and investigation. At this time the things were moving very swiftly, so the Government immediately sanctioned the proposal. Accordingly, on Ist December 1945, Anthropological Survey of India came out of Zoological Survey of India and Dr: B.S. Guha was appointed as the Officer on Special duty in that newly created institution. On 1st August 1946, be became the Director of Anthropological Survey of India and served the institution until death. This Survey of India became solely concerned with anthropological studies and stood as a wing of Government alike the Geological, Zoological, Botanical and Archaeological Survey of India.

Today the Anthropological Survey of India is probably the biggest anthropological research body of its kind in the world that was established by the initiative of Dr.B.S.Guha. He initiated researches in physical anthropology mainly on two distinct lines. First, the osteological studies of historic and prehistoric human remains on the materials excavated by the Archaeological Survey of India. The other significant work was a continuation of the study of the anthropometry of the Indian population that was started at the request of the 1931 Census organizations. Outstanding anthropological works of Dr. B.S.Guha include Reports on Human Remains excavated at Nal (1929), Mohenjodaro (1931 and 1937) and Racial Affinities of the Peoples of India in the Census of India for 1931, Vol. I, Part III (1935), Racial elements in the Population (1944). He obtained anthropometric measurements of 2511 subjects from various parts of undivided India representing thirty-four population groups. The ethnic types of India were classified on that basis.

We should pay a rich tribute to this founder Director of Anthropological Survey of India who had realized the importance of anthropological researches for the country in those days. Not only that he was able to persuade the Government successfully to form a research institution, Dr. Guha had a far-reaching vision and so considered the subject anthropology in an integrated way. He combined the physical and social-cultural anthropology beautifully with linguistics, psychology, biochemistry (food and nutrition) and pre-historic archaeology.

Govind Sadashiv Ghurye (1893 -1983)

Although Govind Sadashiv Ghurye is regarded as father of systematic teaching of Sociology in India, Indian anthropology is also indebted to him. He took his professional training in the University of Cambridge under the supervision of W.H.R. Rivers. and A.C. Haddon. Coming from Cambridge in 1924 he became the first Reader and the first head of the Department of Sociology in Bombay. He served there for thirty-five years and disseminated his valuable thoughts in reputed journals of India and abroad. He enriched both Indian anthropology and sociology.

The first publication of Ghurye 'Caste and Race in India' appeared in 1932 under the 'History and Civilization Series' edited by C.K.Ogden. It became a basic book for the students of Indian sociology and anthropology. In this book he delineated a comprehensive picture of the institution of caste discussing its origin, features, functions, development, etc. The book was revised in 1950 with the change of title. The revised book in the title "Caste and Class in India" portrayed the subtle development in the system, which came into being with numerous changes in the political, economic and social milieu of the country. Prof. Ghurye also examined the position of primitive tribes in India with the question of administration. He whole heartedly supported an assimilation of the backward tracts to the rest of India, politically. Therefore he tried to draw the attention of Indian intellengentia towards the problems that were created by the British Government in handling the so-called aborigines of India.

The book "The Aborigines—'so-called' and their Future" (1943) was an interesting as well as a thought provocating book where Prof. Ghurye succeeded in documenting the problems of aboriginals from the plane of anthropology to that of Sociology, even of politics. He recommended that scheduled tribes should neither be called as 'Aborigines' or the 'Adivasis'. Further, they by themselves should not be treated as a category By and large, they need to be placed together with scheduled castes and ought to be treated as one group of backward classes.

G.S. Ghurye was born in 1893 and passed away in 1983. He could be able to inspire his disciples in his own ideology of integrating the two related disciplines, sociology and social anthropology. M.N.Srinivas performed this responsibility quite effectively. Generations of Indian anthropologists and sociologists are obliged to the enormous contribution of G.S.Ghurye.

Irawati Karve (1905 -1970)

Irawati Karve is a woman anthropologist of India who was born in Burma and educated in Pune. She did B.A. in Philosophy and M.A. in Sociology (1928) from Bombay University before proceeding to Germany for advanced studies. For an outstanding research in anthropology, the Berlin University conferred on her the D.Phil degree in 1930. This marked the onset of her long and distinguished career of anthropological research. Her professional training was accomplished under the supervision of Eugene Fischer in the University of Berlin. She acquired knowledge of both social and physical anthropology.

Coming back to motherland in 1939 Dr. Karve joined the Deccan College Post-graduate and Research Institute of Pune as Head of the Department of Sociology and Anthropology. She presided over the Anthropology Section of the Indian Science Congress in 1939. Her research interests were concentrated on the following aspects :

1. Racial composition of the Indian population
2. Kinship organization in India
3. Origin of Caste
4. Sociological study of the rural and urban communities

She wrote several research papers in different technical journals. Some of these were written in English and others were in Marathi, her mother tounge. The academic proficiency as well as topics of general interest brought her in the limelight of fame and public appreciation. She acquired a wide circle of readership. Irawati Karve also conducted anthropometric studies in Maharashtra (by the financial aid from Emslie Horniman fund) the results of which was published in book-form in 1953. It provided a very useful data to mark a stage in the progress of knowledge about the people in Maharashtra.

The most important books written by Irawati Karve are, 'Kinship Organization in India' (1953) and 'Hindu Society: An Interpretation' (1961). In Kinship Organization in India, she surveyed the kinship organization of India by zones—North, Central, South and Eastern. To interpret the inner integration of Hindu society she illustrated the ancient Hindu mythologies and tried to relate them with modern customs. The latter stands a scholarly treatise in English.The same enterprise was again found in the work 'Yuganta' (1967) which was written in Marathi. It won Sahitya Academy Prize as the best book of that year. In 'Yuganta', Prof. Karve studied the characters and society in Mahabharata. The subject of the book is secular, scientific and anthropological in widest sense.

Dhirendra Nath Majumdar (1903 -1960)

Dhirendra Nath Majuradar was born in Patna in 1903. He received his early education in West Bengal and became graduated from the Arts Faculty of the University of Calcutta in 1922. He also did his M.A. (1924) in Anthropology from the same University where he was placed in first division and secured the first rank among the successful candidates. He conducted first fieldwork in Bihar among the Hos of Kolhan. At this time he got the company of the veteran Indian Ethnographer, Rai Bahadur Sarat Chandra Roy. University of Calcutta awarded him with Premchand Roychand Scholarship in 1926 for the merit of his original research.

D.N.Majumdar was one of the earliest students of anthropology who received their Master's degree from the University of Calcutta, after the establishment of this discipline in 1920. In 1928, Prof. Radhakamal Mukherjee selected him as a lecturer in Primitive Economics in the Department of Economics and Sociology at the University of Lucknow. Among the Indian scholars of that time Sri Majumdar had an in-depth knowledge and abiding interest in Anthropology. His contribution to anthropology was many and various. University of Calcutta again awarded him Mouat Gold Medal in 1929. In 1933 he went to Cambridge. There he studied social anthropology with Prof. T.C. Hodson and physical anthropology with Prof. G.M. Morant. In 1935, he received Ph.D. degree from Cambridge. The subject of his dissertation was cultural change among the Hos. In 1936, he was elected as a fellow of the Royal Anthropological Institute of Great Britain and Ireland. In 1939, he presided over the Anthropology and Archaeology Section of the 26th. Session of the Indian Science Congress held at Lahore. He was a good conversationalist and optimist with firm faith in his vocation.

In 1945, Prof. Majumdar founded the Ethnographic and Folk-culture Society of U.P. The principal aim of this Society was to collect ethnographic data on folk cultures of Uttar Pradesh. Two years later, the Society brought out the reputed journal, 'The Eastern Anthropologist' and D.N.Majumdar became the Editor. From the beginning of his anthropological career, he combined anthropometric surveys with ethnographic investigations. He tried to coordinate between the cultural anthropology and physical anthropology. Although he was not engaged in any investigations in prehistoric site, he kept himself concerned with the major developments in archaeological anthropology and occasionally he lectured on this subject.

In the field of anthropology, Majumdar's vast contribution came mainly in the form of extensive anthropometric and serological surveys, which were carried out among the tribes and castes of Bihar, M.P., U.P., Gujarat and Bengal. He emphasized statistical techniques in the analysis of anthropometric and serological data. This particular method was designated as biometrics method in anthropology. In U.P., he tried to delineate caste hierarchy on the basis of biometry. Although he considered racial factor in the formation of caste structure, but strongly opposed racism and other single factor explanations for the functioning of the caste system in India. Majumdar's study on school-children in the city of Lucknow is a landmark in Indian 'physical anthropology'.

In the field of cultural anthropology, D.N.Majumdar is famous for his critical ethnography. He undertook extensive field works among the Hos of Kolhan in Bihar, The Khasas, The Korwa and

The Tharus of Uttar Pradesh, The Gonds of Baster in M.P. and The Bhils of Gujarat. He was a painstaking and patient field worker being influenced by S.C. Roy and B.Malinowski. He conducted each fieldwork (among a particular tribal area) over several years and observed the people in repeated short and long trips. He was attentive in learning of tribal and folk dialects. In an ethnography he tried to cover all the major aspects of a culture such as economics, kinship, religion, etc. He studied the cultures as a whole, rather than the piecemeal problems. In later years he became interested in investigating the cultural change, but did not neglect the village studies or urban surveys. In physical anthropology, he paid special attention in serology, demography and studies regarding growth of human bodies. Prof. Majumdar was firm believer in applied anthropology. He thought the findings useless unless those had practiced values. The great scholar passed away in 1960.

From 1923 until death, he wrote valuable papers and books on varied aspects in anthropology. These were published mainly in the reputed journals like Man in India, from Ranchi; The Eastern Anthropologist from Lucknow; the Journal of the Gujarat Research Society from Bombay; the journal of Royal Asiatic Society of Bengal from Calcutta, etc. Besides, there were Proceedings of Indian Science Congress, Calcutta; 'An Appraisal of Anthropology today' edited by Sol.Tax et. al published from University of Chicago Press; Hindi Vishva-Bharati edited by Shrinarayan Chaturvedi, published by Educational Publishing Company Ltd. A few of his important books are: Race Elements in Bengal (co-author C.R. Rao), Social Contours of an Industrial City (assisted by N.S.Reddy and S.Bahadur), A village on the Fringe, Race and cultures of India, An Introduction to Social Anthropology (co-author T.N.Madan), Fortunes of primitive Tribes, A tribe in Transition.

Nirmal Kumar Bose (1901-1972)

Professor Nirmal Kumar Bose was one of the leading Indian anthropologists who was born in 1901. Although Calcutta was his birthplace, the school education was taken in Patna, where his father was posted as a Civil Surgeon. He was not at all acquainted with anthropology at the beginning. He passed B.Sc. from the Presidency College of Calcutta with first class honours. In 1921 he was admitted in M.Sc. course under the University of Calcutta; he had a bent in mind for the researches in geology. As a student he was a follower of Gandhiji in all ideas and activities. Therefore, once he took a firm decision to leave the Government College in response to Gandhi's Non-cooperation Movement. He did not care about his academic career and left Calcutta.

After this, he made himself settled in Puri. The elegance of ancient Orissan temple attracted him very much. He met a temple architect named Ram Marasana from whom he knew about the traditional canons of Orissan architecture. He became guide lecturer and started giving lectures to the visitors. At this time, as a matter of chance he met Sri Asutosh Mukherjee,the then Vice-Chancellor of Calcutta University. He persuaded Bose to join the M.Sc. course of anthropology in the University of Calcutta. In 1925 , Nirmal Kumar Bose completed post-graduation in anthropology with brilliant result. He further proceeded to research in the same University. Rai Bahadur Sarat Chandra Roy was a source of inspiration and encouragement for young Bose. The works of Malinowski and some other American diffusionists also attracted him. Moreover, he was conditioned by the ideas of Freud, Marx and Gandhi. In anthropology, he did not want to place himself in any conventional school; rather he declared his identity as 'Social historian'.

N.K.Bose started his work among the Juangs of Orissa in 1927; his painstaking investigations revealed the life and activities of the poor tribals. He restrained himself from the preparation of the stereotyped voluminous monographs and aimed at observing the life of the tribal people in the background of the Indian society and culture as a whole. He acknowledged a number of scholars and writers from whom he got his intellectual stimulation. His basic insight of social science was grown at the time when he proceeded to solve the socio-political problem of the country under the idealism of Gandhi. He wrote two outstanding books on Gandhi philosophy—"Selections from Gandhi" (1934) and "Studies in Gandhism" (1946). The third book was based on his personal experience as gathered during 1946-47 when he got an opportunity to be the private Secretary of Mahatma Gandhi. It is entitled as "My Days with Gandhi" (1953).

Nirmal Kumar Bose was appointed as assistant lecturer in the department of Pre-historic Archaeology of the Calcutta University in 1937. Later, he was given the positions of lecturer and Reader in Human Geography. He served the University of Calcutta from 1937 to 1959. During 1957-58, he visited U.S.A. as a visiting scholar. His intellectual interests and capabilities bestowed him a number of high positions. He was the Director under Government of India for five years, he became the Commissioner for the Scheduled Castes and Scheduled Tribes during the period from 1967 to 1970. The Indian Science Congress Association elected him as President in 1949 for their section of Anthropology and Archaeology. The Asiatic Society offered him two medals—Annandale and S.C.Roy Medal for his outstanding researches in anthropology. At the time of death, he was holding the positions viz. President in Asiatic Society, President of Bangiya Sahitya Parishad and Editor of 'Man in India'.

Prof. Bose's approach to the study and analysis of tribal problems was unique in anthropology. He made himself emerged in tribal welfare during the period from 1953-1969.He had written 40 books and over 700 articles in English as well as Bengali. The subjects, which have been dealt by him, are temple architecture, art, prehistoric archaeology, human geography, urban sociology, geology and travelogues, apart from cultural anthropology and Gandhism. The books which deserve special mention are. Cultural Anthropology (1929), Hindu Samajer Gadan (1949), Peasant life in India (1961), Culture and society in India (1967), Problems of National Integration (1967), Tribal life in India (1971) and Anthropology and some Indian Problems (1972). The notable collection of Essays and travelogues (in Bengali) are, 'Nabin-O-Prachin'(New & Old) and 'Paribrajaker diary' (Diary of a wanderer). He wrote very lucidly without any jargons. He was a good speaker too. The country had lost this great academician in 1972. His creativity, humanity and moral endeavour was beyond criticism.

Mysore Narasimhachar Srinivas (1916-1999)

Professor Mysore Narasimhachar Srinivas was one of the founders of modem sociology and social anthropology in India. He was born in 1916 in the city of Mysore and passed away in 1999 in Bangalore.

Prof. Srinivas studied anthropology and sociology in B.A. course at Mysore. He moved to Bombay for his M.A. and found G.S.Ghurye as a teacher who integrated sociology and social anthropology under the rubric of sociology. Srinivas also secured a L.L.B. degree at this time. But, after this he went to Oxford for Ph .D. degree. In Oxford he got an opportunity to work closely with eminent social anthropologists namely A.R.Radcliffe-Brown and E.E.Evanspritchard. He was appointed there as a lecturer in Indian sociology. In 1951 Srinivas felt homesick and returned from Oxford to join Maharaja Sayajirao University of Baroda. He became the first Professor and Head of the Department of its newly opened department of sociology. Prof. Srinivas devised new syllabi not only for B.A. and M.A. course but he also chalked a new guideline for Ph.D. work. He made two faculty appointments—I.P. Desai was chosen from sociology stream and Y.V.S.Nath from the anthropology stream. In 1959 he moved to Delhi and founded the department of sociology in Delhi University. In this respect he got the cooperation of V.K.R.V.Rao, the then vice-chancellor of Delhi University. There also Srinivas gave four faculty appointments including equal number of persons from each discipline of sociology and anthropology. Andre Beteille and Gouranga Chattopadhya were taken from anthropology while M.S.A. Rao and Savitri Sahani came from sociology. M.N. Srinivas wanted an integration of social anthropology and sociology rather than their mere juxtaposition. He patterned his staff-structure accordingly.

M.N. Srinivas spent twelve years in Delhi University and during this period the academic standard of this University rose so high that it attracted a large number of students from all over India, even from other countries. The focal attention of Srinivas was on Ph.D. students; he established a strong fieldwork tradition through them. The University Grants Commission in 1968 bestowed the crowning glory of this department through its recognition as a centre of Advanced Study. This yielded generous grants for scholarships, fellowships, faculty positions, secretarial staff, and visiting professorship as well as a separate building for this department. When V.K.R.V.Rao established the Institute for social and economic change in Bangalore (1972), Srinivas was invited to take the post of its Joint Director. A brilliant role was performed by Srinivas to make this Institute an outstanding centre for sociological and social anthropological learning in South India. He was also selected as a visiting professor in National Institute of Advanced Studies, which was set up within the campus of the famous Indian Institute of science at Bangalore. This position was held until death. In all places and positions he carried forward the tradition of integrating sociology and social anthropology to fulfil the idealism of his teacher G.S.Ghurye.

Prof. M.N. Srinivas was greatly influenced by the views of eminent sociologists and social anthropologists namely Durkheim, Max Weber, Radcliffe-Brown, Evanspritchard, Raymond firth, Talcott Persons, Merton, Bottomore and others. At the time of course formulation for B.A. and M.A. in Sociology at Baroda as well as in Delhi, he included some important basic books of anthropology. Further, in every course he prescribed texts dealing with wide varieties of society and culture, both traditional and modern. Srinivas believed in close contact between student and teacher. He became a pivotal figure in the emergence of Indian anthropology.

The researches and enquiries of Prof. Srinivas were concentrated mainly on the traditional subjects viz. religion, caste and village. He also took interest in the new subjects like industry, urban community, hospital, etc. He preferred participation observation as important method of fieldwork. His idea about fieldwork has been reflected in the book entitled "The field-worker and the field' (1979) which he wrote with A.M. Shah and E.A. Ramaswamy. Prof. Srinivas not only established and strengthened several academic institutions; he was an active member of several institutions like the University Grants Commission, Indian Council of Social Science Research, Economic and Political weekly, as well as many Committees and Commissions of Government. Although he was very selective in accepting a membership but once it was accepted, he used to work very hard with heart. He blended seriousness with witticism.

Prof. Srinivas forwarded the concepts of Sanskritization (1962) and Westernization (1966) to account for the social change in India. Although these concepts do not affect the social structure but the concepts are very important to analyze the superficial change processes particularly in the later half of nineteenth century and first half of twentieth century. Srinivas owned many honours through prizes, medals, awards, fellowships etc. for his outstanding work at regional, national and international levels. Some of the important books are as follows: "Religion and Society among the Coorgs"(1952), "India's Village"(1955), "Caste in Modern India"(1962), "Social change in Modern India"(1966), "The Remembered Village"(1976) is a reflexive and affectionate return to his original fieldwork. In the later years the work and thought of his writing was changed. He wrote the "Itineraries of an Indian Social Anthropologist" (1973), "My Baroda Days"(1981), "Sociology in Delhi"(1995) and "Reminiscences of a Bangalorian"(1995). Three of these later writings have been reprinted in his book 'Indian Society through Personal Writings' (1996). In the last few years he had been writing his autobiography. Indian Sociology and Social Anthropology, both are indebted to M.N. Srinivas for his critical and diligent work.

Lalita Prasad Vidyarthi (1931 -1985)

Lalita Prasad Vidyarthi was a dynamic anthropologist of India who brought honour and glory to India and Indian anthropology. He was born in 1931 in a multi-caste village near Patna. For holding a middle caste status in the village, from the very childhood he noticed discomfort and restrictions into caste behaviour. Initial schooling of Prof. Vidyarthi was started in that surrounding and he went to Gaya in 1940 when his father joined as Mokhtar (Assistant lawyer) in the Mukhsudpur estate. In

Gaya, Vidyarthi was admitted in a middle school and passed Matric in 1946. He got First Division in Matric. In 1948 he passed I.A. from Patna College but missed first division for a few marks. In 1950 he sat for B.A(Hons) examination from Patna College and again secured first class. In 1951 he acquired the first position while he passed M.A part I in Geography from Patna University. After this, his interest shifted from Geography to anthropology. He did M.A. in Anthropology from the University of Lucknow in 1953. In the same year he joined Bihar University (Now Ranchi University) as the founder lecturer. From 1953 to 1956 he held the particular post.

Thereafter Vidyarthi left the service and accepted the position of Fellow in the University of Chicago for the period from 1957-1958. He received his Doctorate degree from this very University in 1958. Then he came back to his homeland and joined the old workplace. L.P. Vidyarthi was the Reader and the Head of the Department of Bihar University (Now Ranchi University) from 1958 to 1968. In 1968 he was promoted to the post of Professor and continued as Head of the Department as earlier. He was not only a successful teacher, a devoted scientist too. His concept of Third World Anthropology was different from the existing legacy of western and colonial paradigms. He can be said the founding father of Action and Applied anthropology in South Asia. In the middle of 1950s he established an Action Research Unit in the Department of Anthropology at Ranchi University and conducted research in Chotonagpur Region. As a student of D.N. Majumdar, Vidyarthi followed Majumdar's ideology and experimented that in empirical situations.

Apart from the influence of D.N.Majumdar, Prof. Vidyarthi received the intellectual stimulation from the American scholars specially Sol. Tax, Redfield, Milton Singer and Fred Eggan. Because he was close to them while passing his days in United States. Dr. Vidyarthi had a larger vision and a noble mission. He was a man full of gratitude and thoughtfulness. He received a lot of fellowships and awards during the period between 1950 and 1983 from India and abroad. His professional assignments bounded him with a number of reputed bodies of national and international importance. He was elected the President of International Union of anthropological and Ethnological Sciences (IUAES), continually from 1973 to 1978. In 1978 he also became the President of International Congress of Anthropological and Ethnological Sciences (ICAES). Besides, he acted as Vice-President, Chairman, Local Secretary, Member, Director and Advisor of different committees, organizations, seminars and symposia. From 1979 to 1982, he was the Vice-President of the International Social Science Council, UNESCO. In 1983, he became the Dean of the faculty of Social Science.

L.P.Vidyarthi kept notable contribution in different aspects. He was concerned with village studies, studies of sacred complex and pilgrimage sites, applied and action anthropology, scheduled castes, folklore research, urban-industrial anthropology, leadership studies, field work tradition, anthropological theories. He tried hard for socio-economic welfare and advancement of the tribes. The dimensions of work was so extensive and the number of books and papers were so many that after his death a commemorative volume was published in 1986 being edited by M.K.Gautam and A.K.Singh where different learned scholars got an opportunity to discuss the life and work of Prof.Vidyarthi.He wrote twenty one books by himself, in nine books he was the co-author. Thirty-one books were edited and more than 150 substantial articles were wrote by him during the period from 1954 to 1982. However, some of his important works are: Sacred Complex of Hindu Gaya (1961), The Maler: Nature-Man-Spirit Complex in Hill Tribes of Bihar (1963), Cultural Contours of Tribal Bihar (1964), Applied Anthropology in India (1968), Conflict and Tension of Social Trend in India (1968), Tribal Culture of India (1977), Rise of Anthropology in India (1978, two volumes). Trends in World Anthropology (1979), etc.

Prof. Vidyarthi was conferred the distinguishable service award jointly by the association for Anthropological Diplomacy, Politics and Society, and Association of Third World Anthropology in the year 1983. He was the first person from a developing country to receive such an honour. His multidimensional academic works greet him as one of the top anthropologists of the world. We have lost this great academician forever in 1985.

I. PHYSICAL ANTHROPOLOGY

Chapter

HUMAN ANATOMY

PARTS OF A HUMAN SKELETON

The study of human skeleton is essential for understanding the phylogenetic position of man. It is also important for knowing the functions of the body. The bones and cartilages constitute a framework of the body among the vertebrate animals, which we call skeleton. The skeleton performs some very important purpose such as :

1. It gives the body a definite form and shape.
2. It provides support to the body by bearing the weight.
3. It gives protection to the delicate soft organs of the body.
4. It provides space for muscular attachment, which facilitates movement of different parts of the body.
5. The bone cavities, especially the cavities of long bone are filled with marrows. Blood cells are made in the bone marrow.

The skeleton is divided into two parts—*exoskeleton* and *endoskeleton.* The exoskeleton is the external or superficial part of the skeleton whereas the endoskeleton refers to the internal part. The exoskeleton is very insignificant in case of man as it is represented only through nails, enamels of teeth etc. Therefore, the main concern of human skeleton is the endoskeleton. This endoskeleton can be divided into two parts—the axial part and the appendicular part. The axial part of the skeleton is composed of the bones of the head and trunk while the appendicular part includes the limbs.

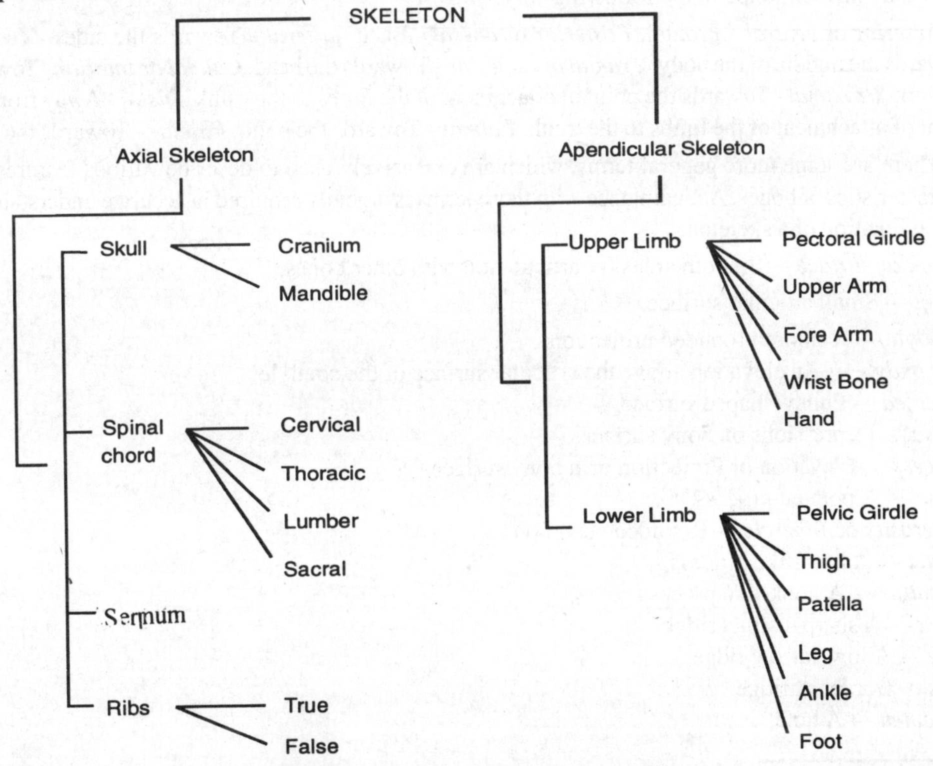

Fig. 3.1. Anatomical Configuration of a Man

After the death of a person, the muscles, ligaments, cartilages etc. gradually detach from the bones. Even the bone marrows are drained and ultimately the dried bones are left. However, a human skeleton is made of 206 bones, which are of different shapes and sizes. Each has different functions too. These bones may be classified into five classes as per their varying shapes and appearances—long bone, short bone, flat bone, pneumatic bone and irregular bone.

Long bones : These bones are made up of a tubular shaft where the ends are known as *epiphysis.* Such bones are located in the limbs such as humerus, ulna, radius, femur, tibia and fibula.

*Short bones :*These bones are strong, compact and capable of limited movement. Some sort of spongy substances make up these bones. For example, carpus and tarsus bones of hands and feet.

Flat bones : These bones consist of two thin layers of compact bones, separated by some spongy substances. Bones of skull and scapula belong to this class.

Pneumatic bones : This kind of bones is comparatively rare in occurrence. When the spongy substances between two layers of flat bones of the skull undergoes absorption, pneumatic bone is formed. The sinuses can be shown as for example.

Irregular bones : These are the bones with peculiar forms. Each of them possesses a unique shape, different from one another. So none of the above mentioned classes include these bones. However, the irregular bones are found to be made of spongy substance and remain enclosed within a layer of compact bone. Bones of the spine are the irregular bones.

Man is an *orthograde* animal according to its anatomical position. Because, the skeleton shows an erect posture with two feet running parallel to each other. The arms hang by the sides with palms facing forward. In this position, some specific terms are especially used to denote the various parts of the body, in contrast to *pronograde* * Position. However, these various parts or directions of the body may be mentioned in the following way:

Anterior or *ventral* - Frontal. *Posterior* or *dorsal* - Back. *Lateral* - Towards the sides. *Medial* - Towards the middle of the body. *Cranial or Superior* - Towards the head. *Caudal or Inferior* - Towards the foot. *Proximal* - Towards the point of attachment of the limbs to the trunk. *Distal* - Away from the point of attachment of the limbs to the trunk. *Palmar* - Towards the palm. *Plantar* - Towards the sole.

There are some more general terms, which are extensively used to describe various features and characteristics of bones. Acquaintance with those terms is urgently required in accurate understanding of the function of a skeleton.

Articular surface — Smooth areas for articulation with other bones.

Facet — Small articular surface.

Condyle — A smooth rounded projection.

Epicondyle — An elevation above the articular surface of the condyle.

Trochlea — Pulley-shaped surface.

Fossa — Depressions on bony surfaces.

Process — Elevation or Projection on a bony surface.

Spine — A pointed process.

Tuberosity or *Tubercle* — Rounded elevation.

Cornu — A horn-like process.

Hamulus — A hook-like process.

Crest — A sharp distinct ridge.

Line — A low narrow ridge.

Lips — Border of ridge.

Foramen — A hole.

---

* The animals that stand or move on four limbs are called pronograde animals. Their anatomical position is referred to as pronograde position.

Canal — Opening of a bony tunnel.

Sulcus — Groove or furrow.

Incisura — A notch.

Hiatus — A gap.

Shaft — A tubular part of a long bone. Epiphysis - The expanded ends of the long bones.

The *axial* and *appendicular, both* the parts of a skeleton show different sections along with different types of bones which may be mentioned in the following way :

I. Axial skeleton

(*i*) *Skull*

(a) Cranium

(b) Mandible

(*ii*) *Spinal Chord or Vertebral Column,* composed of 33 vertebrae.

(a) Cervical includes 7 vertebrae where the first one (Cervical-I) is known as Atlas and second one (Cervical-II) as Axis.

(b) Thoraric section includes 12 vertebrae.

(c) Lumbar section shows 5 vertebrae.

(d) Sacral portion is composed of 5 vertebrae.

(e) Coccygeal or Caudal section is made with 4 vertebrae.

(*iii*) *Sternum*

(*iv*) *Ribs* form with 12 bones.

(a) True ribs are 7 in number. They are connected in front, with Sternum.

(b) False ribs are 5 in number where rib no. 8, 9, and 10 join with the cartilage of True rib no. 7. The rib no. 11 and 12 are free.

II. Appendicular skeleton

(*i*) *Upper limbs* are composed of 5 sections.

(a) Shoulder Girdle (2) consists of Clavicle (2) and Scapula (2).

(b) Humerus is (two) in number.

(c) Radius is also (two) in number.

(d) Ulna is again (two) in number.

(e) Hand (2) is composed of Carpus (8), Metacarpus (5) and Phalanges (14).

(*ii*) *Lower limbs* are composed of 6 sections.

(a) Pelvic Girdle (1) consists of two (2) Innominate bones.

(b) Femur is (two) in number.

(c) Patella is (two) in number.

(d) Tibia is (two) in number.

(e) Fibula is (two) in number.

(f) Foot (2) is composed with Tarsus (7), Metatarsus (5) and Phalanges (14)

I. Axial skeleton

The skull

The skull rests on the uppermost end of the vertebral column. It is divided into two important parts as the *cranium* and the *mandible.* Other features of the skull are as follows:

(a) Cranium is a case for the accommodation of the brain. The term cranium means a skull without mandible.

(b)The cavities for the special sense organs are situated within the cranium.

(c) The frontal portion of the cranium consists of a number of small bones, which not only develop the upper part of the face; they make the upper jaw also. But the lower jaw, mandible has to be taken into account to form a complete face.

(d)Different openings are found on the facial part as the passages of air and food.

(e) Both the jaws provide sockets for the teeth. Upper jaw is fixed as a part of face while the lower jaw (mandible), being a movable part, project an important feature.

The skull is made up of 21+1, i.e. twenty-two bones, where twenty-one bones remain united through the sutures forming the cranium part of the skull. The other *bone* mandible, though form a part of the skull, moves freely. This mandible is connected with the cranium by a *synovial articulation*—the temporo-mandibular joint. Out of 21 bones, eight bones of the cranium make a dome-shaped bony case (calvaria) to contain the brain. The other 13 bones unite through sutures to form the frontal as well as the lower portion of the cranium. Rest part is the mandible, which completes the facial part of the skull. However, the calvaria, the top region of cranium is composed of eight bones in the following way:

Frontal (1) — It forms the front region of the cranium, which means the forehead region.

Parietal (2) — Two separate quadrilateral bones make the sidewalls as well as the roof of the cranium. Each one is irregularly quadrilateral in shape.

Occipital (1) — This single bone makes the posterior and inferior part of the cranium. It means the back and a part of the base of the skull has been formed with occipital bone.

Ethmoid (1) — It is a single bone, irregular and cuboidal in shape. It has been placed at the anterior part of the base of the cranium. This bone makes the lateral walls and roof of the nasal cavity, the medial walls of the orbit and the nasal septum.

Sphenoid (I) — *This* bone looks like a flying-bat and placed in front of the temporal bones and the basal part of the occipital bone, at the base level of the cranium. In fact, this bone represents the skull base, which consists of a centrally placed body with two wings—greater and lesser. For this reason, this bone is irregular and bat-shaped.

Temporal (2) — These bones are placed at the two sides covering the base of the cranium. More specifically, a temporal bone has five parts, such as, the squamous part, petrous region, mastoid process, tympanic parts and the styloid process.

The front as well as lower part of the cranium is the facial part, which is composed of 13 bones in the following way:

Inferior Nasal Conchae or *Turbinated* (2) — These are two bones of nose. Each one forms a curved lamina (Scroll-shaped) to be placed horizontally in the lateral walls of the nasal cavity.

Nasal (2)—These two are small flat bones which are placed side by side, immediate below the frontal bone to form the nasal bridge. These bones bridge across the gap between the two maxillae (superior maxillary).

Lacrimal (2) — These two small bones are situated on the nasal side of the orbit. Therefore, these bones make up the front parts of the medial orbital walls.

Vomer (1) — It is a thin bony plate, which is quadrilateral in shape. It is composed of two fused laminae, situated in the lower part of the nasal septum.

Superior maxillary or Maxilla (2) — These two bones form the whole upper jaw. Each of them is situated below the orbit to form the floor of the orbit. It also forms the greater part of the palate,that means the roof of the mouth as well as the floor and the lateral wall of the nasal cavity. Each one (maxilla) has a body and four processes, namely, frontal, alveolar, zygomatic and palatine.

Zygomatic or Malar (2)—These are two in number. Each malar bone being placed in the upper and lateral parts of the face makes the cheek prominent.

Palatine or *Palatal* (2) — These two bones make up the back part of the palate being situated in-between the maxilla and the pterygoid process of the sphenoid bone, at the posterior part of the nasal cavity. Each palatine bone is composed of a perpendicular plate, a horizontal plate and three processes, namely, the tubercle, orbital, and sphenoidal.

The lowest part of the face, the *mandible is* the most important segment of the skull. It is the strongest bone of the skull.

Inferior maxillary or Mandible (1)—This single bone is the largest, strongest and only movable bone of the skull. It consists of two parts—body and ramus (2). The body or the horizontal part is a curved one. Each ramus-ends or the broad flat bone turns upwards and slightly backwards from the posterior end of the body. The ramus-end has two processes—the coronoid and condyloid. The notch, which separates these two processes, is called the mandibular notch. The joining area of the body with the ramus is known as the angle of the jaw. Besides, the area where the mandible is connected with the cranium by a synovial articulation is called temporo-mandibular joint.

The skull as a whole may be viewed from various angles, which can be termed in the following ways :

(i) Norma verticalis	—	(view from the above)
(ii) Norma frontalis	—	(view from the front)
(iii) Norma occipitalis	—	(view from the back)
(iv) Norma basalis	—	(view from below)
(v) Norma lateralis	—	(view from the side)

These different views may be included in the systematic description of various bones.

Norma Verticalis (view from the above) : The contour line is more or less oval or sometimes nearly circular when it is viewed from the above. Four bones are to be seen from the above, namely, (1) the *frontal bone* with two frontal eminences; (2) the *two parietal bones* with parietal tuberocities, one on each bone; (3) the *occipital bone.*

The bones are united through the sutures, namely, (1) the *coronal suture* is found in between the frontal and the two parietals; (2) the *sagittal suture* lie in between two parietals; (3) the suture between the parietals and the occipital bone is known as the *lambdoid suture.*

The important landmarks are as follows:

(i) The *lambda,* the point of intersection between lambdoid and sagittal sutures.

(ii) The *vertex,* the highest point of the skull. It is situated on the sagittal suture, at the middle region i.e. a few centimeters behind the *bregma.*

(iii) The *bregma,* the meeting point of the coronal and the sagittal sutures.

(iv) The most convex portion of the parietal bone is known as *parietal eminence.* The parietal foramen or minute vascular opening is situated a few centimeters in front of the lambda.

(v) The *obelion* is the point between the two parietal foramina.

Norma Frontalis (view from the front) : It is more or less an oval outline and exhibits the forehead, the orbits, the bony nose, and the prominence of the cheek and the jaws—upper and lower as the mandible.

The Forehead—The forehead is formed by the frontal bone, which is articulated below with the nasal bones on each side of the median plane. The point of intersection of the nasal bones with the frontal bone is called the *nasion* point. The *glabella* point is located above the nasion, in-between the upper line of two eyebrows ridges. The eyebrow ridges or the elevation that extends laterally on each side from the glabella area is known as the *superciliary arch.*

The Orbits—Each orbit has a medial and a lateral wall, a roof, a floor, and a base. The cavity of the orbit is pyramidal in shape. The supra-orbital or superior margin is formed by the frontal bone and a foramen is seen here being known as *supra-orbital foramen.* This superior frontal bone turns backward to create an orbital plate, which gives rise to the greater part of the roof of the corresponding orbit. The lateral margin is made with the frontal and zygomatic bones. The medial margin or wall is developed by the maxilla, lacrimal and frontal bones. The base or inferior margin is formed by the maxilla and the zygomatic bones. A foramen is seen below the inferior margin, known as *infra-orbital foramen.*

The prominence of the Cheek : The cheek becomes prominent due to the presence of the zygomatic bone, which is situated at the lower and lateral side of the orbit. It presents two surfaces—the lateral and the temporal. The lateral surface contributes the lateral wall of the orbit while the temporal surface is located in the temporal fossa. In the level of articulation, a frontal process articulates with the zygomatic process of the frontal bone, and a temporal process articulates with the zygomatic process of the temporal bone.

The Bony Nose : The nasal bones and the superior maxillary form the bony external nose. These bones are found to create a pyriform aperture while the external nose is provided with a cartilagenous framework, by means of fibrous tissue. The nasal aperture is bounded by the maxilla and the nasal bones. The nasal cavity is divided into right and left by the nasal septum. The lateral walls of both the nasal cavities show three or four curled bony plates, known as *conchae.* Again, the median plane of the lower margin of nasal aperture (pyriform) exhibits a nasal spine, which is a sharp bony structure developed by the junction of two maxilla.

The Jaws—Upper and Lower

The upper jaw is known as *maxilla* or *superior maxillary.* These are two bones. Each one shows the following parts.

(i) A body that contains the maxillary sinus;
(ii) A zygomatic process, which extends laterally to articulate with the zygomatic bone;
(iii) A frontal process, which is projected upward to articulate with the frontal bone;
(iv) A palatine process that extends laterally to meet its counterpart from the opposite side; and
(v) An alveolar process, which bears an upper row of teeth.

The lower jaw is known as *mandible* or *inferior maxillary.* The lower row of teeth is carried by the alveolar part of the mandible. A foramen is seen on the mandible near the second premolar tooth, which is known as *mental foramen.*

The following sutures are to be seen wnen the skull is viewed from the front, (a) Fronto-nasal; (b) Fronto-maxillary; (c) Fronto-zugal; (d) Intra-nasal; (e) Naso-maxillary; (f) Inter-maxillary; (g) zygomatico-maxillary, etc.

Norma Occipitalis (view from the back)

The outline of the backside of the skull is roughly pentagonal, which has been formed with the parietal bones, the occipital bone and the mastoid parts of the temporal bones. The most important sutures found from this view are as follows:

(i) The sagittal suture meets the lambdoid suture, where the point of intersection is known as *lambda.*

(ii) The lambdoid suture meets the parieto-mastoid and occipito-mastoid sutures on each side of the skull, where the meeting point is known as the *asterion.*

(iii) The occipito-mastoid suture is important as because it separates the occipital bone from the mastoid part of the temporal bone. The mastoid foramen is a small opening that is frequently found close to this suture.

Besides, the *occipital protuberance* is an important landmark being known as the *external occipital protuberance* (internal occipital protuberance is found inside the occipital bone in the same place). It is a median projection, which is seen on the midway between the lambda and the foramen magnum. The carved ridges form arches laterally from the protuberance, one on each side. These are known as the *nuchal lines.* Nuchal lines can be divided into the *inferior nuchal lines,* the *superior nuchal lines,* and the *supra* or *highest nuchal lines.* The supra nuchal lines denote the upper limit of the neck.

Norma Basalis (view from the below)

A large number of different parts of small bones and foramens are found at the basal region of the skull. For the purpose of description, it can be divided into three separate regions—the *anterior part,* which is formed by the palatine process of the maxilla and the horizontal plates of the palatine bones; the *middle part,* which connects the posterior part with the anterior part and extended upto the anterior margin of the foramen magnum; and *posterior part* goes further back at the base of the skull.

The anterior part

In this part we find the bony palate which forms the roof of the mouth and the floor of the nasal cavity. It is constituted by the palatine process of the maxillae in front and the horizontal plates of the palatine bones in the behind. The special features of this anterior part are as follows:

(i) Inter-maxillary suture;
(ii) Inter-palatine suture;
(iii) Palato-maxillary suture;
(iv) Tubercle of the palatine bone;
(v) Palatine crest;
(vi) Posterior nasal spine;
(vii) Alveolar process;
(viii) Incisive fossa;
(ix) Lateral incisive foramina;
(x) Median incisive foramina;
(xi) Greater palatine foramina;
(xii) Lesser incisive foramina.

The middle part

In this part we find the temporal bone with all its divisions—petrous, mastoid, tympanic plate, styloid process, and squamous bone. The sphenoid bone is clearly identified with its body and three pairs of wings—greater wings, lesser wings and pterygoid processes, but the lesser wings of the sphenoid bone are not seen in the view. The important features of this part are as follows:

(i) Pharyngeal tubercle;
(ii) Tympanic part of the temporal bone;
(iii) Pterygoid process along with pterygoid fossa;
(*iv*) Foramen ovale;
(*v*) Foramen spinosum;
(*vi*) Foramen lacirum;
(vii) Carotid canal.

The posterior part

In this part we find the lower base of Norma basalis. So, it can be said that the lower surface of the skull base is placed posteriorly and includes the occipital bone, which is constituted by four parts around the foramen magnum; (a) a squamous part behind, (b) a lateral part on each side and (c) a basal part in front. The foramen magnum is located in between the mastoid processes. The middle of the anterior margin of the foramen magnum is located as the basion point. The important features of this part are as follows:

(i) Foramen magnum;
(ii) Occipital condyle;
(iii) Styloid process;
(iv) Mastoid process;
(v) Condyler fossa;
(vi) Jugular foramen

Norma Lateralis (view from the side)

In this position, the lower border of the parietal bone is found to articulate with the squamous part of the temporal bone. An arched process extends forward from the lower part of the temporal bone, and finally joins with a corresponding backward extension of zygomatic or mallar bone. This backward extension of the zygomatic bone, as an arch, is known as *zygomatic arch* and the arched process of temporal bone is called as the *zygoma* or *zygomatic process.* The temporal bone articulates with the lower border of the parietal bone and the backward region shows the lambdoid suture by which it comes in contact with the occipital bone. The back and lower part of the temporal bone moves downward to give rise a blunt projection, known as *mastoidprocess.* The important landmarks from this side view are as follows:

(i) Temporal line;
(ii) Temporal fossa;
(iii) Zygomatic arch;
(ix) Styloid process;
(x) Infra-temporal fossa;
(xi) Pterygopalative fossa;

(iv) Pterion;
(v) External auditory meatus;
(vi) Squama of the temporal;
(vii) Mastoid portion of the temporal;
(viii) Mastoid process;
(xii) Squamosal suture;
(xiii) Spheno-temporal suture;
(xiv) Spheno-frontal suture;
(xv) Zygomatico-temporal suture, etc.

Dentition

Teeth appear in different periods in the life of a man. They bear some important clues of significant changes. In man, there are two sets of teeth—deciduous or milk teeth and permanent teeth.

The milk teeth begin to erupt during the first and second years. The other set, the permanent teeth start to replace them from about the sixth year. The deciduous teeth or milk teeth are twenty in number, which are arranged both in the upper and the lower jaws. The arrangements are as follows: Incisors 2/2; Canines 1/1; and Molars 2/2 (arranged both the upper and lower jaws). In case of permanent teeth, the number increases; thirty-two teeth are arranged in the following ways: Incisors 2/2; Canines 1/1; Premolars 2/2; and Molars 3/3 (arranged both the upper and lower jaws). Diagrammatically, the dental formula of permanent teeth can be mentioned as follows:

	Molar	Premolar	Canine	Incisor	Incisor	Canine	Premolar	Molar
Upper Jaw—	3	2	1	2	2	1	2	3
Lower Jaw—	3	2	1	2	2	1	2	3

The anatomy of tooth reveals three different parts—(a) the *crown,* the projected portion beyond the gum, covered by hard white substance known as *enamel;* (b) the *root,* the portion embedded in the sockets of the maxilla or mandible; (c) the *neck,* the constricted portion lies between the root and the crown. The characteristic features of the different teeth are as follows:

(i) Incisors—These are front teeth in each dental arch. The crown of an incisor is chisel-shaped, and this tooth bears a straight single root.

(ii) Canines—These are placed after the incisors. The crown of this tooth is large and conical. The root is usually single.

(iii) Premolars—These teeth show conical crown with two pyramidal cusps. The root is generally single.

(iv) Molars—These are popularly known as grinding teeth. Crowns are large, dome-shaped. All upper molars possess four cusps while the lower molars have five cusps. But each molar exhibits three roots.

Hyoid Bone

This bone is U-shaped, and situated in the ventral floor of the pharynx below the base of the tongue, in front of the epiglottis.

Besides skull, the axial skeleton shows the bones of the trunk, which is composed of the spinal chord or vertebral column, sternum and ribs.

The Vertebral Column

The vertebral column is composed of a series of independent, irregular bones, which are known as vertebrae. Each vertebra is intervened by fibro-cartilaginous disc, called intervertebral disc. The vertebral column bears the body weight and transmits it to the pelvis and the lower extremities. It also provides a protection for the spinal cord and the membrane.

The vertebral column is composed of 33 vertebrae, where each vertebra consists of a body and a vertebral arch. However, all these vertebrae enclose a ring or foramen that is called as vertebral foramen, through which the spinal cord passes. The uppermost or the first vertebra holds the base of the skull, and known as *atlas.* The lowest portion of the vertebral column is situated in between the hipbones. The vertebrae have been classified into five groups according to their nature and position. *Cervical* : These are seven in number and located in the neck region. In size, they are the smallest of all movable vertebrae.

Thoracic : These are placed in the region of thorax and twelve in number. In size, they are larger than those of cervical vertebrae and movable also.
Lumbar : These are five in number, which are found in the lumbar region. These vertebrae are large and massive as well as movable.
Sacral : These are five vertebrae of the pelvis region. All are immovable or fixed vertebrae, fused with one another. A triangular bone is formed by the fusion, known as sacurm.
Coccygeal or *Caudal :* These are usually four in number and found in the tail region. Here the fusion is much prominent than the sacral region. It also gives rise to a small triangular piece of bone, the coccyx. Though the number of caudal vertebrae is normally four, but it may vary between three and five.

In a view from the lateral side, the vertebral column presents four curvatures, namely, *cervical, thoracic, lumbar* and *sacral.* The thoracic and sacral represent second and fourth curvatures and known as *primary curvatures* as they develop during the foetal life. But the first and third, the cervical and lumbar curvatures are called *secondary curvatures;* they develop later on. The secondary curves are concave posteriorly. The cervical curve appears when the child endeavours to hold up its head and to sit upright. The lumbar curve develops when the child begins to walk. The vertebral column is S-shaped as the cervical curve is convex forwards, the thoracic is concave, the lumbar is again convex and the sacral curvature is again concave forwards.

The Sternum

The sternum is a long and flattened bone consisting of three parts. The first part is the longest and is termed as the *manubrium.* The second part is known as the *body* while the last part is termed as the *xyphoid process.* The sternum is placed in the front wall of the chest where the ribs are connected.

The Ribs

The thorax looks like a cage in which some important soft organs, such as the heart, the lungs and the great blood vessels are protected. The greater part of this cage, the thoracic region, contains twelve ribs on each side. These ribs are curved bones, which are connected with the vertebrae posteriorly.

The first seven ribs are connected with the sternum by cartilages, anteriorly. These are known as *true ribs* or *sternal ribs.* Other five ribs are known as *false* or *asternal ribs.* The ribs namely eight, nine, and ten ribs have been joint with the cartilage of the true rib numbering seven. Each of these three ribs joins to the cartilage in such a manner that one goes immediately above another. But remaining eleven and twelve of the false ribs remain free, as their anterior ends do not attach anywhere. These last two ribs are therefore known *as floating ribs.*

II. Appendicular skeleton

The Upper Limbs

The upper limbs or upper extremities are composed with the following bones—the shoulder or the pectoral girdle; the upper arm; the forearm; the wrist; and the hand.

The Shoulder or *Pectoral Girdle*

Each pectoral girdle is composed of two bones—the *clavicle* and the *scapula.* The clavicle or the collarbone is a long bone, which stretches from the top of the sternum to the shoulder and lie on the anterior side, almost horizontally. It is placed at the root of the neck. On the one side it articulates with sternum and on the other side with the acromian process of the scapula. The scapula or the shoulder blade is a large bone, flat and triangular in shape. This bone is placed at the postero-lateral side of the chest wall and connected with the sternum by the clavicle. It also articulates with the humerus. On the back of the scapula a prominent ridge is present, known as spine. The upper and outer part (especially at the lateral side of the bone) shows a cavity for articulation of the humerus. This cavity is known as the glenoid cavity.

The Upper Arm

The upper arm is a single long and strong bone, which articulates with the scapula above and with the bones of forearm below. This bone is known as the humerus. The humerus shows a long shaft and two enlarged ends—upper and lower. The shaft is more or less cylindrical and the upper end is large and rounded to articulate with the glenoid cavity. The lower end is broadened, rather transversely expanded to form pulley-shaped articular surface known as trochlea. The convex articular surface is called the capitulum. This lower part articulates with the bones namely radius and ulna.

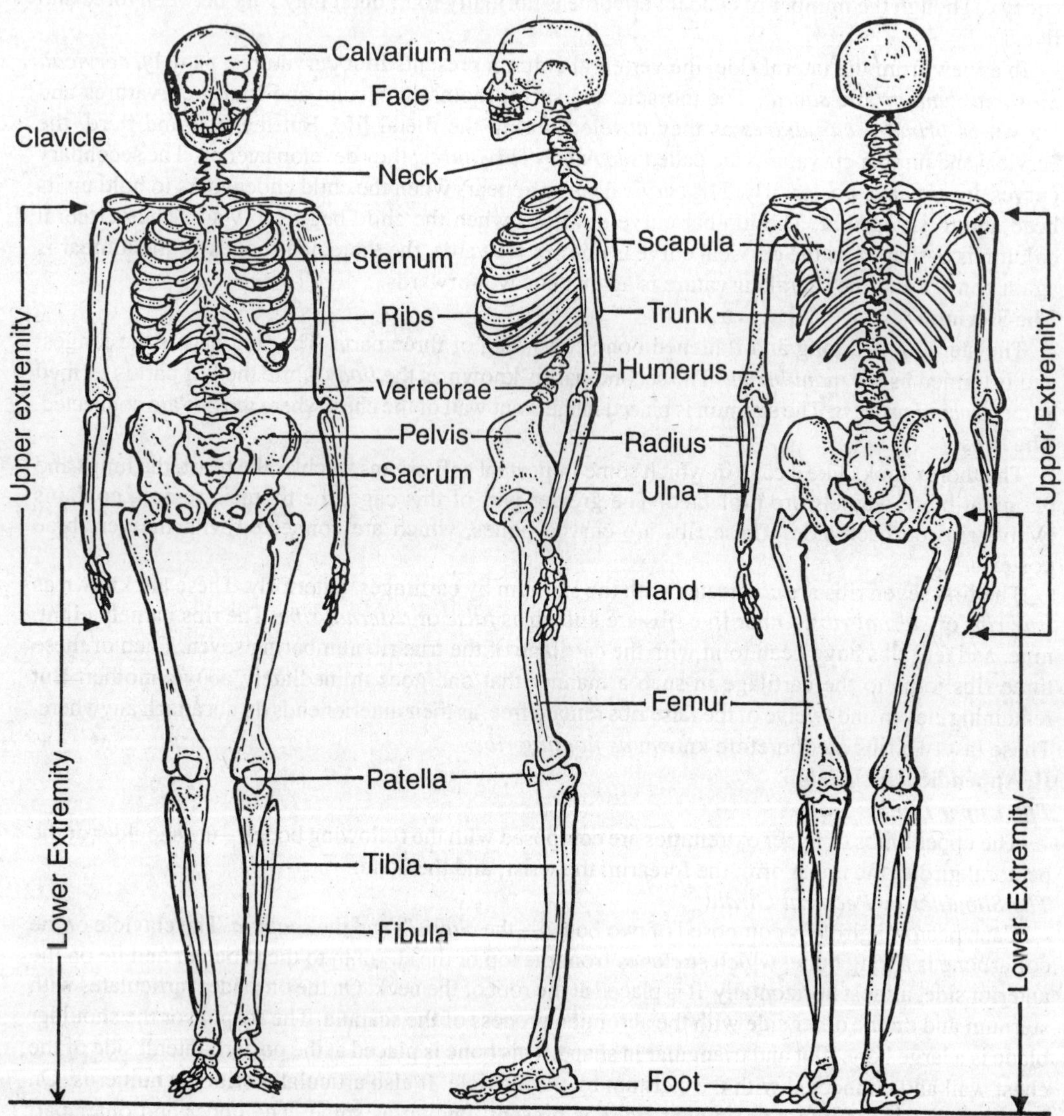

Fig. 3.2. Major Bones In a Human Skeleton

The Forearm

The forearm is composed with two separate long bones—the radius and ulna. The radius is the lateral bone of the forearm, which has two extremities and a shaft. The upper end is disc-shaped that articulates with the humerus and ulna. Besides, the upper end has a neck and tuberosity. The lower end shows a medial and lateral articular surface by which it articulates with the wrist bones. The ulna is situated on the inner side of the forearm. It has a shaft and two ends—upper and lower. The upper end presents a deep notch, the trochlear notch, which exhibits a hinge joint for articulating with the lower end of the humerus. The lower end is slightly enlarged. It has a smooth articular surface and a prominent styloid process.

The Wrist Bones

The wrist or carpus consists of eight irregular-shaped bones, which are arranged, in two rows—the proximal and the distal. The proximal row is made with bones namely the scaphoid, lunate, triquetral, and pisiform. But the distal row consists of the bones e.g. trapezium, trapezoid, capitate, and hamate.

The Hand

The hand consists of 19 bones, which are arranged, in two groups—the metacarpus and the phalanges. The metacarpus includes miniature long bones, which connect fingers with the wrist. These are five in number. Each metacarpal bone comprises of a shaft and two ends—a proximal base and a distal head. The distal end is provided with a large head. The phalanges are the finger bones. Each finger has three phalanges except the thumb, which has only two. Each of these phalanges is provided with a shaft, a head and a base. The number of phalanges is 14 in total.

The Lower Limbs

The lower limbs or lower extremities are composed of the following bones—the pelvic girdle; the thigh; the patella; the leg; the ankle; and the foot.

The Pelvic Girdle : It consists of two irregular shaped bones—the innominate bones. The dorsal part of the innominate bones articulates to the sacral vertebrae. The anterior part is carved around to meet together in the midventral line—the symphysis. Thus, the innominate bones enclose an oval or circular pelvic cavity. Each innominate bone exhibits three fused parts—the pubis, the ischium and the ilium. The ilium is the fan-shaped expanded upper part. It includes the upper part of the acetebulum. The ischium part is situated at the below, behind the lower part of the acetebulum. The ischium and pubis enclose a large foramen, which is known as obturator foramen. The pubis forms the rest part of the innominate bone. These three parts of innominate bone get united at the centre to form the cavity known as the acetebulum where the head of the femur fits. When we observe from the side, the iliac crest is found to form the upper border of the bone (innominate bone). The acetebulum lies on the lateral surface and the obturator foramen is located below, in front of the acetebulum.

The Thigh or *Femur* : The thighbone or the femur is the largest and strongest bone in the skeleton. This bone articulates with acetebulum of pelvic girdle at above, and below with the tibia. This bone is remarkable for its shaft, upper end and the lower end. The upper end consists of a head, a neck and two trochanters—greater and lesser. It shows trochanteric crest, trochanteric fossa and quadrate tubercle. The posterior border of the shaft is formed with a rough ridge, known as linea aspera. The lower end articulates with the tibia and fibula.

The Patella or Knee-cap : This bone is flat and somewhat triangular in shape. Its margins are rounded and their apex is directed downwards. This bone is situated in front of the knee-joint.

The Leg : Each leg bone is consisted with two separate long bones—the tibia and the fibula. Tibia is the larger of the two leg bones. It has an expanded upper end, triangular shaft and a slightly expanded lower end. The upper end or head bears two condyles, medial and lateral. The anterior border of the tibia forms a sharp crest, which is called as shin. The lateral border of the lower end presents a notch, known as fibular notch. Besides, the projected short process of the lower end is termed as the malleolus. The lower end is much smaller than the upper end. However, tibia transmits the body weight from the femur to the foot. Fibula is the slender lateral bone of the leg. It has an upper end, a shaft and a lower end. The fibula is fixed to tibia at both ends.

The Ankle : The ankle consists of seven bones in two rows, known as tarsus. The proximal row is composed of two bones - the talus and the calcaneum. The distal row shows four bones, of which three are the cuneiforms (medial, intermediate and lateral) and the other is the cuboid. The 7th, the navicular is placed between the talus and the medial three bones of the distal row.

The Foot : The foot consists of metatarsus and phalanges. Metatarsus—These bones are five in number. Each bone exhibits a shaft, a head and a base. They are posited in between the toes and ankle. Phalanges—Three small bones are found to be present in each toe, other than the big toe where the number is only two. Thus, phalanges are 14 in number. The phalanges of the toe are much smaller in comparison to those of the fingers.

COMPARISON OF ANATOMY WITH ANTHROPOID APES

Man is the most highly developed animal in the earth. He has grown some special and distinctive modifications that mark him off from other animals—contemporary as well as ancestral types. To find out the nearest relatives of man we must turn to the living apes that seem to have originated from a common ancestral stock. The ape-group is supposed to be very close to human, in spite of vast differences of physical characters that exist between these two groups.

The anthropoid or man-like apes are commonly called as apes and can be divided into two groups—lesser apes and great apes. Gibbons (genus: hylobates) with their sister variety Siamangs* belong to the group of lesser apes whereas the Orangutan (genus: simia), Chimpanzee (genus: pan) and Gorilla (genus: gorilla) belong to the group of great apes. The Gibbon and Orangutan are Asiatic—East Indian in origin while the Chimpanzee and the Gorilla are African. All are the creatures of tropical climate. Among these four, the Gibbons are called acrobat because they have developed an amazing agility in behaviour. Not only that, this type of apes are wholly arboreal, specialized in pendulum swinging. They proceed largely by brachiating. Again, they can run faster than a man on the ground. In standing position Gibbon's fingers touch the ground, as the upper extremities are very long. But the length of leg is almost proportionate to the trunk. With this last character he beats all other anthropoids and stands next to man. The other form of Gibbon named Siamang belong to the genus *Symphalangus* and found in the Island of Sumatra. Since Oligocene epoch the Gibbons seems to have persuaded with independent evolution of traits.

The great apes have been distinguished from the lesser apes for their greater weight. This heavy weight has negated brachiation and brought them down near the ground. Feet have lost their prehensility and forelimbs are found to be used in walking. To accommodate a barrel shaped chest; the shoulder has become broad with stout and massive spines. Tails are lost. The brain shows much more development. In body structure, an Orangutan is smaller than a man, a Chimpanzee has more or less the same bulk as that of a man, and a Gorilla is far more than a man. Among these three, the Orang possesses a little capacity of brachiation and live on the trees. The Gorilla and the Chimpanzee spend much of their time on ground; their arms are relatively long and strong. But this design of the body greatly changes in man. Man walks straight on two legs, the legs have been long as well as strong. None of the great apes can walk or run entirely on two legs; they have to put their knuckle down at the time of walking.

It has been suggested that, although the apes show certain kinds of adaptations towards upright posture, practically, no evolution of apes can go in the direction of man. Man's ancestors probably had many primitive characters that bear more similarity with the monkeys than the apes. The common features lie in the proportion of hands, the development of the thumb, leg and foot muscles, the number and sequence of the eruption of teeth, late obliteration of the cranial sutures, absence of a typical long shelf below the front-jaw *(simian shelf)* etc. The Old World monkeys seem to be more close to the humans than the New World monkeys. They exhibit same number of teeth as in ape and human. Possibly these monkeys gave rise to a progressive branch which advanced through the

* Siamangs are more agile than the Gibbons and they are a little larger also.

Pithecoid or monkey stage and reached the *Hominoid* stage. Evidences have been preserved in the fossil remnants. No surviving primate can therefore be the ancestor of man.

Anatomically a man is fully erect with bi-pedal gait. The heels, together with feet point outward, the arms hang by the sides of the trunk (body) with palms, and the face looks forward. Head is considered as the superior part of the body whereas the inferior part is towards the feet or caudal region (tail). The body-surfaces have been identified as dorsal (backside), ventral (belly side) and lateral (two sides of the body). Anterior part of the body is carried forward when a man walks. In descriptive anatomy this part is synonymous to ventral. Similarly the posterior and dorsal connote the same meaning. The anatomical changes of man can be best illustrated with the forms like Orthograde and Pronograde. Orthograde refers to the position, which is completely upright. The Pronograde denotes_a form, which stands or moves on four legs. The evolutionary change from pronograde to orthograde signifies a position where the trunk has been shifted along with vertebral column. The spines move vertically leaving the former horizontal position. The weight-bearing axis comes closer to the trunk's centre of gravity. Such a shift in the position of the vertebral column has led to the changes in the function of the column, particularly in reference to its supportive or weight bearing capability.

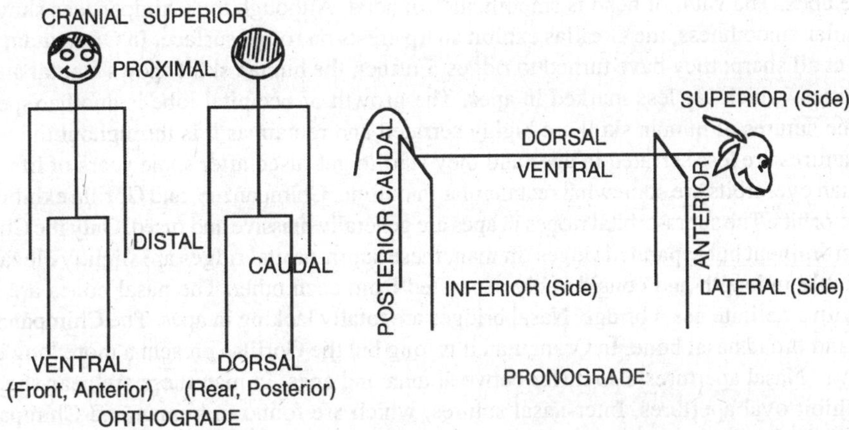

Fig. 3.3. Positional Difference between Orthograde and Pronograde

The vertebral column is not a rigid structure at all. It is like a flexible rod with arch. In case of quadrupedal (pronograde) animals, the highest point of this arch is placed at the middle of the back where the anticlinal vertebra with its spinous process remains more or less perpendicular to the vertebral column. Among the bipedal (orthograde) animals, this anticlinal vertebra is not always clearly identifiable, not even placed strictly at the middle. Though the term bipedal is normally associated with man but some other animals may show bipedalism in different ways and on different occasions. As a matter of fact, there are four kinds of Bi-pedalism—bipedal hopping, bipedal running, bipedal walking and bipedal standing. Bi-pedal running and hopping are found in almost all groups of primate. Bi-pedal walking is found among the men as well as the chimpanzees. But the bipedal standing is strictly restricted to the man. It is only man who can stand erect on two feet and can walk slowly—forward or backward with fully extended legs.

A pronograde animal (quadruped) can not stand on hind legs. It is not only a question of balance; it does not give comfort too. As soon as a quadruped tries to stand on two hind legs, his viscera tends to sag down into the pelvic cavity; it no longer gets the support of ventral body wall. In case of an orthograde animal, the abdominal contents are fixed or suspended in such a way that those do not slip downwards. These facts suggest that the erect posture of man can not be a mere adaptation with a long span of time. Rather it is an outcome of a long process of specialization where considerable genetic changes were involved. Anthropoid apes appeared in between the two forms.

Orthograde posture as well as locomotion again requires a rearrangement of the muscles, especially at the back. In pronograde animals, the muscles those control the forelimbs are attached to the immovable shoulder girdle, pulling the flexible rib cage forward and downward. In orthograde position, as the body weight suspends vertically through the spine, the simple bow-like curvature of spinal column (as found in quadrupedal animals) is altered. Not only it takes S-like shape, the horizontal position becomes vertical where the upper end joins with the cranium and the lower end extends upto the caudal region. It is interesting to note that the muscles require a lesser strength in this position for holding the body erect. However, the concomitant changes in the body that take place with the upright posture are as follows:

***Skull*:**

Skull is of immense importance as a seat of brain. Human brain is not only larger than that of apes, it is much complex in function. The cerebral cortex is unique to man because it generates speech and different other superior mental faculties. The cranial capacity of man ranges from 1300 cc. (cubic centimeter) to 1450 cc. Among the anthropoid apes this capacity is far off .* Human cranium is quite expanded; it is larger than the facial portion. In apes, this proportion is just reverse. As the frontal area of human brain is enlarged, their forehead has been arched and more prominent than the apes. The vault of head is smooth and rounded. Although the Chimpanzees show more or less similar smoothness, the Gorillas exhibit sharp crests on rough surface. In Orangutan, the crests are not at all sharp; they have turned to ridges. Further, the human skull shows frontal and parietal tuberocities, which are less marked in apes. The growth of occipital lobe is another speciality of man. The sutures of human skull are highly serrated and remain as it is throughout the whole life. These sutures are less serrated in apes and they start to get fused after some years of life.

Human eye-orbits are somewhat rectangular in outline. Chimpanzee, and Gorilla exhibit round or oval eye-orbits. The supra-orbital ridges in apes are generally massive and fused. Only the Chimpanzee shows prominent but separated ridges. In man, these supra-orbital ridges are slightly elevated. They are variable in length and considerably separated from each other. The nasal bones are short and broad with a definite nasal bridge. Nasal bridges are totally lacking in apes. The Chimpanzees show a short and broad nasal bone. In Orangutan it is long but the Gorillas present a more long bone with wide base. Nasal apertures also differ between man and apes. In man these are pear-shaped while apes exhibit oval apertures. Inter-nasal sutures, which are found in human and Chimpanzee, are almost nil in Gorilla and Orangutan. Sharp inter-nasal spine in human that separates face from the floor of the nose is not found in apes. In man pre-maxilla can not be differentiated from maxilla but in apes pre-maxilla and maxilla are always separated. The alveolar *prognathism is* usually variable in human skull but it is well marked among the apes. Moreover, the alveolar arch which is parabolic in man becomes broad and U-like in apes.

Mandible

The lower jaw or mandible of man is small, slender and light with a well-developed chin. In apes, it is large, massive and heavy without the trace of a chin. The sigmoid notch in human mandible is deep and narrow, but in apes it is shallow and broad. The ascending *ramus* in man is narrow and high which is broad in apes. M*ylohyoid* ridge as well as groove is well marked in man; they are faintly marked in apes.

The muscles responsible for the movement of the mandible are weak in human. They are strong and well developed in apes. Lighter (weak) muscles indicate that there is more room inside the jaw for tongue. Apart from these, the presence of genial tubercle and absence of simian shelf characterize the human mandible. In apes, these features are just opposite i.e. the simian shelf is present and genial tubercle is absent.

---

* Orangutan's cranial capacity ranges between 365cc. And 425cc. Chipanzee's capacity is within 400cc. To 500cc and Gorilla who shows the largest brain next to man possesses a varying capacity between 400cc and 550cc.

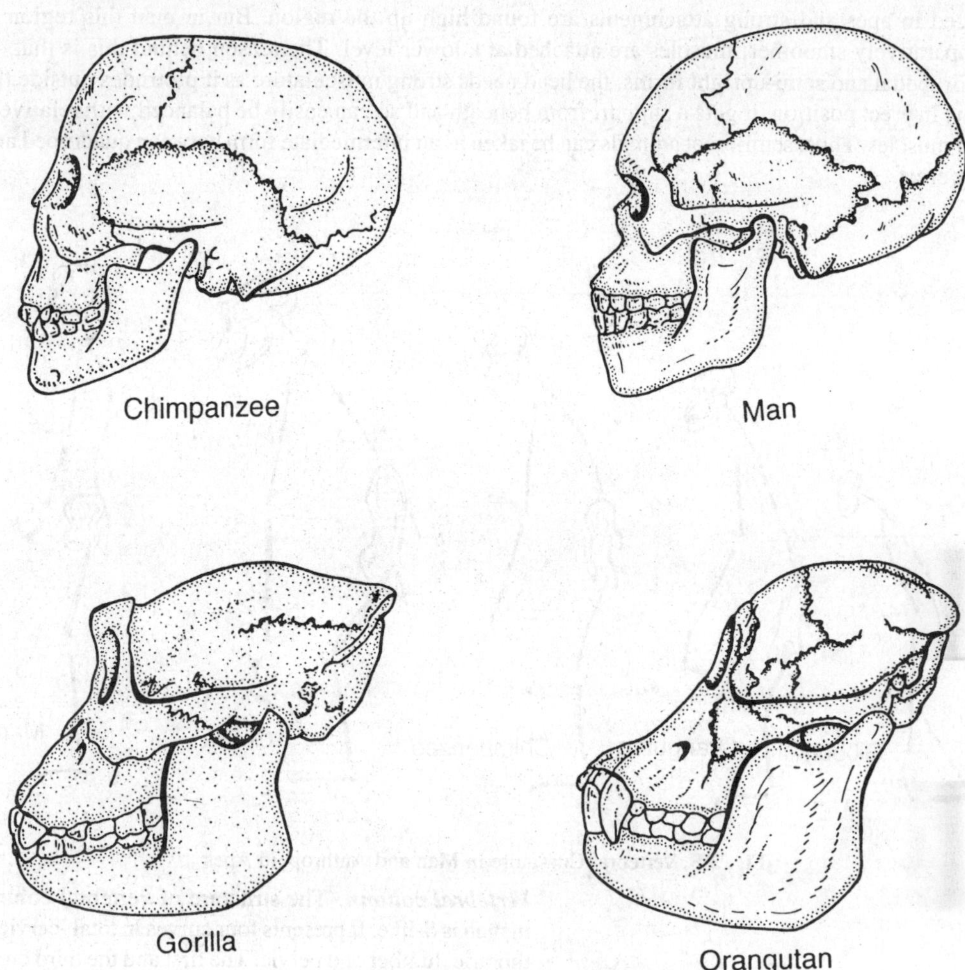

Fig. 3.4. Skulls Showing Resemblance In Man and Great Apes

Teeth

Human teeth are smaller in size than the apes. Canines and incisors are more or less equal in size and no space is provided between two teeth. In apes, the canines are extraordinarily large, pointed and sharp. They are projected beyond the level of other teeth. Not only that, canines of the apes interlock when the jaws are closed. The long upper canine cut against the first lower premolar, and the lower canine passes in front of the upper canine. Therefore, *diastema* is invariably present in apes but human dentition shows no diastema. The roots of the molars are convergent in man and divergent in apes. The chewing motion among men is followed from side to side and also up and down. It is generally known as rotary motion. Among the apes, this chewing motion goes as up and down. The nature of human teeth is suitable for omnivorous diet.

The Linkage between Skull and Body

The position of the foramen magnum has been shifted in man. The skull is no longer suspended but balanced beautifully on the first vertebra (atlas) of the vertebral column. It is a remarkable feature of man in contrast to all quadruped animals. In case of monkeys and apes, the foramen magnum is placed much more posteriorly on the skull. The muscle attachments that facilitate the movement of the skull are located at the nuchal region, on the backside of the skull. This region is rugged in apes and strong attachments are found high up the region. But in man this region is comparatively smoother; muscles are attached at a lower level. The reason behind this is that, in quadrupedal and semi-upright forms, the head needs strong musculature as it protrudes outside the body. In erect position, it gets a support from beneath and so can easily be balanced with relatively light muscles. Thus, semi-erect animals can be taken as an intermediate form between quadruped and fully erect.

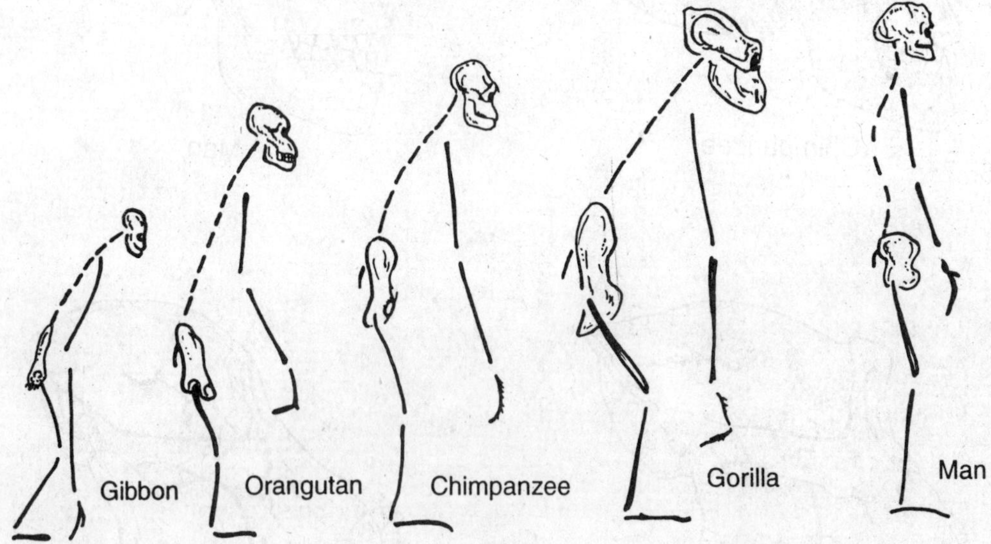

Fig. 3.5. Vertebral Curvature in Man and Anthropoid Apes

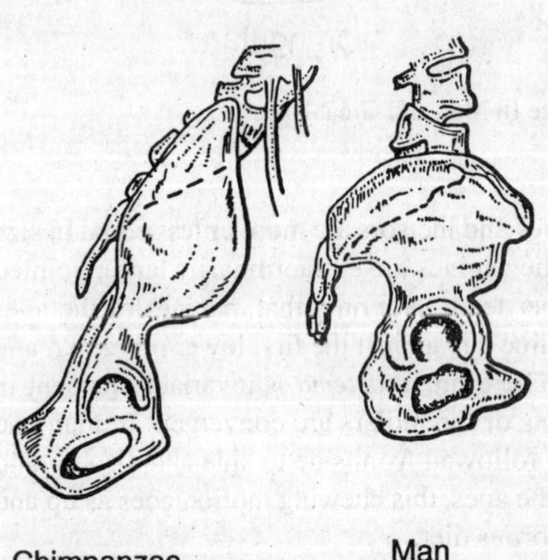

Fig. 3.6. Lumber Curve as Differs in Man and Chimpanzee

Vertebral column: The structure of vertebral column in man is S-like. It presents four curves in total -cervical, thoraric, lumber and pelvic. The first and the third curve are convex while the second and the fourth curves are concave. This arrangement keeps the trunk straight. Of these, the thoracic and the pelvic curves are formed during the foetal life. But the other two, the cervical and the lumber curves develop quite slowly with time as the child tries to sit upright and walk on feet. These two curves are directly related to the change of posture and bipedal locomotion. Monkeys and apes are devoid of these curves. They show only two-curves in their vertebral column—dorsal curve and sacral curve. Therefore, they cannot stand fully erect; their body bends forward. The posture can be termed as semi-erect. They also show some other differences in spine, especially in the cervical region.

Thorax : In human, the shape of the thorax has been altered due to the shift in the direction of gravity, which

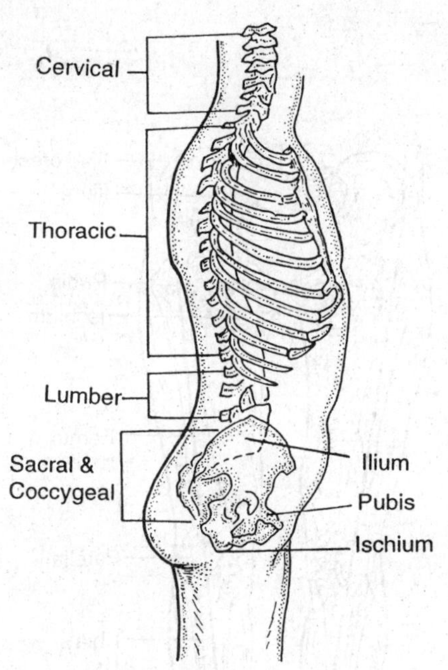

Fig. 3.7. S-Twist Vertebral Column in Human

is a resultant of change in the body axis. However, the thorax in human is barrel-shaped, capable for lateral expansion. Its transverse diameter is greater than the dorso-ventral diameter. The pull of the thoracic contents goes downwards. The lateral movement of the arms is found outward from the body, which tend to broaden the shoulder girdle and increase the length of clavicles (collarbone). In monkeys and apes, the thorax is bulging and it looks comparatively downward. The pectoral muscles in both anthropoids and man cover the outer part of the chest being attached to the upper part of the arm bone and the shoulder blade. It helps in free lateral movements of forelimbs both in anthropoids and man.

Hands

Hands have undergone remarkable changes in course of evolution. Therefore, human hands can be distinguished from any other primate. Man is the only animal who does not employ his hands in locomotion. Rather he uses his hands for various other manual works.

Arms are more elongated in apes than the human. This greater arm length has facilitated in arboreal adaptation; they have been expert in hanging and swinging rather than standing or walking. Since thumbs are sort in size, their precision grip is lesser in comparison to man. Greater size and opposability of thumb has given man great control of hand. He is gifted with both power grip and precision grip. Even the digits are under his control.

Pelvis

Erect posture and bipedal gait of man have made considerable modification in pelvic girdle. The pelvic girdle in mammals is usually consists of two irregular bones known as *innominate bones.* The dorsal portion of each bone articulates with one or two sacral vertebrae and encloses an oval space to form the pelvic cavity. Each of the pelvic halves consists of three fused parts—*Ilium* or the dorsal part that articulates with the *sacrum,* a superior or anterior part called the pubis and an inferior or posterior part called the *ischium.* The pubis and the ischium are united to form the ventral arch (symphysis) where the halves join together in the middle line of the body. The flattened bases of the ischia form the bottom part of the pelvis, the haunch bone, upon which the animal rests its weight in sitting position. As the weight of the head, neck and trunk is borne by pelvis, the ilium bone in human has become short and broad. The angle between the ilium and ischium has got reduced but the transverse diameter of the pelvis has been increased. It gives the pelvis a funnel like appearance with a small orfice directed downward. This type of funneling on the upper part of the pelvis has not occurred in any large primates except man. The pelvis in apes and monkeys is considerably narrow. The whole pelvis in man has been laterally expanded in the form of a basin not only to accommodate the internal organs but also to withstand the pressure exerted by upper girdle, arms and head upon it.

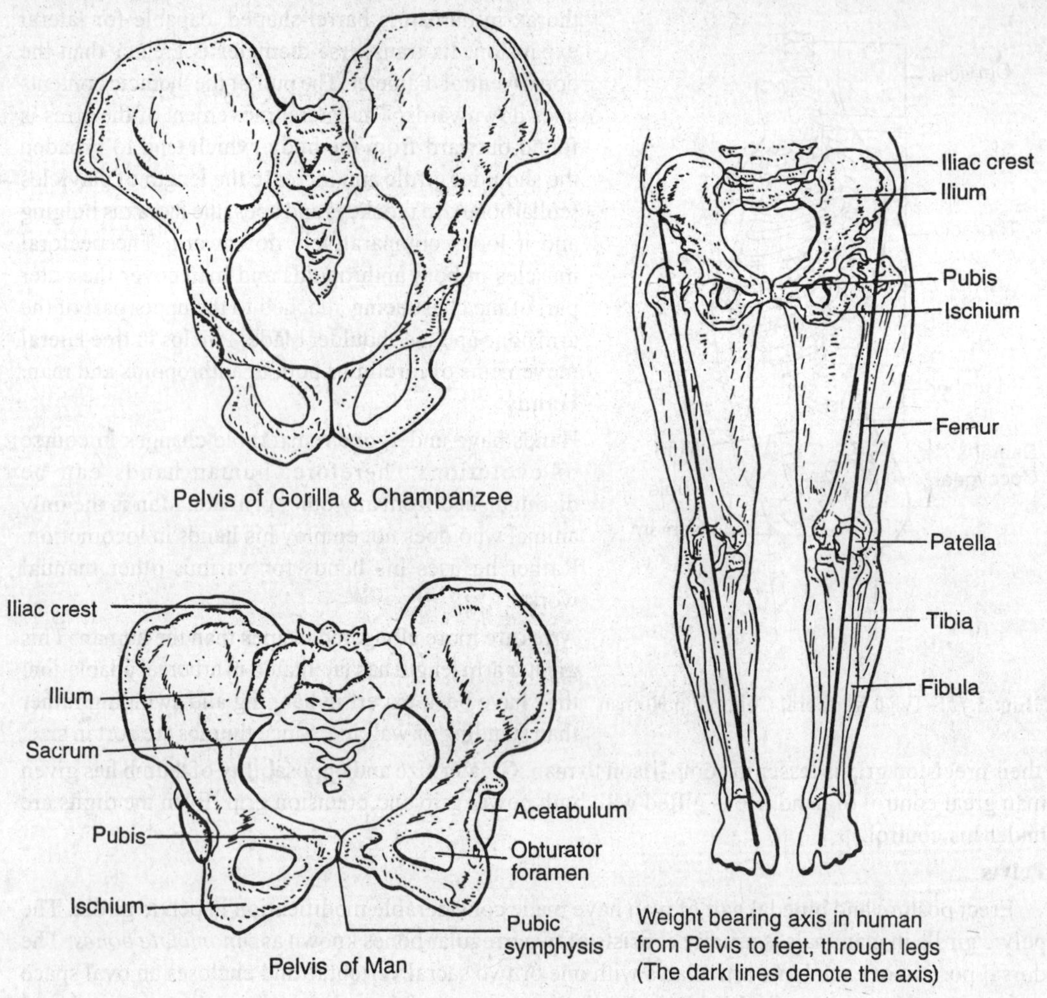

Fig. 3.8. Changes in Pelvis for Erect Posutre

Femur

The hind limbs are attached to the pelvis by ball-and-socket joints where the three portions of the innominate bones the ilium, ischium and pubis unite together. The widening of ilium and the narrowing of angle between ilium and ischium bring the gluteus maximus behind the joint of the hip. Not only that, this gluteus maximus has become the powerful extensor of the leg. The femur renders support to the pelvis from below by articulating its head at acetebulum, which looks downwards, and outward.

The pelvic girdle, shoulder girdle, leg and hand—all these anatomical structures are the parts of locomotor complex, but since most of the weight of the body in upright posture passes through the thigh bones, notable changes are found in the femur. Human femur resembles with the apes in general form. But its shaft is slightly curved. It shows the special features like, the development of linea aspera and popliteal area, a larger angle made by the axis of shaft with the head and neck, larger size of medial condyle than that of the lateral condyle and so on. In fact, the articulation between the bones of leg and foot underwent alterations at knee and ankle for an effective adjustment in erect posture.

Foot

In man, foot is no more a grasping organ; it bears the weight of the body. Structural changes have been obvious for the functional change. For example, the great toe in apes is short in size and opposable to facilitate grasping. In human it has been placed in the same line with other toes. It is the biggest of all digits. In apes, the third toe is longest of all but lateral toes are also developed. They do not show a gradual reduction (in size) as found in man.

Unlike ape, a transverse arch is visible in the foot of man. The medial and lateral arches are also formed. Axis of the foot runs in between the first and second toes. The articular facet of cuneiform bone is flat which articulates with a flat area of first metatarsal bone. But in apes, articular portion of cuneiform bone is convex in shape and it articulates with a concave area of metatarsal bone. In man, the great toe has been firm and rigid for bringing the upright posture.

Within the periphery of our knowledge, the first anthropoid ape was the Propliopithecus, which used to live in Oligocene epoch. There is no doubt that this primitive anthropoid ape was a Gibbon like animal. The Gibbons of present day seem to have originated from Propliopithecus. The fossil ape Pliopithecus was probably a descendant of Propliopithecus and it was almost similar to the modern Gibbon. Therefore the little arboreal anthropoid Gibbon must be considered as the nearest representative to a common ancestor of man as well as all great apes that are found at present. The Gibbonoid stage is the most important stage towards human evolution, especially in reference to locomotor adaptation. The primates like tarsiers, lemurs and monkeys frequently adapted an erect sitting posture in the trees, but primitive gibbonoid apes were found first to modify their bodies for upright posture. Therefore, we can conclude that, in the perspective of evolution, the progress of human line began with the gibbonoid stage (small and primitive anthropoid stage) much before the stage of giant anthropoids was achieved. As man is an erect biped walker, the gibbon is an erect tree climber. It is also an erect biped while on the ground.

The principal modification that took place in the gibbonoid stage of evolution is as follows :

(i) An erect sitting posture at the time of rest.

(ii) The broadening and flattening of the thorax connected with arm locomotion, which altered the gravitational pull.

(iii) A tendency of the skull to be balanced on the vertebral column with a basal articulation in contrast to the former state of suspension with occipital articulation.

With the passing of time the modern gibbons became tremendously specialized in brachiating by the elongation of their arms. They also developed projecting spike-like canine teeth. However, the gibbonoid heritage flourished after the separation of the line towards the giant anthropoid apes. On the other hand, the giant anthropoids evolved from small and primitive anthropoids, gibbon proceeding the Oligocene epoch.

This stage is regarded as the second stage in the wake of modern man and has been termed as giant anthropoid stage, which occurred during the Miocene epoch. In the middle of the Miocene, anthropoids became much advanced in structure. At this time a bifurcation took place among the anthropoids. One line gave rise to orangutan, chimpanzee and gorilla and other line went towards man. In fact, during the lower and middle Miocene, some generalized forms of anthropoid apes belonging to the Dryopithecus family came into existence. A section of them directly progressed towards gorilla and chimpanzee and other remained as an intermediate form between the two. All of them became more powerful and intelligent than the earlier gibbonoid form. Nevertheless, it is found that the principal modifications towards the man were accomplished at the gibbonoid stage, prior to the development of great or giant anthropoids.

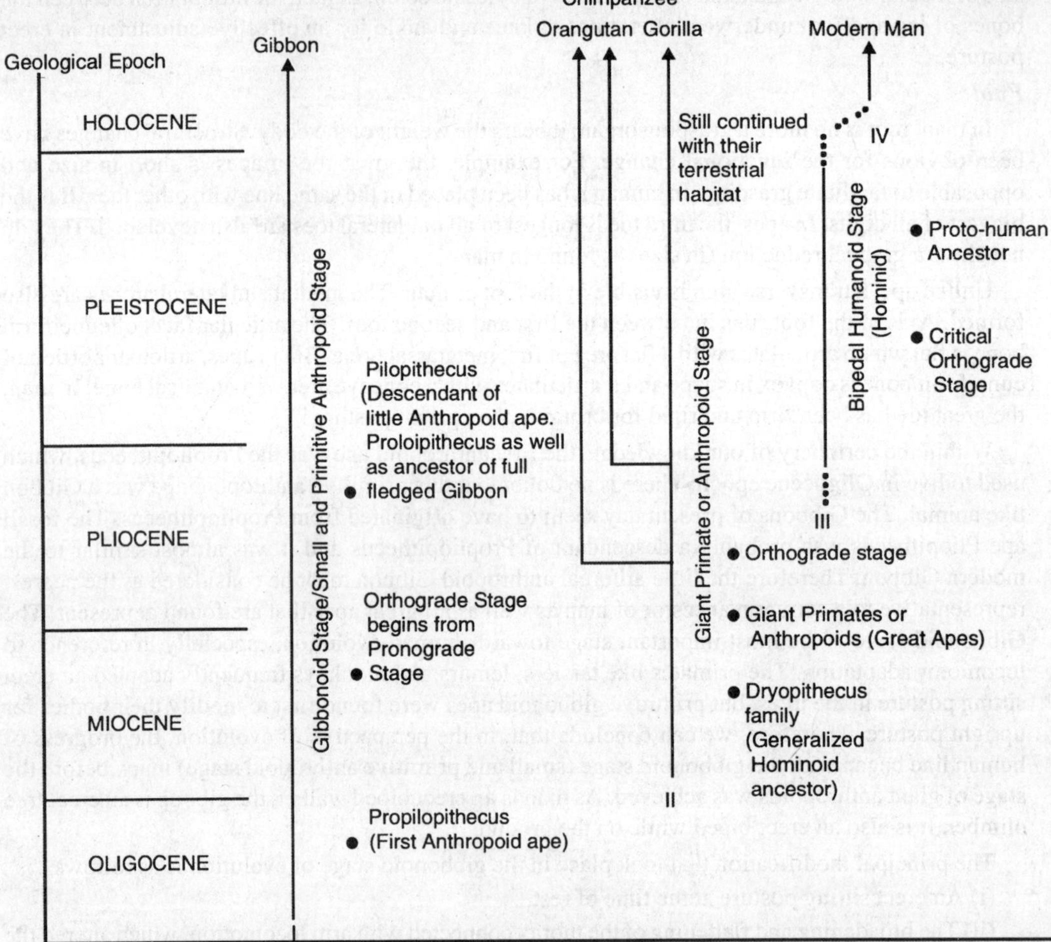

Fig. 3.9. Stages Suggesting Evolutionary Relationship between Man and Apex

The bipedal hominids evolved during Pleistocene epoch as a cousin of the same ancestral stock. They had neither shown any development towards brachiation nor had grown any powerful canine teeth or sharp-cusped molar as noted among the anthropoid apes including Proconsul. Proconsul was a little monkey-like tree-going quadruped creature. Recent palaeontological theories classify Proconsul under the group Dryopithecus, the culminating phase of pongid line that forked about twenty million years ago. Therefore, Proconsul can be taken as ancestral to both ape and man. There was a strong debate among the scholars namely Hooton, Gregory, Keith and Broom regarding the number of splits, which occurred in the evolutionary path. Keith, Broom and Hooton suggested a single split that one branch proceeded upward to produce Orangutan, Chimpanzee and Gorilla and the other branch moved towards the man. Previously Hooton like Gregory advocated a different view. They presumed double splits where orangs first diverged out from the chimpanzee-gorilla and man group and later on man branched off from the chimpanzee-gorilla group. Whatever is the situation, main hominid line was definitely separated before Pleistocene. Modern man had to surpass the Hominid stage for attaining the present anatomical features.

It is therefore clear that man have evolved from a monkey like stage, instead of ape-like stage. Hominid stage is the third stage of development in the way to man. But no fossil record has yet been revealed as the first hominid. It is wholly a matter of conjecture to project the evolutionary line. However, the most primitive but distinctive primitive hominid form, so far as discovered, is the

Pithecanthropus erectus or erect ape-man or fossil ape-man of Java. Pithecanthropus represents a late and unprogressive survival of one of the earliest forms of ground-dwelling human primate. They were transitional between the modern man and the anthropoid apes. These proto-human ancestors of man when took to the ground, they adopted erect posture and different mode of locomotion.

According to Dr. Dudley Morton (1926), the posture is a result of a bio-chemical interaction between organism and gravity. In case of kangaroo and some extinct reptilian forms, a heavy counterbalancing tail helped the animals to retain an erect position. The gorilla, chimpanzee and orang do not possess a tail; rather they show a massive head and chest. Therefore, the weight of their forepart of the body necessitates the support of the forelimbs while moving on the ground. On the other hand, their hind limbs are weak to bear the total weight of the body and the pelvis too is unable to transmit the whole weight to the legs. As a result, these anthropoids bend forward to rest the weight of their frontal part of the body upon the long arms. Because of the greater length of arms, spines can not go parallel to the ground. Biochemical and genetic analysis of the anthropoid apes and man as shown by V. Sarich and A. Wilson (1966) suggest that the gibbons splitted off from the evolutionary path about twelve million years ago, orangs about ten million years ago, and gorilla and chimpanzee about five million years ago. The animals that exhibit more similarity in blood chemistry denote a long history of close relationship and a comparatively recent divergence. Therefore it can be inferred that the gibbon is the most specialized type of small ape which was pioneer to diverge from an ancestral stock. The other great apes and man stand in the following order of increasing specialization: orangutan, chimpanzee or gorilla, and man. Although the line evolving Orang, chimpanzee and gorilla was separated from the line of man much earlier, still man seems to be more primitive than the Orang in adherence to basic configurational plan.

STUDY OF FOSSIL PRIMATES

To study the emergence and development of man, study of fossil primate was found most essential. Scientific minds could not satisfy themselves with the legends and miraculous stories those were in vogue for a long time. So at the beginning of nineteenth century, with the emergence of different branches of science namely Zoology, Palaeontology, Comparative anatomy. Geology, etc. a few energetic scholars went forward in search of man's antiquity as well as relationship with other animals. They tried to remove man from the domain of legends and miraculous stories. The knowledge of zoology and comparative anatomy declared man as the end product of the process of organic evolution, which meant nothing but a transformation of organisms from simple and homogeneous to complex and heterogeneous forms. Palaeontology and geology discovered a lot of bone remains of various extinct animals from different layers of earth. Those unwritten evidences of nature were recorded and analyzed carefully to establish the dim antiquity of man, prior to the age of history. The past history of man was thus hidden in the fossilized remains of plants and animals, the time and sequences of which were interpreted by the geologists. Further, archaeologists tried to establish the emergence of man with the help of stone implements.

Fossils are the hard parts of plants and animal remains, which get infiltrated by mineral substances and preserved in the strata of earth in the form of stone. Fossils are therefore no less genuine than the written record and naturally very useful for reconstructing the past events of ancient days. In tracing the ancestry of man, physical anthropologists have concentrated their attention on the primates, the foremost or highest order of animals. The word 'primate' means 'first in rank or order'. Primates form a distinctive and recognizable group among the mammals that are very close to man. Therefore their study begins with the earliest form of extinct primates that has been preserved as fossils.

The living primates of the modern time belong to four groups in general—Prosimian, Monkey, Ape and Man. A comparative study of the physical features between living primates and the available remains of extinct primates (belonging to different groups and geological periods) reveal the relationship between the two,which in turn help in determining the exact line of evolution. However, it does not imply direct ancestry, rather refers to the taxonomic groups. Isolated fossils are judged in terms of probable degree of variability, which may be shown by the members of a given group in respect to the particular bones in question.

The order primate has been divided into the sub-order, *Prosimii* and *Anthropoidea.* The sub-order Prosimii is lower than the Anthropoidea and it includes only Prosimians. The sub-order Anthropoidea accommodates other three groups — monkey, ape and the Man. In this connection, let us orient ourselves with the terminology as pointed out by Prof. W.E.Le Gross Clark. In the writing, 'The Fossil Evidences of Human Evolution' (1955), he introduced the term 'Simian' in order to refer all primates above in the level of lemuroids and *tarsioids.* But when he grouped all primates together, including the Tree shrews, he used to term 'Prosimian'. He also brought the term 'Hominoid' to designate the apes as well as the man-like creatures. True human forms were then referred with the term 'Hominid'. According to this scheme, the members of the sub-order Prosimii are all called *Prosimians* and the members of the sub-order Anthropoidea, excluding the family Hominidae (representing man) are known as *Hominoids.* Therefore for sketching the evolution of mankind, we have to probe deep into the history of primates. Comparison must be sought especially from the stage of Prosimii.

Primates of Eocene Epoch

Plenty of small prosimians, found from the Eocene deposit have been considered as the earliest primates of evolutionary significance. In fact, these prosimian fossils are found in every period all way up through the Cenozoic Era. Some prosimians with a little change in their bodies continue till the recent days.

These small animals of Eocene first exhibited distinct primate characteristics in contrast to the primate-like creatures of Palaeocene Epoch. Five families among the prosimians were distinguished in Eocene Epoch. They were Adapidae, Anaptomorphidae, Microsyopidae, Omomyidae and Tarsiidae. The family Adapidae was sub-divided into two sub-families namely Adapinae and Notharctinae, each of which had several genera. Possibly ancestors of present day lemurs belonged to the sub-family Adapinae and ancestors of present day lorises belonged to the sub-family Notharctinae. The lemurs, which are at present, found in Malagasy, show such a close resemblance with the ancient lemurs that they are regarded as 'living fossil'. However, Adapis were found abundant in both the Old World and the New World.

The fossil remains of the family Anaptomorphidae were discovered largely from North America. The family was composed of several genera. The snouts of these prosimians were shorter than those belonged to the family adapidae. Dental formula also differed. As they bear many similarities with the present day tarsiers, some of the authorities have taken them as the ancestral form of the living tarsiers. The family Microsyopidae was not frequent in occurrence and included relatively large size animals having similarity with Indridae of Madagascar. The family Omomyidae was widely distributed in Europe, Asia and North America. They suggest a relationship between the Eocene primates of Asia and America. The family Tarsiidae again resembles with the present day tarsiers in certain features.

Some authorities point out that certain animal forms belonging to the order insectivora of the late Cretaceous Epoch have given rise to the prosimians. The family Tupaiidae thus includes tree shrews, which live in South-East Asia and show close resemblance to those animals of Cretaceous Epoch. W.L.Strauss (1956), Le Gros Clark (1962) advocated to include tree shrews under the order primate, since they are the simplest type of earliest prosimians.

Primates of Oligocene Epoch

The fossil evidences of this epoch were relatively less as compared to the previous epoch. Most of them are from Africa. The notable names are Parapithecus, Propliopithecus, Aeolopithecus etc. They hold an intermediate position between prosimians and anthropoids.

Parapithecus-The Parapithecus is known on the basis of jaws and teeth which came from early Oligocene deposit in Fayum, South-west of Cairo in Egypt. The owner of the jaw was a small primate about the size of a little squirrel-monkey of present day. The shape of the jaw resembles to that of the tarsiers. Their teeth show primitive characteristics and catarrhine dental formula i.e. 2:1:2:3, similar to the earliest and most primitive ancestral catarrhine monkey forms. The reduction of premolar teeth suggests shortening of muzzle on jaw. Again, the form of jaw indicates an expanded brain and a short face. The premolars recall Eocene tarsiers while the arrangement of the cusps of the molars are similar to the anthropoid apes. So, the remains may be considered as an intermediate form between prosimians and anthropoids. Some scholars have advocated that the catarrhine monkeys and some higher primates were evolved from Parapithecus. In fact, due to inadequate information, there is a great disagreement among the anthropologists regarding the place of Parapithecus.

Propliopithecus-Propliopithecus was a neighbour of Parapithecus. From the same site at Fayum (Egypt), from the early Oligocene deposit, another incomplete lower jaw with teeth was discovered where the jaw was without the ascending ramii. The comparative analysis of the physical features of the jaws between Parapithecus and Propliopithecus proves that Propliopithecus was much advanced. It can be regarded as a primitive anthropoid ape, closely related to gibbon. The jaw was about the size of a small modern gibbon but deeper, shorter and more pointed. The canines were smaller than the canines of present day gibbon. Some authors are in the opinion that the Parapithecus had given rise to the Propliopithecus.

As different varieties of Prosimians stand as the ancestors of monkeys, the ancestors of Hominoids are found among the primates of Oligocene epoch. Although Propliopithecus is considered as the ancestral form of gibbon, the Aeolopithecus, another variety of Oligocene primate exhibited much more resemblance to modern gibbon for which it was ultimately held that the latter might be the actual ancestor of gibbon.

Primates of Miocene-Pliocene Epoch

From the deposits of Miocene-Pliocene Epoch, several fossil remains belonging to hominoid stage have been discovered in different parts of Asia, Europe and Africa. Some of them are considered as the ancestral forms of the Pleistocene hominids.

Pliopithecus-The Pliopithecus was discovered in Europe (Germany) from early Pliocene or late Miocene deposit. This fossil remain contained only an incomplete mandible. It is a gibbon like animal in respect to dentition. The narrow snout, widely set eyes also resemble a gibbon but some other characters like upper limb, spinal column and general body proportions differs from gibbon. These characters are more similar to a monkey.

Pliopithecus is regarded as a direct descendant of the Egyptian Propliopithecus. Therefore, it is also an ancestral form of modern gibbon. Such evidence reveals that perhaps the gibbon family line was struck out from the main evolutionary line at the beginning of Oligocene Epoch.

Lemnopithecus-This fossil remain was discovered from the lower Miocene deposit in Kenya. It was comprised of a jaw and teeth. The specimen showed much similarity with modern gibbon. Therefore Hopwood, who discovered the fossil, suggested that this ape be evolved from Propliopithecus like Pliopithecus. Although the particular species evolved about ten million years later than the Pliopithecus, it proceeded in the direction of hylobates, the living gibbon of present day.

Dryopithecine- From the Miocene and Pliocene deposits, a good number of primates, classified as Dryopithecines were discovered in Europe, Africa, Middle East, China and India. They were thought to be the common ancestor of living anthropoid apes and man. These animals belonging to Dryopithecine group not only show considerable variation in respective to dental characters and other morphological traits, all of them do not comply with the same geological age. It seems that primates under the group Dryopithecine followed different evolutionary trend to give rise the various apes like gorilla, chimpanzee and orang of today.

Dryopithecines can be broadly divided into two groups—(a) *Dryopithecinae,* a sub-family of Pongidae and (b) *Ramapithecus* assigned to the family Hominidae, *Dryopithecinae* is splitted into the groups, Dryopithecus and Gigantopithecus of which Dryopithecus is again splitted into several variety or genera. *Ramapithecus,* on the other hand show no sub-divisions but a number of types like Brahmapithecus, Kenyapithecus, etc.

Dryopithecus-Most of the Dryopithecus fossil remains included jaw and teeth which have been recovered from the Miocene-Pliocene deposits in different parts of Europe, Africa and Asia (China and the Siwalik hills of India). To trace the evolutionary line, the peculiarity of their dentition is very important which shows five cusps in lower molars. Hence the five-cusped molar pattern is known as 'Dryopithecus pattern'. Out of these five cusps, three are arranged facing the cheek and the other two face the tongue. However, the cusps are very distinct and togetherly they present a form like 'Y' letter of English language. This particular pattern or its modified form is found in the gibbons and other apes and also in man. The incisors of Dryopithecus are smaller and more vertical than the other members belonging to the sub-family Ponginae. But the canines are larger than those of *Hominidae.*

Dryopithecus was possibly an ape-like creature. Since their limbs were very generalized, that favoured neither brachiation nor bipedal locomotion. Different scholars suggest Dryopithecus as a probable ancestor of modern gorilla and chimpanzee.

Proconsul-The fossil finds of Proconsul were first discovered by Hopwood and then by Leaky. They excavated and explored these remains from the early Miocene deposit in Kenya, East Africa and Uganda. On the basis of evidences three species have been identified as *Proconsul africanus, Proconsul major* and *Proconsul nyanzae.*

In size, Proconsul africanus is an intermediate form between the gibbon and the chimpanzee. The supra-orbital torus of the skull is absent, the incisive region of the upper jaw is narrow and the lower jaw is without the simian shelf. The size of the premolar teeth is reduced and the nasal bones are relatively broad. The skull is cercopithecoid in appearance. The mandibular condyle is of hominoid form.

Proconsul major is represented by right side mandible where the second premolar and all molar teeth are present. The massive size of the mandible suggests that the owner of this mandible was almost similar to gorilla in size.

Proconsul nyanzae has been represented by an almost complete mandible of an adult. On the basis of the size of the mandible, it may be assumed that the body size of the animal was more or less similar to that of the present day chimpanzees. In general, the fossil remains of the Proconsuls are regarded as the African varieties of Dryopithecus.

Sivapithecus : This is another variety of Dryopithecus. Its full name is *Dryopithecus sivalensis,* though commonly known as Sivapithecus. The existence was first discovered from the Miocene deposits of the Siwalik Hills in India. The dental character shows a close resemblance with Orangs of today.

Ramapithecus : After careful examination of the fossil findings available from Miocene to early Pliocene period, the two genera among the Dryopithecines that have been distinguished are Ramapithecus and Dryopithecus. Dryopithecus finds have already been discussed. The Ramapithecus show considerable difference from Dryopithecus. The Dryopithecus pattern of molar cusp is absent in Ramapithecus. In contrast to Dryopithecus, the incisors and canines in Ramapithecus are smaller than the molars. Not only that, in many respect Ramapithecus deviate from Dryopithecus and approaches more toward *man.* For example, the upper jaw of Ramapithecus is shortened; it does not protrude in the front but vertically long. The palate is arched. Lewis in 1934 described Ramapithecus as a hominid.

It was quite likely that the Ramapithecus line was diverged from the Dryopithecus lineage in the Miocene Epoch. The separation became complete by the Pliocene Epoch. The size and shape of the face of the Ramapithecus along with the pattern of dentition suggest that they did not use their teeth for defence like the Dryopithecus. Possibly they were biped whose hands were free for hunting and defence. On the basis of these facts some anthropologists have taken Ramapithecus as an ancestor of Homo Sapiens.

Kenyapithecus : Leaky in 1961 discovered an upper jaw with teeth from Kenya, Africa. He named it as Kenyapithecus. In anatomical features Kenyapithecus strongly resembles with Ramapithecus of India. Some scholars prefer to call it as Ramapithecus of East Africa. However, Kenyapithecus is also an ancestral form of Homo Sapiens.

It is probable that, either in Miocene Epoch or earlier, the Dryopithecine lineage was bifurcated into two distinct lines. One continued as the pongid lineage until the emergence of gorilla and chimpanzee. The other line maintained a continuous progress towards Hominidae. In the late Miocene Epoch Ramapithecus evolved as a representative of on-coming hominids.

Gigantopithecus : Between the year 1934 and 1936, the vertebrate palaeontologist G. H. R. Koenigswald collected three man-like molar teeth of three different individuals from an apothecary shop (ancient name of chemist shop) in Hong Kong. The teeth were exceptionally big. The crown of a lower third molar was about six times larger than that of a modern man and two times larger than that of a gorilla. For the gigantic size and appearance, the animal was named as Gigantopithecus.

The mandible of another Gigantopithecus was discovered in 1957 from the Middle Pleistocene deposit of South China. This mandible lacked the left central incisor, the third molar, the ramii and the body behind the second molar. Von Koenigswald declared Gigantopithecus as an ape, rather than an ape-like fossil human. Dr. Pei Wen Chung believed Gigantopithecus as contemporaneous with Peking man. He accepted Von Koenigswald's view. This particular creature probably existed in the world covering a large time span, from Pliocene to Pleistocene epoch.

Oreopithecus : The Oreopithecus fossil remains were discovered in 1870 from brown coal or lignite deposits of early Pliocene date at Juscany, North Italy. In 1872 the remains were described by Geravis and it was accepted as a cercopithecoid monkey. After a long period, a rediscovery was made in 1958 by Dr. Hurzeler. He had pointed out the Oreopithecus as a hominid, instead of a cercopithecoid monkey. Because, almost a complete skeleton which was discovered in 1958 showed a lot of hominid features. For example, bones of the pelvis and legs possessed such characters, which made them capable of bipedal locomotion. The bones of the foot exhibited similarities with man as well as with the apes and some other members of the cercopithecoids. For the association of varied features, Oreopithecus has been a subject matter of great discussion among the scientists.

Primates of Pleistocene Epoch

The Tertiary period ended with Pliocene Epoch and the Quarternary period, specially the Pleistocene Epoch may be regarded as the 'Age of Man' as various fossils remains viz. Ape-like man, early man and Homo Sapiens have been discovered from this time period. In fact, the Hominid evolution was largely accomplished in Pleistocene Epoch. But the particular spot where man first appeared is still a matter of investigation. Some scholars point to North India, some point to Africa and some say absolutely of other regions. However, majority of scholars agree that the earliest forms of man have came from the sites of Far East (from Java to Choukoutien near Peking) and also from Africa and Europe.

Three major stages have been recognized in dealing with hominid evolution. Those are, the stage of *Australopithecines,* the stage of Homo-erectus *(Pithecanthropines)* and at the top, the stage *of Neanderthal* man before getting the evidences of modern man (Homo Sapiens) in Late Pleistocene Epoch.

Australopithecine

This is mainly a group of African fossil remains that have drawn the attention of the scientists as a link between apes and man. According to discoverer Prof. Dart, they are also known as 'Dartians'. It may be mentioned here that the *Australo* means Southern. In fact, the first fossil skull of Australopithecine group was discovered at Taung in Bechuanaland, South Africa by Prof. Raymond Dart in 1924. Gradually many more fossil remains of this sub—family were discovered from different parts of Africa, and also from China, Java and Palestine. A list regarding important finds along with the African sites is given below.

Region	Site	Name
South Africa	Taung	Australopithecus africanus
	Sterkfontein	Plesianthropus transvaalensis
	Kromdraai	Paranthropus robustus
	Makapansgat	Australopithecus prometheus
	Swartkrans	Paranthropus crassidens
	Swartkrans	Telanthropus capensis
East Africa	Olduvai	Zinjanthropus boisei
	Olduvai	Homo habilis
	Garusi	Australopithecus
West Africa	Tchad	Australopithecus

Australopithecus africanus : The fossil remains consisting of an almost complete juvenile skull, probably not more than five years of age, discovered at Taungs in Bechuanaland, South Africa by Prof. Raymond Dart in the year 1924. The skull possessed twenty milk teeth and four permanent first molars in good condition.

The geological age of this specimen is not very certain. According to Broom it belongs to the Middle Miocene or Lower part of Upper Pliocene epoch. But now it has been established by the others that the geological age of the Australopithecus should be the Lower Pleistocene.

A few important characters are mentioned here in the following ways :

1. The skull in size and facial portion closely resembles that of the chimpanzee.
2. The profile of face is concave like an orangutan.
3. The supra orbital ridges are absent and the orbits are circular like those of the orangutan.
4. The premaxilla is well marked as found among the apes.
5. The nasal bones are flat and resembles that of the chimpanzee.
6. The face is slightly prognathous and more or less similar to chimpanzee.
7. The palate is parabolic in shape which resembles that of the man.
8. The teeth are arranged in the human fashion and the canines are small, not like as in apes.
9. The lower jaw is massive and thick which resembles that of the chimpanzee and orangutan.
10. The chin is also absent.
11. The foramen magnum is placed further forward than in the apes which suggests that the head is more well balanced, like the man.

Prof. Dart has estimated its capacity as 520cc. The cranial capacity of the adult species ranges from 518cc to 733cc. This indicates that their brain is larger than that of the chimpanzee. From all parts of view it is clear that the Australopithecus differs from the present apes and it is more man like. Members of this group acquired the walking gait, but their brain and jaws were not very developed. So, Australopithecus is considered as the first stage in hominid evolution.

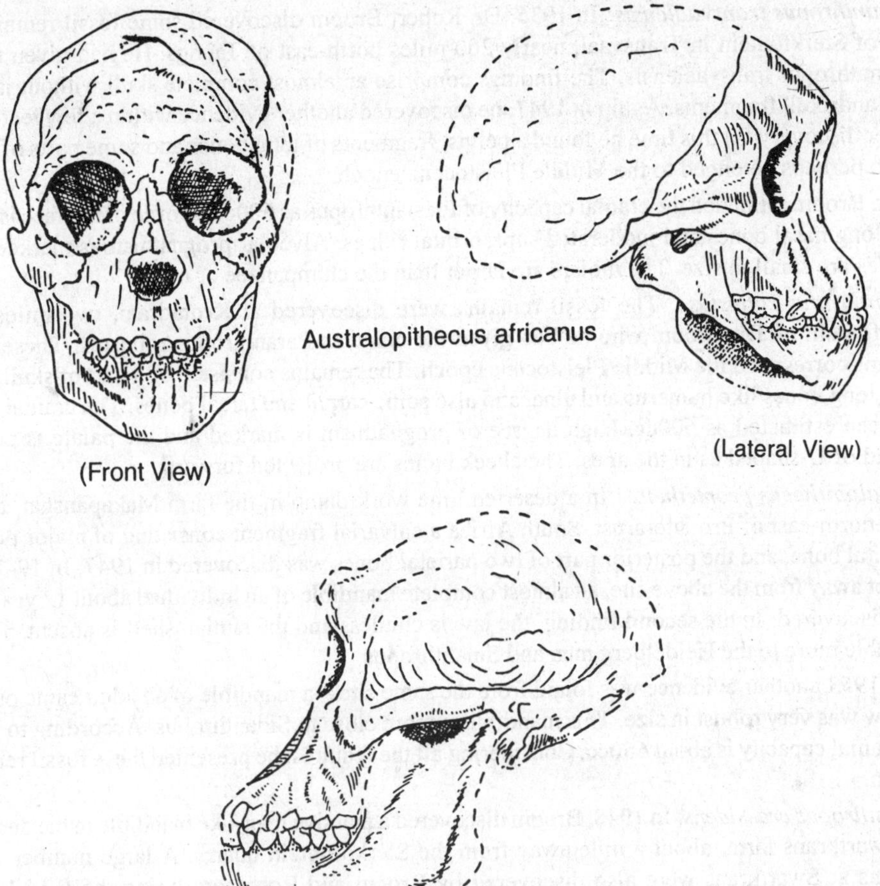

Australopithecus africanus
(Front View)
(Lateral View)
Paranthropus robustus
Plesianthropus transvaalensis
Oreopithecus

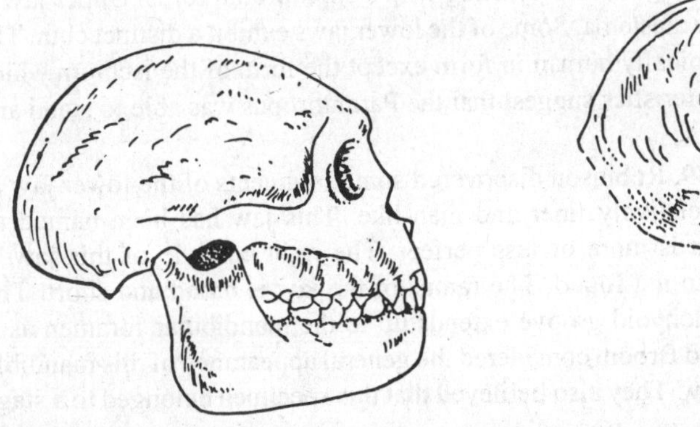

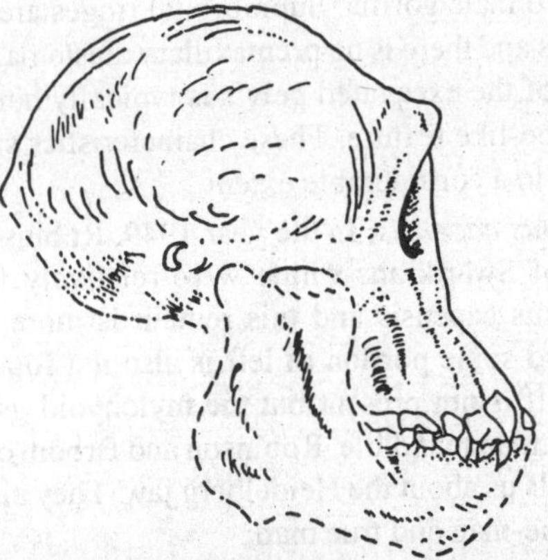

Fig. 3.10. A Few Fossil Hominoids

Plesianthropus transvaalensis: In 1936, Dr. Robert Broom discovered some fossil remains at the cave of Sterkfontein in Transvaal, nearly 200 miles north-east of Taungs. He had given the name Plesianthropus transvaalensis. The findings comprise an almost complete skull without jaw, some teeth and skull fragments. Again in 1947, he discovered another evidence within a few feet from the earlier discovery. At this time he found a pelvis, fragments of long bones and some parts of the skull which perhaps belonged to the Middle Pleistocene epoch.

Dr. Broom estimated the cranial capacity of Plesianthropus as 600cc. Profile of the face is concave with long nasal bones and moderated supraorbital ridges. Alveolar prognathism is marked and the canines are small in size. The molars are larger than the chimpanzee or man.

Paranthropus robustus : The fossil remains were discovered at Kromdraai, two miles east of Sterkfontein by Dr. Broom who he had given the name as Paranthropus robustus. These remains perhaps correspond the Middle Pleistocene epoch. The remains consisted of a part of skull, parts of some long bones like humerus and ulna, and also some carpal and tarsal bones. The cranial capacity has been estimated as 600cc. High degree of prognathism is marked and the palate is parabolic, instead of U-shaped as in the apes. The cheek bones are projected forward.

Australopithecus prometheus : In a deserted lime work damp in the farm Makapansgat, about 13 miles north-east of Protieterarust, South Africa a calvarial fragment consisting of major portion of occipital bone, and the posterior part of two parietal bones was discovered in 1947. In 1948, again, 20 feet away from the above site, an almost complete mandible of an individual about 12 years of age was discovered. In the second finding, the jaw is chinless and the simian shelf is absent. The teeth resemble more to the Heidelberg men and Sinanthropus.

In 1953 another evidence was found from the same site—a mandible of an adult came out where the jaw was very robust in size. This mandible is more close to Sinanthropus. According to Dr. Dart the cranial capacity is about 650cc. Considering all the features, he presented these fossil remains as human.

Paranthropus crassidens: In 1948, Broom discovered a massive man-like mandible in the deposits on the Swartkrans farm, about a mile away from the Sterkofontein quarry. A large number of fossil remains at Swartkrans were also discovered by Broom and Robinson during the year 1949 and 1950. Those fossil remains of Paranthropus include many skulls of which two are almost complete, three complete mandibles, about 300 teeth, the right half of the pelvis and different bones of skeleton.

The teeth pattern, specially the incisor and the canine teeth are typically human in size and form. The premolars and the molars are larger than the modern man. The skull is with a sagittal crest which is similar to male gorilla. Supraorbital ridges are projecting with a supraorbital torus. Upper jaw is prognathus and there is no premaxillary diastema. Some of the lower jaws exhibit a distinct chin. The right half of the excavated pelvis is typically human in form except the form of the ischium which gives an ape-like feature. These characteristics suggest that the Paranthropus was able to stand and walk erect to a considerable extent.

Telanthropus capensis: In the year 1949, Robinson discovered some fragments of the lower jaw at the cave of Swartkrans which were relatively finer and man-like. This jaw has been named as Telanthropus capensis and this remain is more or less perfect. The right condyle of this jaw is missing and some portion of left is also not found. The ramus of the jaw is broad and short. The simian shelf is not present but the mylohyoid groove extends up to the mandibular foramen as it found in human mandible. Robinson and Broom considered the general appearance of this mandible that reminds us about the Heidelberg jaw. They also belileved that this specimen belonged to a stage between ape-man and true man.

Zinjanthropus boisei: The fossil remains were discovered by Mrs. And Dr. L.S.B.Leakey in the year 1959 from a Lower Pleistocene deposit in Olduvai Gorge, Tanganyika Territory, East Africa. The excavated skeletal remains contained an almost complete skull (without the mandible) and the larger part of the tibia. The fossil remains were also associated with several primitive stone tools including nine choppers, one hammerstone, five natural stones and 176 flakes from the same bed along with some other skeletal remains of snake, lizard, frog, rat, mouse, bird, baby pig, antilopes, etc.The age of Zinjanthropus as estimated by Dr. Leakey is about 600,000 years or possibly more. But, it is too much according to some scientists. Two renowned geologists, Evernden and Curtis of the University of California, have attempted for an exact dating by the potassium-argon method. According to them the age of Zinjanthropus is at about 1,750,000 years.

The excavated skull belongs to a youth whose age must be between 16 to 18 years. The sagittal crest is well developed and supraorbital torus is massive. The post-orbital constriction is marked which makes the temporal fossae large. The inter-orbital width is enormous; the face is very long and broad. The zygomatic portion of the maxilla are large. The eye-sockets are wide and set apart. The nasal bones are long and narrow in general and the palate is very high.

The dental characters reveal that the dentition is regular without any premaxillary diastema. The canines and the incisors are not larger than the modern man. The right lateral incisor is somewhat smaller than the left. This condition is occasionally met with in modern man but rarely seen among the anthropoids. The premolars and molars are very large in size. Though the third molar was not fully errupted, still it is evident that the third molar is smaller than the second molar as it is found in man.

The mastoid process is well developed like that of man. The nuchal crest is strongly developed but for its form and low situation it resembles the hominid form. The foramen magnum is situated at the base of the skull more anteriorly towards frontal plane which is seen only in case of Homo sapiens. It suggests that the Zinjanthropus could walk erect. The cranial capacity has been calculated as 530cc which in adult individuals might be 700cc. As per the form of the long bones, the stature has been estimated which is not more than 4 feet 9 inches.

Zinjanthropus is considered as the earliest man who could make tools. Being a tool maker, he continued as a food-gatherer. While gathering vegetables, he used to collect small animals too. The diet consisted of only vegetables as supported by the wear of his teeth. The period when Zinjanthropus lived, the animals were passing through a phase of gigantism. There were pigs as large as rhinos, sheep could stood seven feet at the shoulder. So were the hippos, antelopes, cattle and baboons. Zinjanthropus was unable to deal with those giant forms. Therefore he remained basically as a food-gatherer, instead of a big game hunter. It is not known whether he was capable of articulating speech but it may be assumed that he could express and communicate simple ideas in sound.

Dr. Leakey has taken the measurements on different parts of the skulls of Australopithecus africanus and Zinjanthropus boisei which helps to arrive at a comparative statement in relation to these different fossil remains.

Comparative differences on the skulls of Australopithecus and Zinjanthropus

Measurements	*Australopithecus africanus*	*Zinjanthropus boisei*
1. Maximum Length	147 mm.	174 mm.
2. Maximum Breadth (Supra mastoidal)	120 mm.	138 mm.
3. Maximum Breadth (Inter-temporal)	99 mm.	118 mm.
4. Basion-Bregma Height	105 mm.	99 mm.
5. Bi-orbital Diameter	88 mm.	122 mm.
6. Inter-orbital Diameter	24 mm.	42 mm.
7. Bizygomatic Breadth	131 mm.	188 mm.
8. Upper Face Height	74 mm.	114 mm.

Measurements	*Australopithecus africanus*	*Zinjanthropus boisei*
9. Nasal Length	49 mm.	73 mm.
10. Nasal Breadth	27 mm.	42 mm.
11. Palatal Length	64.6 mm.	84 mm.
12. Palatal Breadth	64.6 mm.	84 mm.

Homo Habilis : The fossil remains of Homo habilis were discovered by Dr. Leakey at Olduvai Gorge in East Africa in the year 1962. The remains are the parts of skull, upper jaw and lower jaw. Although the Zinjanthropus and the Homo habilis are the contemporary representatives, but the Homo habilis is considered as more advanced type. According to Leakey, Tobias and Napier Homo habilis resembles Homo sapiens. But other scientists do not agree with this observation. To them Homo habilis is a variety ofAustralopithecine which may be regarded as a relatively progressive Australopithecus.

Homo erectus

In the stage Homo erectus, the first fossil evidence was discovered from Java in 1891. Many more discoveries were made subsequently from different regions in different times. It is an usual practice to place the new findings under a new name as a new species or with a new generic name. As a result we find several generic names like Pithecanthropus, Sinanthropus and Atlanthropus. As a matter of fact, now most of the scientists have taken a decision to label them under a single species known as Homo erectus. Still, some of them prefer to call the members of this stage by the general term *Pithecanthropine.*

In respect of physical features a wide range of variations are observed among the members who belong to the stage of Homo erectus. The variations are created due to various factors. In a simple way it may be stated that the type is projected on the basis of meagre materials. The materials come through a very long span of time; so it is quite natural that the earlier type will vary from the later one. The geographical variations are also a factor as because the materials were collected from different geographical regions which were situated quite apart from each other. Again, a consideration of age and sex is involved here. Skeletal remains may show a lot of variation for the difference of age and sex.

Pithecanthropus erectus : In 1891, a Dutch Doctor, Eugene Dubois discovered some fossil remains from the village Trinil situated on the bank of river Solo in Java. The skeletal materials consist of a skull-cap, a thigh bone (femur) and three teeth of which one is lower premolar and two are upper molars. The remains were associated with many fossilized plants and animals of different types including rhinoceros, hippopotamus, elephant etc. The pattern of fauna indicates that at one time Java was connected with the main land.

The geological age is supposed to be the Middle Pleistocene epoch. The original remain consisted of the complete top of the cranium or skull cap including the supraorbital region of the frontal bone and the nuchal region of the occipital bone. The length of the skull cap is 18.5cm and the breadth is 13.0cm which gives a cephalic index of 70. So, it seems that they belong to the Dolicocephalic group. The cranial capacity is estimated as 940cc. In case of man it ranges from about 850 to 1700cc. while in apes other than the gibbon it varies between 290 and 650cc. Therefore, in respect of cranial capacity, the Pithecanthropus erectus stands in an intermediate position between apes and man.

The vault is very low and almost like the apes. The cranium is also flattened. The supraorbital ridges are continuous across the frontal bone and fuse in the middle line to form a torus like the apes. The forehead is very narrow and slanting as the apes. The frontal bone shows a slightly marked median keel. The temporal region is not much prominent and the temporal lines are widely separated as it traces in the gibbon, chimpanzee and man. The nuchal plane is inclined more than the man but less than the apes. The main characteristics place it in an intermediate position between chimpanzee and Neanderthal Man. In respect of brain development it also holds an intermediate status between man and large apes.

The teeth are found as of enormous size and each one is larger than corresponding tooth of the man. The roots of the teeth are strong and widely separated as in the apes, but the crowns are more akin to human types.

The femur is a complete one, its length is 45.5cm which suggests that the possessor exhibited an height about 167cm to 170cm approximately. This femur is straight, slender and long. A ridge exists on the femur known as the *linea aspera* which means the powerful extensor muscles were attached on this ridge, essential for erect posture of the individual concerned. In all characters the femur closely resembles with that of a modern man. The name signifies the 'ape man with erect posture'.

A book entitled "Evidence as to Man's Place in Nature" was published in 1863. The book was written by Thomas Huxley, a stern supporter of Darwin. He had illustrated the similarities between the man and the apes. Again, at this time the scientists were looking for 'missing links' between the apes and man. The discovery of Pithecanthropus erectus in 1891 was considered to be the 'missing link'.

Pithecanthropus is a member of Pithecanthropines which has been included under Homo erectus. The Homo erectus represent a stage of hominid evolution beyond Australopithecines.

Sinanthropus pekinensis

In the year 1926, Dr. Anderson and Dr. Zdansky discovered a few bone remains from the village Choukoutien, south-west of Peiping (Peking) in China. Out of these fossil remains, two molars in many respect showed human characters. In 1927, another molar tooth was discovered by Dr. Bohlin. Prof. Davidson Black of Peking Union Medical College, after a careful study of this third kind of molar tooth declared that those belonged to the very earlier types of man and gave the name Sinanthropus pekinensis. Intensive excavations at various sites began following that discovery. As a result further remains were obtained. In 1929, a young Chinese Palaeontologist, Dr. W.C.Pei, discovered a notable complete brain case of Sinanthropus. During the period from 1928 to 1937 different other fossil remains of Sinanthropus were unearthed. These fossil remains were carefully studied by Prof. D.Black till his death in 1934. After that the study was continued by Prof. Franz Weidenreich.

The age of Sinanthropus has been determined by the examination of stratigraphical sequence, the evidences of fauna and available tools. All these circumstantial evidences indicate that the Sinanthropus belonged to a period between Lower Pleistocene to Middle Pleistocene. The animal remains or fauna include a primitive water buffalo, rhinoceros, bears, deer, hyenas, rodents, etc.
Dr. Anderson observed roughly worked pieces of quartz in the strata which were completely foreign material in this locality and no doubt they were brought from outside, by man. However, these quartz pieces were very crude implements like choppers, scrapers etc., made on the chipped core of quartz. Pieces from the antlers of deer were also used as tools. The Sinanthropus fossil remains included three incomplete brain cases, 12 fragments of lower jaws, nearly 50 isolated teeth, a collar-bone, two fragments of long bones, and four ungual phalanges.

In general, the skull is similar to Pithecanthropus, but in many respects it is much advanced. The average maximum length of the skull is 19.4cm, although the range varies from 16.5 to 20.5cm. The maximum breadth of the skull varies from 13.7 to 14.3cm; the cranial index is 72.2. It has a very long head for which it can be technically called as dolichocranial. The vault of the skull is 11.5cm high and the cranial capacity varies from 850cc to 1300cc. with an average of 1075cc. The skull bones of Sinanthropus are very thick and massive. The forehead is receeding. Supraorbital ridges are pronounced and continuous.

A ridge runs down the middle of the skull from front to back at midsagittal plane, known as sagittal crest or median keel. The parietal bones of two sides are more or less flattened. The occipital

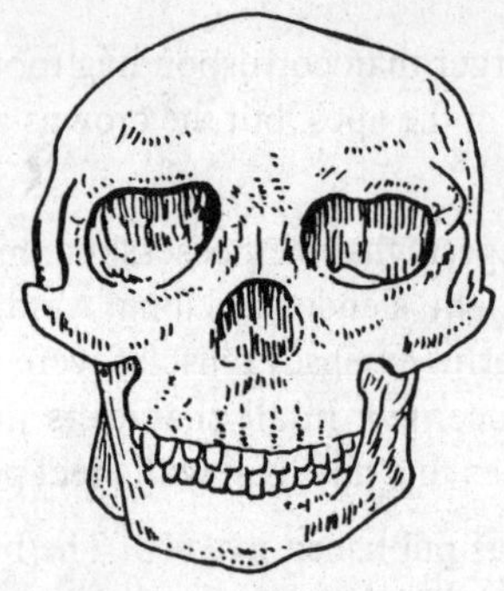
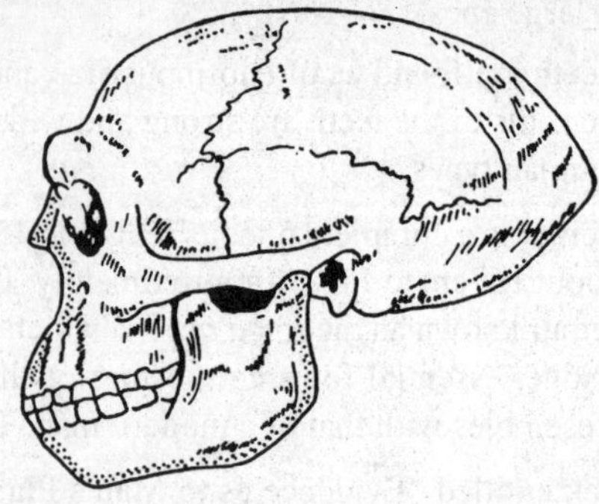

Peking Man (Sinanthropus)

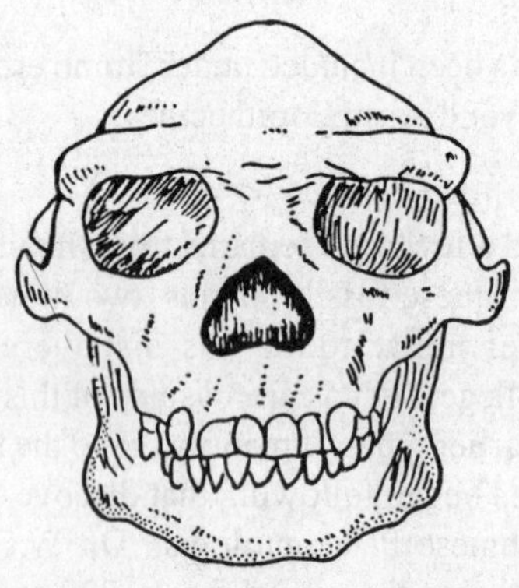
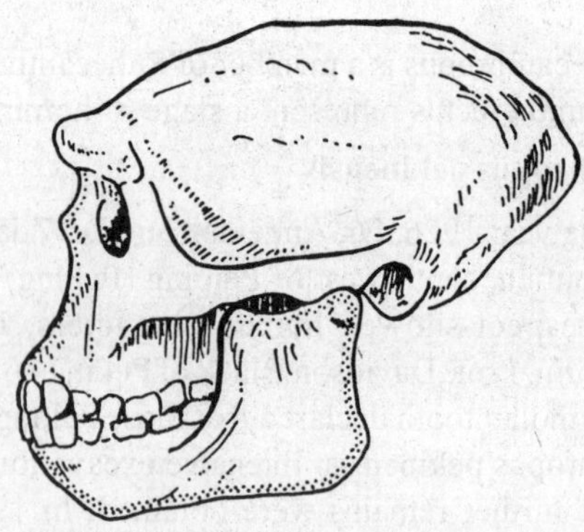

Java Man (Pithecanthropus)

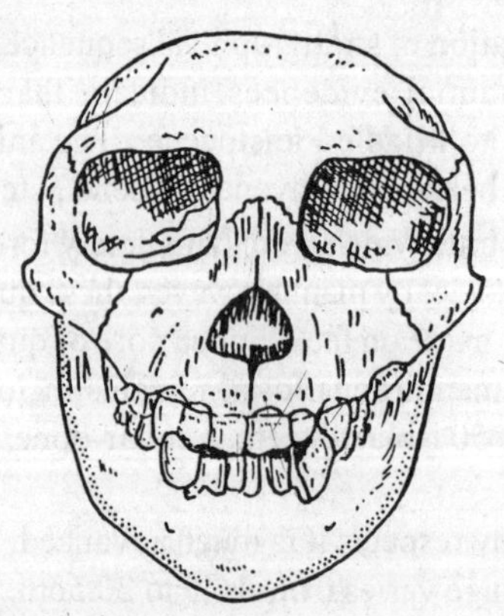
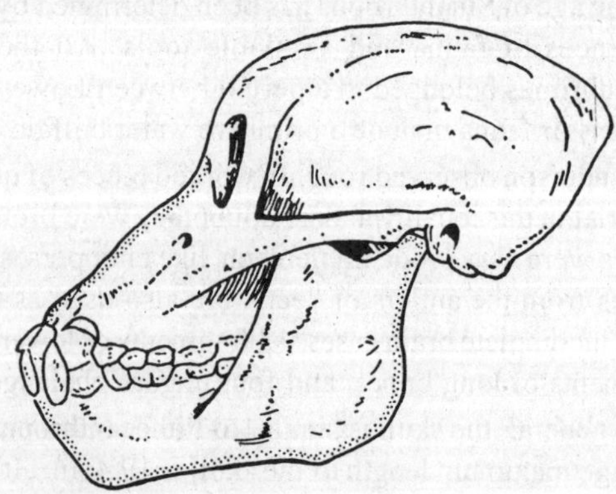

Proconsul

Fig. 3.11. A Sequence of Cranial Development

torus is well-developed and the occipital bone makes abrupt slope while it is arching in case of man. The face is small but projecting. In comparison to length, the face is broad. The nasal bones are broader than the man. The nasal bridge is broad and high and the nasal spine is not present. The cheek bones are quite prominent which resemble to the modern Mongoloids. The orbits are large. The mandible in male is very big, heavy and much greater than the lower jaws of modern man, but the female jaws of fossil remains are comparable to the jaws of modern Mongoloid females. The chin is absent. Thus, a Sinanthropus mandible presents mixed characters of apes and man. All the teeth are large and robust; their roots are longer and enamel is thicker than the modern man. The upper canines are large and they rise above the level of other teeth. But there is no diastema in these teeth. The molars are low and long while in hominids they are high and short.

Sinanthropus resembles so closely to Pithecanthropus that they represent two geographical races of the same Pithcanthropoid stock. Prof. Black said that the Sinanthropus is more advanced than Pithecanthropous, but Prof. Weidenreich was not in favour of accepting this opinion. Rather he suggested, both of them are on the same level of human evolution. According to him, the line of evolution proceeded from Sinanthropus-Pithccanthropus stage through Neanderthaloid forms to modern man. Some other anthropologists think that although the Pithecanthropus and the Sinanthropus are the two different varieties of a same stock, there are some acute differences. Pithecanthropous represents a more primitive form with lower cranial capacity, large frontal sinus, greater flattening of the frontal region, earlier fusion of cranial suture, heavy construction of mandible, presence of diastema in between canine and lateral incisor, etc.

However, Pithecanthropus erectus *(Java man)* and Sinanthropus pekinensis *(Pekin man)* both are the early hominids who lived during the Pleistocene epoch in Asia. A comparison between the two proves the minute differences.

Comparison between the Pithecanthropus and Sinanthropus

Characters	Pithecanthropus	Sinanthropus
1. Size and form of the skull	Small in size and the form is more or less same.	Large in size and the form is same.
2. Maximum length of the skull.	Average 18.5 cm.	Average 19.4 cm.
3. Maximum breadth of the skull	Average 13.0 cm.	Varies between 13.7 to 14.3 cm.
4. Cranial index	70.0 Dolichrocranial	72.2 Dolichrocranial
5. Vault of the skull	10.5 cm (Low)	11.5 cm (High)
6. Cranial Capacity	Average 860cc.	Average 1075cc.
7. Cranial bones	Not so much thick and massive	Too much thick and massive.
8. Forehead	It is receeding and the frontal region is flat instead of bump-like development.	It is receeding and the frontal region shows a bump-like development.
9. Median keel	Present.	Present.
10. Supraorbital region	Heavy and continuous. No furrow is seen.	Heavy and continuous. A distinct furrow separates the forehead from the supraorbital region.
11. Parietal region of the skull.	Flattened	Flattened.
12. Occipital region	Broad and rounded	Comparatively narrow and elongated.
13. Frontal sinus	Large	Very small
14. Palate	Smooth	Rough
15. Mandible	Massive	Massive

Characters	Pithecanthropus	Sinanthropus
16. Lower canines	Smaller	Larger
17. Molars	Larger in size. They increase in size from first to third.	Small in size
18. Diastema and lateral incisors.	Present	Absent.
19. Chin	Absent	Absent.

Wadjak Man

Dr. Eugene Dubois discovered two fossil crania at Wadjak, situated 60 miles south-east of Trinil in Java. He found these in 1889-90 but announced the discovery in 1920. The age goes back to the Pleistocene epoch. The Wadjak skull-I was a female skull with a cranial capacity of 1550cc. But the Wadjak skull-II was the skull of a male with a cranial capacity about 1650cc. Both the skulls are dolichocranial; the forehead is receeding and the supraorbital ridges are prominently marked. The orbits are low and broad. The nasal root is interestingly depressed; it is just below the prominent glabella region. The nasal bones are small, narrow and flat but the nasal aperture is wide. The alveolar projection is seen forwardly. On the whole the Wadjak crania show some striking similarities with the Australian aboriginal skulls.

Homo Soloensis (Solo Man)

Eleven fossil skulls and two tibiae were unearthed in 1931 from Nagandong, the region near Solo river in Central Java. Along with these fossil remains, other findings like the fossil bones of stegodon, elephants, deer, hippopotamus, rhinoceros and recent species of cattle were found. Besides, an axe made of deer antler, a spearhead, several crude stone tools and a few nicely worked bone tools were discovered which occurred in association with the Homo soloensis at the same site. According to geological time scale these fossil remains correspond to the Upper Pleistocene period; culturally the Solo man belong to the Upper Palaeolithic Age. So far as the physical characters are concerned the forehead is low and receeding, the supraorbital ridges are prominent to form a continuous ridge. The mastoid process is well developed. Average cranial capacity is 1100cc. These physical characteristics bring them close to Pithecanthropus than the Sinanthropus. In consideration of the evolutionary line it represents the next stage above the Pithecanthropus but is more primitive than the Neanderthaloids.

Homo heidelbergensis (The Mauer Jaw)

In the year 1907 Dr. Otto Schoetensack discovered a massive lower jaw near the village Mauer, six miles south-east of Heidelberg in Germany. The Mauer sands are very rich in fossil remains. The associated fauna includes the ancient elephant (Elephus antiqus), horse, wild beer, a bison, the etruscan rhinoceros, a sabre-tooth tiger, lion, etc. Geological time of this deposit is Lower Pleistocene. It is probable that the Heidelberg man flourished during the Abbevillian/Chellean cultural period. The discovery of this jaw is much important for the prehistorians because it provides crucial information regarding the Lower Palaeolithic man.

Since the jaw is very large and massive, it gives a very powerful appearance. The ascending ramii are very broad and low, and almost sqrare in shape. The chin is completely absent. The ramii are 5cm in breadth and 6.6cm in height. In modern man this breadth is 3.7cm. The mandibular notch is not very deep. The condyloid process is a large articular surface at a higher level than the coronoid. The coronoid process is blunt, rounded and placed at a lower level than the condyle. The sigmoid notch is shallow. This fossil remain is similar to the jaw of the gibbon. The horizontal ramus is high and massive than in modern man. The lower border is concave as it found in the baboons, not like the apes and man. The angle of the jaw is truncated like that of a orang and also a Neanderthal man. In dentition, the alveolar border looks parabolic instead of U-shaped as it is found in the apes. Diastema is absent and the dental series is regular as well as continuous like that of the modern man.The teeth are of ordinary size and the incisors are quite normal. The canine is small which is not projected beyond the level of other teeth. The premolars are found normal in size. The lower first molars show five cusps. So the dimensions and characters of the molars are much similar to the modern man. It is remarkable that, where the jaw projects simian features, the dentition is altogether human.

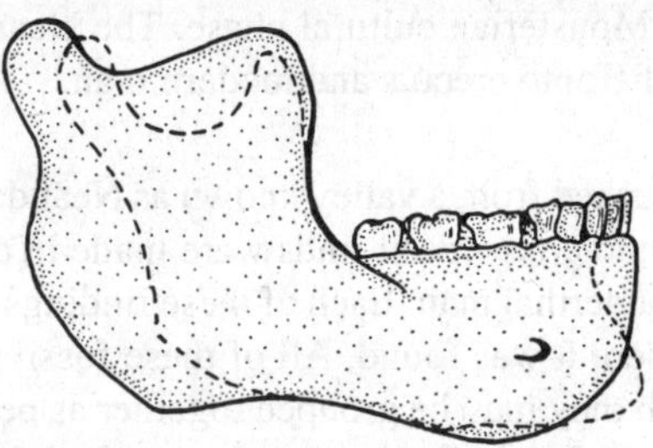

Fig. 3.12. Heidelberg Jaw-compared with the jaw of a modern man (The dotted line denotes the outlines of Homo Sapien's jaw)

Rhodesian Man

In 1921, Mr. T.Zwigelaar unearthed some bone remains during the mining operation at Broken Hill, Northern Rhodesia in South Africa. These remains included a skull that lacked the lower jaw, a portion of upper jaw from a relatively smaller skull, a sacrum, some fragments of a pelvis and portions from femora and a tibia which occurred in association with some stone and bone tools. Besides, broken bones of other animals,which were used as food were also found in fossilized form. The implements and tools closely resemble to those used by the modern Bushmen. Exact geological time period is not known. But, from the evolutionary point of view it may be placed in Upper Pleistocene, though it is not certain.

The skull is long and narrow. The maximum length of the skull is 20.6cm and maximum breadth is 14.5cm. It shows that the skull is dolichocranial. The cranial capacity, according to Dubois is 1400cc while Elliot Smith has estimated it as 1200cc. The supraorbital ridges are prominent in comparison to Neanderthal man and the torus exceeds when it is compared to the gorilla. The forehead is receeding. The occiput is flattened and the mastoid processes are well developed. The face is markedly long and pose close similarity with the Neanderthal man. The nasal aperture is large; the orbits are high and of great size. The palate is large, wide and deep while the teeth maintain the same arrangement as that of modern man. The canine is normal and the third molars are reduced in size.

Some scholars prefer to call Rhodesian man as Neanderthaloid while others like to put them under Homo erecus. Maximum number of physical characteristics of Rhodesian men are similar with Neanderthal man, and for this reason most of the authorities considers the Rhodesian man as a special form of Neanderthal—an African variety of the Neanderthals. Prof. Boule has stated that the Neanderthal man, the Rhodesian man and the modern Australian race possess certain common primitive characters which denote a common origin.

Saldanha Man

The fossil remains as the hominid skull cap and fragment of a lower jaw were discovered by Keith, Jolly and Ronald Singer of the University of Cape Town in the year 1953 from a village near Saldanha Bay which is situated about 90 miles north of Cape Town. These fossil remains suggest that the Saldanha man was very much similar to the Rhodesian man as the frontal bone, brow-ridges, the contours and the form of the skull of both were more or less same.

According to their similarities it can be said that both (Saldanha and Rhodesian) belong to the same group of man. But it is interesting to note that the jaw of Saldanha man, in shape and dimension, is much similar to the Heidelberg Jaw whereas the ramus of the Saldanha mandible is less robust than the Heidelberg jaw. Many other fossil remains of different vertebrates and some stone tools were also found in association with the evidences of Saldanha man. The stone tools comprise of handaxes and other implements of Acheulian type. The existence of Saldanha Man is ascribed to the early part of the Upper Pleistocene period. In the evolutionary sequence they are somewhat older than the Rhodesian man.

Neanderthal

In Hominid evolution, the third stage is represented by the Neanderthal man when the hazy and dim picture of human emergence became much clearer. According to culture context, that was the Middle Palaeolithic Period which corresponds to the Mousterian cultural phase. The Neanderthal stage is an intermediate position between the stages of Homo erectus and modern man.

Neanderthal Man

In 1856, a skull cap and some long bones were unearthed from a valley known as Neanderthal in Germany. Since that time, subsequent discoveries of numerous fossil finds were made in different parts of the world which has been attributed to the Neanderthal man. Each of these findings possess a separate name on the basis of its excavation site where it was found. All of these fossil remains show a good number of common features due to which they may be grouped together as belonging to the Neanderthal race. It is a matter of regret that most of these findings are incomplete. However a list of some important discoveries relating to the Neanderthal man has been furnished below :

Year	Place	Name	Materials
	EUROPE		
1856	Dusseldorf, Germany	Neanderthal	A skull cap and a few long bones.
1914-16	Weimer, Germany	Ehringsdorf-I	Two mandibles and the fragments of skeleton.
1925	Weimer, Germany	Ehringsdorf-II	Skull fragments.
1933	Wurtemberg, Germany	Steinheim	Skull without the lower jaw.
1848	Gibraltar	Gibraltar-I	Skull without the lower jaw.
1926	Gibraltar	Gibraltar-II	Fragments of skull
1899-1905	Croatia, Yugoslavia	Krapina	Skeletal fragments of twenty individuals.
1908	Correze, France	La-Chappelle-aux-Saints	Skeleton
1908	Dordogne, France	La Moustier	Skeleton.
1909-21	Dordogne, France	La Ferrassie	Skeleton of two adults and four children.
1908-21	Charente, France	La Quina	Fragments of many skeletons.
1886	Namur, Belgium	Spy	Two skulls, Fragmentary skeletons.
	ASIA		
1925	Palestine	Galilee	Skull fragments.
1931-32	Palestine	Mt. Carmel	Skeleton fragments of twelve individuals.
1938	Uzbekistan	Baisum	Skull
	AFRICA		
1939	Morocco	Tangiers	Upper Jaw.

After careful observation Hooton has divided the Neanderthal people into two divisions—*Conservative Neanderthals* and *Progressive Neanderthals,* The conservative division shows the classical Neanderthaloid features whereas the progressive division has acquired a few more special characteristic features, which facilitate the division to establish a closer relationship with *Neoanthropic man.* La Chapelle-aux-Saints, La Moustier, La Quina, La Ferrasie are the examples of the conservative division while the Krapina, Ehringsdorf, Steinheim, etc. are the members of the progressive division of Neanderthal people.

The Neanderthal man flourished during Middle Palaeolithic Period, which corresponds to the Mousterian cultural phase. At the end of the Lower Palaeolithic Period the climate began to cool down and continued in the advance of Wurm glaciation.

The flora and fauna that are found in association with the skeletal remains indicate a very moist and too cold climate. The chief animals of this period were the *woolly mammoth* (Elephas primigenius)

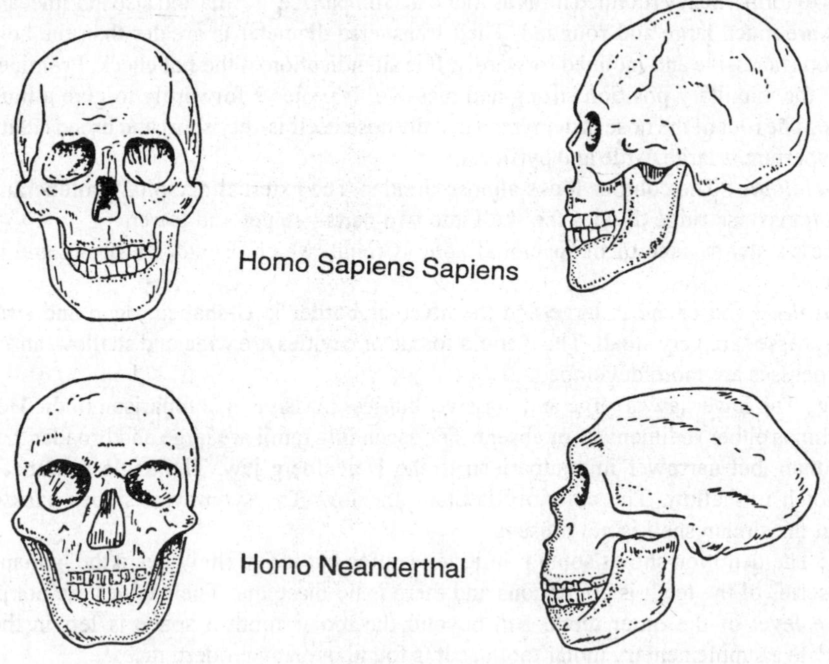

Fig. 3.13. Skulls as different between Neanderthal Man and Modern Man.

and the *woolly rhinoceros* (Rhinoceros antiquitatis). Apart from this typical fauna, the evidences of extinct species of animals like cave-lion, cave-bear etc. were also found.

A development of stone tools was an important phenomenon of this period. The stone tools were made from the flakes of flint and finished by chipping along the edges of one face only. The tool types include points and scrapers of various designs, and similar other artifacts.

A typical Neanderthal man has been represented by the fossil remains known as La Chapelle-aux-Saint that were found from a small cave in the Correze district, France in the year 1908. Excavated fossil finds included a skull with lower jaw, a clavicle, two almost complete humeri, two incomplete radii, some bones of hand, fragments of ilia, two incomplete femora, parts of tibial, several bones of foot, good number of vertebrae and ribs. The associated remains were dressed flints, stone tools as the scrapers and points, skeletal remains of woolly rhinoceros, reindeer, extinct bison, cave-hyena etc. Geological time scale points the people as belonging to Upper Pleistocene where the cultural stage has been depicted as the Mousterian culture.

Physical characters

Skull: General — The skull looks very large and heavy. The bones forming the vault are thick.

The maximum length and breadth of the skull are 20.8 cm and 15.5 cm respectively. The cranial index is 74.5 which indicates the specimen as dolichocranial. But, in fact, the cranial index of the Neanderthal varies between 70 and 76. The cranial capacity is about 1600 cc, which is above the average capacity of modern Englishman. The face is well developed in comparison to the cranium.

Norma Verticalis : The frontal bone is receeding. The skull seems to be much enlarged at the back. The projecting parietal eminences are placed backwardly. The lowness of the vault gave the skull a flattened (Platycephalic) appearance. The temporal fossae are large and the temporal lines rise about halfway up the side of the brain box. The squamosal suture is not so much arched.

Norma Frontalis : The face is long and projected forwardly. The supraorbital ridges immense and continuous to form a heavy rounded torus as found in chimpanzee, gorilla and also in Pithecanthropus. The orbits are much large and rounded. Their transverse diameter is greater than the height. The malar region is massive and inclined forwardly. It is an indication of the flat cheek. Prominent upper jaw makes the maxillary position strong and massive. It projects forwardly to give a muzzle-like appearance. The root of the nose is depressed and the nose itself is very short and broad (Platyrrhine). The nasal aperture is large, wide and pyriform.

Norma Occipitalis : The contour looks almost circular. The external occipital protuberance is not present. A transverse ridge divides the skull into two parts—upper and lower.

The whole outer surface of the occipital bone is rough which provides a strong and muscular impression.

Norma Basalis : The palate is large and the alveolar border is U-shaped, deep and strong. The mastoid processes are very small. The glenoid fossae or cavities are wide and shallow, and the post-glenoid processes are more developed.

Lower Jaw : The lower jaw is large and massive, but less massive in comparison to the Heidelberg jaw. The chin is either rudimentary or absent. The ascending ramii are large and broader than that of a modern man, but narrower in comparison to the Heidelberg jaw. The angle is truncated. The sigmoid notch is shallow. The coronoid is blunt and low. The symphysis slopes backward and inward, but the simian shelf is not present.

Dentition : The dentition shows some primitive features but it definitely bears the human features also. The series of the teeth is continuous and there is no diastema. The canines are not projected beyond the level of the other teeth. But beyond the molar tooth a space is left in the jaw to accommodate a supplementary molar tooth as it is found in many modern races.

The Bones of the Trunk and Limbs :

The strong and stout bones indicate powerful muscular development. The vertebral column is short and massive. The strong ribs indicate large thorax with powerful intercostal muscles. The clavicles are long, slender and arched as in chimpanzee.

The strong and short humerus possesses a large head, which does not differ much from that of a modern man. The radius shows a strong curvature unlike that of a modern man. The forearm is short in relation to the upper arm and the bones of forearm are strong. The hand is almost human in character while the carpals are relatively small and the fingers are comparatively short.

The femur is strong, massive and it has a large head. The shaft of the femur is bowed forwardly as in apes and it does not show any sharp development of linea aspera. The tibia is also very strong and short. The upper head of the tibia is bent back with a backward slope and the knee joint suggests that the Neanderthal man perhaps walked with knees bent.

The foot exhibits some interesting characteristics. The head of the ankle bone is directed inward showing the great toe widely separated from the other digits. In general, the foot does not show much simian features as like the other bones of the limbs. The stature of the La Chapelle-aux-Saints has been calculated by Prof. Boule as 5 feet 1 inch. As a matter of fact, the stature of the classic Neanderthals varies between 5 feet 1 inch and 5 feet 5 inches.

Progressive Neanderthals

A few Neanderthal remains exhibit certain special characteristics and have been known as progressive Neanderthal types. Some of the members of this group need to be explained in order to get an accurate picture.

Krapina : The fossil remains of about twenty individuals in association with some other evidences of Mousterian industry were discovered from Krapina, at Croatia, Yugoslavia in 1899. Unearthed fossil findings were fragmented skeletal remains. Apart from regular Neanderthal features, these fossil remains show some special characters in relation to forehead and the shape of the head (round) which, suggest that the Krapina man is more toward the Neoanthropic man. The Krapina head is brachyccphalic in character.

Steinheim : A skull without the lower jaw, in association with Acheulian industry was discovered from a gravel-pit at Steinheim in Germany in the year 1933. The fossil remains were buried in a deposit of Pleistocene epoch. It seems to have belonged to a young woman whose skull was long and narrow. The cranial index is 70 and the cranial capacity is 1070 cc. The Steinheim skull resembles the conservative Neanderthal type. But, the other characters, such as a lower degree of facial prognathism, presence of canine fossae, well developed squamous portion of the temporal bone and reduction(in size) of wisdom teeth and a few characters in connection with facial and occipital parts, show a similarity with Neoanthropic man.

Mount Carmel: The Mount Carmel skeletons were discovered in the year 1931- 32 from two adjoining caves in Palestine along with the layers which yielded Levalloiso-Mousterian industry. It was the result of Joint Scientific expedition of the American School of Prehistoric Research and the British School of Archaeology in Jerusalem. The first cave is Skuhl from which the fragmentary remains of at least 10 individuals and many other isolated bone fragments were collected. The second cave is Tabun, which revealed a skeleton of an adult female, a mandible of an adult male and many bones and teeth. The Mount Carmel men were tall and the women were short and medium.

The features of the skull suggest that the heads are massive, the foreheads are moderately full and the occiputs are slightly projected beyond the attachment of the neck. The face shows moderate height and the head are of medium height. There are no canine fossae. The chin is almost well developed the teeth are smaller in size in comparison with the typical Neanderthals.

Ehringsdorf : A human lower jaw was discovered from the soil of Ehringsdorf, a village near Weimer in Germany in the year 1014. A German scientist, Schwasbe called it as the *Weimar Jaw.* During the period between 1916 and 1925, some fragments of a shattered skull were discovered in association with the artifacts of Pre-Mousterian and late Acheulian types. The bones of the skull are thin and the vault is high unlike the conservative Neanderthals. The supraorbital ridges are continuous and separated. The length and breadth of the skull are 16.6 cm and 14.5 cm respectively. The cranial index is 74, which categorize the skull as dolichocranial. The cranial capacity has been measured as 1480 cc, which is equal to the average modern man.

End of Neanderthals : Once the Neanderthals were numerous and prominent over Europe. But gradually they became extinct. Some scholars suggest an extermination by more progressive newcomers—the Cro-Magnons, through conflict or deadly war. Others propose that the Neanderthals were extinguished as a result of natural inconveniences like severe climatic condition, food crisis and catastrophic diseases. Third opinion emphasized more on the total absorption by the newcomers. Perhaps more than one factor were responsible for the extinction of the Neanderthals.

Early form of Homo sapiens : Formerly the scientists suggested that the Lower and Middle Pleistocene forms of men were not sapiens. Probably they were the ancestors of Homo sapiens. In this regard the Cro-Magnon type of fossil remains were believed to be the first species of Homo sapiens. But this hypothesis was rejected after the discovery of a few other fossil remains from the earlier layers (in relation to the Cro-Magnon and even of the Neanderthaloids). These fossil remains exhibited a large number of characteristic features of Homo sapiens which are not explainable until we do not suggest the theory of co-existence of the Homo sapiens and the Neanderthaloids. However, in Europe the development of Homo sapiens was parallel to that of the Neanderthaloids. Newly discovered fossil remains have been named as the London Skull, Swanscombe Man, and the Galley Hill skeleton, the Fontechevade Skulls and Boskop Skull. There is a great controversy regarding the fact that whether these above mentioned fossil remains form an early group of Homo sapiens or not. Some of these fossil remains may be discussed in the following ways.

The London Skull : This fossil evidence was discovered in 1925 from Central London. It consists of an occipital, left parietal and a few parts of right parietal bones. It *is* also known as 'Lady of Lloyds' because most probably it belonged to a women of 50 years. Although this skull presents many features of Neanderthal, still it seems to be closer to the Cro-Magnon than the Neanderthal. The cranial capacity of this skull has been calculated as 1260 cc which is 40 cc lesser than the present day English women. However, this London skull comes from the deposits of Upper Pleistocene epoch.

The Swanscombe Man

The fossil remain of the Swanscombe Man is represented by a left parietal and the occipital bones which were discovered at Swanscombe in Kent, England during 1935-36. In the year 1955 the right parietal bone was discovered from the same deposits. Geologically, the period corresponds to the second Inter-glacial or the Middle Pleistocene epoch. A few Geologists also admitted the authenticity of the Swanscombe man. Morant and Le Gros Clark have studied the fossil remains of the Swanscombe skull and declared that this must have belonged to a woman of about 20 years.

In form and features these bones are almost same to those of Modern English women. The cranial capacity is 1325 cc which shows similarity with those of modern skull. Important exception is the remarkable thickness and the great breadth of the occipital bone. Artifacts of Early Middle Acheulian types have been found associated with the Swanscombe man. This suggests an early existence of Neoanthropic man.

The Galley Hill Skeleton

In 1888 the Galley Hill skeleton was discovered from the 100 feet terrace of the Thames River near London. A workman discovered this while he was working in the gravels. The fossil remains were found in a pit, 8 feet down in the ground. The site was situated at a distance about 500 yards from the quarry in which the Swanscombe man was discovered. Earlier it was thought that the Galley Hill man appeared in the second Inter-glacial period; but the recent tests reveal that the Galley Hill man was buried in postglacial times and so its age must not exceed a few thousand years. Some authorities believe that the Galley Hill man represents a primitive variety of Homo sapiens.

The Fontechevade Skulls :

In 1947, Mile G. Henri-Martin discovered some bony remains in the cave of Fontechevade, France. The bony remains consist of two brain cases; one of which is complete, and the other one is fragmentary, These remains are more similar with the modern man in shape and size. The bones are quite thicker. The 'complete skull' is dolichocranial and it resembles with modern man. The second skull is marked for its advanced forehead where the supraorbital ridges are absent. The cranial capacity is estimated as 1450 cc, which resembles that of a modern man. Both the skulls thus show humanoid features.

The remains are found in association with large flakes and the industry has been assigned to a period intermediary between Mousterian and an early period. Some authorities have placed these fossil remains in the Pre-Mousterian period without any hesitation. It may easily be remarked that the forerunners of Neanderthal man may have intermixed with the type of man who represents the Fontechevade skulls. In fact, Fontechevade man highlights the antiquity of Neoanthropic type of man because of their close similarities with Homo sapiens.

The Boskop Skull :

Some fragmentary skeletal remains were discovered from Tranvaal near Boskop in the year 1914. These fossil remains as well as other findings of fossil men from South Africa are of great importance as they unveil the process of raciation in that part of the world. The findings show that the skull was dolichocranial with a remarkable cranial capacity of 1700 cc. The parietal bones are thick. The stature has been estimated as 5 feet 6 inches. In age, the Boskop findings are as old as the Middle Stone Age of Africa. Some other representatives of the Boskop type are Zitzikama woman (1921), Fish Hoek

(1928), Springbon Flats skeleton (1929), Cape Flots (1929), Plettenberg Bay skeleton (1931-32), etc.

Men of Late Pleistocene

A great climatic change appeared just after the Middle Palaeolithic period. The cold climate was replaced by a temperate condition in the Upper Pleistocene, which corresponds to the Upper Palaeolithic period according to prehistoric archaeology. In this period, the fauna, flora and even artefacts became different from those of the Lower Palaeolithic Woolly mammoth and rhinoceros were no more found and the reindeer appeared in abundance. Since the reindeer have played a great role in the material and economic life of the people, the period has been termed as the *Reindeer Age.*

The outstanding feature of the Upper Palaeolithic industry is the use of horn, bone and ivory, besides stones, as the materials for making tools. This new culture is named as the *'Osteo-dontokeratic'* culture. However, the Reindeer Age can be classified into three successive stages—Aurignacian, Solutrean and Magdalenian. A new species of human race with graceful body, straight gait, straight forehead and finer brain appeared in this scenario. They lived in caves and rock-shelters, hunting and fishing were their main economy. The weapons and implements were beautifully made and worked with many skilful techniques. Their high intellectual development and brain capacity left many evidences of their skill and inventive genius in their place of living. Mention may be made about some of their representatives.

The Grimaldi Man

In 1901, Prof. Verneau discovered two skeletons from a cave named Grotte des Enfants in the village of Grimaldi on the Mediterranean coastal region. One of the skeleton was of a woman of about 30 years of age and the other of a boy about 15 years of age. The skeleton of the latter was stained with red ochre and the manner of placement suggested an intentional burial. The cave deposits belonged to Late Pleistocene epoch and the evidence of reindeer bones indicated that the Grimaldi man belonged to the Reindeer Age. The associated artifacts from the same site have been classified as of Aurignacian type. The flints were worked on both the sides in the form of knives, scrapers, gravers, saw, etc. A few ornaments made of bone, shells were also found along with these skeletons.

Both the skulls are long, narrow and high. They are strictly hyperdolichocranial. The cranial index of the female is 68.5 while the boy shows 69.2. The cranial capacity of the female skull is 1265cc, the male (boy) skull is 1455cc. as estimated by Keith. The skulls present an elliptical contour. The forehead is straight and well developed,as well as bulging to some extent. The regions of parietal bosses are flattened. The mastoid processes are small. Although the face is broad, it is narrow below the cheekbones. So the head seems to be disharmonic.

The supraorbital ridges are feebly developed. The orbits are large and rectangular. They are low in relation to their breadth as in the Cro-Magnon type. The nose is broad and it is depressed at the root. The lower border of the nasal aperture is not sharp as in the Negroes. The canine fossae are deep which is a Negroid characteristic. The palate is deep, long and narrow. The alveolar border is U-shaped as in the Negroids and the Australoids. The jaw is very strong with its broad and low ascending ramii and a thick body. The chin is poorly developed and there is a marked alveolar prognathism, which is Negroid feature. The teeth are large and prominent. Upper molars have four well-developed cusps and the lower have five well developed cusps. The dental characters resemble the Australian aborigines. But the majority of the characters of the skull show affinities to the Negroids.

The height of the female has been estimated as 5 feet, and ½ inches. The forearm and the leg are very long in comparison with the upper arm and the thigh respectively. Thus the limb proportions of the Grimaldi man resembles Negroids. Besides, there are other features like the sciatic notch, strong curve of the iliac bones etc., which according to Prof. Verneau suggest the Negroid character. The retroverted tibia, projection of heel bone, and strong bone shaft of the femur also indicate the same.

The Grimaldi man is comparable to the modern Negro. Prof. Sollas and Prof. Boule have pointed out strong resemblance of Grimaldi man with the Bushmen and the Hottentot of South Africa. Some authorities believe that the Cro-Magnon and the Grimaldi were two non-distinct racial groups. Both of them show several distinct characters, but detailed examination indicates a close relationship between the Cro-Magnon and the Grimaldi. This suggests a Negroid admixture in the early population of Europe.

The Cro-Magnon

In the year 1968, a few workman of the Railway unearthed the remains of five human skeletons in association with some dressed flints and a large number of seashells from the village Cro-Magnon in the Dordogne region of southern France. The first information of this discovery was made by a renowned geologist M.Louis Lartet in 1868. These skeletal remains were first studied by Broca and Pruner-Bay and thereafter by Dr. Quatrefeges and Hamy who named these as Cro-Magnon race. Later, several discoveries were also made at different parts of Europe. The geological age of these findings is held as of late Pleistocene which is later than the Grimaldi Negroids.

The Cro-Magnon race not only belonged to the late Pleistocene epoch, the associated implements indicate that these were flourished during Aurignacian stage of the Upper Palaeolithic period. They produced an excellent workmanship on the bone and stone tools, and wonderful drawings of Aurignacian time, Beautiful cave paintings, stone and ivory statuettes of female figure and similar other artistic objects were found in different regions of Europe.

The 'Old Man of Cro-Magnon' is a typical form of this group that has been considered here for description. The skull is large and massive. The length and breadth of the skull are 20.3 cm and 15 cm respectively. The vault is low and cranial index is 73.7. It is dolichocranial. The cranial capacity has been estimated as 1660 cc by Keith and 1590 cc by Boule. It shows marked projection of the parietal bosses, which makes the contour of the cranium pentagonal in form. The forehead is vertical, broad and moderately high. Supraorbital ridges are low and wide. The occiput is bulging and it is flattened at the parieto occipital region. The face is broad and flat. In this way, the head and the face make a disharmonious combination. The orbits are wide and rectangular in form. The zygomatic region is strong, large and prominent. So the cheekbones look strong and prominent. The nasal bones are high; the nose is long and narrow. The maxillary region shows prognathism. The palate is shallow and narrow. Strong lower jaws, but the ascending ramii are not very wide. The sigmoid notch is fairly deep. The chin is prominent with well-marked mental eminance. The teeth are large. According to Boule, the long bone suggests a great height, which are about 5 feet 11½ inches. The muscle imprints are strongly marked on the surface of the bones. The femur is platymeric (the flattening of the thighbone), it is strongly bowed out and possesses a well-developed linea aspera. The tibia is also platynemic (side-to-side flattening of the shinbones). As a whole, the Cro-Magnon types are characterised by robust body with long arms and shins. Moderate brow-ridges on the elevated forehead, high and narrow nose in association with a well-developed chin probably gave the Cro-Magnon a beautiful look.

There are ample evidences to show that the Cro-Magnon people continued to survive through the Neolithic period up to present day. The Cro-Magnon differs from the Grimaldi type in the following points :

1. The Cro-Magnon skull exhibits a pentagonal contour when it is viewed from the above. It is due to marked projection on parietal bosses. This feature goes against the elliptical contour in the Grimaldi.

2. Facial prognathism is less marked in the Cro-Magnon.
3. Less dolichocranial in comparison with the Grimaldi.
4. Excessive long skull in relation to Grimaldi.
5. Leptorrhine nose as against the platyrrhine in Grimaldi.

6. Taller stature than the Grimaldi.
7. Limbs are more proportionate than the Grimaldi.

The Predmost People

The skeletal remains of over 40 individuals were discovered from Predmost in between the year 1890 and 1928. Similar type of skeletal remains was also found from Brunn, a place that was situated about 50 miles at the west of Predmost. A skeleton, a skull and another skeleton was discovered from this spot in 1888,1891 and 1927 respectively. All these fossil remains in association with the artifacts belonged to a period between the late Aurignacian and the early Solutrean.

The height of Predmost man is 5 feet 7 inches and the mean cranial capacity is 1599 cc. This height along with other physical features like the development of supraorbital ridges and preauricular height of the skull suggests the Neanderthal—Cro-Magnon ancestry of the Predmost people. Again, the Predmost skull possesses certain Negroid or Australoid traits, such as flat-sides, long, narrow and high form, prognathism, etc. Some authorities stated that the Predmost man was a contemporary of Neanderthal man, at least at a particular point of time.

The Chancelade Man

In the year 1888, the discovery was made from a rock-shelter near Chancelade in Dordogne, France. The Chancelade man was discovered by two archaeologists named Feaux and Hardy. The skeleton lay on its left side with folded arms. The legs were bent at the knees, just touching the jaws. This is the exact position when it was unearthed. The body was smeared with red ochre. The posture recalls the methods of burial as found among many ancients as well as modern people.

The Chancelade man was discovered from the layers belonging to the Upper Palaeolithic Age. Associated artefacts represent the Magdalenian cultural stage. The reindeer bones were found abundant in these layers. The skull is long and narrow. The length and breadth of the skull is 19.4 cm and 18.75 cm respectively. The cranial index is 70.9 and it is dolichocranial. The cranial capacity has been estimated as 1530 cc by Keith and as 1710 cc by Testut, which exceeds the average of the modern skull. The vault is high and the supraorbital ridges are slightly developed. The vertical forehead is slightly bulging. The parietal bosses are well marked. The mastoid processes are fully developed. Frontalis, a ridge-like elevation is noticed along the sagittal plane of the skull. The face is long and broad but flat in appearance due to the prominence of strongly developed cheekbones. The orbits are large, high and quadrilateral in shape. The nose is long and narrow (leptorrhine). There is no sub-nasal prognathism. The palate is narrow and the alveolar ridge is elliptical in outline. The lower jaw is strong and narrow and it has very broad ascending ramii. The chin is prominent. The skeletal remains suggest that the man was old and must had died between 55 and 65 years. He had a short stature, about 5 feet 1 inch. Strong and massive limb bones also suggest that the man had a strong muscular body.

The upper limbs are longer and the femur is slightly bent. The linea aspera is developed as in Cro-Magnon. The tibia is flattened in a transverse direction. The foot is large. The first mctatarsal of the great toe is distinctly separated from the second toe and this feature resembles with the Neanderthal man.

The physical characters of the Chancelade man are very close to the modrn Eskimo. Beside these physical characters, there are some cultural similarities also. On this ground some authorities are inclined to consider Chancelade as the ancestor of the Eskimo. According to them, at the close of the glacial period, with retreat of the ice, the Chancelade people migrated northward and ultimately arrived at North America. In course of time their descendants have been the modern Eskimos. But, other authorities strongly discarded the view. Hooton declared that the Chancelade origin of the Eskimos could not be accepted.

Again, the Cro-Magnon and the Chancelade men do not differ much in physical characteristics. Still they may be differentiated from one another by some characters in the following ways. The Cro-

Magnon was much taller in comparison with the Chancelade. The face of the Cro-Magnon was disharmonic whereas the Chancelade face was harmonic. The orbits were rectangular in the Cro-Magnon but quadrilateral in the Chancelade.

The Mesolithic Races

A series of culture developed to form a bridge between the Palaeolithic food-gathering and the Neolithic food-producing stages. This transitional stage is known as the Mesolithic period. This period is characterized by use of pygmy stone tools, which are technically known as *microliths*. During this period a new group of Homo sapiens emerged who declared the emergence of first brachycephals in the soil of Europe. The beginning of the Holocene was thus characterised by the introduction of these specialized races. Archaeologically these fossil remains are assigned to the Mesolithic period. The chief representatives of Mesolithic races come from Europe and three important series have been discussed below :

The Muge Man

A large number of skeletal remains have come from a site near Muge on the river Tagus in Portugal. They belonged to late Mesolithic period. The layers of the soil include huge shell refuse and many other artefacts labelled as 'Tardenoisian industry'. Of these skeletons, only nine have been measured and published with scientific details. At first, Paulae Oliveria had studied seriously the characteristic features of the said bone remains. Later, the other scientists strongly discarded the view and published their own views in different journals.

The crania of Muge men are more or less similar; the cranial index ranges from 69 to 82. The cranium box is of medium size with an ovoid contour. The vertex is high but rather flattish on the top of the skull. The occiput is rounded. The frontal bone is sufficiently arched to form a sloping forehead in case of the male crania but the female crania characterize them as mesorrhine. Most of the skull shows a slight alveolar prognathism.

A notable feature is that, here no sex differentiation is possible through scientific measurement. The long bones suggest a height of 160 cm for the males and 152 cm for the females. Regarding the racial affinity of these Muge people, it may be stated that they possibly migrated from a Mediterranean racial homeland. The physical features are identical to the Natufians of Palestine. According to Prof. Boule, they are a variant of the true dolichocephals.

The Teviec Men

The most important skeletal remains of the Mesolithic Age have been unearthed from Teviec in Brittany, France. These people were the coast dwellers who used to live mainly on a diet of fishes and seashells as indicated by the heaps of shells. Twenty-one human skeletons have been found together with artifacts of Azilo-Tardenoisian type. Out of these only 15 skeletons have been studied in detail, of which seven are males and eight are females.

These skulls are slightly larger than the Muge people. The cranial index varies from 72 to 77. The head form is dolicho-mesocephalic. The bones are massive, thick and broad. Vault is high with oval contour. The face is broad and low. The supraorbital ridges are wide. The orbits are low and rectangular. The nose is mesorrhine. The chin is developed and projecting. The height is 150 cm for males and 157 cm for females, respectively. The Teviec men are thus akin to the Europeans. The physical features suggest that these people formed an intermediate group between the Aurignacian - Magdalenian type in one hand and the Mediterranean type on the other, as like the Muge people. In fact, the Aurignacians—Magdalenian group represents gigantic figure whereas the Mediterranieans represent small figures like the Muge people. Therefore the position of the Teviec men is in between these two groups.

The Ofnet Men

The third important series of Mesolithic men in Europe comes from Ofnet cave near Nordlingen, in Bavaria. During the period between 1907 and 1908, R. Schmidt excavated the Ofnet cave where two shallow pits were found with a number of human skulls. All were painted in red ochre. Moreover, in one pit 33 skulls were buried in a neat circle. Of these 19 skulls were of children, 10 of women and four of adult men. But no skeleton was present. The circumstance suggests an act of deliberate burial. Only 20 skulls have been reconstructed and studied later on.

However, the skulls exhibit varying features. They are not homogeneous at all as include only two dolichocranials. Others are the brachycranials and the mesocranials. The cranial index ranges from 70 to 87.

The two dolichocranials show projecting brow ridges, retreating forehead, very broad face and extremely low orbits. Therefore, such people have been described as the Upper Palaeolithic survivals. They are more akin to the Mediterranean type. The mesocranials are smaller and relatively high vaulted. They are considered to be the Magdalenian survivals and more akin to the Teviec type. The brachycranials show long, wide and moderately vaulted skulls with wide faces. Like the Ofnet brachycranials, these people have been referred as the forerunners of the Alpine race. It may, therefore be concluded that the Mesolithic Age, from the close of the Pleistocene to the beginning of Neolithic period, introduced a number of people having distinct racial characteristics, who were the direct ancestors of modern man.

HUMAN PHYSIO – MORPHOLOGY

CELL AND CELL DIVISION

A typical Cell : Its Structure

The body of any living organism is made up of cells. Cells are very minute in size and extremely complicated in structure. Human is no exception.

Each cell is basically a unit of protoplasm, which is said to be the "material basis of life". The *protoplasm is* the combination of cytoplasm and nucleus. The *cytoplasm* is semi-fluid, jelly-like, hyaline substançe; its outer surface is much more thick to form the boundary of the cell. This specialized boundary of the cell is known as cell membrane or *plasma membrane.* The cell membrane controls the in-and-out of various chemical substances. Various living and non-living bodies are found in the cytoplasm. Of the living bodies the *mitochondria* and the *Golgi body* are very important. The mitochondria is the minute, semi-solid body enclosed in a membrane with a complex internal structure. Although the main constituents of mitochondria are protein and fat, it also contains several enzymes, notably oxidative enzyme systems. The golgi body is the local clump of material present

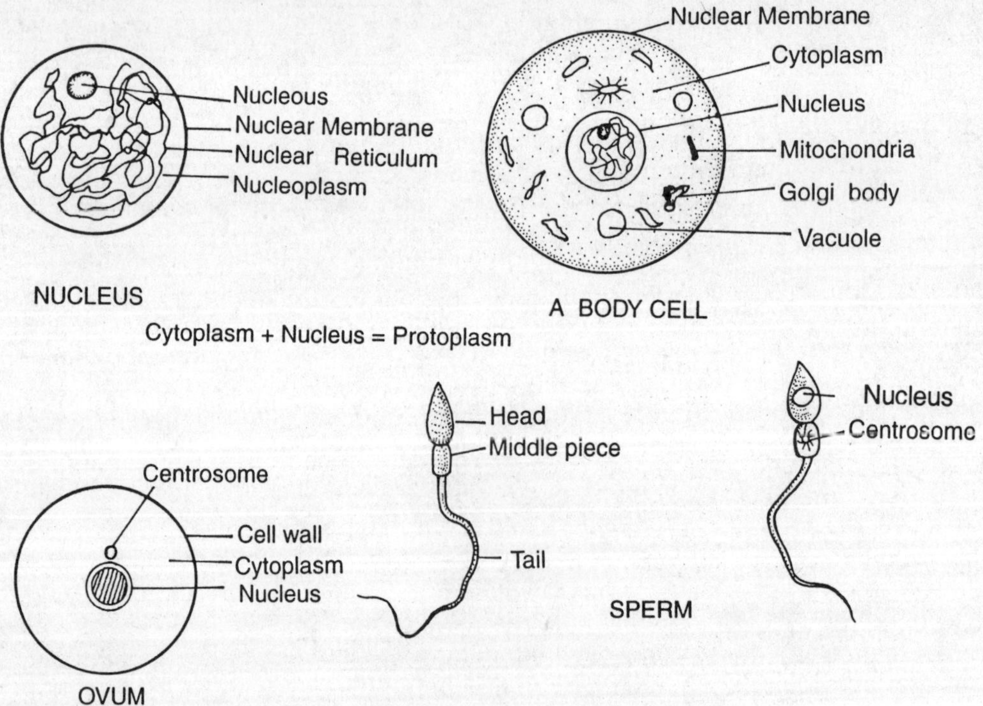

Fig. 4.1. Cells in Human Body

in the cytoplasm. Its functions are not clearly known, but seem to be associated with the formation of secretions. The *vacuole* is important as the fluid-filled space within the cytoplasm of a celi, which controls the cell sap isotonic with cytoplasm by expelling or entering the water from the environment, following the process of osmosis. The *centrosome* is the region of differentiated cytoplasm containing centriole. The *centriole* is the minute granule present in many resting cells, just outside the nuclear membrane.

The *nucleus* is denser than the cytoplasm and it is regarded as the dynamic centre of life. Its shape is absolutely spherical. A membrane, known as *nuclear membrane* surrounds it to keep the core protected. The nuclear cavity remains filled up with a dense jelly-like substance, known as *nucleoplasm.* Some delicate thread-like structures that suspend in the nucleoplasm are called *nuclear reticulum* or chromatin network. The threads are easily stainable by basic dyes; they are made up of a substance, called *chromatin.* Apart from these, one or two small spherical bodies are found in the nucleoplasm, known as *nucleolus.* The nucleus exerts a direct influence on different activities of the cell and plays a great role in transmitting hereditary characters from the parents to offspring.

The cell can be divided into two types - the *somatic cell* or body cell and the *germ cell* or reproductive cell. The somatic cell helps in construction and maintenance of different bodily structures. It can also be sub-divided into nerve cells, muscle cells, etc. in accordance with their nature and function. The germ cell is useful in reproducing new species. It can be sub-divided into *sperms* or male sex cells and *ova* or female sex cells. The *zygote is* also a cell, which is formed by the fusion of an egg or ovum,and a sperm. The human ovum is spherical in shape, and about 1/7th of a millimeter or about 1/175th of an inch in diameter. Though it is the largest of all cells in human body, still not visible in naked eye. In premature stage, a sperm cell is similar to a somatic cell consisting of a nucleus and a mass of cytoplasm. But a mature sperm cell is long and thread-like with an enlarged head formed by the nucleus. It also possesses a long slender tail and a conical middle piece. The tail is used in movement.

A new life starts when a sperm fertilizes an ovum. After the union of the sperm and the ovum, the fertilization takes place when the head and the middle piece of the sperm sink into the egg. The tail of the sperm is left outside. However, the head of the sperm starts functioning as a normal nucleus by absorbing fluid from the cytoplasm of the ovum. Fertilization is thus a process of the nuclear fusion. The sperm loses its identity and the fertilized ovum or egg is called a *zygote,* which begins to divide into two, four, eight and so on, by the process of cell division. From the early stage, some of these cells are set apart to form the *germ cells* while others go to form various body parts, known as *somatic cells.*

The cell division is not a simple process; complicated changes have been noted in the substances of the nucleus. In this regard it is necessary to know about the chromosomes. The *chromosomes* are the slender, rope-like bodies that usually occur in pairs and found in the nucleus. The number of chromosomes remains constant in each species, which generally ranges between 2 to 200. The size and form of chromosomes also vary from species to species. In fact, differences of chromosome constitution mark off one species from another. In case of man, the usual number of chromosomes is 46. There are altogether 23 pairs of chromosome; 22 pairs act as autosomes and the other pair are known as sex-determining chromosome or sex chromosome.

Chromosomes are composed of two kinds of nucleic acids and two main types of proteins. Chromatin is the most important nucleo-protein in chromosomes. The main nucleic acid of the chromosome is Deoxyribonucleic acid or DNA. In 1951, a renowned biologist James Watson with his chemist friend Francis Crick proposed the structure of DNA. The other kind of nucleic acid in the chromosomes is the Ribonucleic acid or RNA, which exists in cytoplasm. DNA samples from different tissues of the same species are all identical in composition. RNA not only differs from DNA; the amount of ribonucleic acid in the chromosomes varies from tissue to tissue. It is found in large quantities in the cytoplasmic particles of the cell. In some cases, the nucleolus also contains large quantities of RNA. It is believed by the scholars that RNA plays an essential role in protein synthesis and carries instruction to the cytoplasm of a cell. In general, the DNA content of an egg is greater than that of a sperm. A variety of suggestions have been made regarding the relationship between the DNA and protein components of the chromosomes but no steady conclusion has yet been achieved. As a matter of fact the nucleus and the cytoplasm exist in a state of symbiosis. The nucleus depends on the cytoplasm for its energy supply and also for the supply of materials out of which new nucleic acid molecules and chromosomal proteins can be synthesized. On the other hand, the cytoplasm is fully dependent on the nucleus for the maintenance of its essential biosynthetic mechanisms.

The Process of Cell-Division

The existence and perpetuation of life depends on the functioning of cells and their divisions. There are two important processes of cell division, which are inextricably linked with each other to maintain the progress of life. These processes are Mitosis and Meiosis. After fertilization, when a sperm cell and an egg cell unite to form a new organism, it is expected to carry double number of chromosomes because both the parents contribute their whole number of chromosomes as appropriate for the respective species. But the process meiosis prevents the doubling of the chromosomes. It reduces the total number of chromosomes into half so that the offspring (newly formed cell) can retain the same number of chromosomes as like parents. On the other hand, mitosis facilitates in forming of two daughter cells from a single cell where each new cell resembles the parent cell in every respect. Unlike meiosis, mitosis involves duplication of chromosomal pairs, which is essential for cellular reproduction, especially during the growth.

The mechanism of Mitosis

A specific process is required for forming two daughter cells out of a mother cell. This process is known as mitosis and it is usually divided into four stages, *prophase*, *metaphase*, *anaphase* and

MITOSIS

(2n) develop [(2n) & (2n)]

Mechanism of *Mitosis*

1. Prophase

2. Metaphase.

3. Anaphase.

4. Telophase.

MEIOSIS

(2n) develop [n & n]

Next develop **[n,n,n,n]**

Mechanism of *Meiosis*

First Meiotic Division.

(Reduction in the Number of Chromosome).

1. Prophase.

i) ***Leptotene;***

ii) ***Zygotene ;***

iii) ***Pachytene;***

iv) ***Diplotene;***

v) ***Diakinesis***

2. Metaphase

3. Anaphase

4. Telophase.

Second Meiotic Division

(Simple Mitosis to develop four daughter cells)

1. Prophase.

2. Metaphase.

3. Anaphase.

4. Telophase.

Figure 4.2. Diagram Representating The Processes of Cell Division

telophase. In some cases it is convenient to designate the transition from prophase to metaphase by the name of premetaphase. However, the whole process of the cycle of mitosis is a continuous one.

Prophase

At the beginning of the prophase stage the cell prepares itself to divide the chromosomes. It is resting stage when the nucleus becomes a little larger. It also involves the condensation of the previously diffuse, invisible or dispersed threads into visible chromosomes. Each chromosome is optically double, which is composed of two identical and closely parallel strands or chromatids lying throughout their length. A chromosome bears a constriction or a narrow region of attachment to the spindle, which is called as *centromere*. Constrictions other than the centromere are sometimes called 'secondary constrictions'. At prophase stage the centromere and secondary constrictions are look alike, but by metaphase this distinction is clear because they develop entirely different relations to the spindle. In this stage (prophase) the outlines of the chromatids present a slightly irregular woolly or hairy appearance. They do not, in general, show a series of granules (chromomeres) as are seen at the meiotic prophase and probably this is a real difference between the two. By the end of prophase the woolly appearance almost disappear and a smooth outline is visible.

Premetaphase

At the end of prophase the nuclear membrane usually disappears and this stage may be defined as the period of spindle formation. During this stage the chromosomes give an impression of struggling and pushing among themselves in order to reach the equator of the developing spindle. The spindle is relatively a solid gelatinous body. Apparently it is composed almost entirely of protein with a very small amount of RNA.

The centrosome seems to act as organiser of the spindle. The centrosome is the differentiated region of cytoplasm containing centriole, which is a minute granule present just outside the nuclear membrane. After disappearance of the nuclear membrane, the centriole is doubled and the two centrioles move apart from each other to form the poles of the spindle. The spindle fibers are probably the bundles of protein molecules. A distinction may be made between continuous fibers and chromosomal fibers. The continuous fibers run from pole to pole whereas the chromosomal fibers connect the centromeres to the poles. In case of bipolar spindles, the equatorial plane is located perpendicular to the long axis, which is in the midway between the poles. As the premetaphase proceeds, the chromosomes arrange themselves with their centromeres, on the equatorial plane of the spindle. It is the most dynamic stage which gives rise to a relatively static Metaphase.

Metaphase:

The nucleolus disappears at the premetaphase and at the end of premetaphase the chromosomes show the maximum degree of condensation. Not only that, at this time the chromosomes acquire smooth outlines. The two chromatids of each chromosome lie parallel or loosely wound around one another, and remain united only at the centromere. They look almost like divided chromosomes but without any separation of the daughter centromeres. Thus, metaphase is a stage during which little or no visible change takes place in the cell; it is a static phase and usually a short one.

Anaphase

The anaphase is marked by the beginning of a separation of the daughter centromeres from one another. The centromeres suddenly split and their halves actively repel one another. As the daughter centromeres move up the spindle towards the opposite poles, they drag the chromatids that are attached to them, away from one another. In fact, the two halves or the identical sets of chromatids move to two opposite poles of the spindle. This movement is autonomous. Thus the mitotic separation of chromatids always starts at the centromere and proceeds along the chromosomes in both directions. After a complete separation of the chromatids from one another, they are called daughter chromosomes.

By this time, the poles of the spindle become further apart than those of the metaphase spindle. The daughter centromeres also get separated from one another and the equatorial region between them seems to be more elongated. As the equatorial part of the spindle elongates considerably, the poles become correspondingly narrowed and eventually the whole thing, appears as a thin bundle of thread-like structures. This thread-like structure between the two separating groups of daughter chromosomes is called the *stem body.*

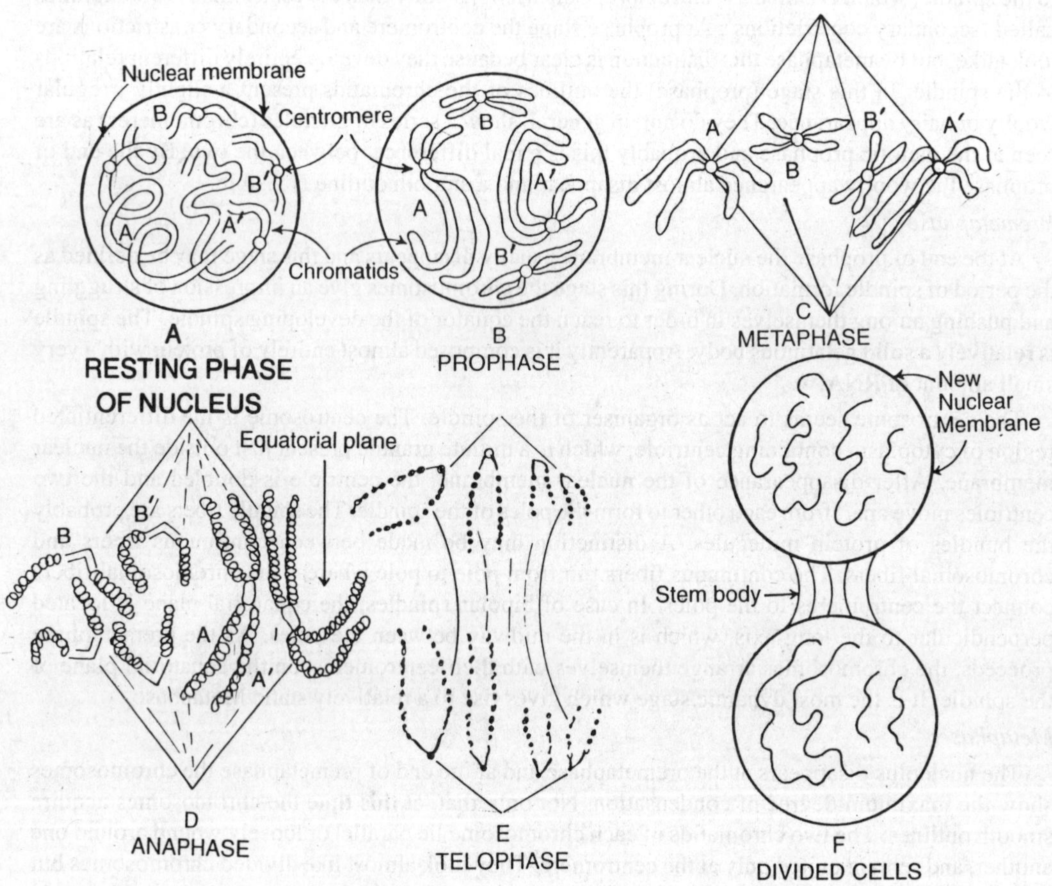

Fig. 4.3. Diagram Showing the Stages of Mitosis

Telophase

As a rule, the two groups of chromatids are now known as the two groups of daughter chromosomes that pass towards opposite poles of the spindle. The polar caps of the spindle seem to be disappearing at the beginning of Telophase, but the stem-body may persist for a long time. The two groups of daughter chromosomes rapidly lose their smooth outlines and undergo de-condensation. An apparently new nuclear membrane encircling the chromosome develops at this stage and nucleolus reappears. But in this telophase stage the chromosome consists of only one chromatid instead of two, and there is only half of the amount of DNA, so that the appearance of prophase and telophase nuclei, in the same tissue is almost always characteristically different. However, ultimately two daughter cells are formed.

Mitotic cell division is very important because the constituents of the chromosomes are never changed and remain identical in all the daughter cells. Thus the daughter nuclei are similar to the mother nucleus. As a result, the process of cell division produces large number of daughter cells, which gradually begin to assume different functions, and in this way the various parts of the body are formed.

The mechanism of Meiosis

The knowledge of heredity is essential in the way to understand the mechanism of meiosis. In fact, meiosis is the antithesis of fertilization. In this process of cell division, the diploid number of chromosomes is reduced to the haploid (half) number. It is a very significant process. Since we know that life occurs as an incidence of fusion between a male and female gametes, in the zygote the number of chromosomes is expected to be double because both the gametes contribute their specific number of chromosomes. Thus, the chromosome number would continue to multiply from generation to generation. Nature's solution to this problem is Meiosis or the reduction division. In higher animal, this process takes place just before the germ cells are formed. The meiosis of the male gametes is referred as *spermatogenesis* while the same mechanism in female gametes is termed as *oogenesis.* However, meiosis involves two types of divisions, occurring one after another. In case of first division, the number of chromosomes is reduced to half but in the second division, the number gets double. Meiosis is therefore a very complicated mechanism of nuclear division, which may be discussed in the following way :

First Meiotic Division

The first 'meiotic division' always begins with a lengthy prophase stage, which differ in much respect from a simple mitotic division. This 'Prophase' show a number of sub-stages, which are known as *leptotene, zygotene, pachytene, diplotene* and *diakinesis.* Next to Prophase, there is a short Premetaphase, which is followed by the Metaphase of the first 'meiotic division'. However, we should start with the sub-stages of Prophase.

Leptotene

It is usually a brief stage when the chromosomes look very elongated and slender. These chromosomes have *a polarized* orientation with all their ends directed towards one small area on one side of the nucleus. This polarization does not seem to affect the centromeres at all. Each chromosome bears a definite number of granules, which form a chain of beads of different sizes known as chromomeres.

Zygotene :

In this stage the homologous chromosomes come together in pairs and become more close throughout their whole length. This phenomenon is called *pairing* or *synapsis.* This process of synapsis usually seems to start at the chromosome ends and then proceeds until it is complete except for some sex chromosomes.

The pairing or synaptic process is undoubtedly facilitated by the polarized orientation of the chromosomes, which brings their ends close together.

Pachytene :

Pachytene begins when synapsis comes to an end, and it ends when the homologous chromosomes begin to move away from each other. The main events that occur during this stage are as follows: a) There is a further condensation of chromosomes so that chromosome pairs become shorter and thicker, b) As a consequence of synapsis between homologous chromosomes the number of observable chromosome is only half (i.e. n) of the somatic chromosome number. In most of the species, this haploid number of chromosome pairs are called bivalents. Each bivalent shows four chromatids or strands. These strands are approximaely equidistant from one another, so that they form a square in cross-section. C) Crossing over between homologous chromosomes takes place during this stage. A small amount (0.3%) of DNA synthesis that takes place during pachytene, is believed to play a crucial role in crossing over. However, Pachytene is usually a long stage.

Diplotene :

The synaptic attraction between the homologous chromosomes suddenly comes to an end and homologous chromosomes of each bivalent begin to move away from each other. This marks the beginning of Diplotene stage. The two homologues of each bivalent appear to be attached with each other at one or more points, known as Chaisma (plural: Chiasmata). It is believed that, initially chiasmata are located at the points of actual crossing over between the homologous chromosomes. As the Diplotene progresses, the chiasmata slowly move towards the ends of the homologous chromosomes. This movement is referred as chiasma terminalization. Further condensation of chromosomes is also visible in this stage, so that they became progressively shorter and thicker.

Diakinesis :

There is no essential difference between late diplotene and diakinesis. In fact, the slender diplotene bivalents gradually turn to thick chromosomes and become very short. They remain sparsely distributed throughout the nucleus, and all loops between the chiasmata lie at right angles to one another. 'Rotation' is usually completed by the beginning of diakinesis, but 'terminalization' may continue right up to first metaphase.

Premetaphase :

In this stage the nuclear membrane disappears and the spindle forms. The nucleolus is also found to disappear and the bivalent chromosomes move towards the equatorial plane while the two centromeres of each bivalent go on opposite sides of the plane.

First Metaphase :

In this stage the chromosomes frequently attach to the spindle somewhere between the equator and the pole. The centromere is attached to the spindle fibers on the equator.

First Anaphase :

The chromosomes now get separated and move to opposite poles. The centromeres do not divide, rather each whole centromere moves in the direction of the nearest pole like the daughter centromeres of an ordinary mitotic division.

First Telophase :

This stage does not differ essentially from that of an ordinary somatic mitosis. Here, two daughter nuclei with a pair of chromatids are formed at the poles. In this way two daughter nuclei contain haploid number of chromosomes.

Second Meiotic Division

Second *Prophase:*

It presents a distinct stage, which is always short and does not include any complication. The spindles become organized rapidly.

Second *Metaphase:*

The number of chromosomes is half the somatic number and the chromatids diverge widely, being only held together *at* the centromeres.

Second *Anaphase:*

The centromeres divide and begin to move towards the poles since the chromatids are quite free from one another.

Second *Telophase:*

This stage does not differ from that of a somatic division, except for the number of chromosomes. Four daughter nuclei are formed, each with haploid number of chromosomes.

It can be concluded that the first meiotic division is causing the reduction in the number of chromosomes but the centromere fails to divide. During the second meiotic division when the centromeres divide, the chromatids do not divide. They simply move apart and form four daughter cells of haploid number. Main importance of this process lies in the maintaining of a constant number of chromosomes in the species. In all sexually reproducing plants and animals, the gametes are haploid (X) in the number of chromosomes. Sexual reproduction is the fusion of a male gamete (X) with a female one (X) and results in the formation of a zygote (2X) from which an offspring is developed. Another importance of meiosis is to help in segregation, assortment and recombination of gene.

SPERMATOGENESIS AND OOGENESIS

The formation of sperms and eggs are referred as *Spermatogenesis* and *Oogenesis,* respectively. Like all other higher animals, in man also, the egg and sperm are developed by the process of meiosis. So a gamete, haploid (X) in number gets diploid (2X) after its fusion with a female one (X). It denotes the formation of a zygote (2X) after fertilization. The body cells of man contain 23 pairs of chromosomes, which are altogether 46 chromosomes. The germ cell contains only one set of the 23 chromosomes. But, at the time of fertilization, a sperm with its 23 chromosomes being fused with an ovum having 23 chromosomes results in the formation of a zygote containing 46 chromosomes.

Maturation of the gamete is the process, which is known as *Gametogenesis,* This gamete is basically a germ cell, which is divided, mitotically in repeated successions to develop into spermatogonia and oogonia, the producers of the sperms and the ova respectively. At some point of time, a spermatogonia stops multiplying by mitosis and this indicates the starting point of meiotic division. The spermatogonia at the beginning is called as primary spermatocyte. This is the stage when the pairs of chromosomes engage themselves in synapsis. The primary spermatocyte divides itself into two secondary spermatocytes by meiotic division. Each of two halves contains haploid number of chromosomes. These secondary spermatocytes divide again mitotically to give rise four spermatids, which are finally grown into sperms. The total process has been termed as *Spermatogenesis.*

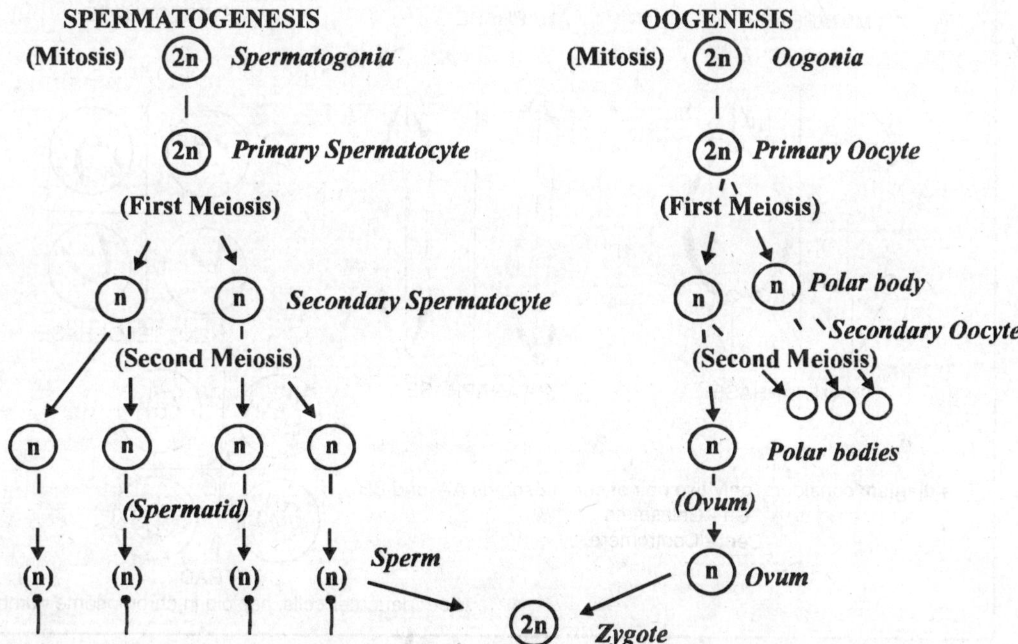

Fig. 4.5. A Schematic Diagram Showing Gametogenesis

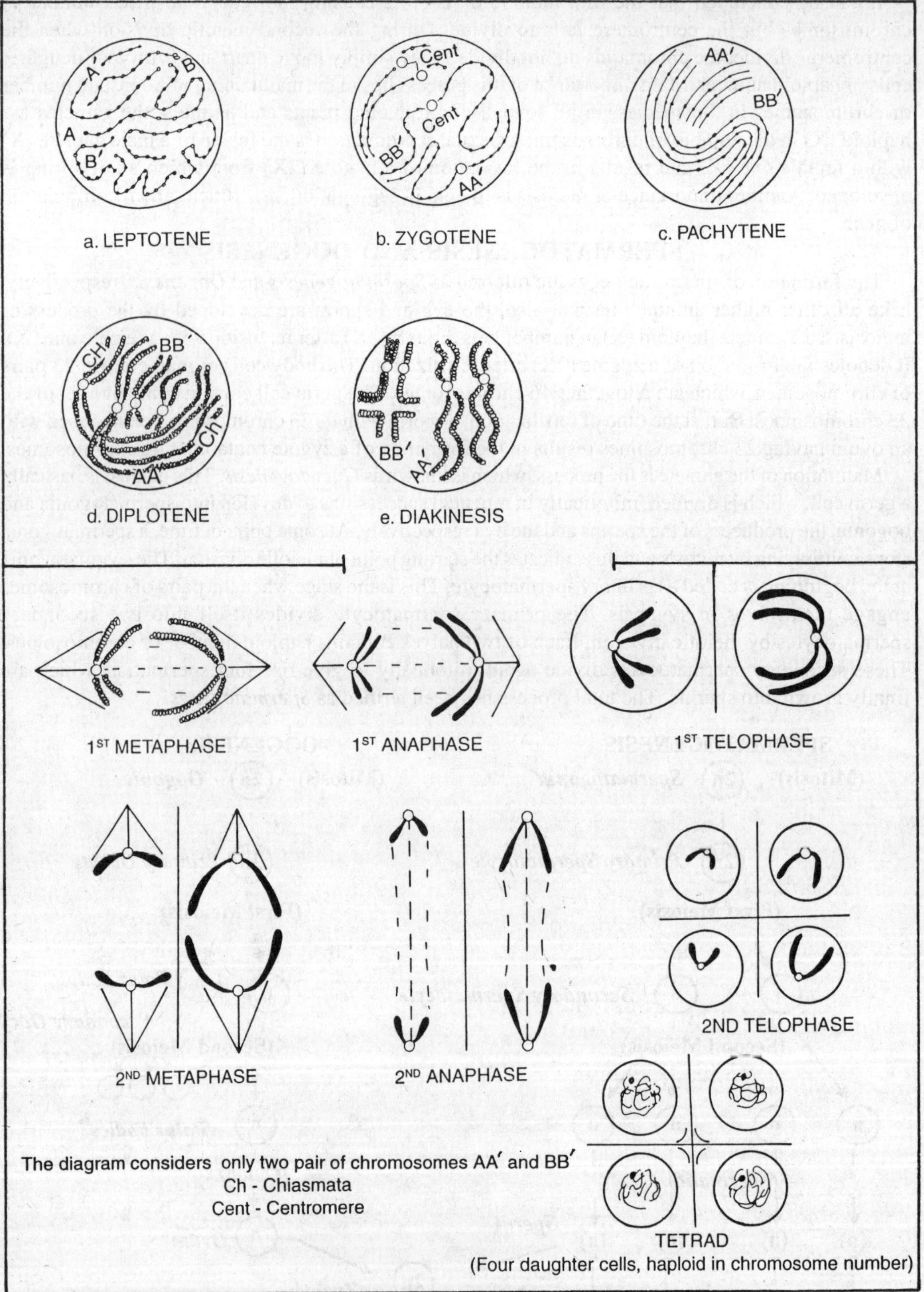

Fig. 4.4. Diagram Showing the Stages of Meiosis

Oogenesis or the formation of ova follows a more or less same process as like Spermatogenesis. For example, germ cell divides mitotically and develops oogonia which gives rise to primary oocyte. In turn, the secondary oocytes develop through meiotic division with one polar body and haploid number of chromosomes. Within the secondary oocyte, two small bodies distinguished - a very small body known as polar body and the other one is oocyte. Each of the secondary oocytes also undergoes an unequal mitotic division to form an ovum and secondary polar body. Therefore, the oogenesis differs from Spermatogenesis in certain features. During the period of growth, a primary oocyte looks much bigger than the primary spermatocyte due to an accumulation of yolk. The primary spermatocyte produces two equal secondary spermatocytes through meiotic division and gradually four spermatids turn to spermatozoa. But, the primary oocyte through meiotic division gives rise to two unequal secondary forms, one is secondary oocyte and other one is polar body. The divisions in case of oogenesis are highly unequal. The ovum receives most of the portion of yolk and cytoplasm to acquire a larger size while the polar body receives only an insignificant portion to form a smaller body. The ovum develops into an embryo after fertilization, but the polar bodies are incapable to take part in fertilization and ultimately they are lost.

This phenomenon offers a fundamental idea on the basic principles of heredity. Subsequent researches in genetics unveil the mechanism of inheritance by which the traits pass down from generation to generation.

COMPARATIVE MORPHOLOGY OF LIVING PRIMATES

Physical anthropologists are much concerned with the order Primate as the animals belonging to this group are very close to man. A comparative study of morphological features discloses this fact. It also reveals the nature and degree of relationship that exists between the man and other primates.

Systematics and Classification

Science can not study the organisms unless the animals are ordered in a systematic manner. Therefore, systematization is essential in the study of living organisms. It deals will all possible aspects of living creatures for putting them into specific kinds of order. A clear and excellent discussion was forwarded by G.G.Simpson (1962), where he stated that the systematics is the study of diversity of animals and of all possible relationships among and between them. The classificatory system that put the animals into groups on the basis of relationship is known as taxonomy. As classifications are the subjects of taxonomies, the organisms are the subjects of classification.

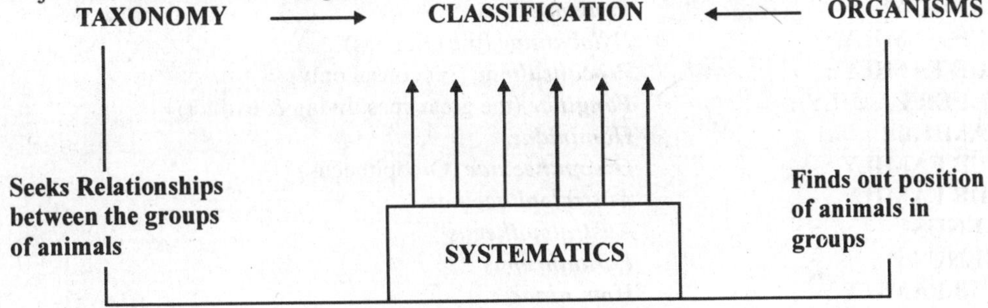

Fig. 4.6. Principles of Systematics

The word taxonomy has been derived from the term 'taxon' which means a category, rather than a group of organisms recognized as a unit. The great Swedish botanist and naturalist Karl Von Linne (popularly known by the Latinized form of his name, Linnaeus), in his famous book 'The systema Naturae' (1735) started to systematize and classify the plants and animals with a binary nomenclature where each organism was designated by a scientific name. It included both a generic and a specific common name. For example, he called the man by the name Homo sapiens - Homo being the name of the genus and Sapiens is the name of the species. From the time of Linnaeus (1707 -1778) to the period of Darwin (1809 - 1882), the typological concept is the basis of biological classification. It governed on the entire animal hierarchy. But gradually that classification was found to be replaced by phylogenetic concept. The phylogenetic classification was basically a statistical analysis of the members of a category, which intends to cover an inbreeding population. Therefore here a variable population have been categorized into species on the ground of genetic homogeneity. Since the typological classification is done on the basis of morphological traits, it is more easy to be determined.

Classification of primates has also been made by various scholars of present day. Classification of Simpson (1945), Buettner - Janusch (1973), and Simon (1972) deserve special mention. In fact, the primates form a distinct and recognizable group and Linnaeus in eighteenth Century first attempted to classify the primates with sub-divisions, scientific and popular names. Actually the difference between the morphological classification and phylogenetic classification is very meagre though the theoretical gap between them is enormous. However, the classification itself is a set of analytical categories by which scientists arrange the data in order to construct a taxonomic system.

MAN AMONG THE PRIMATE

ORDER :	***Primate***
SUB-ORDER :	***Prosimii*** (the *Prosimians* or lower Primates)
FAMILIES :	About 12. Six represent living forms (including some extinct members) : tree-shrews; lemurs, indris & woolly lemurs, aye-aye, lorises & galagos, Tarsius. Other families contain varied early Tertiary forms : tree-shrews with nails; plesiadapids (proto-lemurs); adapids (early lemurs); apatemyids and carpoleslids, both early primate groups not related to living forms.
SUB-ORDER :	***Anthropoidea*** (the higher Primates)
SUPER FAMILY :	***Ceboidea*** (Primates of the New World)
FAMILY :	***Cebidae*** (New World monkeys)
FAMILY :	***Callithricidae*** (marmosets & tamarius)
SUPER FAMILY :	***Cercopithecoidea***
FAMILY :	***Cercopithecidae*** (Old-World monkeys)
SUPER FAMILY :	***Hominoidea***
FAMILY :	***Pongidae***
SUB-FAMILY :	***Hylobatinae*** (the Gibbons)
SUB FAMILY :	***Proconsulinae*** (Proconsul only)
SUPER FAMILY :	***Ponginae*** (the great apes, living & extinct)
FAMILY :	***Hominidae***
SUB FAMILY :	***Oreopithecinae*** (Oreopithecus)
SUB FAMILY :	***Australopithecinae***
GENUS :	***Australopithecus***
GENUS :	***Paranthropus***
SUB FAMILY :	***Homininae***
GENUS :	***Pithecanthropus***
GENUS :	***Homo***
SPECIES :	***Neanderthalensis***
SPECIES :	***Sapiens***

Fig. 4.7. Classification of G.G. Simpson. (It got wide acceptance among the scholars in 1945)

The Concept on Speciation

The concept of species was first developed by Linnaeus as he devised a standardized system of binomial nomenclature as early as 1735. In his system of classification, description of a type specimen was essentially required for the determination of a new species. The species concept has been the most important concept in modern evolutionary biology as because it is fundamental to any discussions of systematics and taxonomy.

Charles Darwin in his books 'Origin of Species' (1859) tried to establish the fact that a new species evolve from a previously existing species by means of natural selection. He also held that the individuals of an organism always struggle among themselves for maintaining own existence; only the individuals with favourable variation get a chance to survive and reproduce. According to Darwin, the favourable traits are selected by nature and the unfavourable traits discontinue. The most controversial aspect of Darwinism was that Darwin could not explain the mechanism by which the new variations arise in an organism. In fact, speciation or the development of a new species depends upon new variations, which never happens suddenly. Not even it is manifested for the factor like mutation. This fact was explained later by the geneticists.

A species may be defined as a population or a group in the population, which show radical difference from others as a consequence of long reproductive isolation. For example, a group being cut off from the parental group when settles in a different geographical area, it adopts the new environment. Certain genetic changes appear among its members with time. Gradually they become so different from the original group that interbreeding between the two groups becomes impossible; a new species is said to be developed. In fact, many species may be distinguishable in a particular place at a particular time because of reproductive gap. This reproductive gap among the two populations of animals indicates an absence of any intermediate form. However, such type of populations or species in the same geographical area is known as 'sympatric species'. When the members of a same species occur in different geographical areas, they are called 'allopatric species'. In the former type, two or more reproductively isolated population are found to occupy the same or overlapping areas,whereas in the latter, individuals belongings to the same species live in non-overlapping or separated geographical regions. Being reporoductively isolated, the sympatric species show distinct characteristic features but the distinction of characteristics is not very clear and complete in case of allopatric species.

It is therefore clear that reproductive isolation is the main criterion for speciation for which a species diverges from its mother population. Most often this is achieved by geographical separation. Because, ecological barrier that often works behind the geographical separation of the zones, create different types of environment enough for molding an isolated species. The most interesting feature is that, if the ecological barrier between two allopatric species can somehow be removed, they behave like sympatric species. On the other hand, if an ecological barrier can be raised between two sympatric species, in the long run, they will exhibit the characteristics of allopatric species.

Speciation is a very slow historical process. It played a vital role in evolution. From the simplest unicellular organism, the utmost development of modem man has taken place by the process of speciation. In the same way, the successful mammals have developed into primates. The evolutionary processes can not go reverse after the differentiation of a species from the parent population. At the same time, the differences of body structure negate the closely related organism to breed. The genetic difference also prevents the reproduction. In these circumstances, if union takes place between the members of two different species, egg will not be fertilized. In case, if the egg gets fertilized, embryo will not survive. Even then, if birth occurs as a matter chance, the offspring will immediately die or it will be infertile. Thus, nature does not favour the mating of two different species and speciation takes the animal kingdom towards novelty.

The Order Primate

Man as a member of order primate share many of his morphological characteristics with the living primates. In general, the primates have shown an increasing ability to adapt varying environmental conditions in the following way :

i) An increasing refinement of hands and feet is found among the primates for which a high degree of manual skill has been manifested. This refinement includes the development of flat nails on the digits of hands and feet (instead of sharp claws), the formation of sensitive tactile pads on fingers and toes, and the development of mobile digits, especially in the thumb and the big toe. Mobility of the digits facilitated an efficient grasping. Although a residual claws known as 'toilet claw' is found in the digit of some primates but in general the primates left the habit of clawing. Grasping was purely an adaptation to arboreal life. Opposability of thumb and big toe enabled the primates to manipulate the objects.

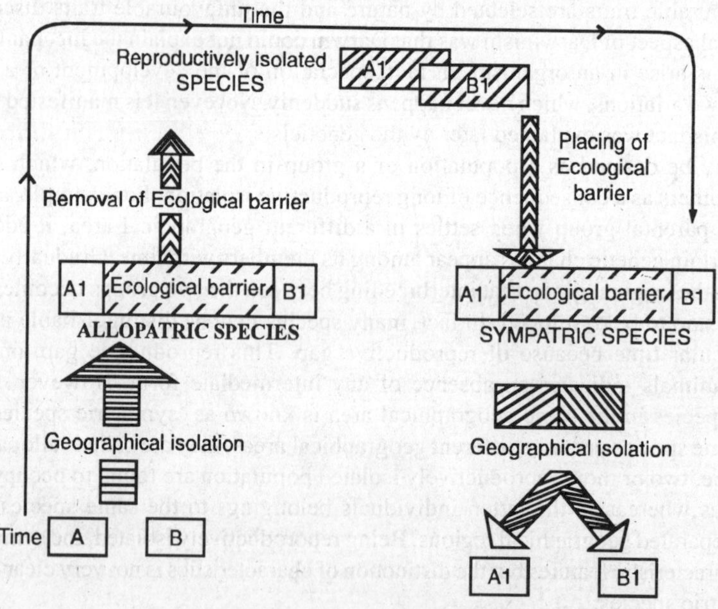

Figure. 4.8. A Hypothetical Model On Speciation

ii) Reorganization of the special senses has been noted among the primates. Specifically the sense of smell (olfactory sense) was reduced which reflected in the flattening of face and reduction of the length of the snout. Reductive smell sense can be correlated with the increased visual sense. In fact, the visual sense has increased throughout the primate evolution. Eyes have been located in the front of face providing a stereoscopic binocular colour vision. Reduction of snout is again related with the increasing cranium size; a continuous development of brain-size has been recorded with special elaboration and differentiation of cerebral cortex. This provided beautiful eye-hand coordination in the development of manual skill. On the whole, a profound change in the morphology of skull structure has been noted.

iii) The next adaptation was the longer gestation period. Most of the primitive primates, say for example, Tupaiformes (Tree shrews) shows a gestation period about 43 to 46 days. But,in case of less primitive form of primates, this gestation period lasts for at least four months, which has been extended to nine months in case of man. Multiple births are rare in majority of the primates. Precisely most of them show less than one birth per female per year in a normal situation. The female sexual cycle varies from a single ovulation per year to as many as thirteen per year in higher primates. They depend on mothers and other adults for a markedly longer period. For instance, the period of post-natal growth ranges from less than one year as found in nocturnal prosimian primates to almost fifteen years in man.

iv) Primates are more social in comparison with other mammals. An increasing complexity as well as quantity of social behaviour is found to be developed in higher primates.

v) Another striking feature is the dentition. Primitive mammalian dentition though does not exist among the primates; still most of the primates show the retention of a simple cusp pattern on their molar teeth. Human is the only exception. However, the dentition of the primates is meant for omnivorous diet.

As per the evolutionary trend, the order primate can be divided into sub-order, infra-order, super-family and family. The morphological comparison of man with other living members of this group seems to be necessary in order to evaluate the position of modern man.

Prosimii and Anthropoidea are the two suborders of the order Primate. Man belongs to the order Anthropoidea that can again be classified into two infra-orders — catarrhini and platyrrhini. Catarrhini bears special significance as the family Hominidae (in which man belongs) is related to it through the super-family Hominoidea. The morphological features can be distinguished in the following ways.

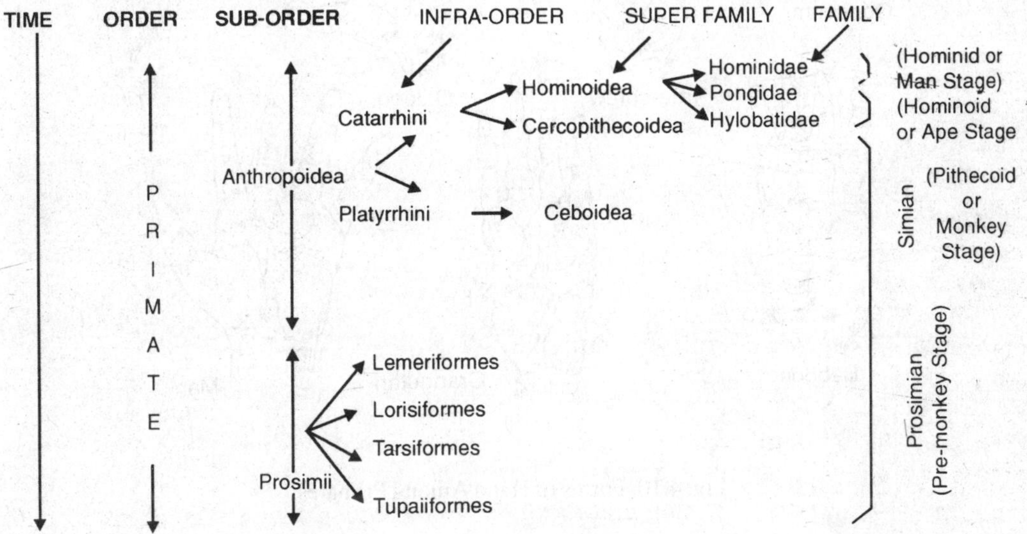

Fig. 4.9. Position of Man Among The Primates

Prosimii and Anthropoidea

1. Prehensile prosimian hands have the ability of grasping things, but their fingers, unlike anthropoids, lack the ability to manipulate the objects.
2. Prosimians have thick and sensitive touch pads below the nails on their fingertips and toes, which are helpful in clinging on the vertical surfaces. But in case of anthropoids the touch pads are smaller and less thick.
3. Most of the prosimians possess single or in some cases double claws which is their grooming claw (commonly known as 'toilet claw'), instead of nails. But this feature is totally absent among the anthropoids.
4. The ears of the prosimians are large and movable for scanning the sound. This provision is lacking among the Anthropoids.
5. Some prosimians have snouts. They possess more developed sense of smell (olfactory sense) than the anthropoids. The snout has gradually been replaced by a smaller nose among the anthropoids.
6. Some of the prosimians are nocturnal in habit and accordingly their eyes are structured. This phenomenon is not found among the anthropoids.
7. Among the prosimians the eyes are placed more or less at the front of the face, but these are set far apart in comparison with the eye-placement of the anthropoids.

8. The brain-to-body ratio, in size and weight, is higher among the anthropoids than the prosimians. But, the olfactory lobes of the brain is larger among the prosimians than the anthropoids.

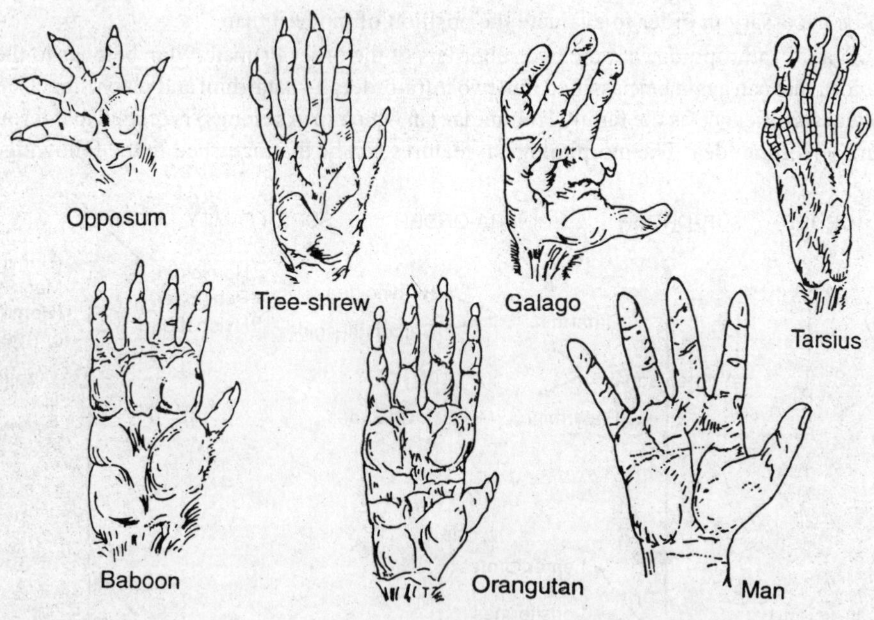

Fig. 4.10. Forms of Hand Among Primates

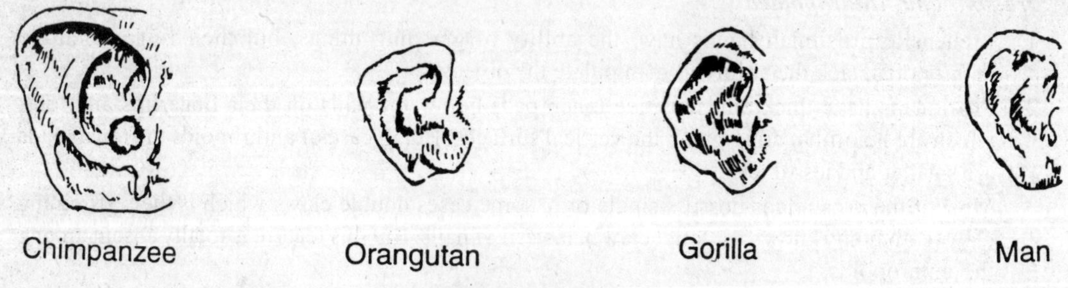

Fig. 4.11. Forms of External Ear

9. Among the prosimians the tear-ducts are present outside the eye-orbit whereas the lachrymal foramen of the anthropoids are situated within the eye-orbits.
10. In dental arrangement, the prosimians show a toothcomb, which is not found in the anthropoids.

Platyrrhini and Catarrhini

The primates that belong to the sub-order Anthropoidea consist of the monkeys, apes and man.

This sub-order is divided into two infraorders, Platyrrhini and Catarrhini. The platyrrhini or New World monkeys differ from catarrhini or Old World monkeys in a number of traits, some of which are mentioned here in the following ways.

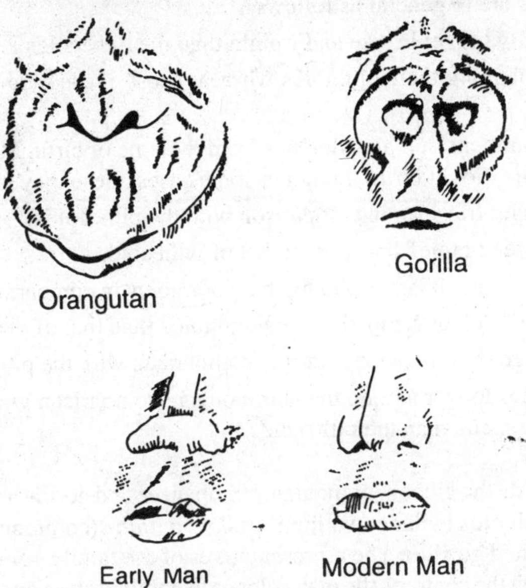

Fig. 4.12 Forms of External Nose

1. New World monkeys are characterized by broad, flat and fleshy noses with widely separated nostrils. Old World monkeys are characterized by narrow nasal septum by which the nostrils are separated. The nose projects and points downwardly.
2. All New World monkeys possess a prehensile tail, which is capable of grasping the branch of a tree. But, the Old World monkeys do not possess a prehensile tail and even in some cases the tails are reduced or absent.
3. In the New World monkeys the cheek pouch is not seen whereas it is found among the Old World monkeys.
4. Ischial callosities are not found in New World monkeys, but present in Old World monkeys.
5. Platyrrhine monkeys have arched nail while catarrhine monkeys are the bearers of flat nails.
6. The pollex and hallux of platyrrhine monkeys are slightly opposable, but in catarrhine monkeys these are capable of independent movement and are completely opposable.
7. Platyrrhine monkeys show a simple stomach, which is not seen in catarrhine monkeys. Among some platyrrhines, it is extremely simple but in others it is sacculated, i.e. divided into series of sacs.
8. Dental formula of New World monkeys is 2:1:3:3 while it is 2:1:2:3 among the Old World monkeys.
9. The palate is short among the New World monkeys. Among the Old World monkeys it is relatively long.
10. The bregma among the Old World monkeys is placed relatively forwardly than that of New World monkeys.

Pithecoid stage and Hominoid stage

The pithecoid or monkey stage includes two super families, *ceboidea* and *cercopithecoidea.* The ceboidea belongs to infra-order platyrrhini and cercopithecoidea belongs to catarrhini. The infra-order catarrhini includes cereopithecoidea, the monkey stage and also the super family hominoidea where we find hominoid or Ape stage along with hominid or Man stage. The major differences between these two stages are in general as follows.

1. Hominoids are usually bigger in size and weight than the pithecoids.
2. The tails are absent among the Hominoids whereas most of the members of pithecoids have tails.
3. Pithecoids show prominent ischial callosities from the time of birth. Most of the hominoids do not show ischial callosities at birth, though in some cases, these may grow after birth.
4. Hominoids have longer life span in comparison with the pithecoids.
5. Hominoids have more complex brains than that of pithecoids.
6. The size of the brain is much larger among the Hominoids in comparison to pithecoids.
7. The cerebral cortex is well developed in the hominoids than that of pithecoids.
8. Hominoids have bigger brain-to-body ratio in comparison with the pithecoids.
9. The gestation period is longer among the hominoids in comparison with the pithecoids.
10. Hominoids are more social than the pithecoids.

Modern Apes and Man

The modern apes include the gibbon, orangutan, chimpanzee and gorilla as man's closest relatives. The modern apes or hominoids include two families, *hylobatidae* (common gibbon) and *pongidae* (chimpanzee, orangutan and gorilla). The representatives of the family pongidae are also known as great apes or giant apes. On the contrary, the man belongs to hominid stage and form family *hominidae.* But it is striking that, both of these stages, hominoid and hominid have been included under one super family hominoidea. Naturally, a comparative account will help us to understand in what respect one differs from the other.

Man walks bipedally with an accurate erect posture; bipedal walks among the apes are occasional with semi erect posture. Again, the upper extremities of apes are greatly elongated than that of a man. The thumbs among the apes are quite short. As a result, they show limited precision grip which is well developed among the man. The size of the brain is much larger as well as developed among the man with complex cortical structure which is not found among any of the apes. Moreover, the man has thc power of articulating speech. These phenomena indicate vast changes in the morphology of man, accompanied by the changes in the structure of skeleton.

All of us know that man is the end product of evolution, which has so far taken place. The anthropoid apes share some similarities as well as differences in morphological features. The structure and functioning of their bodies are to some extent alike. However, the resemblances between man and anthropoid apes are as follows:

1. None of them including man possesses a tail.
2. The vermiform appendix is found in all of these forms
3. All of them exhibit similar blood types.
4. Structure of uterus and placenta are quite similar among them.
5. All of these creatures are omnivorous in food-habit as revealed from the characters of teeth. They chew both meat and herbal food.
6. All of these forms exhibit stereoscopic vision.
7. Each of them possesses opposable thumb in their hands.

Inspite of numerous resemblances the process of change gave man a special position with upright posture and bipedal locomotion. Apes can not be called true biped because they always employ their arms in walking. Differences also exist in terms of size, weight, body form, skin, facial peculiarities, dentition, digital formula, habits, behaviours, etc. For example, gibbons are small arboreal creatures that rarely exceed three feet in height. With a small brain capacity they show a highly developed visual power. Weight of an adult may go upto 50 Kg. The extraordinary length of hands helps them in brachiating. In the ability of swinging from branch to branch of the tree, they resemble with monkeys more closely than the apes. Body is covered with black woolly hair. But the face and ischial callosities do not show any hair. Digital formula is 3>2>4>5>1 for both hands and feet. Considering the above characters, the scholars suggest that the gibbon stand a long way off from the direct line of human ancestry.

The orangutan that lives exclusively in Borneo and Sumatra seems to be closely related to gibbon, instead of other two apes, gorilla and chimpanzee. Orangutans are also arboreal creature but their movement is rather slow than gibbon because of larger size and greater body weight. The height of an adult male is more than four feet and weight varies between 60Kg. and 80Kg. Body is covered with reddish-brown hair. A higher cranial capacity gave this creature a little more intelligence. They are gentle and delightful, capable of making a variety of facial expressions. In fact, they possess a good control over the facial muscles, especially in the area around the mouth and nose. The digital formula of these animals is 3>4>2>5>1. Canines are big and tusk-like. In gibbon, these are big but only pointed. Face of the orangutan is at a glance like man.

The chimpanzee is more man-like than the orangutan. Average height of an adult male is 5 feet and weight is about 125 Kg. They show shaggy hairs all over the body but the hairs are sparsely distributed. Skin colour is variable between black to brown. Cranial capacity is more than a gibbon and an orang. The animals are so tricky that they can be taught many activities involving control and intelligence. They are even capable of using tools. By picking up small grasses and twigs, they often prepare a devise, which is inserted within the nests of the termites to catch the insects for the purpose of eating. Chimpanzees also possess well-developed pointed canine teeth. Like orang they are mostly vegetarian but do not spare ants, meat and fish which come within their reach. Although they have resemblance with man, the anatomical traits are largely similar to gorilla. It has been agreed that modem chimpanzee does not stand in the direct line of man's ancestry. As a matter of fact, the chimpanzee originated during Miocene epoch and pursued on its own evolutionary course along with gorilla (and also orang). Therefore now they are quite apart from man, although once they were the cousins of human ancestors.

Although the chimpanzees in a group often make noise but by nature they are peaceful, in contrast to ferocious and aggressive gorillas. The gorilla is the bulkiest among all primates. An adult male is about five and half feet in average height. Their weight is approximately 200 Kg. Cranial capacity runs more or less parallel to chimpanzee. Although they can brachiate, but spend maximum time on the ground and usually walk on four limbs. During anger they beat on own chest. This is the only time when they stand almost erect. The body is covered with black hair except chest and face. Like orang, none of the gorilla and chimpanzee grows beard and mustache. Face of the gorilla is almost flat with a slightly elevated nasal bone. Shape of the face resembles man while the nose is akin to other apes. Lips are also very thin like apes; no reddish mucous membrane is seen outward as like man. Gorilla is the strongest of all primates. But they never attack others unless they are seriously provoked or badly frightened.

Unlike apes, man possesses longer span of life. The period of growth is also long but rate of growth is quite slow. Arms are shorter than the legs. The body is relatively hairless. Ischial callosity is absent. Tactile hairs are not found. Nose is prominent with elevated bridge and fleshy tip. A median furrow divides the upper lip into two halves. Margins of ears are heavily rolled. Canines are smaller in size and their crowns are placed almost in the same level of other teeth. Foot is arched, both transversely and anterio-posteriorly. The great toe is not opposable; it lies in the same row with other toes but the size has been remarkable for its greatness. Like man all apes show flattened nails on their fingers and toes but their thumbs are opposable. They are all diurnal in habit and vegetarian in diet. All of them live in small family band.

The comparative morphology thus reveals various facts of resemblances and differences by which one can understand the trend of evolution. The great apes seem to be much closer to man than any other animals of the world; they stand next to man in the hierarchy of existing primates. But, the major structural differences between the apes and modern man clearly indicate that, though the apes were the collateral relatives of man in remote past, the immediate ancestors of man were less similar to apes. Rather they possess more similarities with the monkeys as observed in the proportion of hands, in the development of thumb, in the muscular make-up of legs and feet, in the sequence of teeth eruption, in the absence of simian shelf, in the obliteration of cranial sutures etc. Physical anthropologists depend much on the study of comparative morphology to ascertain the position of man among the living primates.

POSITION OF MAN IN THE ANIMAL KINGDOM

All living bodies that use organic food to survive and oxygen for respiration are called animals. They move in response to outside stimuli and feel in their own ways. Man is obviously an animal since he moves and feels.

By examining the animal kingdom we can reach to the conclusion that there is two broad subdivisions - *Protozoa* (unicellular animal) and *Metazoa* (multicellular animal). Higher form of multicellular animals shows bisexual reproduction i.e. an offspring gets chromosomes equally from both the parents. Therefore, the chromosomes are paired in bisexual reproduction where each member of a pair possesses certain differences in the genes. Naturally, the offsprings differ from each other and also from each parent. In the history of animal kingdom, bisexual reproduction is very important as it facilitates the development of more and more complex forms, in course of time. However, the animal kingdom can be classified by the help of taxonomy; special names have been given to the hierarchy of animal groups who went towards higher grades. Each major division or the grades are called as phylum (plural: phyla). In taxonomic hierarchy, *the phylum chordata* holds the uppermost position and includes the *sub-phylum, vertebrates.*

Bilaterally symmetrical animals of higher grades are distinguished by flexible cord placed down on their back at sometime during their life. They form the phylum *chordata.* The cord is called notochord being derived from the Greek word noton, the back. These animals also have a spinal cord, but it is the notochord rather than the spinal cord, which characterizes the phylum chordata. Man belongs to this phylum chordata. Of the chordates, there are many groups who share some features with man, and the sub-phylum vertebrates embraces five classes, such as pisces, amphibia, reptilia, aves and mammalia.

All vertebrates possess well-developed vertebral column, i.e. the spine, to surround and protect the spinal cord. This spinal cord is attached with the brain at one end. Such a structure forms the central nervous system for coordination of the body movement and sensation. The vertebrates also have an internal skeleton of bone and cartilage to surround and protect the organs of sense and the brain which extend upto the limbs. Man shows two pairs of limbs in regular fashion, but in case of snakes and whales evidences are preserved in their ancestral forms.

The mammals are placed in the topmost position among the vertebrates. It is a distinct category of the animals having breasts; mothers nourish their babies with own milk. The females of the mammals not only feed the young after birth; they also establish a social relationship with their offsprings. This trait of establishing social relation is a milestone, which has proved its usefulness in case of man, where we find the transmission of culture, through learning, from generation-to-generation. Besides, there are some other important characteristic features which have been mentioned here in the following ways.

1. The mammals are air-breathing animals having epidermal covering in the form of hair.
2. They are called mammals for the presence of mammary glands, which is used in nourishing the young ones.

3. Presence of sweat and sebaceous glands are noted which undergo fatty-secretion to lubricate the body.

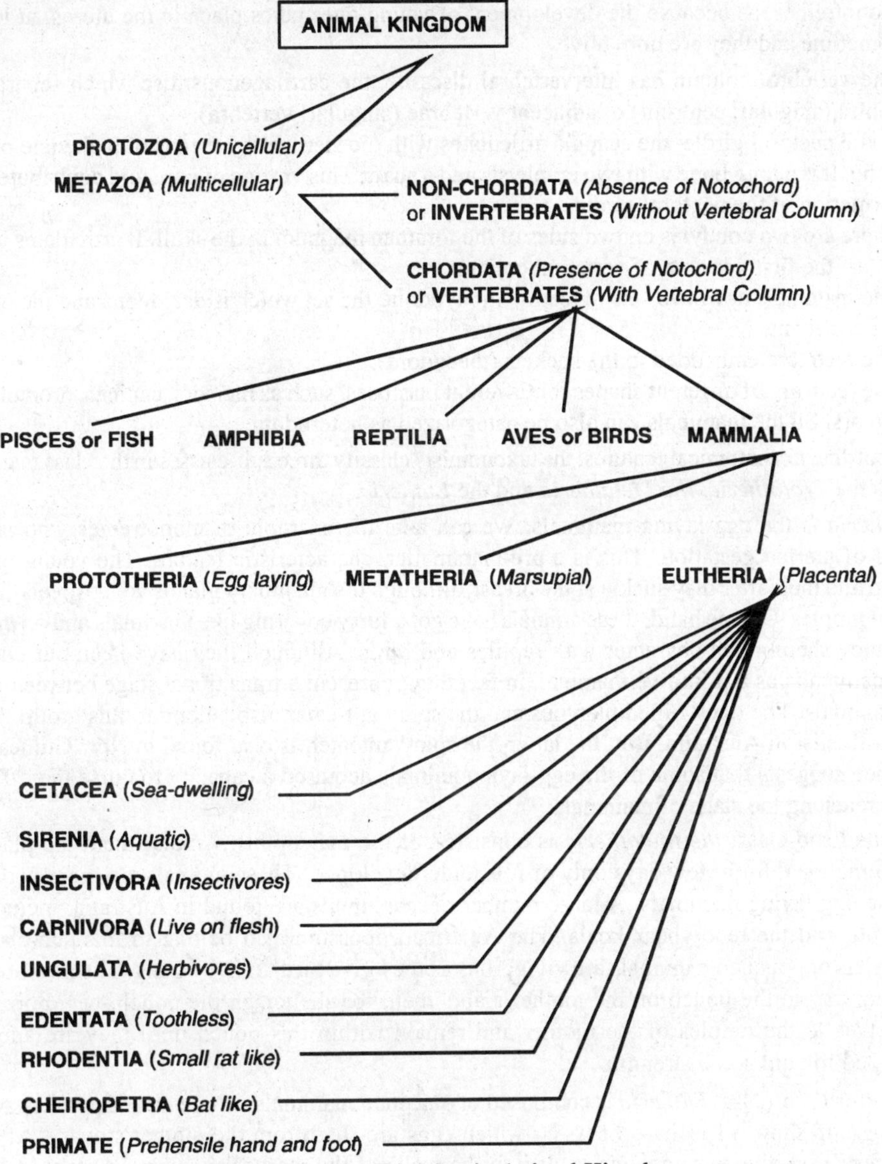

Fig. 4.13. Major Division in Animal Kingdom

4. The body cavity is divided by a partly muscular diaphragm into two parts, an anterior thorax and a posterior abdomen.
5. The lungs are freely suspended within the cavity of the thorax.
6. The heart is completely divided into two parts and four chambers. Each part is consisted with one auricle and one ventricle.
7. The corpus callosum unites the two halves of the cerebrum or cerebral hemispheres.
8. The pons varolii joins the two halves of the cerebellum.
9. The red blood corpuscles are circular in shape and non-nucleated.

10. The mammals are warm blooded or homoithermous, which means that these animals are able to maintain a constant or nearly constant body temperature. They are viviparous, except in monotremes, as because the development of young ones takes place in the uterus, at least for sometime and they are born alive.
11. The vertebral column has intervertebral discs i.e. the cartilagenous disc which separates the centra (singular: centrum) of adjacent vertebrae (singular: vertebra).
12. In the pectoral girdle, the scapula articulates with the sternum by means of a clavicle or collar bone. It is a long bone with two epiphysis and a shaft. This vestige of coracoid contributes to the formation of the pectoral girdle.
13. There are two condyles on two sides of the foramen magnum in the skull. It articulates with the Atlas, the first vertebra of vertebral column.
14. The mammals have two sets of teeth—diphyodont, the set which is deciduous and the other set is permanent.
15. The teeth are embedded in the sockets (thecodont).
16. The teeth are of different shapes for different purposes, such as incisors, canines, premolars and molars. So the mammals can also be categorized as heterodont.

According to anatomical features, the taxonomists classify three sub-classes in the class mammalia, such as the *Prototheria, the Metatheria* and the *Eutheria.*

Prototheria is the egg-laying mammals. We can take the example of monotremes who lay eggs instead of uterine gestation. This is a pre-mammalian characteristic feature. The young ones are hatched and thereafter they suckle at the breast, although the mammary glands are unspecialized and without nipples. On one hand, these animals have got a furry covering like mammals and on the other hand, they share much character with reptiles and birds. Although they have been put under the class mammalia as a primitive mammal, in fact they represent a transitional stage between reptiles and mammals. The duck-billed platypus and the spiny ant-eater also belong to this group. Both of them still exist in Australia. But, the latter, the spiny anteater, is also found in New Guinea. Their existence suggests that some of the egg-laying animals acquired a capacity to nurse the offsprings before reaching the stage of mammals.

Second sub-class, *the Metatheria* is consisted of the marsupials. Among them the placenta is either functional for a few days only or it is underdeveloped. These animals are more widespread than the egg-laying mammals. A large number of marsupials are found in Australia, including the kangaroo, and the teddy-bear koala. The American opossum also belong to this sub-class. The marsupials or pouched mammals are viviparous as they give birth to live young. The new borns find out their way to the pouch on the mother's abdomen, for shelter. In the pouch, new borns attach themselves to the nipples of the mother and remain within this pouch until they are adequately developed to gain some strength.

The third sub-class, *Eutheria* is composed of placental mammals, where man is posited. Animals of this group show a special process by which eggs are shed from the mother's ovary. After this, being fertilized, the egg implants itself into the walls of the mother's womb. In early embryonic stage, a placenta is found to be developed from the wall of the womb. This placenta is a kind of sac which permits interchange of fluids between mother and offspring. However, the placenta helps in long intra-uterine development and gives the birth of large infants. The sub-class, *Eutheria* can farther be classified into nine orders in the following ways.

Cetacea: The sea-dwelling mammals represent the order *cetacea.* The best example is the whale. Their body is fish-like and devoid of hair. The clavicle is absent. The forelimbs are modified into paddle-like structures and hind limbs disappear completely. Naturally, the man is not included in this order.

Sirenia: The members of this order are also aquatic animals. They have porpoise-like body with a covering of scattered hairs. The blunt-nosed sea-mammals like dolphin make up this group. They are herbivorous in food-habit and not nearer to man.

Insectivora: These animals are called insectivores, which means that they live on insects. The examples are the mole, tree shrew, etc., which are usually small in structure and nocturnal in habit. The man is much apart from this order.

Carnivora: These animals live on flesh and the best examples are the dog, lion, cat, tiger, etc. These carnivores are characterized by well-developed cutting and tearing teeth, sharp claw and the absence of clavicle bone. The man is far away from this order.

Ungulata: These are ground dwelling, four-footed hoofed animals, usually having horns. The portion of grinding teeth (molar) is larger in comparison to cutting and tearing teeth and they are generally herbivorous. The examples are the horse, cow, buffalo, etc. As per physical features, the man is widely separated from this order.

Edentata: The mammals of this order show degenerated teeth without having enamel. A few of them are completely toothless. The examples are the anteater, armadillo, etc. In reference to their primitive character, they stand far from the man.

Rhodentia: These animals are mostly small in size, possess a furry body covering and claws in fingers. They also have large, chisel-like incisor teeth. The examples are the rabbit, rat, squirrel, etc. Man can not be placed in this order.

Cheiroptera: These animals are with four-limbs which are found to be modified into wings. The bones have been elongated and make a broad web of skin, which extends upto the hind limbs. Such structural modifications give them an aerial life so that they can fly. The best example is the bat. The man can not be placed in this order too.

After examination of the physical features of the above mentioned eight orders of Eutheria, we can conclude that the man does not belong to these orders. At the same time, it is automatically understood that the man belongs to the remaining order of the placental mammal known as Primate.

Primate: The representatives of this order comprise different forms of animals, which are found to be spread over Asia, Africa, Madagascar, Central and South America. The primate exhibits a large number of characteristic features, of which the most important features have been described here. Such features distinguish them from the other mammals.

But it should be kept in mind that all characteristic features are not exclusive; many of the mammalian characters are retained among the primates. However, we are listing here some characteristics of the primate without differentiation of generalized and specialized characteristic features.

Characteristic features of the Primate

1. They have prehensile (ability to grasp) hands and feet. Apart from generalized limb structures including five fingers and five toes, they have developed this grasping power for better adaptation to arboreal (tree-living) life.
2. The digits (fingers) of both hands and feet are provided with flat nails and sensitive touch pads below the nails, instead of sharp claws of earlier mammals. In case of a few primates, a claw (commonly known as 'toilet claw') is seen on one of the digits. The modification of digits indicates the development of their grasping power towards an arboreal habit.
3. The thumb and great toe are more or less opposable. Either thumb or great toe or both are opposable in varying degrees to enable them in manipulating objects.
4. The clavicles or collarbones are well developed. Such a development facilitates the movement of arms

5. The teeth are specially arranged, suited to an omnivorous diet. At least, once in the life span the primates have three kinds of teeth, whereas the other mammals have specialized teeth for specific purpose, e.g. the canines of carnivores.
6. The grinding molars among the primates show a simple cusp pattern, but other primitive characters of dentition have been lost.
7. Size of the brain has been developed with differentiation of the cerebral cortex. The brain-to-body ratio has also increased in terms of size and weight. A posterior lobe of the brain appears to give rise a higher mental ability.
8. A declination in the sense of smell is noted. The olfactory sense got reduced in association with the flattening face. Protrusion of the snout was also reduced which ultimately turned to a small nose.
9. Bony eye-sockets came into existence with a developed visual sense. The eyes are placed towards the front of the face to permit stereoscopic as well as binocular vision. The eye-orbits are encircled by bony rims.
10. Among the primates, the middle ear is placed within the temporal bone.
11. They show a highly developed and convulated brain covered with much wrinkled cortex.
12. They have simple type of stomach.
13. They have two pectoral mammary glands situated on the thorax. But, some members of lower graded primates possess more than one pair of mammary glands.
14. The testes descend into the scrotum.
15. The penis is pendulous.

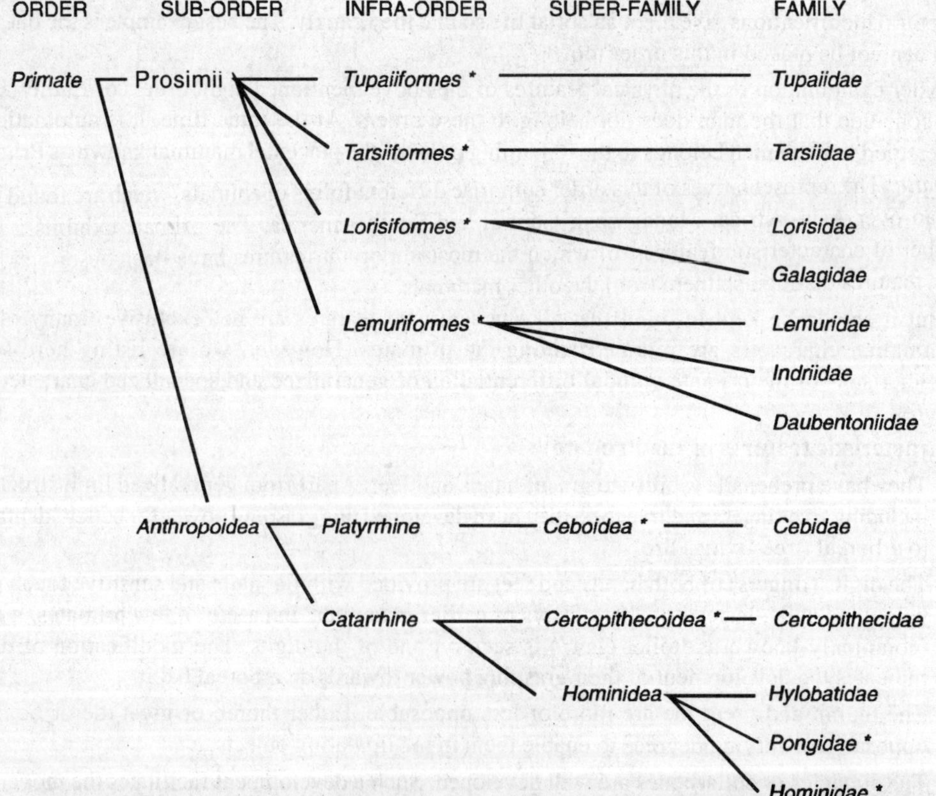

Fig. 4.14. Table Showing Eight Primate Texa (*Each Taxon is indicated by asterisk* *)

16. The natural life span is much longer than that of other animals.
17. The gestation period has been enhanced.
18. Multiple births are not frequent among them. The females have a tendency to give birth only one offspring at a time.
19. The period of postnatal growth is longer. A longer period is taken for the maturity of sexual organs.
20. In comparison with all other mammals, the primates are the most social creatures. Their social behaviours like infant care, vigilance, etc. are complex in nature. Infant primates remain dependent upon their mothers and also other adults for a pretty long period.

Classification of the Primate

The living primates are divided into eight major taxa at various levels as per Linnaeus hierarchy. The term taxon means a category in a classificatory scheme; a group of organisms are recognized as a unit (Taxonomic group). Taxa is the plural form of Taxon. However, the Taxa as shown by Linnaeus have been correlated with four infra-orders, two super-families and two families of the order Primate. Simpson had used an alternative set of names for Taxa but the same animals have been found to be included in each Taxon.

The **order *Primate*** has been classified into two **sub-orders** - the ***Prosimii*** and the ***Anthropoidea***. The Prosimii is further classified into four infra-orders - the *Tupaiiformes* (Tree shrews of Malaysia), the *Tarsiiformes* (Tarsiers of Malaysia and Philipine), the *Lorsiiformes* (Lorises of Africa and Asia), and the *Lemuriformes* (Lemurs of Madagascar). The Tupaiiformes has only one family - the *Tupaiidae;* the Tarsiiformes also shows one family - *the tarsiidae.* The Lorisiformes have two families- the *Lorisidae* and the *Galagidae.* The Lemuriformes have three families - the *Lemuridae,* the *Indriidae* and the *Daubentoniidae.* The sub-order Anthropoidea is classified into two infra-orders and three super-families. The infra-order *Platyrrhine* has one super-family - the *Ceboidea* with one family - the Cebidae. The infra-order *Catarrhine* has two super-families - the *Cercopithecoidea* with one family, the *Cercopithecidae* and another super-family - the *Hominidea* with three families - the *Hylobatidae,* the *Pongidae* and the *Hominidae.* The family, Hylobatidae includes the animals namely gibbon and siamang. The Pongidae possesses three animals - orangutan, chimpanzee and gorilla. But, the family Hominidae presents man as its only member. It may be mentioned here that since the super-family *Hominidea* embraces both the man and the apes, it is also referred as *anthromorpha* while the *term pithecoidea is* applied to the animals belonging to the super-families, Ceboidea and Cercopithecoidea.

Tupaiiformes

On the basis of certain features the Tree Shrews are placed under Tupaiiformes. But there are some controversies regarding the position of the Tupaiiformes in the animal kingdom, as they represent a link with the insectivora from which the Primate came. Three opinions are found in this connection. One, they belong to the Insectivora; two, there should be a separate order for them; three, they are primates. In fact, most of the scholars suggested for classifying the tree shrews with the primate. The studies of W.L.Srtaus in 1956, Le Gros Clark in 1960, and G.G. Simpson in 1962 established that this animal is closely affiliated with the Primates. On this ground the Tupaiiformes has been classified as the infra-order of the order Primate.

Tree Shrews are therefore the most primitive form of living primates. They are found to be distributed all over Southeast Asia. Their distribution covers Indochina, Myanmar (Burma), Thailand, Malay Peninsula, Java, Sumatra, Borneo, Philippines, etc. A considerable number of tree shrews have also been discovered from East Coast of India, Bangladesh, Sikkim, Nicober Islands, etc. The tree shrews are small and extremely active animals with long feathery tails. They resemble with small squirrels. Most of them are arboreal in habit. Some run along the ground and occasionally climb on the low branches and bushes. The brain, in general is, quite complex and approaching towards higher primates.

The visual centers and the eyes are developed like the primates, while the olfactory centers in the brain and olfactory apparatus are small like the primitive insectivores. The dentition is also primate like. The pollex (thumb) and hallux (big toe) are extremely mobile and to some extent opposable.

Some features of the skull show similarities with those of the Lemuriformes. There is an incipient development for the distinct temporal lobe. Besides, the bony structure of the interior eye orbit, the auditory chamber, the palatine bone, etc. show affinity with the Lemuriformes and these features are not present in the Insectivora. The dentition is primitive and the dental formula is 2:1:3:3(upper) and 3:1:3:3 (lower). The molars and premolars are relatively simple. But, three incisors in the lower jaw are typical to the primitive mammalian dentition. However, the third incisor of the lower jaw is much reduced in size and in one genus it has almost disappeared. In seems that the animal is approaching towards primate dentition.

The hands and feet resemble with those of primates. They are able to grasp things relatively well by extremely mobile thumbs and big toes which work by flexion of other digits. This phenomenon may be taken as the beginning of the opposability of hallux and pollex as found in higher primates. Thus, the total morphological features have drawn a line towards the development of prehensile extremities, which led to the evolutionary segregation, and rise of the primates. Again, they possess sharp claws (instead of nails) on the digits to announce their similarity with some other primitive mammals.

Despite different arguments, the total morphological pattern of the Tupaiiformes suggests that they go close to the primates. In fact, these animals are the descendants of a primitive mammal who lie on the way of forming a full-fledged primate.

Tarsiiformes

The Tarsiiformes are commonly known as the Tarsiers. They are found only on islands of the East Indian Archipelago, from the southern Philippines and Celebes on the east to Sumatra on the west.

Tarsiers are tiny animals with very long tails and very long hind limbs. They have a furry covering on the body. The eyes are directed forward and the eye orbits are very large.This give them a startling appearance. As they can rotate their head for 180° , they can look backward without moving the body. These animals lead an arboreal life and show a nocturnal habit.

In skeletal structure, the tibia and fibula are fused together at the lower ends. The tarsus bone of the foot is elongated. The eye-orbits are very large and post-orbital wall is present. The nostrils are widely separated. The foramen magnum is placed farther at the front of the base of the skull, like other prosimians. This phenomenon can be equated to the expansion of the brain. In one way, it facilitated the development of visual centre and on other way it helped in erect position, which is held at the time of hopping. As a matter of fact, the enormous eyes, long hind limbs and the padded fingertips provide these tiny animals a frog-like appearance.

The dentition does not represent any typical form in comparison with other prosimians. The dental formula is 2:1:3:3(upper) and 1:1:3:3(lower). The lower incisors do not form a toothcomb or tooth scraper. The crowns of upper molars are tritubercular, which means a primitive mammalian molar teeth having three cusps. (In evolution of the dentition, these three cusps underwent modification by adding other new cusps.)

Tarsiers are arboreal creature, nocturnal in habit. Most of the time during the day, they sleep clinging to a vertical branch of the tree. They wake up at twilight and hop from tree to tree in search of food, which mainly consists of various kinds of insects, lizards, etc. The thick pads at the tips of the digits allow them to cling on the smooth, vertical surfaces of stem or branch and even to walk along a vertically held sheet of glass.

The second and third digits of the foot show claws like nails. Most of the prosimians have a single grooming claw (or toilet claw) on each foot while the Tarsiers have two such claws. It is important to point out that the Tarsiers are quite different from the Ceboids and Cercopithecoids. The orbital closure of the Tarsiers seems to be a parallel development to that of the Anthropoidea. So, they have been considered as the primitive representatives of the prosimian lineage from which the Anthropoidea has bifurcated. Prominent eyes with bony orbits can be taken as a typical characteristic to build a relationship between the Tarsiers and the Anthropoids. They are the exceptional living remains of the Eocene period.

Lorisiformes

The Lorisiformes is divided into two families, namely *Lorisidae* and *Galagidae.* The animals belonging to both of Lorisidae and Galagidae are distributed in all parts of Africa, south of Sahara, on the islands of Zanzibar off the east coast of Africa and on the other islands off the African mainland, but not in Madagascar. Of the two Asian varieties of Lorisidae, the Slender Loris is found in India, Myanmar, and Sri Lanka but the Slow Loris is found all over the south-east Asia.

The Lorisiformes are the arboreal and nocturnal primates. Their dental formula is 2:1:3:3(upper) and 2:1:3:3(lower) which show a typical prosimian characteristic. The first premolar looks like a canine. Like many other nocturnal animals they have developed the tapetum. This tapetum is nothing but a layer on the choroid of the eye (the pigmented vascular layer next to the retina) that concentrates dim light, so that animal is apt to have available in twilight or moonlight. The tapetum layer is also responsible for the glowing of the eyes in the dark when one flash of bright light falls on it. The lachrymal foramen is placed outside of the eye-orbit.

The hands and feet of the Lorisiformes are prehensile with well-developed digits. They have a single grooming claw (or toilet claw) on each foot for scratching on different parts of the body. The second digits of the feet are therefore equipped with grooming claws while other digits of hands and feet are provided with flat nails. They show a typical primate characteristic for the presence of a single grooming claw. Besides, both pollex (thumb) and hallux (big toe) are opposable and the hands and feet are well adapted for grasping. The tail helps a lot in balancing, as the forelimbs have been shorter than the hind limbs. The body has a furry covering.

In general, the Lorisiformes are small animals, which ranges from large mice to a large rabbit. Their principal food is the insects although they are accustomed in eating tiny plants and animals, whatever is found digestible. So far as the salient features are concerned, the postorbital wall is absent. The palatine bone contributes to the formation of the inner wall of the orbit.

Lemuriformes

All varieties of Lemuriformes live in isolation on the island of Madagascar, the southwest coast of Africa. It may be assumed that their ancestors once arrived in this island and developed into several types by adapting themselves to the local ecological situations. Some occupied bushes,some went on trees and so on. In terms of common name, the animals may be divided into three categories with a wide range of variation in size and colour - the Lemurs, the Indrises and the Aye-Aye. The Lemurs are small in size, but the smallest one, the mouse lemurs (Microcebus) is smaller even than a mouse. They are the smallest living primates. The Indrises (Indri) are large in size. Their, length is about a meter, measuring from head to tail. Third one is the Aye-Aye (Danbentonia). They are the nocturnal animals who attained the size of a cat. They also show long and bushy tail. Ears are large, oblique and without hairs.

Most of the Lemuriformes are diurnal and have prehensile hands and feet. Each extremity is equipped with five well-developed digits. The second digit of each foot is provided with a grooming claw while other digits show only nails. The body is covered with a thick coat of fur. The long bushy tail is though not prehensile, but indispensable for balancing. The eyes are lateral in position with a well-developed binocular vision. Many of them are arboreal and able to leap great distances by swinging. This has been possible because of their well-developed muscles.

The dental formula is 2:1:3:3 (upper) and 2:1:3:3(lower). The first premolar resembles the canine. The incisors and canines are set in the form of a toothcomb. An exception is the Lapi lemur where the upper incisors are not found. In the deciduous dentition the upper incisors remain present, but in the permanent series they do not exist. However, in general the dentition of Lemuriformes is prosimian. They are omnivorous in habit as their food consists of different kinds of insects, fruits, leaves, eggs, etc.

Anthropoidea

The *sub-order Anthropoidea* is classified into two *infra-orders, Platyrrhine* (New World Monkeys) and *Catarrhine* (Old World Monkeys). The Platyrrhine has one super-family, Ceboidea. The Catarrhine consists of two super-families, Cercopithecoidea and Hominoidea.

Ceboidea

The Ceboidea or Platyrrhine monkeys live in the forest regions of Central and South America. These New World Monkeys show a wide range of variation in size as well as in other characteristic features. For example, the Woolly monkeys and the Howler monkeys are the largest while the Pygmy marmosets are the smallest in size.

In general, they are characterized by broad fleshy nose with widely separated nostrils. The teeth are like those of man in structure and number but they have got an extra premolar on each half of both the jaws. The tail is prehensile; it is capable of grasping the branch of tree.

Howler monkeys usually live in the highest branches of the tall trees. They never come down to the ground, not even to drink water. They satisfy their necessity of drinking by sucking the moistures of wet leaves, flowers, fruits and nuts. They move by leaping through trees. The narrow rivers and streams are often crossed in the same way. A big mandible and hypolaryngeal apparatus provide them a fierce and ferocious appearance. The Titis live on the branches of big trees and they also do not go down for drinking water like Howler monkeys. The Squirrel monkeys are small in size like that of the squirrels. They show quadurpedal locomotion though live completely on trees. Their tail is long, but not prehensile. The Uakaris are bald-headed with short tails. The Sakis present profuse hair on the face, and.their tails are not prehensile. Only the monkeys of pithecine group, i.e. the Uakaris and the Sakis, do not possess prehensile tails; their tails are bushy which help in balancing. They are arboreal animals.

The Titis and many other monkeys are diurnal while the Night monkeys (Aotus) are nocturnal. The Spider monkeys (Ateles) have slender body with very long arms and legs. The Woolly monkey's (Logothrix) body is covered with thick woolly furs. Both of these, Spider and Woolly monkeys, possess very long prehensile tails. The dental formula is 2:1:3:3 (upper) and 2:1:3:3 (lower). But,in case of some other group of this super-family, this formula stands as 2:1:3:2 (upper) and 2:1:3:2 (lower). Their food is comprised of the nuts, fruits, buds, leaves, flowers, etc.

Cercopithecoidea

The Cercopithecoidea or Catarrhine monkeys are characterized by narrow nasal septum. The members of this group are chiefly herbivorous and they lead arboreal as well as terrestrial life. The tail is not prehensile and it is either long or short. All of the members have ischial callosities, which means a hard-calloused area on the buttocks. They also have the cheek pouches and their hind limbs are longer than the forelimbs. The Macaques and the Baboons are the typical catarrhine monkeys who show a close resemblance in characters, though distributed in different localities. The Macaques are distributed in Japan, China, Pakistan, India and Southeast Asia. But, the Baboons, are found in Africa, south of Sahara and Arabian Peninsula.

They are well adopted in both the arboreal as well as the terrestrial way of living. During daytime they behave as a terrestrial animal, but in night they sleep on the trees as any other arboreal creature. The size of the males is much bigger than the females. These males are much specialized for defence; they protect themselves as well as their females and youngs. Their hands are capable of precision grip. They build up a socially well-organized group.

Hominoidea

The Super-family Hominoidea is consisted with three families, *Hylobatidae, Pongidae,* and *Hominidae.*

Hylobatedae:

The Gibbons are the animals of this family. They are arboreal creatures. Their slender body is peculiar with extremely long arms; the hands touch the ground when they stand in erect position. The extreme length of upper extremities is directly connected with their mode of locomotion. On the ground they walk in upright position by balancing the body with their long arms. But most of the time they move by swinging from branch to branch of the tree with alternate clutching of hands. They are the perfect brachiator and capable of precision grip. Their habitat is the rain forests of Southeast Asia, including Java, Sumatra, Philippines, etc. They are also found in Northeast India and Myanmar (Burma).

The Gibbons are the smallest of the apes in size. But they show about twelve varieties. The height of the Gibbon does not exceed 3 feet and the body weight of the adult Gibbon varies from 25 to 45 Kg. They show no difference between the male and the female regarding the size of the body. The body is covered with fine woolly hairs, the colour of which is usually white, grey and sometimes black or even a combination of these three colours. Their long and powerful hands are provided with long fingers and short thumb. With the help of these powerful hands, the light and slender body can easily be lifted at the time of travelling through the trees. Therefore a free-swinging and pendulum motion is found. The lower extremities are much shorter in comparison with upper extremities. The big toe is found far apart from the other digits. This wide separation is advantageous for grasping the branches of the tree.

The head is large, elongated and narrow. The brain case is larger in comparison to face and the frontal elevation is low. Brain capacity is about 100 c.c. The face is provided with a flat and broad nose and thin lips. The eye-orbits are pronounced and composed of thick bony rims. The supraorbital torus is not found. Teeth show large, sharp and pointed canines, interlocked at the corners of the mouth. Though, the body is covered with fine woolly hair, the face and ischial callosities are devoid of hair covering. Usually the skin colour of these animals is black. Food consists of fruits, nuts, flowers and leaves. But, the birds, eggs, insects, etc. are also eaten by them often.

Nature of the Gibbons is gentle and pleasant. They live in small families that form with the parents and minor children. The sexes, males and females are very jealous about own sex. So each of them maintains a safe distance from other by using a loud warning hoot; this typical sound is the speciality of Gibbon. Another group of Gibbon known as Symphalangus or Siamang lives in Sumatra. The Siamang is larger in size than the common Gibbon. They possess laryngeal air sacs. Some scholars have concluded that the Siamang represents an intermediate form between the Gibbons and the Giant apes.

Pongidae

This group includes three types of giant apes - Pongo (Orangutan), Pan (Chimpanzee), and Gorilla.

Pongo (Orangutan):

The Orangutans are quite different from the Gibbons in comparison with their bulky and powerful bodies. They are also largely arboreal. Long arms, powerful hands and mobile shoulder help them in brachiation. The forearm is much longer than the upper arm. The legs are short and the feet are long and narrow. The hallux and pollex are very small and opposable. Again, the legs are shorter in length than the hands. On the ground when they walk by using both the hands and feet, they look like an old man who bents because of age. The movement has been slow for their large body size. But at the time of climbing on the trees, they are very swift. At present Orangutans are found only in the islands of Sumatra and Borneo. But the fossil evidences suggest that once they were distributed over a large

chimpanzee. Their body is covered with long, coarse hairs. of reddish-brown colour. The skin colour is usually brownish. No hair is found on face, ear, palms and soles. The face is concave in profile,

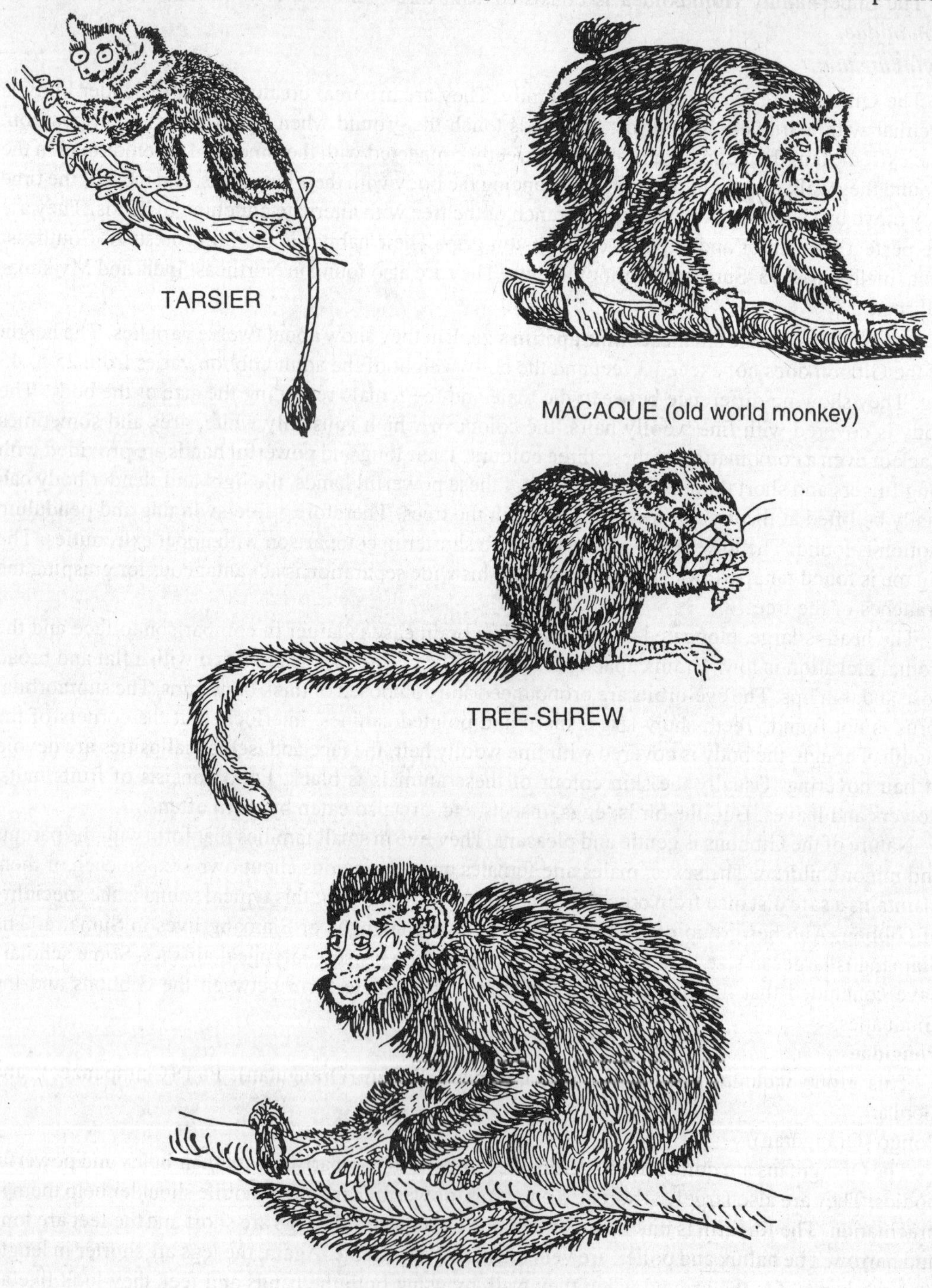

Fig. 4.15. Some Lower Animals Under The Order Primate

i.e. sinognathous. The breasts are situated laterally with the nipples in front of the armpits. Height of an adult Orang crosses 4 feet in erect position while the body weight varies between 60Kg to 80 Kg. The females are much smaller than the males.

The head is large with a high and rounded forehead. The cranium is small in relation to enormous facial portion. The cranial capacity ranges from 365cc to 425cc. Supra orbital torus is absent. The nasal bones are very small in size and have been fused together. The nasal bridge is not elevated; the nasal root is very narrow. The mandible is extremely large and projected forwardly. The jaws are provided with big, sharp and long tusk-like canines, which are interlocked. As because of their arboreal nature they live on trees by building nests on the branches. They like vegetarian diet and their food consist of different kind of leaves, wild fruits, flower, buds, etc.

Pan (Chimpanzee)

As per anatomical features,the Chimpanzees are more of less similar to the gorilla. But, the body of the Chimpanzee is not so well built like Orangutan and the gorilla. Tropical forests of Africa are the homeland of the Chimpanzees. But, their habitats include a variety of environmental settings such as the mountains, rain forests, etc. However, four types of chimpanzee are generally found, each occupying a specific geographical region. They are Pygmy Chimpanzee, Bald-headed Chimpanzee and two kinds of common Chimpanzee.

The body proportion of the Chimpanzees is almost like a man. Average height of an adult male is 5 feet and the weight is 125 kg. Females are a little smaller in size and their average weight is 100 kg. The body, irrespective of male and female, is coated with long and coarse hairs of different colours. The face, hands and feet are devoid of hairs. In case of Bald-headed Chimpanzee, no hair is seen on their heads. They have low-vaulted heads, round in shape. The cranial capacity ranges between 400cc and 500cc. The development of supra-orbital ridges is very poor but these have been continuous to form torus. Nasal bones are fused and the size is very small. The nasal bridge is not elevated at all. But the jaws are well built to be projected forwardly. The canines are also large, sharp and projecting, but smaller in comparison to Orangutan and gorilla. The facial region is small in proportion to the skull. The lips are thin.

The hands are elongated and narrow with long fingers. The thumb is small and opposable. The legs are long and these are larger in comparison with the Orangutan. The foot is also long. The lower extremities are not well equipped for walking upright. The big toe is opposable, so not in the line with other toes. The heel is rudimentary. In erect position, the hands of the Chimpanzee reach up to the level of knee. Often they stand on their legs, but rarely they walk bipedally. On the ground they move quadrupedally with the support of four limbs. However these animals are very expert in climbing and brachiating. They build nests on the trees and during the night they sleep happily in their nest.

The Chimpanzees are quite able to use their hands in different purposes, such as the holding, grooming, building nests and using tools. They are mostly vegetarian, but occasionally eat ants, meat and fish. It may be noted that they have the capability of using tool. They pick up small pieces of grass or twig and prepare some kind of tool. These tools are poked into the nests of termites to catch them for the purpose of eating. The food consists of fruits, vegetable products, eggs, small birds and sometimes rodents also. The animals by nature active and noisy. Though, they live in-groups and move from one place to another in search of food, their groups are not well organized at all and no strong dominance or hierarchy is found.

Gorilla

Among all primates the Gorilla is the largest as well as strongest one. They are found in equatorial regions of Africa, specifically in west central Africa. Two groups of gorillas have been identified - the lowland or coastal gorilla of the west-central coast and the mountain gorilla or eastern gorilla of the highland of central Africa. The habitats range from low rain forests to mountain forests and even sometimes they live on mountain slopes. The height of an adult gorilla varies between 5 feet and 6 feet among the low land gorillas. It is between 5 feet 3 and half inches and 5 feet 10 and half inches in case of mountain gorillas. The average body weight of the adult males varies from 200 to 250 kg.

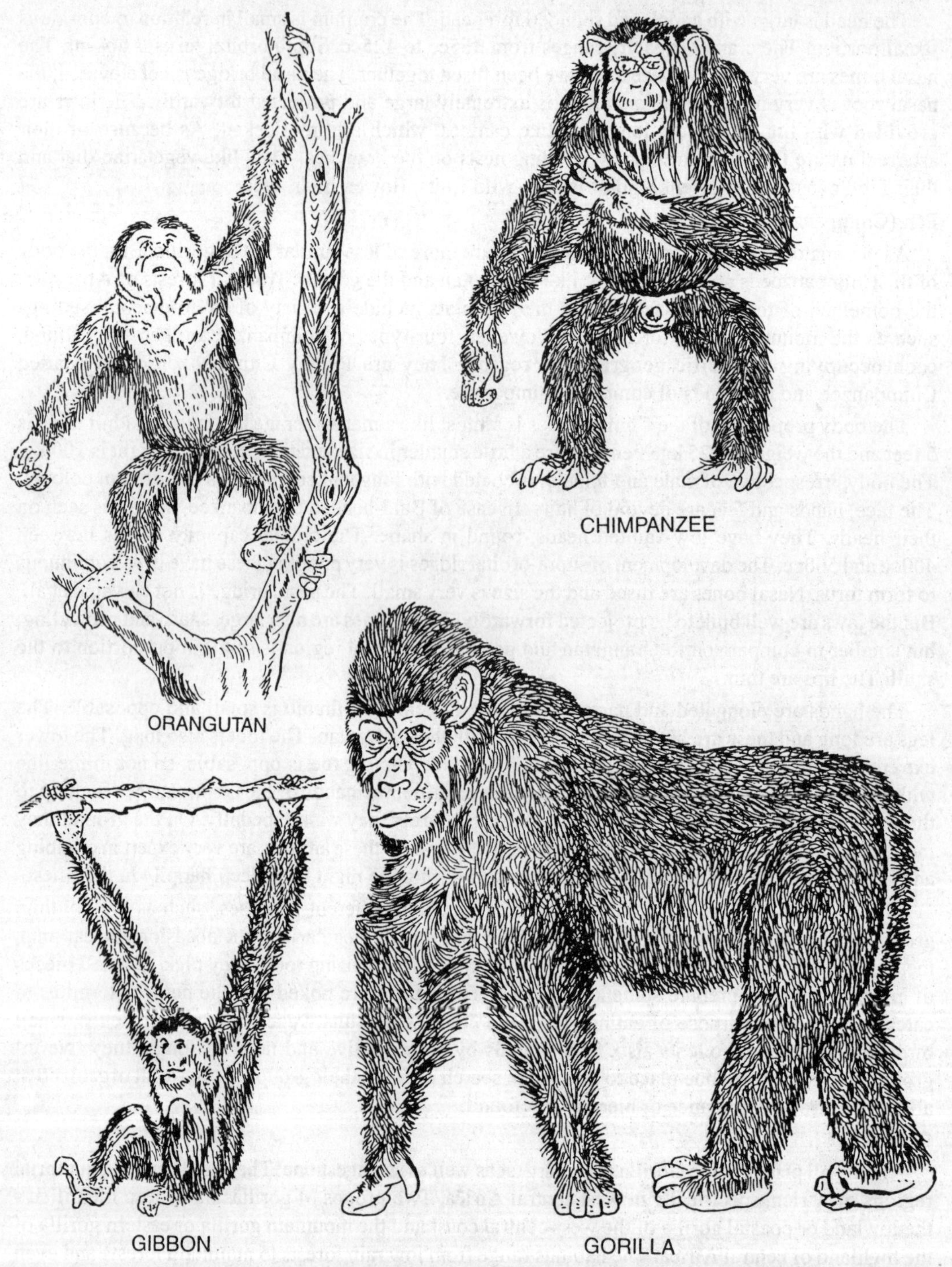

Fig. 4.16. Anthropoid or Man-Like Apes

The body is coated with long black hairs. The face, palms and soles are completely devoid of hairs. Facial portion of the skull exhibits an enormous size while the cranium is a smaller one. The forehead is very low with a prominent sagittal crest. This sagittal crest is weaker in case of female. Average cranial capacity of the males and females are 500cc and 450cc respectively. Supra-orbital ridges are very thick to form a torus. The nasal bridge is slightly elevated while the nasal root is low and flat. Upper jaw is prognathous and the lower jaw is massive. The canines are big, sharp and enormous in size as well as interlocking. The lips are thin and the chin is completely absent. The upper arms are longer than the lower arms, which makes the animal different from a Gibbon, Orangutan and Chimpanzee. The hand is shorter and broader than other apes and the thumb is also well developed. As a whole the hands of a gorilla resemble more to a man than other apes. Though the legs are short, still the feet of the gorilla are like a man in many respects. The heel is more of less well developed and anatomy of the foot indicates that it is less adapted for arboreal prehension. They spend most of their time on the ground, but can climb on the trees. Nests are built on the trees to pass the nights there. Sometimes gorillas walk bipedally but their usual gait is obliquely quadrupedal.

These animals are very gentle and peaceful. But they look ferocious and aggressive when they start chest beating. It is a part of their special behaviour, which is manifested in different occasions. At the time of chest beating they stand almost erect. Though their normal locomotion is not brachiation, still they are called as brachiator. In fact, they are the terrestrial apes who build their nests on the trees only for sleeping. They live in a group of 5 to 30 members where the dominant male member becomes the leader of the group. The females always provide motherly protection to the youngs. They are vegetarian and the food consists of fruits, leaves and different kinds of vegetables.

Hominidae

The family Hominidae is formed with only one group called Homo, which has only one species, the sapiens. This Homo sapiens have been represented by the Man. As a whole, man possesses certain unique characteristics, which helped him to surpass all other animals and to establish a new genus, *Homo*. Within this genus, only the species *sapiens* has been spread and multiplied over the whole world. They have settled in different environmental conditions and acquired distinctive traits.

Homo sapiens

The skull consists of two parts - the cranium and the face. The cranium is a large bony case in which the brain is placed. The brain is complex and its parts are relatively enlarged; the cranial capacity ranges between 1300cc to 1450cc approximately. The forehead is bulging and supra orbital ridges are diminished.

The power of vision not only has increased; a stereoscopic vision has added. The fleshy tip of the nose is made of osteo-cartilagenous framework. The lips are out-rolled, so the mucus membranes are visible. Besides, the upper lip is marked with a median furrow, which is the unique feature of man. In dentition, the canines have been reduced in size and no longer extend beyond the level of other teeth. The nature of teeth indicates that the man deals with a wide range of food. The alveolar border is parabolic and the chin is well developed.

The foramen magnum is situated centrally at the base of the skull for balancing the head on the vertebral column. The arms are shorter than the legs. The shape of foot has also changed, as the big toe is no more opposable. The heels of the feet are well developed. The pelvic region is bowl shaped; its transverse diameter is greater than the anterio-posterior diameter. The chest shows an anterior-posterior tapering. The vertebral column has four curvatures. All these features promote erect posture as well as bipedal locomotion.

Besides, there are other characteristic features unique to man. For example, the body is more or less hairless; articulation of speech, absence of ischial callosities, highly rolled margins of ears etc. These special attributes have given man the top place in the animal kingdom. This unparalleled species under the single genus of the family Hominidae has definitely been emerged from the order Primate through the process of evolution.

Hence we can trace the phylogenetic identification of man in the following way:

Kingdom — *Animal*
Phylum — *Chordata*
Sub-Phylum - *Vertebrata*
Class — *Mammalia*
Sub-class — *Eutheria*
Order — *Primate*
Sub-order — *Anthropoidea*
Infra-order — *Catarrhine*
Super-family — *Hominoidea*
Family — *Hominidae*
Genus — *Homo*
Species — *Sapiens*
Variety (race) — *Caucasoid, Mongoloid, Negroid, etc.*

This means, man's position among the creatures of animal kingdom is very clear and definite. He belongs to the phylum *Chordata* for the possession of spinal cord. The bony segments of spinal cord take him to the sub-phylum *Vertebrate.* The mode of nourishing of the young (by nutrient fluid excreted from the mammary glands of females) proclaims his membership in the class of *Mammals.* Since the embryo develops in the maternal womb, the *sub-class Eutheria* has been assigned to man. Further, this placental animal finds its position in the order *of Primate;* sub-order, *Anthropoidea* and infra-order catarrhine. The family *Hominidae* has been chalked for him under the super-family *Hominoidea.* He got the genus *Homo* for his manhood. But when he first appeared on earth, he was not a perfect man. Man achieved this perfection with the species, *Sapiens.* At present, all living forms of man are regarded as *Homo Sapiens Sapiens.* This is the scientific name to call the man. Here the second sapiens denotes the sub-species as some minor differences exist between the earliest true man and the modern man. The different races as dominant in different parts of world are considered as the varieties of *'Homo Sapiens Sapiens'.*

HUMAN EVOLUTION

Man is a product of evolution. Therefore human evolution is intimately related to the origin of life and its development on the face of earth. It is customary to speak of evolution 'from amoeba to Man', as if the amoeba is the simplest form of life. But in reality, there are several organisms more primitives than amoeba, say for example viruses. The evolution from a self-replicating organic molecule to a protozoan, like amoeba, is the most complex step in evolution, which might have consumed the same extent of time from protozoan to man.

The term evolution was first applied by the English philosopher Herbert Spencer to mean the historical development of life. Since then evolution denotes a change, although the term may be defined in several ways. In the context of man, the biological evolution started with the 'Origin of life'. In the beginning, there was nothing. The first successful formation of protoplasm initiated the life and its continuous development proceeded towards complexity to give rise different life forms of evolved type.

About 10 billion* years after the formation of Universe, the earth was formed. Life on earth appeared far late, nearly three billion years ago. Of the several evolutionary problems, perhaps the origin of life is the most critical, since there is no record concerning it.

Life has been characterized by the capacity of performing certain vital functional activities like metabolism, growth and reproduction. There is no ambiguity regarding this point. But how the first life came on earth is a matter of conjecture. Ancient thinkers speculated that life originated spontaneously from inorganic components of the environment, just after the formation of earth. A series of physio-chemical processes were perhaps responsible behind this creation.

Aristotle (384 BC to 322 BC) was the pioneer in this line of thought and nobody raised any voice against his speculation till seventeenth Century. But in seventeenth Century, an Italian scientist, Francesco Redi (1627 -1697) made an experiment with two pieces of meat. One of the pieces was kept fully covered and the other piece was kept in an open place. After some days he examined both of the pieces very carefully. He noticed that, flies laid eggs on the uncovered piece of meat and so many new flies had born. But the covered piece of meat had not produced any new fly, as there was absolutely no access of flies. Redi tried to establish the fact, that living organisms can not be originated spontaneously from inorganic components. More or less at the same time, Leuwenhock (1632 - 1723) by studying several microorganisms like protozoa, sperm, bacteria etc. under microscope declared that the spontaneous generation was possible for the microorganisms. Later, Louis Pasteur (1822 -1895) also studied much to furnish evidences in support of spontaneous creation.

In fact, scientists of this period were perplexed in finding out how life began spontaneously as a matter of chance. Philosophers, Thinkers and Scientists all had submitted their varied thought and propositions regarding the nature and mechanism of origin of life on earth. Different religions had also put forth different concepts in this connection. However, those various hypotheses and theories have been discussed here systematically, as they came one after another.

* 1 Billion = 1,000,000,000
1 Million = 1,000,000

THEORIES OF BIOLOGICAL EVOLUTION

Theory of Eternity

This is an orthodox theory. It believes that some organisms were there from the very beginning of the Universe. Those organisms are still existing and will be continued in future in addition to some new forms. According to this theory, the original forms are eternal, and they have been preserved automatically. But this view is not at all popular; it is held by a few people only.

Theory of Divine Creation

A Spanish Monk, Father Sudrez (1548 - 1617) proposed this theory. It was based on the Biblical book of Genesis * According to Genesis, of Old Testament of Bible, the world was created by the supernatural power (God) in six natural days. The theory specifies that all creations, including plants, animals and man on earth were created during those six days. Since all species were made individually by god, the theory does not accept the idea of origin of new species from ancestral forms. Life is considered as a vital spirit according to this theory.

The Hebrew and the Christian Church authorities had supported this view for many Centuries. To them, god created Adam and Eve, the two companions of opposite sex about 6,000 years ago, from whom the human beings have descended. Archbishop Ussher (1581 - 1656) pointed out 4004 BC as the exact year for the creation of man. Each and every followers of this theory believed that all creations of god are arranged in a chain where human is posited at the top.

Theory of Spontaneous Origin

The theory contends that life had originated repeatedly from inanimate materials or non-living things in a spontaneous manner. The concept was held by early Greek philosophers like Thales (624 - 547BC), Empedocles (485 - 425BC), Democritus (460 - 370BC), Aristotle (384 - 322BC) and others. Aristotle thought that fireflies originated from morning dew and mice from the moist soil spontaneously. All succeeding Greek philosophers and many scientists shared Aristotle's view till the middle of the seventeenth Century. Louis Pasteur partially accepted this theory.

Theory of Catachysm or Catastrophism

French geologist Georges Cuvier (1769 - 1832) proposed this theory. His observation was based on the fossil remains of varied organisms. According to him, the earth had to face severe natural calamities at different times for which many animal species have been destroyed. But each time when the earth settled after a great Catastrophe, relatively higher forms of animals appeared to replace the situation. Cuvier did not believe in continuous evolution. To him the species never evolved by modification and re-modification; a series of Catastrophes were responsible behind changes where previous sets of living creatures get replaced by new creatures of complex structure. As per his scheme, corals, molluscs and crustaceous appeared in the first phase. Then came the first plants being followed by the fish and reptiles. The birds and mammals appeared thereafter and in the last phase man emerged about five to six thousand years ago.

Theory of Uniformitarianism

This theory was presented by Charles Lyell (1797 - 1837) in his work 'Principles of Geology'. Being a geologist, he could not accept the concept of an unchanging earth. By studying the rocks and geological processes, he came to the conclusion that, at the beginning, some forces were in operation to shape and reshape the earth. Animal forms gradually evolved along with this change. Fossils were the main support for his evidence. This theory on one hand discarded the "theory of Catastrophism" and on the other hand nullified the "theory of divine Creation".

---

* This is the narrative first *book* of Canonical Jewish and Christian scriptures (The first book of Old Testament in Bible).

Theory of Cosmic Origin of life

This theory advocated that the first life seed had been transported through the cosmic particles from other planet. Richter (1865) developed this theory and he was supported by Thomson, Helmholtz (1884), Von Tieghem (1891) and others. According to them the meteorites that travelled through the earth's atmosphere, contained embryos and spores in them; those gradually grew and evolved into different types of organisms. But the concept lacked evidences and interplanetary exchange of viable spores and embryos could hardly be possible in the light of current understandings.

Theory of Cynogen

German scientist Fluger proposed this theory. According to him, the cynogen, a complex chemical compound was developed by sudden reaction between the atmospheric nitrogen and carbon. This cynogen later gave rise to first protein substance, which ultimately created life through various types of chemical synthesis.

Theory of Chemo-synthesis

This theory also recognized a complex type of chemical synthesis. It pointed out different kinds of materials, which in varied natural environment produced a large number of actions and interactions. As a consequence, life developed in a peculiar set up following a complicated situation.

Theory of Virus

Some scientists believed that virus was initially responsible for the emergence of life. The viruses hold a transitional stage between living and non-living. By nature a virus is non-living, but when it reaches to the body cell of the living host, it behaves as living. Therefore, it was thought that such a creature might possess a role in the emergence of life.

Theory of Organic evolution

According to this theory, origin of life must have taken place in this world. First living existence was very minute and in the form of unicellular structure. As the time passed on, most of the unicellular forms were transformed to multicellular forms under the various environmental oscillations. Gradually and gradually simple form of animals was converted to very complex type of animals. As a matter of fact, the geo-environment of the earth underwent a process of continuous change and influenced the animal forms. Complex forms of animals evolved out of the simple forms in a slow and steady way. This process of change has been designated as organic evolution.

The conception of organic evolution maintains its conformity with ancient Hindu religious thought. B.M.Das (1961) wanted to prove this with the example of ten incarnations of Lord Krishna (Dasha avatar).

He mentioned that the first incarnation was a fish (Matsya avatar). He justified his remark by comparing it with the western belief where the life was thought to be originated in water. The second incarnation according to Das was a turtle (Kurma avatar), an amphibian. The next incarnation was a wild pig (Baraha avatar) which represents land-living animals. The fourth incarnation was a mixed form with half man and half animal (Nrisingha avatar). This idea complies with anthropological outlook. All of the anthropologists now agree that the stage before true man was a combination of man and ape. However, the fifth one was a short-statured incarnation (Baman avatar). It indicates the fact that early men were short statured. In this way. Prof. Das described not only the biological evolution, but the cultural evolution too. He also mentioned that Parasurama was defeated by Rama, as Rama possessed bow and arrow, a superior weapon than the axe. The stage corresponded to the food-gathering stage of prehistory and it was followed by a food-producing stage as depicted in the story of Lord Krishna who used to look after the cattle in his childhood and his elder brother Balaram carried a plough most of the time.

In the Christian era, before Darwin, several scientists and philosophers expressed their views regarding the evolution. In this context, Carl Linnaeus (1707 - 1778) made a classic work "Systema Natural" where he described a system of classification involving the plants and animals, known as taxonomy.

CONCEPT	THEORIES		TIME
EVOLUTION	THEORY OF ORGANIC EVOLUTION	SYNTHETIC THEORY	Middle of 20th Century
		MUTATION THEORY	Beginning of 20th Century
		THEORY OF MENDEL	19th Century
		DARWINISM	19th Century
		LAMARCKISM	Early 19th Century
		THEORY OF VIRUS	18th Century
		THEORY OF CHEMOSYNTHESIS	18th Century
		THEORY OF CYNOGEN	18th Century
		THEORY OF COSMIC ORIGIN	18th Century
		THEORY OF UNIFORMITARIANISM	18th Century
		THEORY OF CATASTROPHISM	18th Century
		THEORY OF SPONTANEOUS ORIGIN	18th Century
		THEORY OF DIVINE CREATION	17th Century
		THEORY OF ETERNITY	16th Century
		THEORY OF ARISTOTLE	300 B.C.

Fig. 5.1. Theories In Relation To Human Evolution

He placed man in the order Primate along with apes and monkeys, but he did not suggest any common ancestry for them. Further, he believed that each species was created specially and separately; their position remains unchangeable. In this way, the proposition of Linnaeus was a combination of the Old belief and the new thought.

Mon boddo (1714 -1790) by observing the origin of species, traced the evolution of man from the monkeys. Bonnet (1720 - 1793) also worked on the process of evolution and proposed a 'scale of beings'. His proposition went on an ascending order from the mineral to man. Many more scientists worked with the origin of man. Among them, the contributions of Erasmus, Darwin (1731- 1802), Karl von Baer (1792-1876), Schopenauer (1788 -1860) and Charles Lyell (1797 - 1875) seem to be indispensable for proper understanding of the facts of evolution. Imanuel Kant (1724 - 1804) proposed that the man be descended from the monkey. According to a group of scholars, the expression of Goethe (1749 - 1832) was so meaningful in respect of evolution that he may be regarded as a predecessor of Charles Darwin. Again, another scientist, Malthus (1766 -1834) kept valuable contribution towards formulating the theory of natural selection.

It is justified to trace the history of evolutionary thought from the beginning of nineteenth Century. The first systematic attempt was made by Jean Baptiste Lamarck (1744 - 1829), a French biologist who was an eminent pre-Darwian student of evolution. His theory was published in 1802 in which he proposed the 'inheritance of acquired characters' during the lifetime of the individual. Following Lamarck's proposition, Charles Darwin and Alfred Russell Wallace jointly proposed the theory of the 'Origin of Species' by *Natural Selection.* Charles Darwin's evolutionary theory had its base on the accumulation of small fluctuating variations. He had realized that heredity was an essential factor in the study of evolution, though he did not put much importance to it. August Weismann realized the importance of heredity better than Darwin did. He emphasized on the 'continuity of the germ plasm' and tried to project the transmission of inherited qualities from generation to generation by the germ cells. Hugo de Vries, one of the rediscoverers of Mendel's laws of heredity, announced mutation theory of evolution in 1901. He considered *mutation* (i.e. sudden hereditary changes) as a factor behind evolution. Natural selection found very little or no place in his mutation theory. But, later, the geneticists, biometricians, and palaeontologists revived the faith in *natural selection*. Of these, the most important development took place in the field of genetics; the natural selection was started to be restudied and reinterpreted by the geneticists. Mention may be made of Theodore Dobzhansky and R.B. Goldschmidt, who laid the foundation for the Neo-Darwinian theory. The genetic theory of Natural Selection is therefore referred as Neo-Darwinism. R.S.Fisher, J.B.S.Haldane and Sewall Wright made valuable contribution to the statistical analysis of population and secured own position among the principal proponents of Neo-Darwinism. However, the important theories have been discussed in the following pages.

Theory of Lamarck (Lamarckism)

The French biologist, Jean Baptiste Lamarck (1744 - 1829) spent his early years in military service but when he was stationed at Monaco, he acquired interest in Botany. He also established himself as a distinguished zoologist. His extensive studies on invertebrates formed a base in zoological classification. He was the first scholar to recognize the distinction between invertebrates and vertebrates. But Lamarck's name is usually associated with the 'theory of inheritance of acquired characters'. Of his several writings, mention must be made about three publications relating to the theory of evolution: *Recherches Sur L 'Organization des Corps Vivant* (1802), *Philosophic Zoologique* (1809) and *Historie Naturelle des Animaux sans Vertebrates* (1815 - 1822).

Lamarck expressed the fact that the acquired characters could be inherited. His theory, known as Lamarckism was based on two laws -

i. the law of use and disuse of organs, and

ii. the inheritance of acquired characters.

According to Lamarck, a living body is always influenced by the environmental factors and ultimately this phenomenon initiates an adaptation of organism to its surroundings. As per necessity, some parts of the body may be used more and more. Therefore, those parts tend to show more development or changes in course of time. On the contrary, other parts of the body, which may not be required much, will be weak or demolished due to constant disuse. This change in body structure is reflected in future generations. This means, the characters that are acquired by the use or disuse of different organs can be transmitted to the succeeding generations.

To support his theory Lamarck presented several examples. The most remarkable one is associated with the long neck and high front legs of giraffes. He stated that this animal originally possessed short neck and small front legs. As a herbivorous animal, the forerunners of modern giraffe were acquainted with grass and the leaves of dwarf trees. But following a sudden scarcity of these plants, giraffes had to stretch out their necks to reach the leaves of the tall trees. This stretching affected the muscles and bones of the neck, which started to be modified with time. Not only the neck became longer, the front legs also increased in size. This phenomenon is nothing but an adaptation to the environment, in the way to survival. The modified traits were continued in subsequent generations and eventually all the giraffes got very long necks and well-built long front legs. In another example, he mentioned that the ducks are unable to fly because their wings became weak when they stopped flying. Again, the birds that started to live in an aquatic environment gradually acquired webbed feet through the conquest of survival. Lamarck also cited other examples like limblessness in snake, blindness of moles and certain cave-dwelling forms, acquatic plants with dimorphic leaves (having submerged and aerial leaves), etc. All these changes were held to be cumulative from generation to generation, and also hereditary.

Lamarck's theory had met criticisms from several angles. Although some of his views were admitted by a few scholars, most of the scholars did not accept his theory. The German scientist August Weismann ridiculed the essence of Lamarckism (inheritance or acquired characters) by his experiments, which involved cutting of tails of mice for over twenty generations. All tailless mice in all generations (even in the last generation) produced their offsprings with tails. Therefore he reached to the conclusion that the environmental factors might have an influence on the body cells, but it is not enough to profess a change of reproductive cells. Characters of an organism would not be inherited unless the change could occur in the reproductive cells. However, the proposition of Weismann is popularly known as 'Germ-Plasm theory' as contrary to the theory of Lamarck. According to Weismann the body of an animal is composed of two parts viz. Germplasm (germ cells) and somatoplasm (body cells); only those characters which are located in the germ-plasm will be inherited by the offspring.

The evidence against Lamarckism was also criticized by others, on the ground that cutting of tail is rather mutilation, in which the animal did not participate actively. So some specific cases were required where organisms can actively participate in the activity. In this respect, McDougall (1938) conducted a series of experiments on learning, using white rats. He designed a water tank having two exits, one lighted and the other dark. The lighted exit received electric shock, while the dark exit did not have any arrangement to receive the electric shock. The white rats were dropped into such an experimental tank, and then trained to escape through the dark exit. A number of trials were required for the rat to learn the way to escape from the dark exit. These trials constituted a measure of the speed of learning. The trained rats were bred, and their offsprings were taught the same problem. In this manner, he subjected the rats for experimentation, for forty-five generations. McDougall observed that the number of errors made in learning, the problem decreased progressively from generation after generation. On the basis of this experiment, he concluded that an acquired character (learning or training) is inherited.

Unfortunately, McDougall's experiments met with severe criticism, mainly because the repetition of similar experiments in other laboratories had failed to produce similar results. They could not control the genetic constitution of the experimental rats. Limitation of various other experiments probably initiated the scholars for seeking evidence in favour of Lamarck. A new school of thought in the name of Neo-Lamarckism soon appeared in the scene, which tried to modify the principles of Lamarck in order to make it acceptable to the students of evolution. The foremost position was occupied by Giard (1846 - 1908) of France and Cope (1840 - 1897) of America. However, Neo-Lamarckism was based on the idea of adaptation, integrated with direct and casual relationship between structure-function and environment. The difference between the Lamarckism and Neo-Lamarckism was that, Lamarck believed in direct action of the environment, which, he thought was responsible to achieve final perfection of the individual. But Neo-Lamarckism omitted the very idea. The Neo-Lamarckians argued that a considerable period of time was required for getting the effect of external factors. They also pointed out that if the external factors failed to influence the reproductive cells of the parents, their offsprings would never inherit any of the modifications.

Rapid progress of science in twentieth Century favoured the growth of 'genetics', which supported none of the theories - Lamarckism and Neo-Lamarckism. Still Lamarck deserves appreciation as his proposition helped to open new avenues of thought in the science of evolution.

Theory of Darwin (Darwinism)

Charles Robert Darwin (1809 - 1882) was born as the fifth son of his parents. He had an elementary schooling in Shrewsbury, England. In childhood he took little interest in studies, but showed great interest in hunting birds and shooting dogs. His father and teacher considered him as 'a little below average in intelligence'. Although in school, he showed some interest in mathematics and chemistry, but most of his time was spent in watching the habits of birds, collecting insects and minerals.

In 1825, Darwin was sent to Edinburgh to study medicine, but soon he discontinued the course. After this his father wanted him to be prepared for the post of a clergyman, in the Church of England. So Darwin was sent to Cambridge. While studying at Cambridge, he gained friendship with some distinguished men of science, such as, the botanist Dr. Henslow and the geologist Sedgwick. Dr. Henslow's friendship entirely changed the course of Darwin's life; he nominated Darwin in the position of a young naturalist for the voyage on H.M.S. Beagle (a ship, in which Charles Darwin sailed around the world).

The voyage on the Beagle started on 27th. Dec. 1831 and Darwin visited many Islands in Atlantic ocean, some of the islands in the Pacific ocean including Galapagos islands, many places on the coasts of South America and finally returned after five years on 2nd. Oct. 1936. While on the Beagle, Darwin took notes on the flora, fauna, and the geology of the places visited; and also made extensive collections of living and fossil minerals. All these constituted the basis for his future publications.

Darwin's first publication, Journal of Researches (1839) met with immediate success. In October 1838 he accidentally came across Robert Malthus' essay on population. This essay provided a clue for which Darwin was able to think of the 'struggle for existence' among the animals and plant kingdom. In this respect, he started to collect the data from 1842. The famous geologist of that period, Sir Charles Lyell suggested him to write about the origin of species. In 1858, when Darwin was halfway in his writing, he received a manuscript entitled, "On the tendency of varieties to Depart Indefinitely from the Original type" from Alfred Russell Wallace (1823 - 1913). Wallace requested Darwin to read his essay and to make comments on it. Darwin found that the essay was complete in all respects and contained the essence of his theory of natural selection. Being generous, he decided to withhold his half-completed work, in favour of Wallace. So he wrote to Lyell with a recommendation to publish Wallace's paper at once. But Lyell, being aware of Darwin's strenuous effort since 1842, urged Darwin to write a short abstract of his theory. He wished that Wallace's paper would be published simultaneously with Darwin's abstract. Reluctance of Darwin could not stand against the insistence of Lyell. Thus, in 1859, Wallace's paper and an abstract of Darwin's manuscript together appeared in the *Journal of the Proceedings of the Linnean society.* To start with, Darwin intended to complete his work in four volumes but subsequently he condensed the work into a single volume, entitled 'Origin of Species' which was published in November 1859. The work of Darwin was submitted fifty years after Lamarck and his theory is commonly known as Darwinism. But, the credit went to both the scholars - Darwin and Wallace; the first systematic as well as comprehensive approach in the perspective of evolutionary development was made by them.

Darwin's theory of evolution is based upon four main, rather easily understandable postulates, which may be summarized as follows:

1. *Prodigality of Nature*

All species have a tendency to produce more and more offsprings in order to increase the number of population. For example, a salmon produces 28,000,000 eggs in a single season; a single spawning of an Oyster may yield as many as 114,000,000 eggs; a common roundworm (Ascaris lumbricoides) lays about 70,000,000 eggs in a day. Darwin has even cited examples from slow breeding animals. Elephants appear to be one of the slowest breeders, having a life span of about hundred years. The active breeding age continues from thirty to ninety years, during which a single female may produce six young ones. Taking this estimation into consideration, Darwin calculated that a single pair of elephants, at this rate of reproduction (provided all the descendants survived and reproduced at the same rate) would produce 19,000,000 elephants after 750 years.

All these examples furnish instances of tremendous reproductive potential among all species of organisms. The basic reason behind this huge production is to ensure the survival. Because, in reality we find that, inspite of the rapid reproductive potential, the size of a given species, in a given area, remains relatively constant.

2. *Struggle for Existence*

Above observation led to the conclusion that all the progeny produced by any generation do not complete their life cycle, many of them die during juvenile stages. Darwin therefore proposed his concept of'Struggle for Existence'; the struggle is often generated for the want of enough resource. All individuals can not survive under struggle. According to Darwin, the Struggle for existence may be of different types. It may be a Struggle to overcome adverse environmental conditions (like cold or drought), or to obtain food from a limited source of supply. It may be a fight for occupying a living space, or even to escape from the enemies. However, any of these said situations, evidently leads the members of a group towards competition, in order to meet their requirements. Thus, the nature of struggle may be of three types according to the situations.

i) Intra-specific struggle

When the members of a same species struggle among themselves, the situation is considered as intra-specific struggle. Such a struggle is usually centered round the consumption.

ii) Inter-specific struggle.

The individuals from different species also may go on fighting for survival. An individual from one species may hunt another individual of other species as food. For example, tiger hunt goat and deer; cat hunt rat; lizard hunt cockroach and different small insects; and so on. According to Darwin, in the animal kingdom, a species often stand as prey to other species, which clearly indicates a struggle for existence. Such happenings have been referred as Inter-specific struggle.

iii) Environmental struggle

The environmental struggle is different from the inter-specific or intra-specific struggle. Here individuals irrespective of their species-identity struggle against the environmental hazards like earthquake, flood, drought etc. Those who have greater potentiality for resistance, only they survive.

Darwin believed that the struggle is a continuous phenomenon in the way to survival. It is severe among the members of the same species (intra-specific competition), as they depend on identical requirements of life. The inter-specific competition is though very common, but its frequency is lesser than the intra-specific competition.

3. *Organic variation*

Darwin observed that variation is a universal phenomenon. Except the identical twins, no two organisms are exactly alike. Even the two leaves of a plant or two peas in a pod often show easily recognizable differences. Therefore individuals of a single species must vary from each other. At times, an entire population may exhibit a definite pattern of variation for which it is distinguished from the rest of the species. Such a population showing definite pattern of variation is often referred to as subspecies. Darwin considered these subspecies as incipient species, and he believed that, in course of time, these subspecies would be subjected to further variation to give rise a new species.

Although natural variations are neither advantageous nor disadvantageous to the species concerned, but some variations are considered as favourable and others are unfavourable. In fact, the variations in terms of physiological, structural and behavioural traits play very important role for adaptation in the environment. The new variants are produced continuously but when those variants can not cope up the environment, it is termed as unfavourable variation. Organisms with unfavourable variation easily get defeated in the struggle for survival and in course of time they become eliminated from the world. On the other hand, the new variants that are capable to adopt the pressure of the environment survive long. The new traits of advantageous characteristics pass on to the future generation.

Darwin recognized two main types of variation in nature, viz. Continuous variation and discontinuous variation. By the term continuous variation he wanted to mean small fluctuations of evolutionary significance. It was held as a force for attaining perfection being selected by nature. For example, the long neck of giraffe was evolved out of continuous evolution. Contrary to this, discontinuous variation is mostly large and rare in occurrence. However, they appear suddenly and do not show any graded series. Such discontinuous variations have been regarded as 'sports' by Darwin; to which, Hugo de Vries has given the name 'mutation', at a later period. In the eye of Darwin, discontinuous variation had no evolutionary importance. Darwin draws the example of Dinosaurs. The enormous size and giant stature of Dinosaurs were the result of discontinuous variation. He found negative mode of natural selection behind the extinction of those Dinosaurs.

3. *Natural Selection*

Natural selection is the final outcome of Darwin's evolutionary thought. Individuals differ from each other because of organic variation, which evidently means that some individuals are better adapted to survive under the existing environmental conditions than others. In the struggle for existence, the better-adapted individuals possess a better chance of survival than those who are less adapted. The less adapted individuals therefore get eliminated before reaching maturity and thus a large number of individuals die in the struggle for existence. However, the traits having greater survival value are preserved in the individuals and transmitted to the offsprings, who are supposed to be the progenitors of the next generation. Darwin called this principle, by which preservation of useful variation is brought about, as natural selection. The same principle (natural selection) has been designated by Herbert Spencer as 'survival of the fittest'. In the words of Darwin "the expression often used by Spencer, of the survival of the fittest, is more accurate, and is sometimes equally convenient".

The theme of Darwin's theory may finally be summed up in the following words: The organisms always struggle to maintain their existence as nature decides the survival of the fittest one. Adaptive traits preserved through natural selection gradually change the characteristics of species and thus evolution occurs. The theory of the origin of species by natural selection, though is regarded as a major advancement in evolutionary thought, it lacked the knowledge of heredity, which was essential for the understanding of evolutionary studies. It was really unfortunate that Darwin never came across Mendel's work, who by then, invented the basic principles of heredity. Had Darwin come across Mendel and his work, he would not have to write in the last edition of his 'origin of species' that "the fundamental principles of heredity are still unknown".

The human ancestry was discussed by Darwin in his book, "The Descent of Man' which was published in 1871. He said that life ascended from simplest form of minute organisms to the complex forms through different stages of evolution where man is found at the summit. But, at the time of Darwin very few fossil evidences were discovered; those were insufficient to establish the proposition. This was the first weakness of Darwinism. The second weakness was hidden in the process itself. Darwin wanted to explain heredity by the *'theory of Pangenesis'*, which declared that all parts of the body produce minute particles called pangenes that ultimately get deposited to the sex-cells being carried by blood. Those particles are further carried to the next generation when fertilization takes place and same kinds of organ, cell, tissue etc. are reproduced. However, the theory of Pangenesis, like the Lamarck's principle, accounts for the inheritance of acquired characters. But it too was universally discarded for the lack of evidence. The flaws of Darwin were rectified later, after the development of the science of genetics and the rectified theory was known as Neo-Darwinism.

Mutation theory of Hugo De Vries

Hugo de Vries (1840 - 1935) was a Dutch Botanist, who proposed the third theory of evolution. His 'mutation theory' which appeared in 1901, focussed attention upon the importance of mutation in evolution. In this theory, de Vries declared that evolution is not a slow and gradual process involving accumulation of numerous small changes by natural selection. Conversely, the evolutionary changes appear suddenly and are a result of large jumps, which he designated as mutation.

The publication of de Vries' work raised much controversy among the adherents of Darwinism and the mutationists. The early geneticists extended their wide support to de Vries' theory, mainly because the variations, which they noted in their experiments, conformed to de Vries' observations, but hardly with Darwin's concept. Even eminent geneticists like William Bateson, Thomas Hunt Morgan and others were attracted by this mutation theory. Mutation theory distinguished heritable variations from environmental variations, which Darwin failed to understand in his 'Natural Selection'. As a consequence, in the early years of twentieth Century, Darwin's natural selection was totally rejected in explaining the process of evolution.

Theory of Gregor Mendel

The work of Gregor Mendel virtually remained unknown from 1865 to 1900, until it was rediscovered by three geneticists in 1900, Carl Correns, Hugo de Vries and Eric Von Tschermak.* The real mechanism of mutation was properly understood through the work of Gregor Mendel and the recent discoveries in the field of molecular biology.

De Vries' hypothesis on mutation highlighted chromosomal changes, rather than the changes in the gene themselves. So his mutation theory is considered as out moded on the ground that it did not indicate true mutation. The mutation as understood today are concerned with genes, the discrete units of heredity, which occupy particular loci on the chromosomes. It tells that each gene controls a specific developmental process and responsible for the appearance of specific traits in an organism.

Mendel used the term 'factor', when he described his 'Law of Inheritance'. But in 1900 the term was replaced by the new term 'gene' and a new science gradually developed with the name 'Genetics'. Now it is known that a gene represents a specific segment of the DNA molecule. The product of a gene action, in many cases, is a protein; and the developmental process in a given organism depends on specific kind of proteins produced under the instruction of a particular set of genes. A mutation in a gene often causes corresponding changes in the protein concerned. If mutation occurs in the germ cells of an organism, the change will be inherited by its offspring. Therefore, only those mutations that cause changes in the reproductive cells of the organism are of evolutionary significance. But the structural changes of chromosomes can not be undermined because they often bring considerable effects in the evolution as found in many plants and a few animals like Drosophila, crepis etc.

Although the knowledge of genetics brought a revolution in the field of evolution, Mendel's 'Law of Inheritance' is fundamental in identifying the nature of the offsprings. It explained the basic process of heredity.

Synthetic Theory of Evolution (Neo-Darwinism)

Darwinism in its original form failed to explain satisfactorily the mechanism of evolution and the origin of new species. The inherent drawbacks in the Darwinian ideas were the lack of clarity as to the sources of variation and the nature of heredity. In the middle of twentieth Century, Scientists had come to a consensus to employ all sorts of knowledge - genetic, ecological, geographical, morphological, palaeontological etc. in order to understand the actual mechanism of evolution. Due importance was given to both mutation and natural selection, among other forces of evolution. This led to the emergence of a synthetic theory of evolution, which we also call as Genetical Theory of evolution, or 'Biological theory of Evolution'. Some authors namely David J. Merrell (author of 'Evolution and Genetics') and Edward O Dodson (author of'Evolution : Process and Product") have called this new theory as Neo-Darwinism. But, George Gaylord Simpson and his followers strongly warned against equating the synthetic theory of evolution with 'Neo-Darwinism'. Simpson argued that the synthetic theory had no Darwin. It was not only different from Darwin's; it had drawn its material from a variety of non-Darwinian sources.

After the development of the science of genetics, it has been known that a population shares a common gene pool. Accordingly, the evolution denotes a change of gene -frequency in the gene

* All of these three scientists were basically botanist. Correns was a German, de Vries was a Dutchman and Von Tschermak was an Austrian. They worked independently in three places. All of them reached to same conclusion as like Mendel. Other scientists therefore described the event as the rediscovery of Mendel's law and whole work was recognized by the name of Mendel.

pool of a population over certain span of time. The synthetic theory of evolution does not discard all previous propositions, rather considers them as partially important. Therefore, we find amalgamation of various concepts viz. Natural selection, Mendelian principles, Mutation, population genetics in this theory of evolution. But it is interesting to note that modern genetics does not acknowledge the mutation theory in its original form, as proposed by de Vries. Because that original theory had outrightly rejected the basic concept, 'natural selection' as delivered by Darwin and advocated 'mutation' as the sole force of evolution. However, at present evolution appears to be a complex process involving several complex forces.

EVIDENCES BEHIND BIOLOGICAL EVOLUTION

The idea of biological evolution probably started with Charles Darwin who defined the term evolution as "descent with modification". The word 'descent' refers to the process of origin of new species from an ancestral stock. Again, the word 'modification' introduces an idea of change, which is inherent in evolution. Therefore, both the words jointly mean a species that evolves from its ancestors through changes. This principle is also applicable to man. Man being the highest product in the evolutionary sequence has descended from non-human ancestors. Some more primitive forms may be stated as ancestors of the ancestors. In this way we can reach back to the unicellular form, which formally stand as father of all organisms. To prove this fact, evidences are required for the whole path of biological progress. The biological evolution is exactly same to what we mean by organic evolution.

The organic evolution is concerned with two different types of development, e.g. *ontogeny* and *phylogeny*. The ontogeny refers to the history of development of individual organisms. For example, man, like other organisms, began his life as a single cell, which underwent a complex process of development and finally culminated in a multicellular adult. The ontological changes are of great importance to the embryologists. But a student of evolution primarily deals with the second kind of development, which has been referred to as phylogeny. The phylogeny treats with the evolution of a genetically related group of organisms, in contrast to the development of the individual organism. However, the evidences of organic evolution come from different branches of science, like comparative morphology and anatomy, embryology, palaeontology, physiology, biochemistry, genetics, etc. A few evidences have been cited here.

Morphological Evidences

The structure of animals, both external and internal, provides a source of evidence for evolution. The branch of biology that deals with the form and external structure of animals and plants is called morphology. By comparing organs, musculature and tissues, it can be concluded that man and some other vertebrates have developed from a same stock following evolution. The general similarities between man and higher vertebrates are so close that the medical students, especially beginners often learn the elements of surgery by desecting the dogs, cats, monkeys and other vertebrates, because bodies of those animals can be procured easily. Muscles and tissues of man and ape are usually same in number as well as in function, although the forms are slightly different. Morphological peculiarities of man are mainly counted with the erect posture, stability and solidity of the hind limbs (legs), greater prehensility of the forelimbs (arm), enlargement of the brain, reduction of the face and jaw, and the different type of dentition. In spite of all these facts, no zoologist will hesitate to classify them in the group of primates along with the apes and monkeys, so far as the morphology is concerned.

Anatomical Evidences

The term anatomy is nearly synonymous with morphology and it deals solely with the internal structure of organisms. A comparative study of anatomy reveals the similarities and dissimilarities between man and the other higher primates. The overall skeletal structure and bones are more or less same in man and apes. A layman says, 'man is a descendant from monkeys.' But a student of evolution never agrees with this idea. According to him, as man and monkey share same common anatomical features, they are contemporaries. Probably both of them arose from a common ancestor (an unknown primate, who is not living today) in the remote past. That common ancestor had the potentialities to give rise to man and monkeys, but no known monkey have that potentiality.

Evolution of the organisms can be better demonstrated with the help *of homology* and *analogy.* Homology is the similarity in structure between parts of different organisms due to common origin. Structures having similar genetic basis are said to be homologous. The homologous organs need not be employed for the same purpose. On the other hand, analogy is the similarity in function between anatomical parts of different structure and origin. These structures usually show superficial similarities due to similar function. Structures having similar functions or projecting similar habits are said to be analogous.

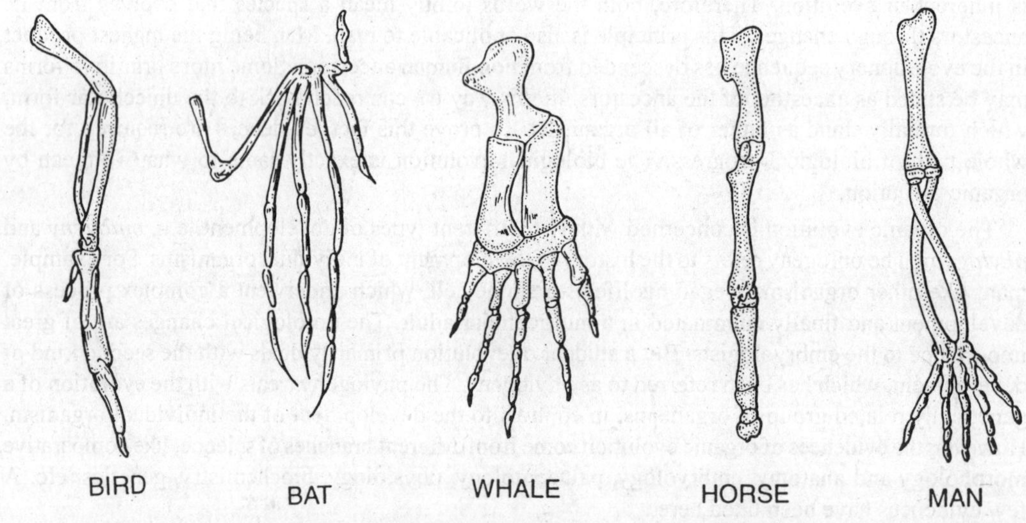

Fig. 5.2. Forelimbs of Vertebrates Showing Homology

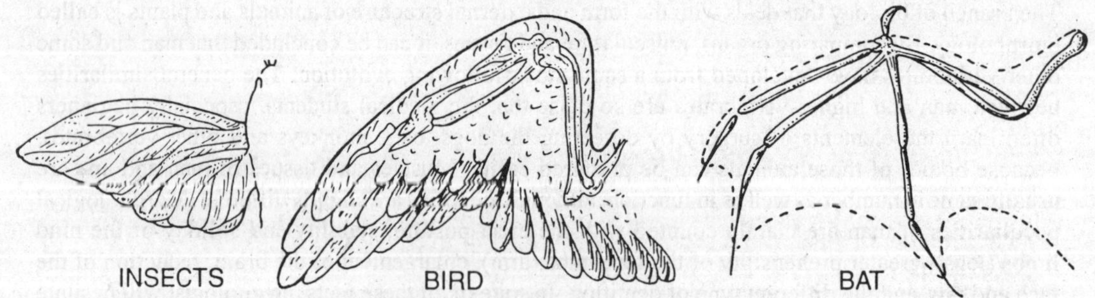

Fig.5.3. Wings Showing Analogy

The forelimb skeleton of vertebrates such as birds, bats, whales, and horses and human are all homologous structures, as these animals possess a similar developmental origin and hereditary basis. But these forelimbs do not look alike as their function differs greatly. In fact, activity of these animals has changed due to environmental influence and the change has been reflected in the adaptability of the forelimbs. Contrary to this, the wings of insects and birds are analogous organs. Although both are used in flight, they differ in their origin and structure. The wings of insects are composed *of chitin* (exoskeleton), and are supported by hollow tubes (veins). These wings are therefore non-living structures, operated by certain muscles present at their bases. Conversely, the wings of birds are supported by a living endoskeleton which remain covered externally by feathres. The supporting endoskeleton has several segments namely humerus, radius and ulna, carpo-metacarpus and the phalanges. Similar analogous structures are found in the fins of fishes and the flippers of the aquatic mammal, whale. Both serve the function of swimming but differ in their skeletal arrangements. Analogical organs may also develop for the purpose of adaptation. Different groups of animals when face same type of environment, they work out more or less similar type of devices for survival.

The vertebrates like fishes, birds, amphibians and mammals are found to possess more or less the same organs in their body e.g. heart, livers, urinary system, nervous system, etc. Now, if we consider any one of these organs, say for example, the heart, we will be able to find out a gradual change in the structure of heart which occur with time. The structure gets gradually complex in higher forms. For instance, a fish has a two-chambered heart. Again, the mammals being developed in form, bear a four-chambered heart. Thus comparative anatomy between the animals yields various evidences of evolution.

Vestigial Evidences

The word vestige means 'a small trace'. In any living organism, the vestigial body parts serve no remarkable purpose; they lie as rudimentary organs or structural residue in living body. Functional analysis of these organs is apparently irrelevant as they show little or no use. But the same functional analysis stands significant to understand the emergence of a new species by evolutionary modification from earlier biological forms. They are able to throw considerable light on the structure of more primitive forms. Therefore a vestige can be an important criterion in tracing organic development. An example may be cited with appendix, which serve a definite digestive function in apes, monkeys and other herbivorous animals. In man the vermiform appendix is not only an useless structure, but also a seat of a dangerous disease called appendicitis. The appendix remains attached to a short portion of the large intestine, termed as caecum. Both the appendix and the caecum are fairly well developed in herbivorous mammals, whose diet consists of abundant cellulose. In these mammals, the appendix and caecum form storage compartments, in which, the mixtures of partly digested food and enzymes stay for a considerable length of time. At this time bacteria act on the cellulose to transform it into easily digestible chemical compounds. Man might have inherited these vestigial structures from his remote ancestors whose diet contained a considerable amount of cellulose.

Another vestigial structure occurs in the inner angle of each eye of man. It appears like a little fold of flesh and is termed as semilunar fold (Plica semilunaris). This structure represents the reduced, movable, third eye-lid, and called nictitating membrane in lower vertebrates, where it is used to clean the eyeball. But no implication is found in the body mechanism of man.

The presence of ear muscles in man is another instance of vestigial structure. In several mammals the external ears are fully movable to hear the sounds coming from several directions. The rudimentary muscles present in the skin of ear of man are incapable of moving the ears; although some individuals have the ability to 'wriggle' their ears. Again, the third molar tooth, which is functional as well as useful for the primates, shows no use in man. It is called 'wisdom tooth' which hardly finds its place in the reduced jaw of man. Usually it erupts much later and sometimes it fails to erupt, as its requirement is minimal in human dentition.

Similar example may be drawn with some specific parts of human skeleton. The spines of the four vertebrae at the lower end of vertebral column fuse together to develop a small round bony structure, known as coccyx or tailbone. This hidden vestigial tailbone indicates that man had tail in certain phases of his development; some of his ancestors definitely possessed this tail. The hidden tail is also seen in the human embryo during its second month of growth. Sometimes a few babies are born with an external coccyx, which has to be corrected surgically by medical practitioners.

The presence and distribution of body hair in man may be regarded as vestigial. Superfluity of hair is found among the anthropoid apes as well as in the pre-hominid forms.

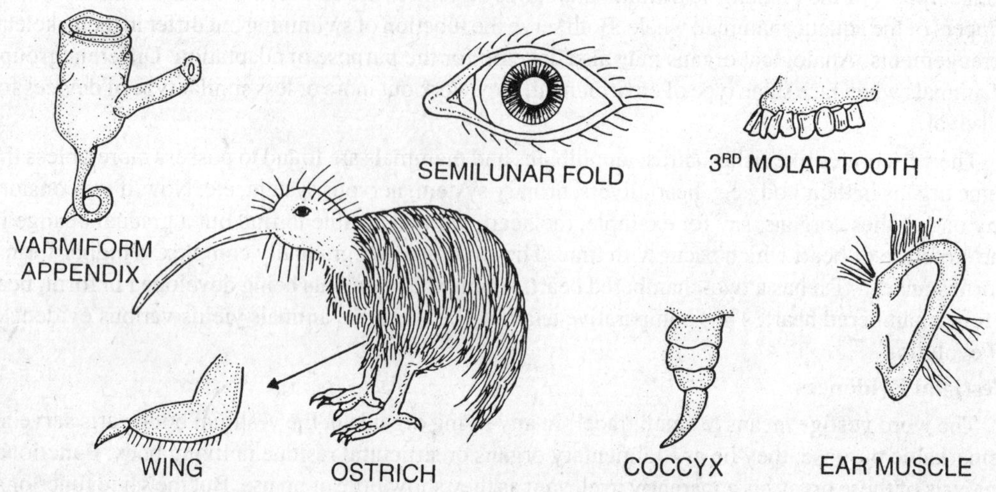

Fig. 5.4. Vestigial Organs of Vertebrates

Dr. Weidersheim gave a list of about hundred vestigial structures in man alone. Such organs are also found in every specialized animal in its anatomy. For example, though snakes are usually limbless in character, vestiges of hind limbs are seen in pythons and boas. In these snakes hind limbs are represented by reduced pelvic girdle, capped by claws. In horse, only the third digit is present; the metacarpal of digit III has greatly enlarged to form the *cannon bone.* Vestigial structures in the leg of horse are found as *splint bones,* which get fused with the sides of cannon bone. Similarly, the vestigial organs are found among the birds. Birds are notable for their flying habits, but a few of them are flightless. For instance, in Kiwi (a bird of New Zealand), the wings are vestigial which remain concealed by the feathers of the body. Again, the animals, which live in perpetual darkness, like deep caves, generally show degenerative changes in their eyes. The cave-dwelling salamander of Central Europe is characterized by complete absence of eyes. Several cave-dwelling fishes are also blind, as their eyes are functionless.

Different vestigial organs are thus formed in the course of evolution due to the loss of their typical function. Consequently these organs show a great reduction in size and as the evolutionary changes continue, they get a chance to be eliminated completely from their possessors.

Embryological Evidcnces

Embryology is a specialized branch of biology, which deals with the formation and development of the embryo. Comparative embryology furnishes several outstanding evidences for evolution. In 1866, the German scientist Ernst Haeckel (1834 — 1919) formulated the 'Biogenetic Law' which is more often known as the 'Theory of Recapituation'. The theory states that 'Ontogeny Recapitulates Phylogeny' which means, the embryos in their development repeat the adult stages of their ancestral forms. The fundamental concept of recapituation stems from Karl Ernst Von Baer (1792 — 1876) who stated "The embryo of a higher species may resemble the embryo of a lower species but never resembles the adult form of that species".

The human embryonic development furnishes several evidences of recapitulation. Each human being starts his life as a single cell, the fertilized egg or zygote, which corresponds to a protozoan ancestor. The zygote divides and becomes a multicellular blastula, which may be comparable to a Colonial flagellate like volvox. The blastula undergoes gastrulation to form a two-layered coelenterate like embryo, which eventually transforms into a triploblastic structure resembling a flatworm. Then chordate features (notochord, dorsal tubular nerve cord, and pharynx specialized for respiration) appear being followed by the development of gillsilts and aortic arches like those of fishes. Next, tetrapoid characters like pentadactyl limb and metanephric kidney make their appearance. This is followed by the development of mammalian, then primate, and finally specific human characters. But, the fact does not mean that the embryonic stage of a fish and human is quite identical. The presence of gill arches in human embryo points out that man's ancestor to distant past had the evidence of gill. Thus embryology serves as a source of evidence in connection with evolution.

Thus, embryos of different animals remain similar at an early stage, including the man. Differentiation among them is possible only at a later stage of embryonic development.

Palaeontological Evidences

Palaeontology is the study of fossil remains, which provides a very reliable information regarding the fauna of particular periods. Any type of remain of living form from the remote past as obtained from the earth may be considered as a fossil. The term fossilization implies the process of fossilizing. Fossil may take any form, and the nature of the fossil record is varied, which may range from an intact mammoth to a simple footprint of a dinosaur.

Any organism may undergo fossilization, but in reality, only a few of them become fossilized. The majority of organisms disintegrate before being preserved for fossilization. Two general types of fossil record have been observed —unaltered remains and altered remains. The unaltered fossils are very rare in nature. The woolly mammoths of Siberia that are preserved in the permanently frozen tundra constitute the examples of this type. Similarly, insects encased in fossil resin (popularly called amber) in the beds of Baltic Shores of Prussia are also the examples of unaltered remains. In altered remains, the fossil show great variations in their constitution, and in most instances the original organic parts become gradually replaced by inorganic salts like calcium carbonate, silica, iron pyrite etc. The fossils may take the form of mould (external impression) or natural casts, or it may be in the form of trails, tracks, footprints, leafprints, burrows, borings, tubes, etc. A brief evaluation of some of the fossil records, especially pertaining to fossil men and pre-hominoid forms present fascinating evidences in support of human evolution. It should be remembered that the Protochordata have not left any fossil record, as they have no bony skeleton. Evolution of horse can be traced beautifully because a complete sequence of fossil horses has been recovered from different geological strata. In case of man, fragments and bones are the mute testimony for the antiquity of man and his early nature.

Evidences from Physiology and Biochemistry

Physiology is the science that studies the activities of organisms. Biochemistry, on the other hand, is concerned with the chemical structure and processes occurring in organisms. In same cases, evidences of physiology and biochemistry have been proved useful in tracing the relationships among certain organisms, which could not be drawn on morphological basis.

Similarities of chemical structure and function are so fundamental that these are essentially similar in large groups of animals and plants. It has been estimated that there are slightly more than hundred chemical elements. All living organisms, plants as well as animals are composed of one or more of these elements, which remain combined in appropriate proportions. Four of them Carbon, Hydrogen, Oxygen and Nitrogen are the most common elements; 99 percent of living things including amoeba, euglena, pea-plant and man are formed in combination of these elements. Again, the abundantly present chemical elements mostly constitute three main organic compounds viz., Carbohydrates, fats and proteins. Carbohydrates such as starches, sugars, cellulose etc. are composed of carbon, hydrogen and oxygen. Fats are similarly constituted of carbon, hydrogen and oxygen. Nitrogen and phosphorus are there, in addition. Protein is composed of carbon, hydrogen, oxygen, nitrogen, sulphur and phosphorus. These three organic compounds combine together in varying ways to form a complex substance called protoplasm. Huxley considered it as 'the physical basis of life'. The protoplasm of all living things—from amoeba to man, has several similar physical and chemical properties. This protoplasm is built into structural units of living organisms called cells. The cells being 'building blocks' of animals and plants exhibit fundamental similarity in all living things.

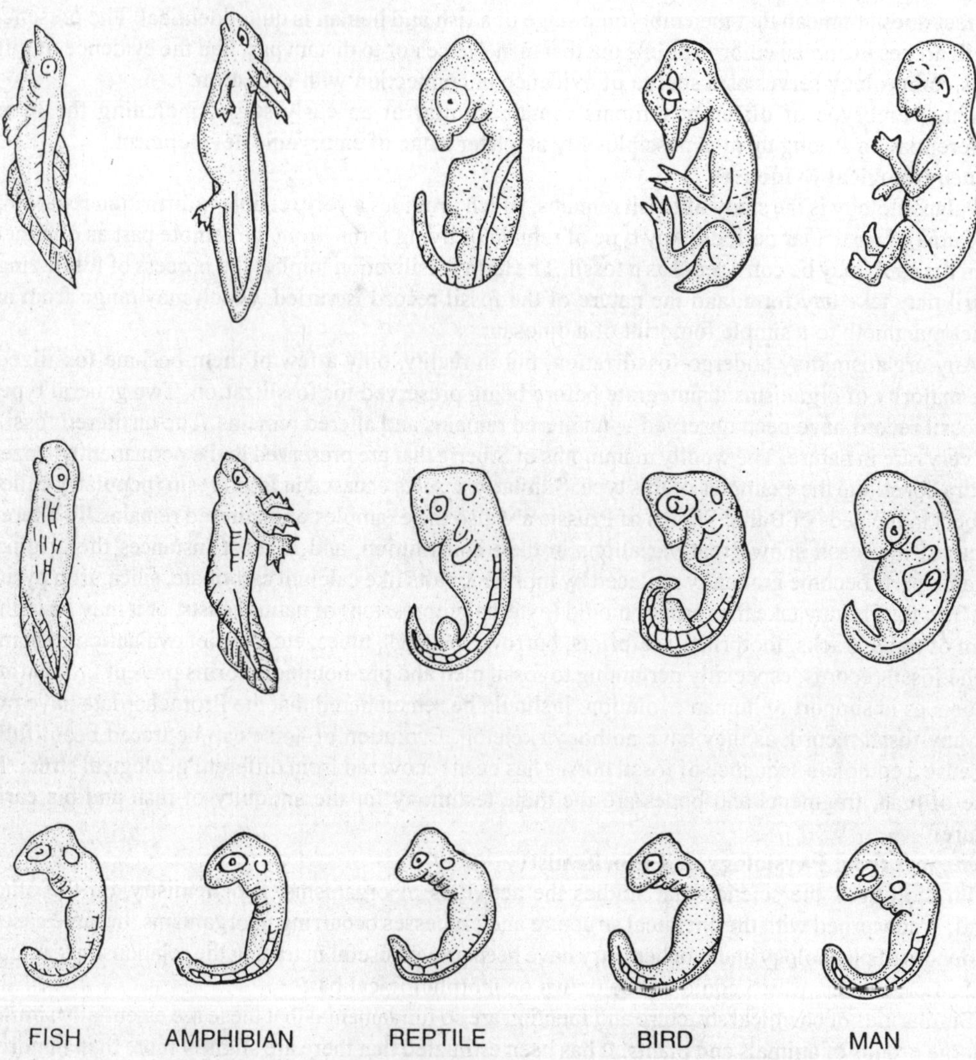

Fig. 5.5. Evidences of Evolution in Embryonic Development

The study of cell physiology has revealed that the process of cell division (mitosis) is essentially similar throughout the plant and animal kingdom. Besides, a large group of animals share some common chemical structure and function of enzymes and hormones in their physiology. The proteolytic enzyme, trypsin, which takes part in protein metabolism, occurs in several groups of animals, from Protozoa to the Mammalia. Similarly, the starch-splitting enzyme, amylase, is found in animals ranging from the Porifera to the Mammalia.

The serology deals with blood serums, and their reactions and properties. Serological tests provide very impressive physiological evidences concerning with evolution. The specialized bio-chemical study, immunology (based on antigen-antibody reaction) reveals the fact that the reaction between antiserum and blood serum of man follow the same course as found in the apes. These serological tests not only confirm the relationship of man to other primates (particularly with anthropoids like chimpanzee), which were originally based upon comparative morphology, but also establish the degree of similarity of the serum proteins among the related species. The comparative serology also helps in resolving difficult taxonomical problems where standard morphological evidences fail to verify the taxonomy of certain aberrant animals.

Evidences from Genetics

Genetics is an important branch of biology, dealing with the science of heredity and variation. The principal fields of genetics from which evidences for evolution are drawn include hybridization, domesticated animal and plant species, gene and chromosome homology, and the nature of genetic (hereditary) material.

The hybridization involves crossing (cross mating) of two genetically unidentical individuals (usually two species) which leads to the production of hybrid progeny. One of the most fascinating example of animal hybrid is the mule, which is a hybrid species, resulting from a cross between an ass and a horse. The mule combines the hardiness of the ass and the sensibility of the horse. But this species is considered as evolutionary dead because the mules are invariably sterile. Such instances are very rare in nature and usually two clearly distinct species on hybridization produce viable and vigorous progeny. The sterility among the offspring individuals is due to genetic incompatibility. Therefore, the genetic study can reveal the closeness between the animal forms in connection to evolutionary consequences.

Domesticated species of animals and plants are of considerable interest to the students of evolution as they provide magnified view of evolution, although in a somewhat distorted manner. In fact, the domesticated animals of today are descendants from wild progenitors. The innumerable breeds or varieties of domesticated species have arisen due to artificial selection by man, but the natural selection also operates side by side. However, the evolutionary significance of domesticated species lies in the fact that they often exhibit considerable changes from their wild counterpart.

Relationships between species can be determined genetically by comparing their chromosomes. The genetic experiments have revealed that the unpaired parts of the chromosomes differ in their genetic contents. The similarities in genetic behaviour between homologous chromosome regions signify the degree of relationship of related species. This is the most reliable evidence in tracing the line of evolution.

The chemistry of chromosomes, from bacteria to man has revealed that they are composed of nucleoproteins, a combination of nucleic acids and proteins. Among the two nucleic acids, DNA and RNA, DNA is the genetic material in most of the organisms. Only in certain viruses, RNA is the genetic material. The available evidences indicate that the nature of nucleic acids is fundamentally same in all organisms—viruses, bacteria, plants and animals. Chemically, both DNA and RNA are very similar, in having sugar-phosphate backbone, and nitrogenous bases, purines and pyrimidines which remain attached to sugar molecules as side chains. The differences between DNA and RNA molecules lie in the sugars, and in one of the pyrimidines. In DNA, the sugar is deoxyribose, while it is ribose in RNA. In DNA, the four nitrogenous bases are adenine and guanine (purine), and cytosine and thymine (pyrimidine). In RNA, the thymine is replaced by uracil. In most organisms (except some RNA viruses), DNA is the genetic material which transmits the hereditary blue print from one generation to the next; RNA appears to take part in protein synthesis. DNA, as a primary hereditary material is capable of undergoing a change, due to genetic phenomenon called mutation. The trait developed as a result of mutation is different from the original, but it is as stable as the original trait. Therefore, mutations account for the hereditary variations, and also form the raw materials of evolution.

STAGES OF EVOLUTION

We have already come to know that evolution allowed the simple living things to become complex. The lower vertebrate animals equipped themselves with various specialized structures (in relation to eye, brain and skeleton) and progressed towards higher forms. At the optimum level they became fully developed to that degree as human life demands. Constant change and adjustment, slowly with time was the pre-requisites for evolution.

For the practical purposes, we should start the story of evolution from the Cenozoic era (Tertiary period) of the geologic time scale, which corresponds, to the 'Age of mammals'. The first mammals were small and simple as known from the fossil remains. At the beginning of the Tertiary period, a group of mammals splitted up into different major lines and primates arose. But no radical change or progress took place at that time, rather retention and moderate improvement of some ancient traits were found. For example, the earliest primates first exhibited the grasping paws where nails developed instead of claws. Five well-separated fingers on toes gave them the ability to rotate the great toe so as to oppose the other digits. At the same time, they got a movable arm by the development of a collarbone and two bones in the forearm.

In 1964, two palaeontologists namely, Van Valen and R.E.Sloan suggested that some very early primates existed in the Late Cretaceous period, about 70 million years ago. In fact, molar teeth of two species belonging to the genus Purgatorius (one obtained from the deposit of Cretaceous epoch and the other from the deposits of Paleocene epoch) were able to show the primate characteristics. Besides, the middle Paleocene deposit of Europe and North America exhibited many remains (fossil) of primates; numerous genera were identified. According to palaeontologists, the earliest primates probably evolved out of the insectivores, a specified order of mammals that subsisted solely on insects. Primates of Paleocene splitted up into various prosimian lines and as a result plenty of small primates emerged out during Eocene epoch. Eocene is therefore called the golden age of the prosimians. Most of them adapted an arboreal life, but some of them also lived in ground like squirrels and rodents.

It is held that the climate of the Cretaceous epoch was uniformly damp and moist. A great environmental shift was found at the end of this period, which extended up to the beginning of Paleocene. This period witnessed a comparatively dry climate for which the vegetation and fauna were bound to change. With the disappearance of swamplands, different kinds of grasses, flowering plants and even the deciduous trees came into the scene. Deciduous plants favoured the growth of different insects and those in turn, attracted the insectivores who increased their number very rapidly with abundant food supply. Since insectivores were very adaptable, they spreaded everywhere—underground, on the ground, above the ground, in water, in woody habitat like bushes, shrubs, vines and trees. The insectivores that took to the trees perhaps contributed to the primate evolution.

In tracing the ancestry of man, we should therefore stretch out our observation back to the Eocene epoch where small ancient prosimians are found to flourish all over the world especially North America, Europe, Africa and Asia. Those early primate potentialities not only gave rise to the lemurs, lorises and tarsiers of present day, they also yielded true monkeys like platyrrhine (broad-nosed) and catarrhine (downward-nose). As the two hemispheres were isolated during Eocene, no primate could move from the Old World to the New World or vice versa. As a consequence, two strains developed side by side, parallel to each other. In South America, New World monkeys such as marmoset, spider monkey, howler monkey, cebus, tity etc. form a group (platyrrhine) who could swing on their tails. The second group of monkeys (catarrhine) developed independently in the Old World. They had no connection with the American monkeys and they were the progenitors of all monkeys in Africa, Europe and Asia. Since these two lines of monkey remained apart from each other, they automatically differed in certain physical features. The New World monkeys are found to possess a flat face with broad nose while the noses of old world monkeys look invariably downward. Again, the New World monkeys retain three premolar teeth out of the ancestral four, on either side of the jaws. The Old World monkeys, on the other hand, had lost two premolars and carry on with two premolars only. This group of monkeys is comparatively larger than the New World monkeys.

A Century before Darwin, it was realized that man exhibits many similarities with monkeys, rather than other animal forms. The great Swedish botanist and Systemist Karl Von Linne' (Linnaeus) placed the surviving primates like great apes, monkeys, lemurs, tarsiers together with men in a single group, under the order Primate. Within this order, New World and Old World forms were also distinguished. The hominoids (the taxonomic group that includes both mans and ape) were found more closely related to the Old World monkeys than the New World monkeys. Although we are not very definite that whether man arose from the same prosimian stock, independently or by repeated splitting, but there is no doubt that catarrhine monkey and hominoid both have been descended from a common ancestral stock. In this sense, the ancient prosimians are the forerunners of man.

In the New World, the primate evolution could not go beyond the monkey stage. As the small monkeys had shown no ability to give rise any higher form. New World obviously lacked the possibility for the origin of man. Contrary to this situation, fossil finds of intermediary ape-man have been recovered from a number of sites in the Old World. The term hominoid refers to those remains. Possibly the first ape-men appeared in later Eocene. From that time onwards, especially from the deposits of Oligocene, a lot of striking fossil jaws and teeth began to be discovered in Europe, Egypt, Burma and possibly in china also.

But the hominids (direct ancestors of human) did not declare their presence until the arrival of the epoch, Pleistocene. Not only that, to get the most evolved structural form, as of Homo Sapiens, hominids had to pass through different stages. It is therefore quite natural to recognize that man emerged from an early primate form and progressed through monkey-hood or ape-hood in order to attain humanity. Evolution created various radiating forms, most of which did not survive. The types, which could adapt the environment, were able to survive. Thus, we can divide the stages of evolution into broad phases: 1. *Prosimii to Hominoid* 2. *Hominoid to Hominid.*

Prosimii to Hominoid

The fossil evidences of early primates that have been unearthed from different strata of Tertiary deposit are mostly the teeth and the fragments of jaw. Only a few cases have been represented by the parts of the skull. Although these evidences are limited in number but they are of immense importance in knowing the antecedent of man.

As previously mentioned, earliest primates come from the middle of the Paleocene deposit. It is really difficult to differentiate the first primates from the other early mammals. Because the bones of primates at this stage was not grown enough to show the discriminating features from that of the mammals. Therefore, only on the basis of dental characters, those remains could be grouped under the order. Primate. Most of them were small prosimians who showed a lot of similarity with modern tree shrews or lemurs. Probably they were distributed in the mountainous regions of North America and France in Europe. Palaeontologists usually categorize the several genera of Paleocene primates into three families, viz., Carpolestide, Phenaco lemuridae and Plesiadapidae. Among these three families, much information is known about the family Plesiadapidae because prosimians belonging to this family had kept greater number of their body fragments in fossilized form. The size of this animal ranged from that of a squirrel to a domestic cat. Among the remarkable features, the claws were flattened, vision was not stereoscopic and the post-orbital wall was absent. Front teeth were large and procumbent like the surviving lemurs (aye-aye lemur) of Madagascar.

In the Eocene epoch, prosimians, specially the lemurs and tarsiers became abundant. They were distributed mostly in North America, Europe and Asia (China). All the prosimians of this time may be grouped into five families namely Adapidae, Anaptomorphidae, Microsyopidae, Omomyidae, and Tarsidae. These porosimians were no doubt primates as they showed same distinctive primate characteristics in their anatomical traits. The brains became slightly larger and the eyes became bigger than their Paleocene forefathers. The enlarged orbits rotated towards the front of the skull.

The position of foramen magnum was shifted forwardly towards the base of the skull which facilitated the skull to be attached with the spine. The snout was also reduced. The forelimbs became shorter in comparison to hind limbs. These facts imply that the said animals were capable to be erect, especially at the time of hopping and sitting like the lemurs and tarsiers of present day. The enlargement of brain was the other important factor that denoted evolution. So far as the dentition is concerned, these animals exhibited only two pairs of incisor teeth, which have been double (four pairs) in case of man. Three pairs of premolars can also be compared with two pairs of human premolars on each side. However, the small prosimians gradually reduced to a few varieties and reached almost at the verge of extinction especially in Europe and America by the end of Eocene epoch. They could not save themselves under a new tide of mammal development; the mammals belonging especially to the group of rodentia and carnivora were their main competitors.

In the old world, many higher primates evolved from prosimians but prosimians themselves could not win in competition. So the number of genera diminished gradually and at present only a few representatives of the family Adapidae and Tarsidae are found alive. The tarsioids have shrunk into one species, spectral tarsier, which live in East Indian Islands. Lemuroids and their relatives, the indris and the aye-aye have been confined to Madagascar and Comoro Islands. American lemuroid sub-family, Notharctinae has kept its survivors (loris and galago) in South East Asia and Sub-Saharan Africa. However, the succeeding epoch Oligocene records an important development of primate. But, unfortunately, the number of fossils obtained from Oligocene level are quite limited. None came from Europe and a few from America. Some others were from Africa, specially Fayum in Egypt. Fayum was not a desert in Oligocene epoch as we find today. Rather it had a tropical rain forest with plenty of rivers and lakes. Its warm congenial climate attracted a large number of primates to settle down. However, the Egyptian primates comprised of a variety like Parapithecus, Propliopithecus, Mocropithecus, Oligopithecus, Aeolopithecus, Aegyptopithecus and so on. Among these, Parapithecus has been ascribed as a likely ancestor of all present day monkeys—old world and new world. In fact, Parapithecus denotes the *pithecoid stage* in primate evolution, next to the *prosimian stage.*

In this context we must consider the *Simians* who came after the *Prosimians.* The term Simian is actually analogous to the primitive ape-monkey form. In the monkey group, the Platyrrhines belonged to anthropoidea; these new world monkeys, for whatever the combination of causes, never went beyond the monkey stage in the primitive evolution. On the contrary, the old world monkey group (Catarrhine) though followed the line of evolution but belonged to the group Simian likewise the platyrrhini. These old world monkeys had undergone some development in a different way and in more specialized direction. On the basis of that reason some authorities like to place them along the man's evolutionary line, in a monkey-like (pithecoid) stage. However, the new development gave rise to a progressive branch which advanced from the pithecoid or monkey stage to the *hominoid stage.* This stage is more akin to man and includes the ancestors of gibbon, the great apes and even man.

Coming back to Parapithecus, we note that this creature is the most primitive form of all known old world monkeys and apes; it holds an intermediate position between prosimian (tarsioids-lemuroids) and anthropoid. Propliopithecus ranks next to it whose jaws and teeth resemble much with the higher primates, especially the gibbon. But, it lacked the long canine tooth as found in the present day gibbon. Nevertheless, there is no disagreement to regard this creature as a remote ancestor of modern gibbon, a forerunner of Hominoidea. Aelopithecus also shows a close resemblance to the modern gibbon. However, it seems that Oligocene offers a critical turning point; primate line has been bifurcated from this time—one became ancestral to the monkey and other to the apes. All Oligocene primates have therefore been categorized under the sub-order 'Anthropoidea', some of whom finally went towards Hominoidea.

If the *anthropoid—humanoid* stock begins to diverge as early as in Oligocene times and if the Propliopithecus is taken as the first anthropoid ape discovered from Fayum (Egypt), the next creature in the evolutionary tree is the Pliopithecus as discovered from early Pliocene or late Miocene deposit in Germany. Pliopithecus resembles more with modern gibbon and this suggests that there must be an ancestral relationship between Propliopithecus and Pliopithecus. However, Pliocene is the last phase of Tertiary period and before that, the Miocene epoch is considered to be blessed with the major evolution of primate. Separation of Gibbon line took place at that time. The line of other anthropoid apes branched off much later during Pliocene from the hominoid ancestral stock. It is probable that orangutan diverged earlier than the chimpanzee - gorilla and that took place in early Pliocene epoch.

Hominoid to Hominid

As aforesaid, the *hominoid stage* is more akin to man but man himself belong to the *hominid stage.* The whole ape-group has been covered by the term hominoid. From the deposits of Miocene-Pliocene epoch several fossil remains of hominoids have come out in different parts of Asia, Europe and Africa. Some of these, namely, Limnopithecus, Epipliopithecus, and Prohylobate are similar to Pliopithecus as discovered from East African Miocene deposit. The Limnopithecus and Epipliopithecus are so close to Pliopithecus that they are thought to have evolved from Propliopithecus. On the other hand, Prohylobate belongs to the family cercopithecidae.

The ape-group of primate is distinguished for some special features. They developed a semi-erect posture and combined it with the habit of brachiation. They also tried for walking on hind-legs at the time of travelling on the ground. This change in the locomotion governed on many skeletal and muscular adjustments in the body and helped in developing new habits. Gradually they became accustomed in arm-swinging and semi-erect walking with two legs. Some changes also took place in the vertebral column as well as other skeletal parts of the body. The partially free hands started to perform other works and the creatures had grown a liking for terrestrial living.

In considering the evolution of apes, we should begin with the Dryopithecine group of primates. Man-like great apes e.g. orang, chimpanzee and gorilla have evolved specifically from this group. The ancestors of man are also linked with this stock. However, Dryopithecines included two different froms of primate—one belongs to a sub-family of Pongidae and other is assigned to the family Hominidae. In fact, the Dryopithecus variety grouped under Pongidae, in course of time, gave rise to the apes of today. On the other hand, Ramapithecus belonging to the family Hominidae is said to be in the direct line of man, though it is not an immediate ancestor. Apart from Dryopithecus and Ramapithecus, many other significant fossil fragments viz. Proconsul, Sivapithecus, Kenyapithecus, Gigantopithecus, Oreopithecus, etc. have been discovered from Miocene—Pliocene deposits. Anthropologists endeavoured to point out their exact position according to the respective anatomical features.

The fossil remains of Ramapithecus recovered from India (Siwalik hills) and Africa, suggest that this type of animals used to live on ground. Their dental characters are similar to the Australopithecines of later period who stand at the base of hominid line. The internal and external changes of the body structure in hominoid stage can be explained as a resultant of climatic and dietetic factors. However, three successive stages have been identified in course of hominid evolution —the *stage of Australopithecus,* the *stage of Homo-erectus* and the *stage of Homo-Neanderthal,* before reaching to the final form, Homo-sapiens.

Anthropologists are in the opinion of placing the Australopithecus under a sub-family of Hominidae, although they are not very sure about whether all of the Australopithecine or one of its members contributed in the development of man. But it is definite that our ancestors maintained a close similarity with some of the members of this group. The next stage to Australopithecus is the stage of Homo-erectus. Simian as well as human characteristics have been blended in this group of fossil remains. The third stage has been represented by the Homo-Neanderthals—an intermediate type between Homo-erectus and modern man. But, in fact, modern man can not be the descendant of Neanderthals on two grounds :

i) The completely evolved modern type of fossil men have been discovered in Western Europe from the layers earlier than those containing Nearderthal remains.
ii) No evolutionary change has been noticed among the Neanderthals with the passage of time. Even the progressive Neanderthals do not show any transition towards the modern man.

These fossil hominids indicate biological evolution in terms of

i) Progressive rounding of brain-case
ii) Increase in brain-size
iii) Reduction in the size of mandible

The majority of the palaeontologists consider the progressive Neanderthals (e.g. Krapina, Steinheim, Ehringsdorf, Mount Carmel etc.) collateral or cousin of the ancestors of Cro-Magnon type. Therefore in no case, the Neanderthals can be the direct ancestors of modern man. They were readily replaced by the onslaught of more progressive new-comers—Cro-magnon or other similar type of population. But, this is not the one or sole reason for the extinction of the Neanderthals. Different other opinions have been expressed in this regard. Some point out an internal conflict; others advocate for catastrophic consequences.

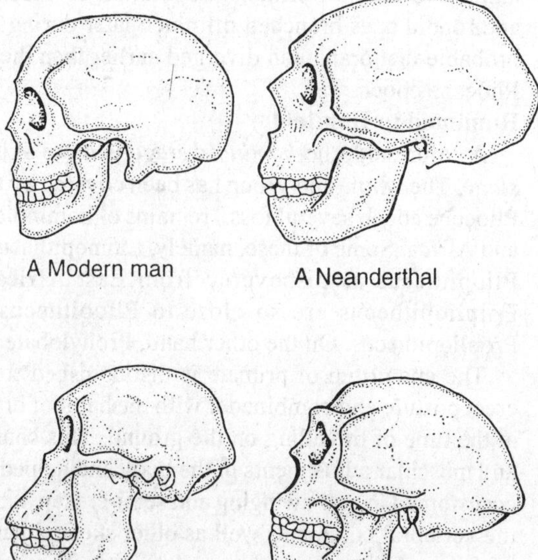

Fig. 5.6. Reconstruction of Skulls in Some Fossil Men (Lateral View)

The Cro-Magnon of late Pleistocene was believed to be the earliest type of Homo—sapiens for a long time. But when a few other fossil remains exhibiting many characteristics of Homo—sapiens (Galley Hill skeleton, Swanscombe man, London skull, Boskop skull and Fontechevade skull) were discovered from the earlier layers than those of Cro-Magnons and even the Neanderthals, the whole model of evolution became defunct. The fact suggests a parallel development of Homo—sapiens and Neanderthal; the newly discovered fossils from Middle Pleistocene deposit are therefore the earliest form of Homo—sapiens. Upper or late Pleistocene men have been considered as Neanthropic men who exhibited a wide diversity of forms like Grimaldi Man, Cro-Magnon man, Chancelade man and so on. The Predmost people also have been included here.

It is therefore very clear that the anthropoid apes, prehominids and modern man, all originated from a common ancestral stock. To trace the antiquity of man we have no alternative other than to study the primates, particularly the stages between prosimii to hominoid that have been unveiled through several fossil forms.

The Birth Place of First Man

Palaeontological evidences suggest that the human evolution did not take place in the new world. Because, in North America all the primates disappeared since Eocene epoch. South of America was only dominated by platyrrhine type of monkeys. Man entered in this world from the old world at the end phase of Pleistocene epoch. Again, Australia being the land of monotremes and marsupials showed no significant development of other mammalian types. On the other hand, in the old world, although the Europe presented many fossil primates and hominoid forms, the actual origin of the hominids possibly took place somewhere in Asia and Africa. Because, the latter two continents exhibited greater number of evidences than that of Europe. Evolutionists differ among themselves in

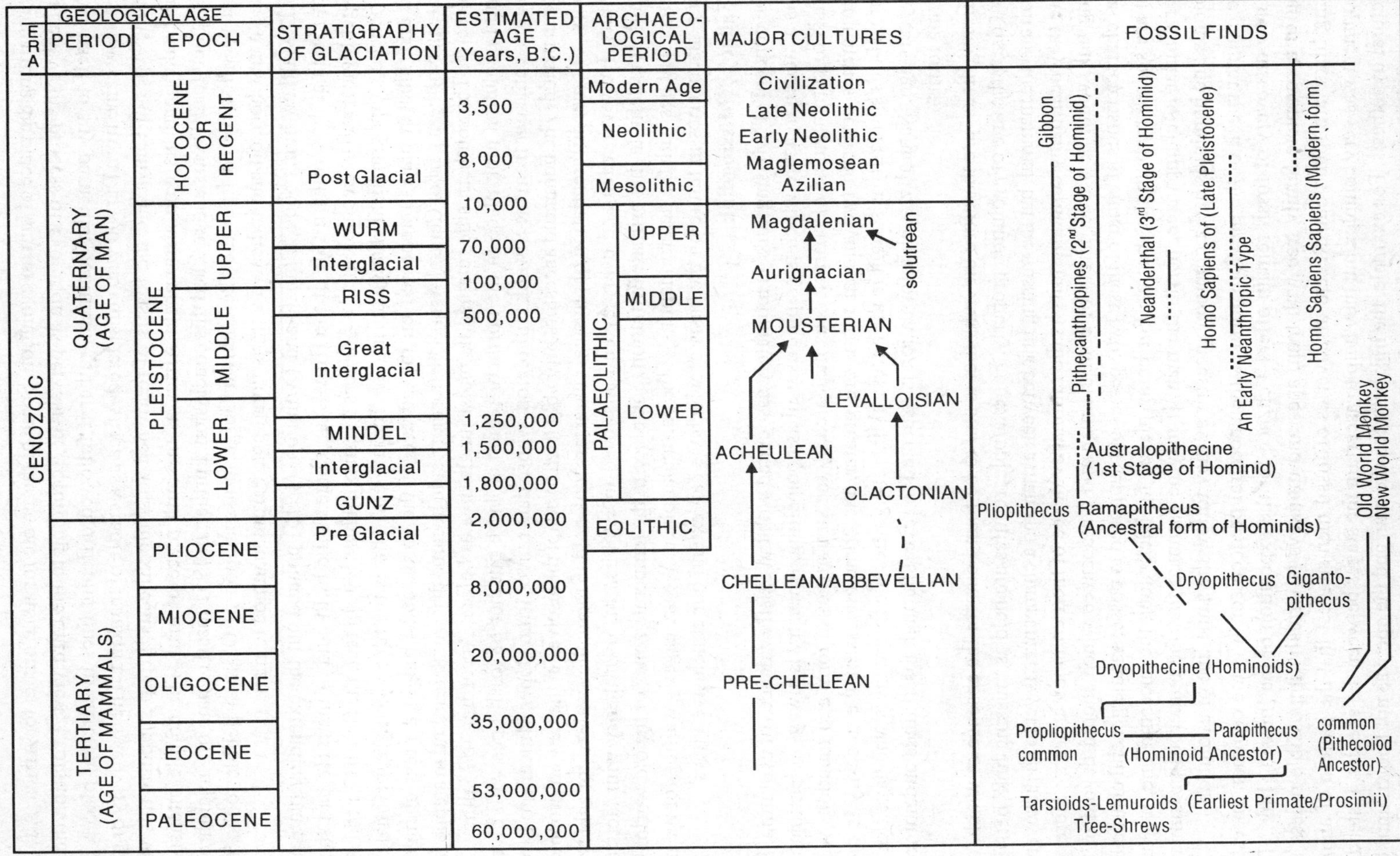

Fig. 5.7. Fossil Evidences In Stratigraphical Sequences

view regarding the place of human evolution. Some of them vote for Asia, others for Africa. A few of them believe the process of evolution as pervasive. According to the third view, evolution did not restrict itself to any particular continent, rather it followed different lines in different places. The oldest view came from Darwin who considered Africa as a cradle of mankind. The discovery of great Miocene Apes in Kenya and the fossils of renowned Australopithecine group supports this view. L.S.B.Leakey and his wife also upheld Africa as the birth place of man. The couple intensively investigated for 30 years and reached this conclusion. Further, in 1951 fossil remain of Zinjanthropus Boisei was discovered from the site Olduvai Gorge in association of Olduvai pebble tools. This opened a new horizon and indicated a transitional phase from man-ape to man.

W.D.Malthew in his book 'Climate and Evolution' (1915) pointed out the extending plateau of Central Asia at the Northern side of the Himalayan range as the birth land of man. He had shown huge geological and palaeontological evidences to establish his view. In fact, the South Central Asia experienced a topographical change during the Tertiary era. This is the time when great Himalayan range and Tibetan plateau appeared on the surface of the earth, and brought a great change of climate in Northern part of Central Asia. The northern side became dry which resulted in steppe land and deserts. The southern side retained its moist jungle area as before. According to Malthew, the hominid evolution took place in the northern region as large number of fossil apes with diverse characters came out from the Siwalik Hill region, whose date ranges from Upper Miocene to Lower Pliocene period. The hominoid ancestors probably tried to adapt themselves in the new environment. In course of time, differentiation of physical characters took hold in them; they walked for achieving humanity. However, each and every evolutionist tried to establish his own theory with facts and findings. It is definite that primate evolution took place in the temperate zones of the old world. But disagreement lies with the determinations of the particular place. We are eagerly waiting for the new striking discoveries in future on the basis of which we can extend our conclusive views.

The Fallacy of Missing-link

The concept of 'Missing-link' is intimately associated with the fossil evidences of evolution. It is held as a hidden treasure which is graphic as well as compelling. Ancestry is always determined on the basis of commonness of characters as indicated by the similarities in form and function. When the similarities between two animal forms are greater, the commonness of the ancestry is more definite. If this similarity is sought in the ancestral line, it signifies the forerunners who have existed some years ago. Naturally, such information is expected to reveal through fossil finds and not from the living forms.

'Missing-links' are therefore the links that is still to be found out, for the absence of which, one can not draw the line of evolution properly. For example, reptiles are held as 'missing link' between the man and the mammal, but the missing link between a man and a chimpanzee is yet to be discovered, despite they bear many common traits as the rational evidence of their common ancestry. Here, the missing link would be a true parent of both the man and chimpanzee which should devoid of the special characteristics of either of these two species but supposed to possess some general characters, which are shared by both the species. For being placed far back along the ancestral line, it would neither similar to chimpanzee, nor to the man. It may be assumed that the numerous prosimians, hominoids and hominids had a common ancestor as they follow the same line of evolution.

We can not speak of the previous form as the ancestral predecessor, it is the common ancestor. Each new discovery of fossil remain offers a chance for the exposition of the 'missing link'. The raising of a complete family tree with man, apes and monkeys is virtually impossible unless the 'missing links' are found out. Recent discoveries of fossil men suggest that the pattern of animal development was very complex but they help to ascertain the stages between the species, instead of conjectural hypothesis. For example, the link between the man and the ape was not known until the famous fossil, Pithecanthropus erectus or the Java Ape-man was discovered.

To get the 'missing link' is not only a matter of chance, there are other hazards too. The main obstacle lies with the formation of the fossils. Since the fossils are formed only under certain conditions, the scope of fossilization is very limited for the remains. If a bone lies on the ground, it decays within a few years. It would not last longer even if it is buried under the earth. But when carnivores drag the bones into the caves, they get a fair chance to become impregnated with hard and insoluble mineral salts as it happened mostly in South Africa. Besides, the bones may fall into a swamp or a lake bed or some other places where they get an opportunity to come in contact with minerals. The minerals usually replace or join the material of the bone and the fossils are eventually formed. Again, a fossil may be totally destroyed for the change of soil-chemistry. In that case, the evidence gets lost. For instance, the Galley Hill area contained no fossils other than one skeleton; the acidity of the ground destroyed the possibility of forming fossils. Similarly, the forested areas usually reserve acid in the soil which annihilates the prospect of fossils, although stone tools may be well preserved in them.

As missing links can be obtained only from fossils, so lack of fossil formation and destruction of fossils often restrain us from knowing the true facts. Darwin considered the continent of Africa as the birth place of mankind because the region was tremendously rich in fossil finds, especially with the apes and hominoid forms. It can be argued that the huge limestone caves of Africa facilitated in trapping of the bones (as fossils), which has not been possible in other places. This suggests that it is not only the abundance of the particular animals, but the scope of fosilization that makes a region visible in connection to evolution. It is also reasonable to think that, for the want of fossil records, some missing links will never be solved out, and Africa may not be the first or single area where evolution of mankind took place.

GENETICS AND HEREDITY

HISTORICAL BACKGROUND OF GENETICS

Man being the most intelligent creature, is very curious about himself. He finds that the children of the families tend to resemble their parents. They inherit the physical and mental traits from either of the parents or both the parents. Moreover, the individuals of a particular human population group show some general similarities among themselves, in spite of tremendous differences. The similarities are produced due to inheritance of paternal characters; some inherent properties of men are biologically transmitted from generation to generation. The science of genetics is an important discipline, which deals with heredity and variation. In fact, genetics is the study of inherited characteristics while heredity is a process that determines the capacity of individual to grow and develop as a result of interaction between the inherited traits and environmental factors. Study of heredity throw sufficient light to understand the origin of man and the significance of races. For example, tigers beget little tigers that inherit the characteristics of their race. On the other hand, genetic study investigates the mechanism of heredity and unveils the internal features of human evolution.

A close observation of different species of animals and plants reveals that a continuous operation of heredity and variation is going on in nature. No two brothers are exactly alike, except some identical twins. This lack of resemblance among the persons related by descent is often referred as variation. The variation denotes a fundamental change in the internal set up of an individual either due to genic differences, which might have originated from mutation or due to environmentally induced differences, which might cause temporary changes in the phenotype. A third source of variation may be due to the hybridity of the parents. For instance, if the parents are heterozygous (or hybrids) for a particular trait, say the eye colour, then it is quite possible to beget offspring having brown and blue eyes. Variation may occur in greater or lesser degree but in most of the cases it has proved useful. Evolution of new forms has been possible for the advantageous variation. Therefore, heredity and variation both are equally important in the study of genetics.

The term genetics came from the Greek word 'gen' which means 'to generate' or to grow or to become something. It was coined by William Bateson in 1905. But the inception of this science occurred or happened earlier than this date. Three botanists, De Vries (Holland), Correns (Germany) and Von Tschermak (Austria) identified the basic principles of genetics in the year 1900. They worked independent of each other, but none of them got the credit. Because, the science of genetics was already born in 1866 through the effort of Gregor Johann Mendel of Brunn, Austria; those botanists rediscovered the work after thirty-four years. According to Webster, the genetics is the branch of biology which deals with heredity and variation among related organisms, largely in their evolutionary aspects". In short, it may be defined as a science dealing with the principles that explain the similarities between parents and their progeny, and the differences among the individuals of a single species. Thus, the genetics has been a science of inheritance and variation.

Inheritance involves the activity of a distinct functional unit known as genes., which get transmitted from parents to progeny and govern the development of the characters of an individual. Character or trait denotes morphological, anatomical, physiological, bio chemical and behavioural features of an organism. The variation of characters is produced for the variation in genes. So, genetics may be redefined as *the science dealing with the structure, organization, transmission and function of genes, and origin of variation in them.*

Different fields of genetics may be considered as the study based on different objects. For example, *Animal genetics* and *Plant genetics* are concerned with the animals and plants, respectively as the objects of study. The *Virus genetics* and *Microbial genetics* consider the viruses and the microbes, respectively as the objects of study. *Human genetics* deals exclusively with the heredity of man. Besides, *Molecular genetics* and *Population genetics,* two specialized fields are usually distinguished from the *Classical genetics.* Molecular genetics mostly deals with the nature and properties of genetic material by considering its physics and chemistry, function, replication, mutation, recombination, regulation and evolution. Population genetics treats with the operation of heredity in natural variable population of freely inbreeding individuals in relation to mathematical and statistical probabilities. However, the *Classical genetics* may be termed as *Cytogenetics* which does not only signify the close relationship between the two allied branches of biology, i.e. *Cytology* (Cell biology) and *Genetics* (science of heredity), but correlates the chemical and genetic organization of chromosomes and their behavioural patterns of inheritance. It (cytogenetics) projects a clear picture about the various aspects of chromosomes and their effects on the development of characters of organisms. At the gross level, cytogenetics and genetics are distinct from each other. But, at the molecular level these are not so sharply defined.

The classical account of inheritance dates back to Hippocrates (400BC) who believed that 'the reproductive material came from all parts of the individual's body and hence that characters were directly handed down to the progeny'. According to Aristotle (350BC) the reproductive materials are actually the nutrient substances, which diverts towards the reproductive path instead of making the parts of body. Though the ideas of Hippocrates and Aristotle concerning heredity differ considerably, still they shared a common view that the inheritance is direct and the substances derived from or intended for specific parts of the individuals are transmitted to the progeny. Several other Greek Philosophers and Biologists, contemporary to Hippocrates and Aristotle, also made observations and speculations regarding the hereditary principles.

Most of the biologists of that period believed that, some primitive organisms originated from non-living matter, i.e. decaying organic matter. The process was called *spontaneous generation.* The doctrine of spontaneous generation was very pervasive until the middle of eighteenth century. William Harvey (1578-1657) first speculated that all animals arise from *eggs* and the semen, which has stimulating effect on the developing process. Dutch physician, Regnier de Graaf (1641-1673) speculated that the characteristics of both the parents are expressed in the progeny in mammals, rather both the parents transmit their hereditary characters to their offspring. Again, a Dutch microscope maker, Anton van Leewenhoek when made a crude hand-made microscope comprising a series of mounted lenses, he observed human sperm cells. He also had observed the sperms of several other animals.

Although it was long known that reproduction takes place by the union or copulation of males with females, but the contribution of male and female, more particularly the facts regarding hereditary materials, were unknown. It was a matter of great debate. However, the concept of Aristotle (that the progeny derived the specific shape from semen while the nutrition was provided by the female) was farther developed during the seventeenth and eighteenth centuries by the discoveries *of sperm* and *ova* in animals, *and pollen* and *eggs* in plants. Basing on William Harvey's speculation, many biologists advanced their ideas and studies, and stated that one of the *sex cells, sperm* or *ova,* is sufficient for performing the function of an organism in miniature form. It has to be developed into adult proportion on receiving appropriate nourishment. Such a theory is known as **Preformation theory**. Long controversies continued during the nineteenth centuries between the *animalculists* or *spermists* and *ovists.* The spermists or animalculists believed that the sperm contained the preformed miniature creature. On the other hand, the ovists confirmed that the ova contained the preformed organism.

Though the Preformation theory is obsolete today, still it provided a sensible concept of reproduction to replace all vague concepts of very early period.

The theory, which subsequently replaces the Preformation theory, is known as ***Epigenesis theory***. This was proposed by Wolff (1738-1794). According to Wolf, neither the sperm nor the ova contained a preformed creature, rather each of them contained undifferentiated material that had the potentiality of developing into an adult organism after fertilization. Epigenesis theory gained an acceptance among the biologists of the nineteenth century. However, at present it is clearly known that both the sperm and the ova are not composed of entirely undifferentiated material, but, on the contrary, they carry the *hereditary material,* the chromosomes, which contain genes.

The French scientist, Jean Baptiste Lamarck (1744-1829) had speculated that the *use and disuse of organs* over a long period of time determine the characteristics of the progeny. He formulated a theory known as **inheritance of acquired characters**. The founder of the modern evolutionary theory, Charles Darwin (1809-1882) suggested that the tissues and cells of an organism threw off minute particles or granules called as *gemmules* which being circulated through the plant and animal bodies, eventually passed to the sex organs to assemble there as gametes. The gemmules of the opposite sex are added during the fertilization, which together lead to the development of a mixture of maternal and paternal organs and tissues in the progeny. The hereditary characteristics are thus believed to be transmitted from the parents to offspring. This mechanism is held as the essence of the doctrine **of Pangenesis**. In his writings, Darwin had mentioned that his theory of 'Pangenesis' is in no way different from the hereditary concept speculated by Hippocrates. A German biologist, August Weismann (1883 - 1914), proposed the **theory of the continuity of the germ-plasm or germ-plasm theory** to explain reproduction in animals. According to this theory, the 'germ plasm' is transmitted to the offspring without any change but a change of the germ plasm (mutation) could lead to a change in inheritance. In fact, the theory distinguished two distinct types of tissues, i.e. somatoplasm and germplasm in the body of an individual. The *somatoplasm* or body tissues are necessary for survival of the individual and proper functioning of the organism, but it does not contribute to sexual reproduction. Naturally, any change. in the somatoplasm can not be transmitted to the next generation. On the other hand, the germplasm produces the gametes, which are the basis of sexual reproduction. The changes in germplasm are easily transmitted to the next generation. Weismann clearly mentioned that acquired characters are not inherited. He had shown that the changes in the characteristics of body do not affect the quality of gametes produced by the germplasm. His conclusive demonstration found the separateness of germplasm and somatoplasm; he disposed off the hypotheses of pangenesis and inheritance of acquired characters.

Before Mendel's experiments on the garden peas (Pisum sativum), John Knight and John Goss, both from England had carried out the breeding experiments in the same plant. John Knight in 1799 made a crossing over between two varieties of peas—unpigmented and pigmented. The unpigmented variety had green stem, white flowers and colourless seed-coats. John Goss in 1824 had experimented the same with green (blue) seeded and yellow (white) seeded plants. His experimental procedure involved the removal of the *stamens* from normally green seeded peas and its pollination happened from stamens of yellow seeded peas. John Knight had also performed reciprocal crosses. The experimental results of both Knight and Goss were quite remarkable, but they did not count the numbers of peas used in their breeding experiments. As a consequence, they had failed to discover the underlying hereditary mechanism, which Gregor Mendel was subsequently able to publish later in 1866. Knight and Goss basically aimed their experiments toward commercial achievement of improved varieties of plants and animals for using them in agriculture and horticulture. They did not seek any information regarding the hereditary mechanisms.

However, Hippocrates' speculations on heredity was not challenged until the final establishment of the basic principles of heredity as proposed by Mendel and published in 1866. But it is a matter of shame that the monumental work of Mendel was largely overlooked for nearly 34 years. From these experiments, *Mendel referred the unit of inheritance, as factor, which we now call gene is particulate and not liquid. The two alleles of a gene do not affect each other in any way, what to say of blending with each other.* He proposed two universally accepted laws of inheritance—the law of segregation and the law of independent assortment with which he explained the dominance of characters. Unfortunately he was not evaluated and his paper lay forgotten in the stacks of many libraries till 1900, until it was 'rediscovered' independently in papers published by three botanists DeVries (Holland), Correns (Germany) and Von Tschennak (Austria) who obtained more or less similar results to those of Mendel. Most likely they came to know of the paper by Mendel through a reference made to it by Focke in 1881 and copied from him by Bailey in 1895. It was really unfortunate that Darwin was not aware of Mendel's work (on heredity) although they were contemporaries. So being ignorant of Mendel's principles, Darwin proposed the theory called ***Pangenesis*** to account for the hereditary transmission of characteristics. If Darwin had knew Mendel and his work, he would not have only appreciated the significance of Mendel's contribution to heredity, the knowledge of heredity would have been utilized to remove one of the major difficulties in explaining the theory of ***Natural Selection***. However, the rediscovery of Mendel's paper initiated a speed of research in the field of genetics.

Mendelian Principles

Gregor Mendel's (1822-1884) experiments on the garden pea *(Pisum sativwn)* are regarded as a great landmark in the study of genetics. He was born in Northern Moravia, Austria—at present a part of Czeckoslovakia. His father had a great fascination for gardening and horticulture; Mendel used to work with him in the orchard and gradually developed an interest towards nature. At the age of 21years, he entered the Augustinian monastery at Brunn (now Brno in Czechoslovakia). After four years, in the year 1847, he was ordained a monk or priest. Rest of his life was spent in that monastery where he made innumerable crosses between varieties of several diverse plant species. In 1851 he entered the University of Vienna, Austria on the recommendation of the authorities of the monastery where he studied all basic sciences including physics, chemistry, mathematics, botany and zoology till 1853. After this he became a teacher at the monastery in 1854 and served there for 12 years. He had presented the facts and figures of his experiments before the Brunn Natural History Society on February 8, 1865. Finally, his 46-paged paper, entitled "***Experiments in Plant Hybridization***", was published in the *Proceedings of the Brunn Society for the Study of Natural Sciences* in the following year, in 1866. The copies of his paper were distributed to libraries in Europe and America, but received no serious attention for 34 years, till the 'rediscovery' by three botanists in 1900. The success of Mendel may be attributed to his brilliant analytical mind, which enabled him to interpret the results of his experiments as well as to define the principle underlying the mechanisms of heredity. In this regard, Mendel deserves to be called as *the father of genetics.*

The principles of Mendelian genetics is based on two 'laws', such as, the *Law of Segregation* and the *Law of Independent Assortment* or *Recombination.* Mendel's success in the hybridization experiments indicates his sound sense of judgement in designing the crosses of garden pea, which were selected only after careful consideration of the suitability of the experiment. He had selected a total of seven pairs of contrasting characters for his final experiments, which are as follows:

1. Length of the plant, either tall or dwarf.
2. Position of the flowers, either axial (distributed along the stem) or terminal (bunched at the top).
3. Colour of the unripe pods, either yellow or green.
4. Shape of unripe pods, either inflated or constricted.
5. Colour of the cotyledons (i.e. nutritive parts of the ripe seeds), either yellow or green.
6. Shape of the seed (outer surface), either rounded and smooth or wrinkled.
7. Colour of the seed coat, either grey or white. (The colour of the flower is often correlated with

the colour of the seed coats; generally the white flowered peas develop seeds of white coat and violet flower produces seed with grey coat.)

In this way, Mendel tested the seven characters individually by carrying out crosses separately for each trait. He considered only one trait at a time.

Law of Segregation (Mendelian Principle I)

The popular belief of Mendel's time was that, the females provided the substance for development of progeny, similar to the soil providing nutrition for plants. Mendel advocated a complete fusion of male and female gametes, where equal contribution of the two parents was necessary for the development of characters in hybrid generation. He also postulated the existence of factors, now known as genes, which are responsible for the growth of various characters. It is these factors that are transmitted from one generation to the next; the characters by themselves can not be transferred. Mendel made a clear distinction between the appearance (Phenotype) and the genetic make up (Genotype), and introduced a concept of dominance and recessiveness. His formulae for determining the different genotypes and phenotypes in different generations brought a dramatic change in thinking about the inheritance pattern of man.

Mendel started his experiment by crossing two pure-bred varieties (strains) of peas, tall and dwarf which were raised separately in the monastery garden. The tall plants were about 6 to 7 feet high, while the dwarf plants were only 9 to 18 inches high. The environmental factors, soil and moisture condition was minutely checked by Mendel to see whether those had any effect on growth. Mendel also noted the results from reciprocal crosses where one parent (Parent A) was used as female in a cross with another parent (Parent B), and again it (Parent A) was used as male with another parent (Parent B). Mendel made himself sure that the male and female parents make equal contribution to the development of characters in progeny; results of reciprocal crosses are always identical.

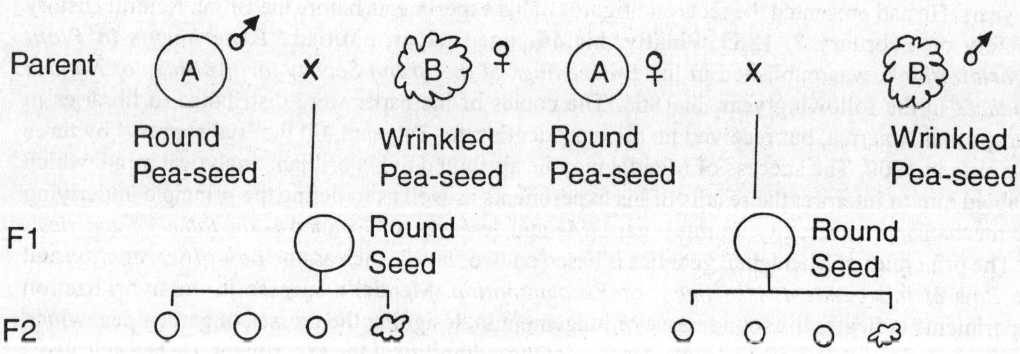

Fig. 6.1. A Reciprocal cross between Mono-hybrids exhibiting Dominant character

Beside the reciprocal cross, Mendel distinguished other crosses as monohybrid, dihibrid, trihybrid and so on. A cross between two parents differing for a single character is termed as monohybrid cross, while those between parents differing for two or three characters are known as dihybrid and trihybrid crosses, respectively.

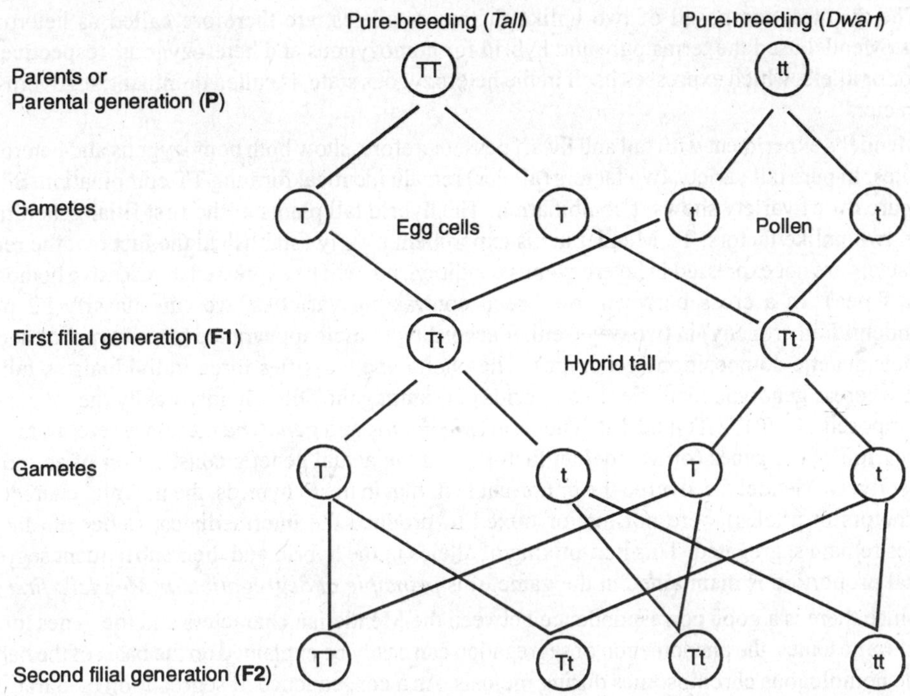

Fig. 6.2. A Mono-hybrid cross showing the Mechanism of Inheritance in second generation

However, in the first experiment Mendel considered only monohybrid cross. One of the two traits, the tall, was expressed in the first generation progeny i.e. all offsprings were found tall resembling one of their parents. Mendel designated this hybrid generation *as first filial* (Fl) *generation* while the cross involving two pure varieties of initial generations was regarded *as parental generation* (P). Mendel further proceeded by arranging a crossing between the tall members of the first generation (Fl). The obtained result was strikingly different. The *second filial generation* (F2) gave rise to a mixed generation of tall and dwarf varieties in the ratio of 3:1. No intermediate form was produced. It is interesting to note that the generation F2 constituted of three-fourth tall plants and one-fourth dwarf plants, although both of their parents were tall. Such a cross in succeeding generations (F3, F4, F5 and so on) also yielded tall and dwarfs varieties in the ratio of 3:1. On the other hand, when dwarf plants were crossed with one another, all offspring appeared as dwarf. In the second generation (F2) also they remain dwarf; no tall plant was produced (Fig. 6.11)

Here, Mendel arranged crossing between one pair of contrasting characters. This type of cross involving two parental strains differing in a single trait has been called as Monohybrid cross. Mendel held each trait as a unit factor or character. Mendelian factors, which determine the characteristic of any organism, are now recognized as genes; a character is produced by the action of a specific gene. Like the two alleles of a gene, Mendelian factors occur in pairs and these are maternal and paternal in origin. Because both male and female of the parents make equal contribution to the development of characters in the progeny. In the first generation (Fl), character of only one of the two parents is expressed. This is known as dominant character. Character of the other parent is not expressed in Fl; such a character is referred as latent or recessive. The dominant character may be of two types. Firstly, it may be pure like that of parent producing progeny with dominant trait only. In this case, factors or alleles are alike and such a zygote is termed as homozygote. Secondly, it may be a hybrid similar to the members of FI generation.

The zygotes, composed of two unlike factors or alleles, are therefore called as heterozygote. Thus, Mendel used the terms pure and hybrid for homozygous and heterozygous, respectively. The factor or allele, which expresses itself in the heterozygous state, is called dominant; it governs on the character.

Mendel's experiment with tall and dwarf peas, therefore, show both homozygous and heterozygous zygotes. In pure tall variety, two factors (alleles) remain identical forming TT combination. Similarly, the pure dwarf variety shows tt combination. The hybrid tall plants of the first filial generation (Fl) carry two unlike factors, Tt. Mendel in his experiment clearly established the fact that the recessive characters are not expressed in heterozygous condition, but will be expressed in recessive homozygotes (dwarf pea). In a cross between true-breed contrasting varieties, we can classify F2 progeny (grandchildren grogeny) in two ways, either according to their appearance (phenotype) or according to their genetic composition (genotype). The phenotype classifies three individuals as tall out of four, whereas genotype identifies two hybrids (Tt) among the talls. Genotypically the F2 generation is composed of 1 TT, 2 Tt, and 1 tt. The terms *phenotype* and *genotype* therefore refer to the visible expression of the genes (or Mendelian factors) and the actual genetic constitution of an individual respectively. Mendel discovered the important fact, that in the Fl hybrids, the parental characteristics (or factors or alleles) were not lost or mixed to produce the intermediates, rather the factors or alleles remain segregated. This non-mixing of alleles in the hybrid and their subsequent segregation (equal proportion is maintained in the gamete) is *principle of Segregation or Mendel's first law.*

Since there is a good correspondence between the Mendelian characters and the genes located in the chromosomes, the phenomenon of segregation can easily be explained on the basis of the behaviour of the homologous chromosomes during meiosis. As a consequence of segregation (separation), the two alleles of a gene get separated and go into different gametes. Therefore, a gamete contains one of the two contrasting characters resulting from allele. When two such gametes fuse together, both the alleles contribute in forming a gene in the zygote. For the manifestation of any character, both the alleles are required. When two alleles of a gene are of same nature; it produces a pure character. But, when the two alleles are not of same nature, a hybrid variety is produced but the character of dominant gene is expressed. Therefore, appearance of an individual does not always conform to its genetic character. In fact, the kinds of alleles of a specific gene or all of its genes, carried by an individual is known as its genotype or genetic constitution while the appearance of the individual with respect to a specific observable character or all of these characters is referred to as phenotype. The concepts of genotype and phenotype was clearly stated by Mendel, though the term themselves were introduced much later. The separation of homologous chromosomes during meiosis may be regarded as the reason for segregation of the two alleles of a gene since the alleles are located in identical positions in the homologous chromosomes.

Law of Independent Assortment **(Mendelian Principle II)**

Studies on the inheritance of only one character at a time enabled Mendel to formulate the concept of gene. The essentials of this concept are :

(i) The development of each character is controlled by gene.

(ii) Each gene exists in two alternative forms, alleles which govern the contrasting forms of a character.

(iii) The two alleles of a gene do not modify each other when exist together in the same cell.

(iv) Each somatic cell has two copies of a gene (identical or distinct alleles), while gametes have only one copy.

(v) The alleles of a gene get separate and pass into different gametes of the hybrid, and

(vi) Genes are the units of inheritance that passed from one generation to the next.

An organism does not essentially have only one character, it has many characteristics with alternative forms. Therefore in second phase Mendel dealt with dihybrid and polyhybrid crossings. He found that the different pairs of characteristic features that came together in the first generation (Fl), separate out in the second generation (F2) (and also other succeeding generation) as a rule. Since each pair of characters is inherited independently of other, Mendel designated this principle as the *Law of Independent Assortment.* The consequences of independent segregation of any two characters can be easily predicted according to the law of probability. The law of probability is the basic concept of the principle of classical genetics. The principle of population genetics is also based on the consideration of law of probability.

In a simple way *probability* is the likelihood of occurrence of an event. An *event* is the occurrence of something. For example, in case of a coin, the occurrence of head or tail in a toss is an event. A *trial* refers to an opportunity provided for an event to occur. So, in a trial with a coin, either head or tail occurs; both can not appear at a time. In this way, under trial, the head of the coin always excludes tail and *vice-versa.* Naturally these events are mutually exclusive. The probability of mutually exclusive events is determined by a simple formula, **P = 1/n**, where **p**= probability of a single mutually exclusive event, and n= the total number of such events. In this example the total number of mutually exclusive events is two, namely head and tail. The probability of getting a head (or tail) in a toss is therefore 1/2. So it is understood that the probability of an event denotes the average frequency of that event. Conversely, the average frequency of an event is its probability. Another category of event can be referred as independent event, where two or more events are closely related but the occurrence of one of the events never affects the probability of the remaining event.

For example, two coins, A and B if tossed, together or separately, the occurrence of the head or tail in case of first coin (A) does not affect the probability of the occurrence of head or tail in the second coin (B). In this way the occurrence of head and tail in coin A is independent of that in coin B. So, they can be called as independent events. As a matter of fact, the probability of occurrence of two or more independent events go together, without being affected of each other.

Sl. No.	Coin A	Coin B	Probability
1	Head	Head	½ X ½ = ¼
2	Head	Tail	½ X ½ = ¼
3	Tail	Head	½ X ½ = ¼
4	Tail	Tail	½ X ½ = ¼

Fig. 6.3. Probability of Head - Tail combination in two tossed coins

In the second series of experiments, Mendel crossed a pure-bred strain of round (smooth) yellow seeds with another pure-bred strain of wrinkled green seeds. He had already established that the genes for both round and yellow were dominant over their respective alleles, wrinkled and green. So in this case, consequences of independent segregation *for seed* shapes and colours were easily predicted. Since the shape and colour of the seed were independent events, their segregation pattern were also independent. Two contrasting seed shapes (round and wrinkled) were mutually exclusive events, just like as other two contrasting forms of seed colour (yellow and green). So probabilities of various character-forms correspond with the frequencies of the forms as appearing in the second generation (F2), in case of monohybrid crosses. (Round 3/4 , wrinkled 1/4, yellow 1/4 , and green ¼). Mendel considered this proportion as a multiple of 3:1 or $(3+1)^2$. In other words, 9:3:3:1 ratio is the consequence of the superimposing of two ratios, one for each character difference.

Segregation for *WRINKLE SEED* (Ww)		Segregation for *GREEN SEED* (Gg)		Independent Segregation for for Seed Shape and colour	
Seed Shape (Frequency = Probability)		Seed Colour (Frequency = Probability)		Phenotype	Frequency = Probability)
Round	¾	Yellow	¾	Round Yellow	¾ X ¾ = 9/16
Wrinkled	¼	Wrinkled	¼	Round Green	¾ X ¼ = 3/16
Ratio, 3:1		Ratio, 3:1		Wrinkled Yellow	¼ X ¾ = 3/16
				Wrinkled Green	¼ X ¼ = 1/16
				[Ratio	9 : 3 : 3 : 1]

Fig. 6.4. Independent Segregation of Probable characters resulting from a Dihybrid cross (in second generation)

However, the phenomenon can be illustrated with the particular experiment as conducted by Mendel. For the dominant alleles of round and yellow traits, the symbols **RR** and **YY** were used. The recessive traits for wrinkle and green were symbolized as rr and yy respectively. The parent generation (**P**) was constituted with two pure varieties of peas — round yellow (**RRYY**) and wrinkled green (**rryy**). When they were crossed, two kinds of gametes having genotype **RY** and **ry** were produced. After fertilization, in the first filial generation (Fl), those gametes gave rise to double hybrids (dihybrid) with the genotype **RrYy**. But all these offsprings were apparently round and yellow for the presence of **R** and **Y** as dominant alleles; **r** and **y** remained only as recessives. Thus, Fl dihybrids were heterozygous with a combination of **RrYy**. Each of them (male and female) produced four kinds of gametes which contained factors (alleles) like **RY**, **Ry**, **rY** and **ry** in equal frequencies. The four gametes of the female possess equal chances to be united with the four gametes of the male. Combinations depend upon the frequencies of probability of meeting. The number of zygotes can be determined by multiplying the four classes of male gametes with four classes of female gametes. Genotypes consisting of 4x4 or 4^2 or 16 individuals are found in F2 generation. This dihybrid cross between round yellow (**RRYY**) and wrinkled green (**rryy**) seed varieties of peas has been represented in the (Fig. 6.5). The checker board of Fig. 6.6 reveals the analysis of 16 F2 individuals where on the average, 9/16 (9 out of 16) combination carry both Rand Y, 3/16 (3 out of 16) carry R but not Y, another 3/16 (3 out of 16) carry Y but not R, and only one of sixteen i.e. 1/16 carries neither R nor Y. As a result, a phenotypic ratio or proportion, 9:3:3:1 is found. This means, 9 round yellow seeds, 3 round green seeds, 3 wrinkled yellow seeds and 1 wrinkled green seed will be available in the second generation (F2).

The ***Law of Independent Assortment*** thus refers to the segregation (or separation) of two (o even more) pairs of alleles independently into the gametes. Mendel observed that the four kinds of male and female gametes, RY, Ry, rY, and ry, were produced out of dihybrids (RrYy) of Fl generation. This happened because each of the paired alleles, (R and r) and (Y and y) segregated or separated from each other quite independently and none of these paired alleles could enter in the same gamete. In other words, R always segregates from r and Y from y. Further, the segregate or separation of alleles belonging to one pair ((e. g. Rr) is always without reference to those belonging to another pair (e.g. Yy) in the gametes (during Meioses). Following the same principle, from a trihybrid (parents differing in three characters) cross, 8 different kinds of gametes are obtained. The 8 male and 8 female gametes would unite randomly to yield a total of 64 zygotic combinations which can be classified into eight different phenotypic classes, having the ratio of 27:9:9:9:3:3:3:1.

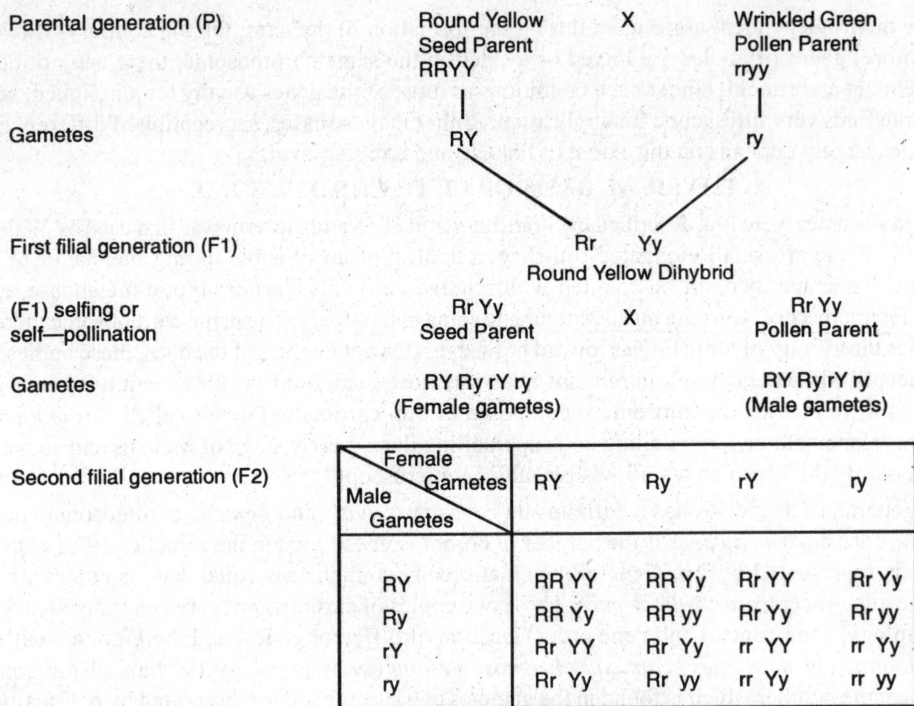

Fig. 6.5. A Dihybrid cross showing the Mechanism of Inheritance (in second generation)

Phenotype class	Genotype class	Number of individuals	Ratio of Phenotypes
Round and Yellow	RR YY RR Yy Rr YY Rr Yy	1 2 2 4	9
Round and Green	RR yy Rr yy	1 2	3
Wrnkled and Yellow	rr YY rr Yy	1 2	3
Wriknled and Green	rr yy	1	1
		16	

Fig. 6.6. A Checker-board showing phenotype, genotype and phenotypic ratio of a Dihybrid cross (in second generation)

Mendel had suggested the general formulae to determine the numbers of different kinds of gametes as perfect to the size of F2 population. A perfect F2 size indicates the minimum number of F2 individuals in which each genotype is expected to occur at least once. Accordingly, in a hybrid, segregating for n number of genes, 2^n different types of gametes (male and female) are expected to be produced. The union of 2^n kinds of male gametes with 2^n different female gametes would yield 4^n possible zygotic conditions, which relates to the size of perfect F2 population.

This second principle of Mendel, the law of independent assortment though is considered as an important landmark of genetics, has its limitation too and so can not be treated as a universal law. It is only valid when two (or more) pairs of alleles situated in different chromosomes become able to pursue their independent assortment during the formation of gametes. On the contrary, if two (or even more) genes or alleles are linked or located in the same chromosome, there can not be any independent assortment. Under such conditions as most of the genes usually remain linked, so this principle finds very little scope for application. Rather in these cases, segregation of different genes or alleles largely depends on the extent of linkage and crossing over.

PHYSICAL BASIS OF LIFE : CHROMOSOME

Chromosomes were first described by Strausburger in 1875 and the term was first used by Waldeyer in 1888. These are small elongated rod-shaped bodies, clearly visible during the stages of cell-division. Their number can be counted with relative ease only during mitotic metaphase, under powerful microscope. Chroma means colour and soma means body. The name chromosome therefore signifies the affinity of these bodies toward basic dyes. On application of the dyes, these bodies stain very deeply while the cytoplasm remains relatively unstained. Studies subsequent to Mendel have clearly shown that the chromosome is composed of thin chromatin threads called *chromanemata.* The chromanemata undergo coiling and supercoiling during early stage of Meiosis and so became progressively thicker to be readily visible under a microscope.

In general, each species has a definite and constant *somatic* and *gametic* chromosome number. *Somatic chromosome number* is the number of chromosomes found in the somatic cell of a species, which is represented by **2n.** Generally, somatic cells contain two equal sets or copies of each chromosome except *sex-chromosomes.* These two copies of chromosomes are generally identical in morphology, gene content and gene order (location of different genes) and therefore are called as *homologous chromosomes. Gametic chromosome number* is precisely the half of the somatic chromosome number which is found in the gametes of a species and is represented by **n**. Usually, the gametes of a species contain one set or one copy of each different chromosome along with a single sex-chromosome from the single pair of sex-chromosomes. Since, a zygote is produced by fusion of one male and one female gamete, it contains 2n chromosomes (somatic number) in which each chromosome is represented by two homologous chromosomes.

The somatic cells are found to contain *diploid number* (**2n**) of chromosome while the germinal cells or gametes contain *haploid number* (**n**) of chromosomes. Homo sapiens or the human beings show 46 chromosomes where 44 are called autosomes and 2 are called sex-chromosomes. *Autosome* - These chromosomes differ from sex-chromosomes; their kind and number remain same in male and female. *Sex-chromosome* - These chromosomes are meant for sex-determination and so differ in kind between the male and female. Since the chromosomes occur in pairs, we have found 22 pairs of autosomes and 1 pair of sex-chromosomes. Therefore, in case of human being, 2n=46, the diploid number and n=23, the haploid number.

It may be mentioned here that a normal human female shows 2 X chromosomes but the male shows 1 X chromosome in each diploid cell, which after reduction division produces two kinds of germ cell or gametes. In case of female, each of the two halves contains X-chromosome. In case of male, one half is with X-chromosome and another half is with Y-chromosome. Thus the number of chromosomes in the somatic cell appear as the diploid number (2n), i.e. 22 pairs of autosome and 1 pair of sex-chromosome, on the other hand, the number of chromosomes in germinal cell appear as the haploid number (n) with 22 autosomes and 1 sex-chromosome, either 1 X-chromosome or 1 Y-chromosome.

Chromosome morphology

Chromosome appearance (i.e. morphology) changes with different phases of cell division. Especially the metaphase stage of mitotic division is most suitable for the study of chromosome morphology. The size of chromosomes, the position of centromeres, the presence of secondary constrictions and above all the shape of the chromosomes are revealed through individual's karyotype. *Karyotype* is nothing but an arrangement of chromosomes in a cell; it determines the chromosome constitution of a cell. So, the number, the types, the size and the shapes of the chromosomes make up the karyotype of a species. Normally all members of a species possess an identical karyotype. The karyotype, is therefore equivalent to the genetic make up *(genome)* of a species where a fixed number of chromosomes are found as distinct from each other in gene content. These chromosomes do make pair and contain a definite number and kind of genes, which follow a definite sequence. Homologous chromosomes contain an identical number and kind of genes in the same sequence.

Each chromosome consists of two chromatids attached to each other at a constriction, the centromere.

Chromatid : In metaphase stage of cell division, each chromosome longitudinally divide into two identical parts. Each of the halves is known as *chromatid.* The two chromatids of a chromosome appear to be 'joined' together at a point called *centromere.* In fact, each chromosome consists of two chromatids attached to each other at a constriction, called the centromere. It is almost universally accepted that the chromatid is the structural and functional unit of chromosomes. It can not be further divisible into smaller subunits unless it is adversely affected to lose its structural integrity and functional capability.

Centromere : The region where the chromatids of a chromosome appear to join with each other during mitotic metaphase is called as *centromere.* It looks like a constriction in the chromosome and termed as *primary constriction.* If there is another constriction in the distal region in the chromosome, then this secondary constriction is called *satellite.* Presence of satellites can be taken as a constant feature to the karyotype of an individual and perhaps of a species. Because in most species, each chromosome is characterized by its length and by the position of the centromere. So, the position of centromeres serves as an important landmark in the identification of different chromosomes of a species. The chromosome is divided into two transverse parts by the centromere; the divided parts are known as *arms.* In general, one arm of a chromosome is longer than the other. However, chromosomes may be divided into four classes on the basis of the position of their centromeres: 1) Metacentric (M); 2) Sub-metacentric (S); 3) Acrocentric (A); and 4) Telocentric (T).

Metacentric chromosome : Chromosomes are called metacentric (M) when centromere is found to be located at the centre of the chromosome, i.e. the two arms on each side of the centromere are almost equal in length.

Submetacentric chromosome : In most of the chromosomes, the centromere is located on one side of the central point. Such a centromere is known as *submedian.* Therefore, the chromosomes with the centromere placed between the median and terminal positions are called as submetacentric (S) chromosome or subterminal chromosome.

Acrocentric chromosome : In some chromosomes, the centromere is located very close to one end of a chromosome. These chromosomes have developed a class called as the acrocentric (A) chromosome. It may be noted that the human chromosomes do not have terminal centromeres, but chromosomes of other species do.

Telocentric chromosomes : These chromosomes are with terminal centromeres and generally such telocentric chromosomes are unstable.

Chromosomes are the most complex nuclear components not only in their morphology, but also in chemical nature. The main chemical constituents are Nucleoproteins, which are made of two kinds of substances - proteins and nucleic acids.

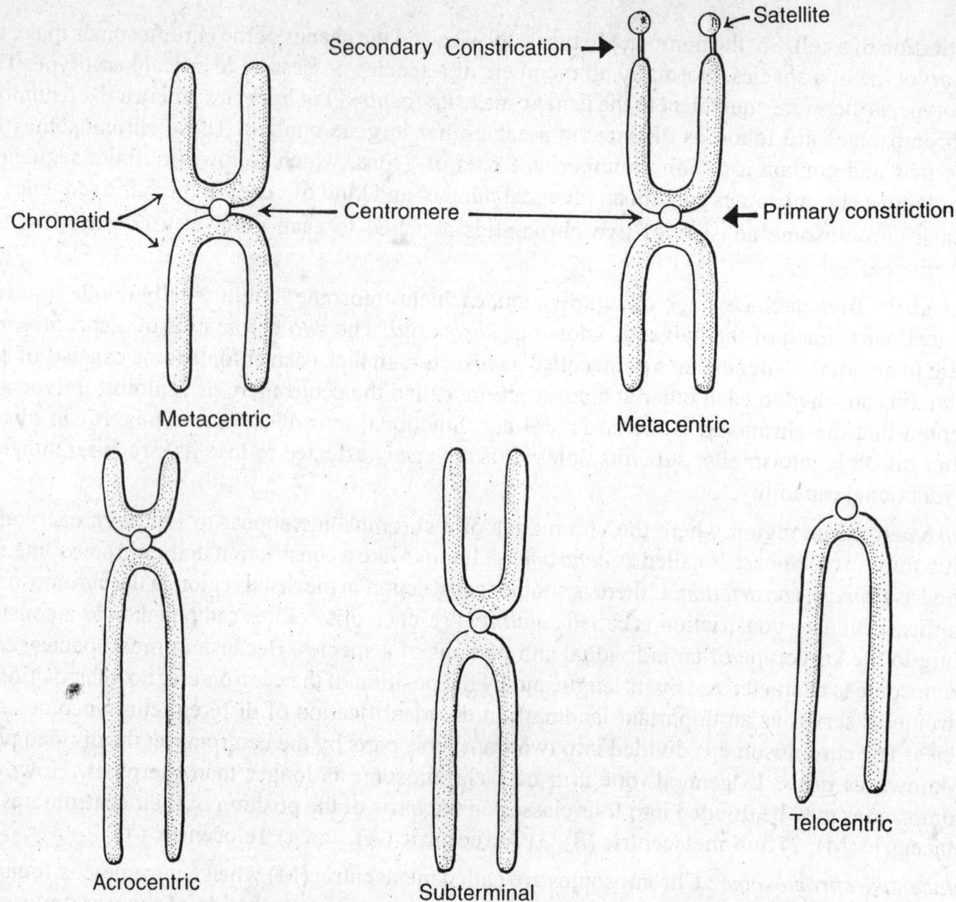

Fig. 6.7. Position of Centromere in Chromosomes

CHEMICAL COMPOSITION OF CHROMOSOME

The two kinds of nucleic acids are termed as deoxyribonucleic acid (DNA) and ribonucleic acid (RNA). It is very difficult to conclude, whether one of these substances (proteins or nucleic acids) contains the genetic material or the genetic material exists in some combinations.

Usually chromatin is composed of DNA, protein and RNA. But their percentage differs from phase to phase. For example, the chemical composition of metaphase chromosomes differs markedly from that of interphase chromatin, which contain relatively less DNA but more protein and RNA than the latter. Again, some bacteria are found with only DNA. Apparently the genetic information are found to be carried by both DNA and RNA; they are referred as principal genetic material. In most of the organisms, the genetic information is coded in the DNA while RNA takes part in the translation of coded information into action. In the absence of DNA, RNA serves all function.

The genetic material must possess certain definite properties, which can be summarized as follows:

a) The genetic material must be extremely stable so that it can withstand any assault from physical and metabolic agents, although occasional changes may occur in it by the process of mutation.

b) This material must be able to replicate itself with great accuracy.

c) This material must be able to direct the process of synthesis of proteins, which carry out the activities within the cell and produce phenotypic expression of the gene.

d) It is also necessary that the different parts of the genetic material should be able to perform different functions. As a matter of fact, there are many thousands of inherited traits in an organism, which by means of genetic control are distinguishable from each other.

However, the genetic material differs from one individual to another as well as from one species to another.

DNA

DNA is made up of four different nucleotides. These nucleotides being units of nucleic acid shows a nitrogenous organic base, sugar and phosphate. Each nucleotide contains a molecule of sugar, specially deoxyribose; a phosphate group, phosphoric acid; and an organic base - a compound formed by carbon, hydrogen, oxygen and nitrogen. The sugar and the phosphate group are identical in each nucleotide. But the organic base becomes different. It forms either with *purine* or with *pyrimidine.* Again, there are two types of purine bases - *adenine* (A) and *guanine* (G). The pyrimidine base can be of three types - *Thymine* (T), *Cytosine* (C) and *Uracil* (U). The four nucleotides that are found in DNA are *deoxyadenylic acid* (or deoxyadenylate), *deoxyguanylic acid (or* deoxyguanylate), *deoxycytidylic acid (or* deoxycytidylate) and *deoxythymidylic acid (on* deoxythymidylate). Thousands and thousands of these nucleotides link together to form one strand of DNA molecule. As a result, *polynucleotide chain* is built up where two nucleotide show a phosphate group in between.

DNA is a long and large molecule having a very high molecular weight. Each molecule is composed of two complementary strands, coiled together in a helix. This structure is called as the Watson-Crick model of DNA. The nucleotides are always found paired - **A** makes pair with **T**, and **G** with **C**.

After careful measurements it has been found that the exact amount of DNA in a single cell remains same in every cell within a single species. The only exception to this is the gametes or germinal cells which contain half of the amount of DNA as present in its somatic cells. This proves that DNA is definitely the genetic material. Besides, the four bases of the genetic alphabet - **A, G, T,** and **C** can produce various combinations and sequences which in turn yield a large variety of genetic messages. So, the genes are the segments of DNA molecule composed by the letters of the genetic alphabet.

RNA

DNA can not synthesize proteins directly, rather it requires an intermediate substance for carrying the information to the active synthesizing -centers of the cells. This translator and activator of the message is another nucleic acid known as *ribonucleic acid* (RNA). The composition of *ribonucleic acid* (RNA) is somewhat similar to that of Deoxyribonucleic *acid* (DNA). The sugar molecule in its backbone possesses ribose instead of deoxyribose. The *phosphate* is the same. Three of the four *bases* are also the same, *adenine, guanine,* and *cytosine.* The fourth *base* is *uracil* (U) rather than *thymine.* So, the RNA alphabet are **A,G,C,** and **U** instead of **A,G,C,** and **T** as in DNA. The *nucleotides* in RNA are *adenylic acid, guanylic acid, cytidylic acid,* and *uridylic acid.* RNA commonly appears as a single strand though double stranded RNA is not absent.

Ribonucleic acid (RNA) is also a large organic molecule composed of a large number of similar units known as nucleotides. Each nucleotide contains a *nitrogenous base, ribose,* and a *phosphate group.* The single strand of RNA shows double helix which remain folded in the middle and twisted around itself.

The actual synthesis of proteins takes place in little particles and that is found in the cytoplasm (the material outside the cell nucleus). The little particles are known as *ribosomes,* which are rich in RNA. Two types of RNA are required for protein synthesis, i.e. messenger RNA and transfer RNA.

The *messenger* RNA (mRNA) is produced by DNA in the nucleus. It is single stranded and synthesized from one strand of DNA in the nucleus in presence of the enzyme RNA *polymerase.*

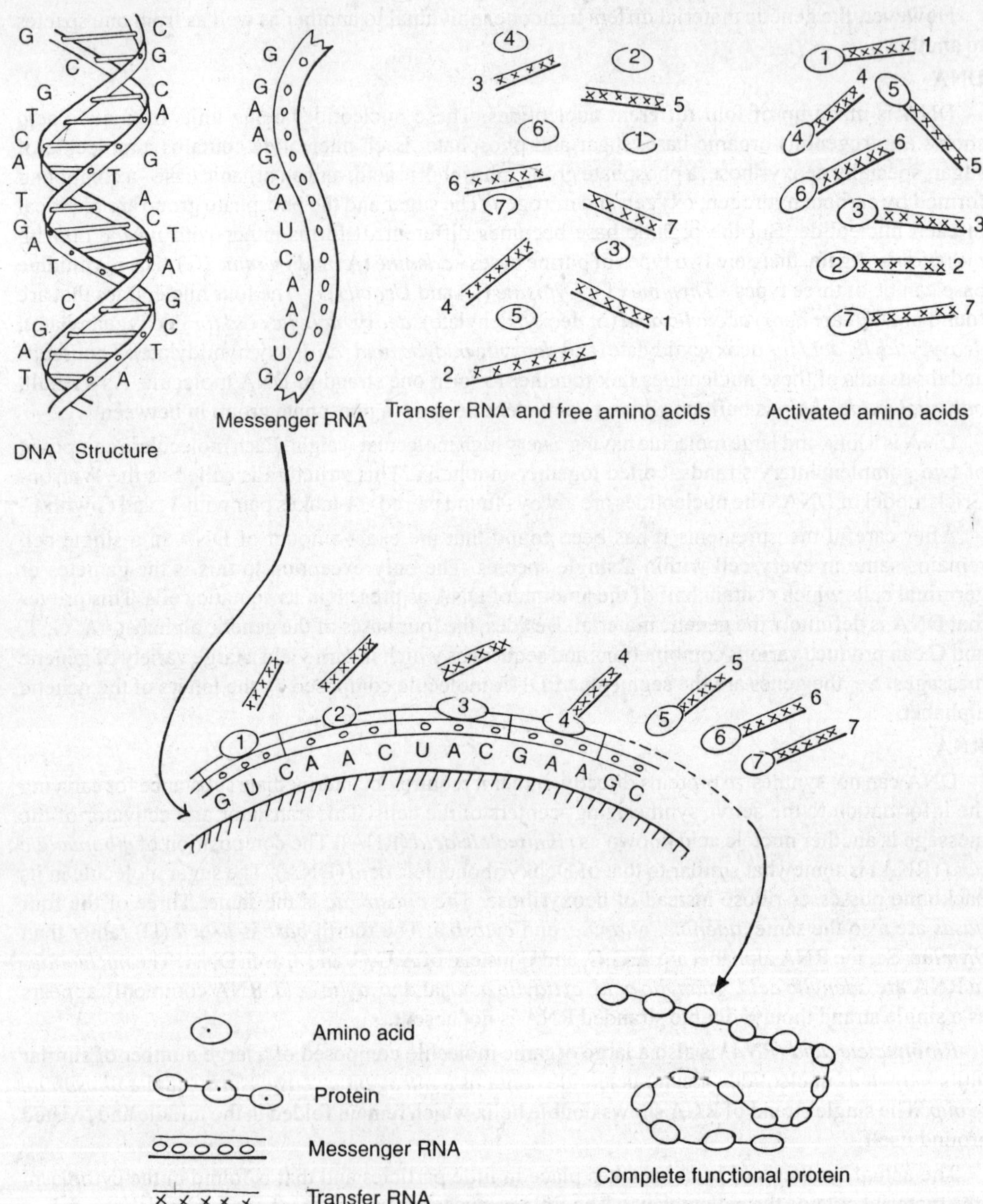

Fig. 6.8. Diagram Showing DNA in the sequence of Amino Acid

This single stranded messenger RNA carrying the blueprint for protein synthesis comes out of the nucleus and goes to the ribosomes and there takes part in protein synthesis. During genetic transcription, the genetic code of DNA is transcribed to the nucleotide sequence of messenger RNA.

Transfer RNA (tRNA) is also synthesized in the nucleus and subsequently transported to cytoplasm. These are small single stranded units and many of these are produced and transported to the cytoplasm to act as conveyors in order to bring appropriate amino acids to the ribosomes as per the code in the mRNA, for protein synthesis.

The hereditary information passes on to the site of protein synthesis from the DNA in the nucleus of the cell. A strand of mRNA is made with the particular sequence of genetic alphabets, organic base that has been imposed by a section of DNA. This strand carries the instructions from the DNA, then leaves the nucleus and enters to the cytoplasm. The instructions are the sequences in which amino acids are to be attached with one another to form a complete, functioning protein. The tRNA is believed to be a relatively small molecule, which is necessary for activation of the amino acids. Each amino acid becomes attached to the end of a tRNA molecule.

This step creates an array of activated amino acids without any sequence. The mRNA becomes associated with activated but unordered amino acids. In fact, the amino acids are put into specific sequence according to the sequence of *bases* that the mRNA brings up to the ribosomes from the DNA. The amino acids are linked into a protein in the sequence, which is ultimately determined by the DNA. The tRNA is dropped off and as a result a complete as well as functioning protein molecule is developed.

It may be concluded that the genes as hereditary units are located in chromosomes. The chromosomes are composed mainly of proteins, DNA, and a small amount of RNA. DNA replication is usually followed by cell division whereas RNA synthesis is always associated with the DNA present in the chromosomes. It means, the RNA synthesis is dependent on DNA. It is also known that different types of RNA (mRNA and tRNA) are complementary to different segments of chromosomal DNA; the information for the synthesis of chromosomal proteins is provided by DNA. All these suggest that the DNA is the most important genetic material while the RNA is next to it.

CHROMOSOMAL ABERRATIONS

In normal course of cell-division cycle, the chromosomes duplicate and segregate in an orderly manner. The sequence of gene loci in the chromosome also maintains an orderly arrangement. But recent genetic research reveals that, in rare cases, genetic variation (i.e. variation in the structure and arrangement of the chromosomes) do occur and bring about certain major clinical disorders through phenotypic changes in organisms. Each chromosomal abnormalities are usually associated with different congenital malformations and diseases.

The gross change or abnormality in chromosome is usually designated as chromosomal aberration. These aberrations may be of two kinds - *Spontaneous* aberration and *Induced* aberration.

i) The naturally occurring structural rearrangements of the chromosomes are called *Spontaneous aberration.* The reason behind such aberrations is not clearly understood. Factors like cosmic radiation, nutritional insufficiencies, and several other environmental factors may hamper the original chromosomal structure or number.

ii) The architectural changes deliberately produced by the use of a physical or chemical agents are called *Induced aberrations. A* variety of agents are able to induce mutations. They also cause breakage in chromosomes and all these result in chromosomal aberrations. Spontaneous aberrations occur very rarely whereas scope of induced aberrations is several times higher than that of spontaneous aberrations.

The geneticists are more concerned with the spontaneous aberrations because these phenomena occur naturally, without any known causal factor. They have grouped these aberrations into two broad categories - Structural aberration and Numerical aberration.

Structural Chromosome Aberration

In this category, aberration alters the chromosome structure but do not involve a change in chromosome number. The mechanics signify chiefly a rearrangement through loss, gain or reallocation of chromosomal segments. However, the structural aberrations of chromosomes can be classified into four common types, which are as follows.

Deletion : The loss of a chromosome segment is known as deletion or deficiency. The deletion of a portion of chromosome is a very rare event. It produces some striking genetic and morphological / physiological consequences. Each deletion gives rise to a distinct set of symptoms which characterizes an abnormality and called as a *syndrome.* A specific deletion in chromosome no. 22 produces a condition, called 'Philadelphia 22'; this is associated with chronic myelogenous leukemia. Again, another deficiency in chromosome no. 5 creates 'Cri-du-chat' (cry-of-cat) syndrome where the individuals produce a characteristic mewing cry like cat during childhood. They also possess some unique facial features and exhibit severe physical as well as mental retardations.

Duplication : The presence of an additional chromosome segment (as compared to that of normal number) in a nucleus is known as duplication. In this process, a segment of a chromosome is added to another chromosome; the extra part of the chromosome constitute duplication when this extra-chromosome segment is located immediately after the normal segment following the same orientation (i.e. the same gene sequence is maintained), it is called *Tandem duplication.* When the gene sequence in the extra-chromosome occur in a reverse order, it is known as *Reverse duplication.* The Reverse duplication is almost same as Tandem duplication, but here the additional segment is inverted in order. For example, the sequence will be **e d c** in place of **c d e**. Sometimes, the additional segment is found to be located in the same chromosome but awav from the normal segment; such cases are termed as Displaced duplication. Another case is the Translocation duplication, when the additiona chromosome segment is found to be translocated into a non-homologous chromosome. In general, duplications do not produce any drastic consequences as like deletion in terms of phenotype and survival. It has been postulated that the increase in DNA content per cell accompanied the process of evolution; the origin of new genes with distinct functions was possible only for the event of duplication.

Inversion : When a segment of a chromosome is found to be oriented in reverse direction, it is called inversion. Two breaks are required within a chromosome to get this situation. The segment rotates in 180°.angle and reinserted between the breaks. As a result, the linear order of the genes becomes exactly opposite, in comparison to its normal homologous segment. Suppose, the normal order of a few genes in a segment of the chromosome is ABCDE (in the original chromosome). If an inversion takes place between B and D (i.e. BCD segment), the order of the genes in the inverted segment will be ADCBE. However, inversion may be of two types - *Paracentric inversion* and *Pericentric inversion.* If the inverted segment does not contain a centromere, it is termed as Paracentric inversion. But if it contains a centromere, it is called Pericentric inversion.

Translocation : Integration of chromosome segment into a non-homologous chromosome is known as translocation. It involves the transfer of a segment of chromosome to a different part of the same chromosome or a different chromosome. There are three basic types of translocation—simple, reciprocal and shift. *Simple translocation* shows an attachment of a small terminal segment of a chromosome to the end of a homologous chromosome or to a non-homologous chromosome. This attachment is not a fusion at all. In *Reciprocal translocation,* the breakage takes place in two non-homologous chromosomes, and is followed by the reunion of broken segment to the wrong partners. *Shift translocation* requires at least three breaks in the chromosome. Among these, Reciprocal, and shift are the most common types.

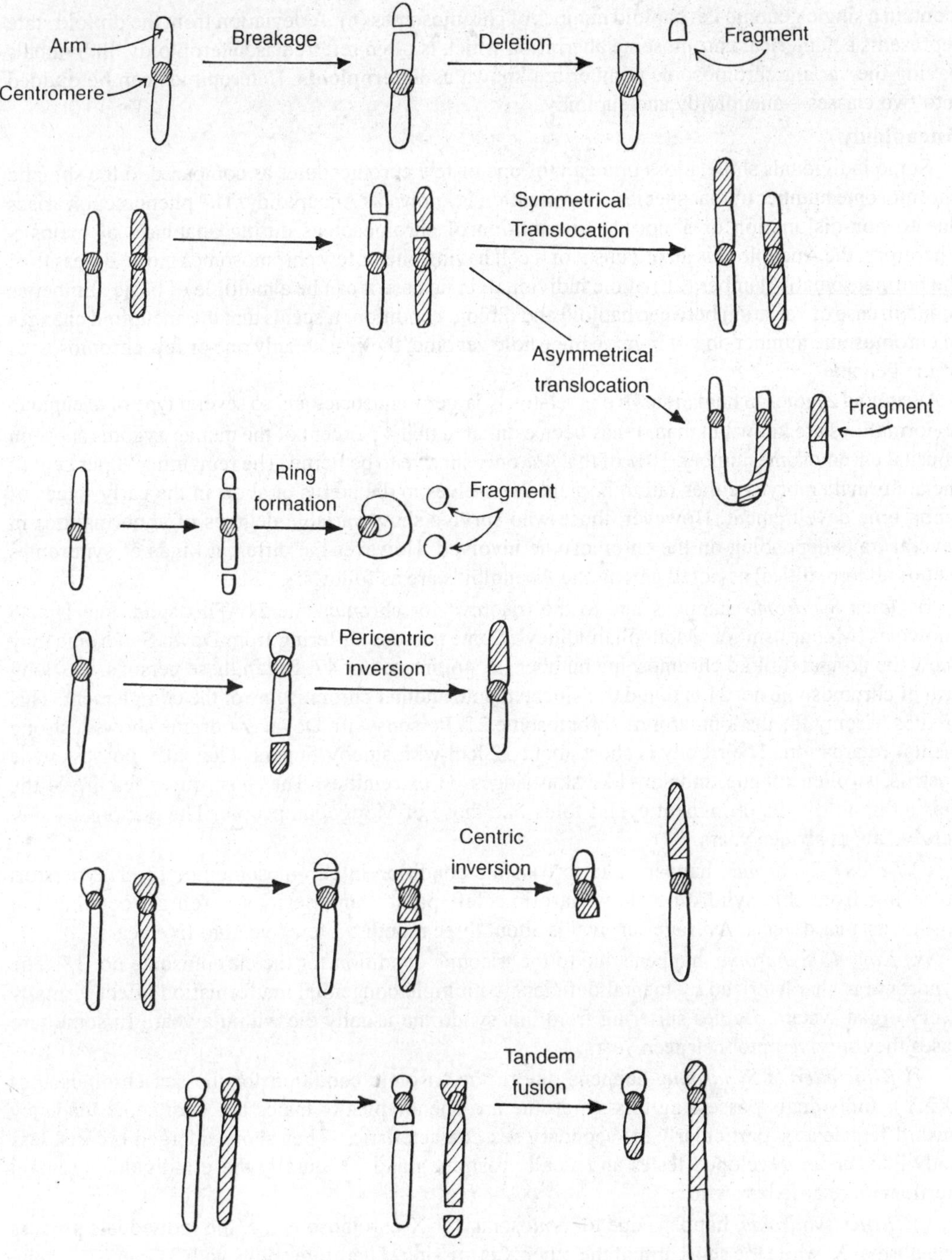

Fig. 6.9. Schematic Representation of Aberration (Chromosome Structure)

Structural Chromosome Aberration

The somatic cells of a diploid organism contain two sets of homologous chromosomes (2n number of chromosomes) or in other words, two copies of the same genome are present. But their gametes

contain a single genome i.e. haploid number of chromosomes (n). A deviation from the diploid state represents a numerical chromosome aberration which is often referred as heteroploidy. Individuals having the variant chromosome number are known as **heteroploids**. Heteroploids can be divided into two classes—aneuploidy and euploidy.

Aneuploidy

Some individuals show a loss or a gain of one or few cnromosomes as compared to the somatic chromosome number of that species. The situation is known as Aneuploidy. The phenomenon arises due to non-disjunction or abnormal distribution of chromosomes during anaphase of meiosis. Therefore, the Aneuploid is an organism or a cell having one or few chromosomes more, or less than the normal somatic number (2n) of the individual; in no case it can be a multiple of basic number as found in case of variation between haploid and diploid condition. It seems that the aneuploid changes in chromosome number do not involve the whole genome; they relate only one or few chromosomes of the genome.

Aneuploid zygotes in human survive in relatively larger frequencies and so several type of aneuploid abnormalities are known in man. It has been estimated that 4 per cent of the numan zygotes show an unusual chromosome number. 10% of that 4% only survive to be borne. The remaining 90per cent of the abnormal embryos either fail to implant themselves in the uterus or abort in the early stages of embryonic development. However, those who survive show variable degrees of abnormalities in several traits depending on the chromosome involved. However the different kinds of syndromes (set of abnormalities) as noted among the Aneuploids are as follows:

i) *Down Syndrome,* happens due to the trisomy* for chromosome 21. This syndrome is also known as 'Mongolism' or 'Mongolian idiocy'. Some persons suffering from Down Syndrome may show the normal diploid chromosome number i.e. 46, instead of 47. But in these persons, the long arm of chromosome no. 21 is found translocated onto another chromosome of the complement. This creates trisomy for the long arm of chromosome 21. Persons with Down Syndrome shows a strong mental retardation. Their body is short about 120cm with stubby fingers. They also possess wide nostrils, swollen tongue, monkey-like skin ridges on extremities. The most queer feature is the epicanthic fold—the prominent eyelid folds like those of Mongolian people. The persons usually survive about sixteen years.

ii) *Patau's Syndrome,* happens due to trisomic condition in chromosome no. 13. The persons suffering from this syndrome show harelip, cleft palate, and serious cerebral, occular and cardiovascular defects. Average survival is about three months; a few live upto five years.

iii) *Edwrd's Syndrome,* happens due to the trisomic condition for the chromosome no. 18. The syndrome is characterized by mental deficiency; multiple congenital malformation affect virtually every organ system. Babies suffering from this syndrome usually die within a year.. In some rare cases they survive upto their teen years.

iv) *Klinefelter 's Syndrome,* happens due to the trisomic condition for the sex chromosomes (XXY). Individuals possessing this syndrome are phenotypically males but with some tendency toward femaleness, particularly in secondary sex characteristics. They show enlarged breasts, less body hair, under developed testes and small prostrate glands. Naturally these individuals remain sterile with retarded growth.

v) *Turner syndrome,* happens due to monosomy for X-chromosome i.e. the individuals possess one normal X, while the short arm of the other X is missing. Therefore adults with Turner Syndrome are females having virtually no ovaries. The secondary sex characters are also poorly developed. Besides, they show short stature, low set ears, webbed neck and a shield-like chest. But it is interesting that these individuals generally do not show any mental retardation.

---

* If the somatic cells of an Organism contain three copies of any one chromosome of the haploid complement the condition is known as *trisomy* and is denoted by $2n + 1$.

Euploidy

Some individuals possess one or more complete genomes in a cell which may be identical with or distinct from each other. Most common types are those in which two copies of the same genome are obtained. Since the basic chromosome number or genomic number is **x**, the above situation is represented as **2x**. This means, all Euploid variations are designated with reference to the Diploid (**2x**) state and not to the somatic chromosome complement (**2n**). Euploidy can be further distinguished into two categories -*Monoploids,* including haploid and *Polyploids. Monoploids* denotes the presence of a single copy of a single genome (**x**) as like the haploids, representing the gametic chromosome number of a species (**n**). On the other hand, presence of more than two genomes in a cell is known aspolyploidy. This means, organisms showing polyploidy possess more than two sets of chromosomes in their nuclei. However, beside monoploids and polyploids, another category known as diploids is found. But diploids do not represent any deviation. Rather, they convey the normal condition of the organisms. The diploid individuals possess two sets of homologous chromosomes - one paternal and one maternal.

LINKAGE AND CROSSING OVER

The role of chromosomes in hereditary mechanism was postulated independently by Boveri and Sulton, soon after the rediscovery of Mendelian principles. They believed that chromosomes carried hereditary unit. In 1903, Walter Sulton put forward his chromosomal theory of inheritance. But later it was discovered that the unit of hereditary material is gene, not the chromosome. The behaviour of genes (located in the chromosome) was first interpreted by Bateson and Punnett (1906), on the basis of their experimental results involving a cross between two varieties of sweet peas (Lathyrus odoratus). This failed to conform with Mendel's principle of independent assortment.

Bateson and Punnet studied the flower colour and pollen shape in pea. They crossed a variety of sweet peas with purple flowers and long pollen grains (PPLL) with another variety of red flower and round pollen grains (ppll). They expected a ratio of 9:3:3:1 from the above cross as two character differences (dihybrid) was considered at a time. But surprisingly, they observed a different gametic ratio. At first they became perplexed with the situation but soon after they found out an affinity between the two dominant genes (P and L) for which they tend to stay together in the progeny. Bateson and Punnet called this situation as coupling phase. In the second generation, the same dominant genes were found to repel each other so that they tend to stay away from one another in the progeny. This situation was referred to as repulsion phase. Subsequently, the hypothesis of coupling and repulsion was replaced by the theory of Linkage and Crossing over.

Linkage

The concept of linkage has its origin in the Demonstration by Thomas Hunt Morgan in 1910. Morgan noticed a peculiar inheritance pattern of the white eye (W) gene of Drosophila malanogaster. He also investigated for the same in similar kinds of plants and animals and explained that the W is located in the sex chromosomes. Subsequently he studied inheritance of several other sex-linked genes. In 1911, he published his findings in the following way:

a) Genes located in the same chromosome tend to stay together during inheritance. (Morgan called this tendency as linkage).

b) Genes are arranged in a linear fashion in the chromosomes.

c) The intensity of linkage between two genes is inversely related to the distance between them in the chromosome. Morgan also proposed that the coupling and repulsion as shown by Bateson and Punnet, are the two aspects of the same phenomenon i.e. linkage. Later studies have confirmed the above conclusions of Morgan.

The linkage can therefore be defined as the tendency of genes (non-allelic genes) localized in the same chromosome to remain together and enter the gametes in greater proportion following the parental combinations. Linked gene does not show independent segregation.

Crossing over

Crossing-over is just opposite to linkage. This concept has also been established by Morgan. It denotes a tendency of the linked genes to enter the gametes in combinations other than parental combinations due to reciprocal exchange of genes at corresponding positions, along the pairs of homologous linkage units (or chromosomes). Therefore, although genes are inherited as a group, it tends to separate in the progeny by genetic recombinations.This exchange of precisely homologous segments between non-sister chromatids of homologous chromosomes is known as *crossing over.* Each event of crossing over produces two recombined chromatids (involved in the process of crossing over) called *crossover chromatids;* and other two chromatids which do not involve them in crossing over is called *non-cross over chromatids.* The cross-over chromatids get new combinations of linked genes while the non cross-over chromatids continue with parental gene combinations. Recombinant gametes produce recombinant phenotypes. This means, new combinations of the gene are required for the production of new characters among the offsprings.

The crossing over or recombination takes place during Meiosis, especially in the phase, *pachytene.* The frequency of cross-over largely depends on the distance between the respective linked genes, situated in the chromosomes. This rate is not at all constant in all pairs of gene. Several experiments on crossing over have shown that, it is not only the distance between the genes, different other factors such as temperature, nutrition, age and sex often influence the incidence of crossing over. In general, the frequency of crossing over is higher in females than that of the males.

Linkage may be classified into two types, complete and incomplete, depending upon the absence or presence of recombinant phenotypes in test-cross progeny. Complete linkage means lack of crossing over as found in male Drosophila. The absence of recombinants does not create new phenotypes. Incomplete linkage shows a very low frequency of recombination, which means that the concerned genes are loosely linked. However, the genes occurring in linear order and organized in linkage groups often change their combination in progeny. Such variations are subjected to Natural selection and form an important factor in human evolution.

SEX-LINKED INHERITANCE

Sexual reproduction is of paramount significance in the biological world as an effective way of generating genetic variability. It involves the fusion of two specialized haploid cells called gametes—a male gamete and a female gamete. In most of the cases, the male and female individuals differ in many characteristic features, which may be grouped into two categories : *Primary* and *Secondary sex characters*. Primary sex characters are related to the gamete producing organs of male and female individuals and depend on the genes present in the zygote. Secondary sex characters are all those characters that show consistent differences between the male and female individuals of a species, such as, genital ducts, genitalia and many other characteristic signs on the body. In human, these secondary sex characters are related to facial hair, mammary gland development, pitch of voice, degree of muscle development, subcutaneous fat deposition etc. At the same time, external administration of human male and female hormones (testosterone and estrogen, respectively) helps to modify the intensity of the features.

We have already known that a genotype is the actual constitution of an organism while a phenotype is the expression of the genotype in the organism. The genotypes are usually described in terms of alleles at a single locus * . The genes, found in several alternative forms are known as allelomorphs or alleles. Naturally, as the chromosomes occur in pairs, an allele is found to be present at the same locus on each chromosome of a homologous pair. This arrangement gives rise to either homozygous or heterozygous conditions according to the nature of allele. As aforesaid, the homozygous individual shows two identical forms of alleles at a particular genetic locus, on each chromosome of a homologous pair. But heterozygous individual always possesses alleles of alternate character at the locus of two homologous chromosomes.

* A genetic locus is a place on the chromosome occupied by a gene.

In dealing with sex-linked inheritance, we are mainly concerned with those genes, which are carried by sex chromosomes or *allosomes*. It may be assumed that each of the sex chromosomes (**X** and **Y** chromosome) consists of two segments where one of the segments remain homologous between the two chromosomes - **X** and **Y**.

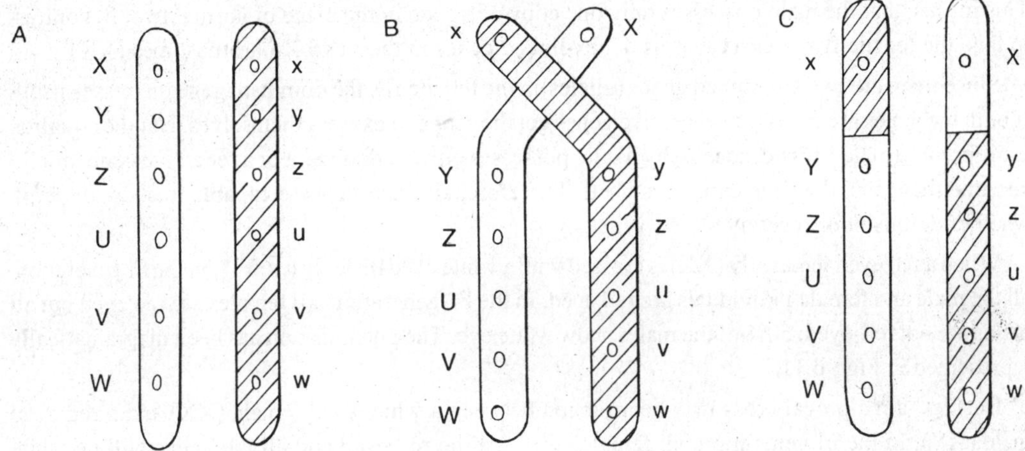

Fig. 6.10. Schematic Representation of Homologous Chromosomes (A) Crossing over (B) Recombination (C)

The genetic materials on homologous segments though have a fair chance to be exchanged during the cell-division, but genes located on this part of **X** and **Y** chromosome seem to have a very little significance and so they are called partially sex-linked. Such traits are sometimes transmitted as simply as autosomal traits, either through **X** chromosome or **Y** chromosome. On the other hand, the non-homologous segments of **X** and **Y** chromosomes are of great significance because true sex-linked genes are located on them. Completely **Y**-linked genes, are called holandric and this type inheritance is extremely rare. In some cases they are considered doubtful also. Rather, the genes on the non-homologous segment of **X** chromosome are the most common types of sex-linked gene and the term sex-linked is usually referred as the **X**-linked traits. Since these genes are mainly carried by the **X**-chromosome, they are commonly called sex-linked genes and the peculiar pattern of inheritance followed by them is termed as sex-linked inheritance. Therefore, it is a particular allele, that being posited in one of the sex determining chromosomes (usually the **X**-chromosome) determine the sex-linked traits.

Sex Linkage in Drosophila

In 1910, when T.H.Morgan was working on the genetics of the fruit fly, Drosophila melanogaster, in his laboratory at Columbia University, he came across a male fly having white eye colour. In Drosophila, the normal eye colour is red; the white colour arose due to mutation. The red eye colour is therefore the wild type and is dominant over its allele for white eye colour. Morgan searched for the pattern of inheritance and discovered the presence of sex-linked gene.

There are four pairs of chromosomes (2n = 8) in Drosophila. Of these, a pair of heteromorphic chromosomes is concerned with sex determination. The male is heterogamatic (XY) and the female is homogametic (XX).

The X-chromosome contains large number of genes and mainly consists of euchromatin. The Y-chromosome is largely inert and has no alleles of any genes, as such, it is composed of heterochromatin. Therefore, in the male fly (XY), the phenotype is determined by the genes present in the X-chromosome. For the presence of a single X-chromosome, male fly can be termed as hemizygous. This means, that the male possesses only one copy of a gene in the place of normal two. In contrast to this, the female fly carries two sets of sex-linked genes in their two X-chromosomes (XX).

Since there are two sex-linked genes (alleles)in the female fly, the dominant gene expresses itself. If both the genes are recessive, recessive genes get the scope to express themselves. But the situation is different in male flies because he by nature possesses either a dominant or a recessive gene (allele) and in either case, the gene expresses itself. Therefore, all male flies are capable to show the trait, whether dominant or recessive.

When a red eyed female fly (XX) is crossed with a white eyed male fly ($\overline{X}$ Y)*, in the Fl generation, all the male and female individuals are red eyed. In the F2 generation, all females are red eyed but all males are not red-eyed. 50% of the males show white eye. The phenomenon has been diagramatically represented in Fig : 6.11.

Besides, a reciprocal cross may be arranged between a white eyed female (XX) and a red eyed male (XY). In the Fl generation, all females flies will be red eyed and all the males will be white eyed. Now if the flies of the Fl generation are intercrossed, in the F2 generation 50% of the female individuals will come out as red eyed. Similarly 50% of the male flies will show red eyes. This means, 50% of both males and females of this generation will be white eyed. (Fig : 6.12)

The inheritance of red-white eye colour pattern in Drosophila thus reveals the peculiar pattern of transmission of sex-linked genes from generation to generation. It is very interesting in the sense that these genes are located in X-chromosome instead of autosomes. Further, the eye colour has nothing to do with the expression of sex; the only criterion of sex linkage is that the gene for eye colour is carried by the X-chromosome.

The characteristic feature of X-linked (sex-linked) inheritance is that, here no transmission of traits (characters) is possible from male to male i.e. father to son. Because, X-chromosome of the male (father) is not transmitted to his sons. rather it passes to daughters. But. the same father can transmit his sex-linked traits to grandsons, through his daughters. Therefore, the phenomenon has been referred as *criss-cross inheritance* in genetics. This criss-cross pattern of inheritance is possible only when the mother is a homozygous recessive $(\overline{XX})$ and the father bears the dominant (XY) gene.

The geneticists usually perform their experiments on standardized organisms, which can be reared under controlled environment. Much of the modern theory of genetics is based on studies of the small fly, Drosophila. Because this animal can be bred in the laboratory in large numbers with relatively little trouble and expense. A sexual generation from egg to egg, may be completed in ten days; a single pair produces hundreds of offsprings. Moreover, the species can be collected from anywhere of the world. The temperate or tropical climate do not impose much variation. Again, as the species exhibit a small number of chromosomes, it is quite easy to observe them.

Sex Linkage in Man

The genes present in the X-chromosome of man follow the same pattern of inheritance as in the

* The line above x i.e $\overline{X}$ indicates the sex linked recessive gene for white eye colour.

sex-linked traits of Drosophila. But the linkage study in man is not at all easy because, in case of man, such study demands some known pedigree to note the inheritance of characters, down through the generations.

A close examination of X and Y chromosomes in man reveals that the majority of the genes for X linked traits are found on the portion of X-chromosome, for which there is no corresponding homologous region in the Y chromosome. The Y chromosome is comparatively small in size and contains very few genes. Moreover, just like Drosophila, man is hemizygous. He has only one X-chromosome. Since there is no allele in the Y chromosome corresponding to the genes contained in the X-chromosome, a single sex-linked recessive gene carried by the male (through the X chromosome) expresses its effect in the phenotype. However, the characteristic features of inheritance for a sex-linked trait can be summarized as follows:

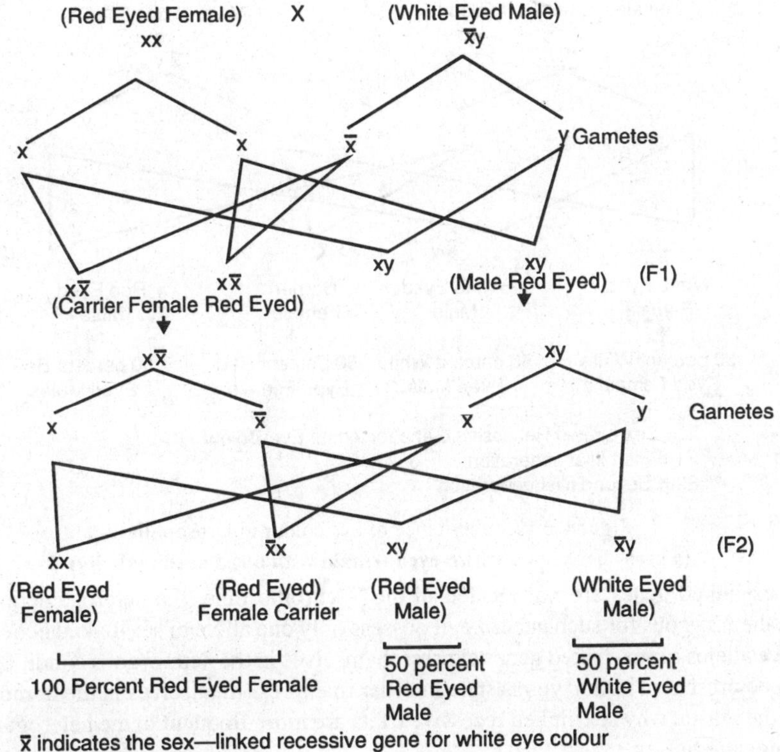

Fig. 6.11. Inheritance of eye colour in Drosophila

(a cross between red-eyed female with white-eyed male fly)

1. Ordinarily, genes governing sex-linked traits are not transmitted from male parents directly to their male progeny. Because individuals receive their X chromosome from their mothers, while their male parents contribute the Y-chromosome.
2. A male transmits his sex-linked genes to all his daughters, since females receive one of their two X-chromosomes from father. These daughters, in turn, transmit this gene to half of their male progeny. Thus, a sex-linked recessive gene is transmitted from a male to its female progeny, and thereafter to half of the male progeny of such females. All sex-linked genes therefore pass from

male to female and then come back to a male of F2 generation (grand-children generation). This *criss-cross pattern of inheritance* is same as found in Drosophila.

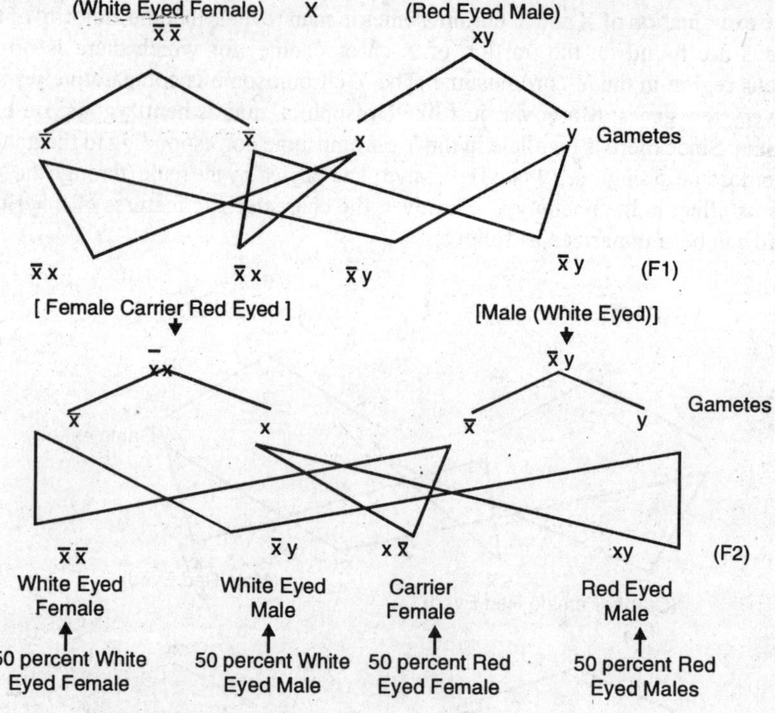

X̄ = Sex-Linked Recessive Gene for White Eye-Colour
F1 = Fisrt filial generation
F2 = Second filial generation

Figure. 6.12. Inheritance of eye colour in Drosophila
(a cross between a white-eyed female with a red eyed male fly)

3. Since sex-linked genes are not located in the Y chromosome, the heterogametic sex (human male) is hemizygous for such genes i.e. it possess only one allele of sex-linked genes. Therefore, recessive alleles of sex-linked genes express themselves in the hemizygous condition, while they have to be present in homozygous state in order to express themselves in the homogametic sex. This is the reason why sex-linked recessive traits are more frequent in the heterogametic than in the homogametic sex.

The cause of sex-linkage can therefore be shown with two reasons:

i) the location of a gene in X chromosome

ii) the absence of its allele in the Y chromosome.

For example, in human, the long arm of Y chromosome is homologous to the short arm of X chromosome. Genes located in that region of X chromosome does not show sex linked inheritance. Rather such genes undergo normal autosomal pattern of inheritance, if their allele is located in Y chromosome. A few genes have been discovered in the non-homologous region of the Y-chromosome, which have no allele in X chromosome. Such genes follow an entirely different pattern of inheritance. For example, the incidence of hypertrichosis (long hair growth in the ears) is found due to Y-linked genes.

In Drosophila, about 150 genes are sex-linked. In human, over 200 genes exhibit sex linkage;

most of these cause genetic diseases. Some well known examples of sex-linked human traits are, hemophilia (inability of blood to clot on exposure to air), colour blindness (inability to perceive one or the other colour, e.g. green colour), optic atrophy (degeneration of the optic nerve), juvenile glaucoma (hardening of eye -ball), juvenile glaucoma (hardening of eye-ball), myopia (near sightedness), defective iris, epidermal cysts, distichiasis (double eyelashes), white occipital lock of hair, mitral stenosis (abnormality of mitral valve in the heart), cystic fibrosis etc. The situations can be explained in the following way: If a male with any of the above mentioned defects marry a normal homozygous female, all their sons will be normal. But their daughters will be the carriers (heterozygous) for the particular disease or defect, although they look almost normal. This condition is produced due to a sex-linked recessive gene. When the daughters (carriers) get married to normal males, 50% of their sons will be perfectly normal while remaining 50% will get the defect. A few concrete instances will help in understanding the situation.

Red-Green Colour-blindness in Man

Perhaps the best known example of sex-linkage in man is the colour blindness. Several types of colour-blindness are known, but the most common type is red-green blindness, which is an X-linked recessive trait. Persons suffering from this colour blindness can not differentiate between the red colour and the green colour. Individually, the red colour blindness is referred *as protanopia* whereas the green colour blindness is known as *deuteranopia.* The genotype for normal vision may be symbolized by (XX), and colour blindness by $(X\overline{X}')$. $\overline{X}$ indicates the sex-linked recessive gene for colour blindness. If a colour blind man $(\overline{X}Y)$ marries a normal woman (XX), in the Fl generation all male progeny (sons) will be normal (XY). The female progeny (daughters) though will show normal phenotype, but genetically they will be heterozygous $(X\overline{X})$. Since these daughters bear the recessive gene of colour blindness, they are the carriers of the trait.

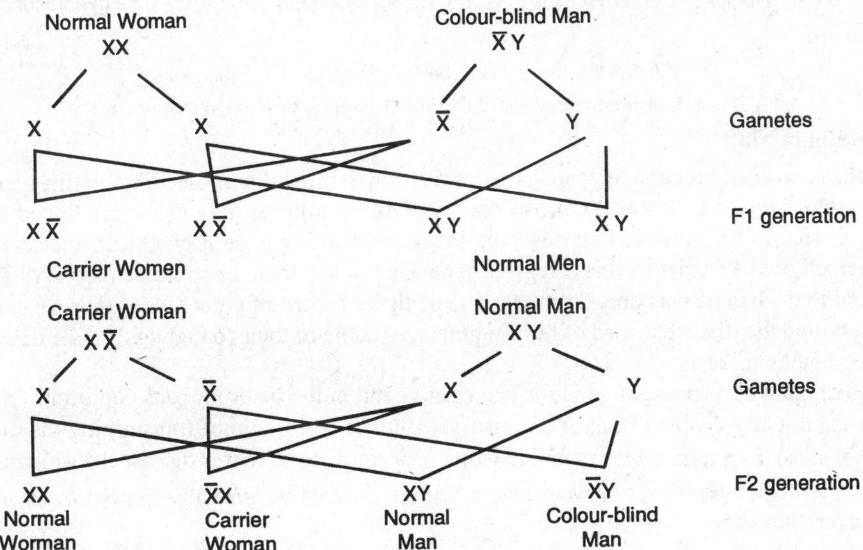

Fig. 6.13. Inheritance Pattern of Colour-Blindness In Human (Situation A)

If such a carrier woman with normal vision (heterozygous for colour blindness) marries a normal man (XY), the following progeny may be expected in the F2 generation: Among the daughters, 50% are normal and 50% are carriers for the diseases; among sons, 50% are colour blind and 50% are with normal vision.

Again, if a carrier woman ($X\overline{X}'$) marries a colour blind man ($\overline{X}Y$), the following progeny will be expected in the Fl generation. Among daughters, 50% are colour blind and 50% are the carriers of this disease; and among the sons, 50% are colour blind and another 50% are with normal vision.

If the colour blind woman ($\overline{XX}$) marries a normal man (XY), the resulting male progeny (sons) in F2 generation will be all colour blind, and the female progeny (daughters) will be all carriers of colourblindness. Thus, the occurrence of colour blind man will be more frequent than the colour-blind woman.

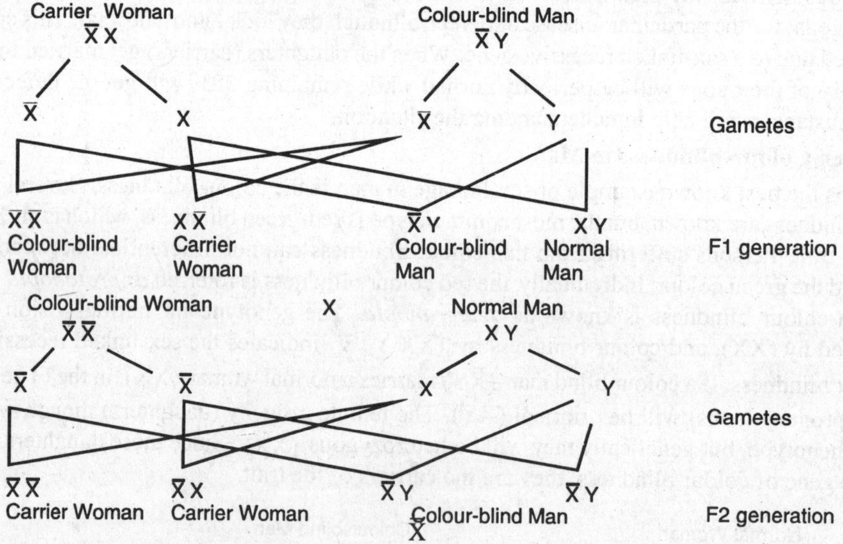

[$\overline{X}$ indicates the Sex-linked recessive gene for colour-blindness]

Fig. 6.14. Inheritance Pattern of Colour-Blindness In Human (Situation B)

Hemophilia in Man

Another sex-linked recessive gene in man is that for Hemophilia or bleeding. In this type, blood fails to clot easily on exposure to air, so that even minor injuries may cause prolonged bleeding leading to death. The trait was first observed in a European Royal Family; Queen Victoria (1819 - 1901) herself was a carrier of this recessive gene. One of her sons, Leopold, Duke of Albany, died of hemophilia, when he was only 31 years. The pedigree record of Queen Victoria reveals that she has transmitted the disease to two of her daughters, as some of their (daughters') male descendants were also hemophiliacs.

A woman may be carrying a gene for hemophilia and in the heterozygous condition ($X\overline{X}$), she does not exhibit any visible effect of the disease. But she is capable of transmitting the disease to 50% of her sons. Hemophilia is very rare in women, because, to have this disease the woman mus be nomozygous ($\overline{XX}$) with the recessive genes. A hemophilic woman normally dies before adolescence due to severe bleeding.

Among the children of a hemophilic man, all the daughters are carriers of hemophilia but none of his sons are affected by the disease (provided the mother of the children is neither hemophilic nor a heterozygous carrier). This is one of the characteristic features of X-linked inheritance, that the father gives X-chromosome to his daughters, but the sons receive no X-chromosome. Hence, there is no scope of transmission ofX-linked traits from father to son, but the grandsons often receive these traits through daughters. Therefore, this trait also follows a criss-cross pattern of inheritance.

If a hemophilia carrier woman marries a normal man, 50% of her daughters will be normal; other

50% will be the carriers. Among the sons, 50% will be normal and other 50% will be hemophiliacs.

Usual frequency of hemophilic male birth is about 1 in 10,000 while hemophilic female birth occurs once in 100,000,000 births.

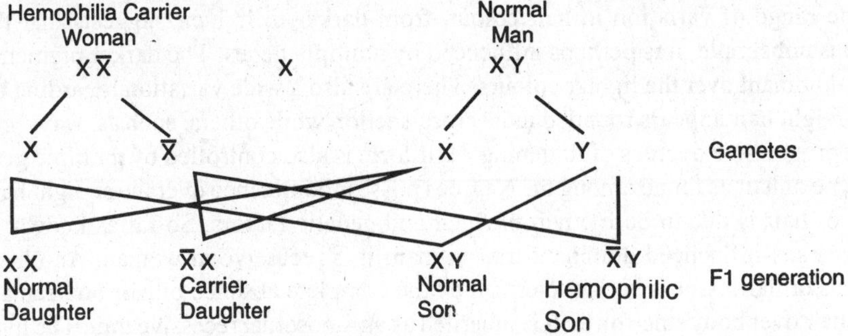

Fig. 6.15. Inheritance of Hemophilia in Man

POPULATION GENETICS

Genetically man is not a standardized organism as it controls and regulates his own environment. The culture and civilization which surrounds him, e.g., home, school, nutrition and several other social factors have influence on the growth and development of a child. So it is very difficult for experimental genetics to use man as the material for genetic studies. Rather, the experimental results obtained through animal and plant sources find universal application in human population. At present all chromosomes in human have been identified and every genetic behaviours have been predicted. Therefore, different inherited characteristics of a population and their mode of inheritance has been disclosed. Mckusick (1975) in his book, 'Mendelian Inheritance in man' had catalogued plenty of inherited traits.

There are also other types of inheritance related to the presence of multiple genes or multiple alleles. In Mendelian pattern of inheritance, clear-cut qualitative differences are produced in genetic types where no blending or gradation of characteristics is found between the parental types. But the inheritance of the characteristics like intelligence, human stature, skin colour etc. are not sharply defined; they show a continuous range of variation in a population, from one extremity to other, being known as continuous traits. As these characteristics mainly differ quantitatively, so it is relatively difficult to study their inheritance in comparison to the qualitative characters. Such characteristics are usually determined by a large number of genes, which do not show clear dominance but are cumulative in their effect in a trait. Members of two or more non-allelic gene pairs producing more or less equal and cumulative effect on a single character or trait is called multiple factors or multiple genes or polygenes. These multiple genes (polygenes) differ from multiple alleles in that the former (i.e. the multiple genes) are non-allelic so that they remain located in different loci and exhibit more or less equal and cumulative effects on a single trait.

According to Mendelian principle, a given locus must be occupied by only two alternative forms of genes called alleles. Without mutation, there can not be more than two alternative forms of genes or alleles in a locus. Recent technical advancements confirm that two or even more kinds of mutation may occur at the same locus, at different times. So, a given locus sometimes may be occupied by more than two kinds of genes or alleles. This presence of more than two kinds of alleles in one locus is called multiple alleles. The dominant relations among multiple alleles differ from one group to another. Most of the alleles express a particular trait in graded series; in some cases entirely different phenotypes are produced. For example, multiple series of coat colour as found in mammals, such as, rabbits, mice, rats, Guinea pig, cats etc. is responsible for multiple alleles. However, some common characters associated with human heredity have been described below which vary from population to population.

Hair

The wide range of variation in hair colour, from darker to lighter, suggests that the mode of inheritance is not simple; it is perhaps influenced by multiple genes. The darker pigmentation types seem to be dominant over the lighter colours. There are also a wide variation regarding the forms of hair. The straight hair appears rounded in its cross-section while others, such as, wavy, curly, frizzly etc. exhibit progressive degrees of flattening. Hair form is also controlled by multiple genes. Frizzly or woolly type of hair as found among the Negroes possess dominance over the straight hair. Baldness or absence of hair is due to both environmental and genetic factors. So far as known, baldness is inherited as a sex-influenced dominant trait in man; it is recessive in woman. Another condition, hypotrichosis or hairlessness is characterized by the complete absence of hair on head and sparsely distributed hair over body since birth. It is inherited as an autosomal recessive trait. The hypertrichosis or excessive presence of hair on body is due to a dominant autosomal gene. Hypertrichosis of the ears is due to a gene present in Y-chromosome.

Eyes

The eye colour follows simple Mendelian inheritance pattern. In general, the *darker pigmentation* of the eyes is dominant over the *lighter pigmentation.* Similarly, the *brown eyes* are dominant over the *blue eyes.* But the wide variations in eye colour have been created by the influence *of modifying genes.*

Myopia or *near-sightedness* (or *short-sightedness) is* a common hereditary defect. It happens due to defect in the curvature of the eye-ball. As a result, the distant objects are focussed in front of the *retina,* which produce only a blurred image upon the *retina.* This condition is inherited as an *autosomal recessive.* Myopia may also appear due to excessive curvature of cornea, being influenced by a *dominant gene.*

The *hyperopia* or *farsightedness* (or *long-sightedness) is* also a hereditary defect. It is an opposite condition where the eye-ball is very small for the curvature of the lens. As a result, the object is focussed behind the retina and the person is not capable of seeing nearby objects clearly. This phenomenon is inherited as a dominant trait.

Astigmatism is another kind of hereditary defect resulting from the unequal curvature of the *cornea.* But the *blindness* may be caused by both hereditary and environmental factors. *Glaucoma* and *Cataract* are two important hereditary defects, which are inherited as *dominant traits.* Besides, *optic atrophy,* (degeneration of optic nerves that leads to complete blindness) is also carried by *sex-linked recessive genes.*

Red-green colour blindness, as stated before, is clearly an action of a *sex-linked recessive gene.* Another defect in the vision is *congenital night blindness,* which is also the resultant of a *sex-linked recessive gene.* The person with this defect, through able to see very well under bright light but can not see clearly in the dimlight. In Some cases night blindness is caused for the deficiency of vitamin-A and those can be corrected by compensating the vitamin-A in the diet.

Ears

The size and shape of the ears exhibit a wide range of variation. Usually, *the free ear lobes* appear to be dominant over *attached ear lobes.*

Deafness is a serious defect, which may happen due to several external factors. But deafness at birth is quite often associated with *muteness.* This *deaf-mute condition* is created due to two *recessive genes.* Another form of deafness, the *otosclerosis* appears as a result of a *dominant gene.* In this case, an abnormal growth of bone is found surrounding the middle ear cavity for which the sound waves move to the ear drum. This condition develops after maturity, at about 30 years of age.

Another kind of deafness begins to manifest after 40 years of age, which is called the *atrophy of the auditory nerve.* An autosomal dominant gene is responsible for this.

Mouth

Harelip (i.e. a cleft in the upper lip extending up toward the nose) is one of the common abnormalities of the lip. This abnormality derives its name from the mouth of the hare. A *recessive gene is* responsible for this trait and this gene has greater penetrance in the male.

The ability to roll the tongue into a U-shape appearance (i.e. the person is tongue roller), is the attribute of a typical *dominant gene.*

The ability to taste a particular chemical, phenylthiocarbamide (PTC), is also due to an autosomal dominant gene. The person who can feel the taste of the chemical is called *tasters.* Inability to taste PTC is inherited as a recessive trait and such persons are designated as *non-tasters.*

Dimples on the cheeks are also inherited as autosomal dominant with differential exparessivity.

Malformation of the Limbs

Polydactyly (or the presence of extra fingers and toes) is inherited as a *dominant trait. Brachydactyly* (or shortening of fingers) also signifies a simple dominant inheritance in man. It results in shortening as well as fusing of the terminal bones of the fingers.

Syndactyly (fusion of two or more digits of the hands or feet) is also a genetic defect. It appears to be inherited as a *sex-linked autosomal recessive trait.*

Arachnodactyly and *brachyphalangy (* extremely long fingers and toes and extremely shortening of fingers, respectively) are found due to autosomal dominant genes.

Long index finger is produced as an influence of sex-linked dominant gene in man and recessive gene in woman. It is very interesting to note that the *functional dominance* of the right hand over the left is due to *a dominant gene* and the left hand over the right is due to a *recessive gene.*

Skin

The skin colour is a *polygenic trait,* which may also be influenced by the environmental factors, such as the sun-light.

Albinism or extreme fair skin is due to the action of a recessive gene.

Another condition *is piebald,* which produces large white spots on the skin of Negroid people appears due to a *dominant gene.* Again, a common type *of skin spotting,* known *as freckles* is inherited as *dominant* trait in certain families.

Usually, the iris of the eyes, the hair and the skin originate from the same embryonic layer. It means, the genes affecting one of these body parts manifests in all three parts. In this way, the blue eyes, blond hair and fair skin go hand in hand, which characterize the *Caucasoid* or white race, while the dark eye, dark hair and dark skin characterize the *Negroid* or black race.

Another condition is *Xeroderma pigmentation.* This condition is characterized by the development of *severe rash* on the skin on exposure to normal sunlight. A *recessive gene* is responsible for this but the child carrying this recessive trait usually dies before reaching maturity.

Another serious defect is the *absence of sweat glands.* This defect results in serious affliction in warm weather unless the body is kept cool either within the bathtub or a swimming pool by immersion. In this connection *Anidrotic ectodermal dysplasia* is one form of this disease which appears to be inherited either as a *sex-linked recessive* or *dominant* trait. Since the mammary glands are the modified sweat glands, then the woman with this disease show no nipples in the mammary gland.

The Skeleton

The stature of the individual is a polygenic trait. But a defect in the single locus may make a person dwarf, even if he inherits many genes for tall stature. Chondrodystrophic dwarfism means a great reduction in the size of the limbs due to the influence of a dominant gene. Persons with this defect possess normal sized head and trunk. Osteo-chondrodystrophy is another kind of dwarfism which results in an irregular development of bones in head and trunk. Sometimes the trunk becomes abnormally shortened, even when the limbs are of normal size. This trait is inherited as sex linked recessive. Osteopsathyrosis (presence of fragile bones) is caused by a typical dominant gene. Sometimes this defect remain associated with otosclerosis (a defect in the ear bone).

Ricket is mainly a disease due to vitamin-D deficiency, but susceptibility to rickets is inherited due to a special dominant gene. Similar to this is Arthritis. Though the disease is mainly concerned with the development of soreness and stiffness of the joints between the bones, some environmental and hereditary factors are associated with it. Susceptibility to Arthritis is inherited as a dominant trait.

Muscles

Pseudohypertropic muscular dystrophy (gradual degeneration of the muscles from childhood leads to death in early teens) is inherited as a sex-linked recessive trait. The incidence is not found among the girls as the boys normally die before reproduction.

Inguinal hernia is characterized by the descending down of the intestine through the opening of the abdominal muscle. The condition is inherited as a dominant trait. Incidence is rare in female sex.

The Nervous system

Intelligence is though very complex as well as variable factor in mankind, the general mental ability of the individual can be measured in terms of Intelligence Quotient (IQ), a test originated by Francis Gallon. The IQ of a person is determined on the basis of his / her performance on standardized intelligence tests suitable to his/her age. Tay-Sachs disease (degeneration of nerves results in blindness. Intellectual capabilities are also lost) is caused by a recessive gene. Heterozygotes behave quite normal. The incidence of this disease is fairly high among the Jews in Central Europe.

Mental Disorder

Amaurotic idiocy is a serious mental defect resulting from autosomal recessive gene. The affected children exhibit a decline in mental ability. Loss of vision, progressive muscular weakness etc *are* the other complications. It leads to death within the age of puberty.

Huntington's chorea affects the voluntary muscles and creates mental retardation at the age of 30. This abnormality is inherited as a simple autosomal dominant. *Schizopherenia is* a common mental problem where the individuals show a tendency to retire from the world of reality. This unusual behaviour is environmentally induced as well as genetic. *Epilepsy* is characterized by unconsciousness and muscular spasms. Epileptic persons are usually subnormal in their mental ability and this defect is inherited as a dominant trait, but the involvement of modifying genes or environmental factor appear to be essential for the expression of epilepsy.

Blood

Anemia is one of the common kinds of defect in the blood. Persons suffering from anemia do not possess adequate amount of haemoglobin in their blood to maintain cellular metabolism. A recessive gene causes low intake of vitamin B12, which indulges the incidence of anemia. Sickle-cell anemia is caused due to the presence of an abnormal haemoglobin in the blood and the affected person suffers from fever and muscular pain. The sickle-cell anemia results from an autosomal recessive gene. The person homozygous for an allele 'S' manifests the disease. The affected persons usually die during childhood.

Thalassemia is also produced due to an abnormal haemoglobin, quite different from'S' allele for sickling haemoglobin, which appears in infancy or childhood. Ovalocytosis i.e. the presence of oval-shaped red blood cells (RBC) in high frequency, is inherited as a dominant trait with varying degree of expressivity. Hemorrhagic nephritis results in the rupture of the capillary walls of the kidneys, causing blood to filter into the urine, appears to be due to a dominant gene. Vericose veins is a condition that appears among the elderly persons. It is characterized by the bulging out of the veins of legs under the skin due to loss of elasticity of the walls of the veins. Heavy weight lifting and standing for long times tend to worsen the condition. This defect is also inherited as an action of ,a dominant gene with limited penetrance.

Hypertension or high blood pressure is influenced by environment as well as inherited as a dominant trait. The symptoms of hypertension are manifested beyond the middle age and this serious abnormality usually invites kidney trouble, heart trouble, apoplectic stroke, etc

Disease

The term disease generally encompasses every kind of human abnormality. Usually non-infectious diseases are created in relation to heredity. Diabetes melltus is one such disease, which results from an endocrine imbalance. A typical recessive gene is responsible for this disease. Affected persons produce inadequate amount of insulin in their pancreas and sugar is not completely metabolized. Excess sugar accumulated in the system comes out in urine. Diabetes insipidus is another disease where excess urine is produced accompanied by abnormal thirst. This disease is due to a dominant gene.

Gout is a condition resulting from defective purine metabolism leading to the production of excessive uric acids. This condition is mostly inherited as a simple autosomal dominant. Besides, several diseases such as asthma, migrane head-ache, hives, eczema, hay fever, colities etc have allergic basis. The tendency to become sensitized is heritable which is inherited in the homozygous condition in early life and heterozygous state after puberty.

Infectious diseases are though manifested as a consequence of invasion of the germs from outside, susceptibility to specific infectious diseases appear to be heritable. For instance, susceptibility to the infection of tuberculosis is believed to be due to a recessive gene. Similarly, different other recessive genes are responsible for the susceptibility to poliomyelitis, diphtheria and scarlet fever etc. Apart from these, defects of chromosomes may create various syndromes as discussed earlier.

Eugenics

Eugenics is the applied aspect of Human genetics. It aims at the improvement of humanity by altering the hereditary qualities in future generations of man. Either physically or mentally, it encourages breeding of those persons presumed to have desirable genes (positive eugenics) and discourages breeding of those others presumed to have undesirable genes (negative eugenics). *Genetic counselling* and *genetic engineering* are the two important aspects of eugenics. Genetic counselling means the act of educating the prospective parents who are either suffering from the diseases or suspected to be heterozygous for some specific genetic disease. It is relatively easy task to identify people suffering from a genetic disease. But the carrier genotypes are very difficult to be known. Only awareness generation programme and counselling may energize the people to go for a genetic checkup before solemnizing a marriage. This is for the good of their future progeny. Once genotypes of two prospective parents is known, it is not at all difficult to work out the frequency or chance of inheriting the diseases among their children. Therefore, genetic counselling and antenatal diagnosis are the only ways to get rid of those unwanted situation. Successful counselling can definitely bring a relief to the possible parents and thus the frequency of genetically defective individuals in the population can be reduced.

Genetic engineering means the manipulation of genetic system within the cell in order to alter the genetic mechanism for the benefit of mankind. The process helps to mend the genetic disorders to make a unit functional. It aims to control the physical, biological and social environment of the individuals especially for the improvement of the phenotype of genetically defective population. The concept of 'genetic engineering' is a recent development towards the scientific creativity.

Although the idea of genetic engineering originated through plant breeding and improvement of crops, it is no more limited to that chore. Its prospect is flourishing day by day. Now it has been so developed that it is going to challenge the law of nature. Since positive eugenics intends at enlarging the proportion of children with most desirable hereditary characteristics, it suggested many new approaches. It, appreciates the plan of creating the sperm banks for preserving the sperms of outstanding men who could serve as fathers of many more children hoping desirable hereditary characteristics. Now, thousands of cases of artificial insemination per year are being conducted to permit women to bear children whose husbands are sterile. The semen of distinguished males could thus be stored in a deep freeze, which would remain functional over 100 years. Similarly it is possible to preserve eggs of females of proved genetic desirability for the future use. The technique of test-tube baby has already acquired enough popularity; extrauterine fertilization has been possible by it. The fertilized eggs or zygote is implanted into a foster mother for further development. This procedure has helped many sterile women to bear children. To control the sex of the offspring, the process amniocentesis * has come into existence. The foetal cell obtained from amniocentesis when subjected to cytological examination, it reveals the sex of the foetus as well as the chromosomal abnormalities, if any. This pregnancy may be continued or terminated according to the situation. Repairing of the defective gene or improving the performance of presently performing genes - both are possible by genetic engineering. Cloning i.e. the exact reproduction of an individual from cellular tissue is also possible by genetic engineering.

As the negative eugenics is concerned with the elimination of undesirable genes from the population (through discouraging breeding among the defectives) normal members of the society get the chance to produce more normal children. Thus, by reducing the progeny of defective individuals, deterioration of human race can be prevented. For example, inmates of mental institutions are never permitted to marry and bear children in most of the civilized countries. This prevents the transmission of defective genes to the future generation, if the mental defects are due to heredity. Consanguineous marriages are also genetically undesirable, as the frequency of the production of defective children is likely to be higher for such marriages. This is because of the simple reason that most of the harmful traits are recessive, so they tend to express in the children of parents who are related very closely. Therefore, most of the societies put some kind of moral and social restrictions on marriages between close relatives. In fact the transmission of serious hereditary defects to the children is threatened in all societies.

Concept of Gene-Pool

All of us know that a *species* consists of a number of individuals and all individuals of the same species constitute a *population.* The individuals of a population may be distributed continuously or may exhibit discontinuous distribution due to some kind of geographical factors. When the population is continuously distributed, quite often, there occur some small groups of interbreeding individuals. Members of each small group inbreed freely occupying a particular territory or area. Mating between individuals located at a considerable distance are expected to be infrequent. However, by virtue of random mating, individuals of each small group are found to share the same *gene-pool.*

Dobzhansky (1951) has referred the total genetic information (encoded in the sum total of genes) in a small inbreeding population as gene pool. In simple terms, a gene-pool consists of all the genes from all individuals of that small group i.e. it is the sum total of all alleles present among the members of a randomly mating population.

The gametes of the all individuals in a small group furnish the pool of genes, which govern on the genes of next generation. The absence of a real isolation between two such small groups has been described as migration whereby addition of some new genes are observed; both the gene-pools get altered. The gene-pool may also change due to the replacement of one generation by another. Any change in the gene-pool constitution affects directly to the respective population group and bring about certain modification of characters in the population.

BLOOD GROUP

The study of blood group has played an important role in physical anthropology, particularly in population genetics. Blood is a special type of tissue in human body, which can be easily obtained from individuals for various types of investigation. It performs basically two functions. Firstly, it

* Puncturing of the amniotic membrane to obtain amniotic fluid and foetal cell.

supplies nourishment to the body by carrying oxygen from lungs to the cellular elements of the tissues in other parts of the body. Secondly, it helps in removing the waste-products (especially carbon-di-oxide) which get accumulated in the body due to metabolic activities. Besides, this particular tissue has the ability to destroy the foreign invadors such as various harmful bacteria that enter from outside.

The blood consists of two types of materials - plasma and blood corpuscles. If we take a sample of blood in a test tube and keep it for a considerable period of time with a anticoagulant, we will find the faintly yellow *fluid, plasma* full of numerous suspended particles known as blood corpuscles. The red cells or red blood corpuscles (RBC) or *Erythrocytes* usually settle at bottom of the tube leaving white cells or white blood corpuscles (WBC) or *Leucocytes* above. But, on the top of the test tube, the maximum portion of faint yellow fluid forms with plasma layer.

The plasma is thus a coagulable fluid where the blood corpuscles remain suspended. It contains three principal proteins like *Albumin, Globulin* and *Fibrinogen.* The fibrinogen has got some

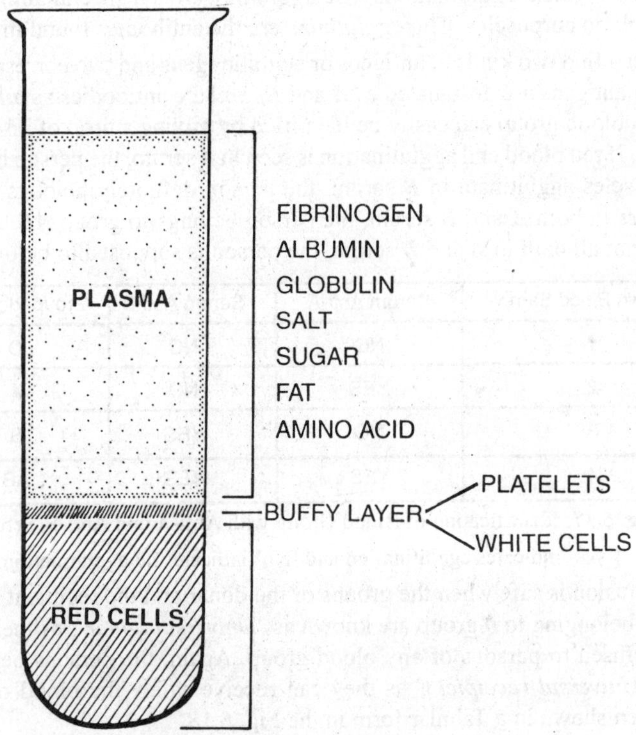

Fig. 6.16 The Components of Blood

special characters so that it causes clotting in whole blood. Besides, many other substances are found to be present in the plasma such as amino acids, sugar, salt, fat, etc. A clear fluid is obtained if coagulating and coagulable elements can be removed from the plasma. The clear fluid is called serum in which fibrin is no longer found. Therefore the equation stands as, *Plasma - Fibrin = Serum.* There are three kinds of blood corpuscles :

i) red blood corpuscles (RBC) or *Erythrocytes* are non-nucleated and look red due to the presence of haemoglobin;

ii) white blood corpuscles (WBC) or *Leucocytes* are nucleated but haemoglobin is not present. It helps in destroying outsiders which enter into the plasma, etc.; and

iii) blood platelets or *Thrombocytes* are nucleated and haemoglobin is absent. It helps to develop thrombin which is acted in various ways in connection with the formation of fibrin.

The A B O System

In 1900 - 1902 Karl Landsteiner divided the human beings into three groups in reference to their blood groups which later on increased into four as shown by Sturli and Decastello. However, the groups are denoted by the letter ***0, A, B,*** and ***AB***. These four blood groups establish the fact that the blood transfusion can react on the serum or plasma of recipient individual causing agglutinations or clumping together of the red blood cells. In fact, for the introduction of certain substances a special reacting substance is created which is called the *antibody* and the foreign substance that has been introduced in red blood cells is termed as *antigen.* When the antigen and the antibody of different groups come in close contact, a physio-chemical reaction is inevitable and that is termed as *agglutination.* For example, if the serum of the person having ***A*** blood group is injected to the person of **B** blood group, the blood cells of **B** blood group person will clump together. The phenomenon is called typical agglutination. The *agglutinogen* is a special antigen existing on the surface of the red blood corpuscles. The *agglutinins* are the antibodies found in the serum.

As a result, we can find two kinds of antigens or agglutinogens and two corresponding antibodies or agglutinins. The antigens are designated as ***A*** and ***B,*** *and* the antibodies as ***anti-A*** and ***anti-B.*** A person of unknown blood group can easily be identified by mixing a drop of his blood with known blood sera ***A*** and ***B***. If red blood cell agglutination is seen in ***A*** serum, the person belongs to group ***B.*** If red blood corpuscles agglutinate in ***B*** serum, the person definitely belongs to group ***A***. If the agglutination occurs in both ***A*** and ***B*** serum, the person belongs to group ***AB*** but if agglutination does not take place at all both in ***A*** and ***B*** serum, the person is supposed to belong under group **O**.

Unknown Blood Sample	Serum Anti-A	Serum Anti-B	Group Identified
1	NO	NO	O
2	YES	NO	A
3	NO	YES	B
4	YES	YES	AB

Fig. 6.17. Identification of Blood Group with Anti-A and Anti-B Serum
['yes' indicates agglutination and 'No' indicates no agglutination.]

The blood transfusion is safe when the groups of the donar and the recipient are known beforehand. The persons belonging to ***0*** group are known as ***'Universal donar'***, as the blood of ***0*** groups can safely be transfused to persons of any blood group. Again, the persons belonging to the ***AB*** group are called ***'Universal receipient'*** as they can receive safely the blood of any group. This relationship has been shown in a Tabular form in the Fig. 6.18.

BLOOD GROUP TYPES	ABLE TO DONATE BLOOD TO PERSONS HAVING THE FOLLOWING BLOOD-GROUPS	ABLE TO RECEIVE BLOOD FROM PERSONS HAVING THE FOLLOWING BLOOD-GROUPS
A	A, AB	A, O
B	B, AB	B, O
AB	AB	A, B, AB, O
O	A, B, O, AB	O

Fig. 6.18. Donars And Recepients (During Blood Transfusion).

The heredity of blood group is now fully known. The three genes in the chromosomes i.e. ***A,B,*** and ***0*** are responsible for the blood group variations in the persons. The allelic genes occupy the same locus on certain chromosomes and contribute in hereditary mechanism. It seems that ***A*** and ***B*** type possess equal expressive power while the ***0*** is recessive to both ***A*** and ***B***. It is also observed that the internal mixtures of the following genes produces genotypically six different combinations in blood groups, but due to recessiveness of ***0*** gene we find phenotypically four blood groups.

Genotypes and Phenotypes of Blood Groups.

Genotypes	*Phenotypes*
OO	O
AA, AO	A
BB, BO	B
AB	AB

During the First World War, it was discovered that the frequency of these classical blood groups varies among the peoples of different racial origin. Since then blood-group study was taken as an important tool by the physical anthropologists. At present, the relative frequencies of four main blood-groups have been universally accepted as an important criterion of race determination. According to Boyd (1950), the racial classification through blood groups has some extra advantage in the study of race which may be mentioned in the following ways :

1) They are inherited following Mendelian principles.
2) They do not change for the differences of climate, food, illness or medical treatment.
3) Their frequency in a population remains very stable.
4) Although they originated very early in the course of man's evolution, but retain their originality.
5) There is a considerable unity between geography and the blood groups distribution.
6) The 'all-or-none' characteristics of the blood group antigens are very useful in differentiating one individual from the other.

It has been found through experiment that antigen ***A*** consists of two distinct antigens - ***A1*** and ***A2***. This means antigen ***A*** *is* separable into antigen - ***A1*** and antigen - ***A2***. Antiserum ***Anti-A1*** reacts with both the antigens - ***A* 1** and ***A2***. Antiserum ***Anti-A2*** is very rare and it reacts only with antigen ***A1***. When the red blood cells are agglutinated by anti-**A1** and anti-**A2** antiserum, then the blood cells classified as ***Al*** and if no such reaction is observed then it is classified as ***A2***. These two blood types ***A1*** and ***A2*** are inherited as separate units. It is noted that ***Al*** is dominant over ***A2*** and also over **0**. ***A2*** is dominant over **0**. So it is clear that instead of three alleles (**ABO** blood types), four alleles have to be considered while dealing with ABO blood group system. It is due to the introduction *of* ***A* 1** and ***A2*** antigens. Now the four alleles stand as ***A* 1, *A2, B*** and **0**, and they take part in the determination of blood groups. As a result we find ten genotypes and six phenotypes as shown in the (Fig. 6.19).

GENOTYPE	PHENOTYPE
OO	O
A1 A1, A1 O	A1
A1 A2, A2 A2, A2 O	A2
B B, B O	B
A1 B	A1 B
A2 B	A2 B

Fig. 6.19. Genotypes and Phenotypes of ABO Blood-Group System with Antigen Al and A2.

Several researchers took interest in searching the relationship between ABO blood group phenotypes and the diseases. Though they could not find any direct correlation, but demonstrated a list of compatibility. The compatibility or incompatibility was decided on the basis of marriage match, according to phenotypes of the parents. The compatible matings were referred as Homospecific while the incompatible matings were called Heterospecific. Various combinations could be possible in both situations in the following way:

Compatible Mating		*Incompatible Mating*	
Male x	Female	Male x	Female
O	O	A	O
O	A	A	B
O	B	B	O
O	AB	B	A
A	A	AB	O
A	AB	AB	A
B	B	AB	A
B	AB		
AB	AB		

Compatible mating results in healthy child-birth whereas incompatible mating often gives birth of still-born babies or the babies die shortly after birth. For example, when a father is of blood-type ***A*** but mother is of blood-type ***0*** or ***B.*** The mother carries *anti-A* antibody in her serum. Now, if the baby gets the blood type ***A,*** all red cells of the baby will be affected by the antibodies, which the mother is carrying. This can be called **ABO** incompatibility. In this way, different other incompatibilities may be created which affect the foetus.

The MNS System

Thirty years after the discovery of **ABO** blood groups, in 1927, Landsteiner and Levine found another blood-group system. This group consisted of **M**, **N**, and **MN** blood groups, depending on two antigens - **M** and **N**. Since these two antigens possess no natural antibodies, they keep no effect in the transfusion of blood. But they are found to be present in the red cells of all human being. When the red cells of a person shows **M**-antigen, his blood-type is designated as **M**. Similarly the presence of **N**-antigen marks the blood type as **N** and when both **M** and **N** are present, the blood type is termed as **MN**. The **M** and **N** antigent are equally dominant; they do not possess dominant - recessive relation. However, this **M - N** factors exist simultaneously along with **A - B - AB - 0** factors, in the same blood but without having any relation to them.

In 1947, Sanger and Race found out another antigen known as **S**. Although this S-antigen differs serologically from **M** and **N** but shows a genetic relation with **M - N** types. As a matter of fact, it occurs especially among the individuals who possess **M**, **N**, or **MN** blood types. Unlike **M** and Nantigen, the **S**-antigen possess an antibody also. Not only that, serologists have been able to discover two antigents as S and s and therefore three blood-types, **SS**, **Ss** and ss can be demarkated. Because of the close affinity, **MN** and Ss systems are considered together. In the combined form they give rise to ten genotypes which are as follows :

ANTIGEN	ANTIBODIES	PHENOTYPE	GENOTYPE
M, N	Anti-M	MS	MSMS, MSMs, MsMS
	Anti-N	Ms	MsMs
S, s	Anti-S	MNS	MsNS, MSNs
	Anti-s	NS	NSNS, NSNs, NsNS
		Ns	NsNs

The Rh System

In 1940 Landsteiner and Weiner discovered the **Rh** factor. They demonstrated that, if the blood of a rhesus monkey is injected to the rabbit, a serum may be obtained. The particular serum from the rabbit agglutinates certain human blood. This new agglutinable factor of the blood has been referred as **Rh** factor. The **Rh** symbol is derived from the word rhesus. The persons who possess this factor are known as **Rh**-positive and the persons who lack it are known as **Rh**-negative. This **Rh** system is independent of all other blood types such as the **ABO** blood system and the MNSs blood systems. The greatest importance of Rh factor lies in connection with pregnancy. In case of **Rh**-negative mother, the **Rh**-positive baby in her womb make a reaction and this phenomenon ultimately affects the baby in the womb who may die before birth. In fact, when a **Rh**-negative mother carries a **Rh**-positive foetus due to her marriage with a **Rh**-positive male, the **Rh** antigen from the **Rh**-positive foetus passes to the placenta and finally to the blood of the mother. This causes the production of an antibody. As the **Rh**-negative mother contain the antigen, this antibody can not do any harm to the blood cells of the mother. But when this antibody passes through the placenta to the foetus who is a **Rh**-positive, immediate reaction takes place to destroy the foetal red cells. This disease is known as *erythroblastosis foetalis*. It is very serious, sometimes fatal for the newborn.

Previously it was thought that ***Rh*** factor represented by the pair ***Rh*** and ***rh*** was inherited according to Mendelian principle and ***Rh*** being dominant to ***rh*** created four genotypes - ***RhRh, Rhrh, rhRh (Rh-*** positive) and ***rhrh*** (***Rh***-negative). But, at present it is found that there are only three ***Rh*** factors like Rh_O, Rh^I, rh^{II}, of which Rh_0 being original is most powerful as well as clinically very significant. These three antigens, Rh_O, Rh^I, rh^{II}, possess theoretically three contrasting factors known as Hr_O, hr^I, hr^{II}, of which only hr^I and hr^{II} are found to exist. This ***Rh -Hr*** system has been worked out totally on the basis of assumption; it can not be demonstrated.

Although at the beginning, the ***Rh***-system was determined by a gene unit causing presence or ibsence of an agglutinogen, but after 1946, eight *types of* ***Rh*** blood were recognized basing on six or more allelic genes. The three antigens of ***Rh*** type determine eight agglutinogens. The eight allelic genes for these agglutinogens are designated as $R_0 R^I$, R^2, R^z, r, r^I, r^{II} **and** r^y. The genes R^Z and r^y are very rarely seen. However, these three elementary antigens and factors, either singly and combined, may give rise to eight phenotypes which are as follows :

Rh_O, rh^I, rh^{II}, $Rh_O rh^I$ (or Rh_1), $Rh_O rh^{II}$(or Rh_2), $rh^I rh^{II}$ ***(Rhy)***, Rh_I, R_2 ***(or Rhz) and rh.***

The symbols have been made further simple at present; ***h*** *is* omitted from all designations. Therefore, the new notations stand as, R_o, r^I, r^{II}, R_I, R_2, R_y, R_z ***and r.***

Again, according to Fisher, the ***Rh*** types have six antigens which are determined by a series of three pairs of alleles. He named these three pairs as ***C, c; D, d;*** and ***E, e.*** A single chromosome can carry either a **C** or **c** gene but can not both. **C** acts as an autosomal dominant to **c**, similarly **D** to **d** and **E** to **e**. In fact, there will be three possible phenotypes ***CC, Cc,*** and ***cc***. The other genes, ***D*** and ***E,*** along with their alleles also behave in the same fashion. Fisher's three-gene hypothesis advocates that these three genes are inherited in a group of three, located on a single chromosome. So here possible combinations are, ***CDE, Cde, cDe*** and so on.

The eight phenotypes according to their short symbols or notation may be projected in the following ways. Weiner's chart corresponds to Fisher's chart.

Weiner's notation	*Fisher's notation*
R_O	***cDe***
r^I	***Cde***
r^{II}	***cdE***
R_1	***CDe***
R_2	***cDE***

R_y	***CdE***
R_z	***CDE***
r	***cde***

Further, Fisher states the relationship between the members of each pair, on the basis of genetic *allelomorphism.* Although an offspring inherits three gene combination from each parent (say for example, cde and CDE), he is not able to transmit both to his child. Only one of them, either cde or CDE is received by his child.

The occurrence of Rh-positive and Rh-negative factors in different populations have some racial significance. The Rh-negative is very rare among the Mongoloids (0.5% to 1.5%). But its frequency is comparatively very high among the white people (about 15%). Again among the Negroes, the Rh-negative factors occurs only in 5 to 8% of the population.

It may be concluded that, in addition to these three commonly known human blood group systems, namely ABO, MNS and Rh, there are several other systems. In anthropological study Rh blood system offer much materials than ABO blood group system.

HUMAN RACE

RACE AND ITS FORMATION

Definition of Race

The race is a term, which has been used to denote a group of persons living in a territory for several generations. In many cases a culturally homogeneous people has been called a race. But an anthropologist looks at it in a different way. They study human race in a purely biological sense and therefore the study of race may be regarded as a branch of science itself. It is a biological phenomenon, which is to be defined in biological terms. The purpose of this study is to trace out the development of modern man and to make a classification. An analysis of biological and environmental factors has been included here in order to understand racial movement with other related problems.

It is now established that all living human beings belong to a single genus and species. *Homo sapiens,* which includes different populations or groups differing in physical features. Though these groups differ from each other in certain traits, they also exhibit a relative commonness of certain hereditary traits. Each of these groups or population constituting the species *Homo sapiens* may be regarded as a race. In fact, the race is a concept, which can be used in various senses.

Eminent anthropologists have put forward their concepts on human race. E.A. Hooton in 1946 explained the race as a "great division of mankind, the members of which, though individually varying are characterized as a group by certain combinations of morphological and metrical features, principally, non-adaptive, which have been derived from their common descent". He made a clear-cut line of distinction between the primary and the secondary races on the basis of their mode of formation. According to him *Primary races* were "differentiated by early geographical and genetic isolation by loss of some genes and fixation of others, by mutations, by inbreeding, and by selection" while *Secondary races* were formed by the "restabilization of blends of two or more primary races". But M.F. Ashley-Montagu in 1960 had provided the genetical clarification of a race as "a population which differs in the frequency of some gene or genes, which is actually exchanging or capable of exchanging genes across whatever boundaries separate it from other populations of the species". In a simple way it may be stated that the race, being the groups or populations, remain more or less reproductively isolated from one another and differ amongst themselves in the relative commonness of certain hereditary traits. These hereditary traits are the resultant of specific distribution of genes responsible for physical characters, which appear, fluctuate and often disappear in course of time by means of geographical and environmental isolation.

The concept of race is therefore nothing but a device of classification where different groups or populations are to be arranged systematically. But in any case, the national, religious, cultural and geographical groups should not be confused with racial groups. For example, the Indians, Germans, Pakistanis, etc. are the national groups, whereas Jews, Buddhists, Christians, etc. form the religious groups. In anthropological perspective the Dravidians or the Aryans are not the races but linguistic groups. In fact, the people under one race should possess a distinctive combination of physical traits and races are distinguished from each other by the relative difference of certain inherited characters, both phenotypically and genetically. Each racial group develops certain characteristics of physiological traits with which it differs from the others. The racial traits often change by mutation and it denotes dynamicity in race. Thus, each race is dynamic, instead of a static group. Stability of a race depends on durability of various genes responsible for various inherited characters. This stability may be achieved by the practice of marrying within the racial group. Any change in any one of the factors depicts a new change in the race as a whole.

Factors Behind the Formation of Racial Groups

We know that a group of organisms possess the same kind and number of chromosomes where each of the chromosomes carries the same number of genes in the same arrangements. Since all human beings more or less look alike and capable of interbreeding as they have been grouped in one species, *sapiens*. It is believed that the modern man, Homo Sapiens Sapiens have evolved from a generalized protohuman form, through successive and parallel species of primitive man. Now, it comprises several varieties or *races* of man who are inhabiting in various parts of the world and show considerable fundamental variation as regards their physical features, due to the variable frequencies of some of their genes. It is assumed that before the close of Palaeolithic Age, all primary divisions of mankind became specialized in their respective habitats by the influence of the surroundings as well as by the chance of variations in relatively isolated populations. Thus, different racial groups came into existence. In fact, the formation of race is a complex process where a number of factors are involved. However, the factors are as follows:

Gene Mutation

Physical characters of an individual may be altered due to sudden spontaneous changes in the structure of a gene and the phenomenon is known as *mutation.* The new mutant gene begins to multiply from generation to generation and develops a new distinctive characteristic in a particular population, provided other conditions are favourable. Such a change is permanent as well as hereditary though occurs seldom. The cause of this change (mutation) is not clearly known, but it is certainly an important process by which different forms of character are produced.

Mutation of a gene producing a particular trait may occur again and again. But unfavourable traits possess a little chance to enter into a permanent gene pool. For example, the hemophilia. For a mutative gene producing hemophilia, a person often meets early death before attaining the age of reproduction. Naturally such a disadvantageous gene does not get a fair chance to spread itself among the population. On the other hand, advantageous mutant genes survive among the population producing novel traits. Thus, mutation offers a high selective advantage as well as a quality of dominance to change the physical character of a population, with time. Increased selective advantage adds more survival values to a gene and as a result, the frequency of occurrence for that gene increases in the population.

Normally, among the human being, one mutation occurs in 40,000 instances. Application of X-ray and some chemicals can enhance the rate of mutation.

Natural Selection

The new bodily characters, which appear as a result of mutation, face a severe competition with old characters. This competition between old and new forms is known as 'Natural selection'. Actually, an advantageous gene multiplies more rapidly than that of a disadvantageous gene and the latter gets eliminated after some time. If we take the example of skin colour, we assume that the earliest men had the light skin colour. Later, the dark skin and the yellow skin have been added as a result of mutation. In the hot moist climatic region of equatorial Africa, the dark skin is advantageous because of its adaptability to that particular environment. So, such a mutant gene finds a great scope to be spread very quickly in tropical areas. But the same dark skin pigment is highly disadvantageous in the cold climate, as nature does not select it. This principle works in case of all traits. The 'Albinism' or the absolute loss of pigmentation in the skin offers no advantage, so has become rare in incidence. The mutant gene responsible for Albino type can not flourish itself in any kind of environment, though it is hereditary in character. However, it is better to say that this adaptability is wholly selected by nature. The genes, only, which possess a survival value in terms of an environment, readily establish themselves and are carried from one generation to other with distinct phenotypical manifestation. This theory of 'Natural Selection' was first propagated by Darwin.

Genetic Drift

Beside the two main biological processes. Gene mutation and Natural Selection,the other important process is the 'accidental' or 'chance fluctuations' of genes, commonly known as 'Genetic drift'. This third factor of race formation works in shaping a particular combination of genes. The gene frequency of a given trait sometimes increases or decreases suddenly without depending on advantageous or disadvantageous position of life; rather it happens by chance being guided by the environmental conditions. The process is more effective for a smaller isolated group (partially or totally isolated) than a larger one. For example, small groups bearing certain special traits often migrate to new territories and drop contact with the ancestral group. In course of time, such a group may split further and further and become isolated to give rise new populations, which are completely different from ancestral population in terms of gene frequency.

Thus, the original genetic composition is altered without an effect of mutation or natural selection. The American Indians were once the people of Asia who lost many traits of their Mongoloid ancestors.

Population Mixture

It is the fourth factor to form the new races. This factor is directly related with hybridization or mixing of different populations. Rise of a new race is followed by the interbreeding of racially distinct populations. The process is known as miscegenation. As a result of miscegenation the genes from different population do not blend together, but the separate gene pools of mixing populations are fused together in different proportions to form a new racial group. The existing racial groups exhibit a population mixture, because in many cases the original races migrated from one area to another and mixed with others. So most of the races have now converted to hybrid races and fail to retain their original traits.

Analysis of factors reveals that migration and isolation are the basic pre-conditions for race formation. Migration breaks the original group; people divided into smaller groups scatter in different directions and become isolated from one another. Here isolation means the separation of a group from all other groups of the same species. So breeding takes place completely or largely within the inhabitants of each isolated group. Isolation is usually brought about by the natural factors like distance, mountain ranges, rivers, forests, seas, deserts, etc. Under the conditions of isolation, gene mutation takes place. Each isolated group shows more or less distinguished characters, different from its mother as well as sister groups, by virtue of mutation. Isolation may be social or geographical. A group may remain isolated within a particular area for a considerable period of time, before moving on to another area. Such isolation is called as geographical isolation, which may even sustain for several Centuries. On the other hand, social isolation negates interbreeding, even when the two populations reside in close proximity. For this reason, where intermarriage between two groups is lesser in probability, the isolation becomes greater.

Usually migration facilitates interbreeding with the inhabitants of the area into which a group is moved. As a result, 'genetic drift' takes place. In subsequent generation, both the groups come closer to each other by borrowing biological traits and each of them try to compensate the deficiencies of the other. Therefore, hybridization has a powerful effect in moulding the populations. It does not merely bring modifications on physical traits and physiological characters, it serves to increase the adaptive qualities of the offspring. Human hybrids often show a greater resistance towards the diseases; an increased index of fertility and intelligence is carried. New gene pools are invariably created when two isolated groups interbreed. But the inbreeding groups are required to remain in isolation; otherwise genetic drift fails to produce distinctive physical characters.

Whatever is the situation, simple migration to a new environment or an admixture of different populations (hybridization), the new genes get multiplied and transmitted from generation to generation by the process of heredity. A new racial group may form with time if there is an absolute isolation. Isolation plays an important role in case of hybridization. As for example, the intermixture of Europeans and Indians has given rise a new race called Anglo-Indian. Further, sexual selection and social selection may be of two significant criteria in formation of new races. The sexual selection denotes the process of selecting marital partner on the basis of personal preference; in course of time the sexually preferred type becomes the dominant variety of population. Social selection is the method of regulating marriage between artificially divided socially approved groups within a population. The mechanism of isolation works in both the cases.

In case of early man, the ancestral stock was probably laid in Pleistocene period. Differentiation among the same species had taken place after their migration into different regions. Genetic changes and climatic adaptation both were responsible for that differentiation; the basic racial types in the world were formed. It seems that in earlier times sexual selection must have played a more vital role than the social selection. Mating between individuals used to occur completely on the basis of personal preferences, instead of social interests. For example, women liked to marry a man with great muscular power or a kind hearted person or a successful hunter and so on. On the other hand a man liked a women for different kinds of body beauty and soft disposition. All these preferences were obviously of relative considerations but were highly operative in the evolution of man. Even at present, the preference of males for fat women in some societies have favoured to increase such body types. In America, recent mating between whites and Negroes, and also the preference of darker Negroes for lighter Negro females tends to lighten the original Negro skin colour.

Race, Versus Culture and Society

Race is a biological concept. Traditional physical anthropology classified the races of mankind on the basis of the relative frequency of different physical traits that are easily discernible in populations. Some distinctive traits, which represent individual variation, tend to cluster in populations and the physical anthropologists try to measure their external manifestations. The modern genetic anthropology has emphasized the counts of gene frequencies in blood types; they are also concerned with some other type of genetically identified characters. This helps them to get rid of the ambiguity caused by the continuous variability of some morphologically measured traits. But it has been found that gene analysis does not reveal any major information, rather it reinforces the traditional racial classification.

In dealing with races of man, we should always remember that all human beings as a member of a single species bear a large number of common characteristics which are greater in number than the differentiating characteristics. Evolution had not proceeded to differentiate the human populations; rather it made the man capable to adapt all sorts of environment. The physical differentiation of man arose to meet the requirements of respective environment. The particular way, in which a people adapt its surrounding environment, is called culture. Culture is chiefly behavioural and cultural factors have played an important role in the physical evolution of man.

If man is held as a unit of a population possessing culture, his way of life including general behaviour is controlled by three factors. Firstly, the instinct of man which is a part of man's biological heritage. Secondly, the personal experience by which he has acquired practical knowledge. Thirdly, the learning where he learns from other members of his own group. By combination of these three factors he controls his environment; create or learn the culture. Therefore, Ralph Linton has rightly defined culture as "a configuration of learned behaviour and results of behaviour whose component elements are shared and transmitted by the members of a particular society". The term society refers to a group of individuals who lead a specific way of life and follow a general behaviour depending on some basic conditions of a common living. A culture can not continue or exist without society.

The fact indicates that the culture is learnt as a member of the society and it is also transmitted through the society. It is also clear that the cultural pattern of man is not determined by the genes but by the society, which he creates to live in. Therefore, concept of culture is quite distinct from the concept of race. The former is a social legacy while the latter is a biological legacy.

The other acceptable fact is that, the physical differences due to heredity have little to do in bringing about differences in culture between two peoples. The cultures, which vary from society to society, do not possess any relationship with race. The environmental factors that act strongly behind the race formation also influence on the culture traits. But their way of action is completely different. Again, the racial traits are changeable like cultural traits, yet the mechanism of dynamicity is not same. So, race and culture can never be synonymous. Some populations of Eastern India are found to have adopted the western culture, which apparently can mislead someone. But in fact, the physical appearance of these people is not identical with the people inhabiting in Western World. Similarity of culture has been evident because of the plasticity (adaptability) of the human behaviour. Genes can not control the attitudes, endowments, capabilities and inherent tendencies of human beings. Therefore, anywhere they can learn to do things that other human beings have done anywhere.

Concept of Racism

The concept of race stands on human biological variations, which can be both external and internal. The external variations include skin colour, hair colour, hair texture, eye colour, stature, body build, nose form and so on. The internal variations are concerned with the susceptibility or resistance to the diseases etc. At present the Homo sapiens sapiens covers a number of races which differ from one another in the relative frequency of certain inherited traits. But originally three basic racial types—the Caucasoid, the Mongoloid and the Negroid inhabited in the earth. Gradually they migrated to different directions and got mixed with each other. The combinations of migration, interbreeding and isolation had diluted the purity of racial types and several minor races were formed.

The number of races in the present world is numerous. Physical mixing of people which started during prehistoric period has reached to its optimum level due to the development of physical communication. Within a few hours, an aeroplane can now take us to the most distant land. Peoples have been much closer to each other and the possibility of biological mixing has enhanced a lot. However, the scientists have acknowledged these phenomena and they are striving to study the variations as well as similarities among the various racial groups. A number of classifications have been put forward. It should be kept in mind that all these classifications are arbitrary. Because each classifier has used his own parameters to trace out the traits. They also have considered the geographical and local races together. The geographical races are the original races, which followed the strict ecological niche. They showed typical distinctive trait frequencies for which each of them was manifested as a unique type. Contrary to this, local races are like interbreeding populations i.e., the ordinary local groups whose members often interbred. Though these racial classifications have found vague, still race, as an explanation of biological variation in human population can never be undermined.

Identification of races or finding out the number of races is quite different from racism. The term racism is concerned with the status of the races; people of a race are considered as superior or inferior of others. According to R.F.Benedict (1940) racism is a doctrine, which says that "one group has the stigmata of superiority and the other has those of inferiority". Probably the idea of racism originated in fifteenth century when some Greek scholars divided the mankind in two groups —the civilized and the barbarians. The famous Greek philosopher, Aristotle also proposed two groups—one group, who is free by nature and the other one who is not free (slave). Later on, Romans announced their superiority over others. As a matter of fact, in the Middle Age, different authorities had presented their hypotheses of superior races, for which the Christians, Nordics and Aryans were thought to be superior. Thus, a concept of superiority versus inferiority was unconsciously nurtured in the mind of the people.

Modern science does not find any race in pure state; all are admixtured racial groups. According to Juan Comas, the process of miscegenation has started since the dawn of human life. This miscegenation (interbreeding between different races) helps to produce a greater somatic and psychic variability. In biological sense, the miscegenation is neither good nor bad. Because, the new genetic combinations which emerge as a result of miscegenation do not produce unfortunate consequences. Rather, sometimes hybrid generations show unusual vigour, as favourable combinations always possess better survival potential. Most of the high civilizations of the world are found to have formed by the people of crossbred group. In this connection we can raise a pertinent question. Whether some races hold a superior capacity for cultural attainment or not. No correlation has yet been made between mental attitudes and racial types. Since no anatomical superiority has been distinguished among the races of modern Homo sapiens sapiens, the issue revolves round the mental processes and psychological aptitudes. Anthropologists failed to achieve satisfactory answer regarding the technological backwardness in some areas of the world and so for a long time they tried to explain the phenomenon as racial inferiority.

Neurological analysis of anatomy and physiochemical functioning of nervous system of different races have recently answered the anthropologists. Controlled psychological experimentation, with or without elimination of cultural and environmental factors suggests that the technological backwardness of some communities may not due to racial inferiority. Although I.Q. test and other aptitude tests reveal some differences in visual, motor and vocational skills but such tests could not eliminate the cultural factor i.e. many skills of intelligence and aptitude have been found to change with changes of cultural environment. All sorts of inferiority may disappear totally with environmental advantages. The technological backwardness therefore can be related to the physical and social environments or historical circumstances. Naturally the superiority and inferiority concept regarding the race is not justified. All races are equally capable of cultural development if the members can associate their innate skill with the required level of cultural experience.

STRATEGY OF RACIAL IDENTIFICATION

It is now clear that all races are fundamentally alike but superficially different. We should therefore know more about the differences.

Different authorities at different times have tried to identify the races of the world in various ways. Some ancient literatures also furnish the accounts of certain groups of population with different physical features. For example, the Sanskrit literature provides the description of dark-complexioned Nishadas (Australoids) as distinguished from the yellow-skin Kiratas (Indo-Mongoloids). The light skin coloured Indo-Aryans also are distinguished. Again, the Chinese people classified the mankind into five groups according to skin colour, as early as 200BC. It may be assumed that the variations of people with regard to physical features were conspicuous from the very beginning. In 1684, a traveler named Bernier traveled the Old World and classified all people whom he met during the journey. This may be considered as the first deliberate attempt towards the classification of human race. But actual credit goes to the Swedish naturalist Carl Linnaeus who first made a full-fledged classification on human species, Homo Sapiens, in the year 1738. Linnaeus categorized all Homo Sapiens into four classes in the following ways:

Homo americanus (American Indian): The people are tenacious, struggle-prone and free. They are mostly ruled by custom.

Homo europaeus: These people are characterized by light skin colour. In nature, they are lively and inventive. They are largely ruled by rites.

Homo asiaticus: These people are tremendously stern, arrogant and stingy. They like to rule on opinion.

Homo afer (African): These people are cunning, slow and negligent in nature. They are full of whims.

The classification of Linnaeus emphasized the mental traits. His all information was based on the observation of culture, instead of physical features. Blumenbach, a notable German Scientist also forwarded a scientific classification of population for the first time in 1775. He classified the world population into five types on the basis of skin colour. They are:

Caucasian or white: The type refers to the white skin-colour people, mainly the Europeans, except Laplanders and Finns. Some people of this type are also found in North Africa and Western Asia.

Mongolian or yellow: This type includes all yellow-skin coloured people like Finns and Laplanders of Europe. The Eskimos of America and some inhabitants of Asia can also be included in this group.

Ethiopian or black: These are mainly black skin-colour people, the inhabitants of Africa. The Caucasians of North Africa are excluded.

American or Red : This type signifies the red skin colour people of America, other than the Eskimoes.

Malayan or brown: This type refers to the brown skin-coloured people, the inhabitants of the Pacific.

Blumenbach called each of these classes as 'race' although the term was first introduced by Buffon, a French scientist.

Pickering in 1848 divided the world population into 11 human races. In 1870 Huxley proposed another classification where five principal races were divided into 14 secondary races. Besides, the scholars like Hackel (1873), Topinard (1885) and Quatrafages (1889) had also attempted to make racial classifications. But their classifications were not at all significant; they are only the modified form of the classification made by Huxley.

As the time passed on, more and more characteristics were added and new methods were implied on racial classification. Gradually a large number of races and sub-races were distinguished. As for example, Denniker (1889) observed 29 races on the basis of hair-form, nose-form and skin colour as subsidiary traits.

The German anthropologist, von Eickstedt made another classification, which was based partly on morphological traits and partly on geographical location. He identified three basic races, such *as, Europoid, Mongoloid* and *Negroid* with 18 'sub-races', 3 'collateral races', 11 'collateral sub-races' and 3 'intermediate races'. In the mean time, scientist Duckworth began to study the human groups on the basis of cephalic index, prognathism and cranial capacity. He found a seven-fold division in the following ways:

(i) Australian; *(ii) African Negro* *(iii) Andamanese;*
(iv) Eurasiatic; *(v) Polynesian;* *(vi) Greenlandish;*
(vii) South African;

The scientist Elliot Smith had also made a six-fold division in the following groups.

(i) Australian; *(ii) Negro;* *(iii) Mongol;*
(iv) Nordic; *(v) Alpine;* *(vi) Mediterranean;*

In 1923 R.B. Dixon used three indices, cephalic index, length-height index and nasal index in order to classify the human races. Next year i.e. in 1924 Haddon employed hair form as a primary character, and stature, cephalic index and nasal index as the secondary traits in his proposition regarding racial classification. In 1931 American anthropologist, Hooton identified three primary races, like white *(Caucasoid), Negroid* and *Mongoloid* with several secondary composite *and* sub-races. This classification was further modified in 1947.

In 1948 French anthropologist Vallois proposed a division of four major races, such as Australoid, Black, Yellow and White where each division consisted with a number of sub-races. Thereafter in 1950, three American anthropologists. Coon, Garn and Birdsell proposed of six stocks, namely, *Negroid, Mongoloid, White, Australoid, American Indian* and *Polynesian* among different 30 races of mankind.

Numerous studies from various regions of the world suggest that hybridization among different groups has taken place since long past. It is a continuous process, which plays a great role in extinction or absorption or formation of new racial groups. The several classifications as proposed by different anthropologists at different times ascertain three major races at the inception. These races are *Caucasoid* (white), *Mongoloid* (yellow) and *Negroid* (black). Some anthropologists are in the opinion of adding one more race; the Australoid and thus four races have been created. It may be mentioned here that the racial classification always honours one of the three distinct categories of findings. The first category of findings includes all variable physical characters as well as some features, which are the resultants of hereditary mechanism. In the second category of findings, genetic processes are dealt, especially in connection of serology. The third category is comprised of rare genetic character.

SOME CONVENTIONAL RACIAL CRITERIA

Racial identification is made on the basis of a number of external and internal physical characteristics. The external physical characteristics are phenotypic characters, which are mostly adaptive in nature, but the internal physical characteristics are genotypic characters, which are strictly hereditary and non-adaptive. Modern genetics has provided much information regarding the bodily characteristics of man. It has found that phenotypic characters may not tally with genotypic characters when recessive genes are present in the body. As a matter of fact, importance should be given on the genotypic characters while studying the racial criteria.

Previous works on this line were chiefly centred round the external physical characters because the former workers believed those characters as hereditary. In reality, they were not so. Although at present, the view of the scientists have broadened and they have been able to overcome the limitations of their understanding, still we find extensive use of phenotypic characters in racial identification. The study of blood groups, dermatoglyphics etc. are the exceptions.

A single character can never be a reliable criterion, so always more than one character is considered for racial classification. On the other hand, adequate number of people is required for the survey. Since some features vary with age and sex of the individuals, it is better to make a comparison among the individuals of same age and sex. W.C.Boyd (1950) had pointed out a number of selective guidelines for choosing racial criterion, which may be cited in the following ways:

(i) The criterion must be objective in order to eliminate an investigator's bias in identifying and classifying the concerned traits.

(ii) The criterion should be non-adaptive to put a guard against natural selection.

(iii) The criterion must be free from environmental modification.

(iv) The criterion should have a little scope of mutation. Because high frequency of mutation may show considerable variation of a trait, time to time in a population.

(v) The criterion should be simple in nature, which can be controlled by a known genetic mechanism.

All these criteria for racial discrimination may be divided into three categories. The first category includes some common variable phenotypic characters whose hereditary mechanism is neither simple, nor perfectly known. For example, the hair colour, skin colour, nose form, facial form etc. Some of these characters can be described after critical visual observations, viz. Hair colour, nasal bridge etc. Others demand instrumental measurements along with visual observation. The second category of criteria considers some physiological characters where genetic process behind the traits is quite well known. A few of these traits are ABO blood group, MN blood group etc. The third category shows some rare genetic traits like colour blindness, some abnormal traits of human blood etc., which occur in low frequency. However, some of the important criteria as considered in racial identification have been illustrated in the following ways:

Skin Colour

A large number of granules are found to persist in the deep layer of the epidermis. These are known as melanin pigment. Varying frequencies of melanin pigments are responsible for the variation of the skin colour. The melanin is said to have some protective power. Continuous exposure to the sun develops more and more melanin pigments, which results in dark skin colour. Naturally white skin possesses lesser number of pigments. But this skin with lesser number of pigments is the best for holding the heat in the body. In fact, the higher concentration of melanin pigments can protect melanoderms from the strong rays of the sun and less concentration of melanin pigments can protect leucoderms from the cold by restoration of heat in the body. It has been noted that ultraviolet rays of sunlight stimulate epidermal cells of the skin to produce more granules or melanin pigments for making the complexion more dark. The process will not operate if the direct sunlight is cut off from the body of the person concerned with the help of any obstruction.

The skin is composed of two layers—the epidermis and the dermis. The skin pigments or melanin are found to be distributed in the epidermis or upper layer where blood supply is absent. The colour of the skin varies to protect against the scorching rays of the sun as well as to restore the heat of the body depending on the environment. But it is a matter of surprise that these different shades of skin colour can not be generalized with geographic and climatic conditions. For example, we find the dark-skinned Negroes of Congo basin, pale yellow-skinned people of Borneo and cinnamon-coloured people of Amazon Valley of South America who live under more or less similar geographical conditions. On the basis of these findings we assume that the earliest men had an intermediate skin colour, instead of white or black. In course of time, probably the genes for skin colour were changed by mutation and brought variations in skin colour according to the environment. Gradually some of the variations were selected by the nature; they outlived the rest. The sexual selection had also played an important role to accelerate the process. Ultimately in different parts of the world specific skin colours developed suited to the particular environment.

The people of the world can be classified into three major groups on the basis of skin colour.

Leucoderms or White skinned people : The Europeans are the best example of this group. But most of the Western Asiatics, North Africans and Polynesians also belong to this group where skin colour varies from pinkish-white to light brown shades. Besides, the brown-skinned people like Hamites, Indo-Dravidian, etc. have also been included to this group.

Xanthoderms or Yellow-skinned people : The Asiatic Mongoloids are the best example of this group. Some Amerinds, Bushmen and Hottentots also show a yellow tinge in their skin colour.

Melanoderms or Black-skinned people : The Negroids are the best representative of this group. Some of them show an absolute dark colour of skin and others have a variable skin ranging from dark chocolate brown to many other shades of dark brown. The Papuans, Melanesians, Pre-Dravidians, etc. are the appropriate examples of this group.

Hair

Hair is the most convenient and striking character in racial classification. This criterion for racial discrimination is very old, as far back as 1820. It provides the basis for a primary classification of present-day races in the world. Studies have been made on different characteristics of hair such as *the form, colour, texture, quantity, cross-section* and *hair whorl.*

Hair Form : The hair can be divided into three types on the basis of hair form.

a) *Leiotrichy* or straight hair : This type of hair form is found among the Mongoloid people. The form is again sub-divided into three types, which are as follows :

i) *Stretched* - This form is usually thick, coarse, straight and stiff.

ii) *Smooth* - This form is thin and soft.

iii) *Flat wavy* - This form of hair has a tendency to become wavy. The wave lengths are usually long.

b) *Cymotrichy* or wavy form of hair: This type of hair form is widely distributed in the places like Western Asia, parts of northeastern Africa, Europe, etc. The people of White race, for example,.

Mediterranean, Alpine, Nordic, etc. show this form of hair. Moreover, this form occurs among the secondary races of hybridized origin. Cymotrichy can be further subdivided in three types, which are as follows:

i) *Broad wavy* - This type shows smaller wavelength which lie on one plane only.

ii) *Narrow wavy* - This type shows narrow waves with strong curvatures. But the waves lie on one plane.

iii) *Curly* - This type shows strong spirals and the curvatures do not lie in one plane.

c) *Ulotrichy* or woolly form of hair. This form of hair shows a considerable range of variation. It is available among the Negroes, Andamanese, Bushman, Papuans, Melanesians, etc. and can be subdivided in the following ways:

i) *Frizzly* - This sub-type shows short and deep waves but not in the form of complete spiral.

ii) *Loose Frizzles* - This sub-type of hair presents circular and flat spirals.

iii) *Thick Frizzles* - This sub-type also shows circular and flat spirals, but in this case the spirals are thickly set.

iv) *Filfil* - In this sub-type, hair remains highly rolled like small knots and appears to be separated by bare spaces on the head. This form of hair is popularly known as peppercorn, which is found among the Bushmen.

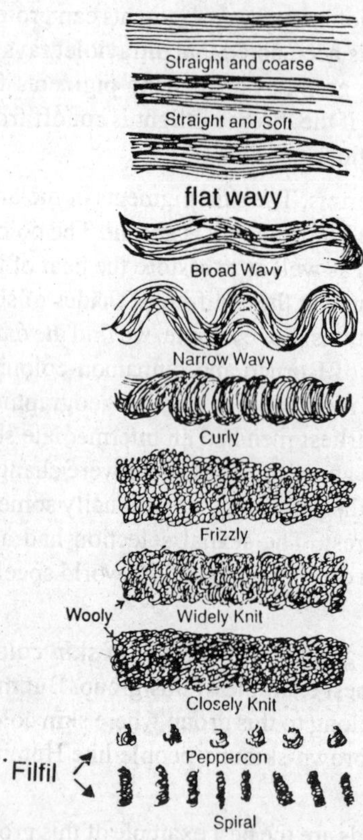

Fig. 7.1. Forms of Hair

In the study of hair form, scholars have tried to find out the relationship between hair form and climatic conditions. Leiotrichous or straight hair is found mostly in dry and cold climate. Ulotrichous or woolly form of hair is easily available in the moist and warm climate. It has also been noted that there are certain groups of people who live in moist and warm climate but show straight hair. The best examples are the American Indian groups of the Amazon basin, the Malays of Java, Borneo and Malaya. The facts suggest that the hair form is probably non-adaptive in character, so it is affected very little by environmental condition.

Hair Colour

The colour of hair shows a wide range of variation among the population of the world. The structure of hair includes a hair-shaft which consists of three parts, namely, the thin unpigmented outer layer, the cortex and the medulla or pith. The colour of hair is obtained for the presence of granular or non-granular pigments. Such pigments are commonly found in the cortex, but sometimes they are also present in the medulla of hair-shaft. Presence of brown or black pigments produces wide variations of black hair. A red-gold pigment is responsible for different golden shades in hair. But the grey colour of the hair appears for the reflection of light from unpigmented portions of the hair-shaft. The environmental influence is not remarkable in case of hair colour.

Dark hair colour is found everywhere in the world. It is very common in occurrence whereas the brown colour of hair is not at all common. Among the Mediterraneans, hair-colour is relatively darker than the Northern Europeans who show a variable colour, ranging from light brown to reddish. Again, some Negro population of Africa and Melanesia show yellow colour of hair. Most of the Oceanic Negroes possess reddish hair in their childhood, which transforms into dark brown colour at the stage of maturity. But this should not be confused with grey hair, which is a sign of old age. During old age, dark colour of hair invariably turns—grey. Because, the medullary spaces in hair shaft increase with age. At the same time absorption of some pigments takes place in the hair-shaft, which brings the change of hair colour - a grey shed is produced. This change is an important common phenomenon.

Texture

Texture of hair may be divided in three categories—*course, medium* and *fine.* Scientifically Garn has studied the shaft of human head hair. He has come to the inference that the thickness of hair shaft varies between 25 $/^{u}$ ($/^{u}$ = mu) to 125 $/^{u}$.An arbitrary classification of head hair has been proposed by him in the following ways:

Fine hair = X = 56$/^{u}$ Medium hair = 57$/^{u}$−84$/^{u}$ Coarse hair = 85$/^{u}$ — X

For example, the Chinese and Japanese people as the members of Mongoloid race show coarse hair. The white men of Caucasoid race possess medium hair.

Quantity

The quantity of hair is usually determined by visual observation and described with the terms like scanty, medium and rich. The same terms are used for head hair, body hair, beard and moustache. Among the Mongoloids and Negroids, the hairs are sparsely distributed on face and body. Caucasoid race or white people show richness of body and facial hair.

Hair whorl

The hair whorl is observed on the occiput of the head. This can also be taken as one of the factor for studying the nature of the hair of an individual. Two types of hair whorl have been distinguished on the basis of direction—*clock-wise* and *anti-clockwise.* Normally a person possesses one whorl though double whorls are also not rare. But three or more whorls are very rarely found. As regard to the inheritance of this character, the *clock-wise hair whorl* dominates over the *anti-clockwise hair whorl.*

Cross-section

In scientific study, the cross-section of human hair can be of two types—*circular* and *oval* or *elliptical.* The circular cross-section is found mainly among the Mongoloid people possessing the straight hair, while oval type of cross-section is prevalent among the Negroid people possessing the Ulotrichy hair form. Recently the cross-section of hair has lost its significance in racial discrimination as studies have shown that both the oval and circular cross-sections may be found in the hair of same head. However, beside the cross-section of hair, the weight of the hair, the longitudinal section of hair and some other characters may also be taken into consideration for studying racial discrimination.

Head Form

The form of head is the most valuable character in racial discrimination. Anthropologists study the head thoroughly as it is relatively independent of environmental fluctuations. To examine the head form, one has to look at it from the above. The head shape presents varieties of contour like *oval, long, pentagonal, narrow,* etc. But such illustrations of head are not at all suitable for statistical treatment. Some accurate measurements are required in this connection. However, measurements on head and skull were first standardized at the Frankfurt Congress, held in 1882. From that time (nineteenth century) up to the present day, the cephalic index is universally measured in the study of the head form.

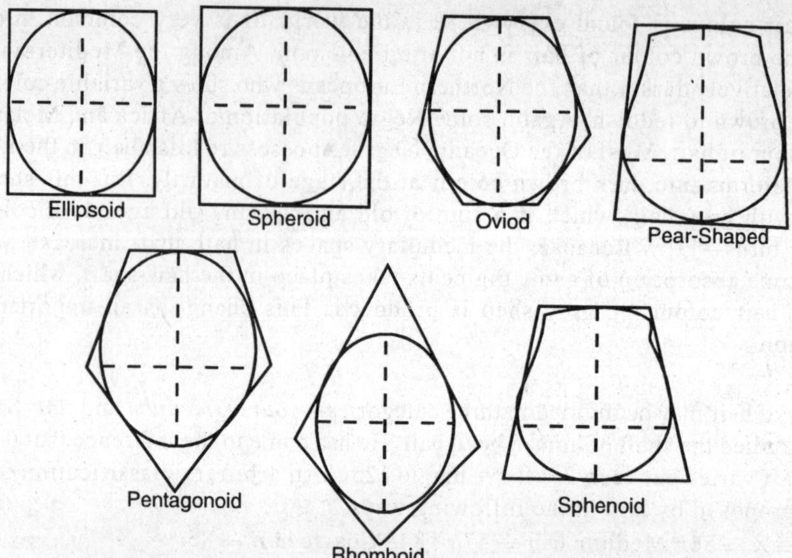

Fig. 7.2. Shapes of Skull as seen from Above (After Sergi)

An eminent Swedish anthropologist Anders Retzins had established this principle to observe the proportion of the breadth to the length of the head, in classification of human race. The ratio of these measurements is called the index and the cephalic index can be illustrated in the following ways,

$$\textit{Length - Breadth Index}\ \text{(Cephalic Index)} = \frac{\textit{Breadth of the Head}}{\textit{Length of the Head}} \times 100$$

The Cephalic index thus indicates the ratio of the maximum breadth to the maximum length, which is multiplied by 100. But it does not provide any idea regarding the contour or real form of the head. However, a classification according to cephalic index has been mentioned here which is known as Martin's classification.

Hyper-dolichocephalic (very long and narrow)	X	—	69.9
Dolichocephalic (long and narrow)	70.0	—	75.9
Mesocephalic (medium)	76.0	—	80.9
Brachycephalic (short or broad)	81.0	—	85.5
Hyperbrachycephalic (very short and broad)	85.6	—	X

The height of head is also required to know for finding out the relations in the length-height and breadth-height indices respectively. The classifications of the head in respect of these two indices have been given by Martin in the following ways :

$$\textit{Length – Height Index} = \frac{\textit{Auricular Head Height}}{\textit{Head Length}} \times 100$$

Ranges :

Chamaeocephal	X	—	57.6
Orthocephal	57.7	—	62.5
Hypsicephal	62.6	—	X

$$\textit{Breadth – Height Index} = \frac{\textit{Auricular Head Height}}{\textit{Head Breadth}} \times 100$$

Ranges :

Tapeinocephal	X	—	78.9
Metricephal	79.0	—	84.9
Acrocephal	85.0	—	X

Our knowledge based on numerous available data from the various parts of Europe suggests that early man of prehistoric period was generally dolichocephalic. The brachycephalic heads developed later, in the Upper Palaeolithic Age of Europe as a result of repeated mutations and other factors. This gradual brachycephaltzation has been described by R.B.dixon as a universal phenomenon. It has also been stated that the development of head shape is controlled by the Mendelian laws of inheritance.

The head form can be modified in many ways with some environmental factors. For example, some tribal folks among the American Indians keeps head under prolonged pressure by a special device for bringing permanent change in the shape of their head. It is an evidence of artificial deformation of the head. Prof. A. Ivanovsky has studied a few people of Southern Russia where malnutrition has brought considerable changes in the form of the head. Franz Boas noticed a good deal of variations in the head form among the children of European immigrants into the United States; the features vary from those of their parents. Dr. H. L. Shapiro also found that the head form becomes shorter and broader among the offsprings of Japanese immigrants who settled in the Hawaiian Islands. Sexual variation of the head-form is the other important factor, which has to be considered in this respect.

It is very difficult to get an accurate head-height measurement in the living subject. There is also a wide range of variations. Unfortunately, the available data on height indices are not adequate to derive any conclusion in a definite way. Further, in many cases, brachycephalic parents are found to produce dolichocephalic offsprings. So this criterion should be used in association with several other reliable criteria for racial discrimination.

Face Form

We generally recognize a man by looking at his face, which possesses many distinguishing characteristic features. The faces are of different shapes as *oval, round, square or pentagonal.* But actual shape of a face can be better understood when it is expressed in terms of relation between the breadth and length.

Face length is measured from nasion to gnathion. The nasion is the upper end of internasal suture where it meets with the frontal bone. Gnathion is the lowest point in the median line of the mandible or the lower jaw, i.e. the gnathion point. The greatest face breadth is obtained by measuring the distance across the cheekbones. It is called Bizygomatic breadth or maximum breadth of Zygomatic Arch. The facial index is found out by dividing the length by its width and thereafter multiplying the quotient by 100.

$$\textit{Morphological Facial Index} = \frac{\textit{Morphological Facial Length}}{\textit{Bizygomatic Breadth}} \times 100$$

A classification of facial indices.

Hyper euryprosopic (Very broad face)	X	—	78.9
Euryprosopic (Broad face)	79.0	—	83.9
Mesoprosopic (Medium face)	84.0	—	87.9
Leptoprosopic (Narrow face)	88.0	—	92.9
Hyper leptoprosopic (Very narrow face)	93.0	—	X

A broad face is usually associated with a brachycephal and similarly a long face is associated with long head or dolichocephal. Though this is a harmonic relation between the head and the face, but not universal in occurrence. The point can be illustrated with the Armenoids who possess long and relatively narrow faces along with relatively short and broad heads. Again, a long head with broad face is available among the Eskimos. These are the examples of disharmonic relation between head and face. In early days, the Cro-Magnon people were in possession of this type of disharmonic face.

Another typical feature of the face is prognathism which can be described as the protrusion of the jaw. When the face does not show any protrusion, it is known as orthognathism. Prognathism is common among the Black races of Africa and Ocenia; it is especially well marked among the Negroes and Australian aborigines. The modern people are generally orthognathous, only a few of them may show a little prognathism.

When the alveolar margins of upper and lower jaw exhibit a projection, it is termed as alveolar prognathism. The forward projection of the facial region is known as the facial prognathism. Apes and monkeys present both of these kinds of prognathism. Among the human population, the Mongoloids and some white people show slight or moderate alveolar prognathism but facial prognathism is almost absent in them.

Although the form of face is considered as an important criterion but it has got a limited scope in racial classification. The facial index and other features vary with the development of age and these are easily affected by the factors like sex, function, etc. For example, the females almost invariably show shorter and comparatively broader faces than the males of the same ethnic groups. The functional variations often reflect on the face. Besides, the hereditary nature of the facial characteristics are not perfectly known which may result in different shapes such as short, broad, long, narrow, etc. So, it is not at all wise to depend on this factor as a racial criterion. It is better to use this factor in association with other reliable factors for racial discrimination.

Nose

The nose presents many interesting features regarding racial discrimination. The proportion of nose can easily be found out by actual measurements, particularly considering its breadth in relation to its length. But some parts of the nose cannot be measured at all; they have to be described in simple ways.

The Nasal Index

The nasal index is expressed as the percentage of the breadth in relation to its length. On the skeleton, the length is measured from the nasion point (where the internasal suture touches the frontal bone) to a point just at the base of the nasal spine. The breadth is the maximum distance on the nasal opening in the skull. In case of living subject, the length is to be taken from nasion to the subnasale where the nasal septum touches the upper lip. The nasal breadth is the highest distance between the two alare or two nasal wings in natural condition. As a natter of fact, the nasal index on skeleton and the nasal index on the living subject never correspond to one another.

$$\textit{Nasal Index} = \frac{\textit{Nasal Breadth}}{\textit{Nasal Length}} \times 100$$

Classification	**(of living)**	**(of skeleton)**
Hyperleptorrhine (Very narrow nose)	X - 54.9	
Leptorrhine (Narrow nose)	55.0 - 69.9	X - 47
Mesorrhine (Medium nose)	70.0 - 84.9	47 - 51
Platyrrhine or *Chamaerrhine* (Broad nose)	85.0 - 99.9	51 - 58
Hyperplatyrrhine (Very broad nose)	100.0 -X	

The Mongoloid people possess short and moderately broad nose; their nasal index takes them to the group of mesorrhine. The Caucasoid people of Asia show the characteristics of leptorrhine nose while the white people of Europe present typical narrow nose. The Negro nose is broad and short for which it has been grouped as platyrrhine. The Australian aborigines also show a clearly marked platyrrhine nose. A. C. Haddon had wrote "speaking generally, the leucoderms are leptorrhine, the Xanthoderms mesorrhine (but the Eskimos are leptorrhine) and the melanoderms platyrrhine".

It may be granted that the nasal index is dependent on environmental conditions. Many anthropologists realize that the natural selection has played an important role in narrowing or widening the nostrils. The people with broad nose and wide nasal aperture generally live under hot moist conditions where they can inhale large quantities of warm and moist air without causing any harm to their respiratory organs. On the contrary the narrower and longer nose is more effective in warming the cold air before it passes on to the lungs. So, the Arctic Eskimos show hyperleptrrhine nose while the equatorial Negroes possess hyperplatyrrhine nose. In India, Aryan-speakers were invaded sometime in the past by leptorrhine. For this reason the country has exhibited some groups of narrow-nosed people in temperate zones. But, the data in connection with the hereditary nature of nose form are inadequate in order to reach at the right conclusion. Therefore it is not wise to attach much importance with the nose as a single point for racial discrimination.

Parts of a Nose

Among the various parts of a nose, the nasal root is marked at the meeting point of the nasal bones with the frontal bone of the skull, which is known as the nasion. Generally the nasal root shows a depression which may be described as shallow, medium, or deep. The two nasal bones join together to form an angle along with their long sides and constitute the nasal bridge. The nasal bridge on living body extends from the below the nasion to the tip of the nose. The nasal bridge is often described as shallow, medium or high. In profile view, the bridge looks straight, concave, convex, or concavo-convex. It may be mentioned here that the primitive people of early period were the representatives of low and broad nasal roots as well as nasal bridges. A few of these characteristics are found to survive among some of the African tribes. The Europeans usually show high and narrow nasal roots as well as nasal bridges while the people of Eastern Asia project an intermediate form of nasal roots and bridges.

There are several other minor features associated with the study of the nose. The tip of the noses can be classified as bluntly rounded or sharply pointed; thick or thin. The horizontal nasal septum can be directed upwards or downwards. Besides, this septum may be straight, concave or convex from the profile view. The wings of the nose or alare may be thin and compressed, broad and flaring or intermediate. The diameter of the nostrils can also be studied as oval or rounded. For example, the Negro nose is short and very broad with medium depressed root, straight or concaves bridge, thick bulbous tip and very thick flaring alare. The nose of Australian aborigines differs from Negro type by more deeply depressed root. The typical Mediterranean people are the representative of leptorrhine who possess a straight nasal bridge of medium height and a narrow root with compressed medium-spread alare. But the nose of Nordic people is higher, longer and narrower with pinched wings. Here also we find inadequate data to draw any conclusive remarks on racial discrimination.

Eye

Sometimes the eyes possess certain definite features in discrimination of the races of man. E. A. Hooton had mentioned only two sharply contrasted varieties of eyes in modern man—the Mongoloid eye and the non-Mongoloid eye. In Mongoloid eyes the palpeberal fissure is oblique and the outer angle of the eye is higher than the inner angle. The eye opening or slit is narrow. External corner of the eye opening is elevated so that slants outward and downward. The inner epicanthus (epicanthic) or complete Mongoloid fold is the characteristic feature, which is found in the population in varying degrees.

The corners of the open portion of the eyes are called *canthus,* which are of two types—inner and outer. Among some specific groups of people the skin-fold hangs over the free edge of the entire upper lid of the eye. It may also extend from outer canthus to the inner canthus. The feature has been described as the Mongoloid fold and found to occur among the Mongoloid people, irrespective of age groups. Young, middle-aged and old persons, everybody exhibit this fold in the same way.

The inner epicanthic fold or the inner epicanthus is the most common variety of all eye-folds. The fold starts on the inner or sometimes medial part of the upper eyelid and covers the free edge of the inner angle of the eye. Sometimes it is found to extend on to the cheek. Although variations are found with age and sex, it is most prominent among the infant, children and women. Another variety of eye-fold is the external epicanthus or external epicanthic fold. This fold starts on the middle portion of the upper eyelid and extends below the outer portion of the upper eye lid, covering the outer corner of the eye. It is frequently observed among elderly male of different non-Mongoloid populations. These are nothing but a slag skin, devoid of elasticity. It appears as a legacy of age factor. Apart from these two kinds of eye-fold, sometimes a skin is found to hang over the central part of the upper eyelid covering the edge of that region; both the inner and outer corners of eye remain uncovered. This type of fold is known as median or cover fold.

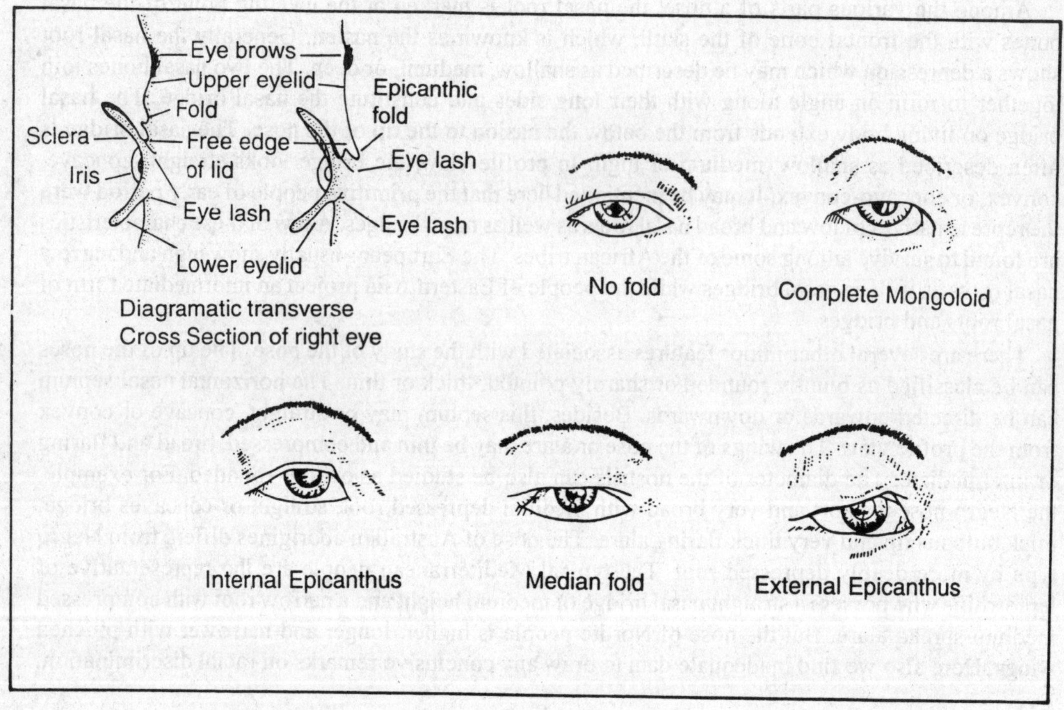

Fig. 7.3. Different Types of Human Eye-fold (Right rye)

Mongoloid eye with inner epicanthus is not only typical to Mongoloid people, all people having Mongoloid admixture show this fold. Therefore inner epicanthic fold possess a great racial significance unlike the other two eye-folds. The non-Mongoloid eye is wide, straight and open; eye-fold is absent. African Negro, especially Bushmen and Hottentot occasionally present this type of fold.

Eye colour depends on the quantity of pigment present in front and back of the iris. The front layer of iris may or may not be pigmented while back part of the iris has a double layer of pigmented cells. This pigmentation is also an important criterion for racial study. It shows a wide range of variation as black, dark brown, light brown, blue brown, and grey brown, blue or grey. Maximum variation has been observed among the people of leucoderms. The Mongoloids and Negroids usually show dark-brown iris.

Stature

This important criterion of racial classification is found quite unsatisfactory as affected greatly by the environmental conditions. An unfavourable environment keeps the people underdeveloped; people show increased stature under improved and favourable environmental set up. This proposition has been justified with the people of Limousion district of France and the Yupa Indians of Venezuela. Shapiro studied Japanese immigrants in Hawaii Island where he observed an increase of stature under different environmental condition. Various research work done by different authorities in this line declare that the stature varies with the differences of occupation, social class, hygienic level and other environmental conditions. Again, environmental and dietetic factors are not solely responsible for the change of stature; the factor of heredity should also be taken into account. In fact, stature is mainly an outcome of hereditary factor but highly influenced by the environmental conditions.

The inheritance of stature follows the general Mendelian rule. It is manifested by the interplay of genes and receives influences from environmental circumstances. In this way stature is a combination of two factors—idiotypical / diotypical (hereditary) factor and paratypical (environmental) factor. A considerable range of variation (within certain limit) is therefore found among the same group of people. For example, a group of short-statured persons may include some tall statured individuals. Again, a group of tall-statured persons often shows some short statured individuals. Knowledge of physiology has revealed that the internal secretions of the glands like thymus, pituitary, etc. have a direct influence on the stature of a person. In addition, different life chances, especially the dietetic pattern keeps an impact. But, as a whole, hereditary factor seems to be more powerful than the environmental factor in determination of stature.

The scales as devised by different authorities are given below :

Topinard's Scale

Pigmy	below 50.0 cm.	— (below 59 inches)
Short	150.0 cm. — 160.0 cm.	(59 inches — 62 inches)
Under Average	160.0 cm — 165.0 cm.	(63 inches — 65 inches)
Above Average	165.0 cm — 170.0 cm.	(63 inches — 67 inches)
High	170.0 cm. and above	(67 inches and above)

According to Topinard the average stature of man is 165 cm. (5 feet 5 inches)

Haddon's scale

Pigmy		below	148.0 cm.
Short	148.0 cm.	—	158.0 cm.
Medium	158.0 cm.	—	168.0 cm.
Tall	168.0 cm.	—	172.0 cm.
Very tall	172.0 cm. and above		

Martin's scale

Pigmy		below	129.9 cm.
Very Short	130.0 cm.	—	149.0 cm.
Short	150.0 cm.	—	159.9 cm.

Below medium	160.0 cm.	—	163.9 cm.
Medium	164.0 cm.	—	166.9 cm.
Above Medium	167.0 cm.	—	169.9 cm.
Tall	170.0 cm.	—	179.9 cm.
Very tall	180.0 cm.	—	199.9 cm.
Giant	200.0 cm. and above		

This classification is largerly used in pragmatic life.

Blood-groups

The blood is a genetically determined factor as proved by the modern scientist .Therefore, it has been an important and reliable criterion for racial discrimination. At present this criterion is extensively used in studying the racial groups, racial movements, racial migration etc. Detail information about the blood groups have been cited in the chapter on genetics.

Dermatoglyphics

Uppermost surface of the human body is covered with hair and sebaceous (oil) glands. Only exception is the palmar and planter regions, which are continuously corrugated with narrow ridges having certain specific pattern. These patterns are known as dermatoglyphics. Literally, dermatoglyphics means skin carve (derma = skin + glyphic = carve). Therefore, it stands as the study of ridge patterns in the skin of the fingers, palms, toes and soles. The patterns are permanent and unchangeable throughout the human life. In fact, the pattern develops early in the stage of foetus formation and remains unchanged upto the final disintegration of the skin. But, it may be worn out or damaged due to some external as well as accidental reasons.

Although the finger patterns are used in personal identification for a fairly long period, the scientific study has started only recently, at the end of the nineteenth century. The anthropologists are more concerned with dermatoglyphics in the context of twin diagnosis, paternity diagnosis, primatology, etc.; they are also trying to establish variations in respect of traits among different human populations. According to Boyd, the dermatoglyphics fulfil many of the conditions for a good racial criterion, as this trait (dermatoglyphics) is non-adaptive as well as resistant to environmental factors. The genetic process behind this trait is not perfectly known but identification of dermatoglyphics is easy without any subjective bias. The ridge patterns can be studied from the different angles applying various methods, but scientists mostly deal with finger patterns and the main line formula of palm.

Finger Patterns

This pattern has been divided into four main types by Henry, as the *arches, loops, true whorls* and *composites.* The composites are the heterogeneous assemblage of patterns. Galton has classified three types of patterns as the *arches, loops* and *whorls.*

A loop is one-side-open ridge pattern, which may be open either to the ulnar side or to the radial side and termed as ulnar or radial loop accordingly. It may be noted here that the classic and widely used notations are A = arches; Lr = Loops radial; Lu = Loops ulnar; and W = Whorls. A number of ridges together makes a pattern when the junction of these ridges shows a triangular island, it is called as *triradius.* In case of arch, triaradius is absent. On the other hand, the whorls possess two triradii and only one triradius is present in loops. In fact, the ridge patterns are identified from the position of triradius.

It is seen that the loops are more frequent than whorls while the arches are found in small numbers.. A gain, the ulnar loops are much frequent than radial loops. Racial variation through fingerprints has been shown in a table.

Population	*whorls (%)*	*Loops (%)*	*Arches (%)*
Mongoloid	40 to 50	50 to 60	1 to 2
Caucasoid	20 to 30	60 to 70	4 to 7
Negroid	30	50 to 60	6 to 7

From the table it has been found that the whorls are most frequent among the Mongoloid population and least among the Caucasoid population. The loops appear more frequently among the Caucasoid

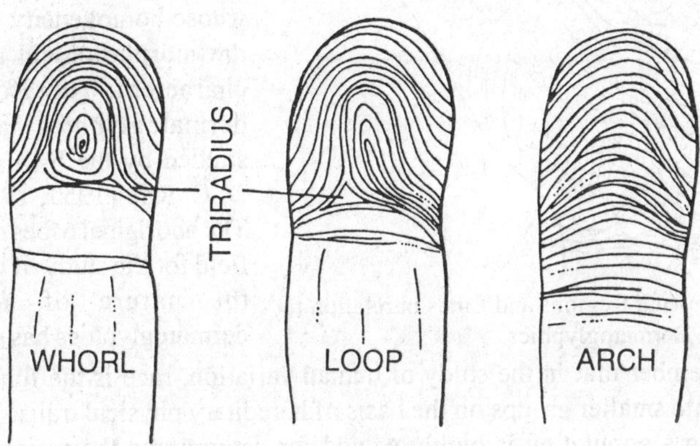

Fig. 7.4. Common Patterns of Dermatoglyphics

people while the Mongoloid and Negroid population show equal frequencies. Lastly, the arches appear with a very small frequency among the Mongoloid people. It is most frequent among the Negroid people. The Caucasoid population exhibits an intermediate stage. However, three indices are usually calculated depending on the frequency distribution of the different finger patterns, which are as follows:

$$\text{Furuhata's Index} = \frac{\text{Whorls}}{\textit{Loops}} \times 100$$

$$\text{Dankmeijer's Index} = \frac{\text{Arches}}{\text{Whorls}} \times 100$$

$$\text{Pattern Intensity Index} = \frac{2 \text{ X Whorls } + \text{Loops}}{n}$$

(n represents the total number of subject)

Palm (Main Line Formula)

Except thumb, at the base of four fingers or digits (II, III, IV, & V), four digital triardii are located which are called as *a, b, c,* and *d* in radio-ulnar sequence. The palm is divided into 13 regions each of which is having a symbol or number for the purpose of making the palmar main lines. The proxima radiant or upper peak of a digital triradius is directed toward the interior side of the palm and this radiant line is fully traced until it ends to the region of the palm by making the symbol of termination. This line is called the palmar main line. Thus, four main lines can be obtained being named as A, B, C, and D. The symbols of the four terminating regions of the four main lines develop the main line formula which can be recorded as in the order of D, C, B, and A It should be mentioned that the mainline A is not usually counted in finding out the population variations. The three formulae that

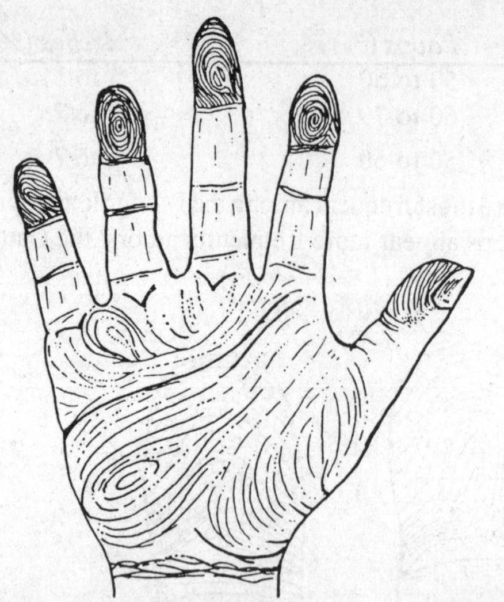

Fig. 7.5. A Palm Showing Regions and Lines in relation to Dermatoglyphics

are mostly observed in man stand as, 11, 9, 7-, 9, 7, 5-, and 7, 5, 5-. According to Wilder, the European formula is 11, 9, 7-, and ; 7, 5, 5-, is the Negro formula.

In India, an anthropologist becomes perplexed by the enormous variations in genetic characters like blood groups, finger or palm prints within the same ethnic strain, whose homogeneity is almost apparent in the morphological and anthropological characters. Therefore, in this country dermatoglyphics has been extensively studied by the S. S. Sarkar (1954,1961), D. C. Rife (1953, 1954,1958) and others. The aboriginal tribes of India offer a unique field for the study of dermatoglyphics. But the nature of sex difference in dermatoglyphics has not been fully known.

We should remember that in the study of human variation, race is the fundamental concept. It divides mankind into smaller groups on the basis of hereditary physical traits. Therefore, the nature of physical traits in a population is highly valued for determining the membership of the people under a race. The observable physical traits are equally important as the measurable physical traits. The earlier physical anthropologists used to depend solely on observations and measurements in classifying the human types in drawing their racial histories. The modern physical anthropologists have not abandoned those traditional kind of observations and measurements, rather they are found to supplement those by the study of blood groups and other bio chemical factors.

RACES OF MAN : THE CLASSIFICATION

Ancient literatures provided us with some descriptions of physical features of human groups. In course of time, more and more characteristics were added; specific methods were devised for the classification of the races. As a consequence, a large number of races and subraces have been distinguished. But it is really troublesome to divide human populations in a clear-cut races as the differentiation of characters between the races are quite arbitrary.

In general, a race is believed to possess a distinctive combination of physical traits which it acquires as a result of inheritance. In 1931, American anthropologist E.A.Hooton recognized three primary races—Caucasoid, Negroid and Mongoloid along with several composite secondary sub-races. But he himself felt the need of modifying this classification in 1947. However, most of the scientists have now agreed to divide the mankind into three principal racial types on the basis of relative commonness of physical characteristics. The principal races, namely Caucasoid, Mongoloid and Negroid have again been subdivided into a number of sub-races. The Caucasoid features are mainly available among the Europeans and their descendants. Mongoloid features are chiefly found in the peoples of Asia and Indonesia. The Indians of North and South America also exhibit some Mongoloid characteristics. But the Negroid features are restricted among the populations of Africa and Melanesia. A few people in America are also found to exhibit some Negroid physical features because they are the descendants of African people who were once taken to America as slaves. Some anthropologists like to introduce Australoid as a separate major racial group because of their peculiarity in physical characters. But, in fact, like the American Indians and South African Bushmen, the Australoids form a minor group and in early times these people were widely distributed in the world. Whatever may be, now we will discuss the physical characters of the various groups and sub-groups in a nut shell.

Caucasoid

This major racial group includes numerous sub-groups with a variety of racial elements. However, the generalized characters are as follows:

Skin colour — Colour of the skin shows a wide range of variation—white, olive, different shades of brown and sometimes dark brown.

Hair - Its form is variable, from flat wavy to different degrees of curliness. Its colour is generally lighter in shades, rarely jet black. Hair texture ranges from medium to fine; rarely it is course. Quantity of body and facial hair is usually moderate or abundant.

Head- Form of head is variable which ranges from dolichocephalic to brachycephalic.

Nose — Form of nose is also variable between leptorrhine to Mesorrhine but platyrrine is never found. Nasal bridge is generally high.

Face - Forehead is high, lips are thin and chin is pronounced. Cheek bones are not at all prominent and facial prognathism is totally absent.

Eye - Colour of the eyes is not black. They show lighter shades.

Stature - Usually tall.

The subgroups that come under this major group are as follows:

1. Mediterranean

The Mediterranean is assumed as the oldest white sub-race, although the exact time of its origin is not known. The people are named after their original habitat—the Mediterranean Shore. In course of time, they migrated to all directions. As a result, now we find these people in Portugal, Spain, France, Italy, Greece, Turkey and some parts of North Africa. Not only that, they are also found as far as in Arabia, Iran, Afganistan, Pakistan and India. Possibly these people were responsible for developing the Neolithic culture in Europe, North Africa, Near-East, South-Eastern Asia, and the region of Upper Nile. They showed a great skill in domestication of plants and animals, in weaving, in pottery, in erection of stone monuments and such other factors during the Neolithic period.

The characteristic physical features of this sub-group are light body build, dark complexion and narrow head form. Three distinct sub-types have been distinguished among the Mediterranean sub-group which have been mentioned below:

a) *Classical Mediterranean* (Basic Mediterranean or Ibero-Insular)

This type is found in the whole of Mediterranean basin and also in Portugal, Spain, France, Germany, Italy, etc. A sporadic distribution of this racial element is observed in Eastern, Central, and North-Western Europe. The same physical features are also visible among the Egyptians of North-Africa, the Berbers of Morocco and in Arabia. The Jewish population of Palestine also shares some of their characteristics. However, physical features of Classical Mediterranean are as follows:

Skin colour : light brown to barnet or tawny white.

Hair : wavy to curly. The colour is black.

Head : head form varies from dolichocephalic to mesocephalic. Cephalic index ranges between 73 to 76.

Nose : usually leptorrhine and straight with medium thickness.

Face : long and oval with pointed chin. Cheek bone is flat and the forehead is slightly high with moderate slope.

Eye : eye colour is usually dark.

Stature : the stature is medium with slender and delicate body-built.

CAUCASOID :

1. **Mediterranean**
   a) *Classical Mediterranean*
   b) *Atlanto Mediterranean*
   c) *Indo-Afghan*
2. **Nordic**
3. **Alpine**
4. **East Baltic**
5. **Dinaric**
6. **Armenoid**
7. **Keltic**
8. **Lapp**
9. **Indo-Dravidian**
10. **Polynesian**
11. **Ainu**

ARCHAIC CAUCASOID or Australoid : (Sub-division of **Caucasoid**)

1. **Australian Aborigines**
2. **Pre-Dravidian or Australoid or Veddoid**

MONGOLOID :

1. **Classical Mongoloid**
2. **The Arctic or Eskimoid**
3. **Indo-Malayan Mongoloid**
   a) *Malay type*
   b) *Indonesian type or Nesiot*
4. **Amerindian or The American Indian**
   a) *Palaeo-Amerind*
   b) *Northern Amerind*
   c) *Neo-Amerind*
   d) *Tehucleche*
   e) *North-West coast Amerind*

NEGROID :

1. **African Negro**
   a) *True Negro*
   b) *Nilotic Negro*
   c) *Bantu*
   d) *Bushman-Hottentot*
   e) *Negrillo (African Pygmy)*
2. **Oceanic Negro**
   a) *Negrito*
      i) *Asiatic Pygmy*
      ii) *Oceanic Pygmy*
   b) *Papuans and Melanesians*
3. **American Negro**

Fig. 7.6. Races of Man In Modern World

b) *Atlanto-Mediterranean* (or Littoral)

This type is distributed in the places like North-Africa, Palestine, Iraq, and the Eastern Balkans. Small numbers of this population are also present in the British Isles, Spain and Portugal.

However, the people are characterised by the following features :

Skin colour : generally dark skin is found.

Hair : the form varies from wavy to curly and the colour is very dark.

Head : form of head ranges from dolichocephalic to mesocephalic.

Nose : usually it is straight with medium height and breadth. The root is usually deep.

Face : long with deep jaws. Cheek bones are prominent. The forehead is retreating with well-developed eye-brow ridges.

Eye : the eye-colour varies from medium-brown to dark-brown.

Stature : though the people are tall, but medium stature also occurs among them. The average height is 170cm. They possess more robust body than the Classic Mediterraneans.

c) *Indo-Afghan* (or Irano-Afghan)

This type are mostly found in Iraq, Iran, Afghanistan, Baluchistan, North-West India and Pakistan. Physical features of this population are as follows:

Skin colour : very light brown.

Hair : the hair on head is wavy in form and black in colour. Body and facial hair is usually abundant.

Head : the form varies from dolichocephalic to mesocephalic. Cephalic index stands between 71 and 77.

Nose : the nose form is leptorrhine with pointed tip. Nasal ridge is straight or convex.

Face : the face is long and narrow.

Eye : the eye-colour is dark.

Stature : the stature is variable, from tall to medium (161 - 178 cm). But most of the people are medium (average 167cm).

2. Nordic

Various opinions have been expressed by different anthropologists regarding the origin and history of Nordic racial type. Some of them said that these peoples were originally 'Aryans'. But most of the modern anthropologists strongly oppose this view .

The people of Scandinavia are considered as the typical representatives of Nordic racial type. This type is well distributed in Scandinavia, Baltic region. Northern Germany, Northern France, some parts of Netherlands and Belgium, and also in British Isles. A sporadic distribution of this population has also been noticed in various parts of Europe. Not only that, they form an important element in the population of United States and some of the British colonies.

The physical features of the sub-group are as follows:

Skin colour : the skin colour is usually pinkish or reddish white.

Hair : the hair on head is slightly wavy and rarely curly in form. The colour is highly variable as reddish-brown, yellow or very light brown, ash-brown, golden brown, etc. The texture ranges from fine to medium. The hair on body and face is either sparse or medium.

Head : the form or head is mesocephalic. It looks finely arched. Cephalic index ranges between 76 and 79.

Nose : the nose is straight and prominent; leptorrhine in form. Nasal root is high and nasal bridge is also high.

Face : the face is long, narrow and its profile is straight. The bones of the face are strongly developed with flat cheek bones. Forehead is more or less vertical with moderately developed eye-brow ridges. The chin is quite prominent.

Eye : the eye-colour is blue or grey.

Lips : both the lips are very thin.

Stature : the stature is tall; average height is about 172 cm.

3. Alpine

Some anthropologists suggest that the Alpines originated in Central Asia. They tried to correlate this sub-group with Asiatic Mongoloid race. The admixture of Alpines with other racial types such as the Nrodic, Mediterranean, etc. are also found.

Concentration of the Alpine population is found in Central and Eastern Europe especially in the region from France to the Urals. They are also found in the Countries like Denmark, Balkan, Norway, Northern Italy, and in the mountains of Asia Minor. A sporadic distribution of this type has been observed in whole of Europe.

The population are characterised by the following physical features :

Skin colour : their complexion is either Olive or Burnet white, or Bronze.

Hair : the hair form is slightly wavy. The colour ranges from medium brown to dark brown; it is rarely black. The texture is medium to fine. The hair on body and face are abundant.

Head : the form is brachycephalic and the cephalic index is 85.

Nose : the nose is either leptorrhine or mesorrhine with straight or slightly convex profile. The tip of the nose is usually short, thick and fleshy. Nasal root is moderately high; the nasal bridge is also moderate in height and width.

Face : the face is broad and short. Its outline is more or less round and oval. The forehead is high. Eyebrow ridges are moderately or strongly developed. The chin is prominent.

Eye : the eyes are straight, dark to medium brown in colour. Occasionally it gets blue.

Lips : both the lips are moderately thick.

Stature : the stature is medium to short; average height is about 165cm. The body shows a strong built.

The *Beaker* : *Folk,* formerly known as "Round - barrow men", "Bronze Age men", etc. are regarded as a cross between Alpines and Nordics. This type of population are found in the region extending from Central Spain to Central Europe, in England and Scotland. They are also found to be sporadically distributed in most of the Central and Western European Countries.

The physical features of this type may be mentioned in the following ways :

Skin colour : the complexion is pinkish or reddish white.

Hair : form is wavy and the colour is dark brown.

Head : the head is large and brachycephalic in form. Cephalic index is 81.

Nose : the nose is long, narrow and prominent on leptorrhine form.

Face : the face is long and narrow. The malar region is prominent. The jaws are large with square chin. The glabella region is high, occiput is flattened and eye-brow ridges are well marked.

Eye : the eye-colour is blue or blue-grey.

Stature : the stature is very tall. The body is muscular and robust.

4. East Baltic

The East Baltic group exhibits an admixture of characters. They have borrowed traits from various sources of which the Nordic and the Alpine influences are dominant. Certain features strongly suggest an Asiatic Mongoloid influence. But the racial history of this population is practically unknown.

The concentration is mainly found in the areas like North-Eastern Germany, Baltic States, Poland, Russia, Finland, etc.

The physical characters have been enumerated in the following ways :

Skin colour : the skin colour is tawny white or creamy white.

Hair : the form of hair is usually straight and sometimes wavy. The texture varies from medium to coarse. The colour is generally ash-brown, rarely reddish. The facial hair is moderate and body hair is scanty.

Head : the form of head is brachycephalic with a flat occiput.

Nose : the nose form is mesorrhine. It is prominent with convex profile. Nasal root is either medium or low. Nasal bridge is moderately high and broad. Nasal wings are also broad where the tip is snubbed.

Face : the face is square in form with prominent cheek. Forehead is high. Angles of the lower jaws are square and the chin is well developed. Eye-brow ridges are moderate.

Eye : the eye-colour is light-blue or grey.

Lips : the lips are medium or thin.

Stature : the stature is extremely variable—usually short and medium predominate. The body is thickly set.

5. Dinaric (Adriatic or Illyrian)

The Dinaric race exhibits both the Nordic and the Armenoid physical features. This population is chiefly distributed in Dinaric Alps region, especially the Yugoslavia, Albania and Austrian Tyrol. A scattered distribution has been noted in Central Europe. In many cases they share some racial elements from Alpine, Atlanto-Mediterranean and possibly Indo-Afghan, apart from Nordic and Armenoid.

These people are characterised by the following physical features:

Skin colour : the complexion varies from lighter burnet to olive shade.

Hair : the form is straight or wavy or curly. The texture is medium. The colour runs from medium to dark brown, sometimes light shades are also found. The hair on body and face are abundant.

Head : the form is brachycephalic with flat occiput region.

Nose : the nose form is leptorrhine. It looks narrow from the profile view. The tip is fleshy in general. The nasal root is high. The nasal bridge is high and narrow.

Face : the form is long and narrow with deeper, heavier and more projecting chin than Armenoid people. Forehead is straight and sloping.

Eye : the eye-colour is brown which varies from light to dark.

Lips : the lips are moderately full and thick.

Stature : the stature is tall; average height is about 172cm.

6. Armenoid

The Armenoids exhibit a considerable amount of Classical Mediterranean, Alpine, Nordic and Indo-Afghan racial elements in them. Recent studies reveal that this particular group is an admixture between the Mediterranean and the Alpine. Dr. Byron has found out the Alpine, Classical Mediterranean and a minor amount of Nordic element in the formation of this race. Asia Minor may be assumed as the earliest known area from which the Armenoid race might have been spread southward to Arabia and India.

The sculpture of a man that has been discovered from the ruins of Mohenjodaro shows Armenoid physical features. Similar type of racial elements have also been projected from among the Babylonians, Assyrians and Hittites. However, Armenoids are concentrated in Turkey, Syria and Palestine. Such type of physical characters have also been noticed among the people of Iraq, Iran and Balkan Countries. The people of Greece, Bulgaria and United States are no exceptions. But, the typical Armenoid representatives are the ancient Hittites.

The physical features have been mentioned here in the following ways :

Skin colour : the complexion is tawny white or olive.

Hair : the hair is usually wavy to curly in form. The texture is coarse to medium. The colour varies from dark brown to black. Abundant hair is found on body and face.

Head : the form of head is brachycephalic with a vertical occiput.

Nose : the nose form is leptorrhine. It is very prominent. Nasal root is high and broad. Its profile is usually convex while the tip is depressed and fleshy.

Face : the face is narrow and elongated. The cheek bones are well developed. The forehead shows high slope where the eye-brow ridges are thick, specially in case of males. The chin shows medium prominence.

Eye : the eye-colour is medium-brown to dark brown.

Lips : the lips are moderately thick and full.

Stature : the stature is medium to tall with an average height of 167cm. The body is well built with a tendency towards obesity.

7. Keltic

This type of people are found in Ireland, Scotland and Wales. They are also sporadically distributed in England and many other parts of Western Europe.

The physical features are as follows:

Skin colour : the complexion is usually pale white.

Hair : the hair form is wavy or curly, rarely straight. The colour varies from medium brown to dark brown and it is rarely black.

Head : the head form is mesocephalic.

Nose : the form is leptorrhine. The nose is long and narrow. The nasal bridge is high. The profile is straight or convex.

Face : the face is long and narrow, cheek bones are compressed and the chin is deep.

Eye : the eye-colour is blue or grey.

Stature : the stature is tall and average height is 172 cm.

8. Lapp

The Lapps are found in Northern Scandinavia, Northern Finland, Sweden, Norway and North-Western region of Russia. This type shows an admixture of various people, e.g. Russians, Fins, Swedes, Norwegians, etc., but they hold some distinct features also, by which they have been identified as a separate ethnic group. Though some of the Lapps are found to be classified with the Mongoloids, the people bear more similarity with the Caucasoids than the Mongoloids. However, the physical features are as follows:

Skin colour : the skin colour varies from greyish yellow to yellowish brown.

Hair : the form is usually straight or slightly wavy. The colour is dark brown or black. The hair on body and face is sparse.

Head : the form of head is brachycephalic; the cephalic index is 84.

Nose : the form is mesorrhine. Nasal profile is concave and the tip is very often snubbed.

Face : the face is moderately broad and remarkably short. The cheekbones are prominent with forward projection. Eye-brow ridges are faintly developed on the narrow forehead. The face shows little or no prognathism.

Eye : the eye-colour is highly variable and an epicanthic fold is occasionally present.

Lips : the lips are moderately thick.

Stature : the stature is short and average height is 159cm.

9. Indo-Dravidian (Dravidian)

The Indo-Dravidian people are distributed in South and Central India. They are predominantly Caucasoid. An admixture of Classical Mediterranean and Australoid (Veddid) is found among these people. The physical features of this type may be mentioned in the following ways :

Skin colour : the skin colour varies from light brown to dark brown.
Hair : the form of hair is slightly wavy; occasionally it tends toward curly. The people possess plentiful hair, black in colour. The hair on body and face is sparse to medium.
Head : the head form is dolichocephalic; the cephalic index is 73.6.
Nose : the nose form is mesorrhine. Nasal root is depressed, nasal bridge is high, and nasal profile is straight. But the tip is to some extent thick.
Face : the face is narrow and medium in length. Little prognathism is seen at times. Forehead is usually rounded and eyebrow ridges are moderately developed.
Eye : the eye-colour is medium to dark brown.
Stature : the stature is medium; average height is 164cm.

10. Polynesian

The Polynesians are a composite race. They originated as the white people but got mixed with the peoples of early Mediterranean, Asiatic Mongoloid and Oceanic Negro. However, the Polynesians are concentrated in Polynesian Islands of the Pacific Ocean namely New Zealand, Friendly Islands, Samoa, Marquesas and Hawaii.

The physical characteristics may be described in the following ways :
Skin colour : the complexion is light brown to yellow brown.
Hair : the form is usually wavy and sometimes straight, but rarely curly. The colour is dark brown to black. The body and facial hair is sparse.
Head : the head form is predominantly brachycephalic; cephalic index is 82.6. Dolichocephalic and mesocephalic forms are also present. But the occiput region is typically flat.
Nose : the nose form is usually mesorrhine. Nasal root is slightly depressed but sometimes it is high. The nose is prominent. Nasal bridge is high and broad. The profile is sometimes convex but mostly it is straight. The nasal tip is thick. Nasal wings are broad.
Face : the face form is long and broad. The cheek bones are prominent. Forehead is high, slightly sloping and narrow. Eye-brow ridges are moderately developed while the chin is well developed.
Eye : the colour of eye varies from medium brown to dark brown. The epicanthic fold is rarely found.
Lips : the lips are moderately thick.
Stature : the stature is tall; average height is about 178cm. The body is muscular and well-built.

11. Ainu

The Ainus represent an ancient racial stock of Japan. Although the population is basically Caucasoid, a considerable amount of Mongoloid admixture is found among them. Besides, they show a close resemblance with the Australian aborigines which has been reflected in a number of physical features. Therefore, some anthropologists like to include these people under the Australoid or Archaic Caucasoid group.

Their concentration is best found in Northern Japan, South Sakhalin and Yezo. The physical characters of this population are as here under.
Skin colour : the skin colour ranges from light brown to brownish white.
Hair : the form of hair is wavy. The colour varies from dark brown to black. The hair on body and face is abundant. For the excessive hair on body and face, these people are often referred as the "Hairy Ainu".
Head : the head form is usually mesocephalic but sometimes it is dolichocephalic. Cephalic index is 76.5.
Nose : the nose form varies from mesorrhine to platyrrhine. The nose is short. Nasal root is slightly depressed and nasal bridge is moderately high. The profile is straight to convex.
Face : the face is short and medium in breadth. The form of face is mesoprosopic and orthognathic. The jaws and the chin are well developed.

Eye : the eyes are horizontal with wide openings where epicanthic fold is usually absent. The colour ranges from medium brown to dark brown.
Lips : the lips are thin.
Stature : the stature is medium to short while the average height is 158cm. They have thick set body.

AUSTRALOID OR ARCHAIC CAUCASOID

The Australoids are found to possess some primitive features that denotes the survivals of some early Caucasoid varieties. This is the reason for which these people are considered as a sub-division of the Caucasoid racial stock.

The Australoids have been classified into two main groups—the Australian aborigines and the Pre-Dravidian or Australoid or Veddoid. These two groups have been further sub-divided into several subtypes.

1. Australian Aborigines

These people suggest an admixture of an archaic Caucasoid type with some Negroids. Certain amount of Oceanic influence has also been noticed in them. The people are concentrated chiefly in Australia.

The physical features may be furnished in the following ways.
Skin colour : the complexion varies from medium brown to dark chocolate brown.
Hair : the form of hair is wavy or curly but rarely straight. The colour ranges from medium brown to black. The hair on body and face is usually abundant.
Head : the head form is dolichocephalic and narrow. Cephalic index is 73.
Nose : the nose form is platyrrhine. It is very broad. Nasal root is markedly depressed. Nasal bridge is moderately high and broad,while the tip is very thick.
Face : the face is short and shows medium to pronounced prognathism. Forehead is receding with prominent glabella region. Eye-brow ridges are extremely large but the chin is usually receding.
Eye : the eye-colour varies from medium to dark brown.
Lips : the lips are medium thick.
Stature : the stature is variable. The average height is about 165cm. The body is medium-built.

2. Pre-Dravidian (or Australoid or Veddoid)

This population is mainly concentrated in South and Central India. The typical representatives of this type are Kadir, Kurumba, Irula, Bhil, Gond, Khond, Oraon, etc.

The physical characters can be mentioned in the following ways :
Skin colour : the skin colour ranges from dark brown to nearly black.
Hair : the form of hair is wavy or curly and the colour is black. The hair on body and face is scanty.
Head : the head form is dolichocephalic and the cephalic index is between 73 and 75.
Nose : the nose form is platyrrhine and it is very broad. Nasal root is depressed and the bridge is moderately high.
Face : the face is short and narrow. The prognathism is moderate. Forehead is sloping with prominent eye-brow ridges. The chin is somewhat receding.
Eye : the eye-colour is dark brown.
Lips : the lips are very often thick.
Stature : the stature is short and average height is 1.57 cm. The body is delicately built.

Veddas

The Veddas of Ceylon is also a typical member of this group. Their physical features are as follows.
Skin colour : the complexion is dark brown.
Hair : the form is wavy or slightly curly. The colour is black, sometimes with reddish tinge.
Head : the head of Veddas is the smallest of all living human groups. The form is dolichocephalic where the cephalic index is 70.5.

Nose : nasal root is depressed and the form is platyrrhine. Nasal bridge is low and broad.
Face : the face is short and broad. Sometimes, little prognathism is visible. Forehead is slightly receding and eye-brow ridges are often pronounced. The chin is also receding.
Eye : the eye-colour is dark brown.
Lips : the lips are medium or thick.
Stature : the stature is short and average height is 152cm. The body is of small size.

Sakai or Sanoi

The Sakai of Southern part of Malay Peninsula is the other distinguished member of the Australoid group. The physical characters are as follows:
Skin colour : the colour varies from yellowish brown to dark brown.
Hair : the form of hair is wavy or curly. The colour is black with reddish tinge.
Head : the form of head is mesocephalic where the cephalic index is 75.
Nose : nose form is mesorrhine, but approaching towards platyrrhine. The nose is flat with medium breadth.
Face : the face is narrow which shows little or no prognathism. The chin is weak.
Eye : the eye-colour is dark brown.
Lips : the lips are of medium thickness.
Stature : the stature is short; average height is 152cm. The body is slender and small.

MONGOLOID

Though the history of Mongoloid people has not yet been known clearly, still it has been stated that this type has originated in the land of Central Asia. From that place the people perhaps spread in different directions.

The Mongoloids can be divided into four main subdivisions on the basis of their geographical distribution. These divisions are as follows: (1) The Classical or Central Mongoloid, (2) The Arctic or Northern Mongoloid or Eskimoids, (3) The Southern or Indo-Malayan Mongoloids; and (4) The American Indians.

The Mongoloid physical features have been characterized in the following ways :
Skin colour : yellow or yellowish brown.
Hair : straight. The body and facial hair is scanty.
Head : usually brachycephalic.
Nose : the nasal root low and the nasal bridge is low to medium. The nasal profile is usually concave or straight.
Face : broad and flat with prominent cheek bones.
Eye : obliquely set with narrow slit-like opening; total epicanthic fold is present.
Stature : the stature is variable.

1. Classical Mongoloid or Central Mongoloid

This type is found to be distributed mainly in Siberia and Amur river district, and sporadically in Northern China, Mongolia and Tibet. The representative groups are, Buriat, Koryak, Goldi, Gilyak, etc. The Tibetans and some other Northern Chinese also present this racial element.

The physical features may be noted in the following ways :
Skin colour : yellow or yellowish brown.
Hair : the form of hair is straight, texture is coarse and colour is black. The hair is sparsely distributed on the body and face.
Head : the form of head is usually brachycephalic with a cephalic index of 85, but mesocephalic and dolichocephalic heads are not uncommon. All forms show a projected occiput region.
Nose : the nasal root is low but without any depression. Nasal bridge is also low and medium in

breadth. The profile is variable—usually straight or concave. The nasal wings are moderately spread.
Face : the face is very broad with square jaws. Forehead is rounded and medium in height. The cheek bones are strongly developed and they are projected laterally as well as forwardly.
Eye : obliquely set with slit-like opening. Typical epicanthic fold is present. The colour is medium to dark brown.
Stature : the stature is variable and the body is well built.

2. The Arctic or Eskimoid

These people are found in Northern Asia, the Arctic coast of North America, Greenland, Labrador and Western Alaska. The representative populations are the Eskimos, Chukchis, Kamtchadales, Yakuts, Samoyedes, etc.

The physical features are as follows:

Skin colour : dark yellow to brownish.
Hair : the form is straight, texture is coarse and the colour is black. The hair on body and face is scanty.
Head : the head form is variable, ranging from brachycephalic to mesocephalic. Eastern Aleutian Eskimo (C.I. 84.6) and, the Kuskokwin Eskimo (C.I. 81.5), both are brachycephalic, but the Greenland Eskimo (C,176.8) and Arctic Eskimo (C.1 78.6) are mesocephalic.
Nose : narrow but prominent.
Face : the face is large, broad and flat with prominent cheek bones.
Eye : the eye-colour is black. In structure they are straight; a complete epicanthic fold is found occasionally.
Stature : the stature is usually short but variable. As for example, the Western Eskimos (168cm) are taller than the Alaskan Eskimos (163.8cm). The body proportion is peculiar with remarkably small hands and feet, large trunks and relatively short legs.

3. Indonesian-Malay Mongoloid

The population is comprised of a large number of Mongoloid peoples who show a considerable admixture of Caucasoid[1] and Negroid elements. Such people are distributed throughout the Southern Asia and known as the Indonesian-Malay Mongoloid racial type. This type is further divided in two groups—Malay type and Indonesian type.

(a) Malay type

This type of people are distributed in Southern China, Indo-China, Burma, Thailand, Malay Peninsula, Dutch East Indies, the Philippines, Japan, etc. The Japanese mostly belong to this Malayan type of racial sub-group (The Mongoloid features appear to be more strong in the Malay type than in the Indonesian type).

The physical features may be discussed in the following ways :

Skin colour . light yellow brown to dark yellow brown.
Hair : the form is straight and the colour is black with occasional reddish tinge.
Head : the head form is brachycephalic with a cephalic index of 85.
Nose : the nose form is mesorrhine and sometimes platyrrhine. Nasal root is slightly depressed and the nasal bridge is low.
Face : it is short and broad with prominent cheek bones.
Eye : the eye-colour usually ranges from medium brown to dark brown, and it is occasionally black. Internal epicanthic fold is present.
Stature : the stature is short and the average height is 158cm.

(b) Indonesian type or Nesiot

This type is found in Southern China, Indo-China, Burma, Thailand, etc. The physical features are as follows:

Skin colour : it varies from light red brown to medium brown.
Hair : the hair form is usually wavy (slight) and the colour is black (sometimes with a reddish tint).
Head : the form of head is mesocephalic; the cephalic index is 78.5. Dolichocephals are not totally absent.
Nose : the form is mesorrhine. But the overall shape is narrow, high and long.
Face : the face is narrower, longer and more oval than the Malay type.
Eye : the colour is black with occasional reddish tint. The internal epicanthic fold is less frequent.
Lips : the lips are thick.
Stature : their height is slightly shorter than the Malay type and it is 155cm. The body is slender in form.

4. The American Indian or The Amerindian

The physical anthropology does not furnish adequate data on American Indians to project the people scientifically under racial classification. More works are still required in order to get a clear-cut idea on their racial characteristics. However, on the basis of present knowledge, it appears that this type is predominantly Mongoloid, but racial elements from Caucasoid, Australoid and Negroid people are also present among them. It is very interesting that the diversified racial elements could not affect the basic homogeneity of physical features. The American Indians are distributed in different areas of North, Middle and South America.

The characteristic features are as follows:
Skin colour : the skin colour varies from yellow brown to red bown.
Hair : the form of hair is straight, sometimes slightly wavy. The texture is coarse. The colour is usually black and rarely dark brown. The hair on body and face is sparsely distributed.
Head : the head form is either dolicho-mesocephalic or brachycephalic. Because of the presence of these two predominating types of head forms in the population, this type is often distinguished as the dolicho-mesocephalic and brachycephalic.
Nose : the nose form is predominantly mesorrhine. The nose is very long. The nasal bridge is usually high with convex profile and the tip is of medium thickness.
Face : the face is broad with typical Mongoloid cheek bones. Forehead is sloping even more than the typical Mongoloids. Eye-brow ridges as well as glabella are strongly elevated. The chin is more prominent than the typical Mongoloids. Shovel-shaped incisors are very common and the face usually exhibits medium prognathism.
Eye : the eye-colour is dark brown to black. Complete Mongoloid fold is practically absent. Internal epicanthic fold is frequently present in women and children. But the external epicanthic fold is a common phenomenon both in male and female.
Lips : the lips are thin.
Stature : the stature is variable.

(a) *Palaeo - Amerind*

1. One group of scholars have designated these people as Lagoa Santa type of Brazil, Ecuador, Orinaco. It is actually an archaic South American race. A few of their living representatives are Botocudo, Buru, etc. Some of the members are found in the Eastern United States and Canada. America shows a sporadic distribution of this type in many of its parts.

The people are dolichocranial. Their skulls are small but with high vault. The faces are long and narrow. The skin colour is more reddish brown than yellowish brown. The wavy hair is black in colour.

(b) Northern Amerind

The North American Indians and the people of the Northern and Eastern Woodlands belong to this group.

The physical features are as follows:

Skin colour : the skin-colour is yellowish brown.

Hair : the form of hair is straight and its colour is black.

Head : the head form is either dolichocephalic or mesocephalic. The cephalic index is within 73 and 75. In the West, these people show a tendency towards brachycephalic.

Nose : the nose form is mesorrhine with straight or convex profile.

Face : the face is oval.

Eye : the eye-colour is medium to dark brown and an external fold is seen.

Stature : the stature is tall and the height varies from 161 to 175 cm.

(c) Neo-Amerind

This type is distributed in South America, Central America and North American Plateau. Their physical features are as follows:

Skin colour : the skin colour is yellow-brown.

Hair : it is straight and black in colour.

Head : the form of head is brachycephalic.

Nose : the mesorrhine nose shows straight or concave profile.

Face : it is broad but shorter than the dolichocephals.

Eye : the eye-colour is black and the external fold is seen.

Stature : the stature ranges from short to tall and the height varies from 155 to 178cm.

(d) Tehucleche

This *type* lives in Patagonia, and probably the Onas of Tierra del Fuego constitute a branch of Tehuelche. The physical characters are as follows:

Skin colour : the skin colour is brownish.

Hair : the form is straight and the colour is black.

Head : the head form is brachycephalic. The cephalic index is 85.

Nose : the mesorrhine nose shows a straight profile.

Face : the face is square and broad.

Eye : the eye-colour is black and an external fold is found.

Stature : the stature is tall and the height varies from 173 to 183cm.

(e) North-West Coast Amerind

Among these people the colour of skin and hair is lighter than any other Northern Amerinds. The stature is medium with long arms and short body. The people are similar to the natives of North East Asia. In this group two sub-types are distinguished:

Northern type and *Southern type* : The Northern type is taller than the Southern type. They live in the North West coast of North America. In Northern type, the people show a concave or straight nose and a broad face with moderate height. But in Southern type, the nose are very often convex and high, and the face is of great height.

NEGROID

This Negroid racial group is mainly divided into two types—African Negro (Ulotrichi Africani as designated by Haddon), and Oceanic Negro (Ulotrichi Orientalis as designated by Haddon).

The generalized characters of this racial type are as follows:

Skin colour : the skin colour ranges from dark brown to black.

Hair : the form of hair is woolly or frizzly and its colour is black. The people show very little body hair and sparse facial hair.

Head : the head form is predominantly dolichocephalic. Their head is round with a protruding occiput region.

Nose : the nose is broad and flat. Nasal root and bridge are usually low and broad.

Face : facial prognathism is often marked. The forehead is rounded with small eye-brow ridges. The chin is rounded and receding.

Eye : the eye-colour is dark brown to black.

Ear : the ear form is usually short, wide with rolled helix and there is little or no lobe.

Lips : the lips are thick and everted.

Stature : the stature is variable.

1. African Negro

This sub-division of the Negroid racial stock is further classified into five sub-divisions—True Negro, Nilotic Negro or Nilote, Bantu-speaking Negroes or 'Bantu', Bushman-Hottentot, and Negrillo.

(a) True Negro

This type of people are distributed in West Africa and in Guinea coast. The physical characteristics are as follows:

Skin colour : the skin colour is dark brown or black.

Hair : the form of hair is woolly and the colour is black.

Head : the head form is dolichocephalic. The cephalic index remains within 73-75.

Nose : the nose form is platyrrhine.

Face : face is prognathous often with a bulging forehead.

Eye : the eye-colour varies from dark brown to black.

Lips : the lips are thick and everted.

Stature : the stature is tall with an average height of 173cm. Their body is well-built with short legs and long-arms.

The typical *Forest Negroes* show slightly different physical characters from that of *True Negroes.* This Forest Negroes live in a region extending from the Senegal River in the West to Sudan, Uganda and Northern Rhodesia. Their physical characters are as follows:

Skin colour : the skin colour is dark brown to black.

Hair : the form of hair is woolly and the colour is black.

Head : the head is dolichocephalic with a cephalic index within 73 - 75.

Nose : the nose form is platyrrhine. The nose is very broad. Nasal root is low with flat nasal bridge.

Face : the face is markedly prognathous, chin is retreating and cheek bones are prominent.

Eye : the eye-colour is dark brown to black.

Lips : the lips are markedly everted.

Stature : their stature is slightly shorter than True Negroes and average height is about 165cm. The face and body are very rough.

(b) Nilotic Negro or Nilotes

Some physical features of the Nilotic Negroes are completely different from the True Negroes. Certain Mediterranean element is responsible for this situation. It is assumed that some prehistoric Mediterranean people moved into Nilotic regions where they got mixed with the Negroid people. Therefore, the Nilotic Negroes of North-Eastern Africa show marked Mediterranean features. Besides, some of the Nilotes as the Shilluk, the Dinka, the Kavirondo and others show certain Hamitic or Ethiopian elements. This phenomena indicates an infusion of Hamitic traits among these peoples.The Nilotes have been referred as Negroids.

Their concentration is found in the regions of Upper Nile Valley and Eastern Sudan. Their physical characters are as follows:

Skin colour : the skin colour varies from very dark to bluish black.

Hair : the form of hair is woolly with black colour.

Head : the head form is dolichocephalic. The cephalic index is 71-74.

Nose : the nose form is platyrrhine, but the degree is lesser than the True Negroes. Nasal bridge is low with broad and low nasal root.

Face : the face is broad and short with less facial prognathism. Forehead is retreating. The chin is better developed than the Forest Negroes.

Eye : the eye-colour is dark brown.

Lips : the lips are thick and everted but a little lesser than in the True Negroes.

Stature : the stature is very tall and average height is 178cm. Legs are long and the figure is slim.

(c) Bantu-Speaking Negroid or 'Bantu'

These Bantu-speaking peoples are essentially Negroes among whom we find an infiltration of Hamitic, Negritto and Bushman-Hottentot elements. A large number of Bantu-speaking peoples of Central and Southern Africa have been included in this group. Different ethnic elements that constitute this group have not yet been clearly defined. But, a wide range of variation in physical characters have been noticed within the group. The physical characters are as follows:

Skin colour : usually dark chocolate, but the colour varies from yellowish-brown to black.

Hair : the hair form is woolly or frizzly and the colour is black.

Head : the head form is typically dolichocephalic, but mesocephalic form is not unusual.

Nose : the nose is generally narrow and more prominent than in the True Negroes.

Face : prognathism is marked. But, the mesocephalic group possesses less marked prognathism with flatter forehead.

Eye : the eye-colour is dark brown.

Stature : the stature is medium or above average. The height is 167cm. while the mesocephalic group invariably shows shorter stature.

(d) Bushman-Hottentot

The Bushman and the Hottentot are more or less same people in terms of physical characters. They show very little differences. But, culturally they are vastly different from each other. The Hottentots are known as the Khoi Khoi and the Bushmen, the Khuai or San. The Bushmen are mainly confined to the Kalahari desert, though previously they occupied a greater part of South Africa. The Hottentots are distributed in South-West Africa.

The physical features may be mentioned in the following ways :

Skin colour : the skin colour is light to brownish yellow in Bushmen and light reddish-yellow in Hottentot.

Hair : the form of hair is pepper-corn i.e. the hair is short and shows a tendency towards coiling—simple coils to spiral knots are found on head; bare spaces are present between them. The colour is black. The hair on body and face is sparse or absent.

Head : the head form is dolichocephalic and high in the Bushmen whereas it is mesocephalic and low in the Hottentots. Parietal bosses are more marked and occiput is less protruding in the Bushmen than in the Hottentots.

Nose : the nose form is platyrrhine with very broad and flat nasal root. The nasal bridge is low and broad. The nasal profile is concave with thick tip.

Face : the face is short and square, and often orthogonathous in the Bushmen. It is more elongated, triangular and somewhat prognathous in the Hottentots. The chin is small and the cheek bones are very prominent. Bulbous forehead shows little developed eye-brow ridges.

Eye : the eyes are often narrow and slanting. The colour ranges from dark brown to black.

Ear : the ears are frequently lobeless.

Lips : the lips are usually thick.

Stature : the Hottentots are slightly taller than the Bushmen. Average height of Bushmen and Hottentots are 145 cm and 160cm respectively. Hands and feet are small. Steatopygia (immense deposit of fat in the buttocks) is more pronounced in the Hottentot women than in the Bushwomen.

(e) Negrillo (African pygmy)

The Negrillo type has been represented by the groups like Akka, BaTwa, BamBute, etc. who live in Equatorial forests of Congo region. Their physical characters are as follows :

Skin colour : the skin colour varies from yellowish light brown to reddish brown, but sometimes it is very dark.

Hair : the form of hair is short, woolly or pepper-corn. The colour of head hair is dark rusty brown, body hair is yellowish, hair under armpits is brown and those are black on pubis.

Head : the head form is mesocephalic with a cephalic index of 79.

Nose : the nose is very broad and flat. The nasal wings are very broad and high.

Face : the face is usually prognathic with weak and narrow chin.

Eye : the eye-colour is dark brown.

Lips : the lips are full but not everted .

Stature : the stature is very short; average height is 136cm. The arms are long and the legs are short with short trunk. Steatopygia is occasionally present in women.

These pygmy statured people are commonly known as the pygmies and they are found to be distributed in different parts of the world such as the Congo region of Equatorial Africa, Malay Peninsula, Sumatra, Andaman Islands, Phillipine, New Guinea, etc. In reference to the geographical position, these population may be grouped into three sub-sections, African pygmy or Negrillo, Oceanic pygmy and Asiatic pygmy. The Oceanic pygmy and Asiatic pygmy, these two types are generally grouped as Negrito.

2. Oceanic Negro

This type of people are mainly concentrated in New Guinea and neighbouring Islands. Their physical characters are as follows :

Skin colour : the skin colour varies from medium to dark brown.

Hair : the form of hair is usually frizzly, rarely curly. The colour is dark brown to black. The hair on body and face is scanty.

Head : the head form is usually dolichocephalic, but sometimes it is brachycephalic.

Nose : the nose is platyrrhine. Nasal bridge is high and broad with depressed nasal root.

Face : the face is less prognathous with small prominent eye-brow ridges.

Eye : the eye-colour is dark brown or black.

Lips : the lips are medium to thick.

Stature : the stature is usually low; the average height is less than 165cm.

This division of Negroids are further divided into two sub-divisions which includes (a) the negritos, both Asiatic and Oceanic, and (b)Papuans and Melanesians.

(a) Negrito

The Andamanese, Semang, Aeta, and Tapiro are the representative groups of the Negrito type. Out of these four, the first three have been grouped as the Asiatic pygmy and the Tapiro is considered as the Oceanic pygmy.

Asiatic pygmy:

Andamanese : these people live in Andaman Islands. Their physical features are as follows:

Skin colour : the skin colour varies from bronze to sooty black.

Hair : the hair form is woolly. The colour is usually black with reddish tinge. The hair on body and face is scanty or absent.

Head : the head form is small, brachycephalic. The cephalic index is 83.

Nose : the nose is straight, sunken at the root.

Face : the face is broad at the malar region but the jaws are not projecting.

Eye : the eye-colour is dark brown.

Lips : the lips are full but not everted.

Stature : the stature is very short. The average height is 148cm.

Semang : these people live in Central region of the Malay Peninsuala and in East Sumatra. Their physical characters are as follows:

Skin colour : the skin colour is dark chocolate brown.

Hair the hair form is woolly and colour is black with reddish tinge. The hair on body and face is scanty.

Head : the form of head is mesocephalic. The cephalilc index is 79.

Nose : the nose is short, flattened and very broad.

Face : the face is round and the upper jaw is slightly projecting.

Eye : the eye-colour is dark brown or black.

Lips : the lips are usually thin.

Stature : the stature is short. The average height is 152cm. Body-built is sturdy.

Aeta: these people live in the Phillipine Islands. The physical characters are as follows:

Skin colour : the skin colour is sooty brown.

Hair : the form of hair is frizzly. The colour is dark brown or black. The hair on body and face is abundant.

Head : the form of head is brachycephalic. The cephalic index is 82.

Nose : the nose is very short, broad and flat.

Face : the face is round or oval.

Eye : the eye-colour is dark brown or black.

Lips : the lips are moderately thick.

Stature : the stature is short, the average height is 146cm.

Oceanic pygmy:

Tapiro : these people are the inhabitants of New Guinea. Their physical characters are as follows:

Skin colour : the skin colour is yellowish brown.

Hair : the form of hair is woolly and the colour is black. The hair on body and facc is abundant.

Head : the form of head is mesocephalic with a cephalic index of 79.5.

Nose : the nose is short, straight and medium.

Face : the face is average.

Eye : the eye-colour is dark brown.

Lips : upper lip is deep and convex.

Stature : the stature is short (average height of male is about 146cm) and the body is muscular.

E. A. Hooton has distinguished two varieties among the Negrito viz., The Infantile Negro and The Adultiform Negro. Some anthropologists suggest a genetical interrelationship among the pygmis of different areas; the branches of an old racial stock possibly diverged towards various geographical directions. But recent studies point out that pygmy is not a race at all. Several environmental factors are responsible for the formation of this physical type. Therefore, the concept of a particular race or a common stock is invalid in reference to the pygmies.

(b) Papuans & Melanesians

Papuans : These people are distributed in New Guinea and other Islands of Melanesia. The physical characters are as follows:

Skin colour : the skin colour is either dark chocolate brown or sooty brown.

Hair : the form of hair is somewhat frizzly with dark brown colour. The body hair, specially facial hair is abundant while the colour often ranges from dark brown to reddish brown.

Head : the head form is typical dolichocephalic.

Nose : the nose is broad with depressed root. The profile is convex with thick tip.

Face : the face shows a high and narrow retreating forehead. It often possesses heavy and continuous eye-brow ridges. The face is prognathic.

Eye : the eye-colour is dark brown.

Lips : the lips are usually thin.

Stature : the stature is variable, but mostly medium. The average height is about 168cm.

Melanesians : these people live in the coastal plains of New Guinea and the neighbouring Islands in Fiji, Admiralty Island, New Caledonia, etc.Their physical characters are as follows:

Skin colour : the skin colour is usually dark chocolate, sometimes copper-colour or very dark.

Hair : the form of hair is usually frizzly, but sometimes it is curly or even wavy. The color is invariably black. The hair on body and face is scanty.

Head : the head form is dolichocephalic in general but mesocephalic and brachycephalic forms are not absent. The cephalic index varies between 67 and 85.

Nose : the nose form is platyrrhine. Nasal root is deeply depressed with straight or concave profile. The nose tip is thick.

Face : the face is average. Forehead is rounded, wider and longer than the Papuans. But the eyebrow ridges are less developed in comparison to Papuans.

Eye : the eye-colour is dark brown or black .

Lips : the lips are not usually thick.

Stature : the stature varies from short to medium.

The American Negroes

This particular type has attracted the attention of many scholars. Because the institution of slavery was found to be continued with them in the United States throughout the first half of the 19th. Century AD. Inter breeding of the people with different ethnic groups such as the African Negroes, the American Indians and the Caucasoid has taken place for a long time. This long process of admixture has resulted in two distinct groups—the North American Caucasoid and the American Negroes.

The North American Caucasoid group combines mainly the traits of European Caucasoid races. Some traits from the American Indians and the Negroes are also found in them. On the other hand, the American Negroes are more and more complex. They include Forest Negro traits, Caucasoid traits as well as the American Indian traits. However, the general physical features of American Negroes may be mentioned in the following ways:

Skin colour : the skin colour varies from olive to dark brown; an increase of Caucasoid trait makes the skin colour more lighter.

Hair : the form of hair is woolly. For the increase of Caucasoid and American Indian mixture, it has become longer. The colour is usually black or dark brown.

Head : the head form is dolichocephalic.

Nose : the nose characters show intermediate features between the Forest Negro and the Caucasoid. But in comparison to Forest Negroes, it is higher and narrower at the root and bridge.

Face : the face is somewhat longer than the Forest Negroes. Prognathism is very little or absent.

Eye : the eye-colour is light, brown or dark brown.

Lips : the lips are medium or thick.

Stature : the stature is variable, but usually tall. These people are taller in comparison to the Negroes of West Africa.

RACIAL ELEMENTS OF INDIA

The physical nature of the contemporary Indians was unknown till the beginning of last century. Because population of India was extremely complex by the continuous penetration of new racial elements from outside, since the time immemorial. However, anthropologists of twentieth century attempted to analyze the ethnic composition of Indian population.

Classification of Sir Herbert Hope Risley (1915)

Sir Herbert Hope Risley tried to classify the Indian population on the basis of anthropometric measurements. He had developed a clear-cut idea about the racial elements of India when he directed the operation of Census for India in 1901. Later, he took the help of anthropometry to affirm his assumptions and published the results in 1915 under the title *'The People of India'*. He identified three principal racial types in India viz. The *Dravidian,* the *Indo-Aryan* and the M*ongoloid.* These three types were again got mixed in varying degrees in different provinces (States). On the whole, Risley distinguished seven different 'physical types' in the Indian population in the following way.

1. The Dravidian type

The stature of these people is short or below medium. The complexion is dark, approaching to black. The hair is similarly dark and plentiful with an occasional tendency to curl. The eye colour is also dark. The head is long and the nose is very broad, sometimes depressed at the root. The people of Dravidian type are distributed in the region from Ceylon to the valley of the Ganges covering the southern part of India,which especially includes the Western Bengal, Tamil Nadu, Andhra Pradesh (Hyderabad), Central India and Chotonagpur. The best example of this type is the Paniyans of Malabar (South India) and the Santals of the Chottanagpur. Risley believed these people as original inhabitants of India who are found to be modified at present by the infiltration of the Aryans, the Scythians and the Mongoloids.

2. The Indo-Aryan type

This type is the most close to the traditional Aryans who colonized India. The people are tall statured with fair complexion, dark eyes, and plentiful hair on face and body. They also possess predominant longhead (dolichocephalic), narrow and long (leptorrhine) nose. The type is confined to Punjab, Rajasthan and Kashmir where the members are known as the Kashmiri Brahmins, Rajputs, Jats and the Khattris.

1. Sir Herbert Hope Risley (1915) :

1. Dravidian
2. Indo-Arvan
3. Mongoloid
4. Aryo-Dravidian
5. Mangolo-Dravidian
6. Scytho-Dravidian
7. Turko-Iranian

II. Giufrida-Ruggeri (1921) :

1. Negrito

2. *Pre-Dravidian or Australoid Veddic*
3. *Dravidian*
4. *Tall Dolichocephal*
5. *Dolichocephalic Aryan*
6. *Brachycephalic Leucoderm*

III. A.C. Haddon (1924) :

1. *The Himalayan Region*
   a) *Indo-Aryan*
   b) *Mongoloid*
2. *The Northern Plains or Hindustan Region*
   Indo-Afghan
3. *The Southern Plateau or The Deccan Region*
   a) *Negrito*
   b) *Pre-Dravidian*
   c) *Dravidian*
   d) *Southern Brachycephals*
   e) *Western Brachycephals*

IV. Fuherer von Eickstedt (1934) :

1. *Wedid or Ancient Indians*
   a) *Gondid*
   b) *Malid*
2. *Melanid or Black Indians*
   a) *South Melanid*
   b) *Kolid*
3. *Indid or New Indians*
   a) *Gracile Indid*
   b) *North Indid*
4. *Palae-Mongoloid*

V. B.S. Guha (1937) :

1. *Negrito*
2. *Proto-Australoid*
3. *Mongoloid*
   a) *Palaeo-Mongoloid*
   i) *Long-Headed*
   ii) *Broad Headed*
   b) *Tibeto-Mongoloid*
4. *Mediterranean*
   a) *Palae-Mediterranean*
   b) *Mediterranean*
   c) *Oriental*
5. *The Western Brachycephals*
   a) *Alpinoid*
   b) *Armenoid*
   c) *Dinaric*
6. *The Nordics*

VI. S.S. Sarkar (1961) :

1. *Dolichocephals*
   - *a) Australoid*
   - *b) Indo-Aryan*
   - *c) Mundari-Speakers*
2. *Mesocephals*
   - *d) Irano-Scythian*
3. *Brachycephals*
   - *e) Far Eastern*
   - *f) Mongolian*

Figure. 7.7. Classification Showing Racial Elements of India

3. The Mongoloid type

The most important characteristic features of this type are broad-head, dark complexion with yellowish tinge and scanty hair on face and body. The stature is usually short or below medium. The nose show a wide range of variation, from fine to broad. The face is typically flat where the eyes are oblique with epicanthic fold. The people of this type are found along the Himalayan region,especially in the regions namely North East Frontier, Nepal and Burma. The best examples are the Kanets of Lahul and Kulu Valleys, Lepchas of Darjeeling and Sikkim, the Limbus, the Murmis and the Gurungs of Nepal and the Bodo of Assam.

4. The Aryo-Dravidian type

This type is known as the Hindustani type. Generally the heads of the people are long with a tendency towards medium. The complexion varies from light brown to black. The nose is usually medium, although the broad nose is not uncommon. But in this case, the broad nose is always broader than the nose of Indo-Aryans. In stature, the people are shorter than the Indo-Aryans who usually show a below average height; i.e. the height ranges from 159cm to 166cm. Thus, the Aryo-Dravidians is differentiated from the Indo-Aryans. The type is considered as an intermixture of the Aryans and the Dravidians in varying proportions. The people of this type are found in Uttar Pradesh, in some parts of Rajasthan and in Bihar.

5. The Mongolo-Dravidian type

This type is known as the Bengali type. The members of this type are characterized by broad and round heads with a tendency towards medium dark complexion and plentiful hair on face. The nose is usually medium with a tendency towards flatness. The stature is also medium but sometimes short. Such people are found in Bengal and Orissa. The notable representatives of this type are the Bengali Brahmins and Bengali Kayasthas. According to Risley this type is not only an admixture of the Mongolians and the Dravidians, some blood strains of Indo-Aryan type are also mixed with it.

6. The Scytho-Dravidian type

The people of this type possess medium to broad head, low to medium stature, fair complexion, and a moderately fine nose, which is not conspicuously long. The hair is scanty on face and body. It is held that the type has been evolved by the intermixture of two distinct racial strains—the Scythians and the Dravidians. Typical example of this type is found in Western India comprising the Maratha Brahmins, the Kunbis and the Coorgs, who are distributed in the tracts of Madhya Pradesh, Maharashtra-Gujrat border region upto the Coorg. The Scythian element is more prominent in higher social groups of these regions while the Dravidian features predominate among the lower social groups in the region. However, this Scytho-Dravidian type is markedly different from the Turko-Iranians in the traits like larger head, flatter face, and higher profile of nose and shorter stature.

7. The Turko-Iranian type

This type is characterized by broad heads and fine to medium nose, which is long and prominent. The stature is fairly tall and the average height of the males varies from 162cm to 172cm. Although the eyes are dark in colour, grey eyes are not uncommon. Complexion of the people is generally fair; plentiful hair is found on face and body. The type includes the inhabitants of Afganisthan, Baluchistan and NorthWest Frontier Provinces (now in Pakistan) who are represented by the Balochis, Brahai, Afghans and some other people of NWFP. In the view of Risley, this type has been formed probably by the fusion of Turki and Persian elements in which the former's features predominate.

Risley's classification faced a considerable criticism from different authorities, especially in respect of the Dravidians, the Scytho-Dravidians and the Mongolo-Dravidians.

Besides, the Indo-Aryans is distributed only in Punjab, Rajputana and the Kashmir Valley according to Risley. But the speakers of Aryan languages actually occupy a vast area in Indian subcontinent, which has not been reflected in his classification. If he had measured the people of Kashmir alone, then he should have placed them in a separate group as they possessed absolutely different physical features. Further, Risley had given much importance in Scythian elements when he discussed about broad-headed people as the Scytho-Dravidian type. In fact, the Scythian invaders stayed so short that they hardly get any opportunity to spread any remarkable influence among ethnic elements of Bombay Presidency where Risley conducted his study.

Risley also stated that the broad-headed elements in Bengal have been influenced by the Mongolian people. But it is difficult to confirm that the brachycephalic elements in Bengal and Gujrat have been derived from the Mongolian element. Although all Mongolian people are brachycephals but the epicanthic fold as a typical Mongolian feature is found only among some people living in Darjeeling and neighbouring districts. This feature is totally absent among the people of other parts of Bengal. Risley also had mentioned a vast area of South India as the land of the Dravidians, but in reality many of these people do not speak Dravidian language and some of them exhibit such physical features which are quite different from the proper Dravidian characters.

Deniker criticized Risley for his conception of Dravidians which are found to occur in two forms—one is long headed, medium stature and narrow nose; and the other one is long headed, short stature and broad nose. These two groups are quite different from each other so far as their physical features, culture and origin are concerned. Among these two types of Dravidian, the second type has been described by some scholars as Pre-Dravidian Again, Risley did not mention anything about the Negrito element in India.

Classification of Giufrida-Ruggeri (1921)

Subsequent to Risley's attempt many other anthropologists tried to analyze Indian people. Giufrida-Ruggeri made the following six-fold ethnic classification for the people of India.

1. *Negrito* : Veddas and some South Indian jungle tribes.
2. *Pre-Dravidian orAustraloid Veddic* : Santals, Oraons, Mundas, etc.
3. *Dravidian* : Tamil and Telegu speaking people.
4. *Tall dolichocephalic element* : Toda.
5. *dolichocephalic Aryan* : (Homo dolichomorphus) : Indo-Afghans, Indo-Iranians, etc.
6. *brachycephalic leucoderm* - (Homo Indo-European brachymorphus) - Armenians, Georgeanus, etc.

In contrast to H.H. Risley, Giufrida-Ruggeri had mentioned the Veddas and some South Indian forest dwelling tribes as possessing the Negrito element. He categorized the Indo-Afghans, Indo-Iranians, etc. under dolichocephalic Aryans. In his consideration, the brachycephalic leucoderms was a type, which included the Armenians and the peoples of Pamir and Georgia. He was inspired by the language based racial classification of his predecessor Sir H. H. Risley. But Giufrida-Ruggeri's classification was too short to denote the enormous variation of physical types that exist among the peoples of India.

Classification of A.C. Haddon (1924)

A.C. Haddon did not agree with Risley and gave his own analysis about the racial elements in India. He divided India into three main geographical regions—(a) Himalayas, (b) the Northern plains or Hindustan and (c) the Southern plateau or the Deccan which is mostly covered by the jungles. He had dealt with each of these three regions, separately, for the reconstruction of racial elements in them.

The Himalayan region

Two principal types are found in this region:

(i) Indo-Aryan : People of this type show tall stature, brown skin colour with varying shades , dolichocephalic head with straight fine leptorrhine nose, well-developed forehead and a long narrow face. This type is represented by the Kanets of Kulu Valley. In Eastern Punjab the Indo-Aryan Kanets exhibit a trace of Tibetan blood.

(ii) Mongoloid : According to Haddon, this type dominates in North Eastern India. In fact, the main racial element of North East Frontier Agency of India is the Mongoloid and the representatives are the Lepcha, Garo, Naga, Khasi, Dafla, etc. who show the Mongoloid features. Haddon identified several racial elements among the tribes of Assam, which have been accumulated due to various invasions at different times. Among these elements the brachycephalic leptorrhine, came from the north and has been converted into Eurasiatic group. The brachycephalic platyrrhine is a variety of Pareoean. Further, the dolichocephalic element has entered from the main land of India into the population of Assam. Some Dravidian elements are also seen. Beside Assam, the people of Nepal, Bhutan, Kashmir and Punjab, show Mongoloid features very prominently.

The Northern plains or Hindustan region

The Indo-Afghan is the predominating type of this region. The people are characterized by dolichocephalic head with straight fine leptorrhine nose, well-developed forehead and a long narrow face. Stature ranges from medium to tall, eyes are dark on light brown complexion. The hair is black and wavy. The representative populations are the Jats and the Rajputs and others. But the places where the members of this type have mixed with the aboriginal people, the admixtured peoples are assumed as the lower caste people.

The Deccan region or Southern plateau

The different racial elements of this region as found by Haddon are as follows:

1. *Negrito* : Some people of this area show Negrito racial strain. Their physical characteristics include medium head, flat nose, flattened occiput, protruding forehead and very dark skin colour. The hair is black and the eyes are brown. The lips are somewhat fleshy and everted. The best representatives are the Kadars of Cochin, the Urallis of Nilgiri Hills and the Pullayans of Palni Hills. On the other hand, Andamanese represent a true Negrito racial element.

2. *Pre-Dravidian* : This is the oldest existing stratum of Indian population. The people are characterized by dolichocephalic heads, short stature, and very dark skin with black hair. The hair form varies from wavy to very curly. The representative populations are Bhils, Gonds, Santals, Oraons, Hos, Mundas, etc.

3. *Dravidian* : The Dravidians are characterized by the dolichocephalic heads, medium stature, brownish black skin colour and the mesorrhine nose. They possess plentiful hair, which are wavy with an occasional tendency to curl. The people of South India speaking Tamil, Malayalam, Telegu, Canarese, etc. belong to this type.

4. *Southern brachycephals* : This type is characterized by Mesocephalic to brachycephalic head and mesorrhine nose. The complexion is brownish black. However, the features are represented by the Paniyans of the Tamil district and the Pavara fishermen of the Tinnevalley coast.

5. *Western Brachycephals* : Haddon had traced a zone of broad-headed people extending from Gujarat to Coorg, along the Western coastal area of India. The people are characterized by brachycephalic heads, almost leptorrhine nose, light brown skin colour and tall stature. Risley had mentioned these people as the Scytho-Dravidians. The best examples are the Nagar Brahmins of Gujarat, the Prabhu, the Maratha of Maharashtra, etc.

Haddon's classification was based mainly on physical characters, artifacts, customs, languages and folk-tales. He justified his own analysis by the help of the evidences. According to his analysis, the oldest people of India must have been the Pre-Dravidians. The Dravidians also lived in India as the original inhabitants at the banks of the Ganges in Western Bengal. The Aryan-speaking people came on this subcontinent in the second millenium BC and spread over the fertile regions of the Punjab. Gradually, they occupied the valleys of the Jamuna and the Ganges. The main drawback of Haddon's analysis was that he did not mention anything regarding the Pre-Aryans of India. The brachycephalic element in India is Alpine in origin as analyzed by Haddon.

Classification of Eickstedt (1934)

Fuherer von Eickstedt had made the German Indian Anthropological Expedition to India during 1926-29. He classified the Indian people in 1934, both from physical and cultural perspectives. Basically he was inspired by the variation in skin colour of the Indian people and suggested four main ethnic elements as constituents of the population in India.

I. *Weddid* or *Ancient Indians:* These are the primitive people living in the forest. Two sub-types are distinguished here.

(a) *Gondid* - These people show dark brown complexion and curly hair. They are totemistic in belief and use mattock. Matriarchal influence is noticed among them. The Oraons, Gonds, Bhils etc. are the best examples of this sub-type.

(b) *Malid*- These people are characterized by curly hair with black-brown colour. Their culture is ancient but now they have been influenced by alien culture. People like Kurumbas, Veddas, etc. represent this sub-type.

II. *Melanid* or *Black Indians:* Racially it is a mixed group, which is divided into two sub-groups.

(a) *South Melanid* : This sub-group is characterized by black-brown skin colour. People live in the Southern most plains of India and possess strong matriarchal influence. The typical example of this group is the Yanaadi.

(b) *Kolid* : This sub-group includes the primitive people characterized by black-brown skin colour who live in the North Deccan forests. They hold strong totemistic beliefs and prominent matriarchal influence. The best examples are the Santals and the Mundas.

III. *Indid or New Indians*: These people are racially more advanced and occupy the open regions of India. They are further sub-divided into two sub-groups.

(a) *Gracile Indid* : This sub-group is characterized by brown skin colour with gracile appearance. The people show strong patriarchal influence as found among the Bengalis.

(b) *North Indid* : This sub-group possesses light brown skin colour. People are patriarchal in nature. The best examples of this type are the Todas and the Rajputs.

IV. *Palae-Mongoloid* : These people show certain incipient Mongoloid characters. The best examples are the Palayan of Wynad.

Eickstedt's classification is regarded as a proper attempt to classify Indian population. Although it was open to severe criticism, but it had a great scope. So, later it was extended with necessary changes and additions. Efforts of B. S. Guha made it more convincing.

Classification of B.S. Guha (1937)

Dr. B. S. Guha's racial classification is based on anthropometric measurements, which were collected during his investigations from 1930 to 1933. Guha traced six major racial strains and nine sub-types among the modern Indian population.

1. The Negrito
2. The Proto-Australoid
3. The Mongoloid
   a) Palaeo-Mongoloid
      i) Long-headed
      ii) Broad-headed
   b) Tibeto-Mongoloid
4. The Mediterranean
   a) Palae-Mediterranean
   b) Mediterranean
   c) Oriental
5. The Western Brachycephals
   a) Alpinoid
   b) Armenoid
   c) Dinaric
6. The Nordics

1. The Negrito

These people are considered as the first comers and the true autochthones of India. They are characterized by dark skin colour, short stature, and frizzly hair with long or short spirals. The head is either small, medium, long or broad with bulbous forehead. The nose is flat and broad. The lips are everted and thick. The best representatives of this type are the Kadars, the Irulas, the Puniyans, etc. of South India.

Such type of characters is also visible among the tribes living in the Rajmahal Hills. In respect of the head form and hair form, the Indian Negrito strain resemble more to the Melanesian Pygmies than to the Andamanese or African Pygmies.

2. The Proto-Australoid

This group is considered as the second oldest racial group in India. The people are characterized by dolichocephalic head, broad and flat nose (platyrrhine nose) which is depressed at the root. They are further short in height, dark brown to nearly black in skin colour. The hair is wavy or curly. Supraorbital ridges are prominent. These features are found among almost all the tribes of the Central and Southern India. The best examples are the Oraons, the Santals, and the Mundas of Chottanagpur region; the Chenchus, the Kurumbas, the Yeruvas and the Badagas of Southern India; and the Bhils, Kols of Central and Western India.

3. The Mongoloid

This type of people is distinguished by scanty growth of hair on face and body. The eyes are obliquely set and show the presence of epicanthic fold. The face is flat with prominent cheekbones and hair is straight. This group can be divided into two sub-groups, such as Palaeo-Mongoloid and the Tibeto-Mongoloid. The former one is further sub-divided as long headed and broad-headed.

In Palaeo-Mongoloid group, especially the longheaded type possesses long head, medium stature, and medium nose. Their cheekbones are prominent and skin colour varies from dark to light brown. The face is short and flat. They are the inhabitants of the sub-Himalayan region; the concentration is most remarkable in Assam and Burma Frontier. The Sema Nagas of Assam and the Limbus of Nepal are the best examples. The other sub-division of palaeo-Mongoloid is the broad headed type who possesses broad head with round face, dark skin colour and medium nose. The eyes are obliquely set and epicanthic fold is more prominent than that of the long-headed type. This type has been identified among the hill tribes of Chittagung, e.g. the chakmas, the Maghs, etc.

Second sub-division of Mongoloid is the Tibeto-Mongoloids who shows no further divisions. Their physical features are characterized by broad and massive head, tall stature, long and flat face, and medium to long nose. The eyes are oblique with marked epicanthic fold. Hair on body and face is almost absent. The skin colour is light brown. The best examples are the Tibetans of Bhutan and Sikkim.

4. The Mediterranean

This group is divided into three distinct racial types, which are as follows :

a) *Palaeo-Mediterranean* : The people are characterized by long head with bulbous forehead, projected occiput with high vault. They also show medium stature, small and broad nose, narrow face and pointed chin. The hair on face and body is scanty. The skin colour is dark. These people probably introduced *megalithic culture* to India. The Dravidian speaking people of South India exhibit the main concentration of this type. The Tamil Brahmins of Madura, Nairs of Cochin, and Telegu Brahmins are the examples.

b) *The Mediterranean* : The features include long head with arched forehead, narrow nose, medium to tall stature and light skin colour. Their chin is well developed, hair colour is dark, eye colour is brownish to dark and the hair on face and body is plentiful. These people live in the regions like Uttar Pradesh, Bombay, Bengal, Malabar, etc. The true types are the Numbudiri Brahmins of Cochin, Brahmins of Allahabad and Bengali Brahmins. It may be assumed that probably this type was responsible for the building up *of Indus Valley civilization.*

c) *The Oriental* : These people resemble the Mediterranean in almost all physical features except the nose, which is long and convex in this case. The best examples are the Punjabi Chattris, the Benia of Rajputana, and the Pathans.

5. The Western Brachycephals

This racial group is divided into three types as given below :

a) *The Alpenoid* : This type shows broad head with rounded occiput, medium stature, prominent nose and rounded face. The hair on face and body is abundant and the skin colour is light. This type is found among the Bania of Gujarat, the Kathi of Kathiawar and the Kayasthas of Bengal,

b) *The dinaric* : This type is characterized by broad head, rounded occiput and high vault. The nose is very long and often convex. The face is long and stature in general is very tall. The skin colour is dark; eye and hair colours are also dark. The representative populations are found in Bengal, Orissa and Coorg. The Brahmins of Bengal and Mysore are the best examples. Both the Alpino and the dinaric people entered into India through Baluchistan, Sind, Gujarat, and Maharashtra. They penetrated Ceylon from Kannada. The presence of this type has been noted in the Indus Valley site, Tinnevalley and Hyderabad.

c) *The Armenoid* : This type shows a resemblance with the Dinarics in physical characters. Only difference is that, among the Dinarics the shape of occiput is much developed and the nose is very prominent. The Parsis of Bombay exhibit typical Armenoid characteristics. The Bengali Vaidyas and Kayasthas sometimes show the features of this type.

6. The Nordics

The people are characterized by long head, protruding occiput and arched forehead. The nose is straight and high bridged. All are tall statured with strong jaw and robust body built. The eye colour is blue or grey. The body colour is fair which is reddish white. This element is scattered in different parts of Northern India, especially in the Punjab and Rajputana. The Kho of chitral, the Red Kaffirs, and the Khatash are some other representatives of this type. The Nordics came from the north, probably from Southeast Russia and Southwest Siberia, thereafter penetrated into India through Central Asia.

Criticism of Guha's Classification

Guha's classification also meets criticism at some points. Firstly, Guha's findings regarding the Negrito element has been opposed by almost all-leading anthropologists. Secondly, Guha tried to prove that all racial elements in India are of foreign origin. Keith strongly opposed this view. Because Keith believed in a racial evolution that had taken place in India and so he took India an evolutionary field of different races. Further, Guha had shown the people of India as Mongoloid and Brachycephalic. He proposed a sweeping distribution of Brachycephals, southwards, round the both ends of the Himalayas, which ultimately extends to the West to spread over the whole of the Deccan. In the northeast and the east, this brachycephalic area is supposed to have spreaded from the Nepal and Bhutan, upto Bengal and Orissa. Dr. Sarkar has strongly opposed the proposition of Guha. In his opinion the brachycephalic population of India does not show a sweeping distribution as has been described by Guha.

Classification of S.S. Sarkar (1961)

Dr. S.S.Sarkar proposed a racial classification based on cephalic index. He suggested six ethnic elements as the main types in the population of India. According to him, India is predominantly a dolichocephalic country, followed by the racial types like mesocephals and brachycephals.

The Dolichocephals.

1. *Australoid* : The Australoids are known by different names, such as, Proto-Australoid, Pre-Dravidian, Nisada and Veddid. The aboriginal people of India exhibit a Veddid or Australoid element in different degrees. Certain tribes of South India, e.g. the Uralis, the Kannikars, Paniyan and other show Australoid features. Sarkar had mentioned that the Australoids are widely distributed throughout India. The features are present among all castes of India although a greatest concentration is found among the lower castes. However, the population is characterized by short stature with dark complexion. Their head is dolichocephalic, the nose is platyrrhine and the hair is wavy.

2. *Indo-Aryan* : The dolichocephalic Indo-Aryans are quite distinct from that of the Australoids. Their physical features denote tall stature and light skin colour. The eye colour is also light, even the hair colour is not so dark as the Australoids. The cranial capacity of Indo-Aryans is higher than the Australoids. Their physique is well built and robust than the Australoids. The best example of Indo-Aryan type is the Baltis of the Hindukush Mountains.

The Indo-Aryan people have frequently met with the people of Indus and the Gangetic Valley of Western India. Therefore, many features of this type predominate in the said region. A sporadic distribution of this type has also been noted in Eastern Bihar, Bengal and Assam. But in the latter areas, the type is confined among the higher castes only.

3. *Mundari-Speakers* : 'The Mundari-speakers' as described by Sarkar are the sturdy, short height people with robust constitution. Other features include a dolichocephalic head, a skin colour lighter than the Australoids. The thick, straight, black hair is more or less similar to those of the Mongoloids.

These people are distributed in the river valleys and plateaus of Eastern and Central India. Chottanagpur plateau, Orissa, and Madhya Pradesh show the highest concentration. They have been migrated from the east and bear some affinities with the Mongoloids.

The Mesocephals

4. *Irano-Scythian* : This type of ethnic element perhaps entered in India from Northwest, almost at the same time during the Indo-Aryan migration. The physical features are characterized by the mesocephalic head and medium stature. These Irano-Scythians are quite different from the dolichocephalic Indo-Aryans, despite certain similarities do exist between them. In Eastern Bihar, Bengal and Assam the dolichocephalic Indo-Aryans have been replaced by the Mesocephalic Irano Scythians. The average cephalic index of Indo-Aryans is 73 while among the Irano-Scythian it ranges between 77 and 79. The Mesocephalic Irano-Scythians appears to be more variable in physical features.

After entering India these people possibly moved southward along the valley of the Indus. Ultimately they reached to Gujarat, Bombay and Maharashtra. Their distribution has been noted upto Northern Mysore, Deccan and further south. In Eastern India this element has been frequently observed among the populations inhabiting in the river valleys of the Narmada and the Son.

The Brachyecephals

5. *Far Eastern* : There is no disagreement that the brachycephalic element came from Central Asia to India during Prehistoric period. The fact is that, the India had a connection with the Islands of Southeast Asia since ancient times and the cultural relationships have been continued till the historical period. A Malayan element is observed in the coastal regions of Chittagung hill tracts. This Malayan strain is quite distinct from other ethnic elements. Dr. Sarkar had defined it in terms of brachycephalic head, short stature, tendency towards obesity and dark skin colour.

6. ***Mongolian*** : This type of people is found in the Northeastern borders of India and the foothills of the Himalayas. The physical characteristics show a predomination of Mongoloid features. The skin colour is yellowish, akin to Mongoloid skin colour. The hair is sparsely distributed on face and body. The eyes invariably present the epicanthic fold. For these typical Mongoloid characters, the people are easily distinguishable from the other populations of India.

Our present knowledge instigate us to conclude that the earliest inhabitants of India were the Australoids who might have received some infiltration of Negrito strains in certain parts of India. The Mongoloid racial strain is also conspicuous in some pockets of India. But this conclusion may not be a final one. More researches are still required to be carried out to solve the problem of racial classification in India.

ANTHROPOMETRY

INTRODUCTION

Background and Scope

Anthropometry being a branch of physical anthropology indicates the measurements of human body, which provides scientific methods and techniques for various measurements and observations on the living human body and the skeleton. Rather it is a device for taking a number of measurements for describing the morphology of man. The measurements maintain their conformity with the anatomical landmarks and usually noted either in a laboratory set up or in a field. The origins of these measurements are very ancient. Long long ago the artists of ancient Egypt and Greece formulated some standard criteria for the human body. But, it was Johann Friedrich Blumenbach (1752 - 1840) who laid the foundations of Craniology in the field of scientific anthropometry. Blumenbach differentiated mankind into different races on the basis of skull-form. His classification is distinguished into three types such as a) square, b) long, and c) laterally compressed. In the same Century, Peter Camper studied the facial form and developed the facial profile angle to find out the extent of prognathism. Again, Charles White developed measurement techniques for the long bones, which has been known as Osteometry. He worked on the upper limbs of the Chimpanzees, Negroes and Europeans. Gradually the development of Anthropometry was accomplished in two different schools, which came to be known as German, and French Schools in Anthropometry. This resulted an atmosphere of confusion and so the need for an international agreement was seriously felt. International Committee for Standardization of Anthropological Techniques was established in London in the year 1932 which embraced the members from 20 different countries of the world. The committee suggested a standardized anatomical nomenclature for using it in definitions and for other specified purposes. The Committee also emphasized the importance of the study of comparative human anatomy for making a correlation of observations on the living, cadaverous and skeletal materials. The Committee is still working continuously towards ultimate perfection.

In India, anthropometry in Physical Anthropology has been greatly influenced by German methods and techniques. The most valuable and recognizing works were done by the scholars namely Ruggeri, Eickstedt, Risley, Guha, Karve, Majumdar, Biswas and others. H.H.Risley was the first ethnographer to collect systematic data on the various Indian populations. The credit of making a systematic study of the racial elements goes to B.S.Guha who collected anthropometric data on a scientific basis and also helped in establishing the study of physical anthropology in India. In 1946, he succeeded in establishing the Anthropological Survey of India and built laboratories for anthropometry and other branches of anthropology especially in physical anthropology. A Committee was formed under the Chairmanship of P. C. Biswas, in the Summer School in Anthropology in 1965 by the sponsorship of University Grant Commission. The objective of this Committee was to select the type and number of measurements as well as to define the measurements with specification of the instruments to be used. It also suggested some measurements as essential for racial study and for growth studies. As a matter of fact, the new avenues are continuously searched out by physical anthropologists. New procedures of measurements as well as new improved instruments are coming into use. New dimensions are also developing in the study of growth and development of mankind, in various populations. The nutritional as well as environmental-effects in connection with the growth and development have been sought in order to relate them with the physical structures and diseases. Studies have increasingly concerned with the dimensions, proportions and shape of man's immediate physical features. It helps in formulating standard sizes for various equipments, which possess an immense value in defence forces, and for other commercial purposes. For example, the efforts have been made to design garments, furnitures, space ships, toothbrushes, etc. to suit effectively with varying shapes and sizes of human body.

The scientific methods and techniques of anthropometry may be subdivided into the following sections:

(a) *Somatometry* means the measurement of the living body including head and face.

(b) *Osteometry* is the measurement of the skeletal parts particularly of the long and short bones.

(c) *Craniometry* indicates the measurement of the skeletal brain-cavity (Neurocranium) and face (Splanchocranium).

(d) Lastly, *Somatoscopy* is being used for observations on the living man, which is also supplemented by somatometry in the study of physical anthropology.

Methods of Anthropometry

Anthropometry is the only science, which measures human body accurately and scientifically. Bodily features differ under different geographical situation and biological environment. Therefore, with the help of anthropometry the form and bodily features of various populations can be expressed quantitatively.

A number of instruments have been developed by anthropologists for taking accurate measurements on the living body and on the skeleton as well. It is necessary to know the limitation and functional efficiency of the instruments, which are specially designed for the purpose of anthropometry. Special attention is required for the system of graduation, which is usually graduated in centimeters and millimeters; sometimes further device of minute calculations is attached to it. The instruments are needed to be handled in a proper way with tender care and it would be better if these instruments are checked before using for taking measurements.

The exact position of the anatomical points or landmarks holds special importance in anthropometry. A few anatomical points are fixed points while others are to be located by trial measurements. First of all the landmarks should be identified on the skeletal parts of the body because they correlate to the definite point on the living body. However, accurate identification of the landmarks, either on the living body or in the skeletal parts demands a thorough and continuous practice.

At the time of operation one should be very careful about the position of the subject who is taken under study. The measurements will be erroneous if the proper position is not maintained. Further, the position of subject varies with the nature of measurements. Caution is required against carelessness, poor lighting and impatience at the time of taking measurements. It is better to repeat all measurements at least twice to minimize the errors in connection with the measurements.

Instruments

Well-acquaintance with the instruments is required for getting accurate measurements. Most of the instruments are found to be manufactured in Switzerland by the supervision of Gneupel and Co. Some are also manufactured in Germany and U.S.A. The reputed companies are Abawerk, G.m.b.H (Germany), Swan Tool and Machine Company and Gilliland Instrument Company (U.S.A.). At present, India is also making these anthropological instruments. Una and Company of New Delhi deserves mention.

1. *Anthropometer Rod* or *Anthropometer of Martin*

It is an important instrument for many of the measurements on the living body. It is a two-metre long rod made of nickelled steel. It consists of four segments of equal parts to form a straight rigid rod. One side of this rod is a fixed sleeve on the top and the rod is graduated in a descending scale towards the bottom (Fig. 8.1). There is another movable sleeve, which can be moved up and down on the rod. These two sleeves or sockets hold two graduated cross bars, each of which is provided with a pointed tip. The main purpose is to take height measurements and the transverse breadth of the body. Besides, this instrument is also helpful in taking long measurements on the various parts of the body from the ground.

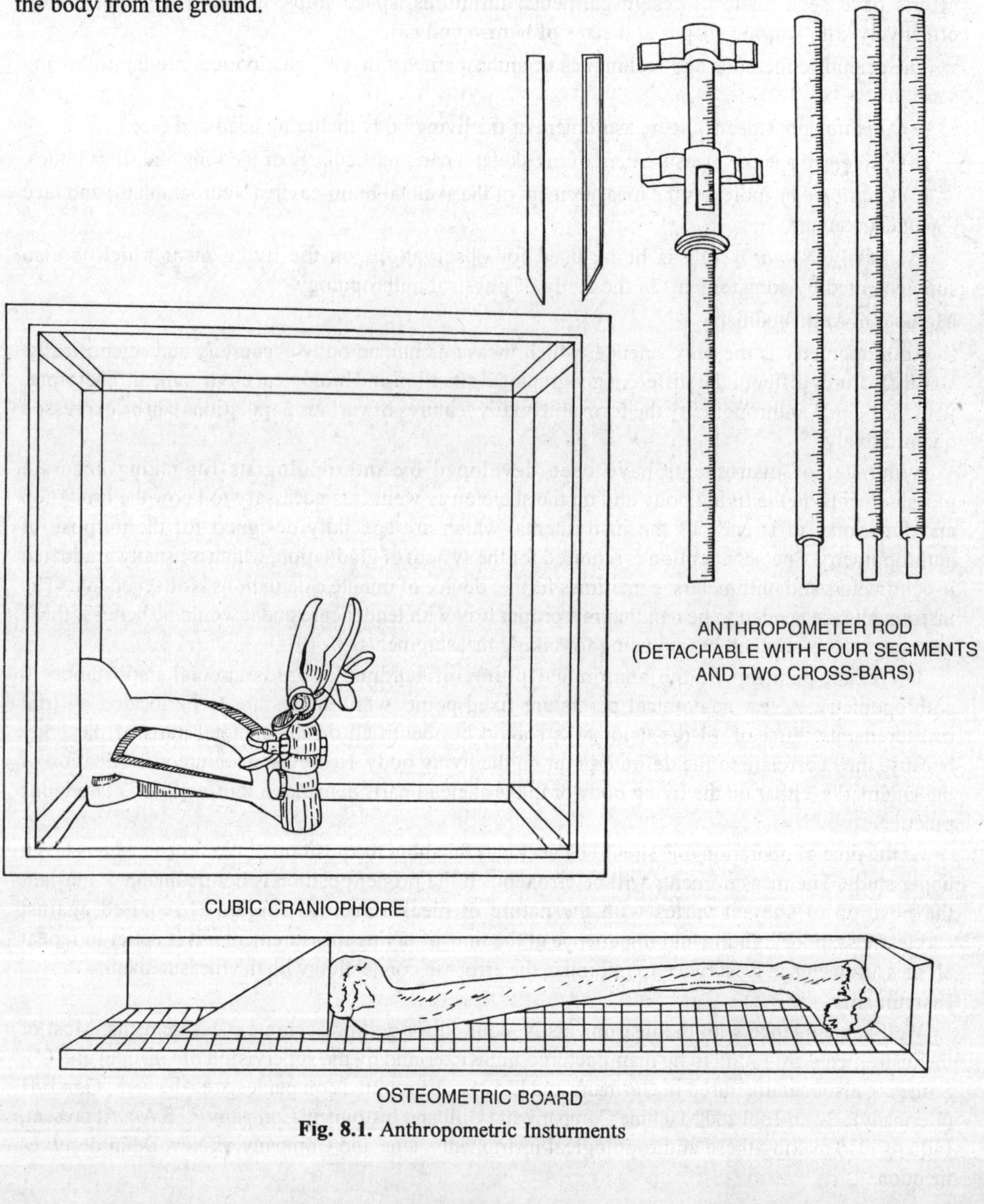

ANTHROPOMETER ROD (DETACHABLE WITH FOUR SEGMENTS AND TWO CROSS-BARS)

CUBIC CRANIOPHORE

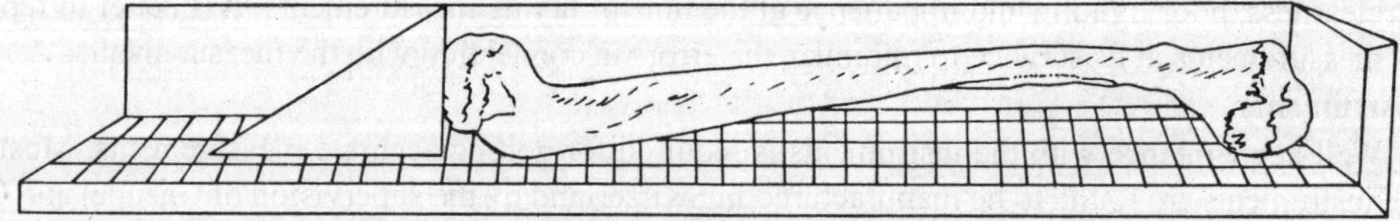

OSTEOMETRIC BOARD

Fig. 8.1. Anthropometric Insturments

Anthropometer rod must be placed in a vertical position by the side of the subject. After locating the landmark by hand, the movable sleeve has to be raised with the help of pointed tip; the cross bar will make the measurement at the upper end of the window of the movable sleeve.

2. *Rod Compass of Martin*

The Rod Compass is a part of anthropometer rod. The upper segment, rather the first segment of anthropometer with its two sleeves and cross bars is known as Rod Compass. It is equivalent to a large sliding caliper and its crossbars have to be adjusted for taking the breadth measurements. During operation, the landmark is located by left hand after holding the rod compass by right hand in horizontal position. The instrument remains in horizontal position and measurement is noted.

3. *Martin's Sliding caliper*

It is a 25cm long straight bar having flattened steel plate graduated in millimeters on its two sides. It has two long arms (12.5cm) of which one is fixed at the top and other one is movable (Fig. 8.1). There is a screw at the middle of the sleeve of the sliding or movable arm. The arms are projected to an equal distance on both the sides of the scale and thereafter they meet at sharp points at one side to have blunted ends on the opposite. The sharp end is used in measuring the skeletons and the blunt end is meant for the measurement on the living body. The scale at the bottom of the bar is useful for depth measurements. But in that case a change is required in the arrangement of the caliper; the movable arm is to be fitted with upside down position.

The bar of the instrument is held by right palm and four fingers. The thumb controls the sliding arm of the Caliper, upward or downward while the left hand traces the landmark for taking measurements.

4. *Martin's Spreading Caliper*

This instrument is mainly used for taking measurements on the living, especially on the head and skeleton. The instrument is composed of two outwardly curved long arms but bounded at one end. The two arms rotate on a screw at the straight ends. Open arm-ends of some instruments are sharp and pointed. These are used in taking the measurements on skeletal parts. Again, some instruments are specially designed with rounded and blunt ends to facilitate the measurements on living body. A straight scale graduated in millimeters is found to be fixed at one of the arms (in the middle of the left curved arm) and the meter scale passes through a flexible socket on the other arm of right hand side. Spreading calipers are available in two sizes, one is of 25cm for taking smaller measurements and another one is of 60cm for taking larger measurements especially on pelvis region. The latter is also known as Pelvimeter.

For collecting measurements, the two arms of the instrument are held by two hands in such a manner that the curved portions remain in between middle finger (below) and the thumb (above). The index fingers should be at the end of the arm and incidentally their tips should touch the end part of the arms. The index fingers are required to locate the landmarks prior to note the measurement.

5. *Tape*

It is useful for measuring the girths of various parts of the body and skeleton. It is made of thin steel, which is graduated in millimeters. Width of the tape is usually about 10 millimeters.

6. *Goniometer*

The instrument is used for taking various angular measurements on the face and skull. It consists of a movable needle, placed on a heavy base and attached to a protractor. A slot exists on the back of the protractor, into which an arrangement of a spring again attaches the instrument with a sliding or spreading caliper. This goniometer is very useful in taking linear measurements.

7. *Cubic Craniophore*

Craniophore is used to orient the skull on a horizontal plane for taking angular measurements.

But this cubic craniophore is used for drawing craniograms by means of diagraph. It is composed of a metallic cube where the length of each side is 30cm. approximately. A jaw is fixed in the centre of the cube. There is an extended arm on one side of it which can hold a skull tightly at the centre of the cube. Metal dish is used in case of fragile skull.

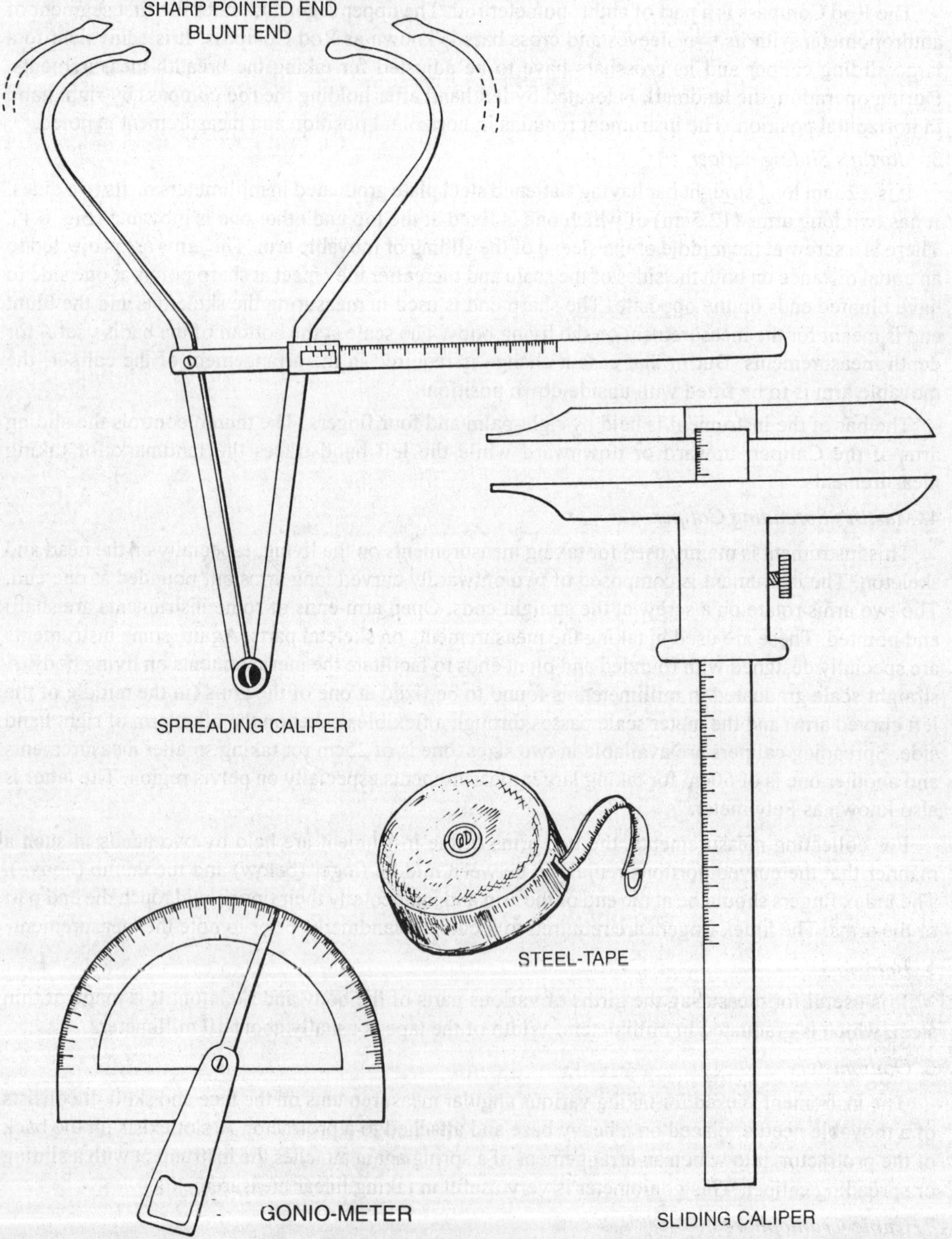

Figure 8.2. Anthropometric Instruments

8. *Martin's diagraph*

This is an instrument for drawing craniograph and it is used along with cubic craniopbore. It has a vertical-graduated bar fixed on a heavy base. Two horizontal needles are fitted with this bar. The lower one carries a pencil-holder exactly vertical in line matching with the point of upper needle. The upper one is curved and pointed which can be moved upwards and downwards. This upper needle can be rotated in any position for the purpose of accurate drawing of the craniographs.

9. Small Martin's Needle : It is used for orienting the skull on a horizontal plane, after fitting into the cubic craniophore.

10. Lens : It is an essential for certain minute somatological observations.

11. Weighing Machine : It is a machine for taking the weight of the living body. Each anthropological laboratory is provided with this machine. A portable machine is essentially required during the fieldwork.

SOMATOMETRY

As said earlier, Somatometry means the measurements of the living body including head and face.

Techniques of Measurements

Measurements should not be taken with shoes on. Minimum number of clothes should be worn at the time of measurements. The subject must stand erect on a level floor against a wall where his back and buttock touches the wall. Again the feet should remain parallel to each other and the heels need to touch the wall. Arms should hang to the maximum; the palms of hands need to touch the thighs. Both the shoulders should be on same plane. The head should rest without any strain in the eye-ear plane or Frankfurt plane, i.e.; tragion and the right orbitable must lie in one plane. All measurements except those concerned with midsagittal plane must be taken on the right side of the body as because it is easier to handle the instruments with right hand. The subject should be asked to seat on a low stool of about 40cm height at the time of taking the measurements on head and face. The head must be straight, looking forward.

Landmarks

Head and Face

1. Alare (**al**): It is the most laterally projected place as a point on the nasal wing. It has to be determined at the time of measuring the nasal breadth.

2. Chelion (**ch**): This point is situated on the mouth-opening line where the lateral margins of the upper and lower lips meet. It denotes the extreme corner points of the mouth.

3. Euryon (**en**) : It is the most laterally placed point on the parietal side of the head. Such a point is only determined by trial method at the time of measuring the maximum head breadth.

4. Frontotemporale (**ft**) : The medial most point on the temporal line—the most anterior and inner point on the linea temporalis or temporal line on the frontal bone. The first finger must be placed on the margin of the anterior and lateral wall of the forehead above the orbits. And then by gradual sliding of the finger on the curved plane of the linea temporalis, the desired point can be located. Generally it lies slightly higher than a tangent which could be drawn on the highest elevation of the upper margins of the eyebrow ridges.

5. *Glabella* (**g**) : It is the most prominent point on the protuberance in the median plane between the two eyebrows. This protuberance is found at the lower forehead above nasal root, which is situated between the eyebrow ridges intersected by mid-sagittal plane and just above the naso-frontal suture.

6. *Gnathion* (**gn**): It is the lowest point on the mandible where the lower margin of the lower jaw is intersected by the mid-sagittal plane. It can be palpated on the lower jaw from the back but placed slightly anterior to chin.

7. *Gonion* (**go**) : **It is the lowest posterior and most lateral point on the angle (formed by the body and the ramus of the mandible) of the lower jaw. The point is located at the lateral side of the angle.**

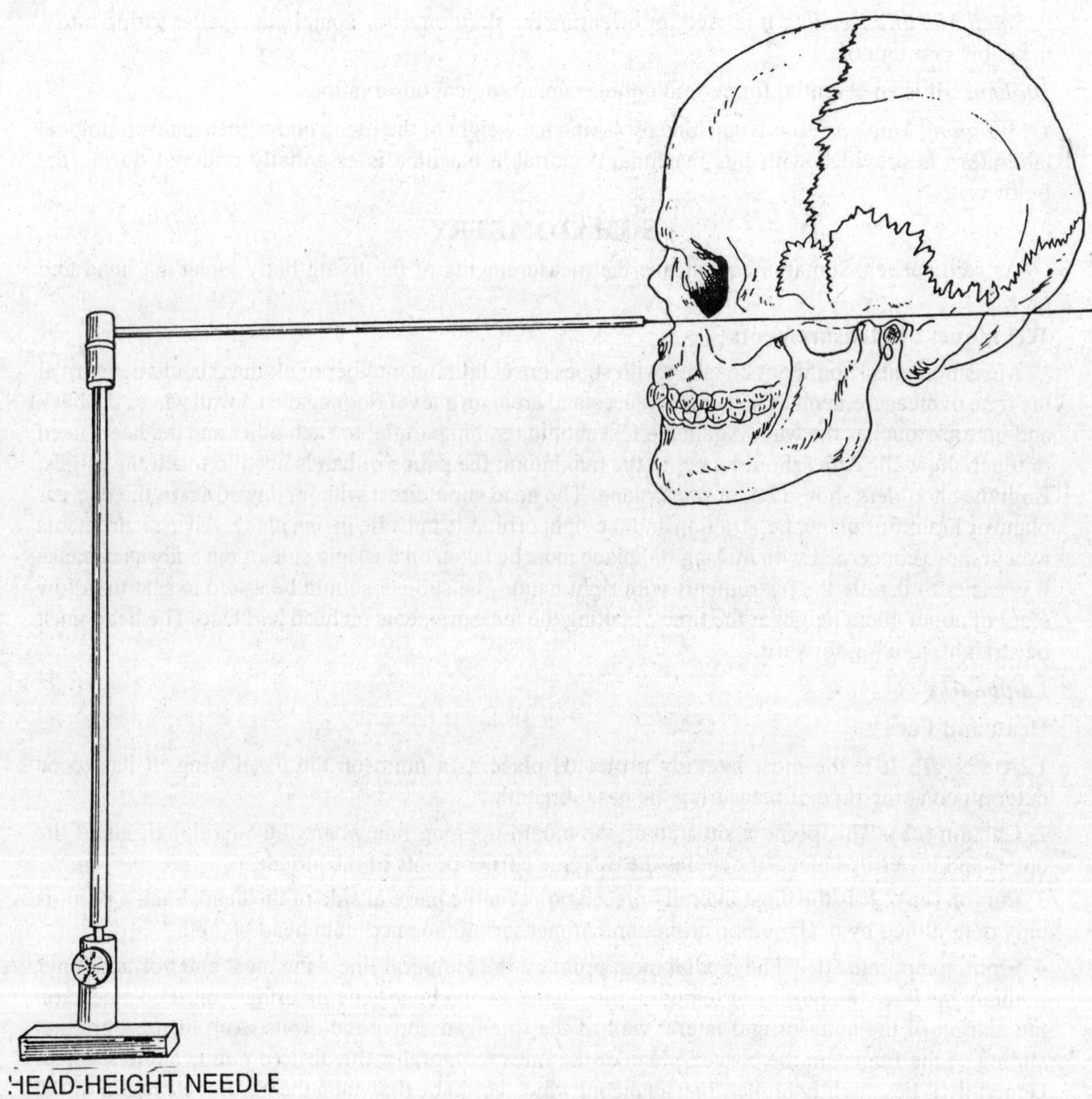

Fig. 8.3. Head—Height Needle To Determine Eye-Ear Plane of a Skull
[Measurement of Eye-Ear plane forms the basis of descriptive cranial geometry. The needle is to be adjusted with a skull in such a way that the top of both ear holes come in the same horizontal plane along with the lower margin of the orbits.]

8. *Inion* (**i**) : It is a point at the meeting region of the superior nuchaeal lines in the mid-sagittal plane. This point can be felt at the junction between the posterior protuberance and the neck, at an extreme posterior position. The tip of this external occipital protuberance is taken as a landmark for measurement on the living body.

9. *Labrale Superior* (**ls**) : This point is located in the mid-sagittal plane and so can be cut by a tangent, placed at the highest elevation of the upper margin of the integumental lip.

10. *Labrale Inferior* (**li**) : This point is located at the lower margin of the lower lip, in the mid-sagittal plane.

11. *Nasion* (**n**): It is a point at the junction where the frontal bone meets the nasal root, being intersected by the mid-sagittal plane. In case of living body this point can not be seen. Therefore it is only felt by means of finger and found at the naso-frontal suture which is located deep in the nasal root. It has to be identified by probing at the depression of the nose. Usually nasion lies in the level of the medial end of the eyebrows, mostly at the lower margins, which is away from the height of the eyebrows.

12. *Opisthocranion* (**op**) : This point is the most posterior point on the back of the head. It corresponds to the mid-sagittal plane on the posterior protuberance.

13. *Orbitale* (**or**) *:* It is the deepest point on the lower border or margin of the eye socket. It can be easily felt through the skin by first finger.

14. *Pronasale* (**prn**): It is the most anteriorly placed point on the tip of the nose when the head is placed in mid-sagittal plane.

15. *Subnasale* (**sn**): It is a point between the lower surface of nasal septum and the upper lip.

16. *Prosthion* (**pr**) : It is a point on the lower margin of the gums of upper jaw at the mid-sagittal plane between the middle of incisors. It can also be defined as the most downward point on the lower border of the gum.

17. *Stomion* (**sto**): It is the point just at the junction of the two lips in the median plane, otherwise the slit of the mouth with close lips as cuts at the mid-sagittal plane.

18. *Subaurale* (**sba**): It is the lowest point on the lower margin of the ear lobe.

19. *Superaurale* (**sa**): It is the highest point on the margin of helix of the ear.

20. *Trichion* (**tr**) or *Crinion:* It is the point on the median plane of the hair limit where the anterior border of the hair on the forehead cuts the mid-sagittal plane.

21. *Tragion* (**t**) : It is a point on the upper margin of tragus where tangents drawn to the anterior and upper margin of this cartilage cut each other. It is on the junction of the anterior-inferior root of the helix with the tragus cartilage. It lies 1-2 mm below the helix spine. Some scholars accept it in the middle of the tragus or at the tip, but others count it at the ear opening or auriculare..

22. *Vertex* (**v**) : It is the highest point on the top of the head in mid-sagittal line when the head is in eye-ear plane. It is not a fixed anatomical point, rather depends on the orientation of the skull.

23. *Zygion* (**zy**): It is the most laterally placed point on the zygomatic arch. These points are to be determined at the time of measuring the bizomatic breadth.

Trunk and Extremities

1. *Acromion* (**a**) : It is the most lateral point on the lateral margin of the acromial process at the scapula when the subject stands in normal position with his arms hanging by the sides. It can be located by palpating the scapular spine with the fingers.

2. *Acropodion* (**ap**): It is the most forwardly placed point of the toe cap, either on the first or the second toe, if the foot is kept in normal position. As a matter of fact, the Acropodion (ap) is a point at the tip of the longest toes.

3. *Dactylion* (**da**): It is the lowest point on the anterior curved top of the middle finger, provided the arm hangs sidewise.

4. *Iliocristale* (**ic**) : It is the most laterally placed point on the iliac crest when the subject stands in normal position. Iliocristale is located on the lateral border at the upper margin of the iliac crest which can be found out by palpating with the fingers from down to upwards along the margin of the crest, to see where it crosses over the upper margin.

5. *Iliospinale anterior* (**isa**) : It is a highest point on superior-anterior iliac spine. It can be located by placing the fingers on the hips and letting the thumb to find out the tip of the bone.

6. *Iliospinale posterior* (**isp**) : This is the most posteriorly placed point on superior posterior iliac crest.

7. *Metacarpale radiale* (***mr***) : It is the most medially placed point on the head of second metacarpal bone, on the stretched palm.

8. *Metacarpale ulnare* (**mu**) : It is the most laterally placed point on the head of the fifth metacarpal bone, on the stretched palm.

9. *Metatarsale fibulare* (**mft**) : This is the most laterally placed point on the head of the fifth metatarsal.

10. *Mesosternale* (**mst**) : This point is located on the anterior border of the sternum where mid-sagittal plane cuts the line joining the articular surface of the fourth rib. The position of the ribs may be counted from the sides.

11. *Metatarsale tibiale* (**mtt**) : This point is the most laterally placed point on the head of the first metatarsal.

12. *Pternion* (**pte**) : It is the hind-most point of the heel when the foot is normally posited.

13. *Radiale* (**r**): It is the highest point on the border of the radius bone i.e. the upper margin of the radiale capitulum when the arm hangs downwardly. This landmark can be determined accurately by pronating and supinating the lower arm. The rotating head of the radius is clearly distinguished from stationary condyle of the humerus.

14. *Spherion* (**sph**) : It is a lowest point on the tip of the medial malleolus of tibia when the subject stands erect. The point is to be identified at the tip of the malleolus.

15. *Stylion* (**sty**): It is a deepest (lowest) point on the styloid process of the radius if the arm hangs sidewise. For locating the exact point, one has to palpate the entire lateral margin of the radius with the thumb-tip.

16. *Tibiale* (**ti**): It is a highest point on the middle of the medial border, which corresponds to the inner glenoidal margin of the head of tibia. The point is very hard to locate when adiposity develops on the kneecap, as found frequently among the females. However, the upper tibiale margin can be located with first finger and thumb by bending the knee. The articular surface of the tibia may be indicated by a horizontal line.

17. *Symphysion* (**sy**) : This point is found on the upper margin of the pubic bone in the mid-sagittal plane. To find out this point, right hand with stretched fingers will slide downwards till the fingers touch the hard bone and it lies on the level of the junction of the penis and abdomen when the skin must not be folded.

Somatometric Measurements

Before collecting the Somatometric measurements it is necessary to keep a record on the subject itself, so that the findings of measurement can be analyzed at any time, in comparison to other findings. However, the information should be collected in the following points:

(1) Name; (2) Sex; (3) Age; (4) Place of Birth; (5) Religion; (6) Caste / Tribe; (7) Gotra / Clan; (8) Occupation; (9) Educational standard; (10) Economical position; (11) Marital status; (12) Number of childern; (13) Physical environment; (14) Common ailments, if any. Date of taking measurements should be furnished in each case.

Head and Face

1. *Maximum Head Length:* This measurement is the straight distance between glabella (g) and

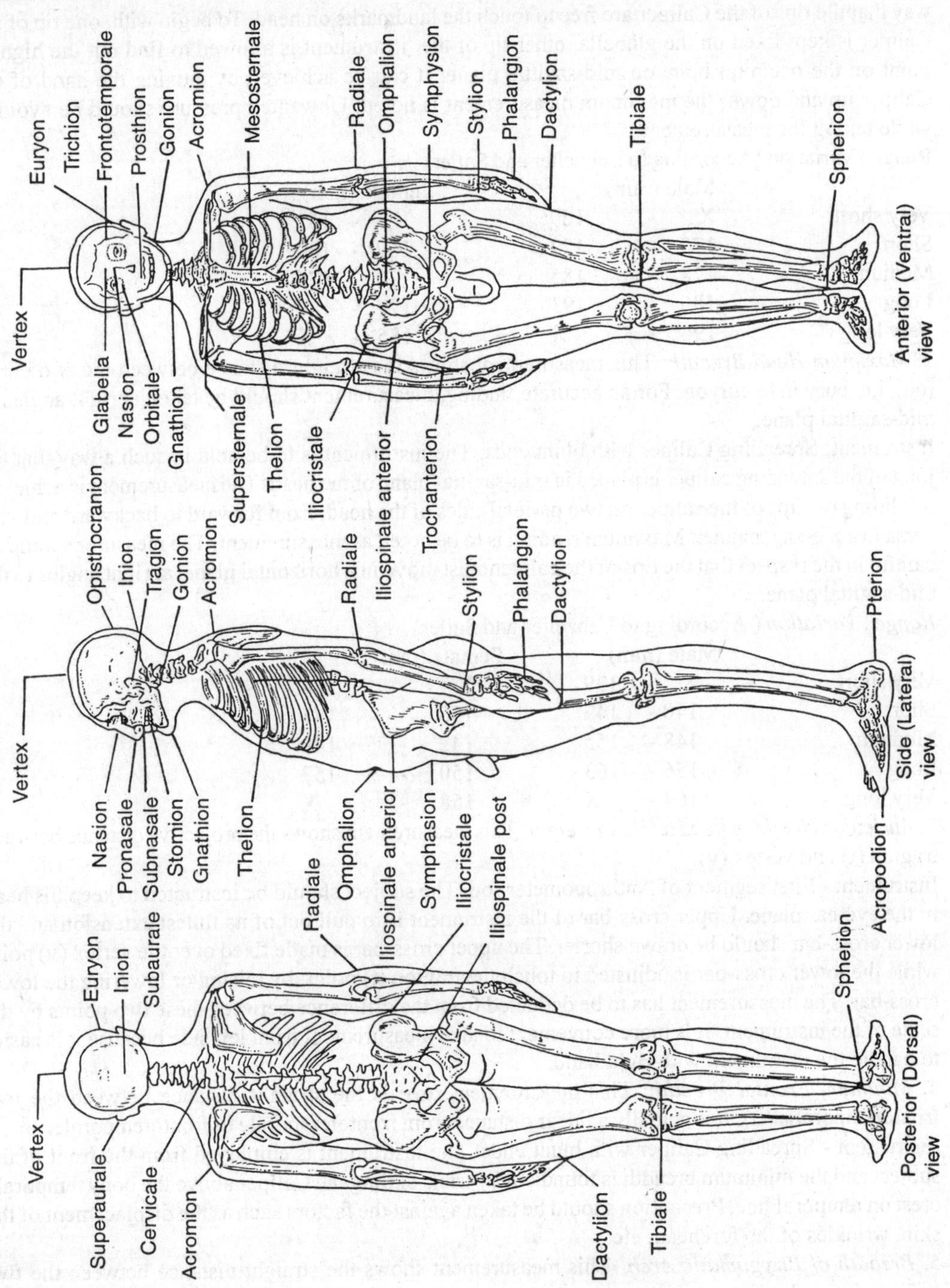

Fig. 8.4. Somatometric Landmarks on Human Figure

opisthocranion (op). The opisthocranion is the most projecting point of occipital bone, which is located on the upper surface of the head in the mid-sagittal plane.

Instrument : Spreading Caliper with blunt ends. The two arms of the instrument are held in such a way that the tips of the Caliper are free to touch the landmarks on head. To begin with, one tip of the Caliper is kept fixed on the glabella, other tip of the instrument is allowed to find out the highest point on the occipital bone on mid-sagittal plane. It can be achieved by moving the hand of the Caliper up and down; the maximum measurement is noted. Unwanted pressure shouid be avoided while taking the measurement.

Range : Variation (According to Lebzelter and Saller)

	Male (mm)	Female (mm)
Very short	X - 169	X - 161
Short	170 - 177	162 - 169
Medium	178 - 185	170 - 176
Long	186 - 193	177 - 184
Very long	194 - X	185 - X

2. *Maximum Head Breadth*: This measurement shows the straight distance between the two eurya (eu), i.e. euryon to euryon. For an accurate finding, measurement should be taken at right angles to mid-sagittal plane.

Instrument: Spreading Caliper with blunt ends. The instrument is to be held in such a way that the joint of the spreading caliper is placed in mid-sagittal plane of the head. The measurement is achieved by sliding the tips of the caliper on two parietal sides of the head, from forward to backward and vice versa in a zig-zag manner. Maximum reading is to be noted as measurement. The measurer should be careful in the respect that the tips of the caliper must move in a horizontal plane, at right angles to the mid-sagittal plane.

Range : *Variation* (According to Lebzelter and Saller)

	Male (mm)	Female (mm)
Very short	X - 139	X - 134
Short	140 - 147	135 - 141
Medium	148 - 155	142 - 149
Long	156 - 163	150 - 157
Very long	164 - X	158 - X

3. *Auricular Height of head or Head Height*: This measurement shows the projective distance between tragion (t) and vertex (v).

Instrument - First segment of Anthropometer rod. The subject should be instructed to keep his head in the eye-ear plane. Upper cross-bar of the instrument is to pull out of its fullest extension and the lower cross-bar should be drawn shorter. The upper cross-bar is made fixed over the vertex (v) point while the lower cross-bar is adjusted to touch the tragion (t) point, by raising or lowering the lower cross-bar. The measurement has to be deducted from the difference between these two points by the scale of the instrument. It is more convenient to take measurement from left side because it is easier to handle the instrument with right hand.

4. *Minimum Frontal Breadth*: This measurement reveals the straight distance between the two frontotemporalia (ft). It is actually a linear distance from frontoteraporale to frontotemporale.

Instrument - Spreading Caliper with blunt ends. The instrument is employed from the front of the subject and the minimum breadth is found out by trial, sliding the Caliper above the bony temporale crest on temporal line. Precaution should be taken against the factors such as the displacement of the skin, wrinkles of the forehead, etc.

5. *Breadth of Bizygomatic Arch:* This measurement shows the straight distance between the two zygia (zy), i.e. the distance from zygion (zy) to zygion (zy). The zygion is the most lateral point on the zygomatic arch.

and the first finger, and applied on the zygomatic arches. Maximum breadth is noted by trial, sliding the hands of the Caliper, backward and forward. This maximum breadth is usually obtained from a place near the ear, not on the cheek. Special attention should be kept in the process of operation, because skin may be displaced at the time of recording the measurement. The joint of the Spreading Caliper must lie on the mid-sagittal plane of the head for a correct reading.

Range : *Variation* (According to Lebzelter and Saller)

	Male (mm)	Female (mm)
Very narrow	X - 127	X - 120
Narrow	128 - 135	121 - 127
Medium	136 - 143	128 - 135
Long	144 - 151	136 - 142
Very long	152 - X	143 - X

6. *Bi - Gonial Breadth:* This measurement shows the straight distance between the two gonia (go).

Instrument : Spreading Caliper with blunt ends. The tips of the Spreading Caliper are held with the thumb and forefingers and two gonion points are searched by index fingers. The measurement is to be taken properly and never on the lower side of gonion.

7. *Morphological Facial Height* or *Total Facial Height:* This measurement denotes the straight distance between nasion (n) and gnathion (gn).

Instrument : Sliding Caliper with blunt ends. The instrument is held in right hand and the nasion point is found out by palpation with the fingers. It appears as a transverse groove above the fronto-nasal suture. After determination of the nasion point, the movable cross-bar has to be placed on the gnathion point for taking measurement.

Range : *Variation* (According to Lebzelter and Saller)

	Male (mm)	Female (mm)
Very low	X - 111	X - 102
Short	112 - 117	103 - 107
Medium	118 - 123	108 - 113
Long	124 - 129	114 - 119
Very long	130 - X	120 - X

8. *Morphological Upper Facial Height:* This measurement shows the straight distance between nasion(n) and prosthion (pr).

Instrument : Sliding caliper with blunt ends. The blunt end of the fixed cross-bar is placed on the nasion point. Then the lower cross-bar is raised against the prosthion point, which is situated at the lowest projection of the gum, between the upper middle incisors. It should be kept in mind that the instrument must be disinfected after every measurement. Normally the instrument is not allowed to touch the flesh part, at the time of taking measurement.

9. *Nasal Height*: This measurement reveals the straight distance between nasion (n) and subnasale (sn).

Instrument : Sliding Caliper with flat arms. The instrument is held by right hand where the lower cross-bar touches the subnasal point, at the lower border of the nasal septum. Upper cross-bar is made fixed against the nasion point. For accurate measurement, the Caliper has to be kept sideways, leaving the middle line of the nasal bridge.

10. *Nasal Breadth:* This measurement shows the straight distance between the two alaria (al). It is the most laterally placed points on the nasal wings.

Instrument : Sliding Caliper with flat arms. Upper cross-bar is placed against the right alare of the nose and the movable cross-bar is adjusted on the other alare. Even a slight pressure on the nasal wings must be avoided for getting accurate measurement.

Range : *Variation* (According to Schlaginhaufan)

	(mm)
Short	X — 24
Below medium	25 — 29
Medium	30 — 34
Above medium	35 — 39
Large	40 — X

10. *Nasal Depth:* This measurement is the projective straight distance between tip of the nose or pronasale and hind-most point of the nasal septum.

Instrument : Sliding Caliper with flat arms. The instrument is held in right hand. Upper cross-bar of the Caliper is placed at the root of the nasal septum and the lower, movable cross-bar is kept on the other side to bring it in touch of the tip of the nose. The main bar of the Caliper should remain parallel to the median line at the time of taking measurement.

11. *Ear Length* or *Physiognomic Ear Length:* This measurement shows the straight distance between superaurale (sa) and subaurale (sba). Superaurale is the highest point on the helix and lowest point of the ear lobe is subaurale.

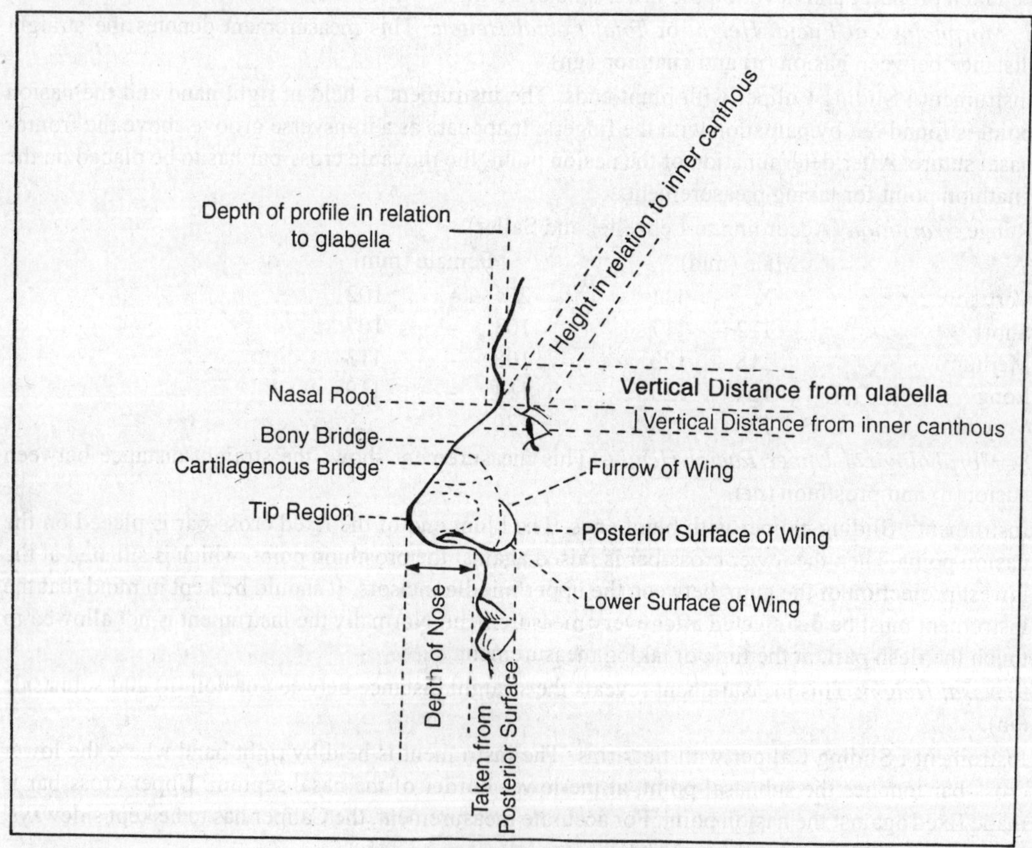

Fig. 8.5. Different Parts of Nose

Instrument : Sliding Caliper with flat arms. Upper cross-bar of the sliding caliper is fixed on the uppermost points of the helix along the longitudinal axis of the ear. The movable cross-bar is adjusted to touch the lowermost point in the same longitudinal axis of the ear lobe. Extra pressure is avoided for correct measurement

12. *Ear Breadth* or *Physiognomic Ear Breadth:* This measurement is the straight distance between the two most lateral points of the ear. One lies at the base of the ear and other lies at the posterior border of the ear cartilage.

Instrument : Sliding Caliper with flat arms. Fixed cross-bar of the instrument is placed at the base of the ear and the movable cross-bar is slided until it touches the most projecting point on the ear cartilage. No pressure should be applied at the time of recording the measurement.

14. *Circumference of the Head* or *Horizontal Circumference of the Head:* This measurement shows the maximum circumference of the head in horizontal position.

Instrument : Steel tape. One end of the tape is held on glabella (g) point by left hand. The other end is held with right hand and taken round following the opisthocranion (op), and again taken back to glabella (g). The tape around the head (g - op - g) should be kept in the same plane and opisthocranion is to be located beforehand. One should be very careful at the time of taking measurement as there is a fair chance to get the tape dislocated.

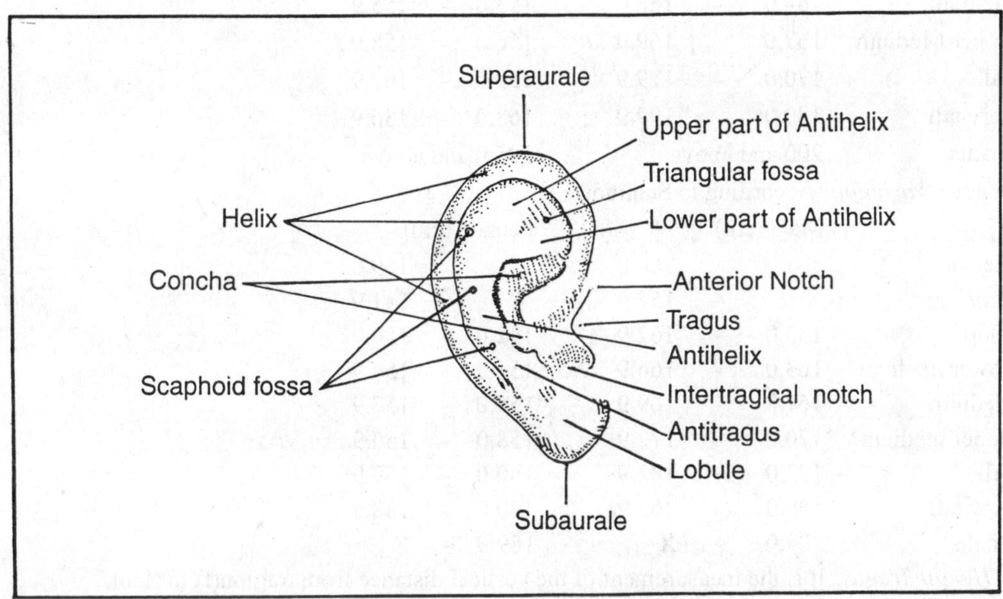

Fig. 8.6. Different Parts of External Ear

15. *Height of Integumental Lips:* This measurement shows the straight distance between labrale superior (1s) and labrale inferior (li).

Instrument : Sliding Caliper with flat arms. Mouth should be closed and the arms of the instrument is to be placed in such a manner that they cut the margins of the integumental lips tangentially while the Caliper position is vertical.

16. *Lip Length:* This measurement reveals the straight distance between two chelion (ch). This is mouth breadth or distance between two corners of the mouth.

Instrument : Sliding Caliper with flat arms. The Caliper need to be placed firmly on the jaws of the subject which must be closed and the facial muscles must be relaxed at the time of taking the measurement.

Trunk and Extremities

Stature or *Height Vertex:* It is the measurement of vertical distance from vertex (v) to floor.

Instrument : Anthropometer. The subject should stand as erect as possible. His arms should hang on the sides, the heels touch one another and the vision of eye must be directed to the horizontal plane as parallel to the floor for keeping the head in natural position.The anthropometer should be kept in vertical position in the median sagittal plane, close to the subject and the moving-cross bar is allowed to touch the vertex (v) lightly. To maintain vertical position of the anthropometer, a goniometer may be attached to its fixed cross-bar. The measurement should be taken carefully.

Range : *Variation* (According to Martin)

	Male (cm)			Female (cm)		
Median			165			154
Pygmies	Under	-	129.9	Under	-	120.9
Very short	130.0	-	149.9	121.0	-	139.9
Short	150.0	-	159.9	140.0	-	148.9
Lower Medium	160.0	-	163.9	150.0	-	152.9
Medium	164.0	-	166.9	153.0	-	155.9
Upper Medium	167.0	-	169.9	156.0	-	158.9
Tall	170.0	-	179.9	159.0	-	167.9
Very tall	180.0	-	199.9	168.0	-	186.9
Giants	200 and above			187.0 and above		

Range - Variation (According to Schmidt)

	Male (cm)			Female (cm)		
Median			169.0			169.0
Very short	X	-	152.9	X	-	141.9
Short	153.0	-	162.9	142.0	-	150.9
Lower medium	163.0	-	166.9	151.0	-	154.9
Medium	167.0	-	169.9	155.0	-	157.9
Upper medium	170.0	-	172.9	158.0	-	159.9
Tall	173.0	-	182.9	160.0	-	169.9
Very tall	183.0	-	203.9	170.0	-	188.9
Giants	204.0	-	X	189.0	-	X

2. *Height Tragus:* It is the measurement of the vertical distance from tragion(t) to floor.

Instrument: Anthropometer. The subject should stand erect maintaining almost the same position as like the previous measurement. The head should be kept to the eye-ear plane. The measurement also finds the head height by subtracting it from the measurement of height vertex.

3. *Shoulder Height* or *Height Acromion:* This measurement shows the vertical distance from acromion (a) to the floor.

Instrument : Anthropometer. The anthropometer should be held in vertical position. The arm of the subject is placed by the side of the body in normal position, the anthropometer should run parallel to the body as it was in previous measurements. The landmark may be located by palpating the point with left hand and the right hand slides the movable cross-bar to touch the acromion (a) when the subject does not shrink or elevate his shoulders.

4. *Elbow Height* or *Height Radiale :* This measurement shows the vertical distance from radiale (r) to floor.

Instrument : Anthropometer. The subject keeps his arms at rest and the position of the subject is same as previous measurements. The movable cross-bar is touched at the radiale point to take the measurement.

5. *Wrist Height* or *Height Stylion:* This measurement shows the vertical distance from stylion (sty) to floor.

Instrument : Anthropometer. Like previous measurements, anthropometer is held by one hand and the other hand is employed to find out the stylion point. Lower sliding cross-bar is used for taking the measurement.

6. *Height of the tip of the Middle Finger* or *Height Dactylion:* This measurement shows the vertical distance from dactylion (da) to the floor.

Instrument : Anthropometer. The subject should stand erect like the previous measurements. His hand must be stretched in the same plane but should not touch the thigh. The arm is required to keep straight instead of curved position and the cross-bar of the anthropometer is allowed to touch the dactylion (da) point for noting the height from the ground.

7. *Height Iliospinale:* This measurement shows the vertical distance from the anterior iliospinale (isa) to the floor.

Instrument : Anthropometer. Left index finger is required to locate iliospinale (isa) point while the right hand holds the anthropometer.

8. *Knee Height* or *Height Tibiale:* This measurement shows the vertical distance between tibiale (ti) and the floor.

Instrument : Anthropometer. Right hand is used for holding the instrument and the tibiale point is located by left hand. The movable cross-bar is used to measure tibiale point.

9. *Height Spherion:* This measurement is the vertical distance between spherion (sph) and the floor.

Instrument : Anthropometer. The anthropometer is to be held in the same manner as like the above measurements. The movable cross-bar is applied on the spherion point to note the measurement from the floor.

10. *Span:* This measurement counts the straight distance between two dactylia (da), from each other when the arms remain full stretched as well as parallel to the floor.

Instrument : Anthropometer. The anthropometer should hold horizontally with zero at the left side of the subject who has been instructed to stretch the arms fully as parallel to the floor. The anthropometer with zero point is placed in such a manner that the tip of the right middle finger touches the zero. Then the rod is carried horizontally at the level of nipples toward the left and the subject is asked to slide the movable socket as far as he can. The reading has to be taken from the inner border of the movable socket.

11. *Arm Length* or *Length of the Upper Limb:* This measurement denotes the straight distance between acromion (a) and dactylion (da), while the arm is hanging downwards parallel to the body.

Instrument : Rod Compass or First two segments of Anthropometer rod. The acromion (a) point is to be located by the fingers of left hand to place the upper fixed bar on it. The movable cross-bar is then raised against the tip of the middle finger or dactylion (da) for taking the measurement.

12. *Length of Upper Arm:* This measurement is the straight distance between acromion(a) and radiale (r).

Instrument : Rod Compass or First Segment of Anthropometer rod. The measurement is to be taken just like the above measurement.

13. *Length of the Forearm:* This measurement is the straight distance between radiale (r) and dactylion (da).

Instrument : Rod Compass or First Segment of Anthropometer rod. The instrument is held in the same way as like the above measurement and the movable cross-bar is placed at the tip of the middle finger while the fixed bar goes on the radiale point.

14. *Length of Hand :* This measurement shows the straight distance between the mid-point of a line joining the two stylion (sty) and dactylion (da) of the middle finger.

Instrument : Sliding Caliper. The extended palm of the subject shows a mid-point of the joining line of two stylion. Then the instrument is applied putting the movable cross-bar on the dactylion point to note the measurement.

15. *Hand Breadth:* This measurement is the straight distance between metacarpale radialis (mr) and metacarpale ulnare (mu).

Instrument : Sliding Caliper. Hand is to be placed on the table to its maximum stretch in a similar manner to above measurement. Then the bars of the Caliper are applied against metacarpale ulnare and metacarpale radiale to find out the measurement.

16. *Shoulder Breadth* or *Biacromial Breadth:* This measurement shows the straight distance between the two acromion (a) points.

Instrument : Rod Compass or First Segment ofAnthropometer rod. The subject should stand with his shoulders straight. The acromion (a) points are located by palpating with the first finger while the other fingers hold the cross-bars of the Rod Compass.

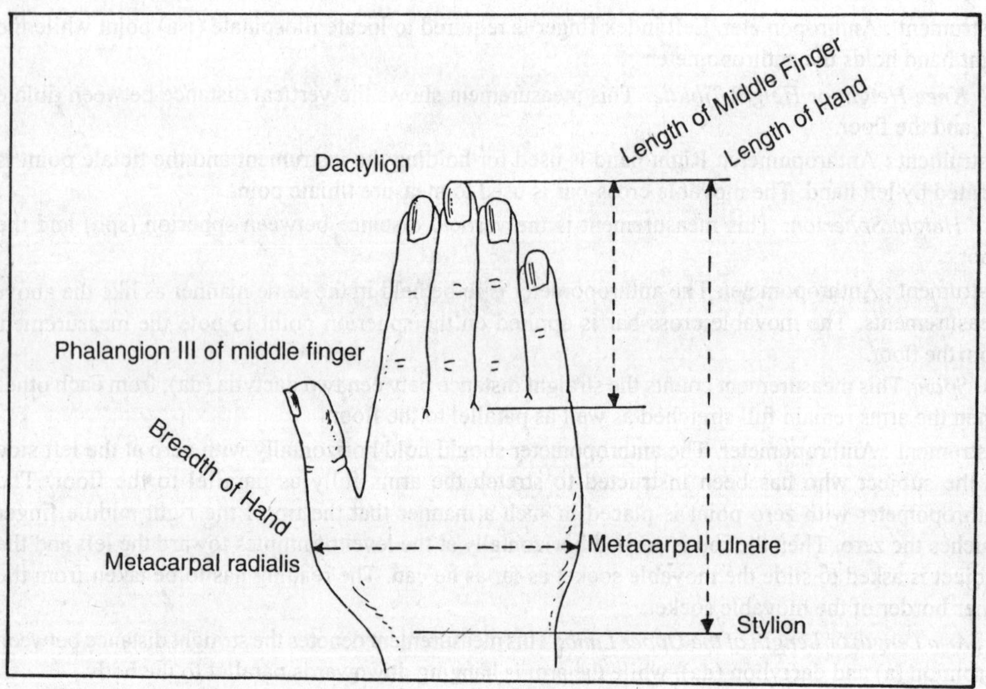

Fig. 8.7. Measurements of Hand

17. *Bi-iliac diameter* or *Bi-Cristal Breadth:* This measurement shows the straight distance between the two iliocristalia (ic).

Instrument : Rod Compass or Anthropometer rod. The difference between the two lateral points on the iliac crest is measured with the help of the cross-bars of Rod Compass. The fingers are used to palpate the most lateral points on the iliac crest. At this time the subject is requested to wear a thin garment for facilitating an accurate measurement.

18. *Chest Breadth* or *Transverse Chest Breadth:* This measurement involves the most laterally placed points on the ribs, at the height of mesosternale (mst).

Instrument : Rod Compass or Anthropometer rod. The subject is directed to stand erect and the measurement is taken when he breaths normally. It is a measurement in between the inhalation state and the state of exhalation. The Rod Compass should be placed horizontally at the nipple level for taking measurement while the arms should be hanged down, slightly away from the side walls of the chest.

19. *Chest Depth* or *Sagittal Chest Depth:* This measurement is the straight distance from mesosternale (mst) to the horizontally placed point in the vertebral column when the arms hang normally on the sides.
Instrument : Spreading Caliper (for long measurement). The instrument must be placed horizontally. This measurement may also be taken with the help of Rod Compass. The subject should be directed to raise the arm slightly for taking the measurement conveniently. The main rod compass is placed at the side of the chest and fixed cross-bar is to be fixed against the chest at the nipple region. Therefore, the movable cross-bar should be adjusted in such a way that it fixes at the back just below the lower edge of the scapula; a slight pressure is given until the cross-bar touches the lower edge of the Shoulder blade. This measurement should be taken very carefully.
20. *Circumference of the Thorax* or *Chest Girth:* This measurement is taken when the subject breadths normally. A measuring tape should be placed horizontally at the level of nipples which passes over the lower scapular angle. The arms may be raised before fixing the tape around the chest and the arms should rest hanging in normal position while taking the measurement.
Instrument : Steel tape. The tape is placed around the chest, covering the landmarks. It should pass on the nipples in front and below the scapular angle at the back. The arms should be in normal position at the time of taking measurement. In case of the female subject, another circumference at the base of xyphoid process, horizontal to the thorax may be taken. The subject should be asked to exhale and inhale. The measurements are taken in both the states accordingly.
21. *Foot Length or Length of Foot:* This measurement denotes the straight distance directly from pternion (pte) to acropodion (ap).
Instrument : Rod Compass or First Segment of Anthropometer rod. The foot is placed on a plane floor in such a position that the weight of the body falls on that. Rod must be placed parallel to the medial region of the foot while the cross-bars rest on the floor. Two cross-bars should be fixed on the two opposite landmarks to take the measurement.

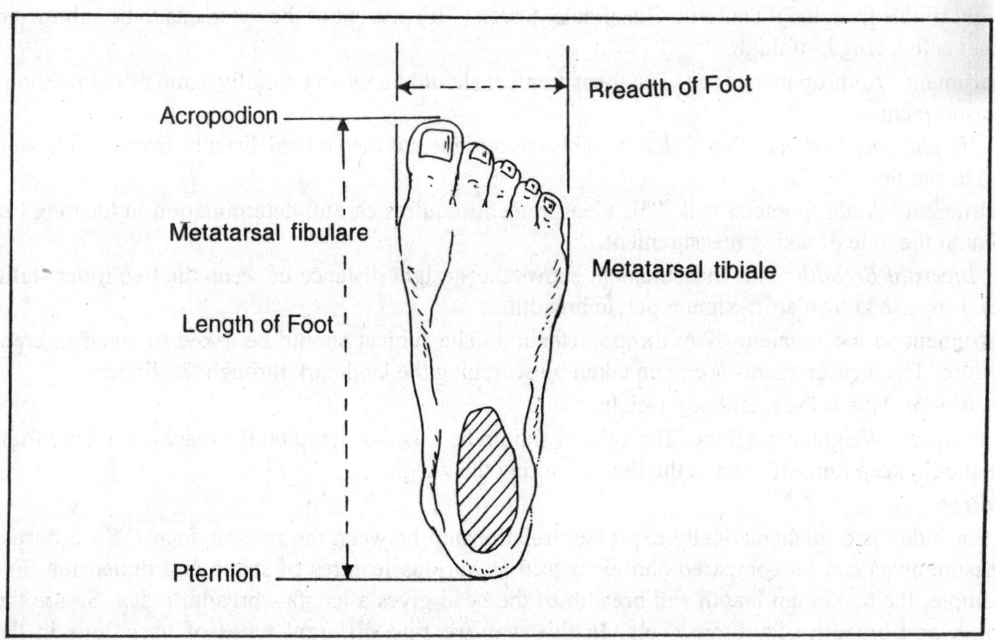

Fig.8.8. Measurements of Foot

22. *Foot Breadth* or *Breadth of Foot:* This measurement shows the straight distance directly between metatarsal tibiale (mtt) and metatarsal fibulare (mtf).

Instrument : Sliding Caliper. Detection of the two landmarks should be made before using the instrument for taking measurement.

23. *Foot Contour:* This is the contour line of foot print.

Instrument : Thin pencil and a flat card board. Contour line of foot produces more clear idea about the foot. The subject is asked to sit on a stool and place the right foot on a clean piece of card board. A 0.5mm pencil with long lead is required to draw a line around the foot. The pencil should be held with 90° angle to the card board. The maximum foot length and foot breadth may be obtained from this contour line. Two different straight lines, one from the pternion (pte) to the middle point of the great toe and the other from the pternion (pte) to the middle point of the second toe should be drawn This forms an angle at the pternion (pte) point, known as the Halux divergence angle.

24. *Sitting Height or Sitting Height Vertex:* This measurement shows a vertical distance from vertex (v) to the sitting surface of the subject when vertebral column is stretched to its maximum.

Instrument : Anthropometer. The subject is asked to sit on a horizontal surface of a stool, preferably 30- 40cm high where his thighs can rest horizontally. The subject should sit keeping his head in eye-ear plane and his body needs to be stretched to the maximum when shoulders should run parallel to the plane and the knees should not be allowed to bend. The measurement should be taken carefully by putting the anthropometer on the sitting surface. The moving sleeve with its extended cross-bar should be adjusted until it touches the vertex.

25. *Length of Lower Leg:* This is a projective measurement that can be obtained by subtracting height spherion from height tibiale.

Instrument : Anthropometric rod. The measurement should be taken carefully. The anthropometer rod must be placed parallel to the medial margin of the foot while the weight of the body should rest mainly on the right foot.

26. *Length of Thigh:* This is also a projective measurement which can be obtained by subtracting height tibiale from height anterior iliospinale. Seven (7%) percent of the value has to be subtracted to get actual length of thigh.

Instrument : Anthropometer rod. The measurement should be taken carefully same as the previous measurement.

27. *Height Symphysion* or *Penal Height:* This measurement is the vertical distance from symphysion (sy) to the floor.

Instrument : Anthropometer rod. This measurement requires careful determination in locating the point at the time of taking measurement.

28. *Bicristal Breadth:* This measurement shows the straight distance between the two iliocristalia (ic). It is also known as maximum pelvic breadth.

Instrument : First segment of Anthropometer rod. The subject should be asked to stand in erect posture. The measurements are to be taken by searching the landmark through the fingers.

29. *Weight.* This is the total body weight.

Instrument : Weighing machine. The subject should be asked to stand on the machine in bare foot. He should keep himself erect at the time of noting the weight.

Indices

An index is a mathematically expressed relationship between the measurements. So different measurements can be compared and expressed as various features of shape and dimension. For example, the maximum length and breadth of the head gives a length - breadth index. So are the length and breadth of the nose, etc. In this way we find different types of variations in the measurements which can produce ratio of certain forms and features. There are some definite formulae to express the measurements accurately, for human body including the head and face. In fact, the measurement ratio is known as an index. In order to find out an index, the shorter measurement is arranged as numerator and the longer as denominator. This gives rise to a fraction which has to be multiplied by 100.

Head and Face : 1. *Length - Breadth Index* or *Cephalic Index:*

$$\frac{\text{Maximum Head Breadth (eu - eu)}}{\text{Maximum Head Length (g - op)}} \times 100$$

Range — Variation (According to Saller)

	Male	Female
Hyperdolichocephalic	X - 70.9	X - 71.9
Dolichocephalic	71.0 - 75.9	72.0 - 76.9
Mesocephalic	76.0 - 80.9	77.0 - 81.9
Brachycephalic	81.0 - 85.4	82.0 - 86.4
Hyperbrachycephalic	85.5 - 90.9	86.5 - 91.9
Ultrabrachycephalic	91.0 - X	92.0 - X

2. *Length — Height Index of Head :*

$$\frac{\text{Auricular Height or Head Height (v - t)}}{\text{Maximum Head Length (g - op)}} \times 100$$

Range — Variation (According to Saller)

Chamaecephalic	X - 57.9
Orthocephalic	58.0 - 62.9
Hypsicephalic	63.0 - X

3. *Breadth—Height Index :*

$$\frac{\text{Auricular Height or Head Height (v - t)}}{\text{Maximum Head Breadth (eu - eu)}} \times 100$$

Range—Variation (According to Martin and Saller)

Tapeiocephalic	X - 78.9
Metriocephalic	79.0 - 84.9
Acrocephalic	85.0 - X

4. *Morphological Facial Index :*

$$\frac{\text{Morphological Facial Height (n - gn)}}{\text{Breadth of Bizygomatic Arch (zy - zy)}} \times 100$$

Range—Variation (According to Marting and Saller)

	Male	Female
Hypereuryproscopic	X - 78.9	X - 76.9
Euryproscopic	79.0 - 83.9	77.0 - 80.9
Mesoprosopic	84.0 - 87.9	81.0 - 84.9
Leptoprosopic	88.0 - 92.9	85.0 - 89.9
Hyperleptoprosopic	93.0 - X	90.0 - X

5. *Jugo—Frontal Index :*

$$\frac{\text{Minimum Frontal Breadth (ft - ft)}}{\text{Breadth of Bizygomatic Arch (zy - zy)}} \times 100$$

Range—Variation

(According to Lundbrog — Linders and Saller)

	Male	Female
Very Narrow	X - 69.9	X - 71.9
Narrow	70.0 - 74.9	72.0 - 76.9
Medium	75.0 - 79.9	77.0 - 81.9
Broad	80.0 - 84.9	82.0 - 86.9
Very broad	85.0 - X	87.0 - X

6. *Jugo—Mandibular Index :*

$$\frac{\text{Bigonial Breadth (go - go)}}{\text{Breadth of Bizygomatic Arch (zy - zy)}} \times 100$$

Range — Variation

(According to Lungbrog — Linders and Saller)

	Male	Female
Very Narrow	X - 69.9	X - 67.9
Narrow	70.0 - 74.9	68.0 - 72.9
Medium	75.0 - 79.9	73.0 - 77.9
Broad	80.0 - 84.9	78.0 - 82.9
Very broad	85.0 - X	83.0 - X

7. *Nasal Index :*

$$\frac{\text{Nasal Breadth (al - al)}}{\text{Nasal Height (n - sn)}} \times 100$$

Range — Variation

(According to Martin and Saller)

Hyperleptorhinae	X - 54.9
Leptorhinae	55.0 - 69.9
Mesorhinae	70.0 - 84.9
Chamaerhinae	85.0 - 99.9
Hyperchamaerhinae	100.0 - X

8. *Nasal Elevation Index :*

$$\frac{\text{Nasal Depth (tip of nose — hind most point of nasal septum)}}{\text{Nasal Breadth (al - al)}} \times 100$$

9. *Physiognomic Ear Index :*

$$\frac{\text{Physiognomic Ear Breadth (lateral posterior margin of helix)}}{\text{Physiognomic Ear Length (sa - sba)}} \times 100$$

10. *Lip Index :*

$$\frac{\text{Height of Integumental Lips (ls - li)}}{\text{Lip Length (ch - ch)}} \times 100$$

Trunk and Extremities :

1. *Biacromial Breadth* or *Shoulder Breadth Index :*

$$\frac{\text{Biacromial Breadth (a - a)}}{\text{Height Vertex (v - floor)}} \times 100$$

Range—Variation
(According to Brugsch)

	Male	Female
Narrow Shoulders	X - 22.0	X - 21.5
Medium Shoulders	22.1 - 23.0	21.6 - 22.5
Broad Shoulders	23.1 - X	22.6 - X

2. *Span — Stature Index :*

$$\frac{\text{Span (da - da)}}{\text{Height Vertex (v - floor)}} \times 100$$

3. *Bicristal Breadth* or *Pelvic Breadth Index :*

$$\frac{\text{Bicristal Breadth (ic - ic)}}{\text{Height Vertex (v - floor)}} \times 100$$

Range—Variation
(According to Brugsch)

	Male	Female
Narrow Pelvic	X - 16.4	X - 17.4
Medium Pelvic	16.5 - 17.4	17.5 - 18.4
Broad Pelvic	17.5 - X	18.5 - X

4. *Pelvis—Shoulder Index :*

$$\frac{\text{Bicristal Breadth (ic - ic)}}{\text{Biacromial Breadth (a - a)}} \times 100$$

5. *Chest Girth Index :*

$$\frac{\text{Chest Girth (circumference of the Thorax)}}{\text{Height Vertex (v - floor)}} \times 100$$

Range—Variation
(According to Martin and Saller)

Narrow chest	X - 50.9
Medium chest	51.0 - 55.9
Broad chest	56.0 - X

6. *Thoracic Index :*

$$\frac{\text{Depth of Thorax (mst-vertebral column)}}{\text{Breadth of Thorax (mst-mst)}} \times 100$$

7. *Inter-Brachial Index :*

$$\frac{\text{Length of Fore Arm (r - da)}}{\text{Length of Upper Arm (a - r)}} \times 100$$

8. *Lower Arm Index :*

$$\frac{\text{Length of Hand (sty - da)}}{\text{Length of Fore Arm (r - da)}} \times 100$$

9. *Length—Breadth Index of Hand* or *Hand Index :*

$$\frac{\text{Hand Breadth (mr - mu)}}{\text{Hand Length (sty - da)}} \times 100$$

Range—Variation

(According to Martin and Saller)

Hyperdolichocheir	X	-	40.9
Dolichocheir	41.0	-	43.9
Mesocheir	44.0	-	46.9
Brachycheir	47.0	-	49.9
Hyperbrachycheir	50.0	-	X

10. *Tibia—Femoral Index :*

$$\frac{\text{Length of Lower Leg (Subtracting from ti — sph)}}{\text{Length of Thigh (Subtracting from isa — ti)}} \times 100$$

11. *Lower Leg-Foot Index :*

$$\frac{\text{Length of Foot (pte - ap)}}{\text{Length of Lower Leg (Subtracting from ti — sph)}} \times 100$$

12. *Femero-Humeral Index :*

$$\frac{\text{Length of Upper Arm (a - r)}}{\text{Length of Thigh (Subtracting from isa — ti)}} \times 100$$

13. *Tibio-Radial Index :*

$$\frac{\text{Length of Fore Arm (r - da)}}{\text{Length of Lower Leg (Subtracting from isa — ti)}} \times 100$$

14. *Foot Index :*

$$\frac{\text{Breadth of Foot (mtt — mtf)}}{\text{Length of Foot (pte — ap)}} \times 100$$

15. *Trunk Index :*

$$\frac{\text{Biacromial Breadth (a - a)}}{\text{Sitting Height Vertex (v - sitting surface)}} \times 100$$

SOMATOSCOPY

The importance of visual observations can not be undermined in the study of somatometry. Although these visual observations are not very accurate, still they are indispensable in somatometry. As a matter of fact, there are a few specific characteristics that can not be put in metrical terms, so such features have to be observed and described with the help of charts and models. Pre-fabricated standard charts are available for comparing skin colour, hair and eye. It is possible to make these observations more accurate by constant careful practice and by consultation with eminent authorities in this field. Some important somatoscopic techniques to study the various parts of the human body are as follows:

1. Body Postures (in erect posture) after Brown :

Type A : The head, trunk and leg are placed in the same axis and in the same plane. The chest is usually well arched, abdomen is flat and buttock projects outwards.

Type B : The head and chest are slightly projected forward; the vertebral column bends backward. The chest is not so high and well arched as in Type A. The buttocks are well developed and strong.

Type C : The anterior abdomen projects outward and the chest is flat. The axis of the leg is almost forward. The buttocks are very developed.

Type D : The head shows a bent forwardly and the abdomen is well arched. The thoracic and buttock regions are strongly arched.

2. Neck

a) *Length* - Long, Medium, or Short.

b) *Adam's Apple Projection* or *Anterior part of the neck* : Marked, or Medial, or Slight.

c) *Size & Volume of the neck* : Slender, or Medium, or Thick.

3. Abdomen

Abdomen : Taut, or Hanging.

4. Body or Trunk

General state of the trunk : Slender, or Medium, or Sturdy, or Plump, or Obese.

5. Upper and Lower Extremities *with Hand and Foot*

a) *Width* : Narrow, or Medium, or Fat.

b) *Symmetry* : Symmetrical, or Asymmetrical.

c) *Defects* : Any muscular defect on elbow or knee. Left hand should be checked as it is less affected by environment.

6. Breasts: The women subject should be taken into consideration who have no children.

a) *Size* : Small, Medium, or Large.

b) *Shape* : Conical, or Intermediate, or Hemispherical.

c) *Symmetry* : Symmetrical, or Asymmetrical.

d) *Nipples* : Small, or Moderate, or Large.

7. Buttocks :

a) *Width* : Narrow, or Medium, or Broad.

b) *Shape* : Conical, or Intermediate, or Flat.

c) *Symmetry* : Symmetrical, or Asymmetrical.

d) *Masculine* or *Feminine type* : Extreme form of femininity is represented in the buttocks of Hottentots tribe of Africa.

8. Hand :

a) *Size* : Small, or Medium, or Long.

b) *Thickness* : Plump, or Medium, or Narrow, or Thin.

c) *Shape* : Conical, or Spatulate, or squarish.

9. Fingers :

a) *Size* : Small, or Medium, or long.

b) *Thickness* : Plump, or Medium, or Narrow, or Thin.

c) *Shape* : Conical, or Spatulate, or Quadrangular, or drum-shaped.

d) *Dexterity of Thumb* : High, or Low.

10. Finger Nails :

a) *Size* : Broad, or Narrow, or long.

b) *Shape* : Flat, or Arched.

c) *Grooves* : Long, or Queer.

11. Foot:

a) *Size* : Broad, or Medium, or Long.

b) *Thickness* : Plump, or Medium, or Narrow, or Thin.

12. Toes :

a) *Size* Small, or Medium, or Long.

b) *Thickness* : Plump, or Medium, or Narrow, or Thin.

c) Shape : conical, or Spatulate, or Quadrangular, or Drum-Shaped.

d) *Dexterity of greater toe* : High or low.

13. Toe Nails :

a) *Size* : Broad, or Narrow, or Long.

b) *Shape* : Flat, or Arched.

c) *Grooves* : Long, or Quer.

14. Skin Colour : The skin colour is affected by use of oil and soaps. Disease, malnutrition and change of climate may also put an effect on the skin colour. Different portions of the body show different skin colour, out of which a few parts must be taken into consideration, such as, the forehead, chest, shoulder blade and inner side of the upper arm. But it is very difficult to determine the exact skin shades in the absence of a standardize skin-colour chart with which the sample can be compared. Various charts have been prepared in this regard by various authorities which help in the study of the skin shades. Among these, the charts made by Luschan, Broca, Fritsch, Huntze, Schultz are famous. The skin-shade of the subject is determined by comparing it with the available shades. For example, Luschan's chart exhibits thirteen colours with varying shades, which have been documented in thirty six plates.

Luschan's Skin Colour Chart :

1,2	—	Yellowish White.
3	—	Carmine White.
4,5	—	Yellowish.
6	—	Light Brown.
7,8	—	Fawn White.
9-11	—	Carmine White.
12,13	—	Pinkish White.
14-29	—	Light Brown to Brown.
30,31	—	Dark Brown.
32	—	Reddish Brown.
33,34	—	Dark Brown.
35	—	Greyish Black.
36	—	Black.

15. Hair : The study of hair may be done in two ways, considering its colour and form. Usually the observations should be made on head and beard, genital and body hair. But at least the hair on head and face are observed.

The hair colour chart of Fischer and Saller is an elaborate one which contain thirty different shades of natural hair colour. The shades are ash blond, light blond blond, dark blond, brown, dark brown, red, reddish brown, etc.

Hair form

1. Leiotrichous or Straight hair
   a) *Stretched* : Thick straight hair.
   b) *Smooth* : Thin straight hair.
   c) *Flat* wavy : Waves with largest radius.
2. Cymotrichous or Wavy hair :
   a) *Broad wave* : Waves with smaller radius.
   b) *Narrow wave* : Strongly curved and short waves.
   c) *Curly* : Waves are deep with broad spirals.
3. Ulotrichous or Woolly hair :
   a) *Frizzly* : Waves with very strong curvature.
   b) *Loose frizzles* : These are flat circular spirals.
   c) *Thick frizzles* : Circular flat spirals thickly set.
   d) *Filfil* : Heavily rolled and known as peppercorn hair.

***Hair texture* :**

Coarse, or medium, or Fine.

***Quantity of hair* :**

Scanty, or Medium, or Thick, or Very Thick, or Rich.

***Hair whorls* :**

a) *Quantity* : Single, or Double, or Multiple.
b) *Direction* : Clockwise, or Anticlockwise.

***Beard and moustache* :**

Quantity : Normal, or Medium, or Thick.

16. *Head and Face* :

Vertex

a) *Curve* : Flat, or Slight, or Medium, or Well-arched.
b) *Shape* : Ovoid, or Pentagonoid, or Elliptical, or Spheroid.

Occiput

a) *Shape* : Barrel-shaped, or Gable-shaped.
b) *Projection* : Flat, or Moderately bulging, or Strongly bulging.

Forehead-

a) *Height* : Low, or Medium, or High.
b) *Breadth* : Narrow, or Medium, or Broad.
c) *Slope* : Absent, or Slight, or Medium, or Pronounced.

Total Face

a) *Height* : Short, or Medium, or Long.
b) *Shape* : Round, or Oval, or Elliptical, or Round, or Square.
c) *Form* : Flat, or Medium, or Arched, or Projected.

17. Eyes :

a) *Eye Opening Axis* : Horizontal, or Slanting, or Oblique.
b) *Direction of Eye* : Outwards, or Inwards.

c) *Eye Folds* : Slight, or Heavy, or Mongoloid Fold, or Epicanthous Fold.

d) *Colour of Iris* : Black Brown, or Dark Brown, or Brown, or Light Brown, or Greenish, or Dark Grey, or Light Grey, or Dark Blue, or Blue, Light Blue, or Crimson Red (Albino).

The colour of iris can be justified by various charts prepared by Bertillon (54 in number of varieties), Martin (16 in number of varieties), Martin and Schultz (20 in number of varieties), Martin and Saller (8 in number of varieties), and Saller (40 in number of varities).

Martin and Schultz Eye Colour Chart:

la	—	2b	Light Blue to Blue
3	—	6	Light Grey to Dark Grey
7	—	8	Greenish
9	—	16	Light Brown to Dark Brown

e) *Colour of Scelera* : Clear, or Speckled, or Yellow, or Dull.

f) *Iris* : It can be measured by means of a lens and can be described by the following terms—Homogeneous, or Rayed, or Zoned, or Speckled, or Diffused

18. Nose:

a) *Nasal Root* : Flat, or Shallow, or Medium, or High, or Deep, or Very Deep.

b) *Nasal Bridge*

i) Size : Short, or Medium, or Long.

ii) Shape from Profile : Concave, or Straight, or Convex.

c) *Tip of Nose*

i) Projection : Upward, or Downward, or Forward.

ii) Profile of tip : Rounded at point, or Fully Rounded, or Flat

d) *Nasal wings or Alare*

i) Thickness : Thin, or Medium, or Thick.

ii) Height : High, or Low.

iii) Bulge : Slight Flat, or Slight, or Strongly Bulging.

e) *Nasal Septum*

i) Size : Short, or Medium, or Long.

ii) Breadth : Thin, or Thick.

iii) Direction : Upwards, or Horizontal, or Downwards.

iv) Profile : Visible, or Non-visible Septum.

f) *Nasal Cavity*

i) Size and Shape : Very narrow, or Narrow, or Long oval, or Short oval, or Round, or Broad, or Very broad.

ii) Length : Small, or Big.

19. Lips

a) *Integumental Upper Lip*

i) Form : Concave, or Straight, or Convex.

ii) Membral Lip : Thin, or Medium, or Thick, or Puffed.

iii) Mentolabial Fold : Low, or Medium, or High.

b) *Integumental Lower Lip* —

i) Form : Straight, or Bulging Downwards.

ii) Membral Lip : Thin, or Medium, or Thick, or Puffed.

iii) Mentolabial Fold : Low, or Medium, or High.

20. Chin

a) Chin Form : Narrow, or Medium, or Broad.

b) Chin Shape : Rectangular, or Square, or Elliptical, or Pointed, or Round.

c) Chin Prominence : Sub-medium, or Medium, or Pronounced.

21. Prognathism

a) Type : Alveolar, or Facial.

b) Size : Slight, or Medium, or Pronounced.

c) Alveolar : Slight, or Medium, or Marked.

d) Facial : Slight, or Medium, or Marked.

22. External Ear

a) Ear Lobe

i) Presence : Absent, or Present.

ii) Size : Long, or Medium, or Short.

iii) Shape : Tounge-shaped, or Triangular,or Square, or Arched.

iv) Thickness : Thin, or Medium, or Thick.

v) Attachment : Free, or Attached.

b) Antihelix Curvature : Weak, or Medium, or strong

c) Antitragus : Weak, or Strong, or Angularly Developed.

d) Helix and Antihelix : Equal, or Narrower, or Broader.

e) Tragus and Antitragus -

i) Size - Small, or Weak, or Medium.

ii) Shape - Long, or Round, or Knob-shaped.

OSTEOMETRY

Osteometry is the measurement of bones especially long bones, which facilitates in comparing the skeletons of different populations. A few special instruments are required for collection of the osteometrical measurements. The most important one is osteometric board; other instruments have been mentioned in connection to craniometry.

Osteometric Board

The osteometric board is a flat rectangular tray with a long and a short vertical wall (See Fig. 8.1). The length of the board is 65cm, breadth is 25cm, and the thickness is about 2.54cm. One longer side and one broader side of this tray remain closed by two vertical wooden pieces about 10cm in height. These vertical walls on the board meet with each other forming a right angle at one corner of the tray. The tray is graduated by metric scales that run on each side for the purpose of measurement. A movable cross-piece is there for determining the measurements. It is an L-shaped wooden piece, which has no attachment with the board, so can be moved freely. At the time of measuring a long bone, this cross-piece is to be placed on the board in such a fashion that one end of it touches the bone. So, when one end of long bone touches the fixed vertical wall of the board and other end is touched by the cross-piece, measurements of the bone can be read out.

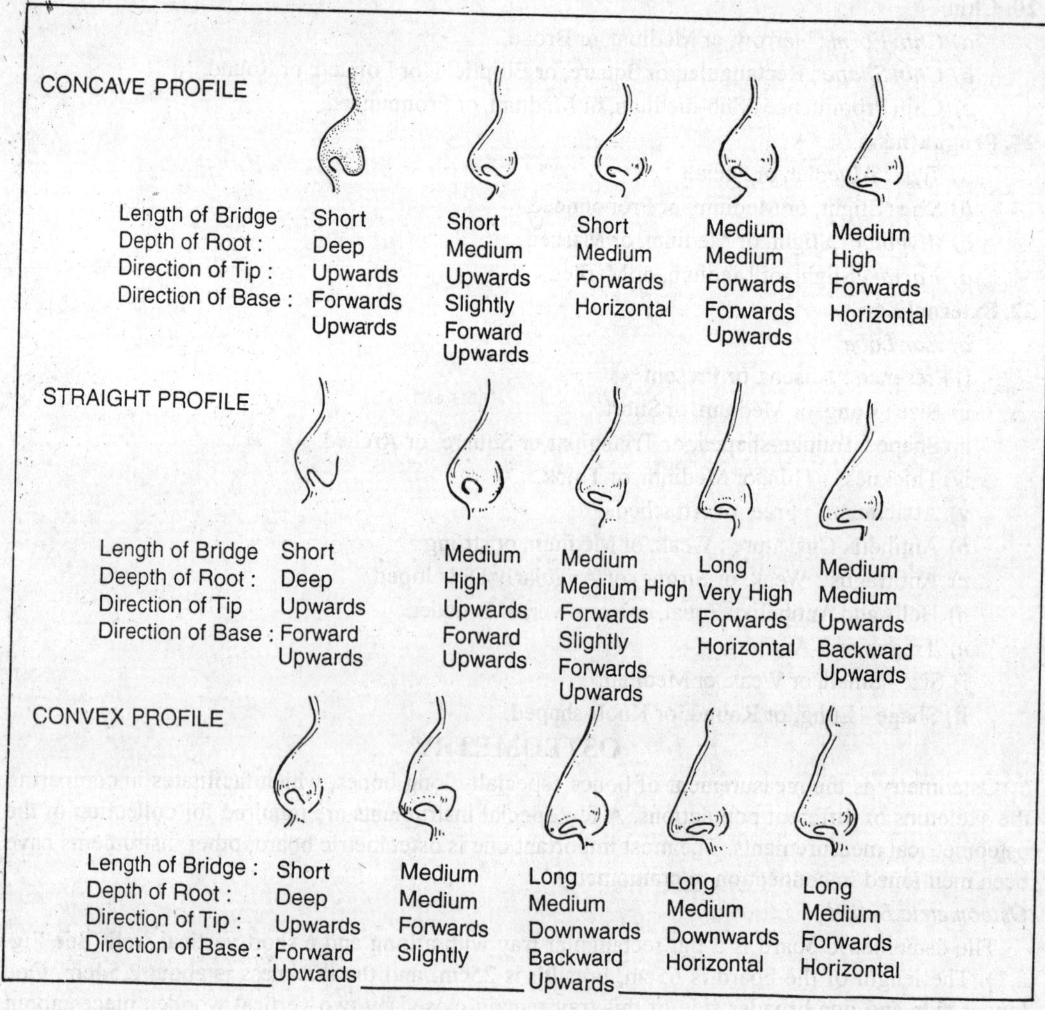

Fig. 8.9. Forms of Nose

Measurements

Humerus :

1. *Maximum Length* :

This measurement is the straight distance between the highest point of the head of the humerus and the deepest point on the trochlea.

Instrument : Osteometric board. The humerus has to be placed in the board in such a way that the head of the humerus touches the vertical wall of the tray at the broader side. In this position, the cross-piece is pressed against the distal end of the bone and the maximum length is recorded from the scale of the board.

2. *Maximum diameter of the Shaft* (in middle)

This measurement is the absolute greatest diameter without regard to the sagittal or transverse plane.

Instrument : Osteometric board and a Sliding Caliper. For determining the highest point in the middle of the shaft, the bone should be pressed longitudinally along the longer side of the vertical wall of the tray and the cross-piece is to be pushed on the middle plane of the shaft. The highest point where the cross-piece touches the shaft is noted with a mark of pencil. Now the greatest diameter can easily be measured with the help of a sliding caliper.

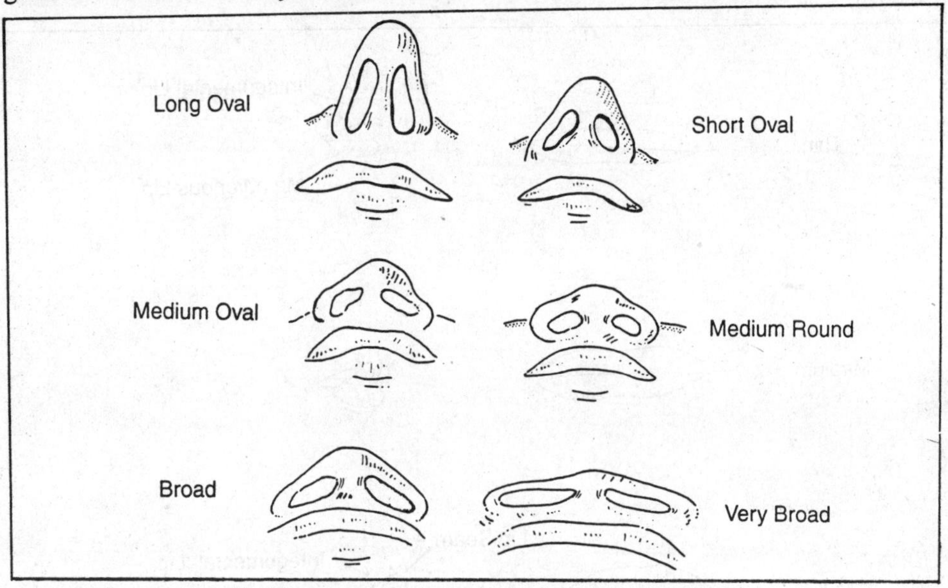

Fig. 8.10. Shape and Size of Nasal Opening

3. *Minimum diameter of the Shaft*
   This is the absolute least diameter in the middle of the shaft.
   Instrument - Sliding Caliper. The thinnest circumference of the shaft has to be determined by trial method. Difference between the two hands of the sliding caliper reveals this minimum measurement.

Radius

1. *Maximum Length*
   This measurement is the greatest distance between the most proximal point on the margin of the head of the radius and the tip of the styloid process.
   Instrument : Osteometric board.
2. *Maximum diameter of the Shaft*
   This measurement denotes the maximum diameter of the shaft.
   Instrument : Osteometric board and Sliding Caliper. At first the highest opposite points on the diameter of the shaft has to be determined by the use of an osteometric board. After this, a sliding caliper measures the difference between those two points.
3. *Minimum diameter of the Shaft*
   This measurement denotes the circumference of the shaft at its thinner most part.
   Instrument : Sliding Caliper. The measurement is to be taken with the help of the sliding caliper by trial method.

Ulna

1. *Maximum Length*
   This measurement denotes the greatest distance between the highest point of the olecranon and the deepest point of the styloid process.

Instrument : Osteometric board. The procedure is same as like the previous measurements concerning with the measurement of maximum length of the bone.

2. *Maximum diameter of the Shaft*

This measurement finds out the maximum diameter of the shaft at the greatest development of the crest.

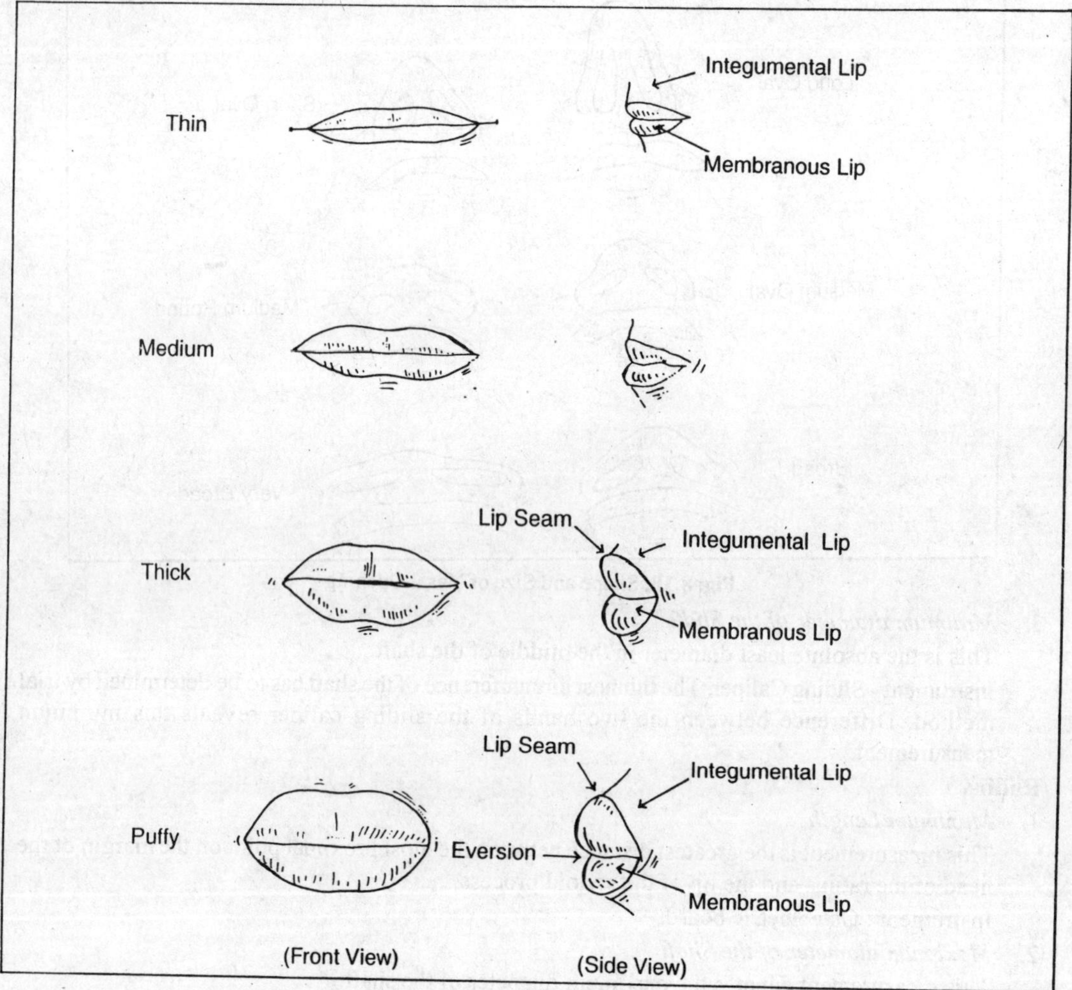

Fig. 8.11. Forms of Membranous Lip

Instrument: Sliding Caliper. This measurement is to be taken in such a way that the two arms of the caliper should lie at the posterior margin and the anterior surface of the bone.

3. *Minimum diameter of the Shaft*

This measurement denotes the least circumference, which is usually found, at the distal end of the bone.

Instrument : Sliding Caliper. The measurement is to be taken with a great caution by using the trial method.

Femur

1. *Maximum Length*

This measurement shows the straight distance between the highest point of the head and the deepest point on the lateral medial condyle.

Instrument : Osteometric board. The femur should be placed with its dorsal side upwards and the procedure of taking measurement is same like other long bones.

2. *Bicondylar Length*

This measurement is the projective distance between the condyles of the femur.

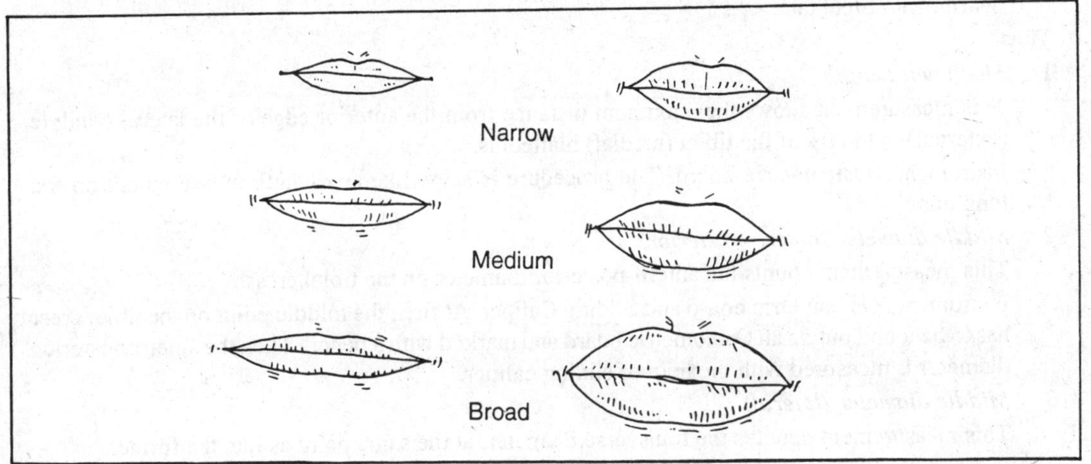

Fig. 8.12. Forms of Mouth Opening

Instrument : Osteometric board. The femur should be placed on the board in such a manner that one of the condyles touches the short vertical wall. Then the movable cross-piece should be applied on the other condyle where the head of the femur is touched tangentially. The readings should be collected *from* the graphical scale on the osteometric board.

3. *Maximum diameter of the Head*

This measurement reveals the greatest distance between the two laterally projected points on the head of the femur.

Instrument : Sliding Caliper. Two arms of the sliding caliper are applied on the two points for taking measurement.

4. *Middle shaft diameter (antero-posterior)*

This measurement denotes the distance between the anterior and posterior surfaces of the femur, especially at the middle of the shaft (at the height of the highest elevation, linea aspera).

Instrument : Sliding Caliper. Since the points can be more or less determined by eye, so arms of the sliding caliper are placed on them very easily to take other measurement.

5. *Middle shaft diameter, lateral*

This measurement also denotes a distance between the two points at middle shaft. So, the measurement is to be taken at the same point as the former one, but it should be taken transversely from one lateral margin to other lateral margin. It means that the measurement is perpendicular to the former measuring plane.

Instrument : Sliding Caliper.

6. *Sub-trochanteric diameter (antero-posterior)*

This measurement shows the anterior-posterior diameter or the sagittal diameter of the shaft below the trochanter.

Instrument : Sliding Caliper.

7. *Sub-trochanteric diameter, lateral*
   This measurement is the transverse diameter at the same point as like the former.
   Instrument : Sliding Caliper.
8. *Circumference of the shaft (at middle)*
   This measurement shows maximum circumference at the middle of the shaft. It indicates the robustness of the shaft.
   Instrument : Steel tape.

Tibia

1. *Maximum Length*
   This measurement shows the maximum distance from the anterior edge of the lateral condyle (external) to the tip of the tibial (medial) malleolus.
   Instrument : Osteometric board. The procedure is same like any length measurement on the long bone.
2. *Middle diameter (antero-posterior)*
   This measurement counts the antero-posterior diameter on the tibial crest.
   Instrument : Osteometric board and Sliding Caliper. At first, the middle point on the tibial crest has to be found out on an Osteometric board and marked with a pencil. Then the antero-posterior diameter is measured with the help of sliding caliper.
3. *Middle diameter (lateral)*
   This measurement denotes the transverse diameter, at the same point as like the former.
   Instrument : Sliding Caliper. The arms of the sliding caliper has to be placed on the same plane of the tibial crest (as stated above) but the lateral points have to be chosen for measurement which goes perpendicular to that of the former.
4. *Antero-posterior nutrient foramen diameter*
   This measurement shows the circumference (i.e. Girth of shaft) of the middle part of the tibia at the level of nutrient foramen.
   Instrument : Tape or Sliding Caliper.
5. *Medio-lateral nutrient foramen diameter*
   This measurement denotes the transverse diameter, at the same point as the former.
   Instrument : Tape or Sliding Caliper.

Fibula

1. *Maximum Length*
   This measurement is the distance between the highest point of the apex of the head and deepest point of the fibular malleolus.
   Instrument : Osteometric board.

Scapula

1. *Maximum Length* or *Morphological Breadth*
   This measurement is the straight distance between the highest point of the superior angle (angle cranialis) and the deepest point of the inferior angle (angle caudalis).
   Instrument : Sliding caliper.
2. *Maximum Breadth* or *Morphological Length*
   This measurement shows the straight distance between the middle point of the glenoid fossa and the point of the vertebral border, midway between the two ridges terminating the spine.
   Instrument : Sliding caliper. It should be noted that the mid-point of the juncture of the spine should be marked prior to the measurement.

3. *Glenoid point height*

   This measurement shows the distance between the inferior angle and the center of the fossa situated near the middle of the glenoid cavity.

   Instrument : Sliding caliper.

4. *Glenoid point breadth*

   The measurement shows the distance between the inferior angle (angle caudalis) and the point on the outer border of the glenoid cavity.

   Instrument : Sliding Caliper.

Sternum

1. *Total Length* : This measurement shows the distance between the suprasternale and the lower margin of the sternum.

   Instrument : Osteometric board or a large Sliding Caliper.

2. *Maximum Breadth* : This measurement denotes the distance between the most laterally placed points on the lateral margins of the sternum.

   Instrument : Sliding Caliper.

Clavicle

1. *Maximum Length* : This measurement shows the distance between the outermost points on the acromion and sternal ends of the bone.

   Instrument : Osteometric board.

Pelvis

1. *Pelvic Height* : This measurement shows the straight distance from the highest point on the iliac crest to the lowest point on the ischial tuberosity.

   Instrument : Spreading Caliper.

2. *Greatest Pelvic Breadth* : This measurement is the straight distance between the most projecting points on the lateral margins of the iliac crest.

   Instrument : Spreading Caliper.

3. *Sagittal diameter of the pelvic inlet* : This measurement shows the straight distance from the most projecting point on the anterior surface of the symphysis to the tip of the most projecting spinous process of sacrum.

   Instrument : Spreading Caliper.

4. *Transverse diameter of Pelvic inlet* : This measurement shows the maximum transverse diameter of the pelvic inlet between Liniae arcuatae.

   Instrument : Sliding Caliper.

5. *Interspinous diameter* : This measurement shows the maximum distance between the antero-lateral margins of the antero-superior spines of the iliac bone.

   Instrument : Sliding Caliper.

6. *External conjugate diameter* : This measurement reveals the distance between the inferior tip of the fifth lumber spine and the antero-inferior margin of the pubic symphysis.

   Instrument : Pelvimeter.

7. *Diagonal conjugate diameter* : This measurement shows the distance from the mid-sagittal point on the antero-superior margin of the sacral promontory up to the postero-inferior margin of the pubic symphysis.

   Instrument : Pelvimeter.

8. *Normal conjugate diameter* : This measurement shows the distance between the postero-superior margin of the symphysis and the lowest point of the third sacral vertebra.

   Instrument : Sliding Caliper.

Sacrum

1. *Maximum Height* : This measurement denotes the distance between the middle of promontory and antero-inferior border of the fifth sacral vertebra.
   Instrument : Sliding Caliper.
2. *Maximum Breadth* : This measurement is the maximum transverse breadth in the level of anterior projection of the articular surface.
   Instrument : Sliding Caliper.

Patella

1. *Maximum Height and Breadth* : The bone is to be placed between the two cross-bars of the sliding caliper in side to side position. Then the bone is moved slowly to find out the maximum measurement.
   Instrument : Sliding Caliper.
2. *Maximum thickness* : In this measurement, the fixed cross-bar of the sliding caliper is put against the anterior surface of the bone, then the movable cross-bar is adjusted against the posterior face, over the thickest part of the bone.
   Instrument : Sliding Caliper.

Osteometrical Indices

1. *Radio-humeral Index* :

$$\frac{\text{Maximum Length of Radius}}{\text{Maximum Length of Humerus}} \times 100$$

Range :

Brachykerkik	X	–	74
Mesotikerkik	75	–	79
Dolichokerkik	80	–	X

2. *Humero-femoral Index* :

$$\frac{\text{Maximum Length of Humerus}}{\text{Bicondylar Length of Femur}} \times 100$$

3. *Tibio-femoral Index* :

$$\frac{\text{Length of Tibia (less spine)}}{\text{Bicondylar Length of Femur}} \times 100$$

Range :

Brachycnemic	X	–	82
Dolichocnemic	83	–	X

4. *Platymeric Index* :

$$\frac{\text{Sub-trocanteric Antero-posterior Diameter of Femur}}{\text{Sub-trocanteric Medio-lateral Diameter of Femur}} \times 100$$

Range :

Hyperplatymeric	X	–	74.9
Platymeric	75.0	–	84.9
Eurymeric	85.0	–	99.9
Stenomeric	100.0	–	X

5. *Pilastric Index :*

$$\frac{\text{Antero-posterior Diameter of middle of shaft of Femur}}{\text{Medio-lateral Diameter of middle of shaft of Femur}} \times 100$$

6. *Platycnemic Index :*

$$\frac{\text{Medio-lateral nutrient foramen Diameter of Tibia}}{\text{Antero-posterior nutrient foramen Diameter of Tibia}} \times 100$$

Range :

Hyperplatycnemic	X	–	54.9
Platycnemic	55.0	–	62.9
Mesocnemic	63.0	–	69.9
Eurycnemic	70.0	–	X

7. *Scapular Index :*

$$\frac{\text{Morphological Length}}{\text{Morphological Breadth}} \times 100$$

8. *Pelvic-breadth height Index :* $\frac{\text{Maximum Pelvic Height}}{\text{Maximum Pelvic Breadth}} \times 100$

9. Pelvic inlet Index : $\frac{\text{Sagittal Diameter of Pelvic Inlet}}{\text{Transverse Diameter of Public Inlet}} \times 100$

Range :

Platypellic	X	–	89.9
Mesopellic	90.0	–	94.9
Dolichopellic	95.0	–	X

10. *Length-breadth sacral Index :*

$$\frac{\text{Maximum Sacral Breadth}}{\text{Maximum Sacral Length}} \times 100$$

Range :

Dolichohieric	X	–	99.9
Subplatyhieric	100.0	–	105.9
Platyhieric	106.0	–	X

Observations

Certain characteristics features are not measurable in numerical figures. Therefore these features may be observed and described. However, some codes against the observations and descriptions are as follows:

Humerus

a) Shape of shaft : Prismatic, lateral prismatic, quadrilateral, plano-convex, intermediary.
b) Supra-condylar process : None, rough, trace.
c) Ridge : Slight, medium, pronounced.

Radius

a) Shape : Prismatic, flexor surface concave, external surface convex, etc.

Ulna

a) Shape : Prismatic, flexor surface concave, quadrilateral.

Femur

a) Shape : Oval, eliptical, prismatic, cylindrical, quadrilateral.
b) Crista hypotrochanterica : Absent, medium, pronounced, very pronounced.
c) Fossa hypotrochanterica : Absent, medium, pronounced, very pronounced.

d) Third trochanter : Absent, medium, pronounced, very pronounced.

e) Linea aspera : Absent, medium, pronounced.

Tibia

a) Shape : Prismatic, lateral prismatic, external surface concave, posterior surface divided into two, posterior surface convex, internal border indistinct, plano-convex.

b) Retroversion of head : Absent, medium, pronounced.

c) Lateral condyle : Convex, concave.

Fibula

a) Shape : Ordinary quadrilateral, approaching prismatic, lateral prismatic, medial surface fluted, lateral surface differentiated into two surfaces, lateral surface fluted, both medial and lateral surfaces fluted.

Scapula

a) Vertebral border : Straight, concave, convex.

b) Shape of acromion process : Sickle, triangular, quadrangular, intermediate.

c) Scapula notch : None, slight, medium, deep, foramen.

Clavicle

a) Strength : Slender, medium, strong, massive.

b) Curvature : Slight, medium, pronounced.

Pelvis

a) Depth and breadth of ischiatic notch : Small, medium, deep.

b) Sub-pubic angle : Small, medium, large.

c) Ischia : Parallel, converging, diverging.

Sacrum

a) Curvature : slight, moderate, pronounced.

Patella

a) Vastus notch : None, moderate, slight.

CRANIOMETRY

The measurements on the skeletal parts can be divided into two broad aspects—*Osteometry* and *Craniometry.* Osteometry is the measurements of long bones, which help us to compare the skeletons of different human populations. We also place the man and others in the phylogenetic order by taking measurements on the primates as well as fossil races. Besides, a variation of range in different human groups in respect to skeletal structures can be determined by means of taking accurate measurements on the bones. In case of craniometry the main objective is to study the form and shape of human skull by means of exact measurements.

The skull is the most important part of the skeleton. Craniometric measurements are divided into three categories, such as,

i) *Neurocranium* (Brain Cavity),

ii) *Splanchonocranium* (Face), and

iii) Other measurements which concern skull as a whole.

Landmarks

1. *Alveolon* (alv) : If a tangent may be drawn to the posterior margins of the alveolar arc, that cuts the mid-sagittal line of the palate on a point. This point is known as Alveolon. It is essential to know this point in finding out the maxillo-alveolar length.
2. *Asterion* (ast): It is the point behind the mastoid process where the parietal, occipital and temporal bones meet. If there is a small bone at this point, all three sutures may be projected in straight line to find out asterion point where they meet.

3. *Auriculare* (au): This point is perpendicular on the ear opening or external auditory meatus touching the root of the zygomatic arc.
4. *Basion* (ba): It is a middle point on the anterior border of the foramen magnum, which is cut by the mid-sagittal plane. It lies exactly opposite to the opisthion (o).
5. *Bregma* (b) : It is a point on the meeting place of the coronal and sagittal sutures. The point may vary but it lies wherever they meet.
6. *Cononale (co):* It is the most laterally placed point on coronal sutures and important for measuring the greatest frontal breadth.
7. *Dacryon* (d): It is the point on the inner wall of the orbit where the frontal, lacrimal and maxillary bones meet.
8. *Ectoconchion* (ect) : It is a point where a line passing parallel to the upper orbital border cuts the lateral orbital margin.
9. *Ectomolare* (ecm) : This is the most laterally placed point on the outer surface of the alveolar margin of the second molar in man but the third molar in case of the apes.
10. Endomolare (enm): This point is located in the middle of the inner margin of the second molar. But in case of the apes, this point is found on third molar.
11. *Euryon* (eu) : This point lies most laterally on the skull. Determination of this point is very essential in noting the maximum cranial breadth.
12. *Frontomalare orbitale* (fmo) : This point is located on the lateral orbital margin where it is cut by zygomatico frontalis suture. More specifically, it is situated on the orbital end of the fronto-jugal suture.
13. *Frontomalare temporals* (fmt) : This point is most laterally placed on zygomatico frontalis suture. It is actually located toward the temporal end of the fronto-jugal suture.
14. *Frontotemporale* (ft) : This is the most projected and inwardly placed point on the superior temporal line.
15. *Glabella* (g) : It is the most prominent point on the frontal bone between the two supra-orbital ridges of the forehead. Its exact location is just above the naso-frontal suture, on the mid-sagittal plane.
16. *Gnathion* (gn) : It is lowest point on the lower margin of the mandible in the mid-sagittal plane.
17. *Gonion* (go) : It is the most downward point in the external border of the mandible where the horizontal and ascending ramii meet.
18. *Infradentale (id):* This point is located between the two middle incisors, where the anterior margin of the alveolar process is cut by the mid-sagittal plane.
19. *Infra-temporalia* (it): This point is the most inward point on the convex infratemporal crest.
20. *Inion* (i): It is the meeting point of the superior nucheal lines in the mid-sagittal plane. The point never lies on the lower end of the external occipital protuberance. In case of weakly developed superior *nuchal* lines one should extend them in the same plane and take the point where they meet.
21. *Lacrimale* (la) : It is a point where anterior lacrimal crest meets with the lacrimofrontalis suture.
22. *Lambda* (1) : It is the meeting point of the sagittal suture with the lambdoid sutures. Sometimes it is not located easily. One should draw two straight lines cutting through these sutures and the point is located wherever these lines meet the mid-sagittal plane.
23. *Mastoiodale* (ms): This point is placed on the tip of the mastoid process, laterally and downwardly.
24. *Maxillofrontale* (mf): It is a point where the anterior lacrimal crest meets the frontomaxillary suture.

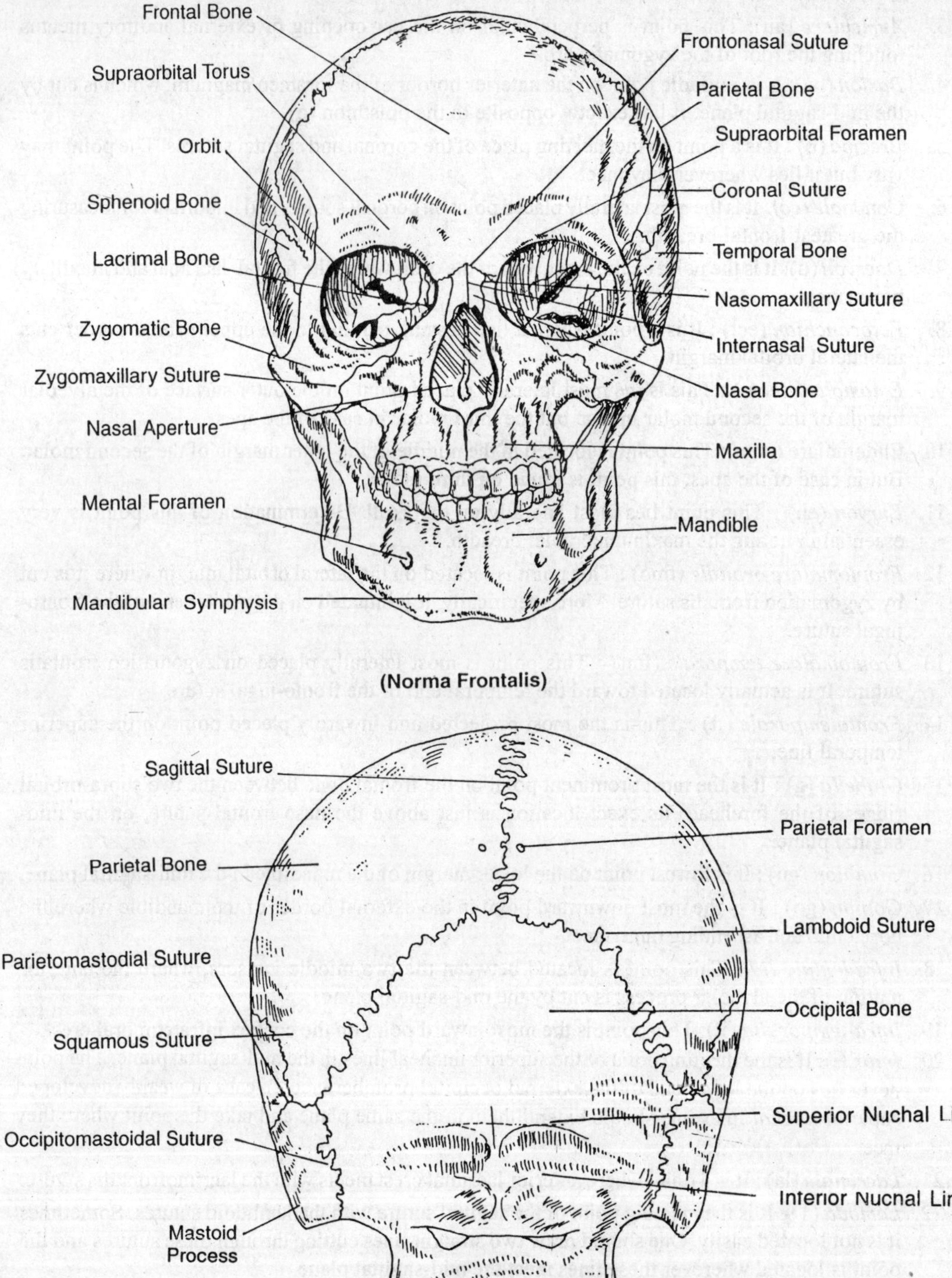

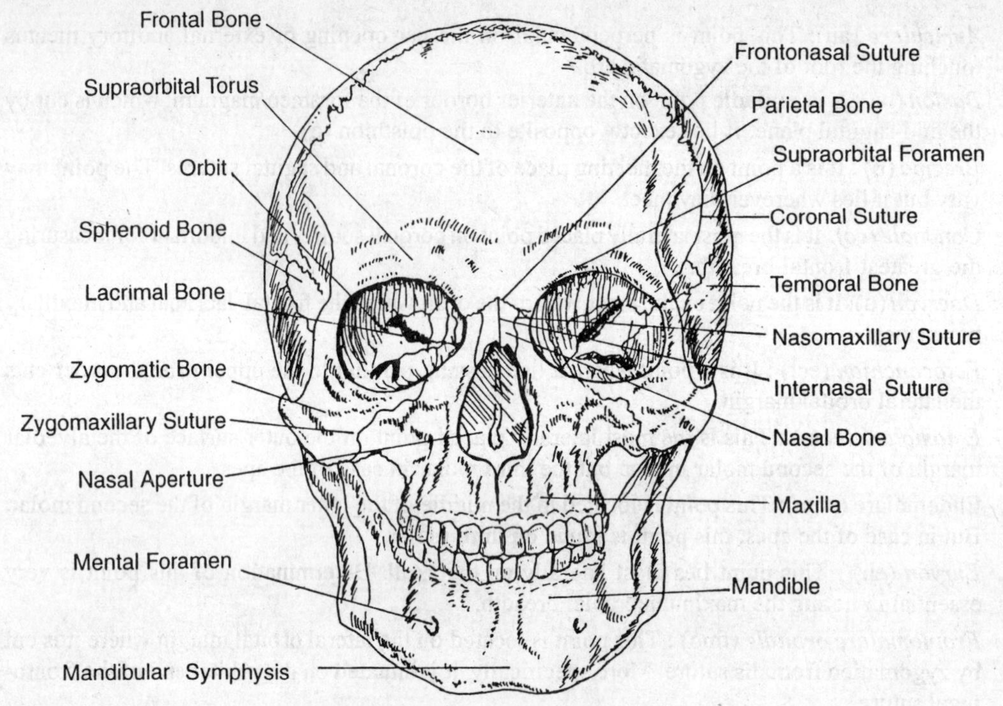

(Norma Frontalis)

Sagittal Suture
Parietal Foramen
Parietal Bone
Lambdoid Suture
Parietomastodial Suture
Occipital Bone
Squamous Suture
Superior Nuchal Line
Occipitomastoidal Suture
Interior Nucnal Line
Mastoid Process

Nomra Occipitalis

Fig. 8.13. Craniometric Landmarks on Human Skull (I)

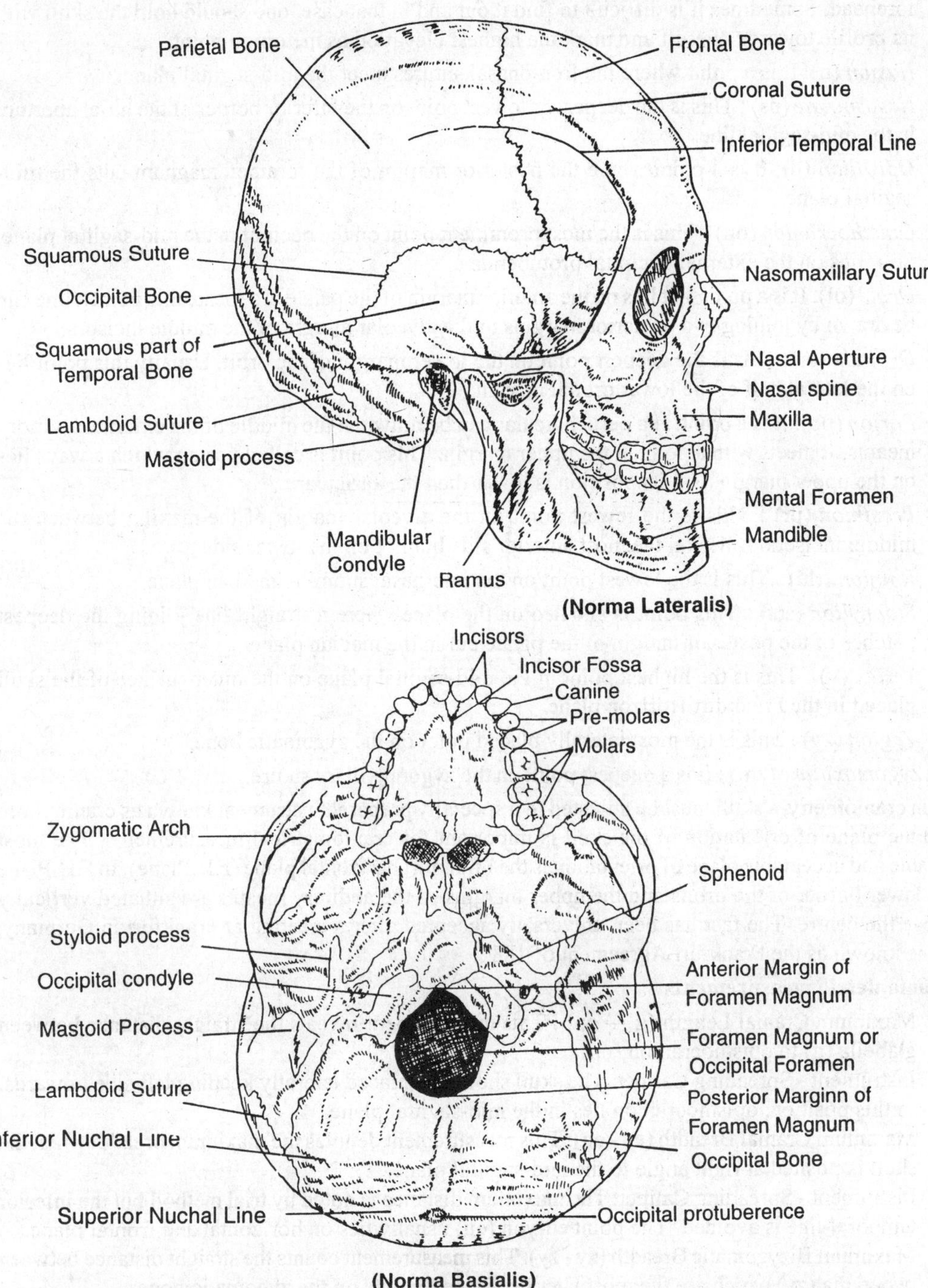

Fig. 8.14. Craniometric Landmarks on Human Skull (II)

25. *Metopion* (m) : This is a crossing-point of the horizontal line and mid-sagittal plane of the forehead. Sometimes it is difficult to find it out and in that case, one should hold the skull with its profile towards oneself and mark the highest elevation as metopion point.
26. *Nasion* (n): It is a point where the frontonasal sutures meet the mid-sagittal plane.
27. *Nasospinale* (ns) : This is the deepest or lowest point on the inferior border of the nasal aperture in the mid-sagittal line.
28. *Opisthion* (o): It is a point where the posterior margin of the foramen magnum cuts the mid-sagittal plane.
29. *Opisthocranion* (op) : This is the most prominent point on the occiput in the mid-sagittal plane. So it lies on the external occipital protuberance.
30. *Orale* (ol): It is a point that lies on the anterior margin of the palate. To locate this point a line can be drawn by joining two posterior margins of the alveolar margin of the middle incisors.
31. *Orbitale* (or): This is the deepest point on the lower margin of the orbit. Usually this point lies on the lateral half of the lower orbital margin.
32. *Porion* (po): It is a point. If a perpendicular can be drawn in the middle of the external auditory meatus, it meets with a point at the upper margin. This point is called porion which always lies on the upper margin. But its position is lower than the auriculare.
33. *Prosthion* (pr) : This is the lowest point on the alveolar margin of the maxilla between the middle incisors. It lies on the most anterior side but not on the lower side.
34. *Rhinion* (rhi) : This is the lowest point on the internasal suture in median plane.
35. *Staphylion* (sta) : This point is situated on the place where a straight line joining the deepest notches of the posterior margin of the palate cut in the median plane.
36. *Vertex* (v) : This is the highest point in the mid-sagittal plane on the outer surface of the skull placed in the Frankfurt Horizon plane.
37. *Zygion* (zy) : This is the most laterally placed point on the zygomatic bone.
38. *Zygomaxillare* (zm) : It is a deepest point on the zygomaxillary suture.

In craniometry, a skull must be adjusted to a specially prepared instrument known as craniophore and the plane of orientation of the skull is important for craniometrical measurements. The most notable and acceptable plane of orientation is the Frankfurt horizontal plane (F.H.Plane). In F.H.Plane the lower border of the orbits and the upper margins of the auditory meatus are situated vertically above the centre. The fact has been universally accepted after a meeting at Frankfurt in Germany, being known as the Frankfurt Agreement of 1882.

Craniometric measurements

1. Maximum Cranial Length (g — op) : This measurement reveals the straight distance between glabella (g) to opisthocranion (op).
   Instrument - Spreading Caliper. The skull should be placed laterally keeping left side upwards. In this position, opisthocranion lies in the mid-sagittal plane.
2. Maximum Cranial Breadth (eu - eu): This measurement denotes the maximum breadth when the skull is posited at right angle to the mid-sagittal plane.
   Instrument - Spreading Caliper. The maximum distance is noted by trial method but the inferior temporal line is avoided. The point euryon (eu) usually lies on horizontal and frontal plane.
3. Maximum Bizygomatic Breadth (zy - zy): This measurement counts the straight distance between two zygia (zy) which are the most laterally placed points on the zygomatic bone.
   Instrument - Spreading Caliper. The skull is placed in norma verticalis position on the table. The measurement should be taken by trial while it shows the maximum.
4. Minimum Frontal Breadth (ft - ft) : This measurement reveals the straight distance between two fronto temporalia (ft).

Instrument — Sliding Caliper. The skull is placed in norma verticalis position on the table. The least distance between the two temporal lines (i.e. temporal crest of the temporal bone) is noted.

5. Basion — Bregma Height (ba - b) : This measurement shows the straight distance between basion (na) and bregma (b).
   Instrument - Spreading Caliper. The skull is placed in norma lateralis position while right side rests on the pad. The measurement must be taken from left end on basion to right end on bregma.
6. Morphological Facial Height (n - gn): This measurement counts the straight distance between nasion (n) to gnathion (gn).
   Instrument - Sliding Caliper. The measurement from nasion (n) to gnathion (gn) point is noted. The mandible is adjusted with the skull carefully and two sets of teeth are pressed against each other at the time of measuring the distance between the landmarks.
7. Upper Facial Height (n - pr) : This measurement shows the straight distance between nasion (n) to prosthion (pr).
   Instrument - Sliding Caliper. The skull has to be placed in such a manner that the occipital region rests on the pad. Adequate attention must be given in finding out the prosthion (pr) point where the movable cross-bar is adjusted.
8. Basion-Nasion Length or Basi-Nasal Length (n - ba) : This measurement is the straight distance between nasion (n) and basion (ba).
   Instrument - Spreading Caliper. The skull should be placed in norma basalis position for taking measurement.
9. Basion-Prosthion Length or Facial Depth (ba - pr): This measurement is the straight distance between basion (ba) and prosthion(pr).
   Instrument - Spreading Caliper. The skull will be placed in norma basalis position for taking measurement.
10. Nasal Height (n - ns) : This measurement is the straight distance between nasion (n) and nasospinale (ns).
   Instrument - Sliding Caliper. The skull is kept in such a manner that the basal region rests on the pad. The fixed point of cross-bar is placed on nasion (n) and the other cross-bar is adjusted against the nasospinale (ns). Measurements should be taken from both the nasal apertures on each side of the spine. The mean measurement stands as the actual measurement.
11. Nasal Breadth: This measurement denotes the maximum breadth of the pyriform aperture.
   Instrument - Sliding Caliper. The measurement should be taken by Sliding Caliper by placing the two ends of the cross-bars on the two farthest points of the nasal aperture.
12. Height of Orbit or Orbital Height: This measurement is the straight distance between the upper and lower margins of the orbital cavity.
   Instrument - Sliding Caliper.
   The cross-bar ends are to be placed on the upper and lower borders in the middle of the orbit. The measurements from both orbits have to be taken.
13. Orbital Breadth (mf- ect): This measurement is the straight distance between ectoconchion (ect) and maxillofrontale (mf).
   Instrument - Sliding Caliper. The cross-bar ends of the Caliper must be placed on the two said points and measurement is noted. For easy convenience the ectoconchion (ect) must be marked before taking measurement.
14. Inter-Orbital breadth (d - d): This measurement is the straight distance between dacryon (d) points.
   Instrument - Sliding Caliper. The two cross-bars of the Caliper are adjusted on two dacryon (d)

points to record the measurement.

15. Bi-Orbital breadth (ect - ect) : This measurement is the straight distance between two ectoconchion (ect).

    Instrument - Sliding Caliper. The two pointed ends of the cross-bars are placed on the middle of the two external borders of the orbits or ectoconchion (ect) points for taking measurements.

16. Palatal Length (pr - alveolar border) : This measurement is the straight distance between prosthion(pr) point to alveolar border.

    Instrument - Sliding Caliper. The fixed cross-bar is placed on the prosthion(pr) point. The other cross-bar of the Caliper is adjusted on the point, tangent at the posterior edges of the alveolar borders.

17. Palatal Breadth (enm - enm) : This measurement denotes the straight distance between the middle of the inner margin of the alveolar border on the second molar, i.e. from the endomolare (enm) to endomolare (enm).

    Instrument - Sliding Caliper. The maximum breadth on the outer margin of the alveolar borders is noted with the help of two cross-bars of Sliding Caliper.

18. Length of Foramen magnum (ba - o) : This measurement is the straight distance between basion (ba) and opisthion (o).

    Instrument - Sliding Caliper. The skull has to be placed in norma basalis position, the points are found to rest on the margins.

19. Breadth of Foramen Magnum : This is another straight measurement on the Foramen Magnum with the help of Sliding Caliper.

    But here two lateral points on the foramen magnum is considered and the maximum transverse diameter is noted.

20. Longitudinal Arc or Nasion - Opisthion Arc (n - o) : This measurement is the arc distance from nasion (n) to opisthion (o) in the mid-sagittal plane.

    Instrument - Steel tape. The zero point of the tape is placed against the nasion (n) by left hand and right hand extends the tape along the sagittal suture, occiput and finally goes to the point opisthion (o). Before taking measurement it should be checked that tape remains on the median line.

21. Transverse Arc (po - b - po): This measurement considers the arc distance from porion (po) to porion (po) over bregma (b).

    Instrument - Steel tape. The zero point of the tape is placed on right porion (po) by the left hand. The right hand extends the tape over the bregma (b), takes it upto the left porion (po) and then the measurement is noted.

22. Horizontal Circumference of Skull (g - op - g): This measurement considers horizontal circumference from glabella (g) to glabella (g), over opisthocranion (op). This is taken in the maximum cranial length plane, at right angle to the mid-sagittal plane.

    Instrument - Steel tape. This measurement is taken in horizontal plane in terms of F.H.Plane. At first, the tape is held on the glabella (g), then it is taken on opisthocranion (op) and again back to glabella (g). The opisthocranion(op) should be marked beforehand. In case of strongly developed supra-orbital arches or occipital protuberance, a cloth tape is required instead of steel tape.

23. Auricular - Bregmatic Height (po - b) : This measurement is the projective distance between left porion (po) and bregma (b).

    Instrument - First Segment of the Anthropometer rod. The skull has to be adjusted in eye-ear plane on the craniophore and the height measurement from porion (po) to bregma (b) is taken.

24. Bi-Condylar Breadth (cdl - cdl) : This measurement denotes a straight distance between two condylia(cdl).
Instrument - Sliding Caliper. The arms of the sliding caliper have to be placed at the most laterally placed points i.e. the two external points of mandibular condyles. The breadth measurement is taken carefully.
25. Symphysial Height or Chin Height (id - gn) : This measurement is the straight distance between infradentale (id) and gnathion (gn).
Instrument - Sliding Caliper. The fixed cross-bar is placed on the gnathion (gn) point of the mandible and the movable one is on the infradentale (id) which is the alveolar border between middle incisor teeth of the mandible. Now the measurement is noted.
26. Bigonial Breadth (go - go) : This measurement is the straight distance between two gonia (go).
Instrument - Sliding Caliper. The mandible is held firmly and two pointed ends of the Caliper is placed against two gonia points for taking measurement.
27. Mandibular Angle or Angle of the Lower Jaw : This measurement is the angular measurement on the point where ramus meets the body while a mandible is placed on a plain surface.
Instrument - Mandibular Goniometer or Mandibulometer. The mandibular goniometer is a special type of instrument used in the measurement of the angle of the lower jaw. The mandible should be placed in natural position on the goniometer and the inclined plane is to be raised slowly until it touches the posterior edges of both the ascending ramii. Inclined plane of the goniometer touches the ramii in a tangent position and the measurement of this angle can be easily found out by the mandibular goniometer.
28. Capacity of Skull: The capacity of cranium box has got special importance in craniometry. Many methods have been used to calculate the cranial capacity of man where variations are noticed in male and female skulls—male skull shows a higher value than the female. The most widespread as well as the best method is to fill the cavity of the cranium with millet or mustard seeds. At first, all openings and small holes are to be plugged with cotton or wax. The mustard seeds or millet seeds are poured through the foramen magnum by means of a funnel. Entire cavity must be filled completely by shaking time to time until the seeds come to the brim and the right thumb is to be used to press the seeds lightly at the foramen magnum. Lastly the seeds are taken out of the cranium and measured in terms of cubic centimeter (cc). For this a volumetric jar is used; reading can be easily found from the scale of the jar. It should be kept in mind that an overall careful management can only yield the correct measurement.

Range — Variation

1. (According to Sarasin)

	Male			Female		
Oligencephalic	X	—	1300 c.c	X	—	1150 c.c.
Euencephalic	1301c.c—		1450 c.c	1151c.c.	—	1300 c.c.
Aristencephalic	1451c.c—		X	1301 c.c.	—	X

2. (According to Flower and Turner)

Microcephalic	X	—	1350 c.c
Mesocephalic	1351 c.c	—	1450 c.c
Megalcephalic	1351 c.c	—	X

3. (According to Sergi)

Microphalic	X	—	1150 c.c.
Elattocephalic	1151 c.c	—	1300 c.c

Oligocephalic	1301 c.c	—	1400 c.c
Metriocephalic	1401 c.c	—	1500 c.c
Megalocephalic	1501 c.c	—	X

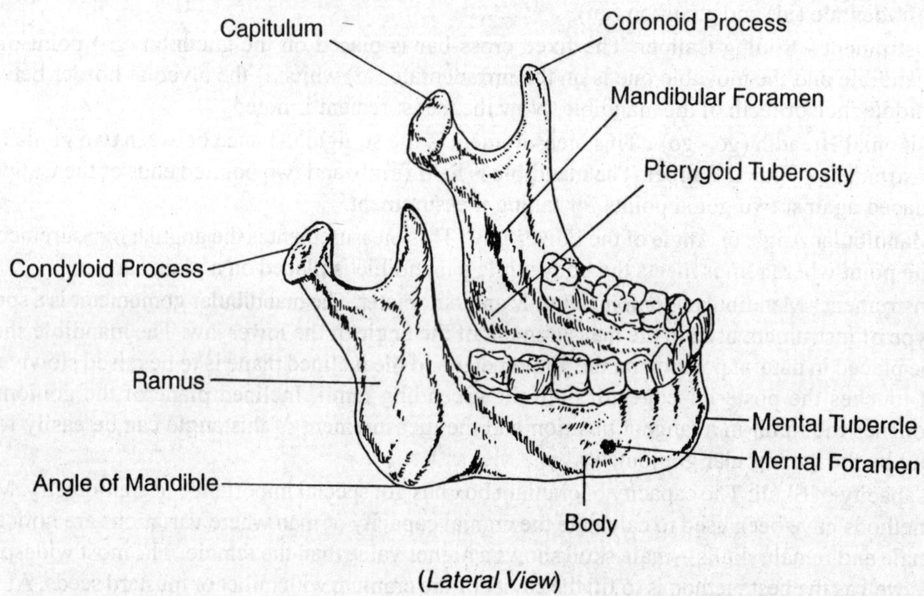

(*Lateral View*)

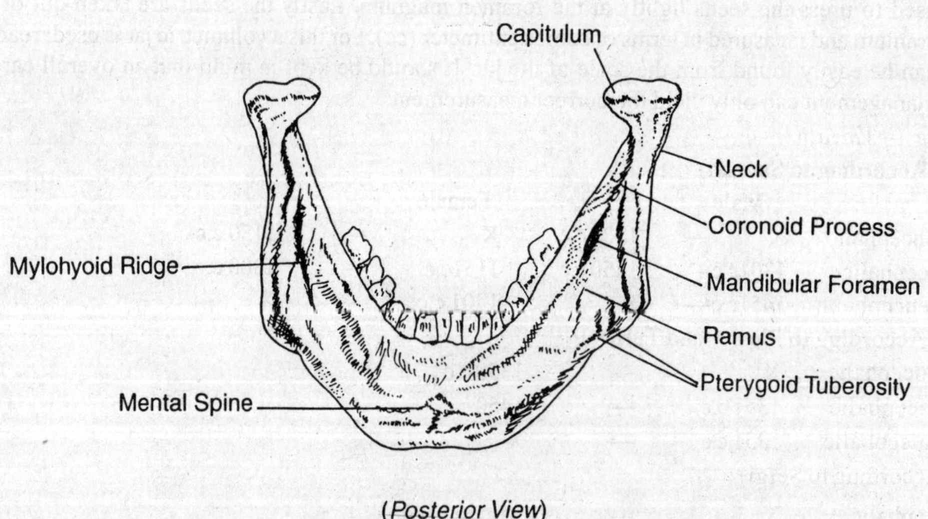

(*Posterior View*)

Fig. 8.15. Craniometric Landmarks on Human Mandible

29. Facial Profile Angle (n — pr / F.H.) : This is the angle that is formed by nasion (n) — prosthion (pr) line with F.H. Plane.
    Instrument — Attachable Goniometer. The prosthion (pr) is the most projecting point of the alveolar process and this angle measures the prognathism of the skull.

Range — Variation
(According to Martin and Saller)

Hyperprognathous	X	—	69.9°
Prognathous	70.0°	—	79.9°
Mesoprognathous	80.0°	—	84.9°
Orthoprognathous	85.0°	—	92.9°
Hyperorthoprognathous	93.0°	—	X

30. Alveolar Profile Angle (ns — pr / FH) : This is the angle measurement where nasospinale (ns) — prosthion (pr) line forms an angle with F.H. Plane.
    Instrument — Attachable Goniometer. This angle measurement is possible only when all bones remain intact in the skull.

Range — Variation
(According to Martin and Saller)

Ultraprognathous	X	—	59.9°
Hyperprognathous	60.0°	—	69.9°
Prognathous	70.0°	—	79.9°
Mesognathous	80.0°	—	84.9°
Orthognathous	85.0°	—	92.9°
Hypergnathous	93.0°	—	X

31. Maxillo-Alveolar Length (pr — alv) : this measurement is the straight distance between prosthion (pr) and alveolon (alv).
    Instrument — Sliding Caliper. The measurement is to be taken carefully from the point alveolon (alv) where the fixed cross-bar of the caliper is placed on the prosthion (pr) at the time of taking measurement.
32. Maxillo-Alveolar Breadth (ecm — ecm) : This measurement is the straight distance between alveolar process i.e. the two ectomolaria (ecm) of the upper jaw.
    Instrument — Sliding Caliper. This measurement is to be taken on the upper jaw to measure the inbetween distance of the ectomoloria (ecm). The point is usually located at the second molar in man and the measurement is to be made by sliding caliper.

INDICES

1. *Cranial Index :*

$$\frac{\text{Maximum Cranial Breadth (eu - eu)}}{\text{Maximum Cranial Length (g - op)}} \times 100$$

Range-Variation
(According to Garson)

Ultradolichocranial	X	—	64.9
Hyperdolichocranial	65.0	—	69.9
Dolichocranial	70.0	—	74.9
Mesocranial	75.0	—	79.9
Brachycranial	80.0	—	84.9
Hyperbrachycranial	85.0	—	89.9
Ultrabrachycranial	90.0	—	X

2. *Length — Height Cranial Index* or *Vertial Index* :

$$\frac{\text{Basion - Bregma Height (ba - b)}}{\text{Maximum Cranial Length (g - op)}} \times 100$$

Range—Variation
(According to Martin and Saller)

Chamaecranial	X	—	69.9
Orthocranial	70.0	—	74.9
Hypsicranial	75.0	—	X

3. *Breadth — Height Index of Cranium* or *Transverse Vertical Index* :

$$\frac{\text{Basion - Bregma Height (ba - b)}}{\text{Maximum Cranial Breadth (eu - eu)}} \times 100$$

Range—Variation
(According to Martin and Saller)

Tapeinocranial	X	—	91.9
Metriocranial	92.0	—	97.9
Acrocranial	98.0	—	X

4. Auriculo — Vertical Index :

$$\frac{\text{Auriculo - Bregmatic Height (po - b)}}{\text{Maximum Cranial Length (g - op)}} \times 100$$

Range—Variation
(According to Martin and Saller)

Chamaecranial	X	—	57.9
Orthocranial	58.0	—	62.9
Hypsicranial	63.0	—	X

5. Facial Index :

$$\frac{\text{Morphological Facial Height (n - gn)}}{\text{Maximum Bizygomatic Breadth (zy - zy)}} \times 100$$

Range—Variation
(According to Martin and Saller)

Hypereurprospoic	Short Face	X	—	79.9
Euryprosopic		80.0	—	84.9
Mesoprosopic	Medium Face	85.0	—	89.9
Leptoprosopic	Long Face	90.0	—	94.9
Hyperleptoprosopic		95.0	—	X

6. Kollmann's Upper Facial Index :

$$\frac{\text{Upper Facial Height (n - pr)}}{\text{Maximum Bizygomatic Breadth (zy - zy)}} \times 100$$

Range—Variation
(According to Martin and Saller)

Hypereuryen]	Short Upper Face	X	—	44.9
Euryen		45.0	—	49.9
Mesen	Middle Upper Face	50.0	—	54.9
Lepten]	Long Upper Face	55.0	—	59.9
Hyperlepten		60.0	—	X

7. Nasal Index :

$$\frac{\text{Nasal Breadth (maximum breadth of pyriform aperture)}}{\text{Nasal Height (n - s)}} \times 100$$

Range—Variation
(According to Martin and Saller)

Leptorhinae or Narrow Nose	X		46.9
Mesorhinae or Medium Nose	47.0	—	50.9
Chamaerhinae of Broad Nose	51.0	—	57.9
Hyperchamaerhinae or Very Broad Nose	58.0	—	X

8. Orbital Index :

$$\frac{\text{Orbital Height (Distance between Upper \& Lower margin of orbit)}}{\text{Orbital Breadth (mf - ect)}} \times 100$$

Range—Variation
(According to Martin and Saller)

Chamaeconch or Low orbit	X	—	75.9
Mesoconch or Medium orbit	76.0	—	84.9
Hypsiconch or Long orbit	85.0	—	X

9. Maxillo-Alveolar Index :

$$\frac{\text{Maxillo-Alveolar Breadth (ecm - ecm)}}{\text{Maxillo-Alveolar Length (pr - alv)}} \times 100$$

Range—Variation
(According to Turner)

Dolichuranic	X	—	109.0
Mesuranic	110.0	—	114.9
Brachyuranic	115.0	—	X

10. Palatal Index :

$$\frac{\text{Palatal Breadth (enm - enm)}}{\text{Palatal Length (pr - alveolar border)}} \times 100$$

Range—Variation
(According to Martin and Saller)

Leptostaphylin	or	Narrow Palate	X	—	79.9

Mesostaphylin	or	Medium Palate	80.0	—	84.9
Brachystaphylin	or	Broad Palate	85.0	—	X

Cranioscopic Observation

Apart from the various measurements, the visual observations help to develop a well-rounded impression about a skull.

Sergi has studied the skull shape in norma verticalis position which can be put in geometrical forms. Maximum length and breadth of the skull as seen in this position have been classified in the following ways, such as,

i) Ellipsoid; ii) Pentagonoid; iii) Rhomboid; iv) Ovoid; v) Sphenoid and ; vi) Pear-shaped.

There are still other information on the skull, which are as follows:

The Vault:

a) Supraorbital ridges : Trace, or Slight, or Moderate, or Pronounced.
b) External occipital protuberance : Absent, or Moderate, or Pronounced, or Doubled.
c) Occipital crest : Absent, or Slight, or Moderate, or Well developed, or Pronounced.
d) Mastoid : Small, or Moderate, or Medium, or Large, or Excessive.
e) Sutures : Serration is None, or Slight, or Medium, or Complex.

The Face :

a) Prognathism — Facial — None, or Slight, or Medium, or Above medium, or Pronounced.
Alveolar — None, or Slight, or Medium, or Above medium, or Pronounced.
b) Orbits — Shape—Round, or Rectangular.
Borders—Sharp, or Blunt.
c) Nasal bones-Narrow, or Medium, or Broad.
d) Palate — Form : Elliptic, or Ovoid, or U-shaped, or Near circular, or Horse-shoe shaped.
Depth : Shallow, or High.

Lower Jaw

a) Size—Small, or Medium, or Large, or Massive.
b) Chin—Form : Pointed, or Rounded, or Square, or Ordinary.
Prominence : Sub-medium, or Medium, or Pronounced.
c) Ramii—Narrow, or Medium, or Broad.

Teeth

a) Number of Teeth—Upper Jaw
Lower Jaw.
b) Size of Teeth—Small, or Medium, or Large.

II. ARCHAEOLOGICAL ANTHROPOLOGY OR PRE-HISTORY

Chapter

ORIGIN OF MAN ON EARTH

FORMATION OF EARTH

The Universe itself was built on the laws of nature. Astronomers estimate that the Universe is existing for some fifteen (15) billion years*. The earth, where we live, is a heavenly body and a member of the solar system. It began as a sort of'dust ball' of small bits of condensed material, collected together by the force of gravity. It was grown out of sun by the pressure of some unknown force. Different theories have been forwarded by different scholars regarding the origin of earth. Among them, the theory forwarded by Sir James Jeans has got the most acceptance. It says that, more than four billion years ago there was neither earth nor the solar system. Instead, there was a giant sun, which faced a violent disruption. Possibly another star, much bigger than the sun, passed by its side with lightning speed which attracted a large portion of sun but could not able to carry the portion, as the speed was very high. Rather, the portion was thrown out into the vast space and ultimately wrecked into pieces. All these pieces, being planets like Earth, Mercury, Venus, Mars, Saturn, Jupiter, etc. began to revolve round the sun, along their respective orbit (Fig. 9.1).

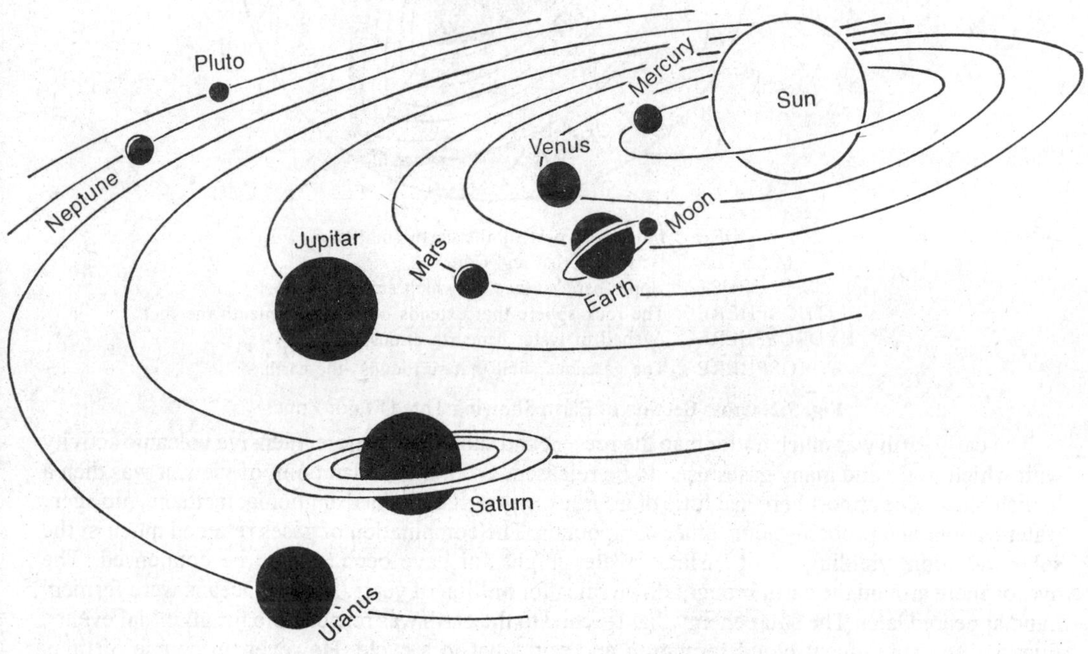

Fig. 9.1. Solar System

* Billion =1000.000,000.

The earth, as 'dust ball' was heated tremendously. Firstly, the colliding particles in it produced heat. Despite much of this was radiated outside, some part of this heat was trapped in the interior part of the earth. Secondly, the interior portion was compressed by the gravity and more heat was generated. Further, the earth contained small amounts of several naturally radioactive elements, which decayed to release energy. For these three reasons, the internal temperature of earth was very high; it existed almost in a molten state. But, in course of time, the ball started to cool down. The melted metallic iron, being very dense, sank into the middle of the earth. As cooling progressed, remaining melt began to crystallize; lighter and low-density minerals floated out towards the surface. Eventually, earth became differentiated into three major componential zones: a large iron-rich core, a thick surrounding mineral mantle and a thin low-density crust at the surface (Fig. 9.2).

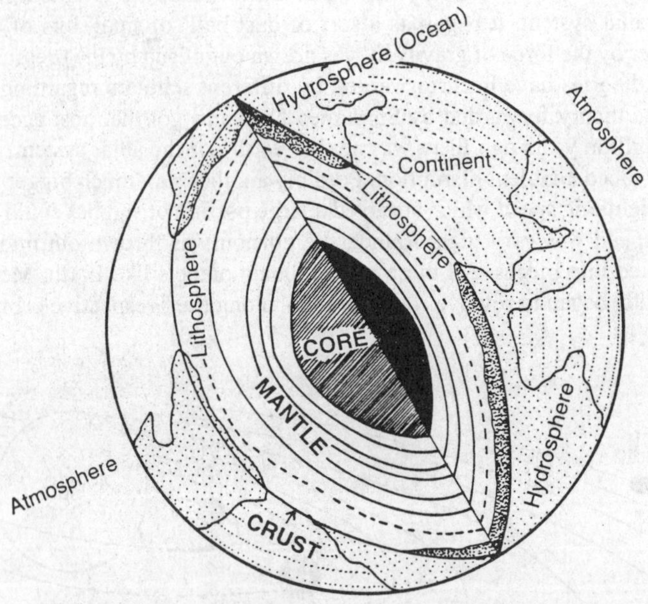

CORE : Innermost part of the earth (molten).
MANTLE : Between Crust and Core.
CRUST : Upper hard surface on which animals walk.
LITHOSPHERE : The rock sphere that extends downward beneath the fcet.
HYDROSPHERE : A shell of water benearth atmosphere.
ATMOSPHERE : The gaseaous shell that surrounds the earth.

Fig. 9.2. Cross Section of Earth Showing Three Major Zones

The early earth was much hotter than the present and subjected to more intensive volcanic activity with which water and many gases used to be released. From the human point of view, it was then a hellish place. The atmosphere had little or no free oxygen. It contained ammonia, methane, nitrogen, water-vapour and probably some other compounds. The combination of gases retained much of the solar radiation, yielding a hot surface. Water might not have been able to be condensed. The hydrosphere around the earth brought down rains for million of years. At first Oceans were formed; Land appeared later. The solar energy that reached to the earth was responsible for all initial events. Evaporation, subsequent cloud formation and rain went in a cycle. However, in course of time, earth's crust became ready for the growth of life. The solar energy permitted the existence of life. The single-celled blue-green algae appeared in large number to modify the atmosphere. Their remains are preserved in rocks, as old as several billion years. They manufactured food by photosynthesis, using sunlight for energy and released oxygen as a byproduct. Gradually enough oxygen was accumulated in the atmosphere to support oxygen-breathing organism.

Organisms were made up largely of Carbon, Oxygen, Hydrogen and Nitrogen. Various combinations of these elements produced four fundamental kinds of organic compounds - Carbohydrates, Fats, Nucleic acids and Proteins. Proteins, the most complex compound among all compounds controlled the combinations of amino acids. Such amino acids formed the basis of life.

A View of the Earth

The study of earth is called Geology. Geologists study the varieties of rock with which the structure of the earth is formed. They are also interested in the composition of earth's interior as well as in the features of earth's surface because they try to discover the age and history of earth.

Man as a surface creature witnessed some of the varied nature and dimensions of universe. They took it as a creation of almighty God and started to speculate about the form of this celestial body. During the fifteenth Century, there was a widespread belief that earth was flat. Christopher Columbus's daring plan to reach the rich Countries of the East by sailing westwards was based on a refusal to accept this popular conception. In fact, certain Greek Philosophers of Fourth Century had rejected the idea of a flat earth. Aristotle (384 - 322 B.C.) was one of the first to reason clearly that earth could not be flat; its shape must necessarily be spherical. He deducted this idea in two ways. Firstly, from the round shadow that is cast on moon during an eclipse and secondly, from the observation that the upper part of an approaching ship appears on the horizon before the lower part. Today the curvature of the earth can be seen clearly from the rockets and satellites.

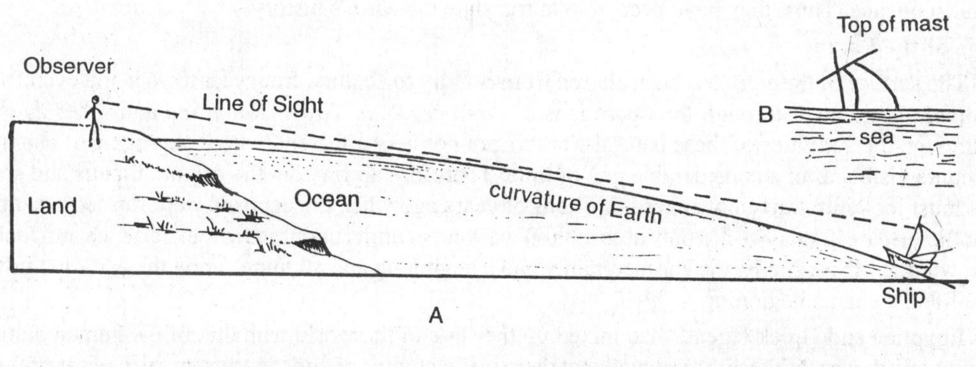

(A) Initial Appearance (B) First visible portion of ship

Fig. 9.3. The Curvature of Earth's Surface

Actually, the earth is not a perfect sphere. It is an oblate spheroid being slightly flattened at the axis ends and bulging out at the equator because of its rotation. The orbit of the revolving earth is slightly elliptical and so its distance from the sun varies with time of a year. Again, our planet earth rotates like a top on an axis, which is 23½ degree inclined from vertical. This inclination is responsible for the change of seasons. When the Northern Hemisphere is tipped towards the sun, the greater heating power of the sunrays produces the season, called summer. The same time stands as winter in Southern Hemisphere; as it remains away from sun. Both summer and winter last for six months.

Most of the surfaces of earth are covered by water. About 30 per cent of the area are *land* whereas *oceans* make up 70 per cent of it. Since about four-fifth of the land-areas are located in Northern hemisphere, most of the Southern Hemisphere is covered by water. The *continents* are low-lying. Their flat areas are called *plains,* higher areas with large summit are *called plateau* and high areas with small summit area are called *mountains.* Three principal kinds of rocks have been recognized —*igneous rock, sedimentary rock* and *metamorphic rock.*

(a) *Igneous rock* was formed out of the molten mass by the process of cooling. Occasionally molten mass came out by volcanic action and gave rise to lava fields and tough igneous rock. This type of rock is usually unlayered except the areas where some of the molten materials had seeped out onto the surface as a lava flow.

(b) *Hardening of sediments formed sedimentary rock.* In that dim past, continuous rain washed down the high areas and flowed towards the ocean basin. Eventually the water driven particles were deposited beneath the sea or other water body. In course of time, mud, sand, clay or lime were hardened and gradually they were compressed to make strata of sedimentary rocks.

(c) *Metamorphic rock* was formed from older sedimentary or igneous rock when they were deeply buried and undergone high pressure or heat, or both. Heat and pressure bring about changes in the original rock. The changes are mostly due to recrystallization of the existing minerals in the rocks. New minerals are rarely formed. But, a new kind of rock appeared.

From the very inception, the atmosphere has been very much active to tear down the elevated portions of the earth and drowning them into the seas with the help of various physical and chemical agents. Naturally the areas around the continents that were lifted above the water went down into the sea by the process of erosion, rather demolition. It happened again and again. Repetition of these phenomenons has been reflected in the layers of sedimentary rocks where each one is distinguishable from others. These layers or strata are the records of earth's history. So, the crust of earth showing different strata follows a chronological order where the layers formed earlier are placed at the bottom and the newest lies at the top. Although, the layers build up through time may not be visible in any one place; geologists try to integrate the layers found in various parts of the world to chalk out the total sequence. Thus, they have been able to translate the earth's history

Age of the Earth

The surface of the earth has been altered from century to century, from year to year and even from moment to moment through the operations *of Streams, Wind, Frost, Rain, Sea* and other agents. Although the activities of these natural workers are not very apparent but the aggregate of changes become visible after a considerable gap of time. Therefore, to find out the origin of lands and seas, we must look into the remote ages—millions of years ago when the deposits were formed one after another. Being human we are only able to think within the limits of our narrow experience and feeble knowledge. Accordingly, the earth certainly had a beginning and all things upon the earth including the life also had a beginning.

Egyptian and Greek legends had mixed up the state of the world with that of the human beings, intimately. It was universally assumed that there was a creator or god; apart from the creator nothing could be possible on earth. People with strong religious conviction had not the slightest doubt that life came into existence due to a dispensation of the Almighty god. The social philosophers conceived the unseen force as god. However, there was no idea about any progressive development from low to high forms of life; it held that the world and its inhabitants were made at the same point of time.

Until Medieval period almost no one was concerned with the age of the earth. The orthodox view as set forth by St. Augustine (AD 354 - 430) declared that all human being in this world was created directly by the god. It preached that men had been descended from a pair of male and female named *Adam* and *Eve.* In the seventeenth century, a date of 4004 BC was widely accepted as the date of birth of Adam and Eve. The significance of layered fossiliferous rocks was not appreciated till then. It was the beginning of nineteenth century when geologists first armed them with the concept of the absolute age of the earth.

In 1899, the physicist John Joly calculated the age of the earth to be about 100 million years. He obtained this result by examining the amount of sodium present in the oceans and by estimating the quantity that gets dissolved in each year being brought down by streams. In this calculation, the age of the oceans was considered more or less same as that of the earth as a whole. The famous physicist

Lord Kelvin calculated the age of earth as twenty 20 to 40 million years from its rate of cooling. Kelvin assumed that the earth had cooled from a molten state with no other source of heat involved. But in 1896 Antoine Henri Becquerel had discovered the radioactivity. Radioactivity has proved itself to be the most precise basis for age determination. Because it is a natural process in which one element is spontaneously transformed into another by disintegration of the central portion or nucleus, of the atom. The oldest rocks of the earth's crust dated by radiocarbon method are now considered to be about 3.3 billion years old. There are still some older rocks beneath these.

Geological Time

The history of earth is subdivided into two parts -*cosmic time* and *geologic time.* The time period, from the massive disruption of sun-star upto the relative completion of the nuclear core of earth and formation of the first ocean is considered as the cosmic time. The period may have lasted about two billion years but bears no direct geological record. The geologic time, on the other hand, has started with the deposition of sedimentary rocks; the oldest stratum was created about some two and half billion years ago.

In 1669, Nicolaus Stenco had set forth two very basic principles that could be applied to sedimentary rocks. The first one is known as the 'Principle of Superposition' which points out the fact that, in an undisturbed pile sediments (or sedimentary rocks), layers on the bottom are deposited first, followed in succession by the layers above them and ending with youngest at the top. The other principle is known as the 'Principles of Original Horizontality'. It is based on the observation that the sediments are commonly deposited in approximately horizontal, flat-lying layers. However, both the ideas were obvious in connection to any case of superimposition. In the late nineteenth century

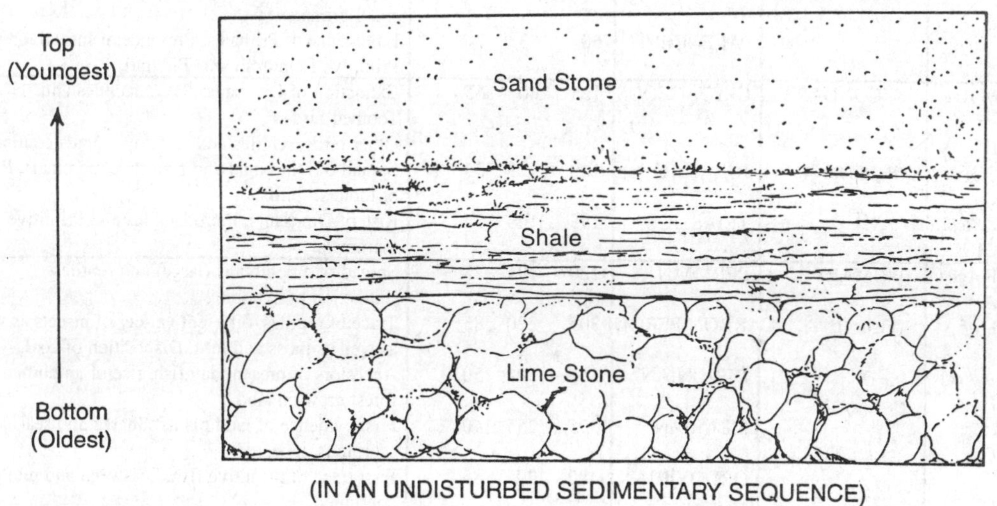

Fig. 9.4. Principle of Superimposition

a number of European scholars including the Italian Giovanni Arduino and the German Abraham G.Werner attempted to arrange the different rocks exposed in their region into a time-sequence. Although this first attempt was oversimplified and incorrect in some ways, it was the basic one. The geologic time scale which we use today is an outgrowth of their work but quite similar to that early scheme.

The largest subdivisions of geologic time are generally recognized as the *Eras*. The smaller subdivisions are called *periods* and the sub divisions of the Period are called the *Epochs*. It is important to realize that the name which are now used for periods were not applied at first to the time divisions but to the sequences of sedimentary rocks located mostly in Europe. Each described a sequence of rocks distinguished by certain fossils and other characteristics. As the study of geology was developed, more and more rocks were found to be studied throughout the Europe and other continents. Inevitably they were further compared to the named systems. Eventually, the correlation of fossils discovered certain assemblages of guide fossils.

However, modern geologist have divided the geological time span into five eras—Archaeozoic, Proterozoic, Palaeozoic, Mesozoic and Cenozoic. Each of these Eras has been named after its most characteristic form of life. Fig. 9.5 presents a simple correlation of outstanding life forms with geological ages.

Geological Age			**Span of Time B.C. (Millions)**			**Characteristic Forms of Animal Life**
ERA	**PERIOD**	**EPOCH**	**From**	**To**	**App. Duration**	
CENOZOIC (*Age of* Mammals)	QUATERNARY (Man)	HOLOCENE	.01	Continuing		Present day Man (Different Races).
		PLEISTOCENE	2		2	Early form of modern man : Pithecantropus, Neanderthal, Cro-Magnon, etc.
	TERTIARY (Mammals)	PLIOCENE	8	2	6	Human precursors : first tool-users. Many existing and extinct mammals.
		MIOCENE	20	8	12	Anthropoid ancestors of Great Apes and Man : Dryopithecus, Sivapithecus, Proconsul, etc.
		OLIGOCENE	35	20	15	Primitive anthropoid apes. Propliopithecus and Parapithecus.
		EOCENE	53	35	18	Variety of modern mammals and abundance of small primates
		PALEOCENE	60	53	7	Emergence of Primates. Presence of Insectivores, Primitive Lemuroid and Tarsioid.
MESOZOIC (*Age of Reptiles*)	SECONDARY (Reptiles)	CRETACEOUS	125	60	65	Extinction of Dinosaurs, Pterodactyles and Toothed birds. Other birds, reptiles and small mammals continued.
		JURASSIC	160	125	35	Spread of Dinosaurs and primitive mammals, Rise of toothed birds.
		TRIASSIC	195	160	35	Rise of Dinosaurs, Pterodactyles and Primitive mammals.
PALAEOZOIC (*Age of Fish*)	PRIMARY (Fishes)	PERMIAN	220	195	25	Spread of amphibians (large) and reptiles. Extinction of Trilobites
		CARBONIFEROUS	305	220	85	Trace of reptiles, different variety of insects as well as well as moss and fern. Disposition of coal.
		DEVONIAN	355	305	50	Ancestors of present day fish, rise of amphibians, forest grew on land.
		SILURIAN	395	355	40	First evidence of land plants. Sharks and sea-scorpions appeared.
		ORDOVICIAN	480	395	85	First trace of primitive fish, Sea-weed and molluscs continued.
		CAMBRIAN	550	480	70	Molluscs, Star fish, Corals and Trilobites is known.
PROTEROZOIC (*Age of Ancient Life*)	(Invertibrates)	PRE-CAMBRIAN	1200	550	650	First multicellular form (Metazoa) appeared - Sponge, Worm, Molluscs, etc.
ARCHAEOZOIC		ARCHEAN	2700	1200	1500	No definite of life. Probably unicellular sea-dwelling forms (Protozoa) appeared.

Fig. 9.5. Evolution of Life in Terms of Geological Record

In the beginning i.e. in the formative phase of earth, the sediments became hardened slowly to give rise the sedimentary or stratified rocks. Evidence of life is completely absent in these earliest rocks *. This is the reason for which these rocks are known as *Azoic i.e.* lifeless rocks. Geologists do not count this period as an era because the particular time exhibits no trace of life at all. The first era is called *Archaeozoic era.* Life, though not very distinct, made its first appearance in Archaeozoic sedimentary deposit as indicated by the presence of carbon in this stratum. It is probable that the first living matters were very tiny as well as soft to make recognizable traces. Archaeozoic era corresponds to the geological epoch *Archaean,* the age of which has been calculated as 2700 million years BC when some unicellular sea-dwelling life forms, protozoa came into existence. It continued for about 1500 million of years. The second era is known as *Proterozoic.* Proterozoic corresponds with *Pre-Cambrian* epoch which started about 1200 million years BC. The first multicellular life forms like *metazoa* sponge, molluscs etc. occurred during this time. This era lasted about 650 million years. In fact, both of these two eras together has been labeled as the 'Age of Ancient Life'.

The third era is called *Palaeozoic, which is divided into six geological epochs such as, Cambrian, Ordovician, Silurian, Devonian, Carboniferous* and *Permian.* The first epoch, *Cambrian* does not yield anything higher than invertebrate animals. The evidence of land-life is not known; only molluscs, starfish and trilobites are found. Therefore, this epoch is often referred as the.'Age of King Crabs'. It started about 550 million years BC and lasted for 70 Million years. The second geological epoch under the *Palaeocene era* is called *Ordovician, which* bears the first traces of vertebrates. Primitive fish, seaweeds and molluscs appeared at this time. The beginning of this period has been dated as 480 million years BC that continued for about 85 million years. The third geological epoch, *Silurian* traces the fishes with bony skeleton. Ancestors of sharks and dogfish were found in this period. First evidence of land-plant stands as a hallmark of this time. The period began about 395 million years BC and continued for 40 million years. *Ordovician* and *Silurian* togetherly represent the 'Age of Mud'. The fourth geological period, *Devonian* shows the ancestors of present day air-breathing lungfish as well as some typical bony fishes. Hence it is the 'Age of Fishes'. Besides, a transitional form of life between fish and four-footed land-dwelling vertebrate (amphibian) was found to appear in this age. Spreading of the forest was the other important feature of this epoch. The epoch started about 355 million years BC and persisted for about 50 million years.The next epoch is the *Carboniferous,* which is marked for Carboniferous deposition. The carboniferous rocks are those from which coal is extracted. Therefore, this period has been referred as the 'Coal Age'. Fish were abundant during this period along with many primitive forms of amphibians. First remains of early reptiles also came from this period as they lived in swamps. Besides, different types of insects, spiders, great forests of ferns and mosses are the evidences of this epoch. The epoch began about 305 million years BC and lasted for 85 million years. The last epoch of *Palaeozoic era* is *Permian.* It is the age of large amphibians and also the land dwelling reptiles. The amphibians and insects spreaded at this time and extinction of trilobites occurred. The epoch started approximately about 220 million years BC and continued for 25 million of years. The last five epochs, starting from Ordovician to Permian, have been categorized as 'Primary Period' to designate the Age of fishes and ancient life.

The fourth era *is Mesozoic* which can be classified into three epochs namely the *Triassic, Jurassic* and *Cretaceous* under the 'Secondary Period'. The period witnessed the evolution of different reptilian forms including the forms of dinosaurs. The Triassic epoch began some 195 million years BC and lasted for 35 million of years. In this epoch, the mammals are traced for the first time in their primitive forms.

Besides, pterodactyle, early crocodile etc, appeared along with the rise of dinosaurs. In the following Jurassic epoch, the reptiles reached to their highest development. Dinosaurs of varying

* The earliest rocks of the world is found in Canada occupying a very modest area, north-west of Lake Superior, in the vicinity of Rainy Lake.

sizes came into existence. Further, the rise of the first bird, particularly the toothed bird has been recorded from this period. The beginning of this epoch can be dated as 160 million years BC that continued for about 35 million years. The last epoch of Mesozoic era is the 'Cretaceous' when reptiles stood dominant with numerous diversifications. Small placental mammals mainly marsupials can be traced along with birds, fishes, sharks and other insectivores. Most remarkable phenomenon of this epoch is the extinction of dinosaurs, pterodactyls and toothed birds. It started about 125 million years BC and lasted for 65 million years. Permian with Triassic is sometimes referred as the

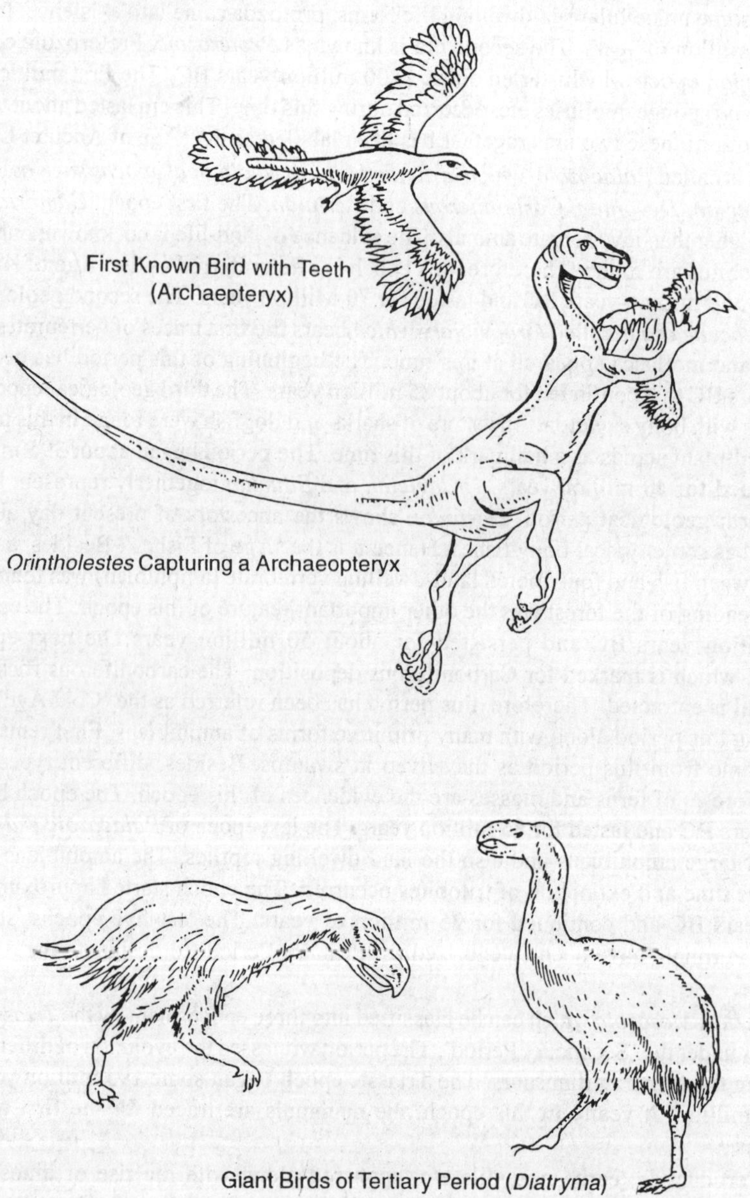

Fig. 9.6. Ancient Birds

'Age of Deserts' for its dry and hot climate. Jurassic is purely the 'Age of the Dinosaurs'. *Cretaceous* is said as the 'Age of Chalk' because in this stratum we get profuse tiny shells accumulated in the form of chalk.

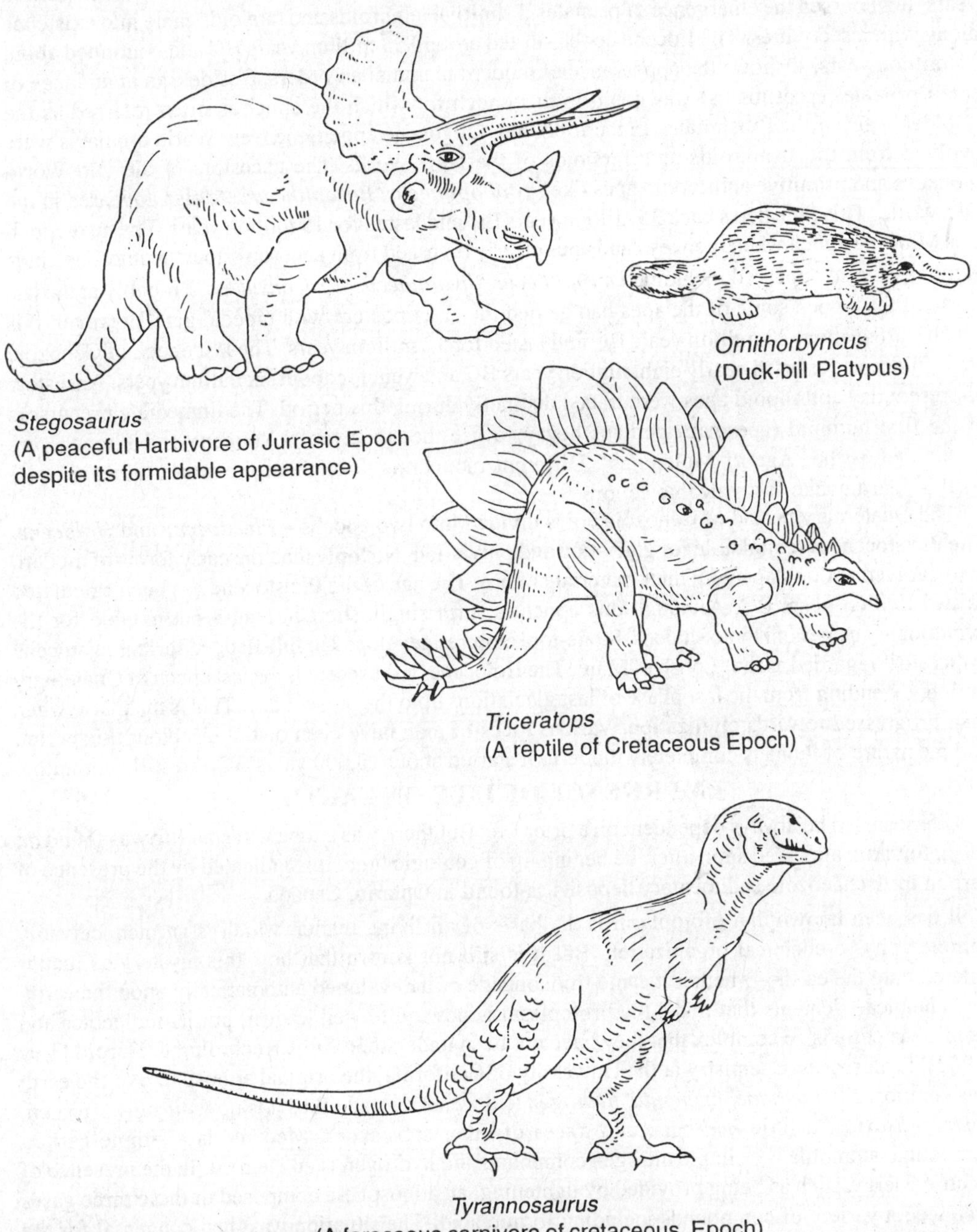

Fig. 9.7. A Few Extinct Animals

The fifth or the final era is known as the *Cenozoic,* which is classified into two major periods — *Tertiary* and *Quaternary.* The Tertiary period is designated as the 'Age of Mammals' and classified into five epochs namely *Paleocene, Eocene, Oligocene, Miocene* and *Pliocene.* The earliest of these, Paleocene dates back about 60 million years BC. Its duration was approximately seven million years. It witnessed the emergence of primates. Primitive lemuroids and tarsioids came into existence along with insectivores. The Eocene epoch started around 53 million years BC and continued about 18 million years. It shows the appearance of modern mammalian orders. Besides, an abundance of small primates (prosimians) was noted in this epoch for which the epoch is often referred as the 'Golden Age' of the Prosimians. In the following Oligocene epoch, the New World monkeys were evolved from the Lemuroids and Tarsioids of the New World. The ancestors of the Old World monkeys and primitive anthropoid apes like *Propliopithecus, Parapithecus* etc also appeared in the Old World. This time dates back 35 million years BC and lasted for 15 million years. The next epoch was Miocene epoch when monkeys and apes clearly diverged from ancestral stock. Numerous kinds of monkeys also appeared including *Dryopithecus, Sivapithecus, Proconsul* etc. Probably at the last phase of this epoch some of the apes had gained the erect posture with bipedal gait. However, this epoch started about 20 million years BC and lasted for 12 million years. The last epoch of Cenozoic era is Pliocene. It started nearly eight million years BC and went for about six million years. Ancestors of present day anthropoid apes were traced distinctly during this period. The immediate precursors of the first hominid (ape-man) perhaps began to use the stone tools. Miocene and Pliocene are considered as the 'Age of Mountains'. Different mountains throughout the world appeared as a result of earthquake and volcanic action.

The Quaternary period of Cenozoic era is divided into two epochs—*Pleistocene* and *Holocene.* The Pleistocene is remarkable for glaciation and pluviation. Not only that, the early forms of modern man evolved at this stage with their material culture. The age of the Pleistocene has been calculated as two million years BC. Although this epoch is surprisingly short in length but notable for the evolutionary stages of Homo sapiens. Stone-tool cultures developed in full-fledged form. Pleistocene is therefore regarded as the 'Epoch of Man'. The Holocene or the recent is the last epoch of Quaternary period, extending from the last phase of last glaciations upto the present day. This is the phase when man progressed towards civilization. Various races of man have been distinguished in this period and the nature of fauna is completely modern. It started about 10,000 years BC and still continuing.

EMERGENCE OF LIFE ON EARTH

Life seems to be always dependent on a prior life. But there was a time when no life was found on earth. First life appeared soon after the beginning of geologic time. It is indicated by the presence of carbon in Archaeozoic sedimentary deposits as found in Ontario, Canada.

It has been known that protoplasm is the basis of all living matter, which is an inconceivably complex physio-chemical organization. But it is still not known that how this mysterious matter entered onto the earth—whether it came from outside or it developed automatically upon the earth. The chemical elements that make the protoplasm is now quite well known, but its molecular and atomic structure is so complex that they have not been understood fully. According to Harold Urey (1952), Professor of Chemistry at the University of California, the original atmosphere of the earth was composed *of methane, ammonia, hydrogen* and *water* instead of, as it is now *nitrogen, oxygen, carbon-di-oxide,* a little *hydrogen* and traces of other substances. Methane is a simple carbon compound, ammonia is a simple nitrogen compound and hydrogen is an element. In the presence of electric energy, such as being provided by lightening, an atmosphere composed of these three gases produced a variety of compounds including formic acid, The situation was then congenial for the formation of amino acids, the chemical building blocks that are found in all forms of life. Actually what happened can not be accurately determined at present, but lightening or ultraviolet energy from the sun could conceivably have effected the first steps of chemical synthesis.

All theories on the natural origin of life postulated that some form of intensive energy—light, heat, radioactivity, ultraviolet radiation or electricity converted simple carbon compounds to chemical compounds with multiple carbon atoms. Laboratory experiments have tried to demonstrate several methods of bringing about this process. It is assumed that the beginning of life was only possible when the several elements that make up protoplasm occurred in proper combination within a suitable environmental condition. Therefore, the whole thing happened as a matter of accident. The elements got a chance to be combined in order to make a unique matter, colloidal in form. The matter began to maintain itself by borrowing energy and materials from its surrounding environment. The process is equivalent to what we now call metabolism. Ultimately it could grow by increasing its mass and could maintain the continuity by the process of reproduction. Above all, as it acquired the capacity to adjust itself in the changing environment, it survived.

Thus, we understand the 'origin of life' but can not point out its occurrence in relation to a particular time. We also understand the kinds of physical-chemical processes that convert the simplest carbon compounds into amino acids and proteins, but we can not make out the sequences of importance. However, the oldest form of recognizable organism is the blue-green algae (cyanobacteria) which first fulfilled the chemical as well as biological requirements of life. Gradually this unicellular protoplasm evolved into differential forms of increasing complexity. It is interesting to note that each successive geological era has been shorter in duration than the preceding era. Moreover, it tends to bring forth more numerous life forms in ever-increasing complexity. Following this principle, the history of living matter shows that the life came into existence in Archaeozoic era and fanned out later with continuous change and progressive development. The speed of evolution has been accelerated with time and at present in the topmost level we find the man who is yet to be developed. Evolution of life will be continued as long as life will exist on earth. But to predict the ultimate form of man or some other higher creature is impossible at this very moment.

To search the animal ancestry and antiquity of man Prehistorians have to depend primarily on geology. In the earlier phases, prehistory was closely related to geology and in fact it was once considered as a branch of Quaternary geology. The stratigraphical method based on the law of superposition has been proved very useful. It tells that in case of deposition, if the earlier deposits meet no subsequent disturbance; the lower beds remain older than the upper beds, the bottom bed being oldest of all. So, the law of superposition can establish the relative ages of the fossils, which are found in a succession of layers, or strata. The older forms of life will be restricted to the lower beds and their descendants will appear in upper bed deposits. Thus, more and more new forms tend to come out from the more recent layers. Around the year 1800, William Smith formulated the Law of Faunal Succession where the basic concept was that the life forms change with time. Old ones disappear and new ones appear in the fossil record, but the same form is never exactly duplicated independently at two different times in history. It is quite likely to find the same type of fossil organism preserved in two separated rocks. Even if the rocks are quite different compositionally or geographically widely separated, they should be of same age. In order to understand a clear stratigraphy, a long vertical section has to be cut on a deposit where one may find various layers arranged one after another from top to bottom. The biological evolution that have taken place since the first emergence of life are stored in the history of earth. By the comparison of strata between different places, different regions, different continents, a complete worldwide chronology according to relative age can be constructed.

The Pleistocene Epoch

The epoch Pleistocene has a great significance in prehistory as the major biological and cultural evolution of man took place during this period. The period is also exceedingly interesting to the geologists as it witnessed profound environmental changes resulting from large-scale glaciations. The change affected the landscape of some areas by the process of erosion and deposition. Geologists are not only provided with the information about the age and extent of glaciations, the epoch also gave them the ideas about long phases of inundations,warm-dry episodes, desiccation and dust storms affecting large tracts of earth and their inhabitants. Repeated migrations of entire flora and fauna were also recorded.

The synonym of Pleistocene is often taken as the 'Great Ice Age'. The term actually is a misnomer as there was no single Ice Age, but a series of Ice Ages differentiated by relatively warmer periods. The succession of climatic fluctuation produced remarkable shift in plant and animal kingdom as well as made considerable alteration in the character of soil and vegetation zones. The features of glaciation are so common in Northern Europe and Northern North America that it arouses curiosity.

Nature of Glaciation

During the last two million years, the great ice sheets were advanced and retreated repeatedly covering 30 percent of the land area of the world. Geologists have been able to recognize four major periods where the large and in depth accumulation of snow occurred due to favourable weather condition. It was certainly more severe than the condition of present-day Greenland and Antarctica; the places still covered with ice sheets. However, the edges of the ice-mass moved outward to cover the great areas of Northern Eurasia and North America. The Ice Age in Northern Hemisphere was always contemporaneous with the changes occurring in Southern Hemisphere.

The glaciation is generally found only in high altitudes or in Polar Regions. But during the Pleistocene epoch, the plain regions were also affected. The blankets of ice engulfed the vast surface of the earth and acted as a scouring agent during the thousands of years of its growth. In many places the ice-sheets were more than hundred meters thick. Today, glacial ice contains about two percent of earth's water. During the periods of maximum glaciation in Pleistocene epoch, a far greater proportion of earth's water was locked up in the form of ice. Sea level became lowered considerably.

Each advance of ice was followed by a retreat of ice. Most of the glacial ice spread out from the great centers of accumulation on land areas surrounding the Arctic Ocean. When the weather became moderate, those great ice fields gradually diminished in size, shrinking backward from their margins. At this time glaciers were created as masses of ice which moved slowly under the influence of gravity. When snow has remained on the ground for a prolonged time and continuously weighed down by addition of more and more snow, the lowest layer slowly changes its position. The compact, crystallized masses of ice from the bottom start to flow slowly downhill being activated by the weight of the overlying mass. A glacier flowing in response to gravity and pressure move forward; it never goes backward. The speed depends upon the degree of slope of the ground. In North Siberia, the rivers flowed down to the Arctic Ocean between cliffs of perpetually frozen ground. Louis Agassiz announced the idea of Ice Age in 1840. Since then we have learnt more about the Ice Age. The accumulated snow of the last Ice Age began to melt back only about eleven thousand years ago, though all of the snow has not melted away. Ten percent of the land area of the world at present is covered by snow. Therefore, we can say that the world is still under the grip of Ice Age; natives of Greenland should readily agree with this opinion.

According to some great pre-historians, glaciations were not the identifying feature of Pleistocene epoch, because a number of glaciations also occurred in the pre-Pleistocene epoch. But Pleistocene glaciations are most remarkable because those persisted for longer periods and were most severe in character.

Glaciers may be of two types—The *Valley glaciers* and the *Continental glaciers.* Valley glaciers are elongated ice streams that form in mountainous areas above the snow line. With the continual addition of more and more snow, they gradually move down the pre-existing valleys and fan out towards the low ground. These glaciers may be from more than 100 miles to less than a quarter mile in length. They follow the original system of stream valleys. One or more tributary or feeder glaciers may join to form wider and thicker main glaciers. In high attitudes same valley glaciers reached the foot of the mountain range. Now such valley glaciers may be seen in the Rocky Mountains like Alaska and the Alps.

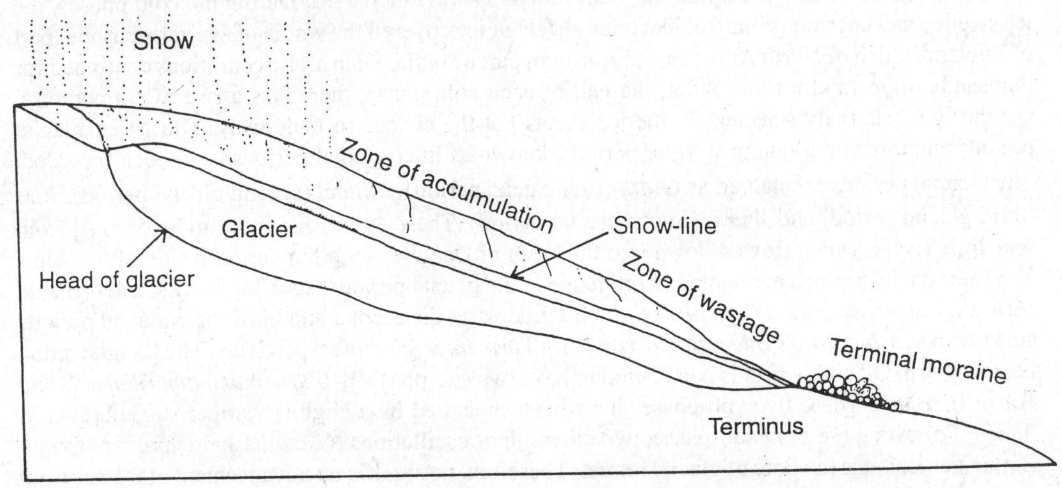

Fig. 9.8. Formation of A Glacier

The Continental glaciers are broad sheets of snow that move radically outward from a central region for their own weight. As they can spread out over the entire continent or large portions of the continents, covering mountain and plains, the name *Continental glacier* has been applied. During Ice Age, about 2,000,000 square miles of Europe and 4,000,000 square miles of North America were glaciated. In North America, three glaciers were identified in Labrador, Greenland and Scandinavia. Glaciations begin where the winter's snowfall exceeds the amount of snow that melts during the remainder of the year. This condition can exist where the mean annual temperature is low enough to prevent melting.

The extensive permanent accumulations of snow from which glaciers originate are called snowfields. Snowfields undergo changes that gradually give rise to glacier ice. Although the exact mechanism of glacial movement is still not well understood, it is quite reasonable to compare the movement to that of a quiet stream of water whose velocity is infinitely less. Glaciers play vigorous erosive role. Small or large, they carry a load of rock debris. Most of this debris remains embedded within the body of glacial ice; some is carried along on the surface. A glacier gains these materials from the surface over which it flows, from adjacent rock masses or from rockslides of higher elevation. It has the ability to pick up soil and loose particles, but not to wear away the solid rocks. Melting glaciers transport and deposit huge quantities of material such as rock fragments, gravel, sand, clay etc. at the front as well as at the margins of the glaciers, which are known as 'glacial drift'. The drift, which shows little or no evidence of stratification, is called 'till'. A *glacial till* contains randomly mixed fragments of all sizes, ranging from great boulders to tiny particles. But the finer part of the debris is carried out beyond the margin of the glacier by melting waters. Sediments of this kind usually show a stratified mass and referred as 'moraine'. However, the rock-fragments available in a drift prove themselves very useful as they indicate the direction of the ice-flow.

Glaciation in Europe

First studies of glaciation were made in Western Europe where there were two great areas of ice formation—Alps and Scandinavia. Following Dr. Albrecht Penck and his colleague Dr. Obermaier who studied on the glaciation of Alpine region in 1909, we have come to know that about two million year ago, some geographical factors worked to upset the balance of environment. After a period of cooling and minor glaciation, there followed a series of swings—from cold period to warm period and again back to cold period. Temperature fell down in the North and south temperate zones, and consequently precipitation (both rain and snow) increased. During the cold phases, the ice swelled and enormous blanket like thick sheets of ice covered the whole of Western Europe, part of England, much of North America and Northern part of India. Such arctic conditions continued for thousands of years at a time. After, the end of each cold phase, there was a rise of temperature; weather got relatively warmer. So the ice-sheets got the chance to melt away. Thus, four glacial periods and three intervening warmer periods, known as Inter-glacial periods have been recorded. The Glacial periods are named as *Günz* (first glacial period), *Mindel* (second glacial period), *Riss* (third glacial period) and *Würm* (fourth glacial period). These names are found to be derived from four little rivers, which flowed down into the basin of Danube along the northern side of the Alps. The Interglacial periods are named in reference to the glacial periods, such as, the first Inter-glacial period is known as *Günz-Mindel Inter-glacial.* Likewise, the second and third Inter-glacial periods are known as *Mindel-Riss Inter-glacial* and *Riss-Würm Inter-glacial* respectively. The last glaciation i.e. the Würm Glacial period is considered to have two sub-phases—*Würm maximum (Würm I)* and *Würm II (Bühl).* These two sub-phases have been separated by a slightly warmer stage known as *Acnen.* Following the *Buhl* sub-phase, two other minor oscillations *(Gschnitz* and *Daun)* are found. But, in general, it is the time of glacial retreat. Therefore, no significant advancement of ice has been noted after this, till now.

All glacial periods were not of same duration and so were the Inter-glacial periods too. Possibly the glacial periods were four to eight times longer than the Inter-glacial periods. Further, all Inter-glacial periods were not equal in length of time, nor they were equally warm. *The Mindel-Riss* Inter-glacial period was perhaps the hottest and longest of all. Modern Radio-metric dating technique (Palaeo-Temperature analysis of deep-sea sediment) reveals that each of the glacial periods lasted about 90,000 years where each Inter-glacial period was continued for about only 10,000 years.

The *Holocene* epoch is considered as belonging to the Post-glacial period. Grahame clark preferred the term 'Neo-thermal period'. He argued that, passing of the glacial conditions was though merely a local phenomenon but a worldwide impact was noticed at the end of the Pleistocene period. With the final retreat of the ice-sheets, the climate began to increase towards normal; no farther advancement of ice-sheet has so far been found. But the remnants of last Glacial period are still visible on one or two parts of the world. For instance, the ice bodies of Greenland and Antarctica has not been totally retreated, although the prevailing temperature is not as high as it was during the Inter-glacial periods. In this condition, nobody can guess whether the climate has permanently settled down or whether another Glacial period is waiting for us in which we will have to enter after the end of the running Inter-glacial phase.

Minor undulations of climate have been identified in each glacial period. Those are termed as stadial and inter-stadial periods where the degree of coolness was differentiated. Stadial periods were tremendously cold but the severity was less pronounced during the Inter-stadial periods. Glaciation affected the North America in a similar way like Europe. Here the Glacial stages have been named as *Nebraskan, Kansan, Illinoian* and *Wisconsin,* which were more or less contemporary to the European glaciation. In India, glaciations, which occurred on Himalayas, followed the same sequence as that of the Europe, but it started a little later. The first glacial period of India corresponds with the second glacial period of Europe.

The cultural evolution that Pleistocene period witnessed can be traced from the lower Palaeolithic culture. This cultural period started from the first glacial period *(Günz)* and carried on upto the third Inter-glacial period *(Riss-Würm).* However, the beginning of Pleistocene epoch is marked by the appearance of a new set of animal forms leveled as 'Villafranchian fauna' in contrast to the warmth loving animals of former epoch. Pliocene. Since the whole landscape was transformed, new types of vegetation appeared with new fauna. Under the exposure of arctic climate and the changed food-habit, the forerunners of man were forced to make an adaptation, which took them in the way to biological evolution. Environmental change stood as a potent agent of biological mutation. It stimulated migration and emphasized isolation. As a result, the Pleistocene epoch experienced the vital stages in the emergence of humanity.

Causes of Glaciation

Pleistocene was an epoch of great climatic fluctuations. Those fluctuations occurred repeatedly at frequent intervals and produced dramatic change in the deposition of land and water. The exact reason behind this situation is not known but attempts have been made to explain it. Two types of theories are found in this context.

I. Geographical Theory

This theory is based on geographical changes that occurred in the earth at the close of the Tertiary period. It advocated a drastic alteration of climatic condition, which was responsible behind the rise of the mountains like the Alps, the Rockies and the Himalayas. As soon as the mountains arose, they created greater areas of high altitude as cold spots, and naturally disarranged the pattern of wind and storm-cycles in the world. The great amount of moisture that was needed to produce the huge ice-sheets was drawn from the moisture of earth and as a consequence the level of oceans were lowered. The actual glaciation was found only in high latitudes or altitudes even during the period of snow deposition as the general lowering of the mean annual temperature rarely exceeds more than a few degrees. This was strikingly different from the conditions of ice-sheets that covered parts of England and extended upto the high mountains of Eastern France, and further into the South, where little or no glacial activity was found.

This interpretation is not fully correct. Because it may explain glaciation in Western Europe but not those which took place in North America and North Asia. Moreover, this theory relating to geographical changes fail to explain the periodicity of extension and contraction of ice-sheets.

II. Astronomical Theories

Many astronomers came forward with their own models of interpretation for glaciation.

(a) The earliest one is known as Milankovitch theory which considered the factors like inclination of Earth's axis, eccentricity of earth's orbit and the longitude of the perihelion (precision of the Equinoxes). It declared that the earth's orbit round the sun is not always circular. It was formerly an elongated one, especially in Pleistocene epoch. Therefore when the North Hemisphere turned away from the sun, it gave rise to extremely cold winter in the North Hemisphere. At the same time, the other hemisphere experienced a hot and short summer. Thus, the variability of the length of the earth's orbit round the sun produced the oscillation of recession and advances of ice. This theory has got certain relevance but it does not explain the Ice Ages in totality. Therefore many scholars have not accepted, the theory. F. E. Zeuner who wrote the book, 'Dating the Past' was in favour of this theory. He claimed that absolute date could be calculated from the curves representing changes of solar radiation on various latitude.

(b) The second astronomical theory was based solely on the position of sun. It said that, since sun is a variable star, the quantity of heat that we receive varies with its position. During the Quaternary period, the position of its poles shifted to some extent, which decreased the amount of heat. As a result, lowering of temperature was noticed in one of the hemispheres.

This theory also failed to explain the occurrence of Ice Ages satisfactorily and therefore remained far from universal acceptance.

(c) The third astronomical theory is based on the thought of Sir George Simpson. It is called Simpson's theory. The theory accounts the rhythmical variation in the solar radiation. Dr. Simpson advocated that the solar radiation is not always constant. They decrease with the appearance 'of 'sun's spots' and increase with their disappearance. When the amount of heat increases, a greater evaporation takes place. Following this principle, during the Pleistocene period, plenty of water from the ocean evaporated to form huge cloud in the sky. The increase of cloud-cover resulted in the decrease of temperature on the earth and resisted the melting of the ice.

On the other hand, when the solar radiation began to decrease, a reversal of the process took place. Formation of less cloud decreased the solar radiation. Naturally the temperature of the earth became high. It facilitated the melting of ice-sheets and the retreat of glaciers.

Although the theory fits well with the Pleistocene Ice Age, it does not explain the early glacial periods particularly that took place in Carboniferous times in South Africa, where there is no evidence of contemporary uplift. Therefore, this view also failed to earn the popularity.

Evidences of Glaciation

The causes of glaciation, however, are not so important in the annals of mankind but the effects are really important. Evidences of the glaciation are still freshly imprinted upon the landscape. There is a dispute with the actual number of glacial periods in Pleistocene epoch. If the evidences of ocean cores (beds of Caribbean and Pacific) is to be relied upon, there were 15 glacial periods before the major four as identified by Dr. Albrecht Penck and Dr. Obermaier. However, the topographical features of Ice Age are as follows :

Iso-static Reaction and Eustatic Movement

We have come to know that great amount of moisture was drawn from the ocean to form the great ice-sheets, which lowered the level of the ocean. The enormous areas of our planet were weighed down under crushing masses of Ice, thousands of feet thick. The weight of the ice on the land depressed some areas, but after the last removal of ice they began to rise slowly for coming back to the original position. This uplifting of the earth due to melting of the ice-mass is known as *Iso-static reaction.* On the other hand, the rise and fall of the general sea level due to melting or forming of the glaciers is known as *Eustatic movement.*

These features have been carefully studied in Scandinavia and in the Great Lakes area. Recent Radiocarbon analysis on the sea-shore objects suggested that in the Neo-thermal period, between C. 12000 and 4000BC, the sea level rose at a rate of approximately three feet a century.

River Terraces

River terraces are topographic platform, benches or steps in the river valley that represent former levels of valley floors and floodplains. Throughout much of the world, beach lines and river terraces were formed during the glacial period. At present they lie above the sea level or river flood plain. Major streams flowing down from the glacial-melt towards the sea made the terraces along the valley sides. This is for the changes in the volume of water in a stream. In colder phases i.e. during the periods of maximum glaciation, the rivers ran slowly with less force. So silt was deposited along the border. Later, during inter-glacial phases, as glaciers melted rapidly, the huge water-flow eroded the banks. More force was applied as volume of water increased. Therefore, silt was alternatively deposited and cut out by new erosion and the terraces as the remains of old flood plains were left behind .

Terrace formation has been extensively studied in Europe both in the Atlantic and Mediterranean areas. Along the East Coast of North America from New Jersey to Florida, seven coast terraces have been found which can be correlated with glacial stages. The highest beach is now about 265 feet above the sea level. In the eastern and southern parts of England raised beaches of Pleistocene time are found. The river Thames in England, for instance, shows a series of terraces. The latest of them corresponding to last inter-glacial period is placed about 25 feet high above the present day river-level. Next to that, there is another terrace, which is about 100 feet high from the present river-level. In this sequence, higher terraces are older than the lower ones. Since the parts of one terrace is considered contemporary with other parts of the same height, terraces at the Thames river valley are regarded as contemporaneous with the same level terraces of river *Seine, Danube, Rhone, Rhine, Somme,* etc. It should be kept in the mind that this correlation is reliable only within a comparatively small area and not for the distant areas. For example, the river terraces in *Burma, Java, India* though formed in the same way but do not follow the same time schedule as the European.

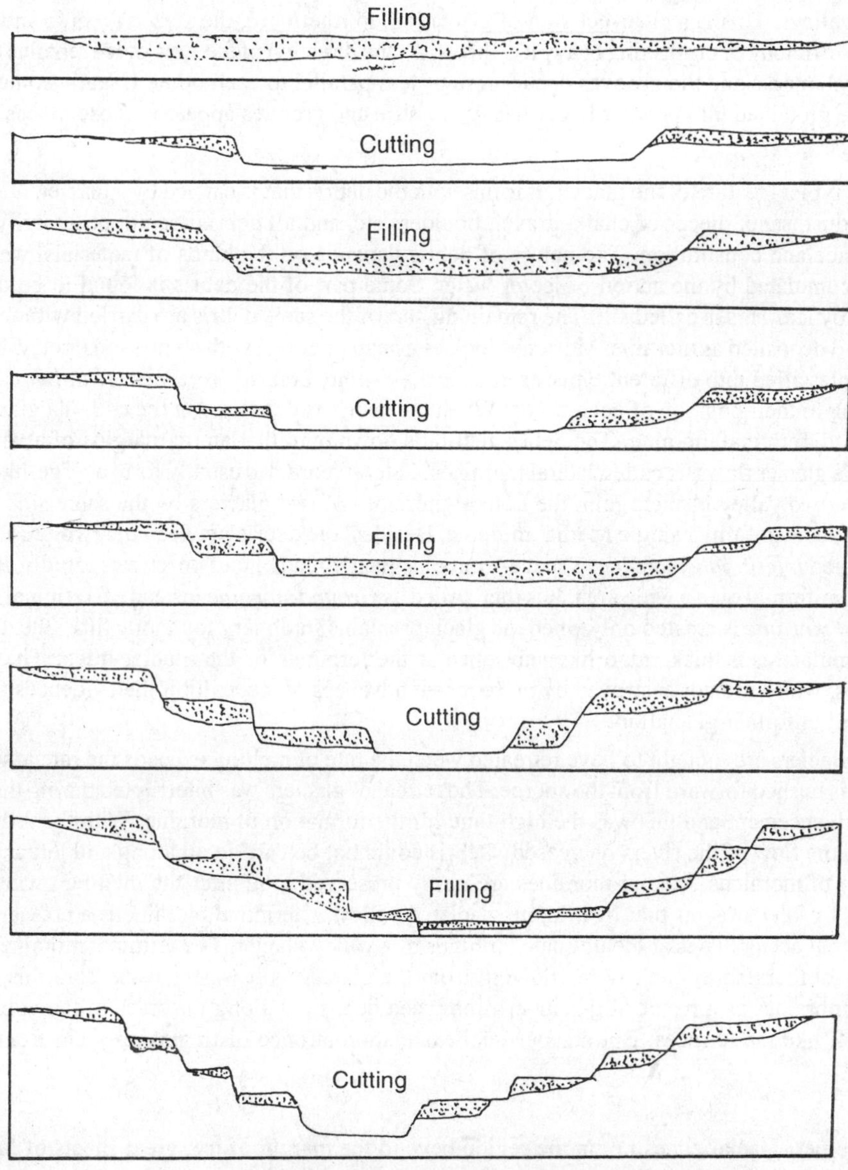

Fig. 9.9. Formation of Terraces through Glacial Action
[Cutting of river-bed took place during the periods of melting ice. Higher terraces are older and contain the relics of early men.]

Evidences of human occupation have often been identified along these beaches or terraces. Not only the stone artifacts, fossil remains and other domestic articles are found to be preserved in older terraces. In India, river terraces exhibiting major human activities include *Soan, Mayurbhanj* and *Narmada valley.*

U- shaped Valley and Scratch marks

Normally a river valley is V-shaped. But the bedrock floor and the valley itself are worn down by the abrasive action of glacier. The moving ice-currents scratch on the rocks over which they pass by. The troughs get enlarged in depth and width. As a result, V-shaped valleys are converted to U-shaped valleys. This is a clear-cut sign of glaciation. Furthermore, the *scratch marks* and grooves bear the testimony of on-rushing heavy ice-sheets. As the direction of the flow of ice remains constant, the scratch marks and the grooves occur more or less parallel to each other. Usually some parts of rocks are grounded into powder under heavy pressure and grooves appear in those places.

Moraine

It is a typical feature of the glacier. It forms with the debris that is carried by a glacier. Debris may contain dust, sand, pieces of chalk, gravel, boulders etc. and all debris are not necessarily alike in appearance and constitution. The nature of debris depends on the kinds of materials, which have been accumulated by the action of ice or water. Some part of the debris is found to be deposited directly by ice. This is called silt. The remaining part of the same debris are carried with the melted water and deposited as moraine. Moraines look as a heap of gravel with some sand or clay. Moraines may be classified into different types as *Terminal moraine, Lateral moraine. Medial moraine, etc.* according to their position of occurrence. When the debris is deposited at the end of a glacier tract, it is called Terminal moraine. The debris that falls down onto the lateral margins of a valley, into which the glacier flows, is called lateral moraines. Lateral moraines usually form a ridge-like deposit but when two valley glaciers join, the Lateral moraines of two glaciers on the same side combine together so as to form a single Medial moraine. Besides, the debris between the two projections of ice is called *interlobate moraine.* The ice-front of a glacier if advances or retreats rapidly, the debris spreads uniformly over a wide area. It is then called as *Ground moraine* instead of Terminal moraine. *Terminal moraine* is created only when the glacier remains stationary for a long time; the drift tends to accumulate as a thick ridge like substance at the terminus of the glacier track. The terminal moraines, which are not destroyed by more recent advances of ice, exhibit the evidences regarding the extent of former glaciation.

The glaciers are thought to have retreated when the rate of melting exceeds the rate at which the new ice is pushed forward from the source. The retreat of glaciers was interrupted during the periods of ice-advancement and this was the high time for the formation of moraine. With the reduction of melting, the flow of the rivers decreased. Debris could not be carried all along and left at places in the form of moraines. Medial moraines are rarely preserved long after the melting owing to their removal by later streams that flow in the glaciated valley. A terminal moraine is a crescent shaped deposit that accumulates at the ultimate terminus of a valley glacier. The terminal moraines may be partially obliterated by melt-water flowing from the glacier as it wastes back from this position. Lateral moraines as a result of glacial erosion when deposited along the sides of the valleys, they may look like the terraces. But a superficial excavation at once distinguishes them from the real terraces.

Loess

In the peri-glacial zones i.e. in the region beyond the margin of ice, great sheets of loess were found to be accumulated. These were the dusts like particles, which used to be blown out by wind from the exposed debris of the glaciated territories. During the inter-glacial phases the surface of loess got weathered, leaving a clear record of climatic fluctuations. In Central North America and

Eastern Europe thick sheets of loess have been noticed which are about a few meters in thickness. Two types of loess can be distinguished in Northwest France during the last glaciation (Würm-I). They are termed as older loess and younger loess as per their age of formation.

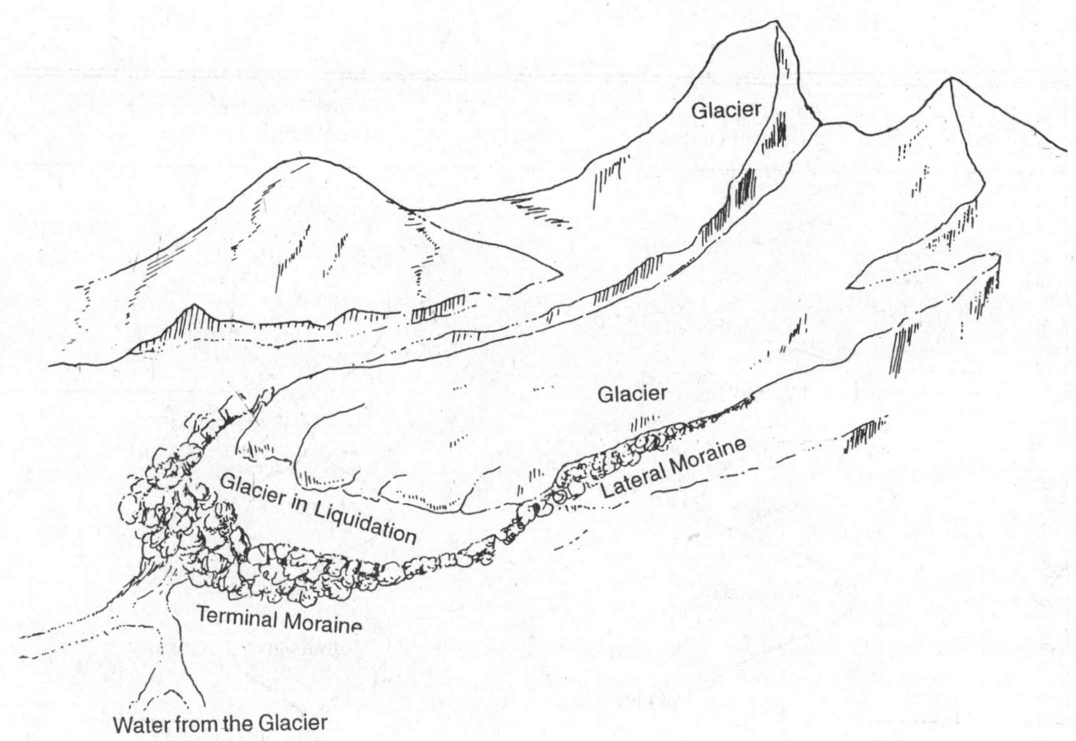

Fig. 9.10. Flow of a Glacier

Pluviation—The counterpart of Glaciation

Many attempts have been made to correlate the pluvials of low altitude with the glacials of high altitude. Although they do not match accurately, still they are considered to be broadly contemporary of each other.

During the Pleistocene epoch, the advances of ice caused the rain-belt to move towards the equator, over the extensive tropical and sub-tropical territories. This climatic shift was expressed as pronounced and prolonged periods of heavy rainfall. Four major phases of rainfall was noted, each of which is called a *pluvial period* and they correspond to the glacial phases. Inter-pluvial phases appeared between two pluvial periods in the same way as inter-glacial periods. Inter-pluvial periods were marked for a comparatively lower rainfall. As a consequence, rise and fall were noted in the water level of lakes and rivers. Periodic activity and inactivity of spring were also noticed. In fact, much of Africa, parts of Asia and North America received tremendous rainfall, more or less in the same time, when glacial periods were in existence.

In many parts of Africa evidences of tectonic movements and volcanic activities are found as a result of earth movement that took place immediately before the commencement of Pleistocene epoch.

Such earth movement was not only responsible for the formation of rift valleys* and upliftment of volcanic cones, it also gave rise to some very high mountains like *Kilimanjaro, Mount Kenya* and

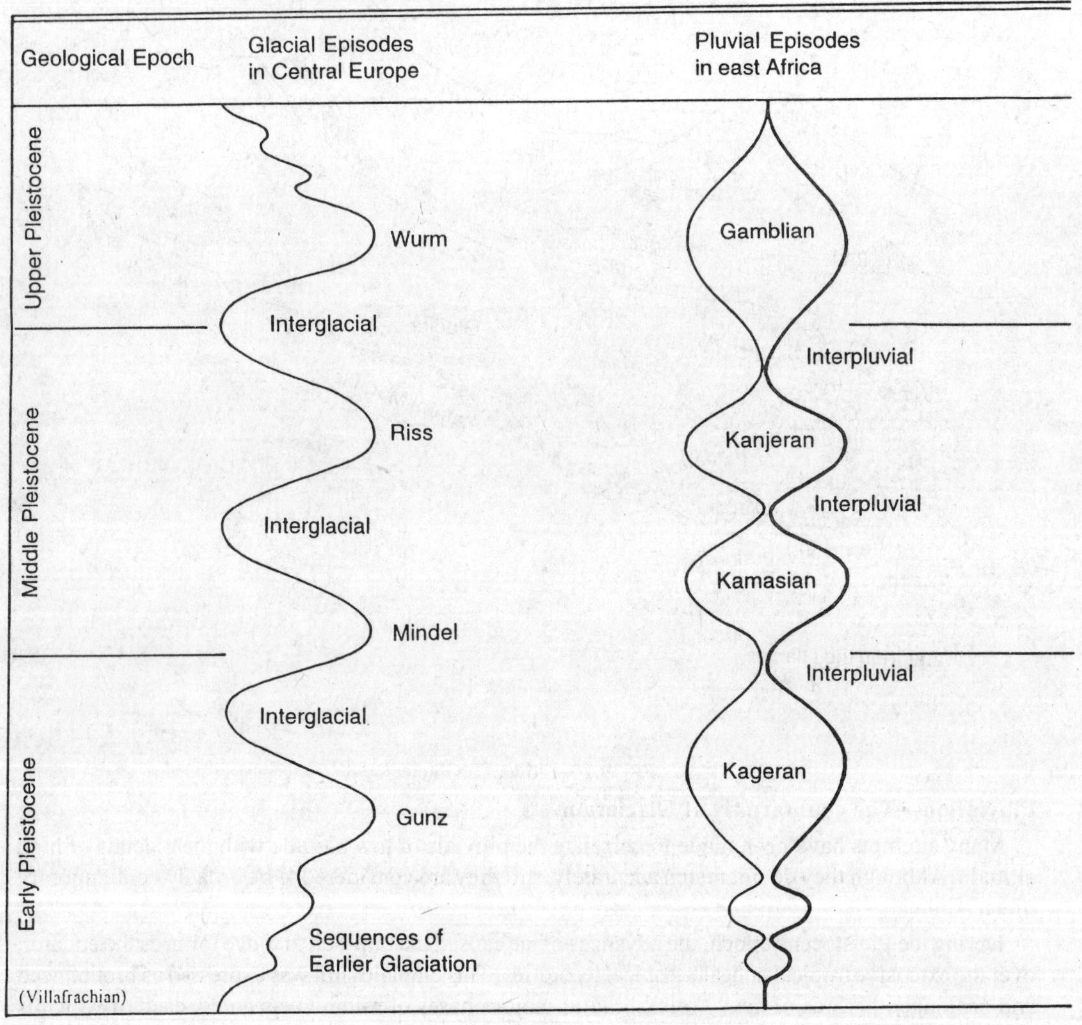

Figure. 9.11. Sequences of Glaciation and Pluviation

---

* When forces working inside the earth cause a block mountain to split, the 'rift' thus made is known as Rift Valley.

the *Ruwenzori Range* which disturbed the atmospheric circulation and increased precipitation around them. Once ice-sheets had formed on the top of those mountains, a high-pressure condition was created and the cyclonic belt moved towards the equator. As an effect of this climatic displacement, rainfall increased along the equator; cloudiness checked the rate of evaporation. Thus, pluvial periods in low altitudes can be explained with the atmospheric change. The root cause was further hidden into the extent of ice-sheets in high altitude.

Although different parts of Africa had witnessed the pluvial episodes, they are most remarkable in reference of East Africa. Distribution of rainfall still varies considerably from area to area in Africa. However, from the earliest to most recent, the pluvial periods have been named as *Kageran, Kamasian, Kanjeran* and *Gamblian.* The Kageran pluvial period of Africa corresponds with the Günz glacial period in Europe. This was the first major change in the environment of Pleistocene. The second major shift appeared with the increase of surface temperature, which was combined, with a general decrease of precipitation. The period was known as the first interglacial (Günz-Mindel interglacial) in Europe and the first interpluvial (Kageran-Kamasian) in Africa. In this way, the second pluvial period (Kamasian), the third pluvial period (Kanjeran) and the fourth pluvial period (Gamblian) correspond with the second glacial period (Mindel), the third glacial period (Riss) and the fourth glacial period (Würm) respectively. Again, the second inter-phivial period (Kamasian-Kanjeran) and the third inter-pluvial period (Kanjeran-Gamblian) correspond to the second inter-glacial period (Mindel-Riss) and third inter-glacial period (Riss-Würm) respectively. Two minor periods of increased rainfall have been identified during the early phase of the post plúvial period. They are known as *Makalian* and *Nakuran.*

In Africa, the evidences of pluvial period have been noted in the following way:

(i) Raised lake and river terraces denote higher rainfall than the present day. Raised beaches are found all around Africa, especially along the Mediterranean upto the Cape. Raised terraces are also found in rift area of East Africa and Vaal River valley in South Africa.

(ii) The study of fossil soils and the chemical alterations such as ferruginization and calcification suggest a different type of climatic condition at the time of its formation. Presence of wind-blown sand in the river valleys indicates an arid condition.

Like the changes of land and sea-levels resulting from the growth and retreat of the ice-sheets, variations in precipitation and temperature (apart from their direct effect) radically altered the conditions for livelihood and the influence went on the vegetation and animal forms. The comparatively arid interpluvial phases were only suitable for human habitation.

It seems that during the final phase of Pleistocene epoch, men became able to spread themselves over the greater part of earth. The early types of men were confined to the warmer parts of the Old World as the remains have been discovered from the region extending between Algeria to the Cape in Africa, from the England and Central Germany in Europe, from the mountains of Northern Iran in Western Asia. India, SouthEast Asia and Indonesia also contributed in this respect. At this time, man possibly acquired his modern form for which the name Homo sapiens stands for. From Northern Eurasia a group spreaded to the New World. Another groups spreaded to Australia and Tasmania and across Polynesia to New Zealand.

BIO-CULTURAL MECHANISM BEHIND THE EVOLUTION OF MAN

Life probably began about millions of years ago in the activation of protoplasm in unicellular form. Since then an unbroken succession has been noted which proceeded towards complexity to give rise different life forms of evolved type. Credit goes to Charles Darwin (1809-1882) and Alfred Wallace (1823-1913) who first comprehensively formulated and systematically explained the biological processes in terms of developmental sequences. In fact, evolution is an ongoing process, which will continue as long as the life will exist. The new forms are continuously coming out and the old forms are disappearing. However, our focus is on man—the most interesting, the most exciting and the most promising of all creatures on the face of the earth.

It is our duty to determine the sequence of events on the universal time scale as well as to know the rate of evolution that differentiate between the anthropoid and human way of life. In course of evolutionary divergence, man is the only animal that gained the capacity to create and sustain the 'culture'. Culture is a new kind of biological adaptation with non-genetic mode of inheritance. This inheritance depends on the symbolic contact and transmission, instead of the fusion of gametes. But it is also true that the fabulous growth of culture would have never been possible unless the biological propensities unique to man were not developed gradually. Culture itself is a product of organic evolution.

Culture gave man an endless capacity to invent and learn; no other animal has made such kind of progress. It is a distinct type of phenomenon that represents the highest level of evolutionary significance. With the blessing of culture man has been able to rule over his immediate environment provided by the nature. The classic definition of culture was formulated by E. B. Tylor in 1871 where he said *"culture is that complex whole which includes knowledge, belief, art, morals, law, custom, and other capabilities and habits acquired by man as a member of society"*. A more contemporary version has been proposed by L. A. White and other founders of modern anthropology.

White's definition of culture is, ".......*an extrasomatic, temporal continuum of things and events dependent upon symboling. Specifically and concretely, culture consists of tools, implements, utensils, clothings, ornaments, customs, institutions, beliefs, rituals, games, works of art, language, etc.*".

Thus culture is the most impressive adaptations achieved by an evolving organism. Although it rests upon and emerges from the psycho-organic mechanism of human body, but it does not belong to the organic structure of man. If we examine the long story of human creativeness, we find that the culture began far back with the first appearance of human species. The growth of culture can be farther traced with the physical developments of man. As a matter of fact, man made numerous inventions and transmitted them from generation to generation as social heritage. Cultural processes enabled man to make adjustments with the diversified nature without going through a biological modification of his organism. Herbert Spencer and A.L.Kroeber designated culture as super-organic. So far as known, in the sequence of evolution, Homo sapiens came last of all with the most remarkable functional complexes such as locomotor adaptation, manual dexterity, visual acquity, social behaviour and intelligence. These are the faculties for which the modern man has succeeded to make significant advancement and his cultural achievement reached at its height.

Locomotor adaptation

It is the most important and fundamental change as the four-footed creatures got erect posture and bipedal locomotion. It involved a number of muscle and bone rearrangements. For example, when the trunk became erect, forelimbs acquired a special significance. With the bipedal gait man became able to walk forward and backward with two fully extended legs. The positions of other two legs (frontal legs) automatically shifted and their functional changes were noted. Change also appeared in the curvature of vertebral column. The spine moved ventrally and the weight-bearing axis came closer to the trunk's center of gravity. The vertebral column is no more a horizontal rigid structure; rather a flexible vertical curved rod. The hands, pelvis, shoulder girdles and legs—all essential anatomical structures are the parts of the locomotor complex.

Manual dexterity

It is another important aspect. With the bi-pedal gait man's hand became a useful organ to perform various functions. Both the hands began to be used in many non-locomotor and non-prehensile activities such as shoving, clubbing, poking with the fingers, etc. But, prehension and precise movements of hand are the unique development in man. All primates have prehensile hands, prehension may be described as special action of hand. When it holds an object, it is called prehensile grip but in other ways, when a hand reaches for an object, it is assumed as prehensile pattern. J. R. Napier divided the prehensile actions of the hand in two kinds—*the power grip* and *the precision grip*. In fact, a hand must hold an object strongly and the power grip produces stability in a kind of clamp, which is formed by partial flexion of the fingers and palm with a counter pressure by the thumb. The precision grip also produces stability when an object is pinched between the flexed fingers and the opposing thumb. The power grip exerts maximum pressure on an object whereas the precision grip is the maximum accuracy of control. However, all prehensile movements of human hand combine the two basic grips, precision and power. For a fine control over the limbs and digits, a voluntary control over the related muscles is required. The thumb becomes completely opposable to other digits when they are flexed.

Visual acquity

A powerful and continuous selection of vision has been evolved as a dominant sense of stereoscopic vision. This stereoscopic vision requires two modifications—one is *osteological* and the other is *neuroanatomical.* In osteological modification, the orbits must rotate toward the front of the skull in a same plane so that the fields of vision of two eyes get overlapped. As a result, the perspective depth of any object can easily be estimated by visual observation. The neuroanatomical modification involves sorting of the optical nerve fibers which cross completely in the brain. This is called total decussation of the optic fibers. It is a mechanism by which all the nerve impulses from the right eye are carried to the left hemisphere of the brain and similarly all the left eye impulses go to the right hemisphere of the brain, vice versa. However, the stereoscopic vision contributes to the visual acuity and therefore a more accurate judgement of distance is possible. Man has developed a high-level precision in striking objects with the help of eye-hand coordination.

Social behaviour and Intelligence

A sharp break between the non-human primates and man has been observed in relation to social behaviour and social group formation. Most likely this difference is due to the development of symbolic and vocal communication in hominid line. As an end product of the whole process of inorganic and organic evolution, man has developed a complex and plastic nervous system which possesses a protracted span of memory for details and has got the ability to use the verbal symbol, language. By virtue of symbolic faculty, the Homo sapiens can store up any sort of information and can pass it on to the next generation.

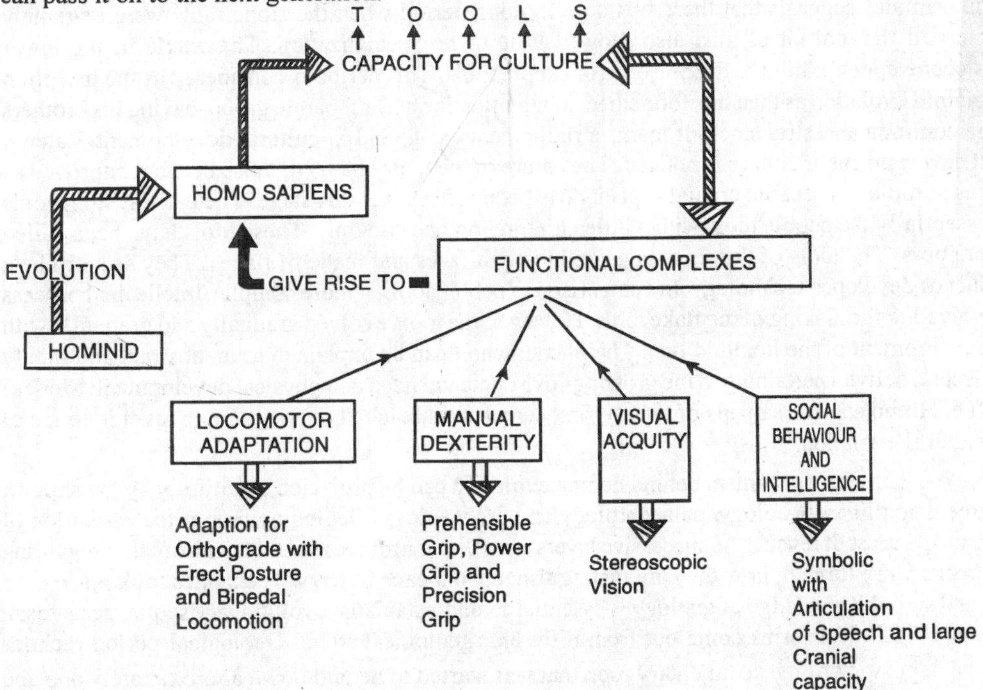

Fig. 9.12. Faculties Producing Bio-Cultural Advancement

Evolution of human behaviour is based upon man's ability to use symbol that in turn is a product of neuro-anatomical evolution. Following the sequence of evolution, we have found that the social behaviour of primates is more complex and more highly structured than those of other mammals. Man has again been separated from them by the development of erect posture, bipedal locomotion and manual dexterity. Moreover, the cerebral cortex has developed highly in capacity and refinement of the visual sense. This cerebral cortex exerts its conscious .control over the muscles. As a result, gregariousness or sociability appeared among the mankind. It suggests that men have come from such a stock that was pre-adapted for symboling and culture. The other important factor, the vocal imitation has been added to it. This is a more explicit and less ambiguous expression that enhanced the rate of communication in a social group. Control of vocalization is also related to erect posture. When the position of larynx changed, the foramen magnum moved towards the base of the skull and the size of mandible became smaller. Naturally the larynx got a long, uninterrupted resonating cavity. In this position, the tongue protrusion favoured in improving the mimicking or matching of low gruff sound or associated grunts. Basing on these developments, gradually, articulation of speech came into existence.

The findings imply that it is the biological development that initiated the uses of symbol. Origin as well as the differentiation of culture was impossible without the faculty of symbolizing. This faculty equipped men not only with intelligence, it qualified them for social behaviour. The social norms are found to be inherited from generation to generation being rooted in the capacity of symbolizing. So the bio-cultural adaptation of men can directly be related to the neuro-anatomical advancement of primates but did not take place in a day. All sensory information, passing through the ears and eyes, reached to the brain and helped to increase the brain-size. This, in turn, enhanced the intelligence as evident in the effectiveness of the manipulative skill.

The prima facie evidence of early culture is the stone tools that bear the evidence of Neuro-anatomical advancement of man. The recognizable tools in association with different fossil remains of the primates suggests that there was a pretty long period when the stone tools were extremely simple and the tool-kit of man also showed little or no specialization. The whole of the lower Pleistocene epoch exhibited the presence of simple tools. This period is considered as the inception of hominid evolution as then the forefathers of man first formed a separate group leaving his brothers of the common ancestral stock. It marks a rudimentary stage in bio-cultural development. Later, a rapid development of culture is noticed. The nature of the tools changed; those became improved as well as complex. Increased cranial capacity has been reflected at this stage. The earliest stone tools are essentially the pebble tools with a little chipping on one end only. These tools have been called as choppers. The tools of the next stage are mostly bifaces and made of flakes. They seem to be a product of developed technology. In comparison to pebble tools, a more complex intellectual process is involved in the making of the flake tools. However, the tools evolved gradually and gradually with the development of the hominid line. The phenomenon can be explained as an abstract and highly intellectual activity pertaining to the growing environmental need and physical development. Modern man i.e. Homo sapiens sapiens of twenty-first century has reached to an optimum level in terms of bio-cultural evolution.

The bio-cultural mechanism behind human evolution can be projected beautifully by the support of three disciplines—geology, palaeontology and archaeology. Geology refers to the formation of earth and its crust. It reveals the successive layers of rock and provides valuable information regarding how rocks were formed, how they are distinguished from each other, why changes took place over time and so on. Knowledge of geology is essential to understand the evolutionary sequence as fossil remains of different forms come out from different geological strata. Geological period reckons with the deposition of the sedimentary rock that was started to be laid down approximately one and half million years ago, before the origin of life. Modern geologists divide the whole span of time upto the recent into various gradations that serve as a scale to estimate the relative age of the fossil remains.

The other discipline palaeontology maintains a close connection with geology. Palaeontology is the study of buried hardened remains or impressions of former animals or plants (fossil). In order to clarify evolutionary relationships, palaeontologists not only search for fossil remains of different animals, they also collect geological information on the succession of climate, environment and plant-animal population. Only a small portion of a bone or a fragment of a skull, even a broken teeth can supply the clue regarding the gross structure of the body of a prehistoric animal; the behavioural pattern can also be deduced. Studying limb bones or parts of the vertebral column, the nature of locomotion can be suggested. Therefore a palaeontologist remains in a position to judge whether the animal was a brachiator or it walked on four feet; whether it had an upright posture and bipedal locomotion or not. Since muscles leave permanent impression on the bones, something could also be said about the muscular attachment. The parts of the spinal bone suggest the size and form of the brain as well as provide some idea about the spinal cord. It has been found that the teeth and the fragments of jaws remain better preserved than the other parts of the body. Teeth give several clues for the reconstruction of the animal. For instance, the size of the animal may be deducted on the basis of teeth. They also provide indications about the type of food and dietary habit that in turn throw some light on the climatic and environmental condition of the particular period when the animal under consideration used to live. Moreover, the presence of a bone of a known animal species in the same deposit along with the fossil fragment of an ancestral man is very important while tracing the human evolution. Because it clearly establishes the age of the particular human remain. Thus, in one hand, the geological structure helps to determine of the age of the fossil animals and on the other hand, fossil animals determine the age of the strata from which the pre-human or human like remain comes out.

Finally, the role of archaeology may be considered which strives to trace human evolution through man-made artifacts. Archaeologists deal with a relatively shorter span of time in contrast to geologists and palaeontologists. The estimation of geologists and palaeontologists are therefore primary and fundamental whereas the estimation of archaeologists is secondary, although the latter seems to be more refined. Since the man-made evidences are proportionately higher than the skeletal remains of human, archaeologists have been able to identify the short stages of hominid development, very accurately. The first hominids proclaimed their humanity by making tools.

Advances in geology, the discovery of more and more fossil finds and the development of modern dating technique have contributed in documenting the age of earth and the evolution of plants and animal forms. Naturally the bio-cultural mechanism behind the evolution of man has nicely been projected.

FOSSIL RECORDS IN SUPPORT OF MAN'S CAPACITY FOR CULTURE

The discovery of fossils not only yields many important information regarding the evolution of body structure in man it also provides an insight to gauge the development of human behaviour. The skull indicates the size of the brain, as the interior surface of the skull is equal to the outer surface of the brain. The brain development corresponds to the mental development. In the same way, the size and structure of the limb bones present some useful facts. A straight thighbone indicates an erect posture of the body. As the arms no longer used for locomotion, they are released for other work. Intelligence is required here to direct the operations by hand. Therefore, it is easy to understand that in manufacturing of articles like tools, weapons etc., a considerable degree of intelligence was involved. In pre-human stage, within a little brain capacity, tools could not be produced. The skill developed with time through certain phases. Cultural evolution, from elementary to advanced forms, maintained an equal rhythm in consonance with the biological evolution. Since an advanced type of tool implies more skill for its maker or user, from a particular manufactured object we can infer the general intention and intelligence of a group of people. Development of intelligence has always been reflected in culture. For example, possession of a good bronze knife indicates that the people knew the art of smelting. The specific purposes of use are revealed from the associated facts.

Skeletal remains of Miocene hominoids suggest that they had an upright posture but had not the brain capacity enough for making tools. These bipedal Proto-hominids emerged in Africa in the late Miocene about fourteen million years ago. They lived there for a very long period, until four to five million years ago. These fossil remains have been classified under the genus *Australopithecus* and not kept our own genus *Homo.* But around two million years ago, when some stone tools began to appear along with the skeletal remains of the *Australopithecines;* some of these Proto-hominids were classified under the genus *Homo.* This group of Australopithecines exhibited better cranial capacity about 750cc which is higher than their earlier counterpart where this capacity ranged from 380cc to 530cc.

The beginning of human culture is actually a little controversial. Palaeontologists suggest a split between monkeys and apes (hominoids) that had occurred before about fifteen million years ago. At the beginning of Miocene (about 22 to 15 million years ago) epoch, there was abundant tropical rain forest in Africa. But afterwards, especially from late Miocene to Pliocene, a drying trend was found which diminished the tropical forests and made the country more or less open. Without intense humidity and rainfall, African rain forests turned grossly to grassland (Savanna); deciduous woodland found in some places. Change of habitat and natural selection favoured the hominoids for terrestrial adaptation. Bi-pedalism developed at this time. But the transformation of an occasional bi-pedal proto-hominid (hominoid) into a completely bi-pedal hominid did not take place overnight. There is a sharp gap between truly human and the pre-human beings according to the evolutionary theory.

True man evolved about 50,000 years ago. To qualify as human, a hominid had to justify himself by work. This criterion is more cultural than biological, although the interrelationship between cultural achievement and biological endowment can not be undermined. There are a number of identifying traits that characterizes a hominid, but the special attributes of humanity are speech and tool - the outputs of a developed brain. Every act of man involves thinking that helps him in making abstraction of concepts, generalization of ideas, anticipation of future consequences and reflection of imagination. All schemes and plans originate in thought; it is a special capacity of human brain. No other animal except man has ever shown the slightest ability to build a culture by accumulation of technology and communication of knowledge. However, the chief attributes, the speech and tool-making propensity worked in the following way:

Speech

Greater brainpower and intelligence allowed man to use symbols in communication that gradually took the form of speech or language. In this context we can remember the definition of language which was put forward by Prof. Herskovits *"Language is a system of arbitrary vocal symbols by which members of a social group co-operate and interact"*. Many animals have been found to make signals towards each other by making different sounds. But they cannot use specific symbols against varied things and actions. Speech helps man to pass down the traditions from one generation to the other.

Tool making

Tool is an object, which is taken from external environment and manipulated by an organism for the fulfillment of some aims. It may be argued that the simple ability of handling tools can not be an exclusive faculty of man because a number of animals in the world are found to use the tools in some way or other. For example, sea otters of California bring boulders up from the sea-bottom to crack the molluscs. Great apes show a manual skill in the manipulation of sticks and strings. They can throw pebbles from a distance for defense. But what they lack is the proper insight and ability of planning. Man uses tools not only with more intelligence and balance; his increased skill and judgement induced him in making tools. Tool being the product of ideas and thought is the most palpable sign of early cultures.

Evolution of brain went side by side with the evolution of hand. A great difference has been noticed between the underdeveloped hand of the anthropoid apes and the human hand. Although the number and arrangement of bones and muscles are same in both, but human hands show special aptitude in performing hundreds of operations which no ape or monkey's hand can imitate. Here precision grip has been added with power grip; it is admirably suitable for other functions that can not be done by any other primates living in trees. This new set of potentialities has opened new possibilities of livelihood in course of evolution. Cultural evolution thus accompanied the biological evolution.

Australopithecine

A group allied to Australopithecus is considered as the basal stage of hominid evolution. All the bi-pedal hominoids have been classified under the genus Australopithecus who had no capacity for making tools. It proves that bi-pedal locomotion preceded the enlargement of brain in the history of biological evolution. Since most of the Australopithecine fossils have been recovered from early Pleistocene deposits of South and East Africa, this region seems to be the main center of hominid evolution.

Recently palaeontologists have pointed out that not all the species but only a particular species of Australopithecines can be regarded as ancestral to the Homo line. They also identified the particular group of Australopithecus with relatively higher cranial capacity and classified them as *Homo Habilis*. *Homo-habilis* fossils are nearly two million years old; they have been identified only from some sites of East Africa and named by Louis Leaky, Phillip Tobias and John Napier. The brain capacity varies from 600cc to 800cc in contrast to other groups of Australopithecus whose brains were not much larger than Chimpanzee as the capacity ranged between 380cc and 530cc. On the other hand, a number of sites in Africa yielded some crude tools of approximately two (2) million years back. Some anthropologists point them as the work of Homo habilis but as those earliest stone tools keep no fossils associated with them, it is impossible to establish the identity of actual makers.

Prof. Clark made a special study on the Australopithecines of South Africa. He found that the difference between a man and an ape, and especially between a hominoid and a hominid is more functional, rather than anatomical. The criteria of humanity go with the ability of speaking and making of tools. Prof. Leaky worked hard in determining the age of Australopithecines because some fossils were found without any evidence of their age. We have already known that a morphologist usually depends on a geologist in placing a fossil in the time scale. Cooperation is also solicited from a palaeontologist. In most of the cases the associated fauna of known date helped to ascertain the age of the Australopithecine. In July 1959 Leaky discovered another hominoid remain with special morphological features from the early Pleistocene layer in Olduvai Gorge, East Africa. In his preliminary report he differentiated the very fossil from other Australopithecus species for its association with some primitive stone tools. He tried to establish the antiquity of the tradition of tool making in reference to the Oldwan pebble tools and thus reinforced the association of technology with the origin of man.

In fact, no tool goes as far back as the Pliocene epoch. Earliest stone tools are crudely-broken pebble tools where primary flaking are found without any sign of secondary retouch. They appeared in association with Villafranchian fauna mainly in Africa, Europe and some places of England. It refers to the Pre-Chellean industry in Europe and Kafuan industry in Africa. The Cromer forest bed in Norfolk, England typically exhibits this kind of tools.

Some progressive anthropologists acknowledge human life towards the close of Tertiary period of the continuing Cenozoic era. They are certain that at this time man started to think seriously to supplement his power of arm for defense as well as for food getting. But since his brain was not much developed beyond the animal stage, his ability was restricted only in utilization of the natural objects placed in front of him. Therefore, he broke a branch of a tree to a convenient length and lifted a piece of stone, convenient to strike on. This is the time when man realized the utility of sharp-edge tools and therefore tried to shape the stones by striking one against another. Such a tool is called *eolith* which delineate the dawn of Stone Age *(Eos* means dawn; *lithic* means stone). To recognize these works on flint is a very difficult task because they present very little trimming on them and no two eoliths look much alike. Therefore such tools have never received the same acceptance as like other tools of early man and have only been granted as artifacts. Some scholars have remarked that eoliths are not at all tools; they are only pebbles whose fragments had been detached by frost or some other natural agency.

In 1910, Prof. Reid Moir of England devoted his time to study the eoliths. According to him, some of these tools are typically pointed like eagle's beak and so he termed them as 'Rostrocarinates'. It is held that these Rostrocarinates have possibly given rise to crude hand-axe of lower Palaeolithic types. Prehistorians are quite confused with eoliths, as no human skeleton has never found with them. Unfortunately very little can be inferred about the life-style of the makers; almost all evidences have been washed away at the time of first great Quaternary inundation.

Homo-Erectus *(Pithecanthropine)*

Apart from the dispute of eoliths, the first major lithic industry belonging to the lower Palaeolithic has been attributed to the *Pithecanthropines* who represent the second stage in hominid evolution (Homo Erectus). Average cranial capacity of the Pithecanthropines was 1000cc. With a noticeable larger brain than Australopithecines, they gained a precise knowledge about the raw materials and within a limit of acquired skill they chalked out an effective way of flint working. The oldest recognized tool, the chopping tool called hand-axe came from them.

The great hand-axe tradition is the first distinct phase to count cultural impact. Evidences have been obtained from Africa, Western Europe and Western Asia. We may start with Africa as the country is called the *Cradle of mankind.* Olduvai Gorge in Tanzania shows a rich assemblage of stone tools. Earliest cultural remnants coming from its Bed-I have been labeled as *Oldwan culture.* As some Australopithecine skeletal remains have been unearthed from this bed, it justifies the possibility that some of the Australopithecines might be the toolmakers. However, the earliest hand-axes which appeared in Bed-II at Olduvai Gorge are most primitive in form and comparable to those available from Europe, labeled as Chellean or Abbevillian* in France. It is basically a core-biface industry. The typical hand-axe or pick of Chellean industry has been called as *Faustkeil* or 'fist Wedge' by the German scholars. But the French scholars described it as *Coup-de-poing,* a blow of the fist or a punch. However, the tools were crudely flaked being struck by stone hammers or being beaten against stone anvils. The irregular rough working edges served the purposes like chopping, cutting, scraping, hitting, picking and prying. The industry corresponds to the first inter-glacial, the Gunz-Mindel inter-glacial of early Pleistocene. The technique employed here is known as Clactonian or 'block on block' technique.

Some tools appeared in the Far East about the same time as Abbevillian in the West. But they were not the hand-axe. China, Burma, Malay and Java witnessed a large number of pebble-tools called chopper. Unlike hand-axe, the chopper shows flaking on only one edge.

The successive beds in Olduvai Gorge, from II to IV, show ten stages in which hand-axes have developed gradually. After the second glacial/pluvial phase, particularly during the period from second to third inter-glacial/inter-pluvial phases, an improvement of technology was noticed both in Europe and Africa. It was actually a specialization of Chellean type although flake-tools became

* Abbevillian is the modern nomenclature for Chellean industry. The older name, Chellean was derived from the site Chelles on Somme Valley, France where the characteristic type of tools were first found. Abbeville is a close spot near Chelles from which a rich assemblage was obtained later.

prominent in the later part of this industry. The implements were flatter, lighter and more oval. Moreover, they were neatly worked with more straight and efficient edges. This particular tool tradition was named after a site at St. Acheul, France. The technique was known as Levalloisian. Speciality of this technique lies in the fact that the flakes have been detached from a tortoise-shaped prepared core. The knocked off flakes are invariably small, thin and neat. Those are further retouched in making of a tool. The remaining part of the used cores is also found to make into tools. Such a technique not only indicated a refinement of hand-axe, it showed the economical use of raw material as well as labour.

Although there is a scope of argument regarding the precise date of Chellean and Acheulian industry, both of this industry belongs to the Lower Palaeolithic. In general, this lower Palaeolithic culture has been ascribed to Homo erectus group whose biological development was enough for making some definite tools. The people were the first and foremost big game hunters who settled in the river valleys like the Thames or the Somme, the Nile, Vaal or Zamberi, the Soan or the Narmada and the Mei Fingnoi or the Irrawady. They were also found to settle by the shore of lakes like those of Olduvai and Olorgesailie in East Central Africa, Karar in Algeria, Torralba in Spain or Hoxne in England. Remains come from open-air sites rather than caves. Sporadic occurrence of charred animal bones from a few sites suggests that the people of that age learnt the use of fire but they were unable to control it entirely for regular use. Evidence of fire is also preserved in the stratified layers. Such layers got discoloured being burnt and mixed with ash and charcoal. Instances come mostly from the spots where the discarded refuses were kept.

The earliest record of using fire thus came from Kenya, East Africa about 1.4 million years ago. But routine usage of fire was established much later, about 50,000 years ago as evidenced from China and Europe. It is held that the spreading of the population from Africa to other cold territories was possible only after they gained sufficient control on fire. Fire helped the people to survive in the freezing winters of temperate regions. However, the industry of lower Palaeolithic being divided into core tool and flake tool tradition reigned in Europe and Africa at the end of the first inter-glacial/inter-pluvial period and in some places continued all the way upto the third inter-glacial/inter-pluvial period. The Acheulian stage lasted relatively longer than the Chellean. A few skeletal remains of *Neanthropic* type i.e. the Swanscombe, the Galley Hill and other skulls have been considered to be of Acheulian period.

Homo-Neanderthal *(Neanderthaloid)*

Anthropologists believe that the Homo Neanderthals have evolved from Homo erectus. This group is recognized as pre-Homo-sapiens and its distribution has been recorded from all over the world specially in Western Europe, Africa, South-east Asia including India and China. They denote the middle Palaeolithic culture.

Neanderthals gave rise to an industry called Mousterian, the name of which was derived from a rock-shelter at Le Moustier in the Dordogne of France. Mousterian industry did not show any core-tool. It shows a preponderance of flake tools. Flake implements of varied shape and size appeared among which points and scrapers were the principal types. The flaking technique was more or less same as in Acheulian industry i.e. a core used to be prepared at the starting point for getting flakes of desired shape and size. The percussion flaking was almost similar to the Levalloisian technique. Only difference is that, not only a number of flakes were obtained by this method, rather here the tool-maker used to continue the process until the last flake came out of the core. Since the whole core was consumed in producing flakes, no core tool was found in this industry. The flakes were nicely retouched thereafter to make specialized tools. Neanderthals' brain capacity has been calculated as 1300cc approximately. With more higher brain function, they realized the utility of lighter tools and made them more effective by finer retouching. These people also started to make various kinds of tools for a variety of purposes.

The Middle Palaeolithic culture characterized by Mousterian industry in Europe and Near East flourished during the last Pleistocene glacial, known as Wurm. The climate drove the people inside the cave for shelter. Therefore, most of the Neanderthal remains and other cultural relic have come from caves and rock-shelters. Regulated use of fire has been noted for the presence of charcoal and ashes in the caves.

Neanderthal people served as a base for further development of mankind. Fully developed man, the Cro-Magnons finds their cultural origin in Neanderthals. Till the onset of last glaciation, the industries retained the lower Palaeolithic culture. People had shown no considerable advancement; no major improvement was noticed in the methods and implements of hunting. But still Neanderthals are significant in two ways. Firstly, in contrast to early men (Pithecanthropines), Neanderthals exhibited some novel adjustment towards nature. They were the pioneers to make a judicial use of fire. They were also the inventors of clothing who utilized the skins of animals. The intelligence of these people helped to extend their range of settlement and so the skeletal remains of Neanderthal men have been recovered from many other places. Flint scrapers as an essential element of this culture indicate that the people often wore the animal skins to get rid of the severe cold, especially when they used to work outside the shelter, under bare nature. The other advancement was noticed in the treatment of dead. The posture of a few skeleton deliver such a look that they had been laid carefully with food and implements. This suggests the beginning of burial and offerings for the dead soul.

Neanderthals occupy an intermediate position between Pithecanthropines and the Cro-Magnon groups of men. They project an evolved form of culture than that of the Pithecanthropines, but differentiate themselves from the more advanced Cro-Magnons. For example, they exhibited only a limited use of bone despite they had huge stock of dry bones after primary consumption. Unlike Upper Palaeolithic men, they did not make any bone implements or weapons. Furthermore, they carried no indication of a developed aesthetic sense. As a matter of fact they practiced no art; no sign of carving and engraving was found from them. No single bored tooth has been observed from this cultural phase, which could suggest the presence of fabricated ornaments to adorn a person.

Cro-Magnon *(Homo sapiens)*

Cro-Magnon man is considered as the earliest specimen of modern man who appeared later than the Neanderthal man. The name has been derived after a rock-shelter in France where this type of specimen was found first. The name Cro-Magnon denotes a race, not a separate species like Neanderthal. The Cro-Magnons contributed to the advanced Palaeolithic culture that flourished both in Europe and Asia, more particularly in northerly parts of the Old World. They represent the Caucasian stock.

Some scholars have designated the upper Palaeolithic industry as *Archaeolithic* industry. At the terminal phase of Pleistocene, a great climatic change was again noticed. The fauna and flora began to be modified with the gradual decrease of cold. Although the severest cold of fourth glaciation was on the Wane, but the period was marked with a brief secondary advance of Ice (Wurm-II or Buhl) and two other minor oscillations (Gschnitz and Daun). However, due to lowering of mean temperature, the main reservoir of snow in the northern continent started to melt and continuous departure of glaciers created massive flood. This in turn caused destruction and displacement on the geomorphology of earth. In many regions namely Egypt, Somaliland, Mesopotamia and India, populations were driven out with the surrounding animals and a long hiatus was created without any trace of man. This hiatus corresponds to the age of Archaeolithic industry in Europe where life could be preserved because devastation affected only a small area of it. Otherwise, everywhere a gap existed at the end of Palaeolithic period. However, this phase of surviving culture in Europe is known as 'Reindeer Age' because the period shows an abundance of a particular deer family in association with other animals like mammoth, woolly rhinoceros, cave-bear, cave-lion, cave-hyena, wild boar etc. Late Stone Age of Africa is comparable to the Upper Palaeolithic of Europe.

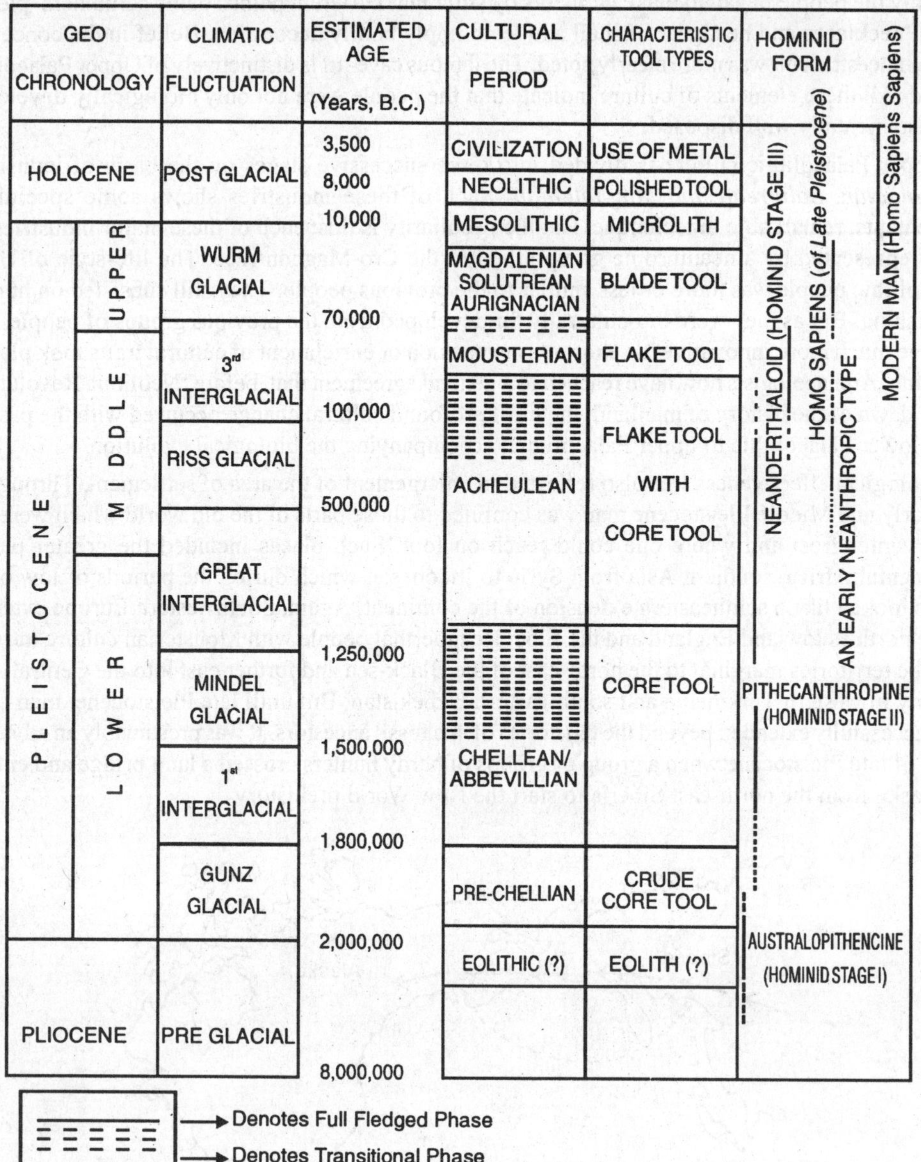

GEO CHRONOLOGY		CLIMATIC FLUCTUATION	ESTIMATED AGE (Years, B.C.)	CULTURAL PERIOD	CHARACTERISTIC TOOL TYPES	HOMINID FORM
HOLOCENE		POST GLACIAL	3,500	CIVILIZATION	USE OF METAL	MODERN MAN (Homo Sapiens Sapiens)
			8,000	NEOLITHIC	POLISHED TOOL	
			10,000	MESOLITHIC	MICROLITH	
PLEISTOCENE	UPPER	WURM GLACIAL		MAGDALENIAN SOLUTREAN AURIGNACIAN	BLADE TOOL	HOMO SAPIENS (of Late Pleistocene)
			70,000	MOUSTERIAN	FLAKE TOOL	AN EARLY NEANTHROPIC TYPE
	MIDDLE	3rd INTERGLACIAL	100,000		FLAKE TOOL	NEANDERTHALOID (HOMINID STAGE III)
		RISS GLACIAL	500,000	ACHEULEAN	WITH CORE TOOL	
		GREAT INTERGLACIAL				
	LOWER	MINDEL GLACIAL	1,250,000		CORE TOOL	PITHECANTHROPINE (HOMINID STAGE II)
		1st INTERGLACIAL	1,500,000	CHELLEAN/ ABBEVILLIAN		
		GUNZ GLACIAL	1,800,000	PRE-CHELLIAN	CRUDE CORE TOOL	
			2,000,000	EOLITHIC (?)	EOLITH (?)	AUSTRALOPITHENCINE (HOMINID STAGE I)
PLIOCENE		PRE GLACIAL	8,000,000			

——— → Denotes Full Fledged Phase

≡≡≡ → Denotes Transitional Phase

Fig. 9.13. A Conceptual Model on Bio-Cultural Evolution of Man

Although the upper Palaeolithic tool making technology appears to have it's root in the Mousterian, (the post-Acheulian tradition), the culture is remarkable for bone tools. A lot of objects made of bone, horn and ivory came into existence. Among the flake implements, blade and burin were the most common types. These were made on pyramidal or cylindrical pieces of flakes, struck off from a prepared flint core. Length of the tools were generally twice than their width. Pressure flaking technique was employed instead of percussion flaking. The tool types suggest that the upper Palaeolithic people became more conscious regarding the shape of the tool. A toolmaker not only adopted a greater physical control over the application of pressure; he also realized the utility of wooden handle and shaft. Presence of needles indicates the knowledge of threading and stitching; possibly the people began to make garments by stitching. Different paints, paint-containers, pierced shells, necklaces and headgear of shell and teeth imply bodily decoration. Belief in the concept of soul or ancestral cult was more clearly noted. The famous cave-art is distinctively of Upper Palaeolithic period. All these elements of culture indicate that the people were not only biologically developed, but also mentally well disposed.

Upper Palaeolithic culture is divided into three successive stages on the basis of industry—*Aurignacian, Solutrean and Magdalenian.* Each of these industries shows some specialized implements, remarkable in technique. But the peculiarity is this; each of these major industries has been represented by a neanthropic group allied to the Cro-Magnon race. The life-style of Upper Palaeolithic people was more or less similar to the previous people. They still subsisted on hunting and fishing. But as they were bio-culturally far developed from the previous groups of people, they became much more innovative; an enormous expansion or enrichment of cultural traits took place at this time. Archaeologists now have reached to a general agreement that, before 'Neolithic Revolution', at the dawn of the history of mankind, the most profound cultural change occurred with the passage from lower Palaeolithic to upper Palaeolithic, accompanying the biological evolution.

Biological effectiveness was also reflected in enlargement of the area of settlement. Throughout the Early and Middle Pleistocene man was confined to those parts of the old world which were free from winter frost and where one could reach on foot. Such places included the greater part of continental Africa, southern Asia from Syria to Indonesia, which during the periods of low ocean levels looked like a southeastern extension of the continent. Again, Southwestern Europe extended to the north as lowland England and the Elbe. Neanderthal people with Mousterian culture intruded into the territories marginal to the north-east of the Black-sea and further east into the Central Asia, notably in western Turkmenia and south-eastern Uzbekistan. But until late Pleistocene, man could not successfully extended beyond the territories of his fossil ancestors. It was presumably an advanced stage of late Pleistocene when a group or groups of hardy hunters crossed a land bridge and entered to Alaska from the north-east Siberia to start the New World prehistory.

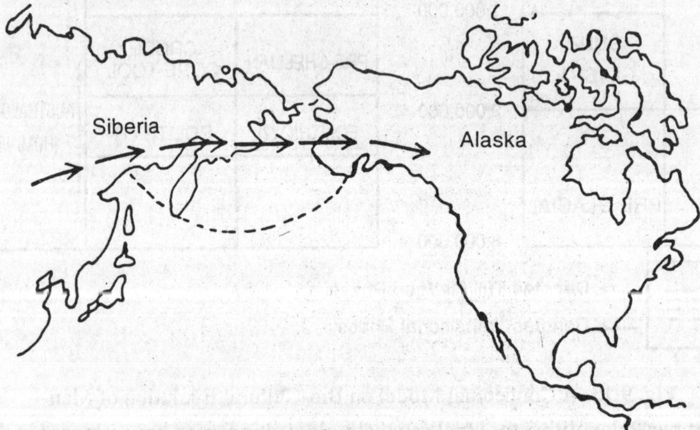

Fig. 9.14. The Ancient Land Bridge (over which Early Men Migrated to the North American Continent)

Evolution got stuck in the New World since the small monkeys arrived there. Only Homo sapiens remains have been revealed from this part of the world. Siberia seems to be the first area, which was occupied by man during the late glacial times. Although the total extent of this territory was vast, the area open to human settlement was very limited in the North for the presence of a great zone of lake and marsh at the south of the northern glaciated zone. Traces of upper Palaeolithic culture have been evident from a few spots only.

It is definite that man stems from an ancestral stock, more primitive than any existing apes. Three major anatomical developments can be pointed out behind the human evolution.

1. The erect posture has been responsible for the rise of culture as an adaptive mechanism. A human level of existence in this sense is more functional than structural. When the forerunners of man got an ability to walk upright, their forelimbs became free to use the tools. In fact, the tree-climbing primates knew no use for tools. Tool using arose in connection with adaptation to life on open ground, away from forest.

2. The technological development has been possible for the enlargement of the brain. Brain capacity increased step by step during hominid development. But this simple criterion is not enough to explain man's capacity for culture—its accumulation and transmission. Because mental propensity depends rather more on the quality of the cortical association areas of the brain than on its size. The power of abstraction and conceptualization form the basis of the systematic manufacture of tools. Capacity for language should also be added to these skills. The brains of the earliest tool-making hominids were functionally advanced for speech. The proto-hominids probably had remained at the stage of occasional tool using for millions of years. Systematic tool making would not have possible until the brain had attained the cerebral complexity, requisite for conceptual thought.

3. The changes in dentition (plan and form of teeth) can be correlated to the changes of ecology. A great change of vegetation not only compelled the man-like apes in terrestrial adaptation, climatic selection also worked on dietary habits. Moreover, the use of tools limited the use of teeth. Many of the functions of teeth was undertaken by the forelimbs and naturally tools became the substitute of teeth.

The man-apes as known from the several fossil skulls and a large number of teeth and jaws showed large canine teeth suitable for carnivorous diet and fighting. Incisors and molars were also very large in them. Because incisors served the purpose of sizing and pulling, and molars were indispensable for grinding of hard foodstuffs. Later, as these functions were replaced by hand and tool, the sizes were diminished

The Australopithecines of early Pleistocene are believed to be slightly aberrant descendants of Pliocene forebears. They were partly carnivorous and possibly used impoverished tools and weapons. Their dentition shows a slightly reduced canines and incisors and non-sectarian premolars and molars. Such dental characteristics are found among all large carnivorous mammals. The initial reduction of incisors and canines that took place in the phase of Australopithecines can be explained as an increasing reliance upon manufactured tools and weapons. The next reduction, in the surface of molar teeth, though appeared with the Pithecanthropines (phase of Homo erectus) but culminated at the phase of Neanderthals. This incidence seems to be related with the acquisition of fire for the purpose of cooking. Subsequent reduction in the size of the teeth and simplification of the pattern was observed when the tool-manufacture techniques got refined and effetive tools were produced by the upper Palaeolithic men. The small canines and incisors are biological symbols of a changed way of life. With increasing control of environment through the use of tools and by widening the nature of diet, man became the most adaptable of all creatures and therefore faced no obstacle in spreading into diversified climatic zone.

Starting from the Pithecanthropines, the producer of the lower Palaeolithic culture, physical evolution of Homo sapiens became complete with the advent of the Cro-Magnons who generated the upper Palaeolithic culture. In between, there were Homo Neanderthals who contributed to the Middle Palaeolithic culture. In fact, the immense and accelerating advances of social-cultural evolution since last 50,000 years could not have taken place without the initial cultural evolution corresponding to the biological evolution as evident from these fossil records.

ECOLOGY THE EARLY MAN

The Concept

From the very beginning man is interested in the surroundings. His knowledge regarding the mysteries of nature and life has developed with experience and power of reasoning. The scientific and technological advancements of recent days although help men to solve some of their problems but many other problems have been created at the same time. Moreover, nature's wealth is gradually diminishing with population explosion and scientific progress. Surroundings seem to be spoilt with filthy air and dirty water. Many old species have gone in the way of extinction. Therefore, the inquisitive scholars form various disciplines have come forward to study the different aspects of environment. Anthropologists became especially interested in the matter of environment because they believe man as a product of interactions between heredity and environment. The phenotypic expressions that are manifested through genes are produced by the interplay of heredity and environmental factors. As a matter of fact, the environment has a great impact on the biological evolution as well as cultural evolution of man. Besides, the variations among the populations of distant geographical regions may sometimes be explained as the differential environmental conditions under which the people are compelled to live.

Ecology is basically concerned with interrelationships between organisms and their environment. The word 'ecology' comes from the Greek word, 'Oikos' which means 'home' or 'habitat' and it combines the word with 'logos', to mean 'the study'. Although the term was coined about 150 years ago but its use was first found in 1865 by German biologist, H. Reiter. Another German biologist, Ernest Haeckel defined it in 1866 as *"the body of knowledge concerning the economy of nature — the investigation of the total relations of the animal, both to its inorganic and organic environment, including above all its friendly and inimical relations with those animals and plants with which it comes directly or indirectly into contact"*. More recently, renowned American ecologist Eugene. P. Odum defined ecology as *"the study of structure and function of nature"* (Odum, 1963). The Indian ecologist R.Misra (1967) defines ecology as *"the study of interactions of form, functions and factors"*. In general, these definitions indicate a wide scope of ecology.

Since ecology focuses on interrelationships of organisms within their environments, it includes the study of both living and non-living environments. When man is held as an organism, apart from himself, all other plants, animals and even the varied materials of the earth make up his environment. So it can not be denied that a distinctive way of life of an organism is dependent on both the living (organic) and non-living (inorganic) constituents of the environment.

We know that the plants provide food for all the other animals including man, directly or indirectly. But the plants also have to depend upon the animals for pollination, dispersal, carbon-di-oxide requirements and other needs. Not only that, the physical environment itself is supported by the organisms in different ways. For example, the plants can modify the temperature and humidity, the animals are able to modify the physico-chemical properties of the soil by trampling or burrowing. It has also been understood that the components of non-living environment are intimately related to each other and so a change in one component always affects the others, considerably.

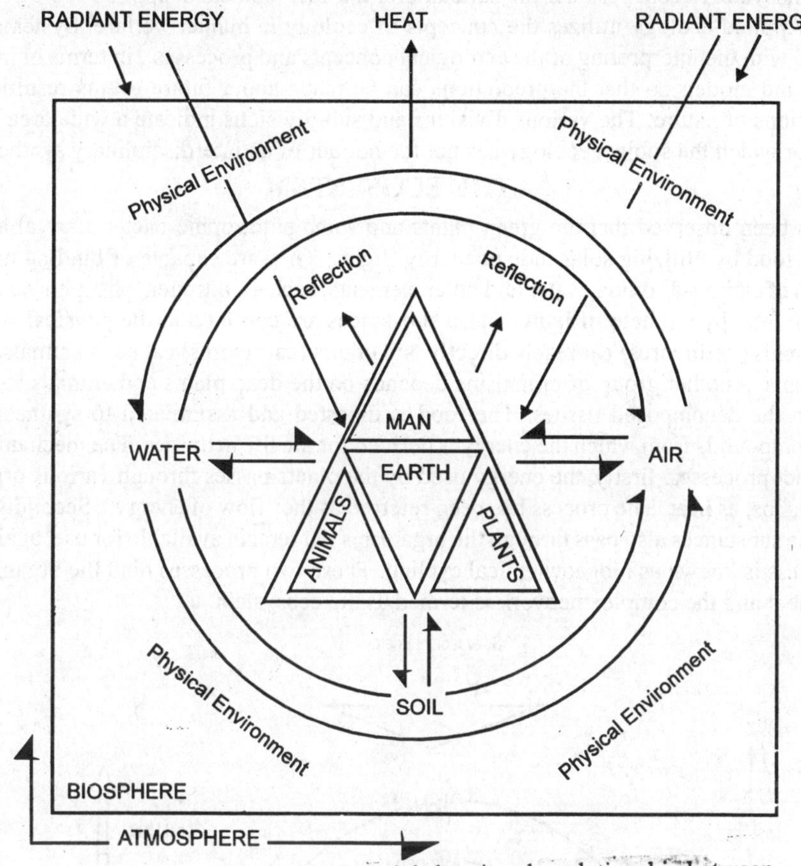

Fig. 10.1 A Conceptual model showing the relationship between Organisms and Environmental Components

For instance, the water in its three forms can modify the light and temperature of the environment. The heat causes the movement of air mass and also increases the rate of evaporation of water from soil and plants. Such a complex relationships of the organisms and the environment, among themselves and with each other give rise to the subject of ecology. Ecological ideas are actually very deep-rooted in human history although the use of the term ecology is comparatively recent. Man became aware of this organism-environment relationship, very slowly in course his exploration for survival. In the late phase of twentieth century, the growth of ecology extended across the continents. The study of man as a biological and cultural entity went under the fold of ecology.

Nowadays ecology is studied in several ways. As a subject, it is divided into various sub-divisions. The most common classification on the branches of ecology is based on the habitat, where the organisms live. Four branches are obtained here such as Terrestrial ecology, Fresh water ecology (Limnology), Marine ecology (Oceanography) and Estuarine ecology. Ecology can also be classified with emphasis on some communities (forest ecology, grassland ecology, etc.) or some taxonomic groups (mammalian ecology, insect ecology, human ecology, etc.). Another aspect in organism-environment relationship leads to the formulation of geographical ecology (explanation of present and past distribution of the organisms), evolutionary ecology (explanation of the environmental causes of speciation and evolution), physiological ecology (physiological responses of organisms to their environment), ethology or Behavioural ecology (concerned with animal behaviour), genecology (to study ecological adaptations in relation to genetic variability), radiation ecology (effects, of radiation on organisms) etc. Recent period is the hallmark for the development of applied ecology and systems ecology. Applied ecology utilizes the concepts of ecology in human welfare. Systems ecology is concerned with the interpreting of the ecological concepts and processes', in terms of mathematical formulae and models so that the predictions can be made about future events resulting from the manipulations of nature. The various divisions and sub-divisions indicate a wide area of scientific enquiry for which the subject ecology has got the honour of an interdisciplinary synthetic science.

THE ECOSYSTEM

It has been observed that the green plants and some autotrophic bacteria are able to prepare their own food by utilizing solar radiant energy (light). They are capable of binding up the simple molecules of carbon-di-oxide, water and other elements, such as, nitrogen, phosphorous, potassium, magnesium etc. by the help of light. These substances are confined to the physical surroundings. Some animals (herbivores) eat plants directly and others (carnivores) eat such animals, which feed on the plants. Another group of organisms depends on the dead plants and animals to obtain their food from the decomposed tissues. The food is digested and assimilated to synthesize different organic compounds from which the energy is derived for the life activities. The mechanism involves two distinct processes. firstly, the energy used by the plants passes through various organisms and finally it is lost as heat. The process has been referred as the "flow of energy". Secondly, along with energy, the substances also pass through the organisms but remain available for use, again and again. This process is known as biogeochemical cycling. These two processes bind the organisms to their environment and the complex network is termed as the ecosystem.

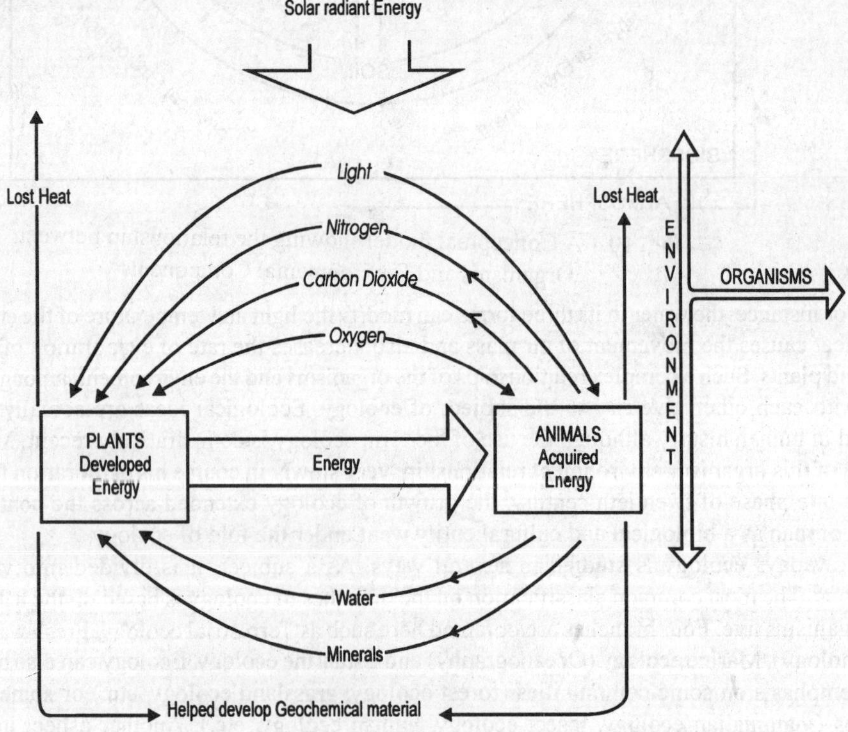

Fig. 10.2. A Diagram Showing Bio-geo-chemical cycling

The term ecosystem was coined byA.G.Tansley(1935) who stated the following words, "*......the more fundamental conception is .....the whole system (in the sense of physics), including not only the organism complex but also the whole complex of physical factors forming what we call the environment of the biome - the habitat factors in the widest sense. Though the organisms may claim out primary interest, when we are trying to think fundamentally we can not separate them from their special environment which they form one physical system. It is the systems so formed which ....are basic units of nature on the face of the earth......... The ecosystems as we may call them are of most various kinds and sizes*". Thus, the ecosystem can be of any size. An ecosystem in a laboratory may be as small as an aquarium. It may range from the size of a lake or forest to the size of the earth. Biosphere (W.Vernadsky 1929) or the Ecosphere (L.C.Cole, 1958) is the usual term which applies to the ecosystem that covers the whole of the earth. In fact, none of the ecosystems in independent, rather all are interdependent in some way or other.

All ecosystems consist of two major components—*biotic* and *abiotic.* The biotic component is comprised the living organisms, whereas the abiotic component includes the physical (non-living) environment. But, both of these components interact very closely to exhibit a definite structural organization. Sometimes, it is very difficult to separate the biotic components from the abiotic components.

Biotic Component

When we consider the biotic components, the organisms are divided into two categories, the *antotrophs* and the *heterotrophs.* The autotrophs can produce their own food. They are the green plants with chlorophyll and certain types of bacteria—*chemosynthetic* and *photosynthetic.* Since these organisms produce food for all other organisms, they are also known as 'Producer'. The heterotrophs depend directly or indirectly on the autotrophs for their food. This type of organisms is further divided into two groups, such as, *Phagotrophs* and *Osmotrophs. The phagotrophs* take food from outside and digest it inside their bodies. They are called consumers. All animals—herbivores (plant eating), carnivores (animal eating) or omnivores (eating all kinds of food) fall in this group. The *osmotrophs* are those organisms who secrete digestive enzymes to break down the food into simpler substances and then absorb the digested food. This group embraces the parasitic and saprophytic bacteria as well as the fungi. They may also be called *Decomposers* because their role has been well documented in the decomposition of the dead organic matter. But the most interesting point is that all of these parasites are not decomposers, rather some of them are consumers (insects and such small animals) who help in the decomposition by breaking down the dead organic matter into small bits. However, the heterotrophs can also be divided into two broad groups as the *biophages* (feeding on living organisms) and the *saprophages* (feeding on dead organisms) Fig. 10.3.

The above mentioned divisions of organisms are principally based on the nature of food which in turn give rise to the trophic structure of the ecosystem. It signifies a step-wise systematic relationships among the living organisms for food, which forms a chain, known as the food chain. Some common food chains can be illustrated as follows:

(*i*) *grass* ⟶ *goat* ⟶ *man*

(*ii*) *seed* ⟶ *rat* ⟶ *cat* ⟶ *hawk*

(*iii*) *algae* ⟶ *zoopl ankton* ⟶ *fish* ⟶ *man*

(*iv*) *algae* ⟶ *insect* ⟶ *frog* ⟶ *snake* ⟶ *peacock*

Such food chains can be traced back to the *Producers* and the position of the organism in the food chain is indicated by the trophic levels. This trophic level is defined as the number of links by which it is separated from the producer; the producers always belong to the first trophic level or the base. A single species can operate more than one trophic level in the ecosystem which means that this species get its food from more than one source. Again, the same species may be eaten by several other species of a higher trophic level; thus, one can find out several food chains linked together in an intersecting manner to develop a network, known as food web. Such a network provides a clear-cut idea about the functioning of the ecosystem.

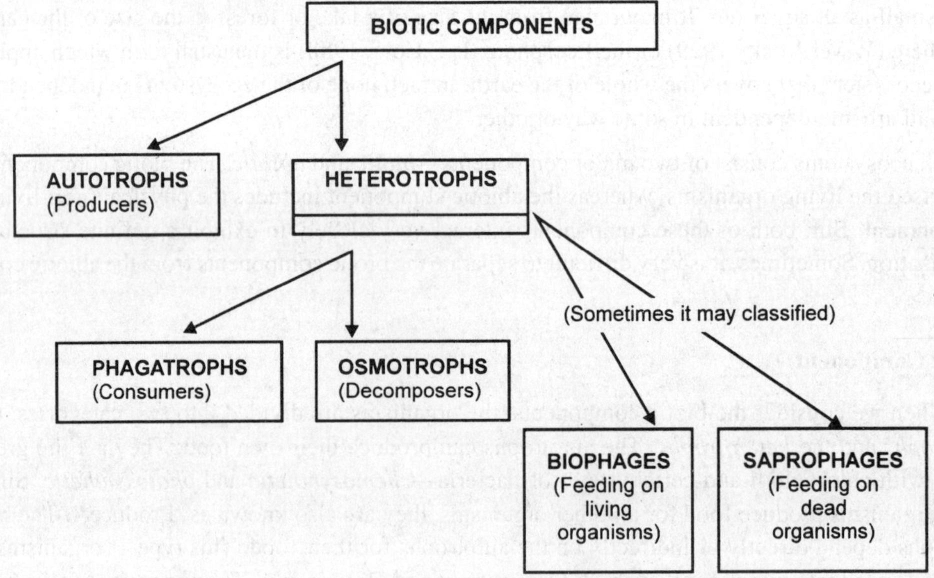

Fig. 10.3. A Classification of the Biotic components based on food

Abiotic Component

Abiotic component of ecosystem refers to the physical environment and its several interacting variables which can be divided into four folds—(*i*) *Lithosphere* which means the solid mineral matter on the earth and the land form as well; (*ii*) *Hydrosphere,* i.e. the water in oceans, lakes, river, ice-caps, etc.; (*iii*) *Atmosphere,* the gaseous mixture in the air; and (iv) the *radiant solar energy.* The position and movement of the earth with its gravitational force are additional components of the environment Fig. 10.4. However, these components create invariability of magnitude and duration of other environmental factors. The energy interacts with rocks, water and gases to produce a complex environment with a large number of identifiable variables such as heat, light, rain, wind, snow, fog, dust, storm, fire, etc. Thus, by the interaction of variables, the environment is created and maintained as a unit where any single component can not be removed or altered without disturbing the other components. Therefore, the environment is a dynamic whole, which remain continuously in a state of flux and also varies in space. The structural analysis of the environment in an ecosystem is urgently required to know the energy gradients and their flow.

Functionally the ecosystem allows the flow of energy and cycling of materials which ensures the stability of the system and continuity of life. The energy needed for all life processes come from solar radiation Fig. 10.5. During photo synthesis, green plants convert light energy to chemical (potential) energy and make it available to other organisms as food. Thus, a continuous flow of energy from sun through organisms maintains the life on earth. Laws of Thermodynamics govern on the transfer and transformation of energy. It says, the energy can never be destroyed, but it is trans

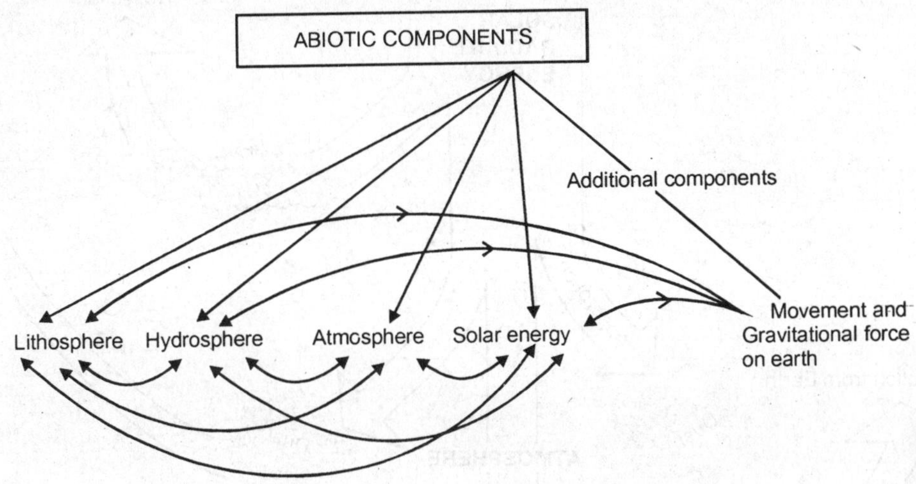

Fig. 10.4. The Abiotic Components and their Interactions

formed into different forms. Therefore, the part of solar radiant energy that is not used in photosynthesis is used in heating of air, water and soil. So some change do occur in cyclic order of nature.

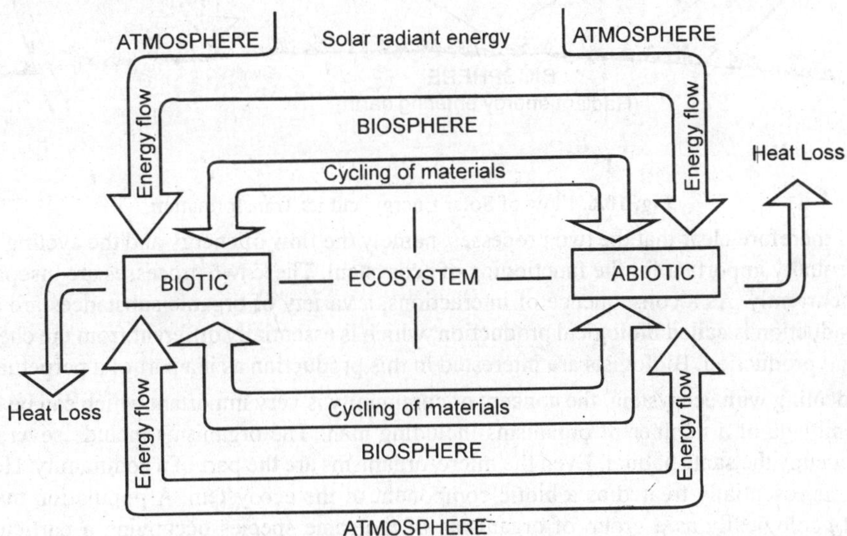

Fig.. 10.5 A Functional Ecosystem (Hypothetical Model)

Ultimately the energy is reflected back to outer space as heat. In fact, a small fraction of available light energy is utilized during photosynthesis and a very little part is stored in animal tissues; the bulk is wasted as heat. The next point is the ratio between the production and assimilation of energy. The small organisms utilize a large part of the assimilated energy for growth while larger organisms cosume a larger part of the assimilated energy for maintenance of the organism (respiration). However, all these mechanisms-the transformation of energy, the food chain, the assimilation etc., are expressed as ecological efficiency.

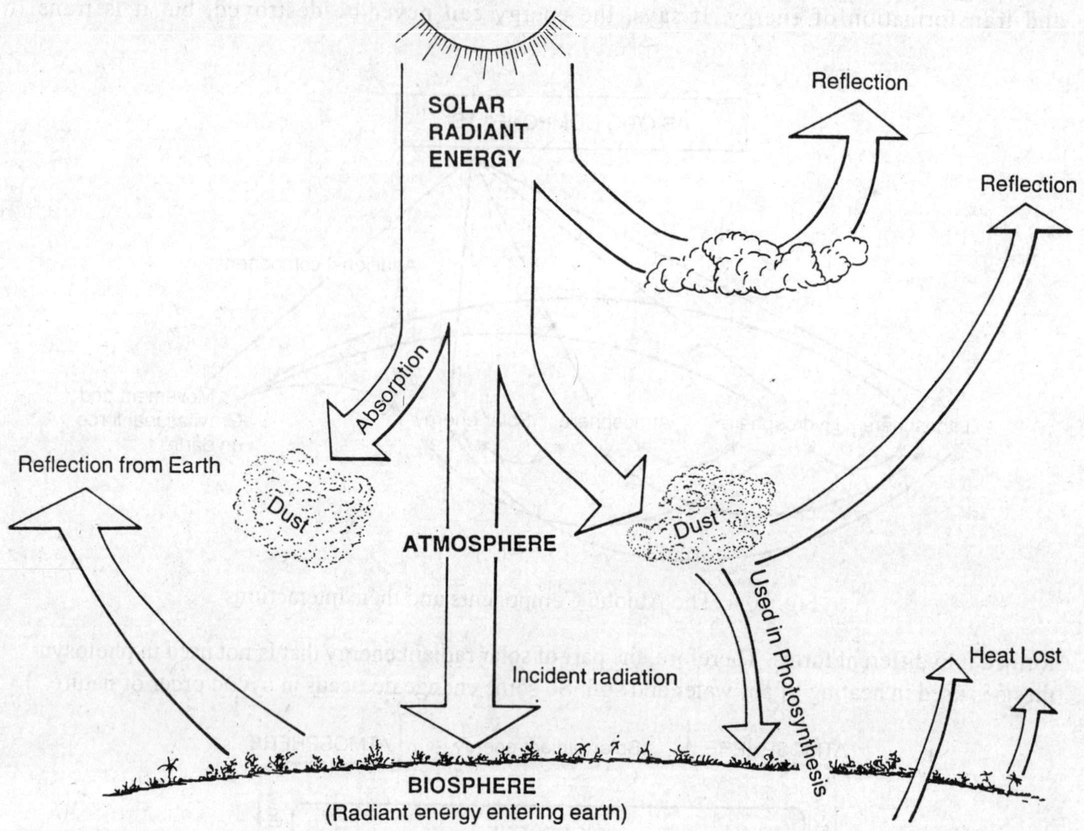

Fig. 10.6. Flow of Solar Energy and its Transformation.

It is therefore clear that the two processes, namely the flow of energy and the cycling of materials are equally important for the functioning of ecosystem. These two processes are inseparable and run concurrently. As a consequence of interactions, a variety of organic substances are produced. This production is called biological production, which is essentially different from the chemical and industrial production. Biologists are interested in this production as it a part of a perpetual process.

In dealing with ecosystem, the concept of community is very important which can be defined as an assemblage of a number of organisms including man. The organisms include several species, which occupy the same habitat. Even the micro-organisms are the part of a community. Here, 'community' is essentially treated as a biotic component of the ecosystem. A population may also be defined ecologically as a group of organisms of the same species occupying a particular space. However, population is the basic unit of the community—an aggregate of individuals of same species who mate among themselves. Naturally the members live in a common area and like to use the same resources of the particular ecosystem. In this manner, by the way of interactions between the organism and its environment, each population develops a style of life, which is designated as the niche or ecological niche of the population.

The niche denotes a specific portion of the habitat occupied by each species. In 1917, J.Grinnell first brought this word. In 1927, C.Elton defined *niche* as an animal's place in the biotic environ-

population in the community. In a simple way, it can be said that the habitat is the address of the species and the niche is its profession. So, several populations with different functions may occupy the same habitat. The ecological niche is referred to the totality of biotic and abiotic factors to which a given population is uniquely adapted.

The concept *of niche* was analyzed by R.H.Whittakar in 1970 as *"functional system of interacting niche differentiated species populations that tend to complement one-another rather than directly competing. in their utilization of the community's space, time, resources and possible kinds of interactions"*. An important term fundamental to the concept of ecological niche is *Biome.* A *biome* is a biotic community characterized by distinct life forms of climax species. The climax species in an integrated unit which acts as an indicator of a specific climate. The growth as well as life forms of climax populations suggests the result of interactions between the organisms and their environment. Though the climax species is a subject of controversial discussion, still each climatic zone represents some particular species as dominants. As a matter of fact, biome has been emphasized on the recognition and classification of natural biotic communities, which can be correlated, with the abiotic environment of that area. The other important concept, apart from Biome and Habitat, is the culture. Culture indicates a man-made environment is contrast to natural environment. Man tend to develop an environment around him by invention as well as learning. The ecological niche appears by the interplay of the *biome, habitat* and *culture.* It keeps a harmonic balance between the biotic and abiotic components in the biosphere of ecosystem. (Fig. 10.7.)

As man is the most superior animal on the earth, he is able to manipulate his own environment. Sometimes he exploits the nature so severely that many life forms, beneficial to his survival are destroyed quite unknowingly. This disturbs the balance of the natural ecosystem and man faces odd consequences. For example, modern techniques of Agriculture and industrial development have brought adverse impact on ecosystem.

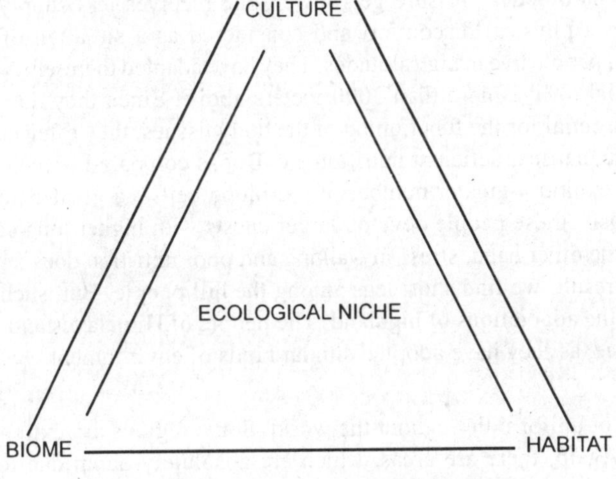

Fig. 10.7. Man in the Ecosystem

ADAPTATION

Relation between ecology and adaptation is very close. The ability of the living organisms to survive in a particular ecological set up is called adaptation. It has been possible for the possession of certain physiological, bio-chemical, genetic and behavioural characteristics. Since an ecosystem changes with time and space, organisms have to adapt themselves with the new environment. Various species of the animal kingdom show the evidences of adaptation. Human are not the exceptions. According to Darwin, evolution itself is based on adaptive selection. The organisms, which can adapt their environment in a better way, are permitted by nature to survive. Therefore, the less-adapted organisms are eliminated readily through death, before the attainment of reproductive age. Thus, adaptation is regarded as a process of modification in the structure and function of an organism for which it can survive and reproduce in a changing ecosystem.

For the alteration of natural environment, not only man, all organisms face a crisis and so they undergo certain changes in their characteristics. Adaptation is thus a process of adjustment, chiefly biological in nature. It takes place under climatic variation. The animals, which fail to adapt, become extinct. Since man is a higher intellectual animal, he has devised a number of ways to minimize the effect of environmental lash. His learned behaviour, culture helps him cope up any sort of adverse climate at any place of the world. In fact, culture represents an intimate adjustment of people in a given environment. Each human culture is a unique one that shows an adaptation of a group of individuals to their local environment. Culture acts as a protective sheath between man and nature; it is a reaction against environment with an inventive brain.

Different types of climatic zone prevail in this world. People inhabit in all sort of extremities. Though they try to combat the situations by means of cultural innovations, still certain biological variation is manifested among the individuals. Environment affects on some biological characteristics for which we find distinguishable differences among the populations, despite they bear major biological characteristics in common. For example, variation in skin colour, hair colour, body build etc., can easily be pointed out as the traits of adaptation.

Environment differs largely for the variation of geo-morphology of the earth. Therefore, places show differences in altitude, temperature, humidity, nutritional constituents and so on. Man have penetrated in all these zones and survived without much difficulty.

High Attitude

The percentage of oxygen in the air is usually twenty-one (21%). It does not change in high altitudes. But as the barometric pressure gets lower, one receives less supply of oxygen in each breath. This leads to a physical discomfort and considered as a situation of oxygen deficiency (hypoxia). Millions of people live in high altitudes. They have adapted themselves in that environment. People are even found to live more than 3000 meters above. Since they do not get the required amount of oxygen essential for the functioning of the body tissues, they often suffer from disturbed sleep, loss of appetite, nausea, deficient nutrition etc. But as compared to the coastal dwellers, the high altitude people exhibit a greater number of red-blood cells—a greater potential for carrying oxygen. Simultaneously, these people develop larger chests with higher lung capacity in course of rapid breathing. On the other hand, stress in walking and poor nutrition does not allow their legs to grow properly. As a result, we find short legs among the hill people. But, such traits are not at all genetical. These are the adaptations of highland. The people of Himalayas and the Andes are alike in physical appearance, as they have adopted similar kinds of environment.

Temperature

Temperature is not uniform throughout the world. Some regions are extremely cold and some are intolerably hot. Again, there are areas, which are absolutely damp due to continuous heavy rainfall. Men have adjusted themselves with all these critical conditions. The body of man has been built in such a way that it can retain an internal temperature about 37 ° C by regulating the atmospheric heat. Therefore, transfer of heat to the body or dissipation of heat from the body is possible according to the situation. For instance, heat is preserved in the body during winter and especially in Polar Regions, on high mountains and in temperate zones. But in hot tropical areas, sweating mechanism keeps the body cool. The skin not only acts as a protective layer: its colour varies with environment. The brownish black pigment of the skin called melanin is found to be present in varying degrees among all populations of the world. More melanin percentage in the skin makes the skin darker. The warm and humid climates usually produce darker skin. It is also an adaptive trait of human body. Because, melanin protects the sensitive inner layers of the skin against the damaging effects of ultra-

Because, melanin protects the sensitive inner layers of the skin against the damaging effects of ultra-violet rays of sun. Since ultraviolet radiation is much more intensive in tropical and sub-tropical areas and also in deserts, people of these areas grow much melanin in their skin. Dark-skinned people in sunny areas are generally saved from sunburn and skin cancer. They remain also resistant to tropical diseases as well. Similarly, the light-skinned people in cold climate show less susceptibility to frostbite than their dark-skinned brothers.

The body-build, height and facial construction, may also be affected by the environment. Loss of heat as well as preservation of heat take place in terms of body surface, which in turn can be equated to the height and weight of an individual. Short and thin individuals are well suited for the warmer parts of the geographical range while the tall and robust persons are quite fit in cooler areas. Although the variation in relation to human face is not much clear to the anthropologists, still it may be noted that the people inhabiting in humid tropical regions tend to show broad, short and flat noses. Negroes are the best examples who poss ss the platyrrhine noses. People inhabiting in low humid areas, irrespective of cold or hot environment, show long and thin noses. In cold regions noses have become narrowest as found among the Arctic Eskimos.

Availability of Nutrients

Survival of man depends on different natural constituents like water, mineral substances, carbo-hydrates, vitamins etc. For maintaining a good health and for proper functioning of various organs of the body, each individual needs a balanced diet where nutrients are proportionately included. But no food is uniformly available throughout the world. Moreover, different food contains different types of nutrients in them. So the nutritional level of people differ with the chances of getting the types of food. For instance,the Eskimos of the Arctic region depends solely on animal protein and fat. They get little or no plant food. On the contrary, in tropical zones, the plant foods are abounding. In the same way, while the temperate regions exhibit varieties of food, the people of desert areas suffer from acute scarcity of food. In fact, nutrition exerts a powerful effect on the rate of growth. Under-nutrition results in slow skeletal growth and delays the sexual maturation. Although, nutri-tional requirements vary with the factors like age, sex, body-size, workload, health condition etc, but in all cases human body possesses the power to adapt with environment. Availability of nutrients in an area is reflected in the stature, weight and general health condition of the people but it does not hinder the process of survival. Human body is found to develop certain automatic mechanism by which it fits itself to any environment prevailing on the earth.

E. Chapple and C. Coon in their book, *Principles of Antnropology* (1942) classified the environ-ment of this earth into eight major zones. They followed the work of a renowned geographer, Preston James.

1. *Dry Land (Desert)*

Such an area shows very little rainfall, ranging between zero to fifteen inches. Naturally vegeta-tion and animal life does not get opportunities to be flourished. Water sources are scanty and irregu-larly distributed. Only a few oases serve the people and animals there. Sometimes rivers originating in mountain areas may provide water to some parts of these deserts and facilitate sedentary garden-ing. The Pueblo Indians have adapted the said environment. Beside, in Australia, Africa and Asia many tribes are found to live under the dry conditions of desert.

2. *Tropical Rain Forest*

This ecological zone shows so heavy rainfall that deep forest develops. Sunlight can not pen-etrate the ground. At the same time, temperature remains very high for which growth of animal life as well as human life has been difficult. A less severity is found in semi-deciduous rain forests where seasonal rain helps the trees to shed their leaves annually. However, we find Amazonian Indians, Congo Negroes, Melanesians and Indonesians to cope up with such an environment. All along the equator, rain forests are found to grow.

3. *Mediterranean Scrub Forest*

The Scrub forest develops in the places where a dry hot summer persists with a mild rainy winter. The situation usually gives rise to a Temperate Zone adjacent to a desert or a sea, but remains shut off by mountains as like the case of California. Broad-leaf trees make this forest rich in small games. The environment is congenial for the growth of human population. Coastal and mountain Indians of California provide the examples.

4. *Temperate Forest*

Temperate forest zones receive fair amount of rain-water. Both the deciduous and evergreen trees are found in this forest. Though natural food plants are limited to this forest, but the games are abundant.

Winter and summer seasons are well demarcated. It presents an ideal type of environment where the primitive existence has been fostered. Vast areas of temperate forest are found in North America and Europe.

5. *Polar Desert and Tundra*

The two extremities of the earth, Arctic and Antarctic regions remain almost ice-capped throughout the year. Naturally these areas are left barren; bushes and sedge grasses are found only in some places. Apparently such places seem to be uninhabitable but in reality about fifty thousands of people live here. For example, the Eskimos, the Chukchis, the Yukaghirs have survived well in the environment of Siberia. The Onas and Yahagans are happily living in Tierra del Fuego.

6. *Boreal Forest*

In the North, the sub-arctic region presents long winters but short summers. Unchecked growth of vegetation has created the forest. Broadleaf trees are rare, rather conifers are abundant. Animals are all furry, suited to cold climate. Many primitive people, namely, Naskapi, Algoaquian, etc. have adapted themselves to this rugged environment.

7. *Grass Land*

In lower latitude, some tropical grassland is found, known as Savanna. In temperate regions also quasi-arid grasslands are visible like Prairies and Steppes. These areas are favourable for the distribution of various types of games. Therefore, such areas embrace a lot of primitive peoples-nomadic hunters as well as pastoralists. Even a few gardener communities are found at the river bottoms. Kirghizes of Turkestan, Mongols of Asia, Masais of East Africa and Plain Indians of North America find these grasslands quite suitable for their livelihood.

8. *Mountain Area*

Mountain areas, except the boreal and polar regions, offer a lot of variant possibilities. Therefore, inspite of difficulties in transportation and communication, such places have attracted a good number of people who gradually adjusted themselves in high altitudes. Incas of Peru, Bhutias and Lepchas of Himalayas etc. are the examples.

STRUGGLE FOR SURVIVAL

Food-gathering

The cravings of hungry stomach initiated man in food-gathering activities. Acquisition of food was an absolute necessity for man to maintain his bodily functions. But man was ill-equipped by nature having no specialized corporeal organ to procure food or fight for survival. Initially he had to depend on the natural environment like other mammalian species. Soon, by dint of higher brain capacity he tried to control his environment. This was purely struggle in the way towards survival. Ultimately man succeeded to cope up the varied ecological settings by making tools. The tools were not only the implements of his response or adjustment to the environment; these also facilitated man's emancipation from the simian way of life and gave birth to culture. Man became successful in exploiting the nature.

In the annals of mankind, the simplest as well as the most primitive culture was predominantly based on the economy of hunting and collecting. Early men used to lead a nomadic life. It was not a matter of choice but a compulsion to them. Their open habitats were usually located at broad river valleys—in terraces or in loessic stations where plenty of natural resources (plants and animals) and raw materials for making tools were available. They preferred flat and open lands as those gave them a broadly open view of their natural enemies. Such topography also helped in free movement. The proximity of water and forest was again important to them. Raw materials for tools were amply provided by the river-borne gravels and pebbles or by the conglomerates scattered along the alluvial tracts. Early men were also found to live in the caves and rock-shelters, especially in cold and rainy phases, when such places were close at hand.

For collecting and gathering of nuts, berries, roots, fruits and tubers practically no tools was required. Bare hands were sufficient in plucking and pulling. Small insects, birds and rodents, and also different kinds of eggs used to be consumed at this time. Those early food-gatherers were not capable to live in all types of environment They preferred the places where resources were bountiful. However, in all circumstances, the food-gatherers used to move in bands. The bands usually consisted of a number of people, cooperative in nature. The society was essentially promiscuous and the population density was very low. Personal belongings were very limited anc perhaps not recognized at all.

The food-gatherers gradually learnt the technique of hunting and became predatory carnivores in their subsistence habits. Although we are unable to pin point the time, when or how the meat-eating habit first took hold in man, but it is one of the most important trait that distinguishes man from his vegetarian anthropoidal relatives. The omnivore man adjusted in all environments by eating everything that was edible. Even he captured the most poisonous plants and animals, eliminating the poison. Like anthropoid apes, man was also inclined to spend his whole time by eating. But in course of time, culture had taught him the lesson to regulate his habits of eating.

About ten thousand years ago all groups of human beings used to lead their life by hunting and gathering. Some of them also took fishing as a subsidiary occupation. Perhaps the first-division of labour between the sexes generated at this stage; the male became the hunter and the female remained as food-gatherer. Many peoples still practise this kind of living. For example, the Australian aborigines, Ihc Eskimos, the Pygmies, the Andaman Islanders, the Onas of Tiarra del Feugo and so on. But number of such communities is very few in present day situations who maintain small population and lead a nomadic life without permanent habitations. These peoples are found to be equipped with minimum number of tools, light in weight.

In almost all hunting situations men have relied upon tools. Because it is extremely difficult for man to let down the enemies or preys without the aid of tool. Throughout the Quaternary period men had to measure their strength against terrible adversities, no matter whether they fought for subsistence or tried defend the life. It is certain that the crude flint weapons alone could not help men to win over the rhinoceros, bisons and other great herbivores on whose flesh they lived. Along with the club, spears, axes and knives, they made different types of snares and traps. The deadfalls and pitfalls stood very effective in capturing the wild ferocious animals. Men also developed the formidable devise of using poison at the tip of the spears. Picks, javelins and spears made of flints, bones and horns were the principal weapons of primitive men in prehistoric days. The hunting techniques differed according to the nature of the games and the situations.

The bow and arrow did not appear until late Palaeolithic or early Neolithic times. It signified a great improvement in the technique of assault, as the projectiles could be thrown from a distance of 400 to 500 meters in order to hit the enemy or beast without giving it a previous alarm. Therefore, at this stage men became able to fight against a lion or a bear with less risk of life. Prehistorians are not sure about the place where bow and arrow was actually invented. Probably, it was invented somewhere the Pacific, in Africa and even in the Americas. The only exception is the Australia where bow and arrow had never reached. The earliest remain of this weapon has been recovered from the Neolithic deposit of Spain about ten thousand years ago, although the evidences of equal type also come from upper Palaeolithic level.

Man, as a hunter certainly did not neglect the abundant fishes in lakes and rivers of that time. But unfortunately no evidence is left to inform us about the methods of early fishing. The artefacts, which were found in association with the skeletal remains of earliest man, do not provide any clue of fishing. Perhaps the practice of fishing appeared a little late when we find the traces of fish-hooks, fish-lines harpoon etc. In certain parts of world, especially near lakes, streams and along the sea-coast, fishing became the principal means of livelihood. Kitchen Midden cultures furnish the proofs. Tiny geometric tools, microliths seem to have been made for fishing. Microliths fitted along the shafts made of bone, horn or ivory served beautifully the purpose of fishing; Harpoons were the most commonly used effective weapon throughout the prehistoric world.

Nets are considered as the most advanced device of fishing. Bas-reliefs as left by the Egyptians of early dynasties show numerous representations on the scenes of fishing where people are found to be engaged in fishing with nets in the river Nile or in the lateral marshes of the Nile Valley. Nets appeared with Neolithic industry; especially those have been recovered from the lake surroundings. Pieces of lightwood possibly served as floats while pierced pebbles or large holed discs of baked clay stood as sinkers. In addition to these nets, a number offish debris have been discovered from Neolithic and pre-Neolithic deposit of Egypt. Similarly Mesolithic deposits of Europe exhibit a large number of fish bones and broken shells. All these suggest that the prehistoric men explored both land and water in order to procure their food. They struggled a lot to sustain their existence.

It is striking that, in the process of evolution, man had grown no specialized bodily equipment suited to any particular mode of life. But he acquired an anatomical privilege and higher intelligence for making extra-bodily equipment—the tool. The ability of making tools not only made the man a social animal, but the most adaptable of all creatures. The antiquity of man is ascertained from the evidence of earliest tools in the known geological strata. Although there is considerable difficulty in identifying the first attempt of making tools, it can safely be concluded that the standardization began with the crude stone hand-axes found in the earlier part of the Pleistocene epoch. It was a long tradition * *of* slowly acquired skill. The ancestors of man who used to live in the Pliocene or Miocene epoch might have used the tools, but they were not possibly the systematic toolmakers. The systematic tool making originated with the coordination of body and mind. Following the perception of tool-making, other cultural activities like making of fire, construction of dwellings and wearing of clothes appeared one by one and these enabled man to meet the challenges of environment and extend his range into all sorts of climatic zone.

The two great discoveries—the making of tools and the use of fire are the achievements of Lower Palaeolithic cultures with which we associate the Pithecanthropines i.e. group forming with Homo erectus, representing the second stage in Hominid evolution.

Tool-making

Prehistoric man fashioned his first tool from stone. The most ancient products of human industry were made of chipped and flaked stone. The common materials *were flint, quartzite, siliceous rock* and quartz. Early men used them according to their availability in a region. For example, in Europe, especially in England and France, all implements have been made on flint. Although normally flint occurs in the chalk deposit, in Palaeolithic and Mesolithic period man obtained this material either from the banks and beds of rivers or from cliffs and sea-beaches. In most of the places, the people preferred flint because of its natural qualities. On one hand, it splits beautifully to produce sharp edged flakes. On the other hand, the material is so resistant that it can only be

*Tradition denotes a set of habits for making one kind of tools for some particular job.

to produce sharp edged flakes. On the other hand, the material is so resistant that it can only be blunted by heavy blow against a hard surface. Therefore, both flake-tools and chipping instruments could be made out of flint. But the places where Archean and Crystalline rocks predominate, flints remain absent. Early men in those places were compelled to choose siliceous rock, quartzite, chert, quartz etc. which showed similar or greater hardness than the flint. But as the said rocks often splitted awkwardly, people began to avoid those and went in search of other rock. Materials like Jade and Obsidian were found out but none of them was suitable. The Jade was extremely hard that the people faced a great trouble in chipping while Obsidian was too fragile to work with. In some regions of the world viz. Mexico, Japan, the Greek Islands of the Mediterranean, Transcaucasia and Armenia, the shiny volcanic rock was found useful, particularly for the production of blades in Neolithic times.

There were various ways to convert a stone into a tool or weapon. The earliest way was to break a lump of rock in halves; the uneven sharp edges were very effective for the purpose of assault. Later, specialization appeared in the production of stone tools. The lump of a stone, whether a pebble, nodule or an angular fragment, required direct blows on its flat surface to detach the flakes. After removing a few flakes from a lump, the lump was found to be used an implement which we call *core tool.* The flakes, which were left as waste products, could again be made into implements after further trimming. This type of secondary work is designated as dressing or retouch and the tools are called flake *tool:* Both the *core and flake* gave rise to certain definite characteristics as a consequence of intentional striking by men. The flat surface of the core where blow was delivered by means of a simple pebble or a hammer stone is known as *striking platform.* Each chip, which was flaked off, shows a swelling on its ventral surface. Just below *the point of impact* (where the blow was struck). The protuberance is called the *bulb of percussion* (positive) which is visible in all flakes detached from the hard rocks. A corresponding cavity on the core is called *negative bulb of percussion.* The 'positive bulb of percussion' is essentially followed by a number of low concentric ripples, termed *as flake-scars.* The parent nodule i.e. the core also shows corresponding hollow ripples. Besides, a large flake detached by a sharp blow gives rise a small scar near the center of percussion, which is called as 'bulbar scar'. The French expression of 'bulbar scar' is *'eraillure'.* It may also be mentioned here that some *'shutter marks '* often diverge from the 'point of impact' as splitting lines. They arise during the process of removal of the flakes. These 'shutter marks' are also known as splits and fissures and remain perpendicular to the *ripple marks.* The ripple marks are nothing but a series of concentric rings around the point of force, which has already been applied at the 'point of impact'. They can be comparable to the ripples in a pond caused by the dropping of a stone in the water (Fig. 10.8).

Technology

From the previous discussion, it is clear that the flaking was essential for any kind of stone work especially in the Old Stone Age. There are various methods of flaking, probably each of them have been used at same time or other, during the Stone Age. They also show a technological development rather then chronological sequences. However, the methods can be classified in the following way:

(*a*) *Direct percussion method*

Flaking by direct blows is called the 'Direct percussion'. It was the most common method of early Stone Age. It is purely *block-on-block* devise. Two different techniques have been distinguished *here–Anvil-Stone technique* and *Stone-hammer technique.*

(*i*) Anvil-stone technique

At the inception of the technology some massive tools were produced by hitting the core against a large, fixed stone called anvil. The technique is called Anvil-stone technique. In the process of working, the anvil stone is to be rested on a block or held against the knee. Sometimes it is held directly by hand. Each blow is delivered obliquely, downwards from the striking platform. As a result, flakes come out from the lower surface of the core: they are usually short but deep and thick in nature. On the other hand, step-like bites are produced on the core, along the edge of the piece of

stone. In a slight modification of this technique, the core is suitably fixed on the anvil and a moving hammer hits the core almost vertically above the anvil. This technique has been called as 'bi-polar technique' *It seems to be widely practised among the Peking man.

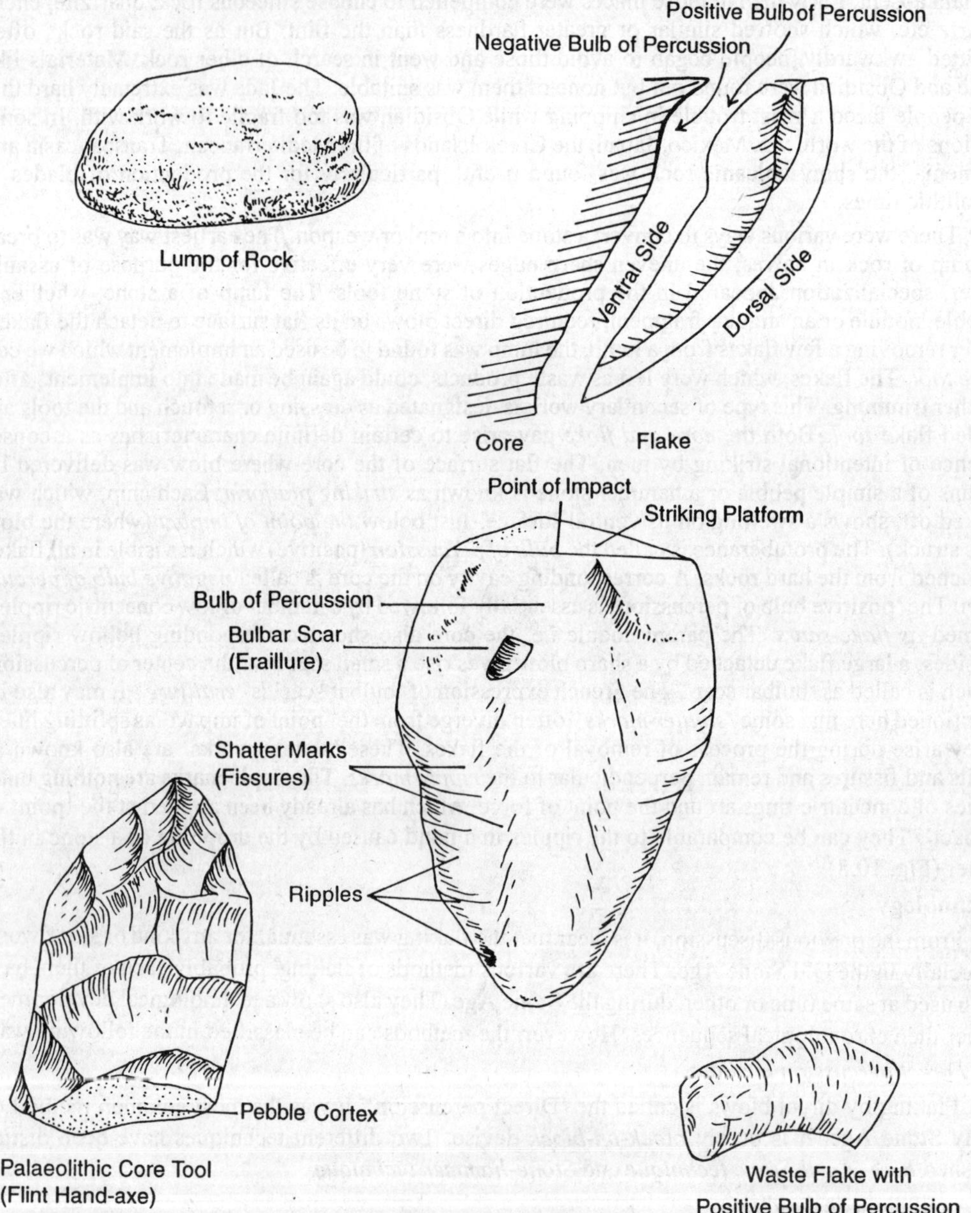

Fig. 10.8. Characteristic Marks in Core and Flake

(ii) Stone-hammer technique

The other technique is known as Stone-hammer technique. Here, the lump of stone which has to be flaked is held by hand. On the other hand, a pebble of suitable size is selected as a striker. The

***In bi-polar technique, flakes are detached without preparing the striking platform.**

lump is striken repeatedly at some particular point. The flakes that come out are short and thick; step-like bites are created along the edge of the core. Gradually different types of hammers came into use by the prehistoric toolmakers. The hammer of soft materials like thick antler roots, long bones, even the hard wood materials were capable of reducing the shattering effects. These hammers touched the core (stone core) at large number of points, so the nature of these scars were quite different from those scars produced by the effect of hammer-stone. In fact, soft-hammers produced the scars of more or less parallel ridges, which were uniformly shallow; the bulb of percussion was no longer conchoidal but diffused. This technique has been known as *Cylinder-hammer* or *Bone-hammer or Soft-hammer technique.*

(iii) Cylinder-hammer technique

The cylinder-hammer technique is obviously a more evolved form of the Stone-hammer technique. It is an effective as well as convenient way of producing smooth skimming flake-scars to make the surface of the tool smooth. The technique involves a powerful blow, especially with a skilful control of finger. The flake-scars with soft bulb of percussion can be produced only by cylinder-hammer technique. It may be mentioned that the early Palaeolithic flake industry, known as *Clactonian* exhibits the evidences of employing the Cylinder-hammer technique.

(b) *Indirect percussion method*

This method shows an indirect flaking technique with the help of an intermediate tool. This intermediate tool is somewhat like a modern chisel, which is placed on the particular point from where a flake is supposed to be detached. The technique requires more and more control of skilful fingers in comparison to 'Direct Percussion' method.

The lump of rock which is used here is fairly larger than that of a usual core. It is held on the left thigh or knee of the toolmaker. The tip of a punch (chisel-like tool) is placed on the chosen point of the core; the blow is given on the butt-end of the punch by means of a hammer. The flake comes out easily. Experiments have shown that the stone-hammers those were employed in the making of these tools, differed in size and weight. Large primary flakes could be produced by the use of a hammer-stone weighing about three pounds but for subsequent dressing, hammer-stones weighing about two or three ounces yielded better result. Therefore, it is clear that the prehistoric men must had kept a series of hammers (with variable weight) in their industry for better production of tools. Sometimes the cores used to be prepared before flaking. Calculated blows along its periphery produced a large number or flakes, similar in size and shape. This prepared core technique is considered marvelous as it involved a previous planning. *Levalloisian flakes* are invariably detached from a prepared core.

Since a punch has been used in producing these flakes, the technique is often referred as *'punching technique'*. It can also be called *'fluting technique'* because at the end of the process, a fluted core is left out. However, plenty of uniformly thin parallel-sided blades (flakes) are obtained in rapid succession.

(c) *Controlled percussion method*

Stone Age toolmakers heavily relied on ' Indirect percussion' method because it produced greater number of flakes and blades without much wastage. Controlled percussion means flaking with pressure which is much-developed device involving both patience and skill. Small flakes are removed from some definite points of the core with the help of a suitable implement (made of stone or bone) but the application of pressure is greatly regulated. The process requires a high-grade precision and so implements with fine cutting edge can be produced. Controlled percussion was especially applied for the trimming of the cutting edge of the tools. The Mousterians and the people of upper Palaeolithic period practised this type of flaking. Many finer stone tools of Neolithic period such as arrowhead, flint daggers, flint knife etc. show pressure flakings on their edge. Even some contemporary primitives are found to shape their stone tools by this technique.

(d) *Grinding and polishing method*

Although the method of Pressure flaking produced a developed form of cutting edge, sharp and refined, a smooth-faced cutting edge was further required for an axe-head or adze-head. Such an edge was proved effective in wood carving, especially for making of boats or dwellings. Therefore, to meet this demand, the method of grinding and polishing made its appearance. The grounded as well as polished axes, adzes and chisels brought revolution in tool making. In Mesolithic stage of culture, this new technology was invented but it flourished in Neolithic stage. This technique was not at all unknown in late Palaeolithic times because a large number of polished bone and ivory artifacts have been revealed from this stage. However, the use of polished stone as axe-heads and adze-heads became conspicuous during the Neolithic Age. Quite surprisingly, polished axe-heads made of stone are still popular among the native population of South Sea Islands.

The finer-grained igneous rock viz. *Basalt* and epidiorite (commonly known as green stone) was found more suitable for grinding and polishing than the flint. So far as the technique is concerned, a flake is to be taken out at first by percussion method. Next to that, the whole surface and especially the cutting edge has to be rubbed on a large slab of wet sandstone or some other hard rock like granite, quartzite etc. This huge stone is called rubbing stone. Sand may be used as an abrasive where the rock is not easily crumbled. A number of rubbing stones have been recovered from Aurignacian, Magdalenian and Azilian industries, but they only served the polishing of bone and ivory and thus helped in making of needles and pins. Stone polishing came in vogue with the Neolithic phase; axes, adzes, gouges, chisels were found to be polished with this technology. Many of these implements were perforated to receive the handles. The boring of the hole for hafting was done by the rotation of circular drill, generally hollow, and worked either by hand or by the aid of a string bow. Wet sand helped a lot in abrasion.

It may be stated that the prehistoric techniques of tool manufacture highlight the flakes. The firsthand chipped flakes were crude in shape and known *as primary flakes.* Subsequent workings were required on the primary flakes to get them as tools. Therefore, many *secondary* and *tertiary flakes* were removed from primary flakes. Thereafter, on the final shape of the specific tool, some minute flaking at the border were also needed to obtain a sharp edge. This type of flakings is termed as *retouching.* Before the invention of pressure-flaking technique, the cutting edge of the tools was made by *step flaking.* Palaeolithic toolmakers knew to control the force of stroke and so a step-like structure was produced at the place of cracking.

A detached piece of stone is always called 'flake' irrespective of its size. When a primary flake gives rise to several other flakes, the original flake gets the status of core in relation to the subsequent (flakes that have been removed later. Such a changing status of the flakes may create morphological confusion in the matter of definition. To solve this problem, we can go beyond the rule and call a piece as flake so long a bulb of percussion (positive) will be identified on it; the number of negative bulb of percussion will not be counted here at all (Fig. 10.9.)

Tool Type

On the basis of technology; stone tools can be classified into three principal types. This classification has been widely accepted by the scholars in this field.

Core tools

The lump of rock or pebble or nodule can be taken to a desired shape by flaking where the detached flakes are totally treated as waste products. So, the nodules or pebbles after elimination of flakes become the tools. Such type of tools is known as *core tool.* Core tools are of two type—single-edged core tools and bifacial core tools.

Single-edged Core tools

The single-edged core tools are often made on pebbles. The primary flakings are found either on one side (in case of chopper) or on both the sides (in case of chopping tools) of the cutting edge. Various kinds of chopper and chopping tools belong to this category.

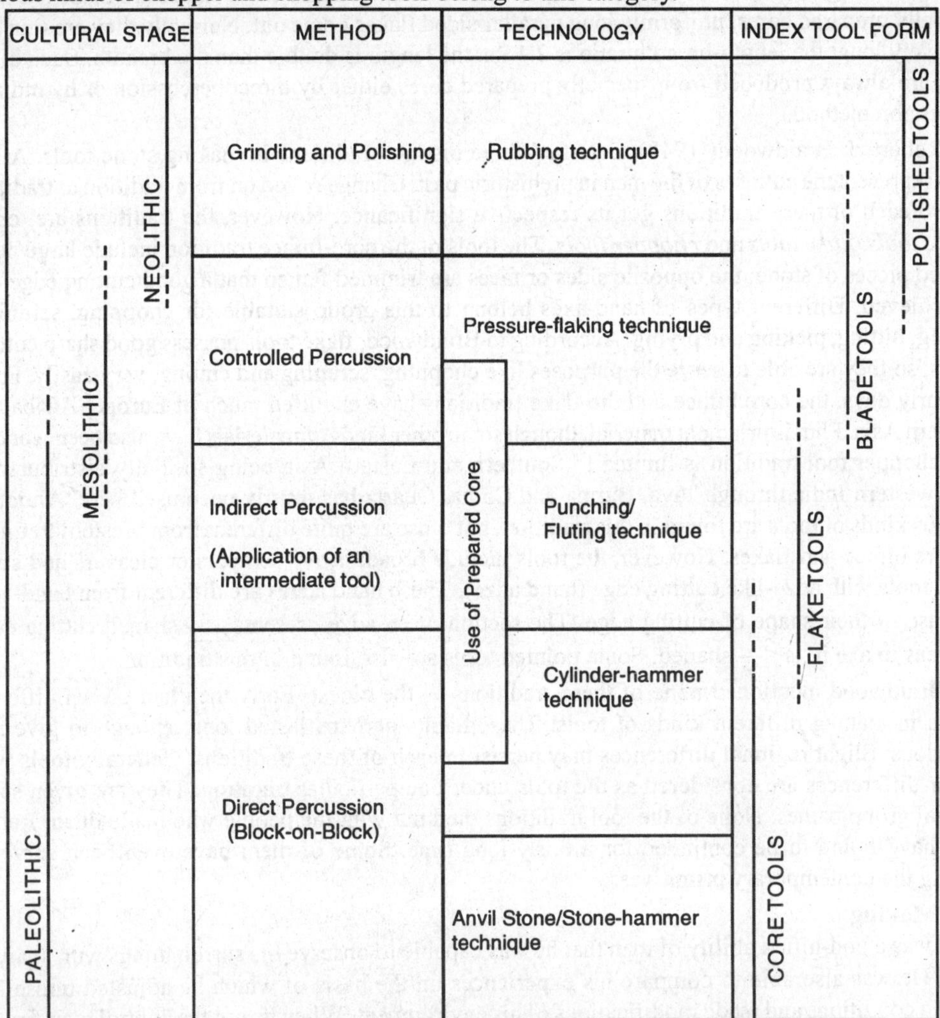

Fig. 10.9. Development of Stone Age Technology

Bifacial Core tools

Bifacial core tools show all round flaking around the periphery. They usually have two faces and a zigzag margin. This special zigzag margin or cutting edge is developed by the intersection of flake-scars that arise from both the faces of the nodule. The most commonly known bifacial core tool is hand-axe.

Flake tools

In case of core tool, the nodule becomes the tool while the flakes are found as left over. Alternatively, when these waste pieces of flakes are utilized as tools, they are called flake tools. The detached flakes may be used as tools with or without further trimming. They generally give rise to elongated or oval artifacts with a *plano-convex section.*

The core is the basic of all. It is the source of flake tools also. But the simple classification of core tool and flake tool may be complicated by the production of large flakes. Because the large flakes are mostly served as the core for further flaking.

Blade tools

On a first glimpse, blade tools are very well made flakes. Since these flakes are detached from specially prepared cores, uniformly long parallel-sided flakes come out. Normally they are thin and relatively long; the length-breadth ratio is 2:1 i.e. the length is double than the breadth. Such blade tools are always produced from specially prepared cores either by direct percussion or by indirect percussion method.

Robert. J. Braidwood(1948) identified three distinct traditions in making stone tools. All of them represent the cultures of the men in prehistoric past. Change rolled on from tradition to tradition and so each of these traditions got its respective significance. However, the traditions are, *core-biface tools, flake tools* and *chopper tools.* The tools of the core-biface tradition include large pear-shaped pieces of stone; the opposite sides or faces are trimmed flat so that a good cutting edge can be achieved. Different types of hand-axes belong to this group suitable for chopping, scraping, cutting, hitting, picking and prying. According to Braidwood, flake tools possess good sharp cutting edges, so they are able to serve the purposes like chopping, scraping and cutting, very easily. From the early days. the core biface and the flake traditions have engulfed much of Europe, Africa and Western Asia. Flint is principal material, though some other kinds of materials have also been worked. The chopper tool tradition is limited to Southern and Eastern Asia being specially distributed in northwestern India through Java, Burma and China. Quartz has mostly been used here. Although various kinds of tools are found in this tradition, but those are quite different from Western tradition of core biface and flakes. However, the tools include broad, heavy scrapers or cleavers and some other tools with adze-like cutting edge (hand adze). These hand adzes are different from hand-axes because of their shape of cutting edge. The section of an adze presents ∠ - shaped cutting edge whereas in axe it. is <— shaped. Some pointed tools are also found in this tradition.

Braidwood mentioned none of these traditions as the oldest. Early men had grown different habits in making different kinds of tools. These habits perhaps lasted long, enough to give rise traditions. Slight regional differences may persist in each of these traditions. Generally tools with minor differences are considered as the tools under one particular tradition. They are given some special group names. None of the tool traditions died out with the people who made them. Rather they have found to be continued for a fairly long time. Some of them have even been survived among the contemporary primitives.

Fire Making

It was a god-gifted ability of man that he was capable to observe his surroundings with analytic mind. He was also able to compare his experiences on the basis of which he adjusted himself to varying conditions and made modifications of his environment. When he got the control over fire, he dominated upon the environment and his cultural progress started. In fact, the using and making of fire provided man with special status and privilege; it raised man above the level of other animals.

Although we do not know exactly where and when the use of fire started, it is probable that man collected it first from the natural sources, out of necessity. Our ancestors definitely could have obtained much knowledge about fire from the incidences like volcanic eruptions or thunderstorms. When a volcano erupts, loud heavy noises come up from deep within the ground. The earth often trembles, too. Fire and smoke blow out from the top of the volcano. Small stones and great rocks leap into the air. Hot ash floats in the surroundings across the sky, which sometimes hides the sun. Rivers of fierce red melted stone, the lava, flow endlessly arising out from the cracks of the volcano. On the other hand, when the thunderstorm blows, continual flashes of lightning evoke terror. Therefore, it is likely that the nerves of the primitive men were not certainly so steady as to experiment with fire under those circumstances. Rather it is the grass-fire, which allured them to experiment with fire. In British East Africa, a vigorous grass-fire is still found which is unable to catch hold the usual riverside or gallery wood, so that the forest does not vanish altogether.

The forest at Anaimalai in South India is often set on fire by friction of bamboo branches. Some

Prehistorians have suggested that the first homeland of mankind was rich in bamboo plantation from where they got the idea of producing fire. During storm the branches often rubbed against one another and fire was created. Early men therefore dared to experiment with bamboo. The Negritoes of Zambales still make fire by rubbing one bamboo across a nick in another. It is probably the original method of getting fire; other ways developed with time.

Somewhere and sometime in lower Palaeolithic, man (Homo erectus) discovered the use of fire but he could not able to attain mastery over it. At this stage he only knew the way to sustain fire but felt greatly insecured before the uncontrolled fire. The controlled use of fire was found in Middle or Upper Palaeolithic i.e. during the reign of Homo Neanderthal, or Homo sapiens. The first suggestive evidence regarding the use of fire came from Africa (Kenya) which seems to be about 1.4 million years old. But the firm and conclusive evidences show a date nearly 500,000 years back; the places were China and Europe. The all rounded global evidences indicate that in late Pliocene, the climate of the world, even in Europe and England was so gentle and warm that the people earlier than Homo erectus (Australopithecine) did not feel like using of fire. In the other way, it can be said that the people of that time had no mental maturity to think about the matter. However, with the realization of the importance of fire, people started to make it according to the requirement. Early men utilized the fire in the following ways:

(i) Fire provided warmth in the chilled environment.

(ii) Fire helped in driving off the wild animals.

(iii) Dark corners of caves were lighted by fire. It also helped men to work in darkness.

(iv) Fire facilitated cooking, it added more taste to food.

(v) Fire was used in hardening the tip of the digging sticks. A sharp fire-hardened stick was equivalent to spear in efficiency.

Hearths of Acheulian time have been recovered from Palestine. But no meat-bone is found burnt. This evidence suggests that perhaps man started cooking a long period after he learnt to keep fire in his home i.e. the place of living. Cooking is directly related to man's ability of making fire. It is supposed that the Upper Palaeolithic people were accustomed to ignite dry fungus with the help of sparks; those sparks were perhaps produced by striking the iron pyrites against quartz or flint. Hearths of Neolithic people were made with the slabs of clay.

The control of fire was a major invention, which helped man to overcome his physical handicap. Many contemporary primitives who live in inaccessible higher altitudes inevitably require fire. They could not survive in the cold regions, especially during the hours of snowfall if they would not know the technique of making fire. The Eskimos, the Siberians and many other North Indian tribals belong to this group. Many advanced group of tribals still requires fire to efface out the darkness. But it is surprising to note that a few tribals like Andamanese and Tasmanians are ignorant of making fire. They borrow the fire from neighbouring tribes and carefully preserve the tended firebrands.

The technique of fire-making may have varied from group to group. The most widely used method is wood-friction. A pointed hardwood stick is usually taken and placed on a piece of dry softwood. The pointed end rests on the piece. In this condition the hard wood is twirled by hand and sparks come out by friction. Some tinder is kindled when the sparks are produced. Among some progressive primitive groups, a bow minimizes the labour of hand. Because, the bow helps to rotate the stick more efficiently as found among the ancient Egyptians and the modern Eskimos. The instrument is known as Bow-drill.

Fire is worshipped in almost all primitives. It is an old practice that can be traced from the beginning of human culture. Since fire appeared as a strange as well as mysterious phenomenon to the early men, they dealt it with awe and reverence. Although we do not know the exact time when fire began to be worshipped but the idea of sacred fire has definitely been associated with the early development of culture. Without fire and tools, man could not have adjusted himself to the changing environment and survived in the struggle for existence.

STONE AGE CULTURE

The study of prehistory is confined to the studies of the climate, geography and specially the cultures, in association with the fauna and flora of the periods. The periods are equally important as they unfold the past history of earth when the ape like man, slowly and gradually, evolved into modern man. The physical environments, especially the climate and geographical condition, have always played an important role in determining the course of evolution where man's ancestors slowly proceeded to acquire a human status. The Pleistocene epoch of Quaternary period witnessed a profound environmental change, which made the biological and cultural evolution of man possible. Although there were a series of climatic fluctuations since the earth was formed, but their magnitude increased a lot during the Pleistocene epoch that resulted into great human evolution and dynamic cultural progression.

The climatic fluctuations of Pleistocene epoch occurred in succession. They were severe as well as long in duration. So major shifts were found in animal and plant kingdom; disposition of land and water was greatly altered. This brought remarkable changes in human migration and settlement pattern. The temperature of earth's surface was lowered in general. That increased the precipitation of moisture in atmosphere either in the form of rain or snow depending on the geographical position. As a consequence, in some areas of earth ice-sheets or snowfields were created *; rivers and lakes were enlarged in other areas. This means that when the ice-sheets were found to be expanded into the temperate zones, the rain belt encroached on the sub-tropical zones. The tropical zones were greatly contracted with the increase of temperature, when the melted water from the snow-fields enlarged the size of the rivers and streams; the regions which had no ice-sheet or snow-field experienced the gradual drying up of the river and lakes and subsequently an arid condition was created. Such climatic oscillations - increase and decrease of the temperature had a direct impact on the lives of plants and animals. Different parts of the world got affected in varying ways in accordance with the nature of the geographical zones. Therefore the entire world of plant and animal had to adjust themselves with the climatic swings. Numerous regional factors also contributed to it. As a result, the overall picture became complicated and difficult for interpretation.

The epoch Pleistocene gave rise to 'glacial periods' in temperate zones of the earth, such as. North and Central Europe (in broad sense) and 'pluvial periods' in tropical and sub-tropical zones, such as, the greater part of the African Continent (in broad sense). There were intervening phases both within the glacial and pluvial periods so that the swing of climatic temperature could be manifested. The first frozen phase is called 'Gunz glaciation' in Europe which roughly corresponds to the 'Kageran pluvial' phase of Africa. A major climatic change occurred immediately after this phase and this was a period of increased surface temperature that corresponded to a period of general decrease in precipitation. This period has been known as first Inter-glacial in Europe and first Inter-pluvial in Africa. Third climatic fluctuation was the second Glaciation, the Mindel glaciation in Europe and the second Pluvial, the Kamasian pluvial in Africa. The fourth climatic shift denotes the

* It is very natural that snow could not form except in high mountains.

second Inter-glacial in Europe and second inter-pluvial in Africa. The fifth climatic change was another return of lowering of the temperature resulting in the formation of huge snowfields. It is known as third Glaciation or Riss glaciation. Its counterpart in Africa, characterized by heavy rainfall is called the Kanjeran pluvial. The change was followed by sixth climatic swing of increased surface temperature on earth; the periods are known as third Inter-glacial and third Inter-pluvial respectively. The seventh climatic swing was another major change—Wurm glaciation in Europe and Gamblian pluvial in Africa. Finally, at the end of the seventh swing, a gradual increase of temperature has been

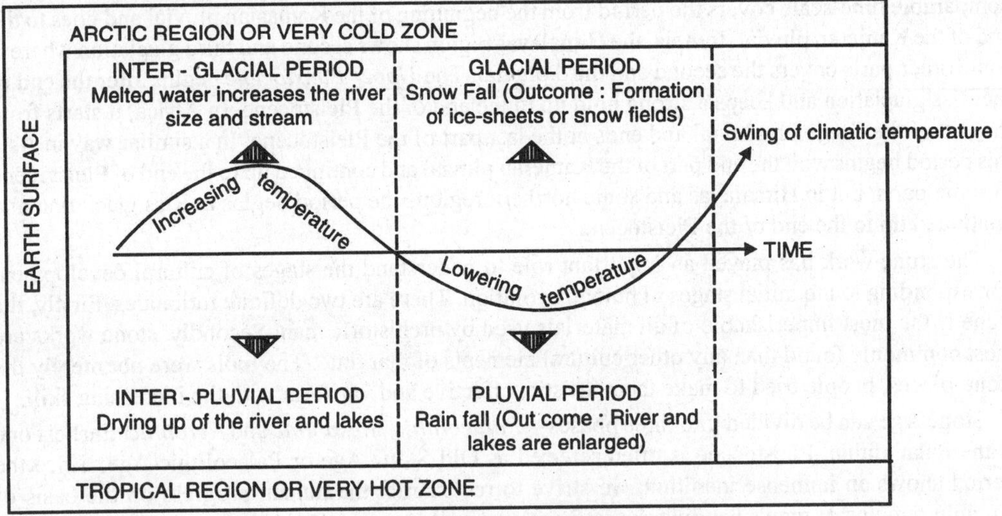

Fig. 11.1. Climatic Fluctuations in Pleistocene Epoch

noted. A relatively warmer and drier condition has engulfed the present-day climate. It may be considered as the eighth climatic cycle indicating the fourth Inter-glacial and fourth Inter-pluvial period. However, in the context of world prehistory, this post-glacial /post-pluvial Neo-thermal period has been referred as the Recent or Holocene period.

The Pleistocene influenced tremendously over the condition of human life. Though we know very little about the cultural inception, it is the very period when the tool users became the toolmakers. The appearance of man-apes, other developed forms towards the man, technological advancement of man—all happened during the Pleistocene epoch. About two million years ago, the beginning of Pleistocene was marked by an assemblage of new mammalian types which were essentially different from the earlier Pliocene types. They were the large mammals including one-toed horse, elephants with special teeth, and certain types of camels and cattle who showed some definite change of form in contrast to their parental genera. However, this animal assemblage has been designated as the 'Villafranchian fauna' and this fauna is a valuable marker at the boundary between the Pliocene and Pleistocene. Most of the animal forms that seemed to be dominant in Pliocene epoch were dropped out in Pleistocene for the sharp change of climate. They were bound to move towards South to avoid the chilled surroundings. The warmer phase though tried to bring these animals again in their old territories, but they failed to cope up with the new climate. Thus, the Villafranchian fauna gave way to the new species in the Middle of the Pleistocene period. This group of fossils who changed their forms is very important in estimating the age of fossil human remains after the Villafranchian time. But, in case of man-apes, it is especially difficult to place them under Villafranchian because they were able to adapt themselves with the climatic fluctuation by using tools. The Villafranchian fauna was restricted to Lower Pleistocene and the tool-users gradually became the toolmakers surpassing the environmental limitations.

The Pleistocene can be roughly divided into three periods, such as the Lower, Middle and Upper Pleistocene. The *Lower Pleistocene* may be taken as the beginning of the Pleistocene and it is continued to the first Inter-glacial period (Gunz-Mindel Inter-glacial) in Europe. In Africa and some parts of Asia, this period is comparable with the duration up to first Inter-pluvial period (Kageran-Kamasian Inter-pluvial). The Himalayan region of Asia has specially shown the final stage of first Inter-glacial period from the beginning of the Pleistocene. The *Middle Pleistocene* starts from the beginning of the Mindel glaciation and sustained up to the end of the Riss glaciation. In Africa, the comparable time-scale covers the period from the beginning of the Kamasian pluvial and goes to the end of the Kanjeran pluvial. In Asia, the Himalayas region shows second and third glaciation whereas some other parts covers the second and third pluvial. The *Upper Pleistocene* begins after the end of the Riss glaciation and keeps a strong hold up to the end of the Pleistocene. In Africa, it starts from the end of the Kanjeran pluvial and ends at the last part of the Pleistocene. In a similar way in Asia this period begins with the end part of the Kanjeran pluvial and continued up to the end of Pleistocene in some parts, but in Himalayas and some northern regions the period begins at Riss glaciation and continued up to the end of the Pleistocene.

The stone work has played an important role to understand the stages of cultural development corresponding to the initial stages of human evolution. There are two definite rationales. Firstly, the stone is the most imperishable of all materials used by prehistoric man. Secondly, stone works are most commonly found than any other cultural elements of that time. The tools were not merely the stone-pieces, people tried to make them artistic, effective and variform with the increasing skill.

Stone Age can be divided into three phases—*Palaeolithic, Mesolithic* and *Neolithic.* Earliest one is the Palaeolithic. Pleistocene is often referred as Old Stone Age or Palaeolithic Age. Since the period shows an immense inanition, we strive to reconstruct the human progress on the basis of tangible remains of man's handiwork that have survived in open-air sites or caves. Position of the sites also have thrown some light on their habitation pattern, though are not sufficient at all. However, in the study of prehistory some terms bear the special meanings; one should study them before reaching to the phase of final interpretation.

Artifact : An artifact is an object that may or may not be a tool. A tool is an implement, which is deliberately made and especially designed by the man for serving a particular need. But, an artifact is anything that shows human workmanship or modifications as a resultant of man's interaction with environment. In this sense, a core, unspecialized flakes and other waste materials, which come out in course of artificial fabrication, have to be regarded as artifact. Each tool is basically an artifact while an artifact is not necessarily a tool. A naturally obtained pebble can not be said an artifact until it is treated by man.

Site: In prehistory, the term site is applied to a specific space where the evidences of human activity have been scattered. The primary evidences are the tools and artefacts. Sites can be divided into three categories (1) *cave or rock-shelter;* (2) *open station and* (3) *settlement site.* The sites are generally revealed by the excavation.

1. Cave or rock-shelter type

This type of sites usually show a number of distinct layers in stratigraphical sequence.

2. Open station type

Such type of sites is particularly rich in showing findings 'in situ' i.e. in original position. They denote a particular cultural stratigraphy.

3. Settlement site

This type of sites are comparatively rare in occurrence. Artefacts are generally found only from the upper Part of excavated area.

An area is to be considered as a site when the entire space yields the artefacts; the artefacts are found in a larger and wider horizontal distribution without a gap.

Assemblage : The term refers a particular gathering or an entire collection of prehistoric artifacts in one occupation zone. The zone corresponds to a single prehistoric or archaeological level and may also be referred as the cultural level. An assemblage is therefore simply a collection of prehistoric artifacts in a distinct level of the site.

Industry: The word implies a manufacturing activity. So any set of artifacts evidently fashioned or used by a single human group of prehistoric days is called an 'industry'. In fact, the assemblage of artifacts of same age in a distinct level of a site is described as an industry of that particular site. Sometimes, a particular site may also be taken as the evidence of several industries. In this regard we can use the expressions like 'stone industry', 'bone industry', 'microlithic industry', etc.

Tradition : This term refers a long practice particularly in connection with technological specialization. If a particular group of tools is found to be continued through time, it may be taken as a tool tradition. Normally a tradition involves the progress as well as modifications on a given set of tool types. The terms like Acheulean, Levalloisian, Aurignacian, etc. indicate the names of different traditions.

Culture : Culture is a term of different expressions but in prehistory, culture means the tradition of a broad period. Actually, it denotes an assemblage of industries made by the people of a same group. Industries that are widely distributed but belong to same age may not be always identical, even when the same group of people is involved in it. But major types of tools remain common indicating a sort of Connection among them. The term 'culture' does not only include the industries but also other factors like art, burial custom, etc. from which we can deduct the life-style as well as minds of the people. In dealing with artifacts or industries we do not concern ourselves with the makers. But in case of culture, we always think about the relics, along with the people who have left those relics. In fact, the succession of human culture is often identified by means of tradition as evidenced by the material remains. For example, when we use the terms like 'Palaeolithic culture' or 'Neolithic culture', we want to grasp the people with their respective work.

Culture Complex : The term is very common and important in prehistoric literature. It means a group of traditions (progression or modification of a particular group of tool types through time) under a single culture as revealed from a cluster of sites within a limited geographical area. For example, the soan Culture-complex in India where a group of tradition has been manifested in a limited geographical area. (Fig. 11.2).

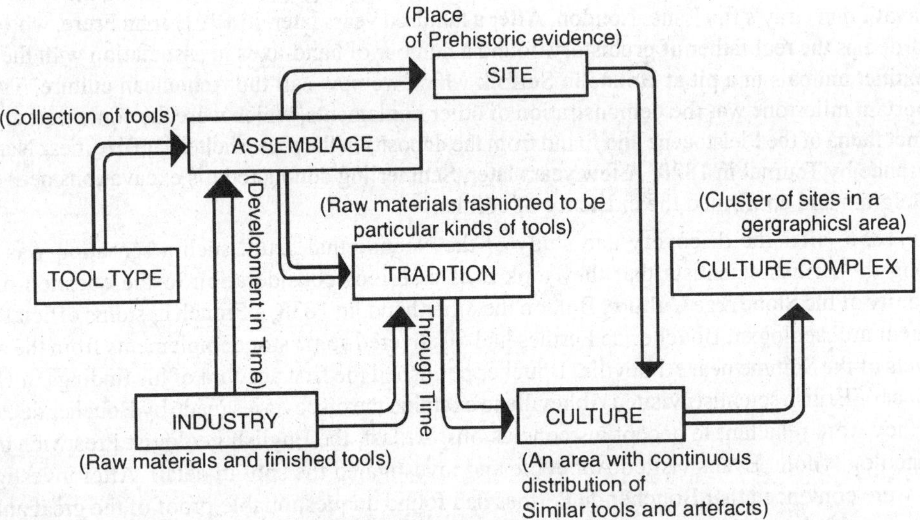

Fig. 11.2 Significance of Tools In Different Perspective

At the beginning of twentieth Century, some British authorities tried to prove the existence of a pre-Palaeolithic period. They called it Eolithic or the 'Dawn of Stone Age'. Evidences include some irregular flints, most of which are blunt and unsuitable for chopping, cutting or scraping. They show only small chips along the edges. However, these flint pieces have been called as Eoliths and considered to have belonged in pre-Pleistocene pre-glacial age, in the level of Pliocene or even before than that in the level of Miocene. But this separate Eolithic period has not been accepted by all prehistorians and there is much doubt about the Eoliths. Majority of the prehistorians argue that the minor chips that are found along the edges of the flints are not actually the signs of flaking by conscious intention, rather they had became chipped off through usage. The pre-human hands manipulated the naturally formed flakes as a tool.

THE DAWN OF THE STONE AGE

The problem in connection with the existence of Tertiary man was first brought forward by a notable French Prehistorian, Abbe' Bourgeois at the Congress at Paris in 1867. He had found a large number of chipped flints from the beds of Upper Oligocene age near the village of Thenay, south of Orleans at France. He claimed that these specimens were man-made. But most of the prehistorians were unable to accept them as the result of human handiwork. After ten years in 1877, Prof. M.Rames exhibited a good number of collections from the Miocene bed. Those specimens also did not get the acceptance of the prehistorians. Prof. M. Boule in 1909, declared these specimens purely as natural products and he was supported by Prof. Abbe' Breuil. Some British specialists grouped these specimens under the term 'eoliths' to refer the dawn of Stone Age. In Eolithic culture the stones have been slightly worked, probably by man. They are so slightly worked that it is very difficult to determine its certainty that whether these stones have been worked at all or not. The term eoliths has been originally coined from Greek word **'eos'** means the Greek goddess of dawn and **'lithos'** means stone, so eos+lithos=eoliths. The eoliths were neither found in association with any human skeleton, nor they showed any clear proof of human handiwork. Still, a good deal of evidence supports the belief that the toolmakers of Eolithic culture were at work in pre-Pleistocene or late Pliocene period. The emergence of eoliths may even happen in the early phase of the Pleistocene epoch.

However, most of the conclusions regarding the eoliths is entirely based on speculation. It requires more data from different aspects along with some evidences of human skeletons to draw a definite inference.

As per the record, the first stone tool was discovered as long ago as 1690 during the course of an excavation in Gray's Inn Lane, London. After a hundred years later, in 1791, John Frere, who is now regarded as the real father of prehistory, found a number of hand-axes in association with the bones of extinct animals in a pit at Hoxne in Suffolk which we now call the Acheulean culture. The next important milestone was the demonstration of other implements, which were the contemporary of an extinct fauna of the Pleistocene and found from the deposits in the rock-shelters at Bize, near Narbonne in France by Tournal in 1828. A few years later, Schmerling conducted his excavations near Lie'ge in Belgium and confirmed the existence of eoliths.

In 1856, after the discovery and study of the Neanderthal skull, such a sensation was grown among the British scientists that they took it as a serious consideration to the question of great antiquity of the Stone Age Culture. But, on the other hand, in 1838, a French customs official and an amateur archaeologist, Boucher de Perthes had discovered some stone implements from the ancient gravels of the Somme near Abbeville. Boucher published the first account of his findings in 1847. A number of British scientists visited Abbeville to examine the discoveries made by Boucher de Perthes. But they were reluctant to accept his conclusions. At last, the English geologist Prestwich and the archaeologist John Evans visited Abbeville and investigated the spot in detail. After investigation, they were convinced that Boucher de Perthes had found unquestionable proof of the great antiquity of man, and a paper was presented before the Royal Society in 1859.

In 1863, the first monograph was published on this subject and that was the famous book by Charles Lyell, the geologist, entitled *Geological Evidence of the Antiquity of Man.* This publication

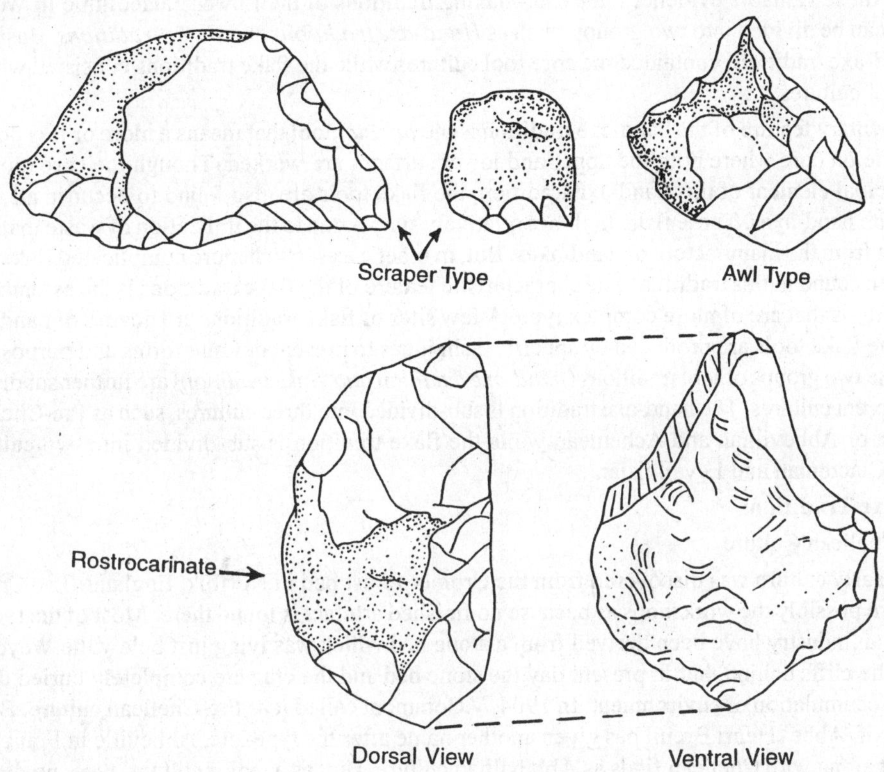

Fig. 11.3. Some Eoliths

was a strong step towards the study of prehistory. From that date, the study of the Stone Age Cultures got a new dimension, which made a quick succession based upon an examination of the stratigraphical evidence. If Eoliths have to be accepted as man-made implements, they must be placed in the Palaeolithic series—the earliest of all Palaeolithic tools.

PALAEOLITHIC

The first or the oldest prehistoric culture is known as Palaeolithic or the Old Stone Age. The term comes from the Greek word **'palaios'** means old and **'lithos'** means stone. Therefore, **palaios+lithos=Palaeolithic**. Although our knowledge regarding Palaeolithic is very meagre and imperfect, still Palaeolithic or old stone age is very important as it provides a clear cut sequence of cultural development throughout the entire Pleistocene period, all over the world. It is considered as a crucial period for all round human evolution; development of cultures can be traced out distinctively in this period. Palaeolithic can be further sub-divided into three phases—*Lower Palaeolithic, Middle Palaeolithic* and *Upper Palaeolithic.*

Lower Paleolithic

The time span of the Lower Palaeolithic was the maximum covering the whole of Lower Pleistocene and bulk of the Middle Pleistocene epoch. During this span many river valleys and terraces were formed. Early men preferred to live near the water supply, as the stone tools are found mainly in or adjacent to the river valleys. Evidence of the earliest stone tools in Western Europe has appeared from the deposits of first Inter-glacial phase in the Lower Pleistocene. Excellent stratigraphic sequences of entire Pleistocene epoch containing Lower Palaeolithic artifacts have been discovered from the Somme Valley in the north of France and the Thames Valley in the south of England. On the basis of those valuable evidences, the tool-making traditions of the Lower Palaeolithic in Western Europe can be divided into two groups, such as *Hand-axe traditions* and *Flake traditions.* Basically, the Hand-axe traditions contained the core tool cultures while the flake traditions consisted with the flake tool cultures.

Elementary feature of the hand-axe tradition is the *bifacial* tool that means a more or less pointed tool made on core where both the upper and lower surfaces are worked. Though the core tools are the principal element of this hand-axe tradition, the flake tools are also found to occur in all levels along with hand-axes. These flake tools are relatively simple due to the utilization of waste materials resulting from the manufacture of hand-axes. But, in other cases, much more complicated flake tools have been found in this tradition. The characteristic feature of the flake traditions is the assemblages of flake tools that are of more complex type. A few sites of flake traditions are devoid of hand-axes where the flake tools are produced by specific techniques to present definite forms and purposes. In fact, these two groups of tool traditions (*hand-axe tradition and flake tradition)* are further subdivided into different cultures. The hand-axe tradition is sub-divided into three cultures, such as Pre-Chellean, Chellean or Abbevillian and Acheulean while the flake tradition is sub-divided into two cultures, such as Clactonian and Levalloisian.

Hand-axe Traditions

1. Pre-Chellean Culture

This early culture was discovered from the Cromer forest bed in Norfolk, England. The Cromer sites were possibly the workshops as because no finished artefact is found there. Most of the tools of Cromerian industry have been derived from a stone bed which was lying just below the Weybourn crag in the cliffs behind, but in present day the stone bed and the crag are completely buried due to natural accumulation of environment. In 1904, V.Commont called it as Pre-Chellean culture. But, in 1929, Prof. Abbe' Henri Breuil has given another name after the type-site, Abbeville in France. He named it along with Chellean finds as Abbevillian culture. But, as a major cultural type, no definite feature could be ascribed for this industry.

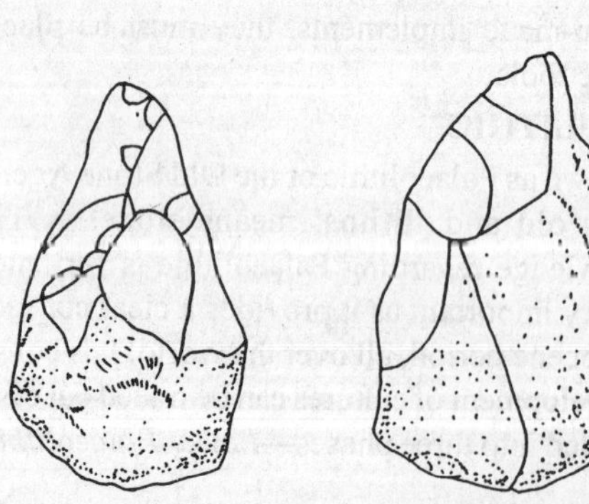

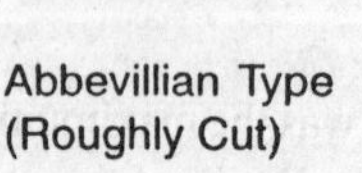

Abbevillian Type
(Roughly Cut)

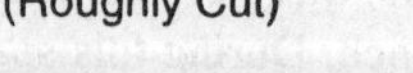

Fig. 11.4. Flint Nodules

Typologically, very crude type of hand-axes, including choppers, discs; scrapers, etc. have been collected as the chief findings. Occasionally more finished tools are found as rare specimens of core-tool type. Primary flakings are evident in these tools but no sign of secondary working has been observed, Essentially, the major findings were the flakes. Although the finished tools are found rarely, but the **pebbly cortex** is significantly present in all Pre-Chellean tools. The primary flakings have been worked out only at the working end. It is definitely an early form of core-tool culture of hand-axe tradition.

The geological age of this culture is the early Pleistocene epoch. During the first glacial period (Gunz), the culture flourished in Western Europe and Africa. Though, no definite representative group has yet been found in association with this industry, still some prehistorians feel that a group alike to Australopithecine might be responsible for this culture.

2. Chellean or Abbevillian culture

It is apparently the oldest tool-making tradition of the core-tool culture in Western Europe. In the gravel terraces of the Somme Valley, a large number of sites have been discovered. Previously this culture was named as the **Chellean**, after the site Chelles on Somme Valley in northern France. The Pleistocene deposits at Chelles are situated on a lower terrace. But in Abbeville the tools are found in situ at the higher terraces at Abbeville. The name **Abbevillian** takes it name from the site Abbeville on Somme Valley in northern France. According to Prof. Breuil, Abbevillian is the combination of Pre-Chellean and Chellean cultures. But, Prof. A.L. Kroebar considered both the Chellean and the Abbevillian as the same culture where the Chellean being the older name and the Abbevillian is more modern nomenclature.

It is mainly a bifacial core-tool culture. E. A. Hoebel, 1958, in his book *Man in the Primitive World*, defined the bifaces as core tools made from the remaining part or heart of a nodule of flint after surface flakes have been removed. The residue of a nodule is taken into a particular shape. The bifaces have greater breadth than thickness and they are tapering to thin edges around the circumference. In this way, two faces i.e. dorsal and ventral sides of the tools are well marked. The characteristic tools are mostly hand-axes or picks; crudest types are found in the Lower Palaeolithic period. German scholars call them 'Faustkeil' or 'fist wedge' while the French have coined the term 'Coup-de-poing', a blow of the fist or a punch. The forms are varied and flaking is generally irregular which produces sinuous cutting edges. In fact, these hand-axes, coarsely flaked with zig-zag margins are probably manufactured either with a stone-hammer or on a stone anvil. So, it is clear that Block-on-Block technique was employed for releasing irregular flakes to manufacture of hand-axes. For the first time, alternate flakings are marked in association with flake tools. But, these flake tools are very crude types with no existence of any definite tool forms. The majority of the flakes are the resultant from the manufacture of hand-axes. Pear-shaped, tongue-shaped, and oval-shaped hand-axes are common types in Abbevillian culture, which is slightly evolved form of the Pre-Chellean culture. The pebbly cortex is slightly present in Abbevillian hand-axes. Besides, discs, scrapers, choppers and also knifes on flakes are found in this culture. It can be accepted as the first tradition of the bifacial core-tools culture. (Fig. 11.5).

The geological age of this culture is the Lower Pleistocene epoch particularly during the time of Gunz-Mindel Inter-glacial (first Inter-glacial period). The distribution is chiefly observed in Western Europe, Africa and Western Asia. The representative population responsible for this culture was an allied group of Homo erectus. It signifies the second stage of Hominid evolution when group alike to early form of Pithecanthropines came into existence.

3. Acheulean culture —

This culture covers the longest time-span of tool-making tradition of the Palaeolithic period. A large number of tools have been discovered from both the Somme and the Thames Valley. The principal stratigraphic development of this culture has occurred in 30-meter deep middle terrace of the Somme Valley at St. Acheul, a suburb of Amiens in France. The tool types are found in the terrace gravel of the Somme. The oldest types relate to the early inter-glacial gravel of the 45-meter high at the terrace of the Valley. Some of the scholars have tried to include the lithic findings occurred

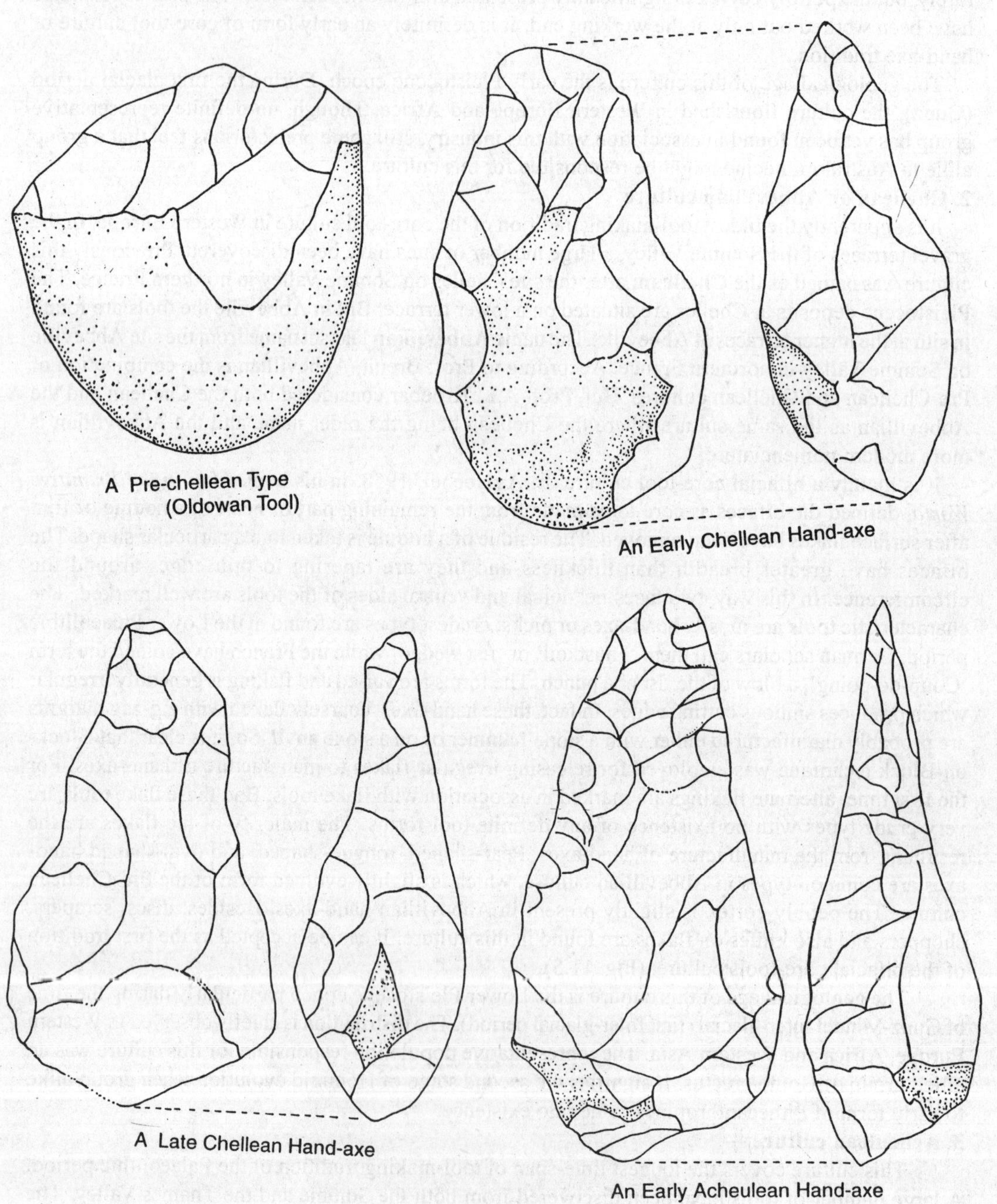

Fig. 11.5. Chell-Acheul Tools

at Suffolk, Swanscombe (Kent in England) in the Thames Valley, Torralba-Ambrona sites in Spain under the Acheulean culture. In general the Acheulean may be divided into Lower, Middle and Upper, though the actual sequence is more complex. In fact, the Acheulean culture is continued with bifacial core-tools and primarily focussed on the manufacture of hand-axes which means the first tradition *(hand-axe tradition)* of the bifacial core-tools culture.

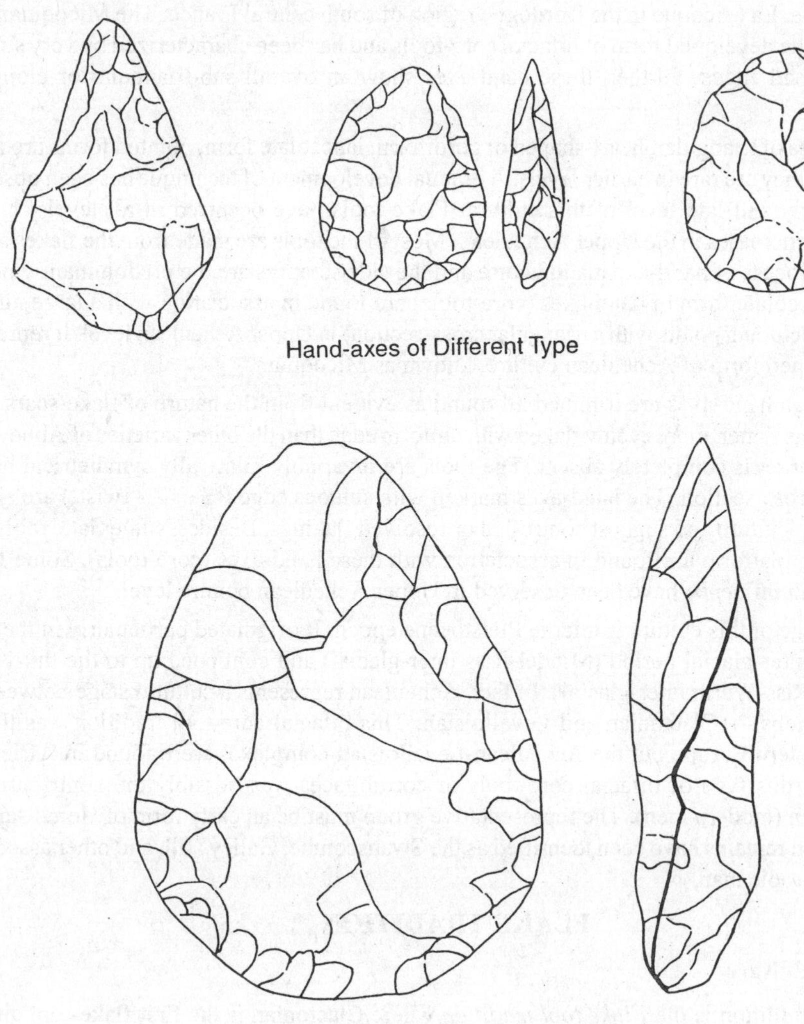

Fig.11.6. Acheulean Tool Types

Lower Acheulean—

This level includes a proportion of roughly made hand-axes, but there is a larger proportion of *ovate* forms of hand-axes than in other levels. Besides, cleavers, the bifacial core-tool with square, or slightly convex, sharp cutting edge at one extremity is found abundant. This type is comparatively rare but found in all Acheulean levels.

Middle Acheulean

This level is marked with *ovate* hand-axes where sinuous edge, known as the **'S-twist'**, stand as characteristic type. The apart from this, fossil human remains have been discovered in direct association with this cultural level.

Upper Acheulean

This is the final level of Acheulean and also known as the *Micoquian.* This name has been accepted after the type-site. La Micoque in the Dordogne region of south central France. The Micoquian type of hand-axes is the developed form of bifacial core-tools and has been characterized by very straight and finely chipped edges. Further, these hand-axes show an overall sub-triangular or elongated form.

The hand-axes of triangular, heart-shaped or cordiform, lanceolate form, pointed forms are found in all levels, but they are rare in earlier levels. A gradual development of technique has been observed between the early and late level of this culture. Flake tools have occurred in all levels, but the number of tools increases in the Upper Acheulean. Most of the tools are made from the flakes struck off during the process of hand-axe manufacture and the side scrapers are the predominant types. In general, the lanceolate form of hand-axes (core-tools) are found in association with a large number of flake-tools (including points with a triangular cross-section) in Upper Acheulean level. It represents the most developed form of Acheulean culture, known as Micoquian.

The Acheulean hand-axes are trimmed all round as evident from the nature of flake-scars. This has made the tools flatter, more evenly flaked with uniform edge than the other varieties of Abbevillian tools; pebbly cortex is completely absent. The tools are invariably bilaterally symmetrical having thin lenticular cross-section. The hand-axes marked with sinuous edge ('S or Z - twist') are typical to the Acheulean culture and suggest controlled or resolved flakings. Besides, some flake tools with facetted striking platform are found in association with these hand-axes (core-tools). Some flakes with Levalloisian influence have been observed in Upper Acheulean culture level.

Geological age of this culture is middle Pleistocene epoch. It originated particularly in the early part of second inter-glacial period (Mindel-Riss inter-glacial) and continued up to the third inter-glacial period (Riss-Wurm inter-glacial). In fact, Acheulean represents a cultural stage between the two flake traditions — Clactonian and Levalloisian. This bifacial core-tool tradition was though confined to Western Europe but the Acheulean-Levalloisian complexes were found in Africa and Asia. However, this type of bifacial core-tools or core-bifaces was possibly the contribution of Neanthropic men (modern men). The representative group must be an early form of Homo sapiens. The fossil human remains have been identified as the Swanscombe, Galley Hill, and other associated types of Neanthropic man.

FLAKE TRADITION

1. Clactonian Culture

The second tradition is the *Flake tool tradition* where Clactonian is the first flake-tool culture. The Clactonian is named after the stratigraphic position of the type-site at Clacton-on-Sea, Essex in England, which is most clearly shown at Swanscombe, Kent, in the Thames Valley. The Clactonian culture is mainly a flake-tradition though some core-tools are found in association with them. These core-tools are the nodules of flint, either alternately worked or flaked along the upper surface of one side as choppers. The Clactonian flakes are rough and struck out unsystematically from the prepared cores. It is suggested that most of these flakes are produced by striking the lump on the edge of an anvil. The manufacture of hand-axes produced some waste-flakes with Clactonian characteristics, but true Clactonian flakes came from the chopper-like cores where numerous flakes with wide, plain striking platforms and dressed edges were produced.

The Clactonian flakes generally exhibited large, massive, unfacetted striking platforms and prominent positive bulbs of percussion. The flake surface formed a wide angle, greater than 90° with the striking platform. This type of flake-tools are mostly crude, but in some cases, actual

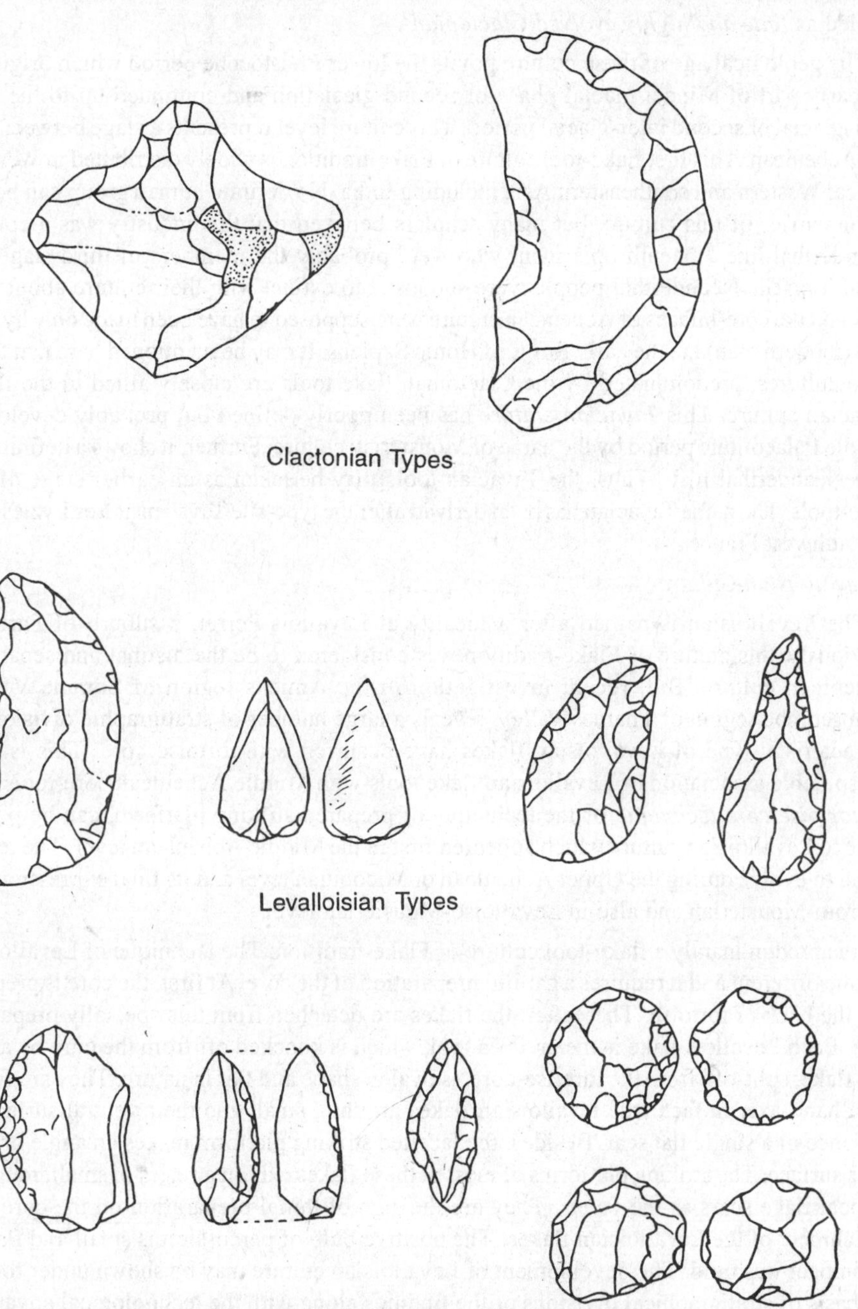

Fig. 11.7. Tools on Flake

retouching or secondary working on the edges of the flakes is also found. Small chopping tools, rough scrapers, discs, knives, blades on flint are the chief findings of this culture. Some flake-tools designed as bill-hook and a type of long flake with concave cutting edge are found here. According to M.C.Burkitt, finely trimmed side scrapers and pointed tools obtained from this stage should be labeled as *'late and highly evolved Clactonian'*.

The geological age of these culture points the lower Pleistocene period which originated during the early part of Mindel glacial phase or second glaciation and continued up to the Mindel-Riss inter-glacial or second inter-glacial period. This culture level represents a stage between Abbevillian and Acheulean. This first flake-tool culture of Flake-tradition is widely distributed in Western Europe, Africa, Western and southeastern Asia including India. No definite human group can be pointed out as the carrier of this culture, but many scholars believed that this industry was a contribution of Neanderthal-like Palaenthropic men, who were probably the members of third stage of hominid evolution. The Neanderthal people were supposed to extinct with their culture about 70000 years back and the core-bifaces of Acheulean culture were supposed to have been made only by Neanthropic man (modern man) i.e. the early forms of Homo Sapiens. It may be mentioned here that the European flake cultures, predominated by the Clactonian flake tools are closely allied to the flake-tools of Tayacian culture. This *Tayacian culture* has been poorly defined but probably developed into the Middle Palaeolithic period by the name of Mousterian culture. Further, it shows a definite association with Neanderthal man. Thus, the Tayacian tools may be taken as an earlier stage of Mousterian flake-tools. The name Tayacian has been derived after the type-site Tayac near Les Eyzies (Dordogne), in Southwest France

2. *Levalloisian Culture*

The Levalloisian is named after a locality at Levallois-Perret, a suburb of Paris in France. Previously, this culture of Flake-tradition was considered to be the distinct and separate from the Acheulean culture. But, recent investigation in the Amiens region of Somme Valley and the Swanscombe region of Thames Valley, reveals a large number of stratigraphic divisions where the characteristic type of Levalloisian flakes have occurred with tortoise core. This is a direct and indisputable association of Levalloisian flake tools with Middle Acheulean core-tools. In this way, the *tortoise core technique* or the technique of prepared striking platform, can be pointed out as basic to Levalloisian culture which appeared first in the Middle Acheulean level. The technique was found to evolve during the Upper Acheulean or Micoquian level and its final expression was arrived in Proto-Mousterian and also in Levalloiso-Mousterian level.

It is predominantly a flake-tool culture of Flake-tradition. The technique of Levalloisian culture is quite different and it requires a careful preparation of the core. At first, the core is prepared to look like the back of tortoise. Thereafter, the flakes are detached from this specially prepared *tortoise-core*. Each Levallois flake is meant for a tool, which is knocked off from the core by a direct blow. The flake, split-off from the tortoise core, is oval in shape and flat in nature. They are finely worked for a hand-axe. In fact, true levalloisian flakes are thin, small and their ventral surfaces show the evidence of a single flat scar. Besides, the facetted striking platform makes an angle of 90° with the flake surface. The striking platforms of most of these flakes exhibit a series of small, roughly parallel, vertical flake scars as the facets. They are the sign of initial preparation on the core before final detachment of the Levalloisian flakes. The positive bulb of percussion is small and flat as because the impact was mild. The development of Levalloisian culture may be shown under four groups on the basis of stratigraphical divisions of the findings along with the technological advancements.

Lower Levalloisian

This phase is characterized by heavy flakes and blades knocked off from tortoise-core or prepared-core. The striking platforms are normally and especially roughly faceted.

Middle Levalloisian

This phase is characterized by smaller, thinner and better retouched flakes than those of the Lower Levalloisian flakes. Numerous blades and rectangular blade-core appeared for the first time in this phase. Another feature is the presence of faceted striking platform. This phase may be labeled as the Proto-Mousterian.

Upper Levalloisian

This phase is characterized by the regular occurrence of the hand-axes, triangular in shape. They occur in association with large and oval flakes of similar size as found in Lower Levalloisian phase. But, Upper Levalloisian large and oval flakes are invariably more thinner and show better workmanship on them. This phase may be termed as early Levalloiso-Mousterian.

Final Levalloisian

In this phase, the retouched blades and triangles are struck off very carefully from the well-prepared cores. Typologically, the tools do not show much difference from the previous phase, but this phase presents a high esteemed workmanship which denote the Final Levalloisian culture and may also be called as the developed Levalloiso-Mousterian.

The geological age of this culture is the middle Pleistocene period. In terms of glacial age the culture is extended between the third glacial (Riss) and third inter-glacial (Riss-Wurm) periods. In fact, Levalloisian appeared as contemporary to Middle Acheulean and merged into the famous Mousterian culture. This flake tradition is found well distributed in Western Europe, Africa, and in India, especially in soanian industries. Although no definite group is held as the carrier of this culture, but an early Neanderthal group seems to be responsible behind it who belonged to the third stage of hominid evolution.

Middle Palaeolothic

The Middle Palaeolithic period is differentiated mainly from the typological point of view where the presence or absence of hand-axes or bifaces is critically important. The *core-tool cultures* have totally been transferred to the *flake-tool cultures* in this level. Therefore, Chellean-Acheulean hand-axes are no more found. Instead, implements have been made on flakes that are knocked off from the nodule.

Both Levalloisian and Mousterian culture were developed on the flake tradition involving a higher technology. Levalloisian culture was started from the Middle Acheulian stage and its developed form is named as 'Proto Mousterian', which became further developed later with the name 'Levalloiso-Mousterian'. This indicates that the Mousterian also emerged from Middle Levalloisian stage. As a matter of fact, evidences of Levalloisian culture come from the open-air sites, whereas the Mousterian had kept its evidences mostly in caves and rock-shelters of South-Western France. Besides, the Mosterian provides the earliest evidence on the regular use of fire and first definite burial has also been discovered from this stage in Western Europe.

Mousterian Culture

The rock-shelter of Le Moustier in the Dordogne area of southwestern France is the type-site of the mousterian culture. It is mainly considered as a flake-tool culture without the presence of hand-axes. This culture shows complete absence of any kind of core-tool. But in early Mousterian levels in France, a large number of small *cordiform* or heart-shaped hand-axes (from the flakes) have been found together with Mousterian points and side-scrapers. These smaller flake hand-axes show at least one flat side along with a sharp, straight, or sweeping curved edge, which is usually retouched on one side only. Though these flake implements are varied in shape and size, but all of them show retouches on one side only. In many sites, the tortoise-core technique for flaking is not found, rather the flakes have been detached by the *discoidal core-technique* (where the prepared core looks like a disc or round in shape) and then retouched. The typical tools of this culture are side-scrapers, points and discs.

The Mousterian tools, in general, show facetted striking platform and secondary workings in the form of step-chippings where the pressure-flaking technique is commonly applied. For the first time, a crude bone-tool industry appeared in the Mousterian stage. Perhaps the Palaeanthropic man in Lower Palaeolithic stage had fashioned bones into tools, but open-air sites were not favourable for preservation of these tools. Mousterian bone tools are mostly made with the broken long bones of animals. Some selected dense bones also bear the traces of use as chopping blocks and compressors or anvils. The typical Mousterian culture was flourished when the climate was very cold. Because this industry was always found confined to caves and rock-shelters. The first time evidence of regular use of fire also supports this fact. The Mousterian culture may be classified into three levels, such as Early, Middle and Late. Early Mousterian tools show immense influence of Levalloisian type as a good number of cordiform flake hand-axes have been discovered from the site at Combe Capelle at France. During the Middle Mousterian, the Levalloisian types of tools has also been discovered but only tool types like side scrapers, points, discs, etc. are found. Late or Upper Mousteian tools became smaller in size. Most of them show monotonous appearance in the form of little side scrapers and points as evident in the rock-shelter of La Quina at the Charente in France. It is interesting that they also had a influence of Clactonian tradition.

Prof. F. Bordes in 1968 classified the Mousterian culture into four levels in the following ways:

Mousterian of Acheulean. It is marked by the hand-axes, large number of side scrapers, backed knives and a type of notched tools.

Typical Mousterian. It shows a sharp decline in the number of hand-axes and knives. The predominance of side scrapers and Levallois flakes were also reduced. This phenomenon is found in the site La Micoque in France.

Charentian Mousterian. It is found in two famous Neanderthal sites at La Quina and La Farrasie in France. The level has been characterized by the strong influence of Clactonian flake tradition and the absence of Levalloisian influence. A good number of side scrapers and notched flakes have appeared in these sites.

Denticulated Mousterian. In this level, hand-axes and backed knives are significantly absent. But, a huge number of tools such as the side-scrapers, end-scrapers, burins, borers and denticulates have appeared.

The geological age of this culture has been ascribed as Middle Pleistocene period . It originated during the later part of third inter-glacial (RISS-WURM) period and continued up to early part of fourth glaciation (WURM). Culturally Mousterian is in between of Lower Palaeolithic and Upper Palaeolithic and usually classed as the Middle Palaeolithic period. The Neanderthal people were definitely responsible for the creation of this culture. But many of the findings still suggest that this transitional phase of culture is the contribution of an early form of Homo-Sapiens.

Upper Palaeolithic

The last part of the Old Stone Age gave rise to the Upper Palaeolithic culture, which covers approximately 1/10th of the time span of entire Paloaeolithic period. During this short span of time, the prehistoric man made his greatest cultural progress. This phase of Palaeolithic period shows diversified and specialized tools made on *blades* by replacement of the hand-axes and flake-tools of earlier cultures. It is also notable that, not only flint and similar rocks were used as tools, bone was also taken as a material for making tools. The ivory and antler were not spared. The culture has been referred as the *Osteodontokeratic culture* for the utilization of bone, teeth and horn at a time. Early man of primitive types disappeared at this cultural stage and the man of modern type came into existence. Earliest man-made dwellings have been discovered from this level. This cultural stage also shows the beginning and flowering of the Palaeolithic art. The *blade-tool tradition* of Upper

Palaeolithic comprises of three cultures mainly—the *Aurignacian,Solutrean* and *Magdalenian,* on the basis of one or more distinctive tool types. But a number of types are common in all cultures of Upper Palaeolithic, they are gravers, end-scrapers, points, etc. Among these, the graver type is very important as it denotes an extensive working on bone and facilitated the development of art.

Although, the three main traditions—*Aurignacian, Solutrean and Magdalenian* form a unit for 'Upper Palaeolithic', with the Aurignacian itself three distinct phases have been recognized—*Chatelperronian* (Lower Aurignacian) at the base, true Aurignacian (or Middle Aurignacian) at the middle and the *Gravettian* (Upper Aurignacian) at the top. The Chatelperronian stage is characterized by the presence of a large carved point with one steeply retouched razor-like edge with blunted back. This tool is known as Chatelperron point, named after the site Chatelperron in France. This phase was limited to the France only. The second phase, the typical Aurignacian or Aurignacian proper represents thicker blades and it was almost pan-European in distribution. In Western Europe the Gravettian supplanted the Aurignacian culture. Most of the authorities believe that this culture had its root in Chatelperronian. The distinctive tool type of this stage was a narrow pointed blade like a pen-knife with blunted back. Some scholars of later period have noticed the existence of another culture known as *Perigordian* between the Mousterian and Aurignacian culture, belonging to the Upper Palaeolithic period. The Perigordian points differ from the Mousterian points by elongated surface flakings. In most of the sites of Southern France Perigordian deposits are placed over the Mousterian deposits. In many sites Perigordian continues until the appearance of *Magdalenian.* Again, in some sites Aurignacian is found interposed between the lower and upper Perigordian. Thus, the Perigordian and the Aurignacian cultural traditions have been closely related in different parts of Western Europe.

BLADE TRADITION

1. Perigordian Culture

This culture is named after a site of Perigord region in southwestern France. The characteristic tools are blades of flint with one edge straight (razor-like). The other side is curved back with steeply retouched edge. The hunters possibly used these blades as knife. The culture reached to its optimum level in the said Perigord region of France.

The Lower Perigordian is the early or the oldest Upper Palaeolithic culture and shows the abundance of large curved points with blunted back which have popularly been known as Chatelperron points. The Upper Perigordian seems to have developed from the Chatelperronian type, showing the straight points with blunted backs. The Upper Perigordian tools include the Gravette points and in addition, this stage notes the first occurrence of small human sculptures. The multi-angle gravers known as *Noailles burins* and tanged points made on blades called as *Font Robert tanged points* are also found in this level. Therefore, it is evident that as synonyms the Lower Perigordians replaces the Chatelperronian and Upper Perigordian appears at the place of Gravettian. The Middle Aurignacian or the true Aurignacian evolves, as it is to mark off the first main cultural tradition of Upper Palaeolithic.

The geological age of Perigordian culture corresponds to the phase of the retreat of Wurm- I glaciation.This cultural stage suggests the appearance of new species of human race, that is the men of Late Pleistocene. This is the first stage in appearance of Homo sapiens. The people of Comb Capelle type and the Grimaldi man seem to be responsible for the Chatelperronian culture, i.e. Early Perigordian. The Predmost people represent the Gravettian culture, i.e. Late or Upper Perigordian. Both of these races belong to Neanthropic race.

2. Aurignacian Culture

This culture is named after the type-site, a rock-shelter known as Aurignac in SouthWest of Toulouse (Haute Garonne) of Southern France. Former Middle Aurignacian is now known only as the Aurignacian. All of the Aurignacian tools including Perigordian types are the usual Upper

Palaeolithic blade tools, e.g. burins, end-scrapers, etc. The leading tool types of this cultural stage include the steep-ended scrapers, nose scrapers, blade artefacts with heavy marginal retouch and

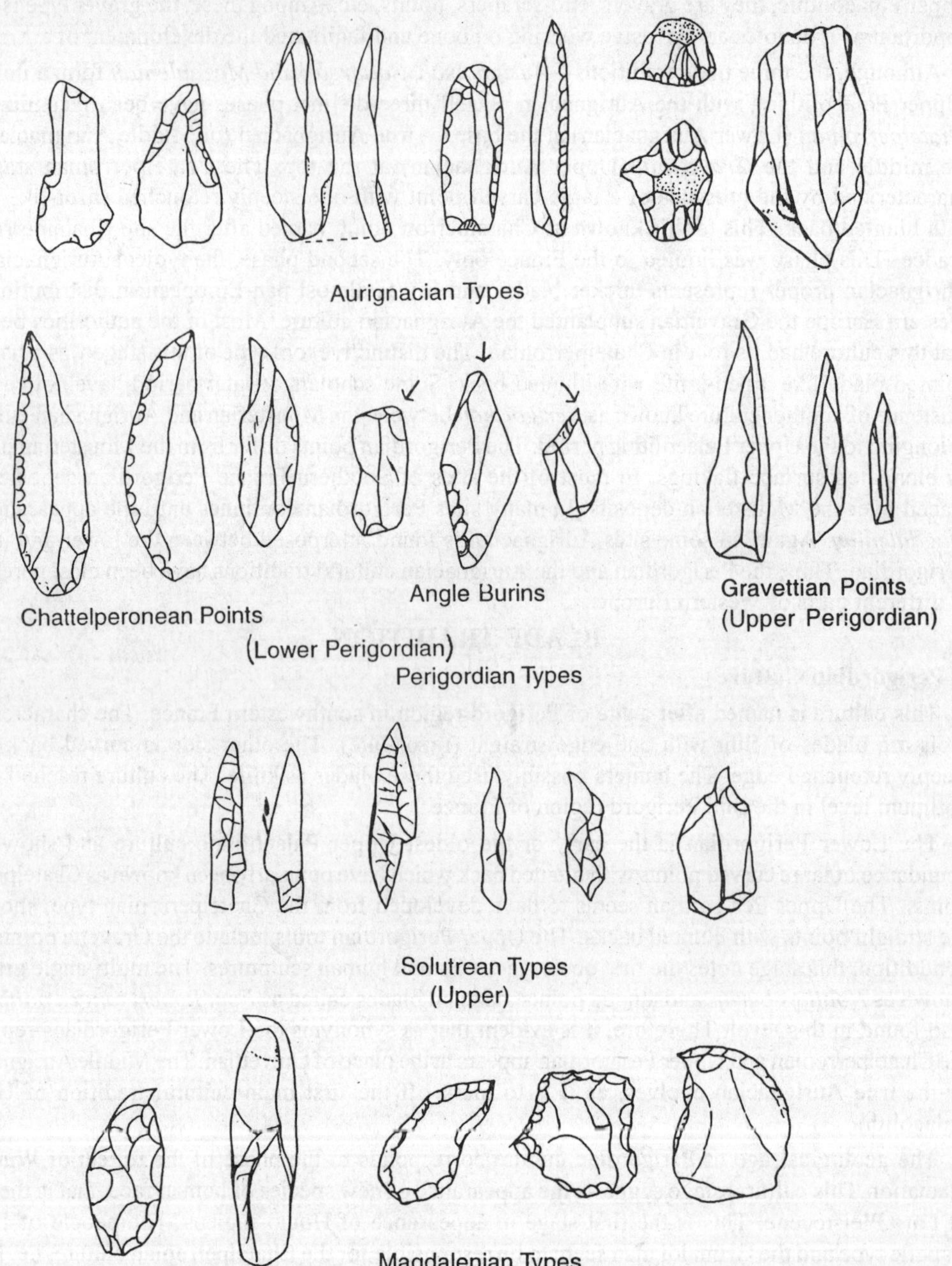

Fig. 11.8. Upper Palaeolithic Tools

split base bone points. The bone was extensively used in the Aurignacian, mainly for javelin points chisels, perforators and arrow straighteners or *batons-de-commandement.* Artefacts of personal adornment for the first time appeared in this stage which probably included necklaces made of pierced teeth and shells, decorated bits of bone, ivory and stone.

Other significant tools are the points, beaked gravers or burins, and keeled scrapers on stone. Besides, novelties in bone tools include split-base lance points, pointed awls or pins. The Arignacian people were artistic and fond of finery as shown by the evidences like decorated articles of ivory with geometric patterns, three dimensional miniature sculptures along with colored engravings and paintings. Along with these, some hollowed reindeer bones were found which seem to be used as tubes to hold the paint.

The geological age of this culture is related to the second part of Upper Pleistocene period and corresponds to the second phase of fourth glacial age i.e. Wurm-II or Bühl. The Aurignacian culture is traceable all over the Europe, but not as a uniform culture. It is also found in Africa and in India, especially in Northern Soan culture. Main responsible representatives of this culture is the Cro-Magnon group of man. In evolutionary line, the Cro-Magnons have been placed as the Men of Late Pleistocene period and they are the first runners of Neanthropic race—the Homo sapiens.

3. Solutrean Culture

This culture is named after the type site located at Solutre near Macon (Saone-et-Loire) in East-Central France. This cultural phase is notable for the finest development of flint workmanship in the Paleolithic period. We have found a high esteem development of lithic industry as the Solutrean toolmakers put much emphasis on the manufacture of their stone tools. The tools are more or less thin and flat, regular and come out as a result of parallel flaking. Though the parallel flaking technique first appears in upper Perigordian level, but it shows its best development in the Solutrean level by the creation of beautiful symmetrical laurel leaf and shouldered points. Along with these characteristic tools, many usual Upper Palaeolithic tools are found to be made on blades such as the gravers or burins, end-scrapers, points, etc. Many of these scrapers also show fine Solutrean chipping technique.

The Solutrean culture is recognized by three fold division in the following ways.

Proto-Solutrean. In this level we have seen the Leaf-points which are much crude where the percussion method was employed. These points are retouched mainly on the upper surface and the bulb on the lower surface has been removed by flat retouch. Usually such a point is shaped by delicate chipping on the lower surface only. These points are commonly known as proto-laurel leaves or proto-Solutrean points.

Typical Solutrean. Early leaf point is rough and thick. But, this level produces thin, regular and skillfully made leaf-points. These are characteristically true bifacial laurel leaves. These tools exhibit the excellence of pressure-flaking technique.

Upper Solutrean. The level produces the leaf-points with constricted sides. The tools include small and beautifully made laurel-leaves. In some cases, one side gets constricted which is characterized as shouldered point in this level.

In general, the Solutrean tools are found as the end-scrapers, side-scrapers, points, gravers or burins, etc. The examples of Solutrean art is rare. But the bone tools are found in many sites. The culture has a limited distribution. An early form is found in the caves of Hungary, Poland and elsewhere in Central Europe. But the main development has been observed in France and in some regions of Spain. According to some Scholars Gravettian and Solutrean have influenced each other until their differentiation was not clear. However, Solutrean is an extraordinary brief period in Western Europe with spotty localization. But its significance can not be ruled out.

The geological age of this culture goes to the second part of Upper Pleistocene period and corresponds to the retreat of the second phase of last glaciation (Würm-II or Bühl). No direct evidence of skeletal remains has yet been found for Solutrean culture. But, other circumstantial evidences suggest that a Neanthropic group allied to Cro-Magnon race was perhaps responsible for this Solutrean culture.

4. Magdalenian Culture

This culture is named after the type-site, a rock-shelter of La Madeleine at Dordogne in southwestern France. The last phase of Palaeolithic period is the Magdalenian culture which is

noted for the wealth of bone and antler tools and especially remarkable for the works of art. The flint industry of the Magdalenian people bore a blade tool tradition and was proved ingenious as well as utilitarian but this flint industry went in a state of gradual decline; the bone industry became more elaborate at this stage and a considerable variety of artifacts made on bone, ivory and reindeer antler

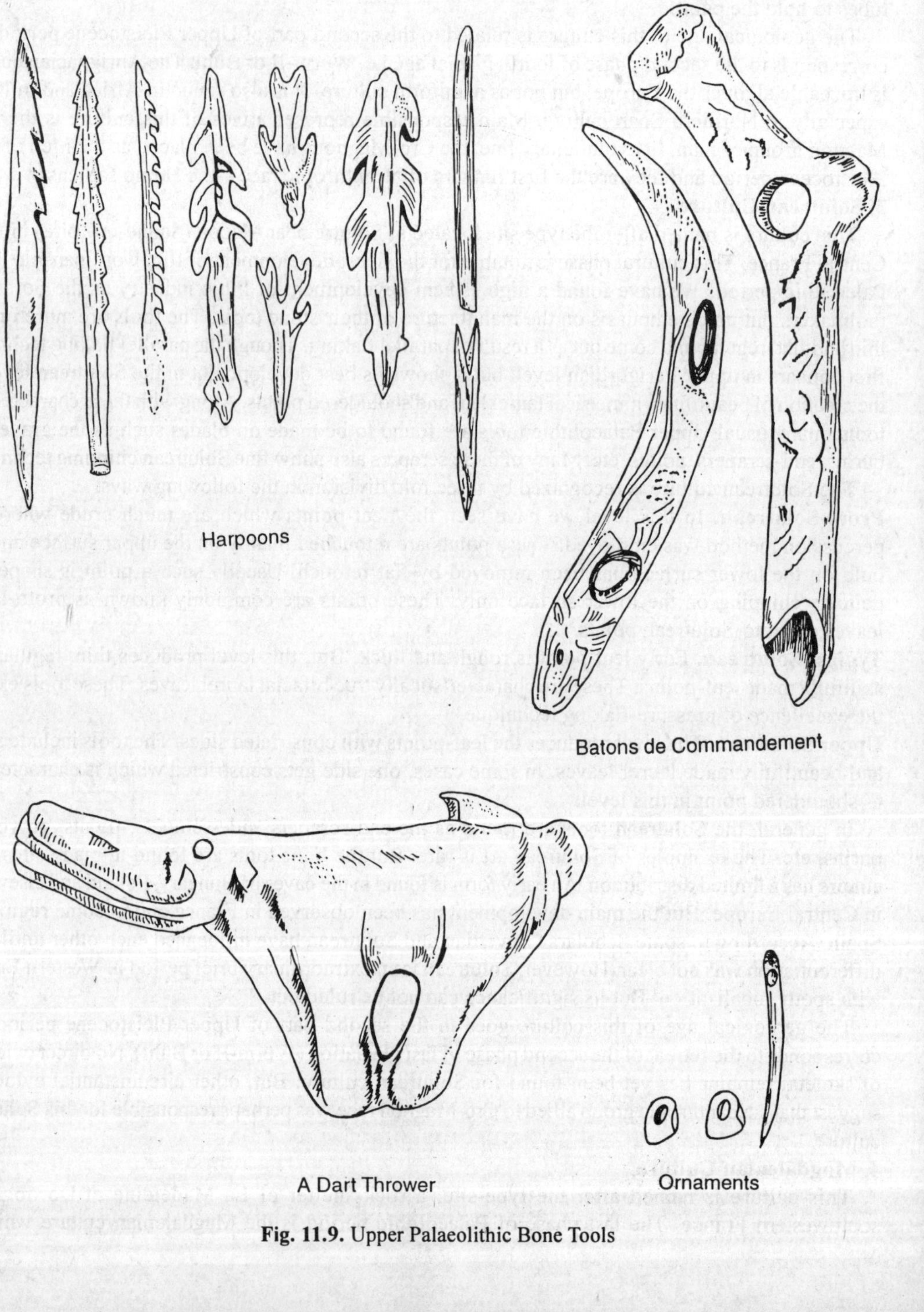

Harpoons

Batons de Commandement

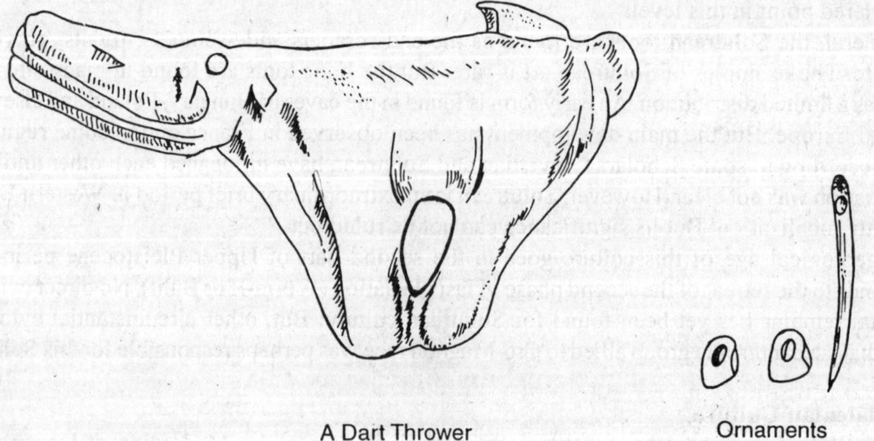

A Dart Thrower

Ornaments

Fig. 11.9. Upper Palaeolithic Bone Tools

were found. They included spearheads with link-shafts, barbed points and harpoons for spearing fish, hammers, etc. Besides, many other bone artefacts of uncertain use were also found. As an example, the baton-de-commandement was first introduced by the Aurignacians; probably it was used as a mending tool for straightening arrow-shaft or spear-shaft. The artifacts like batons-de-commandement were designed with a great artistic skill.

The different stages of the Magdalenian culture have been classified into seven sections in the following ways.

Magdalenian.

In this level, we find transverse burins or gravers. But, in bone tools, the harpoon heads are not traceable. The characteristic tools are made on bone as the javelin points which are flattened, conical with forked bases.

Magdalenian-I.

The flint tools of this level are found as the burins, end-scrapers, star-shaped borers. The bone tools include a large number of bone lance points (slightly convex) and bone needles. First appearance of batons-de-commandement has been noted during this stage.

Magdalenian-II.

The stone tools of this level are comprised of blade-lets, backed-knives, and denticulates. A large number of bone points characterize this level.

Magdalenian-III.

This level also shows the tools like backed-bladelets, triangular points and burins. A large number of bone tools, especially the lance-points continue to occur in this level

Magdalenian-IV.

The flint tools similar to the tools of previous stages appear in this stage. The typical tools of this level are the primitive harpoons made of bone and antler having a single row of lateral barbs.

Magdalenian-V.

The main tools of this level are the harpoons with single row barbs, very long shouldered points with short heads and the gravettian points.

Magdalenian-VI.

The stone tools of this level are the parrot-beaked burins, fattish circular flakes, points and knifepoint. The characteristic tools of this level are the harpoons with double rows of lateral barbs.

In general, the typical magdalenian tools are the long and parallel-sided blade implements. Some tools of this period are found to serve dual purposes — scraper, perforators, double-ended scrapers and scraper burins. It is the richest culture with regard to Upper Palaeolithic art where we have seen most bold outlines. The pigments as used are the black oxide of manganese and red and yellow oxides of iron. They were converted into paint by mixing with some fatty medium. The cave art of the Magdalenian people culminated in the production of polychrome paintings.

The geological age of this culture relates to the final part of Upper Pleistocene epoch and corresponds to the tail phase of last glaciation (Würm). Human skeletal remains of modern type of man have been unearthed from several sites. They are undoubtedly the Neanthropic man and perhaps the representatives of Chancelade group of man who are known as the men of Late Pleistocene.

UPPER PALAEOLITHIC ART

Prehistoric art started its journey and reached to the highest order of proficiency during Late Pleistocene. The Aurignacian culture initiated this interesting phenomena by means of its aesthetic sensitivity in collaboration with the technical proficiency. The Magdalenian culture is the richest stage of Palaeolithic art represented by the men of Late Pleistocene. The usual subject matters of the prehistoric art are the animals of the period. The figures of those animals have been portrayed in paintings, engravings and sculptures. But, the most striking feature of the Palaeolithic art is the absence of a complete scene. The art tradition of the Upper Palaeolithic period can be separated into two categories, as the mural and home art.

Home Art

It includes all types of objects that have been projected in delicate engravings and carvings on stone, bone, antler, and ivory as well as in the round shaped sculptures of that time. Many bone and antler tools, arrow-straightners, spear-throwers, and harpoons are found either engraved or carved with animal forms both in realistic and conventionalized pattern. Not only tools, some flat stones or bone fragments are found decorated with geometric forms in stylized designs. Several engravings of human figures are also included in the home art. All of these objects are movable i.e. they can be removed from one place to another. The French prehistorians have called these mobile art objects as 'Art mobilier'.

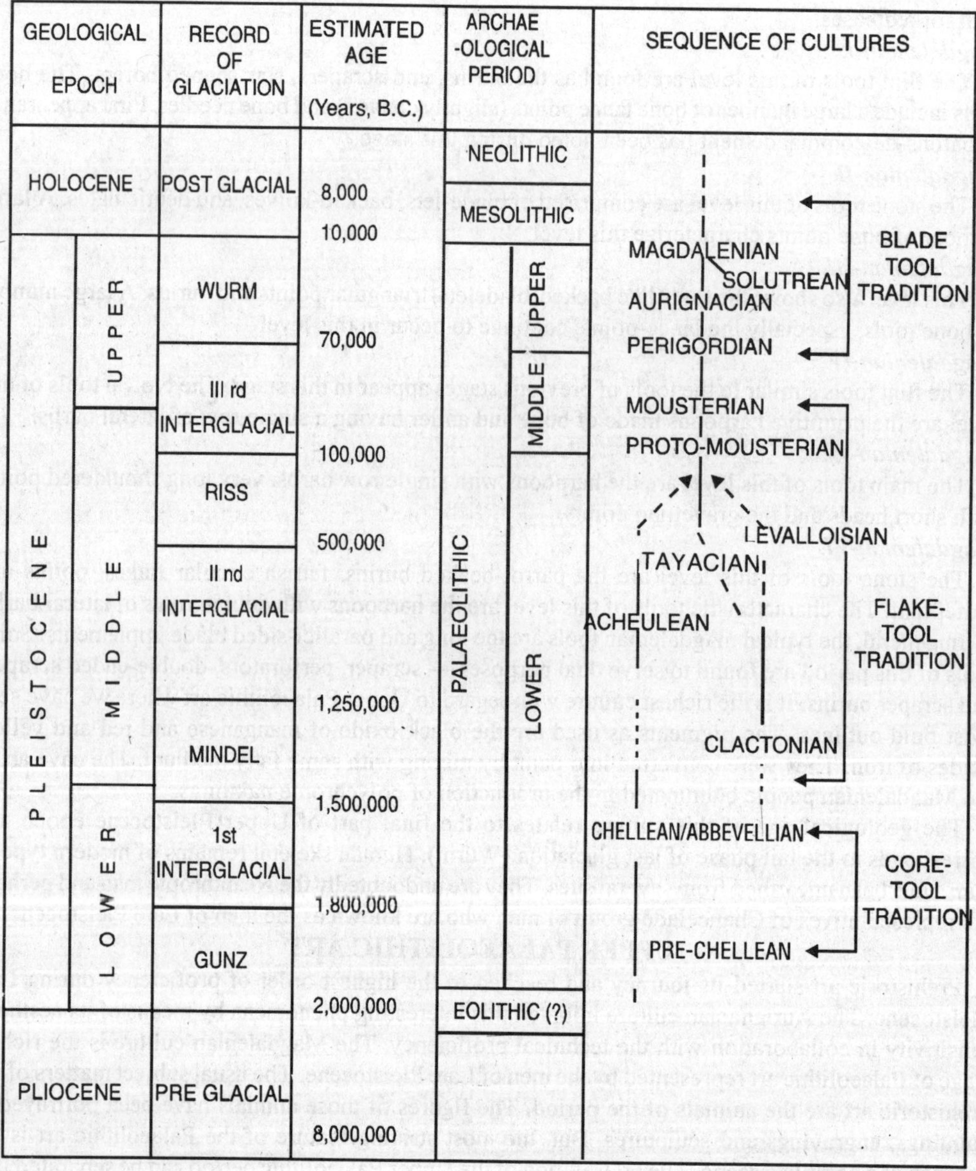

Fig. 11.10. A Conceptual Model Showing European Palaeolithic Culture

A rich collection of home-art materials have been discovered from Aurignacian level in southwestern France, especially at Brassempony, Grotte-du-Pape and Grott-des-Hyenes. They

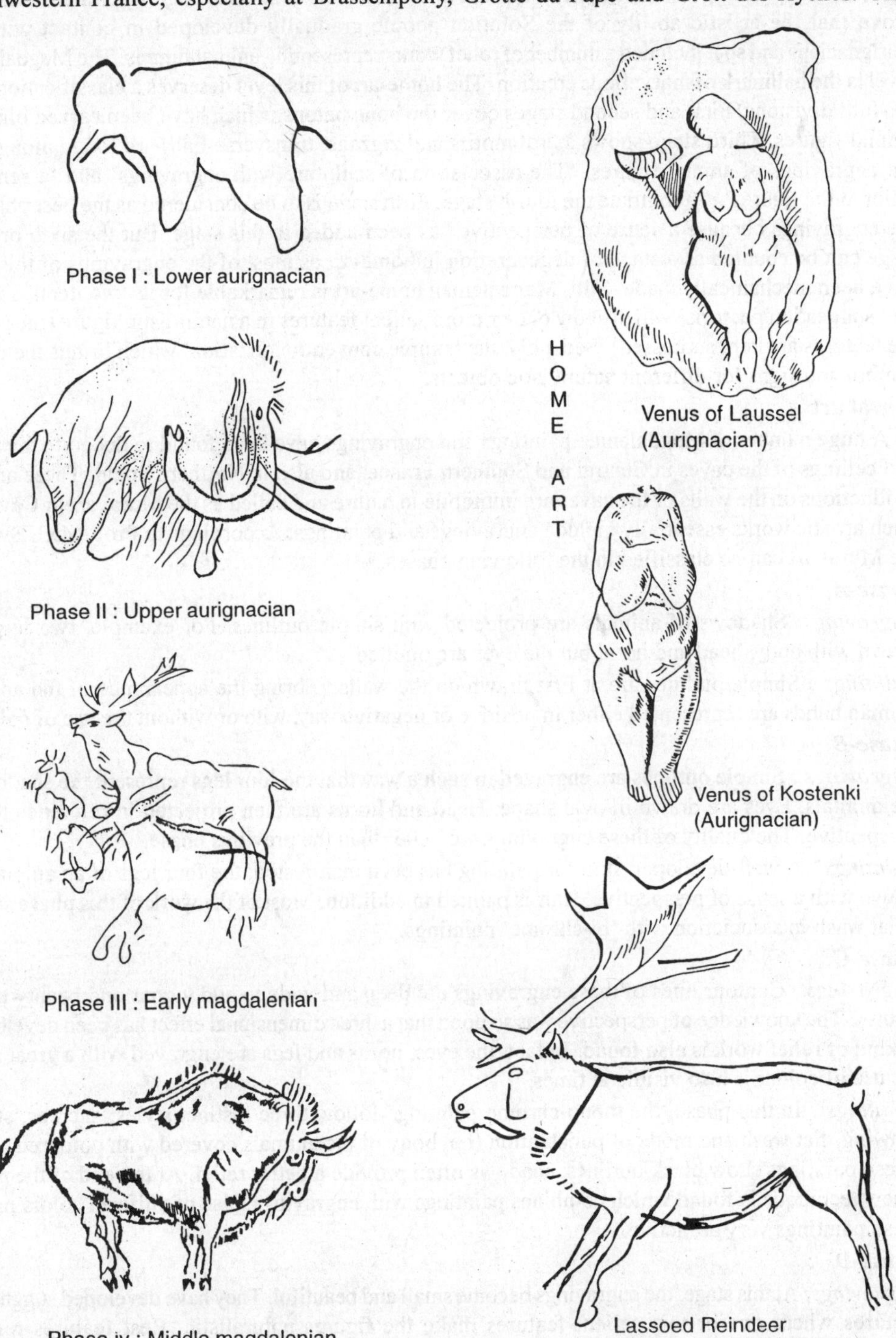

Fig. 11.11 Upper Palaeolithic Art

comprise stone sculptures including the famous 'Venus', carved reliefs and different types of bone-silhouette. Previously it was conceived that the Solutrean level is devoid of art. But, later researches prove that the artistic ability of the Solution people gradually developed in contact with the Aurignacians and so it included a number of relief works representing animal figures. The Magdalenian level is the hallmark of marvellous creation. The home-art of this level deserves a classification with six-fold divisions. First and second stages cover the bone batons, which have been carved out with animal figures. Third stage shows spiral motifs and zigzags; transverse lines are found along with the engravings of animal figures. 'The association of sculpture with engravings' and 'a sense of relief to the figures' differentiate the fourth stage. Fifth stage is to be considered as the best phase of the engravings because a sense of perspective has been added to this stage. But the sixth or final stage can be counted as a stage of degeneration in home-art as most of the engravings of this time have been mechanically made. Still, Magdalenian home-art is remarkable for its two items. Firstly, the 'suggestion pectoris' which show one or more salient features in a naturalistic figure (the rest of the features are kept supressed). Secondly, the 'simple conventionalization' which brings the use of symbol and signs for different naturalistic objects.

Mural art

A huge number of Magdalenian paintings and engravings have been found to decorate the walls and ceilings of the caves in Central and Southern France, and also in Northern Spain. These artistic productions on the walls of the caves are immobile in nature and called as the Mural art or Cave art. Such artistic works essentially include engravings and paintings. Accordings to Prof. M.C. Burkitt, the Mural art can be classified in the following phases.

Phase-A

Engravings: Shadows of animals are projected with simple outlines. For example, two legs are drawn with body, head and horn, but the eyes are omitted.

Paintings : Simple outlines are at first drawn on the walls to bring the appearance of the animal. Human hands are represented either in positive or negative way, with or without the use of colours.

Phase-B

Engravings : Simple outlines are engraved in such a way that the four legs represent the shadow of the animals. Eyes are drawn in oval shape. Head and horns are then projected in reference to the perspective. The quality of these engravings are better than the previous phase.

Paintings : A well-developed skill for painting has been manifested; the four legs of an animal are drawn with a sense of perspective. Hair is painted in addition. Most of the work of this phase shows a flat wash in association with 'bi-chrome' paintings.

Phase-C

Engravings : Contour lines of these engravings are deep and wide to add vigour and beauty to the figures. The knowledge of perspective was so good that a three dimensional effect has been developed. A kind of relief work is also found. In fact, the eyes, horns and legs are engraved with a great skill; the use of colour is also visible at times.

Paintings : In this phase, the mono-chrome paintings follow three distinct styles, such as, stump drawing, flat wash and mode of punctuation (i.e. body of the animals covered with coloured dots). These paintings show black outlines, shadows often provide a better relief. At the end of the phase a new technique is found which combines paintings with engravings. Use of uniform colors makes these paintings very attractive.

Phase-D

Engravings: At this stage, the engravings become small and beautiful. They have developed suggestion pictures where one or two salient features make the figures naturalistic. Rest features remain suppressed. The technique of simple conventionalization also develops during this period, which relates the objects with geometric pattern by the use of signs and symbols.

Paintings : The paintings of this phase are considered as the art par excellence. The drawings are extremely beautiful not only for the three-dimensional effect, but also for the use of perspective. 'Polychrome' paintings characterize this phase.

Finally, it may be considered that phase-A and phase-B belong to the Aurignacian level whereas phase-C and phase-D are the outcome of Magdalenian level. A concluding remark on Palaeolithic art has been made by Prof. E.A. Hoebel in the following way : "Upper Palaeolithic art may be summed up as vividly realistic, an art of flesh and blood, a functional art that aided in survival, eagerly used by a hunting people living in a glacial age".

PALEOLITHIC OF AFRICA

The African Paleolithic is characterized by a variety of stone tool assemblages and industries but these archaic assemblages do not show any distinguishable succession. So far as the environment is concerned, the Pluvial and inter-pluvial phases have dominated in this region, alternately. The earliest evidence of man-made stone tool is better documented in Africa than Europe. Some of the archaic assemblages are purely local while others are much similar to those of Europe. The cultural succession is well established only in certain areas of the continent and not in Africa as a whole.

The earliest African Palaeolithic tradition is *Pebble-tool tradition* i.e. *Kafuan-Villafranchian* industry which corresponds to the first Pluvial, *Kageran.* The hand-axe tradition and flake-tradition of Early Pleistocene though have developed separately, are found to be fused during the Middle Pleistocene period. Many local traditions have made their appearance during this period. However, the climatic oscillations that witnessed the cultural sequences can be differentiated into three geographical areas for better understanding of African pre-history. As the Pleistocene glaciation did not affect Africa, the continent has been demarcated as tropical area where alternate periods of heavy rainfall prevailed with arid condition. No distinct break is evident between Paleolithic and Mesolithic/Neolithic cultural traditions. Further, 'Old Stone Age' has replaced the word Paleolithic here. Similarly the names, 'Middle Stone Age' and 'Late Stone Age' are used to denote, the Mesolithic and Neolithic of Africa.

North Africa

In Algeria of North Africa, a locality is found to yield a series of crudely worked pebble tools in association with undisputed Lower Pleistocene *(Villafranchian)* mammalian assemblage. This is one of the earliest sites to produce implements of definite human manufacture. Besides, the Lower Paleolithic hand-axes of both *Chellean/Abbevillian* and *Acheulean* type along with a variety of flake-tools have been recovered in great numbers from several sites throughout Tunisia, Algeria, Morocco and also in the Sahara region. These are comparable to those of Western Europe. The climate was apparently less arid in comparison to present day situation. But it was definitely warm and wet enough as evidenced by the occurrence of the fossil remains of hippopotamus. Most of the fossil remains available from this region are heavily weathered and patinated. In some cases, the implements are found in direct stratigraphic association with an early fauna, which is similar to that of Europe.

A number of localities have recorded the *Levalloisian* type of flakes. But, the *trueMousterian* sites are extremely scarce in this region. In fact, the *Levalloisian* localities, perhaps the Late *Levalloisian* of the Middle Palaeolithic has been demonstrated by a special development known as '*Aterian culture*'. The type-site *ofAterian culture* is first recognized near the well of Bir-el-Ater, about forty-five miles south of Tebessa in Algeria. The *Aterian culture* has been considered as the North African variation of the Levalloiso-Mousterian culture complex of the Middle Paleolithic period. The tanged points made on flakes with prepared striking platforms that came out of tortoise cores characterize it. These flake tools are retouched by scooping from the sides of the base of pointed flakes. The Aterian points were probably used as arrow or spearheads.

In the upper part of Old Stone Age, the Aterian *culture* was succeeded by the *Oranian culture.* The *Oranian* tradition is composed of two different cultures—the *Oranian* of Western Algeria and the *Oranian* of Morocco. This *Oranian culture,* formerly called as *Ibero-Maurusian,* is found everywhere along the coast, from Northern Tunisia in Western Algeria to the Atlantic seaboard of Morocco. It is to be noted that the *Oranian culture* is quite distinct from the next culture level, *Capsian* belonging to the Middle Stone Age. *Oranian* flint implements are rather crude and a large proportion of bone tools have been included here. A few skeletal remains of Homo Sapiens have been recovered in association with this *Oranian cultures* of North Africa.

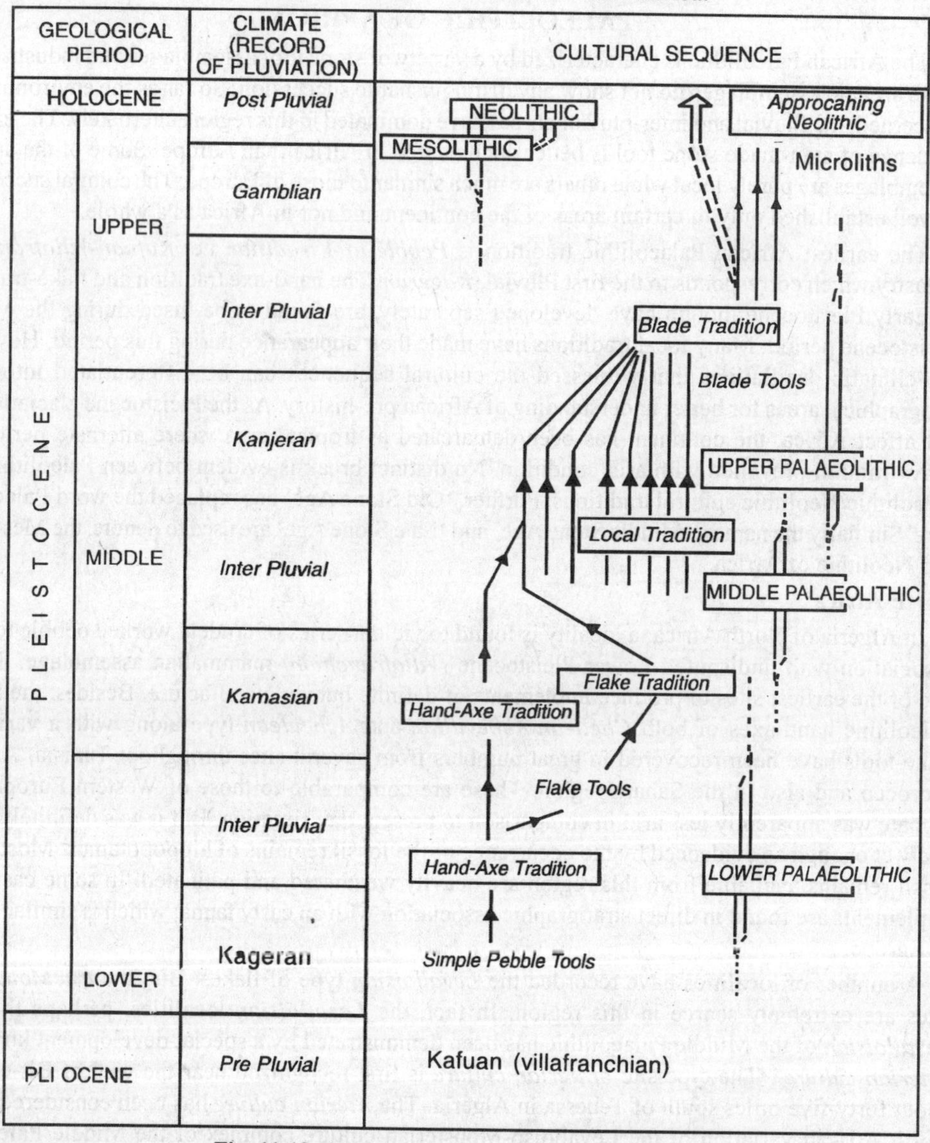

Fig. 11.12. A Model Showing African Prehistory

The Nile Valley of Egypt also demonstrates the presence of Old Stone Age culture. As per archaeological excavation, the Pleistocene terraces of Nile Valley (30 meter in Height) contain typical *Abbevillian,* primitive *Acheulean* and an African version of the *Clactonian.* At the height of 5-meter, it records the developed *Acheulean culture.* The large flakes and cores *of Levalloisian* type have been found from the 9-meter high terrace. The *Capsian* tradition of later period possesses a strong and deep root in the upper Paleolithic.

Artistic activities of North African Upper Paleolithic culture include considerable number of prehistoric rock-carvings and rock-paintings. This art is found to be spread from the Mediterranean coast of North Africa down to the interior of Sahara. A number of male and female figurines have been evident which follow the *silhouette* style and comparable to those of Eastern Spain and also the contemporary tribal art of Bushmen. Some portraits of elephant, buffalo, rhinoceros, ostrich, lion and other extinct animals have also been included under this art. The figures of the domesticated cattle, sheep and goat strongly suggest that the people knew the art of domestication. Again such a depiction of domesticated animals through carvings denotes a phase of Neolithic culture.

GEOLOGICAL PERIOD		CLIMATE (RECORD OF PLUVIATIONS)	ACHAELOGICAL PERIOD		CULTURES
HOLOCENE		POST - PLUVIAL	LATE STONE AGE		NEOLITHIC OF CAPSIAN TRADITION
PLEISTOCENE		POST - PLUVIAL	MIDDLE STONE AGE		CAPSIAN (BLADE TOOL TRADITION)
PLEISTOCENE	UPPER	GAMBLIAN	OLD STONE AGE	UPPER MIDDLE	ORANIAN (BLADE TOOL TRADITION) UPPER ATERIAN (VARIATION OF LEVALLOISIAN-MOUSTERIAN COMPLEX)
PLEISTOCENE	UPPER	INTER PLUVIAL	OLD STONE AGE	UPPER MIDDLE	ATERIAN OR LATE LEVALLOISIAN (LEVALLOISIAN FLAKE TRADITION)
PLEISTOCENE	MIDDLE	KANJERAN	OLD STONE AGE	UPPER MIDDLE	LEVALLOISIAN (FLAKE TRADITION) EARLY LEVALLOISIAN (FLAKE TRADITION)
PLEISTOCENE	MIDDLE	INTER PLUVIAL	OLD STONE AGE	LOWER	UPPER ACHEULIAN (HAND AXE TRADITION WITH FLAKE TOOL TRADITION) AFRICAN MIDDLE ACHEULEAN (HAND AXE TRADITION WITH FLAKE TOOLS VARIATION)
PLEISTOCENE	MIDDLE	KAMASIAN	OLD STONE AGE	LOWER	ACHEULEAN (HAND-AXE TRADITION WITH A VARIETY OF FLAKE TOOLS) TRANSITION FROM CHELLEAN TO ACHEULEAN AFRICAN EARLY CHELLEAN OR ABBEVILLIAN (HAND-AXE TRADITION)
PLEISTOCENE	LOWER	INTER PLUVIAL	OLD STONE AGE	LOWER	OLDOWAN (PEBBLE TOOLS)
PLEISTOCENE	LOWER	KAGERAN	OLD STONE AGE	LOWER	KAFUAN (CRUDELY WORKED PEBBLE TOOLS)
PLI OCENE		PRE-PLUVIAL	EOLITHIC (?)		

Fig. 11.13. Model Showing Stone Age Cultures In North Africa

East Africa

In East Africa, three territories are very important in connection with the prehistoric evidences. They are the Kenya, the Tanganyika and the Uganda. The Kenya gifted the earliest ape skull known from anywhere in the world, which had thrown considerable light on the problems of human ancestry. The earlier discovery of *Pre-Zinjanthropus* from Olduvai Gorge in Northern Tanganyika provided the most striking evidence to the scholars of human evolution; it was regarded as the first tool making hominid discovered from elsewhere. After the remarkable discovery of Zinjanthropus by L.S.B.Leaky the importance of the area was far more increased. The Uganda was the first East African territory to draw the attention of the archaeologists. In 1930 E.J.Wayland worked in this place to find out the cultural succession as well as the climatic change that occurred in the past. Way land identified several distinct cultures of East Africa including *'Kafuan', 'Sangoan'* and M*agosion cultures.* After 1952, W.W.Bishop and L.S.B.Leaky had started their systematic investigation on the lithic cultures of Kenya. The cultural sequences of East Africa were gradually unveiled.

All of the three territories of East Africa—Kenya, Tanganyika and Uganda yielded numerous crude pebble tools which show primary flaking on one end. They come from a bed apparently older than Lower Paleolithic. A great deal of intensive work has been done by Dr. Leaky in this area, who found a series of cultures, broadly designated as *'Kafuan culture'*. However, *Kafuan* is regarded as the earliest human culture of the world, even older than the *Oldowan.* The pebbles under *Oldowan culture* are chipped on two sides along the edge, instead of one side as found in the Kafuan *culture.* The *Oldowan culture* shows a definite *Abbevillian* and *Acheulean* affinity. Moreover, in Uganda, such layers include plenty of typical *Levalloisian* flakes and cores. As per chronology, the *Kafuan* belongs to the first pluvial period, the *Kageran* that again corresponds to the lower part of the Pleistocene epoch.

Thus, the *Abbevillian* and *Acheulean* of East Africa belong to the Middle Pleistocene and are contemporary to the Second or *Kamasian* pluvial period. Next to these, a cultural level called *Kenya Fauresmith* comes forward with well-developed hand-axe and *Mousterian* flakes. On a higher geological level, two distinct cultures are noted as *Kenya Stillbay* and *Kenya Capstan.* This two seem to be contemporary to fourth pluvial period, *Gamblian* under Upper Pleistocene of East Africa.

The lithic industries of East Africa can be broadly divided in the following ways:

The Lower Pleistocene Level.

The culture of this level is known as *Kafuan* and it is probably contemporary to the first pluvial period *(Kageran).* The name has been derived from a site at Kafu Valley in Uganda. Prof. Wayland discovered the pebble tools belonging to this culture in 1919. As said before, crude primary flakings limited to one end only characterize these tools. Significant presence of pebbly cortex has been noted in these tools. In general, the tools are long and broad and can be classified into three sub-divisions.

1. Early Kafuan

This division is characterized by the emergence of long, prepared pebble tools that are undoubtedly oval and flat. Sometimes the phase is also known as *Pre-Oldowan* as the pebbles are roughly chipped to an end and one side only. Signs of secondary retouch are completely absent. Pebble cortex covers 2/3rd of the total length of the tool on the both sides from the butt end.

2. Middle Kafuan

The tools of this division are invariably smaller than those of the earlier division, although the same technique has been followed. Some of the tools show their resemblance to the *Acheulean* tools.

3. Upper Kafuan

The tools of this division can be comparable to the *Levalloisian* tools. These tools are not only smaller in size; they are neat in finishing. A large variety of specialized tools like chisels, points, and scrapers etc. appear at this stage in association with a few small cores.

The Middle Pleistocene Level

This cultural level is contemporaneous to second *(Kamasian)* and third *(Kanjeran)* pluvial period. The culture has been known as *'Oldowan'* after a typical site at Olduvai Gorge. A German entomologist Prof. Kattwinkel in 1911 first noticed this significant 'Old Stone Age site which was thoroughly explored by Dr. Leaky and his wife in 1859. They found some skeletal parts of the earliest hominids in association with a number of pebble tools and flakes. A few broken animal bones were also recovered from this spot which included the remains of birds, amphibians, snakes, lizards, rodents, pigs and antelopes. However, the tools of *Oldowan culture* have been represented by a large number of specimens found in the layers of Lake Deposit. These layers are called beds being formed with a series of variable sediments exposed in Olduvai Gorge, cutting across the Serengeti Plain in Tanzania.

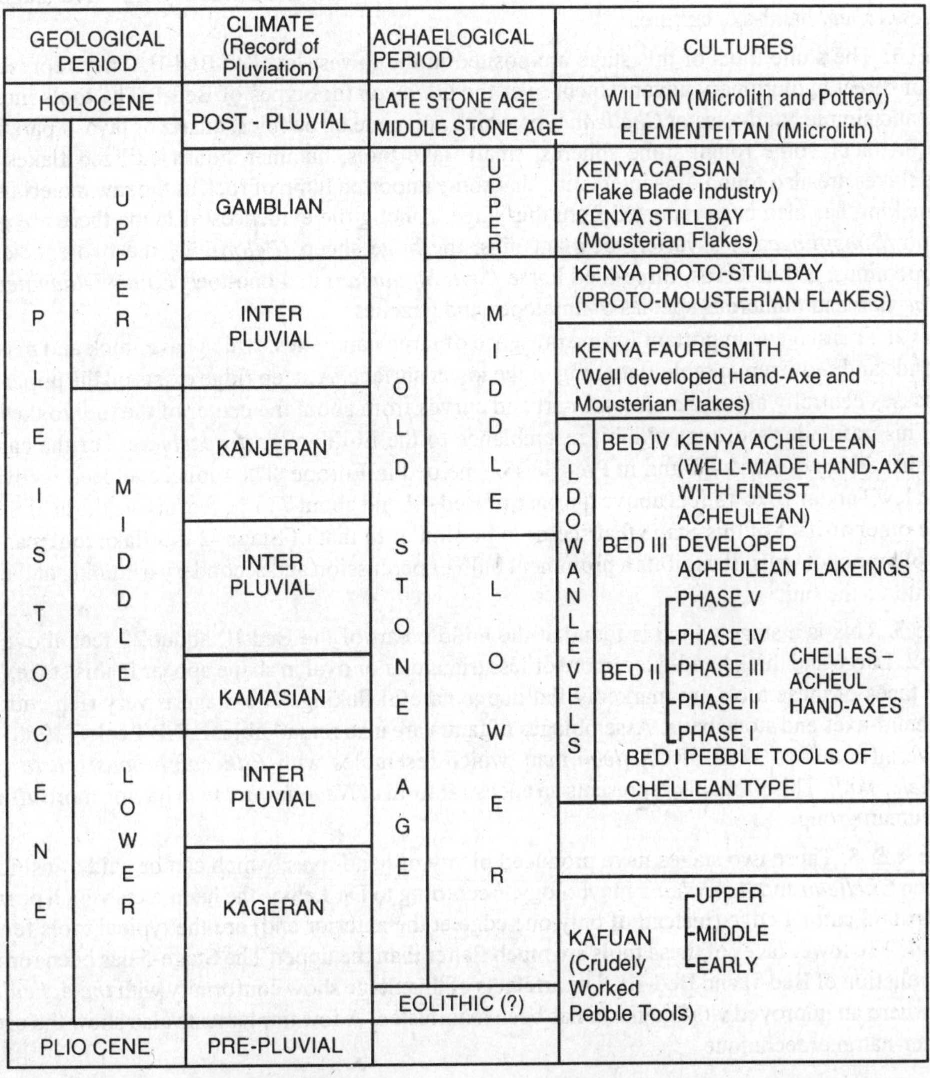

GEOLOGICAL PERIOD		CLIMATE (Record of Pluviation)	ACHAELOGICAL PERIOD		CULTURES
HOLOCENE			LATE STONE AGE		WILTON (Microlith and Pottery)
PLEISTOCENE	UPPER	POST - PLUVIAL	MIDDLE STONE AGE		ELEMENTEITAN (Microlith)
		GAMBLIAN	OLD STONE AGE	UPPER	KENYA CAPSIAN (Flake Blade Industry) KENYA STILLBAY (Mousterian Flakes)
		INTER PLUVIAL			KENYA PROTO-STILLBAY (PROTO-MOUSTERIAN FLAKES)
				MIDDLE	KENYA FAURESMITH (Well developed Hand-Axe and Mousterian Flakes)
	MIDDLE	KANJERAN			OLDOWAN LEVELS: BED IV KENYA ACHEULEAN (WELL-MADE HAND-AXE WITH FINEST ACHEULEAN)
		INTER PLUVIAL			BED III DEVELOPED ACHEULEAN FLAKEINGS
					BED II — PHASE V, PHASE IV, PHASE III, PHASE II, PHASE I — CHELLES – ACHEUL HAND-AXES
		KAMASIAN		LOWER	
	LOWER	INTER PLUVIAL			BED I PEBBLE TOOLS OF CHELLEAN TYPE
		KAGERAN			KAFUAN (Crudely Worked Pebble Tools) — UPPER, MIDDLE, EARLY
			EOLITHIC (?)		
PLIO CENE		PRE-PLUVIAL			

Fig. 11.14. A Model Showing Stone Age Cultures in East Africa

Bed-1 (Olduvai)

Dr. Leaky observed a homogeneous industry in the Bed-I of Olduvai Gorge, which is an evolved form in comparison to the earlier *Kafuan* industry. This is also a pebble-tool culture containing crude choppers and scrapers. As per Potassium-Argon dating, the basal layer of Olduvai Bed-1 is 1.7 million years old. The characteristic tool type in Olduvai Bed—I is crude chopper of varying size, the diameter of which ranges from a Ping-Pong ball to a croquet ball. Anterior flaking makes the cutting edge. Intersection of flake-scars from two-directions result in an irregular jagged cutting edge. These pebble-tools are found on the surface of the basal older gravel and can be compared to the earlier *Acheulean* or *Chellean* types which are thought to have belonged to the period of second glaciation in Europe. It should be kept in mind that this bed of Olduvai has not shown any hand-axe.

Bed-11 (Olduvai)

This bed is quite different from Bed-I. Because, while the Bed-I is composed of volcanic tuffs that ejected into the lake, the Bed-II is consisted of marls (the earthy sediments enriched with lime) and other lake sediments which is about 90 feet thick in places. This bed yields five stages of *Chelles-Acheul* hand-axe culture.

Stage: 1. The stone tools of this stage are posited in the lowest level of Bed-II. They represent a state of direct evolutionary advancement over the *Oldowan* tool types of Bed-I. The tools include crude and simple hand-axes *of Chellean* type which are made on pebbles, quartz or lava. Apart from these artifacts, some rough stone spheres, small flake tools, hammer stones, utilized flakes and waste flakes are also found. It is interesting that some imported lump of rock as the raw materials for tool making has also been reported from this stage. Among the evidences of fauna there are giant baboon *(Simopithecus),* a variety of giant pigs, the huge sheep *(Pelorovis),* the two species of hippopotamus, a rhinoceros, three-toed horse *(Stylohipparion)* and one-toed *Equus Olduvaiensis, Sivatherium* and numerous species of antelopes and gazelles.

Stage: 2. This stage is important for the presence of large hand-axes, which have thick and massive butt-ends and show remarkable flattening of the lower surface. A steep ridge exists on the upper face that passes centrally along the anterior part and curves from about the center of the tool to the butt-end. This particular tool-type exhibit resemblance to the Eolithic '*rostrocarinates*' or the earliest beak-shaped hand-axes as found in Pre-Pleistocene bed in Europe. The tools have been recovered from a level about 10 to 15 feet above the base of Bed - II and about 7 to 12 feet above those of Stage-1. The other artifact of this Stage that seems to be similar to that of Stage -2 is a flake tool made on ivory of hippopotamus. It exhibits a prominent bulb of percussion and secondary trimming at the end opposite to the butt.

Stage: 3. This is a stage which is found at the middle part of the Bed-II, about 20 feet above the Stage-2. Large and thick hand-axes, more or less triangular or oval in shape appear in this stage. The lower faces of these tools are markedly flat due to careful flaking. The Stage is very rich with the thick hand-axes and stone balls. Assemblages of fauna are also remarkable. L.S.B.Leaky (1960) has discovered a human skull of *Chellean* man, which resembles with *Pithecanthropus erectus* and *Steinheim skull.* This probably represents an ancestral form of *Neanderthal man* having more affinity with human group.

Stage: 4 & 5. These two stages have produced plenty of hand-axes, which can be said transitional between *Chellean* and *Acheulean,* in typology. According to Dr. Leaky, the hand-axes which possess an all round cutting edge (instead of only one edge at the anterior end) are the typical tools for the Stage-4. The lower faces of these tools are much flatter than the upper. The Stage-5 has been formed at the junction of Bed-II and Bed-III. The artefacts of this Stage show conformity with *the Acheulean* tools where an improved workmanship has been manifested. A few implements also show the use of cylinder-hammer technique.

Bed - III (Olduvai)

This bed has been formed with the terrestrial or river-borne deposits. It is also known as red-bed for being composed of earth and torrential gravel. Dr. leaky has unearthed 87 isolated stone tools from different levels of Bed-III and classified this bed as Stage-6 to maintain the *Chelles-Acheul* cultural sequence. These tools show a definite technological advancement over the previous Stage and placed at the top of the Bed-II. Artefacts of cleaver type first appear at this stage. So also the stone-balls of the 'bolas' type and some prepared cores. The typical hand-axes of Bed-III introduce a very developed form of cylinder hammer technique, which can be comparable to some of the finest *Acheulean* flaking in Europe.

Bed - IV (Olduvai)

This bed is composed of sand and from the base of this bed a high proportion of large and finely made hand-axes have been discovered. These artifacts are usually much wider than their length in comparison to the earlier ones. Cleavers are very common in occurrence and they are slightly narrower at the cutting edge than the middle portion, which means the types are convergent rather than parallel type. The selection of raw materials seems to have a little influence on technology, because many perfect specimens are found to be produced on coarse-grained quartzite. Dr. Leaky claimed that the makers of these tools had achieved a high level of perfection in their workmanship like the people of Europe where the flint of finest quality was available. He further classified the tools into Stages, from 7 to 11, for keeping up continuity in cultural sequence with other tools that are available from the Olduvai Gorge. A small series of 'S - twist' ovate hand-axes and cleavers in the form of 'V', rather than 'U' stood as typical artifacts of Bed - IV. The uppermost part of Bed - IV are often missing at Olduvai due to erosion. But, a very rich site, known as HK, has exposed the upper part of Bed- IV. In 1951, a small area of it was excavated and 459 artifacts came out which occurred in situ. The artifacts contained a variety of hand-axes and cleavers with blunted cutting edges. It is probably an incident of abrasion, which happened as a consequence of cutting up the hippopotamus carcass. The apparently discarded tools were left in a spot when they turned unserviceable. Plenty of white quartzite as raw materials of these tools has been discovered close at hand and it indicates that the making of new tools was not at all difficult in this Stage. The Western part of the Southern branch of the Olduvai Gorge reveal one cleaver which is comparable to Neolithic axes. If this cleaver could be hafted, it would able to cut down the small trees quite efficiently. Because its edge is found very sharp and straight. Besides, some small and degenerated hand-axes have also been unearthed from this bed.

Like *Oldowan,* the *Fauresmith* culture was flourished in Kenya during the third pluvial *(Kanjeran)* period. This *Fauresmith* culture sequence is held more or less parallel to *Oldowan culture* sequence. Therefore, some scholars have considered Olduvai Bed-III and Bed-IV as Kenya Acheulean. The phase named as *Kenya Fauresmith,* characterized by the appearance *of Mousterian* flakes, denotes an intermediate period between the early and middle part of Old Stone Age. Although the culture *Fauresmith is* commonly referred as *Kenya Fauresmith,* actually the name derives from a site in South Africa, which is situated on the Kavirondo gulf of Lake Victoria, a few miles away from Konam. L.S.B.Leaky in 1932 had discovered the fragments of numerous human skulls in association with some extinct fauna such as *Palaeoloxodon Recki, Simopithecus, Pelorovis,* etc. The neighboring areas of the site yielded several *Acheulean* tools. Hand-axes of lanceolate and ovate forms are conspicuous here as like the *former Acheulean* stage. Many U-shaped choppers showing pebbly-cortex on their butt and a large number of specialized scrapers are found in association with typical *Mousterian* flakes. In general, the tools of this stage are flat, small and neat in finishing. This indicates a development of better craftsmanship.

The Upper Pleistocene Level

This cultural level has been correlated with the fourth pluvial *(Gamblian)* period when we record the disappearance of hand-axes and cleavers. Two distinct cultural strata have been identified in the following way:

A. *Kenya Proto-Stillbay and Kenya Stillbay*

Although *Kenya Stillbay* has evolved from *Kenya Proto-Stillbay,* but they are two distinct industries, based on obsidian. Obsidian is a form of silica which is very hard but can be chipped or flaked quite easily to produce a fractured edge as sharp as a razor. Its colour is usually greenish or grey but the thinnest pieces are really transparent. However, the discs, *Mousterian* flakes points, side-scrapers etc. made on obsidian exhibit high skill of retouching in *Stillbay* than that *of Proto-Stillbay.* Moreover some definite influence from European Upper Palaeolithic, especially from *Solutrean* has been noticed in this industry.

B. Kenya Capsian

The *Capsian* is named after the type site Gafsa (derived from the Latin word Capsa) in Tunisia which is essentially a culture of high plateau. But the *Kenya Capsian* is essentially restricted in Kenya Rift Valley and Northern Tanganyika. It is one of the most controversial of all cultures of East Africa and characterized by a flake-blade industry. Miss Caton Thompson in 1946 has opined that *North Africa Capsian* and the *Kenya Capsian* are related cultures. The *Kenya Capsian* is very close to *Stillbay* industry for the association of blades and bifacial tools in it. The lower level of this culture can be said a 'blade and burin culture' as its chief finds include blades, burins, microliths and bone tools which bear similarity with the European Mesolithic culture. The upper level *of Kenya Capsian* is much widespread and best known among all stone tool cultures of the Kenya Rift. This stage is marked with a steady decrease of large backed blades, burins and end-scrapers. But the microliths, specially the lunates and triangles, are held as characteristic tools. Though a large number of crescent shaped microliths are found along with, they are considered as the variation of backed blade. The most tiny ones, ranging from one inch to less than half inch in length, probably served the function of barbs in harpoons, spears and arrows. For serving this propose, the pigmy tools were definitely mounted in the grooves of a wooden shaft. The presence of bone harpoons, pottery and egg-shell beads strongly suggest that this upper level *of Capsian culture* in Kenya must have continued up to the Post-Pleistocene period.

South Africa

South African sequence of Old Stone Age cultures covers almost the entire region of southern peninsular Africa with northern Rhodesia, Mozambique and Angola. This part of Africa is geologically most ancient and did not face large-scale faulting, rift or volcanic eruptions. The cultural sequence is well-established on the basis of evidences found in the terraces of the Vaal Valley and also from some well-excavated cave sites. The erosional and aggradational phases of the river Vaal, under the spell of climatic changes, have recorded the cultural sequence very beautifully and those have been studied chronologically by the great personalities like J.D.Clerk, H.L.Breuil, and others. However, the stages are as follows :

Pre-Stellenbosch

The basal stage of Old Stone Age in South Africa can be correlated with the European *Chelleo-Acheulean* cultural tradition. Following the sequence of East Africa, here culture begins with a pebble culture of *Kafuan* type. Gradually, it develops into a stage where the tools are similar to *Oldowan culture* in all essential respects. Therefore, *Pre-Stellenbosch* stage is composed *of Kafuan* and *Oldowan* type of industries. The period corresponds to the first pluvial period, *Kageran* and the subsequent inter-pluvial *(Kageran-Kamasian)* period.

Pre-Stellenbosch represents the earliest lithic industries of South Africa. Although the pebble-tools of this stage show resemblance with the tools *of Kafuan* culture and its developed form, the *Oldowan* culture (as found in East Africa), the authenticity of some tools are doubtful because they occur in association of some rolled *Abbevillian/Chellean* hand-axes. Moreover, the hand-axe complex that is found in *Pre-Stellenbosch* is a developed form than the *Oldowan* culture and naturally far more developed in comparison to *Kafuan* culture.

GEOLOGICAL PERIOD		CLIMATE (RECORD OF PLUVIATION)	ACHAELOGICAL PERIOD			CULTURES
HOLOCENE		POST - PLUVIAL	LATE STONE AGE			SMITHFIELD WILTON MAGOSIAN
PLEISTOCENE	UPPER	GAMBLIAN	OLD STONE AGE	UPPER		LOWER TERRACE IV EVOLVED FAURESMITH GRAVEL III STILLBAY (Mousterio-Solutrean Type) GRAVEL II PROTO-STILLBAY
	MIDDLE	INTER PLUVIAL		MIDDLE		GRAVEL I SANGOAN (FAURESMITH)
		KANJERAN				
		INTER PLUVIAL		LOWER	STELLEN BOSCH (South-African Abbeylian Acheulean Stages)	STAGES V Developed Flaking Through Cylinder Hammer Technique) IV Flake Tools of Upper Acheulean Type)
		KAMASIAN				III (Levalloisean Flakes) II (Flake tools of Lower Acheulean Type) I (Tools of Chellean Type)
	LOWER	INTER PLUVIAL			PRE-STELLENBOSCH	OLDOWAN TYPE OF TOOLS
		KAGERAN				KAFUAN TYPE OF TOOLS
PLI OCENE		PRE-PLUVIAL	EOLITHIC (?)			

Fig. 11.15. A Model Showing Stone Age Cultures in South Africa

Stellenbosch Stages'

The *Stellenbosch* stage is higher than the *Pre-Stellenbosch* stage and grossly corresponds to the second pluvial and inter-pluvial period. The type site is located near to the Cape of Good Hope where numerous *Chelleo-Acheulean* types of tools are preserved in sucession. Profound use of hard rocks, especially dolorite, is found in this industry. Local tool-makers preferred the hard rocks and invented the technique of 'prepared core'. The oldest known tortoise cores are found in association with *an Acheulean* industry at Victoria West in Cape Province. The technique is also known as 'Victoria West Technique'. It is used in making tools on hard rocks like hand-axes and cleavers. However, the *Stellenbosch* culture may be classified into five stages :

Stellenbosch-1

The tools of this stage are made mainly on pebbles, boulders or cobbles. Chipping are done from two directions so as to form a sharp cutting edge effective for cutting, chopping and scraping. The striking platform remains obliquely slanting. This technique of flaking is more or less similar to *Clactonian-Chellean* type. Sometimes the tools are flaked all around and the products take a bi-conical shape. The pebbly cortex is placed on a side.

Stellenbosch-II. This stage has been recovered from the old gravel layer -1 of second pluvial period, *Kamasian.* The core tools of this stage do not show any speciality. In case of flake tools, the unfacetted striking platform is found to form a less obtuse angle with the main flake surface. Such a technique of production can be compared to the *Lower Acheulean* of Europe and East Africa.

Stellenbosch-III. This stage is also found in the old gravel layer-I of'*Kamasian* pluvial period. The characteristic feature is the occurance of *Levalloisian* flakes with facetted striking platform. Horizontal flaking are very common in these tools.

Stellenbosch -IV. This stage is found in the old gravel layer-II that deposited during the second inter-pluvial *(Kamasian-Kanjeran)* period. The findings include the flakes with unfacetted striking platform which have been discovered in association with *Upper Acheulean* flake tools with prepared or facetted striking platform. The characteristic tools are cleavers with rectangular and trapezoid cross-section. Such tools are found plenty in this stage.

Stellenbosch -V. This stage is also found in the old gravel layer-II corresponding the *Kamasian-Kanjeran* inter-pluvial period. The characteristic tools are thick and flat core tools having longitudinal flakings all over the surface. Another feature is the application of cylinder-hammer technique in the manufacture of flake tools.

It may be mentioned here that the *Stellenbosch II, III, IV* and V are generally considered as the *South African Acheulean.* The stage-I resembles to *Chellean* of Europe. However, this South African Lower Paleolithic hand-axe complex contain hand-axes, cleavers and crude flakes. Flakes are struck from 'Victoria West' type of cores (hard rock) by means of'Victoria West Technique'.

Sangoan Stage. The *Stellenbosch* is stratigraphically followed by the *Sangoan* which is almost like *Fauresmith* in East Africa characterized by *Levalloisian* flakes. The type site of *Sangoan* industry is located at the Sango Bay of Lake Victoria in Uganda. Since the deposit of *Kanjeran* pluvial period does not yield any industry in South Africa, the *Sangoan* stage occur below the lowest stratum of fourth pluvial *(Gamblian)* deposition. The characteristic tools are discoid cores, large flakes, points, side-scrapers and denticulates. Evolved type of hand-axes and flakes of typical *Levalloisian* are found in plenty. This means that the *Sangoan* has replaced the *Fauresmith* in Uganda.

A lone site of South Africa, particularly in Southern Rhodesia offers a specialized facies of hand-axe industry on pebbles in association with utilized flakes. Most of these tools are prepared on jasper and finegrained silicious rocks. The common types are end-scrapers made by *Levalloisian* technique. Prof. H.L.Breuil is in favour of calling this as a separate stage with the name *Hope Fountain.* In fact, the stage is equivalent to the *Sangoan* stage.

The concluding phase of Old Stone Age in South Africa corresponds with the fourth pluvial period, *Gamblian.* Three stages of cultural succession have been found here.

Proto-Stillbay

The *Sangoan (Fauresmith)* industry develops further to give rise a culture complex of *Mousterian-Levalloisian* tradition. The tools of this stage are hand-axes, 'U'cleavers, flake tools, large circular prepared cores -etc.

Stillbay

The *Proto-Stillbay* evolved into *Stillbay culture.* The pressure flaking technique *of European Mousterian* was completely adopted by the people of Africa at the late phase of Old Stone Age. Flakes came out from a tortoise shaped prepared core. The tools include pressure-flaked points, burins, scrapers, etc. This *Stillbay culture is* also found outside of South Africa, especially in Kenya and Uganda. The shape of the implements as well as the technique of their manufacture suggest that the culture had received the influence of *Solutrean* culture as like Europe.

Evolved Fauresmith

In the deposit of *Gamblian* pluvial period two strata have been identified—Recent gravel stratum and lower Terrace stratum. *Stillbay* and *Proto-Stillbay* cultures have came from the gravel stratum.

But the more recent culture, *Evolved Fauresmith* lies with the lower terrace. The tools of this stage are straight-backed blade-lets, burins and scraper. The tool-types correspond to the Upper Paleolithic and Mesolithic stage of Europe. The *Levalloisian* technique that was exhibited by the *Sangoan or Fauresmith* culture, attained a state of extraordinary perfection at the stage of *Evolved Fauresmith.*

After the *Gamblian* pluvial period when the climate of Africa started to change toward an arid phase, the Late Stone Age industries corresponding to Mesolithic in Europe appeared.

PALEOLITHIC OF ASIA

In Asia, Paleolithic industries have also been recovered from Pleistocene deposits between Ichang and Chungking in the Valley of the Yangtze in South China, in the Mei Fingnoi Valley near Bhankao in Thailand, from the Kota Tampan district in the Perak Valley of Northern Malaya, from Yenangyaung and other sites in the Middle Irrawady in Burma and from the tributaries of the Indus in the Punjab, notably from the Soan Valley.

Paleolithic in India

No skeletal remains of Paleolithic man have been discovered from India though Asia has produced many important evidences of man's early evolution. For example, the Upper Miocene deposits of the Siwalik Hills yielded some remains of fossil apes which seem to have occupied an important place in the evolution of man. Possibly they represent a common stock from which man and the contemporary anthropoid apes have evolved. Robert Bruce Foote collected the first Indian palaeolitł (i.e. the implement which helped the early man to adjust with the environment for survival) from the South India in 1863.

There is no doubt that man emerged during the Pleistocene which was an epoch of great climatic change characterized by phases of repeated cooling and warming. Pleistocene glaciation was a universal phenomenon affecting 30% of the earth's surface. Geological deposits of the lower slopes of the Himalayas, especially the Kashmir and Punjab region show a succession of glacial periods, each indicated by the typical products of glacial action. This can be traced in the structure of the river valleys in the Rawalpindi area where terraces have been formed by the flow of the river cutting into the detritus. The sequence is more or less parallel to that of the Northern Europe. In both the places, man-made tools are found lying on the river terraces being associated with the phenomenon of climatic change.

However, the geological evidence in the Indian Himalayas exhibits four main glacial phases and three inter-glacial phases corresponding to the four main phases of sedimentation and three phases of erosion. The relative duration of these climatic phases was not uniform. The second glacial phase was longer than the first, which is again longer than the third and the last glacial phase was the shortest. Compared with the glacial phases, the inter-glacial phases were longer. Correspondingly, in the adjacent periglacial region (like Punjab plain) formation of river terraces were noted following the phases of aggradation and degradation. Since no artifacts are found in the Kashmir Valley, it is understood that the early men avoided the cold Alpine heights and preferred the open river plains of the Punjab for their habitat. During the subsequent inter-glacial phase, when the climate became relatively warm, animals from Siwalik Hills migrated to the Kashmir Valley. Fossiferous clays near Gulmarg in Kashmir yield the remains of Pine, Oak, and Birch with shells of fresh-water molluscs corresponding to this period. Man perhaps had not entered into this scene as neither his tools nor his bones (as fossil remain) have so far been found. The favourable ecological factors of the area drew the man in the last phase of second glaciation or at the beginning of second inter-glacial as testified by his tools. However, well-defined human industries can be traced in the river terraces and other deposits of Northwest India, particularly in the valley of the river Soan and the Indus, in Poonch, near Jhelum and in the Salt range.

When there were intense glaciations in Kashmir, pluviations occurred in Plains forming wide flood plains and abundant grazing ground for animal life. The environment of early man in the plains of India is intimately connected with the detrital laterite, which is typically a tropical deposit formed through the succession of wet and dry phases. The laterite itself suggests a humid tropical and sub-tropical climate, and the lateritic crust indicates a comparatively dry climate. Still there lie great problems in exact determination of geological age or climatic dating. In Peninsular India, the earliest tools of man come from the Boulder Conglomerate deposits underlying the laterite near Madras. In Central India, in the Upper Narmada Valley (District of Hoshangabad and Jubbulpore) the earliest tools are found to be hidden in the basal boulder conglomerate. Geological age seems to be middle Pleistocene in both the areas.

The earliest fabricated stone tools in India are chipped pebbles like that of the Africa. In Africa, the *Kafuan* pebble tools constitute the oldest lithic industry of early man. One directional flaking or two directional alternate flaking (added later) on river-worn pebble produced zigzag or rugged working edge; the tools were suitable for chopping, cutting and scraping. Archaeologists designated them as chopper or chopping tools. Such tools have came out both from northern and peninsular regions. In peninsular India, these tools are found in association with the hand-axes, the multi-purpose tool meant for cutting, digging, thrusting and many other purposes. Hand-axe suggests tropical forest and woodland environment; it is widespread in the tropical peninsular area. Technologically, these tools are more evolved than the chopping tools as they show multi-directional flaking and symmetry of form. Cleaver is often found associated with hand-axe in peninsular region and denotes *Abbevellio-Acheulean* tradition of Africa. Early man in India, as elsewhere, also made lighter and smaller tools on flake. At first the flakes were detached and worked by a simple technique *(Clactonian)* and later on by a more refined technique of core preparation *(Levalloisian)*. As hand-axe became a characteristic tool of peninsular India, flake tools *employing Levalloisian* technique became an important element of northern Indian tool tradition. The characteristic tools are knife, point, scraper, awl, etc. This technique in peninsular India gave rise to *Acheulean* tradition, which lasted much longer than any other technique that man could devise during Paleolithic Age. The final phase of the Paleolithic in India is still obscure.

Prehistorians have divided Indian Paleolithic culture into two major traditions with distinct geographical features. (A) The *Soan culture* or the chopper chopping tool tradition in the Punjab and (B) *The Madras culture* or Hand-axe tool tradition in peninsular region. Prehistorians have found a constant interaction between Soan and the Madras industries although there is a marked difference between these two industries. The core-tool elements dominate in the south and southeast, while the flake or chopper type is very strong in the North. Moreover, in Central India, a fusion of technology (between two main traditions) has been observed. Certain stone industries from Kurnool in the Deccan and near Mumbai represent a new type of tool tradition, based not on the massive flake but on the slender blade detached from a core.

The Soan Culture

The discovery of stone tools from Soan Valley was first made by Dr. D. N. Wadia and thereafter by Dr. Helmut de Terra in 1928 and 1932 respectively. But the credit of excavation goes to H.De Terra and T.T. Paterson who undertook this important work in 1935. Later, V.D.Krishnaswami, D. Sen and O.Menghin had studied the assemblages of tool.

The river Soan is a tributary of the great river Indus that flows through the city of Rawalpindi in Potwar region. The boulder conglomerate is found on the topmost surface of the river valley. Because when the river was formed in the Himalayan slope, it carried gravel fans and boulder clay with it. The whole Potwar area through which the river Soan has flowed shows periodic loessic deposition, corresponding to the glaciations in Himalayas. Since the whole Potwar plateau was a peri-glacial region, so it is quite natural that the early men inhabiting along the bank of Soan must have witnessed great dust storms. The first traces of men that come from the deposits of second inter-glacial phase are very scanty in number. At the end of the second inter-glacial phase, we get some tools of different character, which further went on change through the subsequent third glacial and inter-glacial periods.

De Terra and Paterson studied all these tools in different phases and discovered five terraces along the river Soan where climatic fluctuations illustrating the environment of man in Pleistocene epoch have been beautifully documented. Five major tool assemblages are depicted here (Fig. 11.16).

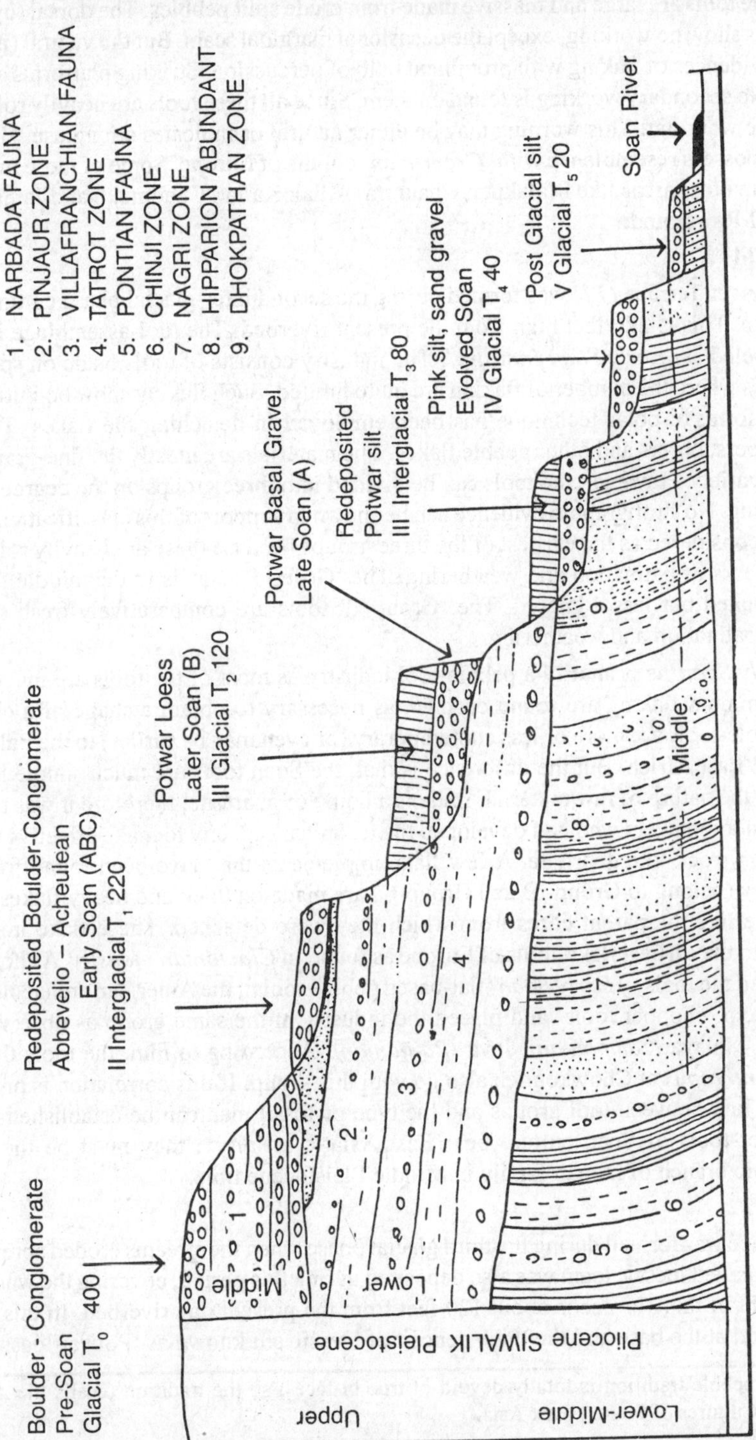

Fig. 11.16 Transverse Section Showing Stone Age Sequence in Soan Valley (After De Terra and Paterson)

Surface of the River (**To**)

The tools that come from the top of the boulder conglomerate constitute the *'Pre-Soan Industry'*. The level can be marked as Terrace ***To*** which is situated about 400 feet above the present riverbed. The available tools are large and massive made from crude split pebbles. The dorsal (upper) surfaces of these tools show no working, except the occasional marginal scars. But the ventral (inner) surfaces reveal the evidences of flaking with prominent bulb of percussion. Striking platforms are simple and unfaceted. No secondary working is found on them. Since all these tools are heavily rolled, therefore the edges are worn out. This worning may be either natural or indicates a long use. However, 'Pre-Soan' tools possess resemblance with *'Cromerian'* eoliths of Europe. Some of these tools have been reported from other areas like Malakpur, Chauntra, Adial, Kallar, Chaomukh and Jammu in Potwar, SouthEast of Rawalpindi.

Terrace- I (T1)

The first river Terrace *(T1)* was formed during the second inter-glacial phase cutting the boulder conglomerate. This is 220 feet high from the present riverbed. The tool-assemblage in this terrace has been labeled as *'Early Soan Industry'*. The industry consists of tools made on split pebbles or pebble halves where the number of flaking are quite limited. Such flaking show no initially prepared striking platform; bi-polar technique has been employed in detaching the flakes. The tool-types include scrapers, borers and other pebble flakes. The materials are mostly the fine-grained quartzite of different variety. However, the tools can be divided into three groups on the degree of patination and weathering. No stratigraphic evidence can be shown in support of this classification. The 'Group A' tools are considered as the earliest of the three groups because these are heavily rolled and show numerous marks of patination and weathering. The 'Group B' stands in the middle. The tools are heavily patinated but less withered. The 'Group C' tools are comparatively fresh showing little evidence of patination and weathering.

The *'Early Soan'* is primarily a pebble-tool industry as most of the tools are made on rounded pebbles. Minimum flaking are found on them as necessary to obtain a shape of a chopper. Such massive pebble-tools are more or less contemporary (or even may be earlier) to the Paleolithic tools of South and East Africa. But the difference is that, the Soan tools are much simple in form being governed by the nature of raw material. India is a home of quartzite; therefore it was really a credit for the Indian toolmakers who had developed the technical capacity to make effective tools out of a hard refractory rock like quartzite. A few flake implements that have been found from the *Early Soan Industry* (belong to Group- B and Group-C) are made on thick and heavy flakes and they are equally crude like the parent cores from which they were detached. These flake implements are comparable to the earliest flake-tools of Europe as found in *Clactonian industry.* Although Paterson has designated the tools *of Early Soan* 'flat-based pebble tools', the American archaeologist Movius has called them 'chopper tools' and placed the industry in the same group as observed in Burma *(Anyathian)*, Malaya *(Tampanian)*, Java *(Pajitanian)*. According to him, the tools that are found with *Pithecanthropus* at Chowkoutien also go with this group. If this correlation is proved correct, the valuable link between tool groups and the type of fossil man can be established more firmly. Thus chopper-tools of Soan culture bear East Asiatic affinity*; they must be the products of palaeanthropic branch of human family in Middle Pleistocene time.

Terrace -II (T2)

This Terrace was formed during the third glacial phase when the streams eroded spreading gravel. Very fine silt called loessic loam was also deposited during this period, covering the whole of Potwar region. Height of this Terrace is about 120 feet from the present day riverbed. In this new terrace, gravel is found at the base and the thick deposit of loessic silt known as 'Potwar loess' is found on

* Early Soan pebble tradition is totally devoid of true bi-faces, so the tradition recalls the *Anyathian* and *Pajitanian* Cultures of South - East Asia.

them. More evolved tool types of early man have been recognized from this level; the Soan industry became developed in technique as the time passed on. The tool-assemblage of Terrace-II *(T 2)* has been named as 'Late Soan Industry'. Typologically and stratigraphically this industry can be divided into two distinct groups - 'Late Soan -A' and 'Late Soan -B'. The tools belonging to 'Late Soan-A' are lain in the basal gravel of Terrace II *(T2),* whereas the tools of'Late Soan-B' are found buried in the thick loessic silt of the same terrace.

The main feature of Late Soan Industry is the appearance of flake tools and cores, which have undergone previous preparation. This technique of detaching flakes from a prepared core may be synonymous with the *Levalloisian* technique. However, from the Late Soan -A we get a number of discoidal cores which are actually flattened circular pebbles, flaked all around the periphery (alternatively from each surface) so as to produce a diamond shaped cross-section with a wavy edge. Sometimes, a patch of cortex at the center is seen. Besides, *Levalloisian* type of core on a flat pebble with faceted platform has been distinguished. Though these cores are not prepared as perfectly as the true *Levalloisian,* but definitely of *Levalloisian* character. Geologically the period goes back to the early part of Pleistocene epoch and corresponds to the *Abbevillian -Acheulean* culture of Europe. The tools of Late Soan -B are made essentially on elongated flakes; majority of the excavated flakes show faceted striking platforms and their upper surfaces are carefully trimmed. According to Paterson, these flakes are similar to the Late *Levalloisian* flakes of Europe.

Terrace-III (T 3)

During third inter-glacial period, the atmospheric temperature increased once again and this released great amount of water from ice caps, which in turn eroded the Potwar basin. Naturally a new Terrace came into existence. It consisted of the gravel covered with the deposition of loessic silt and has been designated as redeposited Potwar silt. This Terrace is situated at 80 feet above the present day riverbed. It is very interesting that no tools of man have came out from this Terrace.

Terrace-IV(T 4)

During the last glacial phase, the weather became extremely cold and humid and there was a deposition of loessic soil. Terrace -IV (T 4) of Potwar region was formed at this time on the river valley of Soan. Height of the Terrace is 40 feet from the present day riverbed. It is composed of loam, sand and gravel. The tool industry that has been recovered from this Terrace is known as *'Evolved Soan'.* In fact, the *'Evolved Soan Industry'* has been reported from two sites, namely, Pindi Gheb and Dhok Pathan. The tools are not much different from the *Early* and *Late Soan,* but technologically they are much developed. The *Levalloisian* technique that appeared in *Late Soan* had shown a further refinement in the *Evolved Soan.* As the flakes of this level are much thinner and slender, they look like a blade. A new type of tool, an awl has been discovered here. It is made on a small round pebble; flakings are found on both surfaces which ultimately help to form a sharp point at one end. However, the tools of this level can be correlated to the tools of Upper Paleolithic as found in Europe and Africa.

Terrace -V(T 5)

This is the lowest Terrace, which show the evidence of post-glacial period. Height of this Terrace is about 20 feet above the present stream.

From the study of different Terraces we can reach to the conclusion that the evolution of man and his tools took place with the succession of environment. The *Soan culture* or *industry* is actually divided into three main phases—*Early Soan, Late Soan* and *Evolved Soan* which originated as early as second inter-glacial period and persisted upto the last interglacial period and beyond, being distributed in North-West India.

The Madras Culture

In South India there are abundant evidences of core culture. Core tool industry of South India is known *as Madras Industry.* Attirampakkam and Vadamadurai are the two important sites in the Kortalayar valley of Madras. The area shows the evidences of pluviations, instead of the evidences of glaciations. The core tools that come out from the second inter-pluvial deposit can be equated with the second inter-glacial phase of the North. But in actual geological sequence it goes into third inter-glacial time or even later. In early 1913, Burkitt and Commiade threw some fresh light on the sequences of Stone Age of India by working out a cycle of pluvial and inter-pluvial stages for peninsular India.

The most common and characteristic tool-type of *'Madras Industry'* is hand-axe. These hand-axes are typically pear-shaped or oval, flaked on both faces to produce a continuous cutting edge. Some pebble tools are also found in association with hand-axes. Technologically the hand-axes are more evolved than the pebble tools as they show multi-directional flaking and symmetrical forms. In technique *the Madras Industry is* almost similar to the Lower Paleolithic industries of South Africa, especially the *Stellenbosch* group, which is also known as *'Abbevellio-Acheulean'*. Not only the tools of these two Continents are typologically similar, similarity is also found in the quality of raw material. Quartz has been used in both the areas. The *'Abbevellio-Acheulean'* tools of India also exhibit a similarity to the West European prototypes. Only difference is that, flint has been used in Europe because this material is very soft to be flaked as well as found abound in their country. However, the hand-axes of India, as in Europe and Africa, show a gradual refinement in form and technique with time. Like Africa, Indian hand-axes are made on both core and flake showing an integration of core-flake technique.

The tool named as cleaver appeared a little later. It is often found associated with *the Acheulean* type of hand-axe. This tool is also standardized into four or five main types with some variations in technique. A cleaver usually suggests a tropical woodland environment where it is used for cutting and shaping of wood and for skinning and flaying of animal carcass. Indian cleavers also resemble to the African proto-type in form and technique. However, the tools of *Madras industry* are also found in other places of India like the valley of river Cauveri and Vaigai, in the West round Mumbai and North of the Narmada, and further North-East as far as the upper reaches of the Son, a tributary of the Ganges.

The sites are generally composed of gravel or boulder conglomerate at the base and a fine lateritic gravel bed at the top. In Attirampakkam, lateritic gravel bed reveals a good number of hand-axes and cleavers. Cleavers show triangular or rectangular cross-sections. The flake tools, which are found in association of core tools, show the mark of retouching and step flaking. In Vadamadurai, findings have been categorized into three groups - first, second and third on the basis of patination as well as typology. The first group is the earliest group where the tools are heavily rolled and show the signs of intensive patination. The core tools include hand-axe, cleaver etc. while the flake tools comprise of scrapers, awl etc. These tools resemble the *Abbevillian-Acheulean* tool types and occur in the deposit of boulder gravel. The second group of tool lies in the lateritic gravel bed just on the top of boulder conglomerate. Most of the tools of this group are pear-shaped and ovate hand-axes. An advancement of typology is indicated with the adoption of *Levalloisian* technique. The third group includes neatly worked hand-axes and cleavers. These tools show little patination and are placed on the top of lateritic bed. The technology went towards more perfection. True *Levalloisian* flakes and cores characterize this group.

The Gudiyan site also reveals early Paleolithic tools *of Acheulean tradition.* In a similar way, the industry is divided into three phases—Phase-I, Phase-II and Phase-III. Similar improvement of typology and technology has been recorded from phase to phase. Basic tools are hand-axe, cleaver, point and awl. The Nevasa site on Praber river valley in Maharashtra is also very significant for early Stone Age Tools. But stratigraphically it succeeds the typical hand-axe culture of Attirampakkam and Vadamadurai. Here hand-axes and cleavers are found in association of scrapers, points, borers and incipient blades. The tool-types suggest that the makers of the tool by this time have been acquainted with cylinder-hammer technique. The age of this industry may be assigned to the late Pleistocene when the Middle Palaeolithic culture flourished.

Therefore, the Nevasa industry is definitely younger than the pure hand-axe culture of Middle Pleistocene period. But the flora and fauna of these two periods (Middle and Late Pleistocene) are more or less same. The Maheshwar site on the river Narmada in Central India and the Nandur Madhmeshwar site on the river Godavari in Western India also reveal the same typological and stratigraphical evidence. The Kurnool caves of South India are no exception. In fact, a developed *Levalloisian* technique have been observed in North-West India particularly from the *Late Soan Industry,* in Gujrat, in Narmada Valley, Madhya Pradesh and also further South in the Deccan and Madras.

With the march of time, the core and flake cultures in India were replaced by blade cultures heralding a new lithic tradition in India toward the close of the Pleistocene. New raw material was tapped for making blade industry. In the Deccan extensive use of cryptocrystallines colloidal silica is found with the names like agate, jasper, chert, chalcedony, flint, etc. The *Levalloisian* technique as a superior technique was gradually able to produce new range of tools - knives, scrapers, burin etc. With the advent of blade technology, the use of multipurpose tools like hand-axes and cleaver was declined; Lower Palaeolithic culture was superseded by Middle Paleolithic culture. In many Paleolithic sites in Gujrat and Madhya Pradesh, particularly in Narmada Valley alluvial tract, an amalgamated culture has been noted comprising of both the Northern Soan element and Southern biface.

Antiquity of Culture in India

The landmass of India presents well-marked peculiarities. The Himalayan ranges bound its North; the sea surrounds South, West and parts of east. In the East, the great overland trade routes between ancient Middle East and China, lies a little apart from India. The NorthWest is arid and semi-desert while the NorthEast is covered with dense tropical forests and mighty rivers. These features made the India a unique land, cut from the rest of the world. The people who once reached to India in ancient days could not able to keep easy communications with the land from which they came from. However, the entrances to India were quite limited. They are as follows:

1. The north-western passage
2. The north-eastern passage
3. The western seacoast

The North Eastern passage was very important for the ancient Indian people as well as culture. The earliest people to come to India were the Mongoloid people who represent a Tibeto-Burman linguistic stock. This stock is found to be scattered from the pacific coast to the center of India today. They penetrated in India along the whole Tibeto-Indian frontier including the Central Himalayas. According to Movius the first phases *of Soan culture* can be related to the western most phase of a widespread Sino-Burmese palaeolithic area. India played an apparently peripheral role to the Central Asia, in terms of culture. As regard to the palaeolithic hand-axe culture of the South, the implements seem to be a little advanced than Europe and Africa; the Old Stone Age culture of India is younger as compared to those of Eurafrica.

MESOLITHIC

The Mesolithic or Middle Stone Age is a part of Holocene (geological Recent) that began at the end of Pleistocene period. It exerted a profound influence on the course of prehistory. A change primarily affected the people of advanced palaeolithic culture occupying the regions subjected to the late glacial or pluvial climate. The time has been grossly counted as 10,000 B.C. In fact, Ice Age ended about 16,000 B.C., but the glaciers in high altitude did not vanish overnight; it took time to melt away.

Mesolithic was a brief period in comparison to Paleolithic. It was a less distinctive period for the use of stone. Moreover, some significant non-lithic innovations were found during Mesolithic. In fact, from the end of the Magdalenian period of Upper Palaeolithic, a rapid change in atmospheric temperature was noticed. The change was so vigorous that it altered the ecological niche and upset the balance of human society. As a result a major readjustment was found in the sphere of culture which we define as Mesolithic. But in some territories, it is difficult to distinguish Mesolithic from Upper Paleolithic (on the basis of cultural ground) because those territories did not witness any sudden change in natural environment.

The most general result that felt in different parts of globe was an universal rise in sea level as an enormous quantity of water from melted glaciers flowed back to the oceans. The evidence has been preserved in the coasts of Scotland and Scandinavia when the sea had reached its maximum height, more than fifty feet than the present day level of shore. However, the change of climate was gradual and discontinuous. The weather in temperate latitude got warmer whereas it became drier in Mediterranean and sub-tropical zones. Such climatic changes altered the plant and wild animal types and affected the human life. Man had to adapt themselves to quite novel conditions. The cultural development of Mesolithic has been best observed in Europe.

Throughout the Upper Paleolithic, the regions such as France, England, Czechoslovakia, etc. were the large tracts of steppe and tundra. Naturally the areas were almost open with shrubs only. For the change of climate, forest was found to invade these areas following a order like birch first, then pines and thereafter oaks and other deciduous trees. On the other hand, in South, forests withered under drought and prairies gradually turned to deserts. The mammoth, woolly rhinoceros, reindeer and bison that were abundant in cold environment withdrew themselves, as they could not cope up the changed environment. Some of them spread eastward and northward; others died out. The new species, which succeeded them, were stag deer, wild boar and urus, the giant wild cattle. The act of hunting became difficult in forest situation and at the same time the new games could not hold up the growing population with enough food. Therefore Mesolithic people began to scatter out from the old territory and diverted their attention towards water creatures like fish, shellfish, waterfowl etc. This shift in the subsistence pattern was mainly responsible for the cultural development of Mesolithic period. The Upper Paleolithic economy of food gathering persisted in Mesolithic but fishing, fowling and collecting were added. At the same time, the base of farming was prepared. Thus, Mesolithic stood as a well-defined stage of terminal food gathering, which can safely be placed between Palaeolithic and Neolithic. The development towards farming served as definite evolutionary link between the two periods. The Mesolithic or the Middle Stone Age is therefore a transitional phase; lying between the Old Stone Age and the New Stone Age, it denotes a period of readjustment.

Most of the Archaeologists believe that the environmental changes induced some European groups to alter their food pattern. There are some other school of thought who argue that the new races of man with new fauna came from the South and the East to crush and absorb the Upper Paleolithic races of Europe. The new races of man were considered as real seafood relishers as their living areas exhibited a store of millions of mussel shells, which perhaps were tossed aside after enjoying the meal. Changes in tool technology were evident at this time. The elegant harpoons of the Upper Paleolithic had become a poor stag-horn copy with eye for lashing to the haft. Flints turned to pygmy tools, microliths.

Microliths are the characteristic tools of Mesolithic Age. These are extremely minute. Some measure only 3/16 inch, or even less in size. The shapes vary greatly but the usual forms are more or less geometrical. Crescent and lozenge shaped microliths are not uncommon at all. All are the tiny tools, which could be attached, joined, or embedded to the wooden or bone handles. Such microliths used to be hafted in rows to act as knives, sows, sickles, spearheads, chisel-edged arrow-points, etc. These new tools involved a new technology as an aid to the new economic life. Mesolithic people invented a wonderful device for killing of the animals—*bow and arrow.* Soon it became a major weapon of the people; they learnt to use chisel-headed arrow points. Blunt-headed wooden arrows were also used for stunning birds and small games without ruining skin. The other tool typical to the Mesolithic period was the stone-axe, meant for woodcarving. It was not like an old hand-axe, rather a true axe with a haft. The haft was parallel to the cutting edge. Prehistorians have named it as '*hewn axe*' to differentiate the type from the previous *Abbevillean-Acheulean* hand-axes as well as from the ground and polished axes of Neolithic.

The most significant developments of Mesolithic were domestication of animal (dog), invention of pottery and the bow. All these three became more important in later period, Neolithic. The dog was perhaps associated with man from the very beginning but it did not serve any utilitarian purpose until the Mesolithic period. A meager evidence of pottery was found in late stage of Mesolithic where its primary function was boiling. The bone-embedded stone arrowheads justify the presence of bow in Europe during Mesolithic. Ertebolle shell mounds specially suggest this. Besides, some small, transverse-edged, projectile points have been discovered from an earlier level of culture, *Tardenoisian*.

Inspite of these developments, Mesolithic showed some apparent degradation. The excellent artistic creation *of Magdalenian* culture disappeared in Mesolithic Age. Instead, some roughly painted pebbles were distinguished.Mesolithic men were contented in daubing a few simple marks on pebbles. It contained some symbols where a few strokes conveyed a particular meaning without being actually a picture. They can be taken as the earliest concept of communication—an intellectual base of writing. Thus, Mesolithic art is nothing but a symbolic representation of oldest hieroglyph. It has been labeled as *'daubbling-art'*.

Mesolithic in Europe

The time span of the Mesolithic period in Europe is considered as roughly between 23000BC. and 12000BC. but some Prehistorians claim that it is in between 15000BC. to 8000BC. The European Middle Stone Age may be divided into two phases, first one is the early phase and second one is the middle and late phase. However, both of these two transitional phases include the following cultures. Phase 1 or Early Phase: *Azilian, Tardenoisian and Asuturian.* Phase 2 or Middle and Late Phase: *Maglemosean, Kitchen Midden or Ertebolle, and Campignion.*

Azilian

It is the oldest transitional Mesolithic culture, which is named after the site at Mas d' Azil (Ariege) in Southern France. The stratigraphic position of *Azilian culture* is clearly established in between the *Final Magdalenian* (below) and the *Tardenoisian* (above). Probably, this culture has evolved directly out of *Magdalenian* and centered in the region of Pyrenees. The main distribution is found in Switzerland, Belgium and Scotland.

The Azilian culture is characterized by two type specimens. Firstly, the flat harpoons made of stag horn with a perforated base and secondly, the painted pebbles with stripes, dots etc. Both of these type specimens have been recovered from most of the sites of *Azilian culture.* In case of painted pebbles, all are river pebbles, rounded and smoothed by natural water action. These pebbles are smeared with red ochre in the form of dashes and dots, wavy lines and their combinations. Such painted pebbles are numerous in number but their significance is still unknown. Among the stone tools there are small blades of flint, which show their resemblance with microliths. These tools include small rounded or thumbnail scrapers, points, backed blades and burins. Besides, some bone antler points and gouges (chisel with curved cutting edge) also have occurred along with a few perforated teeth and ornaments. The fauna of the period is modern and many of those species of animals are still common in Europe. No art has been discovered from this early Mesolithic stage.

Tardenoisian

The people *of theAzilian culture* mostly lived in the mouths of caves or in rock-shelters. The remains have been discovered from a definite stratigraphy in association with the deposits containing Paleolithic industries. With rare exceptions, the *Tardenoisian* deposits have also been found in air sites, usually on sandy soil. The type-site is the La Frdre-en-Tardenois, (Aisne) in NorthEastern France. This *Tardenoisian* culture is closely related and clearly followed from the *Azilian* culture. It is distinguished by the prevalence of pigmy flints. Though the culture seems to be almost pan-European, it is not found in Scandinavia, Northern Russia, Balkan and Italy. The culture specially flourished in sandy thin soil districts where timber remained sparse or lacking.

It is a microlithic culture per excellence where the pigmy flints are chipped to form geometrical shapes, such as the triangles, crescents, trapezoids, transverse arrowheads, and tiny backed blades and points. These are all very small and usually mounted in slotted hafts. Besides, the small flint flakes are also glued into slender bones to produce harpoon heads with sharp cutting teeth instead of barbs. The small pygmy burin or micro-burin is the typical tool of the *Tardenoisian culture,* which is generally regarded as a by-product in the process of microlith manufacture. No axe or painted pebble was found from this culture level. The cave burials are found in this *Tardenoisian* culture that contains complete skeletons along with the microlithic flint tools and bone implements. Some skeletons are kept covered with stag antlers. The physical characteristics of these skeletons represent modern man in all essential respects

Austurian

The type-site is located in Asturia province at North Spain. This culture overlie on the Azilian layers in many rock-shelters and cave sites in the coastal regions of Northern Spain and Portugal. Therefore, and it may be designated as 'Crude coastal culture'. Evidences of this culture come from extensive shell mounds containing the *Asturian* rubbish forming with remains of oysters (edible molluscs), cockles (shell fishes) and limpets (shell fish which sticks tightly to rocks) which were once the principal elements of diet among this shore dwelling population (Fig. 11.17).

The Austurian implements are mostly of pebbles, approaching towards the preparation of rough axes and scrapers of quartzite. All are useful for detaching shellfish from the rocks. The main important tool is the pick hewed from round stone (cobble). Some bone and antler implements are also found in addition. However, the culture is unique, as it exhibits no microlith or blade.

Maglemosean

The sites of this culture are mainly distributed on the shores of inland waters and marshy regions of Baltic area. The term Maglemose has been derived from the Danish word meaning a 'Great Bog'. The culture has been named after a site containing the 'Great Bog' (Maglemose) near Mullerup in Zealand at Denmark. The people of this culture adopted the forest environment where the economy was based on food-collection. (Fig. 11.18). They used to hunt forest animals and birds. Fishing and gathering of wild nuts and berries were also very common. The existence of dog as a domestic animal was noted from this culture. The culture can be demonstrated not only by the game hunt, rather by the use of the axes, wooden implements, dugout boats, paddles etc. An abundance of bone tools are found in association with microliths.

Before the main development *of Maglemosean culture* we have found a *Tanged point culture* to the Southern Baltic area. The chief tool type of that culture was the tanged point, perhaps derived from the Upper Paleolithic culture. However, the *tangedpoint culture* included the microliths, gravers and blade-scrapers of Upper Palaeolithic tradition. Axes were completely absent with exception of a few scattered Antler axes. But the main development of the subsequent *Maglemosean culture* was noted in the axe-culture.

The most important tools *of the Maglemosean culture* are the flint axes where the edges are found to be formed by the intersection of two or more flake scars. Normally, these flint-axes are made out of cores, although a few flake-axes are also found. Stone pebbles with sharpened ends were also used as axes. Perforated antler axes and adzes were not uncommon. Among the bone tools, the bone points, especially barbed bone points, barbless fishhooks, fishing spears, are very important. In fact, many of the stone tools and bone implements suggest the Upper Paleolithic derivation but the typical axe-types are of Mesolithic origin. Especially the hewn-axes are typical for the culture.

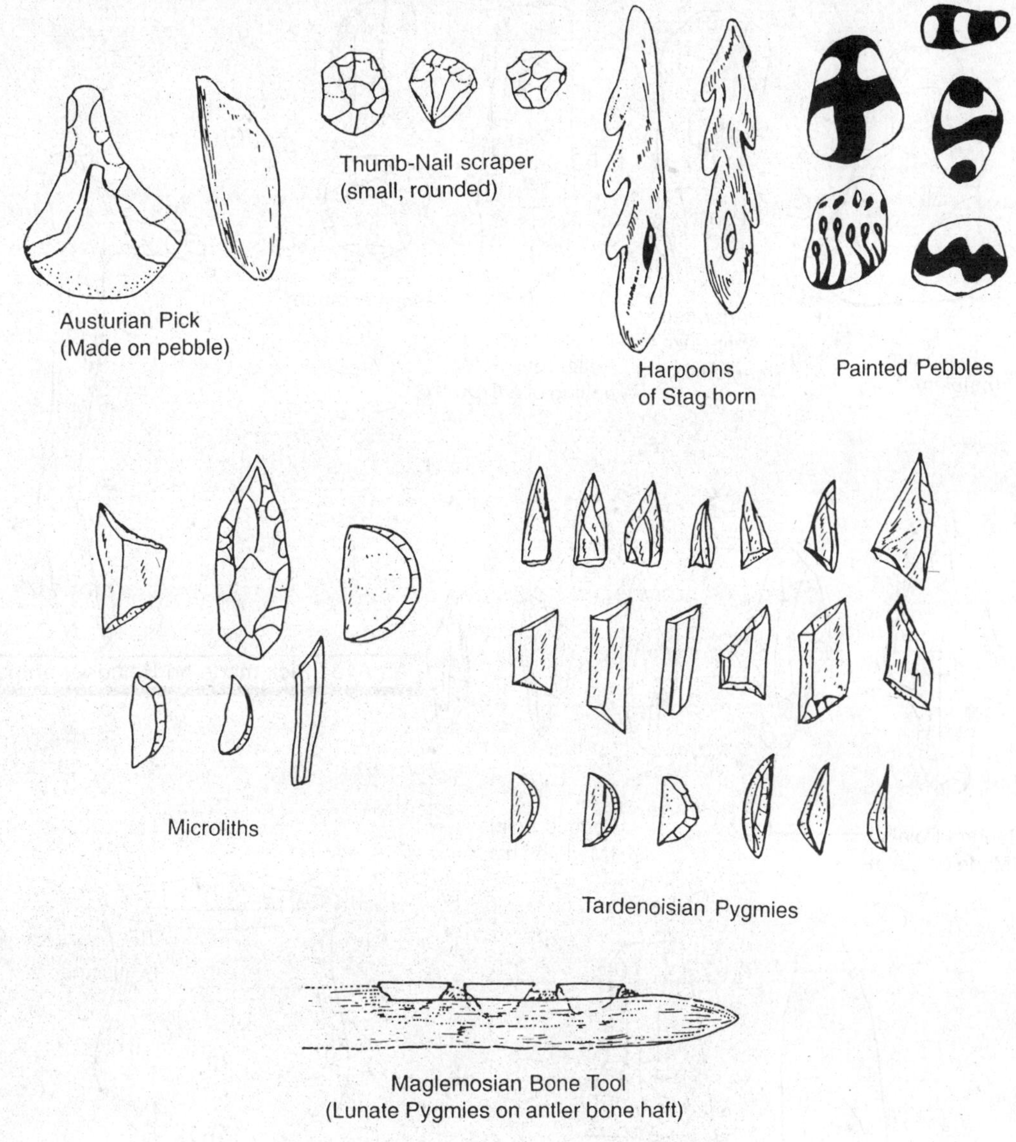

Thumb-Nail scraper (small, rounded)

Austurian Pick (Made on pebble)

Harpoons of Stag horn

Painted Pebbles

Microliths

Tardenoisian Pygmies

Maglemosian Bone Tool (Lunate Pygmies on antler bone haft)

Fig. 11.17. Mesolithic Tools

Ertebolle

Ertebolle culture apparently developed from the late stage of Maglemosean cultural period and it mainly flourished in Denmark. A humid period with heavy forestation was probably responsible for the rise of *Ertebolle culture.* As the condition was difficult for hunting, people turned to molluscs. Hence *refuse mounds of shells have characterized Ertebolle culture.* It is also known as '*Kitchen Midden*' or'*Shell Mound*' culture. However, the culture is found to be distributed especially in Eastern Baltic area in Britain, in Scandinavia and in the interior of Jutland.

The culture is named after its type-site located in France. A large number of middens comprising of shells of edible molluscs have been accumulated in this area. Further, the contemporary coastline in many places closely follows these sites. The deer-antler harpoon has been disappeared and the

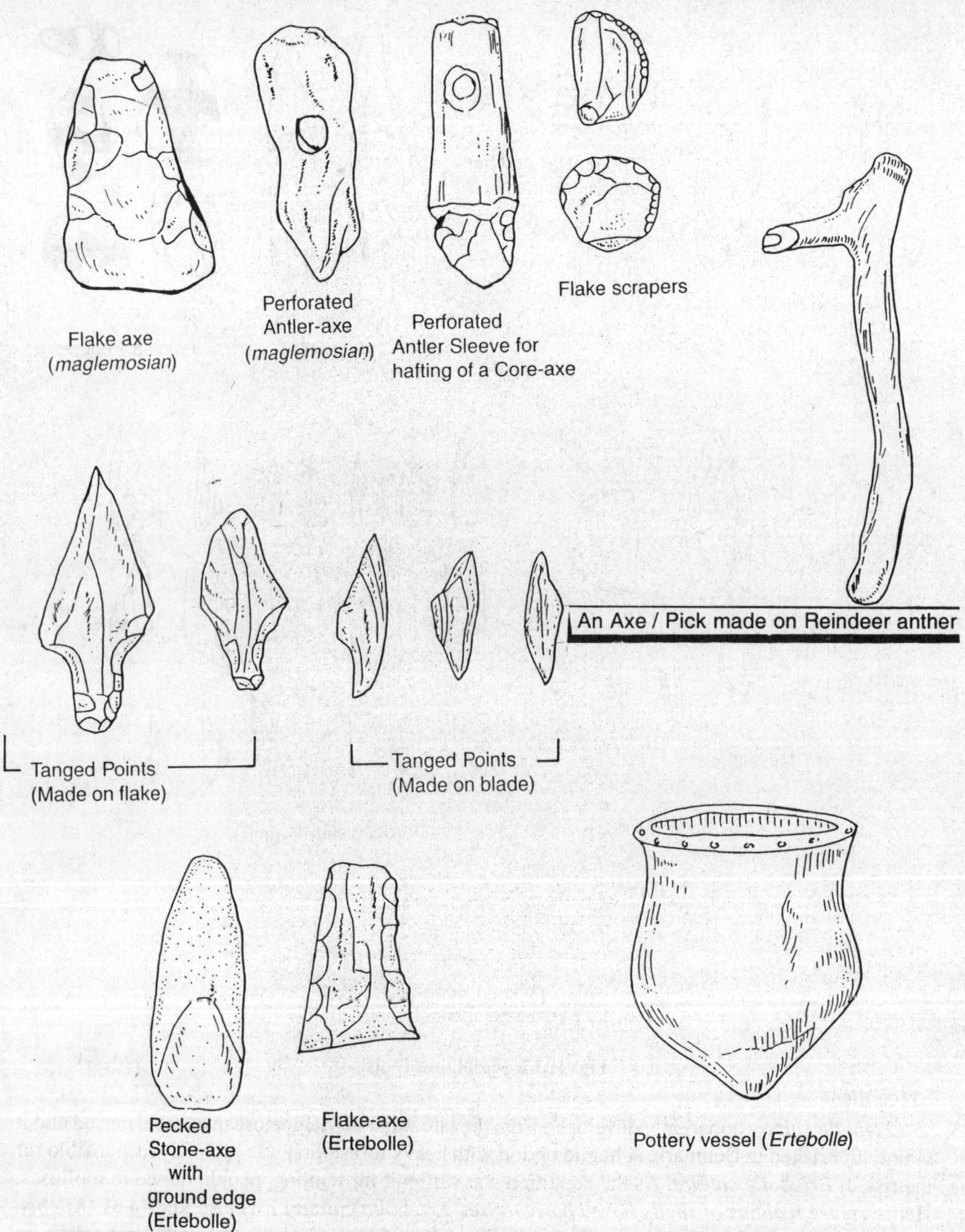

Fig. 11.18. Mesolithic Tools

bone-point bas became rare in this culture. The flake-axes or tranchets of flint are found instead of the core-axes *of Maglemosean* type. But the smooth stone axes exhibit the sign of pecking or polishing for which they are thought to be derived from the pebble as the pebble-axe. Some antler axes with perforation for hafting are also found. Still the most important implement of this culture is the bow. The presence of transverse arrowheads of *Tardenoisian* type suggests the use of the bow. This culture also bear the evidence of domestication. Dog as the first domesticated animal is attributed to *theAzilian,* to some of the *Tardenoisian,* but abundantly found in the *Ertebolle* phase of Mesolithic. Sun baked pottery (coarse cooking pots with rounded or pointed base) have been recovered from several sites *of Ertebolle culture.* These evidences of pottery and the domestication of animal mark off *Frtebolle* as a transitional stage to the Neolithic culture.

Campignian

This culture is more nearer to the early Neolithic owing to the presence of coarse pottery and a few rare examples of domesticated animals. Stratigraphically the *Campignian* is posited before the true Neolithic. The polished stone tools include the campignian axe, the pick and the transverse arrowheads. Besides, there are also rough awls, scrapers, flakes, cores, etc.

The existence *of Campignian culture* can be demonstrated over a large area of NorthWestern Europe. The Mesolithic period as a whole was not progressive. But culture contained the seeds of development for which a sudden change over took place at the arrival of the Neolithic period.

Mesolithic in Near East

Culture of the Mesolithic Europe was an extension of Paleolithic culture. Parallel cultural developments have been noted in the Near East. It also showed an adaptation to the changing environment. The people are the *Natufians,* named after the site Wadi-el-Natuf where their presence was first recognized. The site is actually a strip of land within forty miles of the East Mediterranean between Beirut in the North and the Judaen desert in the South but far away from Heluan, some twenty miles South of Cairo. The early *Natufians* used to live in the open-sites during Mesolithic. Though the economy was primarily based on hunting, the people were expert in fishing as evidenced by the existence of slender bone spearheads (barbed on one edge) and plain fish-hooks. They also had reaping knives where the flint blades were inserted into sockets, in the handles of bone. They also exhibited the production of blade and burin and the use of bone and antler in the same scale as like Upper Paleolithic tradition. Presence of stone pestles and mortars suggest the grinding of cereals. These cereals were naturally produced but harvested by the people. The Mesolithic inhabitants of Palestine contributed to the early development of settled life based on farming. *Natufians* are the earliest Mesolithic people known to us who developed many differences from the Paleolithic food collectors.

Mesolithic in Africa

Since Africa presents a different type of climate in comparison to Europe, no distinct break is evident between Paleolithic and Mesolithic. Further, Mesolithic and Neolithic are overlapped. In North Africa, the Capsian hunters-fishers of Maghreb belonged to the Mesolithic phase until Neolithic impulses passed westward along the northern fringes of the Sahara. They used to live in caves as well as in open-air. For subsistence they largely depended on hunting and on the collection of molluscs. At this time they began to keep domestic animals like sheep which stood as an additional source of food. By examining the caves and open-air middens, it has been found that the people knew the microlith technology; lunate and trapeziform were the common shapes. They used these microliths for various purposes, after fitting them suitably in wooden or bone frame. Different kinds of arrows were evident among these people where microliths armed the arrows. Chisel-ended arrowheads were very common. Blades of reaping-knives or sickles were notably absent in them which suggest that there was no trace of agriculture. But there was polished stone axes and coil-built pottery. Thus, Mesolithic of North Africa did not achieve a separate identity. The entire Paleolithic tradition termed as 'Capsian' (by the French Prehistorians) was continued for a pretty long period till the Neolithic traits diffused from Egypt. The ground axes and the domestication of animals were essentially Neolithic traits. However, the *Capsian culture* in North Africa terminated around 3000BC.

In Post-Pleistocene period (Holocene) East Africa shows a unique culture in Kenya Rift Valley known as *Elementeitan culture.* The important findings of this culture are the occurrence of the lunates, end-scrapers, bone awls etc. The lunates were well made and more symmetrical than the previous level of culture (Kenya Capsian) while the burin became rare. An overall better craftsmanship was noted. This cultural stage represents the Mesolithic phase of East Africa. Another site near the railway station of Khartoum showed a Mesolithic industry with the evidence of pottery. The pottery, which comes from Khartoum, seems to be earliest type of pottery not only for Africa but also for the world. However, the industry contains microlithic tools of different shapes like lunates, crescents, trapezoids, etc. The transverse (chisel ended) arrowheads that are widely found in Mesolithic sites of Europe are also widely distributed in this site at Khartoum. Further, there are long, narrow backed blades merging into crescents and borers, crude scrapers and a few burins. Burins are rare and atypical in occurrence. Small sandstone pebbles, each having a groove around the middle suggest the presence of fishing-line sinkers. But no fishhook is found from this site. Bone harpoon, spear and grooved sinkers led the excavator to infer that those people living at the bank of Nile in Mesolithic times were obviously the fishermen. So far as the pottery is concerned, no complete pots are found. Huge shreds show the wavy-line decoration on them.

In South Africa, the Evolved Fauresmith and stillbay industry of Old Stone Age yields plenty of microliths with small pressure flaked points, small disc cores etc. It corresponds to the Mesolithic phase in South African prehistory.

Mesolithic in India

As in Europe, Africa and several parts of Asia, India also witnessed a great climatic change toward the end of the Paleolithic period. These must have a bearing upon the mode of human living. Although there was widespread archaeological evidence of this environment in the form of microliths, but that was not supported by stratigraphical, geo-chronological and palaeo-biological data. So the microliths carried little cultural significance. Stuart Piggott (1950) stated, "The Indian microlithic industries have not been found at any site in a geological deposit of known age, nor in such stratigraphical relationship to earlier human industries that a connection with the final phases of Paleolithic cultures can be shown". According to Sankalia (1962), in India, the term 'Mesolithic' stands for the Late Stone Age because it does not bear the same stratigraphic, chronological, typological and economic significance as in other parts of the world, particularly like England, Western and Northern Europe, parts of Africa and the Palestine. The term may conventionally be applied in India to denote a cultural stage, which is represented by microlithic industry and not associated with pottery. It also antedates the earliest farming-based village cultures. Robert Bruce Foote had placed these tiny implements (microliths) in the Neolithic period. He coined the term Proto-Neolithic to refer the culture of the regions where microliths had occurred with potsherd. This association according to Foote definitely denotes a higher state of life.

The evidences of'Mesolithic stage' in India are quantitatively and qualitatively richer than the preceding stage of Stone Age, the Paleolithic Age. In certain areas, these industries have survived at least dawn to early historic times showing a transition from Paleolithic industries to a more advanced type of industries. This survival has been possible as most of the Mesolithic sites are associated with the existing landscape that has undergone little or no significant changes. Cultural evidences are thus totally undisturbed, the physical remains of animal and human preserved in Mesolithic deposits enables us to reconstruct the subsistence pattern of the people and environmental setting of the culture. The oldest human skeletal remains yet known from India came from the Mesolithic level. But, is really a matter of great disgrace that no plant remain has yet been found from any

Mesolithic site. A number of living sites of this age that have been discovered mainly from Western and Central India provide us with variety of data regarding the way of the life of the communities.

The Mesolithic sites in distribution cover almost the entire country except a few areas like Indo-gangetic plain, Assam and most of the Western coast of India. In Indo-gangetic plain, their absence can be explained by the lack of primary raw material (stones) for making tools. Assam and Western coast were probably left uninhabited due to very high rainfall and dense vegetation in this area. The regions like Gujrat plains, Marwar, Mewar etc show dense concentrations of sites in contrast to other areas. The vast tract of country between Godavari and Mahanadi has just started to be explored. The bulk of our information comes from the following sites:

1. Akhaj, Valasana, Hirapur, Lunjhanj at Sabarmati Valley in Gujrat,
2. Tinnevalley in Madras,
3. Luni river valley in Rajasthan,
4. Bundelkhand, Baghelkhand, Shivana Valley in M.P.,
5. Bel Pandhari, Nevasa, Suvegaon, Kalegaon in Godavari Basin,
6. Birbhanpur in West Bengal (some sites are also scattered in Bankura, Birbhum, Burdwan and Midnapore Districts),
7. Kurnool and Renigunta in Andhra Pradesh,
8. Nandur Madmeshwar and Malaprabha - Ghataprabha of Maharashtra, etc.

It seems that the Mesolithic people preferred the following environment:

a. *Sand-dune*

In Gujrat and Marwar hundreds of dunes of varying sizes are found on the alluvial plain. Some of them enclose a shallow lake or pond, which were the great sources of getting aquatic creatures. Again, the dunes themselves were covered with thorny scrub bushes; many animals used to live there. Naturally the Mesolithic inhabitants in sandy dune faced no difficulty in collection their food.

b. *Rock-shelter*

The Vindya, Satpura and Kaimur hills of Central India are very rich in caves and rock-shelters. The place was therefore favourite to the Mesolithic people. Not only that, as Central India received ample rainfall, the hills had grown a thick deciduous forest, which provided a variety of plants and games. Some of the rock-shelters have been found to be occupied as early as the Acheulean times.

c. *Alluvial plain*

From early Palaeolithic period man has preferred to live in riverbanks because of the availability of water and games. Numerous Mesolithic sites therefore have been recovered from the alluvial plains. The Birbhanpur site, for example, is located at Damodar's alluvial plain in West Bengal.

d. *Rocky plain*

On Deccan Plateau, many microlithic sites are found. Some are on the hilltops and others are on flat rocky soil. Such occupations must be the seasonal or of short duration, except where there is no river nearby.

e. *Lake-shore*

A few Mesolithic settlements are centered round the shore of the lakes as found in the Gangetic Valley of District Allahabad and Pratapgarh. The settlers perhaps used to get the food supply from the respective lake and the dense primeval forest of the fertile alluvial land.

f. *Coastal environment*

A large number of microlithic sites have been recovered from coasts, for example, from the Salsetle Island and from the teri dune in District Tirunevelli. The inhabitants used to feed upon the marine resources.

Since Mesolithic essentially produced the micro-blades by pressure technique, beautifully fluted cylindrical or conical cores as well as thin parallel-sided blades are common in the sites. The blades are retouched on one or more edges, mostly by steep blunting. Numerous types among the microliths have been noted such as blunted back blades, obliquely truncated blades, points, lunates, triangles, crescents, trapezes etc. Size of these tools varies between 1cm. and 3cm. Though no actual example of hafting is found from India, comparative evidences available from other parts of the world suggest that these microliths must have been hafted on bone and wooden handle or shaft with natural resin. The people perhaps produced a variety of implements and weapons such as arrows, spears, knives, sickles etc. The advantage of these composite tools was that, being made of a number of microlithic components, one of the broken components could easily be replaced without discarding the whole tool. This type of replacement was not possible in case of a single piece tool.

It can be inferred that the Mesolithic people had little material culture. Later, in contact with the metalworking and farming people they acquired a few items of material culture like, pottery, metal tools and stone beads for ornaments. The process of disposing the deads was also unspecialized among them. The bodies used to be buried within the habitation area, whether it is a rock-shelter or an open-air site. At the site Sarai Nahar Rai in the District Pratapgarh all the skeletons are found to be placed in West-east direction keeping the right forehand diagonally across the abdomen. In fact, we are not sure whether the microlith makers of India were the hunters and food-gatherers of Paleolithic tradition or they were the agriculturists. But it seems to be certain that the corresponding period experienced the arrival of new peoples, possibly from the west. This is the reason for which the Upper Palaeolithic blade industry was non-existent in Indian context. Therefore, the microlithic flint industry of India may be considered as contemporaneous with those of early post-glacial date in Europe and Near East or even they may be of a later date. The Upper Paleolithic blade industries of Europe must be ancestral to Indian microliths. However, the microlithic cultures in India had showed a long survival, side by side with the communities making pottery as well as ground stone axes of the Neolithic tradition.

NEOLITHIC

Like Paleolithic, Neolithic is also a Greek word. It means 'New Stone'. The period since the discovery of agriculture to the rise of urban civilization has been bracketed, as New Stone or Neolithic *Age.* This Age lasted approximately from 8000BC to 3000BC. The earlier anthropologists conceived everything on the basis of technological types, so they distinguished two grand divisions in Stone Age, which were different from each other by the technique of stone working. Palaeolithic tools and implements were made solely by chipping. They were very crude in nature. But the Neolithic stone objects were pecked, ground, rubbed and polished; they were far better in finishing as well as in effectiveness. Man was no longer a food-gatherer, he became a food-producer at the advent Neolithic.

Twentieth Century anthropologists defined Neolithic Age in terms of domesticated plants and animals. They also .counted the presence .of pottery and absence of metal as indispensable criteria for a pure Neolithic site. V. Gordon Childe mentioned that the Neolithic had opened an entire new way of life and he termed it as 'Neolithic Revolution'. After the end of Pleistocene Ice Age, first acquisition of domestic plants and animals stood as the greatest achievement of man and therefore the period has been recognized as a great turning point or 'revolution', perhaps comparable in importance only with the 'Industrial Revolution' of recent period. But the origin and early spread of agriculture is still a matter of conjecture and only fragmentary evidences are found on this topic. There is no area in the world where the stages from pure hunting and gathering to farming can be traced step by step in adequate detail. However, the early development of agriculture was rooted in the environmental and cultural conditions at the end of Pleistocene epoch. By this time, the biological development of man was complete as evident in the skeletal remains of modern type of man corresponding to the particular period. It is therefore assumed that the people of this period had grown the capacities for cultural innovation.

Now the question is, how man from being a predator, a parasite on nature turned into a controller of food? We know that the hunting and collecting economy compelled man to lead a nomadic life moving from place to place in search of food. Sources of subsistence were not secured at all. In farming economy by domestication of animals and plants, man became confident about his subsistence. The food-production armed the people to fight against the scarcities of daily life especially in the areas where local resources were fully consumed. It also saved the people from starvation following epidemic or disaster in animal kingdom. The increased availability of food facilitated quick growth of population who settled down peacefully in permanent habitations. Accumulation of food and mechanism of storing made the people relatively carefree and this stimulated the habits of forethought and planning. Within a short period, they become self-sufficient and self-determinant; stability in life was established. Neolithic revolution did not occur overnight, but still it was not a slow and gradual change. Rather it can be said a drastic change which brought a radical difference in the way of life of the people. Neolithic way of life was vastly superior to the Paleolithic. While culture of Paleolithic was carried forward with a greater span of time, Neolithic culture advanced very quickly within a period of a few thousand years.

Possibly it was the women who started the art of cultivation. When man used to go out for hunting, the women would gather wild plants and fruits from the forest. They for the first time noted that the seeds falling on the bare ground grew up into plants from which again seeds could be available. Thereafter, by continuous trial and error, once they really learnt the art of agriculture. If we have to pick out the greatest single change in human history right up to the present, it is the change of 'Neolithic Age' that gave birth of a state of culture in which food was planted and bred instead of hunted and gathered. The event is the most important advance ever taken by man.

A. L. Kroebar (1923) described the food-production as double-barreled activity. According to him, the raising of the plants and rearing of the animals mostly go together, so both of them should be termed as domestication. But it is yet not known that which one of the two, whether cultivation or stockbreeding came first. Archaeologically, the beginning of food-production can not be pinpointed. Because, the difference between the wild and domesticated varieties have been evident later, after passing of a considerable period. However. Neolithic revolution occurred in a number of areas, quite independently. Compared to Europe and United States, much less is known about the areas like Near East and middle America. But wild ancestors of the major domesticated plants and animals were flourished in all of these areas. Not only that, the earliest agricultural hearth has been discovered from the Near East. Recent studies suggest that the domestication of the sheep perhaps started as early as 9000BC.

The earliest experimentation on food-production started in an area extending eastward from the Mediterranean to the Zagros Mountain of Iran, and northward from the Red Sea and the Persian Gulf to Anatolia in Turkey. These experimentation had specially begun from about 10000BC to 8000BC by the Mesolithic hunters dominated in Palestine region whom we call *Natufians.* These first farmers were not the potters. They used to dwell in caves and in open sites but as the primitive farming was developed in them, their settlements gradually became permanent. Several varieties of wheat, barley, lentil, peas were found to be cultivated among them around 8000BC.

Near East and Europe are the best-known areas where developments leading to food-production and settled life have taken place. In Europe, the glaciers began to disappear about 10000BC. A sharp climatic change accompanied with the changing pattern of vegetation and fauna were noted. Some glacier covered zones of Pleistocene became habitable for human and other animals as Ice retreated and temperature rose. On the other hand, the ice-locked water was released and this lifted up the water level in oceans and seas. A number of rivers were inundated by the huge quantity of water. The melting glaciers also created islands, inlets and bays and moved the sea toward inland. Many cold-loving animals like mammoth, woolly rhinoceros, mastodon, sloth etc. disappeared very quickly. Waterways became abound with fish and other aquatic creatures. As the growing forest engulfed the tundra and grasslands, the hunting people faced tremendous difficulty in chasing the new animals in thick forest. Density of animals per square mile also decreased at this time. Therefore it was a compulsion for them to work out a series of specialized adaptation based on new local resources. The people penetrated into every part of the habitable world, except the outer islands of the Pacific and the poor places like Greenland and Baffin Land. They even roamed the plain between Britain and Denmark, which is now under the North Sea. They tried every kinds of available food including acorns i.e. the things that required special treatment for getting them as edible. This is the time when they found many migratory animals within close proximity. Gradually they gained the skill of stockbreeding and realized the importance of naturally grown crops. This initiated them in searching out the places where the crops grew well and ultimately they learnt purposeful growing of plant seeds near their settlement. But it will be totally a wrong interpretation if we project either the congenial environment or the enlargement of habitable territory in favour of subsistence revolution; rather it is the increased exploitative efficiency and adaptability of man that facilitated the great change over from food collection to food-production.

Stockbreeding and cultivation of cereals were the revolutionary steps in man's emancipation from the dependence of external environment. But these two did not necessarily spread together at a time in any region. Further, when man shifted his reliance from roots, nuts and tubers, they developed new types of stone tools. These tools were mainly for pounding or milling so that the husks could be easily removed and the tough kernels could be broken for easy digest. Some tools were also devised especially for capturing new type of fauna namely cattle, red deer, roe deer, pig etc. Thus, Neolithic brought a marked change in tool types. Although the Neolithic people continued to use flake implements made on flint, a quick development of blade tools was found at this time. Blade tools formed the parts of other heavy tools, for example, they were fitted into wooden handles to be used as awl, graver, saw, sickle etc. The people also preferred to grind their stone implements especially those were produced from igneous rock. Wet sandstone was used as abrasive. As a result, the tools of Neolithic period became much tougher and more durable. But they used to take a long time for making, especially when a hole had to be bored through the stone for fastening it with a handle. The new stone-working technique of boring and grinding was proved very important to prepare a kit of woodworking tools including planes, choppers,wedges and chisels. Among them, the war clubs have been differentiated from other kinds of axes, meant for working on wood. Characteristically all Neolithic axes and adzes are hafted with a handle. A polished stone axe-head of Neolithic is commonly known as celt. (Fig. 11.19). Such a celt with wooden handle was extensively used for clearing the wood and chopping the trees. The bows and arrows were widely used in hunting as well as in war. The most remarkable findings of Neolithic deposits are the skillfully made arrowheads and celts. The practice of fishing was improved during Neolithic period. The implements like spear, harpoon etc. were profuse among the stone tools of Neolithic Age.

As a consequence of food-production, population growth was accelerated during Neolithic period. People had settled down in villages and tried to invent certain ways to make the life easier. The important developments that took place in Neolithic are as follows:

Invention of Pottery

Man needed pots for storing food and especially for cooking. Boiling appeared as an important way to make the vegetables and grains edible. Man first learnt to make baskets in Mesolithic by using thin branches of tree. It proved very useful for storing of dry food. They made the first pot by plastering the clay round the baskets. These were hardened or baked by heating them on fire. Primitive men also used wooden scoops and bamboo barrels for boiling. Sometimes they dropped hot rocks into water in a skin-lined hole in the ground. But none of these efforts were very perfect. In Neolithic period, man first learnt to manufacture the pottery. A great range of variation was found in style and manufacture. But it should be remembered that the Neolithic pottery was made without the aid of the wheel * 1 or kiln *2.

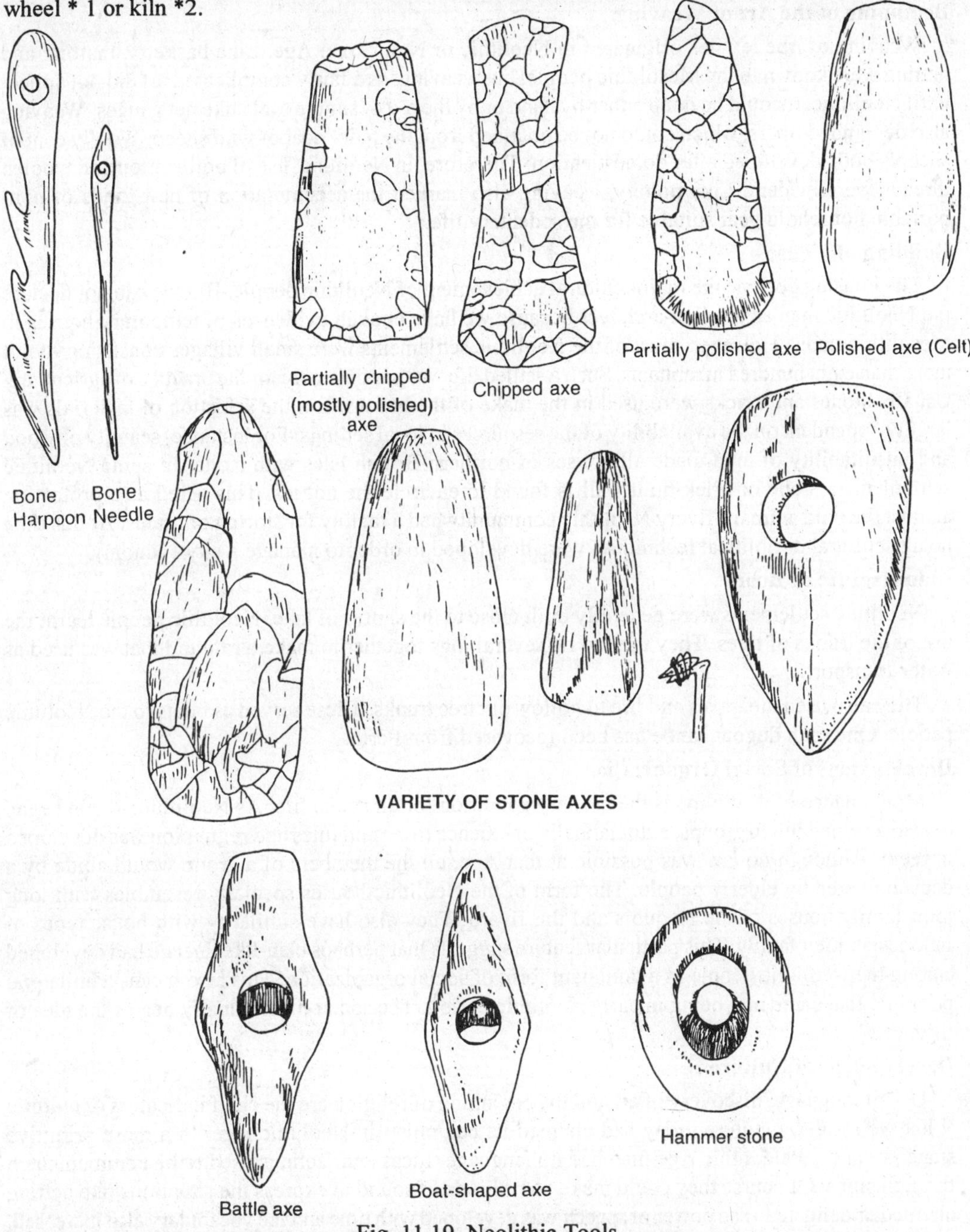

Fig. 11.19. Neolithic Tools

* 1. Wheel was invented in the Old World, particularly in Egypt about 5000 years ago.

*2. All primitive pottery was baked or fired but a kiln was not necessarily used for this purpose.

Pottery is a definite feature of Neolithic. Food collecting people had rejected the clay pots as too burdensome because of their nomadic life, whereas the food-producing people liked them as those were well suited with their sedentary nature of living.

Beginning of the Art of Weaving

Weaving of true textiles originated in Neolithic or New Stone Age. Like basketry, matting and netting were known to the Mesolithic people, but weaving, as a fairly complicated art did not appear until Neolithic. It could be done either by hand or by the aid of simple tools like net gauges. Weaving also demanded fibers, which could not be obtained from the hairy coat of wild sheep. Woolly coat of sheep's body developed after domestication. Therefore, in Neolithic, use of both cotton and woolen threads were evident. Like pottery, weaving also marked the accumulation of new kinds of non-portable household gear suitable for the sedentary life.

Building of Houses

The housing became the major cultural achievement of Neolithic people. But we can not declare the Neolithic man as the first architect because we have enough evidences of temporary houses in Mesolithic times. However, most of the Neolithic settlements were small villages consisting of not more than eight hundred inhabitants. Such a settled life was possible only for the practice of agriculture. Usually stones and bricks were used in the make of the houses. But the selection of materials was largely dependent on the availability of the resources in local settings. For instance, scarcity of wood and unsuitability of mud made all houses of northern British Isles with stone. In some Neolithic settlements a stone or brick-built wall is found to encircle the houses. This acted as a protection against the wild animals. Every Neolithic community had a facility for storing of grains. At this time no agricultural or political techniques were developed in order to manage a city economy.

Manufacture of Boats

Neolithic settlements were generally built close to the shores of lake. Neolithic people learnt the use of the trunks of trees. They used to tie several logs together to make a raft and that was used as water-transport.

They also used their axes and fire to hollow out tree trunks. These served as boats to the Neolithic people. One such dugout canoe has been recovered from Perth.

Development of Social Organization

Maintenance of discipline is the primary condition for all social life. As Neolithic people began to lead a settled life in groups, automatically obedience to certain rules and regulation was developed in them. Although no law was possible at that Age, all the members of a group would abide by a decision taken by elderly people. The form of the Neolithic houses specially resembles with long joint family houses of the Iroquois and the Jivaro. They also have similarity with house types of Indonesian joint family. This particular feature suggests that perhaps clan-like lineage had developed among the Neolithic people as a dominant form of social organization. This also means a unilateral principle that came into being in early Neolithic period. The social life gradually begets the idea of 'property'.

Development of Culture

Use of language, discovery of art and the beginning of religion are the chief indicators of culture. What we mean by culture today had claimed its beginning in Neolithic Age. In a more primitive stage i.e. in the Paleolithic Age men had no language. Ideas and feelings used to be communicated through signs. Of course they could make some kinds of sound to express the situations like getting alarmed or delighted. The power of speech was developed with time and the vocabulary also increased. Although the cave paintings of Upper Paleolithic time have been discovered, the New Stone age paintings are much more developed. Man's love for art has been evident in these paintings. Subjects, forms, techniques of art provide the information about a higher level of culture. A considerable number of clay figurines of women have been found from the deposits of New Stone Age. Probably they are the images of mother god. With the discovery of agriculture people began to worship the land, as 'mother' and they also believed that if the mother could be pleased, she would increase the production of crops. Neolithic Age thus showed a beautiful ad-mixture of art with religion. Apart from the mother earth, different natural forces like sun, rain, storm, etc. were also worshipped aiming at the success in Agriculture.

Neolithic Age can be distinguished into two phases. A stage of incipient farming when people began to experiment with domestication in early post-glacial setting. But within a short period those half-sedentary farmers became a settled peasantry. By producing surplus foodstuffs they made a section of their community free from the responsibility of food production. So some people got the opportunity to become artisans and traders, priests and kings, officials and soldiers in an urban setting. Thus, Neolithic made possible the rise of literate civilizations in the world. A good number of Neolithic sites have been discovered from Asia, Europe and Africa, The New World can not be omitted in this context. Although the record of prehistoric culture in America is much shorter than in Asia, Europe and Africa but the series of discoveries since 1926 establish a modest antiquity of man in America. In fact, the New World contained no primates, except the platyrrhine, (tailed) monkeys which branched off early in the Tertiary and became confined to South America. Throughout the America, there was neither any living anthropoid apes nor any fossil of transitional men or palaeanthropic forms. This indicates that man evolved in Eastern Hemisphere and entered Western Hemisphere after being fairly developed. This entrance occurred in the late phase of prehistory, probably after most of the Pleistocene had passed. The first Americans were the hunter-fishers who gradually developed a farming culture (Neolithic) in America.

NEOLITHIC IN ASIA

A number of centers have been discovered from Asia where ancient farming took place. But the, *'Fertile Crescent'* claims the supreme importance as the birthplace of farming. The agricultural settlements in the Near East occurred along an arc of land (crescent) extending from Palestine, Syria and Cilicia, through the foot-hill zones of Turkey and Northern Iraq, went up to Iran and Capsian shore and Turkestan. The area included the rich soil valleys of the rivers namely *Tigris* and *Euphrates* and came to be known as *'Fertile Crescent'* for the extraordinary fertility of the land. The place stood favourable for the primitive farmers and the focus of farming were located in some particular points within the said territory. Two cereals namely *Triticum dicoccoids and Hordeum* spontaneum, progenitors of Emmer wheat and barley dominated in this zone. Both of them were wild varieties of grass. Besides, there were other varieties too like barley, lentils and peas. The ancestors of one-corn wheat *(Triticum monococcum)* were grown plenty especially in the regions between South Balkan to Armenia. Sheep and cattle fit for domestication were then roving in those regions *. Naturally man proceeded to exploit the organic world before them.

The early historic civilizations of the world were centered on the great rivers—the *Euphrates* and the *Tigris,* the *Nile* and the *Indus.* The antecedent stages witnessed the genesis of farming, which gave rise to permanent settlements, complex social hierarchy and metallurgy, one after another. In most of the cases Neolithic was the contribution of advanced hunter-fishers and food-gatherers of Mesolithic. However, a few early Neolithic sites of Asia have been described below.

Jericho and Jarmo

Excavations of a site in *Jericho* (Tell-es-Sultan) in Jordon and a mound *of Jarmo* in the Kurdish foothills of Northern Iraq have thrown some light on the early stage of farming. *Jericho* yields the traces of agriculture among the *Natufian* hunter-fisher community who lived in the caves of Mount

* Domestication of sheep, cattle or pig did not occur in the same time. Radio-carbon dating suggests that the Domestication of sheep took place as early as 9000BC. Earliest evidence of domesticated cattle was found around 5000BC. Pig was started to be domesticated more lately, only about 2500 years back

Carmel, Palestine about 8000BC. The *Natufians* were actually the Mesolithic hunter and gatherer group of Western Asia. They used to live in rock-shelters but showed a tendency of making settlements at open stations as primitive farming developed among them. They exhibited neither Neolithic celts nor pottery as because they were not true Neolithic people. Instead, they had sickles made of mounted small flints. Repeated cutting of the siliceous grasses *1 made these sickles considerably polished.

The *Natufians* of Near East though maintained their living by hunting and fishing, occasionally they used to harvest wild grasses for grains. The area was rich with natural resources; wild ancestors of wheat, barley and millet were especially abundant. Therefore, the people learnt to utilize these plants. The *Sorghum seems* to be most ancient in use. It is likely that people at this beginning were in search of finding out the spots where the grass like plants yielding cereals grow well. Harvesting formed a part of primitive gathering. As soon as the people realized the usefulness of crop, they exerted a steady pull back. Ultimately they leaned on purposeful growing of those plants near their settlement.

The *Natufians* were more settled than any other Mesolithic communities of the world. By 7000BC we find a substantial development of settlement at *Jericho.* A high wall of tough stone enclosed the town. This *Nautfian culture* of Near East bears the concrete evidence of food cultivation, though in a very incipient form. The first cultivated grains were the wheat and the barley. Later, the millet, rye, flax and bean were added. *Dinkel* gave rise to a new variety of wheat, which earned much popularity among the Neolithic farmers, and it is still grown in some backward areas of Asia Minor. The lowland of Palestinian Valley had an oasis where the Natufians discovered a new relationship with nature by the way to survive under hot, and arid condition. In contrast to *Jericho,* the site *of Sarab* was located in a cool Persian Mountain valley near Kermanshah. Possibly it was the summer encampment of herdsmen-cultivators who had adapted their own pattern of seasonal migration with the flock of wild sheep and goats by the early 7000 BC. The site *of Jericho* and *Sarab* were opposite to each other situated at the opposite ends of Fertile Crescent, therefore revealed extreme environmental conditions. The other contemporary incidences of incipient agriculture and settled village situations have been found out from the sites viz. *Karim-Sahir, Jarmo, etc.* and even from the distant place site like *Catal Huyuk in* South-Western Valley.

The village *Jarmo* was found with incipient farming in the hills of Iraq, above the *Tigris-Euphrates* valley. The settlement was made up of simple houses of packed mud-walls. The grains of wheat and barley were found. Some house hold implements especially the hand-mills furnish the indirect evidence for the use of cereals. Dog was the only animal, which used to be kept. Presence of stone hoes and flint sickle-blades polished with use suggest that the cereals were being reaped, if not necessarily sown. But *Jericho* and *Jarmo* were devoid of any sign of pottery. There was no sign of weaving too. Since these two typical art of Neolithic were lacking among the early farmers of *Jericho ami Jarmo,* Grahame Clark (1961) has designated these farming centers as 'Proto-Neolithic'*2. In 5000BC, the farming villages spread widely in the Near East, all the way from the Fayum Basin along the shores of an ancient lake, at the west of the Nile river, through Palestine and Syria, over to Iraq and Iran. Although the cultural traits in all of these areas were not exactly identical, but there was commonness in the culture.

Syro-Cilicia

A more advanced stage towards civilization has been noted from the Southwest Asia (Asia Minor) which flourished towards the end of 6000BC. Three main foci were distributed in the following ways:

a) The first one was centered on *Cilicia* and *Western Syria* but maintained contacts with *Jericho* of the far South.

---

*1. These grasses contain Silica in varying degree.

* 2. Clark Grahame: *World Prehistory,* Cambridge University Press, 1961, p. 81.

b) The second one was centered on *Northern Iraq* and *Eastern Syria.*

c) The third one was concentrated on Iranian plateau.

The lowermost layer (about 30 feet) of the Syro-Cilician zone exhibits the Neolithic features like pottery and lithic industry. Lithic tools included blades for reaping, knives, polished axes and tanged lance-heads. Impression of a textile garment was also found from a burial at Mersin in Cilicia. An upper layer, 40 feet to 50 feet of the same deposit represents the chalcolithic or the Bronze Age.

Hassuna and Halaf

Traces of broadly similar culture have come out from the basal layer of the great-stratified mound of Tell-Hassuna near Mosul in Northern Iraq. Along with some coarse pottery and remains of cereals, a rich lithic industry has been discovered. The site at Tell-Halaf in the Khabur drainage area of upper Euphrates shows two distinct stages where a Neolithic *pre-Halafian* has preceded the Chalcolithic *Halafian.* The lower *Pre-Halafian* stage yields profuse monochrome wares while the upper stratum, Halafian proper, presents polychrome pottery allied to that of *Ubaid* in Lower Mesopotamia.

Sialk

The area Sialk, between Teheran and Isfahan in Persia reveals well-documented materials comparable in age with the *Hassuna culture.* The lowest layer (**Sialk-I**) is essentially Neolithic showing a form of mixed farming, although a few hammered pins or awls have tried to make it technically chalcolithic. This stage also preserves some traits of Mesolithic as exemplified by the notable presence of microliths, slotted bone (reaping-knife handles) with animal head terminations. But the majority of evidences like painted pottery, ground stone axes, bones of sheep and the implements like sickle and quern suggest that the people of Sialk-I must have belonged to the level of Neolithic culture.

There are a number of areas in Asia, other than Near East where we notice early development of plant and animal domestication. They are mostly located in Southeast Asia and include the sites of India, China, Japan, Indonesia, Philippines etc.

India

Like the greater part of Africa, India lay outside the realm of advanced Palaeolithic culture but the successors of Lower Palaeolithic, after the end of Pleistocene epoch learnt the art of farming and made permanent settlements. A considerable part of Indian subcontinent remained under the control of hunter-fishers whose microlithic flint-work stemmed from the Lower Palaeolithic culture.

The earliest farming communities were confined to Baluchistan and an adjacent area of Sind on the right bank of the lower reaches of the river, Indus. Present day Baluchistan is very arid, unsuitable for living. But the situation was totally different in past, especially during 3000BC, where there an incidence of fair rainfall was noted. The human settlement of this region is therefore likely to go back at least to the beginning of third millennium BC that perhaps maintained a link with the ancient Bronze Age culture of the regions of Far West. A few mounds lack pottery or yield only hand-made wares, but at the site of Kile Gul Muhammed, near Quetta, the majority of the mounds relate to communities using copper tools and turning pottery on wheel.

B.K.Thapar (1974) classified the early farming communities in India into five broad geographical regions. He also added the present day Pakistan as the sixth area. However, the regions *are* as follows:

1. Pakistan, the Indo-Pak sub-continent covers Baluchistan, Swat and the contiguous areas of the upper Sind Valley in Pakistan.

Two phases of Neolithic culture have been arrested in this region. Earlier deposit reveals a pre-ceramic assemblage represented by mud architecture, bone tools like awls or points, chert blades or scrapers, pecked and ground stone objects, a few baskets, and bones of domesticated goats, sheep and oxen. Later phase shows the use of hand-made pottery which are highly burnished and red in colour. Some traits of earlier phase is found to survive. Such a site has been excavated from Sarai Kotla located in Potwar Plateau.

2. North-West region covers Kashmir. Salient features of this Neolithic culture are same with the previous one. Burzzahom is a solitary site in Kashmir, about 6 miles away from Srinagar. H.D Sankalia in his Book 'Prehistory of India' (1977), stated, "Except writing, of which we have no evidence so far, these early inhabitants of the Kashmir Valley had everything essential for living a comfortable life".

3. Eastern region covers Assam, Chittagung and the sub-Himalayan regions including Darjeeling. Here Neolithic culture is characterized by hand-made pottery, grey or brownish in colour. These pots usually possess a cord or basket impression. Ground stone tools include shouldered and round-butt celts, mullers, pestles etc. No direct evidence of domestication (animal or plant) has so far been obtained from this region. Circumstantial evidences suggest that the people used to practise slash and burn cultivation and shifting cultivation. Polished stone axes including 'quadrangular adzes' the typical weapon of Neolithic age have been discovered fiom a large number of sites at Nagahills, Brahmaputra Valley, Khasi-Garo-Kachar hills, Chittagung in Bangladesh and North Bengal.

4. The Chotonagpur plateau with its adjacent Districts namely U.P., Bihar, Orissa, and West Bengal. Neolithic culture of this region is distinguished by a coarse grit-tempered red ware, ground stone tools including rounded butt axes, faceted hoes, chisel, mace-heads (or digging stick weights), pounders and grinding stones. No direct evidence of domestication is found. Occurrence of pounders and grinding stones may imply an acquaintance with husbandry. Many pot-shreds and celts have been discovered from the sites at 'Birbhanpur' on Damodar Valley; 'Bonkati' in Satkahania forest and 'Pandu Rajar Dhipi' on the bank of Ajoy river in the District Burdwan of West Bengal.

5. Mid-eastern region covers the District Saran in Bihar. Neolithic deposits from this region shows a variety of objects. There are, (a) objects of bone and antler consisting of needles, points, borers, pins, tanged and socketed arrowheads, scrapers and also items of personal ornamentation like pendent; (b) ground stone objects include the celts, pestles, querns etc.; (c) microliths include parallel-sided blades, tanged arrowheads, lunates, points and; (d) terracotta objects comprising of figurines. The associated ceramics contain hand-made red-grey, black and red wares, most of which show a burnished surface. Houses are made of mud and daub, Paddy husk impressions on some pieces of burnt clay and the charred grains provide the evidence of cereal cultivation. The 'Bhagat Pahar' sites in Bihar and many other sites from adjacent areas of Purulia exhibit the said materials.

6. Southern region covers the Peninsular India. This region again presents two broad phases of culture, placed one after another. The earlier phase has been best demonstrated in the sites at Shevaroy hills and also at Nagarjuna Konda and Utnur. The principal traits are, hand-made pale-red ware, ground-stone tools, microliths (with restricted or no use of blades) and occasionally the bone implements. Some of the settlements of this phase are represented by ash-mounds which have been proved to be an accumulation of burnt cow-dung, attesting to a pastoral economy.

The later phase has been characterized by a dominant burnished dull-grey ware, ground stone implements including axes, adzes, wedges, chisels, grinding stones, pounders, hammer stone, microliths and parallel-sided blades. Some bone implements, especially the points are also known. Miscellaneous crafts including bead making appeared in this phase. A few painted pottery showing simple linear patterns (in a brownish purple colour over a red surface) have been discovered from this very assemblage. Towards the close of the phase, contact with the Chalcolithic culture of Upper Deccan resulted in the introduction of the use of copper and other concomitant traits. Name of a few significant Neolithic sites in this region are Bellary, Brahmagiri, Sanganakallu., T. Narsipur in Mysore. Typical cell-bearing sites are Nevasa, Chandoli and Nagada in M.P. and Nagarjunkonda in Andhra Pradesh.

Many Neolithic objects have been collected from the surface of the ground. The first discovery of Neolithic objects was made in U.P. by Le Mesurier in1860. Thereafter in 1872 Fraser made some important discoveries in the District Bellary in Mysore. This was again followed up by Foote in many parts of South India. Besides, different important sites have been excavated and explored by Allchin, B.K.Thapar; H.D.Sankalia, M. Wheeler, B.Subba Rao, D.S.Sen, G.S.Roy and others.

China

The other area in the mainland of Southeast Asia is China, which witnessed development of early farming. Today's China runs South up to Indochina and in the North it is extended through Manchuria. But in prehistoric days China was a small region, north and inland, in the Great Bend of Yellow River, Hwang-ho. The northern part of the country enjoyed certain advantages over the South because of an alluvium plain brought down by the river Hwang-ho.. Since the whole of China was relatively ramote from the focal area of the Old World farming, little cultural influence was carried to this region from outside. Another striking feature is that there was a well-marked difference between the environment of South and North China. The South China, beyond the Yangtze valley did not show the same warmth, moisture and heavy vegetation as like the North China. So the habits of daily life as well as temperament of people differed in two of these regions.

Yang-shao and Hsi-yin

In North China, a Neolithic culture has been discovered from the sites namely Yang-shao in Western Honan and Hsi-yin in Southern Shansi, close to the yellow river in the heart of the prehistoric China. In China, Neolithic way of life lasted for several Centuries, possibly up to 1700BC. The material equipment of those people was hand-made pottery with textured surface, woodworking equipments like polished stone knives and projectile.heads. The earliest farmers who laid the foundation of settte life in North China were basically hunting and gathering people like the Danubians of Central Europe. They used to practise agriculture by'slash and burn' technique and maintained livestock largely on forest produces. The main crop was millet and the important domestic animals were the pig and the dog. The village settlement of Pan Pao near Sian in Shensi covered about two and half acres of land where people made mud houses (preceded by pit house) on rounded or rectangular ground plan. The floors and inner faces of the walls were plastered with mud. A clay oven set in the middle of the dwelling helped to keep the interior warm in winter as well as facilitated in cooking. Crops were stored in beehive shaped pits sunk into the loess. Burial remains suggest that the children were often buried inside the room, placing the body into a pot. But adults were buried in cemeteries away from the settlement. Some of these graves were plank-lined. Huge grave goods like dozens of pots and other useful things were provided as a rule. That hunting continued to play a major role in the life of the people has been evidenced by the presence of leaf-shaped and tanged arrowheads made of polished stone or bone. Some of them were even made out of shell or bifacially flaked stone. The other artifacts include stone knife-blades, lunate or oblong in form, polished on either face and along the edge. These knives were often perforated for the fittings of handles. They were simitar in appearance with the knives of Northeast Asia and even with the Eskimos of North America.

The painted pottery culture of Yang-shao was chiefly Neolithic to begin with but it was extended through the Bronze Age upto the coming of iron around 600BC. For the Neolithic part of the painted pottery culture, Kroeber (1923) said, "agriculture is the authentic with millet, possibly also with rice and wheat, and with stock-raising, at first of pigs and dogs, later also of sheep and cattle. Houses were pit-dwelling with roof entrance". The culture of Yang-shao and Hsi-yin was contemporary to each other and both highlighted the Neolithic China.

Japan

Japan is posited far off the East Asia as Britain lies off Western Europe. It is almost like an island, marginal to a continental margin especially from Korea. Japanese civilization was a late development in the history and the elements were infused from the continental mainland. Very little can be said about the earliest traces of human activity on this island. There is no sure evidence of a Palaeolithic period in Japan, but Neolithic remains are abundant. About four thousand Neolithic sites have been discovered from this island. A.L.-Kroeber (1923) had mentioned, "if anything pre-Neolithic is ever definitely discovered in Japan, we might expect it to be post-Pleistocene and Mesolithic". However, there is a difficulty in estimating the accurate date of the beginning of Neolithic. Its end has been noted around AD 200 when iron working was introduced.

Stages of Neolithic in Japan have been identified according to pottery. As the pottery belongs to Mesolithic as well as Neolithic, it is unable to point out the base of the Neolithic. Farming was perhaps an addition of later phase. Archaeologists tried to relate the presence and absence of bones of domestic animals, or sickle (an agricultural implement), or impression of grains with the available pottery in order to reach a tentative dating. Accordingly, pottery development of Japan can be classified into three stages—Jomon, Yayoi and Iwaibe. The first phase of Japanese prehistory is represented by the Jomon culture. It contains hand-made ware where cord markings are commonly imprinted on the outer surface of the pottery. Although this culture has been described as 'Neolithic', there is no evidence of farming, at least until the closing phase of the culture, around 200BC. The only domestic animal of this cultural period is dog. Since the Jomon people preferred to live within the reach of the sea, their middens exhibited huge quantities of discarded shells of mussels, oysters and other remains of shellfish along with plenty of fish bones. Fishing-gear included fishhook with plain or barbed tip, harpoon and spears with detachable heads. The inland games like deer and wild pig was perhaps hunted by arrows tipped with hollow based triangular flint heads flaked on either side. There were also the evidences of grinding stone and mortars by which the wild plants and hard nuts were crushed. The people used to live in the settlements made of trapeziform or circular houses, the floors of which were lowered below the ground. They also exhibited crude stone-axes but fine arrowheads and numerous scrapers worked on flint.

The second stage Yayoi is distinguished by wheel-made pottery. These are less ornamented on the surface but often florid in shape. Imported Chinese bronze is sometimes found in association with Yayoi pottery. Yayoi people were the farmers; they innovated rice cultivation. Tangible evidences of paddy fields, irrigation channels, Carbonized grains, impressions of rice, granaries etc. have been found from this level of culture. Axes, adzes and arrowheads belonging to this culture are made of stones. The culture has continued roughly from 200BC to AD200. The interrelation between the Yayoi and the earlier Jomon culture is very complex. Only in some parts of the country the continuation is found.

The Iwaibe culture of the third stage corresponds to the Iron Age. Here the pottery is wheel-made, grey in colour and unornamented. The florid shape has become more exaggerated than the previous period. By this time Japan is found to enter into Iron Age. The Iron Age of Japan has been dated between AD 200 and AD 700.

The Hidden Neolithic

Down in Southeast Asia lie Indo-china, Burma and Siam and beyond them the islands of the Indies, including the Philippines and Formosa. The area was isolated from the rest of Asia but a distinctive Neolithic culture was found to be developed in this spot. Chronologically it was a late development. The sites in Toukin, Indo-china were first explored. They showed the beginning of domestication with wild rice. The oldest plants like *taros and yams* are also found from this area. Heavy axe-like tools flaked from stone pebbles have been used for clearing the wild vegetation. Presence of grinding stone and pounders indicate the preparation of plant food. The animals which people hunted to supplement their plant food are the mud turtle and numerous varieties of oxen, deer and swine as evident from the nature of bones. The caves and shelter deposits also yield the shells of fresh-water molluscs. On the NorthEast coast of Sumatra, plenty of shell-middens are found. From

Java, Philippines and Sumatra a large number of microlithic industries have been discovered. These industries are comprised mostly of blades with steep retouch; small flakes have also been utilized. Besides, axes aad adzes particularly with lozenge or quadrangular section are found abundant in Philippines, Borneo, Celebes, New Guinea and Melanesia. All these features go in favour of early farming which perhaps aroused .among the Mesolithic people of this area.

The settlement pattern also supports this conclusion. Houses are found to be constructed with heavy timber. They are usually raised up the ground on piles. The solidification of the houses does not mean that the owners never moved from place to place. They used to practise agriculture by the method of 'slash and burn'. These people did not possess much domesticated animals like the West. Dog was absent in this culture. Traces of chicken and pig were found only. For hunting and fishing the people used spears aad blowguns. They had also a variety of nets and multitude of mechanical traps, which exhibit almost a native technology. Evidences of art and craft are very feeble when the Neolithic culture was first founded. Later, the people learnt iron and brass work .as reflected in the making of their weapons and ornaments. But the use of metal was found restricted to the upper level of culture. Basically the people relied on stone tools.

Archaeologists face a problem to find out the starting point of this culture. Many of them have referred this culture as hidden Neolithic of Southeast Asia. The reason is that, the culture was found quite independent in type when compared to the cultures of Western Asia. The art of domestication possibly did not spread from the west to east as there was vast differences in the items of food. However, the nature of Neolithic way of life is entirely different here. Mesolithic fishing people living along the shores of Southeast Asia were to some extent sedentary and such a practice perhaps allowed them to experiment with domestication. They began to plant roots and shoots. But as they did not understand the principle behind life cycle of a plant, so could not relate the seeds with propagation. The sowing of grains was learnt later.

Neolithic in Europe

The Neolithic people of Europe attained a subsistence security greater than their predecessors by learning a number of manipulating activities that involved knowledge as well as perseverance. The rudimeatary subsistence agriculture did not take place within a single small region. The recent review of Dr. A. Whittle (1996) on the latest archaeological evidences of Neolithic Europe suggests that, for Europe as a whole, "Neolithic period began in 7000BC and lasted till 2500BC. But remarkable regional variations were evident. For example, in the Southeast Europe Neolithic Age prevailed between 7000BC and 5000BC, in the Central and West Mediterranean it flourished in 6000BC, in Central Europe it went with 5500BC. But in Northwest Europe, the Neolithic appeared quite late, around 4000BC. Two important Neolithic sites in Europe arc the Danube and the Swiss lake.

Danube

The Danubian culture of Central-eastern Europe took.shape under the influence of East. They grew with farming as well as with domestication of animals like early Neolithic center of South-Western Asia. In Germany and Poland, the Danubian people had built the villages of well-made rectangular houses with timber. Walls of these houses were plastered by mud and roofs were thatched. The people began to make pottery and this practice was carried forward among their descendants. However, the pottery was good in quality and well decorated. For agriculture, Danubians took the help of, 'slash and burn technique'. They exhibited polished stone axes, instead of axes with chipped cutting edges. This polishing of stone tools and the knowledge of agriculture became the striking features of European Neolithic.

The site Tripolje of the Russian Steppe, near Kiev also yielded a late Neolithic culture with the beginning of copper. Polychrome painted pottery is the characteristic feature of this area. It shows an overlapping with Danubian II, as the Danubian culture was spread over a large area.

Swiss Lake

The site is an unusual well-documented underwater preservation in Western Europe. It denotes a. Neolithic phase subsequent to 2500BC. French authors have named it Robenhausian after a pile dwelling zone, standing above the water, extending upto two acres or more. Wheat, barley and two kinds of millets were grown in this spot as found from the evidences. Bones of cattle, pigs, goats and sheep have been recovered in greater number than that of the wild games. Remains of incised pottery are found, ground axes and chisels with or without the wood or antler hafting have also been noted. The perishes are so nicely deposited in the lake mud that the anthropologists of present day can have an intimate glimpse into the domesticity of four thousand years ago The pile dwellings were probably constructed at the end of the Bronze Age.

According to A.E.Hoebel (1958), the birthplace of the Neolithic was the land lying East of the Mediterranean, from which center it spread to east and West. Fanning out into Africa and Europe, it was adopted as well as modified by the Mesolithic folk of Denmark around 2500BC. In fact, with the final withdrawal of the ice-sheets, large tracts of the Old World, especially Northern Europe and Western Asia became sparsely populated. Changing climatic condition imposed successive problems on small hunting groups. They had no alternative other than to adopt the changing environment that took place over much of Europe and North Africa from about 1100BC onwards, in the so-called Mesolithic Age. The axes were probably invented to attack the trees of encroaching forest in the North. The dogs were found to be domesticated to equip the hunting. Trained dogs were able to chase the small games in the forest following the instruction of their masters. The Mesolithic Age was actually a prolongation of the Palaeolithic way of life with little modification. The insecurity of food resisted them to develop the permanent settlements.

Earliest peasant communities in Europe were 'Neolithic' in the sense that they did not have the knowledge of metallurgy and their technology was restricted in the utilization of flint and other stones for making tools. This culture was contemporary to the cultures of Western Asia where the people were found conversant with casting of copper from late 5000BC and later with the casting of Bronze from 4000BC. It is understood that the parts of Southwest Asia were far advanced than Europe in adapting farming. The basic reason behind this phenomenon is the drastic change of weather from the late Pleistocene to Neo-thermal (recent) times, which affected both the regions but in different ways. In Southwest Asia, desiccation of rich hunting.ground brought a challenge to the people so that they were compelled to change their fundamental outlook of exploiting environment. In case of the people of Europe, the gradual increase of temperature though brought some ecological changes, but it offered no definite challenge for survival.

Neolithic in Africa

The Neolithic way of life from its focal region spread to North Africa, southeastern Europe and Inner Asia. At present the Sahara Desert of Africa is continuous with the Arabian Desert keeping the Red Sea in between. The desert region works as a barrier to human beings, whether they are primitive or civilized. But it was more hospitable at some times in the Ice Age. So far as the position, North Africa is cut off from the mainland of Africa and Joined more to Europe as the other shore of the Mediterranean. People bearing Neolithic culture of Western Asia perhaps proceeded along the shore of Mediterranean as a convenient channel. For some time life ran parallel above and below the Mediterranean. The first area to which cereal farming and stock raising were diffused was the Egypt, which in course of time gave birth to the earliest distinctive civilization on Nile Valley.

The land route from Palestine to Lower Egypt exhibits the signs of Neolithic culture. Tasa, Fayum and Merimde are the three remarkable sites where 'pure Neolithic' is found without the presence of metal. The striking feature is that, all of these three spots are more or less similar to that of Near East and Europe, although they appeared after a couple of thousand years.

Fayum and Merimde

The earliest Neolithic community of Africa belongs to an arid phase following a period of heavy rainfall. These people got settled on the Fayum basin of Egypt, on the shore of an ancient lake at the west of the Nile River. Forest, trees and swamps provide the evidences of arid environment. Fayum was not an isolated oasis, but formed a zone of relatively favourable environment extending over a large tract of the Sahara. When the farmers first inhabited this Fayum basin, the water level of the lake stood 180 feet higher than the present day. However, the Fayumis were the hunter-fisher people who adopted a Neolithic way of life, as originally developed in Western Asia. The widespread appearance of barbed bone harpoon heads and winged arrowheads of flint suggest that, it is a parallel culture of French-Sudan. Besides, there are enough evidences of domesticated animals like sheep or goats, cattle and swine. Remains of cultivated emmer and flax have also been obtained. The eqipments for harvesting, storing and grinding of grains are same as found in Western Asia. The coiled basketry, carbonized cereals, wooden reaping knives with flint blades (set in slots) etc resemble with the Natufians of Near East. Numerous blades of axes and adzes made from polished flint and other stones occur with few relics of weaving. Pots of this period are simple in shape and made by hand. Some undecorated flat-based bowls without handle have also been recovered. All these indicate that the economy of the Fayumis was of simple subsistence type and they invented farming later on.

The communities practising almost the same economy was discovered from Merimde-Benisalame near the head of the Delta of the river Nile. It is worthy to note that Merimdians used to bury their deads within own dwellings but without the grave goods. They maintained their livelihood chiefly by hunting and fishing. The food was often supplemented with the produce of farming. Large pots and baskets have been recovered from this site along with barbless fishhooks. These items show a strange similarity with that of the Natufians i.e. the people who lived a long time back in Palestine.

Badari

TheUpper Egypt, notably a region near Badari laid the basis of Early Predynastic culture through simple farming. The Badarian peasants utilized flint and other stones for the purpose of making tools. They cultivated emmer and barley, which were harvested by the reaping knife and stored in large clay pots (silo). The knives were bifacially flaked and they were used with toothed sickles. Among the domesticated animals, there were cattle and sheep (or goat). Evidences of basketry and hand-made pottery accompanied the traces of weaving. Copper appeared in the form of beads, which were hammered to take into shape. The deads used to be buried in a contracted position within oval trench graves inside the dwelling as like the Merimdians. The bodies were generally clad in linen and kept facing to the south. All bodies are therefore found in turning position on its right side where the head is placed towards west. It is peculiar that the Badarians arranged a careful burial for their domestic animals. These bodies were also wrapped in textiles.

Some graves have been discovered from a separate site, Der Tasa where the people provided furniture, distinctive beakers, pots and baskets with the dead bodies. Tasa exhibits a purely Neolithic culture, but for the presence of few pieces of copper—a borer, an awl and two beads, it has been placed under Chalcolithic. Etymologically both the Tasians and the Badarians belong to the 'Pre-dynastic' phase, which preceded the 'historical' dynasty of Pharaohs. The Egyptian Predynastic phase has been dated between 4000BC and 3000BC. It is equivalent to Mesopotamian Ubaid, Uruk, Jemdet and Nasr Predynastic.

NEOLITHIC IN THE NEW WORLD

Recent research in the New World suggests that the food crops on which the American Indians relied prior to European colonization were the domestic varieties of some wild forms. The maize was the most widespread as well as the most important crop; its progenitor flourished in Mexico and perhaps in Central America. There were at least three major areas—Mexico, South America and the eastern United States where the domestication started independently. The earliest domestic form of maize as revealed from Tehuacan and Mexico has been dated as 5000BC. But there were other plants, which used to be domesticated in the New World as early as 7000BC. Such plants include the members of the cucurbit family comprising a variety of bottle gourd, summer squash and pumpkin. People of Meso-America (Mexico and Central America) are often credited for successful planting of maize, beans and squash in the same field. Fertile soil and congenial climate made Meso-America a bountiful zone of unique farming. In South America, the crops started to be cultivated about 2500BC on the Peruvian coast. The items were gourd, squash, lima beans and possibly cotton. A long period, from about 7000BC to 1500BC was required for this part of world to invent agriculture and to facilitate its spreading up to the distant corners of the Continent.

Permanent villages did not appear in Peru until about 2500BC. It was far more late in Meso-America where the date goes to 1500BC. This means, a fully developed agricultural society could not be formed through the cultivation of maize, beans and squash. It helped only in raising the level of subsistence in order to support the formation of settled villages. Domestic animals seem to be economically less important in the New World as compared to the Old World. The animal species are quite different in these two regions. The New World animals include dog, muscovy, duck, turkey, guineapig, alpaca and llama. Among these, the guineapigs, ducks and turkeys were reared for food, alpacas for fur and llamas for transportation of goods. Not only that, Corn arrived in Peru very late about 700BC but in a developed state. The site Huaca Prieta, on the coast of Peru, shows large mounds of 4500 years ago (2500BC). These mounds contain neither fish bone, nor any animal bone. They are also devoid of hunting weapons. Evidences of pottery and weaving suggest a date around 1250BC. The people were no longer dependent on hunting and fishing; they became efficient farmers. The culture therefore represents an well-advanced Neolithic type.

In comparison to Europe and Near East, least is known about the subsistence pattern of Meso-America. The settlement of native Americans is relatively recent. No trace of human occupation has been found in the stratified deposits of Early and Middle Pleistocene. All hominid remains that are found from the New World belong to the Homo Sapiens group The earliest migrants who reached North America were definitely the pre-Neolithic hunters whose descendants succeeded in domesticating New World plants. They also started to domesticate the animals but to a lesser extent. Since the Neolithic way of life was vastly superior to the hunting-gathering way of life, dawn of civilization appeared at the end of Neolithic Age and thus the way to Metal Age was paved. However, the great Neolithic settlement was rooted in the rich soil valley between the rivers Euphrates and Tigris. There were also other spots around it placed within an arc (Fertile Crescent). The area is therefore regarded as the Cradle of farming.

MEGALITHIC CULTURE

In late Neolithic period, the custom of erecting gigantic monuments became popular in North and Western Europe. As the name indicates, *megalith* is a great structure built on large-size stone. (Greek words *megas* means great and *lithos* means stone). They served the funerary or cult purpose. Neolithic farmers paid special attentions in making the burials in contrast to the hunter-fishers of the previous period. The dead bodies used to be buried in a sitting knee-chest position. Arms and legs were bent closely and tightly to form a compact bundle; strips of leather or fiber was possibly used in tying up the whole. Food and personal belongings like weapon, ornaments, stone tools etc. were provided therewith, as a rule. Several theories explain this peculiar position of burial. The first one suggests that the idea of arrangement be copied from nature. The position was almost same as a baby posited in mother's womb, before coming to this world. Since the deceased soul was supposed to enter into a New World, his body was placed abiding nature's universal rule. The second one forwarded a more simple explanation. It said that in this particular position the corpse could be carried to the grave more easily and the grave itself would require less labour in excavation.

From the Neolithic Age, men developed a disposition towards religion. They became the worshippers of sun as well as the dead. The imaginations, being conditioned by wonder and experience took the shape of a belief and the people began to show respect towards unusual incidents that could not be explained with their shallow level of knowledge. At this stage they had grown some elementary skills for manipulative activities. This cultural inventory led the people to move for an incredible toil; they raised huge stones with sheer force to honour the deceased persons. But obviously these erections were not meant for all members in a society; this process perhaps buried the great Chieftains. Some authorities argue that the megaliths are strictly graves, not temples. Others have discarded this view. They said that kings of that period were looked upon as 'Son of God'; so their tombs were built in the form of a sacred temple. However, there is a spirited debate in this regard.

Most of the *megalithic monuments* are made by stone blocks, available in nature. A few of them are slightly shaped; others are rude. The types have been distinguished on the basis of the number of stone blocks and their position of placement. For example:

1. *Menhir (men* = stone + *hir* = long) is made of a large single pillar. It is also known as monolith as only one standing stone is required here.

2. *Cromlech (Crom* = Concave + *lech* = flat stone) is a circle *of Menhirs* i.e. it consists of small pillars arranged in a circular fashion to form a ceremonial ring. When a Cromlech occupies a greater plot of land with more and more stone pillars, it is called *Stonehenge.* Within the Stonehenge, cutting and shaping of great stones with axe often made a semi-circle doorframe. A solstitial orientation is obtained from Cromleches which suggests great annual rites that used to be held there, particularly at the moment when sun's power was optimum. Some other monuments have been found which are similar to Cromlech in function, but different in form. They are called Alignments. If Menhirs are set in a series of rows in a ceremonial plot, they give rise to an Alignment. The most famous Alignments are placed at Carnac in Brittany.

3. *Dolmen (Dol* = table + *men* = rock) is composed of vertical stone blocks which support a roof-slab, like a tabletop. The closed chamber formed by the stone blocks is generally of a room-size where the dead body is interred. All Dolmens are not of same kinds.

a) When the Dolmen is placed under an earth mound, it is called a *Tumulus.* At the terminal phase of Neolithic, multi-chambered Dolmens came into existence where several Dolmens were connected within a single *Tumulus.*

b) Some Dolmens show a long narrow passage up to the main burial chamber like the elongated entrance of Eskimo Igloo connected to the domed ice hut. This type of Dolmen is known as '*Passage grave'*.

c) The other type shows only the passage without the main Dolmen. Here, the principal structure of Dolmen degenerates and the passage reduces to the size of a grave. This structure is termed as *''Hallcist'.* In a much later phase of Neolithic, very small Hallcists suitable for a single individual became larger to accommodate many.

Megaliths are found all over the world. Its main distribution is found in Europe along the Atlantic coast. The countries like Southern Spain, Portugal, Western France, Brittany, Britain, Ireland, southern Scandinavia and Portugal have been pointed out as the area of origin for *Megaliths* in Europe. Brittany is remarkable for *Menhirs;* the long rows of single great standing stones are visible in Carnac, Brittany. England possesses the most impressive ruin of a great *Stonehenge.* Dolmens covered by mounds are found in Cairns of Scotland and Nuraghi of Sardinia. In Japan also, such Dolmens covered with earth have been found. Megaliths are not absent from the New World. But the beliefs associated with huge monuments of Mexico and Peru have not yet been discovered. Moreover, they differ from those prevalent in Europe and Asia though all are derived from the same origin. North Africa presents the evidence of megalithic burial tombs where the *Cromlech* and *Dolmen* are two main forms. But some African forms show a departure from the usual type. In Gambia, *circular*

Tumuli or *Cairns* with two or three upright monoliths have been placed together. They are directed towards the center of the *Cairn*, more or less parallel to a diameter, but parallel to a tangent to the circle.

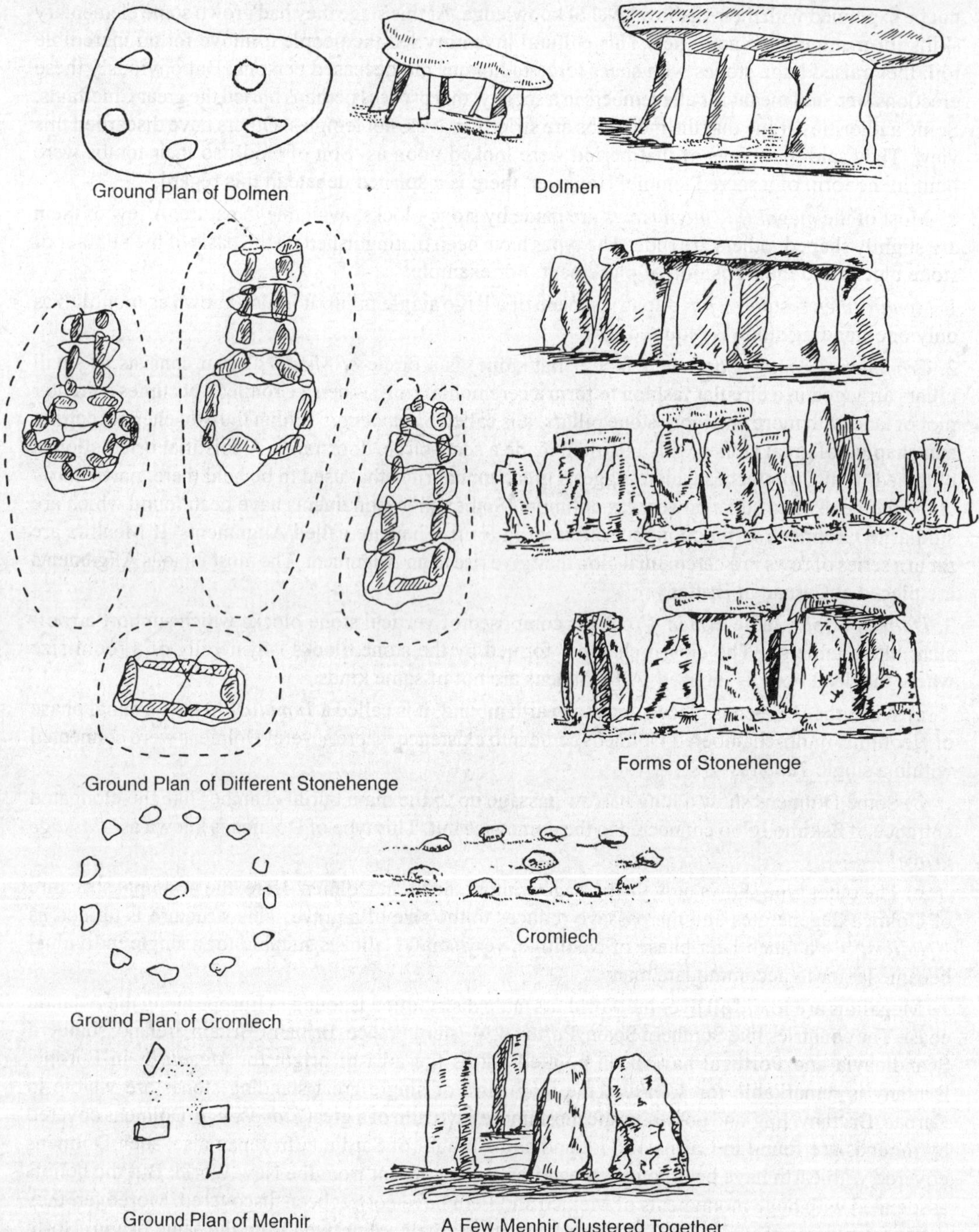

Fig. 11.20. Megalithic Structure

The Megalithic culture in Europe flourished around 2000BC. Both Menhirs and Dolmens occur in Western, Southern and East Asia. Although their connections with Atlantic European ones is very doubtful, but there is no ambiguity regarding the point that Megaliths occurred first in Neolithic time and continued through Bronze Age till quite late in the Iron Age. In Spain, the terminal Neolithic, Early Copper, and Early Bronze Age, each of these periods possesses its typical form of Dolmen. In fact, Megalithic structures arose when the Old Neolithic order was passing out and new social system was going to be introduced by the sea-voyagers.

No distinct break has been observed in the chain of great stone monuments from Peru and Mexico, via Japan, Java and India, to Mesopotamia. They are found to occur in the same fashion in South Sea Islands and in the Lonely Easter Island. Some large black monoliths have also been identified from Pit Cairn Island. All these monuments, with rare exceptions, are the permutation and combination of three elements—a line or circle of standing stones, and a stone chamber like box. The *Stonehenge* is the most elaborate one among all European Megalithic structures. Date of Stonehenge has been considered as a fixed point to calculate the age of the whole system of monuments.

If there was any proof of a common race existing all along the tract of the megaliths, it could be inferred that the monumental masonry was a passion of amusement. The later Neolithic period exhibited a mixture of three different population stocks in varying degree as remarked by Elliot Smith (1915). They were Cro-Magnon, Fur fooz (a brachycephalic type) and the Mediterranean. But the whole culture of megalith builders—the very name of the people, their languages, their customs, except a very few, have left known. They vanished forever as soon as the northern race had gained power and started a career of conquest and destruction. By no means the megalithic tombs suggest the survival of Palaeolithic cave burial.

Megaliths in India

Megaliths are located practically all over India except a few areas like Plains of Punjab, the Indo-Ganga divide, the Ganga Basin, the deserts of Rajasthan and parts of North Gujarat. They are concentrated in Peninsular India especially in the States of Tamil Nadu, Kerala, Karnataka and Andhra Pradesh. Maharashtra came next to them and is followed by Madhya Pradesh, Uttar Pradesh, North East Rajasthan, the Kashmir Valley and Ladakh.

Peninsular India exhibits a variety *of sepulchral* and *commemorative monuments* which are either built of large stones, rude or chiseled or else associated with somewhat homogeneous group of black and red ware, and also an equally homogeneous group of iron tools and weapons. These are mostly collective burials as bones of more than one person have been revealed from them. They occur in cemeteries located in wasteland, often by the side of a river valley or cultivatable fields. It is strange that, there are also a number of burials, which are neither built on large stones nor associated with black and red ware, and iron implements. This means that all burials of Peninsular India do not belong to a particular age.

So far as the megalithic structures are concerned, the *Dolmen* type is found in Bastar State of Madhya Pradeh and Assam. Some *Passage graves* are scattered in Central India and Assam. The *Menhir type* has been reported from Kerala, Cochin, Travancore, Bastar, Chotonagpur, Orissa and Assam. Some contemporary primitives namely Munda of Ranchi and Bhumij of Chotonagpur prefer to erect a single stone (in standing position) on the graves. *The Alignment type* monuments have been recovered from different prehistoric sites of Chotonagpur, Orissa and Assam. The *Cromlech type* is the speciality of Southern India, particularly in Chingleput district of Tamil Nadu and in Kerala. This part of the country also shows some typical types of Megaliths. They are Topi Kal, Hood-stone, Barrow, Cairn, Pit-circle, etc.

Topi kal. This type of megalithic structure looks like an umbrella but without a handle. For the strange shape it is called 'hat stone' or 'umbrella stone'. It is very popular in Kerala.

Hood-stone. This is a large dome-shaped stone whose flat face rests over the ground. It is also found in Kerala.

Barrow. These are simply earth-mounds. Three types have been distinguished as *round barrow, oblong* or *oval barrow* and *long barrow* based on external shape. The whole of South India exhibits this type of monuments.

Cairn. It is more or less like barrow in appearance but made up of stone. This type on monument shows limited occurrence and sometimes found in Chotonagpur among the Oraons.

Pit-circle. It is a kind of lithic structure placed within a pit in the form of a circle. This type is restricted to South India.

The pottery associated with the megalithic monuments of Peninsular India is essentially the black and red ware. Iron objects typical to Indian megaliths include flat celts with two fastened rings, socketed and barbed arrowheads with comparatively long blades, tridents, long swords or lances, spearheads and spikes, wedges, bill-hooks, sickle, hoes etc. Many Bronze objects have come out from the graves of the district Tirunelveli; some of these have been sporadically found in the graves of Sulur, Sanur, Kunnattur etc. Shell objects decorated with linear pattern are more frequent in the graves of the district Chingleput than those of other regions. But the Megaliths of India are not free from problems, like other parts of the world, it possesses some definite ambiguity.

The problems around megaliths in India are seven-fold as furnished below:

1. The spatial distribution of megalithic monuments has not been fully known because the exploration work which started in 1944 is still going on. The findings of exploration, so far as available, indicate megaliths as a product of Southern part of India. But the isolated remains of megaliths in North India namely Rajasthan, U.P., Bihar, Kashmir are not without significance.

2. The different types of megalithic monuments as determined by the surface evidences have only been listed. They are not confirmed or modified later after excavation. An effort of tentative classification came from V.D.Krishnaswami. A unified nomenclature was submitted through his paper on megalithic types of South India that was published in the Journal 'Ancient India', No. 5.

3. The rituals or the process of megalithic burial is not completely known.

4. The particular people, whose culture had been represented in Megaliths, are not yet clearly identified. The skeletal remains and circumstantial evidences though point to the Dravidians as builders of megaliths, but there is no definite proof.

5. The origin of megaliths in India is not clearly determined. Although they occur in profuse and bear close resemblance with those of Europe and Western Asia, the cultural link is still far to be established.

6. The chronology has not been ascertained accurately. The black and red ware of megalithic tombs are found overlapped at the lower end with Neo-Chalcolithic, which has been dated as between 1000BC and 700BC. The upper end is placed with Russet-coated painted and Rouletted wares of AD 1000. The Brahmagiri evidence has sought to establish AD200 as upper limit of the culture and 200BC as the beginning. There is a danger in generalization as some megalithic monuments of India are older than 500BC. Recent C-14 dates have bracketed the megalithic culture of India within 1000BC and AD1000.

7. A comprehensive study on megaliths should be done where present day practice of making megaliths has to be correlated.

All of us know that Aryans brought iron to India. A lot of iron objects have been discovered from the megalithic remains in Peninsular India. This suggests that perhaps the Dravidians were the builders of megaliths who adopted the use of iron from Aryans but met a heavy blow by those invading Aryans. There is counter arguments that Dravidians were slow to adopt the revolutionary means, which gave the Aryans their initial superiority. However, creation of megaliths is still in mystery. We are not very sure whether the Aryans or Dravidians or any other group was in the background. It is expected that this problem will definitely be solved in near future, as the excavations will throw more light through the accumulation of more information.

METAL AGE

The cultural record of man's existence is divided into two great periods - the 'Age of Stone' and the 'Age of metal'. The 'Age of Stone' preceded the 'Age of metal'. The duration of these two periods was not equal. The Stone Age persisted far and far greater period than the Metal Age whereas metal came in use only recently. It was first used in Asia and Egypt about 3500 BC, and in Europe about 2000 BC.Once the use of metal was recognized, the art of writing, the city-life and also an infinite variety of inventions took place which contributed to the rise of human civilization. Metal enhanced man's mastery with stone, wood, bone and other substances.

The 'Age of metal' is sub-divided into two principal ages—Copper Age and the Iron Age. Bronze Age appeared between these two. Towards the end of New Stone Age (Neolithic) man acquired greater knowledge about environment and its resources. With this advanced knowledge he proceeded for metalworking. Metals were neither easy to be found out, nor it was easy to make tools out of them. Because, though some pieces of free copper or meteoric iron was available in nature, most of the metals were embedded in ores, which were difficult to use unless mining and smelting were learnt. In fact, there were four major stages in metalworking:

i) Mining, the discovery and collection of suitable ores.

ii) Smelting, the extraction of pure metal from ore or from impurities.

iii) Alloying, the mixing of different metals.

iv) Forging or Casting, the techniques to shape the metals.

However, Copper Age was the first phase of Metal Age, which continued upto the discovery of iron i.e. before the beginning of the Iron Age. Iron Age dates between 1200 B.C. and 1000 B.C. which means a time of about three thousand years ago from now. Late Copper Age is considered as Bronze Age which marks the Copper Age off from the Iron Age.

Metalworking is a very complex empirical science. Extraction of metal from the ore is the hallmark of metallurgy. It demands an advanced mode of knowledge as well as technical specialization. Although man had learnt the way of making fire, but a higher temperature that was required to melt the metals, was not possible to generate with an ordinary primitive furnace. For this a special blast furnace was required. So, it is assumed that the essentially required apparatus e.g. furnace, crucibles, tongs, etc. had been invented in the 'Age of Metal'. Although a little use of gold and silver were noted in Neolithic period, but the Metal Age is actually counted from the particular time when metal (copper) came into use in a massive scale. Metal Age denotes the large-scale industrial use of metal. Here presence of any metal or the occasional appearance of a metal in precious ornaments is never considered. Advantage of metal is that, when hot it melts and so can be poured into a mould. On cooling, it becomes hard and the edge of the metal implements may be made more sharp and strong than the stone implements.

As writing seems to have come hand in hand with metallurgy, Age of Metal corresponds approximately with the period of history. But written records are available particularly from the

Middle East and around the shores of Mediterranean; for the regions of Northern and Central Europe and also for the remainder parts of the world this record is either lacking or inadequate. Therefore, for those cases stratigraphy is the most accurate as well as reliable method of determining the sequence and dating. Typological arrangements were considered from stratum to stratum, site by site, river valley by river valley, region to region.Tying up with the written records, wherever possible, the final history of Metal Age was reconstructed. Various sequences of culture submitted by the archaeologists in different parts of the world were gathered to draw an over-all picture of the Metal Age in the world. Three major periods—Copper Age, Bronze Age and Iron Age were chalked out.

Copper Age

Copper first appeared in the Old World particularly in Mesopotamia. Some copper tools occurred with the late, improved Neolithic implements in the floor of the Tigris-Euphrates Valley in Mesopotamia. The floor of this valley at the head of the Persian Gulf, had just risen above sea level when the first people came down from the Persian highlands, about 4000 BC with the culture known as Ubaid. Those people began to drain the swamps and founded towns. A Spear-point made of native copper appeared in Al Ubaid III period at Ur of the Chaldea. For the earliest Mesopotamians copper was scarce, still a few more implements were found there. Such findings suggested that copper implements came into use in Mesopotamia between 4000 BC and 3500 BC. The first copper tools from native copper ore were made in the same manner as the stone tools were made in Neolithic Age. Pounding i.e. application of the method of cold hammering drew the shapes. The implements include simple flat axes and daggers. The nature as well as the limited number of the tools suggests that the people of the particular period did not know the art of smelting, casting or molding.

Knowledge of copper came to Egypt and the Asiatic coasts of the Mediterranean from Chaldea. In Egypt, a pin and two beads have been found as belonging to the Badarian period, just after 4000 BC. As we meet with copper at the foundation of early civilization, so it is likely that in later period knowledge of copper spread to the other regions from its center of discovery. As per the deduction, the principal types of copper implements reaching to Egypt and Asiatic coasts of the Mediterranean, advanced further into the Mediterranean islands and thereafter into the Western and perhaps even into the Central and Northern Europe. But by that time the nature of the implements became modified. Neolithic culture of respective region influenced them, although the principal characters were retained. However, copper reached to Gaul simultaneously from the South and the East crossing Black Sea and Aegean civilization. Naturally it took a long time to reach British Isles and Scandinavia. Thus implements made of native copper had been in use for a considerable time in almost all countries including Asia, Europe and Africa, excepting a few countries like Japan, Oceania, etc.

Copper occurs in a native state as pure metal in many parts of the world. It also comes in various ores being mixed with rocks and other metals. But as it was hard for the early people to extract copper from the ores, they avoided using of the ores and liked to work on native copper, which were widely found in the form of small masses and nuggets. Large masses were of limited occurrence. However, the small lumps of copper, scattered on the surface of the ground used to be picked up for making implements. The earliest method of metalworking, the cold hammering, was very simple. The pieces of native copper were hammered in order to flatten them out into various shapes. Neolithic people were the expert stone workers. So their successors came forward to experiment their craftsmanship on the pieces of copper. Since the material was metal, so constant hammering did not produce any chips as like the flint or Chalcedony. Rather they produced different simple stone implements of desired shape. But the principal drawback of this method was that the implements produced in this manner became very brittle. The only remedy was to heat and hammer them in an alternative way while making. The technique is known as annealing.

The next step in primitive metalworking was the learning of the procedure of extraction. It is held that in course of dealing with fire for annealing, once a lump of copper was accidentally thrown into the primitive hearth which was nothing but a stab of clay. Since there was strong fire, the copper nodule melted down. By a coincidence of most unusual kind, some intellectual brain noticed the strange change from solid to liquid. The property of molten copper was known gradually as the people spent their energy and enterprise further on it. But smelting of copper from copper ore was not as simple as the melting of native copper nodules by fire. It requires not only a higher temperature about 1200° C, also involves complicated procedure. Besides, as compounds of copper are different in different ores, so the process of smelting is also different. Let us consider the malachite, the most commonly found ore in the earth consisting of copper carbonate. To separate the copper from it, at first the part of carbon has to be driven out. Copper-oxide will be obtained as a residual deposit. Charcoal have to be added at this stage to provide more carbon which will be able to break the copper-oxide compound into carbon dioxide and carbon-mono-oxide, leaving the pure copper. In the next phase, the metal is allowed to run out into small trenches where it is cooled into ingots of standard size for the purpose of storage, transport and trade. Sometimes another substance, limestone is added to it for refinement. This keeps the metal in a more liquid state and prevents re-oxidization. Primitive copper smelting furnaces were very simple having holes in the ground. However, the extracted copper was possibly worked by the same method of cold hammering at the initial stage of learning.

Invention of the smelting process was a remarkable step in man's mastery over nature and widened the scope of scientific achievements towards high civilization. It was realized that the melted copper would flow into any shape, and the size and outlines of those shapes could be controlled. So in the next step people learned mould casting. By running the molten copper directly from the smelter into the moulds (of desired size and shape), they made different implements and ornaments. This saved the tremendous labour of hammering the ingots. The method has been named as open-mould casting. The first moulds were merely the prepared holes in the ground but soon moulds of clay and stone were introduced.

The next achievement was the making of closed-mould with two or more pieces. This was an evolved form of open-mould, which facilitated full use of the fluid nature of melted copper. The other advantages was, it allowed the metal worker to cast his copper in round shape, instead of entirely fiat or flat in one side. With this trick of casting, people gave up the old shapes, which were merely copies of stone models and began to make truly metallic shapes. Hollow sockets were also made in plenty into which butt ends of axes and spearheads could be firmly mounted. The smiths of Near East were more experienced; they used moulds with multiple pieces for casting more complex objects.

The finest casting technique was the 'lost wax' casting. The most intricate pieces were shaped by this technique. At first, the desired shape has to be carved out of wax. The model is then covered with a coating of clay. One or a few openings are left on the clay-case. When it is put to fire, the wax melts by the heat and eventually the clay wrapped on the model gets hard. The melted wax runs out of the holes. So when the molten metal is poured into the holes, it covers the place of the wax. Finally the clay cast has to be broken to remove the metal in cold state. The shape of each implement achieved by this method is unique because the hand-made model is destroyed in each case and so the same model can not be used repeatedly.

However, it is not yet possible to say that when the smelting of copper began. According to Lauriston Ward, the time was probably during the Jemdet Nasr period, shortly before 3000 BC and the first casting of copper was possibly done in the Nile Valley. However, even before 3000 BC, there was a great burst of metallurgical activity in both Egypt and Mesopotamia. A good number of copper implements and ornaments were obtained from different sites. Some scholars have pointed out that earliest smelted copper Trinkets appeared at Catal Huyuk (Anatolia) in the middle of seventh millennium BC.

The development of industries around the metals started about 3000 BC and therefore the period has been labeled as Metal Age. Copper Age, Bronze Age and Iron Age are the parts of the whole. Prof. Childe termed the period between 4000BC to 3000BC as the period of urban revolution. Because this period had witnessed great discoveries and inventions which brought the people together and made a larger co-operate group. The new trade and techniques demanded specialists and when a man became full-time craftsman (metal worker); the other people came forward to provide him with food, clothing and other necessities of life. Not only this, many other industries came into existence with the rise of mining and metallurgy. To get the raw materials and to circulate the manufactured articles, transport system was also developed. Mode of transaction improved. On one hand, wealth began to be controlled by the officials of the temples as trustee of god and on the other hand, for the sake of keeping records, systems of writing began to develop. The Neolithic people used the animals for meat and milk but the demand of the trade in later period made many animals as beast of burden. The use of horses and camels for the purposes of riding as well as packing has been noted from the copper Age levels in Persia, Egypt and Turkey.

Trade not only started on land but also on water. Picture of boats and sails on Egyptian vases denote the presence of water transport before 3000BC. Instead of raft and canoe (used in fishing and crossing a river), man became more dependent on boats for carrying goods from place to place up the river. Various occupational groups came into existence but all of them were not the producer of food. The non-producer groups gathered in cities and for the difference of wealth and trade, stratification was created within the society. Therefore, in contrast to Neolithic Age, towns and cities were found to be grown in Copper Age.

The appearance of mere copper could not bring an occasion of revolution. It started to change the established order of the things but very slowly. In the early stage tools and weapons were scarce because of the rarity of copper. The forms were copied from the flint implements by using Stone Age methods. When the metallurgy became established in the mining countries, the commercial relations started. Using of stone was reduced, metals became a substitute material of stone. But no where the appearance of copper could change the customs and usage of Neolithic people nor it persisted for a definite span of time. In fact, the non-technological cultural activities - social, political and ideological activities remained unchanged. This is the reason for which the Italian archaeologists liked to designate this phase as Aeneolithic. By the term Aeneolithic they wanted to mean a phase of culture at the dawn of Metal Age (instead of Copper Age). Some other scholars regard the phase as chalcolithic (Greek word chalkos means copper and lithos means stone). The propounders of the term chalcolithic perceived the Copper Age as Stone-Bronze transition where metal was creeping but in a subsidiary fashion. Another name, cyprolithic is also found for the Copper Age as most of the raw copper used in antiquity came from the Isle of Cyprus.

Recent radio-carbon dating suggests that an independent metallurgy in the Balkan started prior to 4000BC. In East Yugoslavia mines with a depth about 20m to 25m have been found. Tools of smelted copper from West Russia are contemporary to the earliest open moulds recovered from Bulgaria and Russia. Both can be dated as 4000BC. But stone continued to be the basic raw material until the spread of bronze.

Bronze Age

Alloying as a higher metal working technique appeared at the end of the Copper Age. Bronze is an alloy of copper and tin. In contrast to copper, the malleable soft metal, bronze possesses a superior quality of hardness. Tin has to be mixed with copper in a proportion of about one-tenth to produce bronze. A higher percentage of tin renders the alloy increasingly brittle. However, bronze predominated between the Copper Age and the Iron Age and therefore this particular period has been referred as Bronze Age.

Several types of bronze were found in the Bronze Age. Everywhere copper was the principal material. A small admixture of tin , phosphorous, arsenic or even sometimes the gold or silver is used with it. Copper-tin mixtures were found most widespread. Some good quality of bronze appeared in the Early Dynastic period in Egypt, close to 3000BC. Some of the prehistorians believe that the entire bronze technique was percolated from outside, probably being cultivated by the people of the mountainous regions of Asia Minor and Armenia who later came to Egypt in search of raw materials. But the fact is far from proved to everyone's satisfaction. Those who believe bronze as invented in South Russia around 3000BC, advocated a rapid spread to East Europe, Near East and North India by 2500BC. Whatever may be, once the invention occurred, it was utilized effectively in different adjacent regions.

Since the copper-bearing strata are much more widely spread over the surface of the globe than those bearing tin, copper was discovered first and then tin. Archaeologists debate as to whether bronze was prepared by measuring out the proportion of the two elements in the metallic state or whether the ores were mixed up before putting them into the furnace. By this hypotheses they wanted to explain the notable differences in the quality of the bronze arising from difference of tin content in the bronze. Not only the tin, a very small proportion of arsenic, antimony or zinc can modify the molecular shape of copper.

Ancient metal workers through their experience understood the varied properties of amalgamated metals. Because they found some copper ores produced a metal that cast better and naturally that particular kind of metal made better and harder implements than others. Thus the early workers became conscious about the nature of the ores. Gradually they understood that copper as a metal is rarely available in the natural state but abundantly found in various ores being mixed with rocks and other metals as copper sulphides, copper oxides, copper carbonates, etc. The metallic copper in ores is formed as a result of a prolonged contact of the outcropping copper veins and lodes with the atmosphere.

As it stated earlier, the most common compound is the copper carbonate (malachite). It was often grounded up to use as a green paint and regarded as the best source of metallic copper. Another kind of copper carbonate, which gave the blue colour, is called azurite. Besides, there are two forms of copper oxide—cuprite and melacouite and many sulpides of copper. As sulphide ores occur in deep veins and hard to find out, early people did not use them much. They have been the chief source of the metal only in recent days.

Early men first picked up the nodules of native copper either to get paint or to make an ornament. Primitive metal workers never went in search of great, concentrated deposits, not even in the period of Bronze Age when industrial use of the copper reached to the peak. However, there were many surface deposits throughout Asia and the mountainous regions of the Near East. In Europe also, they were abundant. But the best sources are believed to be situated in Cyprus, Hungary and Spain. In Africa, the largest number of deposits is scattered in the Katanga region of the Belgian Congo and it is not known that whether they were worked before the Middle Ages.

The Tin which is essentially required for bronze, was not readily available in past. Tin-bearing strata are actually rare as well as limited to a few localities. Tin ores occur in the original deposits in veins and in the form of small crystals in the crystalline rocks known as granulites; it is always an oxide called cassiterite and never found as a native metal. Moreover, as cassiterite looks like a heavy dark sand, it does not seem to be a metallic substance and so very difficult to be identified. Small and moderate deposits occur in Armenia, Syria, North-West Persia and Bengal. Large deposits are only found in Malay Peninsula, in South-East Africa and China. The principal European sources are Bohemia, Spain and the British Isles. In Africa, same old tin mines have been reported from the Northern Nigeria and Transvaal but their age is not certain.

The attrition of rock-matrices and outcropping veins by atmospheric agencies produced alluvial formations in which stream-tin occurs in the form of sand. It is only necessary to wash this alluvium in order to extract the cassiterite. This method was used in the exploitation of tin in Malaya, at Brangka, Perak, etc. places.

The first metallurgists found the beds of copper and tin in a virgin state. They simply dealt with oxides and used a reducing fire of charcoal to separate the metal. This metallurgical process is still utilized in Malay with the primitive furnaces or smelting hearths. Cassiterite is always found in a siliceous gangue, which flakes in the fire. As for the carbonates of copper, the gangue is either calcareous or siliceous; it splits with the heat. From the very early times fire was used for the disaggregation of rocks containing metals.

It is held that man discovered the alloy of these two metals by chance of favouring circumstances, which played an important role in the history of mankind. In the closed moulds, not only the bronze, but any alloy was found suitable. Most copper ores contained other metals such as nickel, lead, phosphorus, antimony, arsenic etc. as natural impurities. Presence of these different metals produced good effects on the casting process and the products. For instance;

(a) The manufactured object if contains arsenic with copper becomes harder than pure copper and provide a more durable cutting edge. It attains a property similar to bronze objects. Warriors with bronze weapons have always found their advantage over the enemies having copper or stone weapons.

(b) When metals with a lower melting point remain mixed with copper ore, the melting point for the whole alloy automatically comes down. This property provided a great convenience to the early metal workers who worked on primitive furnaces. The melting point of pure Copper is about 1085°C whereas for Antimony it is 630°C, for Lead 327°C and for Tin 232°C.

(c) The greatest advantage is found in the process of casting. In a closed mould, it is very difficult to cast copper because explosions or cracking of the moulds often take place. But the presence of other metal with copper minimizes this difficulty.

For successful casting 10 per cent to 12 per cent tin was mixed with 90 per cent to 88 per cent copper. All weapons and tools were made with this percentage. Since the addition of higher tin percentage renders the material increasingly brittle, a content of 30% of tin was used to produce a very fragile white metal which had been very popular as mirror in that olden time.

Bronze Age has been divided into four sub-periods by the early archaeologists. They attempted to classify the objects of each period on the basis of their size, shape, tangs, sockets, blades and ornamentation. It was especially significant in reference to the site of Montelius in France. But more and more excavations throughout the Old World exhibited that the sequences for the development of Bronze Age cultures are not similar in all places. Some types, which occurred earlier in some places, found from a later period in other places. Influences passed forward and backward in the adjacent areas but the main trend went outward from the Near East and the Eastern Mediterranean.

Prof. V. Gordon Childe forwarded a new classification on the basis of usage of the metal. His classification had an academic support as he collected more evidences before presenting the classification. He made three stages like,

(a) When the metal was used for making weapons and ornaments only.

(b) When the metal made tools for using them in various crafts and skilled trades.

(c) When the metal came into use in agriculture and rough work.

The Bronze was seldom used at the beginning for the difficulty of mining, smelting and working. It served as weapon, tool and ornaments of the wealthy people.The mass population, especially in ore-less regions, treated a bronze-piece as treasure. It took several centuries for this new metal to diffuse from the upper classes to farmers and carpenters. Though stone tools were replaced gradually by bronze in Western Asia and parts of Europe, in China it remained confined to the limit of rich and powerful.

Bronze proved itself better for the most purposes than the materials like stone, bone, wood and shell. Different tasks like ploughing the field, felling of trees, harvesting of crop, building of houses became easy as well as quick with the help of hard metal (bronze) tools. Some Bronze implements also stood unquestionable for the domestic uses.

Numerous bronze razors indicate that the shaving of men became quite popular during this period. Vessels had been made of bronze and bronze ornaments for personal adornments also appeared in this time. Elegant bracelets and necklaces were most notable; safety pins out of bronze became plenty for the use of both men and women. A great variety of weapons developed at the same time. Most of them began to outgrow from an old form but with time they took a distinct form. For example, the bronze Celt at the beginning was very simple; wings and flanges were added later and finally the wings had grown so large that they met to form a complete socket. Such a transformation of a Celt to an axe was complete at the last phase of the Bronze Age. In the same way, long swords grew out of the chipped-stone dagger. In the early Bronze Age these swords were very simple in type but in the late Bronze Age they were converted into broad-bladed double-edged sword. Neolithic polished stone axe with the haft hole gave rise to beautiful decorative battle-axe. Bronze spearheads developed out of flint spearheads. Although bronze arrowheads were not absent, but the scarcity of this weapon indicates that bow and arrow lost its popularity.

In the Bronze Age, pottery became a more developed craft of skill; different local styles were evolved. But the decorative art found on jewelry, tool or pottery was not considered as pure art. Megalithic monuments continued to be built in the Bronze Age, but the practice seemed to decline, as burial in cysts became popular. At the late phase, cremation became fashionable. In Bronze Age we further find the sea-going boats which replaced the dugout canoe in order to carry a large number of passengers. Commerce therefore got a chance to be flourished through the river ways.

Weapons made of bronze altered the warfare strategy. But it is sure that the life was not peaceful. The nature as well as the number of weapons suggests that the people preferred face to face combat with hand. Villages were like fortified camps and men used to keep arms with them. However, the intelligence of man lifted them up from their lethal capacities. They invented defensive devices against each killing tool. Therefore, shields, helmets, cuirass etc. were built during this period. Chiefs or kings were in the habit of using the precious bronze armours. Larger settlements were grown in general; in the advanced centers of civilization great palaces of stone and plaster were found to be built.

The use of bronze in almost all countries succeeded the use of pure copper. Neolithic stone implements disappeared slowly. The copper age towns, by the processes of cultural development, invention, diffusion (trade) and migration, gradually turned into Bronze Age cities and new ones were also built. In Mesopotamia, on the agriculturally rich Tigris-Euphrates valley, Sumerian cities grew up in the place of Copper Age settlements like Sumer, Ur, Layash, Erech and Eridu. The Sumerian culture afterwards spread to the North upto Khabur and also to the vicinity of Baghdad. Archaeologists have divided it into two phases, the Uruk and Jemdet Nasr; the names were derived from two important sites. Parallel Bronze Age cities sprang up along the river Nile in Egypt. Bronze Age cities of Egypt are better known than their Copper Age predecessors.

Bronze Age cities grew up more or less at the same time in India on the Indus River valley and its tributaries in the Punjab. Those people knew carts and the potter's wheel and also the bronze making techniques. Some of their techniques were found identical with the Mesopotamian techniques.

The evidences of trade can be traced between Indus and Tigris-Euphrates. Bronze Age influences spread from Egypt and the Syrian coast to Cyprus, Troy and other cities in Asia Minor and also to the Greek islands and mainland where the classic Mediterranean civilizations arose. In the next phase, Bronze Age culture advanced westward through Italy towards Europe. Trade routes were maintained both in land and sea. Thus, Bronze Age spread eastern, western and central Europe and from those areas to Gaul and Britain, and finally to Germany and Scandinavia.

It is difficult to be precise about the date of the inception of bronze industry in different lands and different localities. In the ancient centers like Chaldea and Egypt it existed towards the end of the fifth millennium before the present era; in the Eastern Mediterranean it would be about third millennium before the present era. Perhaps in Gaul it reached around 200BC and in the North of Persia and Caucasus only a thousand years earlier. Nevertheless, all these estimates are merely approximate and unfortunately the documentation is not as perfect as to establish a chronology with absolute certainty.

Copper phase was not a universal phenomenon in prehistory, nor it lasted more than a few centuries. Copper Age appeared largely in those places where the natural distribution of copper nuggets was accidentally found. As soon as the use of copper was established, the discovery of bronze was followed in rapid pace. However, the copper producing zones can be distinguished into two groups. The group in which the metallurgical knowledge had been originated by itself and the other group which gained this knowledge from foreign source. To begin with, the two Americas were ruled out. According to archaeological testimony, neither Algeria, Spain, France, the British Isles and Scandinavia, nor the Central Europe saw the Separation of first copper ingot from its gangue. Europe witnessed a brief and fleeting era of copper. The Aegean islands, Western Asia and, Egypt were left. Among them, Egypt can be expelled from the group of copper-producing countries because this country probably got the knowledge of copper from Asia. The Altai and the Pamirs are equally rich in copper but the antiquity of metallurgy in both the regions does not seem to go very far. In all probability, it was the north of Western Asia where the metallurgic knowledge was first discovered. Thereafter the knowledge, in a rudimentary state, would have gone down into Chaldea with the men who first went there to establish their settlements. Then it would have passed over to Egypt, the Phoenician Coasts and the Aegean islands i.e. the centers from which the knowledge spreaded to Europe.

Indo-china and China were favoured by nature for the discovery of bronze as plenty of cupriferous ore and Stanniferous ore were found there. But we should restrain our pen from drawing any inference until the Central Asia and China are better explored.

Although the social effect of the copper was almost nil, the full-fledged bronze technology influenced the entire culture as well as the structure of the society. As the craft of extracting ores and working with bronze required special skills, the labourers like smiths, artisans and the miners came into being. Each of them assumed a vital role in the society.

Miners and particularly the bronze-founders tended to become a set-off caste with hereditary trade secrets. So class divisions appeared in the society. Not only that, advent of agriculture itself sowed the seeds of social inequality and class-distinctions. Since all soils had not same amount of productive power, it was very likely that some particular families might produce much more crops in the field than others might. Naturally the classification like rich and poor was already in vogue. This inequality increased sharply in the Bronze Age. Different classes like rulers, the nobles, the traders or businessmen, the artisan, the farmers and slaves came into existence. Emergence of cities was found in this Age; the copper Age towns by the process of cultural development grew up into cities.

Apart from the farmers, all other people used to live in the towns. Increase in the production of crops generated surplus food which was enough to foster the town dwellers. The rulers and nobles living in the town accumulated much wealth in their hands and they often used to employ the bronze-smiths to acquire the products such as weapons, armour and various kinds of tools made of bronze. As none of them produced food-grains, they were in the habit of taking crops from the farmers of the village. In lieu, they supplied the farmers some necessary things like tools, sickles, Celts, etc. This period can be said as the germination period of trade and commerce. The exchange of goods was done exclusively by barter. However, the people who lived in town had various occupations. So they had different interest as well.

The necessity of an organization was felt at this time to regulate and control the urban life. This necessity gave the rise to state. With the rise of city-states, the kingdoms became widespread; the political and economic system came into existence. Kings and priests achieved the status like God. These kings along with the priestly class were found to rule the society. Some monumental temples were built to house the kings and priests, and all their works and wealth including the images of Gods and Goddesses. These monuments were built of stone, brick or wood, or by the combinations of these materials, depending on their availability in the locality. The commerce flowed in and out of the so-called temples. Old barter systems was no more found suitable for active and complicated trading, so money economy was invented and great wealth was accumulated within the cities. With the increased scope of leisure and wealth, man's intelligence flourished in different directions; art, architecture, commerce, craft, script etc all developed at a time. Foundation of our modern science was also laid down during this period. The knowledge of arithmetic and geometry facilitated in dealing with weights and measurements; lunar or solar calendars contributed to human attributes. On the whole, Bronze Age brought revolutionary changes in the growth of cities, of states and kings, of social classes, of enduring structures, monuments and writing. All new developments were interconnected with copper-bronze metallurgy.

Bronze Age in Europe

The European Bronze Age poses interesting features. There are no Bronze Age cities in Europe, no known political unification, no writing; even the potter's wheels were not adopted until about 1000BC, when the Bronze Age was ending in Asia. Actually Bronze Age Europe continued like Neolithic Europe with a peasant - culture. But there were communal undertakings, large structures, and presence of chiefs resembling the Megalithic period of late Neolithic. A class of bronze-working smiths obviously appeared at that time but the rest of the people were the farmers. The reason for all these is that the infiltration of bronze took place in Europe after an expectable lag. In fact, bronze crept to Hungary around 1900 BC, Czechoslovakia, Central Germany and Italy around 1800 BC; the Rhine, France and Britain around 1700 BC and lastly to the Baltic Shores .and Scandinavia around 1500 BC.

Iron Age

Iron is the other metal, which gained supreme industrial importance in history. It stood as wonderful material for tools and weapons for its considerable hardness. Its standard was proved higher than the bronze and eventually the materials like stone, bone, wood and shell did not come in comparison. The second advantage was that, it could be found almost everywhere. The sources of iron on earth are as follows:

(i) The earth's surface contains 4 per cent to 5 per cent of iron.

(ii) There are several iron ores, which offer metallic iron in the form of iron oxides, hematite, limonite, magnetite and siderites.

(iii) Small particles of pure iron are found in the basalt rock, though they can not be used for practical purposes.

(iv) The meteorites, which drop on the earth from the sky, also hold a large quantity of iron. The nature of these sources indicated that except from the meteorites, iron is never available in pure state. Even the meteoric iron contains 5% to 25% of nickel for which the implements escape from rust. However, the first iron objects made by man were of meteoric origin. Because, the early men could easily pick up such iron lumps. In quality, this iron is very tough but malleable.

Like bronze, iron made its appearance at different dates in different countries. But the history of iron began with the Hittites of Anatolia in the Near East. By 1000BC it spread out in all directions and became a metal for general use in Western Asia, Egypt, SouthEastern and Central Europe. Therefore, it is seen that iron came into use about two thousand years or more later than bronze and in no country the passage from a bronze to an iron industry did not occur suddenly. Since iron ores were more abundant than that of copper and infinitely greater than the tin ores, a natural question came to the mind that why iron Age appeared so late! There are several views in this respect. Some scholars argue that the people of Near East region knew the working of iron by 3000BC, i.e. as early as bronze. But the iron objects of that era were very few and most of them were made of meteoric iron.

It is still not clear that whether all iron objects came from Near East before 1500BC were made of meteoric iron or some of them were made of iron extracted from ore. If all of them were made of meteoric iron, why people of that place took so much time to learn smelting of iron ores while they made themselves easily versed to smelt copper, lead and several other ores? Even if some of these early iron objects were found to be made of iron extracted from ore, the same question comes again. Why the general use of iron was delayed for about 1500 to 2000 years? This late coming of iron can be partially explained by its high melting point. The melting point of iron is 1530° C which is about 500° C above than that of the copper or gold. Early metal workers were not capable of producing this high temperature; nor they were acquainted with the chemistry of iron. Moreover, for a long time they failed to recognize the spongy mass or wrought iron as a useful metal.

There were other reasons too. The bronze-workers who extended their skill to smelt iron in later period did not foresee any profit at first. Secondly, there was limitation with the softness of pre-steel iron. Thirdly, iron could not supersede the prestige value of bronze for which bronze was considered as a symbol of high rank and treated as a sacred metal.

The smelting technology of iron was much complex than copper. Furthermore, like 'native copper', no native iron could be obtained merely by heating. To obtain metallic iron from the ores, at first it is necessary to get rid of the oxygen or the carbonic acid with which it is generally combined. The silica and other rocks, which form a part of the ore, have to be separated also. For this, the ore must be heated in a charcoal fire. The charcoal takes up the oxygen from the ore (in case of an oxide ore) and passes off as carbon-di-oxide. On the other hand,silica gets combined with a part of the iron and goes off as slag. Rest of the iron is left as a pasty, sponge-like mass, known as 'bloom'. The slag generally contains the ferrous silicate but may show considerable variation for the presence of other rocks in the ore. However, it plays an important part in the technical process because as a scum it floats on the surface of the liquid iron and prevents deoxidizing which takes place in contact with air. The slag is no longer needed after this and so a method of elimination is followed. The spongy mass is heated and hammered repeatedly until the whole slag runs out and iron forms an ingot - a solid mass. The heating and beating of the spongy lump is called forging.

Iron may be obtained in three forms:

(i) The solid mass of iron as available in the above-mentioned process is called 'wrought iron'. It includes a little or no carbon at all. In this state, iron remains relatively soft and further malleable, so can easily be worked.

(ii) If the heat of the furnace increases more than the required temperature, instead of forming a spongy mass, the iron melts completely and flows off. Iron in this state possesses some carbon in it.

(iii) If the percentage of carbon is excessively high, the iron becomes hard but brittle. In this state the iron is called 'cast iron'.

Primitive smelting of iron did not produce the bright liquid needed for casting. Rather it produced a dirty spongy mass with slag in it, which can then be beaten into a desired shape. The fully molten iron for casting, which needs a higher temperature, was first utilized in China, long after the beginning of the Christian era. In between, the process of carburizing (steeling) was invented around 1500BC in the vicinity of the Caucasus mountain, not far from the area where first invention of bronze had taken place. In forging a bar of wrought iron, constant reheating was required. If the smith reheated the bar by thrusting it into the heart of a charcoal fire and if the heat was more than sufficient, the outer surface of the bar would became carburized to a depth proportionate to the length of the treatment and heat of the fire. Therefore the surface of the bar became steel and this hardness increased the efficiency of iron.

The regions namely Near East, India and Far East are still considered rich in iron. India seems to have a tradition of early iron working as it appeared in megalithic graves. Huge forged column upto 42 feet high, weighing about ten tons are found in the North. But unfortunately no archaeological or historical evidence justifies the fact that iron has been in India from before the beginning of Christian era. Iron Age reached its peak in the great classical civilizations of Greece, Rome, China and India. In Europe, Iron Age culture appeared in 650BC where the first phase is known as Hallstalt A or Hallstalt I named after a famous cemetery site in the Austrian Alps. But the arms and implements of Bronze Age continued to be used long after the Hallstalt forms came in. Superiority of the Hallstalt models was recognized and so their forms were copied in copper alloys.

Objects of Iron Age comprises of long, slender daggers and swords made of iron which occur along with lance and bow-arrow. The iron swords and daggers are remarkable for the shape of their hilts, which often have horns, and a conical pommel of characteristic appearance. The celts and scythes are the other propagators of iron industry. Iron objects and implements of a later date have been found from Italy, Spain and Central Europe. They exhibited considerable change and modifications resulting from the influence spread out through Eastern Mediterranean. In the New World, in Oceania, in Polynesia and also among the Tribes of Siberia, the appearance of iron is quite recent.

No compulsive stages have been identified in the succession of stone, copper, bronze or iron. For example, Negro Africa and Japan have skipped the Bronze Age altogether and passed directly from Stone Age to Iron Age. The work of Negro blacksmiths are so great that some scholars think that iron work may have begun in Africa a long time ago. Some old mines are also present there. But it is a matter of disgrace that no archaeological proof exists in this respect. It is also believed that a camel was brought to Northwest Africa from the desert of Sahara by trade about some centuries after the birth of Christ. Iron presumably reached to Negro Africa after this event. Although the most primitive method of securing the metal is still practised among the African Negroes, but nearly all peoples of Negro Africa use iron. Some scholars argue that the Negroes of Africa developed the art of iron-working quite independently.

The discovery of iron opened a new horizon to man. As a hard metal it became very useful in cultivation; agriculture could not make expected headway with the soft metals like copper and bronze. The discovery of iron supplemented the handicaps of the 'copper-bronze' Age. As a matter of fact, iron brought about many changes in the methods of production and increased the yield of crops. As iron was easily available at a cheap price, any peasant could afford iron implements necessary for his agricultural work. Besides, iron led to the manufacture of different vehicles and vessels, which facilitated communication with foreign lands through trade and commerce. As a result, the civilization of Iron Age spread much more widely and over a much vaster area than that of the 'copper-bronze' Age. Types of industries increased in number and workshops became larger, employing more and more men.

The growth was linked up in a chain. Introduction of the cheap iron plough paved the way for extensive cultivation that yielded a higher food-crop. This increase led the population to grow very fast into the hundreds of thousands. As a consequence, cities began to increase in size with many acres of parks and gardens. New empires and trading groups arose. Society became further divided into certain classes as the aristocrats, the middle-class people, the peasants and the slaves. The big landowners were the aristocrats; the middle-class people of those days were comprised of traders or businessmen and the craftsmen. The peasants still used to live and cultivate in the villages with cheap iron implements. The slaves belonged to the lower rung of the social ladder; slavery as a system was widely known in this age. On the other hand, the issues of conflict increased a lot. The need of powerful ruler was felt to prevent the situation and therefore kingship made its appearance.

The invention of true alphabets in the Iron Age brought the possibility of literacy to all classes. In the Bronze Age, the literacy was confined to the priests, officials and clerks only. The arrival of the alphabet instead of hieroglyphic or cuneiform scripts made it possible for all to receive at least elementary education with minimum effort. Side by side, stamped coins of standarad type was invented. This replaced the use of silver-bar, which seemed to be awkward to handle and difficult to manipulate for purposes of fraud. The development of coined money though could not improve the financial condition of the poor but it undoubtedly signified a better medium of exchange useful for minor transactions at the market places. A peasant hereby could earn money by selling his surplus foodstuff and purchase any commodity after his liking. Increased production, transport facilities and invention of coin - all contributed to the improvement of trade and industry. Thus, man marched gradually towards the higher civilization.

Metal Age in India

It is very difficult to say when metal came into use in India. India remained outside the realm of the advanced Palaeolithic culture. The earliest traces of communities based on farming in India were confined to Baluchistan and to the adjacent areas of Sind on the right bank of the river Indus. During the third Millennium BC the environment of Baluchistan was not so arid as it is found at present, so permanent settlement of human being was possible then. From these antecedent stages of the human settlement, a civilization arose in the extreme north of India known as Indus Valley Civilization. It flourished around 2500 BC; it exhibited high degrees of civic discipline that was uniformly distributed over an extensive area and also showed stability over long periods of time. The cultural matrix of Indus Valley Civilization was found to be spread through the Harappa in Punjab, Chanhu-Daro in Nawab-Shah district, Mohenjodaro in Larkana district in Sind, Kalibangan in north Rajasthan and Lothal in Saurashtra. The civilization was earlier than that of the Vedic civilization and some Pre-Aryan people, probably the Dravidians, were responsible for the growth of the ancient culture. The Copper Age covering a time-span from 1800 BC to 1500 BC appeared after the Neolithic period. Its influence was continued further between 1400 BC and 1050 BC.

In Rig-Veda, metal was called as 'aya' but the term did not differentiate between copper and iron. Vedic civilization of the Indo-Aryan flourished in the Indian soil around 1500BC where Rig-Veda was the first of all Vedas. From the vedic source we have came to know that copper was unknown at the beginning of Vedic Age and it *came* into existence especially in Brahmana Age. In fact, the Vedic civilization was totally different from the Indus Valley Civilization. Aryan conquerors were fully acquainted with the use of iron when they came first to settle in Punjab.

The evidences regarding the Copper Age in India came from the surface of Ganges-Yamuna Daub and also from the places like Hyderabad, Nagpur, Madura and Mysore. Early copper objects occurred in small quantities with Neolithic artifacts. Pottery of this period was generally painted. Whether a few of these copper artifacts were locally made or imported, has not been ascertained. Presence of copper objects have been noted from almost everywhere in Deccan, Central India, Rajasthan and Saurashtra.This stage has been designated as Chalcolithic for India. The Nevasa and Daimabad are the two promising sites.

Three hoards of copper objects in Gangetic basin have been excavated from Navdatoli, Chandoli and Khurdi. Other important sites include Bisauli and Rajpur. Bihar and Orissa also exhibit certain sites rich in copper articles. But on the whole, two great zones can be distinguished here. The first one comprises of the upper Ganges and the Yamuna Valley. The other was centered on the Southern plateau of Bihar and extended towards South into Orissa. Both the zones corresponded to the early centers of copper production but possess some differences between them. Unfortunately, there is no record to know the age of these hoards, except the places where they occurred in association with ochre-coloured pottery. The site at Hastinapur is found stratified below the painted grey ware. Gilund culture of the South-East Rajasthan also shows the same instance where some white and red wares were found to occur with copper objects.

The communities occupying the region between the upper Chambal and Godavari basin in Malwan, and the West part of Northern Deccan, seem to have shared certain attributes of Chalcolithic culture.

It is spread all over the Deccan and Mysore plateau. It had its root in purely Neolithic culture of South-East Asia. Recent radio-carbon dating from the Maheswar settlement in the Narmada Valley suggests that the use of copper began in this region around 1500BC and possibly it came to an end about 400BC to 500BC for the spread of iron technology. However, the people used to live in the houses of wood, wattle and daub. Their flint industry resembled to that of the Harappans and it was essentially a blade industry. They had wheel-turned pottery, red in colour and painted in black with geometric designs where representations include either animal or human or floral pattern. Copper was used very economically. It appears that the copper culture principally flourished in North India stretching from the Arabian Sea to the Bay of Bengal. According to Mr. Vincent. A. Smith, a copper Age intervened in North India after the Neolithic period where the people were the original non-Aryan inhabitants of India. Mr. Smith dated the copper implements previous to 1000BC and the most primitive forms may be as old as 1500BC-2000BC.The Gangetic hoard seems to have a much wider distribution and some native people namely Nishadas, Pulindas, Savaras have been spotted as the users of this hoard. In those ancient days they were able to make some beautiful painted pottery along with copper tools and weapons.

A large number of copper ore deposits have been found in Rajasthan, Chotonagpur and Singhbhum in Bihar and Orissa. The finds include tools, weapons and ornaments. Weapons comprise of celts, rings, flat axes, daggers, swords, harpoons, anthropomorphic figures, etc. The largest number of copper implements have been discovered from Gujeria of district Balaghat in Nagpur division of Central Province. But at that time, large scale manufacture of copper implements was virtually impossible for the want of equipments. The old indigenous process of manufacturing copper from pyrites is still continuing in several places of Rajasthan, Sikkim and Nepal. The pyrites after roasting was heated with charcoal and flux in small blast furnaces; the blast is generally provided by hand bellows.

In India bronze implements are scarcely found. The Bronze Age and civilization are reserved for the Indus Valley Civilization proper and its manifestations were found in Punjab, Rajasthan and Sourashtra (Fig. 12.1). Indus people must have possessed the knowledge of smelting and alloying of copper. Since this culture shows many copper implements in association with bronze implements and also as the inhabitants continued the use of long blades of stone, some writers like to include this culture under Chalcolithic. But in fact, it was a definite Bronze Age culture as smelting technology of copper as well as the alloying of bronze were fully discovered by this time.

No evidence of copper age is found in South India. So it is thought that South India had by passed the Copper Age and reached to the Iron Age in third century BC . South Indian Stone axe culture was flourished in the region between Krishna and Godavari river basins as it is clearly stratified in Brahmagiri . The first and foremost primitive agriculturists of this territory could not afford the metal axes for clearing forest or working on timber. Copper was absolutely rare and the pottery was hand-made. Perhaps this Stone Age culture practised by the Mesolithic aborigines of this sub-continent shattered down with the spread of iron-technology during the closing century of the first millennium BC.

Iron was undoubtedly introduced by the Aryans. It replaced the use of copper and bronze. But it did not appear at a time everywhere in India. Its spread over the sub-continent was noted during the later half of the first millennium BC; it was incorporated within the technology of agricultural communities who so far relied on copper and stone. The megalithic monuments such as Dolmens, Cromlechs, Cairns and Menhirs are the only evidences of transition from the copper or Stone Age culture to the Iron Age culture in India. These megalithic monuments are found in Karnataka, Deccan, Central India, Madhya Pradesh,Orissa, Bihar, Assam, Rajasthan, Gujrat and Kashmir. There have been only a few systematic excavations. The monuments at Ranchi show a mixed assortment of finds like polished stone tools, wheel made pottery, copper and bronze objects, copper and gold ornaments and iron slags. Particular date of these monuments have not yet been ascertained. The monuments of Deccan exhibit iron tools and weapons assorted with microlithic flakes of agnate or crystal and carnelian beads. Nevertheless, the present knowledge about the Iron Age in India is so poor that we are not able to pass on the comments in a very distinct way. The transition between the Stone or Copper Age to Iron Age has not been traced properly. The purpose of these monuments is not clear at all although we know that most of them were funerary, connected with the disposal of dead. In Assam, they are connected with fertility rites and ancestor worship. However, it can only be asserted that the use of iron was fairly general by third to fourth century BC in almost all parts of India. Possibly about fourth to sixth century BC iron was first introduced in some parts of northern India. Still there are large pockets - hilly forested region of Central India, Andhra and Assam where iron might not have reached and people are still in a Mesolithic stage. Mesolithic in India mostly bear the features of Late Stone Age.

PERIOD OF TIME (BC)	CULTURAL AGE	AREA OF CONSIDERATION
500 B.C. — 200 B.C.	EPIC AGE (RISE OF MAURA DYNASTY IN MAGADHA)	INDIA AS A WHOLE
1000 B.C. — 500 B.C.	BRAHAMANIC AGE	DO
1500 B.C. — 1000 B.C.	VEDIC AGE	DO
2500 B.C.— 1500 B.C.	BRONZE AGE (INDUS VALLEY CIVILIZATION)	SIND, PUNJAB, RAJASTHAN AND SAURASHTRA
1400 B.C. — 1050 B.C.	NEW CHALCOLITHIC (COPPER + BRONZE)	MYSORE, ANDHRA
1800 B.C. — 1500 B.C.	CHALCOLITHIC	RAJASTHAN, CENTRAL INDIA, SAURASHTRA AND DECCAN
C 2000 B.C.	NEOLITHIC	ANDHRA, KARNATAKA, KASHMIR
C 3500 B.C.	NEOLITHIC	BALUCHISTAN
C 5000 B.C.	MESOLITHIC OR LATE STONE AGE	LANGHNAJ, MYSORE, TINNEVELLEY, BIRBHANPUR
C 125,000 B.C.	MIDDLE STONE AGE OR MIDDLE PALAEOLITHIC	LATE SOAN, NEVASIAN AND OTHER MANIFESTATION IN PUNJAB AND PENINSULAR INDIA
C 500,000 B.C.	EARLY STONE AGE OR EARLY PALAEOLITHIC	SOAN CULTURE OR CHOPPER CHOPPING TOOL CULTURE IN PUNJAB
Disintegration of Indus Valley Civilization started by the end of C. 1700 B.C. (Ĉ = circ		

Fig. 12.1. Chronological Table For Cultural Sequence In Acient India

Distinction between Indus Valley Civilization and Vedic Civilization

Vedic Civilization was cultivated by Aryan people who being originated from an European Stock invaded the autochthons of India, the creator of Indus Valley Civilization in the Indo-Gangetic plain. These invaders were considered as highly intellectual people who gave the world a new language, the Sanskrit and a new religion based on Veda.

Indus Valley Civilization	*Vedic Civilization*
It is a civilization developed by the Dravidian people.	It is a civilization developed by the Aryans.
It was an urban civilization Reflected in city life.	It was mainly the rural civilization, centered round the village.
The people did not know the use of iron. They used the materials like Stone, Copper and Bronze in making weapons.	The people used iron weapons extensively.
The people did not know the use of horse in war.	The people had the cavalry. They introduced horse in India.
Mother cult as well as the cult of Siva were prominent in the religious belief.	People believed in one God.
Bull were of high esteem.	Cows were the high esteem animals.
No social division was found among the people.	Social divisions were conspicuous in the society

Metallurgy in the New World

An apparent independent development of metallurgy following almost similar line has been evident in the New World. American scientists have declared that it holds no connection with the Old World metal industries. Some European scientists who strived to find out the hidden connection, ultimately did not succeed to prove it.

Although the New World smiths knew the use of copper and bronze along with gold, archaeological interpretation has emphasized the artistic aspects of metallurgy rather than the instrumental ones. However, the centers for the development of metallurgy in the New World were the Andean region of South America, Peru and Bolivia. The development of copper and bronze was very remarkable in these regions. At the same time some outstanding work on silver and gold were found which comprised of tiny beads and lacy filigree. Gold working possibly originated somewhere in the North Andes of Columbia or Ecuador about 1500 B.C.

In comparison to different parts of Eastern hemisphere, dating of Peru is not well established. Although copper was invented in South Peru and North Chile before 1000 B.C., but the earliest copper spearheads and digging stick tips did not appear until the Christian era started. Some copper implements of the Tiahuanaco I period have been found in the Lake Titicaca region of the high lands and some implements of the Middle Chimu period from the Peruvian Coast. Tin and bronze appeared earliest in Argentina, about A.D. 400. Tin and bronze implements have been found in Tiahuanaco II period and Late Chimu period.

The finest creation of American smiths came from the area of North Columbia and Lower Central America. Metallurgy was not found in Mesoamerica until A.D. 900. Afterwards simultaneously arose in South Maya area and also on the West Coast. It reached upto the valley of Mexico. Before the Spanish conquest, the Mexican smiths became quite expert in making the tools for agriculture and war. They also produced exquisite jewellaries.

The Inca, the late prehistoric empire was invaded by the Spanish people which had developed many techniques of their own like hammering, embossing, welding, casting and soldering. They used bronze implements like chisels and points in the place of digging sticks. All these techniques were very popular in the Quimbaya area of Columbia, though the exact dating has not been available till now. In Mexico, copper and bronze work were primarily developed by the Aztecs prior to the coming of the Spaniards. But in the North of Mexico metallurgy had never passed beyond the stage of cold-hammering in prehistoric time. Large deposits of native copper still exist in the Great Lake region and many other widely separated areas which have been utilized by the Eskimos, Nora Scoria and even by the Arizona.

As regards the tin, in North America, Cassiterite appeared in the coast of the Pacific ocean. Mexico, presented a special bronze industry though no tin-bearing beds was found there like the Old World.

Although bronze was well developed in parts of the New World and the use of native copper was also widespread, no light was thrown from anywhere to justify an American Iron Age. A few meteoric iron ware used which were made into implements by beating, Non-meteoric iron which was found there belongs to post-Columbian period, probably introduced by the invading Europeans.

In fact, the earliest migrants who reached to North America were pre-Neolithic hunters. Their descendants gradually succeeded in domesticating New World plants and to a lesser extent, the animals as well. About 5000 years or more after the inception of Neolithic Age in the Middle East, it got its way in Middle America. Because it has been found that the major elements of cultural development occurred in the Old World, specially in the Near East and thereafter it diffused to most of the adjacent land masses. During the period from fifteenth to nineteenth Centuries, the Palaeolithic type of hunting culture survived only among the marginal peoples i.e. who inhabited in the Arctic, the Sub-Arctic and Arid regions of North America, the deserts and deep interior forest areas of America, Africa, Asia and Australia.

DOCUMENTARY PROCESS OF CULTURE

Man holds a unique position among all animals in this world because of his potentialities for learning, discovery and invention. He possesses an endless capacity to invent and learn.. This learning takes place by experience; the acquired knowledge passes down from generation to generation for which man has been able to enrich his way of life. The fabulous growth of culture is an achievement of man; no other animal has ever made this kind of progress.

Creativity of man that has been manifested in development of culture was bloomed far back in the history of earth, immediately after the appearance of first man. But we have no written record of that period. The Prehistorians try to unveil the unrecorded part of history by different means e.g. by searching out the surviving specimens of human manufacture (artifacts), by the evidence of the places where they lived, by the structures of the graves in which the dead bodies were buried, etc. Fortunately, the tools, living places and graves give us some important information basing on which we endeavour to build up the story of early men and their succession of cultures. Data are usually gathered from a number of sites in a particular area in order to arrange them in proper sequence, from the earliest to the latest. This chronological arrangement enables the prehistorians to identify various cultural stages to which man belonged. They try to project a total picture of early days by undertaking a comparative study of the development of human culture in different parts of the globe. But the reconstruction of cultural history and the past events are not at all easy. They involve various modes of enquiry and analysis. The field of the prehistorians is distributed worldwide. They divide it into smaller units in terms of area or time. But sometimes, the varied nature of the data in different units makes them perplexed in interpreting the truth. In a particular area, with the passing of time, new forms and techniques are found to develop which continue for sometime.The new ideas may be introduced from outside. As a consequence, the story of human culture changes from time to time. The progress of evolution can be estimated only by means of a comparative study of artifacts available from the geological strata. Here the role of a prehistorian resembles to that of an archaeologists who keeps himself concerned with the antiquities earlier than the historic period.

EXPLORATION AND EXCAVATION

Archaeologists look into the past and go back far beyond the historic period. Systems of writing have developed only about five thousand years back but our early ancestors learnt to make the tools about two million years ago. Archaeological anthropology aims to focus the life-styles of the people during those hundreds of thousand of years and wants to systematize the phases of unrecorded history. They also seek to understand the processes of culture change that occurred in dim past. Unfortunately, the evidences left at different sites are very scarce and extremely hard to be explored. Furthermore, the aid from different other disciplines particularly geology, zoology, botany, geography, chemistry and physics are essentially required.

Exploration Procedure

An archaeologist or a prehistorian can not bulldozes all places indiscriminately for finding out the buried treasures of the ancient world. He possesses specialized knowledge and training in this respect. In fact, identification of a site is required first of all. Sites are often determined accidentally or incidentally through clues, deducted from mythology and folklore. The publications and museum collections of previous archaeological work can also throw hints about the location of possible sites. Sometimes local people of a particular area supply some information. However, a prehistorian selects the site very carefully. Before the final selection, he has to make a preliminary survey in the proposed area. This enables him to get first-hand knowledge about the surroundings of the site and to obtain the surface-collections.

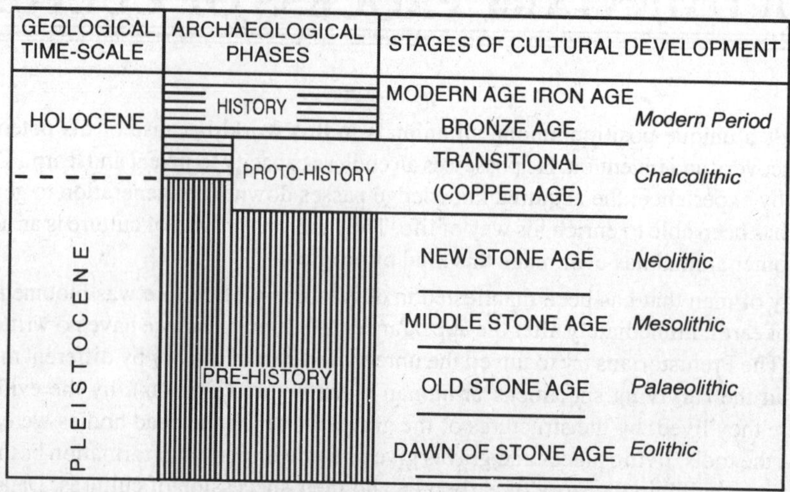

Fig. 13.1. Record of Human Progress

Sites are considered as the 'homes' of the prehistoric men. Excavations of these 'homes' therefore provide valuable information about the people and their culture. Artifacts dropped by the prehistoric men became incorporated in the various strata of the earth. An assemblage of artifacts embedded in a particular geological stratum indicates an industry. So where the prehistorians are capable of finding out the sequence of strata, they also get a sequence of industries. The geological law of superimposition helps a prehistorian immensely. He can also apply this law in analyzing a rock-shelter deposit. The natural floor of a rock-shelter starts to accumulate debris with the occupation of first settlers. For example, the layers from roof and wall break down and fall on the floor, mud and dirt of the atmosphere being carried by the wind enter into the rock-shelter and spread over the matters of the occupants. Everything gets mixed and gradually turns into a hard layer over the floor. With the passing of time, this layer becomes thick. When the successors or a new group chooses the same spot for habitation, a fresh layer of accumulating debris is added on the previous layer. In this way, the cave deposits may show considerable stratification.

On the whole, three types of site have been identified by the prehistorians—(1) *Open station sites* (2) *Cave or rock-shelter sites;* and (3) *Settlement sites.*

Open Station Sites

Most of the sites of Palaeolithic age were the *open station sites* as Palaeolithic men did not know to look for a shelter and making of a shelter was out of their thought. However, such open stations have been found in gravel beds or brick-earth pits. Since they remained open for a long time, the upper layers of the stratigraphy may get twisted or contorted. In South and East England many sites were found where the bottom layers of a gravel pit are evenly bedded but the upper layers are absolutely contorted. Again, some part of a deposit may be destructed by water action when the site is very close to a river. Therefore, an investigator often faces geological problems in identifying open station sites. It is better to collect the implements from different layers of a site. But unless the sites are very rich in tool assemblage, in situ finds are not available. In this case the investigator also faces difficulty in ascertaining the date of the layers.

Cave or Rock-shelter Sites

The *cave* or *rock-shelter sites* are considered as an evolved form of habitation of the prehistoric men. But such rock-shelter sites are relatively less in number in comparison to innumerable open settlement sites in gravel beds.

Usually a rock-shelter or home site in a cave mouth consists of two parts—the shelter and a terrace in front of it. Excavation of both the places is required to yield prehistoric industries. Trial trenches are dug out at first from which an investigator acquires a fair knowledge about the stratigraphy of the place. The investigator may also find industries belonging to different cultures in different layers. Generally the colour and the texture of the deposit vary from layer to layer. Where there is no stratigraphy, the industry remains uniformly distributed from top to bottom. It should also be kept in the mind that the various layers in the rock-shelter are not always horizontal. There may be large boulders or other obstructions over which layers curve. Therefore, an industry obtained in any one of the layer may not be found in the same level throughout the site.

Settlement Sites

No *settlement sites* are found of Palaeolithic age. Rather these are the sites of Neolithic date. Such a site is generally revealed as a mound and a large site appears as an assemblage of mound groups. However, any area showing the indication of past human settlement is considered as a site, regardless of its extent or number of mounds involved. Sometimes the presence of barrows and megalithic structures denote the features of a promising site as those of England. A site may be found in a village or a nearby area, within a forested area, on an abandoned river-channel position or even within a kitchen-midden accumulation. The sites with mound, however small they are, generally visible from a considerable distance. A good number of sites have been located in this way. Some sites are best observed from the air. In Central America, the native chicle-gatherers had brought many sites under the attention of archaeologists. In almost every spot, the local inhabitants remain aware of a site, although they do not realize the importance in terms of archaeology. Many of the local inhabitants, out of curiosity or by chances collect the relics and preserve them in their houses. Therefore, if an archaeologist talks to the local people before survey, he gets more and more fertile information, relevant to his enquiry.

However, the fieldwork starts with survey and surface-collection.

Survey : Survey is the basic pre-requisite of any fieldwork. In exploration, there is no scope of contemplation. Therefore, the objective of survey includes an adequate sampling of sites, a safe coverage of the area, an insurance against the omission of any significant cultural manifestation etc. As soon as a site is selected, it is necessary to point out its location on a map with considerable accuracy. A sketch map also needs to be laid down where the description of the location can be furnished in reference to the local landmarks. For other pertinent information, an index card should be used.

Surface-collection : The surface-collection seems to be necessary from the beginning of the survey. The objective is to get the sample of the materials prior to the excavation. A surface-collection may indicate whether the excavation will yield fruitful result or not. Surface finds like pottery and stone tools often suggest the nature of a site. However, the size of the surface-collection may vary enormously from site to site but not appreciably from one part of the survey area to another. Preliminary site survey and the analysis of surface-collections are the primary activities of exploration.

Excavation Procedure

Excavation comes next to exploration. After the selection of a site, excavation starts from a definite point. A grid plan serves this purpose; the area is divided into a number of squares. The work proceeds in accordance with a previously prepared scale map. The starting point, known as 'datum point' is marked permanently with an object of steel, cement or any other durable material. This is the reference for the future excavators who might come to the particular site in later years. The 'datum point' would make them aware about where the earlier excavation had been done.

The whole area is not dug out. Only some pits are dug being randomly selected on the squares of a grid. Since the sites are often stratified, layers of deposits (natural or human) have to be removed one after another in the same order in which they were accumulated. Objects found in the lower strata are generally older than those, which remain nearer to the surface. The stratification may be disturbed for natural reasons like animal burrowing, frost heaving etc. Again, it may be the case that the former inhabitants have dug out a large hole and piled up the inner dirt on the surface for some reason or other. Therefore, confusion arises; time sequences become difficult to interpret. The other possibility of confusion appears with the members of a community who had started a new settlement next to the old one; the second site though may not overlap the earlier one but placed on the same level. In this case objects from both the sites seems to be of same period.

An excavation may be marine, cave or surface on the basis of its nature. Excavation techniques are varied depending on the type of the site. A cave or a rock-shelter is quite different from a village site; more different is a site, which contains large ceremonial structures. Naturally the techniques of treating them differ. For example, excavation in a village site is done by the skinning or peeling technique. A large square piece of land is taken for digging. Successive horizontal layers are removed unless the deepest part of the occupational debris is touched. The depth of the horizontal layers varies from site to site. If the natural stratigraphy is present, work of the excavator becomes easy; he can adjust his digging. Sometimes a site gets complicated with successive occupations or confusing materials. Presence of too many features embarrasses an excavator to differentiate the cultural groupings.

Prehistorians often find burials from different sites. These are the records of burial customs in ancient world. Egyptian nobility specially attempted to preserve their dead bodies through mummification. Study of different burial customs is one of the important archaeological works because many people throughout the world had the custom of placing various goods in burial with the dead. Perhaps the early men believed in a life after death. Burial offerings aimed to serve the need of a dead person in the next life. From the position of the bodies in burials, from the association of imperishable items and from the skeletal remains, an anthropologist correlates the racial and sub-racial types with the cultural groups. However, the burial mounds of moderate size are excavated by a technique called 'trench technique'. One or more trenches are cut into the mound to know the initial structure of stratigraphy. Then on the vertical profile, successive horizontal slices, from top to the bottom, reveal the features of the mound earth. For caves and rock-shelters, the stripping and trenching techniques are combined. On the whole, the principle remains same. Specimens recovered from different levels are compared to conjectural hypothesis of cultural growth and change.

Not only the man-made objects (artifacts), all other objects like animal bones, plant seeds, shells, ashes, pottery etc. are collected. Records of each stage of excavation are retained through notes and photographs. When an artifact is uncovered, its position is projected in relation to the particular square of the grid; the exact layer of the pit as well as the depth at which it occurs is also noted. Special note is kept regarding original surroundings. Gradually all the artifacts and associated finds are numbered, catalogued and listed in a register. Photographs are taken from the advantageous positions. Finally objects are placed carefully in hard paper boxes or cloth bags, with labels showing their identifying numbers.

Stone implements can readily be picked out. Since bones are more fragile, they have to be detached with penknife. Sometimes the whole block of earth in which they occur has to be cut out. Animal bones and also the remains of plants and pollens help to date a site. If the geographical area is found suitable for tree-ring analysis, logs and beams prove themselves useful. Besides, the bits of charcoal may form an important collection for the purpose of dating. Similarly, the shells found in the sites supply the clue on climate conditions. Pieces of stone and metal, even those which are not tools, may give an idea about the technique of manufacturing the objects. As the organic materials are very delicate, they need special treatment to prevent decay and decomposition.

All the evidences collected from a site though provide the glimpse of a culture of a former living group, but actually they constitute only a small part of total culture. Therefore, such a reconstructed picture may or may not coincide with the prehistoric reality. Still the scholars regard this information enough for first-hand formulation of a rational story. The materials exposed by an excavation are again important because, sometimes they solve a specific problem in cultural history.

An excavation is regarded totally unjustified when it can not contribute to the cultural history. All deposits resulting from human living leave a valuable record not only in the nature of artifacts but also in the association of the artifacts i.e. how they are posited in relation to each other or in relation to the other deposits of the ground. Eminent archaeologist K.C.Chang gave an example of a site where three objects—a pottery beaker, a metal sword and a skeleton appeared together. The beaker and the sword were separated by two feet horizontally and six inches vertically. The beaker was placed near the head and the sword was at the level of the waist of the skeleton. At the first sight they seemed to be the grave goods but Chang suggested some alternative possibilities. According to him, the pottery and the sword belonged to two different strata. Therefore they represented two different time period or settlements. The sword was perhaps responsible for the man's death and that was left thereafter with the body.

Association of objects thus can reveal stories, which may be disturbed by excavation. So when a site is dug, the team should always be alert in this respect so that the site may not be destroyed by the application of commercial steam shovel or by river erosion or by the archaeologist's trowel. From the angle of a prehistorian, a more comprehensive record is available when all evidences of artefacts and objects merge together to draw the line of human evolution; possible stages of biological and cultural evolution are correlated. For this reason, excavation of sites ought to be done with a great precision by the trained personnel, competent for the work. Interpretation of the cultural deposits should be started from the very position in which the materials are laid down. Photographs play a valuable part in the recording and the interpretation of the sites after excavation. These are specially relevant for the burials, structure of the cave paintings and engravings, fire basins, significant soil or debris stratification and of other important finds which can not be drawn out of the site for some reasons or other. Therefore, photographs serve as wonderful visual record especially for the sequences that are found in situ. Good photographs are often useful to jog the memory of a prehistorian when he goes to write the excavation report.

The task of a prehistorian is two-fold. Not only he collects the specimens but also keeps a note in details regarding their occurrence and association with other finds. The usefulness of associated finds can never be undermined. They delineate the picture of an overall environment. Moreover, they appear as a check on the result deducted from the stratigraphy of the earth or typology of the tools. For example, where an industry is obtained as sealed in a layer within a rock-shelter, the associated bones of an extinct fauna proves the antiquity of the industry. Age of the industry must be the same with the Age of that extinct fauna. The authenticity yields no scope of debate. Again, associated finds are often helpful in determination of interchange of objects between two cultures by trade or other ways. If, an industry of doubtful age comes out from an area with some specific object in it, prehistorians stress on those objects to find out the makers; migration or trade is assumed.

PROCEDURE OF DATING

After the collection of artifacts and fossil remains, question of dating comes. Determination of chronology is the most important task in the archaeological anthropology or prehistory. It is supposed that there are two end points in the ladder of the human development. The latest culture complex is posited at the top being represented by the people of America who use European manufactured trade items like glass beads, iron knifes or brass items. In the bottom, there are crude artefacts of natives who utilize those for maintaining existence. It is held that the cultural story is a continuous one from the first to the last stage. The changes in the style and techniques are a general outcome of local development but sometimes they are manifested as outside influences to take part in the local cultural growth.

There are two main approaches of dating—*Relative dating* and *Absolute* or *Chronometric dating.* Relative dating methods are much simpler in form. It can be obtained through the techniques like Stratigraphy, Seriation, Fluorine analysis, Typology, Patination, etc. It provides a sequence of culture but does not present the time frame. Therefore accurate date in terms of specific years is absent here. One may know the age of a particular culture in connection to other cultures i.e. a culture may be termed as old or new in the perspective of associated cultures. But it is impossible to know the actual length of time upto which a culture existed, or the particular time when culture first made its appearance. The precise dating is possible only by absolute chronology.

A. Methods for Relative Chronology

Stratigraphy

This is purely a geological method. Excavation of a stratified site exhibits the cultural levels one after another. The stratum at the bottom is the oldest and the top one is the most up-to-date. Such cultural periods can be understood only in relation to each other. There is no way to ascertain the exact date for any of this culture groups. It is also difficult to know the intervals of time that separated the cultural periods. The spans of years are guessed utilizing the other evidences suggestive of time factor. For instance, prehistorians who have excavated large shell mound sites often consider the number of burials there. They try to determine the probable size of the group by calculating the average death rate and thereafter by relating that figure with the refuge accumulation of the shell mound. Moreover, they estimate the consumption of shell-food per year and then by counting the cubic content of the mound, a possible figure for the age of the mound is derived. Obviously such estimation is not free from error and therefore can not be accurate. Another process of estimation is to compare the cultural unit with other similar units where the length of time has been known. The horizontal extent of a site, the depth to which the materials are present and the areal boundary of cultural division within a geographic space provides the judgement regarding the span of time.

At the beginning, prehistorians in most of the parts of the world would depend on relative chronologies. Sequences used to be placed on a scale of time which, from top to bottom was relative. Such a dating therefore became a subject of change when more reliable dating method was discovered.

Pollen Analysis

Pollen are the microscopic grains containing the male reproductive cells that discharge from the anther of flowers. These pollens are very small in size and only a small part of pollen is released from the flowering plants, but they are quite durable under certain condition. They remain well preserved in peat bogs, lake mud, alpine and desert soil and even within the glacial ice. Since each type of tree has its own recognizable shape of pollen grain, analysis of tree-pollen reflects the tree-composition of a particular area during the period of formation of the deposit. By studying the earlier strata of a peat deposit, one can find out the changes that took place in the flora-pattern from one period to another. The method of pollen analysis is also useful in detecting the climatic change; new kinds of trees replace the old for the change of climate. In North Europe, a great treasure was found in peat deposit. Pollen analysis has been able to reconstruct the stages of forestation in Europe, after the retreat of the glaciers. It is understood that the birch forest appeared first, then came the pine and last of all the oak forest came into existence. Thus, a kind of relative dating is possible with pollen grains.

The method presents a clear-cut picture of ecological conditions to which man had to adjust at different times in past. Density of a forest can be estimated by comparing the number of tree-pollens of a few varieties. Pollens also bear the evidences of human agricultural activities as found in case of some European Neolithic sites. But unfortunately this method has not been much employed in relation to the prehistory in North America.

Fluorine Analysis

Fluorine is a non-metallic element in the form of pale yellow highly reactive gas. Bones (calcium phosphate) lying in the earth absorbs fluorine dissolved in percolating water and results in the formation of stable fluor-apatite. Therefore, fluorine analysis discloses the percentage of fluor-apatite in the specimens and thus can suggest a relative date for the bone materials which belong to the different strata of a deposit. This analysis is based on the principle that the longer a bone will be placed in soil, the more fluorine will be caught in it. The method can not provide an absolute chronological age because the amount of fluorine differs from soil to soil, which gives a differential rate of absorption. But the method have been found quite suitable for the relative dating of bone materials within a particular site. All bones whether of animal or of human lying in the same level exhibit similar fluorine percentage in them. Therefore, the bones acquired from a lower level show more fluorine in them whereas the bones coming from the upper level contain less fluorine. Fluorine test is generally applied to the mixed deposit of human and animal bones. If the quantity of fluor-apatite remain same in both kinds of bone, it is sure that they belong to the same age.

A similar type of technique is the analysis of phosphorous concentration, which works nicely in relation to lake-soil deposit.

Seriation

Pottery is regarded as the best interpretive material by the prehistorians. Because for a pretty long time in history, man did not know the technology of making pottery. But as soon as it became a part of the culture complex, it reflected better than any other single item in showing gradual change of culture. Usually pottery has been found in the sites of agricultural practices; hunting gathering people seldom exhibited this.

Hand-made pottery differs in size, shape, surface-finish and decoration.The style of craftsmanship tend to change over the years. Therefore, it is possible to arrive at a relative chronology basing on the stylistic change in a region where the ceramics had been in use for a long time. This method of dating is known as seriation. It is actually a typological classification of the pottery for the purpose of relative dating. Wherever a fragment of vessel is found in association with the other artifacts, it has become easier to incorporate the findings in the known cultural sequence of that geographical area.

Clay objects are categorized not only according to nature of clay and temper, the shapes and designs are also taken into account. Analysis shows that the pots of a particular site are not always made of local clay. They often come from outside signifying a trade relation with other cultural group. William Flinders Petric, the great Egyptologist forwarded a relative dating sequence for Egyptian tombs on the basis of changes in the form of pots available from some excavated tombs. He suggested an evolution in the design of the pots as the early pots were invariably with handles. Gradually the size of the handles diminished and ultimately nothing were left as handle. Rather, a painted line was found in the place of handle. Renowned anthropologist A.L.Kroeber used this technique in determining the relative age of some sites at southwestern United States, Mexico and Peru.

Pots are easy to break, but the pieces last for an indefinite period. For this reason archaeologists love pottery beyond anything else. They trace tribes, cultures, and also the periods by the styles of pot-making and decoration.

Typological Method

The nature of the artifacts, whether they are crude or fine, gives an idea of relative age. For example, Neolithic tools were more developed than the Palaeolithic tools. Again, Upper Palaeolithic tools were much evolved in form in comparison to the tools of Lower Palaeolithic or Middle Palaeolithic Age. But this method is totally unreliable because it can not yield fruitful results universally. For example, the tools used by the contemporary primitive Australians are as crude as the tools of the Palaeolithic period in Europe. Therefore, such a criterion of determining age is not at all justifiable and if employed, should be dealt with great care.

Age-Area Method

The cultural traits that cover a wide general area and possess greatest distribution are considered as the oldest. This assumption has been proved useful in dealing with closely related forms like the house types. For instance, in the south-west small semi-subterranean houses were distributed over a vast territory and these were greater in number than the large, complex, multi-roomed pueblo structure. But this technique is only suggestive and ought to be verified by stratigraphy.

Patination and Weathering

Exposure of climatic condition or long burial usually brings some change on an artifact. So attempts have been made to deduce the relative age of the artifacts from the pattern and degree of chemical and mechanical weathering. But since there are too many unknown factors to bring about the changes, the method is not fully reliable. It can be used only as a supportive method.

B. Methods for Absolute Chronology

In some parts of the world, nature has contributed in providing a year by year chronology. Although there may be a marginal error, but in most of the cases they help to overcome the limitations of relative dating and accurate dates are obtained. However, the methods *of Absolute Chronology* are as follows:

Dendrochronology

A tree builds up a new layer on its trunk every year. By counting those annual layers one can easily find out the age of the tree when it was cut down. The width of the rings as well as the relative distance between the rings reflects climatic variation over a prolonged period of time. For example, some rings look thick and others are thin. During the years of plenty rainfall, a tree absorbs enough moisture and so the rings (the association of newly formed cells) grow wide. Conversely, in the years of drought, for the want of sufficient water, new cells can not grew plenty and obviously the rings become narrow. It is interesting that all trees of an area get affected in the same manner by the prevailing weather condition, so the ring-sequence follows a general pattern. Suppose, in an area the trees show a sequence of two thin layers at first, then four fat layers, again two thin layers and then one fat layer and so on. By comparing the general tree-ring sequences of two areas, one can differentiate the climatic condition between those two regions. But the method of tree-ring study is not a simple one. It needs trained individuals who are experienced in handling, recording and interpreting the ring sequence and competent to pass judgements on the age of prehistoric specimens.

This method of *dendrochronology* has been applied in southwest United States where the dating touches the time of the Christ. It was an achievement of the astronomer, Dr. A.E.Douglass who was interested in determining the long range effects of sun spots on the weather of the earth. To do so, he utilized the annual growth rings of certain southwestern trees, particularly the brislecone pines in California. The oldest brislecone pine tree has been found alive for 4,950 years. When a Pueblo ruin was excavated in Arizona, the tree ring sequences pointed out the year in which the Pueblo structure was built.

Although *dendrochronology* can not be applied to all kinds of trees or in all kinds of environment, still it is a very promising method for absolute dating. It does not work in the areas like New Zealand where there a little variation is observed in the rate of annual rainfall. However, the method has been successfully employed in England, Germany, Norway, Turkey, Egypt and various parts of United States including Alaska.

Varve Analysis

Varves are the annual layers of sediments that are deposited on the floor of the glacial lake or other relatively quite water bodies. It was discovered by Sir Baron Gerard de Geer, a Sweedish scientist who was concerned with the development of techniques for the interpretation of geological record during the later part of nineteenth Century.

During the summer months, melting of glacial ice took place. Water ran down into the lake. At the advent of winter, the melting stopped and the lake water started to freeze down. During this phase, the finer sand and clay particles that remained in suspension sank to the bottom of the lake. The process went on year after year. Each annual layer (varve) is deposited with a band of coarse sediments followed by another band of finer sediment. However, the varves or annual layers vary in thickness, from half inch to fifteen inches, depending on the degree of warmth and length of summer period. Careful counting of the varves presents a date in terms of years for the period of melting and retreat of the glaciers. Dates about 10,000 years ago from the present have been ascertained by this method for Sweden, Finland and some other Scandinavian sites. This chronology can be compared with the deposition of moraines and other surface features to determine the date of human occupation there.

The method proved its excellence in dating many North European sites of the Late Pleistocene and early recent period. Successive cultural periods in the Baltic and North Sea areas were also determined. But the method can not yield accuracy when applied to the situation of North East of North America because the most recent varve deposits of this area have not yet been connected with the Christian chronology. The other limitation of varve analysis is that, each of these varve may not correspond with the scheme of a year as we think today; the sequence of months might be different in those prehistoric days, contrary to our assumption.

Radio-Carbon Method

Radio-carbon dating is based on the measurement of the decaying rate of radio-active carbon, known as Carbon-14 (C14) in disintegrated products. At present, it is the most valuable as well as widely used method of dating. To know the nature of Carbon-14 we should know that every substances on earth, whether a hard mass, liquid or gas is made up of one or a combination of two or more substances which we call elements. There are 92 elements which occur naturally on earth such as oxygen, carbon, gold, etc. Atom is the smallest part of an element. Each atom has a center, called nucleus which remain surrounded by one or more tiny moving bits named electrons. Within the nucleus, there are more tiny bits named protons and neutrons. The number of proton in the nucleus of an atom is always equal to the number of electrons moving around it. This number stands as 'atomic number' of an element. The 'atomic weight' of an atom equals the number of protons added to the number of neutrons in its nucleus (number or proton + number of neutron). In fact. all atoms of an element bear same atomic number (number of proton). It may be mentioned here that the proton bears positive charge with a minimum mass, electron contains only negative charge and neutron is a neutral mass that maintains the nuclear weight of the atom. The proton in addition with neutron forms a nucleus while electron being devoid of mass contain only negative charge and move around the nucleus. The addition or deduction of electron is responsible for the ionisation of the atom. Again, an addition or deduction of proton changes the 'atomic number' that in turn change the properties of the element, as because the nature of an element is determined by its 'atomic number', in atom level. At the same time the different 'atomic weight' is counted as because they do not possess the same number of neutrons. However, When an atom contains more neutron, the 'atomic weight' increases and the atom is named as isotope. The element carbon has three isotopes basing on the differential number of neutrons : Carbon-12, Carbon-13 and Carbon-14. First two isotopes are more or less stable but Carbon-14 isotope is a radio-active and unstable. Since Carbon-14 is unstable or unsteady, it gives out radio-active rays and changes. This isotope is also rare in comparison to stable isotopes of the element.

In 1931, at the University of Chicago, an unknown radio-active material was detected. In 1941, that was identified as a radio-active atom, Carbon-14 by W.F.Libby, Physicists discovered that the source lay in the high atmosphere where cosmic radiation produces neutrons which strike against nitrogen atoms in the nitrogen sphere of the earth and produce radioactive atoms of Carbon-14. In fact, the neutron bombardment on the nitrogen atom (in the nitrogen sphere) make a reaction. When the neutron strikes on the nucleus of nitrogen atom, a proton is released from the nucleus and external neutron is added on the nucleus of the nitrogen. As a result, the 'atomic number' of nitrogen changes and it gets, equal to the 'atomic number' of carbon. Thus, the nitrogen atom of the nitrogen sphere of the atmosphere becomes converted to the radio-active carbon isotope (C-14). The formula is *Neutron* (bombarded) + *Nitrogen isotope* (N-14) = (a proton of Nitrogen atom is detached and cosmic Neutron is added to the nucleus of the Nitrogen atom. 'Atomic number' of nitrogen is changed, which gets equal to the 'atomic number' of Carbon atom) *Carbon isotope* (C-14).

However, the Carbon-14 atoms created in the air around the earth combine with oxygen to form Carbon-di-oxide and then float down to the ground. Carbon-di-oxide is the substance taken in by every green plants on earth. Animal eat plants and so take Carbon-14 into their bodies. We human beings take Carbon-14 into our bodies when we eat both animals and plants. The quantity of Carbon-14 which is normally present in a living plant or animal body has been estimated. Disintegration may takes place, but it gets always balanced by the intake of fresh Carbon-14. When a living being dies, it stops taking Carbon-14, rather the Carbon-14 which is already present in its body begins to fall apart slowly. Scientists have been able to measure that how much Carbon-14 disappears and how much is left in the object. The half-life of Carbon-14 is 5,730 years*. This means, the quantity of Carbon which one finds at present in a piece of wood will be reduced to half of that amount after 5,730 years. A standard deviation of 300 years has been accepted for this method. Therefore, after mentioning a date we usually write, ± (plus-minus) 300. It means, by adding three hundred years with the given date or by subtracting three hundred years from the given date, one can get a range into which the exact date falls

Radio-carbon method is not absolutely free from disadvantages. The radio-carbon level is generally calculated in terms of present day atmosphere, but we are not sure whether the same level of radio-carbon was maintained in past, thousands of years ago. Further, the changes of earth's magnetic field and solar radiation may have an effect on this method of dating. At the same time, there is a possibility of an object to be spoilt by the action of some other radio-active material; in that case the radio-carbon dating becomes absolutely wrong. Nevertheless, the method has been a precious tool for measuring time. The greatest advantage is, it can cover a longer span of time than any other ordinary method of absolute dating. When it was first developed, the dating limit was only 20,000 to 30,000 years which have been widely extended to cover about 100,000 years at present. Charcoal, the most common material obtained from different sites is regarded as the best organic material for the analysis of Carbon-14 content.

This method of using radio-active carbon unlocks the secrets buried in the Egyptian burial grounds for thousands of years. It also provides surprising information to the scientists about the early men in America. Not only that, the method reveals valuable facts regarding the land that remained covered by the ice-masses for a long time.

---

* Before 1961, the half-life of Carbon-14 was counted as 5,568 years. Now it has been taken as 5,730 years by scientists.

Potassium-Argon method

In principle, this method follows the radio-carbon method. A radio-active form of potassium, potassium-40, is utilized here whose rate of decay has been known. After disintegration, it produces Argon-40 and Calcium-40. Therefore, the ratio of Potassium-Argon may be measured to ascertain the date of minerals and rocks in a deposit; it does not date the fossil specimens directly. However, the method has proved quite useful in dating some hominid fossils as employed in the site, Olduvai Gorge of East Africa where the remains are as old as 1.75 million of years. In fact, this method is able to cover a wide range of time, even far greater than Carbon-14 method. Because the half-life of the Radioactive potassium is 1330 million years. Very old archaeological sites have been dated by this method. It works well in case of the sites which are 500,000 years old or more. Disadvantage of this method lies in the fact that, it can be applicable only to those rocks and sediments, which are rich in potassium. Since these rocks are available only in volcanic areas, the method is restricted to those areas. This is the reason for which potassium-argon dating can not be applied to the sites of South Africa while East Africa yields a very good result.

Fission-Track method

This method of dating employs Uranium 238, which possess a capacity of fission with a half-life of 10^{16} years. The materials like crystal glass that contain Uranium particles are exposed to this dating. The fission spontaneously takes place in the nuclei of Uranium-238 and that causes substantial damage to the lattice for recoiling of the fragments. The damage tracts are visible under the microscope by etching with hydrofluorine acid. Now by calculating the nature, features and number of the fissions, expert scientists make the dating. Since 1960, this method has earned its popularity. It is now widely applied to determine the age of the pieces of volcanic glass where they remain associated with the human remains. It can reveal a date upto five billion years.

Palaeomagnetic method

This method of dating examines the magnetic power of earth. As the magnetic poles of the earth change their direction with time, the fact is employed to decide a date. But the limitation of the method is that it can be applicable only to the objects like fireplace and pottery kilns. When a piece of baked clay cools down it acquires a weak but permanent magnetism. Its direction remains same to the field and the strength continues to be proportional to the intensity of the field. Expert scientists can properly record the direction of the clay at the moment when it cooled down in past. Then by comparing that with the present day direction, the age of an object can be derived. Here the clay necessarily include a kiln, a hearth or a similar thing, which come in contact of fire. The volcanic rocks are also able to show the change in the direction and intensity of geo-magnetic field.

Palaeomagnetic dating method has been employed in some sites at Mexico and Arizona that show dates like 3000 BP. Some Palaeolithic and Mesolithic sites of the Old World have been dated by this method.

Palaeo-Temperature analysis

This method of dating analyses the deep-sea sediments. The sediments contain calcareous remains of small sea animals called foraminifera which are very sensitive to the change of temperature. The frequency of those remains permits an estimation on the temperature of the sea, particularly when they used to live. The deep-sea cores drilled from the beds of Caribbean and Pacific have been studied by Emiliani in 1956 which suggested the dates of glaciations in Pleistocene epoch. According to Emiliani, the varying frequencies of the fossil types in different parts of the cores indicate the alternating warm and cool phases in Pleistocene. He also recorded the fluctuations of radioactive isotopes of oxygen (i.e. the changing ratio of oxygen-16 and oxygen-18 in marine carbonates) to measure the ancient changes in temperature from the fossilized remains in the deep-sea.

The carbon-contained shells of the fossil foraminifera indicate that the each glacial period lasted about 90,000 years followed by an interglacial period, which lasted only about 10,000 years. The last cold phase of the Pleistocene ended about 17,000 years ago as per this evidence. Since undersea conditions alter a little with time and as a steady settling of material takes place on the ocean floor, the Deep-sea cores stand a good means of dating. But, still there are possibilities of error. Howell (1957) stated that the cold periods identified in the sea cores represent merely the subdivisions of last glaciation. But it is sure that the method is able to judge the antiquity of at least 300,000 years.

Racimization

This is the method which involves an analysis on the arrangement of amino acids in organic substances. We know that there are twenty types of amino-acids in all living things which maintain a general configuration for every creatures in the world. But at the death of a organism, the configuration starts to change. The 'left-handed' molecules gradually became 'right-handed'. So by calculating this change, chemists can point out the age of the organic objects. A hominid bone collected from Olduvai Gorge, East Africa reveals a date about 135,000 BP. Some other materials ranging between 40,000 to 100,000 years have been identified by this method as belonging to the period of Neanderthals.

The method of racimization has some limitation too. The specific rate at which amino-acid configuration changes depend on the prevailing temperature of climate. Since the climate varies from place to place, racimization does not yield uniform result throughout the world. Obviously the dating varies for two contemporary objects placed in dissimilar climatic zones.

Obsidian Hydration

Obsidian is a volcanic rock, fragile in nature. It usually contains two percent of water. But when a piece of this rock is broken; it goes on absorbing water until it reaches to a saturation point. This absorption is slow and gradual and the process stops when about 3.5 per cent of water penetrates into the rock. Scientists have calculated the rate of hydration and this help in dating of stone artifacts. As soon as man began to make tools from obsidian, the rocks became chipped and exposed to hydration. Therefore, the extent of hydration can be correlated to the age of the culture and concerned people. This method does not always provide accurate dating because the rate of hydration often differs according to climatic condition and there are many kinds of obsidian rock which show differential rate of water absorption.

Thermoluminescence

Not only the seriation, the method of thermoluminescence can also reveal the age of the pottery. Thermoluminescence is the emitted light from pottery, which can be measured. If the ground-up pottery is heated to about 500° C, some sort of light comes out. The phenomenon results from radioactive influence of the metallic elements like uranium and potassium present in the clay and surrounding soil. It is held that the geological thermoluminescence was driven off at the time of original firing of the pottery. But as the pottery remain further exposed to a steady natural radiation, it again revives the capacity of thermoluminescence. Measuring this thermoluminescence, age of an ancient piece of pottery can be determined.

All the methods grouped under 'absolute dating' may also be leveled as 'chronometric dating' as they are able to point out the dates in terms of number of years; an absolute chronology is obtained from them. Some scholars have again grouped some of these methods as radio-metric method as they base on the principle of radioactivity.

ANALYSIS AND INTERPRETATION OF THE MATERIALS

Analysis and interpretation of the materials are as important as dating. Immediately after the excavation of a site, all kinds of materials are taken to the laboratory for routine work. This is an essential job before the specimens are studied. Some of the items have to be cleaned and repaired.

But thorough washing of the specimen is never recommended as it may spoil the originality. Therefore, certain amount of material in which the specimen occurs may be allowed to go with the specimen. After this, the objects are identified and classified. Each specimen is illustrated with its accurate form and features; reference of the material is also cited. A description of the site and the specific location from which the finding was obtained provide valuable information in understanding the total situation. Nature of associated objects like burials, houses, fire basins, refuse pits etc. give the clue to build up the story. Sometimes settlement patterns are revealed through excavation. In some places only the remnants of houses are found, some other places show walled towns with houses grouped along the streets. Specially demarcated places for ceremonial or civil purposes are even found at some sites. All these data help a prehistorian to reconstruct the cultural life of the people. Artefacts and fossil remains are especially important to him.

Artifacts

Artifacts are the remains of prehistoric past, conceived and made by human. These are used to make inferences on the life-style and food-habit of early man, to note the evolutionary development and adaptations of men to the environment, and also to record their response to competition.

At first, all artifacts are listed in a field-catalogue with respective notes and illustrations. The laboratory work is to identify and classify them in-groups. Groupings are made according to the nature of the material on which the artifacts were manufactured such as stone, bone, shell, clay, wood, etc. They are further grouped according to their forms. This conforms with the identical types as recognized by the geologists and mineralogists. For example, stone artifacts may be classified as flint-objects, steatite objects, etc. Bone objects may be categorized as coming from a particular bone of a particular species e.g. the shoulder blade of a deer or the upper leg-bone of a turkey, etc. Similarly shell artifacts can be grouped as per the scope of their availability. A shell from an ocean is distinguished from a shell belonging to freshwater environment. These can again be classified according to their place of occurrence, which is generally a lake, a riverbed, or a fairly specific locality. Categorization of the artifacts may also take place on the basis of their function. For instance, the stone artifacts may make the groups of axes, knives, scrapers, projectile points, drills and so on. In the same way bone artifacts may be differentiated into the groups of awls, hoes, pendant or beads. For the shell artifacts, groups such as beads, hoes,scrapers, etc. can be made. In this way, the different items like stone, bone, shell, pottery, metal, ivory, wood, etc. can be separated. Not only that, each of the items has to be clearly defined to be placed in an appropriate group.

The nature of the objects and materials in a preliminary way suggest a cultural phase. A more detailed and careful study recognizes the distinctive culture traits in a given area at a particular point of time. The interpretation of culture sequence is essential at this stage as the form of the objects as well as the techniques employed in their change vary greatly as the time passes on. One culture level can be distinguished from other by the type, form, and technological speciality of the artifacts. The situation signifies a varying degree of skill that was employed in making them. Since the peoples belonging to one culture group develop a more or less uniform pattern in the make of their tools, the similarity of technique and skill enables the students of culture to ascertain the line of cultural connection among the sites, despite the difference exists in time and space.

When some stone implements are evident from a site, it is understood that possibly a stone industry was present there. Next consideration is the nature of the implements and their manufacturing technology. Implements may fall either of the two broad divisions—the flaked or chipped type and the polished type, if they are not very small microliths. Suppose, the implements belong to the former group as they satisfy the criteria of that group. Now the finer technologies have to be found out. In case they would have belonged to the polished stone group, the degree of further refinement would have to be searched out. However, a stone, flint or obsidian was chipped and shaped by men for hundreds of thousand years ago. It was an easier process than the shaping stone by polishing.

Therefore, if chipped implements can be seen abundant in a site, it will indicate the activities of an early human group whereas the abundance of polished stone implements suggest a comparatively later date of human culture. The form, size and function of the tools, the workmanship, the particular quality of the stone—all influence a prehistorian to derive his proposition. A Russian investigator, S.A. Semenov tried to make a microscopic analysis of the stone tools especially to understand the wear pattern and striations of the tools in order to find out their function in a better way. Now-a-days computer can help a prehistorian to determine the class of the artifacts by fast calculating the range, mean and standard deviation of the artifacts.

Like the stone tool, pottery is an artifact, which does not perish with time. It is another important indicator of cultural life because making of pottery is related to the higher accomplishment of primitive life. The people of Neolithic period invented this art. It is held that the earliest migrants of North America left Asia before the technique of pottery had spread to Siberia. No sherd is found to be associated with the early prehistoric finds of North America. But in Central America and in the Andean, a few zones with archaic pottery have been identified which are definitely older than the regions of Asia, about a thousand of years. It seems that the pottery was invented independently in the New World. Later, between AD500 and AD700, during the time of Anasazi culture, a group of skilled potters evolved there who worked on ceramic vessels. The other interesting feature is that while Old World (Egypt) potters invented wheel about 5000 years ago, the American Indian potters including the sophisticated Mayas, Incas or Aztecs are still wheelless. In Africa, all people at present possess the knowledge of pottery. Actually the potter's art travelled from Asia to Pacific through Indonesia. Some people very good in pottery making are found in Melanesia and Guam, and Palau and Yap in Western Micronesia. But some other Micronesians and Polynesians have left out the art. As pottery was easy to make and to break also, chances are more to get the potsherds. Changes in pottery have been brought down very slowly and the combinations of features are so distinctive that it is possible to identify vessels or fragments of vessels as belonging to the particular spot in the world where they were made in a specific period of time or to the particular cultural group who used them.

Fossil Remains

The term 'fossil' is derived from the Latin word 'fossilis foderi' which means to dig up. In fact, the fossils are the remnants of animals and plants from past which survive upto the present day being preserved in rocks. When an organic substance like bone or wood gets buried, sometimes a gradual replacement of organic material by mineral matter (silica and calcium carbonate) takes place. These minerals come from earth's crust and formed into rock. The physio-chemical process of change is referred as fossilization.

While the written history of man goes back merely 5000 years, earth demands a history of nearly four thousand thousand (four billion) years. All information about man prior to written history come from the layers of rock, the stratigraphic record. In this respect the prehistorians are indebted to the botanists, zoologists, chemists, physicists and geographers. It is found that the sedimentary rock make up the bulk of layered rocks. They cover about seventy percent of the land surface. In any sequence, the sedimentary rocks lie essentially in their original position; the layers at the bottom are older than the above. These sedimentary rocks are most useful for reconstructing the past events. The fossils, the remains of ancient life are preserved in it.

Most of the fossils represent organisms, which buried quickly in the sediments and had parts that resisted deterioration. Upon rapid burial, the hard parts generally remain intact. The soft parts get dissolved by ground water and leave a mould. In course of time, the hard part being replaced by new minerals gave rise to a fossil. Animals, even without hard parts are occasionally found as fossils where special conditions of soil favour preservation. Fossils used to be collected as curiosities and charms by early men, long before their origin were clearly understood. Amulets made of fossil shells have been found along with the remains of Neanderthal man. The vast fossil records compiled over years exhibit a procession of life forms through time. Although there are many gaps in that record, the overall pattern is very clear. Living organisms, as a whole, have undergone profound changes. Some of the old forms have persisted; others have died out. Still the gradual development of new and more complex forms has been noted within each general group. Therefore, it is a responsibility of the prehistorians to identify the position of the discovered specimens according to the modern scientific knowledge of taxonomy. The relationships between different fossilized forms should be established from evolutionary and ecological point of view. All known species of extinct plants and animals, hominoid and hominid forms have to be taken into account.

On a given site, animal remains may also suggest the extent to which the people had depended[1] hunting and what was the animals that they hunted. Animal bones are subjected to dating, especially in cases of extinct animals when the date of their disappearance are not perfectly known. Nature of the animals varies according to the climatic conditions i.e. animals, which prefer the cold periods do not appear in warm periods. Again, small creatures like bird, rodents and specially molluscs are tremendous sensitive to the change of climate. Therefore, the presence or absence of these characteristic animals gives an impression about the climate of a particular time.

On the other hand, the plant remains including the pollen help to distinguish between wild and domesticated variety of plants. Domesticated crops may show the direction of movement for cultural ideas. For example, corn was first domesticated in United States. But it is now grown by the majority of the Indian groups as the Europeans arrived their land and coined the practice. Ancient carbonized seeds have been recovered from many places. To locate the tiny plant remains at a site; prehistorians use the technique of floatation in which soil samples are deposited in a watery medium. It helps in determining the nature of vegetation that existed in a particular period.

Reconstruction of Cultural History

Study of the skeletal remains along with the nature of flora and fauna deliver an elaborate idea about respective ecological setup. The analysis and classification of artifacts as well as the features of a site offer an objective description to facilitate the placement of the site within the cultural history of the area. Since the adjacent sites often exhibit a high degree of similarity in the material, the closely related sites may be compared either in a relative or absolute time scale to be incorporated in the general culture sequence. Unusual artifacts do not normally appear in closely related sites. But, if appear, such unusual artifacts represent new elements in a culture and so are of critical importance. Not only that, they denote either a contact with neighbouring area or a growing culture-complex.

Prehistorians recognize and name those cultural periods, which offer uniqueness or considerable variation in terms of the life span of the people and the culture content of the complex whole. The periods may be large or small according to their significance. For example. Palaeolithic or the Old Stone Age continued for a greater length of time whereas the culture complexes like Abbevillian, Acheulean, Magdalenian etc. were short-lived. Again, Abbevillian and Magdalenian were included in Palaeolithic, because they resemble with the general features of Palaeolithic but each of them is regarded as a separate unit as each has developed certain special characteristics. Thus, creation of sub-units within a broad cultural period has been justified. Following this principle, the Palaeolithic is divided into Upper, Middle and Lower period and each of these periods is further sub-divided into small divisions. As for instance, Aurignacian, Solutrean and Magdalenian are the three divisions of Upper Palaeolithic. Whenever we find several divisions in one broad cultural phase, the divisions represent the successive culture periods with diverse culture traits. However, the sites and culture units are named as per the owner of the site or after the features of the landscape. For instance, the famous prehistoric culture of the Ohio Valley is called 'Hopewell' after the name of the owner of large and impressive sites, from where this type of material was found out. In Europe, 'Mousterian' was named after a type-site near the town, Le Moustier.

As we find a great range of local variation in the cultures of the contemporary people, variations are observed in the early cultures that developed within the span of any one of the great prehistoric epoch or age. Each human culture represents an intimate adjustment to its particular physical environment and the possibilities for working out a great diversity of customs. Prehistoric investigations shed light on the various social customs of the early human groups. When a series of small sites belonging to one culture show very few artifacts, few burials, very little depth of debris, and little or no evidence of structures, it is inferred that the social group represented by such remains was very small, comprising of a few family units who kept themselves engaged in hunting, fishing or gathering. But when a particular site reveals extensive and deep cultural deposits, evidence of houses, large complex ceremonial buildings, remains of stockade and palisade etc., a different type of social and economic pattern is indicated. It announces the presence of a large and relatively stable group based on agriculture. Each and every item uncovered from a site contributes in building a perception about the very culture. The stockade suggests a defensive mechanism which in turn points out the possibility of an organized social setup where there a war-leader or leaders existed. It is natural that a group having defense must have shown an offensive network. Similarly, presence of ceremonial structures implies special functionaries related to civil and sacred officials. Large and impressive burials along with associated materials provide the evidence of complex mortuary beliefs and practices. Thus, the paintings of Late Palaeolithic caves in Southwestern Europe are not merely a deed of artistic endeavour; they reflect the primitive concept regarding magic and fertility.

Major geological periods are generally dated by radioactivity methods like Carbon-14 analysis, Potassium-argon analysis or by Uranium analysis. Sometimes study of molluscs provides valuable information as found in case of Baltic North Sea area. There are two types of molluscs, fresh-water molluscs and salt-water molluscs. The minor advances and retreats of last glaciation alternately filled the Baltic Sea with fresh water and salty-water. It was evident by the nature of shells preserved in deposits. The first post-glacial water in the Baltic Sea was the fresh melt water from the receding glacier, which was favourable habitat for the first type of molluscs. Later, this water overflowed into the North Sea through a gap in Southern Sweden. Subsequently, the level of the fresh Water Lake was lowered and a particular seashell entered with the salt-water. Again, when the glacier receded, the land level rose and the Baltic area became a body of fresh water. Fresh water molluscs regained a congenial atmosphere to grow with. After this second fresh-water period, once again the water level lowered down for glaciation and salt water entered with the seashells. A prehistorian should know this preliminary geology to recognize when the advice of a trained geologist is required to interpret the association of natural features with the evidence of man's occupation.

PRESERVATION OF THE ARTIFACTS

Preservation is an action in order to prevent, stop or retard the process of deterioration of the specimens. It is sometimes supplemented by the restoration work, which means the treatment of objects with necessary corrections and alterations. The concept of conservation is the ultimate reality, which includes both preservation and restoration.

Now the question is what do we mean by deterioration. Deterioration is the alteration in an object, produced by the interaction between the object and factors of destruction. In dealing with artefacts prehistorians count two factors mainly:

(a) How much an object has been mechanically rolled or weathered.

(b) How far chemical changes have taken place to alter the composition of a specimen.

Stone is the first material that was used by men for making tools, implements and some other objects of daily use. During the interglacial period, when huge ice sheets melted and flowed in the form of rivers, most of the artifacts of the primitive men swept away under the current. These artifacts show the mark of mechanical rolling by the action of water. They were either dragged in a stream or hurled on towards the beach. Whatever may be, for the friction, the edges between the facets got rounded. Some soft stone tools like flint tools became so much eroded that they look like a pebble. More fragile materials e.g. bones rarely survived under such rolling. Therefore, the tools collected from open stations often suffer from severe mechanical erosion unlike the tools obtained from a cave or rock-shelter. In fact, a specimen can not escape from damage when it travels a long way. Other specimens of the same age may show little or no damage when stored in a pit deposit. For this reason, a worn specimen must not always be considered older than a relatively fresh specimen. It is a very relevant point to be remembered. Because an industry is generally examined on typological ground where older tools show a greater degree of weathering than those of the later date.

In case of chemical changes, the surface of the tools, especially of flint tools shows patina. Patina is the resultant of chemical weathering. All the sides of a tool do not necessarily show the same extent of patination. The exposed parts of the tool exhibit greater degree of patination than the unexposed parts. Another type of change, incrustation occurs in certain cases. It is a surface manifestation; not a chemical change. Different close-by materials get fixed on the surface of the tool due to prolonged exposure in nature. Besides, tools of bone and antler undergo special kind of chemical weathering. The organic materials associated with them disappear with time and only the mineral substances persist. Bones in this condition normally become fragile but sometimes when they get fossilized, the organic matter is replaced by minerals and the whole bone becomes stone like, though the cellular structure or the general shape is retained.

Preservation protects the tools from farther damage. So they can be used for some special purposes in future:

i) Prehistorians study them carefully with a long time. They are the indispensable materials of research as valuable inferences in prehistory are drawn on them.

ii) The typo-technology of the tools is often compared with newly discovered tools.

iii) The tools may be utilized as laboratory specimens so that the students of prehistory can get their basic practical orientation.

iv) The tools may serve the curiosity of lay man who tend to acquire a first hand knowledge about the early culture in a museum.

Stone

The stone objects disintegrate at times; the various layers fall apart. Therefore, a prolonged treatment in laboratory is required to restore them back to a sound condition. Some stone objects are found to be covered with microorganisms, which gradually tend to destroy the object. In this case treatment is needed to remove the microorganisms and to cure the object. Besides, stone objects are subjected to various other forms of deterioration. For example, presence of salt can cause damage. Because, salt, by absorbing moisture, turns to salt solution which percolates into the cavities between the particles forming the rock. On evaporation, salt solutions turn into salt crystals, which often appear on the surface of the stone in the form of white efflorescence. Continuous dissolution and crystallization of the salts result in the repetition of strain on stone, the surface of which ultimately turn into powder. Therefore, all stone objects, after initial analysis and interpretation, should be washed thoroughly in repeatedly changed salt-free water to drive out the salts.

Stone sculptures often accumulate dust, dirt and stains. Loose dust can easily be brushed off. Plain water can wash away most types of dirt accretions. Sometimes mild detergent in water produces better result. Acids, howsoever dilute, should never be used to clean the stone objects, except by a trained conservator who understands their action on various types of stone. A deposit of moss or algae may be seen on stone objects, especially on those, which have remained in the open area for sometime. Such a deposit imparts a patchy, green or black appearance to the object. It may also produce pits in the surface of the stone that weaken the structure. A trained conserver is able to remove the moss deposit quite easily by the application of chemicals.

Stone objects often suffer from damage during storage, especially when big and small objects are dumped together. Some objects may break, others get abraded. Therefore, it is wise to store the objects on separate shelves or platforms. These shelves or platforms should be amply padded and strong enough to support the weight, especially in case of stone sculptures.

Clay Material

Clay vessels and other utensils, beads, toys, figurines etc. have been excavated from many sites of ancient culture, in different parts of the world. Clay was certainly an important discovery among man's earliest discoveries of natural materials. Unbaked clay objects are very fragile. They get easily affected by water and so can not be cleaned with any aqueous solution. Such objects should be kept in a dry atmosphere. In case of painted clay objects also, water or humidity gives a detrimental effect. But the clay objects that were once baked can be washed with distilled water to remove the salts present on them. The baked clay objects (produced at any degree of firing) are though more durable than those of unbaked clay objects, they too are fragile, subjected to crack or break. Many such objects break by accident, falling from a shelf or a table and slipping out from hands. Therefore, careful handling is the most necessary prerequisite for this kind of material.

Wood

Wood being an organic material is especially susceptible to deterioration. Man has used wood since the early days of Stone Age. He relied on it in the same scale as he did upon the stone. Naturally he made various kinds of weapons, objects of daily use as well as materials for art and decoration. Although seemingly wood looks hard and durable, but in fact, it is perishable and vulnerable to varied causes of deterioration. It easily falls prey to insects and growth of microorganisms is often found in it. It gets affected by the change of climatic condition. It is combustible too. So preservation of wooden materials demand utmost attention. Great care should be taken in storing them. The oldest wooden tool so far as recovered, is a wooden spear-point about 200,000 years old, found from a spot near London.

Ivory and Bone

The use of Ivory and Bone is recognized since the Upper Palaeolithic period. They have been used in making tools, ornaments, weapon handle, beads, etc. Ivory from elephant tusks was the most common one, but walrus ivory was also recovered from North of Europe. Like other organic materials, both bone and ivory are susceptible to heat and humidity. They expand under high humidity and contract in low humid condition. Therefore, cracks often develop in these materials. But the effect of the climate can be minimized if the objects are coated with a water-repellant substance. Both ivory and bone can be easily stained because they are porous in nature. For the storage of these objects, clean soft tissue papers must be used and finally the objects should be kept on padded shelves or in padded boxes.

Metals

Apart from gold and silver, much copper, bronze and iron objects have been discovered as exhibits of Bronze and Iron Age. Copper and its alloys, like bronze or brass, corrode easily especially when buried in the earth. Corrosion takes place as the soil contains many salts. However, the corrosive layers of its chloride cover the upper surface of the copper object, rendering the metal friable and weak. High humidity accelerates the process of decaying. At present a preservative coating solution is available which can be applied on copper, bronze and brass objects, Since iron objects rusts easily in moist climate, excavated iron objects are often rusted, sometimes to the extent that no or very little metal core is left in the object. If an iron object is already corroded, it should be treated in the laboratory before preservation. Sometimes rusting is so advanced that if it were removed, the whole object or its important portions would be lost. In that case, the rust is also conserved along with the object. Formation of the fresh rust can be prevented, to a large extent, by the application of a water-repellant of its surface. The object is either immersed in wax or wax is applied on it with a brush. Beside wax, some consolidants and varnishes may also be applied on the objects.

Environment plays a major role in the conservation of artefacts. Changes in temperature and elative humidity are particularly important. Since objects remain buried for a long time, they are conditioned by the natural environment. After excavation when they are carried to a new environment, the equilibrium gets distorted and as a result the artifacts may swell, shrink or warp. To avoid all these, perishable objects recovered from an excavation should be kept in a similar environment until they are taken to a conservation laboratory for specialized treatment. If an object comes from a salty ground, that remains liable to much damage. In this case, immediately after the recovery, objects (pottery or organic material) should be washed thoroughly in the salt-free water. This prevents the soil crystallization on the surface of the object for which disruption could have taken place. However, the preserved articles too need protection against light, heat, moisture, mold and insects.

III. SOCIAL CULTURAL ANTHROPOLOGY

Chapter

PRE-HISTORY TO HISTORY

SETTLED LIFE AND EARLY CIVILIZATION

The term civilization has been derived from the Latin word 'city-state' to mean an urban society. A civilization is a larger unit than a culture and in fact it includes a number of cultures. The unity among the cultures is maintained by a general mode of living, which may or may not be an outcome of racial similarity. There are plenty of factors including the climatic condition, which influence the mode of life of the people of diverse cultures in a particular area, at any given point of time. However, the cultures advanced towards civilization when a new relationship was established between man and the animals and plants. At the end of the Pleistocene epoch, approximately twelve thousands of years ago, a great climatic change was found in the world which has been referred by the modern prehistorians as Neo-thermal period. After the passing of glacial condition (Ice Age), the temperature began to rise up very slowly. The vegetation pattern and the faunal characters changed with the transformation of the landscape. At this time, people began to adopt the domestication of plants and animals under special circumstances. The quantitative increase of food and the sure availability stepped up the people towards civilization.

For hundreds of thousands years man had remained at the level of bare subsistence. People wandered from place to place chasing after the games and gathering the fruit-root-tuber. In the post-glacial phase they first felt to be united and so formed groups to carry out the jobs like hunting, fishing and gathering in an organized way. Man's shift from food-collection to food-production was like a revolutionary change in human history, which led to the sedentarism and ultimately laid the foundation of civilization. Villages and towns were made up when man acquired sufficient control over the food-supply. Cultivated fields and the tame herds of cattle and sheep gave the householders plenty of leisure time in their hamlets, which was never known before. Man no longer spent their energies in trapping or tracking wild beasts for food or in searching for berries, wild greens and edible roots. They got the freedom to sit, to think, to experiment and to exchange ideas among themselves. All these were indispensable in producing valuable inventions.

First evidence of such changeover was discovered from the Near East. The Fertile Crescent i.e., the arc of land stretching from western slopes of the Zagro mountains in Iran through southern Turkey and more southward to Israel and the Jordon valley was one of the earliest centers of plant and animal domestication about 10, 000 years back (8000 BC). Wild varieties of wheat and barley were found to grow there at that particular time. The *Natufians* who lived in the caves of Mount Carmel in Palestine and in the neighborhood of Lydda exhibited an excellent control and ultimate mastery on nature through agriculture. Hunting and gathering were the principal means of subsistence for the Natufians. They often used to collect the seeds from the wild plants and store them in front of their rock-shelters within a basin-shaped hollowed-out space in the rock. Possibly women noted the growth of the seeds which dropped by chance on the ground or off shoots which were planted without any definite purpose. Finally they learnt to plant the seeds in the grain fields.

According to English prehistorian V.G.Childe (1950), a drastic change of climate in post-glacial period favoured the domestication of Near East and North Africa. The decline of summer rainfall compelled the people to retreat in some resourceful pockets (oases) in desert. It lessened the availability of wild resources and at the same time energized the people to cultivate grains and domesticate animals. Robert Braidwood criticized Childe's theory on two grounds. Firstly, the climatic changes were not so dramatic as Childe assumed. Secondly, similar climatic changes seemed to have occurred before each inter-glacial period but no food-producing revolution was found as such. Braidwood mentioned that the food-production was not the outcome of the change for climate; rather domestication began only in those regions where plants and animals were found ready for domestication. Accordingly, wild sheep, goats and wild grasses (barley, wheat and millet) grew plentiful in the areas with oldest farming villages.

Neolithic revolution was not only a phenomenon of Near East; it took place in a number of areas of the world, quite independently. Intensive cultivation and animal domestication of late Neolithic-era were responsible for the rapid growth of population in various regions. There was a debate regarding the fact that whether agriculture and sedentary life are complementary to each other or not. Those who opposed this proposition argued that in same regions of world, people began to live in permanent villages before they started cultivation and domestication of plants and animals, whereas in other places, people planted crops without settling down permanently. But recent studies of prehistorians make this point clear by announcing that a culmination of Neolithic sedentary life based on farming always ended in Urbanism.

In the Near East, the early villages were either based on agriculture or on animal husbandry or on a combination of both. Such beginnings were not universal in all places where the major urban centers developed. But it is definite that all these regions had the potency for the practice of agriculture or animal husbandry so that a large population could be clustered together in cities. City life is considered as one of the principal criteria of civilization.

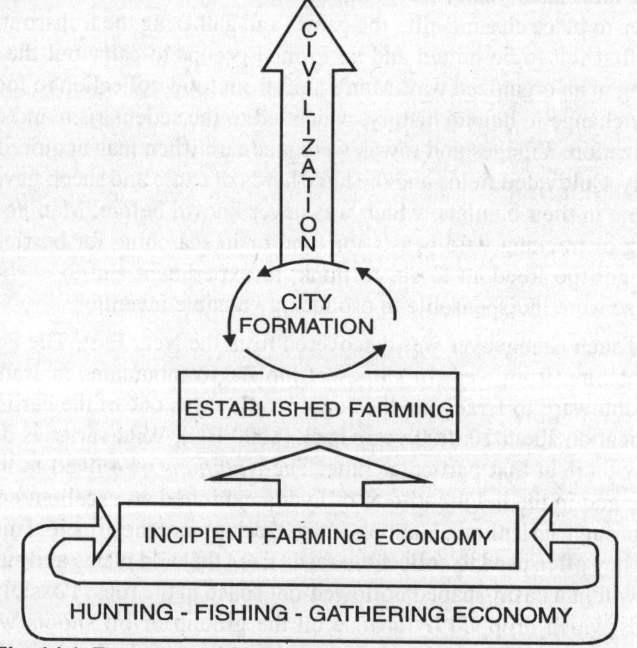

Fig. 14.1. Emergence of Civilization (From Economic point of view)

Rise of Civilization

As a concept, civilization is generally contrasted to the concepts of primitive and folk. Some prehistorians have coined the term 'Metal Age' to denote the time of rising of a civililzation. Because the discovery of the metal and its use was a very important requisite for the development of the civilization. Ploughs-axes-hoes made of metal meant an increase in size and number of agricultural fields. Lumbering and carpentry tools, hardwares improved the standard of the houses and public buildings. Metal spears, daggers and armour made men more efficient in war and enabled them to build great empires. In fact, there are several criteria with which we identify the occurrence of a civilization. For instance, large cities, full-time craft specialists, monumental architectures, great differences in wealth and status, strong and centralized political system etc., are the conventional characteristics of a civilization. The quality and scale of human life change a lot under the influence of civilization. V. G. Childe has prescribed some specific criteria for the development of a civilized way of life.

1. A growing specialization of work, combined with a developing local system within a town for the exchange and distribution of special objects and products.

2. Growth of social integration among a rapidly expanding population in urban centers.

3. The development of trade networks for the exchange of goods and services with nearby urban centers.

4. The invention of rules for the collection and transport to a central urban center of surpluses of food produced by the farmers and herders.

5. The development of many new artistic forms and design to express symbols of the earthly and supernatural power of urban centers.

6. The building of increasingly larger and more complex structures in the form of storehouses, palaces, temples, monuments to heroes and gods; and large scale public irrigation, water supply and drainage system.

7. The growth of political and religious leadership and membership in special organizations. These were inevitable because continuous residence in an urban center began to replace social relations based on family and kinship i.e., personal identification of kinship became weak in an urban center.

8. Invention and development of arithmetic and geometry. These were needed for complex record-keeping, making estimation of value of food and tools, construction of buildings, making of precise calendars and measuring the passage of time.

9. The growth of a small ruling elite or class with different privileges like, access to material goods, travel, special education etc. Thus, the individuals of this ruling class had grown a social difference from other 'citizens' of an urban center or a group of centers.

10. The development of writing as a means of expressing rules for the new forms of social organization in growing urban centers. It specially helped in keeping records of supernatural and political justifications for the new 'citizens' under rule and their associating ruling elites as well.

I. Ancient Civilization in the Old World

Old World is considered as the cradle of civilization. Because here earliest civilization arose in four different centers, independent of each other. It indicated that food produced in huge quantities in the surrounding areas on which the cities were fostered. It also implied transport facilities, which was essentially required to bring the food inside the city and also to the markets. Therefore trade, trade-goods, full-time artisans etc., came into existence. New political inventions with a kind of formal government were inevitable not only to rule the city but also the country that pertained to the city. Thus, different new institutions brought the erosion of the older order of local social organization that used to be maintained on strong kinship, instead of political views. The religion also became affected. Most of the local gods had lost their parochial identity and placed in the extended series of general gods. But it was really strange that such transformation took place only in few spots having high Neolithic culture. The Mesopotamia, the Egypt, the India and the China were the four regions of the Old World, which witnessed the rise of civilization. Among these four, the civilization of china was comparatively recent.

All these ancient civilizations emerged in river valleys. The fertile soil of the valleys developed the possibility of agriculture. In the next phase, irrigation enhanced the productivity and the respective river opened up the avenues of trade and communication.

(i) *Civilization of Mesopotamia*

Archaeologists think that the first city was evolved around 3500 BC in greater Mesopotamia, the area, which is now, shared by Southern Iraq and South-Western Iran. The name 'Mesopotamia' itself means the 'land between the rivers'. The Greeks gave this name to the valley between the rivers, Tigris and Euphrates. However, the settled cultivators of the Near East were the progenitor of this civilization. A number of cities namely Babylon, Akkad, Sumer, Ur, etc., gradually arose along the Tigris-Euphrates river valley. Among those, Sumer became the largest empire covering approximately 10,000 square miles in the lower half of Mesopotamia and had a large number of urban centers. For this, the Mesopotamian civilization is often referred as Sumerian civilization.

The ancient Sumer was a small tract of land on the Tigris - Euphrates delta. It was an unpromising region for the development of a civilization because the area was basically very dry lacking tree, stone and minerals. People used to live in houses made of mud and baked clay. Nevertheless, city life developed there in the third millenium BC. It was possible only for the river Tigris - Euphrates that provided the scope of irrigation to a vast field. The profuse yield not only met up the subsistence, the surplus of the grains was traded for metals, available in favoured regions.

It is interesting to note that the first settlers in this area were not the Summerians but we do not know who were those people and how they were supplanted by the Summerians. However, those people were named as Ubaid people, after the site 'Tell-al-Ubaid' near Ur, which was perhaps inhabited around 4000 BC. The Summerians entered in this region about 3500 BC possibly from Iran and adopted Ubaid pattern of living. All of Sumer was unified under a single government. There was an elaborate system of administration with a set of codified law and specialized personnel. Summerians not only produced the first writings, they also developed different sorts of crafts such as pottery, carpentry, brick-making, jewelry making, leather working, metallurgy, basket making, stone cutting and sculpture. Each city-state of Sumer had a temple, which was not simply an abode of god but was the source of political power.

With the economic specialization, social stratification became conspicuous in the civilization of Mesopotamia. Mesopotamians succeeded best in astronomy. As they were good at mathematics, so became able to develop a complete system of weights and measures. The great world-shaking invention of wheel was made by these people. The wheel not only facilitated the ploughing and transport, it also brought speed in pottery making. Early Mesopotamian metallurgy was the most advanced form of its days anywhere in the world. They used to produce Bronze and for that they had to trade with a few specific places of the world where copper and especially tin ores were abundant. Metal was very expensive in those days. Therefore, its use was primarily held in the hands of the ruling classes and used for the military ends. Metals for everyday purposes were beyond the imagination.

The writings of Sumer were evolved specifically about 3000 BC. They were shorthand pictures, wedge-shaped and called cuneiform. Those scripts used to be written on clay tablets with a sharp stylus or a sharpened bit of wood. After writings, the clay tablets were baked dry and each tablet looked like a page of a book. The priests of the Sumer were in the habit of preserving the temple accounts through clay tablets. A big library of such clay tablets came up in the ruins of Nineveh, a city in Mesopotamia. Henry Rawlinson, a British scholar, after twelve years of hard toil, was able to decipher the Cuneiform scripts. However, the use of this particular script was spread afterwards in the Middle East very widely, among the peoples speaking quite unrelated languages. Summerians also invented decorative seals on which they engraved mythical themes or scenes of everyday life with great precision and delicacy. History of Sumer came to an end around 1750 BC when the civilization of Babylonia began to flourish.

(ii) Civilization of Egypt

In the continent of Africa, the country on both sides of the river Nile is called Egypt. The Nile is one of the largest rivers of the world that flows through the desert of Egypt. Nile is not only a life-giving river of Egypt; it contributed to the birth and development of the Egyptian civilization. Egyptian civilization is, therefore, regarded as the 'gift of the Nile'. As per the archaeological record, Egyptian civilization flourished more or less at the same time of Mesopotamian civilization. Perhaps the rich Neolithic culture that persisted at Fayum in Lower Egypt (on the shore of an ancient lake at the west of Nile), between 6000 BC to 5000 BC paved the way for this civilization. The people of this area though depended on hunting and fishing; they also started the cultivation of wheat and barley. But, the increasing lack of water forced the people to migrate and settle in the valley of the river Nile where annual floods watered the soil to renew its fertility. All the natural resources that Egypt possesses or whatever people have developed there, grew up on the bank of the Nile. If the water of the Nile would not overflow and flood the two banks, it would have been impossible for the people to live therein.

In the Predynastic period, approximately between 4500 BC to 3200 BC, the farming communities spread along the valley of the Nile who used to live in huts made of grass and reed, and produced well-made pottery. In Egypt, the Neolithic culture was succeeded by a sort of Copper Age culture appearing in the valley itself. Early Egyptian civilization sprang up around 3200 BC in different sites like Memphis, Thebes, Karnak, Luxor, Abu Simbel in Nile river valley. Classical civilizations of Greece and Rome look on Egypt as the mother of their civilization.

Around 3200 BC, North and South Egypt were unified by the Semi-legendary King Menes, who was the founder of the first dynasty in United Egypt. Since Egyptians used to designate their King by the term *'Pharaoh'*, Menes is regarded as *'The First Pharaoh'*. However, a sharp culture change marked this period—partly due to introduction of a new political organization and partly for the novel ideas that penetrated from abroad.

Pharaoh was the all-powerful man in Egypt. All the lands belonged to the Pharaoh and his word was the law. The Egyptians worshipped him as a living god. Both the Church and the State were headed by the Pharaoh, though power was delegated to the lesser authorities like priest and Government Officials who acted on his behalf. Naturally the influence of the merchant class in Egypt was much weaker than that of the Sumer. Both the foreign trade and the internal trade were directed under the supervision of the God King. There was no other Egyptian law to regulate these trades.

The expression of the term Pharaoh was "one who lived in a big house". The God-King required heavenly structures to live in. Pyramids, Sphinx, Colossal Royal Tombs were the logical sequences of such a regime. It is strange that those largest stone structures of the world were accomplished without the benefit of any wheel, pulley or crane. Only ramps, rollers and levers were employed in building them.The Pharaohs obviously used to lead a luxurious life and their tombs were crowded with precious stones and metals, ornaments of gold and silver. Even the toilet articles, made of copper were studded with precious gems.

The Egyptians achieved certain other expertise of their own. They invented the technology of preserving the life-less bodies. Such preserved bodies were called *mummies*. The bodies of Pharaohs were essentially made into *mummies* for keeping them in the pyramids. Most of the Egyptian pyramids were built about 5000 years back and they are still the wonders of the world. The other achievement was the solar calendar. Among all gods and goddesses. Sun god *'Ra'* was important to the Egyptians. The people worshipped sun as a source of warmth and they tried to calculate its radiation. Ultimately a solar calendar came from them. The typical Egyptian writing, hieroglyphics also appeared at that time. Hieroglyphic means 'sacred writing'. Writing in hieroglyphic script needed skillfulness and therefore, it became a specialized art, being restricted among a special class of people in the society. However, the scripts used to be written on the leaves of a plant called *papyrus.* Inscriptions were also carved on the walls of the pyramids as well as of the temples. It took long years to decipher the meaning of this script. Champellion, a French professor, discovered the method of reading of the hieroglyphic script in 1823.

The stability of Egyptian culture is perhaps most remarkable as it was extended over a long period of time, almost about 3000 years. This conservation of culture was possible for two reasons. Firstly, for its relatively isolated position and secondly for the restraining influence of the State where the political control, religion and commerce—all were centralized in the hands of a single ruling group. Egyptian civilization was also found to influence many other early civilizations of Old World; important contributions were made in respect to the technologies of making paper, glassware and tanning of leather.

(iii) Civilization of Indus Valley

Indus Valley Civilization is a genuine parallel to the oldest riverine civilizations on the bank of Nile and the Tigris - Euphrates. This civilization was sprouted about 2500 BC and lasted perhaps a thousand years in Northwest on the valley of the river Indus. The two remarkable cities were Mohenjo-daro in the Lakarna district of Sind and Harappa in the Montgomery district of Punjab (Present day Pakistan). Sometimes this civilization is called as Harappa civilization because the first discovered site was Harappa. The Indus plain had a number of self-sufficient village cultures, which gradually developed to form a uniform civilization under favourable condition.

Not much is known about Indus Valley Civilization as compared to the civilizations of Sumer and Egypt. Because the existence of this civilization was unknown until 1920s and by that time the other two civilizations were extensively investigated. Scholars had learnt to read Mesopotamian cuneiform and Egyptian Hieroglyphics along with the abundant texts obtained on clay, papyrus and stone. But they did not find any text from Indus Valley Civilization. Writings there are only found on seals and pottery but the script has not been deciphered.

The territorial expanse of Indus Valley Civilization was more far-flung than the contemporary civilizations of Sumer or Egypt. This was not limited to the banks of the river Indus and its tributaries, but extended far south along the coast of Arabian sea to the present State of Gujrat as far east as Delhi and as far to the north-west as northern Afghanistan. The other noteworthy sites are Manda in Jammu district, Rupar in Punjab, Kalibangan in Ganganagar district of Rajasthan, Lothal in Gujrat, Banawali, Daulatpur and Mitathal in Hissar etc. Credit of this discovery goes to Dr. Rakhaldas Banerjee, an Officer of Archaeological Department of the Government of India who first witnessed a Buddhist stupa at Mohenjo-daro and began to excavate.

Although the Indus Valley is a barren desert at present, it was not so in pre-historic days. Rather it was a seat of ancient agriculture in Indian subcontinent. Like the economy of all earlier Neolithic cultures, the people of the Indus Valley showed subsistence agriculture, including animal domestication. In fact, during Neolithic period, the Baluchisthan and the eastern part of Afghanistan (which make up Indo-Iranian border region) exhibited extreme climates. It was partly a desert area and partly a hilly area, unfavourable for habitation. Therefore, only a few scattered Neolithic sites have been discovered from this area. On the other hand, the adjacent Indus River valley offered a very fertile land for its annual inundation over large areas. Wheat and barley sown at that time became ripen by spring without ploughing or manuring of the ground. This created a dense forest rich in games as well as a wide grassland vegetation, suitable for domestication of animals. Moreover, the alluvium deposition gave the land a higher agricultural potential. The people of Late Neolithic realized this fact and so an entirely new way of life evolved out.

Indus Valley people were the first grower of cotton. They had similarities with Sumer people in crops and domestic animals, style of house building, use of metal ornaments (copper and bronze) and pottery (wheel-made). In contrast to the Egyptian and Sumerian civilization, the Indus Valley Civilization showed the evidence of baked brick. The cities here were not just an agglomeration of houses, they provided the evidence of an excellent and careful town planning with broad roads and beautiful drainage system. The well-planned cities also exhibited large public swimming pools and storehouses (granaries) for the farmers. A municipal Government was perhaps responsible for the maintenance of such cities. But it is remarkable that no pyramids or monumental sculptures were discovered from this civilization. Although no temple was found, presence of typical female figurines carried the religious connotation. Existence of a divine mother is found in the evidences of the seals, terracotta figurines, stone images and amulets. A male god was also worshipped which has been identified with Lord Shiva of later period. Beside the icon of the mother goddess, many other attractive figurines have been revealed from the two principal sites of the Indus Valley (Mohenjo-daro and Harappa). Although they are not very rich and abundant in comparison to the other two old civilizations, still from the statues and figurines, it can be said that the Indus people had made good progress in the art of sculpture.

The Indus people perhaps imported the raw materials for their industry from the places like Rajputana, Baluchisthan and south India. Therefore, they developed good transportation devices in the form of wheeled cart and boat. The people had also made themselves equipped in the knowledge of mathematics, geometry and accountancy. But the span of this civilization was relatively shorter. Recent radiocarbon dating suggests that this civilization arose around 2500 BC and collapsed between 1900 BC and 1500 BC.

(iv) Civilization of China

China, one of the largest countries in the world, is situated in Southeast Asia, on the western coast of the Pacific Ocean. The civilization reached China a number of Centuries later. Like other civilizations, this civilization also followed the same path being connected with the Neolithic revolution. The Yellow River, Hwang Ho of North China contributed here to pass quickly into a Bronze Age with the Shang Dynasty. The Dynasty dated itself from 1766 BC but is thought to have begun about two hundred years later, a little before 1500 BC.

The late succession of this civilization enabled the Chinese to get a few elements from the high civilizations of west. But in early times as china was so far away from the west that she had received only a little stimuli and basic ideas rather than full cultural form, although her Bronze Age civilization projected a similar profile. Chinese culture was able to maintain its own distinctiveness even after receiving foreign influences. One of such influences was the use of iron. China in turn with her invention of paper and printing technology had put the west in debt to her. Writing appeared in this civilization possibly as a native creation based on foreign idea but now has been recognized as the ancestors of modern Chinese writings.

The earliest evidence of cereal cultivation outside the Near East dates from China. Wheat, millet and rice were probably grown in the valley of the river Whang Ho. Of bred animals, there were buffaloes , pigs and dogs, which are still abundant. Hemp and silk fibers were available there, which began to be cut and sewn in order to weave cloth at the dawn of the civilization. The wheel as an instrument of pottery making was arrived among the people. They also had large vessels and ornaments made of copper. The famous sites on the valley of the Yellow River were China, Anyang, Chengchou and Luyang. Here we find the traditional characteristics of a civilization like city life, a system of writing, a state organization, use of bronze *1, monumental constructions *2, class stratifications, a

---

*1. Chinese bronze was imported possibly from Japan, although Japan was not a Cradle of Civilization; she was only a receiver of it.

*2. People did not use stone in building construction, which was plenty in their area. They used bricks.

high division of labour, widespread trade and commercial activities, development of warfare, and an increased knowledge of astronomy and calendrical calculation.

In fact the beginning of Chinese civilization can be traced from a late Neolithic culture termed as *Yang-Shao* or painted pottery culture. Yang-Shao people farmed the rich soil on either side of the middle part of Hwang Ho River, which are the present day provinces namely Honan and Shansi. Their principal crop was millet. People extended their subsistence through hunting and fishing. The pottery was essentially hand-made and produced outside the boundary of the community. *Yang-Shao* culture was developed with the synonym *Lung-shan* culture i.e., the black pottery culture. The late Lung-shan period has been dated as between 4000 B.P and 3500 B.P where settlements grew at a high speed and villages were fortified with high thick earthwalls. At this time, the people learnt wood working technique so that they could build houses and boats to an advanced form. This Lung-shan culture was succeeded in the Yellow River basin after 3500 B.P forming a civilization known as *Shang* or *Yin*. The *Shang Dynasty* flourished between 3500 B.P and 3027 B.P and was overthrown by the *Chou* who ruled from 1027 BC to 256 BC

(v) Civilization of Aegean Sea

It is a civilization of unusual origin. Most of the elements of this civilization were imported. After 3000 BC, this began to grow up with the introduction of metallurgy but without a Neolithic base. Therefore, most of the scholars do not want to mention its name along with other four early civilizations of the Old World.

The Egyptian civilization had posed her effect on Europe. It served as a focus in the west for commerce as well as knowledge,and drew the civilization westward. Joining with Mesopotamia, it stimulated the Eastern Shore and the islands of the Mediterranean.

The Cyclades in the Aegean Sea was not a good place for farmers. But the island was found thickly populated in the Copper and Bronze Ages. The people living there were the producers of copper, marble, obsidian and other materials; they remained engaged in Mediterranean commerce. Importance of the Cyclades was subsided in later times but two other centers e.g., island of Crete and the southern Greek mainland saw the rise of European civilization. The island of Crete was almost equidistant from Asia (Turkey), Africa (Egypt) and Europe (Greece). Since it was posited along the great routes of that time, the civilization was fed by trade. Some writers prefer to call this civilization as *Minoan civilization.* It does not trace a valley-bottom development, but a maritime advancement.

The Cretans were the stockbreeders until sometimes after 3000 BC. Thereafter they opened a copper-using phase. At this time they became very much advanced in the art of navigation, trade and commerce, fine arts and culture etc. They built beautiful cities with good sanitary arrangements. Knossos, the capital of Crete, was a famous city in those days. A female divinity, very similar to the mother goddess, was the most important deity of the Cretans. Knossos had an auditorium where plays were performed regularly and bull fighting was shown. The people of Crete knew the use of gold, silver and various other metals. Byblos, Mycenae etc., are the other important sites under *Cretan civilization.* The cities were numerous who faced on harbours rather than farmlands for their nourishment. The Cretans developed a script in clay tablets, which has not yet fully deciphered. However, around 1400 BC, the civilization got degenerated by the advent of a group of invaders (Greek) from the North. Although the Greeks occupied Crete and demolished the cities, they could not keep themselves immune from the influence of the Cretan civilization. Cretan culture and civilization molded the Greek mode of living.

The different centers of Old World civilizations were not at all identical. Although some patterns may have diffused from one center to another, each civilization was unique with its original culture and style of life. For example, script though found in all these civilizations, but the particular language or writing system was not alike. Again, Egyptian architecture was not similar to the architectural pattern of Mesopotamia or Crete, despite sufficient communications between these centers. Greek visited Egypt but they did not want to imitate the monumental structures of Egypt. They also did not borrow the technology of making mummy. No one could steal the peculiarities or specialties of others. For instance, the metalwork and handicrafts of Indus Valley Civilization were not so good as those of Sumer but the former was more advanced in city planning and sanitation programming. Egyptian's knowledge of medicine was no doubt superior than that of Mesopotamians who were ahead in mathematics and astronomy.

II. Ancient Civilization of the New World

There is a gap between the total cultural history of the two hemispheres, Western and Eastern. Some fifteen or twenty thousand years ago, at the time when the people of Western Europe were adapting their ways of life to the changing environment caused by the retreat of the last Pleistocene glaciers, a band of Asian hunters insearch of games proceeded eastward and once reached to the farthest corner of Siberia. At that time the levels of oceans dropped and the land map of the world changed to some extent. A part of the Bering Sea disappeared and Siberia was joined to Alaska. The hunters crossed a low-lying plain without knowing that the plain would be covered by Bering Sea later on when the glaciers would melt away. They could not even guess that they were entering into a completely new world—the opposite continent. Obviously, the group failed to realize that they had been the first Americans. [vide Fig. 9.14].

The earliest migrants of the new world first settled in Alaska coming from Siberia, crossing the Bering Strait which is about sixty miles wide at its narrowest point. Since it was the closing phase of Ice Age, people could not push them interior into the continent. The whole land was covered by vast ice-sheet. Gradually the ice receded with increasing temperature and facilitated the descendants to move forward upto the extreme limits of South America. Thus, the earliest immigrants to the New World were essentially the hunters and food-gatherers. No hunting-gathering community has ever developed a large population or a civilization. Complex culture, what we call civilization, develops only when the sedentary life comes up. This again depends on the scopes of agriculture and the domestication of plants and animals.

A kind of semi-sedentary life was developed along the Pacific Northwest coast where ocean supplied vast quantities of food. But the seasonal search often shifted the whole group to other areas of natural abundance. In the northeastern part, maize was an important crop but was not as intensively cultivated as in the areas of South. Agriculture as a basic economy was found to develop later in Mexico, Central America and in the Andean. These were the areas where small village populations engaged themselves in intensive cultivation. Agriculture developed in America quite independently of Europe and Asia. But such a transformation of economy occurred much later and, therefore, civilizations in the New World also appeared later, only a few hundred years before the time of the Christ.

Now, the question is regarding the reason of late. In the New World, as in the Old, the earlier stages of agricultural societies did not produce advanced civilizations. They were actually the seeds for the civilization. The difference that marked off America from the Old World, is the process of farming the grains. It has been found that the development of corn from its wild state had undergone a slow process of change as compared to the old-world grains in Near East. In America, a wild variety of maize used to be domesticated. The entire technique of maize cultivation was different from that of millet and wheat cultivation as practiced in the Old World.

There were hundreds of tribal cultures in native America who segregated themselves into ten basic culture zones—six in North America and four in South America. Therefore, the agricultural villages of earlier stage were not only economically self-sufficient, highly provincial also. Civilization could not able to spring out until trade connected the villages; new ware and new ideas were introduced. Intensive agriculture of later period brought a surplus of food, which helped, in braking down the isolation or provincialism of the village, stimulated the thought and ideas, and thus paved the way for civilization. Not in all places, it happened particularly in Meso-America and the Peru.

When the Europeans (Spanish) arrived in the New World in 1519, the Aztecs had grown the control over the most of Mexico and the Incas were ruling over the Peru and adjacent regions. The

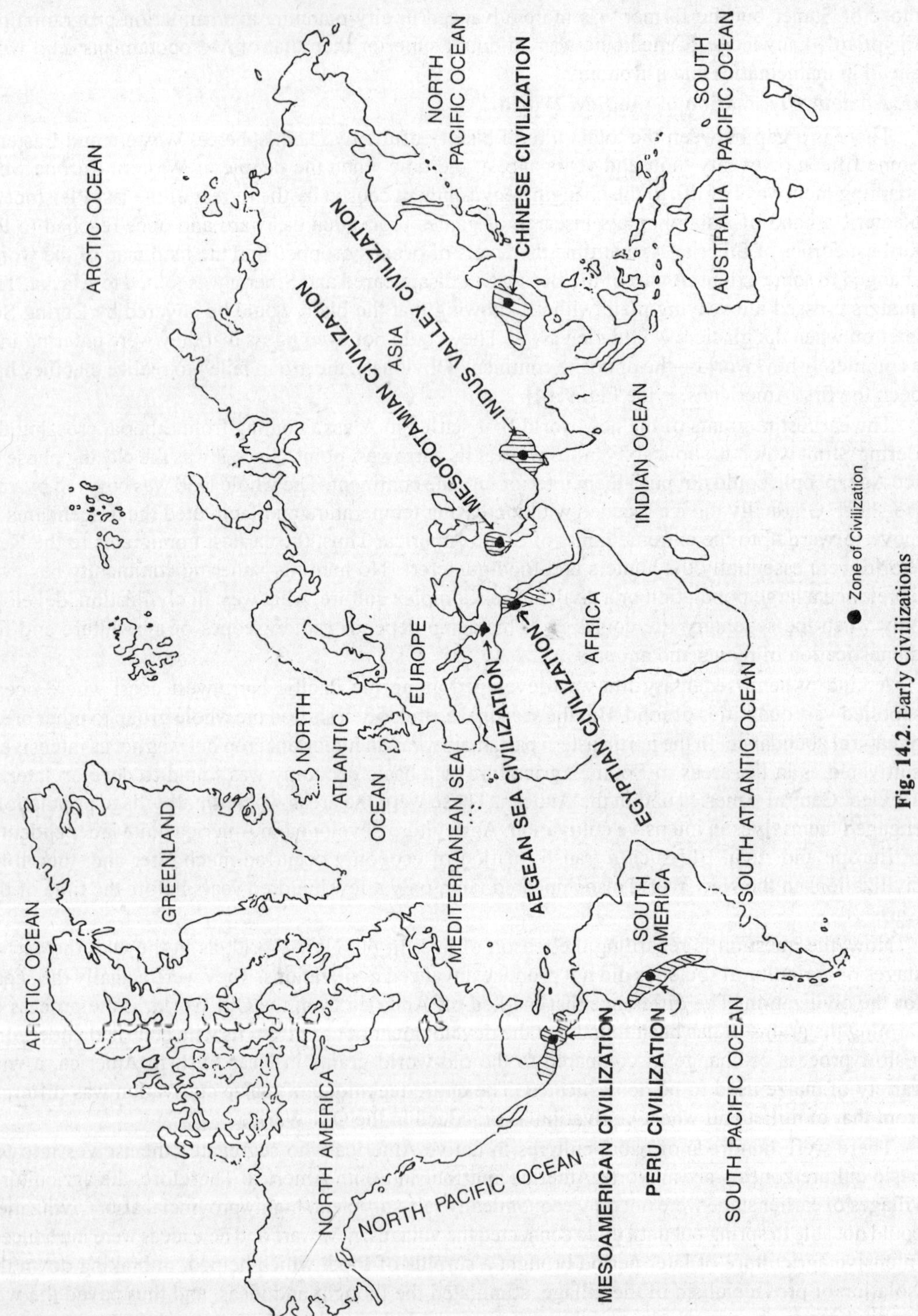

Fig. 14.2. Early Civilizations

Europeans were therefore amazed by the high level of civilization that prevailed in the said two areas. Archaeological research since then has revealed many earlier centres of civilization both in Mexico and Peru, before the periods of Aztec and Inca domination.

(i) Civilization of Meso-America

In North America, the culture area Meso-America consists of some parts of Mexico and Guatemala. It contains the Maya, the Toltec and the Aztec as its most outstanding nationalities. The root of Meso-American civilization is actually extended to Olmec culture which appeared around 1500 BC. The Olmec people relied mostly on fishing than hunting. They used to live in villages and started cultivation of maize. A type of thin-walled pottery was also produced by them. The egalitarian farming community gradually gave rise to the complex culture in North America. The important sites were San Lorenzo, La Venta and Teotihuacan.

The site of San Lorenzo flourished between about 1200 BC and 900 BC. Sixty-five stone monuments were discovered from this site and other ten from surrounding region. The occupation of this site came to an end around 900 BC when stone monuments were deliberately mutilated or buried. But the Olmec culture was continued at the other site, La Venta which persisted for about a few hundreds of years and ended around 400 BC. Still after this, the Olmec culture was found to be alive in the third site at Tres zapotecs.

The Olmec culture also influenced the Maya area as indicated at the site of Izapa on the pacific coastal plain of Chiapas in Southeast Mexico, near Guatemala. The Mayas occupied the core of Central America and during the period of their civilization, the center of development shifted from the highlands of Guatemala in the south to the north—through lowland Guatemala and finally up into Honduras, Yucatan and southern Mexico. Their stone cities first appeared in the lowlands, the Peten, presumably after 300 AD which reached to a climax during the Dark Ages of Europe but disrupted by the arrival of the Spanish after 1000 AD. The Maya culture was characterized by the hieroglyphic writings, position numerals, an elaborate calendar, some pyramidal structures, buildings of lime stone, sculpture, polychrome pottery, etc.

The other site is the Teotihuacan in the highlands of the northeastern section of the valley of Mexico, where a true civilization and a State arose by the end of 200 AD. Although the area was rather dry, it had rich alluvial soil watered by the springs. So it was quite suitable for farming. An adequate food supply fostered the city with a large population. But the city collapsed around 700 AD for the reasons which are not yet clearly understood.

History of Meso-America is peculiar as the center of political power frequently shifted from one region to other or handed over from one tribal group to another. The last group of rulers before the advent of Spanish was Aztecs or Mexicans. Around 1325 AD they settled at the site, Tenochtitlan. They were the contemporaneous people of the Incas in Peru. Aztecs claimed themselves as the successors of the Toltecs. However, the cities and states those emerged in Meso-America were all later than that of Near East. This incidence may be linked to the later emergence of agriculture in the New World. The specialized arts and crafts were found only when the agriculture became able to produce more than enough food for all. Metallurgy, war, religion, architecture, weaving, pottery appeared one after another.

(ii) Civilization of Peru

The Andean civilization flourished along the coast and highlands of Peru. The Andeans had their great development in two departments: the *craftsmanship* and the *politics*. But their religion did not appear to be of same importance as like Mayas or Mexicans. Pottery reached to a high level and it was varied and lively. Andeans invented and used more different techniques of weaving than any other people weave on earth.

However, the initial period in Peru can be dated as 1800 BC to 900 BC which is marked by the first appearance of pottery and woven cloth. Village life was then based on the cultivation of Maize, squash, bean, guavas, chili, etc. The next remarkable period can be bracketed as from 900 BC to 200 BC, when a definite elaborate art style was developed. The style had its reflection on pottery, textile and stone architecture, which have been found, from both the highland and coastal regions of northern and central Peru. The culture has been called as Chavin culture, named after a temple site in the northern highlands of Peru. The first cities came up between 200 BC and 600 AD following a few inter-regional wars. Notable cities were Tiahuanaco and Pucara in the Titicaca basin, Huari in upper Mantaro basin. During the period from 600 AD to 1000 AD, a spread of Tiahuanaco art style has been noticed which reached up to the southern part of Peru. The city of Huari started trade relation with Nazca of the south coast. The entire empire could respond to the orders of a single man. It became not only an efficient and successful system, but at the same time a horrifying one. Despite the city life declined in South Peru, between 1000 AD to 1476 AD a full-fledged city life emerged on the northern coast.

Inca domination started at 1476 AD and continued up to 1534 AD. Incas were very much efficient in military and administrative matters. An emperor was placed at the top of the Inca socio-political hierarchy who used to be considered as a sacred person associated with Sun. The whole Inca tribe was divided into upper and lower classes. Upper class people were treated as nobles who dominated over the lower class people. These people were not only exempted from manual work and service, they had special privileges as to hold land estates, to receive formal education and to marry more than one wife. On the contrary, lower class people were regarded as commoner who enjoyed none of those privileges; moreover they had to till the soil not only for their own but also for the upper class people.

The Inca religion in Peru was state supported. They were not the large-scale human sacrificers as like the Aztecs. Instead, they used sacrifice *llamas* (an animal) in most of the cases. Apart of the realm of art, they achieved a great success in the field of surgery.

In fact, civilizations of both the areas of New World, Meso-America and Peru were flourished around 900 BC. Some of the anthropologists suggest that cultures of these two areas were independent of each other. Others hold an opposite view that there must be a link between these two cultures; especially the Olmec influenced the Chavin culture of Peru. However, these two contemporaneous people, the Incas and the Aztecs extended their empires by constant conquest and assimilation. The Aztecs claimed to be the successors of the Toltecs who contributed in the development of writing and calendrical system of Meso-America. Although the Incas had no writing systems, they developed the use of 'quipus' for keeping records. Besides, Incas were unique in organizing the centralized political authority, which was rudimentary among the Aztecs.

In searching out the genesis of civilization, we usually point out the factors like large fertile valley, abundant crop, growth of population etc., for which the earliest urban centres sprang up. Such an analysis is true but incomplete. Because, it has been evident that, neither the environmental factor nor the population factor, not even a combination of these two can generate cultural complexity. If it were so, civilization would have developed in a large number of places where the above-mentioned preconditions prevailed. Actually, the concentration of people rises and the foundation of the society is built up only when the condition of local culture remains favourable at the particular point of time. The psycho-social attitude of the people is no less important than the limitations and possibilities offered by nature. Once a farming community had grown habits of cooperation and got effective tools and means for greater food-production, the population automatically began to increase. In such condition, if there were no other specific factor to hold the speed down, the small village settlements flourished in the shape of a town and gathered the potency to act as a state in future. As a matter of fact, the surplus food, sooner or later, gave rise to a new and stratified system of wealth. People learnt trading as well as paying. Trading stood essential for the craftsmen because they always did not get their raw materials in the locality. A system of paying was required for them to procure necessary materials from outside. Further, some of the craftsmen made themselves fully involved in the manufacture of goods. Being completely spared from the agricultural work, they also needed money to buy food and other articles of livelihood. Naturally the system of payment came into existence.

The centres of Old World civilization possibly influenced the surrounding peoples as evidenced by the frequent trades that developed between Egypt and Crete, in the period from 3000 BC to 1650 BC. Cretan influence in turn was felt on the mainland of Greece where *the Mycenaean civilization* developed between 1700 BC and 1100 BC. New centres of civilization also developed in India and Far East. For example, cities sprang up in the Ganges valley in India during the period from 1000 BC to 500 BC. Chinese influence spread to Japan where a sophisticated civilization arose by 700 AD. In Southeast Asia a number of regions were influenced by China like Java and Cambodia. However, everywhere a culmination of Neolithic sedentary life based on farming was responsible for the development of urbanism. City life was one of the principal criteria of civilization. Other important criteria were the emergence of a writing system, metallurgy, division of labour, class stratification and a formation of a state which were mostly present in the advanced Old World centres and the New World centres as well.

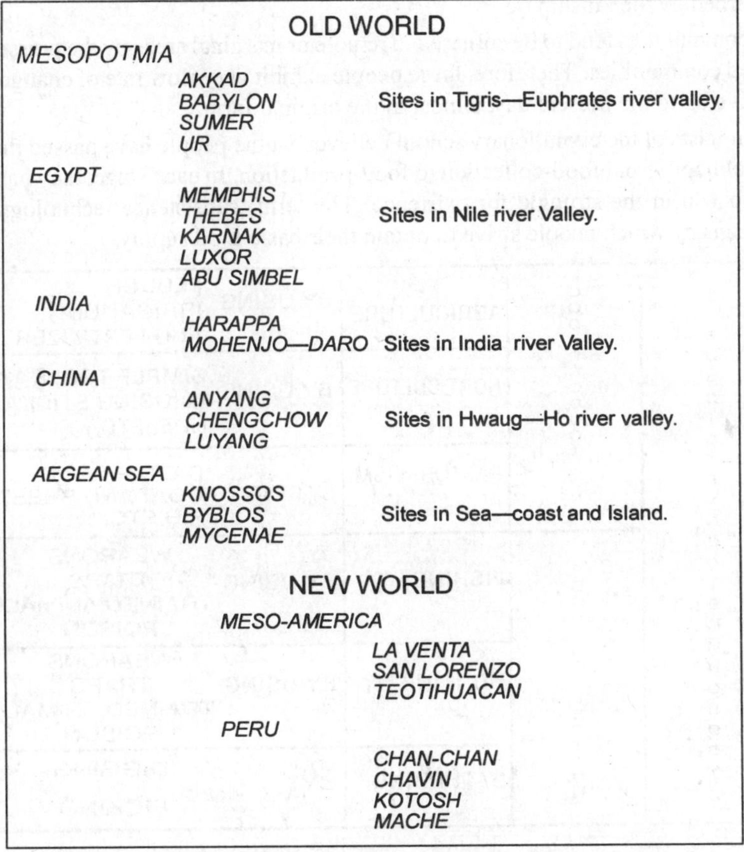

Fig. 14.3. Ancient Civilizations and Their Important Sites

PRIMITIVE SUBSISTENCE TECHNOLOGIES AND MATERIAL CULTURE

Survival was the basic necessity for the early men. They had neither any dwelling nor any clothing. Occasionally they climbed on the trees or entered into the caves for shelter. In winter, they wore leaves or hides of animals to keep themselves warm. Human life was thus fully exposed to the nature.

If we probe deep into the history of mankind, we find that Paleolithic man was the first group of men who became able to depart themselves from the lower animals. They started to explore their surroundings with a relatively developed brain and body structure. But their inventive power was confined only to the production of simple tools, useful for hunting and food gathering. Hunting and gathering are the oldest mode of man's subsistence; many contemporary primitive groups still subsist on them. For example, the Andamanese, the Kadars, the Semangs, the Eskimos, the Australian aborigines and many other African tribes belong to this stage of culture. The economic pursuit of these primitive groups shows a great dependence on the climatic condition in their respective habitat. They share some general characteristics too.

(*i*) The population density remains low for all food-gathering communities. Only those who live in the pacific coast or the Great Plains of America are the exceptions. Because the said regions are very rich in edible plants and animals and gifted with plenty scope of fishing.

(*ii*) The scarcity of the resources often drives the food-gatherers from place to place. But they never cross a specific territory.

(*iii*) The movement of a food-gathering community takes place in a unit group of a few families or a band. But the mechanism of social control remains very loose and the pattern of interaction is usually governed by the kinship tie.

(*iv*) The communities tend to be cornered in remote or marginal areas by the pressure of the larger and advanced communities. Therefore, these people exhibit very slow rate of change and have been able to survive up to the present day, almost in the original form.

Anthropologists of the evolutionary school believe that the people have passed through different stages of livelihood, from food-collection to food-production. In each stage, they had to devise new appliances to win in the struggle for existence. The term 'subsistence technology' refers to the technical means by which people strive to obtain their basic food supply.

SUBSISTENCE TECHNOLOGY	FOOD PRODUCTION	AGRICULTURE	BY USING	PLOUGH IRRIGATION AND FERTILIZER
		HORTICULTURE	BY USING	SIMPLE TOOLS (DIGGING STICK, HOE, ETC.)
		PASTORALISM	BY REARING	CATTLE GOAT AND SHEEP PIG ETC.
	FOOD COLLECTION	FISHING	BY USING	WEAPONS TRAPS TRAINED ANIMAL POISON
		HUNTING	BY USING	WEAPONS TRAPS TRAINED ANIMAL POISON
		GATHERING	BY	DIGGING PICKING

Fig. 14.4. Divisions In Primitive Subsistence Technology

I. Food-Collection

The phase of food-collection announces man's dependence on wild plants and animals for food. Anthropologists have categorized the activities namely food-gathering, hunting and fishing under this stage.

1. *Food-gathering*

It is the earliest phase when men used to collect fruits, roots, seeds, tubers and nuts readily available in the natural set-up. They invented hooked poles for pulling down the cones in order to pick up the nuts. Digging sticks were used for digging up the soil to get the edible roots of the plants. The little creatures namely the beetles, caterpillars, grasshoppers, lizards etc. were also gathered for food. These were usually captured by certain special techniques of hand. Sometimes gum was used in addition.

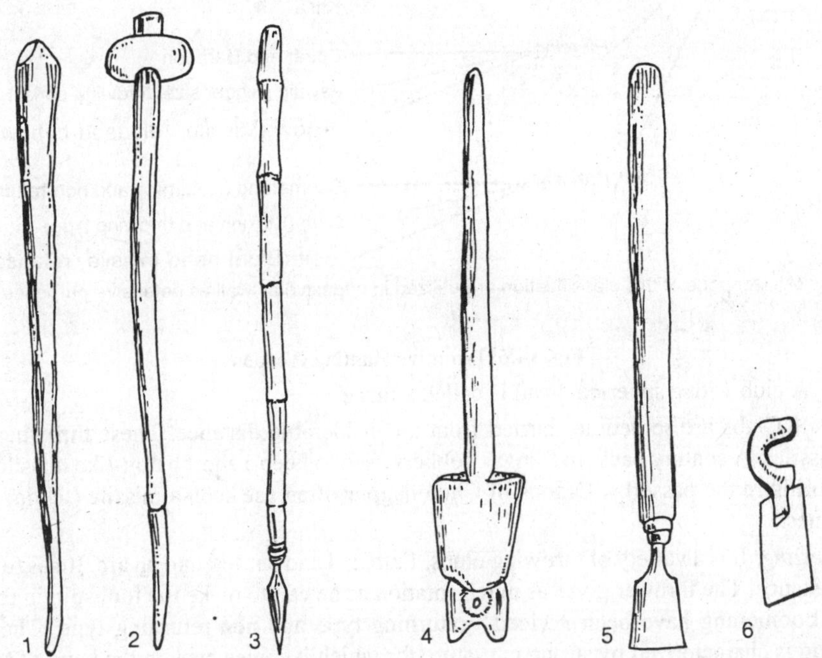

1. Simple digging-stick. **2.** Composite digging - stick. **3.** Digging - stick with ring-stone at the butt end **4.** Digging-stick with broad iron blade. **5.** Digging-stick with a flat-parabolic working end. **6.** Broad-blade digging-stick with a curved handle (spade).

Fig. 14.5. Forms of Digging Stick

2. Hunting

Man is the weakest animal among all big animals. But the intellectual superiority of men has enabled them to devise a large variety of weapons at different time for fighting against ferocious animals and the enemies. The hunting methods as invented by men can be grouped under four categories.

(*i*) Hunting by weapons.

(*ii*) Hunting by trap.

(*iii*) Hunting by tame animals.

(*iv*) Hunting by poisoning.

Hunting by Weapons : Earliest weapon of man was either a branch of a tree or a lump of stone. Gradually the innovative brain taught them to learn the mechanism of making tools. Usually we classify all hunting appliances into three major groups:

(a) Bruising and Crushing Weapons

These are the weapons, which smash down the body of the animals without breaking its skin. Blood shed does not occur normally. Club, mace, boomerang etc., implements belong to this group of weapons.

Club: It is generally made with the branch of a tree. A club may be simple or compound or composite. Simple club is made out of a single piece of wood. In compound club two or more pieces of wood are used. But in composite club pieces are invariably made of different materials. However, a club by nature shows a thick striking end; the other end remains thin for an easy grip at the time of manipulation. The clubs are used to crush the animals or enemies when they are close at hand.

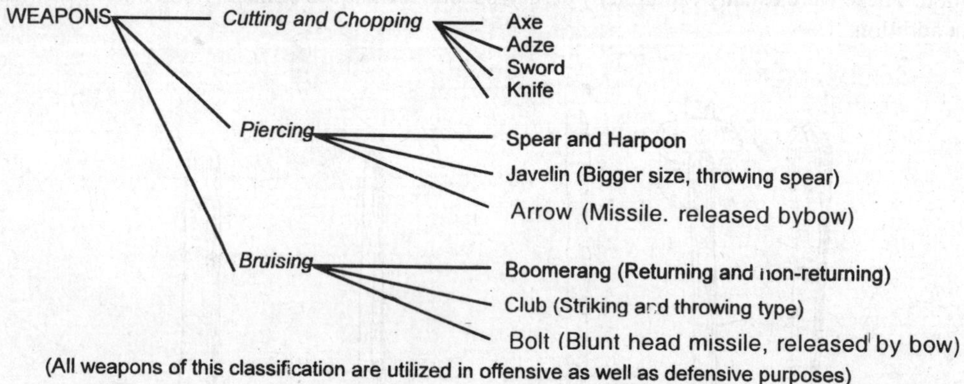

Fig. 14.6. Primitive Hunting Weapons

Mace: A club with a spherical head is called a mace.

Missile : Clubs are sometimes hurled from a considerable distance. These throwing clubs are called missiles. A century back, in Bengal, robbers used to keep a short baton-like missile known as 'pabra' to injure the passerby. Oraons of Chotonagpur often use such a missile (lebda) to capture small games.

Boomerang: It is a variety of throwing clubs, flattened and curved into an arc. It has to be thrown from a distance. The thrower gives as much rotation as he can to make the tool spin in the air. Two types of boomerang have been noticed—returning type and non-returning type. The returning boomerang is characterized by strong curvature for which it comes back in the hand of the thrower following a curved path, when it fails to hit the object. But on striking the prey (usually animal or bird), it drops invariably on the ground. The non-returning boomerang, on the other hand, possesses a lesser curvature and so it never returns back to the thrower, whether or not it hits the object. However, this non-returning variety is extensively used among the Bhils of Rajasthan whereas the use of the returning type is the speciality of the Australian aborigines. In fact, the Australian aborigines are endowed with the credit of inventing this implement.

Sling: It is an appliance to throw small pieces of stone, pellets etc. It consists of a pouch, which holds the missile, and two chords are attached with its horns diagonally opposite to each other. While throwing, the thrower spins it overhead. One of the chords is released with force and the stone flies towards the object. Grass, fiber or leather is used to make a sling. Oraons use bark-fibers, Santals use a type of grass, but the Fizians and the Hawaiians typically use plaited human hair in the making of their slings.

Bolt: It is a blunt octagonal arrow operated by the bow. It can wound a game without any bloodshed. In South America, the Guanacos* are often hunted with bolt and bow.

(*b*) Chopping and Cutting weapons

Such weapons possess sharp edges to create fatal injury. These are usually applied against the objects from a close proximity and by hand.

* A camel like animal whose size is almost like deer.

Axc : It is a weapon for cutting and crushing. Stone-axes without a haft were widely used in Paleolithic period. They have perhaps been evolved into hafted metal axes of present day.

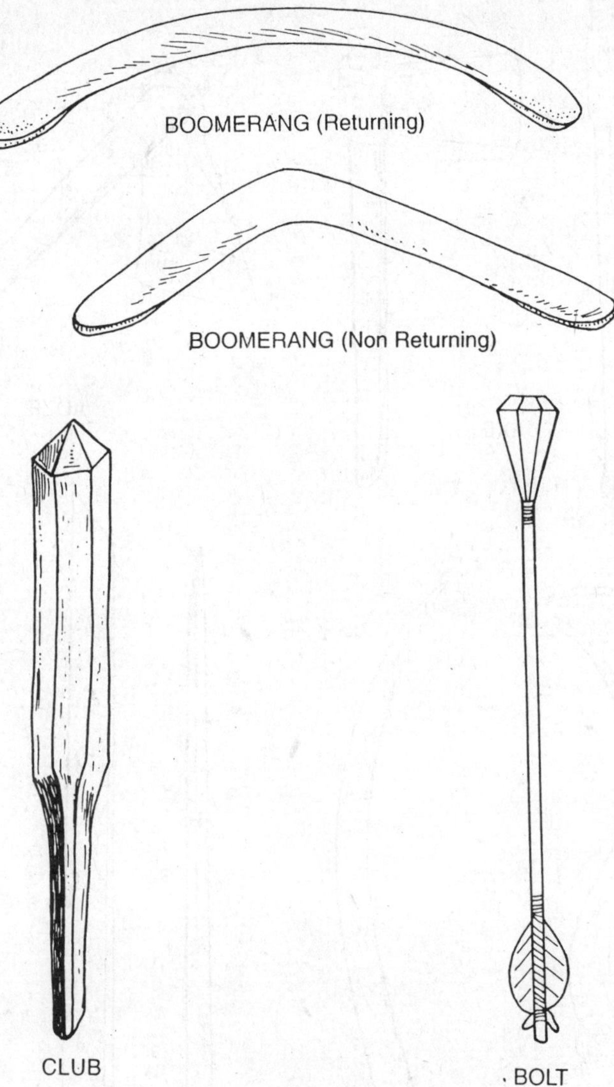

Fig. 14.7. Bruising Weapons

Contemporary primitives use several types of axes, with stone and iron heads. These are not only the implements for wounding and chopping the animals; they are also used in fighting with enemies. Steel-headed battle-axes are common in Europe, Asia and Africa. The 'tangi' used by the Santals and Hos of India is a kind of axe. Axes with double-blades and with a piercing point are not unusual.

Adze: An adze though resembles with an axe, it has some differences too. Unlike axe, its cutting edge forms a right angle with the axis of the shaft. It is a purely domestic tool. Non-literates use this in shaping the Canoes and other wooden objects.

Sword, Dagger and Knife: Sword is a large and long offensive weapon, directly used by hand. It may be single-edged or double-edged. Its tip may be straight or curved, pointed or broad. It is extensively used in Asia, Europe and Africa. Dagger and Knife are the small weapons, probably the precursors of sword. Different materials namely stone, bone and bamboo are employed in the make of these implements. Metal is also popular among some primitives.

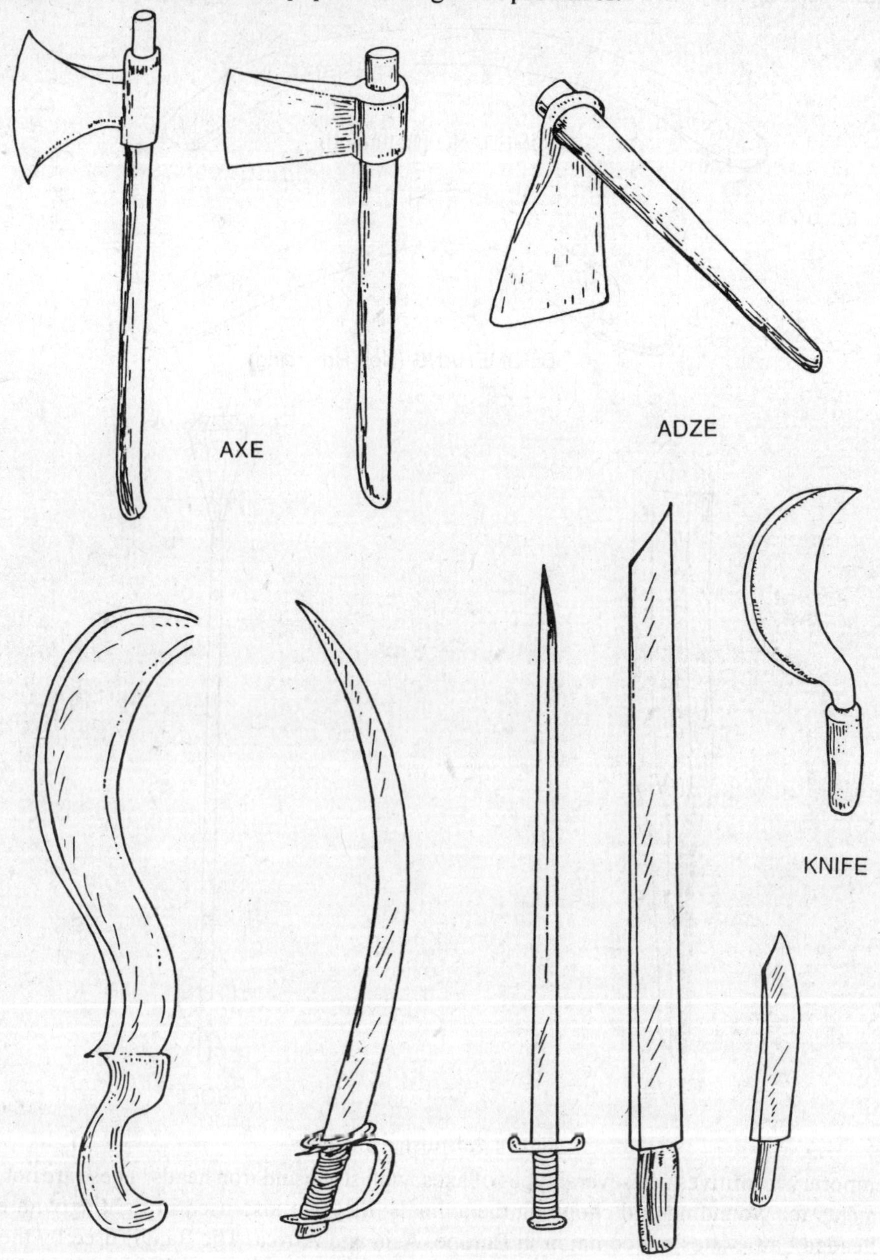

Fig. 14.8. Cutting and Chopping Weapons

(*c*) Piercing Weapons

These weapons are either held in hand to pierce into the body of the games or thrown as missiles from considerable distance.

Arrow: It is a missile with sharp pointed working end, called head. The head is very small indeed and it remains essentially attached with a shaft. Stone, bone and sometimes metals are used to make this head. To release an arrow, bow is essentially required. Various types of bow are found among the primitives made up of wood or bamboo. Some bows show very complicated mechanism.

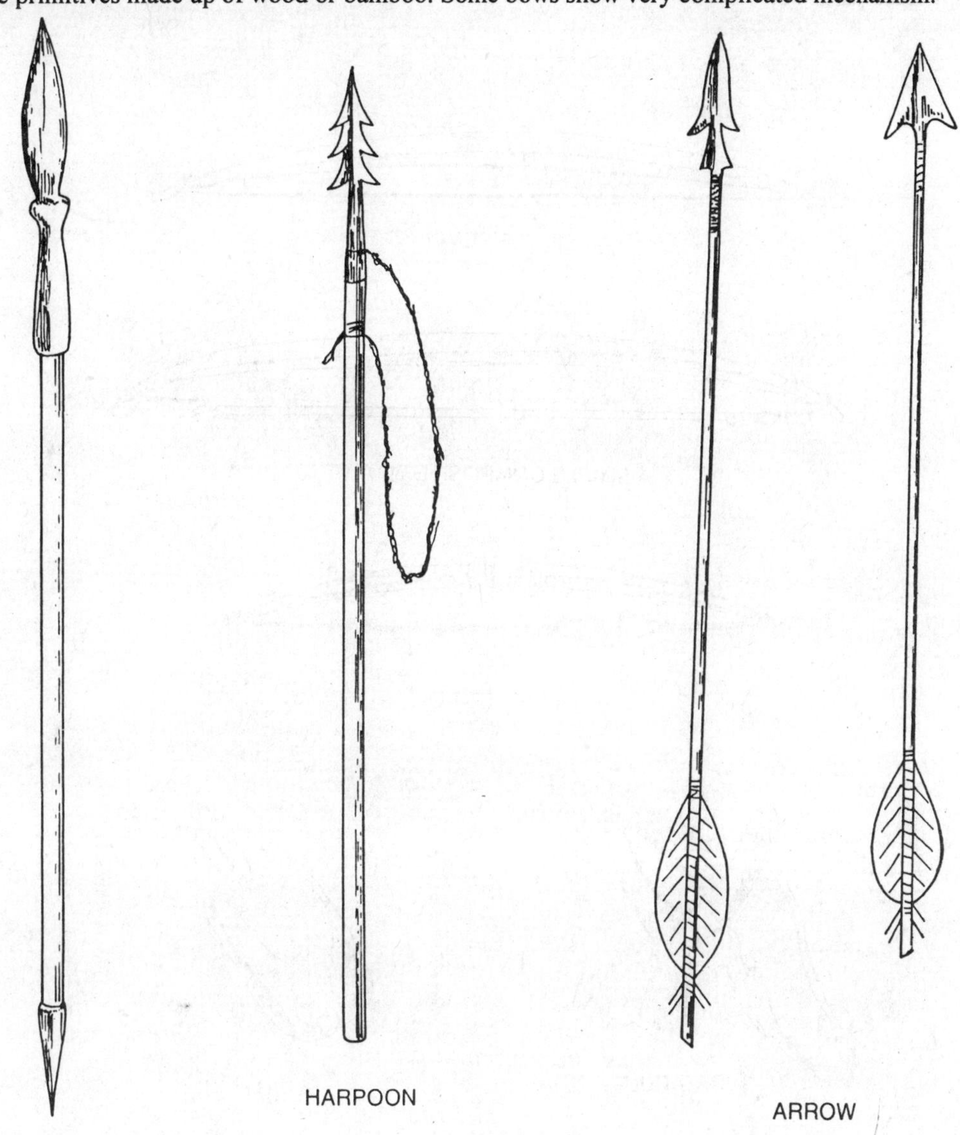

Fig. 14.9. Piercing Weapons

The basic principle in a bow is that, a string has to be tied with two ends of a bamboo split so that the bow gets the shape of an arch. At the time of shooting, the arrow is first adjusted in the middle of the string. Then the string is pulled back and immediately released aiming at the object. The arrow runs by the generated force; least manual labour is involved in this devise. Bow and arrow was invented in the Mesolithic period, but it is still preferred by the non-literates of the world. Australia is the only exception where bow and arrow did not reach at all.

Spear: It is an arrow like weapon with relatively larger size. It can also be thrown by hand towards the game or enemy. The heavy spears with a long handle (shaft), thrusted by bare hand are called 'lance'. The shorter and lighter spears operated as missiles are called 'javelin'. Besides, spears may be of different types basing on structure and building material. It is a favourite weapon of the primitives. The Mishmis make their spears with palmyra wood. Naga spear shows numerous complicated parts.

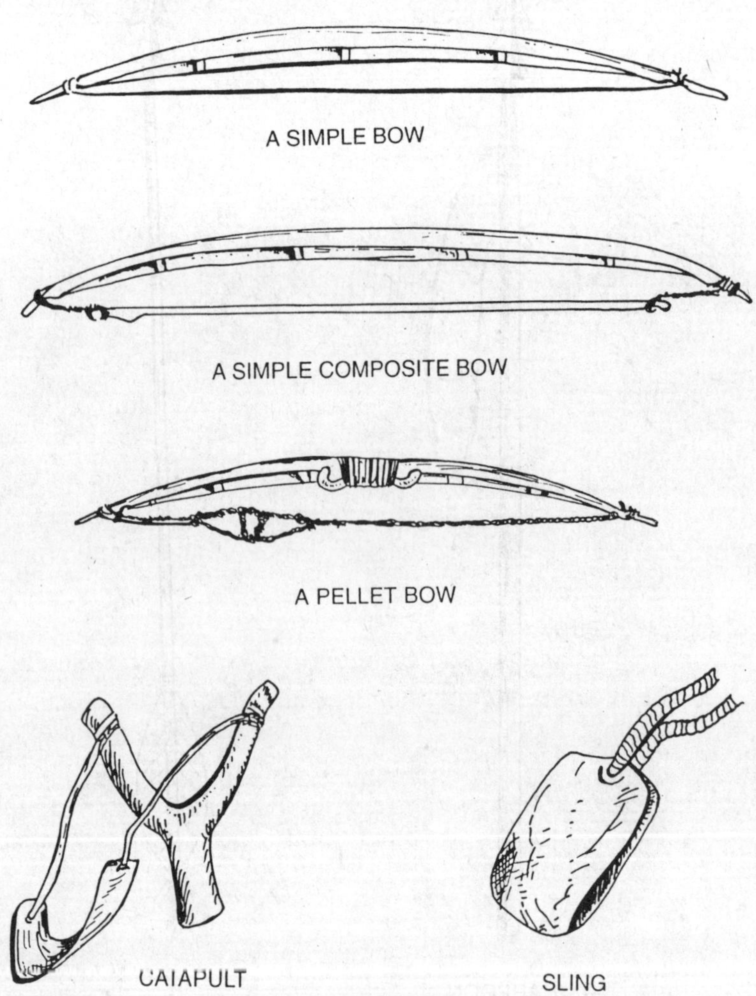

Fig. 14.10. Throwing Devices

Harpoon: It is a modified spear with detachable head. A string keeps the head and shaft connected in such a manner that the head being detached from the shaft is not lost. A harpoon often shows too many barbs, which pierce, into the body of the games to make them absolutely injured. Number of the barbs varies. The Hos frequently use harpoons for hunting.

Hunting by Trap

In hunting, traps and snares were devised to stop the movement of the chased game in order to catch them easily. The traps may be classified into two categories—Manipulative trap and Automatic trap. When the hunter presents himself in the hunting spot for the manipulation of the trap, it is called 'manipulative trap'. But the 'automatic trap' does not need further manipulation of the hunter, after setting the trap.

The most primitive method of trapping is to drive the herds of animals in a dangerous place. Solutrean people used to drive wild horses over steep cliffs. A few of them certainly slipped from the cliffs and the hunters rushed to catch them, dead or alive. In Egypt and East Africa some obstructions (strings) are set up to reduce the speed of the chased games. The Sakais of Sumatra follow the same technique. Another primitive method is to allure the animals with bait so that they enter into a pound.

The Angami Nagas are fond of this method. The same procedure is applied for bird catching in India as well as in Polynesia, Indonesia and East Africa. The third major method of manipulated trapping is sneaking near the places where the particular animals live. A Bushman in Africa goes in the disguise of an ostrich and holds a mutilated head of the bird above himself. In this position he mimics the behaviour of the ostrich until the other ostriches come within his arrow-shot. With the same intention, a Californian Indian puts on a deerskin and antlers, and pretends to graze in the field. The other animals feel free to graze closely. So it becomes easy to shoot them with an arrow.

Several kinds of automatic traps have been distinguished:

(i) Pit-fall trap

A deep pit is dug out on the regular path of the animals and kept covered with leaves. Unscrupulous animals fall into this pit while passing, but they can not escape. Sometimes sharp spikes are also placed vertically into the pit, which injures the animal badly. The Kadars capture the elephants and the Nagas hunt monkeys by this device. The device is also practised by the tribes like Santals, Gonds, etc.

(ii) Dead-fall trap

It is a highly preferred device among the primitives. It consists of a stockade on either side with a crosspiece at the entrance. Bait is kept tied to a stick and supported by the heavy beam overhead. When the bait is pulled, the crosspiece gets loose and the beam falls upon the animal entered in the stockade. Santals, Gonds, Oraons, Baigas often use this trap to catch the small games. Sometimes animals like leopard and tiger are also trapped in this technique.

(iii) Noose-trap

This device is followed by the Maria Gonds in catching hares. A small oval shaped piece of land is selected which is fenced by planting long sticks into the ground. The noose is kept attached to a trigger and set at the entrance of the fencing. When a hare enters into the fence being attracted by the coloured vegetables placed inside, it dislodges the trigger and the noose gets tight. As a result the hare is caught.

(iv) Wheel trap

It is used to capture deer or other fast moving animals. A wheel with big spokes is kept in the way of the animals so that their legs can be stuck in the rotating wheel by slightest touch. A deer can not pull out its leg. In Africa this trap is used in catching the antelopes.

(v) Torsion trap

In this trap spring is attached by torsion. The korkus use it extensively for catching the birds.

(vi) Transfixing trap

This trap uses an arrow or bolt which injures the game when the string of the operating bow is slightly disturbed. The whole thing is set in such a way that it gets disturbed only when the game reaches at the aiming point. Tribal people all over the world are acquainted with this type of trap.

(vii) Cage trap

The trap is actually a cage where the animals enter in search of food. Therefore, baits have to be used in these traps. Three varieties in the cage-trap have been identified.

(*a*) Door trap : It is the cage-trap where the door drops down. The Ao-Nagas and the Lhota Nagas use this trap to imprison the ferocious animals like tigers and leopards. A goat is kept as bait in the large box-shaped trap. When the animal enters into it and pulls the bait, the trigger is automatically released to drop the door. Obviously the animal can not escape.

(*b*) Valve trap : It is a hemi-spherical basket made of bamboo strips. There is a valve, which opens only inwards. The Korkus use this to catch the birds and small games. Food-grains are scattered inside for which the games enter into the basket but fail to come out.

(*c*) Valveless trap : This cage-trap has a thorn-lined door, instead of a valve. As the thorns are directed inward, games can enter into it easily but can never come out. The Baigas show an extensive use of this trap.

Hunting by tame animals

Hunting by tamed animals is a very old practice. From sixth or seventh century AD upto the time of French Revolution, it was a very popular method. The method is still widely used among the primitives of Asia and Europe. In Rajasthan, leopards are tamed to hunt deer. In Central Asia, hooded Falcons and Hawks are trained for hunting. The tribes like Santal, Oraon, Munda, Gond and Khasi highly depend on dogs that sincerely help in hunting of jackel, rabbit, mice, etc. In Malay and Indonesia, deer are hunt by riding on tame horse. The Onas of Tierra del Fuego use dogs to scent out the guanaco, a deer-sized camel. Australian aborigines keep the dogs as scavengers to live in clean surroundings.

Hunting by poisoning

Sometimes poison is used as a substitute of weapon. Hunting by poisoning is especially found in the forest areas of Asia, Africa and America. As long-distance shooting is difficult in the forest, the tribal people for the success of hunting smear poison at the tip of the spear, harpoon or arrow. As a result, a slight scratch on the body of the game makes the death inevitable.

3. *Fishing*

Fish was a supplementary food for the prehistoric men. Due to certain reasons, once the exploitation of terrestrial resources became a problem of the people, therefore, they directed their attention towards aquatic creatures, especially the fish. Evidence of fishing has been obtained from the upper Paleolithic period. Mesolithic people who used to dwell on seacoasts and riverbanks were the extensive consumers of fish.

Fishing was possible only in those areas where fish were abundant. Shellfish were also picked off the rocks at low tide. A fishing economy was often supported either by simple gathering or active hunting or both. Among the contemporary primitives, fishing may be a full time specialization but more often it is a seasonal, part-time or occasional pursuit. For example, the Kwakiutl, Nootka, Tsimshian and some other tribes in west coast of Canada belong to the pre-industrial maritime economics founded upon sea-mammal hunting and seasonal capture of Salmon and Candle- fish. On the other hand, Nuers of southern Sudan spear fish during a brief season in the year when the river Nile is flooded. This is purely an occasional opportunity and the livelihood of Nuers does not at all depend on fishing.

Fish - catching was perhaps started by bare hand. Still the tribal people endeavour to catch fish by hand in shallow level water as well as in soft muddy places. But since the fishes slip away from hand, people thought of various other devices. In fact, the methods of fishing vary from place to place in the world. These can be classified into the following types.

(*i*) Fishing by weapons.
(*ii*) Fishing by traps.
(*iii*) Fishing by nets.
(*iv*) Fishing by tame animals.
(*v*) Fishing by poisoning.
(*vi*) Fishing by screening.

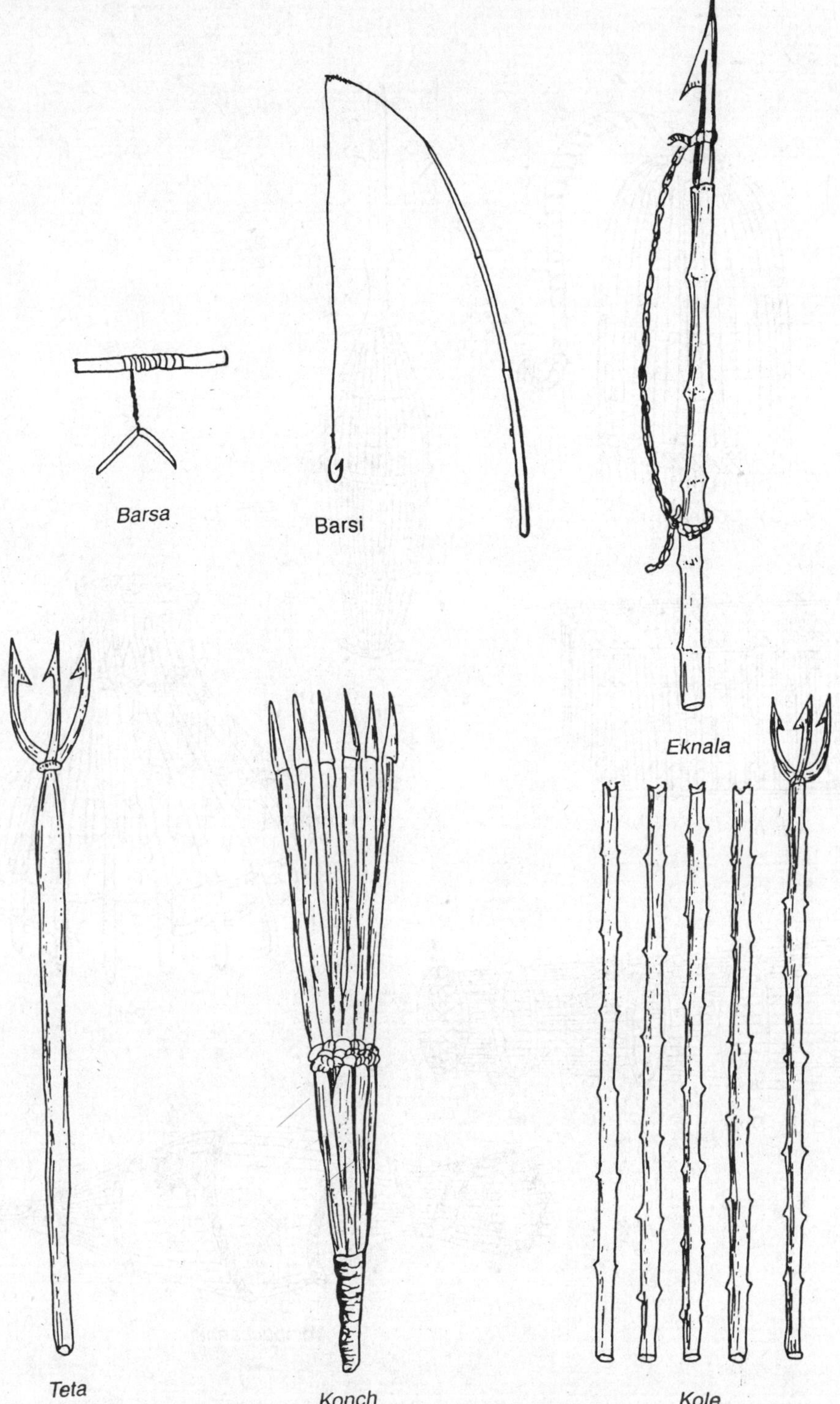

Fig. 14.11. Fishing Implements

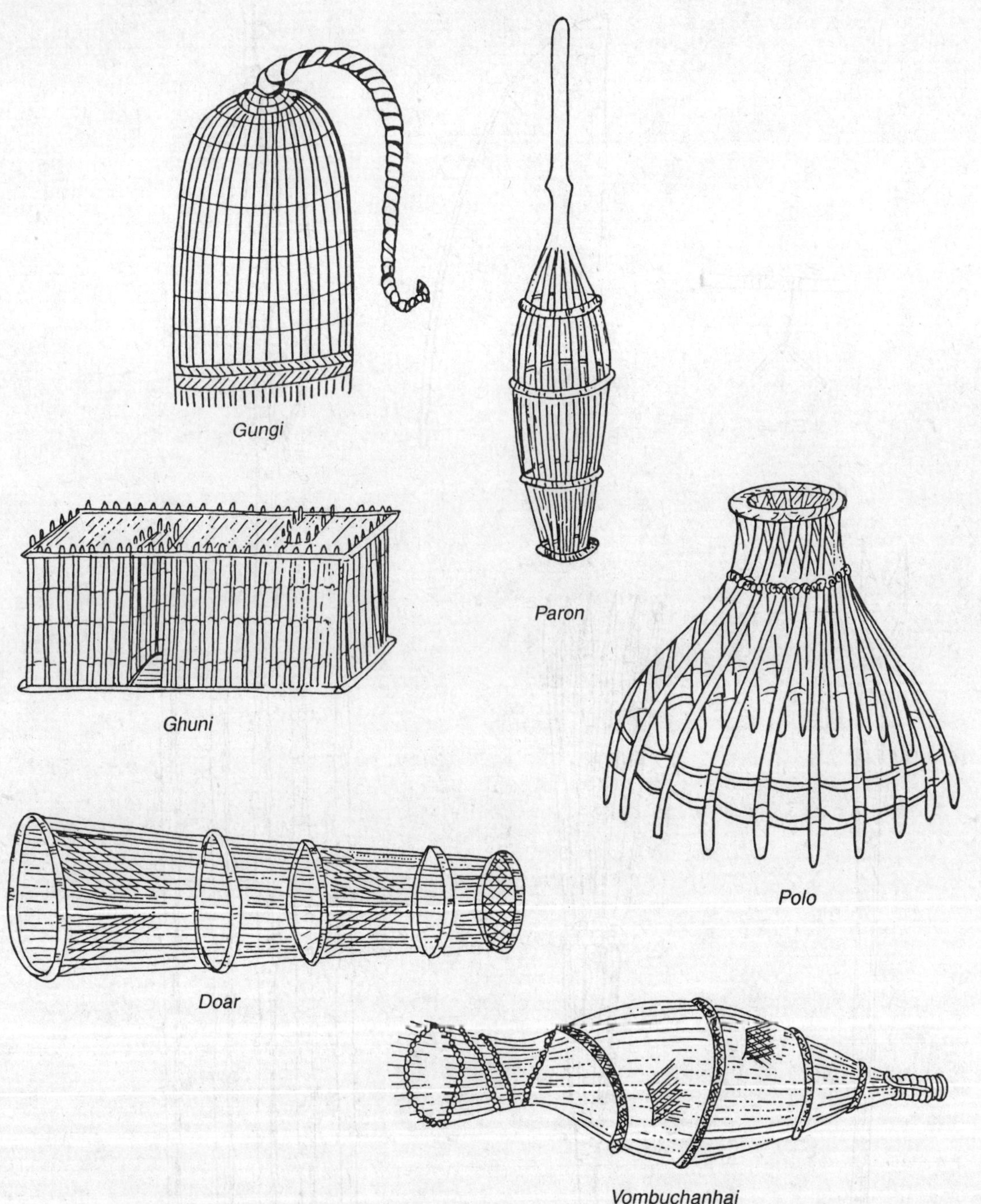

Fig. 14.12. Types of Fishing Trap

Fishing by weapons

Weapons like spear, harpoon, arrow etc., are used in fishing. Such type of fishing is specially suitable in the rivers or streams with very clear water. Arrows are frequently used by the Andamanese of Andaman Islands, Bushmen of South Africa, Santals of Mayurbhanj and Oraons of Ranchi. Use of spear is popular in Bengal. Bunas and Bedias are quite at hand in throwing spears. Harpoons with different types of head (monodent, trident, multiple dent etc.) are very favourite among the tribal people throughout the world.

Fishing is also done by means of hooks and fish-gorges. The technique is to allure a fish to swallow the curved needle to which bait is attached. Tiny creatures like earthworm, grasshopper etc., are provided as bait. As soon as a fish swallows the bait in the hook, it gets entangled and disbalanced. Therefore, it can be turned out easily from water. In Bengal hooks are widely used in catching fish. Spring - gorges are often found to be used in China, Mexico and Africa.

Fishing by trap

Traps came in use as people showed more interest in fishing. The simplest one is polo, the plunge-basket trap of Bengal. Oraon of Chotonagpur and Santals of Santal Parganas use similar kinds of trap. However, traps may be divided into two types: Manipulative and Automatic.

(*i*) *Manipulative trap* : These traps are solely hand-operated, generally cage-like or basket-like in appearance. *Polo; chopa, Muchu,* belongs to this group of traps. These are handy and quite useful in catching small fishes.

(*ii*) Automatic trap : These traps do not require any manipulation by hand. They have to be kept along the currents of water. Some of them are with valves but others go without a valve. Valves are nothing but a trap-door through which fishes can get into the trap easily but cannot come out. Valved traps may show single or double valves, either set in the same direction or in opposite directions. *Baichna, Chaora, Hocha* etc are the traps with single valve. Double valved traps include *Doar, Vombuchanhai, Ghuni,* etc., But *Gungi, Toradang, Kaichu, Ultakukri,* etc., show no valve at all. Besides, an intermediate type has been identified between the valve-less plunge-basket and valved trap. This is called thorn-lined trap because thorns are found to be arranged in the mouth of this trap, which compels fishes to enter into the trap through its constricted neck. *Paron* is such an intermediate trap, used largely in Bangladesh.

Fishing by nets

Appearance of fishing net can be correlated with the knowledge of fabrics. Large-scale fishing has been possible for the net. Nets can be grouped into two broad sections basing on the nature of water-source.

(*a*) Some nets are used in stagnant water like ponds, lakes, bogs and marshy lands. Different types that have been noted are as follows :

(*i*) Hand operated nets: Such nets are conical in shape. The round edge generally has a bamboo-frame. In shallow-level water these nets are very useful to catch small fishes. In Bengal such nets are called as *Chhaknijal, Chant, Chatuni, Ghatgati,* etc., according to the regions. Oraons use this net with the name 'pilni'.

(*ii*) Dip net: It is a variety of hand-operated nets. Three ends of this net are tied with a bamboo pole in such a manner that an acute angle is formed and the net gets a triangular shape. Only one side remains open. Fishermen of Bengal call it as '*vesal jal*'. They dip it in shallow-level water and push forward, keeping the open end at the front. When the net is taken out of water, a large number of fish are caught.

(*iii*) Cast net: It is also a kind of hand-operated net. The speciality lies with its sinkers, which are set along the edges. When the net is cast, it spreads in a circle on water. Such nets are very popular in Bengal with the name *Khepla jal* or *'Jhanki jal.*

(*b*) Some nets are used in running water like stream, river, sea, etc., Several types of these nets are as follows:

(*i*) Seine net: This is a long net provided with wooden floats at the upper edges and sinkers at the lower edges. It is mainly used by the professional fishermen of Bengal and Orissa who call it as *'Bera jal.* The net is usually operated from a moving boat.

(*ii*) Trawl net: This net is typically used for catching the hilsa fish. It is a bag like structure fitted with a bamboo frame. There is a string arrangement with which the mouth of the bag can be closed. A boat is essentially required to draw the net. In Bengal, it is known as *'Shanghla jal'*.

(*iii*) Automatic net: It is a long narrow net with floats at the upper edges. It is kept just stretched across a current or a stream. Fishes swayed by water are entangled in this net. Such nets are often used by the people of Bengal to catch 'Koi' fish. During monsoon they spread the net in water logged paddy fields where plenty of Koi-fish enter.

Fishing by tame animals

Fishing with the help of tame animals is a rare incident. Indian tribes especially tribes of Bengal sometimes train up Otter, a type of aquatic carnivores for driving fish towards the net (dip net) in ponds and canals. In Indonesia and China, Pelicans are trained for fishing in deep-sea water. Cormorant birds are also employed for this purpose.

Fishing by poison

Tribals sometimes utilize poisonous and narcotic plants in fishing. They extract juices from those plants and pour into the water of the stream. They also construct some temporary embankments prior to this operation. This helps them to collect the dead and sense-less fishes. Later on, they release the poisoned water by cutting away the embankments. The Andamanese, the Kukis, the Garo, the Khasis exhibit this method of fishing. The Oraons use 'maona'* for this purpose.

Fishing by screening

The method of screening can yield good results only in the tidal regions. It is a method which does not require confining of water. Only some fences of twigs or reeds are needed to raise closely, across stream or a river. Fishes pass between the reeds or twigs easily during high tide. But they can not escape at the time of low tide. Such fishing is found in some limited areas of Bengal and Orissa.

Since world does not display a uniform physical configuration, the varied climatic condition and vegetation compel the people in choosing different modes of livelihood. The reflection is found in primitive economies that present a great diversity in tools and techniques. Accordingly, we have identified the Bushmen and Eskimo as hunting people; the Nootkas and the Kwakiutl of British Columbia as fishing people. Again, the hunters of Kalahari Desert differ vastly from the hunters of the Arctic coast of North America. For example, the forest and grasslands of Africa offer numerous large herbivores and carnivores throughout the year, whereas the Eskimos of Arctic America starve in the absence of an aquatic animal, seal, which is available only in few months of a year.\ But,in general, excepting a few exceptions, we never find a single or exclusive economy . Collecting, hunting and fishing, none of these elements of simple economy can exist in 'pure' form with any community. Rather, they remain combined in various degrees among different groups of people and often supplement each other.

II. Food-production

Food production is a revolutionary changeover in the history of mankind. People remained no longer dependent on nature; they acquired a great control over certain natural processes by learning domestication of plants and animals. Although the Mesolithic people had shown an aptitude for domestication, but the Neolithic people actually learnt the art. Food-collection was replaced by

* A kind of fruit.

food-production as a predominant mode of subsistence. However, food-production has been classified into two major divisions as horticulture and agriculture. Pastoralism as a transitional phase is identified between the stages of food-collection and food production.

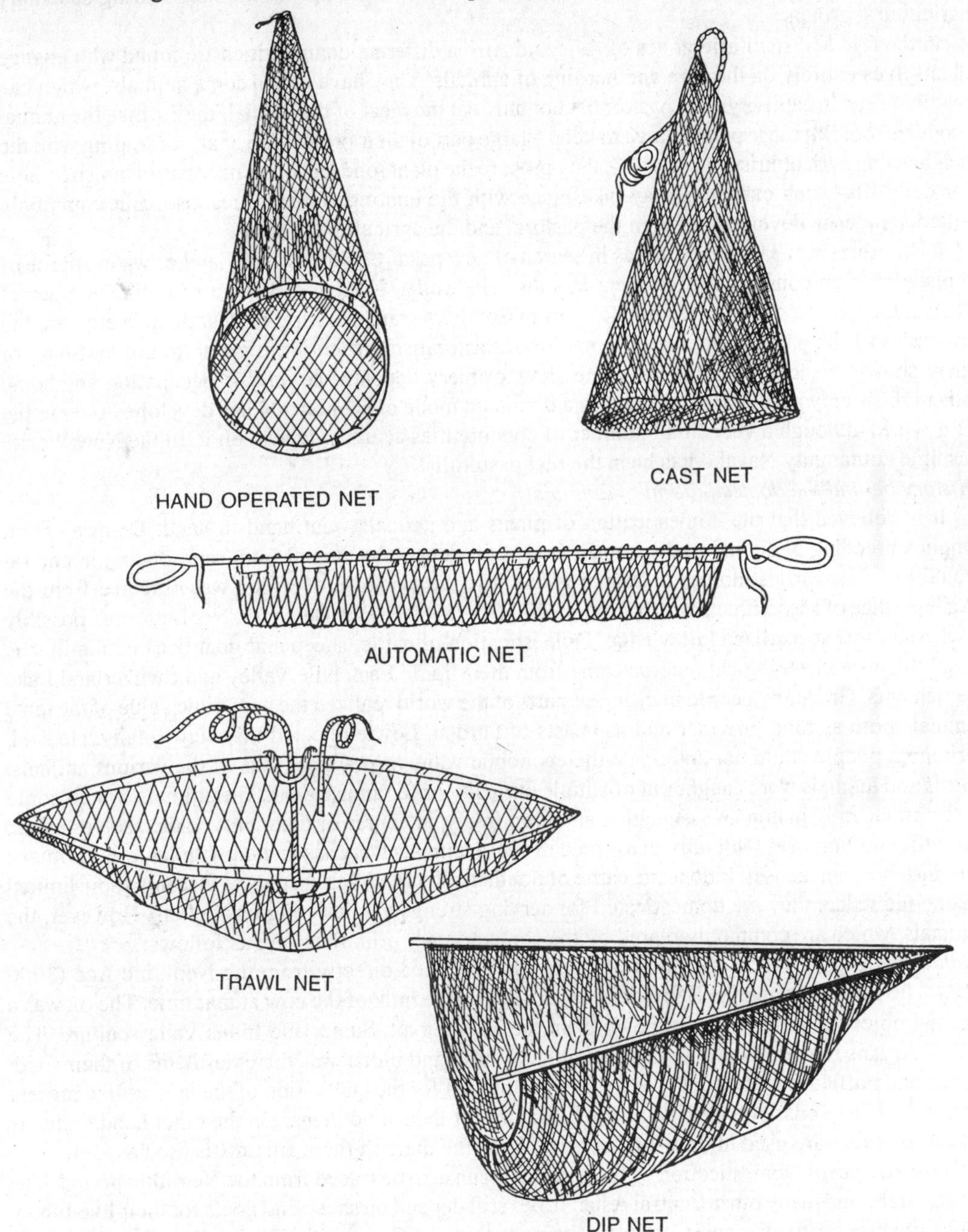

Fig. 14.13. Types of Fishing Net

(i) *Pastoralism*

Pastoralism is a subsistence technology, which is higher than Food-collection and associated with the domestication of animals. It is a type of ecological adaptation that is found in the geographically marginal areas where natural resources permit neither the hunting-fishing nor the agriculture. As a consequence, the inhabitants adapt livestock breeding and lead a nomadic life. But most of these people are bound to maintain a barter relationship with the neighboring sedentary agricultural groups.

In the arid and semi-arid zones of Asia and Africa different communities are found who engage themselves entirely on the care and herding of animals. They have tamed some animals, which can readily breed in captivity. The pastoralist not only eat the meat of the animal, they utilize the animal products too. But these people have to keep a large part of their production apart for trading with the neighboring agriculturists. In this way, they procure the plant food and other necessities indispensable for daily life. Such exchange may take place with the hunting communities also. But commonly interdependence develops between the pastoral and the agricultural groups.

Pastoralists move with their herds in search of new pasture as the grazing land or water source of a place is often consumed after some months. The Toda., Kota and Badaga of India,the Nuer of Africa, the Omaha of Nebrasea, the Basseri of Southern Iran, the Lapp of Scandinavia etc., are the examples of the pastoral people. The nature of Pastoralism differs from group to group. Some of them show advanced Pastoralism, some show primary dependence on this occupation and some others show only partial dependence. As a dominant mode of economy, it has developed only in the Old World although a very small number of communities actually depend on it. In the New World, a single community Navaho has been the real pastoralist.

History behind the domestication of animals

It is believed that the domestication of plants and animals went hand in hand. Domestication implies breeding of animal under human control; it debars the animal from reproducing in natural condition. The earliest domestic animal is the dog, the evidence of which was obtained from the Azilian stage of Mesolithic period (around 9000 BC - 8000 BC) in Denmark. Neolithic men possibly picked up this specialized knowledge. Domestication of cattle, sheep and goat became familiar in Neolithic as archaeological records came from the Middle East, Nile Valley and Switzerland Lake settlements. Gradually people in different parts of the world realized the economic value of the tame animals both as food provider and as beasts of burden. Lowie rejected this view totally. He said, primitive people could not anticipate the economic value or practical uses of the various animals. Birds and animals were caught out of simple curiosity or for amusement. The animals, which could live permanently in human association and bred freely, in course of time, proved themselves useful for different purposes. Not only that, some of those animals were associated with important rituals. For instance, chickens of Indonesia, cattle of Southern Asia and Africa, pigs of Oceania show limited economic value; they are domesticated for serving strange magico-religious beliefs. However, the animals, which are commonly reared by the contemporary primitives, are as follows:

Cattle and Buffalo: Domestication of cattle can be traced directly from the Neolithic Age (3000 BC). The Egyptians and the Babylonians were fed on the milk of the cow at that time. The ox was a special object of veneration as known from the early Egypt, Sumer and Indus Valley culture. The Vedic Aryans regarded cow, as sacred animal as milk and butter was the chief items of their food. Cows and buffaloes were employed to draw the plough for the cultivation of the land as like present day India. The Todas still depend on the buffaloes for their food items. On the other hand, in many places bullocks are used in drawing the carts. Even the dung of these animals is used as fuel.

Goats and Sheep: Domestication of these animals can also be traced from the Neolithic period. The Jews, Arabs and many other Central Asian tribes still depend on sheep and goats for their livelihood. They procure flesh, milk and wool from their pet animals. Some of the Pastoralists regard animal's blood as a source of protein, so they do not hesitate to drink the blood of their animals. Either it is taken directly or with some food.

Pig: Recent archaeological discoveries suggest that pigs have been domesticated first in India and China. Domestication of pig can again be traced back to the Neolithic Age. Pigs are the sources of flesh and milk. They can also be harnessed to the plough or may be used as a carrier. In some places of the world, pig as a religious offering elevates the pride and prestige of the performer. Australian white pigs are now cultivated all over the world. Europe imports pigs from the east.

Donkey, Horse and Camel: These three animals are regarded as transport animals because they are mostly used as carriers throughout the world. Donkey's use is the oldest of all. Possibly it came in use earlier than 3000 BC in Egypt. In Mediterranean region, donkey is still a popular carrier. Horse and camel used to be domesticated in Babylon, about 2300 BC and 1000 BC respectively. Cavalries were very important to the Greeks, Romans, Turks and Mongols as the speed of the horse helped them a lot to win in the conquests. In England and America, even today, the horses are employed in pulling ploughs and common carts. In modern Arabia and Mongolia, horses are extensively used in riding, especially for travelling long distances.

The desert dwellers can not live without camels. This animal was first domesticated by the desert tribes of Arabia, Egypt and India. The entire social-cultural life in the desert revolves round the camel. It provides them with food (flesh) and drink (milk). Its hair and hide are also precious for making rags and garments. Even the dung serves as a beautiful fuel. The two-humped camels have secured a special position in the economy of the Pastoralists of Central Asia. This special variety has proved its usefulness in the agricultural field, in tilling the soil with plough.

Reindeer: Domestication of reindeer is a relatively late Phenomenon, possibly started around 500 AD. Although the places namely North America, Alaska, West of Bering Strait etc., are the home of this animal, but the Eskimos or the American Indians had never domesticated it. Reindeer has been primarily domesticated by the Tungus, the Chukchis and the Lapps. The Tungus ride on them. The Chukchis not only harness the animal to their plough; they also eat its flesh and use them in dragging the sledges. The Lapps utilize the milk of reindeer to prepare cheese, butter, etc.; the flesh is also taken by them. Horns are often used in the manufacture of combs and various other articles of daily use.

Dog: Dog is a sincere animal. It remains very loyal to its master and his property. Therefore, all the people, primitive or advanced like the dog. Its flesh is a delicious food to some tribal people. The Iroquois, the Peruvians, the Polynesians, the Angami Nagas hunt dogs for meat. The Santal, Munda, Oraon and some other tribes of India highly rely upon the dogs in hunting expeditions. The Veddas of Ceylon hunt deer with the help of tame dogs. The Onas chase guanacos, and Hottentots chase the antelopes by dog.The Lapps use dogs to save their herds from the attack of wolves at night. Besides, dogs draw the sledges in Arctic regions. In Germany (Humsburg),carts full of vegetables are often pulled by the dogs to the market. In New Zealand, dog's hair is utilized in making the cloaks.

Bird and Fowl: Domestication of birds and fowls is an ancient practice. Some wild varieties are still domesticated in Assam, Chittagong and Burma. The aboriginal people of Assam and Chotonagpur use fowls not only for food, but also for the purpose of divination and cockfight. In recent time domestication of poultry birds like hen, duck, dove, pigeon, etc., has been popular among the tribals to obtain the flesh and eggs.

(ii) Horticulture

Horticulture is a shallow level cultivation by the application of simple tools and techniques. It involves minimal human muscle power. Use of plough has been denied here. Moreover, neither it approves any improved fertilizer, nor it sanctions the irrigation. Extended land area is not required for this type of cultivation. The principal tools for horticultural are digging stick, hoe, pick and spade. Small portion of land is worked at a time with the help of these hand-operated tools. The digging stick is the most primitive implement which first showed its use in food gathering. A large number of tribal communities of present day still cling to this tool. Extensive use of digging stick has been noted among the Malpaharias of Santal Parganas and the Garos of Assam. It is also prevalent in the tribal belt of Africa, America and Australia.

The oldest digging stick was nothing but a more or less straight branch of a tree, pointed at one end. Its purpose was limited in making of holes in the soil by the pressure of hand. Sometimes the end used to be hardened by fire. An improved digging stick was found later among the Bushmen of

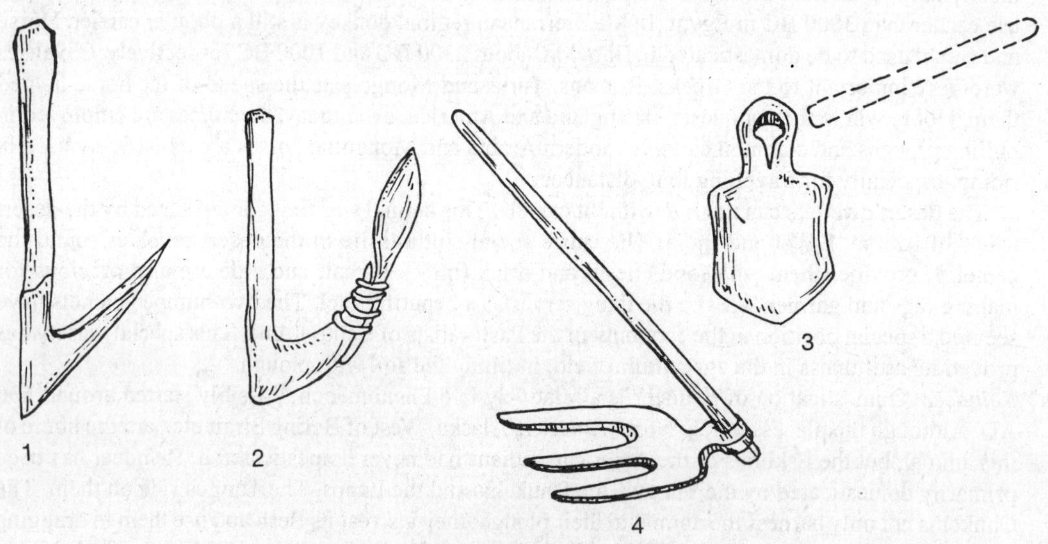

1. Simple hoe with pointed wooden blade (User : Red Kaffir); **2.** Hoe with narrow iron blade (User : Khasi); **3.** Hoe with broad iron blade (User : Oraon); **4.** Hoe with forked blade (user : Lepcha)

Fig. 14.14. Forms of Hoe

South Africa where a horn-sheath was attached to the working end to strengthen the tip and a ring-stone was put at the butt-end to enhance the weight of the thrust. In the same manner, Kharias of Chotonagpur put an iron point at the working end to increase the efficacy of the tool. The Bunas of Bengal use a broad as well as a flat iron-blade at the working end. This is a variety of digging stick, locally known as 'Khonta'. A larger area of soil is cut by it in each push.

The digging stick is believed to have evolved into spade and pick*, both of which were farther developed into a hoe or a plough. In horticulture, a digging stick helps in loosening the soil, a pick is used for breaking up the soil, a spade facilitates in weeding and scraping, and a hoe turns up the soil. All of these implements may or may not be used at a time in a horticultural community. The condition of the soil and topography initiate various modifications on a digging stick. In certain historical phase, after the widening of the working point into a cutting blade, a handle was probably added to it in the position of an acute angle. This gave rise to a hoe out of a simple digging stick. Spade and picks are the intermediary forms.

Horticulture may produce long-growing tree-crops but it never involves permanent cultivation of principal crops. Primitive horticulturists usually led a more sedentary life than the food-collectors. Because they were able to yield more food in a given area unlike that of the collectors, which could support a larger population or densely populated areas. But none of the horticultural societies at present can rely upon their crops for a complete livelihood. They practice hunting and fishing along it. Horticulturist groups may give rise permanent villages and exhibit social differentiation in rudimentary form. Part-time craftsmen and part-time political officials may not be unusual in these societies.

* A digging stick that forms an acute angle with the haft.

Apart from the horticulture, growing of crops with the help of hoe is found in two other types of cultivation—the *Shifting cultivation* and the *Terrace cultivation.*

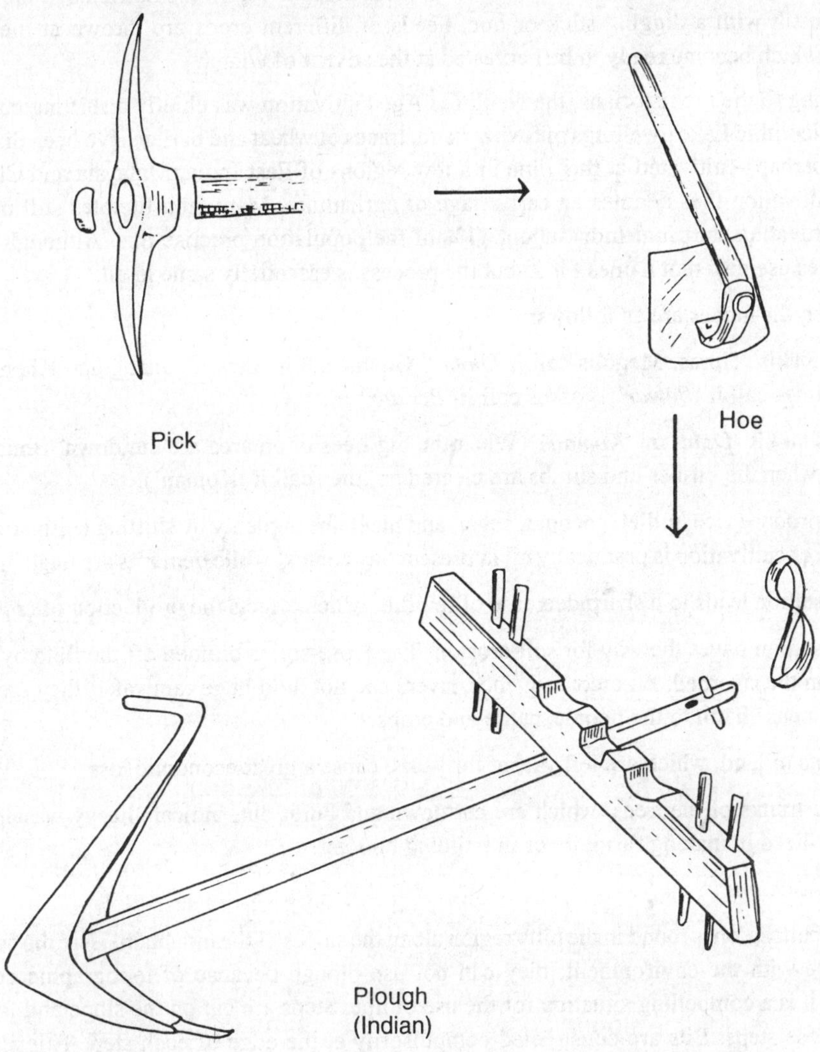

Fig. 14.15. Implements Suggesting a Course of Technological Advancement

Shifting Cultivation

Shifting cultivation is practised especially in tropical and sub-tropical zones. It is a type of horticulture in the sense that here also the fields are cultivated extensively but without the use of a plough, manure or any other technique of irrigation. A plot of land is worked for three or four consecutive years and thereafter the cultivation shifts to some other plot of land. The old plot, which has been abandoned, is left vacant for a fairly long period (about 10-12 years) so that the soil can regain its fertility. However, during this period, the trees, bushes and wild plants are allowed to grow haphazardly. At last, a special technique called *'slash and burn'* technique is applied for clearing up the whole field.

The slash and burn technique has some distinct phases. At first, especially in winter, all the trees and outgrowths are cut down from the base. Those are left in the same spot for several days for drying up under the sun. Finally the dried leaves and branches are set on fire. This continues until ashes are formed. It is believed that the whole process helps to return the nutrients in the soil. The soil is then tilt with a digging stick or hoe. Seeds of different crops are thrown at the onset of monsoon, which become ready to be harvested at the advent of winter.

According to the prehistorians, the Neolithic Age cultivation was chiefly a shifting cultivation. From the Neolithic Lake dwellings of Switzerland, traces of wheat and barley have been discovered. Rice was perhaps cultivated at that time in a few regions of East India, Malaysia and China. The shifting cultivation thus denotes an early stage of agriculture. Many non-literates still prefer this type of cultivation. In tribal India, about 11% of the population practise this. Although different communities use different names for it, but the process is essentially same in all.

However, the names are as follows:

Nagas, Kukis, Tipras, Meghlis call it *'Jhum'*. Gonds call it '*Dahia*'. Juang and Khonds call it *'Podu'*. Baigas call it '*Bewar*'. Korkus call it '*Bendar*'.

Bhuiyas call it *'Dahi'* or *'Koman'*. (When the big trees of an area are cut down, Bhuiya call it 'Dahi' but when the bushes and shrubs are cleared up, they call it 'Koman').

Tribals produce rice, millet, coconut, sugar, and plantains in plenty by shifting cultivation. Merit of this type of cultivation is practically nil in present day context while demerits are huge in number:

(a) Deforestation leads to a sharp decrease of rainfall, which affects the production of crop.

(b) Deforestation paves the way for soil-erosion. The loose-soil is drained off the field by rain and deposited in the riverbed. As a consequence, rivers can not hold huge rainwater; they are flooded causing immense harm to the people, cattle and crops.

(c) The plots of land, which are left vacant for years, cause a great economic loss.

(d) The big trunks of the trees, which are cut down and burnt out, indicate heavy wastage; those could be utilized in making furniture or in building houses.

Terrace cultivation:

Terrace cultivation is found in the hilly region along the slopes of the mountain. Here the cultivators compromise with the environment; they can not use plough because of topographic condition. Therefore, it is a compelling situation for the use of hoe. Steps are cut on the slope and farming is done on those steps. Bits are constructed compulsorily at the edge of each step. This allows the rainwater to stay in the steps since water is urgently needed for cultivation. In case of stiff slopes, although the number of steps increase, but the depth of the steps decrease. The angle of the hoe needs to be very acute, as the soil is generally rocky as well as hard. However, the terrace cultivation is very popular in the places like Assam, Meghalaya, Darjeeling, Nainital, Arunachal Pradesh, Nilgiri hills etc., The tribal people especially Nagas, Kukis, Garos, Khasis, Juangs etc., are in favour of this cultivation.

(iii) Agriculture

It is an intensive as well as a permanent way of cultivation by the help of plough. It also employs different higher techniques like the use of fertilizer and irrigation. Though the Neolithic people were able to produce useful crops like barley, wheat, maize, but full-fledged agriculture started later with the production of various kinds of seeds, roots or shoots which could be deliberately planted or stored for the next season, for the new crop. The technical instruments gradually improved with the improved process of cultivation. The plough showed more efficacy in turning up the soil. The discovery of plough can be equated with the domestication of large animals because all ploughs were universally designed to be drawn by the animals. Continuous furrows on land could be created by this implement.

Two main types of plough have been have identified at present.

(*a*) The simple curved plough found in different parts of India.

(*b*) The quadrilateral plough found in Malay, Indo-china, southern China and other places of Southeast Asia.

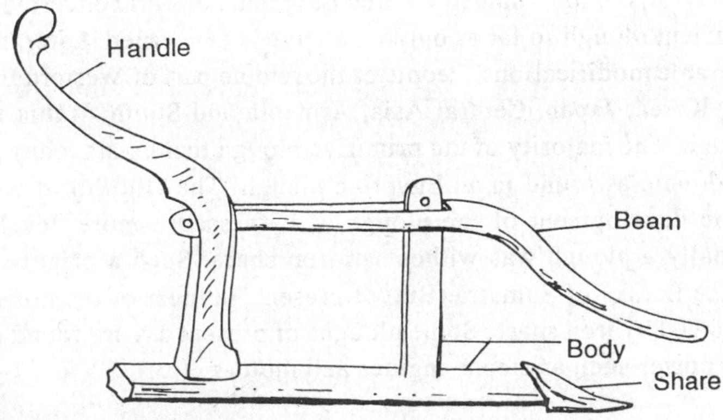

Fig. 14.16. Quadrilateral Plough

A Krumel variety has also been recognized in addition which are nothing but a combination of Indian plough and the quadrilateral plough. The simple curved Indian plough is widely used in Bengal and it is also very popular among the Oraons of Chotonagpur. It is considered to have developed from the hoe, whereas the quadrilateral plough is supposed to have developed out of spade. The third type, being an intermediate type does not confirm its origin. According to some scholars, it is an evolved form of hoe; others suggest a derivation from the spade.

In tracing the ancestry of Indian plough, eminent anthropologist E.B.Tylor found an evolution. He suggested that the haft of the hoe was turned into a beam, the blade was converted to a ploughshare and the main body projected upward to form the body and handle of the plough. The areas, which showed little or no use of hoe, the spade came forward to give rise to a plough. The quadrilateral plough seems to have developed from a spade. The blade of the spade was converted into a ploughshare. The haft turned to form the body and beam of the plough.

The agriculture of the Old World can be distinguished from that of the America and Oceania for its specialized use of plough and hoe. Light implements falling in the category of digging stick are used in America and Oceania. However, the earliest use of hoe was found among the Egyptians in the beginning of Dynastic period (3400 BC). The iron hoe

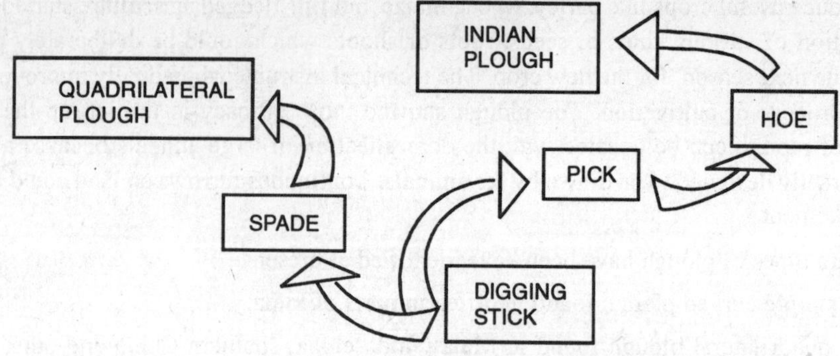

Fig. 14.17. Phases of Technological Advancement From Digging Stick to Plough

possibly developed around 1400 BC as iron working in the Ancient East started at that time. According to Tylor, the long-bladed pointed Egyptian hoe was converted into a plough. That is the most ancient plough so far as our knowledge is concerned. Later, the plough had undergone considerable modifications. People of the remote part of Western Europe, Tribals of Malabar Coast, Korea, Japan, Central Asia, Armenia and South Arabia now know the application of plough. The majority of the primitive plough that exists today, show a single handle instead of double as found in an Egyptian plough. The Old World was not only an early center for the development of the plough, it witnessed a more developed form of agriculture. Originally a plough was without an iron share. Such a primitive type is still surviving among the Bataks of Sumatra. But, at present, in most of the areas, the wooden share has been replaced by iron share. Some ploughs of present day are found to be operated by the mechanical power such as steam engines and motors etc.

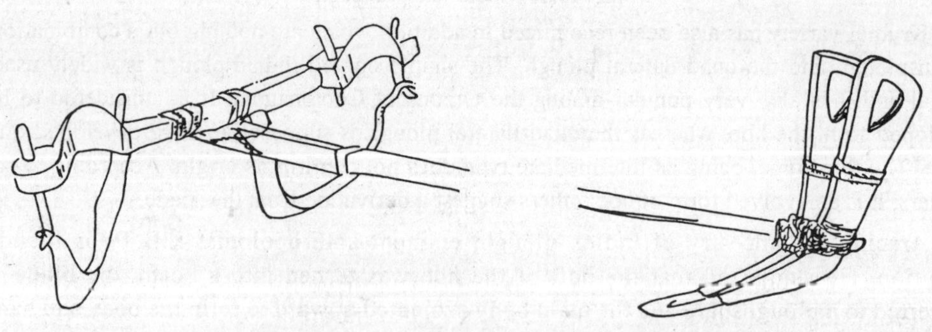

Fig. 14.18. Egyptian Two Handled Plough (An Ancient model)

The breadth and thickness of the ploughshare may differ from place to place basing on the nature of soil. In hilly areas, it gets thick and stout to combat against the hard and rocky soil. In plain, especially in alluvial soil, thin and light blades are quite suitable. However, the operation of agriculture is not limited to the use of plough, a series of specialized implements are involved in it. After tilling of soil, a leveler is needed for levelling the field. It is usually a flat rectangular wooden plank, which is yoked to a pair of animals. At the time of dragging, the ploughman stands on it for putting enough pressure in smashing the clods. Next to this is sowing. The farmer sows the seeds by hand. Sometimes, especially in a large-scale cultivation, a funnel like implement called seed-funnel or seed-drill is used. A few months pass between the planting of seeds and the growing of plants. Irrigation and application of fertilizer are urgently required at this time. A spade may be used for weeding. Finally, for the reaping and harvesting of the seeds, a sickle or scythe is employed. Millet was the first cultivated grain in the Old World. In the New World, its position has been taken by maize.

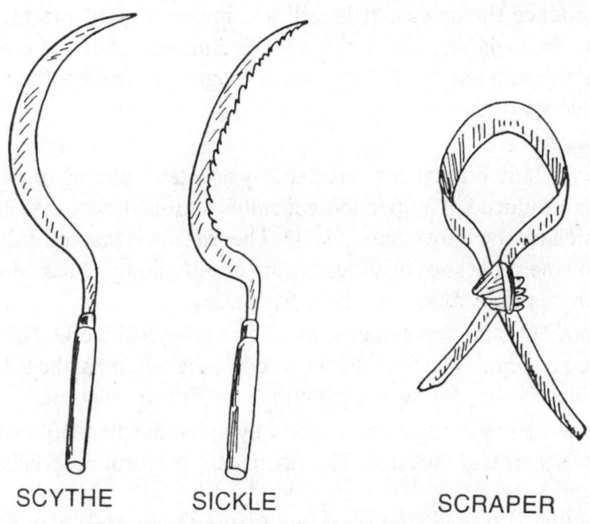

Fig. 14.19. A Few Other Agricultural Implements

The communities practising agriculture show a high degree of craft specialization, complex political organization and large differences in wealth and power. These people are more equipped in food-storage as agriculture is more productive than horticulture. In the history of culture, we find the contribution of agriculture—the towns and cities grew with intensive agriculture. It is also interesting to note that the earliest domestication of animals started more or less at the same time as the cultivation of plants (cereals). The principle underlying these two forms of food production were basically same—the control of reproduction. The bones of the earliest domesticated animals like cattle; sheep, goats and pigs were discovered from the earliest agricultural settlements at Jarmo in Kurdistan, Fayum, Merimde, Badari in Neolithic Egypt.

Apart from the basic subsistence technologies, the other aspects of material culture can be discussed under the following heads.

Fire Making

Fire is the most important material trait, which contributed to the human cultural progress. Its discovery lifted man above the animal level. Suggestive evidence of using fire has been obtained from Kenya of East Africa, which corresponds to Lower Palaeolithic Period. But the confirm evidences that come from Europe and China does not cross 500,000 years. The cave dwellers used to keep fire within their cave not only for warmth and light but also to frighten the wild animals so that they did not dare to come in close proximity. Gradually the early men learnt cooking with the help of fire and gave up the practice of eating raw flesh and plant parts. At this time fire also enabled them to develop the crafts. It is almost certain that neither pottery nor metallurgy could have developed without fire.

Man's acquaintance with fire perhaps had grown in course of experiencing the natural events like the eruption of a stream of lava from a volcano, the ignition of a dry tree by lightning, the sudden break out of fire in the forest from the friction of the trees by storm or violent wind etc. Early men got astonished with those events and later realized the strength of fire. They started to worship the fire due to awe. Gradually they noticed that when two pieces of stone accidentally had knocked together. sparks came out. So they worked out some techniques for producing fire. A few of them are still in vogue among the contemporary primitives.

Percussion or Strike-a-light Technique

This is the earliest technique of fire making. Either flint and stone or flint and iron were stricken with each other to produce the sparks. It is still a common way of producing fire among some primitive tribes of Russia, Belgium, Holland, England, Borneo, Tibet and India. In Southeast Asia, especially in Malay Peninsula and the Philippines, a rough-surfaced bamboo is struck with a piece of porcelain to produce sparks.

Wood Friction Technique

Two wooden sticks of suitable size are selected. By constant rubbing of one piece of wood upon another, wood dust is produced. The friction generates so much heat that the heap of wood dust begins to smolder which can be blown into a flame. The friction is created either by bare hand, or by the help of a bow or by the process of drilling, sawing or ploughing. One of the woods is held at rest on the ground; the other stick is twirled on it from a socket.

(*i*) *Friction by Hand* : It is the fundamental as well as universal method of fire making. Here the wooden stick is rotated by hand. The Santals, the Veddas, the Eskimos, the Papuans, the Californian Indians and many African tribes follow the technique for fire-production.

(*ii*) *Friction by Bow* : Here the twirling is done by some mechanical device. Usually a bow is utilized to produce twists around the stick. The Eskimos, the North Siberians and the Yukagirs use this technique.

(*iii*) *Friction by Drilling* : It is an evolved form of fire making. Here a sharp point is employed to drill the hole. A wooden flywheel is attached to the lower part of the stick. The stick revolves automatically with the winding and rewinding of a string. The Melanesians, the Polynesians, the Chukchis, the Iroquois, the Bhils follow this method even now.

(*iv*) *Friction by Sawing* : A piece of wood is drawn back and forth across another wood- piece as like sawing. Tyior had labelled this technique as 'Stick and groove' type. It is practised widely among the tribes of Australia, Philippines, Burma and India. In Malay, bamboo replaces the wooden pieces.

(*v*) *Friction by Ploughing* : Here a pointed stick is pushed back and forth on a flat piece of wood to produce a groove. The process resembles an act of ploughing. Ploughing is continued till smoke and sparks are generated as an outcome of friction. The Polynesians, the Melanesians, the Papuans and the Negroes practise this.

Fire—Piston Technique

This technique employs a simple instrument consisting of a hollow metallic cylinder, closed at one end. The other end of the cylinder remains open to accomodate a tightly fitted piston. Pressing of the piston upto the bottom of the cylinder compresses the air therein and generates heat to ignite the tinder placed at the closed end. Such a cylinder may also be made up of bamboo, wood or horn. Although this instrument was first invented in Asia, soon it spread to different parts of the world. But gradually its use dwindled down and even became obsolete after 1827 with the introduction of matchsticks, which is now familiar to each and everyone of the world. Nevertheless the fire-piston technique is found to survive among the primitives of Borneo, Java, Malay and Burma.

Lens Technique

Sun's rays if focussed on the dried up objects through a bi-convex lens, fire is produced. This technique has been popular after the invention of lens. A few primitives living close to the advanced societies follow this technique. It is especially true in the context of England.

Cooking

Primitive men first learnt to roast the flesh in fire. Gradually they understood that cooking makes the food delicious in taste and so invented various methods of cooking by trial and error. Some of those early methods still exist among the contemporary primitives of the world. However, we can classify the cooking methods of the Tribals in the following ways :

(*a*) *Cooking without vessels* : Some non-literates simply roast their food because they have no vessels. The Onas hang the slices of flesh from a stick bent towards the fire. Fishes and rats are baked in hot ash. The Nicobarese throw the meat of pig directly on fire. Australians pluck the feathers of a bird and throw it on fire for roasting. Bunas of Bengal present a strange process with the domesticated pig. Before killing it, they feed the animal some unboiled rice and some spices. Then they kill the animal and put that into fire. After sometime when they take out the body from fire, the rice and spices inside the stomach of the pig get considerably cooked. The Bunas thereafter cut the abdomen of the pig and open the stomach by knife in order to get the spicy rice. It forms a very delicious dish. Lodhas of West Bengal steam the meat of the goat or fowl by packing it in a bark of 'sal'-tree; the pack is tied strongly with creepers. Nagas put the meat inside the hollow tube of a green bamboo and bake it in fire.

Among the advanced people of Bengal, we still find some beautiful preparation of fish where fish are baked either in smoke or in steam after wrapping them in a *plantain* leaf. Again, nuts of jackfruit are often put into fire and then eaten up by peeling their cover. All these show the survival of primitive practice.

(*b*) *Cooking in oven* : Some of the tribes are habituated in using oven and they build up ovens in their own way. The Polynesians and the New Zealanders being ignorant of pottery, dig a shallow pit. Inside the pit, stones are lined and the top is covered with a big stone. Fire is put with dried up leaves, and before that green vegetables are kept there to be cooked. The Samoans follow more or less the same technique; only they do not cover the top of the pit.

(*c*) *Cooking in vessel* : Cooking in earthen vessel is widely practiced among the contemporary primitives. Melanesians cook only small quantities of food in earthen vessels upon the ovens. Indonesians use bamboo tubes for cooking, instead of earthen vessels. The same practice also prevails in New Guinea. Many tribes of California and Canada use baskets and vessels made of bark, for cooking. These are made watertight and so the liquids do not flow outside.

(*d*) *Stone boiling* : For the liquid food, the Polynesians use wooden bowl in which after keeping the liquid, hot stones are dropped. North Americans and the people of Kamchatka follow the same technique for heating the liquids.

Storage of Food:

All primitive peoples of present day know the technique of storing food. They store grains, vegetables, fruits, fish and even the flesh.

(*i*) The Plain Indians preserve buffalo-flesh. They dry it up and grind into powder. The powder is then mixed with choke-cherries and packed within raw hide bags. To prevent it from insects and moisture, buffalo tallow is often poured on it.

(*ii*) Pulp of the berries is preserved by making them dried and powdered. Stone dust is mixed with it, hide bags are used in packing.

(*iii*) Corns, beans and squashes need to be stored in a dry state. They are put into jug-shaped pits below the hearth. A foul-smelling plant is generally kept with the corns for protecting them from the insects.

(*iv*) The tribes of Bengal and Assam dry up their surplus fish and cover them with bark.

(*v*) The Maoris preserve their dried sweet potatoes in the communal storage huts. The Polynesians keep the bread-fruit paste in the storage pit, lined with leaves.

Habitation

Habitation is an important element of material culture. It comes under the basic needs of mankind. In the early prehistoric days when men were not able to make their houses used to search the caves or rock-shelters for protection. Caves saved them from the attacks of the beasts, bad weather and prowling enemies. But since caves were few and men were many, they faced a serious problem. Further, the caves were not always situated in a congenial position in respect to water and game. Naturally men strived for making shelters of their own. At the Middle Palaeolithic stage, they became successful in constructing their own shelter. Traces are found from Russia; the Neanderthal men made the pit houses to keep themselves warm. To make this type of dwelling, a hole was dug down, two to three feet under the ground. A roof of poles was then erected over the pit, which was kept covered with bushes and earth.

Among the contemporary primitives, Veddas of Ceylon till now retire to the caves during monsoon. Nomadic tribes build such a dwelling, which can be carried with. The Mongols and the Siberians use a simple type of tent houses. The Kazaks of Central Asia make an elaborate portable shelter known as 'Yurt'. It is a light wooden framework covered with felt. Some Yurts possess wooden doors separated into several rooms. However, no universal house form exists among men. The variations as found among the present day tribal people have been discussed here.

Wind-screen or Wind-break hut

It is the crudest type among all habitations. The Andamanese, the Tasmanians, the Shoshones build this one-wall, roof-less simplest home which gives them protection against gusty wind. Some sticks are erected in a semi-circle. Interlaced boughs or strips of bark or skins of animals are used to cover the gap of the sticks for making the wall. The Onas of Tierra del Fuego live in the worst climate of the world—a climate of bitter cold persist there. Therefore, Onas build simple windbreak hut with guanaco-skin, supported by poles. The Crow-Indians make 'tepee' with a large number of buffalo skin. It is about twenty-five feet high and can accommodate more than twenty people. The Veddas, the Kadirs and the Aruntas also build this sort of dwelling.

Conical hut

A slightly improved form of habitation is conical hut. It shows roof as well as wall but those are not separated. The Lapps, the Eskimos*, the Kharias and the Andamanese exhibit this type of huts. The Beehive hut of the Bushmen can be included here. It looks typically like a beehive; the plinth, roof and wall of this hut are made of bundles of tall grass. The Zulus, the Hottentots, the Birhors and the Koras frequently construct this sort of dwelling. The beehive shaped grass houses of Shoshones is called 'wickiup'.

For getting protection against heat and cold, the Todas of Nilgiri Hills build a half-barrel shaped hut where roof comes down to the ground. The front and the rear parts of the hut have walls. Door remains with the front wall.

Rectangular hut

This is the real form of hut where the plinth, four separate walls and the roof are distinguished. Walls remain firmly fixed with the roof. Rectangular huts can be categorized into some types on the basis of their roof structure.

* Polar Eskimos are nomadic people. But they build permanent houses of stone to which they return after summer's wandering. The snow house, Igloos are the temporary shelters, which can easily be abandoned.

(*i*) *Rectangular hut with flat roof*: Here the ground plan is rectangular in shape and the roof is flat. The Hopis construct these kinds of huts.

HUT ON TREE

WINDSCREEN

Fig. 14.20. Types of House

(*ii*) *Rectangular hut with dome-shaped roof*: Among the Zulus and Hottentots of South Africa, the roofs of the huts are built in the shape of semipheric dome.

BIRHOR ENCAMPMENT IN HAZARIBAGH DISTRICT (BIHAR)

A HUT WITH CONICAL ROOF IN ANDHRA PRADESH

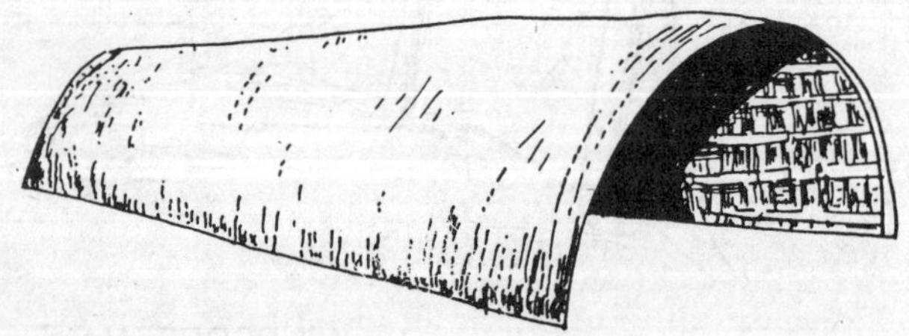

HUT OF TELEGU SPEAKING NOMADIC PEOPLE IN PURI DISTRICT (ORISSA)

Fig. 14.21. Types of House

(*iii*) *Rectangular hut with gable roof*: The ground plan of this hut is though rectangular, the roof is triangular in shape. The Maoris of New Zealand, the Hos of Saraikella and Chirus of Midnapore construct such dwellings. In Oceania and Indonesia the type is widely spread.

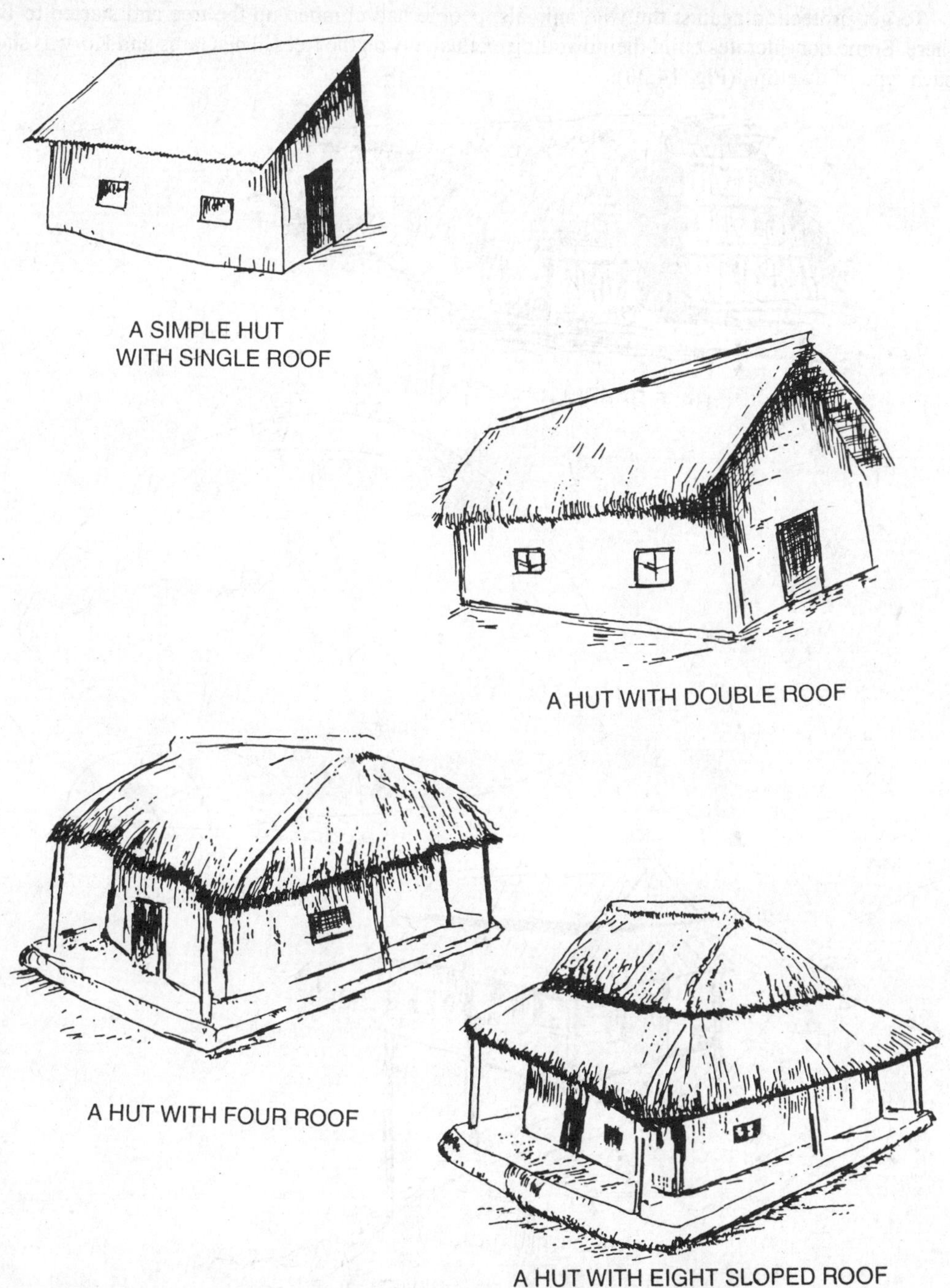

Fig. 14.22. Types of House

Beside these principal types of primitive dwellings, there are some other types too having evolutionary significance.

Tree House

To get protection against the wild animals, people had climbed up the tree and started to live there. Some non-literates build their dwelling exclusively on the trees. The Garos and Korwas show such type of dwelling (Fig. 14.20).

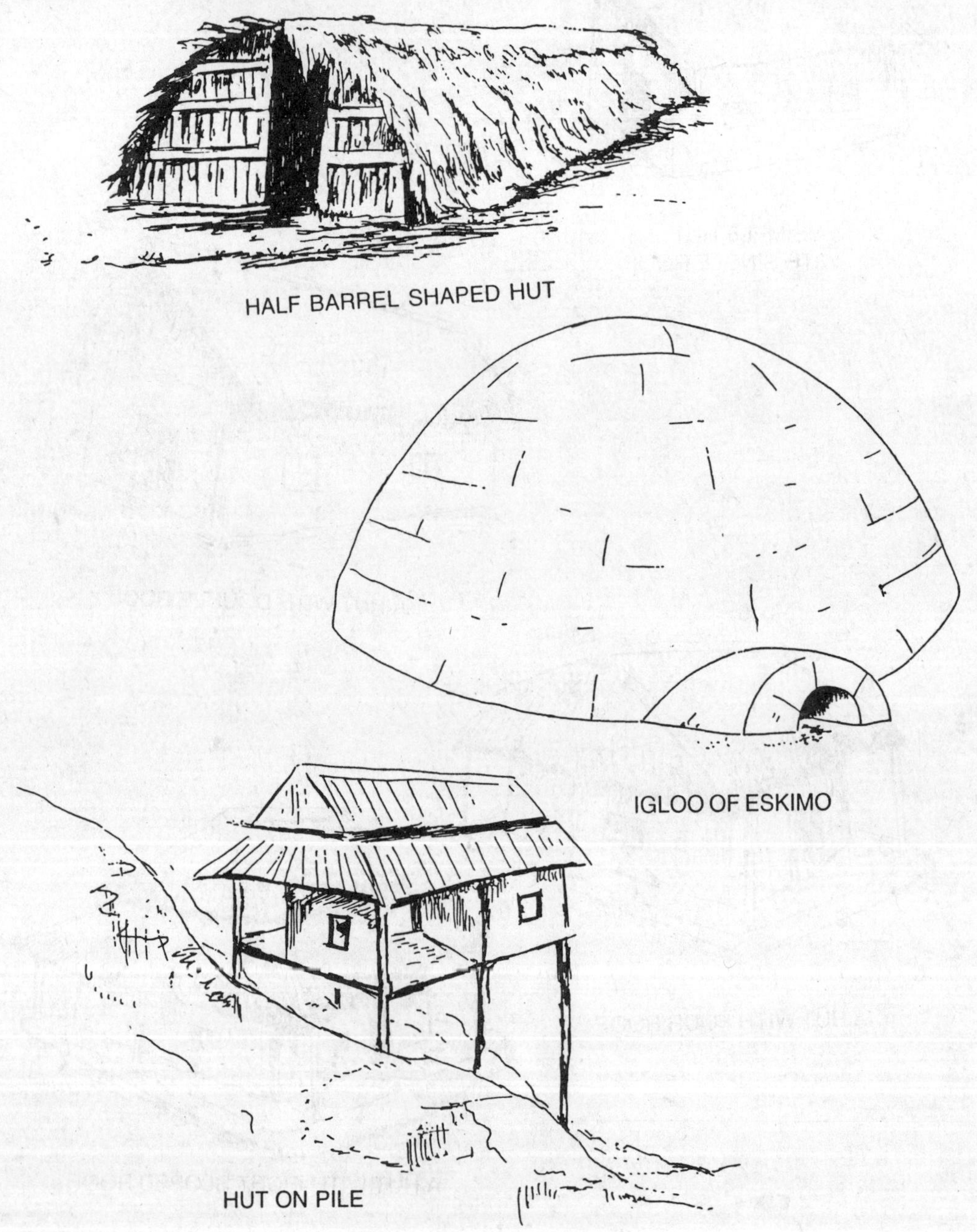

Fig. 14.23. Types of House

Pile House

Houses on piles are very common on marshy lands. A large number of poles are erected straight from the ground and a platform is constructed on them. A hut is finally made on that platform. Such dwellings are prevalent in Borneo and Ireland. In Italy, 'Terramara' settlements are made in the same manner, but on dry land. The Mikirs, the Kukis and a few tribes of Assam make huts on the top of some long posts. Lake dwellings, wherever is found, are essentially the pile dwellings.

Pit House

The most primitive form of dwelling is the pit house, which is still preferred by the peoples of Western North America and North Asia. This house is well suited in the environment of temperate regions where cold winter prevails more or less throughout the year. In tropical areas, such type of house is beyond imagination. The floor of this dwelling is sunk into the ground for some feet and a superstructure is raised on it. Such underground shelters are obviously warmer than those above the ground are, However, these pit dwellings can be traced as a northern-latitude phenomenon of the European Neolithic and later found abundant in Siberia and Western North America.

All cultures have their own attitude of making homes. The settlement pattern varies not only with the environment but also with the pattern of human activity. For instance, a hunting group can not settle in fixed abode, as they have to move frequently from place to place, chasing at the games. So also are the herder groups. These groups have been accustomed to tent like structures. Geographical condition determines the materials for housing. In Assam, Kashmir, Burma and Norway wood is cheap and abundant; houses are, therefore, built in wood. Wooden houses are again found in Japan and in the entire Himalayan region. It gives a protection against falling during earthquake. Tribals of New Mexico, Arizona, south Europe and Spain primarily use stone, as this material is easily available in their place. A few states of India, particularly Uttar Pradesh, Punjab and Rajasthan exhibit plenty of stone houses due to same reason. In Bengal huts are made of mud, straw and bamboo-splits, as such materials are very common in this region.

Clothing

Clothing is considered as a bare necessity to modern man, after food and shelter. Early men of prehistoric days were neither concerned with modesty nor the shame. The impulses towards clothing originated from the want of warmth as well as for the protection of the body against the attacks of insects. There were also sharp thorns and stones in the nature, which often pricked the body of the man. So covering of the whole body or some particular parts of the body were found necessary for the early people. Gradually different other senses developed and motives behind the use of clothing became more pervasive. Observing the life-style of contemporary primitives, we can conclude that the use and the non-use of clothing are purely functional. The motives behind the use of clothing can be related with three factors as mentioned below:

(*a*) *Protection against weather* : Protection against climatic conditions is the first and foremost motive for the use of clothing. Dresses of the primitives vary according to the nature of climate in which they live. The Eskimos of the Arctic region wear leather coats in summer and fur coats in winter. The Tibetans, the Nagas, the Abors and other Himalayan tribes wear woolen dress from head to foot. In the tropical climate, people go with a minimum of clothing. Many African primitives as well as Indian primitives use little coverings on their body. Not only that, they also seek suitable protection against the scorching sun. Tribes like Santal, Oraon and Ho use headgears of different types. The people of Nile use a helmet topped with ostrich feather. In the monsoon ridden areas, people try to protect themselves from rain. In lower deltaic Bengal, the peasants use hats made of palm, coconut and hogla-leaves to get rid of both the sun and rain while working in the field.

Footgear is widely used during travelling. This helps in smooth walking through uneven paths and protects feet from injury. The natives of North America, North Japan and East Siberia put on a special kind of shoes to walk over the snow. In Kashmir, shoes and boots made of straw are used to climb on the snow-capped mountain. Tribes of Chotonagpur make their shoes from the bark of betel-nut tree. On the other hand, the tribes like Naga, Kuki, Khasi etc., use a pack-strap over their head to prevent an injury for carrying burden on head.

(*b*) *A sense of concealment* : The other motive for wearing dress is to save the body from the exposure of evil eyes. Primitive people began to cover some parts of their body to get rid from other's jealousy and this indirectly gave rise to a sense of shame. Among some tribes, unmarried women can go nakedly but the married women are compulsrily clothed. As the married women are considered as the property of their husbands, their figures are kept unseen. The Shilluk women are found to wear leather apron but their men can stay naked. In India, the traditional dress of the Juangs are made out of creeper and leaf girdle; the unprogressive Garos and the Todas still use bark as cloth. However, in the primitive world, coverings are mostly made of gleaming shells, gourds, bark, hide, grass, leaf etc. Those who live close to the advanced peoples use clothes made of cotton, wool and synthetic fibre.

(*c*) *An indication of social distinction* : Social distinctions are often symbolized in special costumes. As ranks differ with status, so the costumes also. Different social classes wear different kinds of dresses. For example, the warrior's class uses special kinds of dress set with weapons. Officers in the royal court wear some specially designed dress. Again, wearing a crown is a symbol of royalty. Although the primitive societies are considered as equalitarian societies, some minor distinction of dress is found which marks off a village headman or a priest from the commoners.

Personal Adornment

A desire for personal adornment developed in the mind of primitive men next to the use of clothing. They tried to display and decorate the body by different means.

(*i*) *Paint* : The practice of painting on body is a common feature among the non-literates. Such a practice can be traced back to the prehistoric time. Women have always been the regular users because they took it with a great cosmetic value. Apart from this, an extensive use of paint has been noted in special occasions, particularly in ritual and social ceremonies. The Santals paint their bodies with numerous colours and go on dancing on the days of feast. The Palayans and Purayans of South India paint their faces with bright colours and dance before the deity. War paint is linked with magical potency; the motive is to influence the enemy and make him defeated.

(*ii*) *Tattoo* : Since the application of paint did not last long, primitive people devised tattooing as a solution. Now it is a widely prevalent practice. From the third or fourth century AD, this practice has been recognized in the primitive world. Tattoos are the permanent marks made on the body, especially on the forehead, nose, breast, neck, arms and palms. The designs include flower, leaf, creeper, coronet, fish, serpent etc. Tatooing became very favourite to the womenfolk. The primary motive is to increase the grace and beauty. It is also supposed to have a protective power against the evil influences created by the occultists.

The technique of tattooing is to puncture the skin with thorn or needle-like sharp objects carrying an indelible dye—commonly the carbon black. Porcupine quill is used by the Bunas of Bengal to incise the skin. The act of tattooing was originated in Polynesia in order to attract the opposite sex. Tribes of Scotland practise this largely. Maoris of New Zealand make tattoo on their face to evoke fear. But to the Abors, it is a mark of dignity. Among the Todas, this practice is confined only to those women who are the mother of one or two children. Tattoo is highly preferred by the Japanese. Negroes and Australian black fellows face difficulty in tattooing as no white dye has yet been discovered which could be applied at the time of incision. Tattoo signifies different meanings to different people.

(*iii*) *Scarification* : This is more or less similar to tattoo. The difference is that, here no particular design is made; only a deep wound is made into flesh which leaves a permanent scar on the skin, afterwards. The practice is more popular among the dark-skinned people. Natives of North Queensland cut their body with a sharp flint. Some tribes treat these scars as identity mark of great warriors. Because, the individuals who really kill a number of enemies in battle possess these scars on their body. The Fizi and the Samoans consider this marks as good as ornamentation. In Central Australia, parallel rows of lines are made on chest and back of the boys as a part of adolescent initiatory rites. These scars convey the status of manhood. Scarification, associated with initiation rite is also found among the Congo Bantus of Africa. But the meaning of such scars is absolutely different among the South American Indians who bear the scars on the body to keep the evil spirits away. Again, to some tribes scarification appears as a remedy of illness. The tribals of Chotonagpur use scar-marks to declare their tribal affiliation.

(*iv*) *Mutilation and Deformation* : In the early stage of human development there was no clear idea about adornment, so many tribes tried to enhance the show of their body even at the cost of major damage on body parts. They used to pierce the nasal septum, the lips or ears for beauty. Some of these practices still exist among the highly civilized peoples of the world. For example, piercing of the ear lobes especially of the ladies is a common practice all over the world. But primitives undertake more vigorous and painful effort for beautification. The Australians, the Melanesians and the Indonesians pull out some of their incisor teeth to look nice. They also pierce nasal septum to push feather, sundry bone or some other ornament through it. The Kadirs chip their incisor teeth to make it sharp and pointed. Some tribes pierce their chin. Chinese women wear very tight iron-shoes to keep the feet small. The Aymaras of South America binds their head strongly to bring it in good shape.

Circumcision and subincision are not the mutilation for adornment; they are mystic status operation having religious significance. Circumcision is the removal of the foreskin of the penis as practiced by the Mohammedan, Abyssinians, and Jews and by some tribes of Africa and Australia. On the other hand, subincision is the slitting of the skin and urethra in females along the length of the male sex organ. It is practised by the Sudanese women and by some Mohammedan women of Sindh.

(*v*) *Ornament* : Ornaments are the objects that adorn the body. They symbolize man's eternal craving for beauty. Both the primitive and civilized people show an inclination towards ornament. The practice is as old as fifteen thousand years. Early men used different natural objects like flowers, leaves, shells, teeth, bones, stones and even grasses for making ornaments. Metals came into use later with the progress of civilization. Contemporary primitives still prefer ornaments made of natural objects. Even the most primitive group of tribals who remain nude, use various kinds of ornaments for extra beauty. Necklaces made of animal teeth and shells are most common among these people. Australian tribes make forehead bands and necklaces with kangaroo teeth. Melanesians and some other primitives those who live near the sea coast extensively use mother of pearl in their ornaments. The tribals of Chotonagpur are expert in making beads out of seed, glass, and stone. Korkus of M.P make bangles and ear rings by grass. Finger and toe rings, bamboo and wooden combs decorated with figures are prevalent among the women of Ho, Munda, Oraon, Santal and other communities. Sometimes, the ornaments are not simply the objects of decoration, they serve the other purposes too. The Abor women wear brass wristlets with spikes for offensive as well as defensive purposes.

It is really difficult to delineate where clothing ends and ornamentation begins. Some scholars strongly believe that clothing has evolved out of ornamentation. However, primitives of present day use diversified materials like grass, straw, bark, hemp, paper, mulberry, leather, cotton, linen etc., for making dress. Most of the tribal men in tropical zone wear a piece of cloth around their waist as girdle. The Maoris, the Zulus and the Bambutes use leafy girdle i.e., a string made of grass or fibre. In India also, the Juangs, the Savaras and the Korwas use leafy girdles. The Melanesians use the barks. East African tribes collect bark from fig tree but the Polynesians collect it from the mulberry tree. In China, the textile material is chiefly hemp among the tribals. Cotton came to be in use about 2700 BC in India. The advanced Indian tribes like Santal, Oraon, Naga, Kukis and others use cotton in their dress. Some of these tribals have learnt the weaving technology. Handlooms are abundant in the states like Assam, Manipur, Nagaland etc., These are also abundant in Pacific Islands.

Primitive Transport

Migration of people and conveyance of cultural traits to distant places have been possible only for the development of transportation. In the early days man was solely dependent on his capacity of locomotion. No way was known to him other than to walk. Heavy loads were carried on hands, head or back while walking. Human intelligence first devised a water transport—a simple float. Although there is no direct evidence of transportation for Palaeolithic, it is held that Upper Palaeolithic people perhaps invented floats or rafts for crossing the small streams. The device was supplementary to swimming; floats helped the swimmers to carry the goods. People of Scandinavia belonging to Maglemose culture of Mesolithic period provided more prominent evidences of water transport. They used dugout canoes made from a single log. On the other hand, people of Finland in Mesolithic period learnt to make the simple land vehicle, sledge.

Primitive transport can be broadly categorized into two groups—land transport and water transport. The first land transport showed the use of domesticated animals. It saved the labour of man and at the same time facilitated the meetings between the peoples of distant areas. Persons or loads could be carried on animals' back. Animals e.g., elephant, horse, bullock, donkey, and camel were trained for this purpose. At the beginning, man walked along with the animal placing the load on its back. Thereafter he learnt riding. The first evidence of riding comes from the herding people of Central Asia. who used to keep both horse and camel. Early Mesopotamian culture showed little or no evidence of horse riding. These people became equipped in horse riding later on, coming in contact with the migrants of Central Asia. After the introduction of wheel, animals were taken to fix with the carts and this set the men free from pushing or pulling of the cart.

Use of animal power in packing and riding is still appreciated by the contemporary tribes. The Peruvians of South America employ llama for the purpose of packing. Lapps use the dog as a pack animal. Reindeers fulfil the same purpose being directed by some European and Asian tribes. In India, Kukis of Manipur utilize the buffaloes for carrying paddy from the field and also for bringing fuels from forest. Since these people live in hilly areas, it is difficult for them to carry the loads along the uneven path. Donkeys are also largely used in hilly areas to carry the burden.

The real comfort as well as speed of transportation was achieved after the invention of wheel*. One-wheel pushcart or wheelbarrow of China is the oldest land vehicle in the world without the utilization of animal power. The earliest wheel was made of solid blocks of wood, which is still used among the Santals of Santal Parganas. Such wheels are also visible in Indore and Burma at present. Old block-wheels are believed to have developed into spoke-wheels of today. The idea of wheel was perhaps generated into the primitive mind by experiencing the circular movement of a bamboo piece or a log. They invented different types of cart and wheel. The wheel is considered as the basis of modern civilization.

To trace the development of water transport we should call back the logs and inflated skins, the earliest and the oldest means of travelling on water. Next invention was the raft, where two or three pieces of wooden logs or bamboo pieces were tied together to increase the surface of the float. It is still preferred by many contemporary primitives for short trips in a river. Eskimo's Kayak, hewn-out trunks of trees used by the Australian aborigines belongs to this group of raft. Besides, rafts made of plantain tree or sola are very common in marshy places of Bengal. Bamboo rafts are useful in rivers. Coracle is a woven bamboo basket, which is used in hilly streams having strong current. Its maximum use is found in the rivers, namely, Godavari and Krishna of South India. Tribal peoples who live in the coastal areas of river and sea largely depend on dugout canoes for carrying people as well as

*1. The wheel was first known from a representation of Sumerian art of 3500 BC. But it is quite interesting that wheel was not used in Egypt till about 1650 BC. It was never discovered in America.

goods. Further improvement on simple canoe is the out-rigger canoes, which was invented by the non-literates of Oceania. Such a canoe essentially possesses two poles for retaining balance. Andamanese extensively use these out-rigger canoes. The double canoe shows two hulls, connected by a platform. These are used for long voyages.

Boats are of more recent date and they are considered to have developed out of canoe. The simplest boats are still preferred by the fishermen communities. It is a small boat where sitting arrangement is made with some horizontal strips of bamboo. Only a man is needed for plying this boat. A better form of boat is relatively big in size and it has a cover on its top to save the passengers from rain and sun. Two persons are necessary for plying this boat. The third type of boat is also plied by two persons—one remains with helm and other remains with oar. There are also other boats as rowed by three persons or four persons respectively.

Poles, paddles, oars and helms are considered as accessories for controlling the water transport. Poles are the earliest appliances to move the canoe in shallow water. For boating in deep-water paddles were devised. Since one end of the paddle is flat, it pushes the water backward and the canoe advances with that force. Use of paddles is generally restricted to the canoe; in boats, oars and helms are mostly used. Both hands and feet are employed to manipulate the paddles, whereas the oars and helms demand only the work of hand.

Large boats are generally used in sea. Such boats were first found in Egypt as early as 3500 BC. As the navigation technique has tremendously been improved at present, some of the primitives have been benefited from it. But the majority of the primitives still revolve within the realm of their simple technology using dugout canoes and simple boats.

From all parts of view, it has been evident that the physical condition of a region directly influences the cultural pattern of the primitive people. People interact with their immediate environment and the reflection appears in their material culture. This material culture is a direct product of overt actions; it consists of tangible objects—the products of technology. Although the primitives show a low degree of technological knowledge and skill, still the study of material culture reveals the basic facts in understanding a culture.

PRIMITIVE ECONOMIC ORGANIZATION

Economics as a subject deals with the production, distribution and consumption of the wealth. Wealth in primitive societies signifies nothing but the supply of food. Whatever may be, some sort of economic activities are found in all societies, which are necessary for the survival of its members. These activities not only satisfy the basic needs like food, clothing and shelter, they also serve the desires for comfort and luxury. The way in which man makes use of his environment (in order to satisfy his needs) gives rise to a variety of economic systems of which primitives' economy is the simplest one.

The primitive economy is almost synonymous to early economy, which was based on a principle of acquisition and consumption. A low degree of technical knowledge, lack of specialization and full of uncertainties characterized such an economy. Under a favourable environment*, people invented simple techniques to explore the resources. Natural conditions of a region restricted or favoured certain kinds of development. Pre-occupation with daily and seasonal food-supply, limitation of transport, difficulties in storage, frequency of hardship and risks of hunger are the other features of primitive economy. In the next stage, for the increase of social complexities and difficulties of movement, man became able to create a kind of secondary environment by employing his intellectual skill and manual labour. Such an effort helped men to produce bulk food, as much as they needed. But, accumulation was difficult and a long term planning was impossible at that time. However, this stage can be related to the production-consumption economics of Neolithic period—the highest

*1. The places where a rich vegetation and animal life was found.

form of pre-historic economy. Still later, with the discovery of metals, the organization of economic life became more improved and complicated, which we call as urban economics.

The economic organizations of the contemporary primitives are more or less similar to the prehistoric days, which are essentially of subsistence type. But they can be classified into two broad categories.

(*i*) Those who follow the production-consumption economics.

(*ii*) Those who belong under production-consumption distribution, economics

Production here is a process of obtaining goods from natural environment e.g., land, water, plants, animals and minerals. It depends on the character of locally available resources and the capacity of people to utilize those. Most of the productions in a primitive community are directed towards the manufacture of consumption goods to serve the immediate needs. Naturally the questions like quality, effectiveness, variety, durability, serviceability etc., do not arise. In fact, the production shows a very low level of technological specialization. Such an economy is incapable of accumulating the goods for technical difficulties of storage and so fails to develop a complex exchange economy. Moreover, as every one produces more or less same range of articles, no demand of goods is observed in the community.

On the other hand, in a relatively developed economy (like agriculture), the rate of production is always higher. It produces much more than one needs. Therefore, the individual or his family can not consume the entire production. This excess production (wealth) demands distribution. The primitives, who surpass the level of production and consumption, exhibit a complex type of economy with production, consumption and distribution. However, the products differ from society to society in accordance with raw materials, technology and human effort employed in producing a culturally desired item.

Most of the primitive economic systems are not acquainted with the use of money, as a medium of exchange. Exchange, whatsoever, is governed by a sense of mutual obligation, sharing and solidarity. They also lack market. People may meet occasionally in a place for simple exchanges, but regular market as an institution is absent, in this economy. The surplus product, if any, are disposed off by lavish feast or by distributing them to the kinsmen or neighbours who are bound to repay in future. These societies show high degree of stability and slow rate of progress. The simplicity as well as uniformity of the techniques contributes in maintaining stability and originality.

Labour in Primitive Economics

Generally a family or a kinship group are engaged in a production. Like all other societies, primitives also show some division of labour i.e., some customary assignment of different kind of work is vested to different kinds of people. Division of labour in a family is found on the basis of age and sex. Certain economic tasks are chalked exclusively for men while others are meant for the women. The men perform usually laborious extensive jobs; women do light works and specially works involving patience. For instance, among the Australian aborigines, the women dig roots and witchetty grubs while men go on hunting. As women often remain handicapped by menstruation and childbirth, so jobs are designed for them in such a way that they can pass a longer time at home. However, women universally engage themselves in cooking and child rearing.

Children can not perform the same kind or volume of work like their parents. They are never given that job, which require much strength. Moreover, age of the particular child is considered before putting a duty to him or her. Girls normally do a variety of work but within the domestic chore. Large tasks are usually accomplished by the cumulative effort of the members belonging to a family or a kinship group.

The social units in a primitive society are very limited. Social relations are, therefore, extremely close, face to face kind. Economic activities are mostly entwined with the social relations. By no means these can be separated from the total social system. Choice and capacity of the sexes as well as the age groups contribute much to the adjustment and readjustment of social relationship, rank and position. The courage of work is derived not only from the simple need of subsistence, but largely from the satisfaction of working together, the pleasures of conviviality and the common interest in the product of work. Specialization of labour is not required in societies with primitive technologies. A man usually possesses a number of skills, ranging from food collecting to canoe making.

Property in Primitive Economics

Property is not a wealth or possession. Rather it is a right to exploit, to use or to enjoy the wealth and possession, which may be movable or immovable, corporeal and incorporeal. However, the conception of property is very old. But it changes with time along with the development of technology, extension of economic network and alteration of ideological set up of the people. Ownership of food, gathered or produced, perhaps generated the idea of property among the primitive people.

A property may be owned either by one individual or by a number of individuals belonging to a social group—kin or clan. When an individual owns a property, it is called private property. Neither the State, nor any other institution has a right to interfere with what is one's own. The tools, implements, clothing, ornaments etc., what a person uses individually, come under private ownership. The children can inherit these goods after the death of the person. But when the members of a group, viz., a lineage, a clan or an association possess more or less similar right on the use of an object or a good, it is called a communal (or joint) property. This is why, the ownership of an agricultural field or a pasture ground is held communally. Children can not inherit such agricultural land or livestock at the death of their father. All members of a society possess an equal access to this property; the property right is vested on the community as a whole.

Most of the primitive groups form their social and economic units unconsciously. The methods of production and distribution are controlled by the local conditions. Since they do not have sufficient means of food-production, they deal the situations of scarcity with cooperation and collaboration. Survival of the group is solicited. The concept of communal property was perhaps promoted by the situation of scarcity. For example, compulsion of natural environment often leads Eskimos to hunt seals or whales in a collaborative way; the hunter of the day distributes the flesh among his neighbours after keeping sufficient food for own family. The pastoral tribe, Kirghiz, shows both private and communal ownership on their pastureland. Private ownership on land arises in winter following the scarcity of grazing ground. The Veddas of Ceylon, who live by hunting and gathering, also show individual as well as communal ownership of property. The individual ownership is revealed in the demarcation of the hunting territory. A trespasser is vehemently resented and punished. On the other hand, collected honey is equally shared among the members of a group. Two or more families often share a cave-shelter or cook food in the same hearth. Distribution is always made in terms of a group and consent of all members is positively required.

Similar instances also come from other tribal groups like Todas, Khasis, Mundas and others. The Todas are the pastoral people who possess the livestock both individually and communally. Sacred herd of the community is sacrificed and their produces are utilized for the general welfare of the tribe, whereas the individually held livestock help in subsistence. Among the Khasis, the cultivated land as a whole is possessed by the clan. As Khasis practise Jhum cultivation; individual families are required clear up their own plot of land before cultivation. Forest produces are held in common. Wood can not be sold out; it has to be used in common. Thus, individual ownership often goes hand in hand with communal ownership. Again, if we consider the Zulus of Africa, we find a tradition of autocracy where a king or a chief is considered as the landowner of the state. He allots the land among his fellowmen for use. Among the Hopis, though the chief owns the land, he has no coercive power; clans are held to be responsible for the distribution of land among the cultivators. Different sorts of ownership rights that prevail among the primitive people suggest that the individualistic and communistic ideas of property are not exclusive of each other; rather they exist side by side.

Land is the basic property in all primitive societies. Farmers, herders and hunters all depend on land. But the notion of the property differs with the use and disposition of this particular valuable. Exhaustion of the soil, warfare and superstition compels the farmers to shift from one land to the other. Herders move from place to place following the abundance or scarcity of fodder and water. Even, the people without domestic animals and agriculture organize themselves into local groups to make a claim on a specific area where no outsider is allowed to utilize the natural resources without their permission. The Australians and the Onas show this type of possession. Multiple possessory right as a queer practice has been reported from New Zealand, Melanesia and West Africa. Under this condition a number of possessors can use the same territory for different purposes. In Melanesia and West Africa, a man is permitted to keep own trees on other's land. Among the Navahos, a man who had abandoned a land after planting a tree may come back later to assert his possession on the tree and he is also supposed to get compensation. In fact, the property ownership does not merely coincide with the political structure of the society; it also depends on the prevailing attitude in a society.

Unlike communal property, individual property can be accumulated or disbursed as per the wish of the individual. It can also be inherited by the successors. But there are certain specific regulations that govern the pattern of inheritance in the primitive society. Accordingly the rules of inheritance may be of following types:

(*i*) Primogeniture or inheritance by the eldest child. It is not a popular practice among the primitives. The royal families in advanced society often show this, especially in connection to the transmission of throne. Still it is found among the Polynesian chiefs where the 'Mana' passes to the first born son. Among the Maritime Chukchis, the eldest son gets the lion's share. In polygynous societies, primogeniture is associated with the superior status of one of the wives. The eldest son of the principal wife of a Masai inherits the largest share of father's property. Maoris of New Zealand shows a special type of primogeniture; if the first child is a daughter, she gets a male name and her status becomes almost like the first-born son.

(*ii*) Ultimogeniture or inheritance by the youngest child is acknowledged by a number of tribes of India, Asia and Africa. Among the Purum Kukis, the entire property goes to the youngest son. In the matrilineal society, Khasi, the youngest daughter is entitled to inherit the whole or bulk of property of her mother.

(*iii*) The inheritance is relation to both property and office is called collateral inheritance. Among the Thongas, at the death of a chief, his eldest son of principal wife becomes a legal heir. But he can not officiate immediately, because as a rule all of his younger brothers avail the chance of ruling, one after another, according to the seniority of age. At last when, the entire brothers die, he get the charge of the office. A more or less similar practice is discernible among the Maori chiefs of New Zealand. In primitive society, collateral inheritance was perhaps devised for an effective utilization of property left by the deceased. Because in most of cases the youngest son remains junior or less experienced at the time of father's death.

Economics among Indian Tribes

Various type of economic organization is prevalent among the primitive societies of modern India. Most of them are found to follow a combination of economy. But the economic level of a tribe is determined on the basis of principal economy of the people. However, various classifications of economic organization have been put forward by different scholars. Thurnwald forwarded an evolutionary scheme of economic development. Adam Smith classified the economic, groups as hunters, pastoralists and agriculturists. Hildebrand's parameters were barter, money and credit.

Grosse pointed out the stages of development in an ascending order like collection economy, cultural nomadic economy, settled village economy, town economy and metropolitan economy. Daryll Forde did not believe in economic stages. He found no single exclusive economy for a tribe; he suggested combinations of economy for the growth of culture. Gordon Childe held the same opinion with regard to prehistoric economies. Forde and Herskovits reached to a consensus that people did not abandon one economy while adopting another. They chalked a five-fold division of economy : collection, hunting, fishing, cultivation and stock raising.

The collecting or food-gathering people presents the oldest known type of economy as it consists of direct utilization or use of the products of nature, without making any alteration in them. Andamanese of Andaman Island, Semang and Sakai of Malayan forest, Negrito in Philippines and many other non-literates belong to this simple stage of economy. But at present none of these communities belong to a 'pure' form of food-gathering economy. They have started woodcutting with hunting, shifting cultivation with chasing, rudimentary domestication with food gathering, and so on. Not only the case of food-gathering communities, absolutely pure form of economy is scarcely found today. Many tribal people of India and of the forests of Asia offer examples of comparatively advanced form of economy. The Santals, Mundas, Oraons, Kols, Kharias etc. though practise agriculture, sometimes go on hunting. The Katkaris though subsist mainly on hunting, practise agriculture too. Again, the nomadic tribes of the pastoral belt, extending from Sahara desert of Africa to the so-called Middle East show a highly developed pastoral culture, but they too practise agriculture occasionally. Bhotia of North U.P. practise both the pastoral and agricultural economics in the same degree. Thus, economics have been almost mixed and complicated for the present day tribals.

Many tribes of central and north eastern region are now accustomed to market economy. They do not hesitate to take the economic surplus to the general market and the village fairs, although their traditional economy still emphasizes barter. Other drastic changes in the mode of economic pursuit that have been discovered in recent decades are the results of industrialization and urbanization. A lot of tribal people are settled as plantation labourers and industrial workers. Some Santals, Khonds and Gonds have migrated to Assam to take up jobs in tea plantation. Central India being very rich in natural ores (coal, iron etc.) gave rise to many industries where the entire unskilled labour force forms with the tribals. Some of the tribal people have also accepted government employment; they are quite happy by constructing roads and working in forest.

Distribution of Goods and Services

Since human society rests on mutual interdependence and reciprocal relationship, goods and services are found to be distributed in all societies, although their degree depends on the level of production. Three types of distributive system have been recognized mainly—reciprocity, redistribution and commercial exchange (Market). The transaction starts with gift giving and proceeds through barter and trade. The ultimate expression is the development of market. Such a distributive system operates both within and between the societies. Further, a society may exhibit all types of transaction at a time; the basic economy is indicated by the predominant type. The value of the goods is determined by their materialistic usefulness or their scarcity in a region. The symbolic quality of certain things is also important in this respect.

A. *Reciprocity*

It denotes giving and taking without the use of money. Reciprocal behaviour is the central theme of this transaction and it is purely individualistic in nature. No political infrastructure is required for this. However, reciprocity can be divided into three sub-types—general reciprocity, balanced reciprocity and negative reciprocity.

(i) General Reciprocity: The giving of gifts is the most elemental form of reciprocity. A pure gift is normally an outright transfer of goods without any expectation of immediate or planed return. Naturally receiving of a presentation does not involve any obligation. Such type of reciprocity is universally found in all families of all societies. Parents seldom meet up the wants of their children expecting a reciprocate behaviour from them. Most of the communities of hunters and gatherers show general reciprocity, following the crisis periods of life. For example, among the Hopi Indians, a family consisting of husband, wife and children forms the basic unit of production and consumption. The family members build their own house, produce food-crop by own labour, weave own clothing, make own tools and weapons, and other necessary items for daily use. Their farming land and the hunting ground are held as a communal property. In case of crop failure or failure in hunting and gathering, when a family becomes unable to sustain itself, the food and other necessities are supplied by the neighbouring families as gifts. The donars expect the same generous behaviour in return, when they will be under similar trouble. Thus, the exchange of goods among the primitive people is highly utilitarian as well as social in purpose. It signifies social cohesion and solidarity between the individuals or groups.

(ii) Balanced Reciprocity: It is an exchange of goods between two parties, either immediately or in the short term. The balanced reciprocity is actually a trade for certain products, which is highly valued or culturally desired. The trade differs from gift-exchange, because it is more concerned with the distribution of goods rather than developing personal relation. Anthropologists have coined the term trade to identify any transaction where one kind of good is exchanged for another.

Trade begins with barter, the direct exchange of goods for goods. Aboriginals of Northern Australia exchange their possession of red ochre with the members of adjacent tribes in lieu of the things like weapon, fish and yam. The Ochre is highly valued as it is used to decorate body, weapons and other implements. Every Australian tribe goes through some form of barter. It is a way of obtaining articles, which are not manufactured within the tribe. For example, at the death of male members, a tribe sent their spears to the neighbouring tribes. Those who accept those sent back some other articles througl the messengers.

The Hopi Indians though belong to a very low level of subsistence economy, often practise barter with several tribes living around them. They give out pinon nuts, mescal, red ochre, shell beads and tanned deer-skin against some farm products and cotton textiles. Similarly, the Aruntas of Australia obtain a few goods, notably a narcotic by trade from the Queenslanders who live about 200 miles away. Productive capacity of Arunta is lesser than the Hopi.

A form of barter known as silent or dumb barter has been noted from Africa. It is a trade between two hostile groups of primitives. The Bantus are primarily a horticultural people who live in large villages with exchangeable surplus. The neighbouring Pygmies are hunters and gatherers who roam in small bands. They like to eat domesticated food-items and use advanced tools and weapons made of iron, which they procure from the Bantus. In return, they supply meat, hides, honey, wild fruits, leaves for roofing the dwellings, and the fibres for mat making. Trading takes place at regular intervals but without involving direct contact. In some agreed-upon places, the Pygmies come to leave their own products and pick up the bunches of plantain and other articles left by the Bantus. Such barter without any face to face contact is also followed by the Reinder Chukchee of Siberia with Maritime Chukchee. Not only between the tribes, barter is also practised among the members of a same tribe; it is an ancient method of transaction, which continues even today.

The Kula Ring of Trobriand Islanders: Kula ring is a complex trading device, basing on the system of barter, whereby the Melanesian communities specialized in producing different goods and artifacts interchange their products. Trobriand Islands lie at the eastern coast of New Guinea and they are separated from each other by sea. Some of these islands are small and rocky, therefore, unable to produce enough food for own inhabitants. But these communities are technologically specialized in certain crafts like pottery making, canoe building, stone and shell cutting etc. Some other islanders produce yam, taro, and pig more than enough. Hence all Trobriand communities unitedly have developed a scheme by which all of them can get a similar chance to enjoy the treasures of the islands—foods and artifacts. Since none of these communities possesses both the raw materials and skills to produce all the commodities, they had to depend on each other for trade with exchangeable surplus.

Members of the communities belong to a ring, called *Kula ring* and they are united by Kula partnerships. Members of each island are linked to the individuals residing on two islands, which lie in the opposite directions to it. Two types of ritual objects move constantly from one community to other, within a close circuit of the ring. Long necklaces made of red shell travel in a clock-wise direction, whereas bracelets of white shell travel in a counter clock-wise direction.

When a man of the community 'A' receives a necklace from his Kula partner of 'B', he must within a reasonable period passes this to his partner in community 'D'. On the other hand, the partner in 'D' after receiving the necklace must give a bracelet to his Kula partner in 'A'. An endless series of rites, ceremonies and formalities are bound up with these exchanges, all of which are very important for the maintenance of Kula relationship. Actually the possession of the ornaments allows a man to visit the home of his trading partner on another island from whom he receives the presents. During the visit exchanges take place between two trading partners.

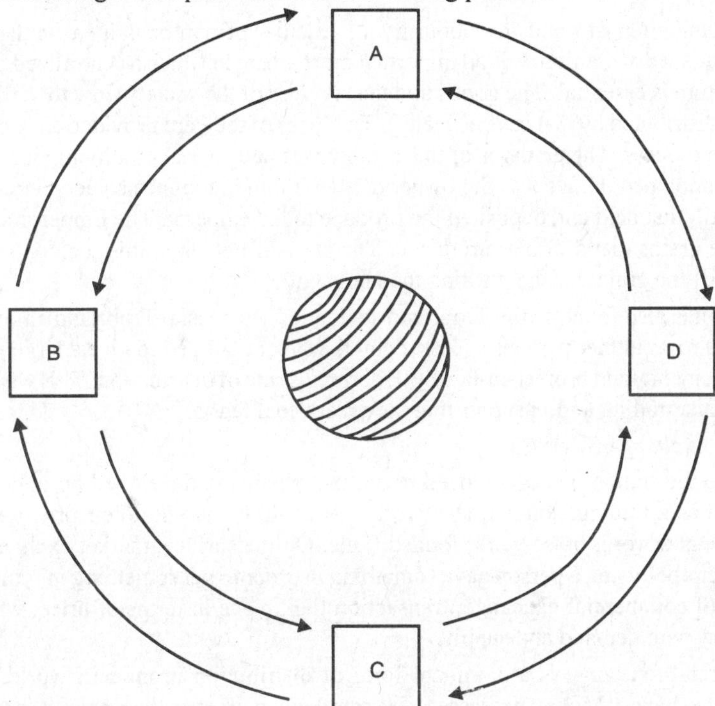

Fig. 14.25. Schematic Diagram of Kula Ring Showing the Flow of Goods

The Aztecs of Mexico show an advanced economy through a system of commerce and trade though the community actually belongs to a lower subsistence economy. They are the horticulturist people having no domesticated animal. Further, as they depend mainly on stone implements, rate of local production is very meager which can not meet up all their needs. Therefore, for fulfillment of certain want, they trade with nearby communities.

(*iii*) *Negative Reciprocity*: It is an attempt to take certain advantages—by getting something against nothing or by giving less value than it deserves. Negative reciprocity is usually practised with strangers or enemies. It is equivalent to simply stealing or cheating in trade.

Reciprocity sometimes equalizes the distribution of goods within a community. In some Melanesian communities of New Guinea, thousands of pigs are sometimes slaughtered and an enormous feast is arranged. This is a cultural practice, though apparently found wasteful. When pig population rise up in alarming numbers, their immediate killing is required for saving the yam and taro patches from destruction. Such a pig feast serves to equalize food consumption. The *Potlatch* is a somewhat similar device for distributing the surplus. It involves a good deal of feasting, gift-giving and institutional exchanges as a way of securing prestige.

Potlatch

It is a strange example of distribution in anthropological literature. American Indian groups of the North West Coast show this. In a potlatch, a chief and his group gives away huge blankets, pieces of copper, canoes, large quantity of food items and other things to their guests. The guests are the prominent persons of the society. The host declares that he has no regard or concern for the economic products, so he is ready to destroy them either through burning, or by throwing them away or giving them to the invitees. Such an action validates or increases the prestige and position of the host. This host chief and his group are supposed to be invited later to other potlatches.

B. *Redistribution*

It is the accumulation of goods or labour by a particular person or in a particular place, for the reason of subsequent distribution. It is an important mechanism in which a centralized control, usually of a political nature is essential. The goods and the services of the society flow towards a centralized point and get redistributed by that central agency. The Incas of the Peru provided an excellent example of redistributive process. The emperor of this society was considered as a living deity as well as the ruler of the common people. He was the owner of everything in the Inca society, including the land. People used to till that land and deposited the produce to the emperor. The Emperor sanctioned only one-tenth of the grains that a farmer produced. The class of nobles in this society was supposed to receive a share of the grains, without tilling the land at all.

Redistribution is also found in the Trobriand society of Melanesia. Trobriand chiefs purposefully marry the sisters of wealthier persons or lesser chiefs who are obliged to give a large portion of their production to their chieftain brother-in-law by Trobriand's rule of distribution. This wealth is expected to be spent in Kula trading and in supporting the great tribal feasts.

C. Commercial Exchange (Market)

To overcome the limitations of a closed trade, the primitives developed an open trade through market. It stood as a trading center where producers could come with their produces like pottery, mats, baskets, leatherwork, brass-work, foodstuff etc. On the surface, market exchanges resembled the balanced reciprocity i.e., a person gave something in order to get something in return. But, soon it became a spot of commercial exchange; transaction took place in terms of price, which was again subjected to vary with demand and supply.

The commercial exchange is a dominant form of distribution in modern world. Money is the medium of this exchange. Money possesses a standard value all over the world; it is non-perishable, transportable and divisible too. So it is considered as an all-purpose medium of exchange in the advanced society. Use of money has also been penetrated into the non-literate societies, but most of them do not utilize this convenient technical device. Therefore, markets being a commercial spot of exchange may not always show the use of money. Tribal people generally remain more satisfied with gift-exchange, barter and other primitive mode of transaction.

In a market economy production unit does not necessarily correspond to the unit of consumption. The size of the consumption unit may vary from a group of household to the whole of a local community. However, markets are widely spread in Africa among the non-literates. The Bakubas of South Western Congo hold markets in every three days. Among the Aztecs of Mexico, local markets are held daily. All the articles in a market have standardized value; the weight and measurements are maintained properly. K. Polanyi (1957) recognized market economies as an important distributive system. It is purely a commercial exchange where the transactions are highly personal. Here, the productions are not directed for simple use, but for gains. It bears no significant link to the social status. Peasant communities throughout the world are accustomed to market economies as they are linked with a central market. According to Polanyi, the role of a market is pervasive. It is not only a place for the distribution of economic goods; it is actually a channel through which the products of the farmers and artisans flow to the ultimate consumer. Market also stands as a center of numerous other social activities.

PRIMITIVE SOCIAL CONTROL

No society exists without a framework of social regulation. It provides an order, which operates among its members who share some common convictions to undergo through a regulated life. Social control is necessary to protect an individual against himself as well as to save the society from chaos. It is the process by which a social order can be established and sustained. In each and every society to control the conduct of the individuals and to compel them to behave in conformity, certain kinds of mechanism are found. They force and restrain the members from doing wrong. These mechanisms can be broadly divided into two categories:

(*i*) Law and Government

(*ii*) Religion.

The former acts as an external means of control, where as the latter exerts an internal control.

Law and Government

The Law and the Government comprise of the political organization. Human life is full of conflict and tensions. To hold the integrity of the group and also to protect the group against the neighbouring communities, law and order is needed. Societies at present entirely depend on political institutions for decision and power. But in the beginning, there was nothing political. No one exercised any general authority to rule or to decide or to negotiate on behalf of the community. There was neither any State nor any Government. Even no legislation was there as like our civilized world. The contemporary primitives still show a more or less same set up. Nowhere, among the non-literates we find any full fledged legal system with the court, the judge, the barristers, the solicitors, the jury, the law books and so on.

Concept of State and Government

The laws are the body of principles, which underlie the activities of a State. A State is a strong Association of people, which has a monopoly of using coercion over its members, within a territory. Government is the authoritative administration of public affairs that exists to perform three vital functions—*legislative, judicial* and *executive* within the territory. R. M. Maclver (1946) defined the government as "the administrative organ of the State, which is constituted by a large body of citizens". He also defined the State as "a community of persons living permanently in a territory under the rule of a government endowed with sovereignty and with the means of enforcing its decisions". Therefore, it is clear that, where there is a government, there must be a State. Sometimes, the term 'State' is used as a synonym of'Society'. J. W. Garner (1951) has pointed out the principal differences between a Society and a State. The State necessarily implies a political organization, which is not implied in Society. The State emerged and developed in society as an inevitable tendency of man's social nature to live with his fellowmen and to establish a proper organization so that he might derive all possible advantage issuing from common actions and direct them to the good life.

Morgan, Maine and others have denied the existence of government in the earliest stage of primitive Society. They had considered this fact in the perspective of simplest societies of the contemporary period. They cited the examples with reference to the societies like Andamanese of Andaman Islands, the Bushmen of Africa, the Yamana of Terra del Fuego, the Eskimos of polar region, the aboriginal people of Australia etc., who live at the lowest rank of subsistence economics and lack any form of organized warfare. Unbridled violence, anarchy, chaos rule in those societies for the absence of government. As there is no scope of civil law, these societies are characterized by the criminal law. This means, some sort of laws prevail there to deal with different criminal offences. But Lowie showed that the view was not correct; absence of authority did not mean a situation of anarchy. Although these societies hardly show any trace of authority outside the family, but when an occasion arises in which collective action is required, a leader is temporarily appointed from among the elders of the tribe. Any man of the community may be chosen for this purpose, but he should be superior of all,either for his bravery or extraordinary performances or quite simply for his scrupulosity. Such a local group can be taken as a starting point for the rise of government.

Nature of the Primitive Government

If it is held that the earliest form of government came out from within the family, the principal authority must be the head of the household. In fact, it is the band, a cluster of associated families who provided the first instance of government for the group. Political organization was formed when a leader or a set of leaders appeared to command on the people and the common people also desired to get their protection to resist the attacks of the hostile groups. The temporary leadership and organization became permanent and more diversified according to the circumstances. For instance, with the settled life in the villages, the sedentary population gave rise to a more complex political life. In North and South Africa, non-literates show the most evolved type of political organization; States have been established with well defined govemment where officials and council exert clear-cut executive, judicial and legislative powers. Similarly, in New York State, the league of Iroquois is a federation of five tribes who are autonomous in local affairs but in the matters affecting the league, a council with fifty representatives is formed. All public opinions are regarded by the council, but final decisions are arrived at by majority vote. The league of Iroquois, thus, represents a non-literate State.

A political unit is usually classified on the basis of its size, composition and other social characteristics. Starting from a small-scale local autonomy, in the way to reach a large-scale regional unification, a number of societies such as band, tribe, chiefdom and state, may be noted in a continuum.

Band : This type of political organization is the simplest of all. The society here is composed of a few small families who essentially lead a nomadic life. Each band is a politically autonomous body embracing the egalitarian hunters-gatherers. Anthropologists suggest that this sort of political unit had characterized the societies in general, before the development of agriculture. However, the bands show a very low population density. Distribution of resources depends solely on reciprocity. Major decisions though come from the headman, but in a very informal way. Headman is usually an elderly person whose personal qualities like skill, good senses or modesty is recognized by the members of band. But he possesses neither any permanent authority nor any extra-ordinary power. Seligman observed that the Veddas lived in a band of five to six families. The band members among the Andamanese as observed by Radcliffe Brown ranged from ten to three hundred persons. The pastoral Bedwins on the other hand, may include upto twenty thousand persons in a band.

Tribe: A tribe comprises of a number of small groups namely clan. These small groups form a larger whole of political integration. Naturally a higher population density is noticed in a tribe. Although a tribe is an egalitarian society like band organization, but the mode of economy is different. The people here are the food-producers and lead a more sedentary life. Shifting hill cultivation and animal husbandry are the general way of food-production. Distribution of goods and services in this society stands on reciprocity. The political organization is informal as well as temporary like that of the band society. No political official serves as a head of the tribal society. Instead, each clan is guided by its clan elders. The wrongdoers are punished and internal disputes of a clan are settled by the effort of these clan elders. But in case of external threat, all small groups in a tribes are united together to resist that threat.

Chiefdom : Chiefdom is a larger entity than a tribe. It has a formal structure of authority, which integrates multi-local political units. Therefore, a number of communities are found under one chiefdom. Chiefdom should have a council headed by a chief. The office of the chief is sometimes hereditary and usually permanent. He possesses the highest social rank as well as authority. Not only the chief, but his family as a whole enjoys greater amount of prestige and privileges in the society. The population density is much higher than the tribe. Social differentiation is pronounced in ranks. Reciprocity works with redistribution. The chiefs redistribute the goods, plan the public labour, supervise the religious ceremonies and direct the military activities. Though these chiefs do not always have the power to compel the people to obey them, but people act in accordance with chiefs' wishes and respect them. Usually the mode of subsistence in chiefdom society is agriculture intensive or extensive. Animal husbandry may or may not accompany.

State: A State is an autonomous political unit having a complex centralized government. It is composed of large number communities, which form cities and towns with high population density. Such a political structure includes a wide range of permanent institutions and keeps monopoly on the use of physical force. It also enforces laws. The main support of the State society is intensive agriculture and animal husbandry. The high productivity results in a commercial market exchange and foreign trade. It also gives rise to social differentiation like class and caste. The earliest states perhaps appeared around 3500 BC in Egypt, Iraq, and India. They arose independently of each other. At present, the Iroquois League provides the best example of a non-literate State. Similar incidence is also found among some highly organized populations of West, North, and East Africa.

E. A. Hoebel (1958) tried to trace the various stages, from simple to complex, as evolved in the traditional political system. He derived his proposition by studying the contemporary communities. He mentioned specially the name of five communities suggesting the form of development.

(*i*) *The Eskimo*: The Eskimos, according to Hoebel, belong to the lowest level of cultural development. The bands in which they live normally does not exceed more than hundred persons. The basis of social organization is bilateral family. No lineage, clan or club is found in them. Naturally rise of government is out of question. A band is usually led by a headman who is considered first among the equals. He does not direct by himself, but his fellow men often calls him to settle the dispute. This leadership status is achieved either for his superior hunting skill or for his proficiency in accomplishing the rituals. Since the society is entirely democratic, no Eskimo can give order to another. As a result people depend on the headman's leadership. But the headman also exercises no legal or judicial authority. For all these reasons, Eskimo society has been considered by some scholars as a state of anarchy with rudimentary laws.

(*ii*) *The Comanche Indian*: The Comanche Indians of North America show a higher development than that of the Eskimos. They have regional chiefs, both for civil and military purposes. Their bands are larger than the Eskimos. The property and status of an individual are counted by the number of horses that one possesses. The society recognizes nine offenses viz., adultery, wife absconding, violation of levirate privileges, homicide*, killing a favourite horse, sorcery, causing another person to commit suicide, failure to fulfil a contract and theft as great crimes. Public opinion and morality plays an important role in dealing with these crimes.

(*iii*) *The Ifugao*: The mountain dwelling headhunter tribe of North America and Philippines show no government. They present a tightly knit bilateral group of kinsmen covering 100,000 individuals. Ifugaos possess a long list of offenses. In cases of trouble, the people go to the 'monkalun' who is considered as a man of highest social class in the community. *Monkalun* intervene the matters but makes no decision or judgement. In fact, he provides the means so that the disputants can reach to a resolution for getting over the conflict. This political system is more organized than the Comanche's political system.

---

* Killing of offenders by the aggrieved kinsmen of dead persons.

(*iv*) *The Cheyenne Indian*: This political system seems to be matured in relation to the Ifugao. It is an well-organized state with a large tribal council where civil chiefs get a ten-year tenure of office. The Cheyennes classify the offences into three categories on the basis of their severity—wrong, sin and crime. They also chalk the degree of punishment, accordingly. But blood revenge, feud and capital punishment are unknown to this community. The people have developed an autocracy.

(*v*) *The Ashanti*: The political system of Ashanti in West Africa provides a link with that of the civilized law. They possess a powerful monarchy. Although a private dispute is ordinarily settled between the two household heads but Ashanti's criminal law is very significant. It serves the social interest in the maintenance of order; the private law has been replaced by the criminal law. According to Hoebel this society is a genuine social advance over the chaos of societies that allow feuding.

The primitive societies those show State organizations are considered as an advanced type. Because State is always a political entity characterized by the presence of a government, in some form or other.

Forms of Government

The different forms of Government as found among the existing societies can be classified in the following way.

i) *Oligarchy:* When the supreme power of the state is vested in the hands of a small, exclusive class.

ii) *Monarchy :* When the supreme power is vested upon a single individual, the King.

iii) *Gerontocracy :* When the government is run solely by the old men of the society.

iv) *Democracy:* When the supreme power rests on a representative group of the common people. Elected representatives from the State form the government.

v) *Theocracy:* When the supreme power is endowed with the priest or other sacred specialist. Such type of government shows a supernaturalistic domination over the government.

The simplest primitive societies essentially show the democracies; all adult males participate in decision making. But a social group always requires some sort of guidance which comes from a leader. The leader acts as a pivot of the community; his directions and declarations are obeyed by all members of the group, in every spheres of life. As a matter of fact, the majority of the people in a community remain ignorant. In danger or crisis when they fail to decide what to do, the leader's guidance help them to pass over the crisis. A leader may be a self-made on group-made man but he should possess some specific qualities:

(a) The person (leader) should have a strong personality.
(b) The person must hold a kind of sympathetic attitude towards others,
(c) The person, should be a good orator or speaker who can mobilize a crowd by his speech.
(d) The person must have a good power of expression, so that the mass can be easily attracted to him.
(e) The person should possess a sound knowledge regarding mass psychology.
(f) The person should be a straightforward, self-less worker.
(g) The person must be conscious about his morality, honesty and virtue.
(h) The person should have patience and adjustment power to qualify himself in any situation.
(i) The person must be thorough and well - informed.
(j) The person should have manifold interests.

However, different types of leadership have been noted in different social situations—from most primitives to most civilised type.

Types of Leadership

(*a*) *The Headman*

In the simplest type of primitive societies a headman is found who exerts some authority over other members. This authority is purely informal in nature, as the communities are invariably democratic, equalitarian, small and autonomous. By and large, every tribal village possess a headman who is either a secular or both secular and sacred head in the village. He is elected or selected by the village elders and remains responsible for the maintenance of law and order in the village. Such a leader earns popularity and confidence by his personality; he holds the group together by solving all disputes and quarrel. The societies in which headman possesses the chief authority are called stateless societies. The Eskimos, the Australian aborigines exhibit this type of leader in their societies.

In some cases, where a number of villages are associated together, existence of a regional headman is found whose court is the highest court of appeal. When a party is aggrieved with the decision of the village headman, he approaches to the regional headman for better justice. Usually, the inter-village disputes are handled as well as disposed off by the effort of regional headman.

(b) The Chieftain

A chieftain is more than a headman for his greater degree of authority and social status. The primitive societies, which have developed 'states', show the chieftains in their society. Function of chieftain varies from society to society. Among the North American Indians two kinds of chiefs are found—the war chief and the peace chief. War chief is the head of military fraternities who functions only in warfare, whereas the peace chiefs are the civil governors to supervise the internal tribal relations. Some criminal cases are also handled by them. Among the Khasis of India, two types of chiefs—sacred chief and secular chief are found. All types of chiefs enjoy high status and many privileges. The position of the chief is sometimes hereditary and generally permanent. But in few cases they are chosen for a limited tenure.

(c) The King

Presence of king is found in advanced primitive societies. In order to bring stability in the administration of government, such a leader is required. Kings usually follow an overall dictatorship and their position is hereditarily achieved. A highly developed kingship is found in Africa.

(d) The Priest King

In certain primitive societies the function of the chief or king is entirely religious. He acts as a priest or a shaman in order to bind the religion and law into a firmly united whole. A headman or a chief or a king is ordinarily considered as the specialist in controlling the actions of man, whereas, a shaman or a priest is the specialist of controlling the action of supernature. Therefore, when the politicians use the religious means for political purposes or when the priests use political means to attain the religious ends, the combination produces the Priest King. Among the Outongs of Java, the headman carries the office of the priest. The chief of the Wabena (Sudan) possesses the religious bearing.

(e) The Talking Chief

In some primitive societies when a chief acquires tremendous power and social status, he employs some mouthpieces who always talk to the public on behalf of the chief. The kings may also use the same devices in some societies. However, the employed persons are called talking chiefs. They perform the hard task to shout among the mass to let them know about chief's decisions. Existence of the Talking Chiefs is frequently found in primitive communities of Africa, For instance, the Ashanti king and Kwakiutl chieftain rarely speak among the public.

(f) The Council

Neither a king nor a chieftain can work without a council under a government. But many primitive societies lack the presence of a king or a chief. Instead, a number of elderly males form a council who take all decisions by joint will. A tribal council is usually governed by a group of headmen. The Tribal Council or the Council of the Tribals is not only prevalent in India, they are found elsewhere in the world. In India, it is specially known as Tribal Panchayet and operated in the village level to control the socio-political sphere of the primitives.

Nature of the Primitive Law

Like any other social institution, law is a part of the society and functionally related to the structure of that society. It is a rule of conduct, essential for the stability of the society. The imposition is done by an authority and accepted by all members of a community. In simple societies, the force of public opinion or a decision arrived at by democratic procedure is impelled on the accused. The community or the council acts as the court. Primitive law stands exclusively on time-honoured customs, usage and beliefs of the people. At times, supernatural fear enhances its force. On the whole, primitive laws are essentially different from the modern laws on the following grounds :

(*i*) *Primitive laws* largely count kinship ties rather than the territorial ties. Kinship bonds are highly emphasized. Both the internal and external problems are dealt with the kin group. Clan elders possess the right to punish the offenders and settle the dispute. They look after the community for the prevention of any deviation of customary law as well as bear the responsibility of organizing the warfare. However, this law may be regarded as a private law, which is totally different from the modern law.

(*ii*) *Primitive laws* coincide with ethical principles and rooted in public opinion. In a homogeneous native settlement, public opinions powerfully influence the life of the people. Since the number of the individuals is limited in such communities, they know each other quite well. Naturally there is no scope of segmentation in the public opinion; common sentiments are shared by all. So, the opinions of the public based on ethical norms stand as exclusive, strong and compelling in the society. These opinions discriminate between 'good' and 'bad', and are expressed in Tribal meetings.

(*iii*) *Primitive laws* do not differentiate between the 'crimes' and 'torts' i.e., between the public and private wrongs as evident in modern jurisprudence. 'Crime' is considered as an offence against the whole community or state, whereas 'tort' is merely a personal wrong. As primitives are not much concerned with the territorial tie, wrongs against the state are hardly recognized. This is the reason for which most of the primitive societies lack specialized personnel and large measures are not taken. Immediate kins of a wrongdoer usually come forward to intervene into the matter. Punishments are chalked by them without the interference of the whole community. In a few rare cases, where harmful effects touch the whole society, collective energy of all members in the society are found to be involved.

Laws in the primitive societies are typically the unwritten laws; there is neither any court nor any codified body of law. Further, these laws vary from society to society. Every society has its own specified customs and conventions, which comply with the norms of the respective society. This means that the individuals in a society are entirely regulated and controlled by its 'sanction'. Sanction is nothing but a social force, which approve or disapprove a mode of behaviour as manifested in the measures like suggestion, persuasion and encouragement. Sometimes it offers a high social appreciation or reward to those who follow the approved pattern of behaviour. Public opinion thus compels a person to behave in conformity with the prescribed usage of the specific society.

The social sanctions can be broadly divided into two categories—*primary* and *secondary*. Primary social sanctions are the immediate sanctions that are obtained by an individual in his society. Primary sanctions are of two types—*positive sanction* i.e., approved behaviour and the *negative sanction*, the disapproved behaviour. Both the positive and negative sanctions can again be divided in two groups—diffused and organized. Some of the social sanctions, whether they are positive or negative, are carried out following a traditional as well as recognized procedure. These come under the 'organized group'. On the other hand, sometimes society expresses the approval or disapproval of the behaviours quite spontaneously. These make up the 'diffused group'.

Every individual in a society has a desire for being praised and he avoids the hatred of his fellows. Again, one aspires to win a reward and avoids a punishment that a community offers or threatens. An individual never wants to be an unworthy, disreputable, dishonourable and discourteous person.

All these factors, ultimately, keep a cummulative general effect on the community at large and help in restoring the equilibrium by providing a collective expression of social approval. Negative sanctions in a primitive society generate from the ethical and supernatural forces because certain acts are considered as displeasing to the god, which make the accused a sinner.

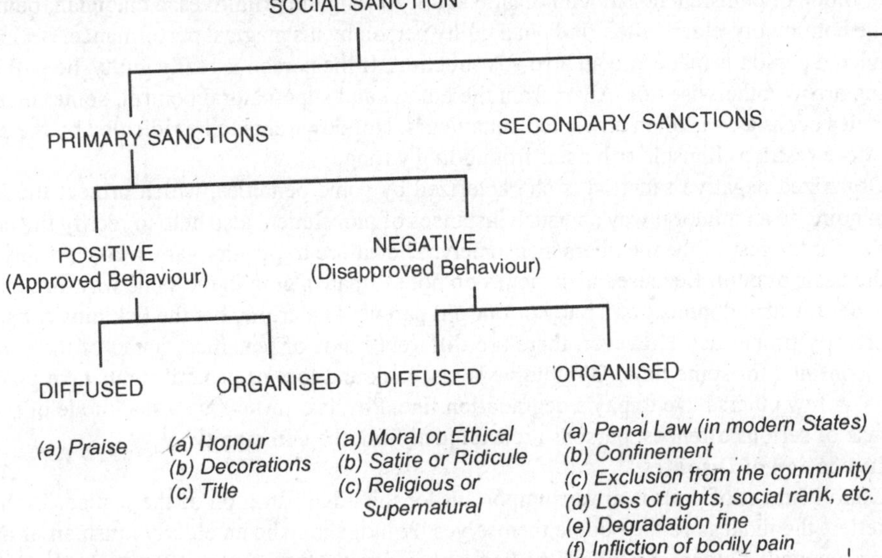

Fig. 14.26. A Model Showing Types of Social Sanction In The Society

Secondary sanctions are concerned more with the actions of persons or groups in their effects on other persons or groups. They include certain procedures acknowledged by the community. Most of the societies sanction a controlled as well as limited act of revenge for the satisfaction of injured person. But this satisfaction should be proportionate to the extent of an injury. Acts like retaliation, duel and indemnification are the socially approved means that can be grouped under secondary sanction.

Forms of *Positive Sanction*

It signifies the appreciation of certain kinds of behaviour. As observance of social standard is always approved by the society, so an individual following the social norms is not only respected, his position in the society is also raised. The common belief is that, ancestor and other spiritual beings are pleased by good conduct. Some believe in reward after death. Ao-Nagas is a head hunting tribe. In their society, the individuals who hunt a greater number of heads achieve a higher position in the society. Many primitive groups spend their surplus food on festive occasions in order to raise their social status and prestige. In diffused type of positive social sanction, an individual is praised for his work; he is respected by his fellow members in the society. In case of organized positive social sanction, special honour is bestowed on a person. He is crowned or glorified in some definite ways. Sometimes the community honours him with specific great title.

Forms of *Negative Sanction*

It is the disapproval for the non-observance of certain social standards. In diffused negative sanction, the force of authority is supernature. This consists of some negative rules, each of which sets a specific form of ideal behaviour, infringement of which definitely evokes the anger of the supernature. This anger in turn harms the transgressor. People of Eddystone Island believe that if a man plucks a fruit from other's tree without the permission of the owner, he will be punished by the ghost. As a result, he may vomit blood or his hand may stick to the fruit. This mode of punishment is automatic and these only works for the breach of a taboo. Among the Nagas, eating of goat's meat is prohibited. They believe that, those who ignore this taboo, their hair become prematurely Grey.

Among the Kamars of M.P., a menstruating woman is considered as a polluted individual. So she is not allowed to enter in the shade of a goddess or into the kitchen. In case of deviation, the supernatural force punishes her as well as the family. The Eskimos never violate the food-taboo because they believe that a particular goddess becomes so angry that she will keep the seals away from them. The other mode of punishment is not automatic at all. Among the primitives, a magician, being directed by the community-elders often find out a guilty person by his magical performance. An image of the convicted person is made and an arrow is shot at it. If the person is really guilty, he will be harmed by the arrow; otherwise not. Apart from the ethical and supernatural control, sometimes satires or ridicules act as a control on unwanted behaviours. Nobody wants to be ridiculed by the society and therefore restrains himself or herself from doing wrong.

Organized negative sanction is characterized by some penalties, which protect the individuals from going to an immoral way and such instances of punishment also help to rectify the behavioural modes for the rest of the members in a society. The nature of penalty varies from society to society on the basis of guilt. Because, all societies do not estimate a guilt in the same way. For example, we consider the abandonment of a baby or our old parents as a crime, but the Eskimos consider this as a perfectly proper act. However, there are different kinds of penalties. Some of the offenders are kept confined for some time. Diminutive social rank and loss of general rights take place in some cases. A few others have to pay a degradation fine. Physical torture is a usual mode of punishment. In case of serious offences, persons are excluded from the community.

Adjudication of Disputes

Establishment of guilt is always important for the administration of the justice. In the simplest societies, the judges are the people themselves. A judge may be an elderly kinsman or a headman. Where council of elders are called into being, they constitute the judges. The authorized persons get together to hear a case and pass the judgement. In most of the cases the accused denies the particular charge against him. Until and unless his innocence is proved, he does not get mercy. Otherwise he is subjected to punishment. Therefore, a judgement essentially comprises of two parts—evidence and punishment. (*a*) *Evidence*

The process by which the innocence or guilt of a person is established is called evidence. In the primitive societies, no judge or prosecutor is found who is expert in cross-examination. Therefore, to get the accurate facts people have to rely on supernatural support. However, the two main ways of getting the evidence are oath and ordeal.

(*i*) *Oath*: It is a promise in the name of God for not to tell a lie. It is said that, if the facts furnished by a person is proved false, the person will be punished by God. This is a technique of compelling one to confess his own guilt voluntarily. Usually a sinner for the fear of supernatural wrath does not conceal his guilt. Among the Oraons and Mundas of Chotonagpur, an individual before producing his evidence is asked to take an oath sitting on a tiger's skin or tiger's jaw. The guilty person never dares to tell a lie for the fear of being killed by a tiger. Among the Maler, the accused touches a sacred axe and utters. 'I will die if I tell a lie'. Santals utter, 'Dharam Dharam' before taking an oath. The results of perjury are dreadful in all primitive communities. Therefore, decision on a dispute is easily derived from the nature of evidence.

(*ii*) *Ordeal*: It is a process of determining the guilt or innocence by submitting the accused to a dangerous or painful test under supernatural control. Among the primitives such tests are usually done with fire, charcoal and water. For instance, among the Oraons, a piece of burning charcoal is placed on the palm of two boys who are suspected of theft. If one of the boys is able to bear the hot charcoal on his palm, he is considered as innocent, whereas the other boy is considered as guilty. Among some tribes of Assam, water is boiled and some pebbles are put into that water. The suspected person is asked to remove the pebbles by bare hand. If he feels pain or unable to do the act, he is declared as the real guilty.

(*b*) *Punishment*

If any one violates the general rule of the tribe, he is punished by the leaders of the community. Punishment corresponds to the organized negative sanction of the society. The several modes of the punishment can be categorized into two principal types—*Trial* and *Weire-guild*, according to the gradation of crimes.

Trial: Among the Oraons, the clans are exogamous. Naturally a member of any clan must chose his mate from other clan. In case of violation of this rule, the man is brought before the village council (Panchayet) for trial. Similarly, since the tribe is endogamous, a man must find his mate from within the tribe (excepting own clan), otherwise the man is punished. Outcasting is a usual punishment. Everybody drop the social relationship with an outcast. Among the boat builder Samoans, a peculiar type of punishment is in vogue. The person who commits a crime is not further allowed to build boats which is his only way of earning bread. Imprisonment as a form of punishment has been noted only from Uganda, in Africa. A more severe form of punishment is the infliction of bodily pain; culprit's arms and one leg are fastened to holes in heavy logs. Although incidence of murder is not uncommon in the primitive societies, but death by hanging is not found. All untoward occurrences e.g. intra-family and village disputes, assault, quarrel over immovable property, theft, divorce, moral and sexual offence etc. are handled and disposed off by the traditional Panchayat.

Weire-guild: It is a punishment where compensation has to be given. The fine may be charged in cash or in kind or in both, depending upon the seriousness of the crime and the capacity of the culprit to pay for it. The Kirghizes pay a fine of one 'Kum' i.e. 100 horses or 100 sheep for killing a common man. The compensation items become seven times higher, if a nobleman is killed. In case of an offence relating to incest, the person is excommunicated (ex. Bitlaha among the Santals). The accused may sometimes be taken back into the community but a community-feast must be arranged by him. The objective behind such punishment is manifold. Firstly, it compensates the party who has incurred a loss. Secondly, it purifies the soul of the guilty and thirdly, it brings the enemies together.

From the above discussion, it has been clear that law is the primary requirement for any society. It consists of a set of principles that permit the use of force to maintain political and social organization within a territory. As a means of social control, it performs four functions : (i) It orders the relationship between the individuals and also between the individuals and social groups. (ii) It allocates and enforces the norms of the community. (iii) It settles the disputes between individuals as well as minimizes the offences in the community. (iv) It regulates the human behaviours against changing conditions of life and makes the culture adaptable.

However, the laws do not change usually; only the interpretations are changed. In primitive societies, since offences are more concerned with the individuals in kin group; this law is chiefly designated as 'Private law'. It differs from the Public law or Criminal law as no specialized officer is found there for the prosecution and punishment of the individuals. Although criminal law is not totally absent in the primitive societies, but they are seldom found. The scope of law expands with the growth of society; the traditional customs and conventions give rise to the sanction of law, with time and force.

RELIGION

Morality or the 'moral code' is often taken as the rules of behaviour in primitive communities. Morality has been sharply distinguished from the 'mores' i.e. moral standard of a group. It is actually a body of principles and rules concerned with the good and evil as manifested by conscience. It remains intimately associated with judgements to stand as a social code. Moral ideas are reflected in the customs and conventions to control the human conduct. For example, certain sentiments arise within a man that ashamed him in doing wrong, condemn him for doing wrong. So the man abstains himself from a sinful act. These strong moral feelings are not the instincts, therefore not universal. They develop out of culture. The specific culture pattern helps an individual to build some moral principles, which are transmitted through various customs, norms, values, prohibitions of the society.

Thus, different cultures produce different moral codes and so the prevailing behaviour in one society may seem to be strange and even repellant to the other.

Religion is a supernaturalism that consists of a system of belief, thought and action. It lies in the core of all primitive and civilized cultures. It acts as an internal controlling force for the society and provides the people with morality. A religion can neither be defined in terms of a particular faith, or in terms of a particular god. In fact, there are a variety of religions and religious ideas. The first and foremost necessity is to examine the nature of supernaturalism.

All religions essentially exhibit a mental attitude towards supernature, which is manifested in belief and rituals. The belief is considered as the static part of the religion while ritual is the dynamic part. Ritual comprises of different actions that aim to establish a connection between the performing individual and the supernatural power. Belief, on the other hand, has no direct impact; it stands as a charter for the rituals and provides a rationale for the same. However, religious attitudes are universal in all known cultures, primitive and modern. They have been associated with the Homo sapiens for last 60,000 years at least, as inferred from the artifacts.

Origin of Religious Belief

General philosophy of the people admits two kinds of ideas—nature and superior to nature i.e. supernature. The concept of nature and supernature is relative in a culture at a particular moment. With the growth of the knowledge some of the supernatural events may seem to be natural. In fact, the difference between the nature and supernature lies in the attitude and realization of people perceived by the help of sense organs. In this context R.H.Lowie quoted the following words: "a sense of something transcending the expected or natural, a sense of the extraordinary, mysterious or supernature". This delineates the features of religious belief, which is essentially mystical and subjective in contrast to the naturalistic belief that is objective and rational in determination of facts.

Anthropologists have considered religion as a product of the evolutionary development of human brain. The capacity of the brain in other animals is so low that it does not permit them to think like man. They never perceive that vastness of universe as they lack sensitive mind and emotional feelings. Therefore, the first religious belief probably came into existence with the origin of first man in early *Palaeolithic* and since then the mystic thought has controlled much of human life until Aristotle, Plato and other Greek philosophers built the foundation of modern scientific outlook, Anthropological inquiry in religion extends as far back as nineteenth century when anthropology emerged as an academic discipline.

There are different theories regarding the origin of religious beliefs. The earliest one was forwarded by E.B.Tylor (1871), where he expressed the view that religion had stemmed from the intellectual speculation about the events like dreams, trances and death. His proposition was three-fold. (a) Religion has been developed out of fear. (b) Though there are great diversities in the forms of religion in the world, the core matter of all religions are same. (c) All religions acknowledge a supernatural power.

Herbert Spencer (1820-1903) thought that religion arose out of ancestor worship. Sir James Frazer (1854-1941) held magic as a source for the development of religion. Most of the scholars of that early period believed that religion evolved by the interplay of the emotions like awe, fear and wonder under the exposure of the nature. But still there were others who took religion as a normal psychological adjustment by which societies built a protective barrier of fantasy against fear. Emile Durkheim (1858-1917) regarded religion as the most primitive of all social phenomena. He found two distinct compartments in supernatural field, which he designated as sacred part and profane part. According to him the sacred part is the religion that refers to gods and deities, and also the sacred performances. Contrary to this are the profane beliefs and practices. Naturally, the second part is not sacred and so it is considered as magic.

B. Malinowski and A.R.Radcliffe-Brown have given functional explanation of primitive religion. Malinowski found religion as associated intimately with various kinds of emotional responses, so he described religion as an adaptation, which dilutes all stress and strain of the individuals. To Radcliffe-Brown survival of a group was more important than that of an individual. Therefore, he suggested social survival by the aid of religion. It is the fear of supernatural punishment for which people adhere to the norms of behaviour in a society. Both the scholars, Malinowski and Radcliffe-Brown derived their explanation following Durkheim's theory of religion. Most of the sociological theories suggest that religion originated from societal needs and symbolizes sacred phenomena. Freud found the root of religion in the unconscious; it provides a therapeutic solution to inner conflicts. Not only Freud, all reputed psychoanalysts agree that religion satisfies some psychological needs of the individual. It relieves anxiety, uncertainty and various tensions. S.F.Nadel (1954) had shown four main functions of religion, which are as follows: (i) Religion has the capacity to supplement the view regarding the world of experience. This is an explanation strictly from the cosmological point of view. (ii) Religion has the capacity to announce and maintain moral values. (iii) Religion has the competence to hold the individuals in a society and to sustain the structure. (iv) Religion provides individuals with specific experiences and stimulation.

In non-literate societies religion is the chief factor that binds the people together. The bond of relatedness is so pervasive that it is manifested in collective activity like law, morality, art, science, political forms etc. According to Durkheim, the origin of religious experience was to be sought in the collective elevation of feeling (euphoria) that occurred when individuals gathered together in a community. Mysterious world of power and force influence different peoples in different ways. Therefore, religion varies in beliefs and content but its flavour remains the same. The individual and the group strive to relate themselves to the unseen universe and ultimately a pantheon of deities is created. People look at them with awe, reverence, fear and wonder; different mysteries move around the orbit of religion.

Concepts in the Development of Religion

Anthropologists tried to trace the evolution of religion from simple to complex forms. Edward Burnett Tylor in his book 'Primitive Religion' (1871) showed this evolution from *animism* to *monotheism, through polytheism.*

Animism

Animism is the earliest concept towards the religion forwarded by Tylor himself. It is a belief in the existence of spiritual beings. Spirits are the ethereal embodiment without real flesh and blood. Although they are non-material, but real enough for those who believe in it. Primitives use different names to refer those spirits—ghost, goblin, genii, trolls, fairy, witch, demon, devil, angel and even god. A spirit does not obey the laws of nature and can transcend matter, time and space. This makes the spirits wonderful and mysterious, and therefore, they have been regarded as supernatural.

Tylor had examined different manifestations of animism among the primitives. Since he was an evolutionist, his interest was not limited only to the forms. He wanted to know why and how the concept of spirit was developed. According to Tylor, all primitive people recognize some sort of subtle invisible matter inside each living body, which initiates the activities of the individual. When the matter leaves the body, all activities of life are stopped and the body becomes still. This unseen subtle matter is regarded as power and termed as soul (anima). As the dead people often come in the dreams with life-like appearance, souls are considered as indestructible and eternal. After leaving the body, they become spirits and begin to dwell in certain places or things but capable of communicating with man. This soul concept is the root of animism

Tylor held that the early men perceived two parts in a man—the bodily self, made of mortal flesh and the spiritual ego, the soul. Although the soul was intangible but formed the vital force of the body. Early men equated this soul with shadow or shade. Warneck and Wundt traced two kinds of souls in man–the body-soul or the fixed soul, and the free-soul or the shadow. The body-soul animates the body and pervades whole of it. Whereas the free-soul being independent of the body can come out of the body and wander in different places. When it departs the body forever, the man dies. During sleep, this soul often goes for wandering and experiences varied things. The dead persons, distant relatives, next-door neighbour, ferocious animals appear in dreams and trances, because the soul interacts with them. However, a man goes on sleeping until the free soul comes back to the body to make him awake. Free-souls can easily be captured and killed by the enemies. Diseases and injuries are created when these souls are affected.

The primitives seem to be much anxious with the ultimately departed immortal body-soul, which can harm the living people. There is hardly any primitive group who does not care this soul. But the ideas regarding the size and shape of this soul differ from region to region. Among the Lakhers of Assam, the soul resembles the original body in appearance and size. The Thadau Kukis of Manipur considers the soul as a minute replica of the individual. The idea is same to their Naga neighbour of Manipur and Naga hills. Sema Nagas view the soul as shadow or reflection of the body. In Chotonagpur and Madhya Pradesh, souls are absolutely responsible for providing the shade and shadow. They Oraons also believe the soul to resemble a shadow but in lighter and more intangible form. The Baigas of M.P. call the shady soul back to the house on tenth day after death with affectionate care. Murias of Baster believe in the same procedure. The Chenchus of Hyderabad know that the souls turn to mortal world immediately after death. Todas of Nilgiri hills hold an opposite view. They think that the souls go to the other world and lead there a life, which is almost similar to this world. The Mundas of Ranchi, after cremation of a corpse, make a small human effigy with grass to represent the soul of the deceased. The Hos, the Murias, the Ao-Nagas, the Purum Kukis, the Baigas follow a more or less similar practice. However, animism has been universally accepted as the essence of primitive religion.

Animatism

A few scholars criticized Tylor's theory of animism. According to them, the theory of Animism was too critical and that did not deal with the emotional content of religion. Prof. K.T.Preuss and Prof. Max Muller postulated a pre-animistic stage where the spirits were not supposed to be only those of the dead man; rather each and every object was considered to have a life, and so all of them were the animate (living) objects. The concept got its popularity in the name 'animatism'.

Max Muller had shown that the earliest form of religion consisted of the worship of various objects in nature. An attitude of awe and reverence worked in the mind of the people regarding the diversified natural objects and phenomena. Being perplexed, they ascribed life to the life-less things and correlated the unseen source of power with God. Prof. Muller tried to support his view with archaeological evidences as found in Egypt and elsewhere.

More recently. Prof. R. R. Merett explained the situation with a specialized form of animistic theory called *'Manaism'*. 'Mana' is a Melanesian term meaning power. According to Prof. Merett, the primitive peoples throughout the world believe in the existence of an impersonal, non-material supernatural power, which belongs to all objects –animate and inanimate. It resides beyond the reach of physical senses but manifested as a physical force. All objects in the nature possess it in greater or lesser degree.

The force *of mana* does not maintain conformity with the regular laws of nature or technological knowledge and skill. Even it does not work always. But when works,it shows an exceptional power to do things which are unusual. It appears all on a sudden as a supernatural attribute of persons and things. Any extraordinary aptitude of man can be explained in terms of mana. Any success in life is therefore, considered as a blessing for mana In Polynesia, the master craftsman is able to show his excellence because he possesses mana. The knowledge of a learned man is reflected for having mana. The weapon with which a mighty warrior wins a battle is also a favour of mana. Whether an outstanding healer or an expert canoeman, every successful man possess the power of mana in the background. The cases of failure are counted as a loss of mana. The Polynesians think that their kings and nobles hold a greater degree of mana than the other classes. Any misfortune in the kingdom like outbreak of epidemic diseases, famine etc., is believed to be caused due to loss of mana on the part of the king.

M. J. Herskovits compared mana with electricity. In his words, "like electricity, mana is impersonal, and like electricity it can be channeled, directed and used to achieve a desired end by one who knows how to manipulate it". Such a power can be diverted from an object to another, from a man or animal to another. Ideas similar to mana are found all over the world, though they are sometimes called in different names. For example, the American Indians recognize a similar power known as 'Orenda', the Iroquois call it 'Wakonda' and the Algonquian Red Indians like the name 'Monitou' .Some Indian tribes namely Santals, Mundas and Hos designate the power as 'Bonga'. This mysterious force influence not only the man and other living beings, it can also act on non-living objects like sun, moon, mountains, rivers, rocks, and seas.

The power of mana is sometimes referred as *'Fetish'*. A fetish is an object like stone, shell, necklace or a piece of carved stone, which is believed to have a power, capable of helping its possessor. The fetish is, therefore, adored, placated, insulted or ill-treated according to its behaviour whether it fulfils or does not fulfil its possessor's wish.

Supenaturalism is less decisive in the simpler societies than in the more complex societies. Simpler people rarely quarrel over doctrinal issues; they stick to the maintenance of rituals.

Components of Primitive Religion

All supernatural beings can be categorized into two broad groups: (i) Those are of non-human origin i.e., the nature gods and spirits. (ii) Those are of human origin i.e., the departed souls like ancestral spirits and ghosts.

The Supernatural beings of both the categories can induce good and evil for men. They have been the causes behind many successes and failures. Diseases, drought, storm, heavy rain, famine, epidemics are also created by them.

Gods and Goddesses occupy an important position in primitive life. The entire universe is departmentalized among gods. These gods and goddesses are usually the self-creators, some of them give birth to other gods. The progenitor gods always hold a prestigious rank. A few of the non-literates believe in a supreme god, creator of the world. This god is regarded as a high god, other gods are considered as lesser gods who help the creator in running the creation. The Maoris of New Zealand depend on three gods—the god of the sea, the god of the forest and the god of the agriculture. They propitiate these gods in turn to get their blessings. Most of the primitives believe that their deities live in the crests of hills. Bandos recognize a goddess who gives rain, when remains pleased. Being neglected the same god begets drought and throws itches or sores to the people.

A large number of primitives in India worship sun as the principal god. The conception of a high god as a creator of the world is prevalent among the tribes of Jharkhand (Santals, Oraons, etc.). Other minor gods remain subordinate to high god. The rank of the spirits is always lower than the gods or goddesses. But they may be malevolent or benevolent to influence on the destiny of man. Commonly the spirits appear in the disguise of animals. They can also take the shape of a rock, lake, mountain, cloud or whirlwind. A lot of spirits are worshipped by the Korwas who are believed to preside over crop, rainfall, cattle etc.

Ancestor spirits and ghosts are supposed to have originated from human beings. All of them are the departed souls. But the ghosts are dreaded, because they are prone to make harms. It is held that the man who dies in an unnatural way becomes a ghost. A ghost may haunt a house and disturb its inhabitants. So they are either avoided or respected by the people. Kannikars of Kerala propitiate the ghosts to get rid of their wrath. The ancestor-spirits, on the other hand, receive higher regards from their descendants. Dead ancestors are often worshipped, as their blessings are indispensable for family prosperity. Almost every Indian tribes practise ancestor worship.

There are some societies where all gods enjoy equal prestige and power i.e., none of the gods are held as superior or inferior in rank. Such a religion is termed as polytheistic religion. Contrary to this, some other societies recognize the supremacy of a single god as a creator of the universe or the dictator of all events. The latter type of religion is called monotheistic religion. Monotheistic religion is generally regarded as a very late development where religion merges with philosophy. Polytheism although indicates a higher level of social development than that of a tribal or village life, it is philosophically less developed than Monotheism. Societies with increasing wealth and organized priest-hood show polytheistic disposition. But religion of primitive people essentially shows the prevalence of animistic beliefs and they are more concerned with evil forces, fears and frustrations. In the history of religion, polytheism first appeared in the advanced cultures of India, Egypt, Mesopotamia and Greece.

Nature of Religious Practices

Religion as a body of belief and practices show a wide variation in religious ideas. The religious practices are also varied. These practices are nothing but the techniques to communicate with the supernature. But they are imperative for the believers who act in accordance with their beliefs. Such practices strengthen the social bonds in a primitive group and denote an added authority towards the customs. However, the practices can be classified into two sections: religious rites and rites of passage.

(*a*) Religious rites

God is the central concept in religion, although the ideas, convictions and emotions vary from society to society. Religious rites aim to appease a god by worship, which can be performed either privately in the home or publicly in the temple. The forms may be different as prayers, offerings, vow celebration or sacrificial performances. Prayer is the simplest of all religious rites where reverence is shown by means of spoken words. It may be a request or a demand or just thanks. Sometimes the process is totally silent or memorized.

Worship with offerings includes various types of goods and food items, which are placed in front of the deity in an elaborate way before propitiation. The vow is a promise of offering something special, to the deity at the fulfillment of certain personal wish. Sacrifice may or may not be associated with vow celebrations. A sacrifice means the giving up of an object, such as food, drink, household goods, life of an animal etc., in the name of a god. The main objective is to influence the god's action. The deities are supposed to grant boons for long life, success in war, cure of diseases etc. Besides, the blessings of gods are also required to keep the people out of danger. Thus, different sorts of fear direct the individuals to follow the path of reverence. Let this fact be justified in reference to some tribal people of India. The great sun god, *Singbonga* among the Mundas is seldom worshipped, as he does not harm anyone. Since this god never gets displeased, no question of appeasing comes. Otherwise, to divert the anger of the god, primitives not only sacrifice the animals, they are often ready to sacrifice the human too. Instances of human-sacrifice are persisting among the Nagas of Assam and Khonds of Orissa till the recent days.

(*b*) Rites of Passage

Rites of passage are completely different from religious rites that comprise of the worship of the nature-gods and different spirits. These special rites are significantly associated with the life cycle of the people, in each and every society. They mark the passing of one phase of life and the entry to another e.g., birth, puberty, marriage, initiation to priesthood, death etc. They are known in English by French equivalent Rites-de-Passage, and popularly known as 'life-crisis' rituals. Arnold Van Gannep first brought this conception to attention. He also distinguished these rites into three types: rites of separation, transition rites and rites of incorporation or aggregation. According to him, rites of separation often get prominence in funeral ceremonies, transition rites stand important in pregnancy, betrothal or initiation, and the rites of incorporation are widely visible in marriage ceremonies.

Puberty rites are frequently performed by the tribal girls at the first signs of menstruation. Sometimes these rites solemnize in ceremonies through the worship of a deity. Among the Uralis of South India, a common tree house is built where all women of the village live during their period of menstruation. Among the Australian aborigines, an initiation rite is chalked out for the boys, which is quite painstaking. A boy has to bear immense physical torture until he is frightened or scratched. His front teeth may even be knocked out. At the end, a ceremony is organized, after which the boyhood passes away. The boy is recognized as a man and he is then allowed to marry.

Mourning rites are universally found among the primitive groups although there are wide variation in performances. The Bagadas of Nilgiri hills dance around the corpse. Such funeral dances are also evident among the Juangs, Savaras, Kols and other tribes of Qrissa. The Mishmis of N.E.F.A. utter prayers in a close-door room for days together; a feast is held thereafter in the honour of the dead person. The Gonds arrange a peculiar ceremony for bringing back the departed soul back. The bereaved relatives go to the side of a river and call out the name of the dead person. After some time, they catch a fish and bring that to home with a belief that the soul would be reborn in the family. The period of pollution differs from one tribe to another. It may be ten days or more.

Significance of Religion

The rites and ceremonies create an atmosphere of benevolence and fellowship. All motives for quarrel and disagreement get eliminated. People are united together and the rejoicing activities energize them; the social sentiments of an organized community are renewed. Religious experiences create such an atmosphere and attitude that human beings are able to regulate their own conduct. Everywhere people have evolved religious systems in which religious behaviour aims to secure similar ends. It bears the testimony to the unity of mankind.

The quality of any particular religion reflects the basic value system of the society. It binds the inter-familial relations, and governs on the economic and political structure of the society. It may include a variety of cults and specialized religious personnel. People try to unload their acute mental pressure under the banner of religion. They seek support and stamina from supernature in the way to struggle for existence.

MAGIC

Magic is the other concept by which supernatural forces can be approached. The term 'magic' has been derived from the French word 'magi' which is used to refer the secret deeds. Unlike religion, the rituals are performed here in order to compel the supernatural power to act in particular ways. It never involves praising or praying, rather it commands on supernature to serve good or evil purposes. Magic and religion are though the opposite approaches but they seem to have always existed together. Sometimes they come so close that they blend together.

Magic corresponds to the concept of profane as advocated by Emile Durkheim. To some scholars magic is more primitive than religion. But other scholars think that there is no reason to ascribe greater antiquity to magic because both religion and magic are the extremely ancient components of human world-view. R. R. Marett and A. A. Goldenweiser pointed them out as two compartments of supernaturalism. The everyday life of pre-literate people was surrounded by unpredictable incidents and terrifying hazards. They had no empirical knowledge to cope with those hazards. Therefore, a quest for supernatural aid appeared in the mind which was further splited into magic and religion. Both magic and religion may be regarded as the tools of adaptation, which have helped man to relieve tensions in different difficult situations.

The valid distinction between religion and magic was first pointed out by Sir James Frazer who is regarded as the founder of the concept of magic. In the book "The Golden Bough', Sir Frazer classified a number of magical formulae which he collected from the primitive peoples all over the world. He wanted to show that all magics work on either of the two fundamental principles: (a) Like produces like (imitative magic). (b) Once in contact, always in contact (contagious magic),

The imitative magic is based on the principle of similarity, whereas the contagious magic is based on the principle of contact. The first principle is derived from the assumption that, like objects and acts has an affinity with each other. So if anything happens to one, the other is readily affected. The assumption for the second principle is that, the things, which were once in close contact, retain an influence over each other, forever.For instance, burning of effigy of an enemy is a common form of imitative magic; by harming the effigy one wants to harm the actual person. Tribal people of Chotonagpur consider thunders as the direct cause of rain. So after a long period of drought, they perform a magical rite on the top of hill. A pig or hen is sacrificed at first and thereafter the people push down a lot of stones, rocks and boulders to create a rumbling noise. The people believe that the rumbling noise like thunder will compel the rain to come immediately. With the same expectation, the Hos also light fire. Smokes reaching on high sky resemble the cloud and the people obviously look for rain.

Contagious magic works in a different way. The non-literates are often frightened to use other's clothing. This is not for hygiene-consciousness, but for the anxiety of being harmed. Clothes are considered as parts of person's body who use them. So evil can be done on them very easily and the user is affected. This is also true for the nail cuttings, hair trimmings, bodily excretions and some other extremely personal belongings. Magicians are in the habit of collecting these things in order to harm an enemy or undesirable person. Sometimes the name of the person is used in contagious magic.

Magic may be good or bad according to the values of the society concerned. The magic, which aims at good ends, is called 'white' magic, whereas the 'black' magic has an evil objective. Both of them are found in all societies but the standards of judgement vary from culture to culture. To counter the 'black' magic, the charms, amulets and spells are widely used. These are the protective devices against devils, spirits and other malevolent forces. Wearing of small objects like charm or talisman protects the believer from much danger. Sir Frazer equated magic with modern science. He said that the cause-effect phenomenon of magic coincides with experimentation-observation of science. Frazer not only reduced his magical principles into laws; he also distinguished a theoretical as well as a practical aspect in magic. Practical magic consists of some positive and negative devices. Positive devices include sorcery, witchcraft etc. while the negative devices contain a number of taboos. However, sorcery and witchcraft both are the symptomatic expressions of social tension and conflict.

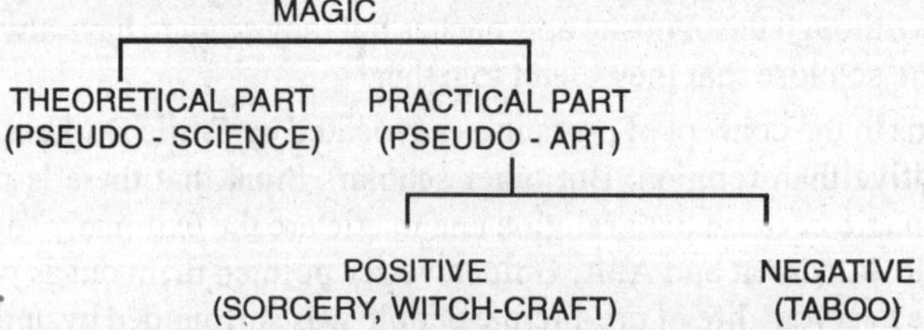

Fig. 14.27. Magic as viewed by Sir James Frazer

Sorcery

Sorcery is a positive magic used for evil purposes. It involves the use of certain materials, objects or medicines for invoking supernatural power to harm people. The materials here are the parts of body on which specific spells are used. It is purely a 'black' magic. Sorcerers are the specialists for sorcery. They are so cunning that sometimes instead of using the tangible relics like hair or nails, they utilize man's shadow or sleeping soul. Primitives are often scared of sorcery, as they know its unpleasant occurrences. In most of the primitive societies, such an act is not socially approved and so treated as a crime against society.

Witchcraft

Witchcraft is a malevolent practice with the help of the spirits. Here ills are carried out by means of thought and emotions alone; tangible objects are not used at all. Therefore, when magic works upon the bewitched, the person begins to suffer but the evidences of witchcraft are not left. The lack of visible evidence makes an accusation harder to prove or disprove. Among the Azande, witchcraft is a part of daily life. Anything that goes wrong—an ache or pain, poor crops, an accident, loss of cattle–everything is attributed as the malevolent act of the personal enemy who has used witchcraft against him. Witchcraft is thus a part and parcel of all economic pursuit, domestic life and community life of the Azande.

A witch by applying a psychic procedure may injure anyone; he may bring the death even. The power arises from a peculiar substance produced inside the body of a witch. Since they keep no external evidence, many innocent persons are often suspected as witches and killed by the mass. Witchcraft and witch-killing both are crimes in present day situation.

Taboo

The Polynesian term 'taboo' means 'forbidden'. All sorts of restrictions came under the taboo. It is the unwritten law of the primitive society. Its function is, therefore, predominantly psychological; being originated in fear, it deals with the forces that are not wholly understood. There are some guidelines regarding the observance and non-observance of taboos. Both are important for a society. Taboos aim to limit an individual to the norms of the society; it avenges itself when violated. This is the reason for which taboo is often referred as a negative magic.

Various kinds of taboo are found in the primitive societies. They have been classified as productive taboo, protective taboo and prohibitive taboo on the basis of the purposes that they serve. Generally the motive behind the protective taboo and the prohibitive taboo remains the same. The supernatural power is supposed to punish the breaker of law, if a taboo is violated. There is some prescribed standard of behaviour, which are not directly controlled, but indirectly supervised by the supernature. Intake of certain food items and adherence to certain acts are never taken for granted. Most of the primitives regard incest as a crime, which deserves punishment from supernature. Moreover, a large number of general taboos are found to be connected with the events of life and death. Attitudes of carelessness and profanity in dealing with supernature are always punished.

Types of Practitioner

Various types of practitioners are found in primitive society who contacts the supernature directly. These practitioners may be part-timers or full-timers, a religious practitioner or a magical performer. The nature of the practitioners varies with the degree of cultural complexity. The more complex is the society, the more diverse type of practitioners are obtained. The cross-cultural research of Michael James Winkeman (1986) suggested that there are four types of religious practitioners in non-literate societies. They are *Shaman, Sorcerer or Witch, Medium* and *Priest.*

Shaman

Societies with subsistence economics show the presence of Shaman. Such societies tend to form with nomadic or semi-nomadic food-collectors. The word Shaman has been derived from the native Siberian tongue, as Siberia is an ancient center of Shamanism. A Shaman is usually a part-time male specialist, proficient in magical rites. He holds a fairly high status in his community. People often call him to cure the diseases. Shamans know sacred songs, pantomime and other specialized formulae to serve the society. Although westerners sometimes call Shaman as Witch doctor, but actually he is more potent and important figure than a Witch-doctor.

The preparatory stage of shamanistic inspiration is painful and lengthy. They not only undergo hard trainings by the older Shamans; they remain far apart from care and distractions of ordinary life. At this time they experience the supernatural beings and the other world. The spirits talk with them and instruct many magical formulae. Thus, they acquire great power to summon a storm, to banish a game or to cure a patient. Two kinds of Shamans are found—the Emotional Shaman and the Steady Shaman. The Emotional Shamans are usually unstable and even epileptic. During the performance of a rite, their physical features get distorted. Muscles are convulsed and eyes are strained. A Shaman at this stage may roll on the ground and brings foams at the mouth. A Siberian Shaman shakes his limbs, sucks up the disease-causing agents and finally gets exhausted. An Ona Shaman behaves wildly. His whole body trembles violently until he falls down on the ground. A Polynesian Shaman, during his performance acquires extra-ordinary physical strength; he eats the food of four adults at a time. An Eskimo Shaman is able to rock and jump a drum on his forehead. A Siberian shaman at this time gathers so much power that he breaks up a metal chain easily with which he is kept tied, The Steady Shamans, on the other hand, are not prone to emotional or epileptic fits. For guidance, they depend on the guardian-spirits that come to them in dreams. A Steady Shaman in California talks to an eagle, a snake, a bear or even to a personified mountain in dream, for boon. Such a Shaman can make himself invisible or can build a bulletproof body. However, both kinds of Shamans with the special skills serve the people in primitive society.

Witch Doctor

The term Witch doctor is sometimes used as a synonym of Shaman and specially refers to the Negroid Shamans of Africa and Melanesia. But actually Witch doctor is a divine personality who exposes the witch. A man either inherits the skill from his parent or learns it from someone else. When somebody suddenly falls ill, it is thought that the person has been bewitched. Therefore, his friends and relatives try to cure him with care, protection and herbal medicines. But when all the efforts go in vain and the symptoms become alarming, there is no way other than to call a witch doctor who knows the remedy of the illness. A witch doctor generally looks awesome as he paints his face and body brightly with clay of different colours. He burns mysterious perfumes around the patient, mutter strange words, twists own body and ultimately cures the patient with his secret power. He knows the medicines for the counter-witchcraft. In some societies like American Indians, the witch doctors are known as medicine man.

Sorcerer or Witch

Sorcerer and Witch both are the malevolent practitioners. So they enjoy a very low socio-economic status in all societies. A Sorcerer and a Witch may be of any sex and usually they are the part-timers. Both of them are dreaded, as they know the way to invoke the supernatural power for causing illness, injury and death. Sorcerers often use different materials for their magic, so when evidences of their malpractice is found, they are killed by the communal vengeance. But in case of the Witchcraft, for the absence of evidences, genuine witches are not always marked out. It is believed that witches possess certain evil substances within their body by which they do harm other people. A magical performance is perhaps responsible for the recognizable changes in the internal organs of the body, which can only be revealed by post-mortem examination. In almost all-primitive societies there are either witch doctors or medicine men or Shamans to act against the evils created by the Sorcerers and Witches.

Medium

Mediums are the part-time religious practitioners and mostly females. They are asked to heal the people while in trance. A medium falls into a hypnotic condition and during that period she is controlled by some spiritual force, external to herself. Different spirits are supposed to communicate with people through the medium.

The process is often referred to as divination, which may be a channel of connection with supernature to get his guidance. Divination often informs a man the source of his misfortunes. The medium usually obtains the guidance through oracle. Primitives believe that most of the misfortunes arise from the practice of Witchcraft.

Priest

Priesthood is a manifestation of developed religion. But it can also be found in the relatively ordered primitive societies where cultures are rich and complex. Priests are usually the full-time male specialists who officiate at public events. They enjoy a very high status in the community; people respect them as they possess the power to reach the gods and goddesses. The priests are also found to organize and maintain some permanent cults. A priest may have mana, but this power lies with the office which he holds and not with him directly as that of a Shaman. Succession of the office is hereditary. The priests have to work in a rigorous structured hierarchy, fixed in a firm set of tradition. Agricultural or pastoral communities that exhibit political integration beyond the community include either Sorcerer or Witch doctor or Medium along with Priest and Shaman.

Distinction between Religion and Magic

Religion and magic are two manifestations of supernaturalism. They are distinguished from one another neither for the goodness of one, nor for the evil of the other. The focus of interest is different in these two approaches; they are fashioned by the attitudes and practices of the believer.

Religion	*Magic*
The Word has been derived from the Latin word 'Religare' denoting Divine communication.	The word has been derived from the French word 'Magi' concerning secret deeds.
It always acknowledges the superiority of supernature.	It does not acknowledge the superiority of supernature.
The priest uses sacred formulae to get the the blessings from supernature.	The magician uses mystic formulae for compelling the supernature to act under his control.
It is a method of submission and reverence.	It is a method of compulsion and command.
Volition of supernatural power is absolute.	No volition of supernatural power is recognised.
Attitudes and behaviours of the performer are devout.	Attitudes and behaviours of the performer are arrogant.
The performances are open to all. No secrecy is found anywhere.	The performances are usually done secretly.
It is mostly communal.	It is extremely personal.
Objective is to appease the supernature in the hope of good result.	Objective is to manipulate the supernature for yielding good or bad result.

PRIMITIVE ART

The search of beauty is an eternal craving of mankind. This urge does not always satisfy the material needs, rather it brings satisfaction to mind and eye. The impulse arises out of an aesthetic sense. Each and every individual responses to it in some way or other. Art is, therefore, a product of deep-rooted human urge and has been integrated with the life of the people as an instinct since the early days of human existence.

The nature and styles of art vary considerably with culture. As culture regulates the patterns of behaviour in a society, it revolves round the specific beliefs and feelings. Therefore, cultural variations in artistic expression are quite natural. As per our modern outlook, we generally want an artist to be original and innovative. But in general, artists are swayed in their own style of creation and sometimes, an ability to lean on traditional pattern is more appreciated than to show originality. Commonly two distinct lines of art are differentiated—pure art and applied art. Applied art aims at some utility rather than searching pure beauty and pleasure. On the other hand, pure art does not bother about utility; they express impulse and rhythm, and manifested in the beauty and pleasure of life. Here the sense of beauty comes from an abstract thought or image as conceived by the respective society. Culture exerts its influence by introducing the specific social values, metaphysical concepts, particular tradition, styles, techniques and also by the material apparatus available therein.

Art is, therefore, nurtured in the society and ends in it. The values and the ideals of society guide an artist. The social genesis of art can never be ignored. Different other factors like geographical condition, climatic set up, economic atmosphere, mood of the artist, medium of working etc., may also influence the creativity of an artist, A person who lives on a high mountain and confined there for the whole life will not be able to visualize the waves of the sea. Similarly, a desert-dweller who has no contact with outer world is unable to present the view of a snowfall or a landslide in his picture. Because artists always externalizes their ideas and perceptions through the art in an objective way.

Nature of Primitive Art

The word 'primitive art' signifies early art which has been gradually evolved with the experience and imagination of the primitives. In a society everybody may not be creative although they have the potentiality of creation. Since art is nothing but an emotional expression, even children spontaneously draw lines when they get materials in hand. They want to express the things, which they have seen or have liked most. Their imagination and experience both work at a time. The object, which they have never found or never heard about, can not be the subject of their drawings.

Anthropologists consider the artistic activities as one of the four basic social activities of human life. But it is a matter of surprise that how art appeared in the dim past, hundreds of thousands years back. What inspiration acted behind those aesthetic activities? The anthropologists, art historians, art critics, philosophers all seek this answer. However, the various pursuits of creativity have been classified in a number of groups like visual art, oral literature, music, dance etc. Among these, the visual art is the oldest as well as a tangible form. It includes drawing, painting, carving, engraving, sculpture, etc. Although, much of the evidences have perished, only a few have survived in the graves, on implements and on the walls of the caves.

Primitive art can not be termed as crude art as some primitive forms of artistic expression are highly complex. They show mature techniques and sophisticated ideologies ranging from naturalism and realism to conventionalized abstraction. For example, Bushmen's art is naturalistic and at the same time full of vitality; Australians' art is abstract and symbolic; Eskimos' art is naturalistic as well as technically sophisticated. What primitive art lacks, is the knowledge of perspective.

The first discovery of primitive art was made in 1879 from a large cave named Altamira near the village Santillana del Mar, about twenty miles west of Santander, Spain. It was a drawing cum painting found on the ceiling of the cave. Many such drawings and paintings were discovered thereafter from the darkest corners of different caves which are definitely of prehistoric period. The pictures include the figures of large and magnificient animals of those days like mammoth, bison, wild bear, wild horse, boar, reindeer, antelope, wooly rhinoceros, cave lion etc. Most of them are the extinct species of Southwest Europe. The forms and the nature of those depicted animals and the radiocarbon dating of the associated fossil remains (including human skeleton) suggest that the paintings are very old. The distribution of the cave art, from the Dordogne of Central France to Spain, indicates that there were some people who lived scatterdly but exhibited some common cultural traits.

Anthropologists relate these paintings to the Aurignacian culture of Upper Palaeolithic period. Numerous marvelous creations were revealed from the caves of Southern France and Spain of Upper Palaeolithic antiquity. But it is really a mystery that why the people went to paint the pictures inside the dark uninhabitable caves. Moreover, a picture was very common in occurrence where a figure of an animal was drawn as pierced by arrow or spear. It seems that the primitive artists though accomplished their work with great pleasure but served some practical purpose as well. The purpose was possibly magical to ensure the increase of animal fertility and success in hunting. The artists chose the dark corners of the caves for maintaining certain kind of secrecy. For light they used lamps with animal fat like the present day Eskimo people. Such ancient art was neither aimless nor a product of leisure. Therefore, in no way they can be defined as 'an art for art's sake'. Rather here the creativity was plunged in usefulness. Thus, ancient art was very close to the life of the primitives; an artist served the demand of the social life.

A Classification on Primitive Art

Primitive art can be sharply distinguished from the modern art and can be classified into two sections—Pre-literate art and Non-literate art. Pre-literate art include those artistic activities which were found in pre-historic and proto-historic periods, whereas the non-literate art is the artistic activities that prevail at present among our primitive counterpart.

(i) Pre-literate Art

So far as we know, the earliest account of artistic skill has been assigned to the people of the later part of the Old Stone Age. Not only the Aurignacian culture, slightly developed Magdalenian culture also exhibited the art products, Those were drawings and paintings in the rock shelters; engravings and carvings on stone, bone, horn and ivory. Colours used by the primitive people were mainly red, yellow and white. Those were natural pigments collected as ochre. Ochre were made into fine powders and mixed with water to get the actual colours. Shades were produced by blending two or more colours proportionately. Besides, the burnt bones provided the black colour. The artists used different grinding bowls of stone for making the colours. Finally, before the application, the colours used to be combined with animal fat and some kind of plant adhesive. Large number of mono or polychrome paintings, realistic as well as symbolic has been discovered as 'Magdalenian art'. A typical female figure carved out of stone was found abundant among the Palaeolithic art objects. Those figurines or Statuettes perhaps represented the fertility cult. If we look critically at the series of drawings and paintings of Palaeolithic period in Europe, starting from Aurignacian to Magdalenian proceeding through Solutrean, we find an increase of technical ability and realism. But conventionalization is very prominent among those arts because early artists liked stability rather than change.

Prehistorian, M.C.Burkitt (1933) classified the Palaeolithic art tradition in Europe into two divisions—Cave art and Home art. The cave arts were confined to uninhabitable caves and were immovable in nature. The motive behind that art was purely magical. Home arts, on the other hand, were found at the entrance of caves or rock-shelters where the prehistoric men used to live. They were chiefly movable in nature and consisted of paintings, engravings, small sculptures (figures of pregnant women, Venus, animal figurines like reindeer, bird, bison, etc.). Stone, bone, horn and ivory were used in production. Some tools and ornaments were also found with decorations and geometric designs on them. Art motifs included different patterns like circles, dots, spirals and entwining lines. According to Burkitt, the purpose of home art was purely ornamental.

In India, no artistic activity has been discovered contemporary to Palaeolithic in Europe. Some rock paintings and engravings were found in Sone Valley; Manikpur and Vijaygarh in U.P.; Singhanpur, Hosangabad and Panchmari in M.P., which can be compared to the standard of Palaeolithic art in Europe, but those are definitely not so ancient. The remains of Indus Valley Civilization, which are considered to be of Bronze or Chalcolithic Age, show certain evidences of artistic ability. We got painted and engraved pottery, beads, jewellaries, brassware, statuettes, seals, etc., where human figures as well as animal figures were depicted.

(ii) Non-literate Art

There is a great diversity in themes and styles of the non-literate art as Adam (1940) and Firth (1951) has pointed out. According to Firth, this diversity arises out of geographical and social differences. The features common to all societies are, (a) the art originated from an inherent aesthetic impulse of the individuals; and (b) the basic configuration of art in all simple societies is more or less same.

The absence of literacy magnified the importance of art in non-literate societies. But the western observers who support the modern art always look down upon the non-literate art. They criticize the primitives as a mentally underdeveloped group with sub-standard knowledge and ideology. Therefore, the media, procedures, techniques and representations of this group are regarded as crude.

A large number of primitive societies are scattered in the world. In Oceania, the primitives are those people who are indigenous to Australia, Melanesia, Polynesia, Micronesia and Malaysia because their cultures were not significantly influenced by the extensions of Asian Civilization into the pacific area. In America, there are several Indian societies in North and South America uninfluenced by the Andean and Meso-American Civilizations. In Sub-Saharan Africa, various indigenous people are found who are neither affected by the extensions of Euro-Asian Civilization nor by the different indigenous Negro States and Kingdoms of Western Sudan, the Guinea coast and the Congo region. In Asia, the Siberian tribes, the Ainu, Central Asian nomads and many other marginal aboriginal peoples of South China and the mainland of South East Asia are considered as primitives who remain quite apart from the great Chinese and Indian Civilizations.

Primitive art has been advanced in technique and sophistication as found in Africa and Australia. African art is said abstract and symbolic. Negro sculpture has reached to a very high level, They use close-grained hardwood and bronze. Most of the African statutes represent dead ancestors. They are not merely the 'works of art'; each of them is a personage, alive with the spirit of the man whom it represents, African masks are the other important forms of art, which express social and religious forces and play essential part in their ceremonies.

Australian arts are not so symbolic but highly stylized. Australian aborigines are exclusively a hunting and food gathering people. They totally depend on nature. Various mythical stories concerning their creation are found to be transmitted from generation to generation by means of art, music and drama. Like all other primitive societies they possess no artist class in their society. Every man, at some time or other, is called upon to act as an artist. He may be asked to engrave on sacred object, to decorate a burial pole or to paint designs on the cave wall, the bare ground or the body of a performer in a ceremony. Art of the Australian aborigines can be classified under three headings—engraving, sculpture and painting. The aboriginal engravers use their skill in decorating the shields, boomerangs and spear throwers; in depicting the mythical stories on sacred objects of wood and stone. Human figures and different other creatures carved on wood and stone are worthy to be praised for their workmanship and artistic quality. Some beautiful cave and .bark paintings have been reported from them. Like the Palaeolithic man of Europe, paintings are made in caves and rock-shelters, but aborigines of Australia do not decorate the rock surfaces, which are far away from daylight. Among these people, art is a living force—an integral part of the daily activities. Tribal magicians try to control the forces of nature; they look for the increase the food supply, punish wrongdoers or destroy enemies. The skill of art is also employed in the communication of tribal beliefs among its members.

Art of the contemporary primitives are largely of decorative type i.e., such an art is used to embellish the artifacts. They may be either formal or representative. Formal types are without a reference of any particular meaning and thought, whereas the representative type tries to portray the designs or figures from some real things. E.R.Leech classified the purpose of primitive art into three groups: (a) Those serving the religious, magical or ceremonial purposes. (b) Those associated with the decoration of personal belongings. (c) Those found with the memorials of dead.

Indian tribal communities of recent period exhibit various types of artistic production. These consist of paintings on different materials, wooden sculptures and engravings, decorations on human body by scarification (tattoo), drawing cum paintings on the walls of the house, doors and floors with certain motif etc. Textiles, basketry and pottery also come within the purview of art evidences of the neo-literates, throughout the India.

The Indians of Tierra del Fuego is regarded as the least developed artists among all primitive communities of contemporary period. They have no interest in art except a few decorations on body and domestic materials. Navaho Indians are just the opposite. They exhibit amazing varieties of art objects—from simple to highly sophisticated forms. There are two schools of thought that deal with the function of art in a primitive society. One of the schools stresses on the propensity of human mind. It believes, man is a born artist who needs no formal training. So art has been carried out for its own sake. The other school holds an opposite view. It says, the early men were essentially practical; each and every work of art was aimed at some utility. Therefore, snakes on the walls of caves were not drawn because of a pleasure in the act; early men believed that by doing so, they could have made the snakes reproduce in greater numbers. In reality both the views are correct in reference to the non-literate cultures. Whatever be the function—practical, magical, religious or decorative, artistic activities relate the individuals to the group. Because, in small communities various aspects of the society remain inter-woven, there is no scope of separating these activities as economic activities, religious activities, artistic activities, etc. Art is, therefore, a part and parcel of primitive life whether they were pre-literates or modern non-literates. Further, the artistic impulse is an innate desire of human mind. Early men perhaps began to indulge those impulses after being technically equipped in producing tools. Gradually they merged into practical purposes of pragmatic life.

Art : A Predecessor of Writing

Art is to writing as thought is to speech. Like art, writing is also a way of communication. Visual symbols are used in writing as like the spoken words in speech. Writing is, therefore, highly symbolic and its origin corresponds to symbolic art and thought.

Very early forms of communication were not the writings, but only the strips of pictures representing some kind of story or event,like a long journey, a battle or even a peace treaty. One had to interpret the significance of the pictures placed in a series, to get the story. It is called a pictogram. A slightly evolved stage was ideogram, which utilized the conventional signs to denote ideas, qualities and actions. But this stage was still a long way back from true writing. The first real step towards writing was phonogram where each picture stood for a monosyllable sound. A number of different pictures, each containing a monosyllable sound were put together to make up a word. But a phonemic system of writing was produced much later.

Writing is immensely important in the cultural history of mankind as it supplemented the memory by storing different incidences of nature and passing them down through ages. Unlike the oral transmission, its field is vast because knowledge being stored in the memory can be transmitted only during the lifetime of an individual; all knowledge dies with the person unless it is communicated to others. The individual may even forget a part of it. But writings can keep all the knowledge alive for the benefit of others, irrespective of time and space. Therefore, we should be grateful to art which helped writing to evolve out as a progressive conventionalization of ‘representative’ art forms. It was possible as the artists shifted their aim and became more concerned to portray the meaning of an object than the image of the object.

SOCIETY

Society is the central concept of social anthropology. It refers to an encompassing network of social interaction and interrelationship and encloses innumerable individuals and phenomena. Social relationships and social behaviours form the essence the society. But any relationship between two persons is not social. For instance, if two persons walk along the same road or travel in the same compartment of a train, independently of each other, their relationship can not be taken into consideration. Because this relationship denotes only co-existence, not sociality. But if they knew each other or exchange greetings among themselves, then only the relationship will be taken as social relationship. Reciprocal recognition (direct or indirect), co-operation and commonness are the characteristic features of every social relationship. A society is, therefore, a population (irrespective of their age and sex) who lives in a given physical proximity and has essentially some common interests.

Society may be looked upon as social process since a series of interactions take place among the individuals. Social process means the operation of social life, the manner in which the existence as well as actions of each living being affect other individuals with which it has relations. No social life is possible without interactions. These interactions are acutally communications, which are conveyed from one person to other or from one generation to the next. Cooperation and conflict are the two elementary forces, which arise out of interaction. Cooperation is the force of working together. Individuals and groups consume their energies for the promotion of common ends or objective. Conflict, on the other hand, is the struggle between the persons or groups for the same goods and services or for recognition. They try to prevent each other from attaining the goal. They do not sometimes bother at all to injure the opposite man or team. Beside these cooperation and conflict, there is another midway force, the competition, which initiates the persons or groups to show their best performance in order to attain the same goods or services or recognition. Dynamicity of social life is found nowhere in the animal kingdom, except among the man. The animals like ant, bee and baboon that are thought as having a rudimentary social life do not stand parallel to man as there no change is observed from the time immemorial.

HISTORICAL BACKGROUND

Ideas about society originated in the thought of Greek and Roman historians who paid their attention to political and military accounts in order to find out the reason behind the rise and fall of the nations in the universe. Socrates, Aristotle (384 - 322 B.C.), Herodotus (485 - 325 B.C), Hippocrates (460 - 357 B.C.), Plato (429 - 347 B.C.) belonged to this group. They postulated a social theory from organismic point of view. Although they did not study the man directly, they analyzed customs religion and political organizations of peoples with didactic interest. In their proposition, universe was an ordered system composed of different parts, which used to function under the natural law. Human society also supposed to function in the same fashion by nature's law. Therefore human beings in a society or state were placed in a natural hierarchy of superior and inferior social relations. According to them change, whatsoever appeared, for the break of harmony of the parts or their activities due to some reason or other.

In the fifth century B.C. the Greek philosopher Empedocles believed that qualities like love, hatred, attraction, repulsion etc. were the basic qualities in the universe which helped in the formation of society. Thereafter, in a gap of hundred years, Heraclitus, another Greek philosopher pointed out 'struggle' as the root cause of social life. But natural history of the species was not discriminated from the traditional history of man until eighteenth century. The actual conception of the society and the main elements of this thought arose much later, from the period between late nineteenth century and early tweaieth century.

We are debtors to the intellectuals like John Locke (1632-1704) who in his essay 'Human Understanding' (1690) stated the role of environment as a basic factor for the development of human thought and action. Baron de Montesqieu (1689-1755) in his book 'Spirit of the Laws' (1748) advocated that the social life and history are governed by natural law. He arrived at this proposition by comparing the actual societies, primitive and modern.

Swiss Jurist J. J. Bachofen (1815-1877), Henry Maine (1822-1888), J. F. Mclennan (1827-1881) attempted to classify societies in their own ways from evolutionary point of view. Auguste Comte (1798-1857) took society as an organic whole. He believed that each branch of knowledge passed through three stages : the theological or fictitious, the metaphysical or abstract and the scientific or positive. He advocated that man should be viewed historically by the investigation of the laws of action and reaction of different parts of the social system. Herbert Spencer (1820-1903) classified worldly objects into three orders: Inorganic (non-living), Organic (living) and Super-organic (society). In organic level, he studied man and in the super-organic level he studied social relationships. Both Spencer and Comte pointed out the constant changing nature of the society.

Karl Marx (1818-1883) was an economic historian, his economic interpretation of history was a classical work. He defined society as an organized multitude of productive forces. To him social changes come out for the changes in productive techniques. Being accompanied with Friedrich Engles (1820-1905), he threw light in the origin of family, private property and the state. They also related property concept with different mode of production.

L. H. Morgan (1818 - 1881) classified the society into strata based on marriage and family. He used the examples of Australian aboriginals to Roman Empire in support of his theory. For the purpose, he also conducted field-tours among the Iroquois of New York State.

Charles Darwin's (1809-1882) epoch making discovery proved biological evolution. He explained the fact that survival of the fittest is found after the struggle for existence. His 'Origin of Species' was published in 1859 and this theory accelerated the speed of social studies of that time. The great social thinkers like McLennan, Maine, Spencer, Marx, Morgan who had already published their views on the evolution of the society, got support from Darwin's theory They considered the society as an organism and related biological processes to social processes.

Later, Emile Durkhiem (1858-1917) held that social facts could not be explained in terms of individual behaviours; collective consciousness or group habits are important to study the society. He further said that the society provided a large number of predefined conventions and norms, which govern on the behaviour of individuals.

L.T. Hobhouse (1864-1929) dealt with the social life of man and its spheres. His theory of social development described the growth of knowledge. He classified the primitive societies in term of economic condition.

Ferdinand Tonnies (1855-1935) put forward the concept of Gemeinschft and Gesellschaft to refer community and society respectively. Elements of human action, which is found in society, ultimately ends in the community of smaller realm.

Thus, gradually the study of society gave rise to the discipline of sociology, (the science of society) which is much older than the discipline of social anthropology. Auguste Comte in France and Herbert Spencer in England were the pioneers in sociology. Comte defined sociology as a science of social phenomena "subject to natural invariable laws, the discovery of which is the object of investigation'. Subsequent sociologists were Emile Durkheim (1858-1917) and Max Weber (1864-1920) from France and Germany respectively, According to Durkheim, the ultimate social reality is the group instead of individual. He said that social life must be analyzed in terms of social facts. These social facts are nothing but the collective ways of acting, thinking and feeling which though seem to originate from the individuals but actually external to them and generated from the group feelings. He wanted to specify the range of external forces and their characteristics from the individual manifestations. Legal codes, religious codes, social statistics, etc. are the social facts as Durkheim had shown. His most important contribution was the concept of the configurafion of social integration *i.e.* solidarity. He classified 'social solidarity' into two types, mechanical and organic. The term mechanical solidarity was applied to the societies where members were found to have a common, shared social experience but they did not necessarily depend on each other to survive. Conversely, the societies with diverse interdependent subdivisions when found to be formally linked into a single system, called organic solidarity, Max Weber, the historian cum sociologist was influenced by Karl Marx. He took individuals as basic unit of the society and classified the societies in terms of social action. Social behaviours were interpreted on the basis of motives and intentions. He further classified the social relationships and divided the nature of authority into three types - traditional, charismatic and bureaucratic.

Social anthropology at the time of its inception was dependent on sociology for its theories but later on it deviated considerably both in field of interest and methods. In fact, it has been related to philosophy, history, etc. at one hand and on the other hand with the biological theory of evolution. However, here we are not again going to discuss the position of social anthropology among the other social sciences, rather we are concerned with the conception of society—how it developed with time and gave rise to the discipline named social anthropology.

It is evident that the major contribution in social anthropology came from Drukheim and Max Weber. The approaches of these two scholars were though opposite in nature (Weber considered individual as the basic unit of the society while to Durkheim the ultimate social reality was the group, not the individual) but both of them examined social phenomena and behaviours for their propositions in contrast to the earlier scholars like Comte, Spencer, Marx, Hobhouse, Tonnies and others who attempted to classify the forms of society using some criteria other than the human social behaviours.

Durkheim followed the British sociologist Herbert Spencer and influenced profoundly the British anthropologist Arthur Reginald Radcliffe. Brown (1881-1955) whose fundamental interest went to study the structure of the society. Radcliffe Brown's concept of 'social structure' was derived from Drukheim's concept of 'social solidarity'. He perceived society as a system where the different parts are interdependent and interrelated. 'Social Structure', therefore, can only be revealed through the study of groups in a society. The relationships between the persons of a group are regulated by a common body of recognized rights and obligations. Social structure is now core content in social anthropology, which is indispensable in understanding of a social relation and a social behaviour.

We should keep the fact in the mind that every human learns to build up social relations with others from the days of infancy. It can be projected not only as the capacity of man for social life, but also an intrinsic need for the society. No human being can make himself a normal person, if he is reared among the animals or in any isolated area. An individual by no means can deny the social influence on him . Individuality and sociality both develop together. In prehistoric time also, man had always lived in society, whatever was the form of that society. The urge for making society was developed as man is social by nature.

SOCIAL GROUPS AND INSTITUTIONS

Human society consists of two parts - social groups and institutions. All standardized social relationships and behaviours can be categorized either into groups or institutions as per their nature.

1. Social Group

A group has been defined by S. F. Nadel (195 1) as "a collection of individuals who stand in a regular and relatively permanent relationship". A group is therefore composed of persons and the persons are always defined according to their mode of action in respect to each other and have the locus in the group.

'Social group'must not be confused with 'society'. Because a society is more parmanent, inclusive and organized than a group and a social group is only a part of the society. Individuals with a common interest or with a common similarity when come together and become organized, a social group is formed. In this sense a society is a group but all groups are not societies. Groups are inevitably smaller in size and even may lack the recognizable structure. They can be divided into two types: primary and secondary. A primary group is a face to face group based on direct personal contact, members deal immediately with each other. The secondary group rest on indirect, secondary or categorical contacts where people may not know each other personally. A family, a local club, a play group, members in a tribal hamlet or a village pocket comes under the primary group. On the contrary, large-scale organizations as developed in our industrial society like business organizations, professional associations, labour unions, etc. are the secondary groups. Besides, due to common environment, some common interests develop among the inhabitants of a common territory, which lead to the formation of other social groups. Such spatial groups are sometimes temporary or permanent. Temporary spatial groups include crowd, mob, audience, etc. whereas the individuals belong to band, village, town, city, etc. form the permanent group. Social groups may also be classified on the basis of blood relationships. For instance, family, clan, moiety, phratry, etc. are the groups where the idea of kindred prevails either in actual or in assumed form (vide. Kinship). Sometimes a social group is the community where a permanent self-contained social group is obtained embracing a totality of ends or purposes. Members of a community find all their social relations inside the group. A community thus possesses a definite structure as well as a system of rules by which the members lead a common life.

Small groups are often made basing upon age, sex or any other specified interest. ie. similarity in religion, recreation, games, education, economy, etc. when give rise to specific social groups, they are called associations. Here people are united to work together and to achieve their purposes. A school, an army, a drama troop, a vocational group, a tracking team, etc. are the examples of association.

Among these different groups, the family and community are regarded as the universal. The former comprises only a few persons but the latter is made up of huge number of persons. Individuals from different groups may also assemble in one group. As for example, a person who is a member of a specific kin group may also be a member of a household group, a lineage group, a clan, a moiety, a tribe and finally a community. Since the domain of social anthropology basically include simple societies, we will remain more concerned about the small communities and they will be the main focus of attention and examination.

II. Institutions

In contrast to social group, institution is a visible aggregate of human beings in interaction. These interactions are necessary for the survival and perpetuation of group. Institution emits certain relatively permanent, organized and structured forms of behaviour which is acceptable to the members of the group. Groups are visible only in institutionalized modes of co-activity. R.R. Marett (1912) said that institution is a convenient word to express all the externals of the life of man in the society. He pointed out two ways by which a given set of institutions could be investigated. He perceived institution as the wheels of the social machine. When they stand still they denote the relationship to each other. But when they move, they indicate the on-going process of custom. In his thought, institutions represent the 'forms of social organization' when they are looked upon as static. Dynamically they are 'parts in movement' to maintain the continuity of social relations. R. M. Mac

Iver (1937) found institution as a "definite organization pursuing some specific interest or pursuing general interests in a specific way".

Gillin and Gillin (1948) defined institution as 'a functional configuration of culture patterns including actions, ideas, attitudes and cultural equipments which possesses a certain performance and which intended to satisfy felt social needs. E.S. Bogardus (1950) defined it as 'standardized set of rules and the machinery and authority for putting the rules into action'. S. F. Nadel (195 1) defined an Institution as 'a standardized mode of co-activity'. Institutions are therefore the 'functional units' based on standardized pattern of social behaviour, recognized and accepted by the society. Traditions and customs are preserved through the institutions. So it is impersonal, formal and concerned with the deeper needs of life. Marriage, religion, law, property, etc. are the institutions in a society.

According to nature, institutions can be divided as religious institution, political institution, economic institution and so on. . Elements of any of these institutions represent relatively self-contained modes of action towards its aims.

The elements work in two ways-.

a) They require each other for the realization of their different aims and thus serve as 'functional unit'.

b) They serve the part of the purposes i.e. part of the actions.

More recently Leeds (1976) has pointed out that the term social institution can be implied to the "forms of standardized action or behaviour linked to a set of complex and interdependent 'norms and roles' and applied to a relatively large proportion of persons with a society or territory'.

Whatever may be the definitions, institution is fairly permanent cluster of social usage. It is more or less endurable, complex and integrated pattern of behaviour by which social control is exerted and basic social desires (needs) are fulfilled. The relationship between grouping and institution is very close. They are interdependent of each other. All social behaviours form and perpetuate by interactions between

(i) grouping and institution

(ii) grouping and grouping

(iii) institution and institution.

For the understanding of social behaviours in a given community, the groupings and institutions may be studied from three angles mainly.

(a) From cohesive point of view *i.e.* to understand the structural and organizational level of a society.

(b) From conflicting point of view *i.e.* to understand the levels of social stratification.

(c) From the view of the social process *i.e.* to understand the changing aspect.

However, we are here neither discussing the social stratification nor the aspects of social change. At first we will deal with structural level, as it is fundamental to understand a society. The main point is this, although social relations and behaviours are common to all societies, there are also some specialized relations and behaviour unique to each society. For example, we Indians consider our whole nation as one society as some unity and commonness prevails among all Indians. At the same time diversified societies have been observed in India which are different from each other by specific criteria. These small societies are usually designated as communities and several communities like Bengali, Muslim, Oriya, Santal, Toda, Khasi etc. are scattered in India. Therefore it can be said that each of these communities or human groups must have some specialized social behaviours over the general social behaviours which provide them with the respective indentification. To understand this social anthropologists use their conceptual tool social structure and social organanization.

SOCIAL STRUCTURE AND SOCIAL ORGANIZATION

Every human group is organized. The individual components do not function independently of each other. Rather they are linked by bonds, the nature of which determines the types of social unit. British sociologist Herbert Spencer used the technical term 'function' for the analysis of society, He perceived close parallels between the human society and biological organism. Further, he generated the idea of function when he conceived human society as a machine. As the interrelated parts of a machine function to keep the machine working, in the same way functional dependence of different parts in a society keep the integrity of the society. Durkheim developed this conception by explaining with social phenomenon. He said that social phenomena constitute the social life to produce harmony within the society. British social anthropologist Bronislaw Malinowski (1884-1942) got inspiration from Durkheim and he related social arangements with biological needs. Institutions were designated as social arrangements by Malinowski as they used to meet the needs of the people and therefore he referred them as functional.

Malinowski's ideas had been granted to the social anthropologists as the 'concept of fimctionalism' but Radcliffe Brown later on modified this concept mixing with his own concept of 'social structure. He felt that the various aspects of social behaviour were responsible for building a society's structure, instead of satisfying the individual needs. He defined the social structure as a total network of its existing relationships and distinguished 'structural-function' from 'function' which Malinowski related with the bio-psychological needs of individuals.

According to S. F. NadeL social structure was "an ordered arrangement of parts which can be treated as transposable, being relatively invariant, while the parts themselves are variable". To Raymond Firth social structure was "an ordered relations of parts to a whole with the arrangement in which the elements of social life are linked together. Those relations must be regarded as built up one upon another, they are a series of varying orders of complexity in some factor of constancy or continuity must be involved in them".

However, social structure refers to the social relations which seem to be of critical importance for the behaviour of the members of the society, so that if such relations were not in operation, the society could not be said to exist in that form. The term 'social organization' was often confused with the term 'social structure' until R. Firth (1951) defined these two terms separately by careful analysis.

Malinowski (1948) used the term social organization and tried to define it in terms of purposive manner in which people acted upon their environment to satisfy their needs. Radcliffe Brown (1952) perceived social organization as the arrangement of 'roles' associated with 'statuses', which ultimately constitute social structure. R. Firth pointed out that the social organization can never be the synonym for social structure as these two concepts promote more or less opposite directions of thought, continuity and change. Society is built on social relations ie. the members of the society carry on their activities which may or may not follow the ideal pattern of behaviour. There are a lot of actions and diversities in behaviour, which arise out of individual choice and decision. The structural study is the study to find out the fundamental social constituents that are revealed in the forms of basic social relations. Those are capable of variation but persist through the process of repetition. Structural elements are therefore like the anatomical framework of human body; they give the shape of the society. Structural study is indispensable to delineate the functions of a society to understand the continuity of social life. On the contrary, social organization is not limited to the ideal pattern of social relations. It also signifies the factors for change *i.e.* the extent to which the social standard deviate as an influence of different external factors. Therefore, if social structure is conceived as a model of social action, the social organization will be the reality. Social organization accounts the reasons of social change. According to R. Firth, a structural analysis alone can not interpret social change. Analysis of the organizational aspect of social action is necessary for the analysis of structural aspect.

Let us discuss the social structure and social organization in respect of primitive society. Khasi society is chiefly matrilineal. They follow matriliny in descent and inheritance. Their pattern of residence is matrilocal; matripotestal authority prevails in the family. Besides, these people are horticulturist. They have a distinct dialect of their own. All such peculiarities gave the Khasi community a typical structure. But, at the same time, the Khasis have their own dress and ornaments, food-habit, pattern of habitation, famliy types, clan, marriage and kinship system, religious belief and practices, political system, educational and recreational units. Study of all these aspects come under organizational study. Since Khasi community is exposed to Christianity and contemporary Indian way of life, a vast change has been found in their old norms and values. Naturally the question comes, how this community is going on with its traditional institutions under the fast waves of change. This aspect demands organizational study. Organizational study is thus the total study of a society including the structural aspect too.

In the context of social anthropology, the concept of social structure and social organization bear the definite meanings. For the intimate and initial understanding of a community one needs to study structural as well as the organizational principles. Moreover, there is a general structural configuration in each and every community, which includes family, marriage and kinship. It is so because, every human society has certain rules concerning sex relations and procreation of children. Institution of marriage plays a very significant role in the formation of family grouping. Types of marriage determine the types of family. Again, the members are bound together by the bond of kinship where sex and age are the two vital points of distinction. A social life is thus organized on the basis of relationships among the individuals. Anthropologists unveil this subtle network with expertise and delicacy.

BASIC STRUCTURE OF A SOCIETY

FAMILY

Famıly is the basis of human society. Although the nature and structure of the family vary from one society to other, but a society without families is not known to us. Relationship between the members of the family is deliberately formed basing on marriage and descent. The members are bound to each other by certain code of norms, right and obligations. The interpersonal relationships within the family make the family an endurable social unit. It is the fundamental social group on which the conception of society rests. In simpler societies, members of a family are found to reside in the same household but in complex societies some of the members may live apart under several circumstances.

The family is not only the basic social group; it is also viewed as an oldest institution of mankind, which has the power to withstand social changes. The biological and social reproductions of the family are indispensable for the society to maintain its continuity.

Origin of the family

At the beginning of human society, there was neither family nor marriage. Only a kind of unregulated animal-like life prevailed. Institutions, namely family and marriage, came in existence after certain stages of social development.

Many anthropological research and speculation have been done to trace out the historical origin of the family. Scholars like L. H. Morgan, J. L. Lubbock, J. G. Frazer and lately R. Briffault were swayed by the evolutionary doctrines of Darwin and Spencer. They tried to project family through unilinear evolution. Their propositions were based on deduction and not on the actual facts. Among them, Lewis Morgan's evolutionary scheme is considered as correct, though not universal. His deductions were logical as well as academic. He advocated a sequence of growth where types of family corresponded with the forms of marriage. According to him society has come up from a stage of sexual promiscuity. He supported this fact with the examples of certain preliterate societies where sex license at some festivals, periodic exchange of wives and lending wife to visitors to show hospitality etc., are still tolerated.

In the book 'Ancient Society' (1851), Morgan had shown five successive stages for the development of family. At the base, there was consanguine family, which formed as a result of group marriage within the members of same generation. Marriage between brother and sister was permitted there. At the second stage, there was the Punaluan family. Although this type of family was an outcome of group marriage, but marriage between brother and sister was forbidden. The Syndyasmian family came at the third stage basing on marriage between a man and a woman. But it was devoid of the norm of exclusive cohabitation; marriage relationship continued till the pleasure of the parties persisted. At the fourth stage, Patriarchial family came to indicate the marriage of a man with several women. Last or final of the stages appeared with monogamous family where the marriage between a single pair was acknowledged with a norm of exclusive cohabitation. Monogamous family according to Morgan resembled with the modern nuclear family. With the same view McLennan and Herbert Muller advocated that the sexual communism prevailed in the early stage of society and group-marriage was only one step higher form of promiscuous condition.

The irony of this stratification is that, some of the stages seem to be purely hypothetical. In fact, no clear traces of early promiscuity or group marriage are found among the most simple people of these days like Andaman Islanders in Andaman, Pygmies of Malay, the aborigines of South Australia and others. Therefore, origin of the family is not clear at all. Early development of the family has been highly speculative in most of the writings though many old records of early man show the evidence of family life. Some of the social scientists argue that there was never a period in human evolution where promiscuous relationship existed between the two sexes. Anthropological evidences in this field are totally in favour of R. H. Lowie who said (1949), 'the social phenomena are fluid. Marriage is the socially approved form of sex relations, which correspond with family, the universal social phenomena'.

Types of family

Variations in the forms of family are quite natural. This depends on the ways of marriage and economic system.

(a) **Monogamous family / Nuclear family**

This type of family is based on monogamous marriage *i.e.*; marriage between a man and a woman. It is the simplest among all types of family as it consists of a man; his wife and children. Here the husband or wife can not remarry till the spouse is alive. The other names of this family are elementary family, basic family, conjugal family, immediate family, primary family, etc. Since this type of family serves as the nucleus of all other types of family, it is also popular in the name of nuclear family. Such families are preponderant in urban and industrial areas. Different tribal groups of India *e.g.* Santal, Lodha, Kharia, Birhor, Chenchu, Khasi, Kadar, etc., show this sort of families in their community. Monogamous family structure is also common among the Australian aboriginals, Polar Eskimos, South African Bushman, etc. This type of family provides more freedom to its members but disadvantage lies in its instability caused by the death of the spouse, divorce, disagreement, etc.

(b) **Joint family / Extended family**

In certain types of family, the nucleus is extended with some closely related kins and the family is called an extended family. Sometimes it is also referred as joint family. According to the handbook, Notes and Queries of Anthropology (1874), a joint family forms when "two or more lineally related kinsfolk of the same sex, their spouses and offsprings occupy a single household and are jointly subject to the same authority or single head". This means that the joint family is a large group, extended upto two, three or more generations with lineally related members, the spouses and children. Such a joint or extended family may rest upon monogamy or polygamy but there must be a number of nuclear families linked together by sibling tie. Usually this kind of families includes parents, grand-parents, grand-children, uncles, aunts and cousins.

Joint families arise and persist as the members carry out their activities in a cohesive manner under the leadership of the eldest person of the household. Cooperation and mutual support are the key words here. Critical economic factors play behind the formation of joint families. Agricultural communities highly favour this type of families to prevent property devolution, to increase more hands in production and to provide the security of family members in accidental cases like death, divorce, disagreement, separation, etc.Nevertheless, the joint family splits and nuclear families come out in different situations. For example, if a young couple move away from parental home or if they begin to handle own familial account and separate budget or if they arrange new establishment for cooking.

Hindu joint family is the typical example of this type of family where the head of the family is a male (Karta) who looks after the affairs of the whole family. By dint of age and experience he exerts his authority over all other members of the family and obtains respect from them. In non-literate societies also, this sort of family exists. For example, the communities like Santal, Munda, Khasi

(Maternal), Oraon, etc., are in favour of this type of families. Outside of India, the Samoans (Paternal), Chiricahua Apaches (Maternal) of America and many other primitive communities exhibit the same.

(c) **Polygamous family**

This type of family is made up of two or more nuclear (monogamous) families affiliated by several marriages. The essential feature is this, one of the spouses remain common to all monogamous families inside a polygamous family. Polygamous families may be of two types—polygynous and polyandrous.

(i) **Polygynous family**

This type of family is based on polygynous form of marriage *i.e.,* where a man marries more than one woman and leads life in the same household with all his wives and children. Such type of families has been noted among the Kulin Brahmins of Bengal and also among the Muslim community. In the context of the tribal groups, Naga of North-East India, Gond and Baiga of Middle India present this type of family. Outside of India, polygynous families are found among the Eskimo tribe, Crow and Hidatsa of North America, and especially among the African Negroes.

Polygynous families in a society often arise following a situation of excess number of women over man. This also indicates an inferior social position of women; men's position remains prestigious because they can afford many wives at a time.

(ii) **Polyandrous family**

This type of family is the resultant of polyandrous form of marriage. Here a woman marries several men and lives together with all husbands and children. Polyandrous type of families is not predominant at all; rather they are confined to small pockets. According to nature, polyandrous families can be divided into two groups—fraternal (adelphic) polyandrous family and non-fraternal polyandrous family. Fraternal polyandrous family is that family where a woman marries two or more brothers. A belt is found in Northern India from Janusar-Bawar to Hindu Kush range through Kangra Valley where the families are chiefly polyandrous. Marquesans of Polynesia also exhibit such type of families. In India, Khasa of Uttar Pradesh, people of Kinnaur, Lahaul and Spiti of Himachal Pradesh, Sinhalese of Sri Lanka, some of the Tibetans and Toda of Nilgiri Hills show this type of families. In non-fraternal polyandrous type of families, the husbands are not essentially the brothers. Such families were once widespread among the Tibetans and Nayers of Kerala. Now-a-days, the Todas sometimes reveal them.

However, the situation arises due to the shortage of women in a society as a consequence of female infanticide or for the excess number of men due to some economic or political reasons.

(d) **Compound family**

This family type refers to a concrete group, which is formed by the agglomeration of nuclear family units or parts of them. Under special circumstances this sort of families form. For example, once due to practice of female infanticide, the number of males in Toda community increased and polyandry became popular. But at present the practice of female infanticide has been totally stopped and so the number of males and females in the community has come to a parallel. Still due to traditional notion, wife of a brother is looked upon as wife of all other brothers in a family. Now, if those brothers marry individually and reside in the same household, then all the brothers with all their wives and children together give rise to the compound family which is normally an unusual phenomenon.

Types of Family

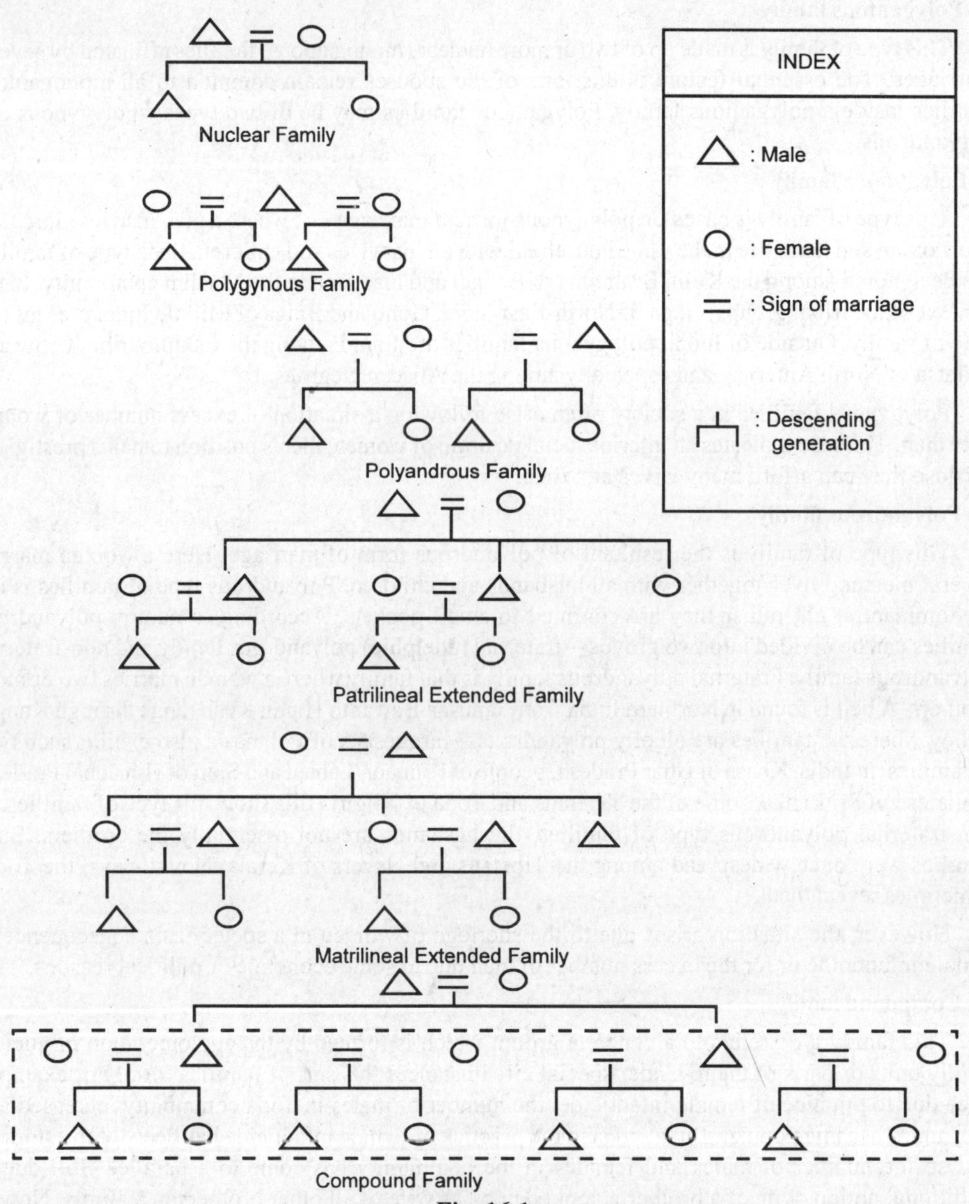

Fig. 16.1

(e) **Composite family**

This is not only an unusual, also an extremely complex form of family. Dieri, a hunting tribe of Australia show this type of families. The male of such family is often compelled to pass long time in a distant forest for procuring livelihood. During that period, a man from the neighbourhood comes to look after the family as a norm. This man establishes marital relationship with the wife and goes back when the actual husband returns. A Dieri woman thus possesses one permanent husband (Tippamalku) and many temporary husbands (pira uru). All the children live in the same house with mother. Such a family is called composite family.

Besides these above-mentioned types of families determined in terms of mates, there are also other diversified criteria basing on which families can be differentiated. For instance, in some society members trace their descent through male line from a common ancestor. Inheritance also passes through the males. Such type of family is known as patrilineal family. Authority rests over the eldest male of the family who is the owner and administrator of the family, property and right. Other persons are subordinated. When the family-head (Karta) dies, the next oldest male or the eldest son of the deceased undertakes the authority. Patrilocal residence is the rule *i.e.*, females go to reside in husband's house after marriage. The whole structure is usually referred as patriarchy or patriarchal society. Opposite to this, there is matriarchal society where authority lies in the hands of the women. Ancestry is traced through female line and property is owned by the women. Matrilocal residence (*i.e.*, husband lives in wife's house) and matrilineal family are the other features. In reality, patriarchy or male dominated society is very common throughout the world, whereas matriarchy in true form does not exist.

In a matrilineal system, descent affiliation is transmitted through female line. Father belongs to some other line of descent. Daughters are highly valued. They do not leave their maternal house after marriage, rather bring their husbands to live with them. Naturally the sons of the family move away to live with their wives. But women of this society are actually supported by the males, typically from mother's side. Mother's brother's role is very prominent in this respect. He is regarded as the closest male matrilineal relative in the parental generation who makes all vital decisions. He stands as a father to the children and takes all responsibility for their upbringing and education. In the long run nieces inherit uncle's property.

However, it can be said that the pattern of authority and rules of descent create two types of family—patrilineal and matrilineal. Patrilineal families are very common; they are found vastly among most of the communities in India. Matrilineal families on the other hand, have been noted from among a few communities like Garos, Khasis and Pnars of the North-East India, Veddas of Ceylon, Nayers, Mappillas, Lakshadweep islanders and some other tribal and non-tribal groups of South India. Outside of India, Iroquois, Zunis, Hopis, Trobriand islanders, etc., show matrilineal families.

Rules of residence further reflect on family types as evident in the following forms.

(*a*) **Patrilocal residence**

The wife goes to reside with her husband in husband's house after the marriage. This type of residence is widely visible in our society. Tribals namely Santal, Munda, Lodha, etc., follow this pattern of residence.

(*b*) **Matrilocal residence**

The husband comes to reside in wife's house after marriage. Khasi, Garo, etc., tribes provide the examples.

(*c*) **Bilocal residence**

Sometimes, the newly married couple is free to decide where they will live, whether with or near the husband's kin, or with or near the wife's kin. Necessity here determines the residence pattern. This pattern of residence is termed as Bilocal residence.

(*d*) **Neolocal residence**

After marriage the couple do not live with or near the close kins of either side. They make a completely separate entity of their own where they reside. This type of residence is called Neolocal residence.

(*e*) **Avunculocal residence**

In some societies, the newly married couple goes to live with wife's uncle (mother's brother). Such Avunculocal residences are found in matrilineal societies. Therefore, occurrence of this type is relatively rare. Still Nayers of Malabar Coast prefer them. Trobriand islanders occasionally like to establish this type of residence.

(f) **Matri-Patrilocal residence**

In certain societies, at first the husband resides with wife in her house. After some time, usually after the birth of the first child, he returns back to his own paternal home with wife. This sort of residence is prevalent among the Chenchus of South India.

Some anthropologists have used the term Virilocal and Uxorilocal to mean the residence 'with the husband's people' and 'with the wife's people', respectively.

Functions of the family

Family as a germinal cell of the society performs certain specific functions, which can be mentioned as follows:

(i) Need for food and shelter is considered as fundamental for child's survival. In a family setting, food as well as shelter is secured for the newborn baby by the effort of elderly members of the family.

(ii) Family serves as a biological unit by providing a common dwelling place for a man and a woman where sexual gratification is possible between them.

(iii) A family regulates the various relationships among its members. It also determines the system of nomenclature and reckoning descent.

(iv) Family projects itself as an economic unit as distinct division of labour is noted among the family members. Men remain engaged in works involving hard labour. Women perform light domestic work and cooking; they also care children upto certain age.

(v) Members of a family are bound to each other by mutual affection and close ties. This emotional congregation is very important; it acts as cementing factor to keep members together. When the children grow and become emancipated, the parents become dependent on them.

(vi) Family behaves as an effective agent in transmission of social heritage. A child begins to learn the customs and tradition of the society gradually from childhood. In a familial setting, the parents and other elders teach the child all the norms and values of the respective society in a very casual way which we call the 'process of socialization'.

(vii) Family helps in handing down the property from one generation to the next. The acquired as well as the self-earned property of the parents go in the hands of the sons or daughters according to the nature of the society. For example, sons of the Santal community inherit the property of father, as it is a patrilineal society. Conversely, the daughters of the Khasi community inherit the property of their mother because the society is matrilineal.

(viii) Family transmits the religious tradition also. Religious beliefs and practices are conveyed from one generation to the other generation through family.

(ix) Family acts as a recreational unit because at times it organizes some recreational activities for the members. Besides, it brings different situations and moods in the family by which the members are entertained.

(x) Family serves as a protective sheath. It protects and cares the sick persons of the family. It guards an individual against emotional troubles. Junior members get their elders as guide or advisor in the major events of their life. It is a dependable shelter where all emotional outbursts are permitted..

Intra - family role and relationship

Some intricate interpersonal relations are normally observed in a family, which exist between the spouse, between parents and the children, and between the siblings. These are of high esteem as important universal activities like economic cooperation, household work, sex, reproduction, child-care, education, protection, etc. are involved in them. Complications appear with the differences of age among the members, for the size of the family, for diversities in cultural norms and for several other factors. Ideal pattern also differs from actual conditions. However, eight types of primary relationships can be picked up from every patriarchal society in the context of nuclear family.

1. Husband-wife relationship

This relationship rests on economic partnership and co-operation, sexual cohabitation and mutual adjustment. It is a joint responsibility for support, care and up bringing of children. Well-defined reciprocal rights are found in the sphere of authority as well as in respect to property deals and divorce decision.

2. Father and son relationship

Father's primary responsibility is to make his son obedient and disciplined from the early days of boyhood. Father provides all sort of protection. Son's duty in turn is to respect and follow his father. At the grown up stage, relationship between the two demands economic cooperation in masculine activities; the son acts mostly under the leadership of father. Obligation of material support is vested on father until the son attains maturity. But the same obligation shifts from father to the son when the father becomes old or incapable.

3. Mother - daughter relationship

It is a parallel relationship as like that between father and son. But economic cooperation or material support does not come within the purview of this relationship. This relationship basically signifies care and protection from mother. In a little grown-up stage a daughter takes the lesson of housework and child-care from mother.

4. Mother - son relationship

Dependency of son is found on mother during the days of infancy. Some disciplines are imposed on him by mother during childhood. As the son grows and mother becomes old, the relationship changes. Old mother is absolutely dependent on son after the death of her husband. She seeks material support and care from son. A life long incest taboo persists between mother and son.

5. Father - daughter relationship

Father's responsibility is to provide protection and material support to daughter until she is married. Economic cooperation and disciplinary instructions are less prominent than father-son relationship. Father keeps no expectation from married daughters. Incest taboo also works here as like mother and son relationship.

6. Elder brother-younger brother relationship

It depends on the difference of age. Usually their relationship is like playmates; they behave like comrades,Where the age difference is greater, elder brother instructs to discipline the young. In the grown up stage, economic cooperation may be found. Generally the younger brother remains under the leadership of the elder.

7. Elder sister-younger sister relationship

It is a parallel relationship as between the two brothers. It also counts the difference of age . Where there is much difference, the elder sister behaves almost like mother. She tries to care and protect her sister.

8. Brother-sister relationship

In the days of childhood they are the playmates. With age comes incest taboo. The interactions become restricted in the adolescent period by the advice of the elders. But affection and cooperation persist between them. Where the difference of age is high, the elder maintains personality and instructs the younger in all matters.

MARRIAGE

Family appears as a result of marriage and it continues through marriage. It is, therefore, a very important institution without which the society could never be sustained. According to Dr. Westermarck, the internationally famous social scientist, marriage is "a more or less durable connection between male and female casting beyond the mere act of propagation till after the birth of the offspring". Marriage enables a child to get a socially recognized father and a mother, R. H. Lowie, the renowned anthropologist said that there were two principal motives behind the marriage, "the universal object of founding a family and the constant need for cooperation in the daily routine of life".

It may be held that family existed in past before the marriage system came in vogue. But we believe that family originated out of biological necessity. The binding force of family tie is embedded in the biological instinct. If we probe deep into the past, we find that men and women in pre-literate society used to live together. They had sex relation with one another and maintained the children in common. The men were the protectors and food gatherers while the women's job was to give birth of children and to nurse them. Due to intra-group conflict and quarrel, gradually the unit with a pair of adult man and woman was sanctioned for living. In course of time that became fortified with custom and law and we got the social institution, 'marriage'.

Marriage is, therefore, a permanent legal union between a man and a woman. It is distinguished from mating which is a temporary union and socially not sanctioned. No evidence of group marriage is found at present. The existing simplest tribal societies like Andamanese, Vedda, Kadar, Chenchu, Birhor, etc., do not show any incidence of group marriage. Although there are ample evidences of pre-marital or extra-marital sex relations in tribal societies, but monogamy stands as the general practice.

Forms of Marriage

At present in human society two forms of marriage is evident

(*a*) Monogamy *i.e.*, marriage of a man with one woman. Monogamous family arises out of this marriage.

(*b*) Polygamy *i.e.*, marriage of a man or woman with two or more mates. Polygamy can be of two types polygyny and polyandry. Polygyny means the marriage of one man with several women. Polygynous families are the outcome of such marriage. Polyandry, on the other hand, denotes the marriage of one woman with several men. Polyandrous families are formed by this marriage.

Polygyny should not be confused with concubinage *i.e.*, where a man cohabits with one or more women who are distinct from his own wife or wives. Since concubinage is recognized by various societies as an accepted institution, and as the customs regulating the relations between a man and his presumptive wives are so varied, it becomes difficult at times to distinguish concubinage from a polygynous marriage especially as regards the secondary wives.

In polygynous societies, jealousy among the co-wives creates problem. To avoid this, sometimes a man marries sisters from a family. Since the sisters grow up together, it is likely that they will be able to lead the life in amity. Matrilocal residence *i.e.*, the man's practice of living in the wife's house also favours polygyny. Such a marriage of a man with two or more sisters at a time is called sororal polygyny. When the co-wives are not sisters, the marriage is termed as non-sororal polygyny. Andamanese, Kanikkar, Urali, etc. tribes show high incidence of sororal polygyny. Besides, among the Totos, Santals, Lodhas, etc., occasional practice of sororal polygyny is evident.

Polyandry like polygyny is of two types—adelphic or fraternal and non-fraternal. When husbands are brothers, it is called fraternal polyandry. Opposite to this is non-fraternal polyandry, when husbands are not brothers. Toda and Khasa tribes are in favour of fraternal polyandry, whereas a few Tibetans and Nayers practise non-fraternal type.

Monogamy is considered as the most modern form of marriage. Human sex-ratio in general is 1:1. This indicates, nature favours monogamy though polygamy is fairly practised in the world. In India, monogamy is found not only among the Hindus and Muslims but also among the Tribals. The marriage act of 1955 has discouraged polygamy by making monogamy a rule for the Hindus. Muslim personal taw though permits polygamy, but other Muslim countries have passed a legislation to eliminate the abuses of polygyny. Tribals follow their own customary law; now monogamy seems to be the most prevalent form of marriage among them. Thus a decline of polygamous form of marriage is gradually expected.

Selection of Mates

In a society one can not marry anyone whom he or she likes. There are certain strict rules and regulations. The first criterion in establishing marriage alliance is the consideration of the group itself.

Exogamy : This is the rule by which a man is not allowed to marry someone from his own social group. Such prohibited union is designated as incest. Incest is often considered as sin. Different scholars had tried to find out the explanation behind this prohibition *i.e.*, how the incest taboo came into operation. S. Freud tried to explain it in terms of Oedipus complex. Malinowski considered the kinsmen analogous to the family members since they were the descendants from a common ancestor; he thought in the same line of Westermarck who held that the sexual desire among the family members used to be treated with horror, so it was prohibited.

In fact, there are some definite reasons for which the practice of exogamy got approval. They are:

(*a*) A conception of blood relation prevails among the members of a group. Therefore, marriage within the group-members is considered as marriage between a brother and a sister.

(*b*) Attraction between a male and female gets lost due to close relationship in a small group.

(*c*) There is a popular idea that a great increase of energy and vigor is possible in the progeny if marriage binds two extremely distant persons who possess no kin relation among them.

Hindus do not select their marriage partners having own gotro-name. It is believed that ‘ gotro’ denotes a large group where members originate from a common ancestor. Similarly we find marriage alliances are not permitted inside a group of a tribe. For instance, clan is a sub-group of tribe that corresponds to ‘gotro’ of Hindus. By nature clan is exogamous. Sometimes exogamy is maintained in the territorial level also. A man can not choose marriage partner from the known girls of his village; he has to marry a girl from other village.

Endogamy: This is the rule, which compels the members of a group to marry within the group. All the tribes and caste groups of India are endogamous. This means a man with a particular tribal identity marries a woman essentially from his own tribal group though with different clan membership. In the same way, persons belonging to a particular caste group always select their marriage partners from the same caste group but with different gotro identification.

The views that support endogamy are as follows :

(*a*) People prefer their own group as members of a group show more or less similar physical characteristics.

(*b*) A group always wants to upkeep their human resource potential in original form. So the members do not want to establish marriage relation with outsiders.

(*c*) Conception of high and low ranks plays among the groups, which resists a high-rank group to develop relationship with a low-rank group.

(*d*) Dissimilarity in religion begets differences in norms and values, beliefs and practices. So people are prone to select their mates from own religious group for maintaining good adjustment.

(*e*) Geographical barriers between two places often discourage the groups to establish marital relationships because of difficulties in access.

Beside these two major rules named exogamy and endogamy, there are other rules also which restrict the marriage on the basis of prohibition and preference.

Prohibition ; An incest taboo is universal feature in the world which means a prohibition on sexual intercourse as found in all societies between the closely related kins like parent and children or between siblings. Sometimes it extends to the cousins. Sigmund Freud proposed that the attraction between a father and his daughter or between a mother and her son gradually raise jealously and hostility towards mother or father respectively. These are the sub-conscious feelings that remain renounced or repressed and result into aversion towards mother or father. However, prohibited relations exist between parent and child of opposite sex, between brother and sister, etc. Incest is a very ancient belief and widely accepted among the primitive people of the world. Incest ban encouraged marriage outside one's own social group and thus helped the members of different groups to form a larger cooperating group.

Prescribed and Preferential Marriage

In many societies, marriage between first cousins is permitted or sought. When a man is compelled to marry a person of a particular category, it is called prescribed marriage. In some societies, there is no compulsion, but people consider certain union as desirable. Such a marriage is known as preferential marriage. In this respect three types of marriage are popular.

1. **Cross-cousin Marriage**

This is the marriage, which occur between the cousins whose parents are brother and sister. A person's cross cousins are therefore, his father's sisters' children in one side, and on the other side, his mother's brothers' children. This type of marriage alliances has been noted among the tribes like Oraon, Toto, Kadar, Gond, Kharia, etc. Sometimes cross-cousin marriage stands obligatory.

Cross-cousin marriage concept was first formulated by Tylor in 1888. Later, by adopting the expressions given by Warner and Radcliffe Brown, two types could be distinguished—Symmetrical and asymmetrical. In symmetrical type of cross-cousin marriage, only one of the two kinds of cross cousin is allowed. For example, Baigas, Garos, Aimol Kukis, and Kadars of India marry father's sisters' daughters (F.S.D.) while Mikirs, Veipei Kukis, Birhors and Todas marry mother's brothers' daughters (M.B.D.). The marriage where the selection of the cousin is not restricted is called as asymmetrical type of cross-cousin marriage. Here a person is free to marry his desired girl either from father's sisters' daughters or from mother's brothers' daughters.

Asymmetrical cross-cousin marriages can be classified into patrilateral and matrilateral form. If the person marries his maternal uncle's daughter, it is called matrilateral cross-cousin marriage. On the other hand, if he takes paternal uncle's daughter as wife, that will be designated as patrilateral cross-cousin marriage. Patrilateral cross-cousin marriages are generally found in patrilineal societies while matrilineal societies exhibit matrilateral cross-cousin marriages. There is a controversy with the point that whether the cross-cousin marriages are really prescribed or preferred. However, they are widespread in all continents of the world except Europe. Hottentots and Bantus of Africa prefer this type very much. Some Australian aborigines prescribe it and some others prohibit it. In Siberia, Gilyaks show matrilateral type when Soyots project the patrilateral type. In India, Veddas and Todas practise cross-cousin marriage as regular phenomena. This type of marriage involves lineage exogamy *i.e.*, cousins belong to different lineages.

2. **Parallel Cousin Marriage**

When marriage takes place between the children of the siblings of the same sex, it is called parallel cousin marriage. The mate may, therefore, come either from one's father's brothers' children or mother's sisters' children. In our country this type of marriage is only found among the Muslim community. Since among the Muslims, women traditionally inherit the property, perhaps to control the property devolution, they preferred parallel cousin marriage. Again, among the Bedwins of Arab, this type of marriage is very popular. Bedwins are the nomadic people. Such marriages help to keep their males within the band and so manpower is protected for fighting and other purposes of defence.

Usually in a community, where cross-cousin marriage is allowed, parallel-cousin marriage is forbidden. A man maintains formal as well as respectful relationship with his female parallel cousins. Joking relationship prevails only among the cross cousins. Significance of joking relationship lies in the possibility of marriage, whereas respectful relationship indicates rigidity of the incest taboo. In most of the societies of the world parallel cousin marriage is prohibited.

3. **Levirate and Sororate**

In many non-literate societies, cultural norms often force individuals to marry the sibling of the deceased spouse. The Latin word 'Levir' means husband's brother. After the death of husband when a woman marries her husband's brother, the custom is designated as levirate. Two types of levirate marriage have been recognized—junior levirate and senior levirate. If the widow remarries her husband's younger brother, it is called junior levirate. Such practice is found among the Santal and Lodha tribes of West Bengal. Senior levirate is just opposite. Here the widow marries the elder brother of her deceased husband. Senior levirate is not widely practised. Kuki tribe follows this. In patrilineal societies, after death, a man's heir is his brother who not only succeeds to his status and responsibilities, also inherits all the possessions of elder brother, including his wife or wives. In this context we should know that, in a few societies son was found to inherit father's all wives, except his own mother. This is known as filial widow inheritance. Example is the medieval Mongols. The Sema Nagas also exhibits the same practice.

The custom by which a man is obliged to marry the sister of her deceased wife is called sororate marriage. The Latin word 'sorror' means sister. Sororate is also of two types—restricted and non-restricted. Restricted sororate is the actual form. But sometimes during the lifetime of wife, the husband marries her sister. That is called non-restricted sororate. The communities like Santal, Toto, etc. show non-restricted sororate type of marriage. Apart from this, in some cases, a man marries several women who are the sisters. This is termed as sororal polygyny. Andamanese, Uralis, Kanikars practice sororal polygyny. Santal, Lodha, Toto of West Bengal also practise the same. High rate of bride price generally leads to difficulties in securing mates, which results in sororate marriage. Younger sister of the deceased girl is supplied to her husband as a compensation for his loss, Levirate and sororate both signify inter-familial obligation and cordiality.

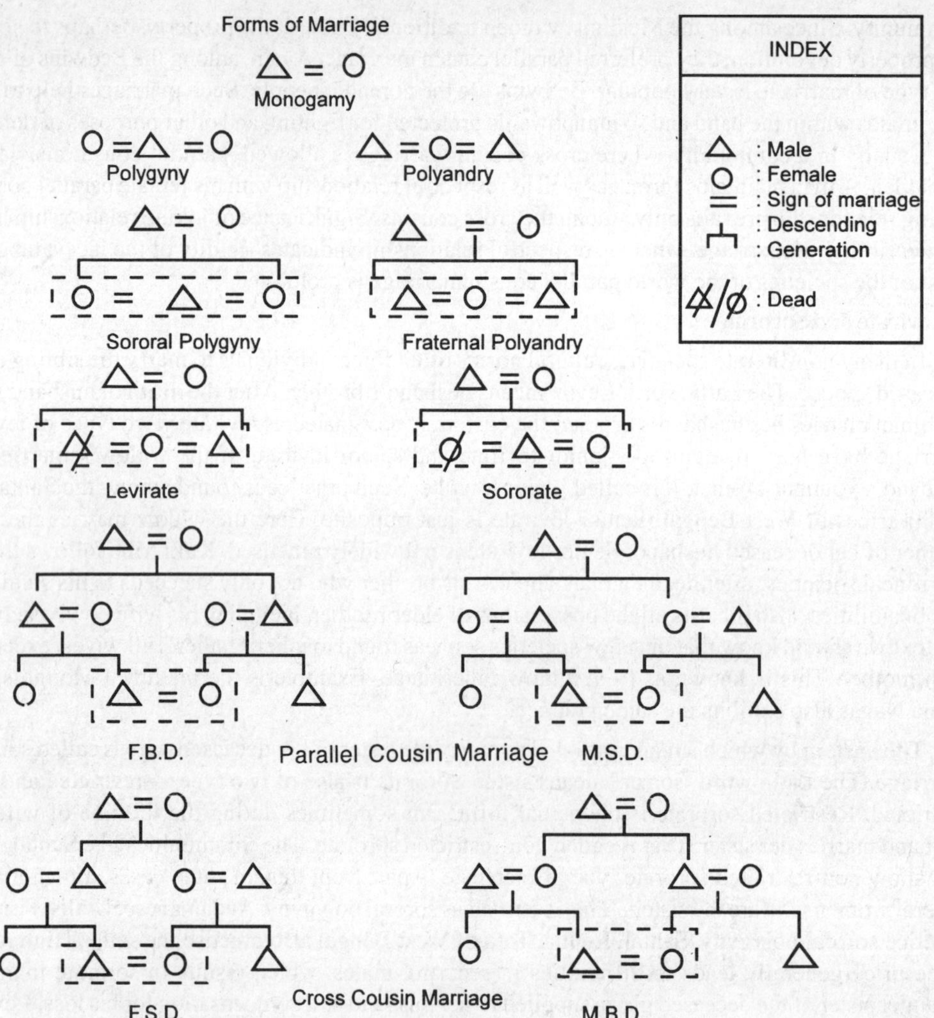

Fig. 16.2

Hypergamy and Hypogamy

In India where caste hierarchy is predominant in social structure, some other restrictions in marriage come up spontaneously. Because each of the caste indicates a particular status, high or low, which is associated with the degree of its ritual purity. Here two types of marriage-match have been taken into consideration.

(*a*) Hypergamy or Anuloma Marriage

This is a situation where an upper caste man marries a lower caste woman. For this unequal match a man does not at all lose his caste status or ritual purity. But his children suffer a lot; they partially lose their hereditary status. But such type marriage had a sanction in ancient India by Hindu social custom.

(*b*) Hypogamy or Pratiloma Marriage

This is a situation where a low caste man is married to an upper-caste woman. Although this type of match was evident occasionally, but failed to get social sanction. Because a woman after such marriage was held as ritually impure for losing her original caste status. According to Hindu tradition, daughters carry father's caste status before marriage and after marriage they are bound to share husband's status.

In ancient India, people were divided broadly into four varnas or castes. Brahmins were placed at the top and then Kshatriya, Vaisya and Sudra, one after another. The rule was like this :

A Brahmin man was permitted to take woman not only from his own caste but also from other lower castes. A Kshatriya man could take a woman from any castes except the Brahmin as it was a higher caste than Kshatriya. A Vaisya man, therefore, was able to take a woman only from Sudra caste, beside his own. In the same manner, a Sudra woman was free to select any man from any caste as her mate. A Vaisya woman was not allowed to marry a Sudra man but she could marry any man from the other three castes. In case of a Kshatriya woman, she was permitted to choose her mate only from the Brahmins, apart from her own caste. She could marry a person neither from Vaisya nor from Sudra, as both the castes held lower rank than her own caste, Kshatriya. Thus, the custom of maintaining purity was so rigid that if a Brahmin woman got married with a Sudra man, their offsprings could avail absolutely no status in the society and used to be known as Chandal, as extremely low and mixed category.

Permissible marriages among Castes

Castes	MAN Select mates from	WOMAN Collect mates from
Brahmin	Brahmin, Kshatriya, Vaisya and Sudra,	Brahmin
Kshatriya	Kshatriya, Vaisya and Sudra	Brahmin and Kshatriya
Vaisya	Vaisya and Sudra	Brahmin, Kshatriya and Vaisya
Sudra	Sudra	Brahmin, Kshatriya, Vaisya and Sudra

Economic Aspect in Marriage

Economic value of marriage is very important as like the rituals of marriage, which varies with culture. Economic transaction takes place either before or after the marriage.

Bride Price: In many non-literate societies, brides' parents are compensated with a payment for the loss of daughter's service in their house. This payment is called bride price, which may be given, in cash or in kind or in both cash and kind. For instance, Oraons of Chota Nagpur take clothes for bride and her relatives. Hos and Mundas take cattle, Ao Nagas take baskets of paddy and a specific dagger. Bride prices symbolize the prestige of woman in her family. Though it is a form of purchase but the purchaser abstains from exerting his absolute right on the purchased. A man neither tortures his wife heavily nor sells the wife, otherwise he obtains various services from her. For any reason, if a girl wants to go back to her parental home, her relatives there try to discourage her. They do not want to get her back to avoid the return of bride price. For this reason, incidence of divorce is rare in the societies where the system of bride price prevails. Such a system is mainly found among the tribes like Santal, Munda, Oraon, Lodha, Gond, Bhil, Chenchu, etc. Sometimes services (manual labour) are provided by the groom to the bride's family as a substitute of bride price. Certain lower castes of Hindu society namely Bagdi, Bauri, Hari, Dom, etc., also take price against bride. The custom of bride price is usually associated with patrilocal residence.

Dowry : It is the payment from bride's side to the groom's side. It indicates a special position of bridegroom in the society. Dowry includes goods or money or sometimes both. Expensive ornaments, clothing, bedding and other useful materials are often offered to the groom or groom's family by the bride's family. In India the custom of dowry is widely prevalent. Usually higher caste Hindus practise this. Muslims and Christians do not lag behind. Recently dowry system has been prevented by legislations but still the customary payment continues. Extortion may be found even after marriage. Bride burning or dowry deaths are often caused by this system. The law stands as a safeguard though seems to have done a little on it.

Age of Marriage

Age of marriage varies from society to society, usually between infancy to adult age- Social norms as well as economic factors regulate the age of marriage in non-literate societies. Pre-marital sexual freedom is generally widely accepted because it is considered as an introductory preparation for marriage. Bi-sexual dormitories are found among some tribes where free mixing between boys and girls is possible. For example, Ghotul of Muria Gond. Pre-marital connections are always overlooked unless they cause pregnancy. But extra-marital sex relations are never tolerated.

Ways of Acquiring a Mate

Marriages among the non-literates are basically a secular contract and so it is necessarily not accompanied with religious solemnization as like that of Hindu marriage. It denotes reciprocal obligation between the two clans or two kin groups, since marriage establishes relationship not only between a man and a woman but also between the relatives of two distant families. Procreation of children is the universal objective of marriage. To an anthropologist, the institution of marriage is of prime importance as it signifies economic cooperation between the couple; security and socialisation of the younger generation. A marriage, however, materializes through economic exchange or mutual service. Sometimes magico-religious and ceremonial performances are involved in marriage. But ways of obtaining a mate is not same in all societies. They are as follows :

(*a*) Marriage by Capture

This is a kind of marriage where a girl is taken away forcibly without her consent. Consent is not taken even from her guardian or near relatives. In all primitive communities, this type of marriage was once in vogue. Manu, the ancient lawmaker of Hindu society referred this type of marriage as 'rakshasa vivaha' and mentioned it as an approved mode of securing a wife among the Hindus. Bhil, Gond, Ho, etc., tribes used to practise this widely. At present, this custom has been modernized anticipating penalty from court. Among the Kharias and Birhors, ceremonial capture takes place instead of real physical capture. The youngman who is in love with a girl, suddenly put vermilion on the forehead of the girl in a public place and runs away. The girl becomes his wife after this. In the Santal community, sometimes this sort of marriage is found to happen. Among the Gonds, got up capture takes place. A mock fight is held between the two parties where the bride weeps and laments ceremonially. Mock capture is also practised in Africa, Melanesia and China. In societies, where a surplus of females is found, groom capture takes place. From the Kambot people of New Guinea such practice has been noted.

(*b*) Marriage by Trial

This is the kind of marriage where a young man has to prove his courage, bravery and physical strength before claiming a girl as wife. Such practice is widespread among the Bhils of Central India.

(*c*) Marriage by Purchase

This is the usual way of obtaining a wife in tribal societies. Here marriage demands a payment for the bride, which is known as bride price. In Vedic age such practice was widespread in Hindu society. Lower castes of Bengal still prefer marriages by purchase. Among the Veiphei Kuki. and RengamaNagas, this practice is quite common. Santal, Oraon, Toto, Lodha, etc., are not exceptions.

(*d*) Marriage by Exchange

It is a modified form of 'marriage by purchase'. Here the bride price is compensated by providing another bride in return. So that payment could not be claimed by any of the two families. In Melanesia and Australia, a man's sister is offered to his wife's brother. The same used to be done among the Kulin Brahmins of Bengal. Bhotias of Almora also show this type of marriage in their community.

(*e*) Marriage by Service

This is the marriage where considerable labour is offered by the bridegroom to the brides' family, prior to marriage. A prospective groom goes to live in the house, which would be his father-in-law's house and works for them. The duration of service varies in different groups of people. It may be a few days or a few months or even a few years. Among the Bunas of Bengal, the groom serves the father-in-law for six to nine months. Veiphei Kukis take this period from two to three years. For Bhils, it stands about seven years. Such a practice is also popular among Aimols, Purum and Chiru Kukis of Manipur, among the Eskimos and the Ainus of Japan. Marriage by service is also associated with matrilocal residence.

(*f*) Marriage by Negotiation

This type of marriage rests on mutual consent of both the parties. The guardians here look for the suitable match and negotiate thereafter. Sometimes go-betweens are recruited. Most of the tribes and castes of India follow this way. Purum Kukis of Assam, Mundas and Hos of Chota Nagpur, Baigas of Madhya Pradesh show maximum rigidity in the process.

(*g*) Marriage by Elopement

The tribes especially who keep dormitories for the youth indulge the adolescent boys and girls to choose their mates. In this circumstance, if parent's consent is not available, elopement takes place. The boy elopes the girl and generally after a considerable period, the couple is received back to the family. Eloped couple when readmitted into the tribe, they have to accept a phase of beating. A grand feast may or may not be arranged thereafter. Among the Oraon, this type of marriage is tremendously popular. Among the Kurnais of South East Australia, about a dozen of girls are eloped at a time.

(*h*) Marriage by Intrusion

This is the marriage, which solemnizes as per the desire of a girl. When a girl is willing to marry a particular person who does not want her, she herself intrudes in his house and begins to stay there without the permission of the members of that house. Naturally they abuse her and the girl has to face different modes of oppression. If she can overcome all those, marriage is sanctioned. Such a strange way of marriage is known as 'marriage by intrusion'. This is observed among the tribes like Birhor and Ho.

(*i*) Marriage on Probation

This is the marriage where a bridegroom is allowed to stay in the bride's house, a few days before the marriage. During this period both the boy and the girl try to know each other very well. If they think that their temperament is compatible to one another, then only they take the decision regarding marriage. Otherwise they separate and for the second situation the boy has to compensate the girl's parents with cash payment. Such a way of getting mate is found in Kuki community.

KINSHIP

Kinship is the method of reckoning relationship. In any society, every normal adult individual belongs to two different nuclear families. The family in which he was born and reared is called the 'family of orientation'. The other family to which he establishes relation through marriage is called the 'family of procreation'. This universal fact of individual membership in two nuclear families gives rise to the kinship system. A kinship system is neither a social group nor does it correspond to an organized aggregation of individuals. It is merely, as the name implies a structured system of different relationships where individuals are bound together by complex interlocking and ramifying ties.

The point of departure for the analysis of kinship is nuclear family. Because each and every individual is trained in the matrix of family. A developing child takes his first lessons from the elders of the family. He learns to behave reciprocally and to establish interpersonal relationships. He responds in particular ways towards his father, his mother, his brothers, and his sisters and expects certain kinds of behaviour in return. His responses remain individualized at first, but get modified in course of time. The personality type is thus moulded as per the expectation of the society. A man becomes acquainted with the traditional norms and values of the society into which he has born and promotes himself as a member of the particular society and culture.

Study of kinship is very useful for understanding the elements of social organization as they serve the mechanism of organizing social activities and coordinating social relations either in a limited sector or of social life or in relation of all social interests. Radcliffe Brown has pointed out that there is a high correlation between social organization and kin terminology. Rivers has mentioned that the particular features of social organization are reflected from the prevailing terminology. Kroeber has shown the role of language in the system of terminology. According to Raymond Firth, kinship is the rod on which one leans throughout the life. In all societies, kinship system offers a great bond of unity by which individuals are linked with each other and a ramifying series of link is established. These are recognized relationships based on actual as well as supposed genealogical ties. Significance of kinship in pre-industrial society is more pervasive and systematic than in modern industrial society. Kinsmen play important roles in social, economic and political spheres.

To get a comprehensive view on kinship system two types of approaches are required. At first, one has to delineate the structure of kinship roles; Secondly, he has to find out associated behaviour to each structure and role in order to know the functional utility of the structure.

I. Structure of Kinship Roles

The bond of kinship embraces a great range of people for which a whole group of related persons is held as an entity. Although it is not a social group but shows the character of groupings as well as institution together. The structure of kinship roles not only reveals the types of relationship but also the specific kinship terms in a society. As per the types of relationship, kins are primarily divided into two groups.

(a) The real or actual kins.

(b) The virtual or artificial kins.

The term virtual kin is applied to the ceremonial kin and social kin, *i.e.*, the kin relations that have been ceremonially established and formed as a result of keen social contact. On the other hand, the real kins are the well-recognized relations formed through blood connection or through marriage. The kins who possess blood relation among them are called consanguineous kin. But the kin relations that develop due to marriage are called affinal kin. For example, in a family, the relationship between the spouses comes under affinal kinship, whereas the relationship between the parent and a child can be marked under consanguineal kinship. An adopted child is treated as one's own biologically produced offspring. So it is also treated as consanguineal kin.

Kinship generally traced from an ego. All persons having relationship with the ego are defined in terms of kinship status. The degree of nearness or remoteness determines this status. When an individual is directly related to the ego, he is taken as primary kin. A father stands a primary consanguineal kin to his children and mother is the primary affinal kin to the father. When a person is related to the ego through a primary kin, he is called the secondary kin. Father's father, father's sister, mother's mother, wife's mother, brother's wife, sister's son, etc., are the secondary kins to the ego. Secondary kins may also be classified into secondary consanguineal kin and secondary affinal kin basing on the nature of tie. Similarly, primary relatives of the secondary kinsmen are ego's tertiary kins. For example, father's sister's husband, wife's brother's son, daughter's husband's sister, etc., belong to the group of tertiary kin. More remote relationships than tertiary kins are designated as distant kins. Both the consanguineal and affinal relatives can be classified into primary kin, secondary kin, tertiary kin and distant kin considering the degree of nearness in the relationship. All the people who are related to an individual through a common ancestor called cognates. Cognates include relatives both from father's and mother's side. When the relatives are related to the ego only through father's line, they are grouped as agnates or patrilineal kins. In the same manner, relatives of mother's side only, are called uterine kin or matrilineal kin.

Kinsmen may extend to an indefinite number of relationship enclosing innumerable distant categories of genealogical connection. For avoiding the complications in the system of nomenclature, every society is found to reduce the total number of kinship and to maintain a framework into which the kinship terms are used. The number of persons in a kin group denotes whether it is a broad range or a narrow range system. Usually, primitive societies are in the favour of a broad-range kinship system while the western kinship system is a narrow range system.

Kinship terms are the terms of addresses, which are used among the kinsmen of a society to designate the kins of different type. There are personal names but almost all societies use some specific kinship terms among the relatives, which are used extensively. An intermediate form is also found between personal names and kinship term. It is called teknonymy. For instance, a person having a son or daughter is often called as "father of——(child's name)', or "mother of —— (child's name)'. This is the combination of parental term and child's name instead of referring the personal name or the particular kinship term. Such terms are very important as they sometimes reflect social phenomena. Study of kinship nomenclature is still a significant method of study, which was first undertaken by Morgan after coining various kinship terms from different parts of the world. He divided the kinship terms into two major divisions and named them as classificatory kinship system and descriptive kinship system. In the classificatory system all relatives were classed in categories following a strict logic. All of them were referred by the same term of designation. For example, a child knows who is his real mother, but the term is also applied to designate his father's second wife, husband's mother, even any other motherly women. In the same way a child may have many fathers other than his own father. Therefore, the descriptive term father or mother may turn to classificatory term; whereas uncle, aunt, cousin, etc., are the classificatory terms in true sense. Because, by the term uncle one wants to mean father's brothers and cousins, mother's brothers and cousins, father's sisters' husbands and mother's sisters' husbands, etc.,a particular category of relatives. The term aunt and cousin also denote the same.

According to Morgan, primitive societies are fond of classificatory kinship terms and in civilized societies descriptive terms are preferred. In a civilized society, a child perceives two different sets of relatives from two different sides—from father's side and from mother's side. He uses special descriptive terms to differentiate each relative in a particular category. For example, father's father and mother's father though belong to the same category but there are specific terms to denote the actual relationship *i.e.*, whether he is father's father or mother's father. Likewise, particular terms are found against mother's mother, father's mother, father's brother, mother's brother, father's sister, mother's sister, and so on. In reality, the words 'classificatory' and 'descriptive' refer to the kinship terms only, not to the whole system of terminology. At the time of exploring a new language, we at first try to infer the meaning of different terms from the objects to which they are applied. Kinship terms are also derived in the same manner.

G. P. Murdock had identified six major systems of terminology on global basis. They are the Eskimo-system, the Hawaiian system, the Iroquois system, the Omaha system, the Crow system and the Sudanese system.

1. The Eskimo system

The system is named as Eskimo system because it is mainly prevalent among the Eskimos, although some Anglo-American cultures also exhibit the same. This system put emphasis on nuclear family. So no other relatives are referred by the terms which are used in the nuclear family setting—father, mother, brother and sister. Next importance is given to the relatives of mother's and father's side. The relatives of these two sides are equally treated according to their category. This means that the same terms *e.g.,* uncle, aunt, cousin, etc., are used for both sides, without differentiating the paternal or maternal line.

2. The Hawaiian system

This system is followed in Hawaii and some other Malayo-Polynesian speaking areas. It is the simplest system where least number of terms is used. In a generation, all relatives of the same sex are referred by one term. For example, mother means all female relatives of both the sides in the category of mother. Similarly, sister means all female cousins of ego from both the sides. Sometimes the system is called as "generational system".

3. The Iroquois system

This system is named after the Iroquois Indian tribe of North America. Father and father's brothers are referred here by the same term. Same principle applies for ego's mother and mother's sisters. But mother's brothers and father's sisters are referred separately by different terms. In case of cross cousins, both the sets are referred with the same term but differentiated by sex. This means, mother's brother's son and father's sister's son belong under the same term. But parallel cousins are invariably referred with different terms and sometimes they are equated with ego's brother and sister under a single term.

4. The Omaha system

This system is though named after Omaha Indian Tribe of North America but also found in some other societies of the world where the patrilineal descent group prevails. Like the Iroquois system, here father and father's brothers are referred by a single term. But the peculiarity of this system is that, here cross cousins of the maternal side merge with the parental generation *i.e.,* all female members are grouped together regardless of their generation. But the members of paternal side show separate terns for their male and female members in different generations. Therefore, mother's brother's children are equated with one's own children whereas, the parallel cousins are considered as siblings.

5. The Crow system

This system is named after the North American Tribe, Crow Indian which show matrilineal descendance. The crow system can be considered as matrilineal equivalent to the patrilineal Omaha system. It specifies the relatives of mother's side instead of father's side. The Crow Indians call father, father's brother and father's sisters' son by one term. Similarly father's sister and father's sisters' daughters are referred under a common term.

6. Sudanese system

This system is rare in occurrence. It is also called as 'descriptive system' because here a large number of terms are used—one for each type of relative for both the father's and mother's side. Each of the cousins is also distinguished.

G. P. Murdock had also tried to classify the kinship terms in three ways.

(*i*) By the mode of use

Basing on the mode of use the kinship terms can be divided into two sections. Some terms are for direct addressing and others are for indirect reference. 'A term of address' is used to call a relative for talking, whereas a 'term of reference' is used to refer a relative for speaking about him or her to a third person. The term of address is the part of the linguistic behaviour that signifies a particular interpersonal relationship. But the term of reference is not a part of relationship, rather it is the designation of a person with a particular kinship status. Sometimes a 'term of address' is same to the 'term of reference'.

(*ii*) **By the linguistic structure**

Here the kinship terms are distinguished as elementary, derivative and descriptive. An elementary term denotes an irreducible word like father, mother, nephew, etc., as in English language, These terms can not be analyzed into components having kinship meanings, therefore, called 'elementary term'. A 'derivative term' is like grand-father, brother-in-law, step daughter, etc., (in English language) *i.e.*, these are the compound words with some lexical elements. A 'descriptive term' is one, which combines two or more elementary terms to denote a specific relative. In Bengali language many such terms are found like *masi-ma, pisi-ma, mami-ma,* etc. Mother's sister is called *masi,* father's sister is *pisi,* mother's brother's wife is called *mami,* and *ma* is the term for calling mother. So two elementary terms have combined in these cases.

(*iii*) **By the range of application**

Here the kinship terms are differentiated into two groups—denotative terms and classificatory terms. Denotative terms are applied to the relatives of a single kinship category, defined by the generation, sex and genealogical connection. Sometimes such a term denotes only one person like father, mother, husband, wife, father-in-law, etc., as in English language. But otherwise it applies to many persons of identical kinship connection. For example. the English terms, brother, sister, daughter, son-in-taw, etc., denote several persons with the same designation. In contrast, a classificatory term is the term that applies to the persons of two or more kinship categories. For instance, in English the term grand-mother stands for both father's mother and mother's mother, uncle may be a brother of any one of the parents or he may be the husband of father's sister or mother's sister. In a similar form, the term cousin includes all collateral relatives of one's own generation and some of the adjacent generation irrespective of their sex, line of genealogical connection or the degree of remoteness.

Kroeber and Lowie advocated six criteria behind terminological discrimination. They are sex, generation, affinity, collaterality, bifurcation and polarity. Biological difference between male and female is reflected in kinship terminology. The facts of reproduction align people automatically in different generations. Affinity arises due to the social phenomena of marriage and incest taboo. Affinal difference is widely recognized in kinship terminology. Kinship terms are found to vary with the relatives of same generation and sex according to the degree of nearness. The criterion of bifurcation applies only to the remote relatives as they are placed on two divergent lines from a common ancestor. Polarity is the criterion which rests on the sociological fact that two persons constitute a social relationship, if live close together. Beside these major criteria, dissimilarity of age and descendance may bring change in kinship terminology. Societies differ markedly with the process of distinguishing the relatives.

II. Associated Behaviour in Kinship Structure

Kinship terms have an intricate relationship with the kinship behaviours in each society. But behaviour towards a particular relative is not same in all societies. Most of these behaviours are culturally patterned and so vary from society to society. The structure of kinship becomes absolutely clear if we analyze the kinship terms in relation to different behaviours. However, the behaviours may be projected at the following, levels.

(*a*) Interpersonal relationship among the kins within the household.

(*b*) Interpersonal relationship outside the household, among the consanguineal kins.

(*c*) Interpersonal relationship outside the household, among the affinal kins.

(*d*) Interpersonal relationship outside the household, among the ceremonial or village kins.

(*e*) Behaviour of the kins, both consanguineal and affinal, at the time of different rites of passage, from birth to death.

(*f*) Behaviours of the kins in secular affairs like economic or religious affairs.

Iravati Karve in her book 'Kinship Organization in India' has pointed out three things that are absolutely necessary for the understanding of any cultural phenomenon in India. Those are, configuration of linguistic regions, the institution of caste and the family organization. In fact, structure of the family provides a basic insight in analyzing the kinship structure. So it will be wise to discuss here a few typical family structures in reference to some known communities of India.

(*i*) Among the Hindu peasants, family includes parents, married sons, and old grand-parents. Father's sisters, unmarried as well as widow may remain there. Normally three generations are covered in a family.

(*ii*) Among the Hindu merchant castes, big-families including married brothers (own or cousin) with their married sons and children are found. At least three generations live together with many nuclear families.

(*iii*) Among the Muslim peasants, parents and the married sons commonly live together. Sometimes married brothers with their married sons and old parents are found (three generations). A notable feature among the Muslims is that, a man generally marries more than one woman and all his wives along their children reside with him in the same household. In this society sometimes marriage occurs between the son and daughter of two brothers.

(*iv*) Among some agricultural tribes namely Santal, Oraon, Munda, etc., the family is normally composed of parents and children. One or two married sons may be present there. Patrilineal patrilocal extended family is a distinctive unit for the economic and ritual activities. Kinsmen gather together following important occasions of life-cycle like child-birth, marriage and funeral.

(*v*) Among the horticulturist tribe Khasi, the family is composed of parents with one or two married daughters. The family compulsorily includes married youngest daughter and old parents.

(*vi*) Among the pastoral tribe Toda, family is composed of several husbands who are brothers, They live with their single or many wives in the same household. Children are obviously included there.

Each of these above mentioned family structure includes certain kins whose specific roles are established by custom. Kinship tie is an adhesive that is essentially required in the formation of family groupings and in holding up a social structure. The relatives as per their respective status undergo certain rules, which regulate their reciprocal attitudes and standard of behaviour. These rules are expressed both through verbal and non-verbal behaviours. Care, affection, protection, respect, cordiality, obedience, avoidance (in physical relation, speech and visual relation), hospitality, joking relation, etc., are observed in the kinship behaviours. For example, parents behave in some definite ways to their children and so also the children. In many societies, brothers and sisters modify and reduce their familiar relation after a certain age.

III. Special Kinship Usage

The study of kinship does not restrict itself with the description of the kins or basis of their classification or general forms of kinship behaviour. There is certain special kinship usage, which hold special significance in respect to the non-literate societies.

(*a*) Avunculate

This is a queer usage found between a mother's brother and his sister's children. Among some matrilineal societies, maternal uncle assumes many of the duties of father as a matter of convention. His nephews and niece remain under his authority. They inherit his property also. Such a relationship exists among the Trobriand Islanders of Melanesia, the Fijians, the African Tribes and the Nayers of South India.

(*b*) Amitate

This kind of usage is more or less similar to the avunculate and found among the patrilineal people. Here father's sister gets great respect and prime importance. She is more than mother to her nephew and exerts her authority on him in many events of life. In fact, it is a social mechanism, which protects father's sisters from falling into neglect, especially in situations when they are driven off from their in-laws house. Polynesian Tonga, Toda of South India etc., communities exhibit this type of kinship usage.

(*c*) Couvade

This is another strange usage of kinship between a husband and his wife. Toda and Khasi community of India can be cited as examples. Here the husband is compelled to undergo an austere life whenever his wife gives birth to a child. He has to maintain a strict diet and to observe a number of taboos along with the wife. Anthropologists regard couvade as a symbolic representation of establishing paternity on the child. Some years back, this particular usage was popular among the'Nayers of South India, Ainus of Japan and also a few communities of China.

(*d*) Avoidance

In most of the societies, the usage of avoidance acts as an incest taboo. A father-in-law avoids his daughter-in-law according to traditional social norm. Same relation prevails between a mother-in-law and her son-in-law and between husband's elder brother and younger brother's wife. This is actually a protective measure against incestuous sexual relation among close relatives who remain in face-to-face contact everyday. Mother-in-law and son-in-law although do not come within the purview of daily relationship, the avoidance between them has been explained in this way that a mother gets displeased by handing over her beloved daughter to a stranger, on the other hand, the son-in-law becomes annoyed for the interference and indirect control of mother-in-law on his wife.

(*e*) Joking relationship

It is just the opposite type of kinship usage in contrast to 'avoidance'. This specially privileged relationship indulges in teasing each other by different kinds of jokes including vulgar sexual jokes. Usually such relationships exist between a man and his wife's younger sisters or between a woman and her husband's younger brothers, between cross cousins, between grand-parents and grand-children. Joking relationships are found in tribal as well as in Hindu society.

Anthropologists consider the kinship terms in a society as extremely functional as well as resistant to change. These terms not only distinguish the relatives but also indicate the form of families, rules of residence, rules of descent and many other important features of a social system.

IV. Rules of Descent

In almost all societies kinship connections are very significant. An individual always possesses certain obligations towards his kinsmen and he also expects the same from his kinsmen. The rules, which affiliate each person to a particular and definable set of kin, are called rules of descent. Such rules vary from society to society. Succession and inheritance is related to this rule of descent. The word 'succession' denotes the transmission of rights and the word 'inheritance' refers the right over the parental property. Usually both the rights go hand in hand. However, the three rules that have been identified regarding the descent are as follows :

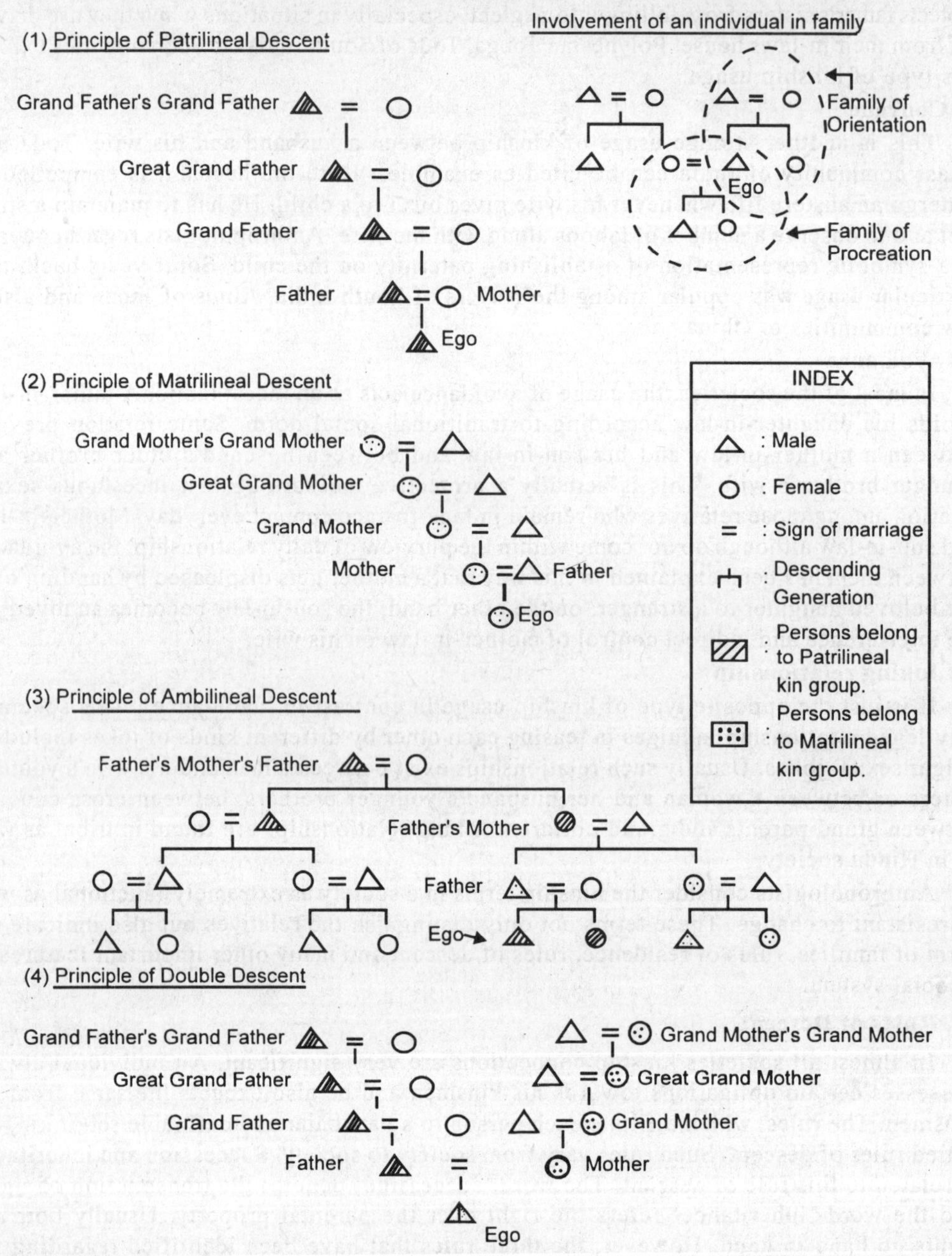

Fig. 16.3

1. Patrilineal descent

When the descent is traced solely through the male line, it is called patrilineal descent. A man's sons and daughters all belong to the same descent group by dint of birth, but it is only the sons who continue the affiliation. Succession and inheritance pass through the male line.

2. **Matrilineal descent**

When the descent is traced solely through the female line, it is called matrilineal descent. At birth, children of both the sexes belong to mother's descent group, but later only the females acquire the succession and inheritance. Therefore, daughters carry the tradition, generation after generation.

3. **Ambilineal descent**

In some societies individuals are free to show their genealogical links either through men or women. Some people of such society are, therefore, connected with the kin-group of father and others with the kin-group of mother. There is no fixed rule to trace the succession and inheritance; any combination of lineal link is possible in such societies.

These three rules may or may not be exclusive. Because, some of the societies entirely go on one rule but in some other societies, two different rules are found to be operated side by side. The societies that exhibit two principles of descent are called Double descent or Duolineal descent. Here an individual claims membership in two distinct descent groups. Therefore, double descent may be of various combinations—patrilineal and matrilineal, patrilineal and ambilineal, or matrilineal and ambilineal. For example, Omahas of North America show patrilineal descent, Chiricahua Apaches of South-West America present matrilineal descent and Samoans of the South Pacific provide the instance of a typical ambilineal descent. Again, the Kapauku Papuans of Western New Guinea is a society where different types of descent group are visible at a time.

4. **Bilateral descent**

The term bilateral means 'two-sided'. There are some societies where no lineal principle operate *i.e.*, individuals in those societies do not relate themselves to a common ancestor. Descent of any particular line is not counted; rather they accept relatives of both father's and mother's side with equal importance. Relatives of two sides (kindred) are counted in reference to ego and include primarily parents, grand-parents, uncles, aunts and first cousins. The nuclear families are always bilateral because spouses come essentially from two different families and the children are related to both of the parents' family. The society of United States is characterized bilateral descent.

It is quite natural that sometimes one may get confused with the principles of bilateral descent and double descent because both use two lines of descent for their manifestation. But the difference is, Bilateral descent does not segregate matrilineal and patrilineal kins, rather combines them in one group. But the double descent although combines two principles (for example, patrilineal and matrilineal) but independently of one another. As a result, individuals hold membership in two distinct social groups, opposite in nature. They, therefore, associate themselves exclusively with one or the other, in different situations and for different purposes. Eskimos of North Alaska show bilateral descent group where kins from at least three generations are recognized, both from father's and mother's sides.

The groups, which obey the principle of unilineal descent, are as follows :

Lineage: A family is bilateral. In contrast to this, a lineage is a unilateral descent group. It is made up of consanguine kins that claim their descent from a common ancestor or ancestress, through known link. A lineage generally includes ancestors of five to six generations in a sequence. Lineages may be of two types—patrilineage and matrilineage. In the former, links are traced exclusively through the male line and in the latter, links are maintained through female line only. If the descent is patrilineal, the child of a legal marriage belongs to his father's lineage. His rank as a noble or a commoner will be determined by the nature of the respective lineage. It may entitle him to become a King or a Chief or a priest. In ordinary cases, one must have a claim on the productive resources of the lineage. In a matrilineal society, every child belongs to the lineage of his or her mother although the authority goes with mother's eldest brother.

The lineage members may or may not share a common residence. The smallest lineage consists of a man and his children. Joint family is also a lineage where members upto three or four generations are available together. Actually members of a lineage form a corporate group who perform the same ritual acts but possess autonomy in everyday affairs. A lineage is always a strict exogamous unit and the ancestor of a lineage is never a mythological or legendary figure.

Clan : A clan is a unilineal kinship group larger than a lineage. Here the members are supposed to be descendant from a common ancestor but the genealogical links are not specified *i.e.*, the members can not demonstrate their actual lineal relationship through a genealogical table. In such condition descent is traced to a mythical ancestor who may be a human or a plant or an animal or even an inanimate object. The term clan, sib and gens indicate the same unilinear kinship group.

Clans are exogamous in nature *i.e.*, marriage partners essentially come from two different clans. Membership in a clan is hereditary. Members of a clan usually remain friendly to each other and help one another following a social need. But sometimes hostile relation between two clans is found.

The particular animal or plant, which remains associated with a clan as group identification, is called totem. According to R. H. Lowie, 'a totem is generally an animal, more rarely a plant, still more rarely a cosmic body or force like sun or wind, which gives its name to a clan and may be otherwise associated with it'. A totem is, therefore, especially significant for the clan. Rivers defined a clan as 'an exogamous division of a tribe, the members of which are tied together by belief in common descent, common possession of a totem or habitation of a common territory'. Clan is found in almost all primitive communities of the world, though not as a universal feature. For example, Santals of India has twelve clans, Lodha tribe has nine clans. Andamanese and Kadar show no evidence of clan. Absence of clan has also been noted from some tribes of America.

Members of a clan regard their respective totem as founding ancestor. They do not always believe that they have directly descended from the totem, it is said that the particular totem has helped or promoted or given some services to their ancestors. Therefore, the members respect totem; they never touch, kill, eat, harm or destroy the totem of reference. For example, a clan among the Santal is named as Hansda. The members of this clan respect duck (local name : Hans) and do not eat the flesh of duck because they believe themselves to be originated from duck. Similarly a clan among Lodhas is Nayek whose totem is sal-fish. Killing or eating of sal-fish is prohibited to this clan. More examples can be cited in this respect. One of the clan among Oraon is named Lakra which means tiger. The clan members never hunt tigers for showing regard to this ancestral animal.

However, clans can be categorized into several types on the basis of its nature of origination.

(a) Patrilineal clan

When a clan is patrilineal in nature, it is called patrilineal clan *i.e.*, all the members are considered to be descendant from a common fore-father through male line. Every child inherits father's clan-name though daughters leave it after their marriage by adopting respective clan-name of their husbands. Santal, Munda, Lodha, Oraon, Bhil, etc., tribes of India exhibit this type of clan.

(b) Matrilineal clan

It is the type of clan where the descent is reckoned from a single ancestress through females. Every child, irrespective of sex acquires their maternal clan-name by birth but sons adopt clan-names of their respective wives after marriage. Garo, Khasi and Nayer tribes of India show this type of clan.

(c) Ancestral clan

Sometimes members of a clan believe that they have been originated from a definite pair of male and female. This type of clan is called an ancestral clan. It is found among the Khasi people of India.

(d) Totemic clan

Instead of a human ancestor, when the members of a clan relate themselves with a particular totem, the clan is designated as a totemic clan. Such clans are frequently found among the primitive communities like Santal, Oraon, Lodha, Kol, Bhil, Gond, Toda, etc.

(e) **Territorial clan**

Sometimes members of a clan identify themselves with a particular territory from where they had been possibly originated. Among the Bison Murias of M.P., clans are named after the villages. Among the Nagas of Assam, Khel is a territorial group, though not a clan.

The term 'sib' is often considered as a synonym of the term 'clan' because it is also a unilateral exogamous group where members believe in a common descent but they may not able to show the link through a genealogical table. Further, this involuntary association is dependent on birth and may be changed through adoption. Sibs do not occur in the lowest stages of culture represented by hunting and the pastoral tribes like Andamanese, Semang, Hottentot, Bushman, Eskimo, etc. Existence of sib has been recorded from the Lhota Nagas of Assam, Bhuiyas of Orissa, Kukis of Manipur, Aruntas of Australia, Bantus and Masais of Africa, etc. The other synonym of clan, 'Gens' corresponds to a patrilineal clan because here kinship is traced absolutely through male line.

Clan is also equivalent to the Bengali term 'gotra'. Members using the same gotra-name do not marry each other. Among the Hindus of India different gotra-names relate with the names of some ancient sages (Kshyap, Sandilya, Gautam, Varatdwaj, etc.) which means people having same gotra-name have been descended from the same fore-father. Gotra is, therefore, patrilineal and it does not possess any totem.

Functions of Clan

1. A clan provides a bond of solidarity among its members. Because the members think themselves to be originated from a common stock. Each and every member feels for others. This creates an atmosphere of protection and security in the group.

2. The men and women of a clan look at their relation as like the relation between brothers and sisters because they are the descendants of a same progenitor. Naturally they do not dare to marry inside the group. Thus a clan regulates and controls the marriage and establishes relationship with other clans of a tribe.

3. For the deviation of any social norm, a clan may punish its member. Since control comes under a disciplinary measure, members are bound to follow the rules and regulations, chalked out by the clan. A person may be excluded from his clan or may face death sentence as per the seriousness of his fault.

4. Clan operates as a Government. It has not only the power to judge on the disputes to maintain peace, different sort of sanctions come through it. For this, tribal councils are sometimes built, in association with clan-heads. Most of the clans possess one headman having the absolute directive and derivative power.

5. A clan is found to control property. For the distribution and caring of agricultural lands, the clan leader performs a vital role among the Oraons. Same is true in case of Khasi and Ao Naga. Among the Todas income from the sacred buffaloes are deposited in the clan account and utilized in solving the common interest of the clan members.

6. Members of a clan are united to co-operate among themselves in various religious and ceremonial occasions. Sometimes clan-leader officiates as a priest. Rectificatory performances are also done under his supervision, in case of any sin.

However, it is held that clans did not appear until societies reached to a sufficiently advanced stages and they disappeared with the development of strong centralized Governments.

A comparative study between Family and Clan

FAMILY	CLAN
It is the smallest social group consisting of parents and their children.	It is a larger social group consisting of numerous families.
It is bi-lateral as counts relatives of both father and mother.	It is unilateral as counts relatives either through father or through mother.
Keen interpersonal relationships exist among the members.	Interpersonal relationships are superficial. Even the members may not know each other.
Structure of the family is very loose and ever changing. Birth, marriage, death, divorce, migration may bring considerable changes in the structure.	Structure of the clan is more or less stable and permanent.
Universally found in all societies.	Not found universally in all societies.
Genealogical relationships can be traced properly within a family.	Genealogical relationships can not be traced accurately among-the members
Members usually share a common residence.	No scope of sharing a common residence.
Members share a common nomenclature.	Members share clan nomenclature.
Exogamy is followed, but in a few societies Parallel cousin marriages are in vogue.	Exogamy is strictly followed.

Phratry

An interrelation between two or more clans makes a phratry. It is, therefore, a larger unilineal descent group than a clan. As in a clan, members of a phratry are not able to demonstrate their genealogical links with the common ancestor, although they strongly believe in such an ancestor. The term phratry has been derived from the Greek w*ord phrater* meaning brother. It is regarded that a few clans historically merged together for some reason or other and developed such an intimate relationship between them that gradually they achieved a common identity where their individual status was forgotten. Phratry is found among the tribes like Aimol Kukis of Manipur, Hopi Indians, Crow Indians, Aztec Indians of America, etc.

A phratry may or may not be exogamous. For example, thirteen clans among Crow Indians are found to be grouped into six nameless phratries, four of which are not strict in the rules for the marriage. Among the Hopi Indians, on the other hand, nine anonymous phratries (each having two to six clans) are found which are exogamous.

Moiety

It is the largest unilateral social group, which results from the splitting of a society into two halves on the basis of descent. The word moiety came from the French word meaning 'half'. Like clan and phratry, the members of each moiety though believe in a common ancestor but can not specify the exact link. Moieties may be exogamous or endogamous. Some of them are agamous too i.e., they do not regulate the factor of the marriage.

The Aimol Kukis of Manipur are divided into two moieties without specific names. Each moiety is further divided into two phratries and each phratry has two clan or sib. Moieties are exogamous and one of them is considered as superior than the other. The superior moiety reserves all the posts of village organization including that of the priest. Both the moieties perform specific religious rites and ceremonies of the tribe, separately. Each has some special performances too. But ceremonies involving higher social-status are performed only by the members of superior moiety. The social structure of Aimol Kukis is thus typical with two moieties, four phratries and eight clans, which can be framed as follows :

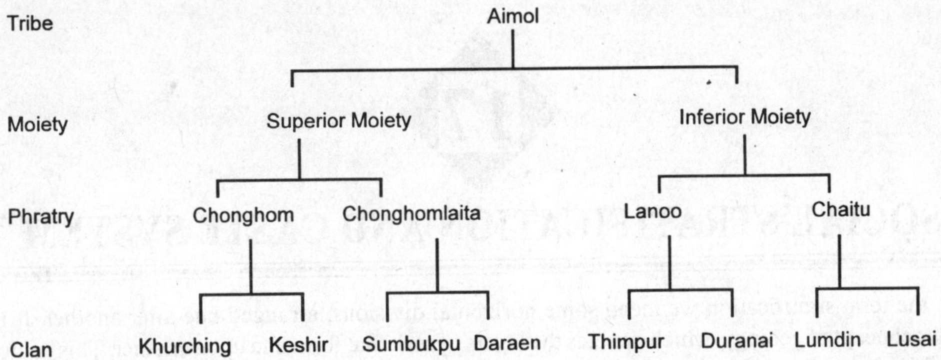

Exogamous moieties are common in Australia and also to some extent in North America. Exogamous moieties have also been noted from the Tlingits of Alaska—Raven and Wolf. Endogamous moieties, Tartharol and Teivaliol are found among Todas of Nilgiri Hills and agamous moieties, Pejuma and Pefuma exist among the Angami Naga of Naga Hills.

SOCIAL STRATIFICATION AND CASTE SYSTEM

By the term stratification we mean some horizontal divisions, arranged one after another. It is a popular concept of geology, which denotes the layers of rock that form one upon another. This analogy is used in the context of human society to differentiate the groups or categories, linked with each other by the relationship of superiority and subordination. In fact, two sorts of divisions, vertical and horizontal are found in a society. Horizontal divisions are qualitatively different form vertical divisions. Since a society contains heterogeneous population, vertical division based on age; sex, occupation etc., determine the existence of various groups or categories in the society. But it does not fulfil the concept of social stratification. The term 'Social Stratification' essentially implies the permanent groups in a society that emerge from social inequality and, therefore, arranged hierarchically as higher and lower groups. The vertical groups are usually less permanent and more varied, whereas the horizontal groups are more consistent, pervading and more deeply embedded in the structure of the society. These horizontal or stratified divisions of the society are generally represented as 'Class', 'Caste', 'Rank' and so on.

NATURE AND REASON FOR STRATIFICATION

There is debate between the sociologists and anthropologists regarding the universality of social stratification. According to sociologists, stratification is a common feature in all-human societies which anthropologists strongly disagree. Anthropologists believe that there are societies, which do not show any differentiation of function and role to produce class identification. To them, the rude simple cultures like the Eskimos, Andaman Islanders, Australian aborigines, Veddas, Bushmen, Semangs etc., are free from social classes. As there are no possibility of accumulation of capital goods, individuals share equal access to all natural resources and none exerts notable political power over others. Therefore, such societies in the eye of anthropologists are unstratified.

The evolutionary theorists treat social stratification as a centralized theme that goes with civilization. Inequality appears when the groups of a society tend to enjoy unequal access to its economic resources. It also relates to the differential access in power and prestige. The anthropologists typify three kinds of societies—egalitarian society, rank society and class society basing on these three conventional parameters.

Egalitarian Society

This type of society is considered unstratified because no special horizontal group is found here having greater or lesser access to the economic resources as well as power and prestige. But an egalitarian society does never mean that such a society is absolutely free from differences in reference to its members. Special abilities or attributes of the individuals like hunting skill, sound health, good perception, physical prowess, attractiveness, intelligence etc., are often praised and that mark out the individuals with special respect. But, for this, no raise of status is observed in the society. These especially talented people are though praised or respected, they do not get any extra privilege from the society. However, the egalitarian society is mainly found among the food-gatherers. The pastoral and horticultural people also exhibit this type of society.

Rank Society

In rank societies all individuals have the same access to economic resources and power, but not to prestige. Rank society is considered as an intermediate form between the egalitarian society and class society; such a society is partly stratified. Usually the agricultural and pastoral societies exhibit the ranks but not compulsorily. Here only a few people of a specific group can avail a particular rank. For example, in Tibet, the shamans are the ranked specialists who mediate between world and supernature. Persons,especially who suffer from some physiological abnormalities like epilepsy is only entitled to get initiation into the secrets of shamanhood. Again, among some of the Australian primitive societies, only the old persons have the eligibility in the position of political leaders. In rank societies, the position of the chiefs is partly hereditary, because it is not always determined solely by genealogical succession. Some other criteria like experience,wisdom, bravery, richness, popularity etc., come into consideration to determine the person's superiority over others.

Class Society

These societies are completely stratified as characterized by the unequal access to the economic resources, power and prestige. The members of each class of this society possess more or less the same life-chances i.e., the individuals of a particular class are exposed towards similar opportunities in terms of obtaining land, animal, money or other economic benefit. For example, correlation can be drawn between social classes and nutritional status, between social classes and personality development etc. Similarly, differences can be projected between the social classes in getting opportunities regarding education, job, justice, etc.

A class society is, therefore, composed of different social classes, which are placed one upon another in a hierarchy of rank and distinction. The ancient Roman society exhibited the classes like slaves, plebeians and patricians. The mediaeval European society was divided into slaves, serfs, free tenants, lesser gentry and nobility. The ancient Hindu society also was split up into four **varnas** —Brahmin, Kshitriya, Vaisya and Sudra.

Class societies are of two types :

1. Open class society, where an individual may have the choice to move up and down along the social ladder. His entry to other classes is open. By fulfilling certain pre-conditions, for example, by acquiring an outstanding degree, by paying an admission fee, by raising economic condition or even by obtaining a royal favour one can change his class-position.

2. Closed class society is that society where there is little or no scope of an individual to shift from his original class position. An individual, for the whole life, has to adhere strictly to the particular class in which he was born. Hindu caste system provides the typical example of closed class..

Thus, a class society is often contrasted with an egalitarian society in which social discriminations are completely absent. An individual may differ from others in rank due to his special aptitude in war, magical performance or oratory that disappear with the death of that person and not transmitted to the descendants. In these societies no group enjoys any distinction of rank except those based on biological factors like age and sex. In fact, property in such groups is so limited that it fails to create significant differences in wealth. According to Marx, a period existed in primitive communism where there was no class and a similar period will also come in future. But most of recorded human history possesses a history of class struggle; conflict and change underlie the factors for development. Marx finally identified the 'production-class' as the most important class in the society, which controls the entire system of production. Economic determinists who followed the Marxian way of thinking emphasized on the objective criteria of a social class. But the American sociologists emphasized more on the subjective nature of class.

The ways in which social classes have originated are many and varied from society to society. In general, it has been admitted that as soon as a quality shared by a large section of the community was held in higher esteem than another was, class distinction arose. Most non-literate communities like the Veddas of Ceylon or the Katkaris of Mumbai with their poverty, simplicity and limited members hardly allow for any class division. Modern classes can be viewed as development from the class structure of a society that prevailed during the Middle Ages where top was occupied by the 'landed gentry'and the slaves were placed at the bottom. However, the specific factors responsible for the growth of social class has been listed as follows:

(a) *Wealth*: With the development of material equipment, wealth began to increase. Since wealth controls the services and goods, it led towards power, which could be transmitted from one generation to the next more easily than the personal qualities. So wealth stood as the most important basis for social stratification.

(b) *Occupation*: Growth of culture led to division of labour, which fostered social differentiation as different degrees of prestige were attached to different occupations. Occupations of exploitation or prowess offered a higher status than that of manual and handicrafts.

(c) *Race*: Difference of race in a community often served as a basis of social stratification as observed among the Negro-white case of U.S.A.

(d) *Conquest*: Invasion and conquest created servile group among the conquered as found among the Sudras of Vedic India.

(e) *Ignorance*: Little communication and ignorance about the condition of other countries often kept the people satisfied with the prevailing condition. The ruling class, therefore, tried to keep the masses ignorant. The religious ideas like the Doctrine of 'karma' of the Hindus or the Divine will of the Christians, thus made the people satisfied with their own status, as found in India and in pre-Revolutionary Russia.

Archaeological evidences and a number of other cultural features suggest that social stratification emerged relatively recently in human history. Possibly about 7500 years back there was no evidence of inequality in the society. Gerhard Lenski equated social stratification with the inequalities of power and privileges in industrial societies. He measured this in terms of the concentration of political power and distribution of income. According to Marshall Sahlins the social stratification developed as the agricultural productivity increased and surpluses were produced. Both the scholars pointed out 'production surplus' as stimulus in the development of stratification. C. K. Meek proposed a different view. He said that the population pressure on resources, particularly in rank societies, is the cause behind the creation of social stratification. There is also a dispute among the anthropologists with the matter that, whether the class societies have really developed from rank societies, or rank societies were formally the class societies, which after gaining equal access to economic resources and power have stood as rank societies having unequal access only to prestige.

Class—Caste Distinction

The main forms of social stratification are reflected in the class-caste distinction—the open versus the closed system. In spite of their differences, these two phenomena of social stratification offer some features in common in relation to their origin, development and function. In fact, they are the agencies of social mobility and social selection as both of them largely decide the position that a man occupies in the society. In 1947, E.T. Hiller in his book "Social Relation and Structure" wrote that, "hierarchical positions become assigned to relatively permanent segments of a society, the social relations so produced are known either as a class or a caste system". This is why both the systems are usually studied in connection with each other.

The Class System

The working criteria of social class may be objective as well as subjective. Economic determinists emphasize on the objective criteria. They relate classes with the instruments of production i.e., with the ownership and non-ownership of land and capital. Max Weber counted the cultural and recreational opportunities, and also the standard of living* along with the economic criterion. American sociologists found a more fluid and blurred nature of classes and thus they emphasized the subjective nature of classes. Mac Iver has forwarded an explicit definition where he perceived class as "a portion of a community, which is marked off from the rest by social status". A social class is, therefore, a category or a group within a society whose members hold a number of distinctive statuses in common which permanently determine their relation to other groups. A social class can exist only with reference to

* Standard of living is considered in terms of consumptuion of economic means.

other social classes. The relative position of classes on the social scale is determined by the degree of prestige attached to their status. In every society, the ruling classes enjoy superior positions than the other classes. The status of domestic servants or public sweepers remains the lowest. The prestige, which a class enjoys and the status derived from it, depends on our evaluation whereby certain qualities are considered as valuable according to prevalent notion of the society. Generally the qualities like knowledge, wealth, purity of race or descent, religion, bravery etc., confer a high degree of prestige on the persons possessing them. This facilitates their admission to the higher strata of the society.

Whenever, hereditary chieftainship or kingship is found, it is expected that all the relatives of the chief or the king enjoy the privileged positions. They are distinguished from other members of the society by special titles or prefixes,and customs peculiar to themselves. This group is usually termed as 'noble' and this term may also be used for those who have similar privileged position, even though they are not related to the chief or the king. Such a group forms the most characteristic example of class.

Ample opportunities are found in this system by which certain individuals may pass from one class to another. So a commoner can be raised to the position of a 'noble' as found in Germany and in England. In Polynesia grading depends on the purity of blood and the people point a noble as an exceptional person who has born out of a union between own brother and sister. In Fiji and Polynesia, the landowners form a definite class distinguished from the chiefs and nobles. In England also this kind of notion is found though not very definite. Nobility of blood, once so esteemed in Western Europe, has now little appeal to the modern man unless it goes together with other qualities.

Certain institutionalized relationships and customary modes of behaviour separate a social class, especially a higher class from others. Distinctive dress and ornaments, speech, types of home and conveyance, ways of recreation, expenditure etc., mark the class. Members of a class share a distinct status in common and obviously there is social distance between two classes. People of each class remain sufficiently aware of this fact and such awareness is termed as class-consciousness. Without the class-consciousness, no dynamics of class action is possible; functional significance of class gets lost. Although each class shows a tendency towards endogamy, inter-class marriages are not forbidden.

The Caste System

Castes are the special forms of social classes where there is no scope of movement for the individuals along the social ladder. Each caste tends to form a close system within itself, making social mobility for its members absolutely impossible. Social position of the persons are determined by birth; one has to bear his caste identity up to the last day of his life. Marital connection outside one's owns caste is prohibited. A caste, therefore, can be defined as an endogamous and hereditary subdivision of an ethnic unit occupying either a superior or inferior rank in comparison to other subdivisions. The whole system is safeguarded by the social law and sanctified by religion. A notion of purity and pollution is associated with it. M.N.Srinivas (1969) has stated that, "the idea of hierarchy is central to caste. The customs, rites and way of life were different among the higher and lower castes".

Caste as a basis of social stratification is unique to India. It is such a peculiar and complex system that no scholar could succeed to define it in totality. Sir H.Risley who undertook the first and authentic work on the people of India had defined caste as " a collection of families or groups of families bearing a common name, claiming common descent from a mythical ancestor, human or divine, professing to follow the same hereditary callings, and regarded by those who are competent to give opinion as forming a single homogeneous community. The name generally denotes or is associated with a specific occupation. A caste is almost invariably endogamous in the sense that a member of the large circle denoted by the common name may not marry outside that circle, but within the circle there are usually a number of smaller circles each of which is also endogamous".

In India, among the Hindus we find the most developed form of caste system. Analogous forms do exist among the Muslims, Christians, Sikhs, Jains and other religious groups in South Asia. Caste like systems has also been observed in the South Asian subcontinent and beyond,from the countries like Japan, Africa, Iran, Polynesia, etc. Other instances come from the Massai, the Somali of East horn, the Jews, the Gypsies and the Burmese. Caste has also been used to describe the system of racial stratification in modern Europe and America. But Indian caste is the best-developed, integrated and consciously framed system that has ever grown up. The system has been with the Hindus for the last 2500 years and has not withered away.

The word caste has been derived from the Latin word 'castus' meaning pure. Portuguese applied the term to the Indian social divisions, which were known as 'jati'. They thought the term quite appropriate to denote the classified groups (jatis) in India, particularly among the Hindus who were always conscious to preserve the purity of blood among them. In fact, Hindu society is composed of about 37,000 discernible endogamous groups, structured in a system, which we designate as caste system. Each distinguishable population of this system is called a jati or a caste. The first official enumeration counted only 2,378 castes as revealed through Indian Census. Rest of the population outside the caste system is composed of different tribal groups and religious communities. Size of those populations is not uniform at all. Some of them comprise of a few hundreds of men while some others show millions of members.

Varna or Jati: A popular debate

Before discussing the nature or origin of caste organization in detail, we have to clarify two interrelated concepts — *Varna* and *jati*, which are very often misunderstood. The confusion is mainly due to the fact that Brahmins as a group called by both, varna and jati whereas no single group is found with the name Sudra. Rather there are many jatis which come under the name Sudra. Again, Manu spoke specifically about four varnas (Brahmin, Kshtriya, Vaisya and Sudra) and about fifty jatis namely Ambastha, chandala, Dravida, Yavana etc. According to Manu, all the jatis were produced by a series of crosses between the members of four varnas at the first hand and afterwards by the crossings between the descendants of the initial unions. Actually, varna and caste are not the same thing. But they are the fundamental groupings, close to each other; one expresses the natural endowment of man and other is concerned with one's nurture and upbringing. Varna refers to an ideal model—a plan or design on which Hindu social organization rests. On the other hand, caste refers to the actual social groups with which people identify themselves.

Varna means colour and has a racial significance as believed by many writers who wrote on it. Varna differentiates between the groups who are within the varna scheme and who are outside of it. The Brahmin, the Kshtriya, the Vaisya and the Sudra—these traditional four groups belong under the varna scheme. The untouchables who form one-fifth of the Indian population as well as the tribal people remain outside the scheme. Within the varna system itself there were two divisions. The first three of the four groups e.g. Brahmin, Kshtriya and the Vaisya were called *Dvijas*, the twice born. Through an initiation ceremony (*upanayan*), the members of those three groups achieved a superior ritual cum spiritual status in order to perform higher rights and duties. Sudras, the fourth group or the last rank in the varna scheme was considered as impure. The members of the Sudra-varna were supposed to be engaged in manual work and were not entitled to go through sacred texts. Varna divisions for human being were based not on the external features or activities, but on the internal state of *Sanskara* * to decide the nature of birth. The Hindus believe that Brahma is the creator of the world; from that supreme soul, the four Varna groups were created. From the mouth of Brahma, Brahmins came out, Kshtriyas came out from his arms, from the thighs the Vaisyas and lastly the Sudras came from the feet. As per the natural properties and normal inner drives of the respective class, men of the Sudra-varna show the aptitude of manual labour for earning bread, men with the aptitude for a little evolved jobs like business or trade are the Vaisya varna people, men with the

* *Sanskara* is the inner qualities or propertes of man based on the nature and function of body, mind and the soul. It is ultimately related with Brahma, the eternal source of energy or the ultimate cause. The idea of *Sanskara* is intimately associated to the concept of 'karma' and 'rebirth' of the Hindus.

aptitude of dealing with power and ruling are the Kshtriya varna people and finally the men with the aptitude of performing religious activities (detached from worldly affairs) are the Brahmin varna people. According to Hindu philosophy a person's social position in this world is determined by his action in previous life. Accordingly the persons who are now born in Brahmin families must have performed good deeds in their earlier lives. Those who are born as Sudras or untouchables must have got punishment for their sinful acts, committed in previous lives.

The term 'Varna' appears in the earliest literature Rig-Veda in its tenth book called 'purusha sukta'. It refers only to the broad categories of the society characterized by differential access of individuals to the spiritual and material privileges. The ranks under varna system are maintained in the same order from the ancient to modern times although the scheme has lost much of its social and legal authority at present. In contrast to *varna*, caste or *jati* refers strictly to occupational groups, which are numerous in number and often internally segmented. Castes can change their identity over time and many new castes are still forming by fusion. The status of a caste, its cultural tradition, its numerical strength may vary from region to region and even from one village to other.

Srinivas differentiated caste from varna in the following way. According to him "the varna scheme refers at the best only to the particular categories of the society and not to its real and effective units", whereas the caste is a "hereditary endogamous, usually localized group, having a traditional association with an occupation and particular position in the local hierarchy of the castes.Relation between castes are governed, among other things, by concepts of pollution and purity and generally maximum commensability occurs within caste". Louis Dumont, the outstanding French scholar who worked on the caste system in India, in his book 'Homo hierarchicus' (1970) has shown that the caste system stands on the fundamental social principle on hierarchy. Both the principles of equality and inequality are present in this system being enforced by heredity, hierarchy and endogamy.

Caste structure has made Indian society stable through the peaceful means of assimilation of the various social groups with diversified customs, religious beliefs, social habits etc. Therefore, even when political systems changed, the social system (caste system) continued without being affected. There are many scholars who highlighted caste as an important and significant institution of India performing a lot of socio-economic functions. For example, it secures employment, education, training, marriage match etc., for the members. It also regulates the behaviour pattern and religious duties* of the individuals.As an effective means of integration, it embraces vast and variegated communities, even the conflicting groups. E. Leach (1960) and L.Dumont (1970) appreciated the caste system for its cooperative and inclusive arrangement where each caste formed an integral part of the local socio-economic arrangement and had its special privileges. J. H. Hutton praised caste as a good institution and pleaded for its reformity.

ORIGIN AND FORMATION OF CASTE SYSTEM

Varna as a functional division of society has been known from the time of Rigveda. But caste has its origin in brahmanical legislation. R. C. Dutta in his book 'History of India' stated, "caste was unknown to the Hindus in Vedic age and was first developed in the Epic age. It divided and disunited the compact body of Aryan Hindus into three hereditary bodies - priests, soldiers and the people". According to Mr. Dutta, it was the formatory phase of caste divisions. Because at this time, except the priest and the soldiers, the mass of the Hindu people were considered under one caste, the Vaisya and this continued through the Epic and succeeding ages.

The origin of caste system is actually not very clear. European writers proposed racial differentiation behind the formation of caste. In their view, colour prejudices among the races in ancient days gave birth to the varna divisions, which gradually turned to caste system. On the eve of Aryan invasion, several races were found on the soil of India. Those included Negritos, Kols and Dravidians. Aryans

* The sacredness of the caste system generates the belief that it has been established with divine sanction. Hence it is everybody's concern to fulfill his caste duties in accordance with the doctrine of *Bhagwat-gita,* the sacred scripture of Hindus

coming to India divided themselves into three classes — *Brahmin*, *Kshtriya* and *Vaisya* on the basis of work. The Brahmins were those who looked after the religion and education. The Kshtriyas formed the worrior class and kept themselves engaged in administering the State. The Vaisyas were the farmers, business-holders and traders. The fourth class, Sudra was added later being consisted of enslaved non-Aryans and the Aryans who had fallen from their high moral standard. The time was the early Vedic period and the stratified system was known as Varna system. Under Varna system, people could change their rank by changing the profession. Intermarriages among the first three ranks were also very common. The system on the whole was very flexible.

From the late Vedic period, the system tended to become more and more rigid. Gradually it became no longer dependent on work but on birth; professions became hereditary. A large number of ranks were created due to mixture and re-mixture of Aryans with the non-Aryans. Each of the ranks now came to be designated as caste. The hierarchy ranged from ritually pure Brahmins to the impure or polluted untouchables. Supremacy of the Brahmins was established and they forwarded legislation for their own benefit. Hindu lawgivers were, therefore, essentially Brahmin, who wanted to keep their place raised from the rest of the people.

The early Vedic religion was very simple. It was made complicated in later period with elaborate rites, rituals and sacrifices; temples and idols came into existence. For the performances of sacred rites and ceremonies, technical knowledge was urgently required and this made the presence of Brahmins important as well as inevitable. The caste system thus acquired its rigidity following the time of Rigveda up to the time of Lord Buddha, around fifth Century B.C. But it began to be losing from the time when Buddhism challenged the caste order. European scholars compared caste system with an iron chain where the members of each rank is bound to the profession of their ancestors and unable to make any improvement for the lifetime.

Indian scholars like S.C.Roy, N.K.Dutta and G.S.Ghurye sought the origin of caste in racial struggle between the fair-skinned Aryans and the dark-skinned non-Aryans. They believed Varna as the root of differentiation in Indian social structure, which became more and more diluted in course of time by racial mixture and hybridization. J.N.Bhattacharya has discussed the formation of sub-castes and additional castes as an on-going process. He observed several causes behind the multiplicity of caste like that of H.H.Risley. However, the usual reasons are:

a) *Migration of people*

If members of a caste leave their original habitat and settle down permanently to another part of country, the cut-off group after sometime develops a new caste distinguished from the parental one. For example, the Rarhi Brahmin and the Varendra Brahmins of Bengal though occupy separate territorial areas, originated from the same stock.

b) *Practice of different occupation*

Each and every caste is associated with a traditional occupation. If one section of a caste adopt a new occupation where most of the members involve themselves with that very occupation, after a considerable time they give birth to a new caste. For instance, 'Sadgops' are those milkmen who had taken agriculture in later period. In the same way, 'Kolu' and 'Tili' have been separated from a caste dealing with oil. At present, 'Kolu' presses the oil and 'Tili' has become oil-merchant.

c) *Change of custom*

Sometimes a section of a caste gets elevated above the other sections of the caste, or becomes degraded below the level of others due to the change of custom. Such a caste generally neglects the established usages of the parental caste or adopts new ceremonial practices. For example, Ajodhya Kurmis of Bihar and Kanaujia Kurmis of U.P. They claim a higher rank than the ordinary Kurmis since they have prohibited the custom of widow-remarriage. Similarly in Bengal, original Kaivarta caste has been splitted into Halika Kaivarta and Jalika Kaivarta. The occupation of the Halika group is agriculture, which is considered as a noble profession in comparison to the occupation of the other, the catching of fish. Therefore, Jalika Kaivartas hold an inferior status than that of Halika.

d) *Inclusion of a part of a tribal group into the Hinduism.*

Usually a large section of a tribe (sometimes a whole tribe even) due to quarrel with its other section or for the want of some benefit enrolls themselves in the rank of Hinduism. They continue either under their own tribal designation or gets a new caste-name but can easily be differentiated from those of the standard caste. The Rajbanshis of North Bengal, the Bhumij of West Bengal and the Gonds of Central India are the examples.

e) *For becoming the follower of one of the modern religious leaders*

This process has evolved a small number of castes. People have been the followers of any renowned religious leader of the time. The Lingayats, original Vaisnavas of Bengal and the Sikhs provide the examples.

f) *For resembling as a national type*

There are a few regional groups who are at present regarded as castes. Such a group preserves a cultural tradition and presents an organization, which is much elaborate than an ordinary tribe. The Newars of Nepal and the Mahratta Kunbis present the examples.

Apart from these, various reasons are responsible for the formation of sub-castes; the actual origin of caste system has been viewed by many eminent scholars. Some of them are as follows:

Theory of Manu

This is the earliest theory. Manu wrote between 200BC to AD200. He postulated that the four original varnas were created from the mouth, arm, thigh and feet of Brahma, creator of the Universe. A large number of castes or jatis were produced later by a series of crosses—at first between the members of four varnas and thereafter between the descendants of those initial unions. For example, the caste 'Nishada' originated by the union between Brahmin and Sudra, 'Chandala' by the union between Sudra and Brahmin, 'Vaidya' by the union between Vaisya and Sudra, and so on.

Some castes were also formed by the process of degradation from the original Varna, on account of non-observance of sacred rites. These castes are called *Vratyas*. Manu categorized Bhrijjakantaka, Avantya, Pushpadha, etc., as *Vratya Brahmins*; Jhalla, Malla, Nata, Karana, Khasa, Dravida, etc. as Vratya Kshtriya; and Sudhanvan, Acharya, Maitra, etc., as *Vratya Vaisyas*.

Theory of Risley

Risley identified two elements, fact and fiction for the growth of caste sentiment. He said, whenever in the history of the world one people had subdued the other, whether by active invasion or by gradual occupation of their territory, the conquerors always have taken the women of the country as wives or concubines but for daughters they have arranged marriages exclusively among their own groups. As a result, if two people are from the same race or colour, complete amalgamation took place. But if they were of different race and colour, progeny ran on a different line. The irregular union between the men of higher race and the women of lower race gave birth of new castes in descending generation. This is found among the Mulattoes (Negro-European), among the half-breeds of Canada, Mexico and South America and also among the Eurasians of India. Once it started in India, the principle was strengthened and extended to all parts of the society. People who speak a different language in a separate district, worship different gods, take different food, observe different custom, etc., were considered as having separate ranks and when intermarriage took place among them, further distinguished caste ranks were formed.

Theory of Senart

Senart said that the caste system in India is the resultant of contact among the different people. As an outcome of interaction between the invaders and the indigenous population, the invaders demanded the ceremonial purity in the performances of their religious rites and beliefs, whereas the indigenous population though introduced their cult, accepted a subordinate position and restrictions on their intercourse with the strangers. Moreover, the offsprings of mixed marriages were not admitted equally with the pure blooded children of the invaders; they form group or castes of lower rank, while the indigenous tribes were admitted to still lower rank in the system. Senart's theory is considered as more reasonable than the theory of Risley.

Theory of Nesfield

Nesfield realized that caste originated in India long after the Aryan invaders had been absorbed in the mass of the native people and by that time all distinction between the two sets of people, Aryan and aboriginal had disappeared. The homogenous population thus formed were divided into seven groups—casteless, the castes connecting with land allied to hunting, fishing, pastoral and agricultural stages, the artisan castes, the trading castes, the serving castes, the priestly castes and the religious castes according to occupational pattern and functions. The rank of high and low solely depended upon the fact that whether the castes represented advanced or backward stage of culture. Thus Nesfield held that occupation is the sole basis of caste and excludes all influences of race and religion for the origin and growth of the system.

It can be concluded that the history of Indian caste is extremely complex and its causes are manifold. For many centuries the social situation of India was in a state of flux;ethnic identities remained diffused. Migration, miscegenation, ecological diversities of the country can be accounted for this kind of situation. It appears that within each ecological and climatic zone, a cluster of communities emerged which established and exercised control over the resources. In course of time, such communities were hierarchically organized or placed in relation to each other depending on their access to resources. Immigrants who came later were also recognized within that local structure.

The *jatis* (castes) being local formations were listed regionally which we have come to know from '*Varna Ratnakar*' written in fourteenth century AD. The major historical text '*Ain-i-Akbari*' also gives us information about the location of communities or dominant communities and artisans in different parts of Mughal India. From 1806 onwards, the listing of caste in India started. It gained nomenclature during the Censuses from 1881 to 1931. Systematic caste-wise enumeration of population is not available in the Census after 1931.

CHARACTERISTICS OF CASTES

Sanction of religion is very important to the caste system. Each caste stands for an entire way of life that is distinctive but at the same time castes of a region may form a single social framework. The belief in the doctrine of *Karma* (quality of action) and rebirth (reincarnation) renders this inequitable structure as a part of the divine order of the Universe. According to Manu, a man should aspire for birth in a higher caste in the next life and that is possible only if he performs his caste-bound duties very carefully. The rights and duties associated with castes undergo stern social rules. Any breach in caste discipline is looked upon as a sin. Thus, each caste has been able to retain its unique function from the time immemorial. The fundamental features of the caste are as follows:

1. *Hereditary Identity*

A person born to a particular caste remains with it for his whole life. He dies in it and his children (except the extraordinary cases of mixed marriage) also belong to it. The caste status is rigidly maintained throughout the life. Neither education, nor wealth can change the caste of a person. Similarly success or failure or disaster—nothing can alter an individual's caste position. This means, nobody can escape from his caste identity, whatever his potentialities or capabilities may be. Caste status is solely determined by birth on which none can exert any control.

2. *Endogamy*

A person of a particular caste is compelled to marry within the network of his own caste. He is not allowed to seek alliances from his above position (*Hypergamy*) or below position *(Hypogamy)* except by breaking the caste rules. A man of higher caste could marry a woman of lower caste according to Manu. But a man of lower caste was never allowed to marry a woman of superior caste position. Hypogamous marriages were also held to be against rule by the *later Smriti** writers.

* A Sanskrit text written later than Vedic Period.

The castes that have been divided into sub-castes are also required to follow endogamy. Among the Hindus, the most ordinary rule of exogamy is the '*Sapinda*' rule by which two sexes are forbidden to marry who have a common ancestor within the six generations counted from the father and four generations from the mother. However violation of the rule of exogamy in relation to a caste or sub-caste (i.e., marrying outside the caste) would mean ostracism and loss of caste purity.

3. *Hereditary Occupation*

Each caste is associated with a particular hereditary occupation, which means all members of a caste are bound to follow the same occupations as chalked for the respective caste. For instance. Brahmins' occupation is restricted to learning, teaching and religion performances. *Kshtriyas* are expected to look after the social order: they are the rulers and warriors. Likewise, the *Vaisyas are* . supposed to produce wealth by engaging themselves as cultivators, artisans and traders. Besides, there are *Tilis in* who are oilmen by profession; *Muchis* are the shoemakers; Chamars are the leather workers; *Lohars* are the blacksmiths; *Sutars* are the carpenters; *Harijans* are the scavengers and so on. In traditional caste system, persons had to follow the prescribed occupation, generation after generation.

4. *Hierarchy*

The popular impression of the hierarchy is a clear-cut one, derived from the idea of *Varna,* where the Brahmins are placed at the top and the *Sudras* at the bottom. These two opposite ends of hierarchy are more or less fixed. Tension arises with the mutual position of a large number of caste groups that appear especially in middle region. In a dispute over rank, each caste cites an evidence of its superiority. Superiority may be claimed in terms of its dietary items, in relation to other caste groups whom it accepts or refuses cooked food and water, with reference to the rituals it performs, the customs it observes, its traditional privileges and disabilities or regarding the myth of origin.

Traditionally agriculture was a common occupation for all castes i.e., most of the people used to practise agriculture in addition to their hereditary occupation. In fact. agriculture included a variety of things—land ownership, tenancy and labour. A potter might be an agriculturist in the monsoon months or a trader of grains for a brief period after the harvest. Similarly, artisans or the people belonging to the serving castes without finding adequate income from the hereditary occupation started to work on land either as tenants or as casual labourers.

All other occupations apart from agriculture are classified as high or low. The occupations practised by the high caste people have been regarded as high standard occupation. Manual work is, therefore, looked upon as low which is frequently practised among the lower caste people. There are still other occupations like swine herding and butchery, that are considered as extremely low.

In traditional social set up a Brahmin always enjoyed supreme rights on land and maintained a high standard of living without manual labour. Peasants and other agricultural castes were found to occupy the middle position of the social ladder. Lower positions got reserved for the field-labourers. The artisans were generally considered as inferior to the farming castes, but their positions used to be more accurately determined in terms of their importance to the village economy. For instance, the castes like potter, weaver, fisherman, oil-presser etc., held a relatively higher rank as their presence were indispensable for the functioning of social economy. On the other hand, castes like barber, washer-man etc., were kept at a considerable distance although their labour were often required in social as well as ceremonial purposes. Besides, between the highest and lowest ranks of social ladder, plenty of other castes were found. In most of the cases their relation to higher or lower ranks remained undefined and the status used to fluctuate from region to region.

Within a single well-defined caste, grades may exist. The '*Kulin*' and '*Moulik*' group among the *Kayasthas** provide the best examples. This happens when one of the sections of a caste devises a

* In Bengal no member of *Kshtriya Varna* is found. Rather *Kayastha* as a caste have been dominant since 5th and 6th century AD.

differentiation of custom or occupation and thereafter claims a superior status by refusing food and marriage alliances from own counterpart. In this way. the shoemakers of Bengal have elevated their status than their brothers working with rawhide.

5. *Purity and Pollution*

The concept of pollution plays a crucial role in maintaining distance between different castes. Occupations adherent to the castes are judged on the basis of their degree of purity and pollution. Sacred occupations revolve round the intellectual, political and production jobs. The profane occupations are scavenging, burning the dead bodies, skinning the dead animals etc., The more profane occupations possess the lower status in the social hierarchy. Thus a lot of gradations are observed along the social ladder for which thousands of small groups have been created, each having a distinctive set of customs and practices. While the higher castes enjoy a number of social and religious privileges, the lower castes remain entangled with a series of.disabilities.

The breaking of pollution rules is considered as sin and such a practice makes a higher caste person impure. Therefore, the sinner has to perform certain purificatory rite in order to regain his normal status. Where the breach of rule is very serious, say for example, when a high caste person eats food cooked by a *Harijan* (low caste) or when a high caste woman undergoes sex relation with a *Harijan,* the offender may be thrown out of the caste irrevocably.

6. *Restriction on Food, Drink and Smoking*

The membership norm of each caste determines what a person may eat and with whom he may sit for a meal. It also prescribes what kind of food should be avoided or not. Elaborate rules govern on the acceptance of cooked food and water from another caste. Food cooked with *ghee,* milk or butter is called 'pakka' food. which may be accepted from the inferior castes. 'Kachcha food, on the other hand, is the food cooked with water. It is accepted only from one's own or equivalent or superior castes. But there are a few exceptions to this general restriction. Food or drink gets sanctified by being offered to a deity in a temple. Such a food is not refused at all even when the cook comes from the lowest rung of the caste hierarchy. For example, the cooks in the famous Jagannath temple at Puri are barbers by caste. Many other significant regional variations have been noted in this respect.

Taboos on vessels are found in some areas. These taboos decide the nature of the vessels (whether they are made of earth, copper or brass) that one should use for eating, drinking or cooking. In the same way, the smoking taboo define the persons (in reference to their caste identity) with whom one may sit to smoke or whose pipe should be avoided for this purpose. In North India 'hukka- offering' is an index of caste equality. Different castes who may smoke on a single "hukka' are considered as equals. For instance, the Lohar (blacksmith) and Khati (carpenter) are allowed to smoke from the same 'hukka' as permitted in the case of a *Jat* and an *Ahir.* The *Nais* (barber) of Bengal, like many other castes have to maintain a separate 'hukka'.

A notable feature in this context is that, in India, the higher castes are strictly vegetarian and teetotaller. They eschew chicken, pork and beef. Self-respecting members of superior castes try to preserve their ceremonial purity by dietary cleanliness. On the other hand, other castes try to avoid or neutralize the potentialities for evil by maintaining the traditional prescription. When two castes contend for superiority, both of them stop accepting cooked food or water from .each other.

7. *Dinstinction in custom, Dress and Speech*

An individual in a caste society lives in a hierarchical world where each caste has a definite culture which is autonomous to some extent. The difference between the cultures is marked in dress, speech, manners, rituals and ways of life. The members of the higher castes are supposed to wear fine clothes and gold ornaments, while the members of the lower castes wear coarse materials and silver ornaments. The speech of the higher castes is refined than the lower castes. The customs and manners of lower castes were far apart from that of the higher castes. Traditionally, the lower caste people were prohibited from imitating the dresses, ornaments and customs of the higher rank and the offenders were punished by the village panchayat.

8. *Rules of Avoidance*

Caste regulates all the customs and rituals regarding birth, marriage and death, and thus determine the status of the caste. Distance between the castes is maintained carefully. Upper castes always try to avoid the lower castes. Purity is held not only through food and drink; the touch, the sight, even the shadow of the lower caste people communicates impurity to the superior caste people. Therefore, where the interaction between the two dissimilar castes is inevitable, the lower caste person is required to keep a minimum distance between himself and the high caste people. In Kerala, a Nayadi had to stand at least 22 metres away from a Nambudri and 13 metres away from a Tiyan, who himself had to keep 10 metres distance from a Nambudri. Inter-caste distance is more prominently marked in South India than in North India. A few years ago. persons from the extremely lower rung of hierarchy were not allowed to use the public road or avail themselves of the public well. They were also forbidden to enter into the temples, to attend the public schools and had to live in a confined state in their own quarters at the outskirts of the village, being tied with some degraded occupations. Thus. various socio-psychological barriers were found to develop between the castes.

Disciplinary Measures in Caste Organization

Each caste behaves as an autonomous social unit. Authority is vested on a board of members called panchayat which controls the code and discipline for the caste and enforces its members to follow the definite rules (regarding religion, etiquette etc.) connected with the caste. The individuals are not only punished for the infringement of the caste rule, penalty is also applied in the cases like adultery, refusal to maintain a wife, immorality and concubine, breaches of any prescribed taboo, killing of a sacred animal (especially the cow), insulting a Brahmin etc. Usually the offenders are punished by fine but in extreme cases ex-communication takes place. Sometimes an offender is re-admitted to caste after having been outcasted for a decade or two. For this re-admission the offender requires to undergo a purificatory ritual. He should also express his regret to the caste assembly and arrange a dinner for his caste members.

Social Dynamics in the Caste System

Castes are the building bricks of the Hindu social structure. They have kept the Hindu society divided in a hierarchical order and resulted in a close linkage between the caste ranking of a person and his social, educational and economic status. This manner of stratification in society enabled the higher caste to serve their deep-rooted vested interests. The priestly caste evolved an elaborate and subtle scheme of scripture, ritual formulae and mythical stories to perpetuate their supermacy and to get hold of the lower castes who remained in bondage for ages. Most of our Shastras (religion scriptures) upholds the four-fold division of Varna and because of this religious sanction, caste system has lasted longer than most other social institutions based on inequality and inequity. Member of lower castes have always suffered from discrimination in all walks of life. Low ritual status paved the way for their social backwardness.

Castes restrtictions have been loosened considerably for urbanization, industrialization, spread of mass education and above all for the introduction of adult franchise after independence. But the system has been able to survive over the centuries because of its inherent resilience and its ability to adjust itself to the ever-changing reality.

MOBILITY IN CASTE SYSTEM

Social mobility can be defined as a movement of an individual or a group from one status category to another status category. This movement may be upward or downward in the social hierarchy, but the possibility of downward movement is seldom considered. Social mobility refers a change of social status, which depends on the factors like occupational position, organizational membership, kinship relations, property ownership, education, wealth and income.

Caste has been viewed as a closed class system, so here social climbing of the individual is virtually impossible. But at present we find certain changes in this system, which have been explained by many scholars with the term mobility. Despite the system has lost much of its force, some old caste practices are still observed. Factors influencing change are many and varied; time to time they have kept repercussions on the caste system. Historically we can suggest three phases while discussing the change in this closed system.

(A) Before British Rule

The caste tradition was first confronted with the rise of Buddhism as Buddhism fostered a religious movement, which went against the growing Brahmanical dictum. The strong hold of Brahmanism lay in *Madhyadesha* i.e. the heart of *Aryavarta* where the Brahmanical rules were originally developed and maintained in strict form. The Brahmanical culture became more and more diluted as it proceeded eastwards towards the provinces like Kashi, Koshala, Videha, Magadha, etc., the birthplace of Buddhism. *Buddha* and his disciples tried to undermine the importance of birth identity and exalted the importance of virtue as a means of salvation. That was the first and major setback on caste philosophy but the system persisted showing tremendous vitality and resilience.

The *Bhakti* movement around twelfth century again opposed the caste ideology. The *Bhakti* saints came from all castes, even from *Harijans* (untouchables). This movement stressed the worth of an individual irrespective of his caste affiliation. The movement was more or less continuous in Indian history. Such a movement originated at different times in different parts of the country. *Basavesvara* developed a new sect, which did not admit caste hierarchy. At about the same time, Ramanuja brought a great reform in *Vaisnavism.* Moreover, there were the attempts of Kabir and Chaitanya (sixth century) to uplift the people belonging to lower and untouchable castes.

In nineteenth century four great movements were found, which among many other things, aimed to abolish the caste system. First of them was the establishment of 'Arya Samaj"(1825) by Swami Dayanand Saraswati in Punjab. It was an association, which considered the conception of caste and sub-caste as artificial barriers among mankind and, therefore, condemned this stratification. Second was the establishment of 'Brahmo Samaj"(1828) in Bengal by Raja Rammohan Roy. This association was opent to all castes; everybody was allowed to hear discourses from Upanishada. Third movement was lodged by Rarakrishna Mission who looked at the system of caste as degeneration in the religion. The fourth movement was theosophical and generated in Chennai (Madras). It worked for the abolition of caste system, particularly since 1893 when Annie Besant became the President of the society. Although these religious reform movements had limitations, they laid a fertile ground for future social movements.

(B) During British Rule

The advent of British profoundly affected the Indian social order. They forwarded a new principle of justice, which pronounced "all men are equal before law". Moreover, this new policy suggested that the nature of wrong should not relate to caste or person committing it, nor against whom it was committed. The whole thing was referred under the Indian Penal code of 1860.

Apart from this new legal policy, the opening of English educational institutions and new economic avenues provided ample opportunities for advancement. These developments brought drastic change on caste system. The Brahmins who so far held the tradition of education and literacy, leaned towards the English education and took administrative jobs in the colonial bureaucracy. It also inspired the lower castes to take education. The other significant feature of this time is the spread of communication, which enabled the lower castes and sub-castes to link together and form a caste association . This association arranged a number of All-India Caste Conferences, which fought for the consolidation of caste in the country by demolishing different barriers among the castes and sub-castes in regional as well as in linguistic levels. The next thrust that weakened the foundation of caste system was the Nationalist movement. It started under the leadership of Mahatma Gandhi. Gandhi campaigned against untouchability and British was compelled to grant special political representation for the untouchables. Dr. B.R.Ambedkar also struggled for the emancipation of untouchables so that they could be able to claim their rights like the upper-caste Hindus.

(C) After British Rule

After the independence of India in 1947, the new constitution was adopted in 1950. It brought measures to abolish untouchability and to prohibit discrimination in public places. Some special positions were reserved for the untouchables, which facilitated their entrance to the higher educational institutions, Government services and the lower houses of Central and State legislation. The ritual and religious basis of a caste was weakened greatly.

Access to higher education, Government employment and urbanization started to cut down the barriers between castes and sub-castes. Restrictions became loose in the choice of marriage. The endogamous circle was widened; many educated people came out of their traditional parochial notion in finding suitable bride and bridegroom for their children. Moreover, industrialization affected the caste principles. The close association between caste and occupation broke down with the expansion of modern education and the urban industrial sectors. Mobility, not only for individuals, but also for the entire caste came to be visible. This incidence reduced the structural and cultural differences among the castes. According to A. Beteille (1969), the divisions based on income, education and occupation became more important at this time than the caste cleavages, for social and economic purposes. Therefore, the individuals who received western education, attained high professional posts and earned large pay packets began to share a common life style, irrespective of their caste identity.

Emergence of a large number of 'caste-free' occupations, expansion of transportation and communication network, influx of population, new democratic conception of equality, scientific outlook, increased facilities of education etc., accelerated the speed of dynamicity and kept a strong impact on the institution of caste. Most of the castes deviated from their traditional functions and inclined towards a fluidity of movement. Consequently, many Brahmins were found to join the service, trade, farming, business etc., giving up their traditional occupation, priesthood. Similarly persons from *Sudra* status were found to become Sanskrit scholars leaving the menial work. Some of them managed the positions of school teachers, clerks, bank-tellers, typists and Government officials. The interpersonal relationships became much easier between the individuals of varying castes. The whole system was modified.

The dissociation between castes and occupation has been more frequent in the towns than the rural areas. It is much pronounced in the big industrial cities. Opportunity of social mobility is, therefore, less prominent in the villages. The untouchables in the remote villages are still confined to filthy trades at starvation wages. But the consolation is that, at present, the Government of India is standing by them with its constitutional guarantee. Any enforcement of disability arising out of 'untouchability' is considered as a punishable offence in accordance with law. Untouchables being designated as Scheduled Castes are now entitled to receive Government scholarships and reservation benefits.

The rural scenario being altered by modern economic forces show an increase of cash-crop production. The grain payments in exchange of services have become unprofitable; mechanized farming has displaced the manual labour. Preference of manufactured goods over the hand made goods, similarly overthrow many persons from their caste-bound jobs. The villagers, therefore, move towards the cities, industrial towns or other prosperous areas for job and better wages. Obviously, the traditional tie between patron and client breaks and interdependency among the castes in a village is greatly hampered. The dominant castes of the village no longer remain dominant; they lose their former power and prestige. On the other hand, the lower castes tend to change their social position by associating themselves with white-colour jobs and also by adopting the Brahmanic way of life. This rise of status, from lower to higher, for a caste is considered as upward mobility of that caste which was first recognised by M.N. Srinivas in 1962. He designated this process as *'Sanskitization'* where the mobility occurs for an entire caste and not for an individual or a particular family. Moreover, the process, by no means, can change the caste status to the extreme. A Sudra can never be a Brahmin by the process of Sanskritization. The position can be raised only to some extent.

In a changing situation of present day, persons belonging to the same caste may be found in different social strata. This is because of their differences in education, occupation and income. So higher caste persons often find lower caste persons in their social circle. It has been impossible to resist the causal contacts between higher and lower castes. Traditional values of the people are changing at a rapid speed. Many of the caste disabilities are no longer tolerated. The people of lower caste groups have challenged religious sanctions towards the social inequalities. New expectations have aroused in them; new awareness and novel urge tend to change their age-old disabilities. In fact, they do not bother for the loss of their ritual status; rather became happy after getting a stable economic source to buy a new status. Further, the special provision of their representation in Parliament and in legislative assemblies have made them self-reliant as well as vocal in claiming different benefits. As a consequence, we find that a new class system is gradually entering into our society to displace the old caste system. This has been composed of different economic categories determined by wealth, education and occupation*[1].

The new status of lower caste people is not always acceptable to the traditionally privileged dominant social group. Therefore, in many areas confrontation is inevitable between the privileged group and the group that is devoid of it. We can conclude that Indian society is passing through a phase of transition. Conflict between the castes will be followed, as long the disparity will be continued. A strong administration is needed to deal with these problems with foresight and tact.

SAFEGUARDS FOR SCHEDULED CASTES

The untouchables who constitute the lowest segment of Hindu social hierarchy have been grouped as 'Scheduled Caste'. The entire history of Indian society tells us the story of discrimination and harassment of lower caste people at the hand of upper castes who used to enjoy greater power and privilege in the society. The sufferings were more rampant in the rural areas than the urban areas. As they remained under utter poverty, so lacked other social, economical, educational and political opportunities. Kathleen Gough in his book 'Criteria of caste ranking in South India' had written, "since the scheduled castes was the lowest caste, they had to serve all castes and no castes served it". In fact, they are mostly employed as labourer in agricultural fields. Sometimes, especially during the construction or repair of the house and at the time of storing of the grains their labour is needed in the house also. But these people are seldom allowed to enter inside of the house. In case, immediately after their departures by sprinkling cow-urine or cow-dung the house is purified*[2]. If they touch the utensils and other domestic articles, the non-inflammable articles are put to fire to let them purified and sprinkling the cow-urine purifies the inflammable articles like clothes. Water and food items touched by scheduled caste are held totally as polluted. So those are not used at all. If such a thing needs to be used, a little part of cow-dung has to be mixed with it. Food is served to these people either in leaves or in broken vessels and that must be done outside the house. In some places, especially in South India, scheduled castes are still not permitted to walk on the main path of the villages and the village merchants do not supply their necessities. A scavenger is required to carry the broom in his hand and he is supposed to shout continuously for announcing his polluted presence.

Mohandas Karamchand Gandhi first realized the role of untouchables in the struggle for power. He succeeded in raising the problem of untouchability on all-India level and incorporated it into his freedom movement programme. He also tried to resolve the discriminatory practices in the Hindu religion by a change of heart within the Hindu community. According to Ambedkar, upper the caste Hindus would never feel the pain of untouchables, as they had never lived as that wretched condition. Therefore, he opposed the activities of any organization started by upper-caste Hindus for the upliftment of untouchables. But he himself founded two political organizations namely 'The Independent Labour Party' (1936) and "The Federation of the Scheduled Castes' (1942) for minority appeal.

---

*1. Special privileges in the form of reservation of seats in academic institutions and employment exchanges provide the key to get education as well as service.

*2. Cow-urine and cow-dung are extremely sacred materials to the Hindus as they regard cow equal to mother. Both the things are used in all sacred purposes, especially to purify the profane objects.

The Constitution of India adopted on 26th Nov. 1949, provided the basic and supreme law in this country that acts as a charter of justice, liberty, equality and fraternity, granted to one and all. It governs almost all aspects of our social life. Its goal is to bring all weaker sections under constitutional purview. Therefore, it has been especially significant to the members of scheduled caste and scheduled tribe population. At the time of first enumeration in 1951, the scheduled castes were about 15% and scheduled tribes were over 6% of the total population.

A large number of lower castes were found who being deprived of their legitimate rights and privileges became unable to stand at their own. Therefore, the constitution declared those castes as scheduled and provided them certain safeguards as under:

1. Abolition of Untouchability

In order to remove the age-old practice of untouchability from the social milieu, Constitution provided the following laws:

1. *Article 17* stated the "untouchability is abolished and its practice in any form is forbidden. The enforcement of any disability arising out 'untouchability' shall be an offence punishable in accordance with law".

2. *Article 13(1)* declared that the 'custom' of untouchability is 'void' as it happens to be inconsistent with the provisions of fundamental rights including the rights under *Article 17* of the Constitution.

3. *Article 15* further wrote that:

A. "The state shall not discriminate against any citizen on grounds only of religion, race, caste, sex, place of birth or any of them".

B. "No citizen shall on grounds only of religion, race, caste, sex, place of birth or any of them, be subject to any disability, liability, restriction or condition with regard to

(*a*) access to shops, public restaurants, hotels, places of public entertainment or

(b) the use of well, tanks, bathing *ghats,* roads, and places of public resort maintained wholly or partly out of State funds or dedicated to the use of the general public".

4. *Article 16* guaranteed equality of opportunity for all the citizens in matters relating to employment or appointment to any office, under the State, irrespective of their religion, race, caste, sex, descent, place of birth, residence or any of them.

5. *Article 19(I)(g)* stated that, all the citizens, including the ex-untouchables, shall have the right to practice any profession, or to carry on any occupation, trade or business.

6. *Article 23(1)* prohibited 'beggar' and other similar forms of forced labour.

7. *Article 25* of the constitution gave them freedom of religion and enables the State for throwing open the Hindu religious institutions of a public character to all classes and sections of Hindus.

8. *Article 29 (2)* guaranteed them the right to education in the same scale with others.

Thus, the total effect of all these *Articles* is to abolish untouchability in all its forms and practices as to achieve social equality. It was mandatory on the part of the State under the *Article* 17 and that should be read with the *Article* 35 (1) of the constitution, to provide the punishment for the various offences of untouchability by appropriate law. Although the Constitution was implemented in 1950, but the law providing punishment for the offences of untouchability was passed in 1955. The Act was originally named as "The Untouchability Offences Act, 1955' (Act No. XXII of 1955) which was amended and renamed in 1976 as "The Protection of Civil Rights Act'.

The abolition of untouchability is a landmark in the socio-religious and politico-legal history of India. As a result, the religious disabilities such as non-access to the temples and holy shrines etc, and social disabilities like non-access to public places, shops and watering places are treated as cognizable offences punishable with imprisonment up to six months and fine up to Rs.500/-.

2. Educational Facilities and Reservations

Due to the problem of untouchability, the scheduled castes remained educationally backward. In order to improve the educational level, the constitution provides some rules, which are as follows.

1. *Article 15* prohibited any discrimination on the ground of religion, caste, race, etc. However, clause (4) *of Article 15* stated, "Nothing in this *Article* or clause (2) *of Article 29* shall prevent the State from making any special provision for the advancement of any socially and educationally backward classes of citizens or for the Scheduled Castes and Scheduled Tribes".

Article 46 stated "the State shall promote with special care educational and economic interests of the weaker sections of the people and in particular of the Scheduled Castes and Scheduled Tribes".

2. The Supreme Court (1964) in a dispute of Chitralekha Vs State of Mysore declared that *"Article 46, 341, 342 and 15(4)* form a group, which has relevance in making of a special provision for advancement of any socially and educationally backward class of citizens in the matter of admissions to Colleges". In this respect, the Union Ministry of Education issued a circular on November 23, 1954 to all the States / Union Territories to provide 20% reservations (15% for Scheduled Castes and 5% for Scheduled Tribes), for the weaker sections of the society in educational institutions. The Government of India continued the pre-independence policy of sanctioning scholarships to the students belonging of Scheduled Castes.

3. Reservation in Employment

Article 16 (1) envisaged, "Equality of opportunity for all citizens in the matters relating to employment or appointment to any office under 'the State' irrespective of their religion, race, caste, sex, descent, place of birth or residence".

Article 16(4) stated, "Nothing in this article shall prevent the State from making any provision for the reservation of appointments or posts in favour of any backward class of citizens which in the opinion of the State, is not adequately represented in the services under state".

Article 335 stated "The claims of the members of the Scheduled Castes and Scheduled Tribes shall be taken into consideration, consistently with the maintenance of efficiency of administration, in the making of appointments to services and posts in connections with the affairs of the Union or the State".

Thus, in order to ameliorate the socio-economic condition of the 'backward classes' specially the Scheduled Castes, at present 15% of the seats has been reserved for Scheduled Castes in case of direct recruitment by open competition. In case of recruitment otherwise than by open competition this reservation is $16\frac{2}{3}\%$ according to a resolution on 25-03-1970.

4. Political Reservation

In order to secure political participation and adequate representation of the Scheduled Castes in the legislature, some provisions have been made in the Constitution:

Article 330 (1) stated "Seats shall be reserved in the House of the people for (a) Scheduled Castes and (b) the number of seats reserved in any State or Union Territory for the Scheduled Castes shall bear, as nearly as may be, the same proportion of the total number of seats allotted to the State or Union Territory in the House of the people, as the population of Scheduled Castes in the States or Union Territory in respect of which seats are reserved, bears to the total population of the State or Union Territory".

Article 332 (I) stated, "Seats shall be reserved for the Scheduled Castes in the legislative Assemblies of the State".

Thus Article 330 (1) and *332 (1)* enabled each and every member of Scheduled Castes to exert his right to contest for a general seat, which every citizen possesses unless he/she is otherwise disqualified.

Article 334 of the constitution pleaded that the political reservations should last for fifty years from the commencement of the constitution i.e., up to 2000 AD.

The castes or communities included in the list of Scheduled Caste under *Article 341* of the Constitution are only eligible to contest in the reserved seat. The President had promulgated the Scheduled Castes Order under *Article 341* in 1950. The Scheduled Castes Order, 1950 was amended once in 1956 and thereafter in 1976. According to the said order, a person must profess to be either a Hindu or a Sikh by religion.

5. Economic Development

The Scheduled Castes who were subjected to untouchability are now mostly the sufferers of acute poverty. In order to uplift their condition, the constitution announced under *Article 88,* "The State shall strive to promote the welfare of the people by securing and protecting as effectively as it may a social order in which justice, social, economic and political, shall inform all the institution of the national life".

Article 39 stated, "The State shall direct its policy towards securing (a) that the citizens, men and women equally, have the right to an adequate means of livelihood; (b) that the ownership and control of the material resources of the community are so distributed as best to subserve the common good; (c) that the operation of the economic system does not result in the concentration of wealth and means of production to the common detriment; (d) that there is equal pay for equal work for both men and women; (e) that the health and strength of the workers, men and women, and the tender age of children are not abused and that citizens are not forced by economic necessity to enter avocation unsuited to their age or strength".

Article 41 stated, "State shall make provision for securing right 'to work' and right to public assistance in case of unemployment, old age, sickness and disablement but within the limits of its economic capacity and development".

Article 42 advocated for just and human conditions of work and maternity relief and *Article 43* provided the right emoluments, wages etc., for the workers. In addition, the *Article 46* pledged 'to promote' their 'economic interests' with 'special care'.

Thus, the Constitution of India made a large number of provisions for safeguarding the rights and special interests of Scheduled Castes in order to lift them up to the level of general mass of the country. The act and orders were categorized into two groups—(*a*) Protective measures and (*b*) Ameliorative or concessional measures.

During the sixth Five-Year Plan (1980-85) for strengthening the economic interest of the Scheduled Castes, a new strategy was built which comprised with the following benefits :

(*a*) the special component plans of the States;

(b) the special central assistance;

(c) Scheduled Castes Development Corporations for the States.

In fact, the various departments of Government for the welfare of Scheduled Castes launched a plethora of programmes. Non-Government Organizations also started to work side by side. Since India is a vast country, so desired social change has not occurred all over the country at the same rate and speed. Even today there are many areas where traditional social structure is found unaltered.

The factors that contributed in breaking the convention of caste can be summed up as follows:

1. *Mass media:* Availability of radio, television, microwave systems, printing press, newspaper, have played great roles in disseminating the information quickly as well as effectively. Weaker sections have been conscious enough about their rights and privileges, declared by the constitution.

2. *Transportation:* Metal roads, extensive railways, waterways and even the airways have minimized the distance at any level. The channel of communication has been expanded.

3. *Mass Education:* Now-a-days education has been made available to all. So the limitation of caste in acquiring education has totally been dissolved.

4. *Demographic Explosion:* Rate of population growth has been so high that traditional social structure undergoes a strain, which hammers on the rigidity of caste system.

5. *Emergence of Metropolitan Cities:* Different big cities like Calcutta (Kolkata), Delhi, Mumbai, Chennai, Kanpur, Bangalore, Ahmedabad, Hyderabad allure the people of villages and they rush to the cities in search of opportunities. Naturally the caste network of a village breaks down.

6. *Legal Provision:* Various provisions of constitution relating to reservation of seats in academic field and Employment schemes as well as the facilities of representation in Parliament and legislative assemblies have gone against the traditional social behaviour. The traditional caste roles are no longer emphasized now.

THE CONCEPT OF 'OBC'

The backward communities in India have been divided into three broad divisions—The Scheduled Castes, the Scheduled Tribes and the OBCs (the other backward classes). At present, the Scheduled Caste comprises 15.6%, the Scheduled Tribes are 7.5% and the other backward classes are 52% of the total Indian population. The first two groups have been defined clearly and enlisted on all-India basis. Untouchability is the very criterion of marking the Scheduled Castes and Scheduled Tribes. Such a criterion has cornered many population groups from the mainstream of the society. But the third group is loosely defined and not listed properly. It is a heterogeneous group as well. For these reasons, the problems of other backward classes (OBCs) stands complicated and remain unresolved while Constitution of India pays sufficient attention to the other two groups.

On 26th November, 1949, our Constitution was enacted. The first Backward Class Commission was appointed on 29th November, 1953 by Presidential Order (under *Article 340* of the Constitution of India) with Kaka Kalelkar as the Chairperson. The Commission was supposed to determine the criteria for 'socially and educationally backward classes' beside the already established categories like Scheduled Castes and Scheduled Tribes. The Commission was also asked to make a list of such classes along with their territorial distribution. However, the report was submitted on 31st March 1955. The Commission enlisted 2399 backward (socially and economically) castes and communities for the entire country and 837 of them were classified as 'most backward'. It also recommended some welfare measures for these 'Other Backward Classes' (OBCs) including certain reservations in Governmental services and educational institutions.

Kaka Kalelkar Commission report was the first national level inquiry of its kind after the adoption of the Constitution. But Central Government did not accept the recommendations of Commission as it was suffering from some grave flaws of methodology and had serious internal contradictions. For instance, it was not clear that how the list of 2399 backward classes was derived. The State-wise lists of backward classes, prepared by the Commission were actually based on the lists made by the Ministry of Education in 1949, for compilation of which Ministry of Education consulted with respective State governments. But such a list was neither checked by field-survey nor tested against the specific criteria evolved by the Commission itself. The report only took the caste a criterion for judging social backwardness. Naturally Governments of India opposed this and emphasized on economic tests.

The Second Backward Classes Commission was appointed on 1-1-1979 under the Chairmanship of B.P.Mandal. This commission adopted a multiple approach. It undertook a socio-educational survey, consulted the Census report of 1961 and also considered the lists of 'Other Backward Classes' (OBCs) notified by various State governments. The report was submitted on 31st Dec., 1980. According to the Mandal Commission's report 'Other Backward classes' included backward Hindu castes and communities (43.8%) and backward non-Hindu communities (8.4%). In fact, all untouchable castes, even those who profess Islam, Christianity or Buddhism i.e., who were not notified as Scheduled Castes came under the fold of 'Other Backward Classes'.

After the recommendation of Mandal Commission, upliftment of 'Other Backward Classes' became a larger national responsibility in the way to remove the mass poverty. On 17-8-90, Union Government issued a Memorandum reserving 27% of the vacancies in civil posts and services for 'Socially and Educationally Backward Classes' (SEBCs). The reservation was also applied to the Public sector undertakings and financial institutions.

The country witnessed a widespread agitation against the reservation policy. Higher caste people argued that there were already 22.5% reservations for the Scheduled Castes and Scheduled Tribes. So the additional 27% reservation for 'Other Backward Classes' would surely bring an enormous strain on the social system. It was also suggested that instead of employment reservation, education might be used as an instrument to bring social change. The rationale behind this proposal was; since social change implies change in attitudes, behaviours, customs, habits, manners, relation and values of people, the people being educated could change the structure of social institutions.

The violation and repression against Backward Classes (including Scheduled Castes and Scheduled Tribes) have been found to increase in rural areas as those people have begun to assert themselves in the society. In Urban areas caste conflict is centered on the issue of reservation because the upper castes have often been victims of this reservation policy. Another strange thing has been noted that being allured by the benefits of reservation and other related opportunities, a section of higher caste groups are trying to establish their identity as Scheduled Caste or Scheduled Tribe by tricky means. Again, some of the real scheduled groups, for getting various facilities and privileges want to continue themselves in the same state of affairs for an indefinite period. Such attempts hinder the socio-economic integration of the exploited groups in the wider national framework.

Although British formulated some socio-legal policies to remove caste disabilities, but they did not think to empower the backward people. Actually they wanted to develop an administrative network capable of exerting social control and extracting revenue. But that was not so easy because of the complex ethno-social situation of the country; the colonial rulers were bewildered with the diversities. The early missionary writings of Charles Grant, Claudius Buchaman, John Shore, Willium Carey and Willium Ward pointed out religion as determinant factor in the formation of social system in India. The Orientalist School further expanded this theme and established the fact that Indian society was differentiated on the basis of religion. Caste was the traditional integrative agency for maintaining social distance as well as social solidarity founded on religion. After independence, the system of universal franchise opened a new phase in the history of caste system in this country. The democratic politics encashed the inter-caste tensions and conflicts for electoral support in the local levels. As a result, caste lost its ritual front but gained more in the political front. The importance of casteism in Indian politics is on the increase. Caste system provided the political leadership with readymade channels of communication and mobilization. Therefore, although a new class system is gradually flourishing, it will be totally unrealistic to assume that the institution of caste will wither away from India in the foreseeable future.

VILLAGE LIFE

In India the study of village communities came into fashion in the decade of 1950. At the end of Second World War, anthropologists shifted their attention from typical monographic study of tribes and concentrated on villages. At this stage a difference was generated between ethnography and social anthropology. It was said that ethnography provided only a descriptive account, whereas the social anthropology offers an analytic view concerned with burning social problems. As a matter of fact, a social anthropologist is always interested in knowing how a society, or a segment within a society works in contrast to an ethnographer who keeps himself engaged in portraying different aspects in a chosen society. Therefore, this change of emphasis, from tribal studies to village studies was explained as a change of approach from comprehensive descriptive to analytic study. Though Indian village was not an autonomous whole like a tribe, it could be treated as an isolable unit of study. The group of Indian anthropologists, who attained maturity in the fifties and sixties, felt that by studying a tribe one would merely run along the periphery rather than to enter into the centre of a large and complex society. Village studies enabled them to overcome the former limitations, they got enough opportunities to get closer to the heart of a complex and changing society.

Each village was a kind of microcosm where anthropologists tested many features of macrocosm. They aimed at solving problems relating to social stratification, social control and the function of religious institution in a village. But Indian village was not purely an isolated and self-contained unit. Actual network of social relations stretched out of a village in every direction; villagers were connected to a wider universe. Anthropologists understood this fact which was valid for both a tribal village and a caste village. However, a tribal village was necessarily smaller and less complex than a non-tribal village. In case of a multi-caste village, the different castes of the village maintained inter-village relationships which might vary in direction and extent, but invariably kept a link with outsiders in terms of various activities like ritual, political, kinship etc. Those multiple relationships between villages were the key to understand the complex microcosm.

Since a peasant village showed a number of connections with the world outside the village, the ritual, political and economic behaviour of the villagers were consistent with one another and this provided a structural explanation for the complex society of India. Furthermore, the technique of the social anthropologists (intensive study by spending a long time in one place) was suited with the apparently small universe of the village. It seemed easy to find out the interconnectedness of different fields of activity by the tradition of long and patient observation. The pattern and ideas that emerged out of such prolonged study could be useful in projecting the nature of complex societies. The village studies thus stood as an important tool in the way to understand a complex society. If we proceed directly to analyze the structure of a complex society, we cannot easily set a feasible boundary as an unit of observation (like the primitive societies) because except the village situation, nowhere these different sets of activities are found to overlap in a significant degree. Therefore, study of village communities became very popular everywhere among the anthropologists.

After 1950, when village studies began to proliferate, influence of British anthropologists toppled down and American anthropologists came to dominate in the Indian scene. They extended their arena of village studies throughout the world, but India provided the most suitable conditions for their work. Some anthropologists of Indian origin also joined with them. The first lot of village studies was almost like village monographs, which offered only a general account of the communities under study. But soon they became thematic embracing more information on village structure where description of castes and caste relationships were prominent. Still at a glance, there were equivalent to tribal monographs*. At present anthropologists have undertaken various other models of study to understand and explain the societies of India, but village studies have not been abandoned. It is still looked on as a small universe although modern villages have articulated with the outside world to a great extent. Economic system of the village has become more closely integrated with the wider economy. Each village has been markedly politicized. Social mobility has increased with population dispersal. As a consequence, many new roles and relations have come into existence as well as many old values, old attitudes and old aspirations have subsided. All recent studies on village community, therefore, emphasize the perspective of wider social, economic and political system. Such studies have brought social anthropology and sociology closer to each other.

CONCEPT OF VILLAGE

The term 'village' usually refers to a consolidated agricultural community. It consists of small clusters of dwellings situated in such a surroundings that is empty of other housing. Although the villages are small settlements, they are economically very important for subsistence. Such type of settlements have existed for over three millennia and found not only in Asia but in Africa, Latin America as well as in some part of Europe.

A village actually stands on the people-land relationship. In a given territory, the residence, landholding and related economic aspects agglomerate; people establish close relationship with each other for easy livelihood. Robert Redfield has defined this localization of people with the features like distinctiveness, homogeneity, smallness, self-sufficiency and last of all he has specified it as "a cradle to grave arrangement". S.C.Dube has perceived the village settlement as a unit of social organization, which represents solidarity different from that of the kin, the caste and the class but plays a vital role as an agency of socialization and social control. According to him caste, religious groups, family, kin group all are the units of social structure, which can not function in isolation. The village is the platform that provides the scope of function and interaction among the individuals as well as the groups.

Anthropologists correlate the emergence of village with man's ability of cultivation. Although the domestication of plants appeared in South west Asia about 10,000 BC, but the first true village came into existence only about three thousand years ago following the situation that people began to lead a settled life with the rise of agricultural economy. Evidences can be shown in reference to Jarmo in Northeast Iraq around 6750 BC. In India, the earliest date of village foundation has been traced back to Vedic times about 2500 BC. At this time, village came out as a small republic—self-sufficient and independent of any foreign relation. The society was made up of families, which gradually extended as village (grama), the clan (vis), the people (jana) and the community (rashtra) in an ascending order. The village stood as a functional unit of social reality being composed of several families sharing a common habitation. According to S.C. Rapson, a village in the Age of RigVeda was comprised of certain number of houses, near each other, for the purposes of mutual defense and that was perhaps surrounded by a hedge or other protective barrier against wild beasts

* S.C.Dube's (1955) *Indian Village, F.G.Baily's* (1957) *Caste and the Economic Frontier, A.Beteille's* (1965) *Caste, Class and Power,* etc. Besides, McKim Marriott's edited book *'Village India'* (1955) and Srinivas's edited book *'India's Villages'* (1960) included a number of village studies contributed by various scholars.

or enemies. In 1891, Sir Henry Maine, the distinguished Victorian historian, having knowledge about the village communities of East and West, described the village as "the least destructible institution of a society, which never willingly surrenders any one of its usage to innovation".

In the post-Vedic literature, more particularly in the epics, we get much clear picture of the villages. It became an important unit of administration headed by a 'Gramini', a leader as well as a chief spokesman of the village. The responsibility of the Gramini was to protect the village and its surroundings upto a radius of two miles. A broad administrative system divided the villages into groups; each with a recognized leader. A group of ten villages went under the leadership of 'Dasa-gramini' and two such groups i.e., twenty villages accumulated under another head called 'Vim satipa'. Thus, a group of hundred villages was recognized under a bigger leader called 'Satagramini' or 'Gram-Satadhyaksha'. The ultimate biggest leader was the 'adhipati' who used to rule over a group of thousand villages. From this time the caste-groups came into consideration and they became the prime factor in forming the cleavages in the villages,

References of Buddhist period (500 BC - 200 BC) shows that during this period the Indian rural economy had been centered round the village communities where the landowners or the peasant proprietors acquired a prominent position. The village organization developed a lot and the villagers became entangled with different sorts of relationships—inter-village and intra-village relationships. Moreover, various other settlement types came into existence. Manu, the Hindu lawmaker of ancient India, mentioned three kinds of settlements—village (grama), town (pura) and city (nagara) of which village was the basic social and cultural unit. From Manu-smrti* we have come to know that, during the particular period, the caste system became tremendously rigid (within the villages) as a protest against Buddhism.

During the Muslim period (1300 AD - 1707 AD) though some changes occurred in Indian society, but those could not alter the basic pattern of economic organization in the village. For example, the new crafts that were introduced, did not permeate in rural areas and remained largely confined to towns and cities. Naturally the age-old rural set-up continued in its own form.

The fundamental change of village life was ushered in British period (1808 - 1947). The new economic and political forces introduced by the British eroded the traditional division of labour based on caste. Practice of hereditary occupation became difficult and so dissociation between caste and occupation started. A sharp decline of traditional handicrafts was noted. New educational system not only facilitated an upward mobility of lower castes, it destroyed the monopolies of higher castes. British administrative policy and welfare schemes at the same time gave a big thrust to village life. As a result, the old institutions of the village lost their function; the quality of life was altered to a great extent. A. S. Altekar explained this change as a social transformation. Despite the above changes, Indian villages still preserve some of its cultural characteristics and social values for which they have appeared as 'timeless' and 'changeless' to the foreign observers.

In nineteenth century, when the foundation of sociology or social anthropology were laid down, a number of surveys were conducted on rural life by the British officials in order to furnish a detailed account of the people and culture of India. Thomas Munro was the first man to formulate a concept of village 'community' in 1806. Sir Charles Metcalfe in 1832 provided the earliest account of the village communities in India. He described a village as an unruinable entity, which form a separate little State in itself and have a profound influence for the preservation of the culture in India. However, the villages were able to retain a semi-autonomous character by performing a number of responsibilities.

* Mani-Smrti is the book written by Manu.

An Outline of a Traditional Village in India

In a traditional Indian village people live close together. They meet each other and interact among themselves more frequently than they do with the people of other villages. They share a common social environment and similar life experiences like storm, flood, epidemic etc. On the other hand, villagers are often swayed by the happy events like bountiful harvest, communal enjoyments etc. A village typically embraces several caste groups, ranging from two to thirty, each of which exhibits its traditional rights and duties. Their privileges and restraints, services or manufactured goods contribute to the total functioning of the community.

A village may include landowning cultivators, tenants, priests, carpenter, potter, blacksmith, barber, washerman, sweeper, labourer, etc., castes. Each of these different rank-holding castes is an endogamous group who restricts their marriage affiliations within own group. The marriage partners are usually chosen from some other villages and kinship ties extend beyond the village. The neighbouring villages, therefore, form, the interdependent groups and the members of a caste in other villages stand as a set of actual or potential kin on whom a man has to depend for different necessities of livelihood. Thus, the kinsmen as well as the village companions constitute a large part of social relations where the life and aspirations of a person remain entwined. When a man goes outside of his village, he has to submit his identity at first as a member of the particular village and thereafter as a member of the particular caste.

Members of a village think for the development of their village in terms of public utilities like the ponds, wells, reservoir, embankment, roads, bridge, grave, pastureland and also other establishment like school, post office, revenue-collection unit etc. But the actual task of establishment and management, maintenance of law and order of a village rest on a headman who is an elder of the village and respected by all members in the village.

In some villages, one caste may be very powerful which is termed as dominant caste. To be a dominant caste, a caste needs to be numerically strongest as well as the largest landholding group. By dint of this position, the very caste acquires a respectable social position and exerts the supreme power in the locality. Therefore, the elders and leaders of that caste form the Panchayat* to rule over the whole village. They decide both the caste and village affairs. Some other numerically large caste group may be present in the village but that can not overpower due to deficient economic strength and social status. For this fact, Brahmins, inspite of their highest ritual status cannot rule in the village when they remain economically weak or numerically small group. But where there is no dominant caste in the village, representatives of Panchayat come from all groups and classes inhabiting the village. Harijan, the lowest rank in caste hierarchy, for their inferior most status in the village, may avail only the task of messenger. Although at present 33% of seats are found to be reserved for women in Panchayat, a few decades back, they had no right of decision making for the mass. Traditional system did not recognize any involvement of women—neither in the village council, nor in the caste council. The caste council is a greater body than the village council. It arose for the preservation of rights, and prevention of oppression and atrocities relating to caste. In a village, number of households belonging to a particular caste group is usually small in number. Therefore, they remain linked up with the similar caste groups of near by villages for administrative purpose.

Solidarity of a village is reflected in the division of labour and interdependence of the caste groups. Ordinarily a village shows several segments (para) where the particular caste-groups or lineage members reside. People prefer to live near close relatives for mutual protection as well as for

* The term 'Panchayat' literally means a group or council of five. In village usage it refers not only to the group that convenes but also to a set of processes for resolving conflict, for redressing transgression and for launching group enterprises (Cohn: 1965).

the easy interchange of food and other things. Members of different castes tend to form small hamlets and, therefore, social distinctiveness is found in different parts of the same village. Inter-caste relations constitute some vertical ties, which may be classified as economic,ritual, political and civic ties of which economic ties are most important. Peasant castes generally preponderate in a village, although there are other castes assigned with various services and crafts. Most of these castes have an access to agriculture. For instance, artisan castes like carpenter, blacksmith, potter, weaver etc., are found to perform agricultural work in addition to their traditional prescribed occupation. Servicing castes of the village like priest, barber, washerman etc., may also possess some agricultural land. The villagers produce most of what they consume. Other demands of life in terms of services and goods are met by division of labour among the castes. The servicing castes serve everyone in the village except Harijan*, the untouchables. Craft-producing castes follow the same thing. Most of the Indian villages are thus self-reliant bodies who do not have to depend on neighbouring villages for procuring their requirements. People rarely cross the village periphery to attend a fair or a pilgrim centre or a weekly market of a nearby village.

The relationships in the village whether it is between a landowner and a tenant, between a master and a servant, or between a creditor and a debtor, all are subsumed under a single category, the patron-client relationship as perceived by M.N.Srinivas in 1955. It is a very crucial as well as highly valued relationship in Indian villages. It brings people of unequal status close to one another as per the traditional norm, usage and social sanction.

The relationship actually forms between a food-producing family and the families, which supply them with goods and services at different necessities. The system is known by various names in different parts of the rural India. In the anthropological and sociological literature it is often called as 'Jajmani system' because it was first described under this name (by William Wiser in 1936) in a village of Uttar Pradesh. The Hindi word 'Jajman' denotes a person to whom a number of clients render their services according to the local custom. The 'Jajmani system' not only bound the different castes of a village together, occasionally it brings the groups of neighbouring villages together without threatening the village solidarity. In fact, the patron-client relationship is a durable relationship, which is inherited through the family or lineage and enforced by the caste system. The neat pattern of alliances on one hand, strengthens the self-sufficiency of the village and on the other hand, foster cohesiveness among the villagers.

In exchange of services and goods, a payment is made by the served class, which may be either in kind or in cash, or in both cash and kind. The essential artisan and servicing castes of the village are paid annually (at the time of harvest) in terms of grain. In some parts of India, they are also provided with food, clothing, fodder and plot of land for residence. On ritual occasions like birth, marriage and death some of these castes are called for special performances. A customary sum of money is paid to them at these times with some gifts in kind. Apart from the life-cycle ceremonies, a number of worships, festivals and fairs of the village require the cooperation of several castes. Even the lower castes may be involved there. For instance, Harijans perform some important duties during a village festival. They beat the drum, carry messages and remove the leaves on which villagers dine. Occasionally a Harijan or a member of any other low caste may cater a religious need to all villagers, including the highest castes. For example, in many parts of India, people relate the outbreak of epidemic diseases (like cholera, smallpox, measles, plague etc.) to the wrath of the village goddess. A member of non-Brahmin caste comes to propitiate in those situations, who may even be a Harijan.

Each caste community possesses its own rights, duties and privileges in the social hierarchy. Generally castes at the top possess special power and privileges which are denied to the lower castes. The lower castes become the tenants, servants, landless labourers, debtor and other clients of higher castes who always try to find out rich and powerful patrons. On the other hand, a tendency is

* Harijans live in a deplorable condition at the outskirts of the village.

found among rich and powerful castes to get more and more clients. However, conflict and cooperation go hand in hand. Economic conflict persists between landowners and tenants, masters and servants or donor or donee. Sometimes, such a relationship turns into exploiter-exploited relationship. The other causes of quarrels are land, wealth and women. Power alignment between individuals or families gives rise to factions. A village may have two or three major factions, which may continue through several generations. But the mechanism of conflict resolution and social mobilization are also rooted in village; a village is able to withstand all sorts of contradictions and continue without disintegration.

TYPE OF VILLAGE

A village in India is usually composed of a main settlement with a number of hamlets around it. As the population increases, some of its residents cross over the periphery for the purpose of dwelling; a number of satellite settlements are created. Moreover, the different castes of the village, cluster together in different hamlets to maintain their individual identity and freeness. As a result, we find several segments in one village like *Bamun-para* (Brahmins' residential area), *Kayet-para* (Kayasthas' residential area), *Kumor-para* (potters' area), *Kolu-tola* (oilmens' area), *Muchi-para* (cobblers' area) and so on. As the cultivation of land is the main occupation of the villagers, they never waste their fertile land. Residences are built on barren lands. Therefore, in most of the cases a village lacks conscious and systematic planning. Huts are built here and there on the basis of individual choice; the villagers only follow the convention of contiguity of lineages or caste-mates. However, a village may belong to a particular tribe or caste or a religious sect. Occupation of all villagers is obviously not uniform. According to N.K.Bose, the most useful and fair means of classifying a village is the consideration of its physical form. Different orders have been distinguished by him for India, where each type is capable of showing a wide range of variation.

In the view of Prof. Bose, the shape of a settlement pattern is determined by the position of its agricultural fields. People want to live within close proximity of own agricultural land for easy supervision. However, he divided the villages broadly into two types.

(i) Villages without an order.

(ii) Villages with some order.

'Villages without an order' is a shapeless cluster where the huts and roads come together without making any specific design. Two sub-types may be distinguished here. (*a*) the massive type and (*b*) small dispersed type. The dispersed type of villages is generally small and shows huts in some discrete clusters. The massive types are often found in western side of West Bengal and Chotonagpur plateau, whereas the small dispersed type of villages are found in the hilly areas, forest areas and also in desert areas.

The villages that are formed with some order show kind of tortuous or irregular roads according to the local requirements. The sub-types are as follows:

(a) *Linear Cluster*—a straight open space is provided between the parallel rows of houses. Santal villages exhibit this sort of assemblage. Coastal districts of Orissa and Andhra Pradesh are the best examples for such type of villages.

(b) *Square* or *rectangular Cluster*—there are some villages where straight roads run parallel or in right angles to one another. The overall shape is either like a square or a rectangle. This type is considered as an evolved type of'linear cluster'. Therefore, such settlements are found in the same areas where linear types are preponderant.

(c) *Isolated homestead*—these are the isolated houses set in the midst of the agricultural fields. The position facilitates constant look-after and watch on the growing crops. When villages are formed with isolated homesteads, a number of villages are treated together under a 'mouza' for the convenience of collecting the rent or tax. Such types of villages are found in northern plains of India as well as the flooded districts of Eastern Uttar Pradesh and Bihar.

Beside the pattern of settlement, villages can also be classified into four types basing on the chief occupational pattern of its inhabitants.

(i) *Farming Village*—Some villages grow entirely on agriculture. That means, the main occupation of the villagers is cultivation. They produce different types of crop for maintaining the livelihood.

(ii) *Non-Farming Village*—This type of villages survive mainly on trading of the agricultural products in the rural environment. Good transport facility in the surrounding area is the first precondition for the development of such villages.

(iii) *Industrial Village*—These villages are usually the spots of small cottage industry like handloom, pottery, metal artifacts, mat weaving etc. Specialized craftsmen produce plenty of articles in the respective village setting.

(iv) *Suburban Village*—These villages grow adjacent to the town. The villagers associate themselves with different activities of the town.

House-type of a village is usually governed by the climate, soil and the materials available in the region. Climatic factor is reflected in the construction of roof, wall, floor, door, and window. For example, a roof has to bear the lashes of wind and rain, so it is often designed in such a way that it can withstand the environmental upheavals. To combat with heavy rainfall, houses are made with sloping roof, which can drain off the water very easily. In hot areas, high roofs are useful to keep the room cool, especially during summer. Similarly, low roofs can keep the rooms comparatively warm in cold areas. Again, in desert areas, walls are built thick to protect the people from scorching heat. The same principle is maintained in hill-stations to keep the cold winds away. Windows are chiefly meant for ventilation; they allow plenty of fresh air and light to come in. Limited number of doors and windows indicate an extreme climatic condition.

People have penetrated to all habitable regions of the globe. The building materials of house obviously differ from place to place keeping conformity with the locally available resources. For instance, Eskimos live in the snow houses called *Igloo.* People along the seacoast make their houses with bamboo. Those who live in the forest of Assam, build houses on stilts. People inhabiting in hills invariably make their houses with wood. It is not only a protective devise against cold; it also saves the houses on uneven surface, from the devastation of earthquake. Where there is plenty of rain and snow in hills, houses are found with inclined roofs. This design prevents rainwater as well as snow from accumulating on the roofs. Throughout the vast region of tropical rain forest and woodland areas, poles and thatched houses made of vegetation prevail. Most of the habitations in Bengal village are mud-built; some well-to-do cultivators only afford to make brick-built houses. Well-ventilated huts of Southern Bengal and coastal Orissa provide the evidences of warm and humid climate. Geography thus guides on the nature of house-types equally as the settlement pattern. Tradition also holds an important factor. But, the distribution of languages or political boundaries does not keep any repercussion in this respect.

S. C. Dube has suggested six criteria for the categorization of villages in India. Those factors are as follows:

(i) Size, population and land area.

(ii) Ethnic composition and caste constitution.

(iii) Structure of authority and power hierarchy.

(iv) Degree of isolation.

(v) Pattern of ownership.

(vi) Local tradition.

According to him each culture-area of India uses its own nomenclature to classify the villages within its territory by size and population. The other factors come side by side. Final governing factor is the tradition, both regional and local. Like dress, speech and manner, the layout of a village and construction pattern of the houses follow the prescribed pattern of the culture-area. Eminent sociologists like A.R.Desai classify the villages into three groups:

(i) The migratory agricultural villages, where people live in fixed abodes for only a few months.

(ii) The semi-permanent agricultural villages, where people reside only for a few years and thereafter migrate. Because, the fertility of the soil gets exhausted by that time.

(iii) The permanent agricultural villages, where people settle permanently and live generation after generation.

The above classification points out a transition from man's nomadic existence to settled village life. However, the sociologists like to divide the villages into two broad groups—*Nucleated villages* and *Dispersed villages* according to the nature of agglomeration of huts. In the nucleated villages the farmers dwell in the village proper, in a cluster. They work on the fields outside the village site. In non-nucleated dispersed villages, the farmers live separately on their respective farms. The social life in the latter is less compact than the former. The village aggregates, in the opinion of sociologists, may also be classified as per their social differentiation, stratification, mobility and land-ownership.

Unity and Diversity in Village Situation

India is a land of villages. Seventy-four percent of its people still lives in villages. Village types differ from place to place, especially in widely separated parts of the country. Again, each village is a distinct entity having a definite pattern and mode of life; a corporate unity for the villagers. Yet the village communities all over the Indian sub-continent possess a number of common features:

(i) The village settlement indicates a body of people who live in a particular piece of land keeping some distance from similar groups with poorly developed roads in between.

(ii) Majority of the people in a village associate themselves with agricultural activity. They depend upon one another in several ways. A vast body of common experience is shared by them while living together.

(iii) Different castes and communities inhabiting in a village remain integrated in its economic, social and ritual pattern, by ties of mutual and reciprocal obligation, sanctioned and sustained by conventions.

(iv) Village social organization represents a solidarity, which surpasses the identity of kin, caste or class, and plays a key role as an agency of socialization and social control.

(v) The important administrative functions in a village are performed by the village council, composed of village elders. It also decides the disputes between the villagers and discusses the matters of common interest like holding of a festival and building of a temple or a road.

(vi) Despite the presence of different groups and factions, a village appears to be an organized and compact whole. The members behave like that of a family at the time of exigencies viz., outbreak of an epidemic, a situation of flood, draught or fire; a fight with neighbouring village etc. The unity of villagers is also observed centering religious occasions, communal feast etc.

(vii) The tie of the caste and kin that often cut across the village, cannot reduce the village unity. Srinivas, used the term 'Horizontal unity' to denote the alliances of caste that go beyond the village It is the opposite concept of'vertical unity' which means the interdependence of castes within village.

(viii) The unity of a village supersedes all sorts of heterogeneity that arise in a population due to diversified occupation, language, interest etc.

The fundamental similarities thus project a general picture of the village all over the country but diversity lies in the internal structure and organization, on account of a variety of factors, namely. size, population, land-area, ethnic composition, caste constitution, pattern of land-ownership, structure of authority and power hierarchy, degree of isolation from urban areas, local tradition etc. A village therefore, presents a unity among the diversities. The people communicate with the larger society to become a part of the modern civilization.

Indian civilization is actually known in villages. The traditional urban society in India was basically similar to village society. Present day alliances are very important to a village system; they are the proper and legitimate factors to understand a village life. The self-sufficiency of the village is only a relative concept and not an absolute one. But Indian village communities restore a strong vitality in them, which enable them to carry on the age-old tradition. Anthropologists have recognized the matter and so they put an emphasis on 'village study' to find out the structure of Indian civilization.

ANTHROPOLOGICAL INVESTIGATION IN INDIAN VILLAGE

Understanding of village community was significant not only for administrative reasons but also for historical insights and ideas. Therefore, anthropologists turned their attention from the microcosms of little tribes to the 'village' communities. The village was taken as a unit of investigation and the findings represented the whole of India. Village communities offered anthropologists a convenient centre for intensive investigation in order to explore overt and covert aspects of human behaviour as a means to study the cultural development. Not only that, 'village study' provided ample scope to perveive the dynamics of change and also peoples's attitude and reactions to change.

The year 1955 witnessed a boon in the study of village communities. In that year, for the first time, four basic books and several papers on Indian Village were published. The studies were done not only by the Indian social scientists, American and British social scientists also came forward. The four important books were S.C.Dube's 'Indian Village', D.N.Majumdar's (ed.) 'Rural Profiles' Mckim Marriott's (ed.) 'Village India' and M.N.Srinivas's (ed.) 'India's Villages'. In the same year, a conference was also held where Iravati Karve was the Chairperson. Prof. Redfield came to participate in that conference. The proceedings of the conference were published as a book with the title, 'Society in India'. Later, the other books namely, 'Caste and the Economic Frontier' written by F.G. Bailey (1957), 'India's Changing Villages' by S.C.Dube (1958), 'Caste and Communication in an Indian Village' by D.N.Majumdar (1958), 'Village Life in Northern India' by Oscar Lewis (1958) appeared one after another. Department of Anthropology and Sociology, in different Universities sanctioned numerous projects on rural research. Planning Commission, Government of India also promoted some rural research projects through different centres.

'Village studies' continued to dominate the study of rural society in India until the early 1960s. After that, a shift in its orientation was found; the status of 'village' began to be evaluated in perspective of Indian society and tradition. A. C. Mayer's work 'Caste and Kinship in Central India' (1960), K.S.Mathur's 'Caste and Ritual in a Malwa Village' (1964), K. Ishawaran's 'Tradition and Economy in Indian Village' (1966) and A. Beteille's 'Caste, Class and Power' (1966) went under this trend of study. The number of research papers in various journals also increased by this time.

The village communities of India evoked profound interest among the foreign scholars because these villages differed widely from the villages of Europe, Russia, Germany and China. So after the Second World War and more accurately after India's independence several American social anthropologists could not resist their temptations to study Indian villages. Special mention may be made about Morris Opler of the Cornell University (1948, 1950); Oscar Lewis of the Illinois University (1951); David Mandelbaum of the California University; Robert Redfield, Milton Singer, Mckim Marriott etc., of the Chicago University who not only stayed in India with their research teams to conduct field-work, but also created an atmosphere of analytic study in Indian villages. The various outlooks,in synthesis,gave birth to some distinct concepts.

Being a disciple of Redfield, Mckim Marriott (1955) applied Redfield's concept of 'Great and Little Tradition' to the complex social situation of India. In 1955, he had studied a village named Kishangarhi in U.P. By modifying the concept of Great and Little Tradition, he brought the twin concepts—Universalization and Parochialisation as operating in the socio-religious system of Indian villages. Marriott defined Universalization as "carrying forward of materials which are already present in the little traditions which it encompasses". Opposite to Universalization, Parochialisation was referred as the "downward devolution of great traditional elements and their integration with little traditional elements. It is a process of localisation". Marriott tried to explain the cultural growth of an indigenous civilization. In his opinion an indigenous civilization is one whose great traditions originate by Universalization and little traditions remain localized by Parochialisation. Milton Singer (1954-1955) also examined the concept of Redfield in the city of Madras (Chennai) with special reference to cultural specialists. Singer further formulated the concept of cultural media, cultural performances and cultural stages in order to understand the processes, unity and continuity of Indian civilization. Following A.L.Kroeber and R. Redfield, Milton Singer conceived Indian civilization as a coherent structure of rural network and urban centres which at the same time act as a medium for the mutual communication of great and little traditions and of other cultural differences between and among tribes and castes. Notions of village, caste, vertical and horizontal solidarity of the castes were also discussed by the eminent anthropologists like Srinivas, Majumdar, Dumont and Pocock.

M.N.Srinivas (1956) had pointed out a process of social change in upward direction, which he termed as Sanskritization. It results only in a positional change without incurring any structural change in the system. Srivinas (1956) also forwarded the concept of' Westernization'; D.N.Majumdar tried to show a reverse process of Sanskritization with the term 'desanskritization' where the Brahmin caste tries to identify them in some matters with other castes. Ishwaran (I960) placed an objection to the term Sanskritization. According to him Brahmanization would be the right word instead of Sanskritization. Again, Prasad advocated the term *Kulinization* in the place of Sanskritization.

The other significant contribution of Srinivas was the concept of 'Dominant Caste'. His paper on the dominance of caste in the village Rampura was published in American Anthropologist (Vol. 61,1959), was appreciated by the anthropologists and sociologists. But S.C.Dube (1961) and B.K.RoyBurman (1978) disagreed with Srinivas over the use and existence of dominant caste. Dube argued that, it is not the caste group but an individual usually dominates in a society. He also mentioned that political power remained concentrated in a few individuals rather than diffused in caste. Moreover, he found the concept difficult in testing in a concrete field situation. B.K.RoyBurman pointed out that instead of a dominant caste group, there might be a number of dominant communities in a particular geographical area. Further, he said that the dimensions of domination might be different for different castes. For example, a community may possess control over the economic resources while another may exert its authority over the social status or may be responsible for the maintenance of law etc. Although Dube and RoyBurman opposed the concept of' Dominant Caste', it was admirably substantiated by Mayer (1958).

The concept of 'Continuum' (Tribe-Caste Continuum) was borrowed from Redfield (1957) as it was reflected in his work 'The Folk Culture of Yucatan' (1941). F.G. Bailey (1960) delineated a systematic interactional mode between tribe and caste as two ideal poles in a linear continuum. He also mentioned that tribe is typically organized on the basis of segmentary solidarity while the caste system is based on organic solidarity. Again, Surajit Sinha, another disciple of Redfield examined some ethnographic materials from the Bhumij village of Birbhum and also from tribes of Bastar (1965). He suggested that the continuum has to be conceived simultaneously on the level of social structure and culture, both as ethnic groups and as local communities. Following Sinha's outline H.N.Banerjee (1969) had made a detailed study on the pattern of tribe-caste continuum among the Korwas of Manbhum.

On the other hand, L. P.Vidyarthi (1961) developed the concept of 'Sacred Complex'; he studied traditional religious places as a dimension of Indian civilization. Marriott and Bernard S. Cohn contributed the concepts of 'Network and Centres' (1958) to study the channels of integration in Indian civilization.

Indian anthropologists got enough inspiration by the work of the European and American anthropologists. In this respect we can mention Peake's work in England's village (1922), Herskovit's work in Haiti's village (1937), and Arensberg's village (1940), Redfield's and Lewis's work in Mexican village (1940), Raymond Firth's effort in fisherman's village in Malayasia etc. But Dumont and Pocock (1957) sought some alternative means for understanding Indian social reality. Indian anthropologist R.Mukherjee (1957) stressed about the inappropriateness of considering the village as a unit of investigation as like Dumont and Pocock. A changing approach perhaps developed at this time; anthropologists emphasized on the limitations of the microcosmic approach. They argued that the articulation of a village to a larger whole had been greatly increased with time. The villagers became accustomed to borrow the services of ritual specialists and artisans of the other village; weekly markets and festivals of neighbouring villages also attracted them, Such intra-village contacts give rise to an interdependent network, which could never be understood if a village was examined as a self-contained autonomous unit.

Multi-village studies were, therefore, undertaken from early 1960s. Policy makers, planners and social engineers provided a warm welcome. At the same time, from 1962, agrarian sector in India began to be advanced under the sway of capitalism. Anthropologist began to ponder over the situation. They have labelled the stages as capitalism, pre-capitalism, semi-feudalism etc. Ramkrishna Mukherjee found a semi-feudal mode of rural Bengal for the exploitation of the peasantry by the landholders. Amit Bhandari conducted a survey in 26 villages of West Bengal in 1970 and observed a more or less same trend as Mukherjee had pointed out. As a matter of fact, no village in India is completely independent at present. They are merely the units in wider social system and the parts of an organized political society. Therefore, anthropologists are bound to widen their field of vision; it goes beyond the boundaries of village proper. A village family typically depends on goods and services provided by the people of the surrounding villages. The social field of the villagers crosses the specific village territory. Villagers often visit the towns or cities for different purposes like economic, social, administrative, judicial etc. As a consequence, the concept of a self-sufficient village looks like a myth. All anthropological investigations at present abide the perspective of a wider social network.

VILLAGE DEVELOPMENT PROGRAMMES

Rich Indian soil from the time immemorial has provided wealth for its inhabitants. It attracted the foreigners from time to time. Pathans, Turks, Mughals and other people invaded India, one after another from tenth century onwards. Similar interest was generated among the Portuguese, Dutch, French and British merchants who came with a view to acquire wealth from India. At the first phase, Muslims ruled over India for a few centuries. They ushered certain changes in the traditional arrangements but that was restricted in the vicinity of urban centres; villages retained their old forms. Indian villages experienced a really severe blow with the advent of British. Rural organizations that existed as self-reliant autocratic communities based on the functioning of the institutions like the joint family, the caste and the village panchayat undergone a profound change.

British brought changes in Indian society basically to serve their own need. They introduced an individualistic as well as a competitive attitude among the people. Moreover, they established two main types of land system—the *Zamindary* and the *Raiyatari* system of land-tenure, which affected the traditional agrarian structure of this sub-continent.

The ancient land system

There is difference of opinion among the historians regarding the exact nature of ancient land tenure system in India. Eminent scholars like Badan-Powell and Radha Kumud Mukherjee believe that private property in land existed even in Vedic period. But others said that the land in ancient India was entirely the property of the king. But both the groups recognized the communal ownership of land in the early village communities. The self-sufficient village, as a centre of agriculture and industry, existed for centuries in India despite political upheavals and dynastic changes. Kingdoms rose and collapsed but the self-sufficient villages survived.

The village population was primarily composed of peasants. The village committee, represented by the village elders, distributed the land among its members. Each family held a plot of land assigned to it and cultivation was done by the collective labour of the families. The system used to be followed from generation to generation. Thus, the peasant families acquired a hereditary right to cultivate as well as to possess the land.

Under the Hindu system, although the kings were entitled to receive a share from peasants' produce (as a protector of the people), they were never the owners of the land. Rather, all of them had their own private land. The share of the kings also differed from time to time in different periods of history. Usually the kings used to demand 1/6th of the gross produces. This share was collected through a hierarchy of officials of whom the 'Deshmukh' or the Chowdhury was the main. By the middle of the seventeenth century the land-system and the revenue administration went in the hands of the Muslim rulers. At this time, the zamindars evolved as a class, out of the Hindu Chowdhuries.

Land Reform in India

The destruction of the self-sufficient 'village republic' took place under the British rule who made the country their colony and exploited colony's raw materials. Lord Cornwallis came to India as the Governor-General in 1786 succeeding Warren Hastings. He introduced Permanent Settlement in the year 1793, which gave special power to Zamindars as proprietors of land. But certain principles were rigorously maintained. Firstly, the right of the zamindars was subjected to regular payment of rent revenue to the Government. Secondly, the amount of land revenue was fixed permanently and no further demand could be made in consequence of any improvement on the estate of the zamindars. Thirdly, the zamindars were to pay to the state ten-elevenths of the revenue received from the Raiyats (tenants) and to keep one-eleventh of it as their remuneration. It should be mentioned that under zamindary tenure, the rights of property in land were conferred upon the native tax-gatherers that had so far never taken any interest in the indigenous culture. In the zamindary areas, the letting and sub-letting of the land was responsible for an extensive growth of non-cultivating rent-receivers creating a chain of intermediaries (tenants, sub-tenants and sub-sub-tenants), whose cumulative burden had to be borne by the actual tiller of the soil. Under Raiyatari system, no intermediary proprietors were recognized; the actual tillers of the soil were vested with a heritable and transferable right of property in their lands. Both the systems were the departures from traditional land system under which land as a private property was not recognized. The new systems made the land a commodity in the market, which could be mortgaged, purchased, leased or partitioned.

Aspects of Change

Village life began to change rapidly. Structural features of a traditional village altered quickly; three important traditional institutions viz., The joint family, the caste system and the village Panchayat became defunct.

The Joint Family—The joint family was the most common form of family organization in India, sanctified in scripture and sanctioned by the secular law. It included several nuclear families covering at least three generations, bounded together by blood tie. Such a joint family stood as a unit of consumers as well as a single producing unit. The members were fed from the same kitchen and the property was held jointly under the trusteeship of the eldest male of the family. The cultivators and artisans of the traditional India necessarily exhibited the joint family system, where collective effort of the family members was required in producing food. A member would never think for individual prospect. But, the new system geared up individualism in such a manner that a man did not want to live subservient to the head of the family; he wanted freedom and selfhood. Further, the land reform measures added impulse towards the splitting of large joint families. A large joint family, a producing unit when divided into separate nuclear families, the small families lost their potency and efficiency for agriculture. First of all, the small units were unable to provide the capital, necessary for maintaining the implements and animals. Shortage of manpower delayed the force and speed of agricultural work. As a result, the part families began to shift from their traditional occupation and scattered in several directions. The towns and cities seemed to them alluring as those offered ample scope of work.

The Caste System—Industries of the village artisans faced a great threat as British compelled the people to use machine made goods, produced outside the village. It not only ruined the self-sufficiency of the village life, also shattered down the central importance of the caste system in the village social structure. The ideal social and religious pattern that kept the caste groups segregated and made them economically interdependent were destroyed. The changing nature of caste adversely affected the village social organization. The whole village drifted away from the ceremonial order within which the reciprocal transactions of caste were organized and reinforced. Therefore, both the significance and rigidity of the caste institution were lost.

The Village Panchayat—The other old institution, the village panchayat was also affected. It lost its autonomous penal power as it was brought under the rule of laws made by the centralized state.

Impoverishment of the rural people gradually deteriorated the agrarian economy. This resulted in a lop-sided, un-balanced position in the national and international economy. On one hand, the low-yield, downfall of the artisans, increasing diminution in the size of the holdings, and on the other hand, the absentee landlords and the resultant group of land-less labourers, appearance of creditors and moneylenders made the Indian rural scene absolutely miserable.

Development after Independence

Rural India has changed beyond recognition during the last 250 years. Autonomy and self-reliance of the village is no more found. Village community has been termed as Peasant society to designate the rural dimension of civilization. Government agencies and several voluntary associations have come forward for ameliorating the conditions of the village, for maximizing the agricultural output and to protect the interest of the villagers. Both the Central and State administration have attempted for the growth of a new social organization in the village level, so that it can cope with the newly grown problems of the villagers. Legislation has passed in provinces to revive the panchayat, the village council. The old councils were arbitrary, conserving agencies whose function was primarily to settle down the disputes of the villagers. But the new panchayats are organized bodies where members are democratically elected by the commoners. It aims to work for the community development rather than conserving solidarity.

From the beginning of the British rule in India, the political and economic decisions that were taken in London and Manchester obviously affected the Indian peasants by destruction of their authentic agrarian economy and by fossilization of age-old social and cultural institution. Independent India was perplexed with the awful situation. It innovated a series of policies and by the method of trial and error it wanted to uplift and nourish the total rural sector. Caste is still a persisting factor in village life, but by no means it can be a prime factor in political decision making. It does not hold the some power as was evident about two hundred years ago.

2nd October, 1952 is considered as a landmark for the inauguration of Community Development projects and Agricultural Extension Services in different parts of India. Since then, new roads are being made and old roads are repaired for the improvement of the total communication system. Five-year plans of Government of India have always been directed at poverty alleviation; they aim to strengthen the rural infrastructure.

TRIBE

The word 'tribe' was picked up by the anthropologists from ordinary usage. In Medieval English, the word tribe conveyed a neutral sense—'a primary aggregate of people claiming descent from a common ancestor'. With the advent of colonization, European anthropologists applied this word to the people who lived in a primitive or barbarous condition in backward areas and did not know the use of writing. Since then the word 'tribe' has been a technical administrative term to denote the aborigines. Different other Indian terms like 'adivasi', 'vanya-jati', 'jana-jati', 'jana-jamity' etc. also bring the same connotation.

Tribes are the indigenous or autochthonous population of Indian subcontinent. Tribal society is often referred as 'primitive society' or 'pre-state society', or 'folk society' or even as 'simple society'. Sometimes, the word 'tribe' is taken as a synonym of the term race but scientifically 'race' carries an entirely different meaning. However, all these terms indicated that the tribals are the backward group in comparison to other advanced social groups. But unfortunately there was no commonly accepted definition of this word. Nobody bothered to offer a precise meaning, as the term for a long time generated no confusion at all. Anthropologists found no trouble to identify and differentiate the tribal groups from the groups of other kind.

The situation was quite easy in Australia, Melanesia and North America where anthropologists conducted their studies at the beginning. But problem arose with India and to some extent also with Africa. Identification of tribes became extremely difficult in India as various types of social groups appeared in association with tribes. Among these diversified social groups, castes created principal hindrance. Many tribes were described as castes and many castes were labelled as tribes as well.

In fact, these two types of communities are different from each other. A tribe is an extremely primitive society, whereas a caste is a part of Hindu society. Except in a few areas, tribal communities tend to retain their pristine tribal characters.

Characteristics of Tribe

Evolutionist thinkers of nineteenth century first attempted to distinguish the tribal societies. They focused on the legal and political institutions. Lewis Morgan (1877) regarded those societies as tribals who exhibited social institutions but lacked political one. Henry Maine, on the other hand, found the distinction in legal terms. In the book 'Ancient Law' (1861), he said that the tribal societies possess their own laws on status rights, rather than the contractual rights. Both Morgan and Maine emphasized the kinship basis of tribal societies in contrast to the territorial foundations of Modern State. The paucity of contractual relation, which Maine identified as a feature of tribal society was widely accepted among the anthropologists of the future generation. Gluckman (1955) and Schapera (1938) made significant contributions on tribal legal institutions. A series of definitions on tribe were forwarded by the anthropologists of the earlier time like Tylor, Perry, Rivers, Lowie and others. But none of those definitions was complete. It was realized that the definition should be based on the empirical characteristics of the particular human groupings found in different parts of the world. Accordingly, features like kinship ties, common territory, one language, joint ownership, one political organization etc., were pointed out as main characteristics of a tribe. But anthropologists did not universally agree with these above-mentioned characteristics. For example, Rivers had not accepted 'common territory' as a vital feature of Tribal organization. He put more importance on the criteria of 'having a single government and acting together for such common purposes as welfare'. To Perry 'a common dialect' and 'a common territory' were the indispensable criteria in labeling a tribe. He said that even the nomadic tribes maintain a specific territorial boundary to roam about. Evans Pritchard also believed in small territory. But he also emphasized the criteria like small population, small-scale social contacts, simple technology, simple economy and little specialization of social function. Redfield added a further criterion—the absence of literature. Kroeber designed a conceptual model where he equated primitives with a small self-containing isolate human group. He also tried to distinguish the primitive society from the peasant society. As per his perception, the primitive society is a whole society; the peasant society is a part society as it possesses an extension in the civilization of that region. Anthropologists thus differed in their opinion. Their inferences were drawn from diverse locations, but each of them had mastery on his respective field situation. This obviously means that the tribal situations vary with the place. Tribal society, therefore, is not an absolute category, although an idealist type of tribal society has been placed at the opposite end of the modern society. Some of the intermediate societies may be more tribal and others are less tribal in nature. In an ideal state, a tribe is a self-contained unit. It makes a whole society in itself; its boundaries demarcate certain limits of interaction in the legal, political, economic and other spheres.

Modern usage of the term 'tribe' suggests that it is a small-scale social grouping (Wilson & Wilson, 1945) that display some form of cultural unity. The members recognize a close affinity towards one another. A compact and self-sufficient economy is found in such groups unlike the modern society. Within the spatial and temporal range of social, legal and political relations, they possess a morality, religion and a world-view of corresponding dimensions. Although tribal societies are supremely ethno-centric, but a tribe is no longer referred in terms of its cultural homogeneity. Field-workers throughout the world have admitted that there is no demarcating wall to differentiate where one 'culture' comes to an end and the other begins. A social group may include a single tribe, or a group of tribes like Baganda, Iroquois etc. It may represent either a culture area or a whole continent as the members share common cultural traits in varying degree. Thus, tribe is purely a technical term based on territorially defined political unity. There is a debate regarding the political qualities of a tribe. Many scholars conceive of tribal societies as being a state of total anarchy. But in reality, many tribal societies exhibit well-established system of government.

Despite so many confusion and ambiguity regarding the group, 'tribe', we should attempt for a workable criterion. Earlier anthropologists had taken tribes not merely as societies of a particular type; they considered the people as belonging to a lower stage of development. In twentieth century when the concept of evolution fell out of fashion, certain emergent characters were chosen carefully to distinguish the primitive societies from the advanced societies. They are as follows:

1. A tribe is a homogeneous ethnic group.
2. A tribe has a sense of identity based on common language and culture.
3. A tribe exhibits a primitive level of technology.
4. A tribe shows no system of writing.
5. A tribe lacks the specialized division of labour.
6. A tribe is associated with a definite territory.
7. A tribe possesses a well-defined political boundary.

According to A. Beteille (1977), a tribe is large enough to be a visible group and small enough to be mobilized for common action. Being cut off from the main stream of Indian life, they usually live in forests and hills and develop a different world-view suited to their particular ecology. Eminent sociologist Duncan G.A. Mitchell (1979) described tribe as a 'socially cohesive unit, associated with a territory, the members of which regard themselves as politically autonomous'. The backwardness of the tribe has been considered in terms of culture and the tribal problems are seen in reference to non-tribals.

Identity of Tribes in India

India is a very complex society but shows an all-pervasive mode of cultural unity. Interactions, interdependence and sharing of certain common symbols are found here, inspite of its multiform diversities. Tribals are not the alien groups; they are the part of the wider civilization. Still they exhibit considerable differences. Each tribe has a traditional identity in terms of territory, which is always referred by its members as a home. They have a common and distinct name. A specific dialect is used by all members of a tribal group. Similarly all the members share an in-group sentiment. These members may or may not be the kin to each other, but kinship operates as a strong associative, regulative and binding principle. Rules of endogamy are strictly followed in a tribe, though the clans and the other sub-divisions are exogamous. The relations of the production are always homogenous. Whatever is the mode of production—hunting, gathering, herding or agriculture, there is no conspicuous separation of social categories. Differential position in the system of production do not create any difference in social position. It is unstratified society.

Tribal economy is basically a non-monetised economy where producers themselves are the consumers. Role of money does not come in existence. The technology is simple as well as unspecialized. Each tribe maintains its specific magic and rituals along with other social, economic and cultural tradition. Politically, they are under the control of the respective State Governments, but each of them exhibits its own tribal council (Panchayat) for local settlements. One of the outstanding features of Indian tribes is dormitory institution. In the absence of educational institutions for boys and girls, dormitories orient the adolescents in different customs related to birth, marriage and death, and also help them in learning of the distinctive moral codes.

Various authorities have described the Indian Tribals by different names. Sir Herbert Risley and Lacey, Mr Elwin and Sri A.V. Thakkar have called them as aboriginals. Sir Baines included them under the categories of 'Hill Tribes' and 'Forest Tribes'. Dr. Hutton called them as primitive Tribes and Dr. Ghurye as 'Backward Hindus'. Dr. R.K,Das and S.R, Das renamed them as 'submerged humanity'. However, the tribal groups are presumed to form the oldest ethnological sector of the Indian national population, the term 'Adivasi' (*Adi* means original and *vasi* means inhabitant) has been applied to them. The constitution of India designates them as 'Scheduled Tribe'.

Indian sub-continent has been the home of the multiple ethnic groups since prehistoric times. The Indian civilization was built on the synthesis of different traditions. There was a time when the whole country was tribal. Local tribal groups gradually converted into castes and ultimately gave rise to caste system. Today, whom we call tribals, are the isolated groups of people who have taken shelter in hardly accessible regions. A few of these groups got a chance to come in contact of caste groups and exhibited a process of absorption. When a tribe became assimilated into the caste society, its tribal identity was lost, but a new status in caste hierarchy (though of a very low level) was achieved. Such a transformation of a tribe into a caste is called Hinduization as caste and Hinduism go together. Ideals of Hinduism are so liberal that it easily accepts the development of a tribe into a caste. It is believed that most of the lower and exterior castes of today were formerly tribes who were gradually incorporated into the caste system.

According to Westermarck, before the coming of Aryans, the savage nation was sub-divided into an infinite number of tribes who had a cruel hatred towards each other. So they indulged no inter-marriages and lived in different territories separated by small arms of rivers or a group of hills. N.K.Dutta also believed this tribal basis of caste. He remarked that, before Aryan invasion there was no possibility of non-Aryan aboriginal groups like Anu, Yadu, Krivi, and Srinjaya to develop into caste. This non-Aryan population of India contained the Pre-Dravidian and Dravidian elements predominantly. Aryan conquerors that came with distinct Varna divisions engulfed all of the Indian aboriginal groups except those who fled away. In respect to the conversion of Tribals, Max Weber said that an Indian tribe gets converted into an Indian Caste when it loses its traditional meaning and significance. From the very ancient time, especially from the time when the Aryan-speaking people came to India, a slow and silent change from tribe to caste had started to take place. Many tribes adopted social customs and religious observances from their Hindu neighbours. Such a practice helped to reduce the differences between two different social systems. However the caste system constantly replenishes itself with new elements and, therefore, the identity of the tribes suffers. Even some communities are found to oscillate between 'caste' and 'tribe' nomenclatures.

The process of tribal absorption is going to be more and more complicated. Most of the tribal groups of the present day have been absorbed into the wider society of India in varying degrees. The process is prolonged as well as continuous. Prof. N.K. Bose in his essay 'Hindu Method of Tribal absorption' (1953), emphasized on absorption and assimilation as two important factors underlying the growth and expansion of Hindu Society. He said that the poor tribals easily came within the fold of more successful productive organization of Hindus. This is the reason for which we find a lot of similarities between tribe and caste in Indian situation. American anthropologists namely Robert Redfield, Milton Singer and Mckim Marriott also studied this interaction pattern with the models of'little' and 'great' tradition.

The tribal societies, as compared to the caste societies, are far underdeveloped and backward in economic, technological and cultural matters. None of the tribes in India is found to maintain a completely separate political boundary. Certain degree of political separateness has been retained in NEFA for its wider structure, but this sort of separateness is not visible in other places. Even the numerically dominant tribes of Chotonagpur, the Santals and Oraons are dispersed. In many cases, the boundaries of states cut across the Tribal divisions. So far as the linguistic boundary is concerned, the Bhils (one of the large tribal groups in India) are found to speak Hindi, In Central and South India, a good number of tribes speak Dravidian language that have close affinity to the languages spoken by the advanced communities of south India. Recently a trend has been very prominent that the Tribals are abandoning their own dialect. The whole of the Tribal India is passing through a critical phase of transition, though all tribes do not show the same degree of change. Dr. A.K. Danda in his book 'Ethnicity of India' (1991) has discussed the on going process of fusion and fission among the population in an elaborate way. He cited a number of cases where the same people are treated as tribes as well as non-tribes in different geographical position. For example, the Gonds are considered as the scheduled tribes in M.P. but not in Maharashtra, except a few selected districts. The Gonds of Chandrapur district (Maharashtra) have been branded as scheduled tribe while the same Gonds in Nagpur district do not belong to the scheduled group. Similarly in West Bengal, the Bhumijs of the District Purulia are though regarded as tribe but they are not tribe in the neighbouring districts like Bankura and Midnapore. Dr. Danda also mentions that, in a few areas where scheduling has been done on the basis of territory, some upper caste Hindus like Brahmins and Kshtriyas have come under the fold of scheduled tribe. The District Kinnore of Himachal Pradesh exhibits such a case. On the same principle, many Muslims in Lakshadweep, many Buddhists in Ladakh and many Christians in Nicobar Islanads bear the identity of scheduled Tribe.

Identity of Tribes is very important for the purpose of enumeration. Different identity of the same tribe, in different places creates serious problems. A number of tribes have been identified with multiple synonyms, which often confuses an enumerator. For instance, Abor and Adi are the same tribe in Arunachal Pradesh; in Mizoram, Mizo is the other name of Kuki; Karbis and Mikirs are the two different names of a single group of people in Assam. In the same way, one can not distinguish between Purum and Chote of Manipur, or Santal and Hor of West Bengal, Orissa and Bihar tract. Immigration sometimes effaces out the generic identity of a tribe. Illustration can be provided with the Bunas of Bengal. They are actually a composite population comprising of different tribes of Chotanagpur area like Santal, Oraon, Munda, Ho, Kharia etc., who came to settle in the plains of Bengal leaving their traditional abode in hilly terrains. Likewise, the Mizos constitute a number of tribes who represent different ethnic groups.

Being exposed towards different religious faiths, many Indian tribes are found to lose their own faith as well as the original system of rites and practices. They identify themselves with the general religious trend prevailing in the respective regions. For instance, some community in the Himalayas has been Buddhists as the region shows an extensive influence of Buddhism. Similarly in certain pockets of West Bengal, Bihar, U.P. and Assam, tribals have been converted to Islam. Again, as Islam prospered along the Indo-Gangetic plain and could not penetrate deep into the peninsular India, in South, we do not find any tribal group who has adopted the Islamic faith. Christianity was brought into India by sea-faring people; its influence is strongest among the tribes near the coast. The tribes namely Meena of Rajasthan, Chaudhury of Gujrat, Deshwali Majhi and Toto of West Bengal, Gond and Dhatra of M.P have moved considerably toward Hinduism in recent decades. The Mizo of Mizoram, Khasi of Meghalaya, Naga of Nagaland, Andamanese and Nicobarese of Andaman and Nicobar Islands, a section of Oraons in Bihar are prone towards Christianity. The Meo and Tadvi Bhil of Rajasthan prefer Islamism to a great extent. The Chakma of Tripura, Bhutia of West Bengal, Mon and Sherdukpen of Arunachal Pradesh and Ladakhi of Jammu and Kashmir are mostly Buddhists.

The identity of an ethnic group is usually valid in reference to a particular point of time and within a specific spatial dimension. Psychic unity and group solidarity are the two fundamental factors that build the social identity. But at present tribal consciousness often surpasses its own tradition and history; it is taking a strange shape in response to other forces. A sort of identity crisis arose in them since the time of the British rule in this country. The situation was further aggravated following the subsequent failures of Indian administrators who were entrusted with the task of ameliorating the ill-fated tribals. Naturally many individuals or groups had learnt to live with contradictions. They have gone through the processes like compartmentalization, assimilation, syncretism (a happy mix, but may not be so rational) and so on. They also had to withstand various domineering forces in the name of progress and development. Therefore, most of tribal societies of today show less continuity and more change. Some positive as well as negative elements have been expressed in the tribal consciousness. Communication channels (improved road, transport, radio, television etc.) and modernizing networks (writing system, literacy, explicit standard of norms etc.) have broadened the world-view of the tribals and they begin to search out commodities and comfort forgetting the old identity.

Classification of Indian Tribes

It is now well accepted that the Indian civilization is the synthesis of several of cultural traditions where tribals hold a definite place. Aside Africa, the largest concentration of tribal population in the world is found in India. To understand the situation we have to orient ourselves in historical as well as proto-historical context. Tribes of India live mostly in the forests, hills, plateaus and different other isolated regions of the country. They are the earliest inhabitants or indigenous populations of this country, who being unable to defend themselves have receded to the remotest corners. Prof. T.C. Das (1960) has called India 'a vast museum of diverse race and cultures'. In 1950, at the time of formulation of the constitution, the tribal groups comprised 5.6% of the total population (as per the enumeration of 1951 census), though were not evenly distributed in any region or state. According to 1991 census, 67,758,000 tribal heads are found to inhabit in different states and Union territories of India, barring a few states (Punjab, Haryana, Jammu and Kashmir) and Union Territories (Delhi, Chandigarh and Pondicherry). The state of Mizoram shows the highest concentration of tribal people (94.75%), whereas Goa shows the least (0.03%). Among the Union territories, Lakshadweep presents the highest number of tribal people (93.15%); lowest number of tribals has been recorded from Andaman and Nicobar Islands(9.54%). On the whole, theses scheduled tribes constitute 8.08% of the total population of the country.

	Total Population ('000)	Scheduled Population ('000)	Tribe (% of total population)
India	8,38,584	67,758	8.08
1. Andhra Pradesh	66,508	4,199	6.31
2. Arunachal	865	550	63.66
3. Assam	22,414	2,874	12.82
4. Bihar	86,374	6,617	7.66
5. Goa	1,170	—	0.03
6. Gujrat	41,310	6,162	14.92
7. Haryana	16,464	—	—
8. Himachal Pradesh	5,171	218	4.22
9. Karnataka	44,977	1,916	4.26
10. Kerala	29,099	321	1.10
11. Madhya Pradesh	66,181	15,399	23.27
12. Maharashtra	78,937	7,318	9.27
13. Manipur	1,837	632	34.41
14. Meghalaya	1,775	1,518	85.53
15. Mizoram	690	654	94.75
16. Nagaland	1,210	1,061	87.70
17. Orissa	31,660	7,032	22.21
18. Punjab	20,282	—	—
19. Rajasthan	44,006	5,475	12.44
20. Sikkim	406	91	22.36
21. Tamil Nadu	55,859	574	1.03
22. Tripura	2,757	853	30.95
23. Uttar Pradesh	1,39,112	288	0.21
24. West Bengal	68,078	3,809	5.60
Union Territories			
1. Andaman and Nicobar Islands	281	27	9.54
2. Chandigarh	642	—	—
3. Dadra and Nagar Haveli	138	109	78.99
4. Daman and Diu	102	12	11.54
5. Delhi	9,421	—	—
6. Lakshadweep	52	48	93.15
7. Pondicherry	808	—	—

[Source : Census "1991", Enemuration started from 9th Feb, 1991 and ended on 28th. Feb., 1991.]

Fig. 17.1. Total Population And Percentage of Scheduled Tribes in India (State wise)

Geographically, India formed a naturally protected region where numerous population infiltrated from outside, one after another, since the days of the Palaeolithic Period. The unique topographical conditions of India attracted the foreigners so keenly that they decided to settle down permanently in this soil. All the strains that entered into this country had been preserved. Indigenous populations were pushed southwards and eastwards into the geographical recesses where for a long time they remained unmolested. At present, all those original peoples of India have come in contact of civilized peoples to a greater or a lesser extent. Many of them have been divided into a large number of sub-tribes, all mutually exclusive, each having exogamous clans with their own customs.

It is a difficult task to categorize the tribals. Different scholars attempted to classify Indian tribes on the basis of different criteria. As a consequence we identify various groups of tribes, distinguished from each other. Classifications have been made on the basis of territorial distribution, racial affinity, linguistic affiliation and occupational specialization of the groups. Even a categorization can be possible as per the degree of culture-contact.

I. *Territorial Distribution*

Though Tribals are scattered all over India but they are chiefly centred into three zones—the NorthEast, the Central and the South.

(i) North-Eastern Zone

It covers the whole of sub-Himalayan region and the mountain valleys on the eastern frontiers of India which merge imperceptibly with those of Burma in the South-east i.e., it covers the areas namely the Assam, the central Khasi and the Garo hills. A large number of tribes agglomerate in Assam, Manipur, and Tripura. The tribal areas of Eastern Kashmir, East Punjab, Himachal Pradesh and Northern U.P. also of come under this zone.

Between Assam and Tibet, the tribes namely Aka, Dafla, Miri, Gurung etc., live. On the West of the Subansiri river lives the Aptanis. The tribes living in the Dehong valley are the Gallong, the Minyong, the Pasi, the Padam and the Pangi. The Mishmis live in the high ranges between the Debong and Lohit rivers. On the Western part, there are Chulikatas and the Belejiyas. The Eastern part includes Digaree and the Meju. Further, there are Khamtis, Singhops and Nagas. The Naga tribe consists of five major groups—the Rangpan and the Konyak live in the Northern part of the Naga hill; the Rengma, the Sema and the Angamis are found at the Western part of the same hill. Besides, Ao Lahota, Phom, Chang, Santam and the Yimstsunger inhabit in the central part of the hill. The Kachas and the Kabuis are found in the Southern side, while the Kalyo-Kengus live in the Eastern portion. The South of the Naga Hills which run through the States of Manipur, Tripura and the Chittagong hill tracts show the Kukis, the Lushais, the Lakhers, the Chins, the Khasis and the Garos. In the sub-Himalayan region, Sikkim and the northern part of Darieeling are dominated by the Lepchas. In U.P. again, a number of tribes, e.g., the Tharus, Bhoksa, Khasa, Korwa, Bijar, Bhuia, Majhi, Cheri, Raji and Kharwar are found.

(ii) Central Zone

This zone consists of plateaus and mountains between Narmada and Godavari. It is actually a mountainous belt that acts as a barrier to divide the North from the Peninsular India. However, the region extends upto the Santal Parganas in the East, Hyderabad in the South and Rajasthan—Gujrat in the West or North West, A large number of important tribes inhabit in this zone from very ancient times. Eastern Ghats and Orissa hills include Savara, Gadaba, Borido, Juang, Kharia, Khond, Bhumij and Bhuiya, The Mundas, Santals, Oraons, Hos and Birhors live in the plateau of Chotonagpur. The Katkaris, Kols and Bhils are found along the Vindhya ranges. Bhils also live in Aravalli hills. Gonds are the largest group that occupy the "Gondwanaland" and extend southwards into Hyderabad and the adjoining States of Kankar and Baster. The Tribes namely Korku, Agaria, Pardhan, Baiga etc. are found around Satpura and Maikal hills. In the hills of the Baster State, the two divisions of the Gond —the Marias and Muria are found.

(iii) South Zone

The zone comprises of the Southern most parts of the Western Ghats stretching from Wynaad to Cape Comorin i.e., the South of the river Krishna. In fact, Hyderabad, Mysore, Coorg, Travancore, Cochin, Andhra and Madras come within this zone. The oldest Tamil literature of the Sangam period furnishes the record of the most ancient inhabitants in this area. People being pushed by the intrusion of more advanced people, took shelter in this area; their successors still survive in the marginal areas.

The Chenchus occupy the area of Nallamalai hills across the river Krishna into the Hyderabad State. The Yeruvas and the Todas live in the lower slopes of Coorg hills along the Western Ghats from the Koraga to South Kanara. The Irulas, the Paniyans and the Kurumbas live along the ranges of Cochin and Travancore whereas, the Kadars, Kannikars,Malvadans and Malakurvans live in the adjacent forest areas.

(iv) Small and Isolated Zone

In addition to these three major geographical zones, mention should be made about another small and isolated zone which embraces Andaman and Nicobar Islands. The Islands are though separated from the mainland, but fall within the political boundaries of India. Some important tribes, such as Jarwa, Onge, Sentineles, Shompen, Andamenese, Nicobarese etc., live there.

II. *Racial Affinity*

Although the skeletal remains of early men are scanty in India, numerous finds of implements show that Indian region has been thickly populated from Paleolithic times. People of the prehistoric days were the ancestors of the existing tribals. But, as various other people entered time to time in this region, a variety of ethnic elements can be traced at present among the vast population of India. Four primary ethnic strains have been distinguished on the basis of physical characteristics—Proto-Australoid, Mongoloid, Negrito and Mediterranean (including Palae-Mediterranean).The first three strains are more conspicuous than the fourth one.

(i) The Proto-Australoid group

Most of the Tribal populations of India, especially of southern, central and partly Northern side show close physical affinity with the Australians. These tribal groups are, therefore, called Proto-Australoid. Earliest ancestors of this group could be traced to Palestine. Some scholars also like to call them as Pre-Dravidian. However, the group is characterized by medium stature, dark brown complexion, curly hair, everted lip, broad and flat nose (depressed at the root), and dolico-mesocephalic head. On the whole, the people exhibit strong muscular and well-built body. The type is represented by the tribes, namely the Gonds of Baster; the Bhils of Rajasthan; the Oraons, Mundas and Santals of Chotonagpur plateau ; the Chenchus, Malayans, Kurumbas and Yeruvas of South India; the Juangs of Orissa, and the Santal, Munda, Kol, Bhumij, Kharia, Ho etc., of Bihar.

(ii) The Mongoloid group

The Mongoloid people came to India from their homes in Northwestern and at present the tribes of Northeast frontier exhibit Mongoloid characters. Their stature in medium, skin colour is Yellowish brown and the face is flat with prominent check bones. The nose is also flat and the eye-slits are oblique with epicanthic fold. The head hair is straight and profuse, but the hair growth is minimal on body and face. Shape of the head is long. All tribes under Mongoloid group show muscular body with great development of calf-muscles. Females are no exception. They are not only healthy and hard-working people, most of them are great mountaineers who carry considerable amount of loads to high altitudes. This group is represented by Naga, Garo, Khasi, Lepcha, Bhotia, Chakma, Mugh, Dalfa, Abor, Mismi etc.

(iii) The Negrito Group

There has been a dispute regarding the existence of Negrito strain in Indian population. Because true Negroid people are found in the Andaman Islands, in New Guinea, in Philippines and also in Malay Peninsula (among the Semangs and the Sakais).0n the mainland of India, the Negrito strain is restricted among some forest tribes of Southern India. Physically they are of short to medium stature and show dark black skin colour, small head with bulbous forehead. The shape of the head is long, the face is short and protruding, the nose is flat and broad, and the lips are thick and everted. Texture of hair is fine but woolly in nature. General body structure is stout and well developed. The Kadars, Irulas, Pulayans, Panyans etc., belong to this group.

(*iv*) *The Mediterranean group*

This type was perhaps responsible for the development of Indus Valley civilization. The people were subsequently dispersed by the pressure of Aryan speaking Vedic invaders. At present this racial strain is mainly traceable in the non-tribal population of India. But the Todas of Nilgiri hills in South India provide a solitary example of this group among all tribal population of India. They show tall stature, fair complexion, long and narrow nose, thin lips, large eyes, wavy hair and plenty of facial and body hair.

III. *Linguistic Affiliation*

Language can be taken as a parameter in tracing the ethnic affinities of the people. During the period from 1905-1931, Dr. Grierson made a linguistic survey in India and found four major language families, such as, Indo-European (Indo-Aryan and Dardic sub-family), Dravidian, Austric and Sino-Tibetan (Tibeto-Burman sub family). The entire population speaking Austric language has been identified as Tribal. Besides, a large population speaking Tibeto-Burman is found among the Tribals. Dravidian speaking tribals are relatively low in number. However, linguistically the tribes of India can be divided into the following groups:

(*i*) *Austric Speaking Group*

Grierson classified the languages of austric family into two divisions—Mundari and Mon-Khemer. The majority of the Tribals inhabiting in central India namely Santal, Munda, Kol, Ho, Bhumij, Lodha, Kharia, Kora, Savara, Juang, Bhil and others speak in Mundari language. The other branch of this language family, the Mon-Khemer is spoken by the Khasis of Meghalaya and Nicobarese of Nicobar Islands.

(*ii*) *Dravidian speaking group*

The Dravidian linguistic group is popular in Central and Southern India. The principal languages of South India, such as. Tamil, Telegu, Malayalam and Kannada belong to this group. The Dravidian speaking Tribals are Gond (speak Gondi), Khond (speaks Khond), Oraon (speaks Kurukh) and Mal Paharia (speaks Malto). The other tribes under this group are Maler, Polia, Saora, Koya, Paniyan, Chenchu, Toda, Kota, Bagada, Irula, Kadar, Malser, Malakurwan etc.

(*iii*) *Tibeto-Burman speaking group*

This language family includes the Tribal languages of various people belonging to the Mongoloid racial element, Naturally it is spoken along the southern slopes of Himalayas, from northern Punjab to Bhutan and also in Northern and Eastern Bengal and in Assam. In fact, the whole northeastern belt of India comes under the fold of this language. The tribes namely Naga, Kuki, Abor, Dafla, Miri, Mikir, Lepcha, Bhotia, Bodo, Garo, Rabha, Toto, Mru, Riang etc., use this language.

There are also some unclassified languages, the speakers of whom are Andamanese, Onge, Jarwa of Andaman Islands. Besides, the tribals who have been greatly acculturated or assimilated with the advanced people, speak the Indo-Aryan language such as Bengali, Hindi, Oriya, Bihari, Assamese, Marathi, Rajasthani, Gujrati etc., according to the influence of the respective region of the neighbourhood. Tribals converted to Christianity are found to speak English quite often.

IV. *Occupational Specialization*

The economic life of Indian tribals is varied. They belong to different stages of economy ranging from food gathering and hunting to settled plough cultivation, through shifting hill cultivation. Though each tribe has its own distinctive type of economy but a tribe is usually found to practise many other occupations as subsidiary occupation. However, the different economies as found among the Indian tribals are as follows:

(*i*) *Food-gathering*

A considerable number of Indian tribes exclusively depend on forest; the entire style of life revolves round the forest. They collect fruits, edible roots, honey etc., from the forest and at the same time go on chasing the animals. In fact, all food-gathering communities practise hunting and fishing as per their opportunities. They are confined to different remote corners of India. For example, the Raji and Soka of Cis-Himalayan region; the Kukis and a section of Nagas in North-eastern Himalayas; The Jarwa, Onge and Sentinelese of Andaman Islands; the Birhor, hill Kharia, Korwa, Juang, hill Maria etc., of Central India. Some of the extremely backward communities of the world are located in Southern India. The tribal concentration of south India includes, the Kadars of Cochin, the Malapantarams of Travancore, the Puliyans of Madras, the Paniyan of Wynad, a section of Chenchus and Yanadis in Andhra, and Aranadans in Nilambar forest of Kerala. These are all food gathering tribes. Some of these tribes, especially those who live on the outskirts of villages, avail the nearby market for selling the forest produces like wood, wild fruits, honey, fodder, gum, wax etc., The Chenchus of Hyderabad, Yanadi of West Chennai, Kurumba of Tamilnadu, Katodi of Baroda etc belong to this group.

(*ii*) *Pastoralism*

There are some tribes in India who subsists on domestication of animals. These people depend on their herds of animals, directly or indirectly for food and other materials. They are found to have adjusted themselves in unusual ecological settings. For example, the Bhotias of Almora and the Todas of Nilgiri hill show the pastoral economy. In both the cases, the social and economic organizations have been built up around the buffaloes. Buffaloes and cows also play an important part in the ritual life of the said Tribals. Milk and milk products are not entirely consumed by the community; a part of it is exchanged to procure other necessities of daily life. Like the Todas, the Bhotiyas are not a purely pastoral people. They show a transition between agricultural and pastoral economy. In the Northwestern Himalayas, the Gujjars, Bakarwals, Gaddis and Jadhs are pastoral communities. They roam with their flocks of sheep, goats, and cattle in search of pastureland on high attitude. In western India, Bharwads, Raisipotras, Raberis are the pastoral people. The Gollas and Kurumbhadis show the same occupation in South India.

(*iii*) *Shifting Hill Cultivation*

In tribal India, shifting hill cultivation is widely prevalent though its name is different from region to region. It is considered as an incipient and ancient type of cultivation. The prehistorians believe that the Neolithic people first discovered this type of cultivation. Here the same plot of land is not cultivated for a long period. The cultivators move from place to place abandoning the old field. As a result, the very field is left unworked for years together. After a few years, the unwanted haphazardly grown shrubs and creepers are cleared. It is held that the cutting, drying and burning of the plants in the same plot of land enhance the fertility of the soil and, therefore, when new crops are cultivated, the production becomes better. Shifting hill cultivation is practised primarily in seven states of India—Arunachal Pradesh, Assam, Manipur, Mizoram, Meghalaya, Nagaland and Tripura. It is partly practised in Bihar, Orissa, Andhra Pradesh, Madhya Pradesh and Kerala. The tribes namely, the Maler, the hill Kharia, the Korwa, the Parahiya, the Savara, the Kuttia Khond, the Baiga, the Maria Gond practise it extensively. There are some communities in M.P., Maharashtra, Karnataka and Sikkim who practise it only marginally. The popular names are Jhum in N.E. Himalayan region, Daya in Northern India, Pady and Dahi in Orissa, Penda and Bewar in M.P, Kurwa and Khallu in Bihar, and Konda paddy in Andhra. The South Indians call it Poduar. In the state of Bastar, its name is Deppa. In Himalayas, it is termed as Khil Kumari in Western Ghats and Walra in Southeast Rajasthan. Still there are other names. The method is most popular among the Mongolian Tribes living in northwestern border of India. As it is a primitive and crude method, no surplus product is obtained.

(iv) Agriculture

Bulk of the tribal people in India has been accustomed with the permanent settle cultivation, although they supplement their economy with the activities like hunting, fishing and gathering. As a matter of fact, at present agriculture has achieved a central place in the economic activity of the tribal people. Most of the tribal groups practise wet cultivation by transplanting method. They know the benefits of irrigation and compost manure, but go on using their traditional primitive skill and primitive implements. The socio-economic and religious life has been complex but well developed. The Santal, Oraon, Munda, Bhumij, Ho, Gond, Baiga, Kharwar, Majhwar and some of the Assam tribes are expert cultivators.

(v) Manual Labour

A substantial portion of the tribals has become landless labourer, who works on others' land for wages. They are paid either in cash or kind, or in terms of share. They possess no right (lease or contract) on land over which they work. On the other hand, they are free from all risks and liabilities in cultivation. These people seasonally migrate to the neighbouring areas for other manual jobs. In the industrial belts their labour is highly appreciated. Thousands of tribal people e.g., Santal, Oraon, Bhumij, Ho etc., have found to be migrated to the industrial towns like Durgapur, Bokaro, Jamshedpur, Rourkella, Bhillai and so on. These tribals are living there permanently selling their manual labour. Besides, many Santals, Khonds and Gonds have gone to Assam for taking up different jobs in the tea plantaions. In Central India the entire tribal belt is rich in natural ores. After the coming of British, those areas have been explored extensively to set up the industries like coal, iron, steel etc. Many tribal groups of Bengal, Bihar, Orissa and M.P. are found to work in those mines and industries. Tribal labour has also been employed in construction of roads, in collection of forest produces (fruit, bark, dye, leaves for bidi making, lac, gum, resin, wax and fodder) and in lumbering.

Most of the tribals who were formerly the landowners have turned to manual labourers owing to the land alienation and continuous exploitation. The total number of agricultural labourers has grown over the years. Some of them have become bonded labourers as like Gond of Rohtas and Korwas of Palamau. Manual labour is a new economic role, which gave the tribals an opportunity to migrate temporarily or permanently in the fields of Punjab; West Bengal, Assam and other areas.

(vi) Handicrafts

Some of the tribals still cling to their traditional craft heritage. They practise craft as a subsidiary occupation. These crafts include mainly basket making (bamboo), tool making (iron and wood), spinning and weaving (cotton and wool). In Assam, cloth is manufactured from cotton threads after dying them beautifully with indigenous vegetables. The Nagas and the Khasis also produce colourful handloom products. Bhotias show a special aptitude in spinning and weaving of wool. The Korwas of U.P., the Agarias of M.P., the Asurs are the well-known iron smelters. Although their techniques are crude but their implements and weapons are quite satisfactory for local use. The Doms and Mahalis of W.B. manufacture different baskets with thin bamboo strips. The Birhors have the monopoly of producing straw-ropes and leaf-cups. The Khonds, Gonds and Savara are at a time engaged in metal working, weaving, cave working and pottery. People living at the North of Brahmaputra, especially Mombas and Sherdukpen make fine bowls, varnished and ornamented, with delicate silverwork. Irulas make bamboo mats and baskets along with ploughshare and wheel. The Tharus of U.P. make household utensils, baskets, musical instruments, weapons, rope and mat. In M.P., the Maria Gonds are mainly occupied in distilling spirits from the products of the forest. The Ghasis prepare gut from the fibrous tissue of the animals. Gujjars of Kashmir and Kinnauris of Himachal Pradesh make sophisticated wooden products. Besides, there are a good number of tribal communities who engage themselves in manufacturing of the items like earthenware, paper pulp, masks, woolen garments etc. In most of the cases, the profit remains marginal but the tribals show a great skill and creativity.

On the whole, Indian Tribal communities that have been grouped into three geographical zones, correspond to the racial as well as linguistic classification. Accordingly, the northeastern group, the central group and the southern group have been identified. This three-fold classification is further strengthened by economic grading. Tribals of Northeastern India are settled agriculturists living mainly on terrace cultivation. Shifting cultivation is prevalent form of economic pursuit in central zone. But the tribals of southern zone are prone towards the food -gathering activities.

The Classification based on Culture Contact

V. Elwin in his writing 'The Aboriginals' (1943) suggested a four-fold classification for the Indian tribes based on the level of their cultural development. He wanted to differentiate the great majority of tribals who had been assimilated (to a greater or lesser degree) into the culture of their neighborhood against the minority group of tribals who are scattered in remote and inaccessible hills and forests. However, the classification is as follows:

(*i*) Tribal communities belonging to class I are still confined to the original forest habitat and follow the old pattern of life. Elwin considered this class of primitives as purest of the pure. The members lead largely a communal life; economically they share one another. They are very simple, innocent and honest. Crime is rare among them. These people are generally shy of strangers. Geographical conditions usually protect these shy tribes from the contact of other people. Hill Marias of Bastar State, Juangs of Keonjhar and Pal-Labara, the Gadabas and Bondos of Orissa, the Baigas and Korwas of Pandaria and Kawaedha, Aranadans of Nilambar forest in Kerala, a section of Chenchus in Andhra belong to this class.

(*ii*) Class II group of tribals are those who have experienced contact with the plains and consequently have undergone changes. This class of primitives are found to live in remote areas, and remain attached to their ancient tradition. But the difference is that, an air of change has percolated in them. Though their village life is still communialistic, individualism has been generated among the people. Axe-cultivation has been ceased and they have proceeded towards agriculture and allied occupations. These people are prone to outside life and, therefore, less simple and less honest than the former group. The Bison-horn Maria, the Binjhwar, the Baiga etc. are included in this class.

(*iii*) Class III comprises of acculturated tribal communities who have migrated to urban or semi-urban areas and got themselves involved with modern industries and vocations. This group of tribals is found to adopt a lot of modern cultural traits as they move out of their native society to join larger plural societies. As a result, they have lost their stronghold on the original culture, religion and social organization. This class of tribal people, therefore, stand in a peculiar state of transition. Some of them have become a part of the lower rung of Hindu society, some others have been described as Christians. In fact, these people being largely affected by external contact suffer from ill feelings. But they have no way out. A large number of Santal, Oraon, Ho and a few other tribals belong to this group.

(*iv*) Tribals in Class-IV are a minority group. They represent the old aristocracy of the country. According to Elwin, tribals of this class have won in the battle of culture contact. Because, inspite of adopting the Hindu faith and acquiring the most advanced style of living, these people have been able to retain their old tribal names and clan identity. They are also found to follow the respective totem rules and observe many elements of tribal religion. The class includes some Gond kings, Naga and Bhil chieftains, Bhuiya landlords, Korku noblemen, wealthy Santal and Oraon leaders, and even highly cultured Mundas. C.B. Mamoria (1957) considered this class as totally assimilated group of tribals within the Indian population.

The members of the first two of the four classes are considered as the real tribes. Because the tribal organizations of those groups are unimpaired, their religions remain alive, the artistic and choreographic traditions are unbroken, and the unaltered mythologies still vitalizes their social life. In order to solve the whole aboriginal problem, Elwin's aim was to raise the people of 1st and IInd classes upto the level of IVth class without letting them to suffer from despair and degradation as like the members of the IIIrd class.

Prof. N.K.Bose used two related criteria—the level of technological development and the degree of geographical and social isolation for classifying the tribes of India.

Tribal Transformation and Its Consequences

The process of transformation starts with small-scale simple adoption. It involves the acquisition of new technical skills, acceptance of new tools and implements, and inclusion of new customs, rites and rituals. A tribe in contact with civilized people acquires a variety of cultural traits. For instance, some tribes in Bihar are found to employ Hindu priests to perform their indigenous rites and ceremonies. Some tribes, especially Mundas, Rajbanshis and Bhils have shown a larger adaptation of Hindu culture traits. As for example, many tribal families of Bengal have chosen typical Hindu names for their children. They also revere tulsi-plant,* abstain from beef, and cremate the dead bodies. Their wives use vermilion marks and iron bangles in a regular way. Similarly, the members of a primitive hill tribe who have moved down to the plains, show a greater tendency of assimilation in contrast to their brothers who are left behind.

The articulation between the tribal economy and the regional or national economy has taken place. A large section of tribal population is found to enter into the greater productive system of the country. This incident, in turn, breaks the homogeneous nature of tribal economy. For the differential acquisition of wealth, the society gets stratified. The changes in socio-economic front gradually transform the entire tribal way of life. The main factor behind this transformation is, therefore, the culture-contact. Not only it has undermined the social solidarity of the tribal groups, it shows far-reaching consequences. It invades the tribal security and introduces various types of discomforts, diseases and vices. However, the channels of culture-contact and the effects of transformation have been listed below:

(*a*) Easy and increased means of communication have opened the avenues of conflict. Such a conflict does not mean the struggle with arms, rather exploitation on culture and material objects. Many tribal areas have been explored to find out mineral resources, forest produces and land for intensive cultivation. Inspite of best intentions, a lot of injustices have been made towards the tribes. Tribals coming in contact with non-tribals try to imitate alien life-style, worldview and ethos.

(*b*) Urban contacts disorganize the primitive social life. The people have not only been exposed to the economic and social-cultural forces of Hindu society, they are also subjected to missionary influence. As a consequence, old myths are forgotten, old Gods are neglected, socially significant simple ceremonies have become complicated. For example, a tribal marriage which was merely an association of local belief and practices has embraced the Sanskritic rituals of Hindus. Many traditional dances have been abandoned due to village politics, rivalry and other social disputes. The dormitories that fostered education as well as recreation in traditional social set up is on a decline, producing tension and sadness in tribal life. High incidence of poor physique, inferiority complex and bitter antagonism have crept up among the tribal people. Interest of the tribal youths has been shifted to other directions causing a total cultural breakdown. Change has also been manifested in the food-habit and dress.

(*c*) Economic and political policies of the British dragged the peace-loving tribals into the orbit of the colonial capitalist profit making system in India. The tribals being uprooted from their traditional habitat and mode of production have reacted in three ways:

(*i*) A section of them have failed to maintain the original tribal structure and social moorings. They have been totally assimilated to the advanced culture of neighbourhood.

(*ii*) Another section have withdrawn themselves from the contact of other groups and transferred to inaccessible regions as a measure of defense. Due to isolation, depopulation (dwindlement of population) has started among some of these tribes e.g., Shompen, Onge, Sauria Paharia etc.

---

* A sacred plant of Hindus.

(*iii*) The third section stands in the mid-way between traditionalism and modernity. Their status has been reduced to a great extent and they are found to serve as bonded labour under the money-lenders, businessmen, contractors and other government officials (i.e., the classes those emerged in Indian society as an outcome of economic and political policies persuaded by the British). Tribals belonging to this section generally work in mines and plantations, on railway and road constructions, and so on. But a few of them, after losing their land and occupation, found no alternative subsistence and so strived to survive by antisocial means. Officially these tribes have been branded as 'Criminal Tribe' (1871). The situation was actually a resultant of acute economic exploitation. Even the nomadic tribes who traditionally secured their livelihood by serving* (periodic) the settled communities, met serious difficulties in continuing their customary life. They took various unfair means to meet up the subsistence and thus went under the fold of 'Criminal tribe' as chalked by the administration. Besides, there were a large number of hill tribes who used to live on hunting and collection of forest products, supplemented by shifting cultivation. When strict rules regarding the denudation of forest were enacted, these tribes found compulsion to come down to the plains but most of them had not succeeded in adapting themselves in settled cultivation, suitable to plain land. The reason was partly the tribal inertia as well as the shyness, and partly the apathy of administration. Dr. Hutton (1946) had pointed out the extra-physical need of mankind in this respect. Considering magic and religion as important tools of adjustment in primitive social set up, Hutton argued that, perhaps the ignorance of appropriate magico-religious ceremony related to the plain-land farming , debarred the hill inhabited tribals to undertake settled agriculture.

(*d*) Religion and magic are the expressions of human experience in a local setting. Religion remains associated with social norms and responsible for their maintenance whereas magic aims at solving problems arising out of various local needs. Primitive religion is a mode of social control and primitive magic is a method of combating the evil forces. Although the tribals in contact with non-tribals have been tolerant towards other religion, many of them under the strong influence of religious reformers (particularly the missionaries) have dared to change own faith. After a prolonged period, a sort of duality appeared in them. Since the religious sanctions behind traditional cultural practices vanishes away, a great turmoil is found in the value system and ultimately the entire social structure is collapsed.

(*e*) A change of dialect has been noted as tribal transformation. Different linguistic traditions have been tampered to a large extent by the culture contact. For passing a long period in close contact of alien people, the original dialect is affected and the individuals acquire proficiency in the speech of neighbouring community. Naturally, the cultural values and norms that is essentially communicated through linguistic symbols gradually get eliminated. However, the impermeable linguistic boundary does not last long. Most of the tribal communities are bilingual at present. Some of them, for instance Bhumij, Lodha, Kora, Mahali and others, in some areas have given up their own traditional language totally and started to speak Bengali with a different accent. Similarly, the Bhils are using Hindi for many years.

(*f*) Another evil effect of contact is the spread of diseases. The improved communication that provided the facilities of trade has also brought new diseases with which the tribals were not acquainted at all. Emigration of labour from tribal areas to plantations and factories, and also the immigration of outsiders into tribal areas introduce the diseases like tuberculosis and veneral diseases. Some of these diseases have been spread like epidemics. The lure of free life unhampered by social control pushes both men and women towards corruption. Indiscriminate mixing of the sexes, drunkenness and immorality have increased a lot.

For the quick spread of some diseases, the missionaries and the philanthropic agencies may be blamed because they collect second hand apparels from the dead and diseased to dole them out

* These tribes used to serve the communities like Marwari, Lakhotia etc., by supplying agricultural implements and utensils, or by providing amusements through acrobatic performances, dances, etc.

among the poors. This is the principal source of getting infection. Again, replacement of their original diet often create the general diseases such as respiratory diseases, dysentery, scabies, ringworm etc.

(*g*) Excessive pressure of exploitation, competitive profit oriented forces of the society have made a few tribal groups desperate, violent and militant. Modernity and other secularity not only haunt their ethnic identity, industrial economy and urban political network have sucked up all their vitality. The accumulated grievance often force them to revolt against administration.

Transformation of tribal life in India started especially after the establishment of colonial rule in India. Although culture contact is the chief factor, but not the sole one. After independence, a good number of welfare and development schemes have been undertaken for the benefit of tribals which definitely play an important role in accelerating the changes. However, the rate of transformation has been different for different tribal groups in various parts of the country.

TRIBAL PROBLEMS AND WELFARE MEASURES

The history of India reveals that the tribal people have always co-existed with the general mass, although circumstances often drove them in the quiet corners of the country. Being confined, the communities achieved a degree of equilibrium with their ecology and reduced to a state of utter ignorance and penury. Rarely they could come out of their habitats and each of them developed a specific way of life with a set of typical thought and behaviour pattern. But the people were happy enough in their own world.

Tribal Administration in India

Since the tribals are the part of Indian society, tribals and non-tribals lived side by side for centuries. Once they were absolutely cut off from the mainstream; later became losers and sufferers by the exploitation of the ruling class. As a consequence, growth of these people was retarded. They fell back in all walks of life. The backwardness of the tribals was, therefore, neither cultural nor social at the root, it arose from isolation. Later, exploitation aggravated the problem. However, the index of backwardness is the underdevelopment which welcomed different problems like poverty, disease, displacement, indebtedness, criminalization and so on. Such problems of tribals were not unknown to the ancient kings of India; the problems varied in size and shape in different phases of history. In general, tribals have always been overburdened with misery and distress. To delineate the accurate picture of tribal problems and their administration, we have to be retrospective in view. Three phases may be chalked out through ages.

I. Pre-British Period

This period records a number of tribals who preferred to live in isolated conditions in order to guard themselves against the constant attacks of the hordes of foreigners who were superior to them both in physical and mental powers. Being cut off from the progressive world, they became backward, downtrodden and poverty-ridden people. The kings and rulers of ancient India were aware of the problems of the tribals but they could not do much because the people were posited far away, out of their reach. In the Mauryan period, the great king Ashoka's inscriptions suggest that he knew the miserable conditions of the tribals. He felt that the welfare of those people should come within the purview of his duties and responsibilities. But very few of the tribals actually got the fruits of his welfare work. A large number of epigraphical and literary evidences stand as witness. In true sense, effective principles of administration were first made available for the tribals during the British period.

II. British Period

When the British rulers established their colonial rule in India, they did not count the tribal population of India who were pushed back into the inhospitable regions of our country by the pressure of invading groups possessing superior power, skill and technology. They were concerned with the rest of the Indian population for the consolidation of their new empire. But within a short period, particularly in the latter half of nineteenth century, they had to face the turbulent hill men, Mal

Pahariyas of Rajmahal hills in Bengal. Those people revolted against the Hindu zamindars in 1772.

Since most of the tribal areas were under or nearer the territories of local feudal or indigenous rulers, the landlords and zamindars in the periphery of tribal habitation exploited the tribals at random. The corrupt officials often conspired with the landlords to force the tribals to vacate their land and this reduced them to the status of serfs. A number of middle class businessmen like traders and moneylenders also showed the habit of taking chances to extract the poor people. As a result, tribals were thrown into inhuman conditions. A great discontent and discomfort was created among them. Innocent tribals became so desperate that they launched violent and militant struggle in some areas.

Lord Cornwallis came to India as governor-general in 1786 succeeding Warren Hastings. In 1793, he established a land tenure system for India with the main objective of creating a class of people with vested interests whose loyalty and support were indispensable for consolidating and standing British domination in India. The new land tenure declared the zamindars as the proprietors of land and this right of ownership became subjected to regular payment of rent-revenue to the Government. In ancient land system, zamindars were nothing but officials under the king. The kings, so long, were entitled to get a share of peasants' produce as a protector of the people, not as the owner of land. But under *zamindary system* as introduced from 1793 zamindars acquired the power to make the land as private property as well as commodity in market. The system though recognized the rights of the cultivators *(Raiyat)* but that was not clearly defined. Naturally the cultivators began to suffer at the hands of zamindars. Upto the middle of nineteenth century (1859), zamindars exercised an authority over the Raiyats and their oppression and exploitation was beyond imagination. Like the peasants, tribals also suffered a lot. In fact, the new rule divided land into smaller fractions and different feudal lords got the possession of land. The traditional economy of the tribals was, therefore, challenged; their old rights on land were lost.

With sympathy for the agrarian discontent, in 1859, British enacted "The Rent of 1859'—the first tenant law. But as this act failed in mitigating the sufferings of poor peasants, the accumulated grievances of the peasants led them toward a movement, which was found to be generated in 1930's. The tribal movements were the counterpart of that peasant movement. But, the agrarian reform was very difficult as the legislative assemblies were dominated by the land-propertied class.

The other great reason behind the tribal agitation was the forest policy of British. Forest always played a vital role in the life of the tribals. A large number of tribal communities still depends on shifting cultivation as their principal means of livelihood. Not only that, forest provides them firewood, different kinds of timber-wood for various purposes. They also collect edible roots, fruits, tubers, leaves etc., to supplement their daily food. Further, forest is a good pastureland where the tribals let their cattle to graze; hay and fodder are obtained from the forest. Plenty of honey is available from the forest, which are generally not consumed by the tribal people. They sell it at the market or exchange it for getting other necessary articles. Wild games of the forest are also hunted to satisfy hunger and some other requirements. Forest supplies them the house-building materials. In biting winter tribals make themselves warm by burning twigs and sticks. In summer, they take rest under the shade of the trees. Different trees, leaves, flowers and forest products are intimately related not only with their economic life, but with the socio-religious life too.

In this circumstance, to resist the environmental deterioration resulting from uncontrolled exploitation of nature, a German expert named Dr. Voelcker was invited. He examined the Indian forest* as well as the agricultural conditions of India thoroughly and suggested for a reform in order to safeguard the biosphere and economic growth. As per his suggestion, the British Government

* Forest keeps a significant impact in maintaining the quality of environment. It conserves the biosphere by producing a stable eco-system. It acts as a storehouse of plants, animals and microorganisms. Human activities and to a much lesser extent natural factors are responsible for causing degradation and deterioration of the forest which endangers the existence of wild life, leading to the extinction of some particular species. Ultimately, a great devastation in environment is possible.

at once designed a policy on forest reservation which came to be known as 'Forest Policy of 1894'. This forest policy affected the tribal way of life very severely. The tribals, who formerly regarded themselves as the lords of the forest, went under the control of forest department sanctioned by the Government. Their traditional right on forest was no longer recognized. Moreover, their free movement inside the forest areas was greatly restricted. Being displaced from the forest homes, some tribal groups met such an embarrassment that they became totally distracted from their social moorings. For example, Lodhas underwent tremendous psychological upset. They did not cooperate with Government and exhibited criminal tendencies. British administrators labelled all tribes of this kind as 'criminal tribes'.* Other tribal groups followed the path of rebellion as their rights on land and other means of livelihood were encroached.

The whole of the nineteenth century witnessed several uprising of tribal people. To handle a series of serious situations, the administrators were compelled to adopt different measures—executive and legislative. At first, British tried to tackle the rebellious tribes with arms; these tribals were considered as dangers before law and order. But soon after they chalked out a definite policy. The administrators bribed the tribal leaders by giving monetary allowance and some retired military officials were encouraged to settle down around the tribal habitation. Military officers seemed to be more effective in suppressing the unrest than the civilian officers. In 1782, Augustus Cleveland, the administrator of Rajmahal hills area suggested to withdraw the particular area from the normal administration. Consequently, local court comprising of local leaders got the power to supervise the civil and penal jurisdiction over the hill tract. The Mal Pahariya was the first tribal group who got released from the clutch of zamindars and allowed to enjoy rent-free land directly from the Government.

With the reference of Mal Pahariyas, under the guidance of Cleveland and his followers, a hill assembly was formed for conducting the affairs of the tribe and it was supposed to maintain its own procedure. Basing on this model, in 1796, a universal regulation for the tribes known as 'Regulation I' was granted by the Government. But as inefficiency and corruption crept in, Regulation I could not be continued. Ultimately it came to an end in 1827 and a new regulation was accepted, whereby the Mal Pahariyas and some other adjacent tribes were brought partially under the jurisdiction of ordinary courts. This new pattern continued from 1827 to 1855. Following the Santal revolt in 1855, a special administration for the affected areas called 'Indian Councils Act' of 1861 come in vogue. Thereafter in 1870, the Parliament gave the power to Governor-general in Council to legalize the regulations. Thus, from 1782 to 1870, different provisions were sanctioned by the Government for suppressing the tribes by the use of physical force. The objective was to keep the tribals apart from the day to day administration; all sorts of contact, communication and participation were denied for them.

In the next phase, the colonial rulers felt the necessity of a special type of administration. Inspite of suspicion and cautious treatment, some special provisions were made for the tribal communities. Accordingly, in 1874, a British Act (XIV) was formulated known as 'The Scheduled District Act' where the tribal areas were specified as 'Scheduled Tracts'. The objective was to exclude the Specific areas from the premises of normal operation where the ordinary laws were expected to be carried out. This Segregation of the tribes into some special areas of reservation was necessary for saving the life and interest of the aboriginal people from critical exploitation.

In 1919, again, another act known as 'the Government of India Act' under section 52(A)/2 was passed which empowered the *Governor-general in Council* to declare any territory as a backward

* In 1871, *one* hundred and fifty-three (153) small communities were branded as criminal tribes who under different restrictions of British became prone towards criminal activities. After independence, the 'Criminal Tribe Act' was repealed in 1952. Since then these communities got rid of the stigma of criminality and treated as de-notified communities.

tract, after obtaining the 'sanction of the Secretary of the State'. The tracts were divided into two categories—wholly excluded areas and partially excluded areas. The tribal areas in different States which were kept totally excluded from the purview of the legislature were called 'wholly excluded area'. But there were some areas of modified exclusion. Those areas were termed as 'partially excluded areas'. For the areas of first category, the power of legislation was solely laid in the hand of the Governor. Neither the central nor the Provincial legislatures was given any power to make the laws applicable. Extensive tribal tracts in the Provinces of Bihar, Bengal, Orissa, U.P., M.P., Mumbai (Bombay) and Chennai (Madras) were declared as 'Partially Excluded Areas', whereas the frontier or border regions in Assam, Laccadive and Minicoy Islands within the Province of Madras, Chittagong hill tracts of Bengal, Lahul in Himachal Pradesh, Spiti in Punjab, Angul in Orissa, etc., were declared as 'Wholly Excluded Areas'. In the mean time, the Census of 1931 also presented an elaborate list of backward groups, labeled as 'tribes'.

On the basis of the recommendation of Simon Commission (1929), in 1935, the other Act of Government of India was passed on. It made more stringent provisions for the special treatment of the tribal areas. The totally excluded areas were placed under the Provincial rule of the Governor to be acted on his discretion and the partially excluded areas went within the fold of Ministerial responsibility. Thus, exclusion was the main policy of British in pre-independent India which served the purpose of keeping the tribals isolated i.e. away from the mainstream of national life. British rulers did not want to face any trouble caused by the tribals.

Inspite of enforced isolation, the administration encouraged the Christian missionaries to move into the tribal areas for their humanitarian work. Two reasons can be furnished against this favour. Firstly, the missionaries wanted to bring a socio-economic transformation by introducing various schemes of health, education and other welfare measures. The influence and assistance of the Missionaries were expected to make the beneficiary groups of tribals faithful and loyal. Secondly, as the activities of missionaries included both conversion and welfare, tribals could be taken out from the path of violence. Moreover, it was also essential to keep the tribals apart from the freedom movement which was radically gaining the ground at that time. The British Government was more interested in collecting the revenues and maintaining the law and order, than to rehabilitate the segregated groups.

III. Post British Period

The administration of the tribal people acquired a new significance after the attainment of independence. The Republic of India realized that the tribal communities had passed through very hard days and , therefore, overburdened with numerous problems. It underestimated the humanitarian approach in tribal welfare. So, for the reconstruction of country as well as to satisfy the members of the community, a new policy was devised with major changes. The principle of charity as recognized by the missionaries was substituted by the spirit of development. The constitution order of 1950 under article 342 defined the scheduled tribes and declared 212 tribes located in fourteen States as 'Scheduled Tribes'. The directive principle of the State policy on tribal welfare, laid down in Article 46 (Part IV) of the constitution declared that, "The State shall promote with special care the educational and economic interests of the weaker sections of the people, and, in particular of the Scheduled Castes and the Scheduled Tribes, and shall protect them from social injustice and all forms of exploitation". Besides, there were many other articles, which pleaded for the general and special welfare of these downtrodden and depressed communities. On the whole, the national policy tried to integrate and level up the outlying underdeveloped tribal groups in a common national social order. The policy of tribals was most explicitly outlined by Jawaharlal Nehru, the first Prime Minister of independent India. He was deeply touched by the problems of the poverty-stricken, ignorant people, those who had undergone a long phase of tremendous exploitation. He found two alternate policies—the open-door policy as taken up by the social workers in which tribals were exposed to all kinds of influences. The other was the closed-door policy as taken up by the colonial government to protect the distinctive and colourful life of the tribals in isolated pockets. Nehru could not absolutely support any of these two policies. Rather he wanted a combination so that the tribals would be integrated into the larger society without losing their distinctive identity. In fact, he emphasized five principles (1957) for the tribal development. Those principles, known as 'Panch Shila', have been furnished as here under.

(*i*) People should develop along the lines of their own genius and we should avoid imposing anything on them. We should also try to encourage in every way their own traditional art and culture.

(*ii*) Tribal rights in lands and forests should be respected.

(*iii*) We should train and build up a team of their own people to do the work of administration and development. Some technical personnel from outside will, no doubt, be needed, especially in the beginning. But we should avoid introducing too many outsiders into tribal territory.

(*iv*) We should not over-administer these areas or overwhelm them with the multiplicity of schemes. We should rather work through and not in rivalry to their own social and cultural institutions.

(*v*) We should judge results not by statistics or the amount of money spent, but by quality of human character that was evolved.

Article 244 (Part VI) empowered the President with the right to declare any area as a scheduled area where a substantial tribal population was found. Provisions for the administration of the scheduled areas and the scheduled tribes were delineated in the fifth schedule, other than those in the State of Assam. Administrative provisions for the Assam tribes were incorporated in the sixth schedule. However, the fifth schedule aimed at safeguarding the interests of the tribes. It tried to resist against exploitations, to protect the traditional way of life and finally to develop the scheduled areas and the scheduled tribes. As a result, some areas in the State like Andhra Pradesh, Bihar, M.P., Maharashtra, Gujrat, Orissa, Punjab and Rajasthan were pointed as scheduled areas. The Census of 1951 enumerated the scheduled tribes of India as 19.1 million i.e., 5.6% of the total population of the country. An office of the commissioner for the scheduled tribes was created to look after he specific needs of the tribes. The Commissioner in early fifties sought the views from different Provinces and States to suggest the prominent characteristics to distinguish the tribal communities. Anthropologically it was important and interesting that the varying criteria like physical characteristics, linguistic affiliation, culture contact, occupational type, territorial distribution were pointed out in identifying the tribal population.

Tribal development programme was taken up as early as 1950's. It can be compared with a school system. The different classes of a school are like the different tribal groups in a country. As the courses of study vary with classes, the problems of the tribals also vary from group to group. The different teachers as found in charge of different classes, the personnel engaged in tribal administration differ with different groups of tribals. Finally, a tribal administration shows its budget and planning at the background as like budget and planning in school administration. The nation visualized a policy of progressive assimilation of the tribal people in the mainstream. Constitution provided special safeguards for the tribal communities for a period of 10 years. Accordingly, a Tribal Welfare Department was instituted in 1951 for the protection and advancement of the scheduled tribes.

In the beginning of First Five-year Plan (1951 - 1955), Rs. 39 crores were allotted. For an overall development of social, economic and cultural life of the tribal communities, the welfare activities were divided into four heads—economic, educational, health and housing. But programmes were not found fully satisfactory as the developmental activities served only the needs of the vocal individuals in the communities. Therefore, in Second Five-year Plan (1956 - 1960), a more pervasive approach was planned which could be implemented through Special Multipurpose Tribal Blocks. Increasing facilities were provided for the tribals throughout the country. Thus ten years passed away. The problems of the more backward tribal communities received the attention of various commissions and study teams. The Scheduled Areas and Scheduled Tribes Commission* (1961), usually known as 'Dhebar Commission' observed that the harmony of tribal life is very important. It recommended for not to impose anything upon the tribals. It also stated that any development programme should aim for the integration of tribals, as they are the members and part of the Indian family. A study team appointed by the Planning Commission headed by P.Shilu Ao, the once Chief Minister of Nagaland probed on the Tribal Development Programme and opined, "the aim of Tribal Welfare Policy should be defined as the progressive advancement, social and economic, of the tribals with a view to affecting their integration with the rest of the community on a footing of equality within a reasonable distance of time. The period has necessarily to vary from tribe to tribe and while it may be five or ten years in case of certain tribes, more particularly the tribes that have already come into contact with the general population by living in the plains. It may be two decades or more in the case of tribals who are still in the primitive food-gathering stage".

The Dhebar Commission divided the Scheduled Tribes into four distinct layers. At the base, they identified a class of tribals who belonged to an extremely under-developed (primitive) stage and needed special consideration. The next layer was formed with those tribals who were in a shifting-cultivation stage. The third layer (counted from the bottom) was comprised of those tribals who had taken to regular agriculture. Finally, the top layer included the most advanced group of tribals those had already been assimilated in the wider society. The commission stressed on the differences of physical environment and advised for special treatment. The shilu Ao team also worked out three principal groups of tribes in India. The tribes living in the remotest corners of the country were found at one of its ends that totally lack the contact of advanced communities. Politically conscious and economically developed communities of the Northeast were found at the other end. In between the two came the tribal communities of Central India who showed varied stages of development. Thus, both the classifications revealed the fact that there was a wide divergence in socio-economic, cultural and technological milieu among the 559 scheduled tribe communities of the country; the extremely backward communities deserved special input.

The Dhebar Commission was in favour of an integrated approach. It strongly advocated that an economic development for the bulk of tribal people can not be achieved until and unless the resources of land, forest, cattle wealth, cottage and village industries are executed on an integrated basis. Accordingly, the Third Five-year Plan (1961-1965) was chalked with a broad strategy of socio-economic development. After this came three annual plans from 1966 to 1968. But all went in vain. The Fourth Five-year Plan (1969-1974) specifically enumerated the problem of scheduled tribes. It said, "the problem of scheduled tribes living in compact areas is essentially that of economic development of the areas and of integrating their economy with that of the country". An expert committee was set up in 1972 under the chairmanship of S.C.Dube for advising on formulation of a new strategy during Fifth Five-year Plan (1974-1979). The Committee reviewed the situation and found that only 40% of the total tribal population had come under the purview of the development programme. Moreover, the great many communities that were lagged behind often gave rise to tension and conflict in the inter-community and intra-community level. Therefore, the Committee suggested a fast and time-bound integrated area development programme suited to the genius of the people. It aimed at elimination of all forms of exploitation ensuring a move towards the goal of equality and justice.

With this new approach to integrated development the concept of 'Tribal Sub-Plan' was adopted in the Fifth Five-year Plan. It was clearly understood that a uniform prescription is neither possible, nor desirable because the multiplicity of the tribal groups and their diversified cultural fabric do not equally respond to a master plan. Therefore, in preparing the 'Tribal Sub-Plan', the following factors were taken into account.

* The Scheduled Areas and Scheduled Tribes Commission was set up under the chairmanship of Sri U.N.Dhebar in 1960.

(*i*) Identification and demarcation of the areas of tribal concentration.

(*ii*) Identification of socio-cultural barriers as well as promoters of change in development.

(*iii*) Assessment of potentialities special problems and felt needs of the tribal areas

(*iv*) Assessment of the resources available for the Sub-Plan.

(*v*) Formulation of sectarian programs.

(*vi*) Designing of suitable administrative set up.

The Sub-Plan was implemented in different States and Union Territories to cover two-third (63.81%) of the total tribal population of the country. 50% to 70% of the State's plan budget was tied up with sectarian programs like power, flood control, large and medium industries, mining and transportation. Integrated Tribal Development Projects (ITDP) were launched within this budget to carry out welfare programmes. 'Special Multipurpose Tribal Blocks' were changed to 'Tribal Development Blocks'. In fact, activities of 'Special Multipurpose Tribal Blocks' were intermingled with the activities of the 'Community Development Project' and so 'Tribal Development Blocks' were worked out to meet up the specific needs of the tribal communities.

Unfortunately, even after the completion of Fifth Five-year Plan, it was found that a universal development for all tribal communities has not been achieved. The socio-economic upliftment of the tribal groups varied considerably from area to area. Therefore, the conceptual framework of sixth Five-year Plan (1980 - 1985) demanded some more clarity. It considered tribal pockets outside the Integrated Tribal Development Project (TTDP) areas, where at least 5000 scheduled tribes were found out of a total population of 10,000. Accordingly the tribals were classified into three categories:

(*a*) Tribals in the areas of tribal concentration.

(*b*) Dispersed tribals in other areas.

(*c*) Primitive tribal communities.

STATE	NAME OF THE GROUP
Andhra Pradesh	*Bondo-Gadaba; Bondo-Poroja; Chenchu; Dongaria Khonda; Gurob-Gadaba; Khond-Poroja; Khonda-Sabaras; Kolam; Konda Reddi; Kutia Khond; Parengi Poroja; Thoti.*
Bihar	*Asur; Birhor; Birjia; Hill Kharia; Korwa; Mal Pahariya; Parhaiya; Sauria Pahariya; Savar.*
Gujrat	*Kathodi; Kolgha; Kotwalia; Padhar; Siddi.*
Madhya Pradesh	*Abujhmaria; Baiga; Bharia; Birhor; Hill Korwa; Kamar; Saharia.*
Maharashtra	*Katkaria; Maria Gond; Kolam.*
Orissa	*Birhor; Bhunjia; Bondo; Didayi; dongaria Kondh; Juang; Kharia; Kutia Kondh; Lanjia Saura; Lodha; Mankidia; Paudi Bhuyan; Soura.*
Rajasthan	*Saharia.*
Tripura	*Reang.*
West Bengal	*Birhor; Lodha; Toto.*
Uttar Pradesh	*Bhoksa; Raji.*
Karnataka	*Jenu Kuruaba; Koraga*
Kerala	*Cholanaickan; kadar; kattunaickan; kurumba; koraga*
Manipur	*Maram.*
Tamil Nadu	*Irula; Kattunaickan; Kota; Kurumba; Panian; Toda.*
Andaman & Nicobar Islands	*Andamanese; Jarwa; Onge; Sentenelese; Shompen.*

Fig. 17.2. Primitive Tribal Groups in India

[These Primitive groups have been identified by the States basing on three norms—pre-agricultural technology, low level literacy and a diminishing population. Initially, at the end of Vth. 5-year Plan, there were 52 communities, which have now been increased to 75. The total population is 2412664 as per 1991 census.]

Three types of development measures were chalked for those three types of areas, viz., *Area oriented; community oriented; and individual family oriented.* The strategy of the Sixth plan was regarded as Modified Area Development Approach (MADA) in contrast to Integrated Tribal Development approach of Fifth Plan.

In 1984, a committee was set up by the planning commission under the chairmanship of Dr.C.H.Hanumantha Rao who recommended decentralization for the Seventh Five-year Plan (1985-1990). Accordingly, Tribal Sub-Plan was extended to all tribal groups of the country. The area-oriented development measures were shifted to family-oriented economic measures where the targeted number of families were directed to be elevated above the poverty line*. However, different pockets of the country were chosen for the implementation of this scheme. Emphasis was given on the formal and non-formal education. It was expected that the education would generate enough confidence among the poor tribals so that they would be able to interact with the outside world quite confidently. The working group also recommended a blending of the elements of fifth and sixth schedule to make the plan highly functional.

The Eighth plan that was proposed for 1990 - 1995 could not take off due to fast changing political situation of the country. Therefore, plans for 1990 -1991 and 1991 -1992 were treated as separate annual plans. The actual Eighth Five-year Plan came to be implemented for the period from 1992 to 1997, which strived to reach every remotest part of the country aiming at all-round socio-economic growth of the tribal people. This Eighth plan stressed on income generation programme with special reference to primitive tribal groups. Five basic parameters viz., Economy, education, health, depopulation and housing were taken as the criteria of selection. The tribal groups who remained engaged in pre-agricultural pursuits like food-gathering, hunting, fishing, etc. (some amount of agriculture may or may not accompany) were labelled as primitive groups. Such groups invariably inhabit in remote inaccessible areas. The examples include the Sauria Paharia of Bihar, Yanadi of Andhra Pradesh, Saora of Orissa, Jarwa and Onge of Andaman and Nicobar Islands. Next to that, the present Ninth Five-year Plan (1997 - 2002) aims at the empowerment of the tribal groups; a separate plan of action has been formulated for the 'primitive tribes' through micro-projects. Seventy-five (75) primitive groups have so far been identified in fifteen (15) States and Union territories, on the basis of pre-agricultural technology and a very low level of literacy.

Plan	***Period***	***Approach***
First Five-year Plan	1951—1955	General approach with a few specific areas of Welfare.
Second Five-year Plan	1956—1960	Approach through special Multi-purpose Tribal Blocks.
Third Five-year Plan	1961 — 1965	Integrated approach in relation to tribal Economics.
Three Annual Plans	1966 1967 1968	A break from Five-year Plans.
Fourth Five-year Plan	1969 — 1974	Integrated approach to link tribal economy with the National Economy.
Fifth Five-year Plan	1974 — 1979	Approach through Tribal Sub-Plan, aiming at all round socio-economic developments of the tribals.
Sixth Five-year Plan	1980 — 1985	Modified Area Development Approach to bring the dispersed tribals under the fold of welfare.
Seventh Five-year Plan	1985 — 1990	Family oriented approach for the removal of poverty.
Two Annual Plans	1990 — 91 1991 — 92	Disturbed situation.
Eighth Five-year Plan	1992 — 1997	Income Generation approach with reference to primitive tribal groups.
Ninth Five-year Plan	1997 — 2002	Micro-project approach for the empowerment of the primitive tribal groups.

Fig. 17.3. Government Plans for Tribal Administration.

* In 1979-1980, the Planning Commission held the poverty line as a figure of Rs.75/-per capita per month.

The Tribal Sub-plan strategy, which evolved during Fifth Five-year Plan, seems to be an important tool in solving the problems of tribals although its area of operation has been changed from time to time. In fact, a steady development of tribals was noted since the time of Fifth Five-year Plan. Different interests of the tribals were protected through legal and administrative support. At the same time, development was promoted through various plan schemes. All these factors helped to raise the standard of living as well as ensured socio-economic integration of the exploited human groups in the wider national framework. The scholars, who are concerned with the planning or effects of planning, have felt the need of a broad-based monitoring system, which is able to evaluate the situation from close proximity putting attention on both the region and the community. To reach to the desired goal we still have to wait for a few years.

Approaches to Tribal Problems

When we plan for tribal welfare, we take notes on varied situations in which tribals live. Their life-styles and capabilities are also considered in order to equip them with the facilities to develop. The objective is to integrate the tribals with the so-called mainstream of society. The problems of the tribals can be viewed in two terms––felt problems and assessed problems. The problems those are felt by the tribals themselves are called felt problems, whereas the assessed problems are those which are assessed and projected by the planners, bureaucrats, policy makers, anthropologists and such others who are the social thinkers. However, here we are concerned with the assessed problems and four major approaches have been differentiated so far, in the field of tribal welfare.

(*i*) Religious approach.
(*ii*) Social Service approach.
(*iii*) Administrative approach.
(*iv*) Anthropological approach.

Religious Approach

Of the said four approaches, the religious approach is the oldest of all. The earliest effort, which has been recognized, is that of the Christian missionaries. These Christian missionaries penetrated into tribal regions mainly with a motive of conversion; they wanted to impose their religious ideas on the simple innocent tribal people. But, at the same time they conducted a lot of welfare activities. A number of schools, hospitals, clubs and libraries were established by them. The attitude was purely religious. Although they wanted to spread Christianity by changing the faith of the tribals, they also endeavoured for an overall development of the unfortunate tribals. With the welfare work they tried to speed up the process of conversion. They actually competed with Hinduism, as there was a general tendency of the tribals to be incorporated into the Hindu society, very slowly. A part of them also went into Muslim society, but a very few of them liked to be included in the Christian society. Those who got integrated in the Hindu society were assimilated with the Scheduled Castes or similar low status groups. Naturally, the Christian missionaries came forward to motivate the tribals directly.

Social Service Approach

Next to the missionaries, a considerable number of voluntary agencies started their social service in tribal areas. In this respect we can name the agencies like Ramakrishna Mission, Bharat Seva-Ashram Sangha, Adivasi Sevak Dal etc., for their substantial humanitarian work. During the last few decades of British rule, these voluntary agencies showed such an over-involvement in this mission that sometimes their objectivity got blurred for the lack of proper understanding about the values and aspirations of the tribal people. Therefore, at times, some well-intentioned measures seemed to be injurious for the people and the approach was obviously discredited.

Administrative Approach

This approach includes 'pre-independence measures' and 'post-independence activities'. For the purpose of administrative advantage, the British government of India demarcated the tribal territories as 'excluded' and 'partially excluded' areas and gave separate political representations to the tribes.

This strategy was bitterly criticized by the nationalistic leaders of independent India. They took it as a 'negative approach of seclusion'. So the exclusion policy was found to be modified after independence. Constitutional safeguards aimed at a 'positive political approach' to uplift the backward communities so that they can be placed in the same plane with other advanced peoples to enjoy equal rights and opportunities of independent India.

Anthropological Approach

Anthropologists believe that the tribals coming in contact with advanced cultures have got the modern outlook of development; people of the civilized world act as change agents. Therefore, they advocate for an ultimate integration of the tribes into the mainstream of Indian life, in order to build up a prosperous nation. Anthropologists differ from the non-anthropologists in their approach. They never dishonour the indigenous way of life and culture of the tribes and want to develop the communities without hampering their original style of life. By dint of their professionalism, they search out the problems specific to the tribals before prescribing a solution. Moreover, they do not like to make a blanket like general policy, which could be applicable to all diversified tribal groups on a common basis. To them, the development programs should conform to the need and desire of the population in concern. Reconstruction of a policy, therefore, would vary with the groups. Special problems of the minority groups need to be considered sympathetically. Further, the anthropologists remain conscious to eliminate the elements that destroy the social solidarity of the tribals along with their freedom.

At present, the missionaries, social workers, administrators, planners and social scientists all have come to the consensus that the tribals are overburdened with innumerable problems as a result of long oppression and neglect. Economic crisis is rampant among them. Therefore, all of the thinkers have jointly recommended a long-term developmental policy.

Policies for Tribal Welfare

The tribal situation of India was first handled by the British during their colonial rule. They planned to isolate the tribals from the main currents of national life. They took the tribals as danger before the administration because whenever they tried to enter into the tribal areas, met encounters and uprisings. Therefore, such regions were cut off from the civil administration. Simple taxes like toll tax, house-tax, hoe-tax were levied. Military officials were kept in charge of those areas, instead of civilian officers. British government made various schemes to keep the tribals under control. Naturally no effort was found in safeguarding the interests of these isolated groups. The educational facilities, medical help etc., were all denied. This non-intervention policy of British was known as 'Laissez faire' policy and it involved an attitude of neglect, which made the poor tribesmen easy victims of exploitation by the civilized people. Traders, moneylenders and corrupt government officials took this opportunity for their ruthless exploitation.

The 'Laissez fair' policy of British government in India had a positive point that it did not put down the traditional customs of the tribals. Even the custom of human sacrifice was allowed to persist among some tribals. Therefore, the tribals got an opportunity to retain their age-old culture, language and way of life. After independence, Indian government became perplexed with tribal situation for policy formulation. Dutta Majumdar (1955) mentioned that a policy relates to "the problem that has been exercising in the minds of thinking persons of India, especially after the attainment of independence, is what should be the place of tribal peoples in the framework of the Indian nation and how they should be developed and brought to a level with the rest of the people—socially, economically, culturally and politically".* However, after the achievement of independence, three alternative policies were put forth:

* N. Dona Majumdar—"The Tribal Problem', in The Adivasis, Publication Division, Ministry of Information and Broadcasting, Government of India, 1955, p.25

(*a*) Policy of Segregation.

(*b*) Policy of Assimilation.

(*c*) Policy of Integration.

Policy of Segregation

This is the policy where some administrators and scholars argued for a separate place of the tribals as like that of the British Policy. They emphasized on the point that the tribals were animist as opposed to Hindus. Not only that, as they had a separate past, that should be carried on in the policy so that the people can retain their distinctiveness. Verrier Elwin and Furer-Haimendorf showed the separateness in terms of religion and other aspects of culture. V.Elwin was a distinguished anthropologist who came as a missionary from England and settled down among the tribal peoples of India to work for them. Gradually he became one of the respected members of the NEFA administration and recommended for a temporary isolation of the tribal groups in some extreme cases. In 1939, he announced his scheme of making 'National Park' with the tribals. He thought that the scheme would keep the tribals away and save them from the interference of the non-tribal mass.

The policy of Segregation was vehemently condemned by different scholars like Ram Mohan Lohia and G. S. Ghurye. So, Elwin in later period was compelled to change his views. He rectified his opinion by saying, "our policy was neither to isolate the tribes nor to freeze their culture and way of life as it is". It was realized that a policy of Segregation might be required in early stages, but can by no means be a permanent policy. Because isolation never leads to progress and advancement, rather brings stagnation and death. From the aborigines of Australia to Arya speaking Khalars and Kati tribes of Rampur Valley of Chitral, everywhere isolation yields the same result. Growth of a civilization always demands contact and intercourse among the people.

Policy of Assimilation

In reaction to conservative policy like Segregation, G. S. Ghurye, a senior sociologist from Bombay (Mumbai), suggested a policy of assimilation in tribal welfare. According to him tribes should not be treated as aborigines or autochthones as they are the backward Hindus. He sought the solution of tribal problem in complete assimilation into the Hindu society. In fact, there are a lot of instances of borrowing the culture traits from others. Such borrowings are quite beneficial when they occur in a natural way in harmony with the cultural setting and psychological make-up of the people.Otherwise they are very dangerous. The sudden and indiscriminate contacts may upset the tribal life. Similarly, the forced measures on unwilling people may bring tragic end as found in the history of the tribal peoples of Australia, Melanesia and U.S.A. In case of India, the tribal folk have a distinct culture; complete assimilation is never possible without injuring them.

The policy of Segregation in the light of 'Laissez faire' policy of British was also opposed by the nationalists of India. Many national leaders like A.V.Thakkar supported tribal movements against the British. Mahatma Gandhi had ordered his workers to go to tribal areas for preparing them so that they could take part in national struggle.

Policy of Integration

According to the eminent Indian anthropologist "Nirmal Kumar Bose, the policy of complete assimilation does not conform to the trends of human history. Because, despite thousands of years of culture-contact and inter-cultural borrowing, Indian society did turn to a homogeneous whole. Still there are heterogeneous cultures like Santal, Bhil, Gond, Oriya, Kashmiri, Telegu etc., therefore, instead of a policy of complete assimilation, a policy of limited assimilation would be more beneficial. It would help to preserve the traditional useful institutions, customs and practices according to the need and desire of the tribals. At the same time new traits could be borrowed and incorporated following the universal rule of change. Further, an integration would be able to discourage the separatist tendencies among the tribals and would bring equal life chances so that the tribals should not suffer from problems. Hence, our policies and programmes of tribal development are supposed to have been based on this approach of integrating tribals with the mainstream in order to bring them in the same level with the rest of the people.

A. V. Thakkar and some other servants of British regime did the pioneering work among the tribes. A vast literature on tribes and castes were published by the administrators like Dalton, Risley and others. There were also the Census officers like Grigson and Hutton. Indian anthropologists namely S. C. Roy and B. S. Guha were no less superiors; they provided a good insight on tribal problems. Besides, there came the contribution of missionaries. V. Elwin's view towards the understanding of tribal problem is still regarded as a landmark, which influenced Mahatma Gandhi as well as Jawaharlal Nehru. The account of the missionaries like Knowles, Wood, Hislop, Nortott, Bodding and Hoffman can not be undermined, although they interpreted the tribal life and aspirations in their own way. National leaders of independent India were highly benefited by those literatures. With the help of planners and academicians, they intensified their efforts to solve tribal problems. Eminent anthropologists, namely, L.P.Vidyarthi, Sachchidananda, P.K.Bhowmick and others have also done excellent work to highlight the tribal life before the administration of Independent India.

In spite of so many efforts, tribal development has been neither uniform nor remarkable in all parts of the country. Manifold reasons can be cited against this sluggishness. Some of them are on the part of the tribals and others rest on the administrative machinery. Management procedures are not always free from lacunae. Lack of proper knowledge hinders in selection of actual area. Shortage of manpower, wrong decision, forceful application may spoil scheme. Lack of coordination among the different branches of an implementing agency or proceeding difficulties for a remote and inaccessible area may also turn down the success of a development project. Further, the loss of zeal, conflict and group dynamics among the implementing personnel may lead to a failure. Usually the tribal people, on account of their specific culture and peculiar psychology show a resistance against the welfare programme. They are quite afraid of novelty as well as want to retain their traditional customs and habits. Therefore, a planner or an implementing agency, without having prior knowledge about the culture and psychology of a group when proceeds for work, meets untoward situations. If the tribal people do not accept his schemes, money spent in those schemes is totally wasted. In recent years, government authorities, voluntary organizations and the planners seek the expertise of the anthropologists to get authentic information regarding a tribal groups. After a long phase of suspect and confusion*, it has been realized that the anthropologists are the only persons who can rightly find out the specific problems of the tribals and point out the solutions in a given situation.

There are other factors too. A large number of schemes related to agriculture, animal husbandry, cottage industry, public health, family planning etc., sanctioned by the government are usually implemented through its various departments. Some benefits are given in the form of grant and others are in the form of subsidy. The Block Extension Supervisors often in order to fulfil their target encourage the tribals to avail the money as grant or loan. Since the loans are not subsidy, that need to be returned in time. Ignorant tribals most of the cases fail to realize this and do not pay back the money. This creates a new set of problems. Again, the rate of subsidy for the same scheme varies with various departments of government and this makes the tribals confused. Different sorts of dissatisfaction, resentment and frustration are generated in them. As a result, the tribals can not keep trust on the government agents.

On the whole, it has been understood that the fundamental basis of tribal life should not be disturbed. The original customs and manners need to be preserved when the exposure for assimilation and integration is favoured. Such a policy would be able to prevent the exploitation of the tribals and

* Initially the assistance of the anthropologists in tribal welfare were not granted with respect, their philosophies were underestimated in most of the cases. This was so because till then there was no adequate and well-established model where anthropologist's knowledge and experience could yield an outstanding result in the welfare of tribal people. Non-anthropologists always differed from anthropologists in opinion.

at the same time would enable them to enjoy the equal rights and opportunities of present day. Success of a project should be evaluated with the ratio of expenditure incurred and the target achieved.

Constitutional Safeguards for Scheduled tribes

India is a Welfare State, committed to the welfare and development of its people in general, and of vulnerable sections in particular. The backwardness of the tribals is not solely responsible for their long isolation from the general society. The other factor is the exploitation by the civilized society. Not only in India, in many other parts of the world, whenever the indigenous people came in contact with so-called civilized people, the results became disastrous. It exposed the innocent and ignorant aboriginals under different types of oppression and exploitation. Tribals dispossessed their land and healthy environment.

During the British rule, the tribal areas were identified specially for the sake of administration. For instance, the Scheduled District Act of 1874 first transferred the Naga Hills District from the jurisdiction of Bengal to Assam. Under the Government of India Act of 1919, this provision was extended to several 'backward tracts' which covered an area of 120 thousand sq. miles, distributed in five of the provinces of British India with a populations over eleven million. The Government of India Act of 1935 erased out the concept of 'backward tracts'. Instead, it considered 'excluded' and 'partially excluded' areas for giving special and separate treatment to the tribes. Accordingly, from 1937, the Naga Hills District, Northeast Frontier tract, Lushai hills and North Cachar hills became 'excluded' areas within the Province of Assam. Many areas in Bihar, Orissa, Bombay and other provinces were recognized as 'partially excluded' areas.

Policy of British government was to keep the tribal's aside, which resulted in their exploitation a the hands of landlords, moneylenders, contractors and also Christian Missionaries who were the supporters of British administration. The tribals lost their land, habitat and milieu,and this non-tribal intervention created many problems like pauperization, causalization and psychological stress and strain. Official as well as illicit felling of forest trees had benefited the outsiders but tribals suffered from the loss of environment. Apart from the problems of forestland and land alienation, other problems of tribals were no less serious. There were curses of tenancy legislation, ceiling on land, fragmentation and consolidation of land-holding and compulsory acquisition of tribal lands for public purposes. Tribals were compelled to incur debts to solve their immediate and pressing economic demands arising out of illness, social obligations, daily necessities etc. Moneylenders and traders made themselves readily available with necessary credits but finally for their unusual demands, tribals had to lose their last penny. The arrangements of advancing loans through government agencies, banking institutions or cooperative societies were not devised for them. British had never thought seriously for the amelioration of their conditions. Not only economically, the poor tribals were deprived in all spheres—social, educational and political. They did not get a fair deal, neither at the hands of government nor from their civilized neighbours. It was only after the attainment of independence that the welfare of tribals in true sense was made under the responsibility of the State.

Constitution of India uttered certain specific provisions for the tribes in order to safeguard their rights and interests as an inhabitant of India. Moreover, it expected a quick upliftment of these unfortunate souls so that they can reach the level of the general mass of the country. Article 342 of the Constitution empowered the President to make initial notification of most deprived tribes in consultation with the State government. Article 366 (25) designated these deprived tribes as Scheduled Tribes. It also furnished a definition, which meant some tribes or tribal communities or parts of or groups within such tribes. Those tribal communities were deemed to be scheduled tribes for the purpose of the constitution. However, the constitution was adopted in 1950. But even after the promulgation of the constitution, five to six years passed to start a full-fledged operation in the favour of Scheduled Tribes. In fact, both the Scheduled Castes and Scheduled Tribes were treated as 'backward classes'—the 'weaker sections of the society'. Apart from the general protection for the weaker sections as stated in Directive Principle under Article 46, the Fifth and Sixth schedules of the Constitution laid down certain special provisions for the tribal areas and recognized the importance of traditional tribal councils so as to have a considerable control of their own affairs and protection of their land and customs.

Fifth Schedule

This schedule is the charter of Scheduled areas as well as Scheduled Tribes. This provision is meant for the scheduled tribes only. Scheduled Castes do not enjoy this. Under this schedule of the constitution, the President is empowered to declare any underdeveloped area having substantial population of the Scheduled Tribes as a Scheduled area. Such areas have been declared in eight States e.g., Andhra Pradesh, Bihar, Gujarat, Madhya Pradesh, Maharashtra, Orissa, Punjab and Rajasthan. Though the scheduled areas are administered as a part of state in which they are situated, the Governor has been given special powers under this schedule. The Governors of the states having scheduled areas are also required to make reports to the President regarding the administration of such areas.

The Fifth Schedule also proposed for an Advisory Council for the tribe. The duty of the Council was to advise on the matters pertaining to the welfare and advancement of the scheduled tribes in the state. Such an advisory council must own its Chairman and 'officers and servants' who will possess the power to initiate any business relevant for the welfare of the scheduled tribes. The paragraph 4 of this fifth schedule empowers the President to direct a state, which has tribes, and to establish an Advisory council for the tribes. All provisions of Fifth Schedule are entirely concerned with the scheduled areas. The objective behind this scheduling of areas was to raise the level of administration in them upto the level of that of the rest of the state in which they are located.

Sixth Schedule

Tribal situation of Assam was very difficult from the beginning (1943). Sixth Schedule of the constitution deals with the administration of the tribal areas of Assam by dividing it into two zones:

A. The autonomous Districts including United Khasi-Jayantia Hills District, the Garo-Hills District, the Mizo District, the North Cachar Hills and the Mikir Hills district.

B. The North East Frontier Agency comprising of North-east Frontier Tract including Balipara Frontier Tract, Tirap Frontier Tract, Abor Hills District and Misinic Hills Districts.

The Schedule advocated for the setting up of a District Council in each autonomous district, and a Regional council in each autonomous region. Both the Districts and Councils were given special powers.

Among other Safeguards, there were,

(*i*) Safeguards for ensuring the political development of the Scheduled Tribes. Article 330, 332 and 334 of the Constitution recommended for the reservation of seats for Scheduled Tribes in the Assemblies as well as in Parliament.

(*ii*) Safeguards for securing adequate representation in the state services. Article 335 advocated for taking into consideration of the claims of Scheduled Tribes, consistently with the maintenance of administration, in the making of appointments to services and posts in connection with the affairs of the Union or a State. Article 16(4) stated that the effects could be given to the above provisions, if necessary, by reserving posts in favour of the Scheduled Tribes.

(*iii*) Safeguards to ensure economic, educational and general development of the Scheduled Tribes. It also pleaded for the raising of the level of the administration of the Scheduled Tribes and tribal areas. Article 6 said, "the state shall promote with special care the educational and economic interests of the weaker sections of the people and, in particular, of the Scheduled Caste and Scheduled Tribe, and shall protect them from social injustice and all forms of exploitation"

Besides, Articles 15, 19,23, and 29 of the Constitution brought protection of the civic rights of Scheduled Tribes for saving them from exploitation.

iv) Safeguards for financial help.

Article 275 of the Indian Constitution announced financial help to the States for the development of scheduled areas and for welfare schemes of Scheduled Tribes under the states.

Article 338 and 339 kept provision for the evaluation of the progress made in the welfare of the Scheduled Tribes.

The list of the Scheduled Tribes was amended in 1976 to cover three hundred tribal groups. In addition, the Government of India accepted the conventions and recommendations made by the International Labour Organization (1957) with respect to tribal and semi-tribal populations. A National Commission for Scheduled Castes and Scheduled Tribes was set up in 1990, by virtue of sixty-fifth amendment of the Constitution. The Commission was supposed to investigate and monitor the matters relating to the safeguards provided for Scheduled Castes and Scheduled Tribes. The Union Government and every State Government were expected to consult the Commission on all major policy matters affecting Scheduled Castes and Scheduled Tribes. These provisions were laid under the Article no. 338. Thus, all constitutional safeguards are more or less same for the Scheduled Castes and Tribes. Because both of them are regarded as the most backward and downtrodden sections of the society.

Tribal Studies in India : An Appraisal

A vast literature is available today on tribal studies as tribal problems were dealt by different researchers, scholars and activists. Analysis of those literatures shows a definite trend upon which all the materials can be divided into three broad categories—*ethnographic studies, developmental studies* and *critical studies.*

The 'ethnographic studies' can be held as denoting the formative phase in the tribal studies. The period may be bracketed as 1774 to 1920. All studies of this phase are essentially ethnographical. Such studies have emerged out of the ruler's need to manage the tribal affairs. After 1920, the ethnographic studies took a new turn to initiate 'development studies'. Because at this time the nationalist movement started which viewed the tribals as a part of the nation and as a result efforts were provided in the development of the tribals. During this phase, many Gandhian workers went into the tribal areas to establish 'ashrams'; they also carried on many constructive activities. The Indian National Congress adopted the 'developmental approach' as a part of its ideology in making a welfare state. The phase continued from 1920 to 1970. Some of the ethnographic studies also appeared in this phase but all of them had an overtone of development.

After 1970, the so-called development model was cracked. Many of the mainstream development projects, while brought the development for non-tribals, pronounced misery for the tribals. At this time, 'critical studies' came out to examine the ill effects of the development model for tribals. The phase is continuing as the critical approach has been highly accepted by the mainstream researchers.

Tribal studies still form the treasure-core of anthropological studies, although the dimension has changed. It provides a basic understanding to the subject. Students of anthropology derive their firsthand knowledge from the tribal field; ethnographic data is tremendously important for them. They need an initial training to collect data, which is the key for their success in future. Tribal field serves as a laboratory where learning takes place by trial and error. An anthropologist in no way can deny the importance of tribal studies.

Tribal Movements in India

Ancestors of present day tribals in India must be of Vedic or even pre-Vedic antiquity. On the basis of Hindu scriptures we can derive a conclusion that Aryans and the autochthonous people had bled and bred together in the past to form the Hindu Civilization. With the passage of time, their age-old hostilities were sublimated into mutual understanding and so peaceful co-existence and amalgamation was possible, But a section of them, the uncompromising autochthones withdrew themselves and went on exile, voluntarily. These tribal groups were orthodox and puritan; they had a great love and respect for freedom and self-identity. So they preferred inhospitable regions in hills and forests to carry on own distinctive style of life.

British colonialism introduced a novel economic system characterized by commercialization of land and forests, well defined property rights, occupational structure, marketing of agricultural surplus etc., in consonance with their own economic system. The ultimate objective was to exploit the socio-economic resources of India for the prosperity of British Empire. Therefore, colonial rulers were found to spread their capitalist network in every nook and corner of the country; the tribals were not spared. Regional rulers and chieftains were either crushed totally or coopted by the British through concessions. The tribals also came in direct conflict with the mercantile capitalism of the British Raj. The tribal feudalism based on communal land relationship collapsed down. On one hand, the hereditary communal rights on land became defunct, and on the other hand, provisions were made in such a way that the tribals could be evicted at any time if their landlord or governmental agents would declare them disobedient, or if they themselves would default in paying the land rents. The tribals were not prepared to face those new challenges and so tensions arose in them. Not only that, the colonial land system gradually set free different forces to encroach upon the tribals; the unfortunate souls were oppressed by both the European and Indian usurpers. The landlords, merchants and moneylenders—none showed pity on them. Even the administrators (who came with a policy to safeguard their interests) and missionaries (who came to equip them with modern educational outlook) were not free from contradictions. They could not improve the material conditions of the tribals. Moreover, as the economic foundation of the life was shattered, it brought rapid changes on superstructure (i.e., religion, language, education, political concepts and value system). The age-old balance between infrastructure (material conditions of life) and super-structure (ideology) got disturbed and the tribals reacted fiercely.

Concept of Social Movement

Social movements originate in collective behaviour. They always possess a special goal which may be viewed as an aspect of social change. The growth of a movement depends on several factors —external or internal. Usually a social system works smoothly when it remains functional. But, when it defuncts,it produces enormous strain in the system for which movement is generated. No social movement is possible without a situation of strain within the social system. External factors also can not be undermined. A socio-political movement is characterized by the well-defined objectives for agitation, an organizational structure, a group ideology and a recognized leadership.

The whole nineteenth century witnessed numerous occasions of riots, rebels, uprisings etc., which occurred against the colonial rule and its exploitative socio-economic network. All these events can be called movement because in each case hundreds or thousands of persons were united to fight against the higher exploitative force. But the nature, form, dimension and magnitude of the movements differed from place to place. In order to interpret the genesis of a social movement, M. C. Paul (1989) discussed three principal theories, which explain the structural conditions as well as motivational forces required for a movement. According to Mr. Paul, social movement is essentially a phenomenon of mass mobilization on specific issues or problems concerning social life. It tries to achieve a definite goal or fulfil certain objectives. Sooner or later, a comprehensive ideology (theory) is developed which is followed by the participants of the movement. All social movements have political implications even if their members do not strive for political power. When a movement accomplishes its goal, it dies off, but the party may stay. However, three major theories on the movement are as follows: (*a*) Strain theory (*b*) Revitalization theory (*c*) Relative Deprivation theory.

The Strain theory was propounded by Neil J. Smelser. It views the social movements in relation to social change. Smelser argued that strain might appear at different levels of norms, values, motivation etc., which affects the structural-functional framework of a social system. Following those situations new mechanisms are readily developed for reintegration. Therefore, movement has been held as an adaptive mechanism to bring out functional equilibrium. The Revitalization theory is a contribution of A.F.C. Wallace, who defined a movement as a "deliberate, organized and conscious effort on the part of members of a society to construct a more satisfying culture". By the term revitalization, Wallace wanted to categorize a number of similar types of movements viz., *Nativistic movement, Revivalistic movement. Cargo Cults movement, Vitalistic movement, Millenarian movement* and *Messianic movement.* Stephen Fuch identified revitalization movement among the Scheduled Castes, Scheduled Tribes and other Backward classes in India.

The third theory, the *Relative Deprivation theory* perceives deprivation as a prime factor behind the movements. It includes two distinct trends:

(*a*) Where relative deprivation is counted in relation to social structure and social mobility. This theory was introduced by R. K. Merton. Later on, W. C. Runciman developed the same concept.

(*b*) Where relative deprivation is counted in relation to social conflict. Karl Marx and Engels developed this concept. Mode of production as the basis of the society was pointed out by Marx. He emphasized on the relations of production into which man entered. He also discovered the elements of conflict and established their relationship with unequal distribution of wealth, property and other means of existence. To Marx and Engels, dissatisfaction arising from deprivation gave rise to social movement.

Nature of Tribal Movement

Tribal movements have been prevalent in India since the days of colonial rule. They are the outcome of severe economic exploitations and inhumane social oppressions. Rebellions of the desperate tribals are found even today, although the character and motivations are different.

During the colonial rule, tribal movements went side by side of the peasant movement in India. Both of these movements were basically directed to resist and drive out the colonial rulers in order to restore the traditional principalities. But since it was a struggle of weak against the strong, in almost all cases, tribals were the ultimate losers. They lost their manpower as well as mental strength. V. Raghavaiah (1971) estimated more than seventy tribals uprising in India during the past two centuries. At the beginning, these movements were rudimentary and unorganized in the form of sporadic attacks. Therefore those were considered as unimportant and marginal. But gradually they gained enough potency to threat the administrators. K. Gough (1974) had classified five kinds of tribal peasant movements in terms of their goal, ideology and method of organization:

(*a*) *Restorative rebellion* aims at driving out the colonial usurpers from the tribal areas to restore the earlier rules and social relationships.

(*b*) *Religious movement* aims at the liberation of a region or an ethnic group under the guidance of charismatic, religious or messianic leaders.

(*c*) *Social Banditary* is a simple form of revolt against oppression and poverty. It is a modest and unrevolutionary protest without any organization or ideology.

(*d*) *Terrorist Vengeance* is a small group action of violence against the landlords, revenue agents, moneylenders and officials. It aims at revenge and justice; but often involves a sense of group-pride. Sometimes, a religious belief is found to be associated with a terrorist act.

(*e*) *Mass Insurrection* is the sudden and dramatic revolts under a single prophetic or charismatic leader. But they invariably lack the ideology of a religious movement. Usually they go through mass boycott to demand justice.

A common political feature was projected in all tribal movements of nineteenth century. It can be described as a militant opposition against the systematic intrusion of colonialism that occurred into the traditional world of tribalism. In each case, this opposition was synonymous to defensive movement. The grievances of the tribals were not only directed towards the ruling British officials, all other supporters like landlords, merchants, moneylenders, etc., were their targets of attack. Because all of these persons were directly responsible for their economic distress and ruin. The violent tribals often killed and robbed the officials; looted and burnt the treasuries and police outposts; plundered the business houses as well as the homes of the landlords, moneylenders and revenue agents. Since they had no specialized weapon, the simple knives, primitive clubs, bows and arrows were only used. The administration in one hand, was completely ignorant about the problems of the aboriginals, on the other hand, were indifferent regarding the protection of these groups.Therefore, they could not rightly cope with tribal upsurges; a number of rebellions appeared one after another. Most of them were rooted in oppression and exploitation of rulers and the neighbouring people. There were also others, which broke out centering some fort of misunderstanding among the rulers and the ruled. The sufferings of the tribals were continued, as they could not win against the strong organized strength of British rulers.

To depict the history of tribal movement in India we should start with *Dhalbhum movement* which took place during 1769 - 74. It was the upraise of Chuars and Bhumijs of Jungle Mahal, Midnapur district in West Bengal who fought against British administration for excessive taxation on land. The *Chakma movement* also started in eighteenth century. It continued from 1776 to 1787 in Chittagong hill tract under the leadership of Damodar Singh. The Chakmas aspired for their traditional freedom and wanted to revive their right over land. The Chuars of Midnapur again revolted with force in 1799, which continued upto 1816. This is known as *Paik movement* in history.

Rebellion of Mal Pahariyas took place in 1778 in Bihar. *Koli uprising in Maharashtra* was found during the period from 1784 - 85. *Kol rebellion* among the Hos of Singhbhum (Bihar) took place in 1832. The famous *Santal rebellion* occurred in 1855 - 56 among the Santals of Birbhum and adjacent districts of West Bengal. Before this, the Santals of Midnapur district went on non-cooperation movement in 1920 - 1921. The Syntengs of Jayantia Hills of Northeast India revolted in 1860 - 62. The Garo tribes of Khasi Hills and Garo Hills also went in the way of rebellion in 1837, 1852, 1857 and 1872. The Lushais and Kukis raided on British camps in 1860, 1871,1888,1889-90 and again in 1892. In Manipur, rebellion occurred in 1891. 'Assam Riot' took place in 1894 and Kacha Nagas revolted in 1880. On the other side, Bhils of Rajasthan, Gujarat and M.P also revolted against the British for occupation on their lands. They fought several times and most remarkable among those were incidents of *Khandesh* in 1817 and 1819. The *Vedchhe movement* (1885 to 1947) which took place in the Surat district of Gujarat also deserves special mention.

Other remarkable rebellions include the *Naikda movement* in Gujarat (1867 - 70), the *Tana Bhagat movement* among the Oraons of Chotonagpur (1895), the *Birsa movement* among the Mundas of Ranchi District in Bihar (1899 - 1900), the *movement* among the Nepalese-Bhutias-Lepchas of Darjeeling District in West Bengal (1920 - 36), the *Bhumij rebellion* of Manbhum (1831- 32), the *Rampa rebellion* among the Reddis in East Godavari District (1879), the *Bastar rebellion* among the Gonds (1910). Besides, there were the records of Khasi uprising, Angami Naga outbursts, Oraon rebellion, etc., which also contribute in the general trend of tribal movement.

A closer look reveals that there were more regional causes for tribal dissatisfaction. For example, the problems in Chotanagpur region were associated with the exploitation by the well-to-do Hindu caste people, the feudal institutions, the massive proselytization and the resultant political awakening and active participation of the people in the political processes. Story of Nagaland was entirely different. Naga was a martial race who had a lust for human sacrifice. When it confronted with a new faith, several new values, new education and new political image grew among the people; the new charismatic leadership coupled with the old psychological orientation of hatred, intolerance and violence was manifested in a rediscovery of native genius. Again, the problem of Bastar land had a new dimension. Diplomatic oscillations of State administration and political bargaining of leaders with the native rulers embarrassed the tribals. At the same time, techno-cultural invasion in the Dandakarnya forest made them psychologically upset.

K. S. Singh (1972) identified two broad trends in the tribal political movements in India. He discovered two broad zones—the Northeastern zone and the middle Indian zone. In the Northeastern region, tribals constitute a majority of the population. British was not able to draw this region completely under its administrative framework, because of difficult geo-political situation. Even before the days of colonialism, this region was not completely integrated into the mainland; people were quite apart from the cultural influences of India. The tribes living in this region acted as a buffer between international and national exposures. Socio-economically they were more or less secure. So the nature of movements in this region was essentially political and secular. All tribes of Northeastern region except Tripura* had never met any threat on their identity; no agrarian or forest-based movement was found in this territory. Therefore, their institutions have remained relatively intact. The occasional agitation, which was noted among these tribals, gradually took them to the path of armed insurgency. But the goal ranged from autonomy to independence.

According to K S. Singh, the tribal politics in Middle India was qualitatively different than the Northeastern region. Industrial economy and the urban political culture had reduced the tribals to minority groups who were scattered in different pockets. Loss of traditional economy made the tribals ferocious and the tribal movement in this part of the country took the agrarian character. These tribal folk participated in the national movement of freedom at a later phase, being called by Mahatma Gandhi and other nationalist leaders. Like non-tribals they fought for driving out the Britishers from India. The national leaders realized the isolated conditions of the unfortunate tribals, so tried to reintegrate them with the main current of Indian life. Gradually tribal movements took a formal shape and much more organized movements were found in post-colonial period which sustained and aimed at bringing about a partial or a total change in the society.

Unlike middle India, political processes in Northeastern region helped a number of small tribes to develop into some larger tribes, not only with new names but also with new ethnic-cum-territorial identity. States were created to accommodate tribal aspirations for autonomy. It surpassed the VI th Schedule model of political autonomy as prescribed by the Constitution. Nagaland enjoys more autonomy than the other states of the Northeast hills. No law can be applied to the Nagas until and unless it is approved by the State Assembly.

The freedom and reform movements of the tribals have been grouped by A. R. N. Srivastava (1986) into three phases.

Phase I: Movements upto 1857 i.e., during the regime of East India Company. These were basically crude agitations, primitive in character. Their aim was to secure some concessions from the rulers.

Phase II: Movements from 1857 upto the beginning of the twentieth century. These were lodged against British economic and political policies in the background of new land revenue system, civil and criminal regulations, and widespread famines. Both tribals and their surrounding non-tribals reacted against the colonial power.

Phase III: Tribal struggles during the first three decades of the twentieth Century. These were mainly economic struggle against the non-tribal intruders—moneylenders, traders, zamindars and administrators who tried to exploit the tribal way of life. During this period, in some parts of the country, tribals joined the groups led by Indian National Congress for taking up scientific issues and struggles launched by Mahatma Gandhi and his associates.

According to Srivastava, all tribal struggles were not politically motivated. Those were close to 'agrarian reform' and 'religious revitalization' type of social movements. But he called the tribals 'the pioneers of nationalistic movement'. Like K. S. Singh, Srivastava also noticed the peculiar disposition of tribes in northeastern region. On the ground of available historical accounts, he classified the tribals of Northeast into three categories :

* Tripura presents a unique case, as tribals are the minority group in their own homeland. A sharp confrontation and conflict was generated with this issue between the ethnic group and the outsiders, which culminated in 1980 through bloody incidents.

(*i*) The areas, which remained almost untouched by the British administration i.e., where traditional way of life persisted and has been persisting virtually undisturbed. The entire Arunachal comes under this category.

(*ii*) The areas, which were in continuous conflict with British. For example, the Naga territory.

(*iii*) The areas, which, after some initial resistance, were brought into British fold. These areas include the regions like the Khasi and Jaintia hills, the Garo hills, the Mikir hills and the Mizo hills. But the traditional chiefs or the leaders still play important role in the decision making process.

Although the British arrived in Northeastern region in 1832, until 1937 the entire hill area of Northeast India was administered as 'Backward Tract' by the British officials. In 1937, a bifurcation in administrative system was found. At this time, the Mizo hills, the Naga Hills, the North Cachar hills and the *Northwestern Frontier Tracts* were branded as 'Excluded areas' and kept outside of ministerial jurisdiction. On the other hand, the Garo hills, the Khasi-Jaintia hills (already occupied by British) and the Mikir hills were marked as 'partially excluded areas'. As a consequence, the hill districts acquired a special identity for the subsequent political developments.

Tribals in Independent India

Most of the tribal movements of post-colonial period claimed regional autonomy and independence. The freedom movement of India provided them a new insight and awareness. As a result, agitation of deprived tribals has turned into separatist movement. Among the notable movements we can name the Nagas, Mizos, Manipuris (Meitheis) who were engaged in persistent actions to establish their independent states. Similarly the Bodo-Kacharis of Brahmaputra Valley of Assam went on in claiming an autonomous state. Strong persisting movement of Chotonagpur was for the creation of an autonomous Jharkhand state. The tribals of Chhattisgarh and Gondwana region in M.P. also launched a movement for a separate state in 1960. In past, voice of Bastar tribes was raised for the same reason. The tribes of the South Gujarat were not the exceptions. But the tribal people of Rajasthan, Bihar, M.P., and Andhra have displayed a political loyalty to the Indian State and showed no separatist tendencies, although they took an active part in agrarian radicalism of 1980's.

The general view on present situation leads us to conclude that the poor tribals had protested several times in past against the Britishers for which a number of great rebellions were recorded in the history of tribals. After independence, they have been peaceful by and large. This is perhaps due to the amelioration of living conditions as favoured by the Constitution.

The study of tribal movements is a recent development in the field of socio-cultural anthropology. Contemporary anthropologists have been interested to study an overall perspective of Indian civilization, as this civilization is the outcome of age-long interactions between indigenous and foreign traditions. The study of tribal unrest has also been taken up to assess the pattern of relationship that exists between the tribal and non-tribal populations of the present day.

A Few Notable Movements

Chuar Movement

The Chuars and Bhumij community of the forested part of Midnapur first protested against the British administration for excessive taxation on their land. The seed of resentment was germinated in 1769 when the landlords of West Midnapur became eager to get independence, following the Maratha invasion. Since 1765, the administration was in the hand of East India Company; the Company had sent lieutenant Farguson to quell them. Although Farguson became overall successful in his task, he could not capture the zamindar of Ghatsila and the king of Dhalbhum. He fought and fought and at last the areas were brought under his control. Some kings and zamidars fled away following the situation. Other kings and zamindars, being defeated, agreed to pay the whole amount of land revenue as demanded by the Company. But,within a short period absconded zamindars were united to resist the British. The Bumijs and Chuars of this region also stood in favour of them. In 1770 they attacked the police force of British. A large number of sepoys were murdered by them. Military force was employed to bring peace in this region.

The peace did not last long. Zamindar Jagannath Dhal revived his power. He made a powerful team with the discontented inhabitants of the locality. In 1773 - 74, they became violent and desperate to drive out the sepoys from their area. The movement was actually a struggle of zamindars against the British, but the contribution of the tribals can not be undermined as they wanted to make themselves free from the grip of alien rulers. At last, Warren Hastings tackled the situation firmly with his wit; he gave back the land-property to Jagannath Dhal. The incident is known as Dhal-bhum rebellion.

The Chuars were again found to be activated in 1799. This phase is known as Paik movement or Chuar movement. The Paik troops of the zamindars were constituted with different tribal people like Bhumij, Kurmi, Kora, Munda, Lodha, Majhi, etc. Traditionally they acted as the guards of the zamindars and in lieu, enjoyed some land-property gifted by the respective zamindar. In British period, especially after the implementation of permanent settlement in 1790, they lost their traditional privilege on land. Moreover, heavy taxes were imposed on them and the price of the salt was raised extraordinarily high. Naturally the Paiks took a rebellious attitude. The queen of Karnagarh not only gave them shelter; she instigated the Paiks towards movement. The Paiks wanted to evacuate the British from their land. During the period between 1799 and 1816, at different times, they fought with different issues. Each time there was a charismatic leader. Chhatrapati Singh led them when they tried to evacuate the British from their land; Achal Singh was with them when they moved against the taxation policy imposed by the British. Subala Singh gave the leadership when they went against high prices of salt. Dubraj Sigh gave them the courage when Marathas attacked them.

Malpahariya Movement

Mal pahariyas of present day Rampurhat (Birbhum district, W.B.) are the descendants of Mal pahariyas of Rajmahal hills. In Mughal period, these people were thought as robbers. In 1770, when British came to rule on them, a famine shattered the whole of rural Bengal. Most of the villages became empty; the cultivable lands turned into forest. But the British troop was not relaxed in revenue collection. Naturally, this attitude made the people furious. This is the time when the Malpahariyas came down to the plains. They started plundering to satisfy their hunger.

The rebellion started in the middle of 1788. The famine-affected peasants of the plains joined their hands with Mal paharias. They began to plunder the offices and the homes of the British officials, traders and moneylenders, at the North of the Birbhum district covering about 100 miles of area. In January 1789, they looted a large market and snatched food from the godown. The frustrated ferocious mass formed teams with three to four hundreds of people and attacked the towns in the neighbourhood. The violence spread even to the Bishnupur district. Being defaulter in revenue payment, the king of Bishnupur was imprisoned by the British. Local people as well as rebellious groups fought against the British rule. In 1790, they captured Rajnagar and most of the Birbhum district. British force faced immense trouble in maintaining peace and defense. It was the first major confrontation of British with the tribals. After this, they began to think seriously about the tackling of these problems.

Santal Movement

The biggest tribal movement in India was the Santal insurrection, which broke out in Eastern India, particularly in Southeast part of the Bhagalpur division in Bihar. It was a broad-based awakening of a tribe that became awe-inspiring even to the British government.

The rebellion has been studied by several competent historians namely W. G. Archer, E. G. Man, W. J. Culshaw, P. O. Bodding, and others. Francis Buchaman's writing on the district Bhagalpur (1810 -11) provides the oldest account from which we have come to know about the conditions of the Santals settled around Dumka sub-division that lay about fifty miles south of the Rajmahal hill tract. At first the region was sparsely populated; the Mal Pahariyas and the Sauria Pahariyas were its first inhabitants. On the recommendation of the then government, the area 'Damin-I-Koh' (Skirts of the hills) was formed in 1832 -33. The whole area included about 1366 square miles. But the hill tract was reserved for the Pahariya people and only 500 square miles at the foot of the hill were allotted to the Santals who came from different directions. The English rulers deputed the laborious Santals in clearing the forest adjacent to the Rajmahal hills. Though the local Pahariyas were greatly affected by this activity, the Santals being pampered by the British dared to clear up the forest and occupied the area permanently as settled agriculturists. Before that, the Santals used to collect forest produces and subsist mainly on hunting and fishing. Sometimes they practised cultivation on hill slopes by employing their traditional *slash* and *burn* method. The main trading crafts were the extraction of oil and the manufacture of lime. However, later, in Damin-I-Koh, most of them became engaged in cultivation.

Sutherland's 'Report on the management of the Rajmahal hills' (1819) focussed on the living conditions of the Santals at the time when he investigated the land tenure system of the Pahariyas of Rajmahal hills. It reveals that the obedience of Santals satisfied the British whose principal interest was to increase land revenue and to ensure trade. In 1838, about three thousands of Santals were reported to live in forty villages at Damin-I-Koh. They used to pay Rs. 2000 per year to the government; their innocence could not apprehend the conspiracy of the alien rulers. Naturally the situation was peaceful and supervision of the government was not very strict on them. In the mean time, flocks of traders penetrated into their areas. As they found the people unguarded, they started to carry away local products like rice grains, different kinds of oil seeds etc. In return, the Santals were paid money, salt, tobacco and cloth. Some moneylenders also came forward to utilize this opportunity. They provided money and food to the tribals when they used to fall in crisis (especially in monsoon). Their motive was to suck them up completely at the time of paying back. They created an asymmetric agreement. The rate of interest ranged from 50% to 500%. Since the tribals were not accustomed to the 'money-economy', the high rates of interest, attachment of land for debt, payment of loan by personal service (which continued even upto third generation), sheer extortion - all these dragged them in a wretched condition. The zamindars (land-holding feudal groups), mahajans (moneylenders), contractors, traders and sepoy (police force) all formed a coterie to snatch their independence and to make them serfs. Gradually the Santals became impatient and restless. They occasionally lodged complain to the British administrators to get rid of their problems. But the administrative power did not come with aids. As a result Santals began to be organized for finding out ways and means to make themselves free from the shackle of feudal confines.

Land became the part and parcel of social, economic and spiritual life as well as the heritage of the Santals. It belonged to the respective persons by whom original clearing of forest were done. The right used to pass to the descendants through the male line. In this circumstance, when some officials received special land grant (as recognition to their service) over the already occupied tribal land, the awardees forced the Santals to vacate the land. The rampant exploitation through land grabbing and ruthless eviction made them entangled in debtness. The whole economic fabric was shattered and hunger drove the poor people in despair. However, the situation could be tackled by the administrative approach but the British administrators were ignorant as well as inexperienced in dealing with the primitive tribes. Moreover, their policy was to exploit the vast socio-economic resources of the native economy. So they did not care about the civic rights of the tribals. Even the judicial system and tribal codes were completely ignored. Further, they lived geographically apart from the tribal people. In the periphery there were police force supported by a number of administrative personnel namely *amlahs, mukhtars,* peons and *barkandazes,* etc., who were nothing but a machinery of extortion with corruption. They allowed the outsider non-tribals to exploit the helpless people. With the passage of time, the tribals realized that the ruling power was in favour of others and it would provide no safeguard for them. Naturally their grievance obeyed no laws.

The Santal resurrection was basically a peasant revolution. Sido and Kanhu, the twin brothers residing at Berhait Valley at the heart of Damin-I-Koh came forward to provide leadership to the Santals. They identified certain external forces (Diku) posing a threat to their existence. Moneylenders, contractors, political infiltrators and other exploitators were grouped under this category. Like all Santals, Sido and Kanhu were poor and hard-working peasants. Common people had a great faith in them, as they were simple, straightforward, honest and sincere. In addition, Sido had a great inclination toward religion for which people used to visit him for religious discourses. Therefore, an impact of religion was pronounced in bringing the Santal population together in order to revolt against oppression. Sido and Kanhu declared that the god had appreciated the attempt of revolution. Thus, it was easy for Sido and Kanhu to organize the agitated Santals.

An ultimatum was issued against government regarding the oppressions and misrule. At the same time a leaflet in Kayathi language was circulated where the objectives were clearly furnished. The leaflet also cited the instances where the British physically tortured the innocent tribals. The whole of Bhagalpur and Purnia came under this circulation. On 30th. June, 1855, about ten thousand tribals were assembled in a village named Bhagnadihi. They sent memorandums to the Government, departmental administrator, District Magistrates of Bhagalpur and Birbhum, O.C. of Dihi and Tikri police station and also to different zamindars and other responsible functionaries. They wanted a permanent solution of their problems and expected answers within fifteen days. After this meeting, about thirty thousand Santals marched towards Calcutta. This was the first mass upsurge in India.

Government turned a deaf ear to the request of the Santals. Therefore, the peaceful protest suddenly turned into violence. The Santals often gathered in thousands and proceeded to plunder the nearby accessible markets, murdered the police officers, traders, contractors and moneylenders. They did not excuse even the minor government officials and other innocents who fell across their path. The government was alarmed and sent troops to encounter the hordes of Santals. In August 1855, the Santals got defeated as the Bhagnadihi village was burnt to disorganize the actionists. British took strong steps in nearby areas also. Many died, many were widowed, and a large number of children became orphan. Still the courage of the Santals could not be resisted. They went forward in Southwest direction and reorganized themselves. In the mean time, on 10th. Nov. 1855, military rule was declared. Massive killing of Santals took place and various measures were taken.

The movement subsided for the time being but the seeds of discontent were not thrown away. Though the mad plan of Sido and Kanhu snatched the lives of a large number of Santals and caused great sufferings in terms of house, cattle, food, belonging and everything, it did not decrease the spirit of fighting. The people fought till the last moment of their life. In the second week of February 1856, Sido died in a confrontation and Kanhu was caught by the English. Kanhu was severely condemned by the British as a murderer and so was hanged on a *mahua* tree in the field of Jhilimili.

Still after this, the turmoil continued and British government had to exert hard to put down the revolt. In fact, the Santal insurrection of 1855 was the prelude to the nationwide Sepoy mutiny of 1857. Many social movements of present day take their impetus from the glorious struggle led by Sido and Kanhu. The official account of this rebellion has been furnished in Mc Pherson's 'settlement report'. The report said that the peaceful development of the district was temporarily interrupted by the Santal rebellion, which broke out in the Damin-I-Koh on the last day of June 1855. According to Mc Pherson, the outbreak was quite unexpected. It perplexed the British administrators and compelled them to change their outlook. However, some special reform measures were taken later. The Santals got a special non-regulation district for them known as Santal Parganas. Upto 1911, the area was under Bengal, but at present it belongs to Bihar. It became a place where Santals could protect their tradition and indigenous activities from disintegration and ruin.

30th June is a memorable day to all valiant Santals. Each year they celebrate this day as 'Hul Dibash'. The charismatic leaders, Sido and Kanhu are considered as the apostles of the insurrection.

It should be kept in mind that the feudal oppression did not come overnight, nor the administrative onslaught appeared for a day. Naturally, the resultant insurrection was an outcome of long drawn organizational preparedness. Santals of district Maldah, W.B. were restive during 1929 - 32 centering a crop failure. Their wrath fell on the non-tribal landlords and they became rebellious against British administrators. Jitu Santal, the champion of the movement lost his life at Adina mosque. This is a less publicized Santal upsurge, known as *Adina Movement* in history.

Munda Movement

Raibahadur S. C. ROY, who spent his lifetime among the various tribal communities of Chotonagpur, found the Mundas as permanent plough cultivators in the uplands. This group of tribals maintained a distinct identity in regard to both land right and social customs.

In 1765, Shah Alam handed over the financial administration of three States—Bengal, Bihar and Orissa, to the East India Company. Chotonagpur was a part of Bihar. The company claimed huge sum of land revenue from Chotonagpur, which was very difficult to be paid by the villagers. Since they could not afford it, the amount of arrears increased day by day. Under the burden of heavy taxation, some people revolted in Tamar Pargana in 1789. East India Company immediately sent troops to crush down the agitative mass, but the fire of discontent continued to smolder till 1795. Trouble broke out again during the period from 1796 to 1798, covering a larger area. Not only the Tamar Pargana, the Rahe and Silli Parganas were also included. In addition, at the turn of eighteenth century, Company imposed stamp and custom duties on Chotonagpur. Therefore, the resentment became more acute and the local king was unable to collect the revenue in time. With this issue, in 1806, company compelled the king to appoint police force for maintaining order and peace. But instead of peace, people rose in revolt, which spread various places. In 1819 - 20, rise of two leaders. *Sale Rudu* and *Konta Munda* was found whose lives ended in company's jail. The oppression and exploitation continued under new administration.

Flames of revolt were lit again in 1832 against the factors, namely, the enhancement of tax, infiltration of outsiders and the evacuation policy of the British administration. Meanwhile, a major change was noted in the life-style of the inhabitants of Chotonagpur. Rights of the Munda cultivators over the land was restricted by the interventions of the British; Protestant and Catholic missionaries came to serve them, In fact, the simple-minded Mundas failed to recognize the real source of their sufferings. The Missionaries worked hard to free the forest tribes of Chotonagpur from superstitions. They taught the tribals the ways to claim better life chances as a human being. As a consequence, in 1858, after the great Sepoy Mutiny of 1857, the Government proceeded to collect information on land ownership in Chotonagpur. At this time, they passed 'Bhuinhari Act' (1869) to protect the ancient rights of Munda related to the land. But the Mundas did not get any benefit out of this Act because they had lost most of their land prior to the enactment. Moreover, the Act had many flaws and the Mundas, for the want of education, were cheated in various ways.

With the assistance of the Christian Missionaries, though the land rights had made some progress, but the Mundas understood that the freedom, which they used to enjoy in the past, would never return again. At this phase, they also realized the forms of exploitation and neglect as made by the Government. During the period from 1889 - 90, they attempted once again to get rid of oppressions, either through wit or the weaponry. But it was not so easy. After a long effort, in 1899 -1900, the Mundas got a real leader, Birsa Munda. Under his leadership, the Mundas again dared to raise their arms to drive out all outsiders away from Chotonagpur. But like Rudu and Konta Munda, Birsa Munda also died in jail. Many of his followers were hanged as well as suffered from long imprisonment.

Mundas of Chotonagpur even today consider Birsa Munda as their hero because his protest could haunt the British in real sense. The Government thereafter started to think seriously about the protection of the Munda peasantry. A series of new laws were made which released the tribals from the clutch of the *Jaigirdar* and *Thikadar* class. The Mundas regard Birsa as a prophet cum rebel. He was claimed to be the 'Dharti Aba', the father of the world. However, this charismatic leader Birsa was a pupil of the German Missionary School at Chaibasha. At the time of revolt in 1899, he instituted a new religion by combining the elements of Christianity and Hinduism. For instance, he propagated Christian monotheism and at the same time thought that wearing of a sacred thread is essential for maintaining ritual purity.

The movement led by Birsa was basically peasant movement, although many other reasons assembled at its origin. Birsa assured the people to deliver a 'Munda Raj' which would be free from all sorts of exploitation and social oppression, However, the movement of Moundas, with religious flavour but politics gradually led to armed clashes on several occasions—with the colonial rulers, landlords and moneylenders in 1789, 1796 and 1832.

Tana Bhagat Movement

Tana Bhagat movement was developed among the Oraons of Chotonagpur. It emerged as a result of the spread of brahmanical influences.

The Oraons, like Mundas, caught Hindu influence very easily which came from different directions—The Gaya and the Sahabad districts of Bihar, the Raipur and Bilaspur districts of M.P., and the Sambalpur and Gangpur districts of Orissa. Not only the influence came from different directions; it was operated in several ways. The most interesting feature is that, the Hindus did not propagate their religion; Oraons adopted the Hindu ways and customs in their own accord.

'Bhagatism' as a religious movement sprang up around 1914 in the Western part of Chotonagpur where Oraons fell under acute deprivation and oppressions made by the local zamindars and policemen. An Oraon leader, Jatra by name, proclaimed that in a vision he had received a revelation of God for the fellow Oraons. His message readily influenced all the Oraons of the region; they joined together under his leadership. Jatra Oraon lived in the village Beparinwatoli under Bishnupur police station of Gumla subdivision. In 1914, he was only 25 years old but his advice was strong enough like that of an elderly priest.

Jatra Oraon said, he had got the order of Dharma (the principal god, Sun) that all Oraons should give up the worship of spirits (ghost) and should stop the practice of animal sacrifice. Instead, they should lead an ascetic life and should abstain from non-vegetarian food, wine, tobacco. Group songs and dances were also prohibited. Further, they should not pay rent to the oppressive zamindars and should stop working for the aliens. Acting upon his order, people refused to work for landlord and disobeyed the rules and regulations imposed by the British rulers.

The followers of Jatra Oraon were called 'Tana Bhagat'. The word 'Tana' means 'pulling together'. Jatra Oraon tried to pull all Oraons under one faith. Tana Bhagat movement is also called 'Kurukh Dharam' movement as it aimed at worshipping the god in its pure form. The devotees of 'Kurukh'* religion began to practise strict ritual purity. In every village, within their own community, wherever they got anything impure or auspicious, they sought to pull (Tana) that out. The final objective was to drive away the misfortunes by the act of praying and singing hymns collectively to god. However, the Tana Bhagat movement helped to develop a campaign of hatred and revolt against all outsiders and oppressors including the Christian Missionaries.

Although Tana Bhagat movement was launched with a socio-religious ethics, but soon it acquired political overtones. The followers of Tana Bhagat participated in the non-cooperation movement led by Mahatma Gandhi. Some of them actively participated in the Congress Sessions outside Bihar. On their persuasion, Congress Government of Bihar passed a special bill known as 'The Bhagat Agricultural Land Restoration Act' with a view to return their lost lands. As a matter of fact, the right over these lands was ceased by the British Government when the tribals refused to pay the rents. At present Oraons enjoys a better situation. Tana Bhagat political sufferers are getting a monthly stipend from the Government.

* Kurukh is the other name of Oraon.

The movement with its obsessive approach of purity made considerable upheaval among the Oraons. As a result, numerous changes appeared in the life-style of this particular people. Most of the social rites and sacraments were altered as well as purified. Oraons, became free from the exploitations of other groups, mainly the landlords and the traders. Naturally, the community was saved from the conditions of dire poverty.

Naga Movement

Population of Naga hills had a long history of discontent and unrest. Their problem originated from the socio-political processes viz., Proselytization and imposition of Western values and education. A conflict appeared between tradition and modernity. After the Second World War, especially when the democratic institutions were introduced in their area, the problem came to the scene. A transformation of native genius took place, which led to the national as well as emotional alienation.

All Naga tribes depend directly on forest produces and games. The Angami and Chakhesang tribes inhabiting the south part of the hill are the expert terrace cultivators. But other sections of Nagas in northern and eastern side practise shifting cultivation. Traditionally they had no unified system of administration. Different groups used to identify themselves in different names like Konyak, Ao, Angami, Sema, etc., but the neighbouring people like Ahoms and Manipuris used a single term 'Naga' to denote all those groups. However, the groups had a strong feeling of hostility with one another. Every Naga village was a republic; a socio-economically self-sufficient unit with its own government. When British came in India with their new administration, Nagas were afraid of losing their customary ownership on hills. They assumed a great danger from the emerging socio-political situation, which was inclined to encroach upon their cultural autonomy. So they wanted to put safeguards against their customary socio-economic rights and moved for autonomy.

First Naga outburst was found at Kohima in 1879. It was a raid upon a British administrative centre that was built in 1876. Their resentment was against the interference of alien people and also against the amalgamation, as British wanted to treat them along the people of plains. However, at the beginning, they had to accept the British rule; the Naga hill district was established in 1881. Although there was least interference in the life of the Nagas, they still felt insecure. Aiming at solidarity and unity, a club was established in 1918 at Kohima. Its branch was made at 'Mokokchung . The club was comprised of emerging Naga elites who submitted a memorandum to the Simon Commission when the members of the Commission went to visit Kohima.

If we trace the development of Naga movement, we find that at different times, the Nagas moved with different ideologies and intention. Leadership changed from person to person. But it is universal that from the inception, Naga movement has been led by English-educated modernized Nagas. Rapid Christianization brought a wave of formal education in Naga hills, which played an important role in the formation of the movement. But factions often developed in course of the movement, among the followers of charismatic personalities. The movement acquired different dimension (in form and content) at different stages, but its basic character remained unaltered. Religion (Christianity) and education were the two important factors, which acted behind Naga unrest. Kohima and Mokokchung townships have been identified as the epicenters of this movement.

At present Nagaland is a full-fledged state in India where Nagas enjoy more autonomy than the other states. The fifth schedule of our constitution provided them special provisions. No laws can be applicable to the Nagas unless that is approved by the State Assembly. Naga movement gave rise to a new consciousness—a sense of solidarity. But even after the formation of the separate state, all sections of Nagas could not come to a consensus and an underground tension continued, particularly among those who still lived in isolated condition within the forest. Some more time passed on to decline that tension. The deviated group gradually realized the futility of violence. Besides, there were continuous efforts of Naga churches, peace councils and government of India. All of them wanted to bring peace among the extremist Nagas. Nagaland as a state received a lot of attention from Central Government. Several developmental schemes were implemented for the upliftment and amelioration of the people in Nagaland.

The principal movement of Naga tribe occurred in Nagaland, but there were other movements too. For example, the three groups of Nagas in Manipur, the Zemei, Liangmei, and Rongmei went on rebellion. This movement is known as *Zeliangrong movement* in history (1917 - 19). In 1891, with the British conquest in Manipur, the hill tribes came under the direct administration of British, They were compelled to pay house-tax to the government. This was never paid earlier when the people were under the control of Manipur king. The economic life of the poverty-stricken tribals was greatly affected by the new tax. They were loyal to the authorities, but still got no sympathy. Obviously their grievances bursted out against the colonial rulers.

Khasi Movement

Very little is known about the ancient history of the Khasis because of the paucity of historical writings. Two prominent Khasi rulers, Siem (chief) of Sutanga and Siem of Khyrim are only known who established their kingdom in the plains of Sylhet. These chiefs of Khasi were hostile with the people of the plains who probably drove them to the hills and compelled them to live in isolated condition. The acquisition of the Diwani of Bengal by the East India Company in 1765 ushered the first contact with the Khasi people. East India Company realized the rich mineral potentialities and trade facilities of Khasi territory. They wanted to bring the hills of the Khasis under their control and, therefore, tried to make friendship with the people through some developmental work. The honest natured Khasis could not anticipate this cunning motive at the first hand. Naturally they became trapped and exploited at the hand of the English people. David Scott, one of the English officials in 1824 made diplomacy with the Khasis. He promised the chief, Ram Singh of Sutanga for military help in the fight of the Khasis against the Burmese. In this respect an agreement was also made and the Khasis gladly accepted the English in their territory. David Scott also planned an economic blockade to exclude the Khasi traders from the frontier markets, for which the sale and purchase of the products dwindled down and Khasi trade was paralyzed. Sylhet became already a part of British possession since the Company took the acquisition of Diwani of Bengal (1765). Treaty of Yandaboo was signed in 1826 (between East India Company and Tirot Singh, the chief of Nongkhlaw) which brought the two connecting valleys of Khasi hills under the control of British. David Scott was so successful in exploitation that later British appointed him as a Commissioner of Assam, where he served till his death in 1831.

Soon after the signing of the Treaty in good faith, the Khasis realized that they had made a great blunder. As the Treaty was written in English language, they could not make out the flaws in it. The terms of Treaty were not equally favourable for both the parties; it was an agreement between a superior and an inferior power where the Khasis had been the losers. The first Khasi chief who gave an encounter was Bor Manick, the ruler of Shillong. In 1828, Bor Manick marched to Doomoreah to seize the collected revenue by the revenue officials of the East India Company. This was the premier challenge of the Khasis and it prepared the ground on which the Brutal Nongkhlow Massacre happened on 4th April 1829. In order to crush the English garrisons, the Khasi warriors attacked with bows and arrows, spears and shields. The incident also inspired the neighbouring chiefs to be united in order to resist the British.

Since 1924, the relationship between the English and the Khasi states was regulated again and again by treaties, engagements and negotiations. Treaties were signed and executed to secure the friendship as well as good will of the Khasi rulers and also to get a firm hold of the hills. In 1857, W. J. Allen prepared a report on the administration of the Khasi and the Jaintia hills. Another report was submitted by Montagu Chelmsford in 1917. Different such reports came at different times and the policy of British (towards this Native State) changed time to time from 'Non-Intervention' to

'Subordinate Isolation'. India's independence affected the political situation of Khasi and Jayantia hills. The Khasis were brought closer to the Indian scene. On 7th September 1949, Dr. Ambedkar being the chairperson of'Drafting Committee' made an amendment of the sixth schedule to protect the traditional rights and privileges of Khasi people. The Constitution was discussed from 17th November 1949 and finally passed by the Assembly on 26 November, 1949.

The influence of British was felt on the political and economic fields, but they could not bring any major change in the Khasi way of life. The metamorphosis of Khasi society actually began with the arrival as well as work of the Western Christian missionaries. The contribution of the missionaries was found mainly in the sphere of health and education. They also tried for the development of Khasi language by giving it a script. This script showed a similarity with Roman script. However, the efforts of the missionaries subserved their primary aim of evangelization. Under the influence of the missionaries the Khasis were benefited a lot and they adopted the values of the missionaries. As a result, a new dimension was added to the tribal ethos and identity, and at the same time a vacuum was created in the socio-cultural system of the Khasis. A majority of them became critical to their indigenous values and customs. They looked down upon these as obsolete and redundant, and therefore, started to hate their own identity as primitive and backward. This process was supported by British rulers both overtly and covertly. Naturally, the Khasi converts with new ideology began to dominate in the socio-economic and political fields. It created a sense of deprivation among some traditional Khasi intellectuals. They reacted against the situation by resisting not only British, but also the Christian missionaries and Khasi converts. This consciousness arising out of deprivation of structural condition revived a nativistic ideology. They urged to unite among themselves to hammer on alien power and control. A strong effort gave birth to an organization called Ka Seng Khasi on 23rd November, 1899 at Mawkhar, Shillong. This Seng Khasi spearheaded a revivalistic movement for the preservation of Khasi religion as well as Khasi identity.

Bodo-Kachari Movement

The Bodo-Kacharis constitute the largest ethnic group among the tribal population of the State of Assam. But Assam embraces also other tribal peoples like Mizo and Nagas and even many non-Assamese of Burma and East Pakistan. Among all these people, the Bodos are the most educationally depressed and economically backward community. If we trace the Bodo-Kachari movement from the very beginning, we find that the nature of the movement has changed from time to time with the changing issues and leadership.

The first recognized movement has been known as Brahma movement in history, which was generated among the Bodo-Kacharis of Goalpara district in later half of nineteenth century. Kalicharan Mech, subsequently called as Gurudev Kalicharan Brahmachari was the leader of the Brahma movement. He acquired certain lessons of education for which other people respected him very much. He propagated Brahmaism where he said that a Jyoti (light) emanating from sun is capable of dispelling all darkness and may take the people to the lord Brahma. The members of the Bodo community, initiated in Brahma faith were known with the title Brahma. But, for that, they had to submit a petition to the Deputy Commissioner of Goalpara. Jamadar Brahma was the first disciple of Gurudev Kalicharan Brahmachari. A sizeable number of Bodo-Kacharis accepted the 'Brahma faith' and this changed their outlook of life. The followers took upon themselves the task of reforming the whole community. So they went forward to open schools, to stop various social ill-practices like bride price, brewing, drinking of rice-beer etc. They also strived for ameliorating the economic ills of the masses by launching co-operative movement and by articulating with the government regarding various socio-economic issues. They made the socio-religious observances and rituals very popular by minimizing their expenses. The most remarkable activity was their representation before Simon Commission for reserving the seats for Bodos in the legislative assembly and jobs under government. All these activities helped to form a tribal league in 1930's. The socio-economic and political problems of the Bodo-Kacharis were projected through the league. The college going Bodos were also united in an association called 'Bodo Chhatra Sammelan'.

Brahma movement brought about a renaissance among the Bodo-Kacharis of pre-independence era. It facilitated the emergence of Bodo elites and the impact went on economical and political field. For instance, the prohibition of practices like drinking and brewing of rice-beer resulted in saving of agricultural produces in poor families. Again, the expansion of literacy among the Bodos helped in producing creative and reformatory writings (in Bodo language). Further, an 'All Bodo Sahitya Sabha' was formed in 1952 for the cultural and linguistic development. The 'Bodo Sahitya Sabha' and 'Bodo Chhatra Sammelan', later, contributed to the Udayachal movement that started under the leadership of Plains Tribal Council of Assam (PTCA). The objective of this second movement was to raise the cultural, political and economic status of the Bodo community.

The genesis of the Udayachal movement was rooted in the conflicts between Bodo and non-Bodo elites, concerning the economic and political power. From the beginning of the twentieth century the incidence of land alienation took an alarming look. Emigrants from the East Bengal (present day Bangladesh) infiltrated into their area and occupied many prosperous villages and fertile areas. Bodos tried to evict those immigrants but failed in their attempt for poor economic standing; most of them already had lost their land through mortgage and litigation. In the mean time, the number of Bodo elites were increased considerably who could raise their voice. But the number of non-tribal population of Assam was still much greater in number who were better equipped as well as better trained than the Bodo-Kacharis. However, the Bodo leaders identified certain political leaders as a source of their misfortunes. Because those leaders, instead of providing protection to the Bodo, made them socio-economic victim. They permitted non-tribal traders and moneylenders to penetrate into the Bodo areas. Those intruders not only created adverse socio-economic situations, they were largely responsible for the spreading of diseases like tuberculosis and syphilis. They also dragged the Bodos towards the addiction of opium. Therefore, feelings of deprivation, distrust, discrimination and injustice played in their minds and they appealed to the government seeking protection. Letters were sent to newspapers and representations were made in this respect. They also cautioned that the delay of government in introducing measures to safeguard their interest might have frustrated them, so they could be involved in desperate action. In 1972, they put forward a memorandum to the Prime Minister objecting to the step-motherly treatment of administration. They thought that, the administration being dominated by the Assamese-speaking people wanted to up-root the Bodo-Kacharis in a planned way. They did not tolerate their second class citizenship in the state.

The difference between the two movements is that, the earlier Bodo leaders strived for improving the material conditions of the people by placing their demands in Parliament and State Assembly. They emphasized on the facilities, such as, education, medicine, irrigation etc. They also claimed the linguistic right so that the Bodo language could be used in schools of Assam. However, their demand was met by the government—the language was introduced in the primary schools as well as in the secondary schools in 1963 and 1968 respectively. But, the leaders and elites of later Bodo movement demanded for establishing a separate state, Udayachal. They were inspired by the action of other tribal groups in the composite state of Assam who won their distinct political identity along with the preservation of traditional cultural identity. In fact, the creation of the states of Meghalaya, Mizoram, Nagaland and Manipur generated a strong resentment among the Bodos. But their demand for a separate Bodoland is still unresolved.

Nebula Movement

This movement took place among the hill people of the Darjeeling district in West Bengal. Originally the people of Nepal, Bhutan, Sikkim and Tibet were hostile to each other. But since the advent of British rule, many of them were found to migrate in Darjeeling who began to live there in amity.

The administration of British was extended to Darjeeling in 1835. At that time the district was inhabited by the Lepcha autochthones and the Bhotiya migrants. In 1839, Campbell became the Superintendent of Darjeeling and encouraged the immigration of the Nepalese peasantry. As a result, within ten years, an influx of people was noted who came mainly from Nepal and partly from Sikkim, Bhutan and also from South of the district. Nepalese formed the majority, so the identity of Lepchas and Bhotiyas were threatened. British kept this district under the 'Non-Regulation Area' because they had already met a few serious ethnic rebellions in different parts of India. The district Darjeeling did not get the right for representation in the Old Legislative Council that was constituted under Government of India Act in 1919. The administration was vested solely upon the Governor-in-Council. Even after the promotion of Government of India Act in 1935, the district was continued to be treated as a 'Partially Excluded Area'.

However, with the growth of the Nepali speaking population, the linguistic situation of Darjeeling was changed a lot. Nepalese as a dominant group demanded for the recognition of their language. Nepali intellectuals began to publish magazines in their own language. Some schoolbooks were also compiled and published in Nepali language. Inspite of adapting the policy of segregation, some people of the district became conscious about the socio-economic growth of their area. As early as 1906, the Gorkha Samity was established with a library, theatre hall and social welfare unit. The activities of this association aroused a sense of ethnocentrism and cultural chauvinism among the Nepalese. In 1920 - 21, under the leadership of Bahadur Rai, another social union was established for the hill people. The members of this union went forward to generate the urge of independence among the people of the district. They wanted to follow the path of Gandhiji as they got acquainted with Gandhi in 1921 during his visit to Darjeeling. The alert British officials, in order to prevent the growing influence of Indian National Congress, declared Darjeeling as 'Partially Excluded Area'. But, by this time, the other problems grew serious. The Lepcha and the Bhotiya leaders could not accept Nepali language as a medium of instruction. Their opposition gave rise to a situation of tension and enmity.

The British administration assumed that the situation was gradually going out of their hands. They could not further depend on the Gorkha leaders who strived for the political power. So they employed Ladenla, a Bhotiya official to develop solidarity among the hill tribes. Ladenla met with six hundred hill people in a cinema hall at Darjeeling on 23rd. December 1934. In that union, he preached an ideology of regional integration in opposition to ethnocentrism and Ne-bu-la was born. Although the Nepali-speaking members formed the majority, Ne-bu-la included the Lepcha and Bhotiya members in the same plane. Their objectives—the employment assistance, social welfare and general upliftment attracted the people of other communities like Sherpa, Limbus, Rais, Gurung, etc. The Nebula movement brought a renaissance among the hill people of Darjeeling. A unity as well as democratic consciousness developed in them.

Since the leaders of the Nebula movement were working under the British Government in different capacities, the common people identified them as agents of the British Government. Instead of confidence, it aroused suspicion among the mass. Still Nebula was found to function smoothly; there was no major issue of conflict. But when government allocated a seat for them in the Provincial Legislative Council (for the three sub-divisions of the district Darieeling), chaos appeared over the choice of the candidate. As the Nepalese were the majority and as they formed a league (Gorkha League), first of all, they aspired for the political position. Ladenla also died suddenly. The Nebula movement automatically perished with time.

ETHNOGRAPHIC NOTE ON A FEW TRIBES

Toda

Introduction

The Toda is a pastoral tribe who live in the Nilgiri hills of Southern India. The geographical position as well as the climatic condition of the area is very congenial for the livelihood of this tribe.

The Todas live there from immemorable past in association with four other tribes namely Badaga, Kota, Kurumba and Irula. So far as the physical features are concerned, the people possess tall stature, fair complexion, long and narrow nose, long head with black wavy hair. These features place them under the Caucasoid racial stock; other groups of the same area belong to Proto-Australoid stock. Ethnographers and physical anthropologists hold a special interest in Todas as the group exhibits pastoralism as well as a tradition of polyandrous marriage. They show no similarity with any other Indian tribes, physically or socially. Sm. Kamaladevi Chattopadhyay in her book 'Tradition in India' (1978) had stated, "The Todas have been a puzzle to many anthropologists and sociologists for their physiognomy, temperament, costume, habitation, way of life, etc., none of which bears comparison with any other Indian tribe". The language in which the Todas speak is very close to Tamil.

The word Toda has been derived from the name 'Tundra'—the sacred tree of the Todas. Numerical strength of Toda population recently shows a sharp decline. Government of India is very anxious for them. For increasing the number of heads, some important measures have been taken by the Government.

Material Culture

The Todas present a classic example of pastoral economy. They know neither hunting, nor agriculture, and rear only buffaloes. From the milk of the buffaloes various products, such as ghee, cheese, butter, curd, etc., are made. These products are partially consumed by themselves. The rest is sold or exchanged with the neighbouring tribal communities. In lieu of milk products, the Todas procure different other things necessary for daily life. Socio-economically a symbiotic relationship is followed among the tribes of Nilgiri hills. The Todas enjoy the highest social status *.

Each family remains engaged in caring huge number of buffaloes. The males of the house take the animals to the field regularly in the morning. The dairy work is absolutely laid on them. Females are not permitted to enter into the dairy house because of a taboo. They keep themselves involved totally in the household work, like, rearing of the children, fetching of the drinking water, collecting of fuels from forest, and so on. Formerly, the males used to cook, but now this task has been transferred to the females of the house. But, the processing of milk, like boiling, churning, curdling etc., are exclusively done by the males.

Milk is indispensable in the socio-religious life of the tribe. Milking is done twice a day—in the early morning and in the evening. The buffaloes are of two types. Some are regarded as the ordinary type; others are sacred. The ordinary buffaloes are maintained for subsistence requirement. Therefore, those are owned by the individual families. On the other hand, sacred buffaloes are kept in special sheds. Milk from them is offered to gods and some of these buffaloes are also immolated during worship. It is interesting that all sacred buffaloes are held as the property of the clan.

The Todas are purely vegetarians. Their favourite dish is rice, boiled in milk (jagari). They also prefer curd, churned milk and plain milk. Further, they take vegetables and green leaves in their principal meals. Recently the tribe showed some change in their food habit. Since they are trying to subsidize their pastoral economy by taking up agricultural activities, bread (made of wheat) and pulse have been added to their diet. Occasionally, they eat meat of deer, but other meat is forbidden for them. Meat of the sacrificed buffaloes is considered as sacred, so they take this during two annual festivals.

Both the males and females of the Toda community are addicted to liquor. They do not make this liquor at home, rather buy from the market. The habit of smoking also prevails among both the sexes. Besides, the people like to use snuff in nostrils and mouth.

* In the social hierarchy, the Badagas are placed next to the Todas. Then come the Kotas and the Irulas.

The markets of the Todas are actually the homes of the neighbouring people. Agriculturist Badaga supplies them paddy and other agricultural produces, in exchange of milk and milk products. The Kotas are basically artisans. They supply various utensils made of clay and iron. Again, Irula and Kurumba, the hunting-gathering groups bring different forest products like honey, fruit, timber, vegetables etc. All of them are interested in getting back milk and milk-made food. Modernization has helped the Todas to sell their milk through cooperative in order to earn cash money.

A Toda village (Mund or Aras) generally consists of ten to twelve huts situated on the hill slopes. The huts are of two types. The first type is represented with half-barrel shaped long huts measuring about 15 feet X 12 feet. The roof and walls can not be differentiated in this type of dwelling. The inner structure of the hut is made up of bamboo-splits or canes, over which dry leaves and branches of the tree are placed. Sometimes mud is used as plastering medium. Such a hut shows no window at all. A single low door is present at the front. Inside of the hut is obviously dark and stuffy. Usually, a hut is divided into two distinct portions—the inside room is used as a workshop where the females do not get an entry. The outside room is meant for living and other household work, including cooking. Cattle shed is separately constructed; the main dairy-house may or may not be attached to it. The second type of hut is not barrel-shaped. It is circular in shape and made of stone. Here also, the roof and the walls can not be distinguished. The roof is very high and constricted gradually from the rounded wall. This type of hut is used to keep the sacred buffaloes.

Earthen pots and containers are used in kitchen. Wooden plates, leaf-cups and iron pans are not uncommon. In dairy work, big cauldrons made of iron are used along with wooden laddles and earthen urn. Today aluminium buckets has come into use.

The Todas are simple people. They do not exhibit much grandeur in their dress. The males use a long strip of white loin cloth which has to be thrown over the shoulder, after covering the waist. This is the traditional garment of the Todas. Sometimes they use colourful cloth for covering the upper part of their body. Women use long thick cloth covering almost the entire body. Nowadays, females have been found to use stitched garments. These are colourful blouses having shirt like lower part. Children peculiarly shave their head keeping locks of hair in front as well as in the back. Women like to decorate their bodies with tattoo marks. They keep long hair in plait. Moreover, they are fond of ornaments. Earring, nose-ring, nose-pin, finger-ring and chain around neck are preferred which are usually made on silver, copper and iron. The Toda women are expert needle-workers. They can stitch fine designs on cotton clothes and woolen shawls, especially with red and black threads. Recently their embroideries have attracted the tourists and the community has been economically benefited.

The Todas use the implements like axe, hoe and knife in making and repairing their huts. Although they were traditionally accustomed in rubbing wood pieces for making fire, at present they have learnt the use of matchbox and the stick. But it is a pity that the people lack the concept of personal hygiene. They take bath rarely, twice or thrice in a month. At the same time, to get rid of dry and chill climate, they smear the molten fat from buffaloes over their bodies.

Social Organization

Like any other tribe, Todas are strictly endogamous i.e., marriage alliances occur exclusively within the tribe. The tribe in principally divided into two subdivisions called 'moiety'—Tartharol and Teivaliol. Each of these two moieties is again endogamous. The members of Tartharol consider themselves superior than the Teivaliol. Therefore, the former remains in charge of the sacred buffaloes, whereas the workers come from the Teivaliol to look-after the herds.

Each moiety is sub-divided into a number of clans. The Tarther possesses twelve clans while the Teivali contains six clans. Each clan is exogamous, patrilineal and territorial in nature. A number of villages are found to be grouped under a clan where the names of the villages are derived from the names of the leaders of the respective village. Each clan further shows the existence of two sub-clans — Kudr and Polm. The Kudr regulates the ceremonial aspect of the clan, whereas the Polm keeps an account of finance. However, no clan ever encroaches the pastureland of other clan.

A sub-clan is composed of a number of families. The Toda families show some queer characteristics. They exhibit polyandrous type of families. A woman with her multiple husbands and children usually form this type of family. The husbands may or may not be the brothers. If the husbands are brothers, the family is called 'fraternal polyandrous family'. Besides, compound families are found among the Todas. In this type, apart from a common wife, the husbands keep their individual wives. All of them continue to live together with their children. Obviously, children of such families are known after their mothers; sociological fatherhood is more important than the biological fatherhood. The kinship system of Toda emphasizes on classificatory terms i.e., each term of designation denotes a number of relatives in the same rank, belonging to a particular sex. For example, all male relatives equivalent to the status of one's own father are called with the term 'father'.

Marriage proposals for girls come at the age of five or six. Father of the boy blesses his daughter-in-law with a new piece of cloth. But the girl stays with her parents until she gets young (15-16 years). Since Todas practise 'fraternal polyandry', wife of a brother is regarded as the wife of all brothers. Once the community used to practise female infanticide. Therefore, the number of females became gradually less than the number of males. As a reason of disparity in the ratio between male and female, polyandry got a sanction in the society. But the evil practice of female infanticide was ultimately stopped due to intervention of British administration. The number of females has again started to increase. Now monogamy is more or less possible in this society. Still Todas go on practising polyandry as a traditional custom. At the same time, availability of more females has urged the males to keep wives on individual basis and such practise has given rise to the compound families. Social system has been more complicated.

Both type of cross-cousin marriage are in vogue in Toda society. In all cases of marriage, bride price is compulsorily paid. Usually brothers share this price. A barren woman or a quarrel-some woman can be divorced. In this case, a buffalo has to be given to wife as compensation. After divorce, both the partners are free to marry again.

Authority in a family rests on the eldest male. Descent is patrilineal. Property passes from father to legal sons according to the rule of inheritance. Females have no right on property. However, three types of property are recognized among the Todas—Personal property, Family property and communal property. *Personal property* includes the clothes and ornaments. *Family property* comprises of hut, utensils and ordinary buffaloes. On the other hand, the homestead, pasture ground and sacred buffaloes owned by the clan come under the *Communal property.* At the time of inheritance of the family property, the eldest as well as the youngest son get the privilege; both of them own one buffalo extra.

Administration of Toda village rests on a council (Noym). There is no separate leader, except the five members of the council. Of these five members, three come from Tarther moiety; one comes from Teivali moiety and the other comes from neighbouring tribe Badaga. The representation of Tarther is greater than the Teivali on account of its superiority. Although theft and robbery are unknown to Todas, the other disputes like criminal offences and unsocial activities are dealt by the council. Besides, it organizes and conducts the communal ceremonies.

Cremation is the tradition among the Todas. The death body is covered with a new cloth and carried to the cremation ground. Food and ornaments are provided with the body. Funeral rites are of two types—Green and Dry. The Green funeral is performed immediately after death. In the Green funeral, a few (generally two) buffaloes are immolated with a belief that those buffaloes will serve the diseased soul in the other world. However, after cremation, pieces of charred bones and hair are collected; those are kept preserved for the Dry funeral, which is to be held after a few months. This second funeral rite or Dry funeral is usually performed collectively, where the chief-mourners of different dead persons join together. According to Toda belief, a soul may rest in peace only after the Dry funeral.

The Todas are found to worship different gods and goddesses along with a large number of spirits. Of them, Tiekirzi is the most powerful goddess—the creator of earth and mankind. Onn, the brother of Tiekirzi is the most important god who saves the Todas from all dangers. Numerous ghosts and evil spirits are considered to be responsible for bringing unnatural death, diseases, epidemic and loss of milk in buffaloes. The people have devised different rites, witchcraft and sorcery to get rid of them. Besides, Todas have a faith in transmigration of soul. Recent census report suggests that a good number of people of this community have turned towards Christianity leaving their traditional animistic belief.

Chenchu

Introduction

The Chenchu is a food-gathering tribe of Andhra Pradesh. Main concentration of this tribe is found in the districts of Mahabubnagar, Nalgonda, Prakasm, Guntur and Kurnool, at the north of Nallamalai hills. Some Chenchus also live in adjacent states like Orissa, Tamilnadu and Karnatak. The word Chenchu means a person living under a tree (Chettu). Another version relates the term with the habit of eating mice. However, in physical features the Chenchus resemble the Proto-Australoid group. They show short stature, long and narrow head, round or oval facial profile, dark complexion, wavy hair, short nose and profuse body hair. The people talk in Chenchu language with Telegu accent.

Material Culture

By tradition, the Chenchus are forest dwelling tribe. They used to roam in dense forest searching fruits, roots, tubers and honey. They also liked to collect tobacco leaves, tamarind, mahua-flower and different other kinds of leaves. The leaves were made into leaf-cups and leaf-plates, which had a great demand in local market. Even today the Chenchus earn a lot of money by selling the cups and plates made of leaf. These people also utilize Mahua-flower in making liquor, which is also sellable in the market. Tobacco leaves are made into cigarettes, which are generally consumed by themselves. Honey is collected directly from large beehives after driving the bees with smoke. Honey is so favourite to them that one can eat a honeycomb entirely including the wax as well as the nascent bees.

In forest the Chenchus usually move in groups. In course of gathering, they also hunt small animals like deer, wild boar, rabbit, wild cock, rat, squirrel and birds. Females accompany the males in hunting. Hunting is mainly done with bow and arrow. Since the people do not possess much skill in hunting, they take the help of traps. Herbal poison is used in fishing. Recently, this tribal community has shown their interest in agriculture. Some of them are cultivating the land; many of them work as agricultural labourer. Forest and road contractors have also been successful in employing many Chenchus as porters, on the basis of daily wage. The people are quite satisfied with these novel jobs.

Most of the Chenchus still depend largely on forest produces. Different kinds of roots, tubers, stems and leaves are boiled to be edible. During summer, when roots and fruits are not available easily, they go on hunting. The flesh of the hunted animals is simply roasted in fire before eating. A few years back, the Chenchus preferred to eat the flesh of monkey but now the practice has been abandoned by rule. On the other hand, they have been accustomed in taking boiled rice along with the gruel. Sometimes, they are found to prepare hand-made breads from millet. Smoking of tobacco and drinking of liquor are the part of their culture.

The Chenchus migrate from one place to another after every couple of years. They are expert hut builders; a hut may be built within three to four hours. These huts are small and conical in shape, made of branches and leaves of the wild trees. The walls and the roof are not distinguished, ventilation is lacking, and one has to stoop down in entering into these huts. Ten to twelve closely set huts usually make a village. The Chenchus prefer to live on hill slopes or high ridges as this position is favourable for watching the enemies from a long distance. In making a settlement, the people primarily search for a source of drinking water. As a food-gatherer, they possess digging stick, spade and hoe. The essential weapons for hunting are bow and arrow, axe and spears. All these tools show iron blades or iron heads. Household utensils are usually made of clay and wood, but cups and plates made of leaf are quite useful. Recently in contact with urban people, they have learnt the use of aluminium containers and cooking pans. Baskets made of bamboo splits and leaf mats are common in every household. The Chenchus, who have started agriculture, keep agricultural implements in addition. Some Chenchus domesticate animals like sheep, goat, poultry bird, etc. Not only they drink milk and eat eggs, a part of the products are sold as well. Besides, dogs are domesticated in each family for the purpose of hunting. These people, in general, are in the habit of rearing up different animals; they often sell the animals in market at a good price.

In olden days, the Chenchus were not much concerned with dress. They used to wear barks of the trees or skins of the animals. In a little advanced stage they had stitched the leaves in order to make a dress. But at present everyone of the community is using clothes, imitating the persons of the neighbouring communities. But still the general dress of the males is just a string around their waist where a piece of cloth hangs from front to back so that the genital organs can be covered. During winter, they use another piece of cloth to wrap the upper part of the body. Women use coarse saris but the children demand for modern dresses as they find in their surroundings. The Chenchus irrespective of sex, like to wear ornaments. The males use earrings and finger rings made of brass and bell metal. The women use a variety of ornaments manufactured of brass, copper and silver. The ornaments made of beads are equally popular. Young girls are fond of colourful glass bangles. Tattoo marks of numerous designs cover the whole body of a Chenchu woman. Foreheads and hands seem to be especially suitable for these marks. The long hair of the women are regularly combed with coconut oil and made into buns.

Although the Chenchus are not good at handicrafts, they often prepare baskets of different sizes with bamboo splits. The musical instrument known as *Jhapeta is* made by the Chenchus themselves. It is nothing but a circular bamboo frame tightened by the sheepskin. Besides, they manufacture a kind of bamboo flute and an instrument resembling the Hawine guitar *(Kineri)*. All these instruments show engraved pictures on them. But the people still use the pieces of flint for producing fire.

Social Organization

The Chenchus are divided into some local groups for the purpose of peaceful hunting. None of them enter into the territory of others. Rather they move from place to place in their own territory and make the temporary dwellings. Existence of four groups have been so far identified :

(i) Konda Chenchus are found in Kurnool and Mahabubnagar districts.

(ii) Koya Chenchus are found in Bhadrachalam district.

(iii) Ura Chenchus are scattered in towns and cities and live exceptionally in close contact of advanced groups.

(iv) Daseri Chenchus are confined mainly in Kumool district.

Not only the tribe Chenchu, but also all its regional groups are endogamous in nature. Ura Chenchu and Konda Chenchu have recently started marriage among themselves. However, each regional group is divided into a number of clans. C.V. Furer-Haimendorf had recorded twenty-six clans among the Chenchus such as Marepalli, Madla, Thokala, Nimalla, Chirugala, Nallapthula, Iravala, Palicherla, Udutaluri, etc. Each of these clans is exogamous, patrilineal and associated with a totem. Totems are generally named after the animals and plants. They are regarded as very sacred and so are often worshipped.

A clan is further divided into a number of families. Most of the families are of nuclear type, based on monogamy. Although polygyny is not restricted but it is rare in occurrence. The community never exhibits a joint or an extended family. Married sons with their wife and children live in separate huts within the domicile of the father. Father as the head of the family exercises his control over others. Thus, the families are patrilocal and patrilineal. The kinship system is classificatory as well as bilateral in type. The Affinal kins are though distinguished from the Consanguineal kins, the Chenchus count both the father and mother line in determining the descent. A large number of persons are designated by a single term. Such terms are applied to the persons of same age and sex who are more or less equal in status.

Marriages take place either by negotiation or by elopement. Cross-cousin marriages are preferred in the community. The wedding procedure is very simple in the case of marriages that occur after the permission of guardians. Both the bride and the bridegroom sit on a mat in the house of the bride. The corners of their clothes are tied tightly as a symbol of union. The invited relatives and neighbours present themselves in that spot to bless the couple for their conjugal life. Bride price is also paid at this time. When guardians do not give consent for a marriage alliance, the boy elopes the girl. They go to an another village and spend some time there. After a few months or a year, they come back to the original village. By that time, the anger of the guardians is subsided and the couple is accepted in the family. Widows of Chenchu society have a sanction for second marriage. In case of maladjustment, either of the spouses may want divorce. Children of the divorced couple stay with mother until they are adult. Thereafter they go back to their father.

Chenchu society is strictly patriarchal. Father's property is inherited by the sons. Distribution is generally equal among the sons, but sometimes elder son gets a little more. Two types of property are recognized—personal and communal. Personal property includes clothes, cattle, household implements, weapons, ornaments, homestead land with the dwelling, etc. Communal property comprises of hunting and collecting ground, which is enjoyed communally. Although women get nothing of their father's property, they inherit mother's ornaments, solely.

Administration of a Chenchu village rests on a community council known as 'Kul panchayat'. It is formed with the representatives of different clans. The head of the council is called 'Peddamanchi' who is helped by a minister. However, the village council remains responsible for settling all types of the quarrels and disputes of the village. It also look-after other important social problems relating to inheritance, divorce, custody of children, etc. At present, the function of the community council has been diminished to a great extent.

In Chenchu language, the soul is termed as 'jivam'. When a soul leaves the body (in order to meet the god), the person dies. The dead bodies are usually buried according to tradition. But recently a trend has been developed towards cremation. Whether it is a burial place or a cremation ground, the area is different for different clans. Neighbours and relatives take the corpse for disposal. Two types of funeral rites have been observed among the Chenchus—*Chinnadilal* and *Peddadinal.* Chinnadilal is performed three days after the death, whereas Peddadinal is performed fifteen days after the death. Chinnadilal is specially meant for children and unmarried persons; Peddadinal is observed in case of married persons. A period of pollution is maintained until the funeral rite is performed. The pollution gets over with a purificatory bath. A grand feast is arranged at this time to entertain the relatives and the neighbours.

Chenchus believe in a number of deities and spirits. The prime god is Bhagaban-taru who lives in the sky and look-after the Chenchus in all their work. Thunder and rain are his weapons. The other important deity is 'Garelamai Sama' who is the goddess of the forest, benevolent in nature. She silently watches the people and saves them from danger, particularly in the midst of forest. Therefore, the Chenchus always worship this goddess before going for hunting or honey-collection. Besides, there are Potsamma, the god who rules on the diseases like pox and measles; Gangamma, the deity of water; Mayasamma, the deity who protects the Chenchus against enemies. The Chenchus have also been found to adopt some deities and ceremonies of Hindus. Magicians and sorcerers are present in this society from the very beginning. Divination is practised occasionally.

The Chenchus drew the attention of anthropologists and social scientists since 1932. Steps were taken again and again to settle this hunting - gathering tribe in the plain land. British government in the district of Kurnool (1932) and Nizam's government in the district of Mahabubnagar had tried for the socio-economic development of this tribe. The Chenchus were encouraged in agriculture; subsidy was given to them for procuring land, cattle, plough etc. But traditional Chenchu mentality did not accept the new ways of life. They had gone back to their original environmental set up ignoring the facilities of colony. The relationship of Chenchus with forest was disturbed since 1952 because forest plan was implemented at this time without considering the conditions of Chenchus. Restrictions came on them, which resisted their free movement in homeland. Under acute socio-psychological pressure the honest kind-hearted Chenchus turned into mischievous criminal tribe. The criminal activities increased day by day. Some agencies began to approach them with different welfare programmes. After a prolonged probing, a gradual change of attitude was noticed in them. They are now accepting different educational schemes, modern medicines and family planning devices. A section of the Chenchus has been supplied with commodities through cooperative stores.

The cultural invasion among the Chenchus was actually started about two millennium years ago when the country was first invaded by the outsiders. Different kings invaded this country at different times. The Pallavas, Vishnukundalinis, Kadambas, Cholas, Kakatiyas, Reddy kings, Vijaynagar kings, the Mughal kings and the Nizams are notable among them. Following these invasions, the Chenchus became more and more defensive but some outside traits percolated among them quite silently. This percolation took place mainly in the sphere of religion. According to V.N.V.K. Sastry *1, many Chenchu gods have been replaced by the Sanskritic gods, in course of time.

Andamanese

Introduction

The Andamanese is a dying tribe who lives in Andaman Islands. The Andaman Islands are a part of a chain of Islands that stretch in the South of the mainland of India, between Cape Negrais (Mayanmar) and Achin Head (Sumatra). In fact, Andaman Islands comprise of the Great Andaman, Little Andaman and a number of smaller islands. The Andamanese is one of the four primitive tribes *2 of Andaman Island who still live in a food-gathering stage. So far as the physical features are concerned, the people show very dark complexion, frizzled hair, thick lips and flat nose. Body built is short and stout with illformed but slender limbs. The people belong to the Negrito racial stock. But they are the descendants of Palaeo-Melanesians, instead of the Negroes of Africa. The blood-group frequencies of Andamanese do not at all resemble with the Negroes. Linguistically they exhibit a separate family having no apparent affinity to any other family of language.

Although the Andamanese represent one of the ancient surviving cultures of the world, the people are friendly enough. During 1788 - 89, when the British administrators tried to colonize Andaman Islands, they saw the Andamanese as most docile of all inhabitants. The population of Andamanese gradually dwindled down. Census enumeration shows that there was 625 heads in 1901, which had turned to only 28 in 1981. The number of females are just one-third of the males. The people have reached almost at the verge of extinction.

During the period from 1906 to 1908, when Radcliffe-Brown conducted his research work among the Andaman Islanders, he found more or less 625 population as shown in the Census of 1901. In his book 'The Andaman Islanders' (1922) he remarked, "during the last fifty years the number of Andamanese have been greatly diminished". It was known that in 1857, there were twelve groups ofAndamanese with a population of approximately 5000, who used to live in different parts of Andaman. Radcliffe -Brown explained the decline as a resultant of European occupation in the

*1. V.N.V.K.Sasoy—'Chenchu—Food Gatherers and Hunters of Nallamal forest: Problems of Transition' in Man & Life, Vol. 19, No. 1 - 2,1993, pp. 83 - 96.

*2. The other three tribes of Andaman are Onge, Jarwa and Sentinelese.

island which introduced the diseases like syphilis among the Andamanese as early as 1870. Besides, in 1877 an epidemic of measles broke out in Andaman. The death rate among friendly Andamanese increased enormously following that incident. Also the birth rate became alarmingly low since then. Now a birth is not only a rare incidence, moreover it has been very difficult for the babies to survive. The diminution of population being combined with some other causes has altered the original mode of life of the Andamanese. The people are found to adopt the customs of other neighbouring population leaving their own. The dialect has also been diluted. Different languages in corrupt form have penetrated in them.

Material Culture

The Andamanese are basically the hunters and gatherers. As the islands are covered with a large number of rivers, and as the sea is near at land, the people show the habit of fishing. Sea and forest have played vital roles in the pragmatic life of Andamanese. As a shrinking community, Andamanese are at present looked after by Andaman Administration in Strait Island. This administrative body has recently encouraged them in horticulture and domestication of animals. The people have shown no hesitation in accepting the new ways of life as Forest Reservation Act had already restricted their movement in the forest. However, all members of the community, irrespective of age and sex, take part in economic activities. They have been trained in plantation, agricultural work and domestication *1.

From the account of Radcliffe-Brown we came to know that Andamanese maintained strict division of labour. The male members of the community kept themselves engaged in construction of hut and canoe; in manufacturing of implements and weapons. They also used to go for hunting, fishing and honey collection. Fishing or turtle hunting might keep them away from home for consecutive days. Women and children would have gone for local gathering. They used to collect fruits, flowers, edible root, leaves and shrubs from the forest. Besides, women had to fetch water, manage the cooking and other domestic jobs. Different kinds of ornaments were made by them and they were very expert in tattooing.

Present day Andamanese also show a great cooperation between the sexes. Wives help their husbands in all economic activities. Traditionally the people are prone to sea fishing, so they have been the great boat-builders. But they do not make rafts like other tribes. As rafts are not at all suitable in rash sea, canoes of different sizes are made by the Andamanese. However, the young generation shows a diverging interest; their attention has been shifted to other activities. Household goods include, canoe, canoe making tools, fishing nets, wooden implements, earthen pots and baskets of different type. The pots and utensils are generally decorated in colour. Colours are obtained naturally from ores. The aesthetic sense of Andamanese is found to be reflected in various decorations. They are also fond of music; music and dance form an indispensable part in all communal feasts and festivals.

Bow and arrow are the favourite hunting implements of Andamanese. They use these from time immemorial. They had learnt to make spears, particularly in the period when they started to employ the dogs for the purpose of hunting. For turtle and large fishes they still prefer arrows and harpoons with a float. Nets are used to catch small fishes; these are thrown either from the shore or from the canoe. Andamanese know the methods of poisoning or stupefying fish in pools and rivers. Digging-stick is an effective implement in digging the roots, just like a hooked pole for tearing the fruit from a tall tree. Adze is sometimes used for obtaining molluscs and for cutting honeycomb. Use of string or thread is very popular among the Andamanese. They make it from the fibre of Anadendron, Gnetum and Hibiscus scandens as noted by Radcliffe-Brown. From the account of Radcliffe-Brown we also know that the people were expert mat-makers. Simple form of mat making was known to

*1. In 1858, before the introduction of dogs, Andamanese had no domestic animals. Now they breed pigs and fowls in captivity.

each and every person. The framework of a mat was made with bamboo or cane strips. Strips of same size were placed parallel to each other. Inhabitants of Little Andaman used bamboo-mats to sleep on them. Traditional huts of the Andamanese were also made with mats. For making such a hut, at first, a strong mat had to be made with strong palm leaves. That was placed on a rafter and tied with strips of cane. Even the hunting shelters of Andamanese at the top of the trees often utilized these mats.

The Andamanese used to make two kinds of huts—ordinary or temporary hut and communal hut. Ordinary huts were meant for a single family, whereas the communal huts were like a camp to accommodate a number of families. At present, the diminishing population of Andamanese shows no encampments or village settlements. Rather, after their settlement in Strait Island (1969 - 70), they live in a colony of cemented huts where the roof is made of corrugated sheet. Now they do not have to search an open space surrounded by forest to keep their huts secured from the gust of wind.

The Andamanese are basically non-vegetarian. They eat flesh of various animals like deer, pig, fox, rat, snake, turtle, bird etc. Fish and shells are also favourite to them. But their principal diet is often supplemented by fruit, roots, tubers, honey, etc. They do not leave even insects. Roasting, baking, frying and boiling are the devices that they employ for processing the food items. There are certain food taboo for pregnant women; pork, turtle and honey are forbidden to them. Administration provides the people with free ration but most of the Andamanese are in the habit of selling them back to the outsiders; they buy opium with that money.

Coming in contact with modern civilization, the Andamanese have learnt the use of cloth. At times, Administration itself supplies them some urban type of dresses. But they do not like such body coverings. In the olden days, the males of the community used to wear a rope girdle or a strip of bark around their waist. A rope-girdle included a number of fine strips of cane or simple leaves running parallel to each other, being tied with the rope. Women used the same girdle but usually with a shoulder strap. The shoulder straps were nothing but a belt-like bark, which lay over the shoulder, crossing along the chest and passing beneath the breasts. However, the preference of ornaments has not changed at all. Both men and women wear different ornaments made of shell, bone, seed and pieces of bamboo. A simple string made of fibre is often found around the neck or around the leg, just below the knee.

Social Organization

Formerly the Andamanese were differentiated into a number of local groups where each of the groups possessed their individual right over certain recognized area, for hunting. In 1857, twelve such groups were in existence covering north, middle and south of Andaman. The population was about five thousand. A sharp decline of the population was noted from 1857 to 1901, and also from 1901 to 1981. Five thousand heads of 1857 were reduced to only 28. Naturally the local organizations broke down and the living members of different groups have been united together. No clan exists among these people.

Traditionally, the society of Andamanese is patriarchal. Father is the head of a family. Old men and women are respected by all. Formerly, the marriages took place according to guardians' selection. Partners used to come necessarily out of the group. But now there is no scope of choice or selection as the society has shrunk a lot. Marriage is settled for a boy and a girl immediately after attaining the puberty. The community as well as administration of Strait Island wants to produce more children as early as possible. The childbirth has been very difficult and at the same time mortality has been pronounced. Most of the children do not survive more than two years. However, during pregnancy, husband takes care of his wife in all respect. Since children are very precious for the community, every member of the society comes forward to nurse the baby and the mother. At the time of delivery, an elderly woman of the community assists the husband.

Although Andamanese form a very primitive social group, but their sense of generosity and hospitality is very remarkable. This is understood from the modes of cooperation, inter-group and intra-group relationships. At present, the community is guided by a chief. Both the chief and his wife enjoy some special privileges in the community. In olden days, a chief had to acquire his higher social position by dint of his extraordinary performance in hunting and fishing. Nowadays the social set up has changed; the community has been reduced to a minority group. As a result, chief's power and position has been dwindled down. But chief's wife plays a motherly role, so the unmarried girls of the community accept her authority.

The Andamanese do not lament over the death of a person, especially if he is an old person. They try to dispose off the dead body within eighteen hours. Burying of the body is a traditional practice of the community. Formerly, the people had a complex system of belief and practices, customs and traditions. But, for a long time, they are out of touch of those age-old practices. Rather, in contact with the advanced people, the diminishing community has received new impetus for survival where the original customs and beliefs have been abandoned. The Strait Island, the modern home of Andamanese shows no communal hut; no communal dance, feast or festival is held there. But the deity 'Pulga' is still worshipped in every household because he is considered as the central force of existence. Unproductive economy and barrenness of women are the main problems of present day Andamanese. The simple people blame on the evil spirits. They believe that the various evil spirits are responsible for various sorts of unwanted phenomena. A lot of superstitions are associated with fire and the evil spirits.

In 1976, L.P.Vidyarthi *1 had called Andamanese as a dying community. In the same year R.S.Mann *2 had pointed out some crucial reasons for demographic decline of this population. V.S.Upadhaya (1987) *3 tried to explain the factor of gradual extinction with the accumulation of harmful recessive genes in isolation. However, immediate research and action programmes have been suggested by the scholars for the amelioration of the situation. Voluntary Organizations with the involvement of anthropologists, social workers, doctors and administrators may play a vital role in saving this rapidly vanishing population.

Birhor

Introduction

The Birhor is mainly a nomadic tribe but majority of them has settled down. They are distributed in the hilly areas of the State of Bihar, Orissa, Madhya Pradesh and West Bengal. The main locus is Bihar where the people are found in the Districts of Hazaribag, Gamla, Ranchi, Giridih, Chatra, Kodarma and Lohardaga. The word Birhor is composed of two *Mundari* words—*Bir* and *Hor. Bir* means forest and *Hor* means man. Therefore, the particular word *Birhor* means the 'people of the forest'. The whole Chotonagpur plateau shows the presence of Birhors who live within dense forests.

Ethnically Birhor belong to Proto-Australoid group. The people show short stature, long head, wavy hair, dark skin colour and broad nose. They talk in Austro-Asiatic language. In the pattern of ABO blood-group distribution, Birhors resemble majority of tribes of Chotonagpur. On the basis of temperament and movement, the Birhors can be divided into two distinct divisions (*i*) Uthlu Birhor or nomadic Birhor and (*ii*) Jaghi Birhor or settled Birhors. Recently another type of Birhors has been distinguished in addition. This type is called Basalu or semi-nomadic Birhors.

The Uthlu Birhors are chiefly nomadic. They roam from forest to forest in search of food. Hunting and gathering is their prime occupation. The Jaghi Birhors live in one place for generations. They show a completely different life and culture from that of the Uthlus. Both the sections of Birhor

*1. L. P. Vidyarthi- 'Development Plans of the Tribes of Andaman and Nicobar Islands' in *Journal of Social Research,* Vol. XIX, No. 2, 1976, pp. 74-85.

*2. R. S. Mann - 'Depopulation among Andaman Aborigines—An analysis' in *Journal of Social Research,* Vol. XIX, No. 2,1976, pp. 37-50.

*3. V. S.Upadhaya- 'Retrieval from the Precipice: A case study of the Negritos of Andaman' in *Journal of Social Research,* Vol. XXX, No. I &2, pp. 67-90.

consider themselves superior than the other. So intermarriage between these two groups are totally forbidden. According to the myth of Birhors, in the origin there were two brothers. One of them could not succeed in crossing the forest while the other crossed it, leaving behind the brother. The third section, Basalu has been bifurcated from the Uthlus. This section includes those Birhors who have settled in a place about a decade. Occasional migration is noticed in this group.

Material Culture

The most striking feature of the tribe Birhor is that these people never form any permanent tie with land, even when they settle in one place. Forming small bands called 'Tanda', they move in quest of food and subsist on forest products and game. The economy revolves round the forest. Traditionally Birhors capture the games by hunting and trapping. Small games like hare, squirrel, porcupine, fowls, small deer, pig, etc., are trapped near agricultural fields of villages. They are also skilled in trapping the monkeys; their skins are sold against cash money to the people of Ghasi caste who need this skin for making their musical drum, 'madal'. Birhors usually avoid dangerous animals or big games for which they have no use. Hare trapping is very economic to Birhors. Hare's meat is of great demand in the locality. Maximum cash money is earned from the selling of hare. A hunting party comprises of eight to fifteen persons including male, female and children (in the age group of 8 to 12 years). Both hunting and trapping require a large manpower. Not only that, children get training in the operation. Although the operations are simple, they involve a lot of patience and experience. One has to guess the place of different animals by recognizing their footprints. Sometimes, several attempts from morning to night go in vain; no single game can be collected. This kind of failure for a few successive days brings chaos and starvation in the community. The hunters have to change their spot. Usually, a group never explores one forest regularly.

Beside hunting and trapping, Birhors collect wild fruits, flowers, roots, tubers, vegetables, mushrooms etc. round the year. In the rainy season, when there is a shortage of game, they have to depend solely on plant food. Food-habit of the Birhor thus depends on the natural availability of food. The people take two principal meals in a day. Lunch is termed as 'Seta Ka!wa Joma' while the Dinner is called 'Ayub Pandam'. Children's breakfast is specially made with ant-egg (how). These eggs are collected from the top of the big trees in the jungle where the ants make their nest. However, the eggs are boiled with salt and served to the children. The adults like to take them as a pickle, after cooking with ginger-pepper-chili and salt. In Birhor community, use of edible oil is very limited. Foods are mainly boiled and baked. Cooking with oil seldom takes place, only when the people get surplus money in their hand for going to the market and to buy oil and spices. But the same Birhors show an extensive use of country liquor called 'Haria'. The males drink it so much that they often skip a meal. The females drink it reasonably. Children or junior boys and girls are the exceptions; the habit of drinking does not form in them.

The Birhors make temporary huts with different kinds of wild leaves. These huts are small and more or less triangular in shape. The huts are constructed in such way that rainwater can not penetrate inside the room. Still the Birhors change their place of abode with the change of season and convenience, at least four times in a year. The huts are generally put up on the slopes of the hill at the fringe of a forest or outskirts of the cultivatable zone of a village.

Beside hunting and gathering, the Birhors also keep themselves engaged in honey collection and mat making. These are the significant sources of their income. The Birhors know specialized technique for extracting the honey from the hives. The quality and quantity of honey varies with the size of bees and beehives. Following the behaviour of the bees they detect the location of the honeycomb. Honey is sold in the weekly market of the neighbouring villages. Birhors are expert mat-makers. They prepare the mat for own use, but surplus is sold out. The rope fibres are collected from different types of creepers and barks of the trees. Trapping nets are also made with these fibres. Among different types of fibres namely, *Mahulan, Sisi, Udal, Chihar, Talhech, Beri, Hensa,* etc., the 'Mahulan' fibre is considered as 'the best' and known as 'chop'. Next to it, are Sisi, Udal, Chihar and Talhech. Both men and women of the Birhor community go to the forest to bring the material for making rope with which traps and mats are manufactured. Chihar is found abundant in the forest; they grow on strong trees. These creepers are thick in nature. So the outer peel is removed and inner part is taken out. After drying these materials in the sun, the actual act of rope making is started. No special implement is needed for this work. The Birhors are skilled enough to make the rope with bare hand.

Some of the Birhors, during their collections of the material for the rope also collect wood from the forest which they sell afterwards at a good price. Those Birhors who have been interested in cultivation by the encouragement of Government, usually cultivate lac as it seems to them more economic. Government is also trying to encourage them in domestication of animals by providing the animals like cow, pig, goat and hen. Some Birhor by instinct go on fishing, though enough fish are not caught. Many of them have accepted the jobs involving labour as the traditional type of occupations do not help in maintaining the family.

The poor Birhors do not have much tools and implements. They have different kinds of self-made nets for fishing and trapping of animals. Besides, they have hunting club, axes, bow-arrow and some long (10 to 12 feet long) bamboo sticks. These sticks are required to keep the net straight. Such a stick also facilitate in pulling when a tangent piece is fixed at its tip. Flute is the favourite musical instrument of Birhors. These people can not spent much on dress, but they have love for ornaments. The males wrap a piece of cloth around the waist and walk bare-foot. The females use sari and blouse but those are usually torn. The females wear plastic bangles and some other ornaments of cheap metal around the neck. Earrings are used only on special occasions like marriage or religious ceremonies. The wealthy Birhors possess some silver ornaments. There is a traditional belief that wearing of silver ornaments keeps a woman safe from the evil spirits and black magic. But most of the family for the reason of poverty can not afford silver ornaments. Rather they keep white flowers on buns as a substitute of ornamentation. The practice of tattooing is not found among the present generation.

The Birhors possess a great knowledge about indigenous herbal medicine. Therefore they are considered as the most efficient herbal doctors, in their locality. Many tribal and non-tribal communities around them are highly benefited.

Social Organisation

The Birhors live in small bands (Tanda) which normally consist of two to twenty-four families belonging to one or more clans. The clans are totemic in nature. Some of them are Aiyen, Kachua, Induar, Bhagwar, Nankari, Sillwari and Kanhar. A clan includes a number of lineages. The people use the term Birhor as their surname. Like some other communities, the house or hut of the Birhors (Kumbha) may or may not correspond to the concept of family. The number of families in a band is determined by the number of cooking fires there. The society is patriarchal; the families are patrilineal and patrilocal in nature. Generally a family consists of father, mother and their children. When the children grow up they are no longer allowed to sleep in the parents' hut. They go to the youth dormitory (Giti-ora) or a separate Tanda meant for them. S. C. Roy (1921) explained the function of Giti-ora as a center for moral and intellectual training. As a forest people, the Birhors take sex natur and inevitable, so dormitory of the youngs is considered as the safe place both for training as well as enjoyment of pre-marital sexual relations. Still there is some restriction on free mixing of the two sexes. In the dormitory (Giti-ora), the young boys and girls are put under the supervision of an old widow. They spend their time with fun. Riddles are propounded and solved; myths, folk tales and traditional fables are narrated and memorized; various songs are sung and dances are performed.

Although the people of a band are bound together by kinship ties but the biological union of the families are not always very strong. When a band includes people belonging to only one lineage, identity of each member can be traced through genealogical link. Marriage alliances among the members of the same clan are never accepted. But the clan possesses no functional significance beyond marriage. Kinship terminology reveals the behaviour pattern of the society. All Birhors use the same terms of address for persons who stand in identical genealogical relationships. The system is classificatory in type. Kinship relations are also classified into joking or respect relationship. Moreover, there are relations of avoidance. It is the moral duty of all closed kins to socialize a child, apart from the role of his own parents. Only legitimate children acquire social status in the society.

The life cycle of the Birhors are eventful with rituals based on tradition. During childbirth, an old lady of the community comes to act as mid-wife (kusrain). Generally a new hut (kumbha) is constructed for this purpose. As soon as the birth is announced, the ladies of the band start beating of plates (Thalis). The Birhors think that, if they do not make noise at the childbirth, the child may turn deaf. However, the midwife cuts the naval cord (Buka) with an arrowhead and buries it in the ground near the door of the hut. After this breast-feeding starts. First solid food is given to the child when he becomes nine months old. Infants are never given a single drop of water. A control on thirst is grown in them from the very beginning, as water is not always available in the forest situation. The Birhors do not leave a baby alone upto the age of one year. They believe that the evil spirits may cause harm to a helpless child.

In Birhor community, bride price is in vogue. The people are fond of monogamy but they do not negate widow-marriage. The Marriages occur after attaining of puberty. The rate of bride price varies with the activeness of the bride. The procedure of marriage ceremony is somewhat peculiar and interesting. They do not require the performance of any priest. In the house of bride (Jumi), both the parties (bride's and groom's party) sit in two groups. The bride and the bridegroom are then called. Both of them immediately run into the forest and unite there secretly. During this period both the party enjoy in the house. Some of them sing marriage songs and utter special rhymes. In the mean time bride and bridegroom come back. The bride is clothed with a new sari brought by the groom's house. Wild flowers are attached with her bun. The groom applies vermilion on the head and forehead of the bride. In olden days, the groom used to cut the tip of his small finger and put that blood to the forehead and parting of the bride. A sumptuous feast is held with meat and country liquor, 'Haria'. The meat-dish is typically prepared either with monkey's or pig's flesh. Next day the groom goes back to his own house with the bride. For marriage, the Birhors strictly maintain clan exogamy. Most of the marriages of present day are arranged-marriage (Sadar Bapla). But the society traditionally had a sanction for ten types of marriages—'Napam Bapla'(Love Marriage), 'Udara-Udari Bapla' (Marriage of Intrusion), 'Balo Bapla' (Capture of a groom), 'Sipundar Bapla' (Capture of a bride), 'Hirum Bapla' (Second marriage at the presence of Wife), 'Sanga Bapla' (Widow/Widower remarriage), 'Kiring joa Bapla' (Marriage for debt), 'Golhat Bapla' (Marriage of Exchange) and 'Bengkhari Bapla' (Marriage as economic help towards the bride). In case of widow or widower's remarriage, the marriage ceremony gets further simplified.

In Birhor society death is considered as the effect of supernatural power. They believe that the soul (Hapram) does not leave the household if death ceremonies are inadequately performed. Therefore, a series of rituals are performed by the bereaved family members. Traditionally the Birhors liked to bury the dead. But in contact with Hindu culture they have learnt to cremate the bodies. After cremation, they immerse the ash in any river or pond. Pollution period is observed for ten days. In case of old men and women, their sons put drinking water in the mouth of the corpse and the door of the hut is kept open so that the soul gets no hindrance at the time of escaping. When an infant dies, the body is kept in an earthen pot and buried in a corner of burial grave (mashani). The funeral ceremony at the 10th day of death is followed by a feast where vegetable dishes are served at the presence of Naya. Naya is the Priest in Birhor society.

The Birhors believe in ancestral spirits who intervene in the affairs of the living. Therefore, in a settlement, a hut is often marked off as the home of the ancestral spirits. Whenever one sits for a meal or a drink, he or she offers a part of food towards ancestral spirit, at first. Besides, there are plenty of gods and goddesses. The goddesses dominate in the Birhor pantheon. The unlettered Birhors distinguish between creator and destroyer. Benevolent spirits are appeased as goddesses, whereas the malevolent spirits are feared. The former is locally known as 'Bhanga Kani' and the latter is called 'Agobh Bhugivo'. However, four main groups of deities have been recognized. They are Kando bonga, Ora bonga, Hapram bonga and Tanda bonga. Among these, Hapram bonga indicates the ancestor spirits. Each band is headed by a sacred specialist (Naya) who performs all religious worships and hunting rituals of the band. The post of the 'Naya' is hereditary and the person needs proper training from his childhood. Sorcerers are also found in Birhor community. A sorcerer identifies different spirits as responsible for particular diseases. The simple Birhors celebrates a number of festivals e.g., Sarhul, Karma, Bonjha, Newa, Jam, etc. These festivals are colourful with song, dance and feast; they play valuable role in the socio-religious life of the people.

The life style of the Birhors was first publicized by the British administrators like E.T. Dalton (1872), L.R.Forbes (1872)* and H. H. Risley (1891). Risley in his 'Peoples of India' referred Birhors as wood man. They are actually a community of handful persons (approximately four thousand in the State of Bihar) who are standing on the crossroad of extinction. After independence, Government has thought seriously to save this community through perspective plan considering the eco-system, felt need and cultural values of the people. Now most of the Birhors are exposed to health, education and economic programmes of Government. The index of life has been a little elevated.

Lodha

Introduction

The Lodhas are one of the primitive tribes in West Bengal who are now known as ex-criminal tribe. They mainly live in the Districts of Medinipore (under Jhargram sub-division), Bankura and Purulia. Some Lodhas are also found in the Mayurbhanj District of Orissa, and Singhbhum District of Bihar. In all regions, they have chosen a habitat inside or adjacent to a forest. The people feel proud to present their identity as 'Savara'. Savara is a generic term that was used extensively in ancient literature to denote the forest dwelling tribes. Physical features of these people show medium stature, dark brown skin colour,. black curly hair and long head. Probably they belong to the Proto-Australoid group. The folk tales of Lodha suggest that their homeland be in Madhya Pradesh. However, the community possesses no distinct dialect of their own. They speak in regional language.

Previously the Lodhas used to live deep in the forest and satisfied with the forest economy like hunting of animals and gathering of edible forest resources. They developed an intimate relationship with forest. Due to the implementation of the rigid forest laws, the people had lost their monopoly on forest produces and became totally cornered. In this critical situation, being deprived of traditional rights, some of the tribals migrated to the other part of the District in search of job. The other group became revengeful; they made themselves entangled with various anti-social activities. In 1871, the people were labelled as 'criminal tribe' by the colonial power and put under the 'Criminal Tribe Act'. The British Government took not only Lodhas but also some other communities as perpetual criminals. It enacted the law to separate the children of these tribes from their born-criminal parents. After independence, Indian Government gave a fresh look to the problem of these unfortunate tribes and the Act repealed with the effect from 1st, September 1952. Since then, the criminal groups have been reckoned as Denotified community or 'Bimukta jati'. One hundred and fifty-three (153) communities were made liberated from the stringent law.

Material Culture

At present the Lodhas have associated themselves with agriculture and manual labour. They no longer depend on hunting or collection of food. Rapid deforestation and reservation of forest have debarred them from procuring forest products and hunting animals. But Government has given them

* L.R. Forbes—Report on the Ryotwaree Settlement of the Government Farms, Palamau, Calcutta, 1872.

some agricultural land. The Lodhas have no apathy for cultivating them. They cultivate the land, as well as work as agricultural labourer in other's field. Nowadays the Lodhas are found to live very close to other tribal communities like Santal, Munda, Oraon etc. So it is not at all difficult for them to work in other's field. By cultivation they produce paddy, millet, vegetables, sugarcane, cotton, tobacco etc. They know the usefulness of manure in order to enhance the fertility of the soil. But in places, Lodhas find no opportunity to associate themselves with land. There they work as ordinary labourer against daily wage. This manual labour is sold in various spheres of work, like road and railway construction, in tea-plantation and mine. Not only the males, but the females as well as children render their labour to earn a daily bread. They also involve themselves in past-time activities. In rainy season, they practise fishing. Honey, snakes and cocoons are occasionally collected from the forest by selling of which the Lodhas acquire cash money. Tortoise, snail and fishes are also caught. Government has not only drawn their attention towards cultivation, they have also been encouraged in domestication of animals.Cattle, sheep, goat, cat, dog, poultry birds are found in most of the Lodha houses. But since the people are very poor, so all families can not afford them.

Rice is the staple food for Lodhas. They prefer to eat stale rice with a pinch of salt and green chilies. It brings a soothing effect in body and mind. Fermented rice is used in preparing the country liquor, 'Haria'. Liquor from Mahua-flower is also favourite to them. Two meals are taken in a day but drinks are used very often. Both the sexes are in the habit of drinking. Children are not prohibited at all. A curry made of potato, arum or any other leafy vegetable is taken with cooked rice. Sometimes, this vegetable dish is replaced by a non-vegetable dish. Fish, toad, snake, lizard, tortoise and fowl are very delicious to the people. In cooking, they rarely put oil; boiling and roasting are the common processes.

Household goods are limited among the Lodhas. For cooking and serving they use earthen pots, iron cauldron, aluminium bowls, wooden ladles and spoon. Cup and plates made of leaf are commonly used in taking food and drink. Among the implements, there are hoe, spade, axe, etc. Bow, arrow and bolt are also found. Arrow and bolt are specially used in hunting being hurled by a bow. The cultivating groups possess agricultural implements, namely plough, leveler, scraper etc., in addition. Many Lodhas are now using cots for sleeping. These cots are made of coconut fiber-rope; the rectangular frames of which are manufactured with bamboo or wood. The Lodhas also make fishing traps of different size and shape.

Since the Lodhas no more live in deep forest, they have been social and friendly with other communities. Sometimes they make village exclusively for themselves, sometimes live in association of other populations. The villages, which are entirely inhabited by the Lodhas, show no definite plan in making houses. Huts remain scattered haphazardly. But each hut has a definite rectangular ground plan on which the framework of bamboo-pole is constructed. Walls are made of mud and the roof is thatched with straw. In design, the roof shows either two or four slopes. Four-sloped huts are more frequent. All huts consist of a single room, which is used for multiple purposes. Lodha huts possess a single door but no window at all. Each Lodha village exhibits a sacred grove meant for the deity 'Baram'. The place is locally known as 'Baram Than'. Besides, the shrine of two important goddess, Chandi and Sitala are found in all Lodha villages. Some Lodha villages, those are situated in close to Hindu villages show altars with sacred basil plant.

Almost all Lodhas of present day are acquainted with modern dresses. But they can not afford these always because of poor economic condition. The males still go on using a small strip of cloth that has to be passed between the legs to tie with waist belt. This is the traditional dress called 'Kopni'. Some of the Lodhas wear coarse dhoti and a short shirt. The female wears coarse sari without much design on them. They do not use any under-garment. The small children mostly stay without dress; only at the time of going to the market places or in festive occasions they feel to wear. Despite acute poverty, the Lodhas are fond of ornaments. As the metal ornaments, especially those made of brass and silver are very expensive, they utilize wood and bamboo for this purpose. They also like tattoo marks on their bodies, irrespective of sex. The various ornaments, used by women are neck-chain, bangle, necklace, armlet, waist-chain, anklet, earring, finger-ring, nose-pin, etc. Property of Lodhas is very few indeed. It includes ornaments, clothes, weapons and utensils. A small plot of cultivatable land and a few domestic animals have recently been added to the property.

Social Organization

The Lodha is an endogamous tribe. It is divided into nine patrilineal clans. Clans are exogamous in nature and each is associated with a particular totem. They names of the clan are Bhakta, Mallik, Kotal, Nayek, Digar, Dandapat, Paramanik, Ari and Bhuiya. A totem is very sacred for the respective clan. Members of a clan show regard to their particular totem in every possible way. For example, Chirka Alu (a kind of yam available in the forest) is revered by the Bhakta clan; tiger is sacred to Dandapat; and moon is holy to Kotal clan. In the same way, Porpoise (a kind of sea-creature, like dolphin), Sal-fish, Manik (a kind of bird), Chanda-fish, Grasshopper and Makar (a kind of shark) are respected as totems.

Since a totem is considered as the first ancestor of the very group, neglect towards the totemic object signifies disrespect to the forefathers, and an injury to it indicates impairing of one's own relatives. Although the present generation of Lodhas can not provide a clear-cut clarification of totem, but totems are still revered by heart.

Each clan is divided into a number of families. Families are usually nuclear in type consisting of husband, wife and their unmarried children. Father plays a dominant role in the family by managing the family affairs and performing duties and responsibilities to other members. Mother remains responsible for the domestic duties. Children obey their parents. Adult sons are supposed to help the father in economic activities. After marriage a son constructs a separate hut and lives there with his wife and children. Polygynous families or extended families are rare in occurrence. Kinship system of the Lodhas can be called a classificatory type. For example, a single term is used in calling, father, father's brothers, father's sisters husbands and mother's brothers. But the Lodhas also possess a number of denotative and descriptive terms in their kinship terminology. The clan name and the family surname are continued through male line. After marriage, the daughters adopt the clan name and surname of their husbands.

The life cycle is full of events and rituals, which starts with birth and ends in death. Economic hardship cannot provide ease to the pregnant women. At the advanced stage, they have to go to the paddy field and to the market with loads on head. Certain taboos are imposed on them at this time. For instance, a pregnant woman is not allowed to touch a basket trap full of fish; sewing of a torn cloth with a needle is also forbidden. Similarly, visit to a cremation ground is not recommended. It is believed that these acts invariably draw the attention of a ghost or evil spirit on the baby in the womb, so that the life of the foetus may be endangered. However, when the labour pain starts, a midwife (Hadi) is called for delivery. Umbilical cord is traditionally cut with a bamboo split but at present use of blade has been more popular. The midwife put the placenta in an earthen pot to bury it in the southwest corner of the room. The purificatory rite (Ekusia) is performed on the twenty-first day after the childbirth. In this rite, the barber cuts the nails of the mother and the newborn. The next rite is the rice-feeding ceremony, which is observed after six months, on an auspicious Tuesday. A priest (deheri) is called and a feast (Kshir) is arranged.

Monogamy is the rule in the Lodha society. Polygamy, especially polygyny is rarely practised. Cross cousin and sororate are allowed as preferential marriage. A marriage is generally negotiated and solemnized in the house of the bride. The groom has to pay the bride price. In case of love marriage, though bride price is not paid, presenting of some clothes towards bride's family is essentially required. Marriage by exchange and marriage by service prevail side by side. No song or dance is found to be associated in the marriage performance of the Lodhas. On the other hand, both male and female possess the right of divorce. Adultery, indolence, barrenness, maladjustment etc., are the reasons of divorce. If the girl is found to be guilty, her family has to pay back the bride price.

All disputes regarding marriage are settled by the council of elders (panchayat) in the village. Remarriage (sanga) requires a prior permission of the village headman and some royalty has to be paid to the village council. As a matter of fact, the village council plays a very important role in the life of the Lodhas. There are three principal posts of the office-bearers—the Mukhiya i.e., the headman, the Dakua i.e., the messenger and the Paramanik i.e., the cook. Posts of Mukhiya and Dakua are hereditary. Paramanik cooks for the whole village in ceremonies but his post is not hereditary. Not only the disputes relating to marriage, all kinds of disputes in the village are resolved by the effort of the village council. To ascertain guilt, the council often takes the help of oath and ordeal. Offenders are punished by physical assault or cash-fine. For very serious cases, ex-communication (social boycott) is prescribed

Dead bodies are disposed either by cremation or by burial. Lodhas discriminate between normal death and accidental death. In case of normal death, the body is cremated on the bank of a pond or a river. But, in case of unnatural death, the corpse is thrown upon a burial ground or in a forest. The funeral customs are also different for these two different cases. In case of normal death, a pollution period is maintained for ten days. Purificatory bath is taken by the family members when the pollution period gets over. A post-funeral rite (Sradh) is also performed, followed by a feast. The rites and performances associated with death seem to have received a strong influence from Aryan culture.

Supernaturalism is intimately associated with Lodha society. Both animistic and animatistic conceptions prevail in this society. The people believe in different malevolent and benevolent spirits who dwell in forest, river and hills. A number of rites are performed to appease the benevolent forces and to encounter the evil spirits. Lodhas also worship some gods and goddesses of Hindu pantheon. The highest god is known as 'Bhagwan' who controls the supernatural world. Next to him is Basumata, the goddess who provides food, water and shelter. During marriage ceremony this goddess is essentially propitiated, because she acts as a witness of the union. Basumata is also worshipped in other time. Apart from Basumata, Baram, the protecting god of Lodhas lives in woods. He is worshipped specifically on two days—on the last day of the month Chitra (March - April) during the festival named 'Jhetel', and on the last day of the month Pous (Dec - Jan). Moreover, he is invoked in each major festival.

Life of the Lodhas is full of worship and festivals. Some have fixed dates but others are performed according to convenience. A few of them are restricted to the families where heads of the families propitiate instead of a priest. Others are communal i.e., those are performed in the shrines by the priest. For conducting the worship and festivals, each village requires a priest (deheri). A priest is usually respected by all members of the community and he is always accompanied by an assistant whose duty is to arrange everything before commencement of worships. The assistant is locally known as Talia or Chharidar. Before any communal worship, the priest employs Talia and Mukhiya to raise the fund. Worships of the Lodhas often involve animal sacrifice. All immolation is done by a specialized person called 'Hantakar' who comes from a particular family, generation after generation.

The medicine man of the Lodha community is known as Gunni. The Gunni collects different kinds of herbs, as he knows the remedial powers in them. Diseases caused by ghost and evil spirits are treated by him. He also practises divination when an epidemic breaks out. But, like priest, the post of the Gunni is not hereditary. The other important personality of the community is Byakra. He is a spirit-possessed person who can forecast all future events.

The recreational life of the Lodhas includes a variety of things. It may be an evening chat or hearing of popular folk tales. It may also mean participation in games, dance or music programmes. By nature the people are lively and merry-making, but poverty has dampened most of their spirits.

After repealing of Criminal Tribal Act in 1952, the Central as well as the State Government had tried to ameliorate the condition of Lodhas in order to uplift their wretched socio-economic condition. Prof. P.K.Bhowmick had continually studied on a section of Lodhas in the District Medinipore of West Bengal. Since 1963, he was trying to rehabilitate these poor people. His Action Research Project in 'Bidisha' has been quite fruitful in changing the criminal mentality of the people. Dr. Bhowmick initially found that the Lodhas of Medinipore were constituted with one-third of the total number of criminals of the District. They were convicted of theft, robbery, burglary etc. Local police used to harass and torture them after every crime in adjacent area, without considering whether they are really guilty or not. However, the welfare measures were taken by the Government as early as 1953. Lodha families were provided with land, hut, cattle and cultivable implements at free of cost. But the scheme failed totally within two years; Lodhas remained unchanged. The situation was reviewed by expert administrators and the anthropologists who emphasized on socio-cultural and psychological aspects. Accordingly, the rehabilitation schemes were again revived in 1959 - 60. Later, the welfare schemes for Lodhas were tied up with Integrated Tribal Development Project. At present, the Lodhas have been able to turn back from their old ideology and practice. Formerly, they had no education, no land, no skill, no art and craft to make an alternative way of livelihood. Now they are organized enough to accept the programmes of education, cultivation, domestication and so on. They do not further indulge themselves in anti-social activities. A great psychological change has been induced in them.

Garo

Introduction

The Garo is a tribe of Northeast India who shows a tradition of' shifting hill cultivation'. The main concentration of this tribe is found in the Garo hills of Meghalaya. More than four lakhs of Garos inhabit in the State of Meghalaya itself. Some large groups are also found in the contiguous regions like Kamrup, Goalpara and Khasi hills in Assam. Here the numerical strength of the population crosses four figures. A sizeable population also lives in Bangladesh (District Mymensingh), West Bengal (District Jalpaiguri and Coch-Bihar), Nagaland, Mizoram and Tripura. This accumulation of population can be enumerated as 15,000 approximately.

The Garos call themselves *achik-mande (achik* denotes hill and *mande* denotes man) meaning hill-men, just as the Lushais (another hill tribe of Assam) who call themselves *Mizo (Mi* means man, and *zo* means hill). The original home of the Garos is not known. But the people assume Tibet as their homeland, A legend in this respect has persisted among the Garos for generations. Major Playfair (1909) in his monograph on Garos had pointed out certain linguistic resemblances between the Tibetans and the Garos. He also had referred about the reverence which the Garos, like the Tibetans, have for gongs. Both the groups attach the same value to the Yak's tail, although the animal never inhabited in their locality . But such scrappy pieces of evidence are not enough for establishing a historical connection of the Garos with Tibet. It is more probable that like most of the plains tribals of Assam, the Garos moved into their present habitat through the Northeastern routes from China and Upper Burma. This movement was a part of the great Mongolian influx that took place into this part of India in prehistoric times.

The Garos belong to the linguistic stock called the Tibeto-Burman. They possess close linguistic affinity with other tribal groups inhabiting in the adjacent areas. Grierson, in his Linguistic Survey of India (1905 - 1931) brought all tribal languages of this area under a single group called 'Bodo'. As regard to the physical features, the Garos are short statured, long-headed and broad-nosed people having a broad facial profile. The forehead is not receding; nose is remarkable as it lacks prominence. The skin colour is usually fair with a yellow tint. The people show more or less similar physical features like the other tribals inhabiting the plains of Assam, particularly the Kacharis. However, the Garos can be divided into two groups—the hill dwelling Garos and the plains dwelling Garos as per the nature of habitation.

Material Culture

From the immemorial past,the Garo is a cultivator group. They practise 'Jhum' method of cultivation like most of the hill tribes of North-eastern region. An extensive area covered by valuable trees is chosen each year. The trees are cut down and after allowing sometime for the timber to dry, the entire area is destroyed by fire. This is the primitive way of manuring the land. However, seeds are broadcasted in such fertile field and similar treatment is given to another area in the following year. As land in the possession of a particular village is limited in area, the villagers have to go back to the same land after some years (6-7years) and naturally the trees which have grown in the mean time are destroyed once again. Plains dwelling Garos are little more advanced than their counterpart. They have adopted the modern method of cultivation from the neighbouring communities. This permanent wet cultivation is very effective in producing crops each year, in the same plot of land. The principal crop is the paddy altough millet, maize, vegetables, ginger, and indigo etc., are grown extensively. The other important crop, next to the rice is cotton, abundant cultivation of which gives the Garos an economic advantage over all other hill tribes of this locality. Largest part of the wealth comes from this product. Maize, potato, orange and tobacco are also the cash crops. The people have learnt to cultivate cashew nut, topioca and lac, in addition.

The Garos are expert hunters. They also go on fishing, especially in the monsoon and other off-seasons (in relation to agriculture). The hilly forest provides them ample opportunity to hunt as well as to gather forest products. By selling honey and fuel, they make money. Bow and arrow are the chief weapons for hunting. The people also know the use of traps. Different kinds of traps are used in fishing but fishing by poisoning is a popular method. Domesticated animals of Garos include pig, goat, sheep, duck and hen. Cattle are rare among the 'Hill Garos' but the Garos of the Plains often keep cattle in their possession. Dog and cat as pet animals are found in every household of both Hill Garo and Plain Garo. Nowadays some Garos are working as day-labourers and some others have taken up the service.

Rice as a staple food is taken three times a day. Maize and millet are also very popular. Boiled grain is generally accompanied by gruel and vegetable dishes. So far as the animal food is concerned, the Garos eat almost everything—goat, pig, fowl, duck, dog, cat, snake, lizard, bat and even the flying white-ant. Elephant's flesh is highly priced. Nothing is practically left by the Garos; only a line of avoidance is drawn for tiger's flesh. The people not only like fish; they have a special preference for the dried fish (nakam). The pork and beef are also relished by them. The Garos never drink animal milk because they compare it with urine. Like any other hill tribe, drinking of liquor forms an indispensable part of life. This liquor is not distilled, rather brewed out of rice, maize and millet. Apart from its daily use, drinking in profuse quantity is noticed during social feasts and religious ceremonies. Smoking and drinking are the habitual practices among the both sexes.

The settlements of the Garos obviously show a water source near at hand. Construction of hut differs in hills and plains. In hill, some wood or bamboo splits are erected from the ground in such a manner that a rectangular platform can be made a little above. The hut is actually made on this platform. Walls of such huts are made of wood or bamboo splits while the roofs are made of straw, grass or leaves. In the plains, huts are directly made on the ground. These huts are totally made of mud where roofs are thatched and sloped. None of the huts show any window, but each hut possesses two doors—one in the front and one at the back. The settlement of the Garos do not exhibit any definite plan; huts are usually huddled together one side or both sides of the village path.

A hut or a dwelling is composed of three distinct chambers, each having a typical name and meant for a specific purpose. The chamber at the front is known as Nokra. This part of the dwelling is used for keeping fuel, cattle, grains and other necessary household articles. The middle chamber is called Nokgachi. It is a comparatively big space, separated into two halves. One half is very sacred; it is the abode of household deities. The other half is used for cooking and making of liquor. When relatives come from outside, this portion of room is rearranged for their sleeping. The innermost chamber, known as Nokaring, is the bedroom of Garos. It has an adjacent verandah behind, where the leisure time can be spent through recreational activities. Each family of the Garos possesses a storehouse but that is situated far apart from the main dwelling. Raw foodstuff especially grains are stored in that house. All storehouses of the village are usually clustered in one place and protected by the collective effort of the villagers. Besides, each family is found to make a temporary hut in the midst of agricultural field for the supervision of the growing crop. Such temporary huts may sometimes be built on trees to get rid of wild animals. All Garo villages centrally exhibit an open space for communal worship. Further, a bachelor's dormitory known as Ghumghar is invariably present at the entrance of each village, which serves as training cum guardhouse of the village.

The traditional dress of Garo male is called *gando.* It consists of a loincloth, six to seven feet long and about six inches wide. It has to be passed between the two thighs and its ends are knotted with a string tied around the waist. A type of sleeve-less shirt is sometimes worn to cover the upper part of the body but more commonly it is left bare. In winter, wrapper is taken in addition. A headgear usually blue or white in colour is also used. Traditional women's apparel is called *riking.* It is a cloth about a foot and half long and about a foot wide, which goes round the waist, A blouse or a wrapper is worn in the upper part of the body. But at present both the males and females of the Garos have adopted modern dresses in contact with neighbouring populations. Both the sexes also wear neck chains and earrings. Women wear various other ornaments like bangle, chain, nose-pin, finger-ring etc., made of brass, bronze, iron and beads. Some well-to-do-Garos use gold ornaments, but the use of silver ornaments is rare in them. The hairstyle of women is very attractive. They keep a comb in their hair at the base of the plaits or bun. Tattoo-marks are very favourite to them.

On festive occasions, the Garos go on dancing with the rhythm of music. At this time men and women wear loincloths studded with beads (cowries). Their head-dresses too, show the rows of beads and they also stick feathers (hornbill) in hair. Musical instruments include 'singa.' of buffalo horn, bamboo flutes, drums, brass gongs and cymbals. Music is an integral part of all religious ceremonies and social functions. The men dance with the swords and shields in hand. Women dance together or separately. Besides drinking and dancing, the Garos hardly have any other form of communal amusements. Womenfolk weave cloth, make mats with bamboo and cane while men build dugouts, boats and rudimentary implements of metal.

Household goods are limited among the Garos. Hoe and digging stick are the most important tools. Cooking utensils are mostly made of clay. Hollow bamboo with a closed node at one side is used for the purpose of drinking water and liquor. Water in household is stored in hard dry coverings of gourd. Baskets of different sizes and shapes made of bamboo are found in all households. The Garos are specialized in making a type of umbrella with cane leaves. But unfortunately no successful cottage industry has been developed yet.

Social Organization

As said before, the Garo community is divided into two broad groups on the basis of their place of settlement. The Hill-dwelling Garo and the Plain-dwelling Garo. The Hill dwelling Garos are divided into twelve regional groups called Akaui, Chisak, Dual, Machi, Matganchi, Coach, Atiagrass, Abeng, Chibock, Ruga, Ganching and Atang. Similarly, the regional groups of the Plain-dwelling Garos have been recognized as Momin, Marak and Sangma, These regional divisions are locally known as Chatchi. Each regional group (chatchi) is again divided into a number of clans (machang). As Garo is a matrilineal tribe, clans are reckoned through mothers. But like other tribes, the clans of the Garos are associated with totem, which are usually non-human objects like animal, bird, plant etc. Some of clans even trace their origin from inanimate objects. So far as the marriage rule is concerned, the regional groups are endogamous in nature but the clans are strictly exogamous. Each clan is composed of a few sub-clans, which in turn give rise to a number of lineages (mahari). The families under one lineage usually inhabit in a village.

Family as a basic unit of the society may be of two types among the Garos—the Nuclear and the Extended type. In extended type of families, the married daughters live in their parents' family with respective husband and children. Matrilineal residence is the characteristic feature of this society.

Although mother holds the key position in the family, father is not neglected at all. He exerts his authority considerably. A sharp division of labour is found in these families. Economic responsibility totally goes with the males of the family. Females engage themselves in domestic chores. They perform the duty of cooking, rearing of children, fetching of drinking water, making of liquor, collection of fuel etc. All-important decisions of the family are taken by them.

Monogamy is the rule but sometimes polygamy is found. A man can marry maximum three times but he should obtain permission from the first wife (zik mamong) and his second wife must be the elder sister of the first wife. Some of them are found to marry their widow mother-in-law. The marriage rules are designed in such a way that a person can keep the whole property of his in-law's house under own control. The enlightened section of the Garos has abandoned the practice of marrying mother-in-law, although they inherit the said property. The custom of bride price is not found in Garo society. Both the couple is entitled to get divorce on the ground of adultery, physical defects etc. Kinship terminology is of classificatory type. Terms of address are common for father, father's brothers, mother's brothers, and mother's sisters' husbands and father's sisters' husbands. In the same manner, female relatives in the status of mother are designated by a common term.

Property of the Garos can be put into two categories—private and communal. All land belongs to the village community and not any individual. It is the village headman (nokma) who is regarded as the guardian of all land (agricultural/Jhum field and pasture field) of the village. Land can be sold by a nokma only with the prior permission of his wife. Private property includes homestead land, hut, personal dress and ornaments, cooking utensils, weapons, implements and other household goods. Private property is only inherited by the successors. The law of inheritance is based on the principle of descent that passes down from mother to daughters. It is strictly restricted to the female line. When a woman possesses no female issue, the entire property goes to another woman of the same machong or motherhood.

The Garos believe in soul and rebirth. Traditionally they burn the dead bodies. Before cremation some typical performances are done; foodstuffs are offered for the soul. But in case of accidental deaths, they bury the bodies. Pollution period is observed for a few days. On its completion, a purificatory bath is taken by the chief-mourner and other close relatives. A feast is also held. The simple people show their regard to various deities*, although the religion (sangsarek) stands on animatism. The spirits are highly estimated for the benefit as well as the harm. All natural forces like thunder, lightning, rain, wind and earthquakes are believed to be controlled by the spirits. So the people try to propitiate them with sacrifices of birds and animals. Sacrifices form an essential part in all religious ceremonies. Spirits are involved not only for good harvest, their blessings are also needed to get rid of different troubles, particularly of those which remain associated with illness, birth, marriage and death. The magical practices like witchcraft and sorcery are prevalent among these people.

A large number of Hill-dwelling Garos have been converted to Christianity. This conversion has brought considerable changes in the life and culture of the Garos. On the other hand, many plains dwelling Garos have adapted Hinduism in rudimentary form. In course of maintaining a socio-economic link with neighbours (Hajong, Rabha, Koch and Bengali), the people have been influenced by them.

The traditional village council is known as Laskar. This is constituted with the village elders. The council intervenes in all matters of dispute and takes the ultimate decision; the village-elders share their views with the headman (Nokma). After the establishment of British administration in Northeast, special power was put to council so that the members began to serve as rural police as well as honorary Magistrates. Only criminal cases of a serious nature used to be heard by the Government

* Rokimi is the deity of agriculture, Gitchipang is the deity of festival, and Michittata is the deity of the village and so on.

Magistrate at the headquarter. Therefore, all simple disputes were decided on oral evidences and judgment was passed by the village elders. The cases for which the oral evidences were not available, oath and ordeal were followed. In 1952, an autonomous District Council was established for the Garos. Although the village headman (nokma) is found to retain his traditional post, his power has been reduced to a great extent.

Khasi

Introduction

The Khasi is a matrilocal community who lives mainly in the Khasi and Jaintia hills in the State of Meghalaya. Some Khasis are found in Garo hills, North Cachar hills and also along the northern and southern slopes down to the Brahmaputra and Surma Valleys. Not only matriliny, some other typical characters mark off the Khasis from the other tribes of Northeast India. They gave rise to a well-developed State organization while many other tribes of this region could not go beyond a simple form of local government. Secondly, the economy of the Khasi people show some advanced features, unlike their immediate neighbours. Inspite of numerous contacts with the surrounding populations, the Khasis maintain a unique socio-economic system based on trading. But they are totally different from the Nagas and the Mizos in the East and the Adis or the Miris in the West.

The original home of the Khasis is not known exactly. The great linguist Sir George Grierson had stated that the dialect of the Khasis is Mon-Khmer which belong to the Austro-Asiatic language family. The people have settled in the hills of North-east India being migrated from South-east Asia. They are the only people in India who speak *Mon-Khmer* language. P.R.T. Gurdon (1903) had discussed the findings of Mr. Logan who through his study of various vocabularies tried to demonstrate the linguistic relationship of Khasi with other people outside the India namely the Mons or Talaings of Pegu and Tenasserim in Burma, the Khmers of Cambodia and the majority of the inhabitants in Assam. According to Gurdon, Pater Schmidt found a similarity of language between the Khasis and Nicobarese; the inhabitants of the Malay Peninsula also use the dialects similar to the Khasi dialect. Barch (1967) said that Khasi is a generic name given to the people of Khasi and Jaintia hills. In search of Khasi identity one finds that only two tribal groups of India—the Naga and the Khasi are still using the 'singular shoulder-headed celt' made of iron as like those celts found in Malay Peninsula in 1875. The Nagas call it *mo-gyu,* but Khasis call it *moh-khiw* or *mo-khu.* The Khasis have never taken to plough like their neighbours. However, from this fact again a connection can be deducted between the Khasis and the people of Malay Peninsula, particularly in reference to the culture of Old Stone Age. Once the Ho-Mundas of chotonagpur also used this kind of celt. The second interesting similarity is found in the practice of erecting the monumental stones, monoliths. This ancient Khasi custom has also been observed only among the Ho-Mundas ofchotonagpur. In fact, the burial customs of these two people are almost similar. Both of them cremate their dead bodies and gather unclaimed bones to put them into stone cairns. The Khasis call it mawshieng (bone-stone) and both the people erect monolith to perpetuate the memory of the deceased ancestors. Similarity of the cultural elements suggests that these groups have originated from a common source, even if they were not related to one another.

So far as the physical features are concerned, the Khasis are short statured, fair-skinned people. Head is oblong with round or oval face. Nose is of moderate breadth.

Material Culture

The system of agriculture among the Khasis is comparatively elaborate and carefully adjusted to the productive power of the soil. Formerly they used to practise cultivation by slash-and-burn method, but for the devastating effect of deforestation now they have been compelled to change their method of cultivation. Like the olden days, still the principal crop is paddy. At present paddy is grown in terraced and well-irrigated fields of the flattish valleys with which the central plateau abounds. Such fields are found along the northern margin of the districts. The rest of their crops e.g., potato, millet, arum, ginger, turmeric, chili etc., are also grown on hillsides for local consumption. Besides, potatoes, pineapples and oranges are grown in huge quantities. These crops are sold to the neighbours in the plains. Khasi community has got the advantage of good soil and heavy rainfall, which facilitates in surplus production. Khasi economy is chiefly a market economy. They sell different produces and earn money. Betel nuts and betel leaves also grow plenty in their locality along with bay leaves (Tejpata or Cinnamonum tamala). These items too are exported by the Khasis. Moreover, the Khasi country is very rich in mineral wealth. Both the hills, Khasi and Jayantia, show ores of lime-stone, sand-stone, coal, iron, silimanite, fire clay, corundum, ceramic clay, quartz and gold. Although the quality of coal is not so good, but it is of great demand to the rural iron-smelters who use waist-high blast furnace and hand-driven pump-bellows in primitive fashion. Largest number of coalfields are located in Cherapoonji and Lakadong. Iron is the other material that is profuse in Khasi and Jayantia hills. As the quality of this iron was very good, iron industry of the Khasis once earned much fame in the country. The practice has been dwindled down in recent days because the hand made products could not win in the competition with the mill-made products. However, small quantities of iron are still exported outside. This iron is sought after for the manufacture of hoes and waits (a form of heavy knife for the general use). Next to iron we can name the vast store of limestone which exists in the southern part of Khasi kingdom and continues to be an important article of manufacture.

Formerly the Khasis had a special love for hunting. They used arrows and spears for this purpose. They were also in the habit of setting different kinds of traps. For example, sometimes a spring-gun was kept hidden in the way of deer in such a manner that as soon as a deer would touch the gun by foot, the trigger would be released and the arrow would pierce the animal with a great force. Formerly, the fishing used to be entirely done by poisoning. Water of a hill-stream was confined with dam and some herbal poison was put to it. At present they use various appliances for fishing. Fish and pork are the favourite food items of the people. They take these with rice. Different vegetables and pulses also supplement their diet. Like any other tribals, the Khasis are fond of country-made liquor. By fermenting the rice or millet they get a type of liquor which is largely used as beverage. The rice-beer is also indispensable in a number of religious ceremonies.

The village of the Khasis are not build-up on the hill-tops. The people choose a convenient area where they find a source of water very easily and where their huts can be protected from the lash of the strong wind which blow across their country during the monsoon. Among the Khasis land ultimately belongs to the people. The chief may have some land and property of his own but he holds no extra privilege over the land by virtue of being a chief. He can not levy any tax on the people for the use of land. Again the lands that are owned by the families and clans are considered to be owned by the State as a whole. Houses are strongly built with the bamboo poles and the planks of wood. They are simple in design and rectangular in shape. Household goods are many including utensils, furniture, baskets, musical instruments etc. Since the people live in close contact with advanced people, a few modern equipments like transistor, tape-recorder; guitar, etc., have been found among them. But in the kitchen they still use the earthen pots made by hand. Most of the elderly women of the community know how to make pottery. Pots are generally painted black. People are at hand in basketry and cane work. They are also skilled in cotton and silk textiles. Indigenous vegetable dyes are used for a variety of beautiful colours.

The physical beauty of Khasi women excels their sisters both in hills and plains. Ethnologically they are different from their neighbouring hill tribes who had come from Northeast. Rather, they share customs and habits with Malaysians. Women are fond of ornaments more than men are. The people in general are cheerful in disposition and have a special love for music. Music permeates in their blood in such a way that on all occasions of life they sing; starting from a lullaby to a love tune or from a marching tune to a harvest tune. Whether in the field or round the fireplace, in joy or sorrow, a Khasi sings. There are marriage songs, festival songs, religious songs and even the mourning songs. Music is core to the Khasi life. But most of the Khasi songs have adopted Welsh tunes for over a long period. Apart from songs and music, there are three past-time games—hunting, fishing and archery, which amuse the Khasi males of today.

At present the Khasis are highly exposed to modern education. Many of them are working in offices. Some hold the positions like college professors, school teachers, doctors, nurses, administrators, merchants and businessmen. In all walks of life, Khasi women have come froward and qualified themselves. But the most interesting feature is to note that a Khasi, being a doctor, teacher or government officer does not hesitate to take turn at the hoe in his potato garden.

Social Organization

The main feature of Khasi society is that, it is a matrilineal society. The traditional social, cultural, ethical, political and economic ideals of the Khasis are interwoven with the origin of the Khasi race. D. Herbert (1903) remarked that, "The Khasis are a singularly progressive and intelligent race with democratic tendencies". The tribe is divided into four main sub-groups.

(i) The *Khynriam Khasi* or *Upland Khasi* distributed in the Shillong Plateau of the Khasi hills.

(ii) The *Phar Khasi* of the Jaintia hills.

(iii) The *War Khasis* inhabiting in the West and South slopes of district Meghalaya and

(iv) The *Bhoi Khasi* settled in the low lands.

Beside these four main sub-groups, two other groups—the *Hadem* and the *Lynngam* have been recognized recently. The Hadems are localized in Jaintia hills while the Lynngams are found in the Khasi hills. Each of these sub-groups maintains territorial endogamy very strictly and, therefore, they are considered as sub-tribes. Sub-tribes are again sub-divided into a number of clans (Kur), lineages and families in their respective territories. Some of the important clans are Lingodon, Kharkongor, Diengdoh, Sohkhlet, Marbaniang, Shiemlieh etc. Members of a clan trace their descent from a root ancestress and all clans are bound together by strict ties of religion, ancestor worship and funeral rites. The clan or kur is the nucleus of Khasi life. Alt institutions - social, cultural and political revolve round it.

The Khasis have their own theory of origin. They believe that originally the hill connected the heaven to earth. On the side of the heaven there were seven huts where men lived. But due to increase of sin, the heaven got separated from earth and men were left to themselves. However, clans of Khasis are reckoned through mothers. Position of women is very high in this society. Mother is regarded as custodian of family rites and religious performances. When a clan settles down in a place, the members make up the village. A village may assimilate the members of other clan. who for some reasons or other settle down in this village.

Choosing of mate from one's own clan is regarded as a great sin. Therefore, a Khasi never breaks this rule. But they are permitted to marry the children of maternal uncle or children of father's own sisters. To the Khasi, marriage is more than a contract; it is a sacred bond between a man and woman, ordained by god. After marriage, a man becomes a member of his wife's clan. He goes to live with wife in her house. Income of a son-in-law is solely handed over to mother-in-law for the maintenance of the family. Therefore, all sons of the family leave their family of orientation after marriage. A Khasi family is essentially ruled by a grandmother who resides with her married daughters and grandchildren. When a family grows excessively large, it splits up into sections. New lineages are thus created, but the original ancestress of the clan is never forgotten.

Mother's elder brother enjoys a special authority in Khasi society. The whole management of a house remains under his control. Divorce is so common and frequent among the Khasis that the children very often do not know the actual identity of the father. They always stay with mother. The event of divorce can not throw them away from their mother's family. The ancestral property passes down through the female line. The youngest daughter inherits the largest share of the property; remainder is divided among other daughters. The sons have no right on the property except his self-acquired wealth. Mother's residential house usually goes to the youngest daughter.

The Khasi State is founded on religion. But there is no temple, mosque, church or pagoda. The Khasis never symbolizes their gods by means of images; their worship is offered towards the spirits only. The people also propitiate the dead ancestors. They believe in transmigration of souls. All religious rites and ceremonies are usually performed by the clan through a priest (Lyngdoh). Individual households also possess own small domestic shrines just like a village that show a sacred grove. The people utter prayers and show their gratitude, but there are no fixed time for it. Above all, the Khasis recognize a supreme power (spirit) as the creator of the world. Some years back, the people used to practise divination with breaking of eggs. The practice of human sacrifice was also found. With the spread of Christianity and modern education, those customs have disappeared. The rites associated with birth, marriage and death were very strict in former times, which is facing crisis today. Due to Westernization a section of Khasi undervalues their traditional norms and practices, whereas another section is continually striving to revive those customs. Specially, an organization of the Khasis known as Ka Seng Khasi has spearheaded a revivalistic movement for the preservation of the Khasi religion and Khasi identity. The British rule exposed the Khasis to a materially advanced culture. Christian missionaries had contributed a great lot in education. Altough the majority of the Khasis are Christians but there are small groups of Hindus and Muslims too. The enlightened Khasis especially live in towns like Shillong, Jowai and Cherapoonji.

The most remarkable feature of tribal life in Northeast hills is the state organization. Democratic spirit has been strongly reflected in the indigenous administrative set-up. The Khasis are no less exception. This tribal group had developed a three-tier system of political organization. State came into being through voluntary association of villages and groups of villages. At the base there were traditional village councils (Dorbar Shonong); each council was in charge of looking after the social affairs of the respective village. The elders of the village constituted this 'Dorbar'. In the next level, several villages joined together to form the 'Dorbar Raid'; its duty was to handle the disputes between the villages. The chief (Siem) being posited in the apex was supposed to be assisted by the two 'Dorbars'. All judgements and verdicts followed the prevailing tradition. Some important families of Khasi community had the potency to send a chief. But the selection was vested in the hand of a body represented through priestly (Lyngdoh) clans. Both the Siem and Lyngdoh used to be elected by democratic principles. They were the instruments to carry out the decisions of 'Dorbars' and not permitted to act of their own.

In small States, the office of the Siem and Lyngdoh are often combined in one person. Some specific factors led to the emergence of Khasi states under the Siem. Firstly, as the villages grew in size and population, people thought to make them united under one common administration. Secondly, there was a taboo, which resisted the people to administer a justice or to pass a verdict by themselves. Naturally, necessity of an alternate personality (Siem) was felt. Thirdly, there was a popular belief that god wanted special person (Siem) endowed with a super-power to decide right or wrong. Fourthly, as Khasi states prospered with strength, power and influence, chances of foreign invasion increased. Fortification and other measures of defence also demanded the command of a chief. However, such a three-tier political system prevailed from unrecorded time. It gave enough room to the British administrators as well as the researchers to analyze the characteristics of Khasi State and to cite its differences or similarities with other republics existing in other parts of the world.

With the ascendancy of British, certain changes were introduced. The greater part of the Khasi and Jayantia hills (which were under the control of native chiefs) formed a subsidiary alliance with the British power. The tone and the colour of administration underwent a metamorphosis. Although the people continued to govern themselves through their elected leaders in the local councils (Dorbars), the chiefs (Siems) were compelled to submit themselves to the paramount power. No revenue could be collected from the common people, but the chiefs were committed for a regular payment. Half of the profits from the natural resources of the state (mines, minerals, different forest produces) were required to be given to the British government.

morality as the state is founded on religion. Justice is administered keeping conformity with the norms and tradition of the people. The Khasi rules are very severe with the lawbreakers and do not want any repetition of the crime. Punishment of a murderer is death through beating. Adultery is punished by life-imprisonment. In case of robbery or theft, punishment is designed in such a way that the culprit remembers it upto the last day of his life. Usually a culprit is ordered to sit on a bamboo platform under which hot chilies are burnt. All petty offences are thus dealt in local council; great crimes are finalized at the court of central jurisdiction. Since 1952, the area has been administered under the Sixth Schedule of the Constitution of India. The attitudes of Khasis towards the development programmes of the government are positive.

Munda

Introduction

Like Santals, Oraons and Nagas, the Mundas are a numerically dominant group of India. The members of Munda community refer themselves as *Horo,* the superior man. The term also serves a functional designation. The people believe themselves as the descendants of the Supreme god (Sing Bonga), so they declare themselves as the head or ruler of the area. The main concentration of the Mundas is found in the Chotonagpur plateau in Bihar. They are also spread over the adjoining states of West Bengal, Orissa, Madhya Pradesh and Assam. A section of Mundas has been discovered from Tripura too. Although brief accounts of these people were furnished by E. T. Dalton (Descriptive Ethnology of Bengal: 1872), H.H.Risley (Tribes and Castes of Bengal: 1891), J. Hoffman (Munda Poetry, Music and Dances: 1907) and others, the pioneer ethnographic account on Mundas came from S.C.Roy in 1912. Sri Roy studied the Munda community of Chotonagpur quite extensively. The Mundas are now a well-studied tribal community having one encyclopedia of their own. This encyclopedia is known as Encyclopedia Mundarica (1930) which is completed in sixteen volumes. Besides, there is an appendix volume filled up with illustrations. This voluminous encyclopedia, authored by John Baptist Hoffmann *et al* (1857 - 1928), has been recognized as an authentic account of Munda land and its people. It covers a number of aspects—customs and festivals, songs and dances, rites and rituals, etc.

Racially, the Mundas are the Australoid people. They are so ancient that they could lent their name to Austro-Asiatic language, Mundari or Kolarian which belong to one of the four language families in India. S.C.Roy tried to trace out their original homeland on the basis of mythology. He said that the Mundas were originally the inhabitants of Northwest India who later migrated to the chotonagpur plateau and settled there by clearing forest. Sachchidananda (1979) also advocated a similar view. Considering the tradition of Mundas, he commented that the people came from Southern part of North India to modern Bundelkhand and Central India, over Rajasthan and Northwestern India. Finally they reached to Chotonagpur area. Some other scholars argue that these ancient Mundari speaking population came from South east Asia through North east region and gradually spread over a large part of the country, especially in Eastern and Central regions. In fact, the Mundas pervaded whole of the country. The Savaras of the south and the Korkus of the west are the Mundari speaking communities. The language of Badagas in Nilgiri hills seems to have been influenced by Mundari language. Again, the Punjabi language shows many Munda words. Munda place names are also discernible in Goa and Northwest Himalayas. Similarly, the Vedic Sanskrit exhibits some Munda terms. The name Munda itself is of Sanskritic origin. In ancient days these people were called as Nisadas.

The present address of Mundas is Chotonagpur plateau, under Khunti sub-division of Ranchi District. They are living there since 1765. The particular time corresponds to the reign of Nagabansi Maharaja (king) in Chotonagpur. The history of Mundas can be divided into five distinct phases:

I. At the first phase, the Mundas settled in Chotonagpur by clearing up the forest. A village named Khunt-Katti was established. The Mundas, who came to reside in this village, became totally

named Khunt-Katti was established. The Mundas, who came to reside in this village, became totally isolated from outside influence and formed an honorific group.

II. In the second phase of history, the Mundas went under the control of Nagabansi Kings. This led to the formation of a state with the character of feudalism. The tribal polity based on joint ownership of land was distorted at this time.

III. A further breakdown of traditional socio-economic system was noted in the third phase. A Nagabansi king was defeated by Mughal and so he was taken to the court of Akbar as a prisoner. Later, being released when the king came back to Chotonagpur, brought with him the Mughal style of administration. Some Hindus and Muslims started to work as 'Jaigirdars'. Many businessmen, moneylenders and religious specialists came to live in that spot. The area grew as a small urban centre of trade and commercial activities. Various Hindu temples were built. The entire way of life proceeded towards change. The Mundas were swayed in the new wave.

IV. The British Colonial rule did not affect the feudal pattern of land rule but the Munda country was brought under direct administrative control of East India Company. Growing trend of Sanskritization and Missionary contacts brought an adverse effect on traditional social system.

V. The colonial period witnessed a number of reactionary movements in Munda country. After independence of India, movements were found in various parts of Munda country. Not only that, rampant urbanization, industrialization and social welfare measures have altered their age-old bearings.

So far as the physical features are concerned, the Mundas are the short and stout people with an average height of 160 cm. The skin colour is dark-brown. Head is usually long with wide face. Nose is medium in size and broad. Lips are thick. Head hair is wavy. In ABO Blood group system, they show an average frequency of O ranging between 55% to 58%.

Material Culture

Munda villages are situated on highland among the hills and hillocks. Three to four huts remain together centering a courtyard. This quadrangular plan is called 'racha'. Each household possesses a kitchen garden at its back. The compound used as kitchen garden is known as 'Bakri'. Vegetables are grown in this garden and a manure-pit is found in one of its corner. Huts are made exclusively of mud where the roofs are thatched. At the entrance of the household, there is invariably a cowshed called 'Unri Ora'. A special sleeping room, 'Giti Ora' is a part of house but found outside of the homestead. The kitchen is called 'Mandi Ora' which is usually a separate room within the homestead. A kitchen must show two parts—*Sare* and *Ading.* The former is used for general cooking, whereas the latter is extremely sacred meant for ancestor worship.

Agriculture is the primary occupation of the Mundas. Agricultural implements are many and varied like plough and yoke, hoe, ladder, sickle, garden spade etc. Irrigational devices are known to these settled agriculturists. The principal crop is rice. Millet and maize are also grown with different kinds of common vegetables e.g., tomato, brinjal, onion, cauliflower, cabbage, gourd, chilli etc. Incidence of hunting is very rare among these people, but fishing continues as a subsidiary occupation where the ecological condition permits. Nets and basket traps are the common fishing implements. Most of the Mundas are small and marginal farmers. Many of them work just as labourers under wage system, both in agricultural and non-agricultural sectors. Since a considerable section of Mundas has been brought under the educational system, a few of them have secured respectable service under government as well as private organizations.

Rice is the staple food of Mundas. It is taken three times a day. The first meal is taken in the morning where stale rice is taken with salt and chilli. The second meal is taken during mid-day, where hot rice is eaten with pulse and vegetables. Sometimes fish is added to this meal. The third meal or the evening meal is made ready with hot rice. It is often accompanied by pulse and vegetable curry. The people like pork but abstain from beef. Though an egg is not forbidden, but it is not taken at all. Even the milk of the cow is not taken except during illness.

Home-brewn liquor, called 'Haria' is a part of culture. It is prepared out of fermented rice. Rice soaked in water is left for a few days; some yeast (bakhar) is added to it. Both male and female drink the liquor regularly. A large-scale use of this liquor is found in social and religious ceremonies. The Mundas feel free to offer it to their deities. Another type of liquor is made from Mahua-flower, which is not so popular in use. But that too is indispensable for some rare occasions. Habit of smoking is prevalent among the males. Tobacco wrapped in a 'sal' leaf serves this purpose. Besides, the common 'bidi' is largely available in the local shops.

The dress of the Mundas is very simple. The males wear a piece of loincloth or a coarse 'dhuti'. Women usually use a coarse sari. Ready-made modern garments like blouse, shirt, trousers etc., are preferred by the young generation. Footwears are specially used while travelling a long distance. Females love ornaments. Silver bangles, earrings, finger-rings, neck-chains are used widely. Necklaces made of plastic-beads, gold-like imitation jewellaries are also popular among the young girls.

Household goods of Mundas consist of earthen vessels, iron and aluminium utensils. People usually sleep on the floor of the room by spreading gunny bags or mats made of palm-leaves. In some houses, raised fixed rectangular platforms made of bamboo-splits are found to serve as cot. Small wooden seats (pira or pidi) are used in every household. The well-to-do families can only afford wooden chairs, benches and stools. Broomsticks out of palm leaves, grass or bamboo strips are very common in occurrence; they are meant for sweeping. Small burner-lights (kupi) are used in evenings to light the room. Among musical instruments, two kinds of drums—*Dama* (drum with a metal-case) and *Dumang* (drum with an earthen-case) are very popular. Flute made up of bamboo is extensively used by the people. Well-to-do families generally afford to keep all these instruments.

Social Organization

Risley (1891) had observed thirteen sub-tribes among the Mundas. But the number of clans (Kili/Gotar) differs in the writings of different researchers. Some well-known clans are Hansda, Kachchap, Lang, Tuli, etc. All clans are totemic in nature and each is divided into several lineages (Khunt or Bansha). Each lineage is further divided into sub-lineages. Families are usually patrilineal, extended in type; but nucleus families are also very common. Some families are polygamous where wives live together in one household with a common husband. Both classificatory and denotative types of kinship terminology are prevalent in this society. Because of close contact with neighbouring population, kinship terminologies include some Hindi as well as Bengali terms, according to the nature of exposure.

Marriage rules of the Mundas follow sub-group endogamy and clan exogamy. Lineages are, therefore, all exogamous. A marriage solemnizes in bride's house. Elaborate marriage ceremonies, as observed by S.C.Roy, are not found at present. But the pre-marriage ritual, 'betrothal' is still very important where boy's guardians pay a visit to bride's house for finalization of the marriage proposal. This custom is known as 'marang-para'. The system of bride price continues with its traditional implication. On the day of marriage, a wedding march (barat) is arranged from groom's house to bride's house, which is often accompanied by a musical troop. For performing the rituals, a priest is essentially needed. Next day, the bride goes to groom's house and there also a ceremony (chuman) is performed. Divorce is permitted in this society, but no compensation has to be paid. Moreover, different sorts of other marriages like polygyny, junior levirate, junior sororate and cross-cousin marriages have a special sanction.

Pregnancy brings restriction of food and work. On the fifth month of pregnancy, a ritual named 'Sutam dharam' is performed. The birth is handled by a midwife (Dai) of Hari community. Pollution is observed for eight days. On the ninth day, a purificatory ceremony, 'Nai Narta' is arranged. Local barber and washerman attend this ceremony. Name giving is done through another ceremony (Sakhi).

No separate rice-gining ceremony is held among these people. The ear-piercing ceremony (Tukai lutu) is optional which may or may not be performed before the age of puberty.

Death is looked as a departure of soul (Jiu) from the body. It is believed to be caused either by natural agencies or by spirits. Cremation and burial, both are practised by the people. Some rituals are essentially associated with disposal of the dead bodies. Pollution period continues for nine days. During purificatory ceremony, barber and washerman of the village are called. It is peculier that, nowadays, a Brahmin priest is also employed for performing these rituals.

The Mundas are basically the worshippers of nature. They also propitiate their dead ancestors. The supreme deity, 'Sing Bonga' is held as the source of all power and energy. A large number of village gods and household deities are worshipped in addition. The people also recognize the existence of different malevolent and benevolent spirits. They practise black magic, divination and witchcraft. The festivals like Sarhul, Karam, Jitia, Manda, Sohorai etc., are observed in the community. Numerous indigenous songs are composed for these different festivals. It is notable that certain Brahmanical rites and practices have been penetrated among these people in varying degrees.

The traditional village council of the Mundas was composed of five heads, selected from different clans. Although its importance has been dwindled down, still it forms essential machinery of social control at the village level. Formerly, there was another body to embrace a group of Munda villages. But that traditional system has been largely replaced or superimposed by the present day statutory Panchayat body.

The Mundas of Chotonagpur are one of the pioneers who revolted against the pressure of colonialism with agrarian issues. The Tamar resurrection of 1819-1820 marks the inception of Munda movement. Later, the Mundas were organized under the leadership of Birsa and protested against zamindars in order to expel them from own territory. Birsa is revered as a god by the present day Mundas. Birsa movement is the best-known millennial movement, which kept a tremendous impact; the foundation of British rule in Chotonagpur was shaken for some time. Since 1915, the Mundas were striving for autonomy in Chotonagpur. They took an active part in Jharkhand movement. However, the Mundas can be said a progressive tribe. They strongly hold their tribal identity but not isolated at all. As they live very close to the advanced people, process of Sanskritization operates in them very highly. They also involve themselves in patron-client relationship with many serving and artisan communities (like Lohar, Ghasi, Thakur, Gosain etc.) of the neighbourhood. Annual payments are made to communities by the Mundas for their service throughout the year.

CULTURE

Meaning and Definitions

The term culture stands for the sum total of human behaviour, verbal and non-verbal and its products—material and non-material. The word 'culture' connotes different meanings in different situations. For instance, we commonly use this in humanistic sense to designate manner, taste and intellectual development. It is said that a man is quite cultural or some people have more culture than others, or some human products are more cultural, such as, visual art, music, literature etc. Poetic expression of culture denotes sweetness and light. To a historian, culture simply signifies social development in artistic and intellectual fields. Modern biological usage of 'culture' is restricted to the medium of growth where a mould or a bacterium may be grown artificially. In Sanskrit language, the synonym of culture is 'Sanskriti' which has been derived from the Sanskrit term 'Sanskar', the meaning of which is "the refinement of soul" by various acts, through a series of births and rebirths. The term 'culture', thus has been used in different context for a long time, but its exact meaning is still vague.

Etymologically, the word is linked with words like 'cultivate', 'cultivation' and 'agriculture'. In seventeenth century, its meaning was applied to human development. This metamorphical meaning was further developed in eighteenth century to give rise to a general term. In Germany, the word was started to be spelt as 'cultur'. But soon it was converted to 'kultur'. During nineteenth and twentieth centuries, the term became very distinct and got entangled with different meanings in different spheres.

In anthropological usage, the concept of culture is not only broad and comprehensive; it is very significant as well as scientific. It does not limit itself with specific field of knowledge, rather encompasses the whole range of human activities. Each and every manifestations of human existence e.g., tools, weapons, utensils, clothing, ornaments, customs, beliefs, institutions, rituals, games, work of art, etc., are included in culture; the numerous behaviours that govern the way of life in any society (economic behaviour, social behaviour, political behaviour, religious behaviour, educational behaviour, recreational behaviour, etc.) are also covered by culture. In fact, culture helps man to satisfy certain inborn needs so that he learns to survive in an environment. In other words, this is the ability for which man copes with his environment and not forced to submit himself in it, like other animals of the world. Culture, as a product of body and mind, marks off man from other higher primates. World shows different cultures with time and space; each of them is noteworthy in its own way. Each person represents a particular culture in which he or she belongs. Difference between the groups of people is explained by the dissimilarities in their culture.

Heredity and learning are the two chief factors that ascertain the behaviour of organism. In man, heredity controls the instinctive behaviours like sucking, swallowing, yawning, etc. These behaviours come down by the biological process of transmission. But, the behaviours like speaking of a language (English, Bengali or any other language), wearing of a dress etc., are learnt from others through the social process of transmission. The degree of learning is determined by the individual capacity. Instances of social learning are found to some extent in other animals, but the absence of speech invariably limits their power of acquisition and communication. Culture is, therefore, unique to human. Being an essentially learned behaviour, it gets transmitted from generation to generation by a complex mechanism. Human beings enjoys the longest childhood among all animals, which also indicates a great dependence on learned behaviour.

Different scholars tried to define culture as per their own conceptions. The earliest definition was delivered by the British scholar E.B.Tylor in 1871. He introduced the term 'culture' in anthropology, borrowing the German word 'Kultur'. According to him, culture is the complex whole which includes the knowledge, belief, art, moral, law, custom and all other capabilities and habits acquired by man as a member of society. The weakness of this definition was that it did not distinguish social organization and social institutions from the general concept of culture. Such an inclusive use of the term culture was continued by Boas (1896,1940), Malinowski (1927, 1931) and other ethnologists. B. Malinowski stated that "culture comprises inherited artifacts, goods, technical processes, ideas, habits and values". He regarded that these things are interrelated and form a total way of life, which secures for an individual the satisfaction of his bio-psychic drives, fulfillment of other wants and cravings that ultimately provides with freedom. F. Boas's conception was pluralistic as well as relativistic. He emphasized on the plurality of local cultures as functioning and organized whole. Radcliffe-Brown (1952) held society as a system; its function contributed to culture in the way of maintaining relationships among the components. The view of Radcliffe-Brown was much similar to that of Malinowski.

After Malinowski and Radcliffe-Brown, most of the anthropologists accepted the significance of functional interelatedness in a culture. They wanted to demonstrate the same, empirically. Mead (1935, 1938), Kroeber (1923, 1948, 1952), and Benedict (1934) perceived culture as the central concept of social science. Redfield (1930) defined culture as "sum total of conventional meaning embodied in artifacts, social structure and symbol". Linton (1940) pointed out, "A culture is the configuration of learned behaviour and the results of behaviour whose component elements are shared and transmitted by the members of the society". According to White (1943, 1949, 1959), culture was an extrasomatic temporal continuum of things and events, depending upon symbolism. N.K.Bose (1949) conceived culture as a crystallized phase of human behaviour. To Kroeber (1948) culture was the mass of learned and transmitted motor reactions, habits, techniques, ideas and values and also the behaviour they induce. Herskovits (1955) stated culture as man-made part of environment. Kroeber with C.Kluckhohn (1952) defined culture as "patterns, explicit and implicit, of and for behaviour acquired and transmitted by symbols, constituting the distinctive achievement of human groups, including their embodiments in artifacts".

In 1963, Kroeber and Kluckhohn found that there was hundreds of definition for culture. Both of them reviewed those and formulated a summary which has been accepted by the social scientists. They believed, "culture is not behaviour, nor the investigation of behaviour in all its concrete completeness. Part of culture consists in norms or standards of behaviour. Still other part consists in ideologies justifying or realizing certain selected ways of behaviour. Finally every culture includes broad general principles of selectivity and ordering in terms of which patterns of and for and about behaviour in very varied areas of culture content are reducible or parsimonious generalization". In the book, 'culture' Kroeber and Kluckhohn furnished several definitions collected from the writings of anthropologists, sociologists, psychologists and others. They also classified them into four groups:

(a) Descriptive or enumerative definitions, which listed the content of culture but did not specify the criteria of selection.

(b) Historical definitions, which emphasized social heritage or tradition.

(c) Normative definitions, which stressed the social rules or socially patterned behaviour.

(d) Psychological definitions, which laid importance on learning and psychic bearings.

The definition types were not mutually exclusive; still there was no contradiction among them. In fact, they referred two separate but related ideas—(*i*) real things and events, specific to human activities, and (*ii*) a concept or an abstraction from real phenomena.

The usage of the term culture though was adopted by the anthropologists of nineteenth century, later in twentieth century, it spread to other fields of social science with profound impact. Goodenough (1957, 1961) identified two different meanings. The first meaning indicates 'the pattern of life within a community' i.e., it denotes the regularly occurring activities along with the existing material and social arrangements. The second meaning signifies an 'organized system of knowledge and belief, whereby a people structure their experience and perceptions, formulate acts, and choose between alterations'. This is the inner configuration of culture. Therefore, a culture represents both the realms of phenomena and ideas.

The general agreement about culture is that it is learned; it allows man to adapt himself to the natural and social setting, which is greatly variable and manifested in institutions, thought pattern, and material objects. Culture is actually a conceptual tool by which we try to distinguish the customary ways of thinking and behaviour of the communities. Within each and every society some particular ways are noted through which needs of its members get satisfied. Culture, thus, makes headway by creating a definite form of ideas, institutions, utensils, language, tools, services and sentiments. It is, therefore, the man-made environment that found in all societies into which men are trained.

Imposing upon the physical environment, culture sprouted very simply at the beginning of human history. Gradually a series of transformation took place and a variety of cultural patterns were produced. At present there are numerous cultures—some are simple, others are sophisticated as well as complex. The wide range of culture projects various life-activities of people. Language as a vehicle helps culture to be transmitted from one generation to other. It is never inherited like genetic traits. The entire knowledge and pattern of behaviour flows from one person to another and this is how culture passes on. It can change and grow slowly. Today none of the anthropologists treat culture as a totally closed system. Because many solated tribes of hunters and gatherers are found to keep the relation with the neighboring group, which in course of time influences the respective culture.

Ways to look at the culture

Ethnocentrism

At the time of analyzing cultural pattern, a person should not be egocentric. Because, the person whose vision is limited to own needs and desires, fails to submit effective analysis while dealing with other people. To judge other people's way of life in terms of own cultural viewpoint is called ethnocentrism. Generally the cultural codes lie beneath the everyday behaviour in a hidden state. If an anthropologist is unable to recognize that, whole of his operation will be unsuccessful.

Cultural Relativism

When anthropologists go to study others cultures; sometimes they can not share alien customs or ideas ethically. A society's problems and opportunities should be understood or evaluated in terms of the standards or values of the respective society. Consciousness about cultural relativity allows anthropologists to cope with the situations of diverse cultural practices. For instance, many foreign scholars who had observed religious immolations in India, are found to criticize the practice instead of seeking inner significance, which is associated with the deep-rooted cultural values of Indians.

Nature and Conf guration

If we consider a particular cultural structure, we find that it is an organized whole made up of interrelated parts. Each part has its specific form and function, but they are not independent of each other. Therefore, a change in one part affects the other parts in varying degrees and ultimately the entire system is influenced. The components of culture are considered as sub-systems as the culture itself is conceived as a system. A culture may be compared to a human body where functioning of its components can be equated to different kinds of activities in human body.

Interrelatedness of the components in a culture is actually an old concept. It was realized by the social philosophers of eighteenth century who first attempted to draw links between different human phenomena like law, custom, political institution and so on. But the idea did not develop scientifically until twentieth century. Malinowski (1939) in defining culture regarded all of its parts (related to each other) in a systematic way. He stated that the items of culture aim to serve the basic needs of the members in a group. According to Malinowski, basic needs and cultural responses are directly related to one another, in all cultures. He had shown a number of references to explain this fact. For example, an agricultural tool or an implement for food production,or a technique related to food preparation finally serves the need of hunger. Similarly, other basic needs like shelter, safety, clothes, health, reproduction etc, are met through different other cultural responses.

The smallest isolated unit in a given culture is known as element. A cultural element may be a pattern of behaviour or a material product of such behaviour. When a few of these elements are combined together in a meaningful way, a greater unit is formed which we call a cultural trait. Some of the cultural traits together give rise to a trait-complex. A number of trait-complexes finally build up the total cultural configuration.

A culture can, therefore, be divided into segments on the basis of trait-complexes. A trait-complex can be broken into smaller units called traits and each trait is the accumulation of elements. If agriculture can be taken as a trait in culture, ways of rice production i.e., ploughing, leveling, weeding, harvesting etc., are the different elements. A trait may be an assemblage of certain actions or sentiment, or attitudes or objects. Trait-complex is the outcome of the interrelation of culture traits. Culture traits can be carried by the migrating people and generally a trait, especially if it is a material object, can travel a long distance. But for a non-material trait, the capacity of travelling is lesser.

Despite very strong individual differences, the members of a particular society project some common beliefs, attitudes, values and behaviours. This is the general agreement of any cultural system. The distinctive way of living in a society represents a generalization of behaviour so that the conduct of its members can be regulated. It does not disturb the personal habit system of any individual but makes up the norms for living together. As a result, members of a society respond to the situations more or less in the same way. Anthropologists seek to discover the typical customs and range of acceptable behaviour that constitute the pattern of a particular culture.

Norms of behaviour or types of performance are not same in all cultures. For instance, when two Europeans meet together, they shake their hands. But when two Bengalis meet, they greet each other by a typical posture, 'Namaskar'. This is for the difference of culture pattern. Here, the term pattern refers to a specific set of behaviours, which characterizes the culture. Each and every culture has its own pattern.

A culture again shows two types of patterns in it, opposite in nature—*ideal pattern* and *actual pattern.* Members of a society recognize some idealistic modes of behaviour as right kind of behaviour in particular situations. These ideal behaviours are expected from all members of the society. But in reality, behaviours deviate a lot from the ideal model. Therefore, actual patterns are rarely similar to the ideal pattern and such disagreement between ideal and actual pattern is discernible in every culture.

A.L.Kroeber distinguished these two aspects of a culture as *ethos* and *eidos*. Ethos is the values of a culture, which make the culture distinct and separate from other cultures. Eidos give the formal appearance of a culture with all its constituents. C.Kluckhohn also differentiated the explicit and the implicit elements of a culture. To him, everything in a people's life could not be perceived through sensory observation. The elements, which can be identified by the aid of eye and ear, are considered as explicit or overt items of a culture. The other elements like motivations and impulses, which can not be identified by sense organs, are the implicit or covert items of a culture.

R. Benedict believed in the integration of a culture by its content. She said, a culture is made up of many patterns. Further, she had shown that the patterns of a culture are harmonious so that they could be bound together in some consistent way. According to her, the harmony gave the culture a particular style—a unique configuration. M. E. Opler suggested a number of summative principles in culture called themes. Themes are the general motivations responsible for the various sets of behaviour types in a society. A theme is known by its expression and corresponds to the idea of pattern or configuration of a culture.

Our daily activities are controlled by the patterns or theme of culture. Opler's concept of theme is considered more elastic than Benedict's concept of pattern but both are the keys for understanding a specific cultural structure — a particular way of people's living. Each culture possesses its own theme or pattern distributed in different levels (material, social and religious levels) which varies from culture to culture.

Apart from the two main cultural patterns (ideal and behavioural) in a society, there are also some other patterns. No society is entirely homogeneous. Internal sub-groupings have been formed on the basis of age, sex, and social, economic or occupational specialization. Each of these sub-groups presents a characteristic type of behaviour, applicable exclusively for its own members. The most interesting but difficult problem of social science is to determine how patterns of culture change with time. It also endeavours to discover the extent to which the behaviours of the people are regulated by the individual will or motivated by the patterns of conduct that prevail in their society.

Aspects of Culture

It is well established that the building of culture was possible only for a few special god-gifted abilities, such as, erect posture, bipedal locomotion, stereoscopic vision, higher brain capacity, precision grip and power of speech. From the very beginning men have used these special capabilities to solve their problems and to make their life smoother. They were not aware about what we now designate as culture but the individuals unconsciously developed a boundary of culture around themselves.

The social environment of any human group plunges into culture. It is the culture, which guides the people—what to eat and how to eat, what to be done and how to be done, how to talk and how to think. Human behaviours are grossly determined by culture as the individuals are born, raised and live in it. A person can never be free from cultural influences. These phenomena have been explained by Leslie White as 'cultural determinism'. According to White, culture is the matrix, which is governed by its own laws of growth and operation. Neither human biology, nor human psychology can analyze the principle of its reality.

Every human society, literate or non-literate has a distinctive culture, which governs the behaviour of its members. It consists not only of things or events, which we can observe, count or measure; it also comprises of shared ideas and meanings. Some specific practices, beliefs and behaviours are shared with family members, relatives and friends because individuals of a group think and act in more or less similar way. The culture of any society also represent an adaptation or an adjustment to the conditions of life as people live in it, including the physical, social and supernatural environment. However, the behaviours are mostly learnt through imitation and expressed by the aid of language. Although the culture is the source for many ideas and perception, an individual does not remain totally limited to his cultural codes. As a dynamic entity he sometimes modify them in certain ways. This introduces change in culture over time and we recognize the universality of culture with dynamicity. Economy, climate and topography contribute to the dynamic nature of culture. However, each culture possesses some definite fields or aspects into which the behaviours of its men can be classified. They correspond to some major dimensions of human experience and are listed as follows:

(a) Economic Aspect of Culture

There is no society without the methods of production, distribution and consumption. There is some form of exchange, which may or may not be followed in terms of money. All these constitute the economy. Anthropologists study the pattern of behaviour around the economy because economic behaviour is considered as the basic of all types of behaviour of man. The total cultural life evolves from economy. Moreover, a change in economic life keeps certain repercussion on all other aspects of life. According to Karl Marx, economy is the infrastructure and the cultural life is the superstructure. Among the non-literates, economic behaviour is simple but obvious. In these societies gift exchange, barter, trade etc., are the general mode of economic transaction. Gift exchange is the ceremonial exchange where passing of presents is associated with an idea of getting future return. Such ritualized gift giving has been reported from Kwakiutl, Haida and some other tribes of North West Coast of North America. Trobriand Islanders of Melanesia, people of North Australia also practise this. A custom of funerary gifts is in vogue in West Africa. Barter is the direct exchange of a good against another good. It is the most primitive way of transaction. Hundreds of years ago, when money did not come in use, people used to give out their things for the things, which they wanted. Barter system of exchange is still practised between the Chukchis of Siberia and the Alaskans, between the Congo Pygmies and the Bantus of Africa. Such a primitive practice is also evident in California, Malaysia, New Guinea and other places. However, at present, the most prevalent form of exchange in non-literate societies is trading. It is an exchange considering the direct value of goods, usually in terms of money.

Economy often holds a dual meaning in non-literate societies. On one hand, it satisfies th material need of the people, on the other hand, it fulfils people's desire of prestige. For example, during the potlatch of Kwakiutl, thousands of valuable blankets are burnt, canoes are broken and even slaves are killed in order to announce a chief's prestige. Prestige is determined by the rate of expenditure, not by the amount of savings. Economy thus regulates the general character of the social, political and spiritual processes of life and, therefore, role of economy in shaping the life of the primitive people can not be denied. It is held that the concept of property and market arose from economy.

(b) Social Aspect of Culture

Men are called social animal, as they can not live without society. A great variety of behaviour that entangle with social life are found to be manifested in day to day activities of the people. Relationships, formed between individuals and groups, subsequently give rise to a form of network, which we designate as social structure or social organization. According to R. Firth, social organization is the dynamic phase of social structure, "the continuity principle of society". Different groups and institutions are enclosed in this sphere and so various types of behaviour originate. Most of the behaviours of the people conform to the social norm. They decide the factors like, how men and women seek their mates, how primary grouping proliferate into broader units, why non-relationship categories form basing on age and free association, and so on. Such behaviours also explain how the social rules are transmitted from generation to generation and cultural consistency is maintained.

In all human societies, men and women form attachments and set up groupings, which transcend the lines of kinship. But social behaviours essentially vary from culture to culture in accordance to the respective norms of the society. Non-literate societies are not different in this concern.

(c) Political Aspect of Culture

For ordering the human relations, a power system is necessary which is termed as political structure. In each and every society, whether it is modern or primitive, a kind of political structure is found. The associated behaviours of this structure are known as political behaviour. Generally a political system operates within a territorial framework. It maintains relation with other groups by diplomacy and warfare. Within the group, it controls the conduct of the members by exercising laws.

Primitive political organization stands as an organization for war and peace. Physical force is applied either for maintaining the existing order or to attack outsiders. Punishment is often given to the offenders who go against the law. Even in the most backward societies, where government is not found in full-fledged form, some kinds of laws are practised which are said to be customary. Customary laws have been noted from most of the native African political systems. Ashanti political structure shows an extreme degree of complexity among the non-literate societies. Here five territorial divisions are found under a permanent chief who is again guided by a group of elder advisors having military power. Thus, the types of political structure, the techniques of settling the affairs, dealings methods of various crimes and offences etc., come under the political behaviours being an important aspect of life.

(d) Religious Aspect of Culture

The invisible powers that rule behind the universe are admitted to all societies and the ways in which they are approached have been the indispensable part in the routine of living. People try to harmonize themselves with those powers. Different beliefs and practices have developed in this respect, which vary from one culture to other. Starting from the animistic and animatistic beliefs, the ideas about spirit, ghost, magic, gods and goddesses, all project these forces of supernature. Religious attitudes have been formed accordingly. Gods and ghosts are the opposite attributes—for good and ill. People appease the gods, the benefic forces, by prayers and offerings. These are the ways to get their blessings. But the malefic forces are not revered; they are tried to be taken under control by uttering magical formulae and making some specific performances. Besides, to ascertain and interpret the will of the gods, several other techniques are followed by the people. Divination is one such technique by which decision of supernature is sought. Religion implies the emotional response of mankind towards the great invisible powers and religious behaviours have been integrated in cultural life. From the simplest society to most complex society, people of all social strata exhibit certain kind of religious behaviour, which may be in the form of individual or group activity.

(e) Educational Aspect of Culture

The next aspect of culture considers the educational behaviour of the people i.e., the behaviours, which originate from the educational activities in a society. Education is the process, which moulds the individuals as per the expectation of a society. As like the training in schools and colleges, or through religious discourses, culture itself takes a substantial part to train up their young members. Cultural traditions that are acquired by each and every individual at own initiatives generally go beyond the formal schooling. Teaching and learning techniques are silent here. Both the literate and non-literate people are exposed to this silent educational system. It is an in-built mechanism in every culture. It starts with the informal teachings by the elders in the family. Children are prone to imitate the activities of grown-up persons at different situations. They also observe various performances and ceremonies. All these gradually help them to assimilate the underlying values and norms. Moreover, the popular stories, parables, proverbs, riddles of the society teach them the ideals to discriminate between good and bad, right or wrong. Such contribution of culture can no way be overlooked; rather they are especially valuable among the non-literates. Because, the non-literate peoples, being devoid of the formal courses of learning, acquire each and every practical lesson from daily experiences. The silent exposure of education offered by culture is indispensable for their physical and mental development.

(f) Recreational Aspect of Culture

Search for recreation is a universal human instinct. In every culture some of the behaviours are identified around recreation. They vary from simple to complex but obviously required for the relaxation of the individuals, to cast off the monotony of daily life. Some aesthetic sense are usually involved in these behaviours that denote the level of mental excellence of the people. Art, drama, music, dance etc., are the tangible expressions of recreational aspect of culture.

The above mentioned aspects or fields of culture can be further systematized into three levels on the pattern of interaction.

(*i*) *The Material level,* where interaction is found between man and nature. This is the category where we can place the material objects like the tools of production, means of subsistence, materials for shelter, instrument of offence and defence, etc.

(*ii*) *The Social level,* where interaction takes place between two or more individuals. This level includes all sorts of interpersonal relationships as reflected in marriage, family, kinship, etc.

(*iii*) *The Religious level,* where interaction takes place between man and supernature. This level is, therefore, composed of intangible objects like idea, belief, knowledge, rite, etc.

Among these three levels, the material level is considered as the primary level and it is placed at the base; the whole of a culture depends upon it for the mechanical means of adjustment. The social level is subsidiary to material level. The third importance is ascribed to the religious level, which is totally abstract in nature. However, the influences of interaction are not equal in all directions, although they are always integrated.

Culture and Personality

By the word personality we mean the psycho-biological world of an individual which gives one a definite self-hood. Since the behaviours of the individuals in a society are regulated largely by culture, it influences the personality also. Difference of culture is responsible for the difference of personality norms. Cultural traits, both material and non-material contribute to the personality formation. Self-hood develops only in human being. Animals like dog and chimpanzee though have some sort of personality in the form of individual trait but do not possess a self.

Culture is an integrated system of learned behaviour. When an individual borns in a society, he is automatically born into a culture. Although the individual can not perceive the culture directly, he is exposed to the social influences that play upon his physical and mental structure. Gradually he adapts the pattern of behaviour sanctioned by the group; culture moulds him as per the expectation of the society.

The process by which a child acquires the culture content along with personality structure and selfhood is called socialization. He experiences social interactions primarily with parents and other closed kins. Thus he learns various social behaviours i.e., what to do, how to do and why to do in different social situations. The moulding of an individual by the particular culture starts unconsciously from the early days of infancy and there is no way to escape it. But it is also true that each child is dynamic and unique personality. Each of them is inclined to his individualistic behaviour type. For example, a boy may be docile or aggressive, reflexive or impulsive, intense or relaxed, introvert or extrovert as per his own mood and motive. The cultural determinants act on the individual peculiarities and set a pattern within a given society delimiting the normal behaviour. The kind of speech, the form of tools, practice of religion etc., are absolutely determined by the culture. Culture builds a generalized personality although differences in temperament and variation in individual experiences prevent the cultural standardization of personality.

The way in which a cultural tradition is formed in a society and how it changes progressively aroused the interest of the anthropologists. They tried to utilize the notion of personality to refer the characteristic behaviours and the ways of thinking and feeling. The beginning of the concept of 'culture and personality' may be linked with the great American linguist and anthropologist Edward Sapir (1884 - 1939) who being influenced by German psychologists criticized the contemporary concept of culture. In 1934, he argued that 'the more fully one tries to understand a culture, the more it seems to take on the characteristics of a personality organization'. The study of the development of personality was the best way according to Sapir to understand a culture. However, most of the earlier anthropologists believed that a typical personality structure prevails among the people of a given society, as the members share the same culture. They also believed that the differences between the cultures are created for the differences of personality types. Later, the typical personality types were conceptualized alternatively as configurational personality, character structure, basic or modal personality and national character.

Ruth Benedict in her books 'Patterns of Culture' (1934) pointed out that the cultures are not only patterned, they can be characterized in terms of different personality types. She believed that each culture is organized around a central ethos; people share certain fundamental psychological attributes in an integrated manner. She tried to analyse cultures from configurational point of view.

Margaret Mead, a student of Benedict followed a more or less similar trend of thought. She studied three different societies in New Guinea and noted that those societies were different from each other psychologically. Not only that, she found some differences in the psychology of male and female in each of those three societies. Mead arrived at a conclusion that, it was not the biology but the culture that make the personality differences between the two sexes. In her book, 'Coming to Age in Samoa' (1928), she had shown how certain child-rearing practices produce typical character structure among the adults.

The psychoanalyst Abram Kardiner in the book 'The Individual and His Society' (1939) stated that the social experiences of the family (specially during child-rearing phase) and subsistence techniques often put forward a basic personality structure, which are reflected in the institutions like religion, ideology and politics. Since Kardiner was basically a psychoanalyst, his interest was focussed on the psycho-dynamics of personality and culture. Therefore, he examined the effects of social institutions upon personality and the effects of personality upon the social institutions, *vice-versa.* Ralph Linton in his book, 'Cultural Background of Personality' (1945) supported the same view. Cora Dubois in his work, "The People of Alor' (1944) used the term 'modal personality' to point out the basic personality i.e., the most common types of patterned behaviour by the members of a society. In fact, culture produced a notion of 'common personality' that had been extended to delineate 'national characters' during the Second World War in order to depict the nature of the natives—whether they were friendly or unfriendly in attitudes.

Melville. J. Herskovits in his book, 'Man and His Works' (1947) tried to relate culture and individual in terms of enculturation. Enculturation is the process by which an individual learns the forms of conduct acceptable to his group. He seldom comes into conflict with other members of the society by thought, values and act. As a consequence, he tends to be moulded into a right kind of person as desirable to the group. According to Herskovits complete success is never achieved. Some persons are influenced easily than others; some resist the enculturative discipline. Still, by and large, most of them reach to a common consensus of belief and behaviour so that the elements of culture constitute the matrix within which personality structure of individual develops.

E.A.Hoebel in the book, 'Man in the Primitive World' (1958) defined personality as the sum of integrated behaviour traits and he perceived it as analogous to the culture of a society. According to him, the bodily constitution, the physical environment and the culture combine to produce the personality structure of an individual. Enculturation encompasses the whole universe of the individual, so that he learns to internalize the norms of the particular culture. The process involves selection as well as elimination of multitude kinds of behaviour that an individual indulges.

It is, therefore, clear that man is the product of cultural milieu in which he is born. Heredity provides him with certain generic human faculties at birth. In course of time he learns different things and his personality takes a definite shape. It is the culture, which conditions him according to its expectations. The studies on culture and personality flourished in 1930's by the initiative of American anthropologists. But this trend of study did not last long. After 1960's the popularity of such study dwindled down. They not only became unfashionable but unreputable too. Still, the concept of personality is regarded as a cardinal concept in anthropology.

Theories on Culture

Anthropology is a product of scientific developments that took place in the western world. Western civilization has got a base on the scientific, political, moral, religious, philosophical and aesthetic formulation of the Greek and Roman people to which social scientists revived their interest during the Renaissance. Developmentalists began to document their data in order to demonstrate socio-cultural progress of man. This accelerated the emergence of socio-cultural anthropology.

From the very beginning social-cultural anthropologists were overwhelmed with the problems of growth of culture and cultural parallels. They tried to deal with these problems and proceeded through three principal approaches. These approaches have been regarded as theories of culture. They are as follows :

(*i*) The theory of Evolution or Evolutionism.

(*ii*) The theory of diffusion or Diffusionism.

(*iii*) The theory of Structural-functional analysis or Structural Functionalism.

***The Theory of Evolution* (Evolutionism)**

The word 'evolution' describes the process of qualitative change. Evolutionism is the scholarly activity of describing, understanding and explaining of this process. The theory was basically drawn from the sociologists and appeared at the formative phase of anthropology i.e., the middle part of the nineteenth century when Darwin and Spencer went forward to explain evolution behind all phenomena.

Evolutionism has been granted as the guiding principle of anthropology. The first attempt to define evolution was made by Herbert Spencer. In his book 'First Principles' (1862) he defined evolution as 'a change from an indefinite, incoherent homogeneity, through continuous differentiation and integration'. Since the definition was not all pervasive, later Spencer had to modify it. In the perspective of evolutionary bias, Edward. B. Tylor (1832 - 1917) and Lewis Henry Morgan (1818 - 1881) devoted themselves wholly to study the evolution of human society and culture.

Tylor is called the father of modern anthropology for his contribution to the concept of culture. In the book 'Primitive Culture' (1871) he first defined culture or civilization. Tylor postulated that the culture was evolved from the simple to complex, and all societies had passed through same stages of development. Tylor attempted to show this in the context of religion. He explained cultural parallels as the psychic unity of mankind. Tylor pioneered the doctrine that culture was learned, not biologically determined. He eliminated the consideration of race from the study of culture.

The other proponent, Morgan also advocated a uniform and progressive cultural evolution. He believed that human societies had evolved from lower to higher types through successive stages. The stages were supposed to stand in the same order in all parts of the world as the mental processes stood almost similar among the peoples. In the book 'Ancient Society' (1877) Morgan demonstrated human progress through the stages like Savagery (lower stage). Barbarism (middle stage) and Civilization (final stage). He speculated that the institution of family was evolved through five distinct forms and tried to correlate them with the evolutionary consequences of marriage regulation. Morgan also considered the evolution of technology, property, kinship and political structure.

Beside Tylor and Morgan, Haddon and Levi Bruhl also tried to show the evolutionary sequences. Haddon traced it through art forms. He recognized an early realistic form, which reached to an abstract symbolic form through certain geometric forms. Levi-Bruhl's subject was logic. He showed its evolution from a primitive to modern form. Karl Marx took sufficient interest in Morgan's evolutionism because it supported his own materialistic conception of history. However, the conjectural reconstruction of culture gained enough popularity in nineteenth century. Anthropologists of this period exhibited a high degree of ethnocentrism. They had a blind faith in the idea of advancement of society and culture; they judged other cultures in terms of their own model. We can utter the names of Bachofen, McLennan, Lang, Frazer, Westermarck and Brinton in this connection. However, evolutionism that arose in 1860's began to decline around 1900. It remained under eclipse nearly half a century and again began to assert itself from 1940's. The nineteenth century concept of evolution is known as *Unilinear* as it emphasized one straight-line course of development. Some scholars like to designate it as parallelism because it occurred in all societies in the same manner, inspite they were quite apart and isolated from each other.

***The Theory of Diffusion* (Diffusionism)**

Diffusionism or the theory of diffusion is the second approach on which social-cultural anthropology rests. Evolutionists tried to explain culture in terms of internal growth but the diffusionist's explanation was different. They considered the factors like accepting and adopting of cultural traits from outside. They were especially interested in the external growth of culture.

Early evolutionists although made significant contribution in understanding the cultural processes, their method of comparative analysis was unjudicially used. Further, they were ignorant about the cultural borrowing. Most of the evolutionists were armchair theorists who depended simply on the accounts furnished by travelers and missionaries. The uncritical ethnocentric attitude did not count the perspective of time and space in studying man and culture.

According to the diffusionists cultural traits are often borrowed. The total culture may not emerge into similar forms in different societies as spontaneous growth is found in each society, which involves certain potentialities as well as common human nature. The theory of diffusion emphasizes on the process of borrowing and incorporation for which the traits travel from one culture to another and the diffused traits may undergo all sorts of change. The study of Diffusionism have been conducted by three schools which are as follows:

(i) British School of Diffusionism

Here the main spokesmen were Grafton Elliot Smith (1871 - 1937), William James Perry (1887 -1949) and W.H.R.Rivers (1864 - 1922). Elliot Smith was an anatomist who had a keen interest in archaeological work. In course of his excavation at Cairo in Egypt, he found some pyramidal structure similar to English monuments of Europe. He came to the conclusion that the English monuments were the crude imitations of Egyptian pyramids. Moreover, he regarded Egypt as a source of all higher culture. In the book, 'In the Beginning: the Origin of Civilization' (1932) he forwarded the following propositions:

(a) Man was basically unimaginative and uninventive. So culture arose only at one favourable geographical setting.

(b) Such favourable circumstance existed only in Egypt.

(c) Civilization was naturally diluted when it spread from place to place. For example, the cult of sun worship diffused from Egypt. Sun worship was also found in Maya and Mexico.

(d) In 4000 BC, civilization originated in Egypt and then spread in other directions with several modifications.

Perry supported Smith's view. Both Smith and Perry insisted on the universal spreading of culture from Egypt. They designated the other people outside the influence of Egyptian civilization as 'ape-like natural men' because those people neither possessed arts and crafts, nor had the clothing, ornaments, agricultural implements, marriage customs or any elaborate rituals. But the people were satisfied with the making of hunting implements.

Rivers was originally a medical practitioner and his interest in anthropology was primarily concerned with psychology. He believed in field investigation and explained the differences between the Melanesian and the Polynesian cultures in terms of successive waves of migration. His classical work was 'The Todas' (1906) where his interest went with the spread of culture. He also developed an important technique named 'genealogy' for recording the kinship nomenclatures. Like Smith and Perry, Rivers also failed to build up a competent group of disciples to carry on his culture-historical interpretations. As a result, by 1930, diffusion studies were declined in England, which advocated independent parallel evolution of any culture trait in two widely separated areas was rarely possible; people prefer to borrow inventions from other culture rather than developing own ideas.

(ii) Continental or German School of Diffusionism

Fredrick Ratzel (1844 - 1904), Fritz Graebner (1877 - 1934) and Jesuit Wilhelm Schmidt (1868 -1954) were the main proponents of this school. Sometimes this school is also called as Kulturkreise School as it suggested the concept of culture-complex or culture circle. It says that a number of different culture-complexes (Kulturkreise) developed at various times in several parts of the world and later on those were diffused. According to continental diffusionists, diffusion is a continuous process and different strata of diffused-in culture traits may be found in a culture.

Ratzel was basically a geographer who became the founder of Anthropogeography. In his book 'Ethnology' (1929) he said that the development of universe was not uniform. Therefore, a group of people with simple technology might have nurtured advanced social structure and might have practised a complex form of worship. Ratzel observed the African culture as richly developed along the East Coast of Atlantic and so to him Atlantic coastal zone contained the most ancient elements. The term Kulturkreise was coined by Leo Frobenius (1873 - 1838) who followed Ratzel's observation and searched out the origin of African culture in East Indies.

Graebner was a historian who was inspired by the researches of Frobenius. He invented some methodological principles in 1911 and delineated two basic criteria as 'Criterion of form' and 'Criterion quantity'. By 'Criterion of form' Graebner wanted to state the similarities between two cultural elements resulting from diffusion. On the other hand, the historical relationship between the two cultures was stated by 'Criterion quantity'. This relationship was manifested by the similarity in the number of its elements. Schmidt attempted to reconstruct the cultural history of the whole world. Like Graebner, he tried to show the emergence of ancient cultural traditions and their spatial ranges. He pointed out four main 'grades' of Kulturkreise namely primitive, primary, secondary and tertiary. Each grade was distinguished in different layers. For example, the primitive grade included the culture of hunting and gathering peoples of the world. This grade was again classified into three divisions (Kries).

(a) The central or exogamous kries corresponded to the culture of Africa and Asia. It was featured with exogamous hordes and monogamous families.

(b) The Arctic kries corresponded to the culture of the Somoyeds, Eskimo and Algonkians. The features were the exogamous clans and sexual equality.

(c) The Anarctic kries corresponded to the cultures of Southern Australians, Bushmen and Tasmanians as featured by exogamous clans and sex totems. In the same way, the other grades, primary, secondary and tertiary were also classified into sub-grades. Schmidt was assisted by another scholar named Jesuit Wilhelm Koppers.

The diffusionists of Continental or German School of Diffusionism depended solely on the evidences of material culture; they did not count the diffusion of social institutions. They provided some important insight to the culture processes, particularly in reference to the influence of environment on culture. They also showed the role of complex centers in disseminating culture. However, these diffusionists were primarily concerned with the discovery of the facts related to diffusion.

(iii) American School of Diffusionism

This school of Diffusionism was chiefly conducted by Clark Wissler (1870 - 1947) and Alfred Louis Kroeber (1876 - 1960). Both of them were the students of Franz Boas (1858 - 1942), the leading opponent of evolutionism in early twentieth century. Boas rejected the evolutionary theory of nineteenth century, which propagated the idea of universal laws behind all human cultures. He believed in 'Historical Particularism' which said, "each culture is unique and that similarities between cultures are best explained by the assumption that cultural exchange has taken place". He also stressed the importance of collecting data through a detail recording of culture with description.

Students of Boas namely Alfred Kroeber, Robert Lowie, Edward Sapir, Melvillie Herskovits, Ruth Benedict, Clark Wissler, E. Adamson Hoebel, Margaret Mead, all followed the views of Boas very strictly. Boas divided the study into four phases:

(a) Facts of the situation must be described at first.

(b) Analytical study from particular to general must be done.

(c) Dynamics of culture contact ought to be sought.

(d) Psychic make-up of the individual should be considered.

Clark Wissler not only developed the basic viewpoints of Boas; he defined the restricted area (culture area as called by Boas) more precisely. To him, each culture area was consisted of a certain set of culture-complexes and a central point of dispersal was located in them. Non-material traits of culture could not travel long distances without getting diluted on the way. Long distances, especially mountains, oceans, arid deserts were the obstacles for culture diffusion. Discontinuous diffusion might take place owing to migration, unusual changes in fringe areas and for strong individuality. Wissler plotted various culture areas in America in terms of technological, artistic and institutional features. He said that a culture area might have included different distinct populations. He also suggested a 'culture center' for each of the culture-area from which the various culture traits spread out. He also adopted 'Age and Area' theory of naturalist, which was applied in anthropology by N.C.Nelson in 1919.

Koreber revised Wissler's model of culture-area and formulated a scheme of seven 'grand areas'. He did not approve 'culture center' as argued by Wissler. Further, he divided and analyzed the culture areas in terms of culture content. He also talked of geographical determinants. According to Kroeber, culture areas often resemble with each other despite they have differences between them. Immense cultural similarity is visible when communication between two areas is frequent. So marked cultural difference is produced for limited communication. This is the reason for which people of the neighbouring area exhibit close resemblance. Regarding the point of origin of characteristic traits in 'culture area' Kroeber stated the center 'as a locus of superior productivity'. In his book 'Anthropology' (1923), he presented a diagrammatic representation on accumulation and diffusion of culture traits.

However, the aforesaid three schools of diffusion assumed that, most of the aspects of the high civilization possibly emerged in culture centers (one or few); they got diffused to the rest part of the world under special circumstances. Distance and resistance are the two forces that check the process of diffusion. However, the diffusionists had dealt with the culture history, culture growth, culture parallels, transmission of culture elements etc. Diffusion is very important to understand the phenomena of culture change and stability. Approaches of diffusionists became greatly popular in a period between late nineteenth century and early twentieth century.

***The Theory of Structural Functional Analysis (* Structural Functionalism)**

Both evolutionism and diffusionism strived to explain culture and culture parallelism. They could not throw any light on the on-going process of culture. No psychological clarifications were made on the internal structure of culture, which included customs, values, beliefs etc. Although diffusionist's anthropology still holds a little ground in America and in England, but actually both the evolutionism and diffusionism are outmoded as they provide only history. Anthropologists began to raise questions regarding the inner constituents of a culture. They became interested in finding out the facts that whether there is any relation between different parts of a culture or not. Structural—functional approach answered all these questions.

The theory of Structural – functional analysis or Structural Functionalism rests on the propositions of two great anthropologists—Bronislaw Malinowski (1884 - 1942) and Arthur Reginald Radcliffe-Brown (1881 - 1955). In the background, there was one French Sociologist (Emile Durkheim) whose concept of 'social solidarity' led Malinowski to see the society as an integrated group. Malinowski also used Tylor's definition of culture, but to Malinowski culture was an instrument, which enabled man to maintain his biopsychic survival. He assumed that nothing is loose within a culture; all cultural traits serve the needs of individuals in a society i.e., the function of a cultural trait lies in its ability to fulfil certain basic or derived needs of the members of a group. Malinowski passed a very long period among the natives of Trobriand Island. He observed their culture minutely and came to the conclusion that each aspect of culture, whether it is economic organization or social organization or religion or language, is rooted in the needs of the people. He noticed an intimate interrelationship among the various aspects of a culture for which self-sufficiency of a culture has been secured. For the same reason when change touches an aspect of culture, it is carried on other aspects too. As a result, the whole culture undergoes a phase of transformation.

The needs, which Malinowski correlated with cultural responses, are as follows:

Basic Needs	Cultural Responses
1. Metabolism	1. Commissariat.
2. Reproduction	2. Kinship
3. Bodily Comfort	3. Shelter.
4. Safety	4. Protection.
5. Movements	5. Activities.
6. Growth	6. Training.
7. Health	7. Hygiene

Malinowski's theory though had both merits and demerits; it laid the foundation for a functional theory of culture. The concept is regarded as Functionalism in anthropology. Later, Radcliffe-Brown added a dimension of structure in it; he showed how functions could be analyzed in the context of structure. Unlike Malinowski, Radcliffe-Brown directed his attention to study the survival of society. So he put much emphasis on the society itself rather than the individuals in a society. He dealt with the social structure in reference to the interfunctioning interdependent parts of it.

Radcliffe-Brown was trained under A.C.Haddon and W.H.R. Rivers. In 1906, Rivers had sent him to Andaman Islands to reconstruct the cultural history of the Islanders. He studied the myths, ceremonies, festivals and customs of the people in depth, for the absence of written records. Thereafter he analyzed the data from Structural-functional point of view. Since Radcliffe-Brown received an intellectual stimulation from Durkheim, he considered Durkheim's concept of'function' as an appropriate method for scientific study of the society. He did not pay any attention to the problems of social change and believed in the continuity of social structure. According to him, social structure can never be destroyed by the changes. Some changes may appear in its structure but without any breach of continuity. Finally Radcliffe-Brown compared the social structure to that of an organic structure where the structure is derived from the function of its parts.

Structural-functional approach speaks the following points:

(i) Everything has a form that determines its purpose.

(ii) All things are in a state of flux, cycling between integration and disintegration.

(iii) Every form has a structure where parts are of different orders of importance.

(iv) Each part has to contribute to the proper functioning of the total system through acculturation of its own potential.

(v) The general trend of every system is to balance the harmony, as each and every part fulfils its potential activities as per to its purposive form.

(vi) Any change in any part of the system will disturb the flow of proper action and results in disharmony. It ultimately ends in decay.

(vii) The parts are usually connected by strong exertions. So the disequilibrium, which is produced for any change, is soon vanished and a harmonious state is re-established.

R. Linton (1936) distinguished between the form and function of a culture. According to him, form is the organization of the contents of a whole culture. The form of a culture-trait or a culture-complex can be perceived by the senses and it is possible to describe them objectively. But the function is the activity or operation that has to be understood from the relationship of the traits within a culture. Sometimes it is also reflected in the relationships of things outside the society and culture. However, in finding out the functional relations of cultural phenomena, study of cultural variables and epiphenoma are important rather than the organic constants like physiological necessities or psychological imperatives.

Neo - Evolutionism

The evolutionary approach did not die out with nineteenth century; it went just into eclipse. Actually, Boasian reaction had simmered down its popularity for about fifty years. Again, anthropologists revived their interest in the evolutionary process of culture. One of the vehement critics was R.H.Lowie who stated, "evolution is far from dead and our duty is merely to define it with greater precision" (1917). However, the idea of evolution, in later period was followed up mainly by the anthropologists like V.G.Childe (1892 - 1957), J.H.Steward and L-A.White. In 1930's, L.A. White criticized 'Historical Particularism', the concept forwarded by Boas. The resurgence of cultural evolutionism was found from 1943, the year in which White's 'Energy and the Evolution of culture' appeared in the journal, 'American anthropologist'. Among the three proponents of Neo-evolutionism, L.A. White is considered as the most outspoken anthropologist.

Neo-evolutionism is no doubt an advanced concept than the traditional or classical or unilinear theory of evolution. Neo-evolutionists dealt with the cultures as a whole, than a particular culture. They argued that the evolution was essentially divergent as the cultures of the world did not pass through similar stages of development; development of the cultures depended on the bio-cultural balance on the ecological niche. To them, evolutionary stages were appreciable only in estimating the growth of culture, but not for the reconstruction of culture.

Julian Steward (1955) classified evolutionary thoughts into three schools—unilinear, universal and multilinear. In his opinion, ideas of Tylor and Morgan should be grouped under unilinear approach because those scholars advocated a similar line of cultural development for all societies of the world, throughout the history. Steward coined the term 'universal evolution' to label the ideas of L.A.White and V.Gordon Childe who tried to project the evolution of culture as a process of Thermo-dynamics. It is the increase of energy and technology for which the culture progressed i.e., culture = Energy X Technology. White emphasized on the harnessing of energy as a measure of cultural development, whereas Childe stressed on the technological revolution. But finally both the scholars related the fact with human advancement in social and mental plane. As a support of his proposition, White cited the examples of Negro-Africa and Near East where evolution did not follow the same sequence but went on in the same direction. Gordon Childe had drawn the evidences from the sequences of cultural evolution that occurred in the emergence of the high cultures of Peru, Mesoamerica, Mesopotamia, Egypt and China. He was able to show that the development followed a more or less similar course.

J. Steward designated himself as a multilinear evolutionist though the term 'multilinear evolution' was not coined by him. By the theory of multilinear evolution Steward condemned the approach of Universal evolutionists (L.A. White and V.G. Childe) for their inability to explain the specific instances of cultural evolution. He himself searched for the parallels of limited occurrence, rather than universals to establish the causal principles that account both for parallel as well as divergent evolution. Since Steward was basically an ecologist, his focus was on the techno-environmental and techno-economic criteria. He classified the societies or socio-cultural integration basing on their level of ecological adjustment to the environment. He sought for specific cultural similarities and differences. But inspite all of his efforts, multilinear evolution failed to yield a wide scope. It could not develop a comprehensive set of evolutionary principles to cover the growth of culture, from earliest prehistoric days upto the present.

Marshall Sahlins and Elman Service were the students as well as the colleagues of White and Steward. They tried to combine the views of Steward and White and so recognized two different kinds of evolution, specific and general. Actually they wanted to mean that the evolution always move in two directions at a time. On one hand, it creates diversity by the adaptive mechanism; some new forms appear as distinct from the old. This divergence and adaptation has been referred as "specific evolution". On the other hand, the "general evolution' refers to the general progress of human society for which the higher forms arise from the lower forms by suppressing it. Thus Shalins and Service found out both parallelism and convergence in the cultural evolution.

The latest theory of evolution is known as 'Theory of differential evolution' as proposed by R.L.Carneiro and S.F.Tobias. Although the theory is recent, but the concept is very old. E. B. Tylor as early as 1871 identified different spheres of human life as intellectual, moral, political etc., which went on continual progress, but their advancement did not put forward with equal steps. A.E.Hooton in the field of hominid evolution also noted that all parts of the human body had not been evolved in the same scale. Carneiro applied the same concept in cultural development. His term 'differential evolution' was analogous to Hooton's term 'asymmetrical evolution'. By using this term he wanted to show a differential rate of evolution which occurred in various spheres of culture, both inter-societal and intra-societal. The societies did not only evolve following variable degree of development, differences are also found in respect to different spheres of the same culture. This concept of differential evolution forms a part of Neo-evolutionism.

Evolutionism, whether it is looked upon as unilinear or multilinear, are neither contradictory nor the separate views. Neo-evolutionists made a refinement over the traditional theory of evolution and it appears that the Neo-evolutionism is more like a methodology than a theory. The reasons are as follows:

(a) Neo-evolutionism is based on the assumption that the significant regularities in culture change do occur and it is concerned with the determination of cultural laws.

(b) Neo-evolutionism deals with the form, function and sequence of culture, which have the empirical validity. Therefore, only after searching the broadest regularities and differences carefully, we can talk about evolution. One must examine the wide spectrum of comparative data before arriving at any conclusion. Hence it is perfectly a method.

(c) A theory always stands on concreteness and specificity. Neo-evolutionism must not be a theory in true sense as it lacks concreteness and specificity. For instance, migration as a factor of diffusion may bring new traits in an area. Neo-evolutionism, therefore, fails to satisfy the problem of concreteness and specificity.

(d) Nature of culture is dynamic i.e., culture changes with time and space. Neo-evolutionism can not solve the problems of time and space to deduce a cross-cultural similarity.

For all these drawbacks, the Neo-evolutionism remains far from being a theory. It is regarded as an essential methodology.

Structuralism

Next to A.R.Radcliffe Brown we can name the French anthropologist Claude Levi-Strauss whose approach of cultural analysis has been known as Structuralism in Anthropology. Levi-Strauss was primarily concerned with the psychological factors and processes in order to explain the various cultural practices. He considered art, rituals and patterns of daily life as the manifestations of culture and tried to relate those surface representations with the underlying structure of human mind. In doing this he stressed on the cognitive process i.e., he wanted to understand how people perceive and classify the worldly objects around them. According to him, human mind undergoes a process of dualism and so he made a distinction between conscious and unconscious levels. He also proposed two structural models, mechanical model and statistical model to understand the patterns of interaction and thought in a society. By 'mechanical model' he referred to the idealistic and normative behavioural patterns in a society, which facilitates in integration and cohesion of the members. In contrast to this, the 'statistical model' indicates the real behaviours, which exist among the members of the society; these offer a scope for alternatives and contradictions leading towards pluralism. Thus, Levi-Strauss introduced a concept a scale for measuring the daily life phenomena.

There were also other Structuralists in France as well as in Britain who became influenced by C.Levi-Strauss. Edmund Leach was one of them. He accepted the concept of ideal behaviour and actual behaviour of Levi-Strauss at the time of dealing with social phenomena. But he was in the favour of comparative generalization, which he derived partly from the work of Radcliffe-Brown, and partly from the work of Levi-Strauss. Aside Leach, the other British Structuralists were Rodney Needham and Mary Douglas who did not follow Levi-Strauss's principle in heart and soul, but applied structural analysis in studying particular societies and particular social institutions.

Levi-Strauss's approach acquired greatest significance in linguistics amongst all social sciences. Methodology of structural linguists was benefited a lot. This structural approach could also be related to ethnoscience as ethnographic data invariably involved a description of the culture (of a specific society) from which the underlying rules of thought of the particular people may be projected.

Language and Culture

Language is exclusively a human achievement. It enables man to share the experiences and thoughts of his fellows and to re-create his personal experiences for their benefit. A society that lacks language can no means assure the continuity of behaviour and learning necessary for the creation of culture.

Language is not something that is genetically inherited; rather it is an art that has to be learnt. It forms a system of communication where different sounds are put together in a meaningful way. In the animal kingdom though some animals are found to communicate among themselves by their own system of signaling, language is unique to man. for example, birds often pass message through distinctive sounds, squirrels cry in danger to make their fellows alert. The peculiar body movement of bees conveys their discovery of food resource. Similarly, the body-odour is a vehicle of information in ant society. All living primates use more or less these three means—sound, body movement and odour for communicating with each other. The great apes are also very efficient in communication. They make various types of call for various purposes, but possess no linguistic capacity as such. In fact, animal sounds are fixed and instinctive that vary within a limited range. These sounds are never similar to a language. Therefore, it is the language, which separates the human being from other animals by facilitating accumulation and circulation of knowledge. Without language there would be no way to re-create the past experiences or to pass the knowledge on others.

Language evolves from symbol. Behaviours of man can be classified into two types—non-symbolic behaviour and symbolic behaviour. A man yawns, coughs, blinks, stretches, scratches himself, cries out in pain, shrinks with fear, bursts into anger and so on. All these are non-symbolic behaviours and are not peculiar to man. Therefore, such behaviours are shared not only with other primates, with many other animal species as well. On the other hand, the symbolic behaviours are restricted to the mankind itself. Various kinds of symbols are used in communication. For example, men use amulets, confess sin, make laws, observe codes of behaviour, explain dreams, classify relatives in specific categories etc. with the help of symbols. Without symbol there would be no culture; man would merely be an animal. Sturtevant in his book 'An introduction to Linguistic Science' (1947) defined language as 'a system of arbitrary vocal symbols by which members of a social group cooperate and Interact'. Ability of using symbols transformed our anthropoid ancestors into human. No culture or civilization could have been generated and perpetuated without symbols.

Symbols in Speech

The most important form of symbolic expression is articulate speech. Without speech we could not get any political, economic or military organization. No codes of etiquette or ethics, no laws, no science, no theology, no literature, no music or game, could be formed. We would belong almost on an ape level. Paraphernalia of rites and rituals would also become meaningless without an articulate speech. The articulate speech thus facilitates in communication of ideas and feelings from one individual to other. It also transmits the accumulated knowledge from generation to generation and helps to preserve the tradition.

Language and speech are almost synonymous. Both depend on our ability to make abstractions, to assign symbols on them and to manipulate the symbols quickly. An infinite number of combinations and recombination of meanings are possible here. It is the man who only can spontaneously combine the symbols for building up new sentences to speak. He is also able to make out the meanings of those sentences, which have never encountered before. Contrary to this, a gorilla or a chimpanzee, being closest to man (in terms of body structure) can not combine the symbols to produce new messages, despite their symbolic capacity is very high.

Man possess language as long as he possess culture. The tool-making propensity of man has been correlated with language. Tools and speech, the two distinctive features of mankind are thought to be inseparably connected with each other, from physiological point of view. If we look at the animal kingdom, we will find that the brain of no animal, except man, is adequate in making symbols against different objects and actions. The animals use only certain signals to express a few emotions. Those signals are in no way equivalent to symbols. Further, the animals do not possess the ability to associate the symbols with sound. This limitation of brain debarred the animals, including apes, from making tools. In fact, the continuous and cumulative development of human culture would have never been possible without the aid of language.

***The Structure of Languages* (Structural Linguistics)**

Each and every language abides some definite pattern of speech, which may or may not follow the formal 'rules of grammar'. An unconscious rule based on sound makes up the essential structure of a language, Children during the days of their infancy learn this, hearing from the elderly persons.

The speech structure is, therefore, based on the sound elements, which constitute the phonology, morphology and syntax. Phonology means the patterning of sounds. Human vocal tract is able to make a wide range of different sounds. But no language uses all those sounds. There are some typical sounds for each language; other sounds seem to be difficult for the members for lack of use. The study of the total range of sounds in a language is called phonology or phonetics. The phonemes are the smallest sound unit, which can be conventionally classified into two basic categories: vowels and consonants. Vowels are the sounds, which come out of the vocal cord without any significant interference. On the other hand consonants are the sounds that are released only after getting some obstruction by some parts of the mouth like tongue, lip, etc. Generally the number of the phonemes varies from twenty to thirty in the languages. Analysis of phonemic structure is inevitable for linguistic research.

The second aspect of language is morphology. This is the principle related with the structure of the words. It denotes how sounds are arranged in sequence and how the different words are formed. The smallest unit with a specific meaning is called a morph. One or more morphs with similar meaning make up a morpheme. Although neither a morph nor a morpheme is a word but some words can be build up on a single morph or morpheme. A number of prefix, suffix, lexicon, gender signs etc., may also take part in the formation of words. The third aspect of language is syntax. It is the principle on which the words are arranged in a sentence to convey a definite meaning. In fact, morphology and syntax together constitute a language's grammar. Speakers of a language follow the implicit rules of syntax but as they are not consciously aware of it, so can not readily explain how they have constructed it. A linguist tries to make the rules explicit through the analysis of morphology and syntax.

The study of words and sentences (i.e., grammar) is different from the analysis of sounds in a language. The sound system or phonology is though an important concern of linguists, but stands meaningless without Semantics. The knowledge of semantics is essentially required to derive the linguistic meanings. These meanings are carried through the words and expressed in the sentences. However, the anthropologists often use language as a model to get the perception of culture i.e., how a culture encodes the view of the world and how communications are embedded in a social system.

French anthropologist Claude Levi-Strauss, the great proponent in Structuralism, pointed out the sound system of a language as a conceptual model for understanding a culture. He stressed on the cognition i.e., the processes of thinking and memorization. Finally he dealt with the ways in which people perceive and classify the things in the world around them. Edmund Leach was highly stimulated by Levi-Strauss's view. However, the idea of structural linguistics reshaped the methods of analysis in many social sciences. N. Troubetzkoy advocated four basic operations in structural linguistics.

(a) It studies the unconscious phenomena, instead of conscious linguistic structure.

(b) It treats the terms in relation to one another, not as independent entities.

(c) It projects phonemics as a complete system.

(d) It aims at discovering some general laws either by induction or deduction.

The most widely explored method in anthropology for semantic analysis is called componential analysis. It is an important tool for kinship studies in primitive societies. Kenneth. L. Pike (1964) contributed to methodology by providing etic and emic dichotomy. Linguistic researches of present day unveil the deeper problems of ethnology and social anthropology.

The Origin of the Language **(Historical Linguistics)**

Since the spoken words leave no remains for the archaeologists, the origin of language is difficult to be found out. Possibly it was evolved with the Homo sapiens in the Stone Age. But there is a debate regarding the possession of fully developed language by the Neanderthals. At present there are four to five thousand languages in the world of which more than two thousands are only spoken form, without a system of writing.

At the outset, there were some popular theories with amusing names, associated with speculative origin of speech.

(a) *The bow-wow theory*—It said that the speech had been imitated from the barking of dogs and other animals.

(b) *The pooh -pooh theory*—It said that the speech had been derived from the automatic emission of painful feelings or sensations.

(c) *The dingdong theory*—It expressed that there was a mystic harmony between sound and sense. Speech developed gradually by their correlation.

(d) *The yo-he-ho theory*—It advocated that the words came into existence as a result of strong muscular action of mouth. They relieved the body by letting the breadth and a sound was produced automatically.

(e) *The gestural theory*—It said that, men used their tongue in the same rhythm with gesture and posture and ultimately words came out.

(f) *The Tarara-boom-de-ay theory*—It spoke that the speech was evolved through the expression of joy. Primitive men when succeeded in hunting a great mammoth, they were spontaneously filled with joy and made different sounds, which paved the way for speech.

None of these theories could be accepted as entirely satisfactory and so abandoned by the serious scholars. But perhaps each of those theories contributed some elements, which altogether was responsible for the origin of the speech.

However, the languages, inspite of their inherent stability, undergo change. Historical linguistics has pointed out a hypothetical language at the origin and discovered some methods of reconstructing the course of change. One of these methods is the internal analysis of homologous language. When two or more languages show considerable resemblance in words and grammatical forms, it is held that they have come from a common ancestral language. By comparing presumably related languages, linguists can reconstruct the various features of the ancestral one. Such a reconstructed language is called proto-language.

Languages may appear similar for other reasons too. Contact between two language communities often leads one language to borrow from other. A large number of words, especially the names of new cultural items, are borrowed from the other language. Conquest and colonization show an extensive borrowing. As prolonged contact often results in greater resemblance between the languages, isolation produces divergence in a language. Two groups of people speaking same language, sometimes drop communication with each other and become separate, either physically or socially. With the passage of time they show changes in phonology, morphology and syntax. If this separation continues, two dialects of the same language give rise to two distinct families.

The other method for the reconstruction of language- evolution goes with the searching of the sequence of documents. But it is possible only in case of written languages. Writing is the graphic representation of speech. Probably it developed from pictographic drawings. Most of the modern systems of writing are alphabetic where each symbol denotes a speech sound. It has been found that the fundamental vocabularies of a language change at a given rate with time. This rate of change has been calculated for a number of historic (written) languages. Usually 19% of the vocabularies change in a thousand of years. Therefore, when two related languages show a difference of 38% on their basic vocabularies, it is understood that the people of a language-group got separated in two groups, at least two thousand years ago. Such a device in estimating the time of separation between two related languages is called glotto-chronology.

Interdependence between Language and Culture

A language comes into existence when two or more individuals agree to attach same meaning to the same sound. The meanings denote the symbols, which may have certain kinds of physical form. A symbol may be a material object or a colour or a typical sound. It may also be an odour, a motion or a taste. There is a subtle difference between language and speech. Language is the conceptual code, the system of knowledge that enables a person to produce and understand speech. Speech is the actual behaviour—the utterance of words or sounds by a person. The language is, therefore, more than the speech and it is a peculiar means of expression that can not stand alone. It depends on culture and explains a culture. The symbolic values of a language help in understanding of the tangible elements of culture.

The languages usually tend to preserve their existence in terms of territory or racial population. This does not mean that a language is biologically inherited. As a mater of fact, no inherent relationship is visible between race and language. But, war or trade or migration when brings one racial element in contact with another, the original language becomes altered or diluted. For example, the Negroes who entered America from Africa wholly adopted the new language. Today sixteen million American Negroes go on speaking English whose forefathers did know nothing but African Negro language. Languages are neither the product of geographical condition, nor the climatic condition; they are entirely social. All words and speech forms are learnt directly from one individual to another. All languages are sufficiently flexible (expandable) to equip the speakers in communicating with varied things of life.

The speech and the culture often form a unit as opposed to race. A decisive change in the structure and content of a language seems to be impossible without some changes in culture. Both the language and culture influence each other. Since anthropology aims to derive a holistic knowledge about men, language as a cultural phenomenon comes under the purview of anthropologists. They remain concerned with those aspects of language, which come across the ethnological or ethnographic study. Nature and meaning of language is sought in relation to every aspect of culture.

Language can not be observed or studied directly. To describe a language, one has to observe the speech behaviour of the people, who use the particular language. Because, the pattern of speech is related to the pattern of behaviour and communication procedure. Construction as well as utterance of sentences, both is equally important in a language. Each language shows an obvious and clear-cut connection with culture through vocabulary. As a culture increases in complexity, so does the vocabulary associated with a language. Linguistic change is a part and parcel of cultural change. Without language, an individual or a group of individuals finds no way of planning their activities.

Neither any explanation of actions, nor any direction of actions is possible without a language. Therefore, language must be as old as the oldest cultural artifacts. It began more or less at the same time when culture began and has been continued ever since.

Culture Change

Culture is essentially dynamic. No culture, howsoever simple or isolated, can be completely static. It changes with time. The term, culture change, denotes modifications in the elements and pattern of a cultural system. As culture consists of learned pattern of behaviour and belief, changes often take place along the alterations of human need. New ideas, beliefs, values give rise to a new set of attitude and behaviour.

The dynamic nature of culture can be shown from any sphere of human life. The simplest example is our grandparents whose speech, dress and manners are somewhat different from us. Old books, newspapers and photographs also yield similar evidences. The difference is much conspicuous as time passes on. Change may also be noticed in the variation of an accepted pattern of design or in a new method of preparing food. The trend of fashion, style of art, even custom change with time.

From the writings of the earlier social theorists like Durkheim and Radcliffe-Brown it has been known that human society is based upon a organic solidarity so that the customs, institutions and other social activities remain integrated to maintain the social order. Social and cultural systems cannot maintain equilibrium when they face any force externally or internally. This is the root cause of culture change.

Social change is generally differentiated from the culture change because by the former term we mean the changes, which occur in the structure of social relationship. Therefore, such changes are only evident in the context of social role and relations or in the relationship between group and institutions. Usually social and cultural change remain closely related. Sometimes cultural change precedes the social change or precipitate social change. However, the records of prehistory and history show that patterns of culture in human societies have been advanced through constant change although the rate and type of change are not same in all societies. We can cite examples of two dramatic changes—the Neolithic revolution of prehistoric days and the Industrial revolution of recent centuries. Both the changes were sudden and drastic involving the alterations on the subsistence level through techno- economic thrust.

Any change in social-cultural system is generated with the variation of ideas, values and beliefs held by its members. All individuals are the potential sources of new ideas and each of them has the capacity to conceive the new ideas. Therefore, a new idea or behaviour may originate within the society or it may be borrowed from outside,or may be imposed on. Whether the change is internally produced or externally induced or deliberately planned, it involves social and cultural elements but the processes are more or less similar.

Factors of Culture Change

Anthropologists favour basically a Marxist or Neo-Marxist interpretation on the factors influencing the culture change. For instance L.A. White in 1949 declared that 'the amount of energy harnessed per capita per year' is responsible for culture change. J. Steward in 1955 identified 'adaptation of a culture to its environment' as a primary factor. Other scholars identified the importance of 'religious ideology' (Max Weber, 1922), 'cultural theme' (Opler, 1945), 'cultural focus' (Herskovits, 1955), etc. Geertz (1957) had pointed out inherent incongruities and tensions in social-cultural system as the main factor of change. G.P.Murdock in 1949 stated that "social organization is a semi-independent system comparable in many respects to language and similarly characterized by an internal dynamics of its own". Other theories came from Hallowell (1955) and Wallace (1961) who advocated the psychological aspect in explaining culture change.

Evolution is a change whatever slow it is; any evolutionary change occurring within a society may be a factor of critical importance. Murdock (1949) had used 'evolution' to designate the 'processes of orderly adaptive change'. His explanation was, when the hunting and food gathering people having simple culture had started domestication, their food supply became certain and regular. This new technology increased the population growth and so posed critical problems. The need for division and specialization of labour, social control and management of surplus food led to adaptive changes in cultural pattern. Similarly, according to him, if for any reason, a society shifts from a matrilocal residence to patrilocal residence, adaptive changes necessarily develop to embrace the total range of culture. However, two major types of evolutionary changes have been distinguished:

(i) General or macro-evolutionary changes. These changes take place in culture influencing the totality of culture. They appear as a natural process.

(ii) Specific or micro-evolutionary changes. These changes take place in particular socio-cultural system. The sources of change are varied, both external and internal.

Source of Culture Change

(a) *Change of Environment*

Since culture is an adaptive process, it always adjusts with a new and changing environment. The situations like major geographic or climatic alteration in an area, unlimited population growth within a territory, sudden change in subsistence technique, etc., may have an important role in culture change. For example, an environment may undergo great climatic undulations or a people may abandon their old habitat (or may even be driven out) for some reason or other. Now, if the new home experiences drastic change of environment or if the people have to change their old method of subsistence, the existing culture will be altered profoundly, irrespective of the fact that their new situation is inferior, superior or worse. The transition from Palaeolithic to Mesolithic in Northern Europe was thus associated with great climatic change. The change of environment at the end of last glacial period also allowed forests to grow in the place of tundra, steppe and grassland. The vegetation and fauna changed simultaneously.

Increased population in a given area similarly leads to culture change. To ensure the food-supply for an influx of population, as we know, the prehistoric men could not depend merely on hunting and gathering technology. They started domestication of animals and plants, which ultimately culminated in Neolithic revolution. The situation brought another kind of culture change; a partly urbanized civilization developed.

(b) *Dicsovery and Invention*

These refer to the innovative processes that may originate from inside or outside of a society. Discoveries and inventions do not compulsorily beget changes. But, if the people of a society accept a discovery or invention, and use it in a regular way, the culture change is inevitable. Significant inventions and scientific discoveries of modern times have been listed as steam locomotion, X-ray, radio, television, telephone, computer, bactericidal medicines and so on. The new objects, new technical processes and new knowledge are the important sources of change for all societies and cultures. There are some innovations, which are unconscious, but others are intentional. The innovations of prehistoric times were mostly unconscious. For instance, fire was invented in a casual way through playful activities of primitive men. Friction between two stones produced fire all on a sudden. In the same way, very little consciousness played in the making of the first stone tool or in finding out the least way of cooking.

Though all human individuals are the potential sources of new ideas, innovations do not take place frequently. Moreover, there is a question of acceptance. All innovation or discovery do not always satisfy a social need. For this reason, differential rate of innovation is observed between two societies. Social factors also regulate the course of innovation.

(c) *Diffusion*

Diffusion is the spread of culture from one individual to another and from one society to another. Cultural elements are diffused or borrowed by this process. Any sort of contact between two societies with different cultural pattern imposes change on both societies. Such a process of culture modification includes new items, new technical devices, new ideas, new belief etc.

The term acculturation is sometimes used in the place of diffusion to mean the significant changes that take place in a group by the act of borrowing. But actually these two terms connote different meanings.

The term 'diffusion' points out the cultural traits or complexes that diffuse from one culture to another. Conversely, acculturation is the result of culture contact where the total way of life changes due to the influence of alien culture. Further, acculturation takes place in one direction only. It may or may not lead to assimilation. Redfield conceived acculturation as "the modification of indigenous life under the influences from the white man's world". Linton, Herskovits, Hallowell and Beats kept important contribution in relation to acculturation. There are also two other concepts, *transculturation* and *contra-acculturation*. When mutual exchange of culture traits and complexes occurs between two societies, instead of one way borrowing, it is called transculturation. Contra-acculturation involves the idea of a dominant and a recessive culture. The dominant culture breaks down at first and thereafter develops a reaction to recover the loss of its individuality. Another term, *enculturation* was coined by Herskovits to denote the process by which a man adapts himself to his culture and learns to fulfil some essential function related to his status and role. The process starts from the first learning day of the life and continues until death.

Contact between cultures may take place in various ways, such as,

(*i*) Contact Through Trade and Commerce

Trade and commerce facilitate contacts between different peoples and places. In the city of Kolkata, people from different states come for trading. They form groups and settle at different corners of the city. Due to close contact, cultural traits often percolate from one group to the other.

(*ii*) Contact Through Industry

Large industries bring various cultural groups together. For example, the steel industry at Tatanagar has provided the livelihood of many Behari, Marwari and tribal people (Ho, Santal, etc.). Therefore, the people with diverse cultures have got the chance to come in close proximity. In tea gardens, plain-land tribals are working side by side with hill-area tribals. It is very difficult for them to maintain their cultural distinctiveness.

(*iii*) Contact Through Preaching of Religion

The Christian missionaries often go to different areas for spreading their religion. As a result, contact takes place between the peoples having different religious faiths. Indigenous tribal cultures are often influenced.

(*iv*) Contact Through Conquest

Sometimes contact is established after conquest. For instance, British conquered India and spread their own culture. Not only that, at different periods of Indian history, the Vedic Aryans, the Sakas, the Huns, the Pathans and the Moghuls came from outside. They settled in India and spread their cultures.

(*v*) Contact Through Education

Different educational institutions help in establishment of contact between people of diverse cultural groups. Such type of contacts is mainly found among the advanced people. At present, social workers make contact with backward people for welfare purposes.

Models to study the Culture Change

The study of culture change essentially requires a comparative framework so that the sequences of change, whether it is in individual behaviour or in social structure or in stages of cultural development, can be compared minutely in order to reach to an accurate and systematic statement. Comparison is possible in two ways:

(a) Through macroscopic model

It studies the change involving a long time span like the analysis of cultural evolution in human societies, from lower Palaeolithic upto the twentieth century. L.A. White (1949) and Childe (1951) utilized this model in their study of cultural evolution. Both of them dealt with the trends and stages in reference to the culture of mankind as a whole. R. Redfield (1941) also followed this model in order to study the social and cultural change. He compared four communities in Yucatan—a tribal village, a peasant village, a town and a city. Using the variables like degree of isolation and degree of homogeneity in culture patterns, he was able to show the change that occurred over a relatively long time.

(b) Through microscopic model

It studies the changes, which take place within short periods of time like the dissolution of joint families in recent days. Such studies are less extensive in scope. Murdock (1949) and Leach (1954) followed this model to analyze the change in social organization. Murdock proposed that a change in social system began with the modification of the rule of residence, which leads to other changes in kinship structure and terminology. Leach's model of study is specially known as an 'oscillation model' because he had shown a regular oscillation between two polar types of political value systems in Kachin society. Spindler and Goldsmidt (1952) studied the changes in individual and group behaviours by comparative analysis. Finally a 'genetic model' was proposed by Rommey (1957) where he emphasized on linguistic, archaeological and ethnographic information controlling the geographical and historical factors. This particular model extracted the idea of Sapir (1916) and Eggan (1954) who held the view that genetically related tribes are determined for their sharing of common traits in physical type and for using of related languages.

A Concept on Cultural lag

'Cultural lag' means maladjustment in culture. The concept was first introduced by W.F.Ogburn in 1922 through his book 'Social Change'. Since then it has been placed in the terminology of social sciences. The idea behind this proposition was, the various parts of modern culture do not change at the same rate; some parts change more rapidly than others. As there is a correlation and interdependence of parts, a rapid change in one part of culture requires immediate readjustments in other parts of the same culture as for instance, industry and education are correlated. So a change in industry needs an immediate adjustment in the educational sector, through certain other change. Otherwise cultural lag will appear.

The extent of lag varies according to the nature of cultural material and may exist for a considerable number of years. Usually a sudden change in one part of culture is evident due to some discovery, invention etc. The maladjustment is the resultant of strain as the balance of the ordered structure breaks down for the time being. A cultural lag may affect an individual or a group. Anthropologists often face the problems of cultural lag while imparting development programme in the communities.

FIELD WORK

As we have discussed the nature of anthropology (vide. Introduction), it is now very clear that anthropology should be called a science. Every science stands on a scientific method of investigation and research by which it looks at the world. Anthropology also possesses its own scientific method to look at the world of man. Social-cultural anthropology is purely a social science. Like physical science it does not require a close-door laboratory, it accumulates the data from open field.

As science establishes the relation between cause and effect, it involves that stages like experiment, observation and inference. But in many cases experimentation is neither possible nor desirable. For instance, we can take the subject astronomy—one of the most accurate sciences, which deals with the celestial bodies like stars, planets and comets. We are not able to experiment with the planets, so in this case experimentation has to be replaced by some kind of observations. One should observe the occurrence of the phenomenon and their repetition with verification. Hence, a discipline, which develops either through experimentation or observation, can be called as science; experiment or keen observation is the basic prerequisite of a science. Social-cultural Anthropology is a non-experimental observational science. The physical and biological scientists essentially need a laboratory room for conducting their experiments, but to the anthropologists laboratory is the field. They carry out observations through fieldwork. It is a unique method of unusual kind.

Development of The Concept of Field Work

At present, although anthropology has been divided into several sub-disciplines where considerable use of laboratory is found, but the discipline itself emerged from the field. The development of scholarship came after the Europeans who had been travelling to distant corners of the globe for about four hundred years, since the 'Age of Columbus'. The explorers in course of their voyage visited different peoples and diverse customs. Soon, the missionaries, traders and government officials followed them and wrote lot of reports on what they had seen.The facts were invaluable to make the other Europeans aware about the variegated human life on earth. Many European thinkers became interested about the non-European queer cultures and gradually 'study of man' was initiated basing on the accounts of travellers, missionaries and government personnel.

The anthropologists of nineteenth century were totally involved in exploring the variety of human cultures but they were quite apart from the rigorous life of the actual field. Sitting on a comfortable armchair, they simply looked into the accounts served by the other people. Nevertheless, certain theories on human behaviour were built at that phase.

The value of the fieldwork was realized at the beginning of twentieth century when the outlook of the anthropologists changed. It was understood that an anthropologist should face the situation by himself in order to get accurate and relevant data. So many anthropologists of this time made themselves engaged with the groups of aborigines. These anthropologists started to live among the alien people for a long period leaving his own family, friends and the whole of the cultural setting. This practice helped them to gather sum-total knowledge about a foreign culture. E. B. Tylor was the first scholar who emphasized the need of direct data-collection in anthropology, but F. Boas was the pioneer to begin with this practice.

In the book 'History of Anthropology' (1934), A. C. Haddon divided anthropologists into two distinct groups:

(a) The workers in the field who collect the facts and (b) the arm-chair workers who 'welcome them into coherent hypotheses'. Haddon himself successfully combined the two. Nineteenth century anthropologists were not the data-gatherers. The earliest attempt of professional data gathering, as mentioned earlier, was made in America by Franz Boas. He conducted Jessup North Pacific Expedition in 1897. The second attempt of fieldwork was done in England under joint leadership of Haddon, Rivers and Seligman during the period from 1898 to 1899. It is known as Cambridge Expedition to Torres Straits. In these expeditions, the research teams were comprised of anthropologists who went to a distant place for contacting the natives. Each member of the team was supposed to concentrate on one or more aspects of the native life viz. Material culture, property, kinship, totem, taboo, etc. Reports were written separately on each aspect of the tribe as if no connection existed among them. These field works essentially had some drawbacks. Since the visits were very brief, the investigators got little scope to be acquainted with the natives. Further, the language was a bar to them. Interpreters could not make out the meaning as elaborately as the investigators wanted. On the whole, the short-term visit, compartmentalized view, the lack of acquaintance as well as the difference of language hindered the anthropologists in getting a clear picture in totality.

The most outstanding fieldwork tradition in anthropology was developed by B.Malinowski. He believed that the various aspects in the life of a people were interrelated, so he never studied a society in piece-meal manner. Malinowski also stressed on fieldwork as primary way of anthropological data gathering. In the book 'The Agronauts of the Western Pacific' (1922) he wrote about the necessity for inclusion of a statement of field procedure. He wanted to include the same in a report where the results of the same fieldwork were furnished. According to Malinowski, a cultural anthropologist must "possess real scientific aims and know the values and criteria of modern ethnography. ... .He has to apply a number of special methods of collecting, manipulating and fixing his evidence". During the fieldwork, Malinowski at first began to collect data living with his local friends, the white people. But soon he realized that there must be some flaws in his technique of collecting data. So immediately he had cut off the company of white people to set up a camp in a native village. Living in a village, he became able to observe the habits, attitudes, customs, behaviours etc., of the native people, which w«s never possible to achieve through the questions or from documents. Thereafter, Malinowski established participation as an important technique of fieldwork.

Next to Malinowski, we can put the name of A.R.Radcliffe-Brown who did an extensive fieldwork in Andaman Islands. Radcliffe-Brown was the first pupil of W.H.R.Rivers in social anthropology. In 1906, Rivers had sent him among the Andaman Islanders for reconstructing the cultural history of those people. Instead of conjectural reconstruction, he studied the myths, ceremonies, festivals and customs of the people in depth and analyzed those from structural-functional point of view. In fact, the proper anthropological fieldwork was exhibited in succession by the efforts of Boas, Haddon, Malinowski and Radcliffe-Brown during the first quarter of the twentieth century. They often undertook painstaking work to learn as well as to record the social-cultural life of the natives.

In the context of field work H.J.Herskovits (1942) has advised, "see as much as you can, participate whenever you are permitted to do so, and compound your experiences by discussing them formally with natives as widely as you are able". It is the rule that an anthropologist has to qualify himself according to the situation and scope. We can illustrate this point against the background of the field-report of E.E.Evans-Pritchard—what he experienced with the Nuer and the Azande of East Africa. The Nuers treated him as an equal but the Azandes, as superior. Nuer people forced him to be a member of their community, whereas the Azandes compelled him to live outside of the community.

Evans-Pritchard carried out his fieldwork in Sudan and Kenya from 1926 to 1938. He delineated the procedure of anthropological research into three stages—the fieldwork stage, the stage of comparison, and the stage of inference. The third stage was the inference stage where the general tendencies and functional relationships common to all human societies are supposed to be discovered. Evans-Pritchard did not share Radcliffe-Brown's theoretical concept and he was in opposition to comparative method in anthropology. But he put the whole emphasis on the fieldwork.

The early fieldworkers tried to understand how all the parts of a society fit together to make a working whole. They were different from armchair anthropologists namely Maine, Fustel de Coulanges and others not only in ideology but in the tedious process of collecting the raw data. Fieldworking anthropologists were in the habit of filling their notebooks with details of what they saw and heard, and those unprecedented ethnographic activities resulted into interesting monographs. As a matter of fact, a cultural anthropologist has to live and work in two worlds. Field is his laboratory room where he is supposed to enter for different observations. The conventional field of the socio-cultural anthropology is outside the realm of main stream life. Because the aboriginals live in far off places, particularly hills and forests under miserable conditions. Anthropologists often face various difficulties like physical danger, disease, lack of transport, food, etc., but they tackled all these untoward situations with patience and perseverance. In field, the data is recorded by observation and interview. It is very difficult for the anthropologists to shift their focus from the small, simple non-literate society to complex modern world. All anthropologists, irrespective of their connection with University, Government, Industry, Hospital or any other arena, follow the general method and approach of fieldwork.

Since social anthropology is an empirical discipline, it languishes for the absence of a deep respect on facts and for a loose attention to their observation and description. A self-indulgent attitude may produce a disastrous effect. In fact, intensive fieldwork is a very difficult undertaking; it demands physical, mental and moral preparation of the field-worker. Because different kinds of restrictions, in relation to food, dress, behaviour, movement, etc., are usually imposed on a fieldworker as essential preconditions of fieldwork.

Importance of Fieldwork

It was not until the second World War that anthropologists reached to a maturity which brought a general recognition for tremendous importance of the empirical knowledge about the complexities of human relations distinguishing one group from another. The anthropologists strived to understand the basic and elementary forms of interpersonal and inter-group relations of peoples as well as their institutions, grouping and associations, no matter whether they are rudimentary or not. It was realized that the sensitive finer points of the life need to be revealed for resolving any sort of conflicts—psychological, social or political. The procedure of anthropologists is still experimental and variable but it has now been extended further beyond the stage of casual observation among the isolated individuals.

Classical anthropological works have been undertaken in small, isolated primitive communities. The reason behind the selection of a small population is that, such a closed society offers an in-depth, micro-level study at the cost of very little expense. As the members of the group live within a small area, all of them remain exposed to the same natural forces and enjoy similar life chances. Their way of life is, therefore, relatively stable in comparison to complex societies. An anthropologist can easily qualify himself to a small setting by knowing each and every person. He can be totally involved into the particular social system.

Staying a prolonged period in a small community, an anthropologist gets the total picture of a society as well as its culture. Since he does not possess an instrument to measure the social roles, positions and inter-personal relations, his data are basically qualitative in nature. A few of the anthropologists, for certain aspects, sometimes rely on quantification. For example, G.P. Murdock used statistical techniques in analyzing the basic data on kinship. Similarly, A.L.Kroeber, Harold Driver and others applied statistics to survey the cultural elements in California. The sensitive social-cultural anthropologists think that there is no harm in counting heads, items and other countable materials in anthropological research. To them, quantification is necessary and unavoidable according to situations. For example, sample surveys or census survey as essentially based on enumeration, therefore, no way is left here to escape from quantification. However, anthropology, specially the social-cultural anthropology, is strictly a field discipline. No other professional social scientist, inspite of dealing with the same field, does show the patience and perseverance to study a small population extensively.

Fieldwork is the part of training in the subject social-cultural anthropology. Every anthropologist should undergo this training in course of his preliminary study. It enables a student to go beyond the horizons of his own society and perceive an alien culture with subjectivity. Learning about two different societies gives a student a comparative view i.e. he acquires competency to estimate the similarity or dissimilarity between any two societies or cultures. Since many factors in non-literate societies are regarded as more or less constant, trainee anthropologists are free to study a few variables in details with a hope for establishing connections among the various parts of the society. The students of anthropology thus not only gain first-hand knowledge about a culture, they also understand how different parts of a culture remain interwoven within a whole.

Recently the non-literate cultures have started to disappear or change at a very high speed. Anthropologists are aware about the importance of recording the life-ways of numerous tribes as a part of human profile. They also realize that they should not waste literally a minute in writing the general books where each year witnesses the extinction of aboriginal cultures that had not yet been described. This is the professional responsibility of the anthropologists to record each and every type of primitive culture before they lose their originality.

The full-length reports of non-literate cultures,which anthropologists study afresh, get the status of monographs and can be utilized by the students in their training. The already established theories and techniques are usually applied in the training. A trainee anthropologist learns a lot through the field training just like a physician who gains practical experience through his internship. Field training provides the junior anthropologists with an insight for which they proceed to study other peoples' way of life and culture; it is an invaluable experience to them. What an anthropologist tries to understand for himself, he tries to teach that to his successors.

The variety of communication media in the modern world has desegregated the majority of non-literate societies and the technological advancements of the developed people have brought rapid social change in these societies. As the time goes on, the number of the primitive societies are diminishing. Not only that, several new problems are cropping up where contribution of anthropologists seem to be indispensable for bringing out the solutions. Usually a problem in simple society can not be dissociated from other aspects of the society and different kinds of values are associated with every action, so it is not possible for a lay man to treat these problems. Anthropologists are essentially required for achieving this end. The expertise of an anthropologist is derived from his fieldwork; the training as well as professional application rest on fieldwork. To know a people and their social facts, an anthropologist spends his time with them. The value system of a people can not be understood through mere questions without mixing with the people. A primitive society is, therefore, closest to the laboratory conditions for the students of social-cultural anthropology.

Pre Requisites of Field Work

Selection of Venue

Selection of a venue is very important prerequisite of fieldwork. By tradition anthropologists select a homogeneous ethnic group where a specific sense of identity is maintained on the basis of common language and culture. Such a group also reveals a primitive level of subsistence technology and lack highly specialized division of labour. Naturally an anthropologist feels delighted when he finds a group where he alone is able to handle all the people and phenomena in a closed circle. Social anthropology deals with the whole of a social life, not with the isolated problems like sociology. Social-cultural anthropologists, therefore, believe that the understanding of a part of a social life is absolutely impossible without the reference of the total social life.

Self Preparation

Every fieldwork whether it is a training or research involves certain kinds of preparation stage. A.L.Epstein (ed.) in his book "The Craft of Social Anthropology" (1969) has pointed out "preparation for fieldwork has come to be seen as an essential part of the training of students in the subject and fieldwork itself is a unique and necessary experience". It is nothing but an orientation of the investigator before leaving for the field. This preparation is generally of two folds—*Intellectual preparation* and *Material preparation.* The 'intellectual preparation' chiefly means the library work. Since an investigator plans to study a tribal group, he needs a prior introduction about the group. Various literatures are available in the library from which one can build his preliminary ideas on the particular group. The investigator also needs the information about the area to which the particular group is located. Usually such groups are cut-off from the main land and situated in remote corners, within the forest or hilly tract. Naturally the investigator should have grown a first-hand knowledge about the field-locality before reaching there. Besides, various other categories of literature are available in the library, which relate developmental work, culture change, acculturation etc., as the communities have already been affected by them. The investigator should consult them according to his specific purpose. Thus, the philosophies of change, processes of change, motivations and theoretical models acquaint the investigator with the implications of change on different aspects of life. He should also know the various field methods and techniques as used by the anthropologists in their fieldwork. Once 'Notes and Queries on Anthropology'was the main guide for the students in collecting data. Now there are a good number of books on anthropological methods. However, an intellectual preparation provides the glimpse on the people, area and the techniques of work.

The other kind preparation is the 'material preparation'. For the sake of research or training, an investigator has to stay long time out of the home, in the field. So he needs some essential things to use. These things include the chief requirements of daily life, items that are typically used by the investigator and the materials essential for the survey. Therefore, the bag of the investigator becomes filled up with a variety of items like bedding, dry food, medicines, water-bottle, torch, paper, note-book, pen and pencil, camera, tape-recorder, etc. At this time the investigator also needs to select and reserve his place of living (field camp). A prior information about the area enables an investigator to be fully equipped while he is in the field. Otherwise, he remains overburdened with his own problems and can not concentrate properly on the work. It is a foolish idea to seek comfort in the field, but carrying necessary articles often saves the investigator from embarrassment.

Along with material preparation an investigator has to make himself 'mentally prepared' to face the odds i.e., what he has not seen in his own society. He is also supposed to maintain a double standard of norm, value and ethics to win the field. In field, he should forget his own social status, values and sentiments. Rather he has to attune himself with the manner and etiquette of the particular field as a successful fieldworker. He ought to be satisfied with the available accommodation. Local people may come forward to provide a space for living when they cordially accept the newcomer. This happens only where the investigator can satisfy them with an amiable behaviour. Ideally a field camp need to be set up in the midst of the community so that one gets more and more opportunity to be acquainted with the local people.

Establishment of Rapport

Objective of the fieldwork is to get a reliable and verifiable information. An effective and fruitful fieldwork depends on the genuine acceptance of the investigator by the village community. Until and unless the investigator is accepted by the community, no real information is obtained. Misleading answers can not form the body of data. Therefore, an investigator should devote a reasonable time for creating rapport in order to win the confidence of the people. Cooperation can be sought only after the establishment of friendship. At the first step, an investigator is supposed to find out some key persons or contact persons in the community where he intends to work. At this stage, he does not wish to know each and every person of the community, rather picks up some influential persons who possess decision making capacity or hold leadership qualities. Because such persons can help the investigator to proceed smoothly into the field. These persons are generally regarded in the society. So when they introduce the investigator with the members of the community, the commoners are bound to cooperate with the stranger. In the second step, the investigator has to interact with the people very freely and frankly. He is required to explain the purpose of his visit. At the same time, he must assure the people against any sort of harm caused by him.

An investigator should start with friendly greetings as per the respective cultural pattern. This is the key to his entry-point. Use of pen and pencil is not recommended at the initial phase. Because a noting may make the people suspicious which hamper in building of an easy relationship. It should be remembered that the human mind is highly sensitive. An anthropologist should win the people with love and sympathy, instead of wit and sophistication.

The investigator should also make himself prepared to face any sort of untoward incident. He is supposed to tackle the situations with intelligence and common sense. It is always better to bear an identity card. Sometimes the people may get ferocious as they think the investigator an agent of government who has come either to exploit them or to impose an exorbitant tax upon them. The investigator can not apprehend the type of wild conjecture that a community would make about him. While working in a minority religious group, it is not at all unlikely that the group may associate the investigator with some anti-religious organization. But it is interesting to note that the success is often built upon these initial troubles. Once rapport is established, the field is gained forever.

An anthropologist should not deviate from his code of ethics. He must not betray the simple people. Therefore, if any secret of the society comes out which the people do not want to disclose to others, an anthropologist must restrain himself from announcing that publicly. Sometimes an investigator suffers from culture shock where he faces entirely opposite cultural values, norms and practices. From the very beginning he should be mentally prepared for bearing such a shock. Otherwise he will be biased and these biases will resist him in making rapport as well as proper interpretation of facts.

Rapport is the most difficult but the most interesting stage in fieldwork. A good rapport brings the investigator so close to the people that they provide him not only with all required information, they try to protect him in all ways. With time they really forget that the person is an outsider, rather take him as their own kin or a wellwisher. Naturally, the people further do not feel shy in allowing the investigator to watch all of their socio-cultural activities. At this time the investigator must be very cautious about himself. He may participate in eating, dancing, drinking or going together. But he should not violate any rule or norm. A slightest vulgarity may spoil his whole fieldwork. He should also lead a well-regulated life in the field camp. He may appear to be a laughing stock at the first sight, but gradually his generosity, keenness and sincerity reserves a place for him in the alien society. One may keep some petty things like cigarettes, betel-leaf, toffees etc., for distributing in the field. But that should be distributed only in close situations. It is a common experience of the anthropologists that they step down in a .field like a stranger and face abuses as well as suspect of the local people, but come back as a king causing heart-break to many of the villagers. Tears rolls down from the eyes of the people as they think themselves departed from their closed kin. This can be taken as an index of a good fieldworker.

Use of Interpreter

An investigator has to employ an interpreter where he is not familiar with the language of the people. The interpreter should necessarily be a member of the community who knows the local language as well as the language of the investigator. Such an interpreter also helps in creation of rapport. It should be borne in mind that a fieldwork yields the best result if it is conducted through native language and if the conversation can be made direct between the investigator and the informant. Because, the interpreter can not always explain the things as clearly as the investigator wants. Although his presence seems to be very useful but gaps are often created in the meanings while conveyed through the interpreter. For the lack of proper understanding, the analysis of the investigator may turn imperfect.

Selection of Informant

The other important aspect of the fieldwork is the selection of good informants. The selection must be made very carefully, as everybody in the field can not be good for information. An informant should be intelligent as well as prompt in answering questions. He should also be experienced in the topic on which the investigator wants information. Usually an investigator wants information on different topics. Data on each topic must be derived from such a person who possesses a direct knowledge on that topic. For instance, information regarding the agricultural work must be taken from a person who practises agriculture every year. Data on marriage should be collected from a person who has been recently married. Similar rule is to be followed for collecting information on death customs. The persons who know least about a topic can not satisfy the investigator. At the same time their information remain fabricated with imagination. So this type of information do not have any face value; the wrong is understood when they are cross-checked with the same information provided by the others. For this reason, a young man of the community who takes minimum interest in religion can not be able to supply reliable data on religious practices. It is better to ask a religious specialist of the community or some other persons closely associated with such activities. Similarly it is wise to collect data on food and childbirth from the women; they are able to submit better detail in comparison to man.

Choosing of right persons for right purpose is the special aptitude of anthropologists. They know whom to choose and why to choose. Some people are found in all societies who are well informed but so biased that they show a tendency to suppress or distort the facts. Again, other people are there who can not narrate a thing in a systematic manner and so big gulfs are often found in their expression. These people do not have the patience to listen the investigator about what he exactly wants; they generally go on talking in their own way. Their answers are, therefore, inappropriate and most of the time moves along the periphery of the topic. The third category of people continuously suffers from distrust. Naturally they speak superficially under some sort of psychological pressure. As a consequence the investigator gets irrational and misleading answers.

It is the sole responsibility of the investigator not only to select the appropriate persons, but also to train them as per his need so that they can answer systematically and precisely to all queries. Again, it is necessary for an investigator to prepare a list of four to five reliable informants for different topics where each group should have a chief informant. Generally the main information is collected from the chief informant of each group. Different topics are covered in this way. Later, data on each topic is verified by the panel informants of the respective groups. Corrections are made, if they seem to be necessary.

Data Collection

Data are collected at the last phase of fieldwork. They are largely based on our sense-observations. The word 'observation', in the context of data collection, includes all forms of sense perceptions used in recording responses as they impinge upon our senses. But response is not a datum. "A datum is what is observed, is manifest or phenotypical"* Anthropologists look for the facts by observing the people carefully; they try to understand the themes of every behaviour. Being an observer, one

* Galtung, Johan: Theory and Methods of Social Research, Allen and Unwin, London, 1967, p. 27.

should keep himself aloof from the group-struggle of the community. He is supposed to be an objective observer free from all bias and inhibitions. But there are certain limitations for him. No confidential information can be gathered in this way. Moreover, there are certain phenomena in the society for which the time and places are not fixed. Obviously, the observer faces a great trouble in identifying them. Again, a few phenomena may co-occur at a time in different places. The observer can not present himself simultaneously in both the places. Above all, societies may have some special activities, which are not open for general observation; even all members of the community are not permitted to watch them. To overcome these limitations, anthropologists are compelled to take the help of other method. The method of'interview' has proved its efficacy in eliminating the limitations of 'observation'.

At present, most of the anthropologists supplement their 'observation' by the methods of'interview'. The collected data is purely qualitative in nature. Whether it is a training or research, an anthropologist begins with a broad area of interest and gradually plunges into the field. His aim is to find out the principles, how and in what ways the various aspects in the life of a people are operated. He also wishes to know how these different aspects remain interrelated. In course of study he eliminates as many variables as possible. Among all social sciences, anthropological fieldwork is a unique enterprise. No other empirical study concerned with society offers such a scope of in-depth observation and interview. The collected data in anthropological fieldwork is completely free from statistical error. Since anthropology is a science of sensitive mind, the fieldwork for data gathering initiates a psychological stimulation among the people who expect material help and solution of heir problems. The investigators should make up their mind in such a way that they must not tell a e or give talse hopes to the natives in order to achieve the target. But in reality, some of the workers, especially many foreigners go to the field in a hurry and want to squeeze the maximum information within a short period by distributing money, gifts and other attractive items. Some other workers go in disguise of government officials and create a grave atmosphere in order to get the data very easily. All these practices go against the professional ethics of fieldworkers. An ideal fieldworker though collects data for his own benefit but also tries to do some good for the respective people. In doing so a field is never spoiled after his work, rather when other workers penetrate into the same field after a few years, the people come spontaneously to cooperate them. Otherwise a fieldworker himself feels uneasy and does not dare to handle the same field twice.

Methods and Procedure of Fieldwork

The fieldwork is always conducted with a definite objective. It is needed for research or training. Fieldworkers generally gather data either to test a hypothesis or to generate a hypothesis. Hypothesis testing is called 'deductive research' and opposite to that hypothesis formulation is called 'inductive research'.

All modern sciences are rooted in observation. Every scientist considers observation as the most important as well as primary factor. The scientific method of any discipline involves two stages:

(i) Collection of data

(ii) Analysis and interpretation of the data.

The data for social-cultural anthropology are collected in two ways :

(*a*) *Textually* i.e., from the texts, particularly from various records and reports. Record is primarily concerned with a translation that takes place in the very moment, whereas report is written after the event has taken place. Textual records and reports are the documents intended to convey instructions regarding transactions or to aid the memory of the person involved in transaction. They are often collected from the books and manuscripts, letters, newspapers, diaries, public documents etc.

(*b*) *Contextualiy i.e.*, directly from the field through observation. Contextual data can again be of two types—*Primary data* and *Secondary data.* Primary data are those which are collected afresh for the first time and thus happen to be original in character. Eyewitness events are taken into account.

On the other hand, Secondary data are those, which have already been collected by someone else and such data, are generally processed data. Hearsay evidences are also counted here. In social-cultural anthropology, as the field is equal to the laboratory room, bulk of data comes from the field.

Method of social-cultural anthropology is chiefly observation, supported by interview. An anthropologist wishes to study something about which he knows practically nothing. Sometimes he selects such an arena where different misconceptions prevail. His nature of approach is purely exploratory.

Observation

Observation is the oldest method of scientific investigation. The first knowledge of man about the universe bloomed with his power of observation. In fact, observation means the examination of phenomena or a thing without altering it. An observer can not impose anything on the phenomena; he watches the natural movement objectively. Observation may be of two types—intensive observation and extensive observation. Intensive observation denotes an in-depth observation on a small closed community. Collected information is expressed in the form of qualitative description. Extensive observation is just opposite to intensive observation. Here observation is applied on a large field. The quantitative data are usually collected by using schedules and sorted out by applying statistical method. The phenomena are expressed in the form of numerical data. However, the social-cultural anthropologists prefer intensive observation or micro-level study. This type of observation is totally informal and can be achieved by personal shorthand devices. Practice helps to develop the power of observation.

The two wings of *intensive observation* are participant observation and non-participant observation. Anthropologists depend more on 'participant observation'. So this method has been erroneously identified with anthropological approach. For 'participant observation', an observer has to participate actively in the group under his study. He does not necessarily carry out all the activities as followed by the members of that group, but his presence, as an active member is valued in the group. Actual purpose of this participation is generally kept concealed. Contrary to this, the non-participant observation suggests a direct observation without any involvement. It is simply the watching and noting of the phenomena as they occur in nature with reference to their cause and effect or mutual relationship. It is possible to study the village plan, habitation pattern, flora, fauna, water sources etc., by this method.

Human behaviour is neither stable nor tangible. Pattern of behaviour or interactions between individuals change in every moment. Each individual possesses his own individuality and there are also the factors like age, sex, social position and so on, which influence on the behaviour of a person. This variability of human behaviour poses a great problem to an observer. Further, as the object of observation and the observer belong to the same category, both remain sensitive to each other and liable to influence one another. Therefore, a student or a researcher of social-cultural anthropology often faces a changing situation. What one observes now, may not be the same when he is away from the scene. His very presence may also bring substantial difference in the observed behaviour.

It is, therefore, clear that the personal bias of the observer and different types of other personal experience may also affect the events and processes that are sought to be observed. Moreover, there is an obvious impossibility of studying everything by participant observation. A fieldworker can never participate fully in the life of an unknown community for some practical reasons. Another fundamental difficulty with 'participant observation' is that it lacks standardization. What field-worker experiences in a field can not be absolutely identical for a second investigator. Two persons may not record the same facts from the same field. The irony is that, it is almost impossible for an outsider to be a genuine 'participant observer' in all ways. Say for example, an anthropologist never qualifies himself as a criminal while studying a gang of criminals. Thus, a thorough and careful observation by participation in all social situations is virtually an impossible task. This drawback has to be supplemented by 'interview'.

Regarding the utility of the method 'observation', it can be said that it eliminates the fallacies which enter in the question-answer approach—the 'interview'. It has no room for suspicion and speculation. The phenomena are best known as they are studied directly in real situations. It provides a fieldworker with perfect information. The behaviours under hypothetical conditions are contrasted from the actual behaviours in genuine life situations. 'Observation' offers no scope of errors as arising from misunderstanding and misinterpretation of words or from the bias verbal report submitted by a person. A 'Participant observer' generally acquires more popularity in a community than a 'non-participant observer'. Because he gets an ample scope of mixing with the people and does not stay at the periphery like a non-participant observer. Different sorts of appreciation inspire and encourage the natives to explain the significance of various social customs before an outsider.

'Participant observation' may be of two types according to the nature of participation. The investigator plays the role of either 'participant observer' or 'observer participant'. When the observer is external of a group to start with and afterwards involves himself in the group, we call him as 'participant observer'. But sometimes the observer is internal to the group and practises introspection to study the social reality i.e., he observes the other members of his own group. This type of observer is referred as 'observer-participant'. In social-cultural anthropology all fieldworkers including the researchers and applied anthropologists belong to either of these two categories.

Interview

An 'interview' may be defined as face-to-face meeting of persons on some particular points. The purpose can vary widely to include, as for example, a meeting to arrange for a course of action, to collect information or to select persons etc. Normally it involves putting of relevant questions to the interviewee and recording his response. Set questions are asked and answers are recorded in a standardized form. An interview, without any purpose, is not an interview at all but an idle conversation. In social-cultural anthropology, an investigator is mostly a stranger in the native situation, so he has to ask various things which he does not know. However, the person starts with a working hypothesis. It is the preliminary plan of work based mainly on textual information. It has to be tested in the field and modifications may be made as per the objective of fieldwork. In the words of George A. Lundberg "A hypothesis is a tentative generalization, the validity of which remains to be tested. In its most elementary stage the hypothesis may be a hunch, guess, imaginative data which becomes the basis for action or investigation". An anthropologist generally conducts a structured interview and uses a schedule. He never uses a closed questionnaire as like the sociologists.

Difference between Schedules and Questionnaire

The schedule is the form containing questions or blank tables, which are to be filled by the investigator after getting information from the informants. A questionnaire, on the other hand, is a form prepared and distributed by the investigator to secure responses to certain questions; the respondents themselves fill the form. The schedule contains direct questions as well as questions in tabular form, whereas the questionnaire is made up of only questions; it is totally devoid of any table. Purpose of questionnaire is to collect information from the respondents who are scattered over a vast area. It provides reliable and dependable data. Questions are mainly of two types—*open-end questions* and *closed-end questions,* although there are other types of questions like, pictorial questions, dichotomous questions, leading questions, ambiguous questions, ranking items of questions, etc. The open-end questions are highly valued in anthropological investigations; the sociological investigation has a preference on closed-end questions.

Apparently, the schedule and the questionnaire methods are similar. But there are some subtle differences. The schedule is the direct method of data collection. In questionnaire, the data are collected indirectly through circumstances. In schedule, the questions are set in a very short form but questionnaire often shows lengthy questions to bring detailed information. The information through schedule is more reliable than the questionnaire method. The schedule method can cover only a limited area, whereas the questionnaire method covers a wide area. Moreover, in the questionnaire method, the questions are formulated on the basis of the cultural and educational background of the respondents. In a word, it is respondent-oriented. But the schedule method is just opposite to this. Here an investigator selects his informants in a random way and the informants are given some sort of freedom in answering the questions.

Rules of Interview

It has already been known that the success of interview depends on the selection of suitable informants. An informant does not suit for all kinds of information. The age and sex of the informants are also taken into account. For instance, in the questions regarding the history of village settlement, youngs would never be the good informers; elderly persons of the village can tell a lot about this topic. But, the young informants are more reliable than the old persons for providing information on play activities. Therefore, it is a question of decision that whom to be selected to get information with greatest reliability. The informants should be selected without trial and error.

No hard and fast rule can be laid down regarding the procedure of interview. Several factors may affect an interview. The conditions for successful operation are:

(i) The place of interview should be free from any disturbance and distraction. All places are not favourable for putting all types of questions.

(ii) Privacy should be maintained if the informant desires for it. Because most of the informants, howsoever simple and cooperative may be, do not want to disclose the personal matters of their life in presence of others. So all places are not congenial for interview.

(iii) Questions should be short, interesting and clear to hear. Answers should not be limited to 'yes' or 'no'; the informant should speak out of his own mind.

(iv) There should not be any language barrier for the informant. The interview must be conducted in the language of the informant and the questions must be framed with simple words so that the informant may follow them easily.

(v) Bullying questions (like those of policeman) or tricky questions (like those of lawyers) should be avoided. Instead, the questions should be asked in a sweet and reassuring tone. The investigator should be frank and straight, rather than cunning.

(vi) In every step, the confidence of the informant needs to be generated. The investigator must show his interest in everything that an interviewee whishes to tell him. But, on the whole investigator is expected to have the ability to control the course of interview in accordance with the objective of the study.

(vii) The investigator should have the expertise to note the reactions of the informant that is expressed through his talk, gestures and facial expressions.

(viii) The investigator is supposed to talk and put questions in a friendly way. He should not argue with the informant, nor supposed to show any kind of authority over him. Any sort of unwanted advice or moral instruction should also be avoided on the part of investigator.

(ix) Informant's personality is to be respected and his viewpoint must be appreciated. Extracting information against his will is a malpractice.

(x) The investigator should cultivate the qualities like honesty, politeness, sensitivity and good sense of humour. These will help to remove the stumbling blocks from one's path.

(xi) The investigator must not be impatient or frustrated. He should keep in mind that his valuable time and efforts are the keys to success.

(xii) The time and place of the interview must be pre-planned. It is always better to make a prior appointment with the informant. Because arrival of investigator during the time of important domestic work or agricultural work may not at all please the informant. He may feel annoyed or psychologically disturbed. So his answers may go far from the reality.

(xiii) The convenience of informants must not be forgotten and the investigator should be very punctual.

Success of an interview depends on the personality and skill of the interviewer. No amount of preliminary preparation can be the substitute of tact and spontaneous insight in a concrete situation. In every step, the investigator has to judge the informants, whether they get the actual meaning of his questions or not. He himself should also be careful in interpreting the answers. Extraction of proper meaning is very important in both ways. Sometimes a situation or the nature of an information debars the interviewer in using the formal schedule. Writing down of the answers may be misinterpreted by the simple-minded native people. Therefore, memorizing of both questions and answers yields better results. But absolute reliance on memory may have the possibility to gather inaccurate facts.

The types of interview generally vary as structured or unstructured, individual interview or group interview etc. The structured interviews involve the use of pre-determined questions and highly standardized techniques of recording. Standardization ensures that all informants should reply the same questions; the questions bear the same meaning to all informants. The interviewer may repeat the questions if the reply is not to the point. The use of open-ended questions is mostly found in structured interview. This means that the available answers are not limited with short expressions like 'yes' or 'no' as found in closed questionnaire. Such questions also offer a scope of free discussion so that the informants can present their ideas and feelings very clearly. On the other hand, the structured pattern does not allow the informants to go out of the tract, while in discussion. This type of interview has also the names like 'controlled interview', 'guided interview' and 'directed interview'. Contrary to this, the unstructured interview (uncontrolled, unguided or undirected) is the less systematic way of interview. It involves relatively lesser standardization of relevant techniques and operations. It does not follow a system of pre-determined questions; rather it is an informal way of talking where questions devėlop spontaneously with the course of discussion. The informants are urged to describe different incidents of their life; the interviewer picks up the necessary materials from their narration. The interviewer possesses the freedom to waive a question and informants gets the freedom to talk on events that seem to them significant. This type of interview furnishes a free-flowing account on the personal or social contexts of beliefs and feelings. In case of 'individual interview', a single person is called upon for the discussion. Such an interview is generally held for collecting personal as well as confidential information. In 'group interview' two or more individuals are interviewed at a time. Its utility lies in accumulating information on group life. The other advantage is that the procedure is less expensive and less time consuming.

Techniques of Interview

The techniques are usually the parts of a broad method. *'Observation'* and *'Interview'* are two broad methods of data-collection. Each of this method consists of a number of techniques. Techniques of interview can be classified into two distinct types on the basis of nature and scope of investigation —*techniques for documentary interview* and *techniques for special interview.*

Techniques for documentary interview

The techniques, which are devised mainly for the documentation of real facts, are called *'Documentary Interview Techniques'*. Four different techniques have been distinguished here, namely, Survey with Census; Narration or Description of facts; Case history recording; and Representation through genealogical table. Each of these techniques follows a special procedure typical to its own.

(i) Census survey technique

The fieldwork usually begins with this technique. The community or the area under investigation is surveyed in order to count the people therein. The investigator moves from door to door to get him introduced with the families and during this visit he collects the primary information of the respective family. A schedule is essentially used for this purpose. The technique is very useful in gathering a basic data about a people.

(ii) *Description or Narrative technique*

In this techniquc the informant is required to narrate certain facts on the basis of his experience, no particular case is investigated. As narration differs with individuals, the content and pattern of analysis vary from man to man due to the variation of individual psychology. The social-cultural anthropologists normally avoid this technique but its use becomes inevitable in collecting information on religion. The reason is that the religious attitudes stand on ideas, beliefs and things which occurred in long past; one has to hear them from certain sources where the total matter has been visualized.

(iii) *Case history (Concrete) technique*

This is the technique where a person (informant) is allowed to state his own experience in reference to the recent past. In this type of investigation a large number of concrete cases are collected and analyzed in order to draw a conclusion. The technique has proved its worthiness in collecting data on a particular topic, particularly for the cases like birth, marriage and death. It is the most convenient way of data collection.

(iv) *Genealogical technique*

In this technique, the pedigree of the informant is traced. The genealogical knowledge plays an important role among the non-literate peoples; ancestry can be traced back to several generations with the help of standardized symbols and charts. A large number of collaterals are known by the name. This method helps to gather huge information relating to social structures and other institutions of human group.

The functional value of the kinship terminology has been identified by the anthropologists. Genealogical data are, therefore, extremely valuable in understanding the nature of clan, the regulation of marriage, inheritance pattern of property, succession of cheiftainship etc. The technique was first employed by W.H.R.Rivers in his Torres Straits expedition. He stated, 'the Genealogical method makes it possible to investigate abstract problems on a purely concrete basis. It is even possible by its means to formulate laws regulating the lives of people which they have probably never formulated themselves".*

Techniques for special interview

Documentary interview techniques may not always yield the correct data. It is, therefore, necessary to check the available data for final representation. Some special techniques that are employed for the final verifications of the data are called 'Special Interview Technique'. These techniques have been chalked basically on two principles—*Interview in depth* and *Repeated* (panel) *interview.*

(i) *Interview in Depth*

In a general way, an informant is interviewed only once, although the clarifications may be sought in a second interview. Under the special technique like 'Interview in depth', an unusually long period is consumed for interview. Such interviews (in-depth) may be of two kinds—single interview in depth and multiple interviews in depth.

(a) Single interview in deptn—Here the informant is interviewed for a long time but in a single sitting, the interview gets completed. Such an interview is required for a deeper understanding of opinions and attitudes of the subject. The whole personality of the subject may be disclosed by this technique. Two important types of it are 'Focussed interview' and 'Clinical interview'.

---

* W.H.R.Rivers—A Genealogical method of collecting social and vital statistics, J.R.A.I., Vol.30,1900.

Focussed interview : The interview, which focuses on a typical incident, is called the 'focussed interview'. It involves an interrogation to illuminate the factors or stimulus acting upon informants, the results of which are reflected on their behaviour. However, the central theme of this interview revolves round a particular situation and the questions are common to all people selected for the interview. Usually the subjects of focussed interview are, adult education, dowry system, family planning, reading of newspaper, listening to the radio etc. The aim is to study the repercussion of a particular situation in depth and to analyze the effects of communication. The 'focussed interviews' are used effectively in the development of hypothesis about the aspects of specific experience that may expectedly lead to changes in attitude on the part of those who are exposed to experience.

Clinical interview : This type of interview is quite similar to the 'focussed interview' but the difference is that the clinical interview is concerned with broad underlying feelings or motivations, or with the course of individual's life experience, rather than with the effects of specific experience, as found in focussed interview. It resembles the method of questioning as a doctor asks a sick person in order to diagnose him. Here the investigator does not at all help the informant to express himself and not even he guides the informant in choosing the topic. The procedure is mostly utilized in social psychology and psychotherapy. Its use in social-cultural anthropology is not popular at all.

(b) Multiple interviews in depth—Here the informant is interviewed not only over a fairly long period, he has to submit numerous interviews at different dates. There are three types of 'multiple interview'—the memoir interview, the interview with prisoners and the psychoanalysis. Among these, the 'memoir interview' is considered important for the social-cultural anthropologists. They take the help of this sub-technique while verifying the facts. On the other hand, psychoanalysis is purely a therapeutic technique for treating the patients with mental disorders and personality troubles. Thirdly this technique can also be applied on the criminals of the prison who do not normally want to disclose their motivations of crimes. In later two cases, the investigator notes the statements, which deviate with time.

Memoir interview: It is a technique where in different sittings an informant is allowed to exercise his memory to describe the events which he experienced a long time back. Once the technique was very popular among the social-cultural anthropologists. They used to collect different sorts of data in this way. For example, an old agriculturist could be asked for submitting the data on agriculture, as he was once actively engaged in agriculture. Formerly this technique was also widely used to show the aspects of change. The procedure was to obtain permission from a person to see him regularly over a period of time. After getting the permission, the investigator was supposed to meet him frequently to ask the questions on the events in which he had been involved. To a certain extent, the investigator used to play the role of a memoir writer. Anthropologists of present day have abandoned this particular technique of interview.

(*ii*) *Repeated* (Panel) *Interview*

The technique is almost similar to that of the previous one i.e., 'Interview in depth'. But the difference is that here a group of persons are interviewed instead of a single person. The persons of the group are subjected to repeated interview, at various intervals. A panel is essentially made with an idea that a set of questions can be applied on the same people at regular intervals, throughout the period of inquiry.

It is essential to know the methodology of fieldwork, but practical work in the field is more effective than theoretical knowledge.Still before going to the actual field one should orient him with the theoretical framework of existing methods. It is a matter of regret that the established methods and techniques have been poorly described by the eminent anthropologists. However, a devoted anthropologist may evolve certain new and effective techniques while working in field. An increasing attention has to be paid on the technical problems of methodology.

Analysis and Report Writing

After the data are collected, the fieldworker turns his focus of attention on their analysis and interpretation. Analysis of data involves a number of closely related operations that are performed with the purpose of summarizing the collected data and organizing them in such a manner that they would yield answer to the questions in the light of hypothesis. The dividing line between analysis and interpretation of data is very difficult to draw. Because these two processes often merge imperceptibly. As analysis refers organizing of data in a particular manner; it is the interpretive ideas that govern this task.

Analysis and interpretation vary from study to study on the basis of the problem chosen by the fieldworker (researcher). The objective of the field-researchers is to study how people behave in specific organizations, communities or circumstances. He concludes whether anyone would behave similarly in all situations. If not, he demonstrates the factors that appear as resistant in a given situation. Usually a research aims at achieving a wider validity. Such validity can be established by gathering and analyzing the field-data. Similarities between persons and situations can be revealed by a number of social science researches. But in most of the cases, huge data are obtained from the field where analysis follows a completely mechanical procedure. In fact analysis is a tough task which determines the validity of a research. It demands an appropriate combination of data and reading them against the instruction for verification or falsification of hypothesis. Moreover, integrated operation is needed for the analysis of data, which involves the following steps:

(i) Classification or establishment of the categories for data.

(ii) Application of categories to raw data through coding (symbols).

(iii) The tabulation of data.

(iv) Statistical analysis of data.

(v) Inference about casual relations among the variables.

The above mentioned steps of operation combine to form the preparational part for the analysis of data. The main research report should contain the findings that need to be communicated to others. The prerequisite to communicability is that one must be clear about the persons for whom the given report is intended to be written. Therefore, a report must contain the following points in order to supply the necessary information to all social scientists of present as well as future days.

(a) Statement of the problem with which the study is concerned.

(b) The research procedure.

(i) The study design including the selection of venue, the universe and the nature of sample.

(ii) The techniques of data collection.

(iii) The methods of representation.

(c) The results of major findings.

(d) The implications of the research results for theory and practice.

Interpretation is the core of the research for the simple reason that the usefulness and utility of research findings depends on it. The task of interpretation is not an easy job; rather it requires a great skill and dexterity on the part of the researcher. It is an art, which can be learnt only through practice and experience. It should be remembered that even if the data are properly collected and analyzed, wrong interpretation may lead to inaccurate conclusions. However, interpretations are reflected in final report; writing of report is the last step in a research study.

Research reports vary greatly in length and type. But all of them are the product of slow, painstaking, accurate inductive work. The usual steps involved in writing reports are:

(i) Logical analysis of the subject matter.

(ii) Preparation of the final outline.

(iii) Preparation of rough draft.

(iv) Rewriting and polishing.

(v) Preparation of final bibliography and

(vi) Writing of the final draft.

It has become customary to conclude the report with a very brief summary, which rests on the concise presentation of the research problem and methodology. The major findings and major conclusions are also included here. At the end of the report, appendices should be enlisted in respect of all technical data such as questionnaire, sample information, mathematical derivations and the like ones. Bibliography of sources consulted should also be furnished at the end of the report. Index is also considered as essential part of good report and as such must be prepared and appended at the end after Bibliography. A judicious use of statistics in research reports is often considered a virtue for it contributes a great deal toward the clarification and simplification of the material and research results. Statistics are usually presented in the form of tables, charts, bars and line-graphs and pictograms. Statistical presentations should be neat and attractive to look into problem at hand. Good photographs can be provided with the report, because they are often worth more than a thousand words.

By nature anthropological research comes under the social science research but it goes in its own way. Any sort of anthropological fieldwork is based upon participant observation, which differs from the other methods of data collection. The fieldworkers or researchers merge fully into the lives of people, by being or becoming members of the group. As they do not stand at the periphery merely as observer, so they get ample scope to know the people and the society. Their attention is concentrated on each and every phenomena of the society; they try to link up all phenomena and behaviours in a society. The data of an anthropologist is, therefore, purely qualitative in nature. Field-notes are often recorded in the memory.

HUMAN GROWTH AND DEVELOPMENT

INTRODUCTION OF HUMAN GROWTH AND DEVELOPMENT

The fundamental tenets of anthropology are not restricted only to the analyses of past, present and future of the mankind, it also explains the developmental phases that the individuals experience in course of their journey in life.

Change is the characteristic of life; from the moment of conception till death human beings undergo a lot of changes. These changes represent growth, development or decay that comes automatically with time. How a seed i.e. a fertilized human cell turns into an adult individual with varied potentiality, what typical phases have to be passed over, why constraints appear in between and how those can be minimized, the necessary preconditions of normal healthy biological development, behavioural propensities associated with environmental dynamics, etc are the pertinent scientific questions that come under the purview of human development. The structure of needs and path of psycho-social development of the children do not always follow the same channel. The impacts of genetic and environmental factors are also added to social experience. In fact, the pattern of growing up, aging and maturity - all developmental tasks[1] across the life span depends upon both the internal and external stimuli. Hence, the scientific knowledge about human development enables us to describe and explain how children grow as adult individuals and what are the other changes that appear with time; what creates the wonderful range of diversity among the people and why in spite of diversity in ethnicity, culture, gender, race, age, and ability, a regulated pattern of development can be traced from conception to late life in a sequential order.

A basic overall plan for growth is shared by all members of any animal species. Contrary to the other animals, primate group experiences a special evolutionary step which has been much more special in case of human beings. Human growth and development can be explained both generic and phylogenetic terms. Since Homo sapiens is the highest form of creature in the animal kingdom, certain directions of development and maturity are obvious for them from phylogenetic point of view. All human beings all over the world are therefore alike in physiological characteristics; they have the same potential to grow and develop although the capability and concern vary from individual to individual.

With the changes of outlook recently Human development has been designated as Life Span Development which scientifically examines the factors of growth, change and stability that occur all along the life-span of human beings.

Development means change – a kind of systematic progression from one state to another. This change is irreversible and permanent having a lasting effect on the individual. It helps individuals to reach to a certain fixed point and to attain maturity.

1. Each phase of development has some characteristic traits. These common traits can be listed as developmental tasks.

CONCEPT OF GROWTH AND DEVELOPMENT

The terms 'Growth' and 'Development' are actually two different terms having separate meanings but most of the time they are used interchangeably. This creates a lot of confusion. Terminologically 'Growth' means increase and enlargement of the body or some parts of the body i.e. the cases where the body becomes heavier, larger, longer, etc. It is a change in the organism which can be observed and measured in the physical sense denoting an increase in size, length, height, weight, proportion, etc. Therefore, all changes in quantitative form which could be objectively observed and measured come into the domain of growth. On the other hand, development refers to the qualitative changes. These are the changes in kind, structure or organization - a complex process of integrating many structures and functions, which may not be observable or measureable outwardly. Thus, 'growth' can be one of the components of developmental process or 'development' in its quantitative aspect can be termed as growth.

'Growth' and 'Development' both are actually incremental processes. Both refer to a dynamic process of aging and maturation. Although interdependent and interrelated to each other, they are never synonymous. As per Elizabeth B Hurlock (1977), in most places 'growth' and 'development' goes together; in reality they are different but never takes place alone. For example, when we refer a growing child, we want to mean that the child is showing an increase in size and structure. Not only the child becomes larger outwardly, the size and structure of his internal organs and the brain also increase .The growth of the brain cannot be measured directly but as a result of the growth of the brain, the child acquires a greater capacity for learning, remembering and reasoning. The overall situation tells us that the child has grown mentally as well as physically; so we are able to conclude that 'development' is a qualitative and integral change occurring at physical and mental plane by improving the efficiency or functional ability of an individual.

In the words of Hurlock, 'Development means a progressive series of changes that occur in an orderly predictable pattern as a result of maturation and experience'. She projected development as a progressive series of orderly, coherent changes. Here the term 'progressive' signifies directional changes which are essentially forward moving rather than backward. The other two terms, 'orderly' and 'coherent' suggest that there is a definite relationship among all changes taking place; between those that proceeded and those followed them. Development is thus both a process and a product of transforming and transferring, from one stage to another, monitored on time scale. It is essentially based on systematic laws of progression and complexity, and embraces both quantitative and qualitative changes.

It is a wonder as well as a mystery that the development of every human being starts with the formation of a single cell at the time of conception. (Although a general sequence of development is followed but there are individual differences in timing and specific expression of developmental changes.) All individuals possess the potential to change throughout their lives. Rapid and dramatic changes occur in early life - the period between prenatal period and the onset of adulthood. Each year a child undergoes a series of changes. His height and weight increases; hair grows in length, teeth appear at some point of time, body proportion and dimensions change due to constant growth. Similarly for the brain development, the mental development takes place. Although the size of the head or brain does not increase remarkably, immense change is noticed in the level of thinking, in power of speech, in memory and intelligence. The functions of the glands in the body and the reflex mechanisms do also change with time. Mentally, the child becomes different at each successive stages of growth. This can be viewed as development. Early development is more important than development of later period because, early childhood is characterized by plasticity and the child remains most malleable during this very period. Growth usually stops at maturity but development in some form may continue throughout the life. Human beings show lifelong capacity for change and they can influence their own development.

Comparison between Growth & Development

GROWTH	DEVELOPMENT
1. Growth is a part of developmental process.	1. Development is a wider and comprehensive term. It signifies an overall change in the individual.
2. Growth denotes distinct change in the physical or biological areas only.	2. Development denotes the increasing functional ability of an individual and integration of changes in different spheres like physical, mental, emotional, moral, social, etc.
3. Growth is directional; the changes follow forward motion.	3. Development is multi-directional. It shows a transition from lower to higher plane of activity.
4. The changes produced by growth are subjected to measurement; one can measure them objectively and accurately.	4. As development signifies improved function and behaviour, changes are difficult to be measured by numbers. Those can be assessed by observation and other means.
5. Growth is a compartmental manifestation – external or internal.	5. Development is a holistic manifestation.
6. Growth implies quantitative changes.	6. Development implies qualitative changes.
7. Growth stops after attaining the peak maturational level.	7. Development is a lifelong process.
8. Growth is not affected by learning.	8. Learning and experience has a lot of impact on development.
9 Growth need not necessarily lead to development. For example, a child may grow by becoming fat, but this growth does not bring any functional improvement of body mechanism.. So it can be said that growth may or may not always beget development.	9. Development is integrative, sequential as well as progressive. Development is possible without growth as we see in case of a few children who do not gain height or weight. Although they do not grow in size at all, they experience functional improvement in physical, social, emotional or intellectual fields.
10. Growth cannot be changed or modified from time to time.	10. Development can be changed and modified from time to time.

Individuals change both inwardly and outwardly. Since human being is a part of animal kingdom, their growth and development should be traced in both ontogenic and phylogenic terms. We have already searched the relationship of man with other animals and discussed the development in phylogenic terms {chapter: 4 & 5}. But in discussing development in ontogenic term, we must be careful about the human qualities which are dramatically different from other animals. Man is more delicate and adaptive than other animals. Again, they are not passive recipient of outward influences. They actively shape their own material environment and have the ability to bounce back. Further, they can create ideas and follow the same, which continuously give rise to a social environment what we call culture in other word.

According to Papalia et al, the reason for the complexity of human development is the growth and change that occur in different aspects of the self. The aspects are intertwined and each aspect of development affects the others. The domains of development as described by Papalia are as follows:

- Physical development, which includes growth of the body and brain, sensory capacities, motor skills and general health.
- Cognitive development, which includes change and stability in mental activities like learning, attention, memory, language, thinking, reasoning and creativity.
- Psycho-social development, which includes change and stability in emotions, personality and social relationships.

BASIC PRINCIPLES OF DEVELOPMENT

Individual is a dynamic organism and the development is subject to many influences. It proceeds with certain regulated principle. As per R.M. Learner, individual development 'is characterized by an increase of complexity of organization – i.e. the emergence of new structural and functional properties and competencies at all levels of analyses as a consequence of horizontal and vertical coactions among its parts, including organism-environment actions.'

Three guiding principles mainly work behind human growth and development; each indicating certain specific direction. These principles primarily build the conceptual framework for the study of human development.

1. Simple to Complex Development

This principle advocates that the development is never haphazard; it progresses along logical lines which are similar for all human beings. Individuals are bound to pass through simple to complex forms in course of their development. This means that maturation always follows a preordained course. The specific milestones of development may vary from individual to individual because there is no standard age for sitting, standing or walking except the order which will remain same for all. For example, one baby sits at six months, another baby cannot sit even at the end of ten months, but both of them learn to sit with support before they could sit alone. In the same way all babies must stand before they start walking.

2. Top to Bottom development

This principle is also known as Cephalo-caudal principle based on head to foot development. It means that the development proceeds gradually from upper to lower part of the body. The structural and functional developments first occur in head, then in the trunk and lastly in the legs. So an infant learns to turn his head and lift it up before he gains control over his trunk or legs. For the same principle babies use their hands to grasp prior to their effort of walking. Coordination of arms always precedes coordination of legs. Even in the fetus, embryo head, brain and eyes are found to develop earlier than the other parts.

3. Inner to Outer Development

This principle is also called Proximo-distal principle based on near to far development. Here development proceeds from the central axis of the body outward towards the periphery or the extremities. It is side-wise development. Here the spinal cord develops before outer parts of the body. Babies first acquire the ability to move their shoulder, then the arm and hand, and finally the fingers. This means that the finger muscles develop last of all to enable fine motor dexterity.

These principles signify a typical pattern and orderly process in growth and development. We can therefore safely predict the norm that how most of the children develop at the same rate and at about the same time. Although there are individual differences in children's personalities, activity levels, and timing of developmental milestones, the principles of development are unique and universal.

There are also many other important principles that works behind development. They explain the basic process of development in the following way.

1. Development is a continuous process

Development is a continuous process right from the day of conception in mother's womb. Each stage of development influences the succeeding stages. Foundation of one stage is generally built upon the preceding stage. As a child grows, he constantly acquires and adds new skills for further achievement; one stage of development ushers the way for the next stage of development. In the context of motor development we find a baby lifts and turns his head before acquiring the skill of turning over. Infants can move their limbs (arms and legs) before attaining the skill of precise grasping. Mastery of climbing stairs involves increasing skills from holding on to walking alone; more maturation is still needed to walk up and down the stairs with alternating feet.

Sometimes development slows down, at times accelerated. Children born prematurely may lag behind for about a year or so, but later the same children catch up to the norm and show development in the normal path. An orderly pattern is maintained both in prenatal and postnatal development. Therefore, development can be said continuous, directional and irreversible.

2. Development follows an orderly sequence

Every species or organism on this earth follows a specific order and pattern of development. The same is found in case of human being. Each stage of development leads to the next. The rate and speed of development may vary in individual cases but the pattern of the sequence remains same for all. For example, in a fetus, head and trunk appear first before the limb buds are formed. Then the limb buds gradually give rise to arms and fingers. Similarly, babies strive to stand before they walk; cooing is followed by babbling; even children learn to draw circles before they can make squares. A child from a poor family and a child from an affluent family, both follow the same pattern of development, although the latter may develop at a faster rate due to different facilities available at home. On the whole, it may be stated that development may be delayed but the sequence remain unaltered. The orderly pattern is discernible in all areas of development like physical, mental, motor, speech, etc.

3. Development proceeds from general to specific patterns of functioning

It has been observed that general activity always precedes specific activity in course of development. The early responses of the baby remain very general in nature which gradually is replaced with specific ones. For example, babies wave their arms in general; random movements are noted as long as they are not capable of making specific responses like reaching for an object held before them. Similarly, babies first learn to grasp an object with the whole hand prior to manipulating the same with thumb and fingers. This growth pattern reveals the fact that infant's first motor movements are very generalized, undirected, and reflexive and in the early years they are totally unable to control their finer muscles.

The leg movements in the same manner start with kicking and it continues until the babies are able to creep toward an object. Growth thus occurs from gross muscle (large) movements to more refined (small) muscle movements; finer muscles are required for subtle manipulation of tiny objects (with fingers). In all areas of development this change over from generalized to specific function is established. If we consider the emotional development, we will find that the earliest emotional responses of the new born are generally diffused excitement which slowly gives rise to specific responses like fear, anger, joy, surprise, hatred, etc. The same incident is noticed in case of language development. In early stages of language development children are prone to use some known words indiscriminately for any type situation or thing, but with time the vocabulary increases and they learn to use correct specific words.

4. The rate of development varies for different areas

It has been known that the nature of development is similar to all people throughout the world and all aspects of development are interrelated. But it proceeds at different rates for different parts of the same body. For this reason human growth is called asynchronous.

Developments of mental and physical traits are though thought to be continuous but they do not go hand in hand. The brain, for example, grows at a very rapid rate during prenatal and early postnatal stages and shows more or less full weight several years before pubertal changes commence. Conversely, during the early years of life only very limited increment is noticed on the genital organs and related system. In fact, all systems and organs of the human body thus grow on their own rate and reaches to maturity on own time scale.

There are ample of instances which exhibit the differential rate of structural functional development in human body. During the prenatal period and throughout the infancy, upper part of the body develops faster than the lower limbs. This imbalance of the body proportion is corrected in late childhood when legs lengthen at a higher speed than the trunk. The heart, liver and digestive organs grow slowly in childhood, but rapidly during the early years of adolescence. The feet and hands reach their maximum limit at the onset of adolescence but the face and shoulders show slow development. In the same way, memory for concrete facts and objects develops more quickly than memory for abstract and theoretical material. Again, sometimes, one aspect of development may slow down the growth and development of another area which seems that only one area is developing. For example, growth of language usually gets slower when children learn to walk. Because, children cannot learn two new things simultaneously at a same pace; all their energy is spent in acquiring a new skill.

5. Each phase of development has characteristic traits.

Children do not necessarily pass through the various developmental stages at the same age because of varying developmental rates. But since developmental phases follow a relatively stable sequence, each phase can be demarcated with some typical characteristic traits. For example, by the age of one and a half years, a child is expected to walk independently. In the same way, certain physical development like changes in body proportion, appearance of body hair, teeth, etc. are likely to occur at certain phases. These specific routine changes are normal and known as developmental tasks. Normally children of an age group show some listed physical change or behaviour at each phase of development. These are called 'characteristic behaviour'. 'Problem behaviour' may appear for not fulfilling the 'characteristic behaviour' in time. In general, there is a powerful impulse to grow and mature which is spontaneously expressed in each phase and a child acquires the joy of unfolding new abilities.

6. There are Individual Differences in Development

Although the pattern of development is same for all children, they may not follow the smooth, step by step mode and specific rate, and naturally do not reach same point of development at the same age. The developmental timetable is characteristically unique for each child. The individual differences arise because each child is controlled by a typical combination of hereditary endowment and environmental influence.

Every person is biologically and genetically different from others. This is even true in case of identical twins; no two individuals can ever exhibit identical hereditary endowments or similar environmental experiences. Therefore, it can be inferred that the individual differences are generated from interacting influences of heredity and environment. In fact, there is a range of ages for any developmental task to take place. Some children may talk at ten months while others take a few months more to start talking. Some children are quite active in the childhood, while others remain passive. This does not mean that a passive child will be less intelligent as an adult. Comparing one child's progress with or against another child is a wrong validation.

Rates of development, as mentioned earlier, are never uniform within a child itself. A child's intellectual development may progress faster than his emotional or social development. Sometimes motor development becomes unusually slower. Some changes occur in a flash, others take weeks, months, or years to reveal.

7. Early development is more critical than later development

Early childhood is characterized by plasticity and the child remains most malleable during this period. Attitudes, habits and patterns of behaviour, which are established during early years determine to a large extent that how successfully a person is going to adjust in his future life. Early foundation is very important because that tend to persist in later life. The first two years of life are most critical. The origin of human competence is specially embedded in a critical period of time between eight to eighteen months. What a child experiences during this time span remain preserved for his future competence; this is a vulnerable time than any other time before or after. Experiences of this time are usually carried forward and are reflected in the behaviour of a child as he grows older and older.

8. Development is a product of maturity and learning

Maturation is the unfolding of traits i.e. characteristics that are considered as intrinsic hidden potentiality of an individual resulting from hereditary endowment. Conversely, learning is the adoption of a new behaviour or modification of previous behaviour that can be cultivated by constant practice, exercise and conscious effort of the individual. Children acquire competency in learning by using their hereditary resources if they get ample opportunity to learn. Learning usually takes place through imitation, identification or training under different condition of motivation.

All behavioral manifestations are the product of learning. Postnatal maturation and learning are closely interrelated; one is influenced by the other. Development as a whole depends on the interplay of heredity and the socio-cultural forces of the environment. For instance, a child may possess a special mechanical aptitude by dint of his hereditary potential but that will not be expressed properly if his environment does not offer chances where he can play and manipulate with different mechanical instruments. Similarly, a child with superior neuro-muscular organization is genetically potent for high quality musical performance. But if this child remains deprived of opportunities for practice and systematic training, he will never be able to show up his skill to the optimum level. Sometimes learning is merely repetition of an act by imitation or identification; few people show their excellence to copy from others within very short time without much effort. This is their inherent capacity – a blessing from genetic makeup. Some learning depends on deliberate training that is selected, directed and purposive in nature. Thorough practice can provide certain ability for all learners but the genius is produced only when one holds good genetic predisposition. Since heredity and environment are different for different people, it seems obvious that individuals will encounter dissimilar factors which make them different from others. On the whole, a stimulating environment is required where varied experiences allow a child to develop his potential.

9. All development is interrelated and the child develops as a unified whole.

Development could be split into separate domains like physical, cognitive, emotional, social, moral, etc. but the domains are interrelated and interdependent. It means that there is no water-tight compartment; the domains mutually influence one another and finally an integrated approach is needed to comprehend the total developmental process.

To understand a child and his development, we generally break the study into manageable pieces as the growth of body and the growth of mind. Further, the different components of development are considered - motor development, reflex development, language development, emotional development, personality development, etc which are interwoven nicely to signify an integrated whole. A small maladjustment in one area of development could disturb the other areas and can lead to serious problem. There are no individual whose bodily functions, mental abilities or personality organization lack unification at any phase of development. So to study human growth and development we must recognize the totality and regard the child as unified whole. Coordinated, coherent and continuous growth is the precondition of development. The earlier developments are integrated to form a pattern and then move forward for developing another scheme of new pattern, and thus passing through numerous diversified systematic changes, an organism becomes a unified whole and fulfils the course of life.

10. There are social expectations for every developmental period

Every cultural group expects its members to master certain essential skills and behaviours at various periods during the life-span. Robert J. Havighurst labelled them as Developmental tasks. Developmental tasks are nothing but the developmentally appropriate tasks that a child should be doing at a given stage in life. People of all ages are aware of these social expectations in form of developmental tasks. Parents and teachers know at what ages children are capable of mastering which pattern of behaviour for making good social adjustments. On the contrary, young children also understand what people say to them and ask them to do. They realize approval and disapproval of their behaviour in the society. However, social expectations largely determine the pattern of learning in a particular social milieu.

11. Every area of development shows potential hazards

Each area of development is associated with certain developmental hazard. Some of them are environmental in origin, others originate from within. Regardless of the origin, hazards inevitably affect the physical, psychological and social adjustment of a child. They can alter the pattern and speed of development; forward movement may be stopped and a regression towards a lower stage may be evident. Advance awareness of such hazards helps to prevent an adverse situation or alleviate the underneath cause.

An understanding of the principles of development helps us to plan appropriate activities and stimulating and enriching experiences for children, and provides a basis for understanding how to encourage and support young children's learning.

AREAS OF DEVELOPMENT

A huge amount of physical growth and development occurs during the first year of a baby's life. These early stages of development are critical in laying the foundation for the baby's future. Different scholars studied growth and development from various perspectives and each of them forwarded a scheme to understand the dimensions. The key areas of development as shown by the scholars are as follows:

1. Physical development
2. Motor development
3. Language development
4. Cognitive development
5. Emotional development
6. Social development
7. Moral development
8. Personality development

AREAS OF CHILD DEVELOPMENT	
Physical	Total biological changes.
Motor	Muscles actions for body movement.
Language	Communicating wants, needs and understandings.
Cognitive	Capacity for logic, reasoning and creative thought.
Emotional	Expression of feelings to others.
Social	Building productive relationships in society.
Moral	Distinguishing right from wrong, change of behaviour by understanding (conscience).
Personality	Developing a sense of self, uniqueness and worth.

Compared to other species, human beings experience a prolonged period of physical growth and development; a child's body changes continuously and dramatically until it reaches the mature adult state. The changes proceed according to a carefully regulated and controlled course of growth plan. Although birth is generally considered as beginning point of life, physical development of a child actually starts from the fetal stage. Both the prenatal and the postnatal progression are important to the babies. Before birth blood circulation, breathing, nourishment, elimination of waste matters, body temperature regulation, etc are supported through mother's body, but soon after birth the baby becomes a separate entity who learns to manage all sorts of body function independently. His physical growth and development turns to a new mode for adjustment to the new life outside of mother's womb.

- **Physical development**

Physical development refers to a process which brings overall changes in the body in terms of size, body composition, body proportion, brain structure, sensory capacities, reflex mechanism, functioning of glands and many other body systems. It accounts both external and internal structure and mechanisms, but first get manifested in the external dimensions of the body. The body size (height and weight) increases very fast during the first few months. A normal newborn is 18 to 22 inches in length and his weight is within 2.5 to 3.5 Kg. Immediately after birth babies lose about 10% of their body weight due to the loss of body fluids which is regained after a very short period. The average birth weight doubles at the completion of fourth month; it is tripled by one year and quadruplets by the second year. Similar pattern is maintained for the height. Height increases by about 10 to 12 inches during first year, by about 5 inches in second year and 3 to 4 inches in third year. On an average the height of the baby at four months is 23 to 24 inches and at one year 28 to 30 inches and by two years 32 to 34 inches. At this time head remains 1/4th of the body length. Up to 10 years, boys are slightly taller and heavier than girls. Between 11 and 12 years the rate of growth for the girls is faster, but after 13 years boys grow faster again and soon they get taller and heavier than girls. The height of a child depends more on heredity, while her weight, though also influenced by heredity, depends more on environmental factors like nutrition, diet, living standard, family surroundings, etc. The children of tall parents are so expected to be tall, while the children of fat parents may not necessarily be fat. Average height and weight range of a population varies with the geographical area and children are normally labelled as tall, short, thin, fat, etc as per their respective position on the scheduled range.

Remarkable bodily growth continues up to the first 20 years of life in humans. Growth is rhythmic and not regular. Studies of growth reveals that growth comes in waves and there are four distinctive periods of growth of which two periods are specially characterized by rapid immense growth called growth spurts. First growth spurt occurs during the period between birth and 2 years and the second one is found at the puberty. A rapid growth spurt is noticed among the girls in the age between 10 and 14 and among the boys in the age between 12 and 14. This is followed by a period of slow growth until 18 to 20 years, when full adult height is attained. Apart from increase in height and weight there are a number of other changes. The form and proportion of body changes as the children grow from infancy to adolescence; they progress from the chubbiness of infancy to the long-legged, more slender body of childhood. By adolescence, they become more slender and thin. As we know that the different parts of the body grow at different rates, some parts of the body attain mature proportions earlier than others. More or less by sixteen years of age, different parts of the body assume their mature proportions.

During infancy, the head of the infant remain much larger in comparison to the rest of the body. It constitutes 1/4th of the total body length as compared to 1/8th in the case of adults. The head grows in width up to age of 3 years, but continues to grow in length until eighteen years of age. This growth pattern is the same for boys and girls, though boys' heads are slightly larger than girls' heads at every age. The nose remains small and somewhat flat during the first few years of life. Gradually it becomes larger and achieves its mature size by fourteen years of age The mouth, chin and entire lower part of child's face do not grow considerably (as compared with the upper part) before the onset of puberty. The forehead and the eye-balls reach their mature size as the child approaches puberty.

At the age of six, the trunk increases twice in length and width as compared to its size at birth. The child looks to be slim until puberty, then again the body widens out. The length of the arms and hands increases about 60% to 75% in the period between the birth and 2 years. By the age of 8 years the arms become 50% larger in comparison to what was found in 2 years; at this time arms seems to be thin but acquire an adult look. The legs grow at a slower pace than the arms. By 6 years of age a child's legs become equal about half of the body length. This ratio remains constant for the rest of the life. It should be remembered that head growth slows down, while the trunk and limb growth increases.

Age	Boys		Girls	
	Height (cm)	Weight(kg)	Height (cm)	Weight (kg)
6 years	108.5	16.3	107.4	16.0
7 years	113.9	18.0	112.8	17.6
8 years	119.3	19.7	118.2	19.4
9 years	123.7	21.5	122.9	21.3
10 years	128.4	23.5	128.4	23.6
11 years	133.4	25.9	133.6	26.4
12 years	138.3	28.5	139.2	29.8
13 years	144.6	32.1	143.9	33.3
14 years	150.1	35.7	147.5	36.8

Source : *Indian Council of Medical Research, New Delhi. Growth and physical development of Indian infants and children (1972). Technical report series No. 19.*

Growth of muscles and bones are also important in accounting physical development. The bones of an infant are soft and flexible as they are composed mainly of cartilages. As the child grows, the bones become broader as well as longer. This process of hardening of the bones is known as Ossification. Ossification begins early in the first year and ends during puberty; its rate is different for different areas of the body. Bones not only increase in size, they also increase in number. At birth, infants show approximately 270 bones. This number increases to about 350 in puberty. Girls are more advanced than boys in bone development since birth. By 6 years they are about 1 year ahead of boys and by 9 years, one and a half years advanced. Though the bones of an infant do not fracture easily, they are very susceptible to deformities because of their soft and spongy nature. Dietary deficiency or hormonal deficiency can retard or delay ossification; bowed legs and other skeletal deformities may occur. Any kind of bone deformities in childhood may persist throughout the life of the child.

Increase in weight at all periods of growth is due to the development and addition to the weight of the bones, muscle and fatty tissue. Muscles regulate the functioning of the vital body organs like the heart, the digestive system, etc. They are also responsible for strength and co-ordination of activities. Muscle fibers present at birth are much undeveloped in form; they develop in length, breadth and thickness as the child grows. The large muscles develop more rapidly than small ones and determines the motor ability that progress from broad to precise. Fatty tissues develop rapidly during the early years of childhood and so increase of weight at this time are caused mainly by fat tissues. On the contrary, rapid development of the muscle tissues are found only in the early adolescent years. This is why the chubby little babies quickly turn into slim and slender adolescents.

Early sensory capacities appear with the sense of perceptions like touch, smell, taste, hearing and sight. Perception is the interpretation of sensory input; perceptual ability is an integral part of infant growth. Although the developmental stages of perception are difficult to define, touch is probably the first sense to develop in the infants; all body parts remain sensitive to touch. This sensitivity develops during the first five days and increases tremendously for several months thereafter. Smell and taste which are well-developed at birth continue to improve with time. Hearing starts in the womb; fetus can respond to the sounds. But immediately after birth, for a day or two, hearing seems to be impaired for entering of the fluid into the inner ears. As soon as it dries up, efficiency of hearing comes back. Vision is the least developed sense at birth. By the age of three months, babies are able to see things clearly and distinctly with discrimination of colour. More acute vision develops a few months later at the age of 5 or 6 months; this is called 'binocular vision' where both the eyes simultaneously can focus on an object and allows the perception of depth and distance.

Teeth take years to form. Process of teeth development can be traced from the fetal stage, particularly when the fetus is six weeks old. But no presence is noticed even at the birth. The first tooth usually erupts between 4 and 12 months of age, on an average by 7 months. By the age of 2½ year children possess 20 teeth. These teeth are temporary and called the' milk teeth'. Teeth appear in the same sequence in all children but the ages at which they erupt are extremely variable. Girls seem to be a little advanced here. Around 6 year of age, most children begin to lose their milk teeth and their permanent teeth start erupting. By the age of 12 years all the 20 milk teeth get replaced by 'permanent teeth', which are considered as durable. By the age of 13 most of the children show 28 teeth. The last 4 teeth are called the wisdom teeth; they erupt late, if at all, between the age of 21 and 25 years.

Growth of nervous system is very fast before birth. By the age 3 to 4 years, the number and size of nerve cells (neuron) keep increasing. Thereafter these cells grow at a relatively slower rate. Human brain contains billions of neurons. Although neurons are present at birth, they are not linked. After birth, the links between the neurons develop rapidly. These links, or connections, are called synapses. "Brain wiring" occurs as new links form. The larger the number of synapses, the greater is the number of messages that can pass through the brain. These links facilitate child's interaction with the world. They influence the ability of a child to learn, solve problems, get along with others, and control emotions. Brain weight at birth is generally 350 grams which grows into 1400 grams approximately in adults.

- **Motor development**

The word motor refers to muscular movements. A child in the process of growing up begins to gain specific muscle control; this enables him to do some voluntary movements. Physical development is therefore intimately related to motor development. Motor development refers to the development of control over the body movements involving muscle action. The child's capacity to sit, crawl, stand, walk, run, hold an object, climb stairs, write, draw, etc. are included under motor development. Development of motor abilities and skills begins before birth. Fetal muscles are developed enough to move its arms and legs; at the end of fourth month of pregnancy mothers can feel this movement at times. A few weeks after birth, motor development progresses at a higher speed. But due to neurological immaturity of the baby, most movements are still random and uncoordinated; they also cover larger areas of the body. Motor development can be classified under two heads (i) **Gross motor development** and (ii) **Fine motor development**. Gross motor development involves large muscles and signifies improved skills specially in the muscles of arms and legs so that a child indulge himself in running, jumping, hopping, cycling, skipping, etc. On the other hand, Fine-motor development involves the small muscles of the hands and fingers to show the skills like precision in grasping, holding, cutting, drawing, brushing, painting, sewing, etc. i.e. the activities which require fine motor control.

The development of the motor skills is determined both by maturation and learning opportunities. When the central nervous system, bones and muscles are mature enough and the environment provides the right opportunities, babies start to explore new things and gain new experiences. Skills proceed from simple to complex – from gross action to specific action following the two basic principles namely Cephalo-caudal and Proximo-distal. Fine muscles are activated later than the larger muscles. The whole process is so neat, orderly, automatic and smooth that it is believed to be genetically programmed.

Large Muscle Activities (Gross motor development)

Large Muscle Activities are noted in walking. Walking enables a child to move around more efficiently and as it frees hands, the child starts to explore and manipulate the things around him. Delay in walking may be an indication of some kind of problem with the child, related to his physical, mental or socio-emotional well-being. The average age for unassisted walking is between 13 and 14 months. Some children are always exceptional; they start walking a little earlier or later than this. The ability of walking develops in a series of stages and with the passage of time the strength, speed and coordination advances. By the age of 6, most of the children show mastery in the basic motor skills namely running, jumping, climbing, pushing, pulling, throwing, grasping, etc. They can balance themselves on a rail, on a wall or on a chalk mark on the floor. By 7 or 8 years the children enjoy strenuous physical activity involving the large muscles.

Fine Muscle Activities (Fine motor development)

By 5 or 6 years of age children are though found to coördinate fingers and hands in simple precision activities (such as writing, sewing, craft work, etc.), but at this time their finer motor skills develop in a rudimentary form. Considerable progress is noted in the age between 6 and 10. Improvement continues not only during middle childhood but also in late childhood and adolescence. There is a great variation in the pattern of learning fine motor skills. A child learns to do those things quickly for which, he possesses the ability, get opportunity to practice and receive encouragement for accomplishment. The direction of motor skills development is the same for girls and boys, but generally boys are found to be better in all physical activities as compared to girls. This is due to the fact that boys have larger muscles and more strength than girls. Besides, there are cultural reasons, as boys are usually encouraged to participate in sports and games and do other tasks that involve a lot of physical power. Girls, on the other hand, are discouraged to play muscle games and pushed into such activities that involve fine motor skills. On the whole, there are fixed ages for acquiring skills and competencies for each task whether it involves gross motor skill or finer motor skill. Environmental obstacles, physical or mental handicaps may hamper the normal pattern of motor development.

As children grow from infancy into middle childhood, they gain more and better control over their muscles. In the initial years of childhood, a child gains control over his gross or large muscles, which enables him to control gross movements involving large areas of the body, e.g. walking, running, jumping, etc. Later, the child gains control over his smaller muscles and that enables him to obtain skills like throwing and catching, writing, sewing, etc. The child will therefore walk before he runs; throw the ball before he is able to catch it and scribble before he is able to write.

Developmental Milestone

The attainment of different skills and abilities in babies are commonly known as developmental milestones. Child development experts have carried out a lot of research on young children to work out a standard i.e. what most children can do at different ages and the rate at which they grow. From this research, milestones of development have been identified. The selection of right time for each activity, which is hemmed in sequence are therefore called milestones in development. For example, a baby can move its body from side to back at 2 months and from back to side at 3 months. At 4 months, it can turn over completely. The baby hitches or moves in a sitting position around six months, crawls and creeps around 8 months, pulls the body up and stands by 10 months with support. He learns to stand without support by 11 months and thereafter strives to walk. Initially he walks with support by 12 months and without support around 14 months. His hands and palm can scoop up an object (called palmer scoop) around 5 months. Around 9 months the baby can use his fingers to pick up fine objects. To use the fingers, adequate control over finger muscles is required and that is obtained only after neuro-muscular maturation and eye-hand coordination. The motor skills of babyhood are integrated later; they are of great importance to the baby for developing personality.

Again, a child becomes mature for learning the skill of writing only by the age of about six to seven years. Expecting a child to start writing at a younger age is likely to create a lot of pressure on hand and arm muscles which can be harmful afterwards. Thus, 'milestones of development' refers to a set of functional skills or age-specific tasks that most children can do at a certain age range. Although each milestone corresponds to an age level, but the actual point of time when a normally developing child reaches the milestone can vary quite a bit.

Milestones for Motor Development (during first two years)

Head Control		Trunk Control		Arm & Hand Control		Leg Control	
Age	**Activity**	**Age**	**Activity**	**Age**	**Activity**	**Age**	**Activity**
1 month	Can lift the head, Simple eye movement	1 month	—	2 weeks	—	1 month	—
2 month	Raises head lying on stomach, Horizontal eye movement	2 month	Turns body from side to back	1 month	Thumb sucking	2 month	Move slightly
3 month	Vertical eye movement, Control on lip muscles (to smile)	3 month	Turns body from back to side	2 month	Diffuse movement	3 month	Move frequently
4 month	Circular eye movement	4 month	Turn body completely (roll over)	3 month	move frequently	4 month	move vigorously
5 month	Raises head lying on back	5 month	Pull body to sitting position	4 month	Reach and grasp	5 month	kicking
6 month	Holds head in sitting position	6 month	Sit with support	5 month	Hold objects what is given	6 month	Hitching (backward movement in sitting position)
		8-9 month	Sit without support (for several minutes)	6 month	Grasp dangling object	7 month	Crawling (lying on stomach, pushes the body forward by pulling with arms and pushing with legs)
				7 month	Put objects to mouth	8 month	Crawling continues

				8 **month**	Pick up objects (fingers used with opposed thumb)	9 month	Creeping (pushing the body forward on hands and knees)
				10 month	Steady picking	10 month	Stand with support
				14 month	**Scribble** starts	11 month	Stand without support
						12 month	Walk with support
						14 month	Walk without support
						16 month	Walk up stairs
						22 month	Jump in place
						25 month	Walk on tip toe

The first three months of a child's life are a time of great wonder. Major developmental milestones at this age are centered on exploring the basic senses and learning more about the body and the environment. To the pediatricians milestones are of immense importance; these are used in a scale to check child's growth and development status. Developmental biologists and researchers also deal with them with paramount interest as they wanted to control harmful deviations and depict the future trend of human development.

At birth, babies also display a variety of motor reflexes as necessity for survival. Some of these reflexes gradually disappear before the first year, while others that are more useful for later life become stronger and better coordinated. The following table gives a list of the major reflexes displayed by the newborn.

SOME NEWBORN REFLEXES				
REFLEX	**STIMULATION**	**RESPONSE**	**AGE OF DISAPPEARANCE**	**ADAPTIVE FUNCTION**
Rooting	Stroke cheek near Corner of mouth	Head turn towards source of stimulation.	3 weeks (becomes voluntary head turning at this time.)	Helps infant to find the nipple
Sucking	Place finger in infant's mouth	Infant sucks finger rhymically.	permanent	Permits feeding.
Swimming	Place infant's face Down in water.	Baby paddles and kicks in swimming motion.	4-6 months	Helps infant survive if dropped into body of water.
Eyeblink	Shine bright light at infant's eyes or clap hands near head.	Infant closes eyelids quickly.	Permanent	Protects infant from strong stimulation.

Withdrawal	Prick sole of foot with a pin	Foot withdraws with flexion of knee and hip.	weakens after 10 days	Protects infant from unpleasant tactile stimulation
Babinski	Stroke sole of foot from Toe toward heel.	Toes fan out and curl as foot twists in.	8-12 months	Unknown.
Moro	Hold infant horizontally on Back and let head drop slightly, or produce a sudden loud sound Against surface supporting infant.	Infant makes an embracing motion by arching back, extending legs, throwing arms outward and then bringing them in towards body.	6 months	In human evolutionary past, may have helped infant cling to mother.
Palmer grasp	Place finger in infant's hand and press against palm	Infant grasps adult's finger	3-4 months	Prepares infant for voluntary grasping.
Tonic neck	While baby lies on back, turn head to ones side.	Infant assumes a 'fencing position', one arm is extended in front of eyes on side to which head is turned, other arm is flexed.	4 months	May prepare infant for voluntary reaching.
Body righting	Rotate shoulder or hips.	Rest of body turns in same direction.	12 months	Supports postural control.
Stepping	Hold infant under arms and permit bare feet to touch a flat surface.	Infant leaves one foot after another in stepping response.	2 months	Prepares infant for voluntary walking.

Sources: *knobloch & pasamanick,1974; prechtl & beintema,1965.*

• Language Development

Language is the means by which we communicate our thoughts, feelings, needs and wants with one another. It is our most human characteristic, essential to all human relationships. There are many different ways that people use language to communicate, like speaking, writing, gesturing, sign language, facial expression, pantomime, art and so on. Of these, speaking is the most effective and widely used form. Speech utilizes words or sounds to convey meanings. Speech development is therefore considered as prelude to language development. All sounds created by a child are not necessarily 'speech'. Speech is a complex motor- mental skill originated from the combination of diverse factors - the maturation of the vocal cord is necessary in association with several muscles in one hand and on the other hand it involves putting of appropriate meanings to the sounds produced. So until a child gains enough control over the neuromuscular mechanism, clear and distinct sounds cannot be produced. Further, the child should learn to associate proper meaning with the specific sounds. Voice and speech is therefore intimately related to be expressed as language. Voice is the sound that is produced when air from the lungs is pushed between vocal folds of larynx, causing them to vibrate. Speech is the talking skill facilitated by the muscle actions of the tongue, lips, jaw, and vocal tract in response to recognizable sounds.

The ability to communicate begins early in a newborn's life and like other areas of development, follows predictable pattern. Most of the children cannot utter first word until they are 12 to 15 months old. There are two distinct criteria to determine a sound as speech. Firstly, the child must know the meaning of the word and secondly he should pronounce that word in such way that others can follow the meaning. Words are the set of sounds which are assembled and combined in the sentences maintaining grammatical rules of construction. Children have to discover its underlying structure and use it meticulously at time of communication. Infants must hear language to learn it. What a child hears during the first three years of life will determine his or her adult vocabulary. Children at this age have an incredible capacity for learning language. On the contrary, infants and toddlers who hear fewer words develop smaller vocabularies.

Children acquire language through a series of identifiable stages. The first stage is called the pre-linguistic stage (0 to 12 Months) when infants learn to use language and try to communicate with others but quite inefficiently. During this period they use three crude forms of sound for drawing attention of others - crying, cooing and babbling. These are called pre-speech forms. Of these, babbling is most important as it constructs the basis of real speech.

Early sounds are explosive in character. These are made by the child from the 1st day of their existence. Most popular among them is crying which conveys the needs of the child; other sounds of this time are grunt of pain or disgust, squeals of delight, yawns, sneezes, sighs, coughing, growling, etc. These are caused by chance movement of vocal apparatus. Many of these early sounds gradually disappear as the vocal mechanism is organized. Cooing, a high–pitched sound appears after a couple of months which is a sign of pleasant emotion. Cooings soon turn into babbling which is a slightly advanced form - utterance of vowels combined with consonants such as ma-ma, da-da, na-na, ga-ga, etc. Later, child becomes expert to repeat these sounds by stringing together like ma-ma-ma-ma, da-da-da-da, etc. Babbling is regarded as 'play-speech' full of pleasure. It develops around six months of age and continues throughout the babyhood; its peak period lies around eight months. Babbling is not only important for the purpose of enjoyment; this verbal practice makes the foundation of real speech. Babies try to babble for expressing themselves into a conversation. The socializing value of babbling is also great.

Like many other skills, speech can be learnt by trial and error, by imitation and by training. The child primarily begins to acquire those words that stand for people, objects and events. Children learn to understand their native language long before they speak it. In order to develop good language skill a child must have a rich language environment i.e. the child should have opportunity to listen to any standard language; he should also have exposure to different kinds of stimulating material which can help him to build up a sound vocabulary and good verbal expression. Moreover, in learning to speak, a child has to accomplish four major tasks:

1. Comprehending the speech of others.
2. Pronouncing the words correctly so that others can apprehend.
3. Building a vocabulary - both general and special. A 2-year old child usually keeps 200 to 300 words in his stock.
4. Combining words successfully into sentences. Children begin to frame sentences from about 12 to 18 months of age. At first they use monosyllable words, then bi-syllable words, and finally complex sentences. Before two years, the baby speaks with words made of two syllables in a sentence having no grammar.

Both receptive and expressive communication start improving as babbling of babyhood and crying get largely reduced. Normal speech development gains significant strides where they learn proper pronunciation, making of sentences (even though with poor grammar) and building of vocabulary. Also the content of speech takes a turn. From talking about self, self interests and self needs the child moves on to socialized speech around six years wherein others and their concerns are spoken of. Amount of talking determines the proficiency in speaking. Every year, as the children grow older, they learn more and more different words; the stock of special words also increases. But all children may not follow the ideal pattern. Language development depends largely on participation of children in social interactions.

Milestones of Language Development

Age	Activity
0-3 Months	• Recognizes different tones of voices. • Coos and gurgles when content. • Cries to show hunger, tiredness and distress. • Smiles in response to others' faces. • Recognizes carer's voice.
6 Months	• Babbles and coos. • Babbles consist of short sounds e.g. 'da da, ma ma'. • Laughs, chuckles and squeals. • Cries to show distress. • Begin to understand emotion in parent or carer's voice. • Begin to enjoy music and rhymes accompanied by actions.
9 Months	• Babbling continues. • Begin to recognize own name. • Imitate simple words. • Pointing begins. This is often accompanied by a sound or the beginnings of a word. This demonstrates an increasing awareness that words are associated with people and objects. • Babbling begins to reflect the intonation of speech. • May understand simple, single words e.g. bye bye.
12 Months	• Babbling becomes more tuneful and inventive. • Strings vowels and consonants together to make repetitive sounds. • Use gestures to ask for things. • Enjoy games. • Understand more than they can say. • Begin to respond to simple instructions e.g. 'come here', 'clap your hands'.
15 Months	• Have about ten words that their carers can understand . • Words are used to mean more than one thing depending on the intonation the baby uses. • Pointing is accompanied by a single word.

18 Months	• Two words are put together e.g. 'bye bye dog'. • Telegraphic speech appears, with children using key words in a grammatical way e.g. 'dada come'. • Vocabulary increases with children learning 10-30 words in a month. • Repeat words and sentences. • Use language to name belongings and point out.
2 Years	• Quickly learns new words. • Use plurals e.g. 'dogs.' • Starts to use nagatives e.g. 'there no cats'. • Both active and passive vocabularies continue to increase. • Sentences become longer although they tend to be in telegraphic speech. • Questions are asked frequently, What? And Why?
3 Years	• Speech is understood by strangers. • Sentences contain four or more words. • Imitates adult speech patterns accurately. • Knows and understands nursery rnymes. • Enjoys asking questions. • Talk to themselves during play. • Pronouns are usually used correctly. • Rhymes and melody are attractive.
4 Years	• Vocabulary is now extensive. • Longer and more complex sentences are used. • Are able to narrate long stories including sequence of events. • Play involves running commentaries. • Can use language to share,take turns, argue, collaborate, etc. • Begin to describe how other people feel. • Questioning is at its peak. • Speech is fully intelligible with few, minor incorrect uses.
5 Years	• Sentences are usually correctly structured although incorrect grammar may still be used. • Pronunciation may still be childish. • Have a wide vocabulary and can use it appropriately. • Vocabulary can include shapes, colours, numbers, etc. • Questions become more precise. • Offer opinions in discussion.
6 Years	• Understands 13,000 words. • Understands opposites. • Classifies according to form, colour and use. • Uses all pronouns correctly.

7 Years	• Understand 20,000-26,000 words. • Understands time intervals and seasons of the year. • Is aware of mistakes in other peoples' speech.
8 Years	• Form complex and compound sentences much more easily and exhibit few lapses in grammar. • Carry on meaningful conversations with adult speakers and follow fairly complex instructions with little or no repetition. • Able to read age appropriate texts with ease and begin to demonstrate competence with writing simple compositions. • Have acquired various social amenities in common usage, such as 'please' and 'thank you' and will know when and where to use them.

• Cognitive Development

Early childhood is not only a period of amazing physical growth; it is also a time of remarkable mental development. Cognitive development refers to the growth and change in the mental capacity that begins in infancy and reaches maturity in adolescence. It involves many abilities such as attention, perception, memory, thinking, reasoning, problem solving and intelligence. Gleitman in 1981 defined the term cognitive development as an intellectual growth that begins at birth and continues through adulthood. He also mentioned that the children begin learning from the moment they are born by looking, listening, and interacting with people and objects – cognitive development is the development of the learning structures and systems in the brain.

It was once believed that infants lacked the ability to think or form complex ideas and remained without cognition until they learned language. It is now well known that babies are aware of their surroundings and interested in exploration from the time they are born. Soon after birth, babies begin to learn; they gather, sort, and process information from around them, using the database to develop perception and thinking skills. In fact, a neonate is able to see and hear. But the capacity to look at an object carefully or with interest (attention) develops at a later age. A child tries to recognize the shapes (triangle, rectangle, cylindrical, etc) and sizes (big, small, etc) of objects before he gets the conception of colours (red, blue, yellow, green, etc), numbers (1, 2, 3, etc) and alphabets (A, B, C, etc). The child starts to keep all his learning into memory i.e. the child acquires a new ability of storing the first hand learned materials in the brain for future use. The full span of memory gradually develops with age. The infant also attains the power of generalization as well as the power of discrimination. For the power of generalization, a child can react warmly to a lady who looks like his mother. Conversely, for the power of discrimination, he becomes successful to distinguish between stranger faces and familiar faces. Again, on the basis of similarities and differences the child classifies objects and events into various categories, which is called categorization.

Although senses can function at birth, but those are not well at all as in later life. Neonates experience heat, cold, pain, etc. and reacts to these. After a short period they can distinguish between certain sounds, smells and tastes. Some more time is required to distinguish between dark and light or to recognize colours. Babies cannot ask about the things because they do not have 'words for things'. They want to capture the world around them through the senses. Each and every experience is totally new and they begin to register the similarities among repeated experiences of the same kinds of things. This builds their knowledge base. A great deal of knowledge building goes on even before children are able to speak and to exploit language. The knowledge building is actually an ability to organize the world around them into different categories and can be regarded as one of the basic building blocks of cognitive development.

Cognitive development cannot be understood without knowing the basics of how the categories and concepts are framed by the children. Infants primarily learn to group things together on the basis of perceptual features. These features include colour, shape, texture and so on. Initial perceptual categories are formed through the co-occurrence of certain visible features. On the basis of perceptual similarities, infants of 3- 4 months of age can recognize that certain sets of entities can be put into categories. Categories and concepts are thus fundamental building blocks of cognition. Perceptual categories appear basing on visible features of objects. Conceptual categories take a broader range of features; it includes the function of objects and other more abstract features. Conceptual categories can also incorporate information that comes from other people, such as names and descriptions for things. Categorizing and concept formation, both the skills come as a consequence of the accumulating experience and knowledge, and increasing levels of abstraction on the part of the child. Infants seem to have some impressive general capacities to build categories and then develop concepts, and thereafter use those in flexible ways.

There are a few definite steps in cognitive development. The first step is concept learning. The process of 'grouping together' means putting similar things in one category (concept learning).) For example, all objects made of wood and having four legs can be labelled as furniture. Therefore, the concept of furniture includes chair, table, almirah, bed, etc. At a later age the child begins to reason what is right what is wrong. He solves simple and complex problems of daily life (problem solving). He also develops the skill to perform goal-oriented behaviour, to go through logical thinking and thereby to manipulate his environment effectively (intelligence). At the pre-primary stage the child's thinking is largely guided by what he sees rather than based on concepts of reasoning. It is only at the primary stage that the thinking becomes more logical. But even at this stage the child is able to think logically only with respect to concrete objects and events. Abstract thinking comes still later after the age of eleven or twelve years. For this reason, at the primary stage when children have to learn basic concepts they should be taught through play and activities and not through the 'chalk and talk' method. Talking, reading, singing or playing music to infants and toddlers is congenial for quick cognitive growth. Children learn from the sounds. They try to imitate the words as well as the facial expressions. This stimulation is also important for the language development.

Cognitive development of the children can therefore be enhanced by playing with them, singing to them, reading to them, planning stimulating activities for them, and encouraging them to explore their environment. Babies are like sponges; they absorb everything from their environment. Cognitive progress enables a growing child to memorize a poem, imagine how to solve a math problem, come up with a creative strategy, or string together meaningfully connected sentences. Human cognition thus involves the building up of an internal, mental structure that has a relation to corresponding structure in the outside world. These mental structures lie at the core of psychological functioning. In fact, Language and thought both is the product of cognitive development. These two skills are closely related because they involve planning, remembering, and problem solving. As children mature and gain experience with their world, both skills develop simultaneously.

Milestones of Cognitive Development

AGE	ACTIVITY
From Birth to 3 Months	• See objects more clearly within a distance of 13 inches • Focus on moving objects, including the faces of caregivers • Tell between sweet, salty, bitter and sour tastes • Detect differences in pitch and volume • See all colours in the human visual spectrum • Respond to their environment with facial expressions

From 3 to 6 Months	• Recognize familiar faces • Respond to the facial expressions of other people • Recognize and react to familiar sounds • Begin to imitate facial expressions
From 6 to 9 Months	• Understand the differences between animate and inanimate objects • Tell the differences between pictures depicting different numbers of objects • Utilize the relative size of an object to determine how far away it is • Gaze longer at "impossible" things, such as an object suspended in midair
From 9 to 12 Months	• Understand the concept of 'object permanence', the idea that an object continues to exist even though it cannot be seen • Imitate gestures and some basic actions • Respond with gestures and sounds • Like looking at picture book. • Manipulate objects by turning them over, trying to put one object into another, etc
From 1 Year to 2 Years	• Understand and respond to words • Identify objects that are similar • Tell the difference between "Me" and "You" • Imitate the actions and language of adults • Can point out familiar objects and people in a picture book • Learn through exploration
From 2 to 3 Years	• Sort objects by category (i.e., animals, flowers, trees, etc.) • Stack rings on a peg from largest to smallest • Imitate more complex adult actions (playing house, pretending to do laundry, etc.) • Identify their own reflection in the mirror by name • Respond to simple directions from parents and caregivers • Name objects in a picture book
From 3 to 4 Years	• Demonstrate awareness of the past and present • Actively seek answers to questions • Learn by observing and listening to instructions • Organize objects by size and shape • Understand how to group and match object according to colour • Have a longer attention span of around 5 to 15 minutes • Asks "why" questions to gain information
From 4 to 5 Years	• Rhyme • Name and identify many colours • Draw the shape of a person • Count to five • Tell where they live • Draw pictures that they often name and describe

• **Emotional Development**

Emotion is a subjective feeling, a cognitive interpretation, a physical reaction, and a behavioral expression. Emotional development refers to the development of emotions in children. Foundation is laid in psychosocial development i.e. the development and expression of various emotions tied intimately with brain maturation and cognitive learning. It involves varieties of feelings and their manifestations in stages and circumstances, which affects personal and social adjustments of the children. Severe stress or early abuse can damage a child's emotional development.

All of us have abilities to identify feelings of others and manage our own impulses that arise occasionally in our life like happiness, elation, sadness, fear, anger, jealousy, trust, pride, confidence, humour and so on. These are all emotions. These emotions are elicited by a wide range of stimuli coming from various objects, situations and people. But differences exist between the emotions of children and adults. Emotional expressions start as a useful pre-speech form of communication. Newborns show their emotions only by smiling and crying, in pleasant and unpleasant situations. The former is characterized by a relaxing of the body and the latter by a tensing of the body. For instance, when the babies are physically uncomfortable, or want of something, they cry. Conversely, when feel sociable, they smile. At this stage, there is no other way left for them to communicate their needs. The outstanding characteristic of the babies' emotional makeup is the complete absence of gradations of responses showing different degrees of intensity. Whatever the stimulus, the resultant emotions is intense and sudden.

However, emotions are quite simple at birth and remain completely undifferentiated in form. This continues for about a month. As time goes by, particularly after a couple of months, babies respond more and more; emotional responses become less random and more differentiated. Babyhood emotions markedly differ from those of older children and adolescents and definitely from those of adults. It should be remembered that emotions are more easily conditioned in babyhood than in later ages. Since emotions are mostly conditioned by the earliest social experiences of a child and guided by cultural and environmental factors, many scholars are in favour of grouping social development and emotional development together.

Like other areas of development, emotional development follows a typical pattern. The first sign of emotional behaviour in infants is general excitement due to strong stimulation. The general excitement becomes differentiated into distress and delight by about three months of age. All of us know that infants cry intensely at the time of hunger and pain. They also cry if left alone in a room. This shows a state of distress. Otherwise, they remain delighted and make pleasant sounds by cooing and gurgling when rocked, patted or played with. By about two years, a child starts displaying of other emotions like fear, disgust, anger, jealously, joy, elation, affection, etc. By the age of five, they are able to suppress the expressions of shame, anxiety, disgust, disappointment, anger, envy, jealousy, joy, elation and so on. Infants show their displeasure by screaming and crying but in a little grown up stage displeasure is shown by throwing things, banging head against the wall or lying flat on the ground. Much later reactions include running away, locking oneself in a room or verbalizing abusive words.

The typical fear response in babies is cry. The commonly observed fears among small children are loud noises, animals, dark room, sudden displacement, strange persons, places, and objects. As they grow older, the type of fear changes and usually concentrated on the supernatural objects, imaginary creatures associated with dark places, death or injury. They are even afraid of failing, or of being ridiculed. Fear responses are generally curbed by social pressure with the passage of time and sometimes children consciously withdraw themselves from the feared objects.

Anger is another frequently expressed emotion in childhood. Small children gets angry when their demands are not fulfilled, when somebody interfere in their play, when toys do not work in the way as they want, or if people do not give much attention to them. In older children, constant nagging and finding faults, teasing, lecturing, making unfavourable comparisons with other children often lead to anger. Unlike younger children who exhibit their anger by throwing things or through temper tantrums, the older children exhibit their anger by being withdrawn to themselves or threatening to run away.

Most childhood jealousies generally originate from the conditions that are prevalent in the home environment. Jealousy in children is triggered by the birth of another child in the family. This is called 'sibling jealousy'. This is actually a feeling of threat in anticipation of losing a central position in the home and particularly love and affection of the parents. In older children, jealousy may arise when parents make comparisons among siblings and show favour to a child who happens to be obedient or intelligent, lovable or sick or handicapped. The jealousy developed in the home is carried over to school when teachers and classmates make comparisons among friends. Children also feel jealous when they are deprived of material possessions. Most of these jealous behaviours indicate an underlying structure of insecurity and uncertainty. The jealous responses in younger children are more direct and aggressive while older children express them in a more indirect way by being sarcastic or critical of others.

Among the infants, the emotions of joy, happiness and delight come from their physical well-being. In older children, it is tagged with the success - successful achievement of goals or point of strength that they set for themselves. Joyful expressions range from a quiet, self-satisfied contentment to laughter. They even show their joy by jumping up and down, clapping their hands, hugging and kissing the person, animal, or object that has given rise to their joy. Emotional development may vary in frequency, intensity and duration among the children. Variations are produced according to children's physical state, intellectual state and environmental conditions. Healthy children tend to be less emotional than those who are poor in health. Children who are intelligent tend to have better control over their emotional expressions. The status of satisfaction or deprivation of socio-emotional needs also influence the pattern of emotional responses.

Milestones of Early Emotional Development

Approximate Age	Expression of fundamental Emotion
At Birth or soon after	• Neonatal smile (reflexive smile - birth to one month; not occur in response to stimuli; usually during sleep) • Distress (in response to pain) • Disgust (in response to unpleasant taste or smell)
2 to 3 months	• Social smile - responding to external stimuli; usually responding to a face; infants show preferences for other human faces
3 to 4 months	• anger • surprise • sadness
4 to 7 months	• fear (separation anxiety, stranger anxiety, etc)
7 to 12 months	• Self awareness • Shame • Jealousy • Shyness • Pride • Guilt • Contempt
1 year to 3 years	• Temper tantrum • Self-conscious emotions appear but depend on monitoring and encouragement of adults. • Vocabulary of words for 'talking about feeling' expands. • Empathy appears

3 to 6 years	• Self-conscious emotions links to self –evaluation. • Understanding of causes and consequences, so emotion regulating strategies develop. • Behavioural signs of emotion improve in accuracy and complexity. • As language develops, empathy becomes more reflective.
7 to 11 years	• Self-conscious emotions become integrated with inner standards of excellence and good behaviour. • Emotional self-regulation becomes more internal and are adjusted to situational demands. • Empathy increases as emotional understanding increases.

Emotions are especially intense during early childhood. Small children are easily aroused to emotional outbursts, but display disequilibrium in emotional behaviour. Emotionality of this age is psychological rather than physiological in origin. Further, most of the emotions are transitory and short lived - a child can cry intensely at one moment and on the other moment he may laugh happily. This momentary shift of emotion is due to lack of understanding of the situation; children do not mature intellectually at this stage. As they grow older, their emotions become more persistent. However, Children's emotions are transparent. One can easily read their mental state following the style of behaviour i.e. whether a child is happy or not. If a child remain tense, that can be readily known from his restlessness, crying, nail biting, thumb sucking, etc. Numerous outbursts of temper occur, and the children suffer from anxiety and feelings of frustration. Girls often dissolve into tears, whereas boys are more likely to express their annoyances or anxieties by being sullen or sulky. Older children acquire a strong incentive to learn to control their emotional expressions because of peer pressure and a desire for approval and acceptance. Characteristically, emotional expressions in late childhood are pleasant ones.

Emotional tension heightens in adolescence resulting from sharp physical and glandular changes that take place at this time. Adolescent emotions are often intense, uncontrolled and seemingly irrational as the time is considered as a 'period of storm and stress'. No stereotyped behaviour pattern is observed due to response of some specific stimulation as found in children. The hormones, the growth spurt and the reproductive maturity all these are not merely physical for they also have an emotional impact. The emotional pattern of the adolescent is therefore called 'heightened emotionality' wherein the person is irritable, moody, and feels intensely. However, maturity sets in adolescence makes way for adulthood and the person learns to adjust appropriately to the social situations. The peer group influence increases. Adolescents begin to notice and take interest in the opposite sex. Making friends and adjusting to new social situations in educational institution or searching for career are learnt during this time. Great deal of interest is shown in personal grooming, looks and clothes. Adolescents also ponder over several philosophical issues and try to find a suitable answer to questions such as 'Who am I? What is the purpose of life?' etc. The search for identity takes a meaningful turn, the adolescents becomes able to adjust well in the society and can develop a positive self-concept.

• Social Development

Social development refers to the way that a child learns to interact with others and develops social relationships at home and outside. As an infant, strong ties are established between the child and the intimate care giving person, generally mother, father or any other family member. Sometimes an appointed person may also take this place. However, the quality of relationship between parents and the child determines the social skills of the child in adult life. The trust which the child gains through his early experiences at home and the independence which he develops in exploring things around him is carried for the later life. Further, this is influenced by the knowledge component of the child, the sex identity and concept of self.

Beginning with a social smile babies learn to respond to the social environment. At first they understand the immediately observable aspects like physical appearance of the people around and the common pattern of social behaviour. Later, they grasp the psychological processes e.g. perceptions, intentions, motivations, attitudes, beliefs, abilities of the people. Social thinking becomes organized and integrated as the children grow further older and relationship with peers as well as with visitors of the family are established. Peer sociability begins in infancy with isolated touches, smiles and vocalizations. By the age of two, Children tend to bind with one another, seeks friendship, attention and approval of peers, and thus group behaviour begins during preschool years.; The horizon of the school children is more wide and sensitive. Young children are likely to be more cooperative and friendly than competitive and hostile. Group loyalties are formed near adolescent age. By interaction with the members of same age and sex group, a child develops an image about himself or herself, called self-concept.

The socialization process (i.e. the process by which members of society shape the behaviours of immature individuals in order to grow them as competent and contributing participants of the said society) is also important in this context. Peers serve as socialization agents for one another by reinforcing and modeling a wide variety of social behaviour. Peers also play a dominant role on developing self concept by giving feedback to a friend i.e. what kind of person he or she is and the kinds of behaviour for which he or she will be accepted or rejected by others, etc. This is actually a stage of prosocial behaviour. The growing child has to adapt to a number of newer context - the school, the neighbourhood and the peer group. Through social development, a child learns to accept others points of view and the ability to work in a group. Learning of qualities like sharing, co-operation, waiting for the turn, respecting and listening to elders, etc. are also parts of the social development process. Further, developing empathy and understanding the needs of others can be included in the area of social development.

Milestones for social development

Approximate Age	Social behaviour
Birth to 1 month	Helpless a social No interest in people.
2 to 3 month	Visually fixates at a face, smiles at a face, may be soothed by rocking.
3 to 4 month	Recognizes his mother. Distinguishes between familiar persons and strangers, no longer smiles indiscriminately. Expects feeding, dressing, and bathing.
4 to 6 month	Wants to be picked up by anyone who approaches them. React differently to scolding and smiling faces. Tries to attract the attention of other baby.
7 to 9 month	Protests separation from mother. Attempts to imitate speech, simple acts and gestures.
10 to 12 month	Responsive to own name. Wave bye-bye. understands "no-no!"Gives and takes objects. May be afraid of strangers
1 to 2 year	Obeys limited commands. Repeats a few words. Interested in his mirror image. Feeds himself.
2 to 3 years	Talks, copies parents' actions. Dependent, clinging and possessive about toys, enjoys playing with another child. Resists parental demands. Gives orders. Inability to make decisions.
3 to 4 years	Likes to share, Cooperative play with other children, goes to nursery school, Imitates parents. Beginning of identification with same-sex parent, practices sex-role activities. Intense curiosity and interest in other children's bodies. Dresses and undresses with assistance. Attends own toilet needs.
4 to 5 years	Prefers to play with other children and becomes competitive. Prefers sex-appropriate activities. Dresses and undresses independently.

Foundation of later social behaviour is laid in childhood because of two reasons. Firstly, the type of behaviour the children face in social situations affects on their gross personal and social adjustments. Secondly, once established, they tend to be persistent for the whole life. Whether a man is jovial or ill tempered, introvert or extrovert, shaky or dashy - the kind of character in a person are build up basing on early social experiences. As a baby the special bond exists between the mother and the child; infants use their mother as a secure base to explore the environment. They listen to their parents' and try to get parents approval and attention in all situation. Since they remain strongly attached to and totally dependent on parents, cannot tolerate short separation from their parents. These attachment behaviours are often called dependence. Although attachment and dependency appear to be the basis of social development; dependency behaviour changes greatly during the development process. In general, frequency of dependent behaviour gradually decrease with age specially when attachment extends to other adults (teacher) and peer group. This period witnesses a transition, from interactions that are basically 'self-oriented' to interactions that are 'other oriented'. Thus the period is marked as the formative period of self-concept - the image of self is established in children; they conceptualize who they are and how each of them differs from others. At this time they also develop an increasing awareness of another important aspect of the self-concept, namely, sex-role identity. Apart from the physical factors, the other factors that mould sex-role identity are imitation and identification. A little boy who observes his father engaged in rough and tough play, imitates that behaviour while a little girl who observes her elder sister playing quietly with dolls imitates her. Similarly the little boy who identifies with his father is likely to incorporate his father's personality into his own, whereas girls may try to identify with their mothers.

A significant factor in the social development is the opportunity for social interactions (among siblings and other peer group.) This depends on the family constellation and the attitude of the parents towards children. Each family provides a unique set of experiences for the development of social attitudes, values and convictions. The alliances and rivalries of the elders, their different complex interactions are interpreted by the children. So eventually when they move from the family circle to broader circle, the experiences at home influence on the development of the other relationships. It has been found that the child who is valued and accepted by the family and who has a good sibling relationship can adjust in a better way with the peer group. Because that child has learnt to accept, work and cooperate with others, even with those who hold opposite opinions and convictions.

The group of peers in adolescence period provides a sense of security and belongingness. Perhaps at the end of his period, self realization comes. Children become mature for the social life. They can evaluate their own strengths and weaknesses; attitudes and values are solidified in a clearer light. They are able to make proper judgments for the world around them. Both social skills and emotional maturity are needed to forge relationships and relate to others.

• **Moral Development**

Morality means the conformity to the moral code of the social group. The term is derived from the latin word 'mores', which means manners, customs and folkways. Moral codes are therefore the rules to which members of a given culture become accustomed over a period of time. It denotes a standard of conduct for the particular group. Moral development refers the learning of moral concept and moral behaviour by the children. It includes acquisition of values such as honesty, integrity, sincerity, righteousness, conformity to group standards, and so on that a society expects from the children to imbibe and practice. But moral behaviour should never be confused with ethical behaviour, which is a more inclusive concept. Ethics or ethical behaviour is applicable for the whole of human conduct in general, rather than any particular group. This also indicates that moral standard may vary group to group.

Morality is a purely human quality. It has a vital emotional component, for its powerful effective reactions that cause us to empathize on others distress or to feel guilty on our own fault. Secondly, it has a cognitive component, for which we develop the concepts of justice and fairness with broadening social experience. Thirdly, it has a behavioural component, for the adoption of societal prescriptions for good conduct or norms. In every social group certain acts or behaviours are considered either right or wrong. However, learning to behave in a socially approved manner is a gradual and long process.

Babies are viewed as amoral, meaning that they do not have an understanding of right and wrong behaviour; it is through social learning experiences (learned over many years) that children come to understand morals. By the time children enter school, they are expected to be able to distinguish right from wrong in simple situations. Naturally at this stage they are supposed to know the difference between good and bad behaviour. To be a moral person a person should learn to follow the laws, customs, and rules of the society. Even growing children are expected to behave as per rules set by parents and others in authority; they learn these through trial and error method by direct teaching, experience and identification. Thereafter they develop moral concepts or the abstract verbal form.

Moral development in early childhood proceeds on a low level. During the preschool years children do not have their own concept of morality and possess no clear understanding of social rules, except for the regulations laid down by the adults. The reason is that young children's intellectual development does not reach to that point where they can learn or apply abstract principles of right and wrong. They merely learn how to act and do not ask about the reason behind. In this stage of moral development, children obey rules automatically, without using reason or judgment, and they regard adults in authority as omnipotent. In a little grown up stage, they judge all acts as right or wrong in terms of their physical consequences, rather than in terms of the motivations in the background. They also learn that they are expected to follow these rules, failing which they will be punished or will not get social acceptance .When early childhood comes to an end, habits of obedience are found to be established through constant practice, for those who live in a disciplined atmosphere.

Discipline is the perfect way of teaching moral behaviour to the children approved by the social group. Three elements are included here – (i) rules and laws which serve as guidelines for approved behaviour, (ii) punishment for willful violation of rules and laws, and (iii) rewards for behaviour or attempts to behave in a socially approved way. During the early childhood years, major emphasis is laid on the educational aspects of discipline; punishment is given only when there is strong evidence that children know what is expected of them, but they willfully violate those expectations. To increase young children's motivations towards socially approved behaviour rewards are given; rewards serve purpose of reinforcing the motivations. Gradually, the child begins conforming to the social expectations in order to gain rewards. This maturity in moral behaviour is closely linked with maturity if social and emotional development as well.

Milestones for moral development

Approximate age	Moral Internalization	Moral Construction
1–2 years	• Amoral	
3–6 years	• Wide variety of pro-social acts begins.	• Right behaviour satisfies child's own needs. • Assisted by the adults for self control.
7–11 years	• Stereotyped approval-focused orientation.	• Stereotyped images of good and bad. • Concern for approval to justify behaviour. • Generation of self - control strategies.
12–adulthood	• Internalization of societal Norms continues.	• Self control strategies expands. • Moral self regulation continues to improve.

Moral codes develop from generalized moral concepts (ethics). In late childhood, the code of conduct and morality learnt at home is extended to the social group with which the older children identify themselves. In fact, children make a conscious choice about their peer group. Discipline also helps in this process. Praise, rewards, punishment and consistent application of rules enable the child to develop moral behaviour. Development of conscience is found in late childhood.

Conscience is a conditioned anxiety response to certain kinds of situations and actions; it is built up by associating certain acts with punishment. As a result a feeling of guilt is internalized and children avoid those acts what they know as wrong to avoid punishment. The pattern of morality further changes during Adolescence. Adolescents no longer accept a moral code in an unquestioning way which is handed down to them by parents, teachers or even their contemporaries. They develop a conscience of their own and so want to build new moral codes by modifying the old concepts of right and wrong on the basis of their mature practical experiences. Another important change takes place in adolescence. Since parents and teachers cannot watch adolescents as closely as they did for young children, adolescents are expected to assume responsibility for control over their own behaviour. Morality is thus rooted in internal control and not depends on external agencies for its mature manifestations.

• Personality Development

Personality is not a specific attribute; it is the quality of individual's total behaviour. It is made up of many components – objective and subjective. Objective components are observable and measurable like body size, body features, aptitude and talents, habit and action pattern, etc. Subjective components are those features in a person that are hard to be measured and studied. Those are motives, aspirations, ideas, feelings, attitudes, etc. Personality is gradually structured in a person by the combination of different traits with self concept. It refers to the enduring characteristics that differentiate one person from another and remain more or less stable over the life span.

The potentials of personality development are present in human being at birth. As the time passes on, personality of an individual is formed by the constant interplay of temperament and environment i.e. the adult personality structure is built up. The core of the personality is the self concept, which remains fundamentally same all throughout. If we trace the development we will see that the personality pattern begins to take shape in early childhood. Parents, siblings and other relatives who constitute the social world of young children, initially help in building the base of the personality. Their feelings about the child, their attitude and behaviour with the child, all are important in framing the self concept in a child. As early childhood progresses, the attitude of the peers and their way of treating keep a strong impact on child's self concept. Since the environment of young children is limited to a large extent to their homes and companions, parents and peers take a leading role in conditioning the personality. We can also mention the process of child rearing in this context.

Children are born with characteristic temperamental differences that are reflected in activity rates and sensitivities. Individual differences are apparent at birth and are shown in responses to food intake, in crying, in motor activities, and especially, in sleep. A disturbed prenatal environment can also alter the foundation of personality. For example, if the mother is subjected to severe or prolonged stress, her mental condition may cause a modification of the newborn infant's behaviour pattern. There are ample of evidences which show that infants who are separated from their mothers after birth do not make good adjustment to postnatal life as like those infants who remain with their mothers. Again, children who remain under strict, authoritarian discipline accompanied by frequent reprimand and corporal punishment tend to build up resentment against all persons in authority.

Where the parental aspirations are unrealistically high, children are doomed to failure. Regardless of how children react, failure leaves an indelible mark on their self-concepts and leads to feelings of inferiority and inadequacy.

Genetic studies of the persistence of personality traits over a period of years have revealed that patterns established early lifc remain almost unchanged as the child grows older. Each child in a family learns to play a specific role, but difference arises with the child-training process used by parents in different cultural background. Personality traits are strengthened or weakened depending on interaction with environment which is called as quantitative changes. The roots of personality are found in babyhood; attitudes of parents, siblings and relatives who stay with the child are generally mirrored in the child and the child accepts those as 'self'. Later, the peer group influence is also added to it which may reinforce and establish or contradict and damage the previous influence of the family on the child. In early adolescence, both the boys and girls seem to be well aware about good or bad traits in them because by this time their self concept becomes more stable and consistent giving rise to self esteem. They earnestly take attempts to improve their personality.

Adolescence is the threshold of adulthood. At this time they begin to concentrate on those behaviour which are associated with the adult status. Development of an Integrated Personality is found as a result. The different aspects of development, discussed above, do not occur in isolation, rather they are inter-related and to some extent interdependent. Motor development makes the child mobile, increases his range of activity and helps him to explore the surroundings. This further helps in cognitive development. Language development also helps the child to communicate in a better way; it allows the child to interact with more people. Such interaction leaves a significant impact on the child's social, emotional and moral development. If all these developments are adequate and at the right time the child feels confident of himself in later life i.e. the personality develops.

Milestones for Personality Development

Age	Features
0-2 months	• No sense of self as an individual.
2-3 months	• Discover they can cause things to happen.
up to 6 months	• Discover their physical self; distinguish between the self and the world around them. • Learn that they can act upon other people and objects.
6-12 months	• Realize that they and other people are separate beings with different perspectives.
5 years	• Comprehend the basic physical differences between males and females and accept their constancy.
6 to 8 years	• Friends will become increasingly important, and he will value their opinion of him.
9 to 11 years	• Develop self-concept with both positive and negative aspects.
12 to 15	• Mood swings. • Affection toward his parents also tends to decrease. • Desires for independence emerge.
16 to 18	• Struggles with parents decrease coinciding with an increase in independence.

Stages of Growth
(Life span conception)

Developmental changes occur covering the whole life-span from conception to death. It gives a complete picture of growth and decline at different periods during the entire life-span. Although the concept of periods of development is a social construction, life shows systematic progression in a definite order through a series of phases; step by step a child move closer to some form of adult status. Developmental changes enable people to adapt to the environment in which they live. Nature of these changes is mainly biological and psychological. They may be rapid or slow, inclining or declining, but require constant readjustment on the part of all individuals.

Initial years of development are very important in the life of most of the species because more developmental changes take place during this time than in any other period later. It has already been mentioned that in humans, physical growth is greater and extensive in the first year of life than in any other single year. Similarly, changes involving social interactions, the acquisition and use of language, memory and reasoning abilities, and virtually all other areas of human functioning are greatest during childhood. Also the events and experiences of the early years have been found to effect strongly on individual's later development.

The stages of development may be categorized grossly in two main types:

- **Prenatal development**
- **Postnatal development.**

Prenatal development includes all the changes that take place in the womb of the mother. Therefore it is also called "intra-uterine development" where the uterus is the environment. Postnatal development on the other hand, refers to all the stages that follow after the birth till the very end of life. Prenatal period begins at conception, or fertilization, when the genetic material from a male sex cell (sperm) unites with the female sex cell (ovum) to form a single cell, called a zygote. Zygote is the fertilized ovum (egg); .life begins with this single cell which cannot even be seen with eyes. The human ovum at conception is a homogeneous mass and about 0.1 mm in diameter. The zygote receives 23 chromosomes from the mother and 23 from the father, and these 46 chromosomes replicate over and over as the zygote reproduces itself through mitosis. The length of prenatal period often varies in length, on an average it lasts for 280 days. Prenatal development is orderly and predictable.

Prenatal development covers the period from fertilization to birth. It comprises of three distinct stages:

(1) The period of the zygote (from fertilization to end of two weeks).

(2) Period of the embryo (2 weeks to 2 lunar months).

(3) Period of the fetus (end of 2 months till birth).

The period of zygote

It lasts approximately two weeks. Thirty-six hours after fertilization, the zygote starts it first duplication by the process of mitosis. Although the first cell division takes many hours, each successive division occurs with increasing rapidity. The mass of cells (60-70 cells together by the end of 4th day) called 'blastocyst' move down the fallopian tube into the uterus. For 4-5 days it floats freely in the uterine cavity. The size of zygote at this time is like a pinhead because it receives no source of nourishment and kept alive only by the yolk in the ovum. Now the cells begin to specialize - it divides many times and separates into an outer and an inner layer. The outer layer gets differentiated into placenta, umbilical cord, and amniotic sac while the inner layer develops into a new human being.

Approximately 10th day after fertilization, the cells of the zygote become sticky and attach to the wall of the uterus firmly. This process is called implantation - the zygote is totally buried in the uterine wall. The period of the zygote thus ends after two weeks of fertilization to give rise the period of embryo. By this time a woman first suspects that she may be pregnant as misses her menstrual period; the prenatal development is well under way.

The period of Embryo

It extends from 2 weeks to 2 months. During this time the organism is like a miniature human being. Cell differentiation takes place quickly; most dramatic changes of prenatal period are observed at this time. All major internal and external structure of the body first appears at this time. The embryo can turn itself within the uterus and the heart beat can be heard.

To be very specific, in the third week, the inner cell mass (embryonic disk) differentiates into three layers of cells – Ectoderm, Mesoderm and Endoderm from which all tissues and body organs emerge. The ectoderm gives rise to the nervous system, outer skin, hair and sweat glands; the mesoderm creates deep layers of skin, muscles, skeleton, circulatory system and also a variety of internal organs and connective tissue; and endoderm develops into digestive system, lungs, urinary tract and internal organs. A primitive heart which began to form earlier, by the end of the third week connects itself to the vessels and starts to beat so as to form a cardio-vascular system; it is the first functional organ system of the body.

Around the fourth week, the embryo looks something like a tube of about 0.1 inch long. At the end of fourth week, the embryo assumes a curved form; the upper and lower limbs begin to form as tiny buds. The embryo's body changes less in the fifth week, but the head and brain develop rapidly. The upper limbs take the full form, but the lower limbs just appear and look like small paddles. By the sixth week, the head continues to grow rapidly and differentiation in the limbs namely elbows, fingers, and wrists become recognizable. It is also possible to discern the ears and eyes. The lower limbs develop fully in the seventh week.

By the eighth week the growing embryo takes a child-like appearance. Almost half of the embryo consists of the head. Eyes remain open for about a week before the growth of eyelids. The eyes, ears, nose, jaws, toes and fingers are easily distinguishable and a tail like structure (an extension of the spinal column) which was present before, disappear at this time. All internal and external organs are visible in rudimentary form. This period is of immense importance as environment begins to affect the development of cells.

The period of Fetus

This is the longest prenatal phase which extends from the beginning of the third month till birth. At the end of the embryonic stage, the newly differentiated organs, muscles, and nervous system start to get organized and interconnected. Three major support systems of carrying fetus - the amniotic sac, the placenta, and the umbilical cord become strengthened. The amniotic sac is a watertight membrane filled with fluid. As the embryo grows, the amniotic sac comes to surround it. It acts as a cushion to support the growing embryo within the uterus and also provides with an environment of constant temperature. The placenta is the organ by which the mother and embryo can exchange materials. After implantation, placenta is formed by the contribution of both the mother's tissue and the embryo's tissue; its villi push into the uterine wall. Each villus contains fine blood vessels to send nutrients and oxygen to the embryo and to collect waste products from there (finally to be excreted through mother's body). The semi permeable placental membrane facilitates this exchange. Placenta remains connected to the developing organism by umbilical cord, which also houses the blood vessels but blood cannot pass between the mother and the fetus. As said above, only oxygen and nutrients pass from the mother's blood to fetus through villi, and waste products of the fetus are eliminated.

The body proportions increase as growth continue. By 5th month all the internal organs assume actual proportions. Nerve cells that are present from the third week, increase greatly in number between 2nd to 4th month and the nervous system is found to develop. Fetus's appearance changes drastically. Among the external changes, the head grows first than the other parts of the body. The skin which has been transparent begins to thicken during the third month. The fetus's eyes move from the sides of the head to the front. Nails appear on fingers and toes by the fourth month, and pads also appear at the ends of fingers that uniquely identify the individual for life. Head hair begins to grow. In the third month, the kidneys begin to excrete urine into the surrounding amniotic fluid, which is freshened by the mother's body every three hours. Sexual development becomes apparent in males by the end of this month with the appearance of external sexual organs. In females the oocytes form the outer covering of the ovaries. Fallopian tubes, uterus, and vagina develop and the external labia become discernible. A bone structure begins to support a more erect posture by six months.

Internally, by three months, the brain assumes the basic organization; nerve cells grow and establish connection with brain as well as external environment. The Fetal activity begins in the third month when the fetus is capable of wiggling the toes, and swallowing; but the mother feels none of this. The fetus also appears to become sensitive to environmental stimulation for it moves its whole body in response to touch stimulus. By the fourth month the eyes are sensitive to light through the lids, and by the fifth month, a loud noise may activate the fetus. During this same month the fetus swims effortlessly. The fetus is now capable of kicking and turning, and may begin to display rhythms of sleep and activity. By the seventh month, brain connections are sufficient for the fetus to exhibit a sucking reflex when the lips are touched.

From seven months of age, the fetus has a slightly better chance of survival outside the mother's body. The brain is sufficiently developed to provide partial regulation of breathing, swallowing, and maintaining body temperature. However, a baby born after only seven months of development will need to be provided with extra oxygen, will have to take food in very small amounts and will have to live for several weeks in an incubator for temperature control. In the eighth month, fat appears under the skin, and although the digestive system is still too immature to extract adequate nutrients from food, the fetus begins to store maternal nutrients in its body. But even a baby born at eight months is susceptible to infection. By the eighth month, the mother's body starts contributing disease-fighting antibodies to the fetus that she has developed through her own exposure to foreign bodies. This process is not complete until nine months of fetal age and is very important, because these antibodies help to protect babies from infection until around six months of postnatal age, when they can produce their own in substantial amounts. On completion of 9 months or 280 days, the fetus gets ready for birth. It is surprising that in the beginning of third month a fetus weigh only 5 grams and 2 inches in length, which turns to 3 to 4 kg in weight and about 20 inches in length at the time of birth. The size at birth actually depends on a number of factors like race, sex, parents' body constitution, mother's health and nutrition and so on.

Postnatal Period is commonly divided into the following age periods:

1. Neonatal - Birth to 2 weeks
2. Infancy or babyhood - 2 weeks to 2 years
3. Childhood
   (*i*) Early childhood (preschool years). - 2 to 6 years
   (*ii*) Middle Childhood (school years) - 6 to 12 years
4. Adolescence - 12 to 18 years
5. Adulthood
   (*i*) Early adulthood - 18 to 40 Years
   (*ii*) Middle adulthood - Early 40s to early 60s
   (*iii*) Late adulthood - Early 60s and above.

• Neonatal stage

Neonatal stage begins with birth and ends when the infant is approximately two weeks old. It is the shortest of all developmental period. During this period babies adjust themselves physiologically to the new environment. After birth, within a few days, the babies lose about 10% of their body weight because of a loss of fluids. They tend regain this weight from 5th day and usually go back to their birth weight by 10th to 14th day. New babies show a large head measuring 1/4th of the body length and a receding chin. The skull bones are not yet fused; a very soft location is usual above the forehead at the junction of frontal and temporal bones, known as 'bregma'. This bone joint grows properly and hardens within first month of life.

The newborns are independent entities; hence growth is oriented towards functional state of life. In fact, the adjustment to life outside uterine walls starts with the detachment of the umbilical cord from the naval. This is the time when babies have to make some major body adjustments. For example, they start a new mode of breathing following the cut of umbilical cord. With the birth-cry the lungs are inflated and respiration begins. The rate of respiration at this time ranges from forty to forty-five per minute. By the end of the first week of life, the rate normally drops down to approximately thirty-five per minute which is more stable than it was at first. Secondly, babies feel cold for being transferred from a higher temperature level (100F in the mother's womb) to a lower temperature level (60-70F in the delivery room). Thirdly, no more they receive nourishment through umbilical cord, and starts sucking and swallowing of breast-milk. Last adjustment is for the organs of elimination. Elimination of waste begins a few hours after birth, usually within an hour after feeding. Neonatal sleep is broken by short waking periods which occur every two or three hours, with fewer and shorter waking periods during the night than during the day. The prenatal environment, length of the gestation period, kind of birth (normal or other), postnatal care and parental attitude mostly influence on the adjustments of newborns in the postnatal environment.

Newborns are physically weak and dependent of others for care and protection. Their sensory capacities remain well developed at this time. They show some useful reflexive behaviour. Reflexes can be defined as inherited responses towards stimulation in specific parts of the newborn's body. According to Baron (2001), the reflexes namely blinking, rooting, sucking, tonic neck, Moro, babinski, grasping, and stepping are shown by the babies at birth.

• Infancy or babyhood

Infancy lasts from two weeks to two years of life. It is the true foundation period of life characterized by rapid change in bodily systems, psychological development and neuromuscular organization. However, growth takes place mainly for addition of more cells or increase in the protoplasm. When the anabolic processes exceed catabolic processes, an increase in size, shape and weight is noticed. All these signify the infant stage.

During infancy growth is so rapid that more than 50 per cent of birth length and 200 per cent of birth weight increase at this time. The cells become larger in size. The cervical and lumber curvatures of the spinal column appear as the baby begins to straighten the head and tries to sit up and to stand.

Infancy is hazardous period. These hazards are physical and psychological in nature. Physically, it is hazardous because illness and accidents of this time are very serious often leading to permanent disabilities or death. Babies also face difficulties in making necessary radical adjustments to a totally new and different environmental conditions i.e. difficulties appear to execute respiratory, digestive and vascular functions which become a threat for the infants. High infant mortality rate is the evidence for this statement. Psychologically, infancy is the time when many behaviour patterns, many attitudes, and many forms of emotional expressions get established. Biological and psychological growth brings change in the physical capacities and intellectual development. Since behaviour pattern, interest and attitudes are established during infancy, serious psychological hazards may come if poor foundations are laid at this time.

It is a stage of decreasing dependency and increasing individuality. Individuality is shown in appearance and in patterns of behaviour. Even identical multiple births express this individuality. Infancy or babyhood is the beginning period of socialization; babies learn social behaviour under the supervision of elders. They also become conscious about their sex-role from the very period; sex-appropriate trainings are provided for them. During this period the babies learn to speak; hence able to communicate with the immediate social world.

• Childhood

Childhood ordinarily spans from the end of infancy up to the start of adolescence. Infants therefore attain childhood before reaching adolescence. It is a period of relatively steady progress in growth and maturation as well as it shows rapid progress in neuromuscular or motor development. Childhood is often divided into two sub periods - early childhood and middle childhood.

(i) Early childhood

Early childhood extends from 2 to 6 years. This stage is also known as pre-school stage or toy age. Some scholars like to call it as a period of eruption of milk teeth. Psychologists label this age as 'exploratory' or 'questioning age'.

Physical development proceeds at a slow rate in early childhood; the physiological habits started in the babyhood stage get confirmed in this stage. But the brain continues to grow rapidly, achieving 90 per cent of its full weight by the age of 5 years. Hand preference (whether left handed or right handed) is established by the age of 4. The children of this age need about 12 hours of sleep a day to function efficiently. The development of speech as well as the power of comprehension is very fast at this stage. Vocabulary is acquired at a rapid rate and children use those words to ask questions about things and people. They learn about numbers, colours, shapes and the reasons for everyday events.

This period is also the skill acquiring period as the child easily repeats and picks up skills. The emotional development follows a particular pattern according to intelligence, gender, family background and child rearing practices experienced by the individual. Play is important for the overall development of the child. Play can be influenced by the motor skills acquired by the child, their popularity among other group-mates and the socioeconomic status of their families. Parents, companions and different family relationships perform important roles in socialization and in developing the self-concept in the child. Highest contribution comes from parents who consciously want to influence children's behaviour through discipline and teachings.

Physical hazards of early childhood include illnesses, accidents, obesity, left-handedness, etc. Psychological hazards come with unsocial content of speech, inability to establish empathetic behaviour, failure to learn social adjustments due to lack of guidance. Children at this stage are not expected to be moral, but their behaviour needs to conform to the moral standard of the social group. So they are taught to feel guilt and shame towards the undesirable behaviour; different sorts of reward and light punishment are chalked for them to convince the gap between the socially approved behaviour and wrong behaviour respectively. Some special interests are found to grow at this time in children. They may show curiosity in myths, religious beliefs and observances, even think about the mysteries like birth and death. They often get surprised with their own body mechanisms and hold an egocentric attitude.

(ii) Middle childhood

Middle childhood extends from the age 6 to age 12 years which corresponds to the school years. It is the period of eruption of permanent teeth, though not all erupt. It is also considered as troublesome age or the pre-gang age by some scholars.

The growth of this period is generally influenced by health, nutrition, immunization, sex and intelligence. Overall growth rate remains slow but steady. The growth in legs is most pronounced feature of this period and accounts for most of the increase in height. Children become expert in motor skills like throwing and running with a great perfection. The skills developed during this period can be categorized into four groups: self-help skills, social-help skills, school skills, and play skills. All areas of speech - pronunciation, vocabulary, and sentence structure improve rapidly. Comprehension and reading ability are also found to increase.

Emotional expression of middle childhood is intense although basic emotional pattern is almost similar to that of early childhood. In this period children are better equipped at recognizing emotions of others and empathizing with them. Older children learn to control the overt expressions of their emotions and to use emotional catharsis to clear their systems which are usually caused by social pressures. Their thinking develops rapidly and store of information grows at a fast pace.

During this period children develop closer ties with their peers. They compare their abilities, appearance and characteristics with those of their peers. Boys show more extensive peer relationship than the girls. Older boys are interested in activities with their peers and want to belong to a group or gang. These children often reject parental standards, develop antagonistic attitude towards persons of the opposite sex. There is rapid understanding of concepts as a result of intelligence and learning opportunities.

Children during this period develop moral codes influenced by moral standards of the groups they belong to. Still misdemeanors are found among some of them. This can be explained as ignorance i.e. either the children do not understand what is expected from them or they deliberately break the governing rules. The interests of older children are broader than those of younger children and include many new subjects like clothes, human body, sex, school, future vocation, status symbol and autonomy. Sex-role typing in middle childhood influences children's appearance, behaviour, aspirations, achievements, interests, attitude towards opposite sex persons, and self evaluation. The self concept and self esteem which develops greatly in middle childhood links to cognitive, emotional, social, and personality development.

- **Adolescence**

Adolescence is the threshold of adulthood. It extends from twelve years to eighteen years i.e. it starts from the time of puberty and continues up to the onset of adulthood. This is the period when the individual becomes sexually mature; the hormonal influences play a leading role in order to attain sexual maturity. This period in the life-span of the individual is very important, as it is a transitional phase between childhood and adulthood; a time of huge changes. During adolescence puberty is a short phase, which occurs at different ages for boys and girls. The average age in girls it is thirteen and is marked by the onset of menarche or first menstruation. In boys it comes approximately at the age of fourteen years and is determined by the nocturnal emissions. Puberty is marked by an acceleration of growth which is known as adolescence growth spurt. The adolescence spurt is a constant phenomenon and occurs in all children, but varies in intensity and duration from one child to another. Puberty can be equated to the functioning of the pituitary gland; it is the time when the pituitary gland at the base of the brain begins to send signals to the sex glands to increase of hormones.

Adolescence is therefore intimately linked to the secretion of sex hormones both in boys and girls. The male sex hormone is testosterone secreted by the gonad, testes. The female sex hormone is oestrogen secreted by ovary. The gonads, ovary and testes are stimulated by pituitary hormones, called Follicle stimulating hormone (FSH) and Luteinizing hormone (LH). Together, these are called gonadotrophins. The pituitary hormones, FSH and LH, in turn are controlled by hormones of specialized region of brain, called hypothalamus. Hypothalamus secretes gonadotrophin releasing hormone (GnRH). Under appropriate internal and external message GnRH is released from hypothalamus that has effect on pituitary causing release of gonadotrophins. Gonadotrophins act on gonads and effect release of testosterone and oestrogen in males and females respectively.

Differentiation in primary and secondary sexual characteristics marks the adolescence period. There are also other changes like maturation of reproductive organs, sharp growth in height and weight, rapid muscular and skeletal development, alteration on a variety of physiological functions. Men become considerably larger, acquire broader shoulders and a deeper larynx; women enlarge their pelvic diameter and deposit fat in various strategic places, including the breast. Changes in puberty are also influenced by hereditary factors and various environmental factors like nutrition, health, emotional stress, etc. Puberty changes affect physical well-being, attitudes and behaviour of a child.

Adolescence is not only a period of great physical changes, it also ushers a lot of psychological, social and emotional changes that leads to legal maturity. A great change is noticed in interest and capacity. Behaviour pattern and value system also changes; adolescents are capable of abstract reasoning, sophisticated moral judging, and realistic planning for future. Early adolescence is called a Gang Age because peer group influence is tremendous at this time. They tend to spend most of the time out of home being closed to friends. This is the time when adolescents hate wise counselling of parents and elders, rather put importance on friends' experience. Their attitudes, speech, interest, apparel, behaviour – everything depends on the nature of friend circle. If friends are addicted to alcohol, drugs, tobacco or other narcotics, an adolescent is keen to imitate that. Another new feature of this time is to take ample interest in opposite sex; they want to follow the gender roles demanded on them by the society.

The developmental task of adolescence calls for overcoming childish attitudes and behaviour. Some adolescents reach maturity within short time showing excellent mastery on developmental tasks that to be carried on in adulthood. The universal interests of adolescences are recreational interests, personal and social interests, educational interests, vocational and religious interests and interests in status symbols. It is a period of heightened emotionality; its intensity depends on the degree of physical as well as psychological changes. Adolescent emotionality can be attributed mainly to the fact that boys and girls come under social pressures and face new conditions for which they received little preparation during childhood. They are now supposed to make adjustments to new patterns of behaviour in order to fulfil new social expectations. Early adolescent emotions are often intense, uncontrolled and seemingly irrational, but improvement is noticed with each passing year.

In the late adolescence, peer group influence begins to wane. At this time ego develop in them and they look for self identity. No longer had they belonged to large group activities, number of companions gets limited, and they want to belong to selective social group and indulge themselves in various formal social activities. Social insight and social competencies increase. This change of outlook and interest is determined by sex, intelligence, social environment and cultural background in which they live.

There are changes in morality in this period shifting from specific moral concepts to generalized moral concepts of right and wrong and there is a control on their behaviour by the development of conscience. Adolescence is also projected as a period of 'storm and stress'. Relationship with family members is usually strained; they feel that parents are unable to understand them. Strong desire for independence leads to many clashes with parents and other adults who are in authority. Stress is also associated with educational and vocational issues. Adolescents generally aspire unrealistically high; therefore often fail to reach their goal and suffer from dissatisfaction. Without achievement neither success and satisfaction nor social recognition and prestige can be earned. Overall personality level shifts in adolescence. There are constant attempts of improving personality. This comes from within of an individual. An adolescent sets ideals, takes an account of own strength and weakness, improves self-concept, and tries to reinforce favourable behaviour for an honourable acceptance in society.

• Adulthood

The term 'adult' means 'grown to full size and strength' or 'matured'. Adults are, therefore, individuals who have completed their growth and are ready to assume their status in society as full grown individuals. Adulthood is a period of adjustments to new patterns of life and new social expectations. An adult is expected to play new roles, such as that of spouse, parent, and breadwinner and to develop new attitudes, interests, and values in keeping with these new roles.

Adulthood is the longest period of the life span – a period of more or less stability. Childhood and adolescence are the periods of 'growing up'; adulthood is the time for 'settling down'. It begins at about 19 or 20 years and lasts up to 65 years approximately. Changes in adulthood are more gradual and less dramatic than in childhood. This period is further subdivided into three periods – early adulthood, middle adulthood and late adulthood. During the long period of adulthood, certain physical and psychological changes occur at predictable times.

(*i*) *Early Adulthood*

Early adulthood extends from age 18 to 40 years. It is popularly known as 'settling down age' because at this time human beings have to assume various responsibilities to settle down in broader life. Young adults remain at the peak of their strength, energy and endurance. By this time their body functions are fully developed. With higher physiological and psychological capacities they enter into a new age called the reproductive age. It is a time of commitments and independency. This period is crucial because until now most boys and girls have had someone by their side like parents, teachers, friends or others who help them in making adjustments in the environment. But, as adults, they are expected to make all adjustments for themselves.

Parenthood is probably the most important phase which comes in the lives of most young adults. In the beginning of adulthood, most men and women prepare themselves for marriage, parenthood, and jobs. The early adult years pose many new problems which are categorically different from the problems experienced in the earlier years of life. Once individuals decide upon the pattern of life which they believe will meet their needs, they develop another sets of behaviour, attitudes and values which will tend to be characteristically theirs for the remainder of their lives.

Social activities in this period are mostly reduced because of family, vocational and professional responsibilities. This is the time when individuals have to plan their professional life and settle in an occupation. Social mobility in men most of the time is due to their own efforts and hard work. Vocational adjustment can be very demanding, which includes selection of a vocation, settling down in their occupation and adjusting to respective work environment. The vocational adjustment can be seen as per achievements made by the individuals, frequent changes in their jobs and the job satisfaction experienced by the individuals.

Family adjustment in this period can also be very difficult. There are many changes in the family roles and responsibility. Marital adjustment is one of the most difficult adjustments which young adults have to make. During the first one or two years of marriage, the couple normally makes major adjustments to each other, to members of their families, and to their friends. While these adjustments are being made, emotional tension may arise and this is understandably a very significant period. After adjusting to each other, their families, and friends, they have to look for parenthood.

In some cases like lack of preparation for marriage, early marriage, unpractical and too much of romantic ideas of marriage, and a lot of role changes can cause many problems after marriage. Parenthood automatically brings certain changes in attitudes, values, roles and responsibility. Women have to make a lot of changes for adjusting to the new family, especially if it is a joint family and also at the arrival of the children. Many other factors control the adjustment to parenthood, like attitudes towards pregnancy and desire for parenthood, age of the parents, sex of the children, spacing between children, parental expectation and the children's temperament. On the other hand, marital faithfulness and trust establishes the foundation in marriage. Communication between the partners is the key area.

Success in marriage can be judged by several criteria like husband- wife compatibility, cordial parent-child relations, good adjustment with in-laws, ability to cope up with differences, good handling of finances, we feeling, etc. Single parenthood among men and women has now been more acceptable than in the earlier times.

(ii) Middle adulthood

Middle adulthood or middle age is an intermediary stage between adulthood and old age. It begins at 40 and extends to 60 years. It is characterized by different achievements - professional and otherwise. This time of life is evaluated by introspection. It is also called Empty Nest period as children often leave home. Many observers view this period as a time of stress, which is commonly referred as 'Middle age' crises.

The developmental tasks of the middle aged adult are centered on success in career, adjustment in marriage and satisfaction with children. At work an individual attains great achievements and experiences a climax. Depending on the foundations, the marriage may be shaken up or strengthened. The relationship with children assumes a new dimension as they too start early adult life.

Physical and psychological decline is observed in this period. Middle age can be a difficult time and successful adjustment to this stage depends on the essential base laid down in the earlier stages. How well a person has mastered on his civic responsibilities in the early years of adulthood will influence his degree of success during middle age and this will also determine that how happy he will then be as well as during the closing years of his life. It is a period of achievements and evaluations. This age has been taken as the dreaded period of transition and stress.

Sometimes Middle adulthood or middle age reveals as a time of boredom and emptiness as children have grown up and moved out for higher studies or employment, or married off. The individuals suddenly feel lonely and do not know how to use their free time. Physical changes are drastic in appearance, physiological functioning and sexuality reduces to a great extent. Adjusting to altering situations becomes very difficult for most of the individuals. In women the menopause sets in for reduction in estrogen and that cause psychological stress. Physiological and psychological changes are also found among the males which affect their attitude, behaviour and self-evaluation. Now a day, a person seems to be more successful in adjusting to middle age if he can hide his physical signs of aging and be involved with ways that keep him young and trim.

Mental decline starts in the middle age. The interests in religion are also on the increase during this age. Most of the people in this age group show changes in the pattern of recreational activities. Although an interest continues in physical recreation, but that involves fewer persons; individuals at this time are more adult oriented than family oriented. Further, social interests and activities are greatly modified by social-class status, sex, and marital status.

(iii) Late Adulthood (Old Age)

Late adulthood or old age begins at 60 and extends till death. Physical and psychological decline fastens up in this period and the period is known as 'senescence'- a time of growing old or advent of ageing. Decline comes partly from physical and partly from psychological factors. The physical cause of decline is the change in the body cells, not due to a specific disease, but for degeneration. Psychological causes of decline include unfavourable attitudes towards self, other people, work, or life in general. This can lead to senility, just as changes in the brain tissue can. Individuals who have no sustaining interests after retiring from work are likely to become more depressed or disorganized.

There are individual differences in the effects of ageing. People age differently because they have different hereditary endowments, different socio-economic and educational backgrounds, and different patterns of living. These differences are apparent among members of the same sex, but they are even more apparent when men and women are compared. Ageing takes place at different rates for the two sexes. How an individual copes with the strain and stresses of everyday life will also affect the rate of decline. In modern times, medical techniques, cautious clothing and grooming make many men and women to look, act, and feel as they did when they were much younger.

The physical changes first mark by changes in appearance. Different internal systems change and therefore physiological malfunction, sensory and sexual changes are noticed. The changes in motor capacities signify changes in strength and speed; more time is needed at this time to learn new things or to cope up with new set up. There are various causes for the changes in the mental abilities. Most important among these are lack of environmental stimulation and lack of motivation to be mentally alert. Changes in interests are also caused by the reasons like deterioration in health and economic status, change in residence and marital status and change in values. Leisure time increases and loneliness appears. Bent of mind goes towards religion, perhaps due to the concern over death.

The employment opportunity for older workers are restricted by compulsory retirement, hiring practices, pension plan, social attitudes, sex of the workers and the kind of work. Retirement causes change in roles, interests, values and overall life pattern. The pattern of family life established in early adulthood starts to change with the onset of middle age. People who felt happily married in early adulthood may find their marriage become more satisfying to them as they grow older. With time mutual interests develop; when children grow up and leave home, partners come closer together. Illness or retirement on the part of the husband may make the wife feel useful again, as she was before. Satisfaction with marriage among older people also increases if their children are successful and happily married, and if they can maintain good relationships with their grandchildren, even if the contacts with them are infrequent.

The most common living arrangement for the elderly in our society is that an elderly couple can live either alone or with their married son and family. The same applies for an elderly widow or widower. Recently, 'homes for the elders' have been introduced in our society; many helpless elders are found to live in those homes. Certain problems in adjustment specific to this stage are physical and economic dependency on others, difficulty in establishing new contacts, etc. Old people are supposed to behave in a mature manner to their own children who by this time have become adults having different interest and outlook. Common physical hazards at this stage are diseases, physical handicaps, malnutrition, accidents and sexual deprivation. Reduction in income and loneliness add fuel to fire.

Meeting social and civic obligations is difficult for many older people as their health fails. Sooner or later, most old people have to adjust to the death of a spouse. It may also necessitate changes in living arrangements. As grown up children become increasingly involved in their own vocational and family affairs, the elderly can count less and less of their companionship. This means that they must establish affiliations with members of their own age group if they are to avoid loneliness.

It is usually expected that old people will play a decreasingly less active role in social and community affairs as well as in the business and professional worlds. Because of unfavourable social attitudes, few rewards are associated with old age roles, no matter how successfully they are carried out. At times, from a feeling of uselessness and unwanted, elderly people suffer from inferiority and resentment. These feelings are not at all conducive to good personal and social adjustments. With poor self-concept, ageing people are expected to adjust with decreasing strength and gradually failing health.

In the present world of increasing life expectancy there should be a call for revisions in the roles of the senior people. They should be helped to evaluate their shortcomings from a different dimension. Attitude of the community must be modified towards the elderly people. They need to be encouraged to contribute in the society afresh with appraisal. New interests have to be grown among them so that they can spend their time in a productive manner. It is an established fact that in spite of an overall deterioration of physical and mental skill, knowledge and wisdom greatly increase in old age.

Senescence

After the active phase of the life-span, declining process starts resulting in old age or senescence. During this period many molecular and cellular changes occur. Organismic changes are also included here. These changes are measurable and can be explained, but these do not exhibit any specific pattern or well-defined sequence. In some individuals the changes are fast; in others slow. Individuals are characterized by several senescent characteristics e.g. graying of hair, loss of strength, reduction in sensory capabilities, poor homeostatic mechanisms, reduced resistance of body against diseases, cardiovascular irregularity etc.

The term 'senescence' is used when talking about the changes which occur during the period of obvious functional decline in the later year of life-span. Some people prefer to use the term 'ageing' for the same process and period. Actually ageing means simply growing older; any change related to age can be designated as aging. Thus, the onset of puberty may be described as an ageing change, but that is definitely not a senescent change.

According to Comfort (1960), "Ageing is an increased liability to die, or an increasing loss of vigour, with increasing chronological age, with the passage of the life cycle."

According to Strechler (1962), "Senescence is the changes which occur generally in the post-reproductive period and which result in a decreased survival capacity on the part of the individual organism."

According to Maynard Smith (1962), "Ageing processes are those which render individuals more susceptible as they grow older to various factors, intrinsic or extrinsic, which may cause death."

From the above definitions it is evident that senescence has at least three cardinal characteristics. (*i*) The changes that occur during ageing are quite harmful; they increase the chances that an animal will die i.e. ageing involves a decrease in the ability of an animal to cope with its environment. (*ii*) Death is the ultimate result of ageing. It is sudden, but the process of ageing involves a progressive increase in the probability of dying. (*iii*) The process involved in aging is common to all members of a species having universal inescapable consequence of getting older. Changes are,

1. Decline in metabolic efficiency.
2. Decrease in the power of replacing worn out old cells and repairing the damaged tissues, organs and organ systems.
3. Deterioration in the structure and function of body.

On the whole, ageing and senescence are fundamental intrinsic properties of living organisms.

There are few theories to explain the process of ageing:

1. Ageing takes place due to interaction of the genetic material and the environment; it is expressed as a product nature and nurture reaction.
2. Ageing is caused by accumulation of some harmful products of metabolism in the cells and in intercellular spaces. In other words the internal environment of cells undergoes adverse changes resulting in ageing.
3. Ageing is due to intrinsic property of the genetic material (DNA) i.e. the program already exists in the body as a predetermined property of life.
4. Wear and tear occurs in the tissues of the body for continuous and constant usage which is not replenished in the long run.
5. Aging is a resultant of non-stop metabolic activity. Very high rate of metabolic activity advances old age; its slow rate enhances the life-span.

Visible changes due to ageing are morphological and physiological. Because of ageing the tissues do not renew and as a result cells show senile involution. The memory declines; aged persons need more time to learn and to react. The speed of conduction in motor nerves also shows a decline. There is reduction of density in long bones and vertebrae, and therefore, height and sitting height show decrease. Muscles become weak and bones get brittle. Bone marrow does not produce as many new cells as they could before when the body was young. Vital capacity and muscle tone declines. A stooping body with dried wrinkled skin is often found in old persons. Cells cannot retain water causing the skin dry. Less volume of blood and less urine is formed for the same reason. Heart's efficiency to pump blood diminishes. Brain and kidneys receive smaller quantities of blood. Eye sight fall short. Dental decay and disorders may cause chewing difficulties and speech problem. Hair becomes thin and turns gray. The number of taste buds in the tongue gradually diminishes with age. Inactivation of certain useful enzymes of the body and production of some defective proteins at the same time increase defects of the DNA structure. Some cells of the brain accumulate worn out cell pigments. This signifies that the cells in different parts of the body lose their efficacy. The intercellular proteins especially collagens undergo a marked change in their constitution. These proteins which are permeable, flexible and easily soluble in younger person become less permeable, rigid and insoluble as age advances. Such property changes affect gaseous exchanges and expulsion of nitrogenous wastes. Gerontologists have worked out on a large number of changes in cells and in extracellular substances of different tissues.

Life expectancy has been increased dramatically since 1900. The number and the proportion of elderly people have increased a lot. Although above 60 is the traditional concept for late adulthood, many old people are found belonging to 80s and 90s. Social scientists specialized in gerontology[2] advocate three groups among the older adults e.g.' young old' (from 65 to 74 years), 'old old' (from 75 to 84 years) and 'oldest old'(from 85 and above). Researches prove that those who belong to the 'young old' group usually remain healthy, vigorous and active. Those who belong to 'old old' group are not that sort of active like the people of previous group. Finally, those who crosses 85 years, are really infirm having difficulty to manage own livelihood. This is actually a classification of older people by functional age. Functional age may not always proceed in the same pace as Chronological age. This means, a person at the age of 80 may maintain such a good health that he may look functionally as a young old. Again, it may be the case that the chronological age of a person is low, but his biological age is much advanced. Biological age remain lower than chronological age for those who mature late.

FACTORS AFFECTING GROWTH AND DEVELOPMENT

There are two basic determinants of development – heredity and environment. A complicated mixture of hereditary and environmental influences probably acts throughout the whole period of growth. This is the great 'nature-nurture' debate which puzzles the scientists. The prenatal period sets the stage for the rest of a child's development throughout life. Adjustments to postnatal life depend on prenatal environment, the type of birth and experiences associated with it, length of the gestation period, parental attitudes and postnatal care. A healthy prenatal environment contributes to good adjustments in postnatal life. We should therefore take into consideration of both the stages – Prenatal and post natal and search for factors which are detrimental or conducive for growth and development.

Heredity (Genetic Foundation)

New life begins by the union of a male sex cell (XY) and a female sex cell (XX). These sex cells are available in the reproductive organ - gonads. Male sex cells are known as spermatozoa (spermatozoon); these are produced in the male gonads called testes. On the other hand, female sex cells known as ova (ovum) are produced in the female gonads called ovaries. Both the male and female sex cells are similar in terms of their content. Each mature sex cell holds 23 chromosomes containing genes, the true carriers of heredity. In the light of modern scientific discovery, it can be

2. It is the study of the aged and aging processes.

said that each chromosome consists of long, double stranded molecules of chemical substances called deoxyribonucleic acid or DNA[3]. DNA carries biochemical instruction and determines all hereditary characteristics.

Normally all body cells of men and women caries 23 pairs of chromosomes i.e. 46 chromosomes; half comes from father and the other half comes from mother. Among these, 22 pairs are autosomes have no relation to sexual expression; rest one pair, XX in females and XY in males take part in sex determination. When an ovum (X) is fertilized by a sperm carrying (Y)[4] chromosome, the zygote is formed with XY combination i.e. denoting a male. But when an ovum (X) is fertilized with a sperm carrying (X) chromosome, the zygote is formed with XX combination and a female is produced. Thus, the genetic endowment is the contribution from both the parents. Genetic transmission not only determines the sex of the individual, it dictates on genotypes and phenotypes (See also Chapter No. 6). Expression of different physical and mental characteristics as well as capacities is controlled by the interactions of several genes, dominant and recessive in character. Genetic configuration of a person is entirely a matter of chance; it cannot be controlled by any means. There is no way to know which sets of genes, whether from the maternal or paternal side, will be passed on to the child.

Heredity affects and limits the later development beyond which no individual can go. If prenatal and postnatal conditions are favourable, and if individuals have strong motivation, they can develop their inherited physical and mental traits to their maximum potential, but can never alter them. Biologically speaking heredity is the sum total of traits potentially present in the fertilized ovum (Combination of sperm cell & egg cell), for which off-springs resemble to their parents and grand-parents. It greatly influences different aspects of growth and development i.e. height, weight and structure of the body, eye colour, hair colour, hair texture, intelligence, aptitudes, instincts and so on.

Gene depends for its expression firstly on the internal environment created by the interactions of all genes, and secondly on the external environment. Determination of body size is certainly a complicated affair involving many genes, yet a disturbance in a single gene or group of genes may produce widespread and drastic damaging effect. The height, weight, body-build of a child is controlled by both the genetic and environmental forces, especially by their interaction. The gene based control is found on dental maturation and skeletal maturation; the timing of ossification is partly influenced by genetic factors and partly by environmental factors.

Genetic influences continue throughout the entire period of growth. Chromosomal abnormalities strongly suggest genetic command on growth. Genetic factors play leading role to create differences between male and female patterns of growth. They keep influence on the rate of growth; both the early matures and the late matures are within their grip. It gives us an understanding of how genetic inheritance of disease occurs and affects a person's development and health experience. Genetic materials passed to the developing infant remain responsible for some birth defects or developmental disorders. The National Institute of Child Health and Human Development has identified several classes of developmental disorders with strong linkages to the transfer of genetic material; neurological problems like Down Syndrome, metabolic disorders such as hypothyroidism, degenerative conditions including muscular dystrophy, etc can be linked to genetic disposition. In the same way, behind the cognitive development specific genetic factors work along with other biological mechanism, as is seen in many genetic causes of mental retardation.

---

3. Watson and Crick in 1953 discovered the chemical structure of DNA molecule and unlocked the genetic code. Genes are segments of DNA along the length of the molecule. It gives a string like look. Genes pass on from parents to offspring with specific genetic instructions.

4. The Y chromosome contains the gene for maleness.

Environment

Environment plays an important role in human life. Environmental factors refer to external substances or conditions that stimulate human growth and development. Therefore, a person's environment consists of the sum total of the stimulations which he receives from outside. Environment may be of different types such as physical environment, social environment and psychological environment.

Physical environment is composed with all outer physical surroundings both in-animate and animate which are manipulated by the person in order to ensure food, clothing and shelter. Geographical conditions e.g. weather and climate comes under physical environment having considerable impact on a growing child. Social environment is constituted by the society. Individuals, groups, institutions, laws, customs, religion – everything keep an impact on the child and his behaviour is regulated. Psychological environment is rooted in individual's reaction with an object or event. Love, affection and fellow feeling attitude strengthen human bond with one another. Growth and development cannot ignore these stimulations of environment by any means.

Environmental Influence in Prenatal Stage

The physical environment of a child during prenatal phase is undoubtedly the intrauterine environment. The child is closely linked to this environment up to his or her birth.

There are several environmental agents (factors) that may cause damage during the prenatal period and affect the later development. These factors are collectively referred as Teratogens. A teratogen is a toxic agent, which can cause a birth defect or malformation. It could be a chemical, virus, drug, alcohol use, radiation, maternal infection or a disease present in the mother who could increase the chance for the baby to be born with a birth defect. Sometimes a medication necessary for health is also a teratogen: thyroid medication, blood thinners, and lithium are just a few examples. About 4 to 5 per cent of birth defects are caused by exposure to a teratogen.

The harmful effects of Teratogens depend on the factors like amount and length of exposure, combination of few such agents, genetic constitution of the developing baby, age of mother, age of embryo or fetus at the time of exposure. They can deter the normal growth, can bring defect in bodily structure, can create different type of abnormalities and behavioural disorders and even can pose psychological problems. At the zygote stage, the organism is rarely influenced by the teratogen. Maximum susceptibility to teratogen is likely in embryonic and fetal stage. In fact, after fertilization, when the fertilized egg is connected to the uterus, a common blood supply exists between the mother and the embryo. If any agent dissolves in the mother's blood at this time, that can reach to the developing fetus. Teratogens are therefore thought to have the ability to affect the fetus from two weeks onwards after conception.

The commonly found Teratogens are as follows:

- Drugs (Some prescription drugs and Illegal drugs)
- Hormones
- Narcotics
- Nicotine
- Caffeine
- Alcohol
- Radiation

Other harmful factors include maternal factors such as maternal disease, mother's nutrition, mother's age and physical activity, incompatibility to blood types, emotional stress of the mother, etc. Further, growth and development may be arrested for the Lack of medical care (due to poverty or isolation or other reasons), for the Lack of social support, and even for the lack of appropriate information.

Drugs

Gross anatomical defects of limbs may be caused for taking a drug called Thalidomide by the mother during pregnancy. Respiratory troubles in children are seen if mother has taken barbiturate drug during labour period to ease pain and distress. Effects of anesthetic drugs go to the sensory-motor function of the newborn causing damage to the muscular, visual and neural function. Besides, many other drugs, apparently seem to be non-harmful, e.g. some antibiotics, analgesics, anticoagulants, tranquilizers, vitamins, etc can create birth defects if administered during pregnancy. Some women are in the habit of taking wide variety of drugs, they continue this habit during pregnancy and naturally the severe effects fall onto the developing embryo and fetus.

Hormones

There are several hormones which doctors had used injudiciously and indiscriminately to the women. They include birth control pills, progestin, diethylstilbestrol (DES), androgen, synthetic estrogen, etc. Among these, diethylstilbestrol (DES) was most widely used as a drug to prevent miscarriage. Since back kick of taking a drug during pregnancy do not always show up immediately, doctors could not got the view of its disastrous effect. It has been found that some daughters of those women (who had taken diethylstilbestrol (DES) during their pregnancy) mostly have developed a rare form of vaginal or cervical cancer reaching at their puberty and others run higher risks of miscarriage or premature delivery.

Narcotics

Use of narcotics is continuously increasing in women. Chronic user of Heroin and methadone, are always at risk for a variety of problems namely, premature birth, physical malformation, respiratory distress, and increased mortality at birth. Infants exposed prenatally to either heroin or methadone become physiologically addicted; they also show the symptoms of drug withdrawal e.g. restlessness, irritability, tremors, vomiting, fever, shrill cry, convulsions, etc. Cocaine and marijuana used by the pregnant women similarly produce harmful effect in children. Low birth weight, small head circumference, sometimes acute lymphoblastic leukemia, a childhood cancer may occur. Nervous system of the infants is also highly affected. Women addicted to morphine and codeine is likely to bear premature addicted babies with profuse neurological problems.

Nicotine

Cigarette smoking has surely an adverse effect on fetus. Heart rate of fetus is often accelerated. Vigorous smoking by pregnant women increases the incidences of death; an apparently healthy infant may unexpectedly die. Otherwise, smoking during pregnancy often retards physical growth of the fetus, lowers the newborns' birth weight and their subsequent resistance to illness. It may also enhance the chances of spontaneous abortion.

Caffeine

Although no clear evidence regarding the adverse effect of Caffeine on the fetus has been shown, but heavy intake of coffee, tea, cola or chocolate may increase the risk of miscarriage. So a pregnant woman should restrain herself from the food and beverages containing caffeine.

Alcohol

Heavy alcohol consumption by the mother during pregnancy usually produces severe as well as complex set of handicaps for the children. Fetal alcohol syndrome (FAS) is produced if maternal consumption exceeds 3 ounces of absolute alcohol per drink, and if she takes 4 to 5 such drinks per day. Due to Fetal alcohol syndrome (FAS) children suffer from slow prenatal and postnatal growth, undergo facial and bodily malformation, congenital heart disease and disorders of central nervous system. The central nervous system (i.e. baby's brain and spine) is the most sensitive area to teratogens throughout the entire pregnancy. As a teratogen, alcohol affects this central nervous system. At any time during the pregnancy, it has the potential to cause birth defects and health problems in the baby. Even smaller doses of alcohol, one ounce or more a day, increase the possibility of low birth weight, developmental delay, difficulties in breathing or sucking, and spontaneous abortion. This is why alcohol consumption should be avoided during pregnancy.

Radiation

The other potential source of birth defect is radiation or X-ray of mother during pregnancy. For a long time we know that radiation can cause gene mutation. It can induce greatest damage to the developing embryo and fetus, especially during early pregnancy. So any radiation exposure should be avoided by the expectant mothers, whether it is for diagnostic reasons, or for treatment of pelvic cancer. Sometimes radiation may be indirectly received from occupational hazards or from atomic energy sources; pregnant women must remain conscious of that. Apart from spontaneous abortion, radiation can give rise to structural malformations, brain damage, microcephaly, increased susceptibility to certain forms of cancer, shortened life span and even mutant genes, effects of which can be felt through generations. Frequent X-ray treatment is never recommended for carrying mothers.

Children exposed prenatally to heavy metals like lead, mercury, etc always show lower intelligence and poor level of performances. These children are generally subjected to all sorts of childhood illness. Mothers who deal with pesticides during pregnancy face the same problem with their children. High levels of lead exposure have been linked to premature birth, low birth weight, brain damage and a wide variety of physical defects. Ionizing radiation is most dangerous as it can cause mutation damaging the DNA in ovary and sperm. Miscarriage, slow physical growth, under developed brain and malformations of skeleton, etc are the other evil effects. A child may appear normal at birth, but the possibility of later problems cannot be ruled out.

Infectious diseases

Disease producing microorganisms are toxic to the embryo and so they cause abortion rather than malformations. In early pregnancy, placenta acts as a barrier against harmful agents but later it allows many substances to reach to the fetus. Antibodies produced by the mother to combat against infectious diseases are transmitted to the fetus for producing immunity at birth. Mothers contracting infections therefore pass on to the developing child and create serious damages. Viral diseases namely Rubella (German measles), chicken pox, hepatitis, etc are very dangerous during embryonic and early fetal period. If mother suffers from Rubella or German measles in early pregnancy, the child is likely to undergo deafness, mental retardation or heart trouble; viral infections may also produce deaf-mutism, cardiac lesions, low intellect in children. The Harpes viruses are found to be associated with a variety of congenital malformations. Mothers who suffer from Syphilis or other sexually transmitted diseases normally face miscarriage. If survive, children become extremely week and mentally deficient.

Among the bacterial and parasitic infections, the most common one is Toxoplasmosis, which is acquired for eating raw and uncooked meat (or for the contact with infected cats). Toxoplasmosis damages eye and central nervous system of the developing child in the womb. Besides, there is Acquired immune deficiency syndrome (AIDS), the most threatened disease of today caused by human immune deficiency virus (HIV). This virus undermines effective functioning of the immune system; like other viruses it can travel across the placental barrier to the fetus and so offspring of the mother who carry this virus have every possibility of contracting the disease.

Malnutrition

Mother's nutrition keeps an effect on the growing fetus. Adequate and well balanced diet is necessary for all women during pregnancy. Because, the food supply of growing fetus is linked to his mother; different nutrients of food comes through mother's blood stream, via the semi permeable membranes of the placenta and the umbilical cord. The women who are ill and undernourished show a higher incidence of premature birth and mortality rate. Several other complications like anemia, toxemia, prolonged labour, threatened and actual miscarriages, etc are found among them. The babies born to such mother often suffer from the diseases like pneumonia, rickets, cold, bronchitis, etc after birth. Prenatal malnutrition can pose severe consequences for the development of central nervous system. Although these consequences are difficult to demonstrate clinically, but brain cells are surely reduced for poor nutrition. Weight deficits are also recorded. The higher the nutritional deficiency, the greater is the loss of brain weight, especially if malnutrition occurs in last three months of gestation period. Some cases of mental retardation (MR) is also said to be associated with maternal malnutrition in prenatal stage.

Expectant mothers who gain 10-12 Kg weight during their pregnancy are less likely to miscarry or bear babies who are stillborn or whose weight are dangerously low at birth. Care must be given to mother's diet by including vitamins and minerals (such as calcium, phosphorus, sodium, potassium, iron, etc) with proteins, fats and carbohydrates. Vegetables, green leaves, seasonal fruits, pulses and cereals in addition to milk, eggs and meat or fish provide a balanced adequate diet for the baby and the mother.

Age of the mother & physical activities

Maternal age is very important for the health and well being of children. Older pregnant women are prone to suffer various birth complications on account of diseases like diabetes, high blood pressure and severe bleeding. Naturally there is greater likelihood of miscarriage, premature delivery, still birth and other birth defects including retarded fetal growth. Older mothers show greater susceptibility to chromosomal abnormalities; those who are over 35 (at the time of pregnancy) are more likely to give birth to a Down syndrome baby. With age, reproductive organs also get weak and fertility tends to decline. Having a child at an older age poses high risk to both mother and child. Children of advanced age mothers generally become less intelligent and may exhibit severe morbidity which cannot be treated by nutrition or medical care.

Age between 20 and 35 is the ideal for childbearing. Adolescent or teenage mothers usually experience an increased incidence of pregnancy and delivery complications. This is for the immaturity of reproductive organs. Such pregnancy endangers physical health of both mother and the developing child. Nutritional requirements remain very high in teenage; teenager girls therefore need more nutrition at this time for their own growth.

It is well known that exercise is good for health. Normal work and regular exercise prevents constipation and improves respiration, circulation, muscle tone and skin elasticity which keeps women comfortable during pregnancy. So these are preferred for safer delivery. A healthy expectant mother can continue her routine practices like jogging, cycling, swimming, etc. But heavy exercise can endanger the developing fetus. When the mother is tired and over worked the fetal activity increases. If somehow exhaustion goes beyond limits, it can cause still birth or bring irritability in the child.

Incompatibility of Blood Types

Difference of blood composition between the fetus and the mother may lead to biochemical incompatibility. This is commonly created by the Rh factor, a protein substance found in the blood of most people. It should be remembered that majority of people are Rh-positive while only some show Rh- negative blood group. When an Rh-positive man is married to an Rh-negative woman, adverse situation is created for the offspring. If the baby carries Rh-positive blood, mother's blood tend to produce antibodies against 'foreign 'positive Rh factor. These antibodies are formed in mother's circulatory system and pass on to the fetus through placenta. Consequently, the red blood cells of the fetus are destroyed and that prevents supply of oxygen to the fetus. However, it gives rise to erythroblastosis leading to the death of the child in most of the cases. If survives, the child is likely to suffer from anemia, jaundice, deafness, heart defects, etc. Its extreme result may be cerebral palsy or mental retardation.

Usually the first born Rh-positive baby of an Rh-negative mother is not affected by the blood compatibility. With each successive pregnancy, the risk becomes greater. A preventive control for Rh problem has been developed recently. The blood of the newborn is tested immediately after birth (Blood sample is collected from the umbilical cord). If it is seen that an Rh-positive child has been born to an Rh-negative mother, a vaccine is pushed into mother's body within three days. This starts to destroy baby's Rh-positive cells before the mother's body begins to produce antibodies. This also serve as a protective shield for all other children born later. Children already affected by Rh incompatibility have to receive blood transfusion as long as they will survive.

Emotional stress

Although there is no direct connection between maternal and fetal nervous system, mothers emotional state can influence fetal development. When a mother is emotionally upset due to stress, tension or anxiety, blood supply to the fetus is not adequate and growth is hindered. This can result in premature delivery, still born baby or an irritable child.

There is a specific mechanism by which stress affects the developing baby. Different emotions of mother like rage, fear, anxiety, etc activates the autonomic nervous system and thereby liberate certain chemicals namely, acetylcholine and epinephrine into her bloodstream. Again, in this condition endocrine glands secrete adrenal hormones as a defensive response. As a result, the brain, the heart, the voluntary muscles of the arms, legs and trunk are stimulated; cell metabolism changes. Stress hormones in turn change the normal blood composition and altered substances containing stress elements are transmitted to the fetus through placenta. So whenever mother is under intense stress, the developing baby is also stressed. Whatever is the cause of mother's tension, her anxiety and depression is carried to fetus increasing the fetal heart rate and other activity level. This is not only a damaging factor for the fetal growth, but also harmful for future psychological adjustment of the baby. Prolonged and intense emotional stress during pregnancy may lead to catastrophic events in the life of the child. Such children show a wide range of problems like, low birth weight, hyperactivity, irritability, difficulty in eating and concentrating in studies, excessive bowl movements, gas pain, sleep disturbance, cleft-palate, etc. Marital discord, domestic violence, economic insecurity, etc are usually the reasons of maternal stress.

Delivery Complication

Prenatal development comes to an end with the onset of the birth process. There were five kinds of birth each with its distinctive characteristics. These are Natural or spontaneous birth, Breech birth, Transverse birth, Instrument birth and Caesarean birth. Sometimes the fetus may undergo difficulties and complications during delivery which can affect the normal intelligence and future development of the baby. Most crucial of these complications is the lack of oxygen supply; the situation is referred as anoxia. Anoxia is very dangerous to baby irrespective of the fact, whether the birth is normal and spontaneous or assisted. Usually in case of complications, special assistance is required to ensure good health and development of the new born.

Environmental Influence in Postnatal Stage

After birth, the environment may exert either a positive or negative effect on growth. There is no more mothers influence in postnatal stage. The child is now an independently growing organism. The environment changes from intrauterine environment to general environment. The factors which are important here are as follows:

- Nutrition
- Hormones & Exercise
- Disease & Injury
- Climate and Season
- Affection & Stimulation
- Socio-economic & Cultural Factors

Nutrition

Proper growth and health of a child stands on good nourishment. Good nourishment again depends on food habit. A balanced diet - combination of proteins, carbohydrates and fat along with necessary vitamins and minerals can promote growth with appropriate developmental milestones. This also ensures that the physical growth is not stunted.

An adequate supply of calories is essential for normal growth, although the need varies with the phases of development. Nine 'essential' amino acids have been claimed to be significant for growth and absence of any one of them will result in disordered or stunted growth. Besides, role of minerals and vitamins are immense. For instance, zinc plays a part in protein synthesis and it is a constituent of certain enzymes; a deficiency of zinc therefore causes stunting, interference with sexual development and falling out of hair. Iodine is needed for the manufacture of the thyroid hormones. Bone cannot grow properly without an adequate supply of calcium, phosphorus and other inorganic constituents such as magnesium and manganese. Iron is required for the production of haemoglobin. Among the vitamins, Vitamin A helps in vision. Vitamin C is said to control the formation of intercellular substance in bone. Vitamin D helps in mineralization of bones as well as calcium-phosphate metabolism; its deficiency is the cause of rickets.

Nutritional status of a child should be judged with care. Neither under-nutrition nor over-nutrition is good for health. Both are treated as malnutrition as keep adverse effect on the structural and functional development of the child. Just like physical development, a child's cognitive development can be hampered by poor nutrition. Undernourished children may have cognitive deficits and later be diagnosed with learning disabilities or other intellectual impairments. During childhood malnutrition delays growth; in the years proceeding adolescence it delays the appearance of the adolescent spurt. Growth studies have demonstrated that malnutrition may cause serious impairment for growth; majority of malnourished children fail to achieve their full genetic potential of body growth. In underdeveloped countries, malnutrition plays a major role in inhibiting the growth process.

In developing countries, the most common nutritional problem results from a diet containing very low amount of protein. The condition is referred as protein- calorie under-nutrition, which appear in two forms of dietary diseases – marasmus and kwashiorkor. The disease marasmus is manifested in the first year of life. It originates from maternal malnutrition and thereafter the baby does not get sufficient breast milk or supplementary feeding and naturally undergo severe weight loss, muscular atrophy, and decrease of subcutaneous fat. On the other hand, kwashiorkor is not a product of starvation. It occurs after weaning, in the age between one year and three years. These children lack minimum protein in their food and show a peculiar type of swollen abdomen with fatty liver. The body remains extremely lean accompanied by swelling of face and limbs.

In affluent countries, overweight and obesity are serious nutritional problems related to maladaptive eating patterns and physical under-activity during childhood. High-calorie food with large amount of sugar and fat, and very few beneficial nutrients calls various types of diseases in near future. Moreover, the psychological and social consequences of obesity are very hard to tolerate both in adolescence and youth. The physical features shatter down the self-esteem, self-concept, and confidence; it creates only stress and brings trouble in relationships with family, friends and partners.

Hormones & Exercise

There are a number of endocrine glands inside the human body which are ductless glands. These are situated in some specific parts of the body and secrete chemical substances directly into the bloodstream. The secretions or chemical substances are called hormones which are regarded as growth promoting substance. Most of the hormones secreted by the endocrine glands play a significant role in regulating the pattern of growth and development as per instructions of the genes.

The most important hormone controlling growth from birth up to adolescence is somatotrophin – a growth hormone. This is in fact a polypeptide secreted by the pituitary. It contributes to the growth of bones and thereby increases the height of persons. Growth hormone controls the rate at which growth takes place and works until the steroid- induced adolescent spurt is started. Its administration causes the amino acids to be incorporated into tissues to form a new protein. It is also responsible for an overall growth in most of tissues including brain. Thyroid hormone plays a vital role throughout the period of growth. The activity of the thyroid is judged by the basal metabolic rate which decreases gradually from birth to adolescence. But the children who suffer from hypothyroidism show a delayed growth; skeletal maturity, dental maturity and growth of the brain all are affected. Individuals born with pituitary tumors or those who develop pituitary tumors in childhood may experience different type growth patterns. Tumors on the pituitary gland usually cause to release larger quantity of human growth hormone, resulting in 'gigantism'- an unusual type of excessive growth, opposite to 'dwarfism'.

During adolescence a new phase of growth occurs under the control of steroid hormones secreted by the adrenals and gonads. The gonads of both sexes secrete estrogens in small quantities from the time of birth onwards. At puberty the estrogen level rise sharply in girls and to a much more limited extent in boys; the sex differences is possibly due to an inhibitory hormone secreted by the seminiferous tubules of the testicle. Testosterone, produced by the testicle, is important in stimulating growth and useful for the greater growth of muscle. Gonadotrophins are responsible for the growth of both, ovaries and testis, and later on secrete amounts of estrogens and testosterone that facilitates the growth and development of secondary sex characters. The body constitution and structural growth of girls are different from boys. The boys in general are taller and stronger than the girls but Girls show rapid physical growth in adolescence and excel boys.

All hormones have the power to raise or lower the activity level of the body or certain organs of the body. The gland pancreas secretes pancreatic juice, not into the blood, but into the intestine. There it acts upon food and plays an important part in digestion. This pancreas also discharges into the blood, a substance called insulin. This is carried to the muscles by the help of blood and enables them to use sugar as a fuel to add more strength. If the pancreas fails to produce the secretions, the organism lapses to the unfavorable conditions of growth and development. Similarly, the adrenal glands (posited very close to kidneys) work. These make secretion of adrenaline, a very powerful hormone responsible for strong rapid heart-beat and release of stored sugar from liver; it is also found to control the blood pressure. Although growth is determined primarily by the amount of growth hormone produced by the human body, the production of growth hormone may be altered under circumstances.

The functional activities of the child come in the fold of exercise of the body. Play is a good exercise for growth and development in childhood. Repeated play and rest build the strength of the muscle. Muscular strength increases mainly due to better circulation and oxygen supply. The brain also develops through activity. A child does not play or engages himself in various activities knowing that it will help him in growing. Willingness in playing is natural and the urge comes deliberately in growing children.

Disease & Injury

Physical growth can be affected by illness or disease. Nourished children generally become less affected by common childhood illnesses like measles, chicken pox, viral fever, etc. But in malnourished children such diseases take a serious form. First of all, poor health always suppresses the immune system and these children remain more susceptible to infections diseases. Secondly, diseases by interacting with malnutrition produce detrimental effects for which growth is retarded. Illness, conversely, can influence the nutritional status of the children by reducing their appetite and limiting the absorption of nutrients from food. As a result, a perpetual vicious cycle is created with diseases and under-nutrition, and children of developing countries are affected most.

A number of infections and infestations are found to hamper nutrition metabolism, particularly the protein utilization, which is essential for growth. Parasitic infections decrease the velocity of growth. Because, the parasites (tapeworm, mosquito, etc) draw substantial quantity of nutrients from the host body for fulfilling own need. In gastro-intestinal infections, the pathogenic organisms which enter into the body consume the nutrients disturbing the physical growth. Recurrent and persistent diarrhea leads to growth impairment. However, children who develop a severe illness in infancy or early childhood are more likely to demonstrate inhibited growth pattern. The integrated nature of growth and maturation is largely maintained by a constant interaction of genes, hormones, nutrients and other factors. Certain drugs, for example, the attention deficit hyperactivity disorder drug Ritalin has a growth suppressing effect on children; most of the cases administered with this drug have shown shorter and lighter physique in comparison to their peers of the same age.

On the whole, a clean environment free of germs reduces child's risk of illness and disease. Diseases are found to interact primarily with under nutrition to affect physical and mental health. Chronic illnesses generally have more ability to impair growth and development than the acute illness. Besides, there are the causes of injury. Any serious head injury may cause brain damage and affect mental development of a child. Fracture of the bone may hamper the skeletal growth.

Climate and Season

Body builds of people around the world can be related to the governing climate of the specific area. It is believed that the climatic factors in the origin had created the races when people could not cross their geographical boundary. In fact, climate has a great influence in determination of height, weight, complexion, features and body constitution. People of tropical area and the people of arctic region can be easily differentiated by their contrasting physical features. This distinction is apparent since childhood and it is held as results of evolutionary adaptation to climatic condition over generations. A European child will be white and tall; his or her hair colour, eye colour, facial structure will be governed by the characteristics of the same race. Some features develop to withstand the climate. For example, the short and compact body of the Eskimos are meant for heat conservation while greater body surface of African Negroes facilitates heat loss.

Seasonal influence can be noticed to some extent on human growth and development. Considerable seasonal variation is visible on velocities of growth in the temperate regions of the world. Children grow faster in summer than in the winter months of the year. Growth in height is very fast in spring and growth in weight is fastest in autumn. This is true for all ages, including adolescence. The mechanism of this seasonal effect is not known clearly; probably variations in hormone secretion are also involved here. Another important clue is to be noted that, infections and infestations are very common in hot and humid climate.

Climate though seems to have a very minor effect on overall rate of growth in man, it has been suggested that each major race of mankind varies in stature according to the climates in which they live. However, many studies have been undertaken to note seasonal variation of growth in mankind.

Affection & Stimulation

Children need parental affection as stimulation for normal growth and development. 'Failure to thrive' and 'Deprivation dwarfism' are two growth disorders resulting from the lack of maternal attention and affection. "Failure to thrive (FTT) is a non–specific term applied to infants and young children who are failing to grow in a normal fashion." (Smith, 1997) Most often, failure to thrive involves children under the age of three years. A diagnosis of 'failure to thrive' is not only based on the measurement of a child's weight and height on a given point of time, but is based on a pattern or trend of growth and that is consistently less than expected for a particular child. Loss of fat and muscle tissue gives the baby a wasted appearance; cognitive, social, and behavioral deficits are also significant there. Passivity and apathy develops towards the physical environment and the child exhibit little smiling and vocalization. The other disorder, the Deprivation dwarfism appears late between 2 and 15 years of age. As the name indicates, these children present very short stature. But weight remains proportionate to height and they do not seem malnourished. This disorder occurs mainly for extreme low level of growth hormone; emotional deprivation affects hypothalamic-pituitary communication and reduces growth hormone secretion. As a result, growth is inhibited. If these children can be removed from adverse social situation to emotionally adequate environment, the levels of growth hormone is quickly restored. But, if the treatment is delayed, development remains hampered and dwarfism can be a permanent feature. Maternal nurturing difficulties are generally associated with growth disorders which are again rooted in poverty and family disorganization. Different social problems like unemployment, job instability, crowded living condition, inadequate birth spacing, divorce, etc keep parents under acute stress, so they cannot meet up the emotional requirements of their children.

Children from broken home or orphanage do not grow and develop to an optimum level. There are many negative factors like unstable family, insecurity, sibling rivalry, loss of parents, inadequate schooling, etc. which may affect growth and development. Emotional deprivation resulting from rejection by family members leads to personality disorder. Emotional dissatisfaction is detrimental towards healthy personality development. Favourable interpersonal relationship is a good source of stimulation, congenial for better personal and social adjustment.

Socio-economic & Cultural Factors

The physical growth of human beings is definitely controlled by socio-economic factors. Children of different socio-economic levels differ among themselves in average body size and structure; this is true for all ages of both the sex. The higher socio-economic groups are always more advanced. They show a superior nutritional status associated with fewer infections. Their position is therefore better than their coevals in the lower socio-economic groups along the course to maturity. Reason is the life-style factors such as diet, exercise, recreation, comfort, access to employment-income- education, safety, security, satisfaction, healthy home and neighbourhood, etc which are all time adequate for them.

Size of family and home condition exerts profound influence on the rate of growth. In a large family with limited income children do not get proper nutrition. As a consequence the growth is delayed or stunted. If there are many children at the same time, all children do not get proper attention and individual care, and therefore show differential rate in growth. These children usually show light body and low stature. It has been observed that, only child develops more rapidly, both physically and intellectually than other children who have siblings. Children of affluent families may show a tendency towards obesity but the other facets of development like cognitive, linguistic, emotional, social, moral, etc remain absolutely normal. Urbanization also has a positive effect on growth. Better child care is supported by sufficient food supply, appropriate health and sanitation services, and a higher level of educational facility.

If we consider the factor of culture, we will find that culture differs from ethnic group to ethnic group. Each ethnic group has its own culture which shows unique food habit, typical method of child rearing and infant feeding, and different values and belief systems (e.g. taboo, totem, etc) attached to food items and thereby affect growth and development of the children. Further, each ethnic group follows some adaptations in different geographical areas and maintains a specific gene pool. Body structures keep a reflection of culture from childhood.

Secular trend in human growth

Secular trend is the trend that is observed on a span of several generations. Secular trend in human growth is the progressive pattern of change (over successive generations) in the age at which physical and behavioural milestones are achieved. It has been found that overall economic conditions of world have improved in last 100 years. Children of industrialized countries have acquired a larger body and taller stature more quickly than before; they went to the path of sexual maturity more rapidly than their counterparts in other underprivileged countries. This trend has been documented in most of the European countries, especially in Japan, North America, and Australia. The phenomenon is referred as "the secular trend" in growth. Recently, this secular trend has also been reported from some developing countries.

Although different hereditary and environmental factors activate during the prenatal and postnatal period, growth is the result of the concerted effect of a complex network of many regulatory factors with varying interactions. Growth potential is found to be modulated by a variety of social factors. Urbanization is one of them. The factor of urbanization is recently added here; it keeps a positive effect on growth being associated with better conditions of living. Optimal growth can only be achieved when all positive factors operate in harmony. Historical evidence suggests that secular increases[5] in growth do not occur in all countries. Since most of the under developed counuries are stricken by poverty[6], famine and disease; secular decreases are found there instead of secular increase.

METHODOLOGY FOR GROWTH STUDIES

Human development is the scientific study of processes of change and stability. It seeks to describe, explain, predict, and modify development. Ways of studying human development are still evolving, making use of advanced technologies.

Like all other scientific research, growth studies or developmental research begins with a hypothesis[7]. Scholars are interested in developmental changes which are systematic, coherent and organized. They want to know the various facets of adaptability, how a growing individual deals with ever-changing internal and external conditions of existence. Fundamental theories appeared as a result of the previous research. In areas where there is little or no existing theory, study starts with new research question. Based on this question or hypothesis, researcher selects his research strategy which is specifically an approach with selective methods.

Approaches in Research Design

Researchers use different designs and methods to study human development. Two basic kinds of research design are called, Correlational and Experimental.

Correlational Research

A researcher usually works with a specific problem to which he has to find out a solution. The clearer is his conception of the problem or question, the better will be the possibility of getting the solution. The question is based on the hypothesis which itself is derived from provisional conjecture or from established facts. This hypothesis is either proved or disproved by conducting research.

A Correlational study is an attempt to find out a relationship between two variables (factors that change). Variables may include characteristics, attitudes, behaviours, or events. The goal of correlational research is to determine whether or not a relationship exists between two variables, and if a relationship does exist, the number of commonalities in that relationship. Relationship between variables has to be found out as they happen to occur, without researcher's intervention. Correlation allows predicting one variable on the basis of another, the directions of which may be positive or negative. In a positive correlation, the values of the variables increase or decrease (co-vary) together. In a negative correlation, one variable increases as the other variable decreases. In a non-existent correlation, there is no relationship between variables. Correlational study suggests possible causes for outcomes, but never permits to draw conclusions on cause and effect.

---

5. It denotes Increase in height and weight. Average increase has been estimated as1 c.m. in height and 0.5 k.g. in weight per decade between 1880–1950 in the 5-7 years age group. For adolescent group, the same data increases to 2.5 c.m. and 7 k.g. per decade. Along with a fast reduction is noted in the age of menarche.
6. People who have less than 60 per cent of the income that an average person expects may be considered to be at risk of poverty and social exclusion.
7. A hypothesis is a testable statement that reflects what the researcher expects to find in a study.

Experimental Research

In contrast to correlational study, experiment is absolutely a controlled procedure, where the researcher manipulates an independent variable in order to see its effects on the dependent variables. So Independent variable is the factor which the researcher manipulates and dependent variable is the other factor which he observes. Dependent variable is expected to be influenced by the independent variable. In experimental studies, researcher takes special caution to control for unknown characteristics of subjects which can reduce the validity of research. Experimental research leads to conclusions regarding causation.

A number of factors can affect the outcome of any type of experimental research. For example, investigators face the challenge of finding samples that are random and representative of the population being studied. Additionally, researchers must guard against experimenter bias, in which their expectations about what should or should not happen in the study. Researchers should also control extraneous variables, such as room temperature or noise level, that may interfere with the results of the experiment. Only when experimenters carefully control extraneous variables can they draw valid conclusions about the effects of specific variables on other variables.

Experiments can be performed in the psychological laboratory or in the classroom or outside the class-room under rigidly controlled conditions. Two types of experiments are possible here - Laboratory experiments which involve high levels of control on the variables and Field experiments, based on rare opportunities to get the subjects in natural settings. Experimental researches with children may sometimes stand unethical because of their strong manipulative and exploitative character. Because, children are usually more vulnerable than adults to physical and psychological harm; immaturity may bring a threat to their psyche as research respondents. Although the studies, in which children are observed, questioned or tested psychologically possess least potential for harm. But occasionally some sentimental issues or embarrassing topics raised by the interviewer, like death of a parent or divorce can cause anxiety, stress and feelings of insecurity in children. Therefore, to check the situation, special ethical guidelines have been developed by the initiative of federal government, research associations, funding agencies and the like who guarded against stress provoking matters and thereby exposing the children under puzzled condition.

However, developmental research scientists are interested to find out how development takes place throughout the whole lifespan of individuals. To study the age effects, they have devised some special type of correlational design, often referred as Quasi-experiment. Quasi-experiments are like experiments (natural experiment) but lack the control base of true experiments. That is, the age cannot be contolled. Two **approaches** are very popular here.

(*i*) **Longitudinal study**

(*ii*) **Cross-sectional study**

Longitudinal study describes the age changes and the cross-sectional study describe the age differences. To overcome the limitations of longitudinal and cross-sectional studies, another study called Sequential study is designed.

(*iii*) **Sequential Study,** where individuals in a cross-sectional sample are tested more than once over a specified period of time. This is to determine differences in each age cohort over a period of time.

Besides, there are other two forms:

(*iv*) **Cross-cultural study**, where groups of different cultures are studied.

(*v*) **Co-twin study** is the fourth approach, which is purely a controlled experiment with children.

The first three approaches are contrasted with the experimental approach. The Longitudinal, Cross-sectional or cross-cultural studies may be thought of as "experiments done by nature." The investigator, for one reason or another, cannot manipulate any of the variables and must be content to identify important factors and observe relationships between them. However, true experimentation can often be done with children. Experiments with children, like all experiments, involve the manipulation of the independent variable, the measurement of a dependent variable, and the control of all other variables.

Longitudinal Study (Observe one group at different times)

In this approach the same group of individuals is studied over an extended period of time. It involves re-examining of the same people at intervals. Data is first collected at the commencement of the study, and then additional data is gathered from time to time all through the length of the study. In some cases, longitudinal studies can last several decades. It assesses changes in one or more persons. Generally one single characteristic e.g. vocabulary size, IQ level, height, aggressiveness, etc are measured. The researcher may also look into several aspects at a time to find out the interrelationship among the factors. In fact, the researcher follows the same group of subjects through the various stages of development that are measured. Among important longitudinal studies we can mention Barkeley Growth Study of 1928, which was conducted in the University of California at Berkeley. A massive data base was collected over a long span of time; the study provided important insight into the course and continuity of development after studying delicate issues like mental characteristics, behaviours and experiences along with physical growth, intelligence, personality, etc. Even child rearing practices were considered in the background.

Longitudinal studies are usually conducted on relatively small groups of individuals, which mean that it is usually difficult to relate the results to larger populations. Another problem that is associated with this study is that the participants often quit from the study, which shrinks the size of the sample. This reduces the amount of data even further. A special type of longitudinal study is known as retrospective study which involves looking back over a period of time. For example, the medical records of past years may be studied to establish the existence of a trend.

Longitudinal investigation can examine the growth and development of Cohort[8], but findings based on one cohort may not be applicable to a different cohort, universally. For instance, children's IQ may be affected by the quality of schooling from one decade to another or by the generational changes in parental consciousness regarding the importance of stimulating children's intellectual abilities. Since longitudinal study enquires the same people repeatedly, investigators can easily track the data about each individual concerned, but people enlisted for the study in course of time may lose their interest to cooperate with or may move away from the locality as the case of migrant population. Another problem may arise with this study that the whole project may not be possible to handle by a single investigator.

Terman and Oden (1959) studied a group of gifted children from their early school years through middle age. They found that on the average the children who talked early, walked early, were physically superior, and were typically social leaders as well. By the age of 35, many of them were listed in such books as American Men of Science and Who's Who.

Advantages of longitudinal study:

- Sensitive to individual patterns of change.
- Permits analysis of development on each case or each group.
- Permits the study of growth increments.
- Provides an opportunity to analyze relationships between maturational and experimental processes.
- Creates an opportunity to study the effects of cultural and environmental changes on behaviour and personality.

---

8. Cohort is the group of people who share a similar experience for growing up in the same place in a given period of time i.e. exposed to a particular set of cultural condition in the same historical era.

Disadvantages of longitudinal study:

- Enormous amount of time is involved.
- Tremendous expensive to carry out.
- Follow-up study is involved owing to great length is covered.
- Since data is extensive, cumbersome to handle.
- Sometimes participants drop out of the study, shrinking the sample size and decreasing the amount of data to be collected.
- Sometimes fill-in gaps are required by retrospective reports.

Cross-sectional Study (Compare groups that differ in age or background)

In cross-sectional study, people of different ages are assessed. This type of research aims to compare developmental levels at various ages or backgrounds. Many children at different ages are studied in groups according to their age, and the results on the same sets of measures are compared for the groups. For example, the approximate age at which an infant can be expected to roll over, creep, crawl, pull himself up to a standing position, and walk unaided- all these can be determined by observing the behaviour of groups of children from birth until the age of about 15 months. The procedure is to make the criteria of the group at first and then selection of the subjects; thus a group of one- month-old infants, another group of two-month olds, and similarly different group of babies at every month of age can be selected for a cross-sectional research design. Comparing the results of development of the children in each group the researcher can spot the appropriate time of any skill development.

This kind of study provides information about differences in development among different age-group, rather than features changing with age in the same person. Again, a researcher may measure or observe a group of young adults and compare this data with information gathered about a group of elderly participants. Gesell and his co-worker developed norms for four aspects of human growth-motor behaviour, language behaviour, adaptive behaviour, and personal-social behaviour. They studied large numbers of children in each age group, ranging from birth through adolescence, to determine the approximate age at which each step in the growth process normally occurs.

A cross-sectional study may also compare people from different economic backgrounds. If the reading ability of six-year-olds were measured in low, middle, and high-income families, one would have a "cross-section" of reading ability at that age for the various income groups in a community.

Advantages of cross-sectional study:

- It can be done relatively quickly, so saves time and energy.
- Gives a picture of typical characteristics at different ages.
- Relatively inexpensive to carry out.
- The whole project can be handled by a single investigator.

Disadvantages of cross-sectional study:

- It does not consider individual differences; deals with group averages.
- Provides only an approximate representation of developmental process.
- It cannot eliminate cohort or generational influence.
- It does not take into consideration the cultural and environmental changes occurring over time.

Cross Cultural Studies (Compare groups from different cultures)

Another important way of gathering data on human development is the cross- cultural method, which may sometimes be thought as a special kind of cross-sectional study. People differ culturally to the extent that their customs, roles, and other learned behaviours that are passed on from generation to generation are different. cultural standards of one society cannot be applicable to other societies ;

what may be normal or acceptable for one group may be abnormal or unacceptable for another group. Sensitivity to others' norms, folkways, values, mores, attitudes, customs, and practices necessitates knowledge of other societies and cultures. It is therefore impossible to investigate the effects of certain variables, because they do not appear in the same way in all societies. Cradling, for instance, is not practiced in most Western countries. Yet the practice, in which infants are bound firmly to a board and kept moving for most of their first year of life, is of interest to the psychologists who study motor development. Dennis and Dennis (1940) found a way to study the effects of such enforced physical restriction. They compared the age of walking in Hopi children who had been cradled with those who had never been cradled, and found that the average age of walking was not affected by cradling.

Cross-cultural research reveals the variations that exist across different groups of people, Landauer and Whiting (1964) studied the height of adult men as a function of the degree of stressful treatment of infant boys. Their data were taken from observations of eighty non-literate societies. They found that the average man in societies in which baby boys were ritually scarred, circumcised, or otherwise stressed, was taller than in societies where infants were not stressed.

Co-Twin Studies (Differences between identical twins are not caused by heredity)

Developmental psychologists who want to rule out the effects of heredity in their investigations often use the co-twin study as a method of research. Co-twin studies typically compare identical twins who have been reared apart or who have been given different kinds of training. Hilgard's work in 'training digit memory in a pair of identical twins' (Hilgard, 1933) was an example of a co-twin study. Hilgard trained one of the twins to remember digits in the first year, then trained the other twin in the second year, and then compared their performance of the twins for digit memory through frequent memory tests. He found that although both twins profited from the training, the twin trained later did better than the twin trained in the first year. But, both the twins lost their achievement gains after training was ended. Hilgard designed the same training at different developmental stages to manipulate the time of training. The dependent variable in an experiment is what is measured. Hilgard measured the performance on digit memory tasks. So the dependent variable stood as the performance on these tasks. Time of training was therefore the independent variable.

An important characteristic of experiments is that all other variables are held constant, as far as possible. Ideally, all experimental subjects should have identical experiences, apart from the differences the experimenter produces by manipulating the independent variable. Then we can be sure that the changes we measure in the dependent variable were produced by changes we made in the independent variable. Because controlling other variables is crucial, experimenters have often used co-twin studies in developmental psychology. Identical twins have identical heredity and usually a very similar environment. The use of twins gives control over important genetic variables that could be controlled no other way.

Geseli and Thompson (1929) studied the stair-climbing ability of two identical twins. One was given six weeks of training while the other was not. Their data indicate that training for this motor ability is not necessary at all. Motor ability is naturally achieved in course of development.

Types of Approaches

Mode of Research	Characteristics
Longitudinal	The same group of subjects be measured repeatedly and the measurements are compared.
Cross- sectional	Groups at different stages or from different backgrounds are compared.
Cross-cultural	Groups from different societies are compared.
Experimental	The independent variable is manipulated, the dependent variable is measured, and all other variables are held constant.

The Longitudinal and cross-sectional designs are classical approaches, uniquely suitable for studying human growth and development. The purpose of both the approaches is to arrive at establishing norms and trends in development.

- Jones and Conrad (1933) administered intelligence tests to all residents in a small New England town. They grouped the subjects' test results by age and then compared the average score per age. Test results indicated age trends in various aspects of intellectual ability.
- Several investigators have studied the growth of Individuals In height and weight from birth to adulthood. Nicolson and Hanley (1953) measured each boy's height and weight at various age intervals. They found that some boys undergo periods of rapid growth as early as age 10, others as late as age 17.
- Psychologists are often interested in the influence of parent models. In the child's development of self-control. Mischel (1953) examined preferences for larger, delayed rewards as opposed to small, immediate rewards among children of the Trinidadian Blacks, Grenadian Blacks, and Trinidadlan Indians. Anthropological data indicate that Trinidadian Blacks are more Impulsive, self-indulgent, and less likely to provide for the future than the Grenadian Blacks or Trlnidadian Indians. Trlnidadian Black children were found to show the greatest preference for immediate rewards.
- Fantz et al. (1962) used fixation preference to measure the visual acuity of infants. They presented various stimulus patterns to infants and measured the length of time each infant looked at each pattern (fixation preference). Data from this study indicate that infants have surprisingly good visual acuity.
- In a study of language development, Young (1941) studied children from well-to-do families and poor families. He measured the rate of language development for children in higher socio- economic families and compared this rate to the rate of development of language in children from poorer families.
- Palmer (1930) was interested in daily weight fluctuations. He weighed three pre-school children at approximately one hour intervals during the day for periods of three to five days. These measurements indicated that there are several consistent daily fluctuations in a child's weight.

Difference between longitudinal and Cross-sectional approaches

Features	**Iongitudinal**	**Cross-sectional**
Method of study	Same group is tested over the years repeatedly.	Different groups are tested belonging to different developmental stages or age levels at a particular time.
Cost and time	It is expensive, and takes years to arrive at a generalization.	Less time taking and comparatively less expensive.
Interpretation of results	One has to wait for years till the entire data are collected.	Interpretation is done as soon as the data are collected, since data are gathered quickly.
Quality of data	Shows individual growth and change over a long span of time. So there is possibility of sample mortality.	Large data are collected and norms can be established. No possibility of sample loss since data is collected in one attempt.
Professionals involved	Many observers and expert researcher are needed for the study.	Relatively few persons under a researcher can collect the information.

There are several popular **Methods** under these two approaches:

(*i*) Observation
(*ii*) Interview
(*iii*) Questionnaire
(*iv*) Case Study
(*v*) Standardized test
(*vi*) Clinical
(*vii*) Baby Biography
(*viii*) Sociometry

The methods were deviced for understanding, recording and interpreting the behaviour of the children with accuracy. They range from incidental and subjective nature to well designed and objective procedures. Methods are different in nature to suit in different situations; ultimate aim is to understand, control, and predict the course of development.

Observation Method

Observation is the most commonly used method in behavioural sciences. It implies collection of information by own observation, without interviewing the respondent. To avoid any kind of distortion scientists choose to observe subjects' behaviour directly through observational method. Observation can be of two types-Naturalistic observation and Laboratory observation. In Naturalistic observation the researcher or investigator observe people's behaviour in real world situations, like pre-school, nursing home, or any institutional settings. The strength of this method is that it allows researchers to view behaviour as it really happens in a natural setting. The drawback is that the researcher cannot control outside variables that might impact behaviour. Conversely, in Laboratory observation, researcher himself designs the setting to observe a particular behaviour. The benefit is that the situation remains fully under control. But the drawbacks are that the setting can be unnatural and the subjects are aware that they are being studied. Both of these factors can have an impact on results of the study.

In either research method, observer records participants' behaviour within an environment. Observational method, especially Naturalistic observation reduces the possibility of subjects giving misleading accounts of their experiences, not taking the study seriously, being unable to remember details, or feeling too embarrassed to disclose everything that happened. But individuals who agree to be observed and monitored may function differently as respondents who do not want to be observed and monitored.

Observation has become a scientific tool and useful method of data collection for the researchers, especially for formulating a research problem. This method can be planned and recorded; it is subjected to checking for validity and reliability. It is one of the oldest methods of child study. It deals with overt behaviour of persons in appropriate situation. The method can be designated as 'measurement without instrument'. The information that is obtained is related only to what is currently happening. The method is neither complicated by the past behaviour nor by the future intentions. To increase the accuracy of observation, cameras and other electronic devices may be used. These help to record children's movements and speech more accurately.

Advantages:

- The method can be used for all people irrespective of age. But more useful to study small and shy children.
- No special tool or equipment is needed.
- It can be used anywhere – in a workshop, in a classroom, in a park, etc.
- The method can be used for individual study as well as group study.
- Current information can be obtained.
- Subjective bias is eliminated.

Disadvantages:

- There lies great scope for personal prejudice and bias of the observer.
- Records may not be written with cent per cent accuracy.
- Limited information is available.
- It reveals the account of overt behaviour only and cannot reach to the covert behaviour.

Interview Method

Interviewing is a key method of data collection and used extensively in social sciences and educational research. This method of collecting data involves presentation of oral verbal stimuli and reply in terms of oral verbal responses in a face to face contact. It ranges from highly structural to unstructural flexible questioning. It is more effective than observation as it penetrates deep into the personal life to unveil attitudes, opinion, motivation, etc. i.e. discovers how individuals think and feel about a topic and why they hold certain opinions. It is a tool by which the sociological and psychological measurements are possible. A problem with interview is that the memory and accuracy of the interviewers may be faulty and they may lack proper training. On the other hand, information from the respondents may not be always accurate. Some respondents may forget when and how certain events took place; others consciously or unconsciously distort their reply. Some others hanker for acceptability and exaggerate their answers to gain prominence. Finally, wording or framing of a question may not be very specific which is a lacuna on the part of investigator.

There are several kinds of interviews. Hitchcock (1989) lists nine types: structured interview, survey interview, counselling interview, diary interview, life history interview, ethnographic interview, informal/unstructured interview, and conversations. Cohen & Manion (1994), prefers to group interviews into four kinds, including the structured interview, the unstructured interview, the non-directive interview, and the focused interview. However, interview method has superiority in terms of creating rapport; it allows researchers to observe the hidden part of verbal communication, such as their use of gestures.

Advantages:

- It is the most dynamic way to understand an individual; useful to obtain detailed information about personal feelings, perceptions and opinions.
- It investigates issues in an in depth way.
- Easy to conduct.
- Confidential information may be obtained.
- Greater flexibility is available i.e. it suits in many situations.
- Usually achieves a high response rate.

Disadvantages:

- Expensive in terms labour, money and time.
- Danger of subjective bias.
- Needs trained and competent interviewer.
- Some respondents are not easily approachable.
- It may suffer from language handicap; free expression may be restricted.

Questionnaire Method

Questionnaire is a systematic compilation of questions. It is a device for securing certain answers to specific questions; a form full of printed questions, set in a definite order, is provided to the respondents. Respondents are expected to fill up those forms by themselves, either in presence of investigator or in their leisure time. This method of data collection is quite popular particularly for big enquiries.

The difference between questionnaire and interview is that Questionnaire contains written questions to which participants respond in writing, often by checking or circling responses. In Interview also, a series of questions are delivered but presented orally by the interviewer and respondents also answer them orally. Both questionnaires and interviews can be highly structured, but it is common for interviews to be more open-ended, allowing the participant to provide detailed answers. Open-ended questions do not provide choices for the participants to select; rather, they must formulate an answer in their own words. Closed-ended items ask participants to choose among discrete categories and select which one best reflects their opinion or situation. Questions with ordered choices are common on questionnaires and are often similar to the individual items in a personality inventory or a summated attitude scale.

Respondents, who are not easily approachable for interview, can be reached conveniently by this method. In cases, where the subject is unable to fill out a questionnaire, with very young children for example, questions may be placed through a structured interview. Care should be taken in creating a questionnaire. Questionnaires generally comprise a combination of open and closed questions, providing balance between depth and authenticity of information, and fixed-option data which are more easily quantifiable. Each type has advantages.

1. For research exploring feelings, attitudes or types of behaviour; and where resources are plentiful, open-ended questions are preferable.
2. For demographic or performance data, and where time, subject or topic sensitivity, objectivity and ease of scoring and analysis are important, closed questions are more practicable.

Questionnaires are cost effective as compared to face-to-face interviews and particularly, when the number of research questions increase. This is especially true for studies involving large sample sizes and large geographic areas. Questionnaires are easy to analyze. Data entry and tabulation can be easily done with many computer software packages. Questionnaires reduce bias. There is uniform question presentation and no middle-man bias. The researcher's own opinions do not influence the respondent to answer questions in a certain manner. There are no verbal or visual clues to influence the respondent.

General advantage with Questionnaire method:

- It is economical; saves the time and energy of the investigator; a quick way to collect a great deal of information.
- By the use of large samples the results can be made more valid and reliable.
- Respondents get the scope to read the questions repeatedly for fuller understandings.

General disadvantage with Questionnaire method:

- Filling a lengthy form may be tedious for respondents.
- There is a possibility of getting ambiguous reply.
- It is difficult to know whether the willing respondents are truly representative or not.

Case Study Method

This is a comprehensive study method for a social unit. Case study method is a popular form of qualitative analysis and involves careful and complete observation of a social unit, be that a person, a family, a cultural group, an institution, or an entire community. It studies one single case in depth providing a detail picture of that case. If the case is a person, it highlights on every corner of his personality and behaviour. For this reason, case study method stands most useful in the study of human growth and development.

It is a study of depth rather than a study of breadth. While findings of such study provide a great deal of information about a specific person, the results are often difficult to generalize to larger populations. For this reason, case studies are most often used in clinical research or other cases where certain aspects of the subject's life cannot be reproduced or duplicated.

In case-study method, an investigator usually studies an individual who has a rare or unusual condition or who has responded favourably to a new treatment. Case studies are typically clinical in scope. The investigator - often a physician, psychologist, social worker, counsellor, or educator interviews the subject, obtains background records, and administers questionnaires to acquire quantifiable data on the subject. A comprehensive case study can last months or years. Throughout the duration of the case study, the researcher documents the condition, treatment, and effects in relation to each patient and summarizes all of this information in individual case reports. Although case studies are valuable for obtaining useful information about individuals and rare conditions, they tend to focus on the pathology i.e. the characteristics and effects of a particular disease, and therefore applicable only to individuals with similar conditions rather than to the general population. Growth, development or change in a child can easily be noted. The method helps to understand children having physical, intellectual or emotional difficulties and strives to solve those problems. A child's learning style and coping style is readily determined which facilitates offering of appropriate guidance.

Advantages:

- A thorough intensive investigation is possible.
- This method enables the researcher to trace the natural history of a social unit.
- It can furnish a real and enlightened record of personal experiences which reveals inner cravings, tensions, and motivations of an individual that drives towards action.

Disadvantages:

- The method is time consuming and expensive.
- Case situations are seldom comparable.
- Such study never provides the knowledge of impersonal, universal, non-ethical, impractical aspects.
- It may reflect observer bias.

Standardized test

The standardized test is a test that is administered and scored in a consistent or 'standard' manner and can be applied to a large number of cases. Standardized tests are designed in such a way that the questions, conditions for administering, scoring procedures, and interpretations remain same and can be administered and scored in a predetermined, standard manner. These tests are used in psychology, as well as in everyday life, to measure intelligence, aptitude, achievement, personality, attitudes and interests. Attempts are made to standardize tests in order to eliminate biases that may result, consciously or unconsciously, from varied administration of the test.

The opposite of a standardized test is a non-standardized test. Non-standardized testing gives significantly different tests to different test takers, or gives the same test under significantly different condition or evaluates them differently. Standardized tests are considered as fairer than non-standardized tests. The consistency of Standardized tests permits more reliable comparison of outcomes across all test takers. Because, this kind of test can be administered to a group of subjects under exactly the same experimental conditions and scored in exactly the same way. In standardized test, number of items, method of construction, method of interpretation, all are controlled. Further, there are many other steps like developing working concept, assembling of test items, review and verification of test items, construction of pilot test, etc none of which can be neglected.

In the study of human growth and development, standardized tests are used to generate a carefully selected sample of behaviour. Psychological attributes of the people like intelligence, self esteem, etc are checked with different growth phases. Tests come in different forms. Some involve open-ended situations with standard stimuli, others involve structured situations where the range of possible response is narrow and the answers come as either right or wrong. Psychological or psychometric tests consist of a series of questions that the individual who is being tested, answers.

These answers are then analyzed carefully to reveal information about the interviewee. (For instance, Intelligence tests reveal IQ level). Tests are useful in sorting out the differences between children in a particular setting but could be biased against particular individuals or groups. However, any psychological test should be uniform, objective and interpretable.

Advantages:

- Scientific, accurate and free from bias.
- Quite economical.
- Since the questions and answers are set and the scoring is objective, it allows for easier interpretation of data.

Disadvantages:

- The simplicity of scoring and administration may lead to errors.
- It can greatly affect the validity and reliability of the assessment, if not properly standardized.
- Lack of experience and skill may lead to wrong interpretation.

Clinical Method

Clinical method is directed towards solving of unusual behaviour. The clinical set up or environment is associated with health care and treatment of the individuals who come to the clinical psychologists for advice and treatment of their physical and mental disorders. Since the method is used to investigate the root cause of a problem or an exceptional behaviour, it is applicable to an individual case which has some problem. It provides insight into the adjustment problem, reveals underlying causes of misbehaviour. Accordingly, treatment and therapeutic measures are suggested; attempts are taken to rehabilitate the persons in their environment.

Although clinical studies are carried out to understand the stress and emotional problems of the children, but this method is also used with normal and well adjusted youngsters. It is very useful to study the derivatives of growth and development. Help of certain specialized techniques are taken and the method involves some definite steps.

1. Preparation of case history.
2. Study of the environment.
3. Direct observation of the individual during interview or play.
4. Psychological examination with help of certain tests and technique.

Clinical method is actually an outgrowth of psychoanalytic theory which treats every child as a unique individual. The aim is to obtain a clearer understanding of child's psychological functioning and experiences. It exhibits a wide range of data about one subject by combining observation and test scores with flexible questioning. By questioning, a psychologist wants to be acquainted with the inner mental makeup and covert thinking of the individual which is usually tallied with test result and supplemented by observation. Clinical psychologists have an understanding of childhood and adolescent problems. They are trained to work with both adults and children and have specialized experience in all types of developmental behaviour and many types of psychological difficulties which influence children and families. So they can render help to young people and their families in a number of ways like,

- Help to develop problem solving skills.
- Help in increasing ability to manage existing psychological problems.
- Help to cope up with stress.
- Encourage in maintaining good relationships with family and neighbourhood.

Advantages:

- Discloses details of an individual pattern of behaviour by intensive study.
- Offers important insight into the processes of development.

Disadvantages:

- Collection and recording of information may be unsystematic and subjective.
- Diagnosis and treatment both become useless if the clinical researcher is not technically proficient or immature.

Biographical Method

Biographical method means the collection of Baby Biographies. It is one of the oldest method used in the field of child psychology to study the regular development of a child with his changing behaviour. Baby Biographies are considered as earliest source of information regarding infants' development, originated in late nineteenth century or early twentieth century as an attempt to record a single child's behaviour. Parents used to record the day to day activities and physical progress of their own children very systematically through narrative description. Baby biographies are therefore like diaries written by the enthusiastic and enlightened parents. The first published record, 'The Biography of a Baby' was written by Milicent Shinn who studied the growth of her young niece during first year of life. One of the most famous Baby biographers was Charles Darwin who wrote information about his own son which was published in 1877. The most influencing Baby biographer was J. Piaget who observed his three children and provided extensive information on children's intellectual development. These are the personal documents - records of intensive observation on the sequential growth of children.

Baby biography or diaries on children was a unique tool in throwing light upon many important and vital aspects of personality of an individual. Not only did they enlist the personal history, but also provided delicate reflection of emotions and feelings of the respective child. When the field was new and methods of study was under formulation, the biographical method was of immense importance; insights originated from the rich potential of Baby biography. This legacy has been drawn to contemporary method of naturalistic observation, where development of individual children is followed over time.

Sociometry

Sociometry is the method for describing and measuring social relationships among individuals in a group. It shows the degree of relatedness among people. Measurement of relatedness can be useful not only in the assessment of behaviour within groups, but also for interventions to bring about positive change and for determining the extent of change. It is a powerful tool for assessing dynamics and development in groups devoted to therapy or training.

Jacob Levy Moreno coined the term *sociometry* and conducted the first long-range sociometric study from 1932-38 at the New York State Training School for Girls in Hudson, New York. He noticed that in groups and communities people are attracted to or draw away from others and these choices create patterns of interaction. He discovered when people chose whom they interacted with; they have a higher level of satisfaction in being together, a greater sense of belonging, and achieve their purpose. Moreno himself defined sociometry as "the mathematical study of psychological properties of populations, the experimental technique of and the results obtained by application of quantitative methods" (Moreno, 1953). Moreno came to the conclusion, "Choices are fundamental facts in all ongoing human relations, choices of people and choices of things. It is immaterial whether the motivations are known to the chooser or not; it is immaterial whether [the choices] are inarticulate or highly expressive, whether rational or irrational. They do not require any special justification as long as they are spontaneous and true to the self of the chooser. They are facts of the first existential order."

Sociometric methods have been developed over time, with the principle that the investigator is an active group member. Material generated from the group, belongs to the group members. When this information is taken away by the researcher and managers to make decisions about the group, this is no longer sociometry. Group members can create new patterns for themselves, and enhance authentic companionship and greater mutuality amongst people in the group. All groups have thus intricate networks, based on both socio-emotional choices, and psycho-social networks. Sociometry enables us to enter into the world of interpersonal choices, attractions and rejections, and their effects. This method assists in exploring and improving relationship dynamics.

Sociometric tests can contribute in the field of human growth and development by weighing psycho-social networks - choice, attraction, repulsion among a group of children, whether in class or in play group. It also can be used in a family setting to review the adjustment pattern with the family members and to work out rehabilitative measures. Sociometric assessments are most often used to determine eligibility for special education and for intervention for adaptive behaviours or socio-emotional problems. Children identified with special education needs, such as learning problems, mental retardation, attention deficit disorders, and autism spectrum disorders, etc may benefit from assessment and intervention toward enhancing their social skills. In the general education, children who are shy, rejected, or engage in bullying or aggressive behaviours or who simply have limited social skills may also be benefitted. Not only the therapeutic measures can be availed, this method also helps to clean the path of normal development.

Sociometry is a new technique for studying the underlying motives of children. It tells us, peer-rating is sometimes more useful than rating by the superiors and thus adds another dimension to the understanding of peer relationship.

In many situations the researchers are found use a combination of methods in observing children behaviour and to make the study more meaningful, objective, comprehensive and reliable.

APPENDIX

SUGGESTED READING

Allchin, R and B. Allchin (1983) : *The Rise of Civilization in India and Pakistan*; Select Book Service Syndicate, New Delhi.

Allchin, R and B. Allchin (1993) : *Birth of Indian Civilization*; Pengiun Books, London.

Ashley Montagu, M.F. (1951) : *An Introduction to Physical Anthropology* ; Second Edition, Charles C. Thomas, Illinois, Chicago.

Ashley Montagu, M.F. (1960) : *Human Heredity*; The American Library, New York

Ashley Montagu, M.F. (1960) : *(A) Handbook of Anthropometry*; Chalres. C Thomas, Springfield.

Ashley Montagu, M.F. (1962) : *Man : His First Million Years*; (Fifth revised print), New American Library, New York.

Ashley Montagu, M.F. (1972) : *Culture and the Evolution of Man*; Oxford University Press, New York.

Baden-Powell, B.M. (1957) : *The Indian Village Community*; Araf Press, New Haven.

Bailey, F.G. (1957) : *Caste and Economic Frontier*, Manchester, New York.

Bailey F.G. (1960) : *Tribe, Caste & Nation*; Manchester University Press, New York.

Banerjee, H.N. (1994) : *Introducing Social and Cultural Anthropology*; K.K. Publication, Calcutta.

Bansal, I.J.S. and P. Singhal (ed.) (1983) : *Anthropology in Indian Context*; *Today and Tomorrows :* Printers and Publishers, New Delhi.

Barnouw, V. (1989) : *Introduction to Physical Anthropology and Archaeology;* Fifth edition, The Dorsey Press, Illinois, Chicago.

Basam, A.L. (ed.) (1975) : *(A) Cultural History of India*; Clarendon Press, Oxford.

Basu, M.N. (1957) : *Material Existence of Man*; Dipika Publishing, Calcutta.

Basu, M.N. (1961) : *Field Methods in Anthropology and other Social Sciences*; Calcutta.

Basu Roy (1993) : Kalighat — Its Impact on Social-cultural Life of Hindus, Gyan Publishing House, New Delhi.

Braidwood, R.J. (1961) : *Prehistoric Man*; Chicago Natural History Museum (Popular Series, Anthropology, No. 37), Chicago.

Beals, R.L., H. Hoizer and A.R. Beals. (1977) : *An Introduction to Anthropology*; Fifth edition, Macmillan Publishing Co., Inc., New York.

Bean, R.S. (1932) : *The Races of Man;* The University Society, Inc. New York.

Beattie, J. (1964) : *Other Cultures*; Routledge & Kegan Paul, London.

Beteille, A (1966) : *Caste. Class & Power*; Oxford University Press, Indian Branch, Bombay.

Beteille, A (1967) : *Castes Old & New*; Asia Publishing House, Bombay.

Beteille, A and T.N. Madan (1975) : *Encounter and Experience : Personal Account of Field-work*; Vikas Publishing House, Delhi.

Benedict, R (1942) : *Race and Racism*; Routledge & Kegan Paul, London

Benedict, R (1946) : *Chrysanthemum and the Sword*; Houghton Mifflin, Boston.

Benedict, R (1959) : *Patterns of Culture*; (revised edition) Mentor Books, New American Library, New York.

Bews, J.W. (1935) : *Human Ecology*; Oxford University Press, New York.

Bhattacharya, A (1979) : *(A) History of Ancient India*; Arnold Publishers, New Delhi.

Bhattacharya, D.K. (1972) : *Prehostaric Archaeology*: Hindustan Publishing Corporation, New Delhi.

Bhattacharya, J.N. (1896) : *Hindu Castes & Sects*: Thacker Spink & Co., Calcutta.

Bhowmick, P.K. (1963) : *The Lodhas of W.B. : A Socio-Economic Study*; Puthi-Pustak, Calcutta.

Bhowmick, P.K. (1980) : *Some Aspects of Indian Anthropology*; Subarnarekha, Calcutta.

Bhowmick, P.K. (1982) : *Approach to Tribal Development*; Inter-India Publication, New Delhi.

Bidney, D. (1953) : *Theoretical Anthropology*; Columbia University Press, Columbia.

Boas, F. (1940) : *Race, Language and Culture*; Macmillan, New York.

Boas, F. (1955) : *Primitive Art*; Constable & Co., London.

Boas, F. (1962) : *Anthropology and Modern Life*; Norton, New York.

Boas, F. (1965) : *The Mind of Primitive Man*: Collier-Macmillan, New York.

Boas, F. (1972) : "*Anthropology*" in *International Encyclopedia of Social Sciences*; Vol. II, pp. 99-108.

Boas, F (ed.) (1938) : *General Anthropology*: D.C. Heath & Co., New York.

Bordes, F. (1968) : *The Old Stone Age*; London.

Boyd, W.C. (1958) : *Races and People*; Abelard-Schuman, London.

Bose, N.K. (1953) : *Cultural Anthropology and Other Essays;* Indian Associated Publishing Co., Calcutta.

Bose, N.K. (1967) : *Culture & Society in India*, Asia Publishing House, New Delhi.

Bose, N.K. (1972) : *Some Indian Tribes*; National Book Trust, Delhi.

Bose, N.K. (1972) : *Anthropology and Some Indian Problems*; Institute of Social Research, Calcutta.

Bose, N.K (1976) : *The Structure of Hindu Society,* translated from the original book '*Hindu Samajer Garan*' (in Bengali); Sangam Books, New Delhi.

Bottomore, T.B. (1979) : *Sociology*; Second edition; Blackie & Son, New Delhi.

Brace, C.L. (1863) : *(The) Races of the Old World : A manual of Ethnology*; John Murray, London.

Brace, C.L. (1991) : *(The) Stages of Human Evolution*; Fourth editon, Prentice-Hall, New Jersey.

Brace, C.L. and M.F. Ashley Montagu (1965) : *Man's Evolution: An Introduction to Physical Anthropology*; Macmillan, New York.

Breuil, H. and R. Lantier (1965) : *The Men of the Old Stone Age*; George G. Harrap & Co., Ltd., London.

Broom, R. (1950) : *Finding Missing Link*; Watts, London.

Buettner-Janusch, J (1969) : *Origins of the Man*; Wiley Eastern Pvt. Ltd., New Delhi.

Burkitt, M.C. (1929) : *Our Early Ancestors*; Cambridge University Press, London.

Burkitt, M.C. (1969) : *(The) Old Stone Age*; Fourth edition, Bowes & Bowes Publishers Ltd., London.

Butler, J.A.V. (1959) : *Inside the Living Cell*; Basic Books, Inc., New York.

Carneiro, R.L. and Stephen *F. Tobias* (1963) : *"The Application of Scale Analysis to the Study of Cultural Evolution"* in Transactions of New York Academy of Sciences, Series 2, Vol. 26, pp. 196-207.

Chapple, D.E. and C.S. Coon (1942) : *Principles of Anthropology*; Henry Hold & Co., New York.

Chattopadhaya, K. (1978) : *Tribalism in India*; Vikas Publishing House, New Delhi.

Chaudhary, B. (1982) : *Tribal Development in India : Problem & Prospects;* Inter-India Publications, Delhi.

Childe, V.G. (1936) : *Man Makes himself*; Mentor Books, New American Library, New York.

Childe, V.G. (1942) : *What Happened in History*; Penguin Books, London.

Childe, V.G. (1952) : *Social Evolution*; Watts & Co., London.

Childe, V.G. (1957) : *The Dawn of European Civilization*; Routledge & Kegan Paul, London.

Childe, V.G. (1957) : *The Dawn of Indian Civilization*; Sixth edition. Routledge & Kegan Paul, London.

Clark, Grahame (1947) : *Archaeology & Society*; Second edition, Methuen, London.

Clark, Grahame (1961) : *World Prehistory*; Cambridge University Press, London.

Clark, Grahame (1967) : *The Stone Age Hunters*; Thames & Hudson, London.

Clark, G. and S. Piggott (1965) : *Prehistoric Societies*; Hutchinson, London.

Clark, J.G.D. (1935) : *The Mesolithic Settlement of Northern Europe*; Cambridge, London.

Clark, J.G.D. (1952) : *Prehistoric Europe*; Methuen & Co., London.

Clark, J.G.D. (1970) : *The Prehistoric Africa*; Praeger, New York.

Clark, W.E. le Gros, (1950) : *History of the Primates*; Second edition, British Museum (Natural History), London.

Clark, W.E. le Gros, (1955) : *The Fossil Evidence for Human Evolution*; University of Chicago Press, Chicago.

Clifford, J. and P. Fred. (1986) : *Physical Anthropology & Archaeology*; Alfred Knoff, New York.

Cohn, E.C. (1956) : *Elements of Genetics*; McGraw Hill Books Co., Inc. New York.

Cole, J.B. (1988) : *Anthropology for the Nineties*; Collier-Macmillan Publishers, London.

Cole, S (1963) : *The Neolithic Revolution*; British Museum, London.

Cole, S (1965) : *(The) Prehistory of East Africa*; New American Library, U.S.A.

Comte, A (1983) : *The Positive Philosophy of East Africa*; Translated (from French) and condensed by H. Martineau, Kegan Paul, London.

Coon, C.S. (1936) : *Races of Europe*; The Macmillan Co., New York.

Coon, C.S. (1963) : *The Origin of Races*; Jonathan Cape, London.

Dalton, E.T. (1872) : *Descriptive Ethnology of Bengal*; Govt. Printing Press, Calcutta.

Danda, A.K. (1991) : *Ethnicity in India*; Inter-India Publication, New Delhi.

Danniel, G. (1950) : *A Hundred Years of Archaeology*; Duckworth, London.

Darwin, C. (1859) : *The Origin of Species*; Modern Library, New York.

Darwin, C. (1871) : *The Descent of Man*; Volumes I & II, John Murray, London.

Datta, K.K. (1940) : *The Santal Insurrection of 1855-57*; Calcutta University Press, Calcutta.

Datta-Majumdar, N. (1955) : "*The Tribal Problem*" in *The Adivasis*; Publication Divisions, Ministry of Information & Broadcasting, Govt. of India, New Delhi.

Desai, A.R. (1969) : *Rural Sociology in India*; Vols. I, II and III, Popular Prakasan, Bombay.

Dixon, R.B. (1989) : *The Racial History of Man*; Swati Publications, Delhi.

Dobzhansky, T. (1951) : *Genetics and the Origin of Species*; Third edition, Columbia University Press, New York.

Dobzhansky, T. (1965) : *Heredity and the Nature of Man*; Allen & Unwin,

London.

Dobzhansky, T. (1973) : *Genetic Diversity and Human Equality*; Basic Books, New York.

Dube, S.C. (1952) : *Anthropology*; Chetna, Hyderabad.

Dube, S.C. (1955) : *Indian Village*; Cornell University Press, Ithaca, New York.

Dube, S.C. (1958) : *India's Changing Villages*; Routledge & Kegan Paul, London.

Dube, S.C. (1990) : *Indian Society*; National Book Trust, India.

Dumont, I (1972) : *Homo Hierarchichus*; *The Caste System and Its Implications*; Granada Publishing House Ltd. London.

Dumont, L. and D. Pocock (ed.) (1957) : *Contribution to Indian Sociology*; Institute of Social Anthropology, Oxford.

Dunn, L.C. and T. Dobzhansky (1952) : *Heredity, Race and Society*; Mentor Books, New American Library, New York.

Durkheim, E. (1938) : *The Rules of Sociological Methods*; Translated version, G.E.G. Catlin (ed.), University of Chicago Press, Chicago.

Durkhein, E. (1961) : *The Elementary Forms of the Religion Life*; Translated version, J.W. Swain, Collier-Macmillan, New York.

Duverger, M (1964) : *Introduction to Social Sciences*; George Allen & Unwin Ltd., London.

Elwin, V. (1959) : *A Philosophy for NEFA*; Sachin Roy, Shillong.

Elwin, V. (ed.) (1959) : *India's North-East Frontier in the Nineteenth Century*; Oxford University Press, London.

Ember, C.R. and M. Ember (1993) : *Anthropology*; Sixth edition, Prentice-Hall of India, New Delhi.

Engles, F (1940) : *Dialectics of Nature*; Translated and Edited by C. Dutta, International Publishers, New York.

Epstein A.L. (ed.) (1969) : *The Craft of Social Anthropology*; Associated Book Publishers Ltd., London.

Evans-Pritchard, E.E. (1937) : *Witchcraft, Oracles & Magic among the Azande*; Clarendon Press, Oxford.

Evans-Pritchard, E.E. (1940) : *The Nuer*, Clarendon Press, London.

Evans-Pritchard, E.E. (1951) : *Social Anthropology*; Cohen & West, London.

Evans-Pritchard, E.E. (1964) : *Social Anthropology & Other Essays*; Free Press, New York.

Evans-Pritchard, E.E. (1965) : *Theories of Primitive Religion*; The Clarendon Press, London.

Fagan J.J. (1965) : *View of the Earth*; Holt, Rinenart & Winston, New York

Firth, R. (1956) : *Elements of Social Organisation*; Second edition, Watts & Co., London.

Firth, R. (1958) : *Human Types: An Introduction to Social Anthropology*; New American Library, New York.

Firth, R. (1964) : *We, the Tikopia*; Allen & Unwin, London.

Firth, R. (1996) : *Religion : A Humanist Interpretation*; Routledge, London

Flint, R.F. (1957) : *Glacial & Prehistoric Geology*; Wiley, New York.

Frazer, J.E. (1946) : *The Anatomy of the Human Skeleton*; J.A. Churchill Ltd., London.

Frazer, J.G. (1922) : *The Goldden Bough*; Abridged edition, Macmillan & Co. Ltd., New York.

Freud, S. (1943) : *(A) General Introduction to Psycho-analysis*; Translated version, Garden City Publishing Co., New York.

Forde, C.D. (1961) : *Habitat Economy & Society*; Fourteenth reprint, Methuen & Co., London.

Foster, G.M. (1969) : *Applied Anthropology*; Little Brown, Boston.

Foster, G.M. (1973) : *Traditional Culture & Impact of Technological Change*; Second Edition, Allied Publishers Ltd., New Delhi.

Foote, R.B. (1868) : *Foote Collections of Indian Prehistoric & Protohistoric Antiquities*; Supt. of Govt. Press, Madras.

Gannep, A.V. (1960) : *The Rites of Passage*; Translated from the original book, *Les Rites de Passage of 1909,* Routledge & Kegan Paul Ltd., London.

Ghurye, G.S. (1961) : *Caste, Class & Occupation*; Popular Prakashan, Bombay.

Ghurye, G.S. (1979) : *Caste and Race in India*; Popular Prakashan, Bombay.

Ghurye, G.S. (1995) : *The Scheduled Tribes;* Third edition, Popular Prakasan, Bombay.

Giri, H (1998) : *The Khasis under British Rule*; Regency Publication, New Delhi.

Gluckman, M (1963) : *Institutions of Primitive Society*; Oxford University Press, London.

Grey, H. (1958) : *Anatomy, Descriptive & Applied*; Edited by T.B. Johnson, D.V. Davis & F. Davis, Longman, London..

Grierson, G.A. (1967) : *A Linguistic Survey of India*; Vols. I,II,III & IV, revised Edition, Motilal Banarsidas, Delhi.

Guha, B.S. (1935) : *The Racial Affinities of the People in India*; in the Census of India, 1931, India.

Gurdon, P.R.T. (1975) : *(The) Khasis*; Cosmo Publications, Delhi.

Haddon, A.C. (1924) : *Races of Man & Their Distribution*; Milner & Co., London.

Haddon, A.C. (1945) : *History of Anthropology*; Watt & Co., London

Halbar, B.G. and (1991) : *Relevance of Anthropology*; Rawat Publication,

C.G. Hussain Khan (ed) New Delhi.

Harris, M. (1968) : *The Rise of Anthropological Theory*; Thomas Y. Crowelt, New York.

Herskovits, M.J. (1955) : *Cultural Anthropology: an abridged revision of Man & His Works*; Oxford & I.B.H. Publishing Co., New Delhi.

Hoebel, E.A. (1958) : *Man in the Primitive World*; *Second edition*, McGraw-Hill Book Co., New York.

Hoebel, E.A. etal (1955) : *Readings in anthropology*; McGraw-Hill Book Co., New York.

Honigmann, J.J. (1967) : *Personality in Culture*; Harper & Brothers, New York.

Honigmann, J.J. (ed.) (1997) : *Handbook of Social & Cultural Anthropology*; Vols. I & II, Rawat Publications, New Delhi.

Hooton, E.A. (1965) : *Up from the Ape*; Motilal Banarsidas, Delhi.

Howells, W. (1945) : *Mankind So Far*; Doran & Co., Inc., New York.

Howells, W. (1956) : *Mankind in the Beginning*; G. Bell & Sons Ltd. London.

Howells, W. (1960) : *Mankind in the Making*; Secker & Warbug, London.

Howells, W. (1961) : *The Emergence of Man*: Random House, Inc., New York.

Howells, W. (1963) : *Back of History*; Revised edition, Doubleday & Co., Inc., New York.

Hughey, M.W. (ed.) (1998) : *New Tribalism : The Resurgence of Race and Ethnicity*; New York University Press, New York.

Hutton, J.H. (1963) : *Caste in India*; Cambridge University Press, London.

Huxley, J. (1941) : *Man Stands Alone*; Harpar & Bros., New York.

Huxley, J. (1964) : *Evolution : The Modern Synthesis*; Harpar & Row, New York.

Indian Council of Social Science Research (1985-86) : *Survey of Research in Sociology and Social Anthropology in 1969-1979,* Vol. I, II & III, Satvahan, Delhi.

Kardiner, A (1939) : The Individual & His Society, Columbia University Press, New York.

Kapadia, K.M. (1966) : *Marriage & Family in India*; Oxford University Press, Bombay

Karve, I. (1953) : *Kinship Organization in India*; Deccan College Monograph Series, Deccan College Post Graduate and Research Institute, Poona.

Keith, Sir. A. (1925) : *The Antiquity of Man*; J.B. Lippincott Co., Philadelphia.

Kessing, R.M. and F.M. Kessing (1971) : *New Perspective in Cultural Anthropology*; Holt, Rinehart & Winston, New York.

Kluckhohn, C. (1950) : *Mirror for Man*; G.G. Harrap & Co., London.

Koenigswald, G.H.R. Von (1956) : *Meeting Prehistoric Man*; Thames & Hudson, London.

Kohler, W. (1925) : *The Mentality of Apes*; Harcourt Brace & Co., New York.

Koshambi, D.D. (1965) : *The Culture & Civilization of Ancient India*; Routledge Kegan Paul, London.

Kothari, C.R. (1993) : *Research Methodology: Methods & Techniques*; Fourth reprint, Wiley Eastern Limited, New Delhi.

Kroeber, A.L. (1952) : *The Nature of Culture*; University of Chicago Press, Chicago.

Kroeber, A.L. (1954) : *Method & Perspective in Anthropology*; University of Minnesota Press, New York.

Kroeber, A.L. (1976) : *Anthropology*; Revised edition, Third Indian Reprint, Oxford & I.B.H. Publishing Co., New Delhi.

Krober, A.L. and C. Kluckhohn (1952) : *Culture, a critical view of Concepts and Definitions*; Harvard University, Cambridge.

Laskar, G.W. & R.N. Tyzzer. (1982) : *Physical Anthropology*; Third edition, Holt, Rinehart & Winston, Inc., London, Chicago, etc.

Leaky, L.S.B. (1953) : *Adam's Ancestors*; Methuen & Co., Ltd., London.

Leaky, R.E. (1981) : *The Making of Mankind*; Dulton, New York.

Levi-Bruhl, L. (1923) : *The Soul of the Primitive*; Allen & Unwin, London.

Levi-Strauss, C. (1967) : *Totemism*; Merlin Press, London.

Levi-Strauss, C. (1967) : *The Scope of Anthropology*; Cape, London.

Levi-Strauss, C. (1968) : *The Savage Mind*; Weidenfeld & Nicolson, London.

Levi-Strauss, C. (1968) : *Structural Anthropology*; Penguin Books, London.

Lewis, J. (1969) : *Anthropology Made Simple;* W.H. Allen, London.

Linton, R. (1945) : *The Cultural Background of Personality*; Appleton, New York.

Linton, R. (1965) : *The Study of Man*; Owen, London.

Lowie, R.H. (1947) : *An Introduction to Cultural Anthropology*; Rinehart & Co., London

Lowie, R.H. (1947) : *An Introduction to Cultural Anthropology*, Rinehart & Co., London.

Lowie, R.H. (1950) : *Social Organization*; Routledge & Kegan Paul, London.

Lowie, R.H. (1960) : *Primitive Society*; Routledge & Kegan Paul, London.

Lowie, R.H. (1960) : *Primitive Religion*; Peter Owen Ltd. London.

Lowie, R.H. (1966) : *Culture & Ethnology*; Basic Books, Inc., London.

Madan, T.N. and G. Sarana (ed.) (1962) : *Indian Anthropology*: Asia Publishing House, Bombay.

Mair, L. (1957) : *Studies in Applied Anthropology*; University of London Press, London.

Majumdar, D.N. (1950) : *Affairs of a Tribe : A Study of Tribal Dynamics*; Universal Publishers, Lucknow.

Majumdar, D.N. (1958) : *Caste & Communication in an Indian Village*; Asia Publishing House, Bombay.

Majumdar, D.N. (1958) : *Race and Culture of India*; Asia Publishing House, Bombay

Majumdar, D.N. and T.N. Madan (1957) : *An Introduction to Social Anthropology*; Asia Publishing House, Bombay.

Majumdar, R.C. (1962) : *The History & Culture of the Indian People*; Bhartiya Vidya Bhawan, Vol. III.

Malinowski, B. (1926) : *Crime & Custom in Savage Society*; Routledge & Kegan Paul, London.

Malinowski, B. (1927) : *Sex & Repression in Savage Society*; Kegan Paul, London.

Malinowski, B. (1945) : *Dynamics of Social Change*; Yale University Press, New Heaven.

Malinowski, B. (1954) : *Magic, Science & Religion*; Doubleday, Garden City New York.

Mamoria, C. B. (1957) : *Tribal Demograhy in India*, Kitab Mahal, Allababad.

Mandelbaum , D. G. (1970) : *Society in India;* Popular Prakasan, Bombay.

Mann, R. S. (1980) : *Experimetation & Handicaps in Tribal Development;* Rawat Publication, Jaipur.

Marett, R. R. (1909) : *The Threshold of Religion*; Methuen, London.

Marett, R. R. (1912) : *Anthropology*; Holt, New York.

Marriot, M. (1955) : *Village India*; University of Chicago Press, Chicago

Mead (1950) : *Sex & Temporament in three Primitive Societies*; Mentor, New York.

Mead, M. (1960) : *Coming of Age in Samoa;* Third edition, Morrow, New York

Mehta, P.C. (1994) : *Voluntary Organization & Tribal Development*; Shiva Publishers & Distributers, Udaipur.

Merton R.K. (1957) : *Social Theory & Social Structure;* Free Press, New York.

Mitra, P. (1923) : *Prehistoric India;* Calcutta University Press, Calcutta.

Montagomery, C. W. (1993) : *Fundamentals of Geology*; Second edition, Wm.C. Brown Publishers, U.S.A.

Morgan J. D. (1990) : *Outline of Prehistory*; Deep & Deep Publication, New Delhi.

Morgan L. H. (1964) : *Ancient Society,* Harvard University Press; Cambridge.

Moore, W.E. (1955) : Economy & Society, Dobleday, Garden City, New Delhi.

Murdock, G. P. (1934) : *Our Primitive Contemporaries,* Macmillan Co., New York.

Murdock, G. P. (1949) : *Social Structure;* Macmillan Co., New York.

Nadel, S. F. (1951) : *The Foundation of Social Anthropology*; Free Press, New York.

Nadel, S.F. (1953) : *Anthropology & Modern Life*; Australia National University, Canberra.

Napier, J. R. and P. H. Napier. (1967) : *A Handbook of Living Primates*;. AcademicPress,

New York.

Nell, J. V. & W, J, Schull (1954) : *Human Heredity*; University of Chicago Press, Chicago.

Neogi, P. (1979) : *Copper in Ancient India,,* Janaki Prakashan, Patna.

Oakley, K. P. (1975) : *Man the Tool Maker*; Sixth edition, Trustees of the British Museum (Natural -History), London.

Ogburn, W. F. (1950) : *Social Change;* Viking, New York.

Osburn, H. F. (1924) : *Men of the Old Stone Age;* Charles Scribners Sons, New York.

Osman Hill, W. C. (1957) : *Man as an Animal*; Hutchinson University Library, London.

Pelto, P. J. & G. H. Pelto. (1978) : *Anthropology Research*; Second edition, Cambridge University Press, New York.

Penniman, T. K. (1952) : *(A) Hundred years of Anthropology*; Second edition; The Macmillan Co., New York.

Piddington, R. (1960) : *An Introduction to Social Anthropology*; *Vols.* I & II, Oliver & Boyd, London.

Piggott, S. (1949) : *Approach to Archaeology;* Adam & Charles Black, London.

Piggott, S. (1950) : *Prehistoric India to1000 B.C,* ; Middlesex, Penguin Books, Harmondsworth.

Piggott, S. (ed) (1961) : *(The) Dawn of Civilization;* Thames & Hudson, London.

Playfair, A (1909) : *The Garos*; David & Nutt, London.

Polanyi K, C. M. Arensberg H.W. Pearson (ed.) (1957) : *Trade & Market in the Early Empires*; Free Press, New York.

Publication Division, Govt. of India (1969) : *India : Social Structure*; Ministry of Information & Broadcasting, Patiala House, New Delhi.

Radcliffe-Brown, A.R. (1922) : *The Andaman Islanders*; Cambridge University Press, Cambridge.

Radcliffe-Brown, A.R. (1952) : *Structure & Function in Primitive Society*; Cohen and West, London.

Raghavaiah, V.R. (1971) : *Tribal Revolts*; Andhra Rastra Adimjati Sevak Sangha, Nellore, A.P.

Raha, M.K. & P.C. Koomar (ed) (1989) : *Tribal India*; Gyan Publishing House, New Delhi.

Rao, M.S.A. (ed.) (1978) : *Social Movements in India*; Vols. I & II, Manohar Publications, New Delhi.

Redfield, R. (1942) : *The Folk culture of Yucantan*; University of Chicago Press, Chicago.

Redfield, R. (1960) : *The Little Community and Peasant Society & Culture*; University of Chicago Press, Chicago.

Redfield, R. (1971) : *Primitive World & Its Transformations*; Cornell University Press, New York.

Rickard, T.A. (1932) : Man & Metals, MicGraw-Hill, New York.

Risley, H.H. (1915) : *Peoples of India;* Reprint, Thacker Spink & Co., Calcutta.

Risley, H.H. (1981) : *The Tribes & Castes of Bengal*; Reprint, Firma

K.L.M., Calcutta.

Rivers, W.H.R. (1906) : *The Todas,* Macmillan & Co., London.

Rivers, W.H.R. (1914) : *Kinships & Social Organization*; Constable & Co., London.

Rivers, W.H.R. (1924) : *Social Organization*; Kegan Paul, London.

Rogers, J.J. (1?93) : *A History of the Earth*; Cambridge University Press, Cambridge.

Roy, S.C. (1950) : *The Mundas & Their Country*; Reprint, Asia Publishing House, Bombay.

Royal Anthropological Association (1929) : *Notes & Querries on Anthropology*; Fifth edition, Routledge & Kegan Paul Ltd., London.

Roy Burman, B.K. (1981) : *Strategy of Tribal Development : An Anthropological Approach in Tribal Development & Its Administration*; Concept Publishing Co., New Delhi.

Sachhidanand (1979) : *The Changing Mundas*; Concept Publishing Co., New Delhi.

Sahay, B.N. (joint ed.) (1980) : *Aspects of Social Anthropology in India*; Edited by, L.P. Vidyarthi, B.N. Sahay & P.K. Dutta, Classical Publication, New Delhi.

Sahay, K.N. (1999) : *Social Anthropology in India*; Commonwealth Publications, New Delhi.

Sahu, C. (1998) : *Primitive Tribes in India*; Sarup & Sons, New Delhi.

Sahlin, M.D. (1960) : *Evolution & Culture*; University of Michigan Press, Michigan.

Sahlin, M.D. (1972) : *Stone Age Economics*, Aldine, Chicago.

Sapir, E. (1949) : *Language: An Introduction to the Study of Speech*; Harcourt Brace Javanovich, New York.

Sarkar, R.M. (1954) : *Fundamentals of Physical Anthropology*; Post-graduate Book Mart, Calcutta.

Sarkar, S.S. (1954) : *The Aboriginal Races of India*; Bookland Pvt., Ltd. Calcutta.

Sarkar, S.S. (1957) : *A Lobaratory Manual of Somatology*; Scientific Book Agency, Calcutta.

Sarkar, S.S. (1957) : *"Race & Race Movements of India"* in Cultural Heritage of India, Vol. I, Calcutta.

Seligman, C.G. (1930) : *Races of Africa*; Thornton Butterworth Ltd. London.

Sen, D. & A.K. Ghosh (ed) (1966) : *Studies in Prehistory;* Robert Bruce Foote Memorial Volume, Firma K.LM. Calcutta.

Service, E.R. (1962) : *Primitive Social Organization*; Random House, New York.

Service, E.R. (1978) : *Profile in Ethnology*; Harper & Row, New York.

Shapiro, H.L. (1939) : *Migration & Environment*; Oxford University Press, London.

Shapiro, H.L. (ed.) (1960) : *Man, Culture & Society*; Oxford University Press, London.

Sharma, B.A.V. et al (1983) : *Research Methods in Social Sciences*; Sterling

Publishers Pvt. Ltd., New Delhi.

Sills, David. L. (ed.) (1960) : *International Encyclopedia of Social Sciences*; Reprint edition, The Macmillin Co. & the Free Press, New York.

Simpson, G.G. (1971) : *The Meaning of Evolution*; Bantam, New York.

Singer, M. (1972) : *When a Great Tradition Modernizes*; Praeger Publishers, New York.

Singer, M & B.S. Cohn (ed.) (1996) : *Structure & Change in Indian Society*; Rawat Publication, New Delhi.

Singh, I.P. & M.K. Bhasin (1968) : *Anthropometry*; Kamla-Raj Enterprises, Delhi

Singh, K.S. (1998) : *Indias Communities*; Vols. I, II & III, Oxford University Press, Delhi.

Singh, K.S. (ed.) (1972) : *Tribal Situation in India*; Indian Institute of Advanced Studies, Simla.

Singh, K.S. (ed.) (1982) : *Tribal Movements in India*, Vols. I & II, Monohar Publication, New Delhi.

Singh, K.S. (ed.) (1991) : *The History of the Anthropological Survey of India*; Anthropological Survey of India, Calcutta.

Singh, K.S. (ed.) (1992) : *Ethnicity; Caste & People;* Monohar Publications, New Delhi.

Singh, Y. (1977) : *Social Stratification & Change in India*; Monohar Publication, New Delhi.

Singh, Y (2000) : *Culture Change in India,* Rawat Publication, Jaipur.

Srinivas, M.N. (1952) : *Religion & Society among the Coorgs of South India*; Oxford University Press,Oxford.

Srinivas, M.N. (1955) : *India's Village*; Asia Publishing House, Bombay.

Srinivas, M.N. (1958) : *Method in Social Anthropology*; University of Chicago Press, Chicago.

Srinivas, M.N. (1962) : *Caste in Modern India*; Asia Publishing House, Bombay.

Srinivas, M.N. (1966) : *Social Change in Modern India*; Allied Publishers, Bombay.

Stern, Curt. (1973) : *Principle of Human Genetics*; Third edition, W.H. Freeman Company Publishers, Francisco.

Steward, J.H. (1955) : *Theory of Culture Change*; University of Illinois Press, Urbana.

Stibbe, E.P. (1938) : *An Introduction to Physical Anthropology*; London

Sturtevant, E.H. (1947) : *An Introduction to Linguistic Science*; Yale University Press, New Haven.

Sullivan, L.R. (1928) : Essentials of Anthropometry, American Museum (Natural History), New York.

Tax, S. (ed.) (1962) : *Anthropology Today*; University of Chicago Press, U.S.A.

Tax, S. (ed.) (1965) : *Horizons of Anthropology*; John Dickens & Co. Ltd., Great Britain.

Thapar, B.K. (1970) : *"History of Domesticated Animals"* in HUMAN EVOLUTION, Vol. II, Academic Press Inc., London.

Thapar. R (ed) (1977) : *Tribe, Caste & Religion in India*; Macmillan India Ltd., New Delhi.

Thurnwald, R. (1932) : *Economics in Primitive Communities*; Oxford University Press, London.

Titiev, M (1964) : *Introduction to Cultral Anthropology*; Holt, Rinehart & Winston, New York.

Tylor, E.B. (1924) : *Primitive Culture*; Reprint, Volumes I & II, Estes & Lauriat, Boston.

Tylor, E.B. (1930) : *Anthropology : An Introduction to the Study of Man and Civilization,* Walts, London.

Upadhyay, H.C. (1991) : *Scheduled Castes & Scheduled Tribes in India;* Anmol Publication, Delhi.

Vidyarthi, L.P. (1968) : *Applied Physical Anthropology in India;* Kitab Mahal, Allahabad.

Vidyarthi, L.P. (1978) : *Rise in Anthropology in India* ;Volumes I & II, Concept Publishing Co., Delhi.

Vidyarthi, L.P. (1979) : *Rise of World Anthropology*; Concept Publishing Co., Delhi.

Vidyarthi, L.P. (ed.) (1986) : *Tribal Development & Its Administration*; Second edition, Concept Publishing Co., New Delhi.

Weber, M. (1930) : *The Protestant Ethic & the Spirit of Capitalism;* Translated by T. Persons, Scribnar, New York.

Weber, M. (1947) : *The Theory of Social & Economic Organization;* William Hodge, London.

Westermark, E.A. (1901) : *The History of Human Marriage*; Reprint, Vols.I, II & III, Macmillan, London.

Wheeler, Sir, M. (1959) : *Early India & Pakistan to Ashoka*; Frederick A., Praeger, New York.

Wheeler, Sir, M. (1966) : *Civilization of the Indus Valley & Beyond*; Thames & Hudson, London.

White, L.A. (1949) : *The Science of Culture*; Farrar, Straus & Cudahy, New York.

Whittle, A. (1996) : *Europe in the Neolithic*; Cambridge University Press, Great Britain.

Wilkinsin, T.S. & P.L. Bhandarkar (1979) : *Methodology & Techniques of Social Research*; Himalaya Publishing House, Bombay.

Young, P.V. (1975) : *Scientific Social Survey & Research*; Revised Edition, Prentice-Hall of India, New Delhi.

Zamora, M.D. (1991) : *Anthropology: An Overview & Other Essays;* Reliance Publishing House, New Delhi.

Zamora, M.D. (1993) : *Perspectives on Cultural Change & Development*; Reliance Publishing House, New Delhi.

Zeuner, F.E. (1959) : *The Pleistocene Period*; Hutchinsen & Co., London.

Zeuner, F.E. (1964) : *Dating the Past*; 4th Revised edition, Methuen & Co. Ltd., London.

NOTES

NOTES

NOTES

NOTES